Practice Problems *for the* Civil Engineering PE Exam

A Companion to the *Civil Engineering Reference Manual*

Tenth Edition

Michael R. Lindeburg, PE

Professional Publications, Inc. • Belmont, CA

How to Locate Errata and Other Updates for This Book

At Professional Publications, we do our best to bring you error-free books. But when errors do occur, we want to make sure that you know about them so they cause as little confusion as possible.

A current list of known errata and other updates for this book is available on the PPI website at **www.ppi2pass.com/errata**. We update the errata page as often as necessary, so check in regularly. You will also find instructions for submitting suspected errata. We are grateful to every reader who takes the time to help us improve the quality of our books by pointing out an error.

PRACTICE PROBLEMS FOR THE CIVIL ENGINEERING PE EXAM: A COMPANION TO THE CIVIL ENGINEERING REFERENCE MANUAL
Tenth Edition

Current printing of this edition: 1

Printing History

edition number	printing number	update
9	1	New edition. Copyright update.
9	2	Minor corrections.
10	1	New edition. Copyright update.

Printed in the United States of America

Professional Publications, Inc.
1250 Fifth Avenue, Belmont, CA 94002
(650) 593-9119
www.ppi2pass.com

Library of Congress Cataloging-in-Publication Data
Lindeburg, Michael R.
Practice problems for the civil engineering PE exam : a companion to the Civil engineering reference manual / Michael R. Lindeburg.-- 10th ed.
p. cm.
Includes bibliographical references and index.
ISBN-13: 978-1-59126-048-6
ISBN-10: 1-59126-048-5
1. Civil engineering--Problems, exercises, etc. 2. Civil engineering--Examinations--Study guides. I. Lindeburg, Michael R. Civil engineering reference manual. II. Title.

TA159.L578 2005
624'.076--dc22

2005054365

Topics

Background and Support
Water Resources
Environmental
Geotechnical
Structural
Transportation
Systems, Mgmt., and Professional

Where do I find help solving these Practice Problems?

Practice Problems for the Civil Engineering PE Exam presents complete, step-by-step solutions for more than 540 problems like those found on the Civil PE exam. You can find all the background information, including charts and tables of data, that you need to solve these problems in the *Civil Engineering Reference Manual*.

The *Civil Engineering Reference Manual* may be purchased from Professional Publications (800-426-1178 or www.ppi2pass.com) or from your favorite bookstore.

Table of Contents

Topic VI: Transportation

Topic VII: Systems, Management, and Professional

Preface to the Tenth Edition

When I wrote the first edition of my first book more than 25 years ago, I naively considered my job to be done. Ten editions of this book, *Practice Problems for the Civil Engineering PE Exam*, have taught me otherwise: improvements and corrections are inevitable. Altogether, this tenth edition has about 60 more pages than the ninth edition. All of the problems and solutions from the previous edition have been retained. An additional 105 problems were added to chapters that heretofore had none.

I redrafted the problems and their solutions to be consistent with the newest structural and transportation codes and standards referenced by the civil PE exam. For example, the structural codes have been updated to include ACI 318-02, ACI 530-02 and 530.1-02, AISC/LRFD 2001, AASHTO (Green Book) 2002, ASCE 7-02, NDS 2001, and IBC 2003. Many of the concrete problems were given alternate solutions so that you could use either ACI 318 Ch. 9 strength design methods or legacy methods, both of which are now part of the exam. So regardless of whether you choose to study strength design or allowable stress design methods, you will be supported.

Some of the problem statements have been changed to include additional information in order to more closely simulate the exam and to eliminate the need to make assumptions that might get you off track. Suggestions for improvements from readers of previous editions have been incorporated.

I've tried to anticipate your questions by adding strategic commentary throughout many solutions. My daily dose of fan mail (as I like to view the correspondence) brings questions that show me how to improve logic and flow. In these solutions, you won't find any professor chalkboard talk such as "the interested reader may prove the expression in his/her own time," or "it is patently obvious what the next seven steps should be, so we'll skip them." Although I might leave out some interim steps (e.g., omitting the steps to solve for the roots of a quadratic or higher-order equation, which is something your calculator can do), for the most part the solutions are complete down to the algebraic manipulations.

The first seven editions of this book were written with a singular purpose: to serve as solution manuals for the end-of-chapter problems in the corresponding editions of the *Civil Engineering Reference Manual*. Starting with the eighth edition, this book essentially became an independent collection of solved practice problems that you could take with you anywhere.

The chapters in this edition continue to parallel those in the *Civil Engineering Reference Manual*. This book contains numerous references to the equations, tables, and figures in the *Civil Engineering Reference Manual*. This book uses the same nomenclature, data, and methods. Though independent, these two books intentionally complement each other.

Even if you don't have the *Civil Engineering Reference Manual* with you while you review this book's solutions, you should be able to follow the solution steps. Except where the intended equation is obvious (e.g., $A = \pi r^2$ and $v = Q/A$), the actual equation is given in variable format before the numbers are plugged in.

As usual, I've already begun the next edition of this book. Even before the next edition arrives, however, you can keep up with changes to this book by logging onto the Professional Publications' web site at www.ppi2pass.com/errata. I'll do my best to make *Practice Problems for the Civil Engineering PE Exam* and the *Civil Engineering Reference Manual* the most useful books in your library.

Michael R. Lindeburg, PE

Acknowledgments

So, who helped me with this new edition? Wow. Where do I start?

For many of the chapters in previous editions, I asked experts from industry and academia to review, to update, to reorganize, and to write original material. These subject matter experts also incorporated new topics into these chapters according to their own experience. This edition shares a common developmental heritage with nine previous editions. Looking at the acknowledgements in those books, there are hundreds and hundreds to whom I owe gratitude.

Many of the revisions needed to bring forth this edition were made by a team of structural and transportation experts. S. K. Ghosh, PhD, of S. K. Ghosh Associates, Inc., Palatine, Ill., revised the concrete chapters. William Wood, PhD, PE, director of the School of Engineering Technology at Youngstown State University in Ohio, attended to the masonry revisions. Alan Williams, PhD, SE, a senior engineer with the California Department of Transportation, revised the sections on steel. Similar work on the traffic and transportation chapters was completed by Norman Voigt, MS, PE, a licensed consulting engineer and surveyor in Pittsburgh. All of these contributors applied themselves to the new edition with the diligence and attention to quality that you've come to expect from PPI.

Within PPI, a number of staffers rode shotgun on this new edition. Project editor Sean Sullivan did the majority of the editing and proofreading. Typesetters Kate Hayes and Miriam Hanes formatted the raw text into the pages that you see. Illustrators Amy Schwertman and Tom Bergstrom revised my figures to be clear, consistent with PPI's style guide, and less silly. Sarah Hubbard and Cathy Schrott managed the editorial and production departments, respectively. As Editor in Chief of Professional Publications, I got to check, review, and revise everyone's contributions, but few changes were needed.

Elizabeth, my wife, continues to bring me tea throughout the day while I'm working and writing. That routine hasn't changed in 25 years, although she brings me much less pie with the tea these days. And there still isn't a day I sit down to write that a wee one doesn't come bounding up the stairs to jump on my lap. These days, however, the wee one is only my cat, not one of my children.

If you put all of the acknowledgements of previous editions side by side, you'd see how my wife and family contributed and survived during the previous 25 years of authorship. The first editions mentioned a wife, and then two children. The children grew older and had a more significant presence in my acknowledgments (and in our lives). Within the last five years, they went off to college, and I wrote they no longer had to endure a father who pecked away so much at a keyboard, or who brought boxes of manuscripts to proofread on every family vacation they took together, even those to far distant lands.

For this edition, I'm proud as punch to report that Jenny, my older daughter, has finished a master's degree at George Washington University, and that Katie is just about finished at UC Santa Barbara. Both have chosen courses of study that put them squarely on the path to help humanity. Both of them are attracted to men with technical backgrounds, both have my sense of humor, and both talk frequently about the books they are going to write some day. These girls are the best books that Elizabeth and I ever created.

And, as kind of a private joke between her and me, I will continue to include this secret message to Jenny in each edition of this book until she figures out how to decode it. [Jenny] y92 e8e 8 w5q6 wqh3 24858ht r49j 5y3 eq4i3w5 j94h8ht y974w 59 5y3 233 y974w 9r 5y3 h8ty5? 8 yqf3 j6 3oe3w5 eq7ty534 u3hh6 59 5yqhi r94 5yq5. y34 w9dd34 04qd58d3 i305 j6 go99e d84d7oq58ht 2y8o3 y34 dyq5534 qhe u9i3w 59ww3e g35233h 974 529 d9j07534 h99iw 8h j6 9rr8d3 i305 j6 w08485w w9q48ht.

So, there you go. A zillion people have helped with this new edition. That's a lot of people for me to thank. Even thanking all of them would require a staff. And then I'd have to thank the staff. Ah, such is the life of authors. We're never finished.

Thank you, people!

Michael R. Lindeburg, PE

Suggested References for the Exam

Structural

Building Code Requirements for Structural Concrete (ACI 318) *with Commentary* (ACI 318R). American Concrete Institute.

Notes on ACI 318: *Building Code Requirements for Structural Concrete with Design Applications*. Portland Cement Association

Manual of Steel Construction, Allowable Stress Design. American Institute of Steel Construction, Inc.

LRFD Manual of Steel Construction: Vol. 1, "Structural Members, Specifications, and Codes." American Institute of Steel Construction (AISC) (only if you intend to work in LRFD)

Design Handbook: Beams, One-Way Slabs, Brackets, Footings, Pile Caps, Columns, Two-Way Slabs (ACI SP-017). American Concrete Institute

Building Code Requirements for Masonry Structures, ACI 530, *and Specifications for Masonry Structures*, ACI 530.1. American Concrete Institute

National Design Specification for Wood Construction. American Forest and Paper Association

Standard Specifications for Highway Bridges, AASHTO

International Building Code. International Code Council

PCI Design Handbook. Precast/Prestressed Concrete Institute

Minimum Design Standards for Buildings and Other Structures, ASCE 74. American Society of Civil Engineers

Transportation

Policy on Geometric Design of Highways and Streets ("AASHTO Green Book"). American Association of State Highway and Transportation Officials (AASHTO).

Highway Capacity Manual. Transportation Research Board/National Research Council

Manual of Uniform Traffic Control Devices—For Streets and Highways (MUTCD), D6.1. U.S. Department of Transportation, Federal Highway Administration (FHA), Traffic Control Systems Divisions, Office of Traffic Operations

Roadside Design Guide. AASHTO

Thickness Design for Concrete Highway and Street Pavements. Portland Cement Association

AASHTO Guide for Design of Pavement Structures, AASHTO. Volumes 1 and 2

The Asphalt Handbook, Manual MS-4. The Asphalt Institute

Design and Control of Concrete Mixtures. Portland Cement Association

Traffic Engineering Handbook. Institute of Transportation Engineers

Trip Generation. Institute of Transportation Engineers

Water Resources

Handbook of Hydraulics: For the Solution of Hydraulic Engineering Questions. Brater, King, Lindell, and Wei

Urban Hydrology for Small Watersheds (Technical Release TR-55). United States Department of Agriculture, Natural Resources Conservation Service) (previously, Soil Conservation Service)

Environmental

Wastewater Engineering: Treatment, Disposal, and Reuse (Metcalf & Eddy). George Tchobanoglous and Franklin L. Burton

Recommended Standard for Wastewater Facilities ("10 States' Standards"). Health Education Services, Health Resources, Inc.

Standard Methods for the Examination of Water and Wastewater. A joint publication of the American Public Health Association (APHA), the American Water Works Association (AWWA), and the Water Pollution Control Federation (WPCF)

Geotechnical

NAVFAC Design Manuals DM 7.1 and 7.2. Department of the Navy, Naval Facilities Engineering Command

Codes Used to Prepare This Book

The information that was used to write and update this book was the most current at the time. However, as with engineering practice itself, the PE examination is not always based on the most current codes or cutting-edge technology. Similarly, codes, standards, and regulations adopted by state and local agencies often lag issuance by several years. Thus, the codes that are current, used by you in practice, and tested on the exam can all be different.

PPI lists on its website the dates and editions of the codes, standards, and regulations on which NCEES has announced the PE exams are based. It is your responsibility to find out which codes will be tested on your exam. In the meantime, here are the codes that have been incorporated into this edition.[1]

STRUCTURAL DESIGN STANDARDS

AASHTO: *Standard Specifications for Highway Bridges*, Seventeenth ed., 2002, American Association of State Highway and Transportation Officials, Washington, DC

ACI 318: *Building Code Requirements for Structural Concrete*, 2002, American Concrete Institute, Farmington Hills, MI

ACI 530: *Building Code Requirements for Masonry Structures*, 2002, and ACI 530.1: *Specifications for Masonry Structures*, 2002, American Concrete Institute, Detroit, MI

AISC/ASD: *Manual of Steel Construction, Allowable Stress Design*, Ninth ed., 1989, American Institute of Steel Construction, Inc., Chicago, IL

AISC/LRFD: *Manual of Steel Construction Load and Resistance Factor Design*, Second ed., 2001, American Institute of Steel Construction, Inc., Chicago, IL

ASCE 7: *Minimum Design Standards for Buildings and Other Structures*, 2002, American Society of Civil Engineers, New York, NY

IBC: *International Building Code*, 2003 ed., International Code Council, Inc., Falls Church, VA

NDS: *National Design Specification for Wood Construction*, 2001 ASD ed., and *National Design Specification Supplement*, 2001 ASD ed., American Forest and Paper Association, Washington, DC

PCI: *PCI Design Handbook*, Fifth ed., 1999, Precast/Prestressed Concrete Institute, Chicago, IL

TRANSPORTATION DESIGN STANDARDS

AASHTO: *AASHTO Guide for Design of Pavement Structures*, 1993, American Association of State Highway and Transportation Officials, Washington, DC

AASHTO: *A Policy on Geometric Design of Highways and Streets*, Fourth ed., 2001, American Association of State Highway and Transportation Officials, Washington, DC

AASHTO: *Roadside Design Guide*, 2002 ed., American Association of State Highway and Transportation Officials, Washington, DC

AI: *The Asphalt Handbook* (MS-4), 1989, Asphalt Institute, College Park, MD

HCM: *Highway Capacity Manual*, Fourth ed., U.S. Customary version, 2000, Transportation Research Board, National Research Council, Washington, DC

MUTCD: *Manual on Uniform Traffic Control Devices*, 2003, U.S. Dept. of Transportation, Federal Highway Administration, Washington, DC

PCA: *Design and Control of Concrete Mixtures*, Fourteenth ed., 2002, Portland Cement Association, Skokie, IL

ITE: *Traffic Engineering Handbook*, Fifth ed., 1999, Institute of Transportation Engineers, Washington, DC

[1]Although any edition can be used to learn the subject, the exam is "edition sensitive." The code versions, editions, or years that are tested on the exam are posted on Professional Publications' website, www.ppi2pass.com.

How to Use This Book

This book is primarily a companion to the *Civil Engineering Reference Manual.* As a tool for preparing for an engineering licensing exam, there are a few, but not very many, ways to use it.

Before it grew to behemoth size, I envisioned this book being taken to work, on the bus or train, on business trips, and on weekend pleasure getaways to the beach. I figured that it would "carry" a lot easier than the big *Civil Engineering Reference Manual*, which I knew before I started to write it was going to be a big "mutha." Though I don't think you'll be taking this book on any backpacking trips, you might still find yourself working problems in the cafeteria during your lunch break.

The big issue is whether you really work the practice problems or not. Some people think they can read a problem statement, think about it for about ten seconds, read the solution, and then say "Yes, that's what I was thinking of, and that's what I would have done." Sadly, these people find out too late that the human brain doesn't learn very efficiently in that manner. Under pressure, they find they know and remember too little. For real learning, you have to spend some time with a stubby pencil.

There are so many places where you can get messed up solving a problem. Maybe it's in the use of your calculator, like pushing log instead of ln, or forgetting to set the angle to radians instead of degrees, and so on. Maybe it's rusty math. What is $\ln e(x)$, anyway? How do you factor a polynomial? Maybe it's in finding the data needed or the proper unit conversion. Maybe it's just trying to find out if that proprietary building code equation expects L to be in feet or inches. Conquering these things takes time—more time than you may want to spend when time is at a premium. And unfortunately, most people have to make a mistake once so that they don't make it again.

Even if you do decide to get your hands dirty and actually work the problems (as opposed to skimming through them), you'll have to decide how much reliance you place on the published solutions. It's tempting to turn to a solution when you get slowed down by details or stumped by the subject material. You'll probably want to maximize the number of problems you solve by spending as little time as you can on each problem. After all, optimization is the engineering way. However, I want you to struggle a little bit more than that—not because I want to see you suffer, but because the optimization is in getting you through your licensing exam. We're not trying to find the optimally minimum study time.

Studying a new subject is analogous to using a machete to cut a path through a dense jungle. By doing the work, you develop pathways that weren't there before. It's a lot different than just looking at the route on a map. You actually get nowhere by looking at a map. But cut that path once, and you're in business until the jungle overgrowth closes in again.

So, do the problems. All of them. Don't look at the answers until you've sweated a little. And, let's not have any whining. Please.

1 Systems of Units

PRACTICE PROBLEMS

1. Convert 250°F to degrees Celsius.

(A) 115°C
(B) 121°C
(C) 124°C
(D) 420°C

2. Convert the Stefan-Boltzmann constant (0.1713×10^{-8} Btu/ft^2-hr-°R^4) from English to SI units.

(A) 5.14×10^{-10} W/m^2·K^4
(B) 0.95×10^{-8} W/m^2·K^4
(C) 5.67×10^{-8} W/m^2·K^4
(D) 7.33×10^{-6} W/m^2·K^4

3. How many U.S. tons (2000 lbm per ton) of coal with a heating value of 13,000 Btu/lbm must be burned to provide as much energy as a complete nuclear conversion of 1 g of its mass? (Hint: Use Einstein's equation: $E = mc^2$.)

(A) 1.7 tons
(B) 14 tons
(C) 779 tons
(D) 3300 tons

SOLUTIONS

1. The conversion to degrees Celsius is given by

$$\begin{aligned} {}^\circ\text{C} &= \left(\tfrac{5}{9}\right)(^\circ\text{F} - 32^\circ\text{F}) \\ &= \left(\tfrac{5}{9}\right)(250^\circ\text{F} - 32^\circ\text{F}) \\ &= \left(\tfrac{5}{9}\right)(218^\circ\text{F}) \\ &= \boxed{121.1^\circ\text{C} \quad (121^\circ\text{C})} \end{aligned}$$

The answer is (B).

2. In U.S. customary units, the Stefan-Boltzmann constant, σ, is 0.1713×10^{-8} Btu/hr-ft^2-°R^4.

Use the following conversion factors.

$$\begin{aligned} 1 \text{ Btu/hr} &= 0.2931 \text{ W} \\ 1 \text{ ft} &= 0.3048 \text{ m} \\ \Delta T_{^\circ\text{R}} &= \frac{9}{5}\Delta T_\text{K} \\ \Delta T_\text{K} &= \frac{5}{9}\Delta T_{^\circ\text{R}} \end{aligned}$$

Performing the conversion gives

$$\begin{aligned} \sigma &= \left(0.1713 \times 10^{-8} \frac{\text{Btu}}{\text{hr-ft}^2\text{-}^\circ\text{R}^4}\right)\left(0.2931 \frac{\text{W}}{\frac{\text{Btu}}{\text{hr}}}\right) \\ &\quad \times \left(\frac{1 \text{ ft}}{0.3048 \text{ m}}\right)^2 \left(\frac{1^\circ\text{R}}{\frac{5}{9}\text{K}}\right)^4 \\ &= \boxed{5.67 \times 10^{-8} \text{ W/m}^2\cdot\text{K}^4} \end{aligned}$$

The answer is (C).

3. The energy produced from the nuclear conversion of any quantity of mass is given as

$$E = mc^2$$

The speed of light, c, is 3×10^8 m/s.

For a mass of 1 g (0.001 kg),

$$\begin{aligned} E &= mc^2 \\ &= (0.001\ \text{kg})\left(3\times 10^8\ \frac{\text{m}}{\text{s}}\right)^2 \\ &= 9\times 10^{13}\ \text{J} \end{aligned}$$

Convert to U.S. customary units with the conversion 1 Btu = 1055 J.

$$\begin{aligned} E &= (9\times 10^{13}\ \text{J})\left(\frac{1\ \text{Btu}}{1055\ \text{J}}\right) \\ &= 8.53\times 10^{10}\ \text{Btu} \end{aligned}$$

The number of tons of 13,000 Btu/lbm coal is

$$\frac{8.53\times 10^{10}\ \text{Btu}}{\left(13{,}000\ \frac{\text{Btu}}{\text{lbm}}\right)\left(2000\ \frac{\text{lbm}}{\text{ton}}\right)} = \boxed{3281\ \text{tons}\ (3300\ \text{tons})}$$

The answer is (D).

2 Engineering Drawing Practice

PRACTICE PROBLEMS

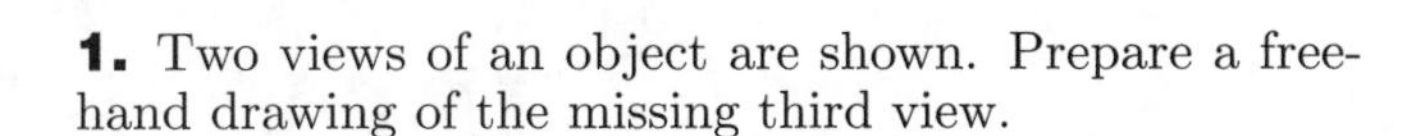

1. Two views of an object are shown. Prepare a freehand drawing of the missing third view.

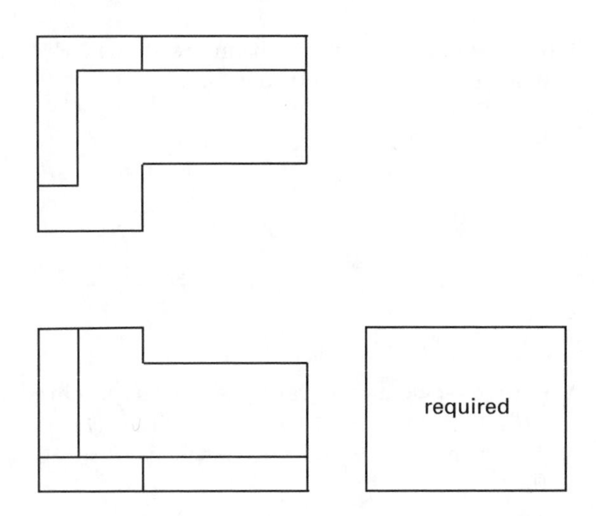

2. Two views of an object are shown. Prepare a freehand drawing of the missing third view.

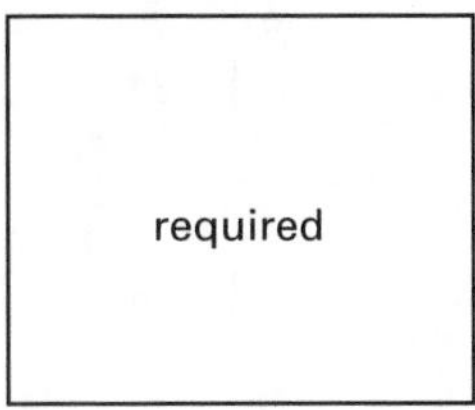

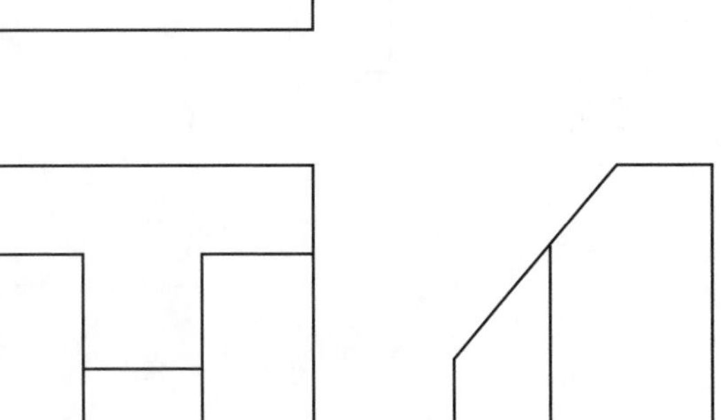

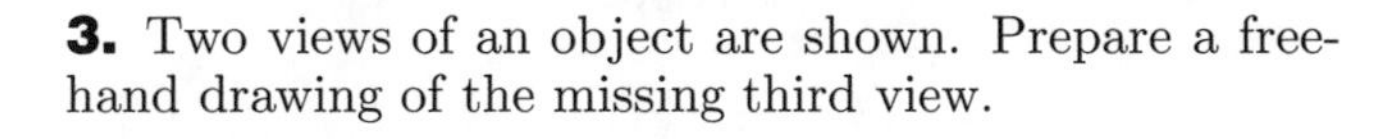

3. Two views of an object are shown. Prepare a freehand drawing of the missing third view.

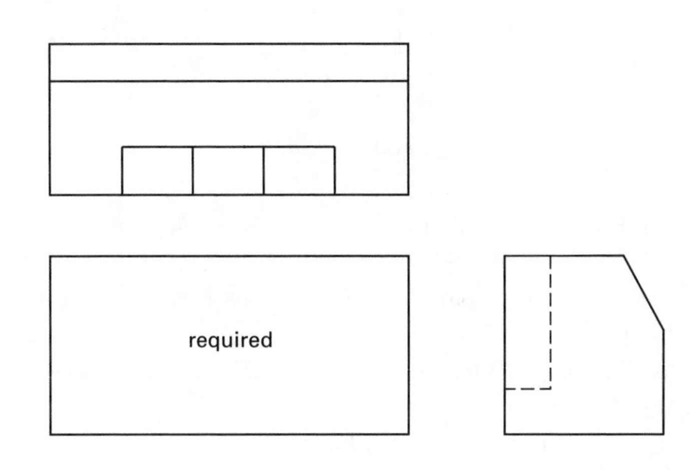

4. A pictorial sketch of an object is shown. Prepare a freehand three-view orthographic drawing set.

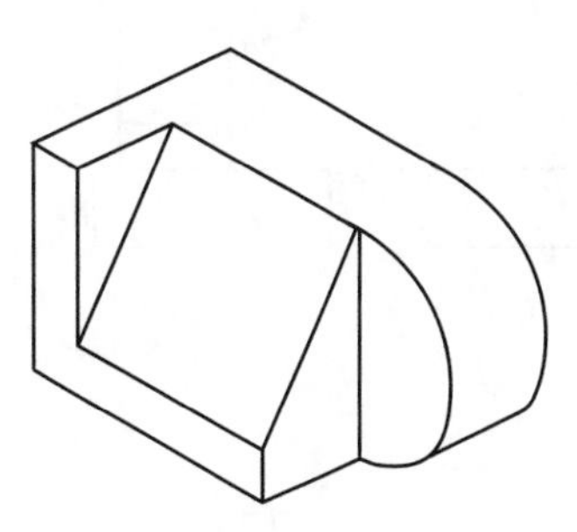

5. Plan and elevation views of an object are shown.

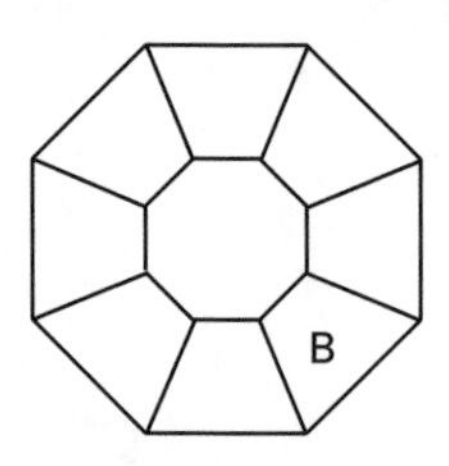

plan view

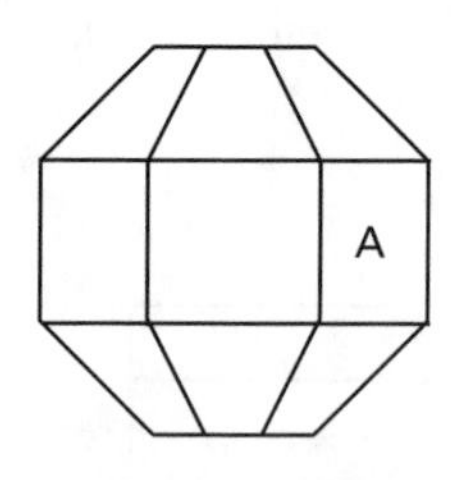

elevation view

(a) Prepare a freehand drawing of the view that shows surface A in true shape.

(b) Referring to part (a), what is this view of surface A called?

(c) Prepare a freehand drawing of the view that shows surface B in true shape.

(d) Referring to part (c), what is the view of surface B called?

6. Prepare a freehand isometric drawing of the object shown. Hint: Lay off horizontal and vertical (i.e., isometric) lines along the isometric axes shown.

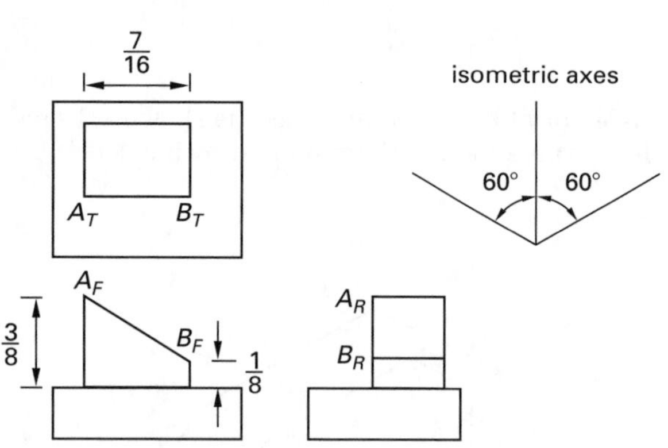
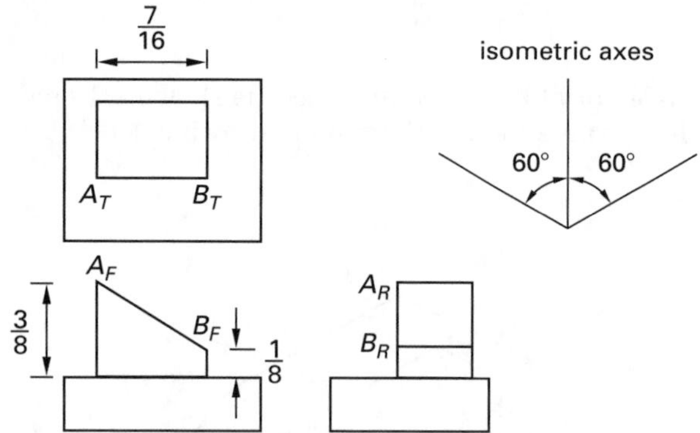

7. Two lines in an oblique position are shown. Prepare a freehand right auxiliary normal view.

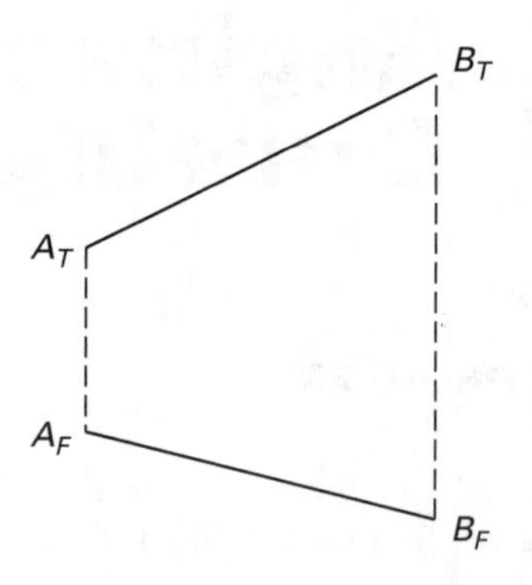

8. Two lines in an oblique position are shown. Prepare a freehand read auxiliary normal view.

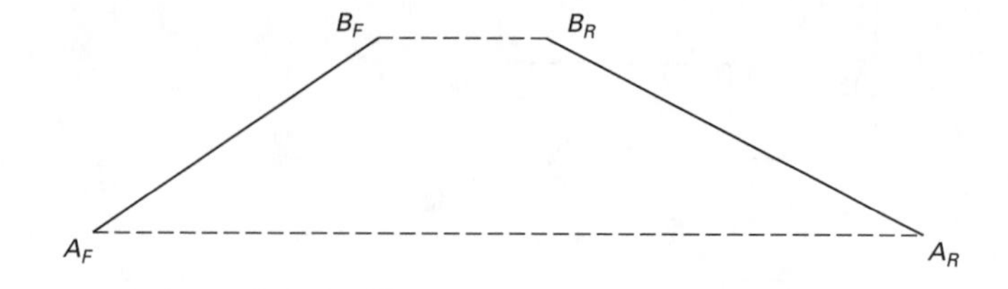

9. A parallel-scale nomograph has been prepared to solve the equation $D = 1.075\sqrt{WH}$. Use the nomograph to estimate the value of D when $H = 10$ and $W = 40$.

$D = 1.075\sqrt{WH}$

W: 100, 80, 60, 40, 30, 20, 10

D: 80, 60, 40, 30, 20, 10

H: 50, 40, 30, 20, 10

SOLUTIONS

1.

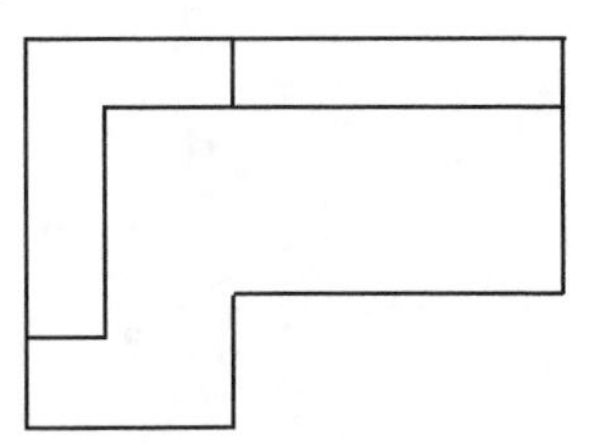

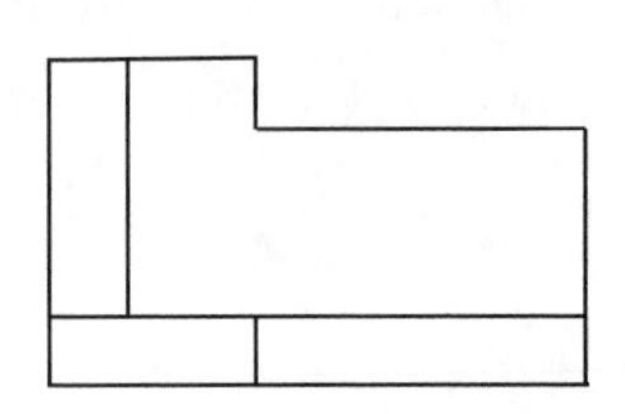

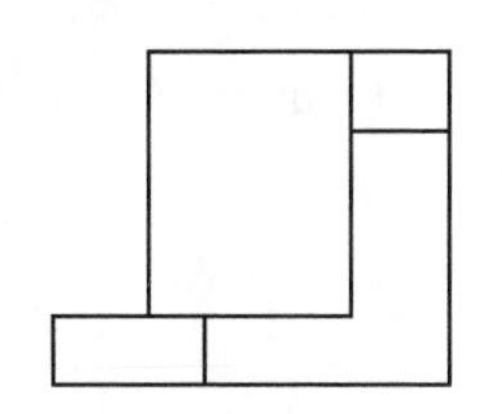

2.

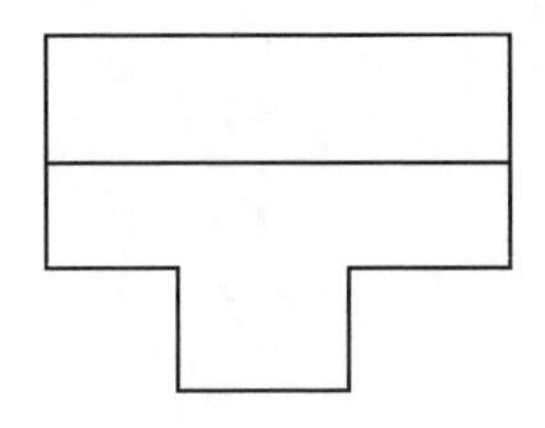

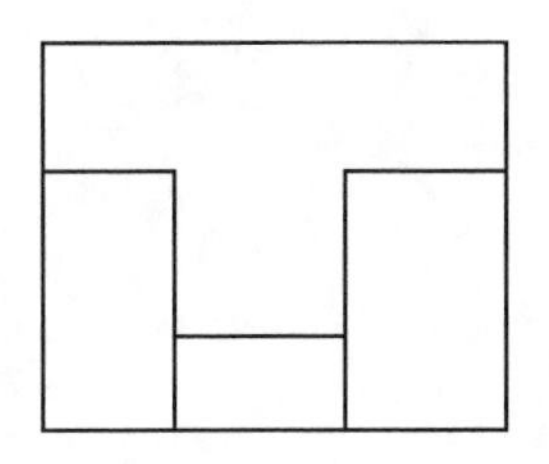

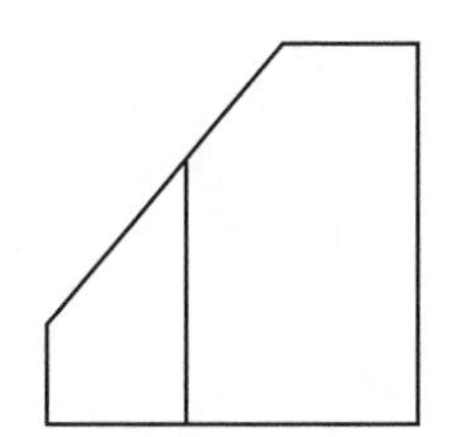

3.

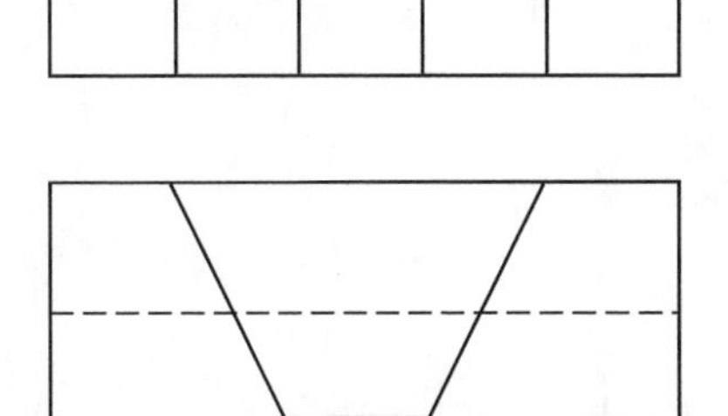

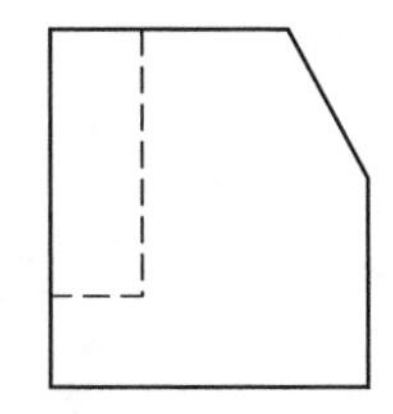

4.

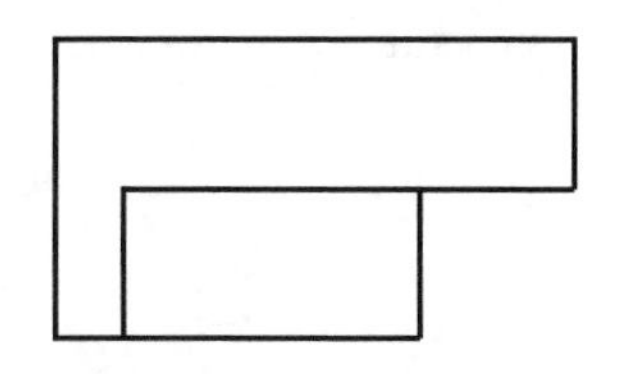

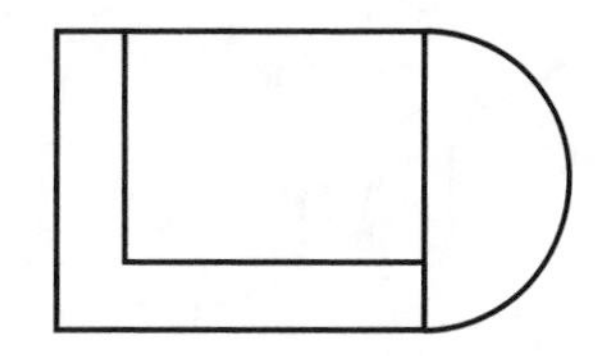

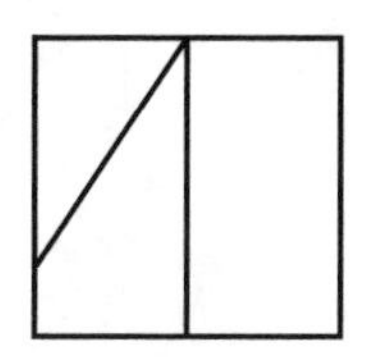

5. (a)

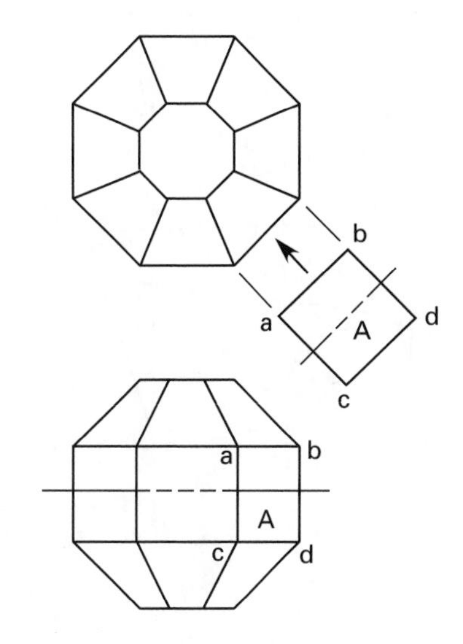

(b) An elevation auxiliary view shows surface A in true shape.

(c)

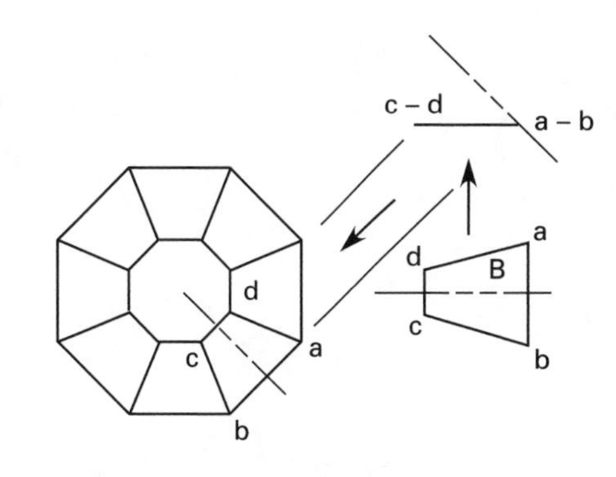

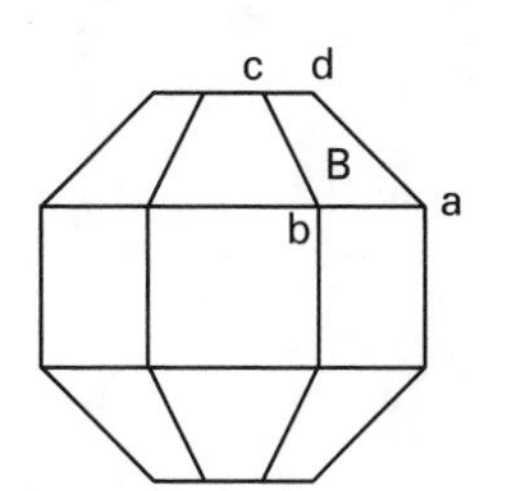

(d) An oblique view shows surface B in true shape.

6.

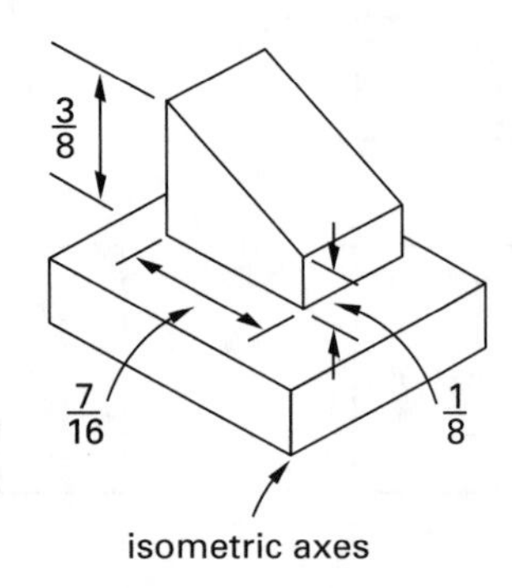

7.

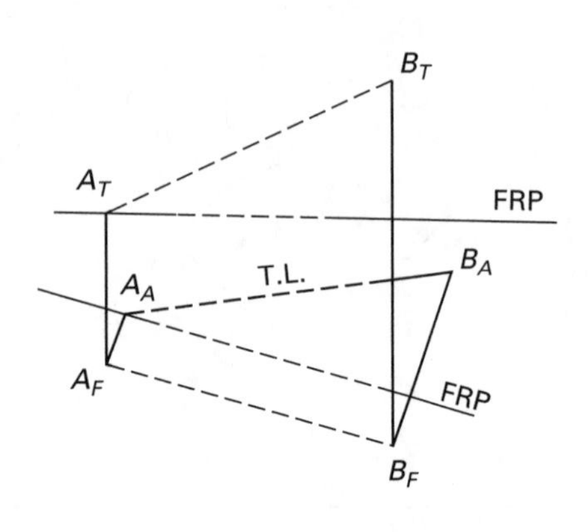

8.

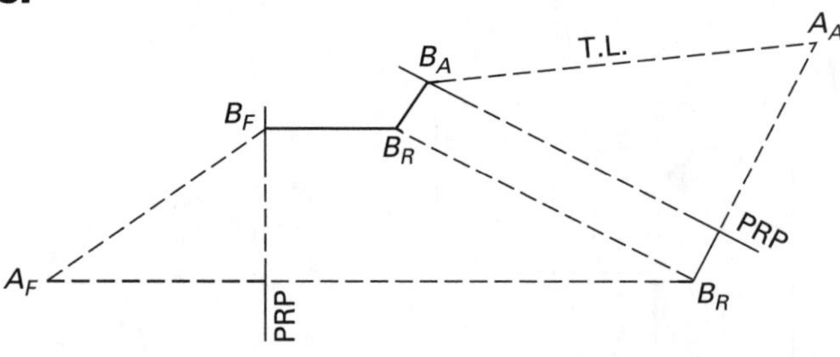

9. A straight line through the points $H = 10$ and $W = 40$ intersects the D-scale at approximately 21.5.

3 Algebra

PRACTICE PROBLEMS

Series

1. Calculate the following sum.

$$\sum_{j=1}^{5}\left((j+1)^2-1\right)$$

(A) 15
(B) 24
(C) 35
(D) 85

Logarithms

2. If every 0.1 sec a quantity increases by 0.1% of its current value, calculate the doubling time.

(A) 14 sec
(B) 69 sec
(C) 690 sec
(D) 69,000 sec

SOLUTIONS

1. Let $S_n = (j+1)^2 - 1$.

For $j = 1$,

$$S_1 = (1+1)^2 - 1 = 3$$

For $j = 2$,

$$S_2 = (2+1)^2 - 1 = 8$$

For $j = 3$,

$$S_3 = (3+1)^2 - 1 = 15$$

For $j = 4$,

$$S_4 = (4+1)^2 - 1 = 24$$

For $j = 5$,

$$S_5 = (5+1)^2 - 1 = 35$$

Substituting the above expressions gives

$$\begin{aligned}\sum_{j=1}^{5}\left((j+1)^2-1\right) &= \sum_{j=1}^{5} S_j \\ &= S_1 + S_2 + S_3 + S_4 + S_5 \\ &= 3 + 8 + 15 + 24 + 35 \\ &= \boxed{85}\end{aligned}$$

The answer is (D).

2. Let n represent the number of elapsed periods of 0.1 sec, and let y_n represent the amount present after n periods.

y_0 represents the initial quantity.

$$\begin{aligned}y_1 &= 1.001y_0 \\ y_2 &= 1.001y_1 = (1.001)\big((1.001)y_0\big) = (1.001)^2 y_0\end{aligned}$$

Therefore, by deduction,

$$y_n = (1.001)^n y_0$$

The expression for a doubling of the original quantity is

$$2y_0 = y_n$$

Substitute for y_n.

$$\begin{aligned} 2y_0 &= (1.001)^n y_0 \\ 2 &= (1.001)^n \end{aligned}$$

Take the logarithm of both sides.

$$\begin{aligned} \log(2) &= \log(1.001)^n \\ &= n \log(1.001) \end{aligned}$$

Solve for n.

$$n = \frac{\log(2)}{\log(1.001)} = 693.5$$

Since each period is 0.1 sec, the time is given by

$$\begin{aligned} t &= n(0.1 \text{ sec}) \\ &= (693.5)(0.1 \text{ sec}) = \boxed{69.35 \text{ sec}} \end{aligned}$$

The answer is (B).

4 Linear Algebra

PRACTICE PROBLEMS

Determinants

1. What is the determinant of matrix $\mathbf{A}$?

$$\mathbf{A} = \begin{bmatrix} 8 & 2 & 0 & 0 \\ 2 & 8 & 2 & 0 \\ 0 & 2 & 8 & 2 \\ 0 & 0 & 2 & 4 \end{bmatrix}$$

(A) 459
(B) 832
(C) 1552
(D) 1776

Simultaneous Linear Equations

2. Use Cramer's rule to solve for the values of x, y, and z that simultaneously satisfy the following equations.

$$x + y = -4$$
$$x + z - 1 = 0$$
$$2z - y + 3x = 4$$

(A) $(x, y, z) = (3, 2, 1)$
(B) $(x, y, z) = (-3, -1, 2)$
(C) $(x, y, z) = (3, -1, -3)$
(D) $(x, y, z) = (-1, -3, 2)$

SOLUTIONS

1. Expand by cofactors of the first row since there are two zeros in that row.

$$D = 8\begin{vmatrix} 8 & 2 & 0 \\ 2 & 8 & 2 \\ 0 & 2 & 4 \end{vmatrix} - 2\begin{vmatrix} 2 & 0 & 0 \\ 2 & 8 & 2 \\ 0 & 2 & 4 \end{vmatrix} + 0 - 0$$

by first row:

$$\begin{vmatrix} 8 & 2 & 0 \\ 2 & 8 & 2 \\ 0 & 2 & 4 \end{vmatrix} = (8)[(8)(4) - (2)(2)] - (2)[(2)(4) - (2)(0)] = (8)(28) - (2)(8) = 208$$

by first row:

$$\begin{vmatrix} 2 & 0 & 0 \\ 2 & 8 & 2 \\ 0 & 2 & 4 \end{vmatrix} = (2)[(8)(4) - (2)(2)] = 56$$

$$D = (8)(208) - (2)(56) = \boxed{1552}$$

The answer is (C).

2. Rearrange the equations.

$$\begin{aligned} x + y \quad &= -4 \\ x \quad + z &= 1 \\ 3x - y + 2z &= 4 \end{aligned}$$

Write the set of equations in matrix form: $\mathbf{AX} = \mathbf{B}$.

$$\begin{bmatrix} 1 & 1 & 0 \\ 1 & 0 & 1 \\ 3 & -1 & 2 \end{bmatrix} \begin{bmatrix} x \\ y \\ z \end{bmatrix} = \begin{bmatrix} -4 \\ 1 \\ 4 \end{bmatrix}$$

Find the determinant of the matrix $\mathbf{A}$.

$$\begin{aligned} |\mathbf{A}| &= \begin{vmatrix} 1 & 1 & 0 \\ 1 & 0 & 1 \\ 3 & -1 & 2 \end{vmatrix} \\ &= 1\begin{vmatrix} 0 & 1 \\ -1 & 2 \end{vmatrix} - 1\begin{vmatrix} 1 & 0 \\ -1 & 2 \end{vmatrix} + 3\begin{vmatrix} 1 & 0 \\ 0 & 1 \end{vmatrix} \\ &= (1)\big((0)(2) - (1)(-1)\big) \\ &\quad - (1)\big((1)(2) - (-1)(0)\big) \\ &\quad + (3)\big((1)(1) - (0)(0)\big) \\ &= (1)(1) - (1)(2) + (3)(1) \\ &= 1 - 2 + 3 \\ &= 2 \end{aligned}$$

Find the determinant of the substitutional matrix $\mathbf{A_1}$.

$$\begin{aligned}|\mathbf{A_1}| &= \begin{vmatrix} -4 & 1 & 0 \\ 1 & 0 & 1 \\ 4 & -1 & 2 \end{vmatrix} \\ &= -4\begin{vmatrix} 0 & 1 \\ -1 & 2 \end{vmatrix} - 1\begin{vmatrix} 1 & 0 \\ -1 & 2 \end{vmatrix} + 4\begin{vmatrix} 1 & 0 \\ 0 & 1 \end{vmatrix} \\ &= (-4)\big((0)(2) - (1)(-1)\big) \\ &\quad - (1)\big((1)(2) - (-1)(0)\big) \\ &\quad + (4)\big((1)(1) - (0)(0)\big) \\ &= (-4)(1) - (1)(2) + (4)(1) \\ &= -4 - 2 + 4 \\ &= -2\end{aligned}$$

Find the determinant of the substitutional matrix $\mathbf{A_2}$.

$$\begin{aligned}|\mathbf{A_2}| &= \begin{vmatrix} 1 & -4 & 0 \\ 1 & 1 & 1 \\ 3 & 4 & 2 \end{vmatrix} \\ &= 1\begin{vmatrix} 1 & 1 \\ 4 & 2 \end{vmatrix} - 1\begin{vmatrix} -4 & 0 \\ 4 & 2 \end{vmatrix} + 3\begin{vmatrix} -4 & 0 \\ 1 & 1 \end{vmatrix} \\ &= (1)\big((1)(2) - (4)(1)\big) \\ &\quad - (1)\big((-4)(2) - (4)(0)\big) \\ &\quad + (3)\big((-4)(1) - (1)(0)\big) \\ &= (1)(-2) - (1)(-8) + (3)(-4) \\ &= -2 + 8 - 12 \\ &= -6\end{aligned}$$

Find the determinant of the substitutional matrix $\mathbf{A_3}$.

$$\begin{aligned}|\mathbf{A_3}| &= \begin{vmatrix} 1 & 1 & -4 \\ 1 & 0 & 1 \\ 3 & -1 & 4 \end{vmatrix} \\ &= 1\begin{vmatrix} 0 & 1 \\ -1 & 4 \end{vmatrix} - 1\begin{vmatrix} 1 & -4 \\ -1 & 4 \end{vmatrix} + 3\begin{vmatrix} 1 & -4 \\ 0 & 1 \end{vmatrix} \\ &= (1)\big((0)(4) - (-1)(1)\big) \\ &\quad - (1)\big((1)(4) - (-1)(-4)\big) \\ &\quad + (3)\big((1)(1) - (0)(-4)\big) \\ &= (1)(1) - (1)(0) + (3)(1) \\ &= 1 - 0 + 3 \\ &= 4\end{aligned}$$

Use Cramer's rule.

$$x = \frac{|\mathbf{A_1}|}{|\mathbf{A}|} = \frac{-2}{2} = \boxed{-1}$$

$$y = \frac{|\mathbf{A_2}|}{|\mathbf{A}|} = \frac{-6}{2} = \boxed{-3}$$

$$z = \frac{|\mathbf{A_3}|}{|\mathbf{A}|} = \frac{4}{2} = \boxed{2}$$

The answer is (D).

5 Vectors

PRACTICE PROBLEMS

1. Calculate the dot products for the following vector pairs.

(a) $\mathbf{V}_1 = 2\mathbf{i} + 3\mathbf{j}; \mathbf{V}_2 = 5\mathbf{i} - 2\mathbf{j}$

(b) $\mathbf{V}_1 = 1\mathbf{i} + 4\mathbf{j}; \mathbf{V}_2 = 9\mathbf{i} - 3\mathbf{j}$

(c) $\mathbf{V}_1 = 7\mathbf{i} - 3\mathbf{j}; \mathbf{V}_2 = 3\mathbf{i} + 4\mathbf{j}$

(d) $\mathbf{V}_1 = 2\mathbf{i} - 3\mathbf{j} + 6\mathbf{k}; \mathbf{V}_2 = 8\mathbf{i} + 2\mathbf{j} - 3\mathbf{k}$

(e) $\mathbf{V}_1 = 6\mathbf{i} + 2\mathbf{j} + 3\mathbf{k}; \mathbf{V}_2 = \mathbf{i} + \mathbf{k}$

2. What is the angle between the vectors in Probs. 1(a), 1(b), and 1(c)?

3. Calculate the cross products for each of the five vector pairs in Prob. 1.

SOLUTIONS

1. (a)

$$\begin{aligned}\mathbf{V}_1 \cdot \mathbf{V}_2 &= \mathbf{V}_{1x}\mathbf{V}_{2x} + \mathbf{V}_{1y}\mathbf{V}_{2y} \\ &= (2)(5) + (3)(-2) \\ &= \boxed{4}\end{aligned}$$

(b)

$$\begin{aligned}\mathbf{V}_1 \cdot \mathbf{V}_2 &= (1)(9) + (4)(-3) \\ &= \boxed{-3}\end{aligned}$$

(c)

$$\begin{aligned}\mathbf{V}_1 \cdot \mathbf{V}_2 &= (7)(3) + (-3)(4) \\ &= \boxed{9}\end{aligned}$$

(d)

$$\begin{aligned}\mathbf{V}_1 \cdot \mathbf{V}_2 &= \mathbf{V}_{1x}\mathbf{V}_{2x} + \mathbf{V}_{1y}\mathbf{V}_{2y} + \mathbf{V}_{1z}\mathbf{V}_{2z} \\ &= (2)(8) + (-3)(2) + (6)(-3) \\ &= \boxed{-8}\end{aligned}$$

(e)

$$\begin{aligned}\mathbf{V}_1 \cdot \mathbf{V}_2 &= (6)(1) + (2)(0) + (3)(1) \\ &= \boxed{9}\end{aligned}$$

2. (a)

$$\begin{aligned}\cos\phi &= \frac{\mathbf{V}_1 \cdot \mathbf{V}_2}{|\mathbf{V}_1||\mathbf{V}_2|} \\ &= \frac{4}{\left(\sqrt{2^2 + 3^2}\right)\left(\sqrt{5^2 + (-2)^2}\right)} = 0.206\end{aligned}$$

$$\phi = \cos^{-1}(0.206) = \boxed{78.1^\circ}$$

(b)

$$\begin{aligned}\cos\phi &= \frac{\mathbf{V}_1 \cdot \mathbf{V}_2}{|\mathbf{V}_1||\mathbf{V}_2|} \\ &= \frac{-3}{\left(\sqrt{1^2 + 4^2}\right)\left(\sqrt{9^2 + (-3)^2}\right)} \\ &= -0.077\end{aligned}$$

$$\phi = \boxed{94.4^\circ}$$

(c)

$$\begin{aligned}\cos\phi &= \frac{\mathbf{V}_1 \cdot \mathbf{V}_2}{|\mathbf{V}_1||\mathbf{V}_2|} \\ &= \frac{9}{(\sqrt{7^2 + (-3)^2})(\sqrt{3^2 + 4^2})} = 0.236\end{aligned}$$

$$\phi = \boxed{76.3^\circ}$$

3. (a)

$$\mathbf{V}_1 \times \mathbf{V}_2 = \begin{vmatrix} \mathbf{i} & \mathbf{V}_{1x} & \mathbf{V}_{2x} \\ \mathbf{j} & \mathbf{V}_{1y} & \mathbf{V}_{2y} \\ \mathbf{k} & \mathbf{V}_{1z} & \mathbf{V}_{2z} \end{vmatrix} = \begin{vmatrix} \mathbf{i} & 2 & 5 \\ \mathbf{j} & 3 & -2 \\ \mathbf{k} & 0 & 0 \end{vmatrix}$$

Expand by the third row.

$$\mathbf{V}_1 \times \mathbf{V}_2 = \mathbf{k}\begin{vmatrix} 2 & 5 \\ 3 & -2 \end{vmatrix} = \boxed{-19\mathbf{k}}$$

(b)

$$\mathbf{V}_1 \times \mathbf{V}_2 = \begin{vmatrix} \mathbf{i} & 1 & 9 \\ \mathbf{j} & 4 & -3 \\ \mathbf{k} & 0 & 0 \end{vmatrix}$$

Expand by the third row.

$$\mathbf{V}_1 \times \mathbf{V}_2 = \mathbf{k}\begin{vmatrix} 1 & 9 \\ 4 & -3 \end{vmatrix} = \boxed{-39\mathbf{k}}$$

(c)

$$\mathbf{V}_1 \times \mathbf{V}_2 = \begin{vmatrix} \mathbf{i} & 7 & 3 \\ \mathbf{j} & -3 & 4 \\ \mathbf{k} & 0 & 0 \end{vmatrix}$$

Expand by the third row.

$$\mathbf{V}_1 \times \mathbf{V}_2 = \mathbf{k}\begin{vmatrix} 7 & 3 \\ -3 & 4 \end{vmatrix} = \boxed{37\mathbf{k}}$$

(d)

$$\mathbf{V}_1 \times \mathbf{V}_2 = \begin{vmatrix} \mathbf{i} & 2 & 8 \\ \mathbf{j} & -3 & 2 \\ \mathbf{k} & 6 & -3 \end{vmatrix}$$

Expand by the first column.

$$\mathbf{V}_1 \times \mathbf{V}_2 = \mathbf{i}\begin{vmatrix} -3 & 2 \\ 6 & -3 \end{vmatrix} - \mathbf{j}\begin{vmatrix} 2 & 8 \\ 6 & -3 \end{vmatrix} + \mathbf{k}\begin{vmatrix} 2 & 8 \\ -3 & 2 \end{vmatrix} = \boxed{-3\mathbf{i} + 54\mathbf{j} + 28\mathbf{k}}$$

(e)

$$\mathbf{V}_1 \times \mathbf{V}_2 = \begin{vmatrix} \mathbf{i} & 6 & 1 \\ \mathbf{j} & 2 & 0 \\ \mathbf{k} & 3 & 1 \end{vmatrix}$$

Expand by the second row.

$$\mathbf{V}_1 \times \mathbf{V}_2 = -\mathbf{j}\begin{vmatrix} 6 & 1 \\ 3 & 1 \end{vmatrix} + (2)\begin{vmatrix} \mathbf{i} & 1 \\ \mathbf{k} & 1 \end{vmatrix} = \boxed{2\mathbf{i} - 3\mathbf{j} - 2\mathbf{k}}$$

6 Trigonometry

PRACTICE PROBLEMS

1. A 5 lbm (5 kg) block sits on a 20° incline without slipping. (a) Draw the freebody with respect to axes parallel and perpendicular to the surface of the incline. (b) Determine the magnitude of the frictional force on the block.

(A) 1.71 lbf (16.8 N)
(B) 3.35 lbf (32.9 N)
(C) 4.70 lbf (46.1 N)
(D) 5.00 lbf (49.1 N)

SOLUTIONS

1. (a)

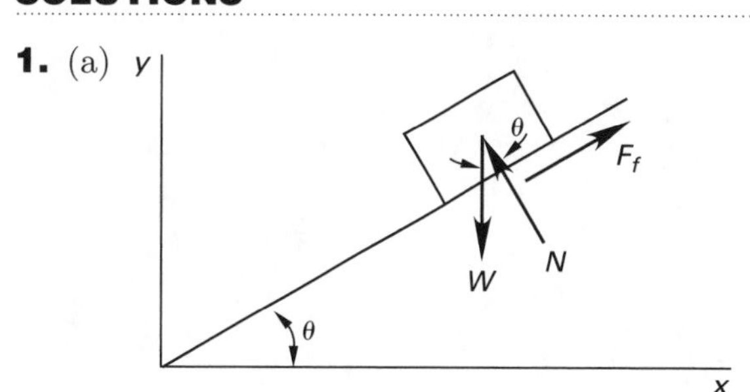

Customary U.S. Solution

(b) The mass of the block is $m = 5$ lbm.

The angle of inclination is $\theta = 20°$. The weight is

$$\begin{aligned} W &= \frac{mg}{g_c} \\ &= \frac{(5\ \text{lbm})\left(32.2\ \frac{\text{ft}}{\text{sec}^2}\right)}{32.2\ \frac{\text{lbm-ft}}{\text{lbf-sec}^2}} \\ &= 5\ \text{lbf} \end{aligned}$$

The frictional force is

$$\begin{aligned} F_f &= W\sin\theta \\ &= (5\ \text{lbf})(\sin 20°) \\ &= \boxed{1.71\ \text{lbf}} \end{aligned}$$

The answer is (A).

SI Solution

(b) The mass of the block is $m = 5$ kg.

The angle of inclination is $\theta = 20°$. The gravitational force is

$$\begin{aligned} W &= mg \\ &= (5\ \text{kg})\left(9.81\ \frac{\text{m}}{\text{s}^2}\right) \\ &= 49.1\ \text{N} \end{aligned}$$

The frictional force is

$$\begin{aligned} F_f &= W\sin\theta \\ &= (49.1\ \text{N})(\sin 20°) \\ &= \boxed{16.8\ \text{N}} \end{aligned}$$

The answer is (A).

7 Analytic Geometry

PRACTICE PROBLEMS

1. The diameter of a sphere and the base of a cone are equal. What percentage of that diameter must the cone's height be so that both volumes are equal?

(A) 133%
(B) 150%
(C) 166%
(D) 200%

SOLUTIONS

1. Let d be the diameter of the sphere and the base of the cone.

The volume of the sphere is given by

$$\begin{aligned} V_{\text{sphere}} &= \left(\tfrac{4}{3}\right)\pi r^3 = \left(\tfrac{4}{3}\right)\pi\left(\frac{d}{2}\right)^3 \\ &= \left(\frac{\pi}{6}\right)d^3 \end{aligned}$$

The volume of the circular cone is given by

$$\begin{aligned} V_{\text{cone}} &= \left(\tfrac{1}{3}\right)\pi r^2 h = \left(\tfrac{1}{3}\right)\pi\left(\frac{d}{2}\right)^2 h \\ &= \left(\frac{\pi}{12}\right)d^2 h \end{aligned}$$

Since the volume of the sphere and cone are equal,

$$\begin{aligned} V_{\text{cone}} &= V_{\text{sphere}} \\ \left(\frac{\pi}{12}\right)d^2 h &= \left(\frac{\pi}{6}\right)d^3 \\ h &= 2d \end{aligned}$$

The height of the cone must be 200% of the diameter.

The answer is (D).

8 Differential Calculus

PRACTICE PROBLEMS

1. What are the values of a, b, and c in the following expression such that $n(\infty) = 100$, $n(0) = 10$, and $dn(0)/dt = 0.5$?

$$n(t) = \frac{a}{1 + be^{ct}}$$

(A) $a = 10,\ b = 9,\ c = 1$
(B) $a = 100,\ b = 10,\ c = 1.5$
(C) $a = 100,\ b = 9,\ c = -0.056$
(D) $a = 1000,\ b = 10,\ c = 0.056$

2. Find all minima, maxima, and inflection points for

$$y = x^3 - 9x^2 - 3$$

(A) maximum at $x = 0$
inflection at $x = 3$
minimum at $x = 6$
(B) maximum at $x = 0$
inflection at $x = -3$
minimum at $x = -6$
(C) minimum at $x = 0$
inflection at $x = 3$
maximum at $x = 6$
(D) minimum at $x = 3$
inflection at $x = 0$
maximum at $x = -3$

SOLUTIONS

1. If c is positive, then $n(\infty) = 0$, which is contrary to the data given. Therefore, $c \leq 0$. If $c = 0$, then $n(\infty) = a/(1 + b) = 100$, which is possible depending on a and b. However, $n(0) = a/(1+b)$ would also equal 100, which is contrary to the given data. Therefore, $c \neq 0$.

c must be less than 0.

Since $c < 0$, then $n(\infty) = a$, so $\boxed{a = 100.}$

Applying the condition $t = 0$ gives

$$n(0) = \frac{a}{1+b} = 10$$

Since $a = 100$,

$$\begin{aligned} n(0) &= \frac{100}{1+b} \\ 100 &= (10)(1+b) \\ 10 &= 1 + b \\ &\boxed{b = 9} \end{aligned}$$

Substitute the results for a and b into the expression.

$$n(t) = \frac{100}{1 + 9e^{ct}}$$

Take the first derivative.

$$\frac{d}{dt}n(t) = \left(\frac{100}{(1 + 9e^{ct})^2}\right)(-9ce^{ct})$$

Apply the initial condition.

$$\begin{aligned} \frac{d}{dt}n(0) = \left(\frac{100}{(1 + 9e^{c(0)})^2}\right)\left(-9ce^{c(0)}\right) &= 0.5 \\ \left(\frac{100}{(1+9)^2}\right)(-9c) &= 0.5 \\ (1)(-9c) &= 0.5 \\ c &= \frac{-0.5}{9} \\ &= \boxed{-0.0556} \end{aligned}$$

The answer is (C).

Substitute the terms a, b, and c into the expression.

$$n(t) = \frac{100}{1 + 9e^{-0.0556t}}$$

2. Determine the critical points by taking the first derivative of the function and setting it equal to zero.

$$\begin{aligned}\frac{dy}{dx} &= 3x^2 - 18x = 3x(x-6)\\ 3x(x-6) &= 0\\ x(x-6) &= 0\end{aligned}$$

The critical points are located at $x = 0$ and $x = 6$.

Determine the inflection points by setting the second derivative equal to zero. Take the second derivative.

$$\begin{aligned}\frac{d^2y}{dx^2} &= \left(\frac{d}{dx}\right)\left(\frac{dy}{dx}\right) = \frac{d}{dx}(3x^2 - 18x)\\ &= 6x - 18\end{aligned}$$

Set the second derivative equal to zero.

$$\begin{aligned}\frac{d^2y}{dx^2} &= 0 = 6x - 18 = (6)(x-3)\\ (6)(x-3) &= 0\\ x - 3 &= 0\\ x &= 3\end{aligned}$$

This inflection point is at $x = 3$.

Determine the local maximum and minimum by substituting the critical points into the expression for the second derivative.

At the critical point $x = 0$,

$$\begin{aligned}\left.\frac{d^2y}{dx^2}\right|_{x=0} &= (6)(x-3) = (6)(0-3)\\ &= -18\end{aligned}$$

Since $-18 < 0$, $x = 0$ is a local maximum.

At the critical point $x = 6$,

$$\begin{aligned}\left.\frac{d^2y}{dx^2}\right|_{x=6} &= (6)(x-3) = (6)(6-3)\\ &= 18\end{aligned}$$

Since $18 > 0$, $x = 6$ is a local minimum.

The answer is (A).

Integral Calculus

PRACTICE PROBLEMS

1. Find the integrals.

(a) $$\int \sqrt{1-x}\,dx$$

(b) $$\int \frac{x}{x^2+1}\,dx$$

(c) $$\int \frac{x^2}{x^2+x-6}\,dx$$

2. Calculate the definite integrals.

(a) $$\int_1^3 (x^2+4x)\,dx$$

(b) $$\int_{-2}^2 (x^3+1)\,dx$$

(c) $$\int_1^2 (4x^3-3x^2)\,dx$$

3. Find the area bounded by $x=1$, $x=3$, $y+x+1=0$, and $y=6x-x^2$.

4. Find a_0 for the two waveforms shown.

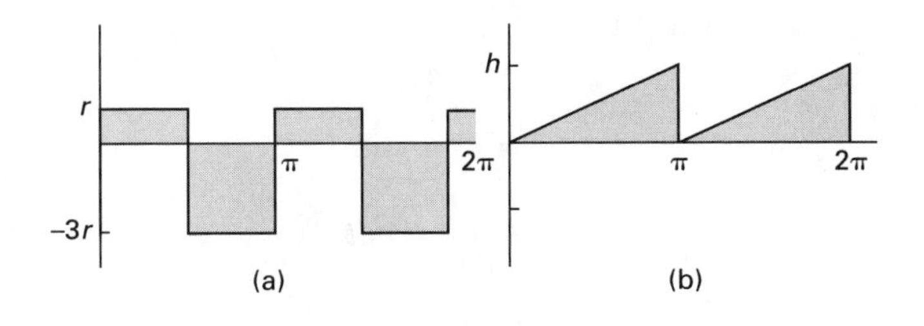

5. For the two waveforms shown, determine if their Fourier series is of type A, B, or C.

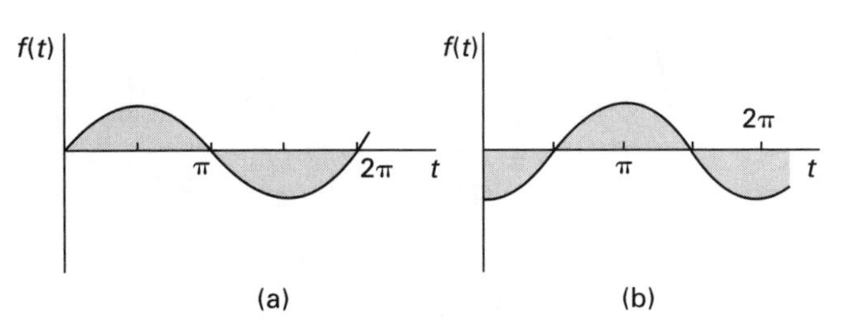

$$\begin{aligned}\text{type A: } f(t) &= a_0 + a_2\cos 2t + b_2\sin 2t \\ &\quad + a_4\cos 4t + b_4\sin 4t + \cdots \\ \text{type B: } f(t) &= a_0 + b_1\sin t + b_2\sin 2t + b_3\sin 3t + \cdots \\ \text{type C: } f(t) &= a_0 + a_1\cos t + a_2\cos 2t + a_3\cos 3t + \cdots\end{aligned}$$

SOLUTIONS

1. (a)

$$\int \sqrt{1-x}\,dx = \int (1-x)^{\frac{1}{2}}dx$$

$$= \boxed{\left(-\frac{2}{3}\right)(1-x)^{\frac{3}{2}} + C}$$

(b)

$$\int \frac{x}{x^2+1}dx = \frac{1}{2}\int \frac{2x}{x^2+1}dx$$

$$= \boxed{\frac{1}{2}\ln\left|(x^2+1)\right| + C}$$

(c)

$$\frac{x^2}{x^2+x-6} = 1 - \frac{x-6}{x^2+x-6}$$

$$= 1 - \frac{x-6}{(x+3)(x-2)}$$

$$= 1 - \frac{\frac{9}{5}}{x+3} + \frac{\frac{4}{5}}{x-2}$$

$$\int \frac{x^2}{x^2+x-6}dx = \int \left(1 - \frac{\frac{9}{5}}{x+3} + \frac{\frac{4}{5}}{x-2}\right)dx$$

$$= \int dx - \int \frac{\frac{9}{5}}{x+3}dx + \int \frac{\frac{4}{5}}{x-2}dx$$

$$= \boxed{x - \frac{9}{5}\ln\left|(x+3)\right| + \frac{4}{5}\ln\left|(x-2)\right| + C}$$

2. (a) $\int_1^3 (x^2+4x)dx = \left[\frac{x^3}{3} + 2x^2\right]_1^3 = \boxed{24\frac{2}{3}}$

(b) $\int_{-2}^2 (x^3+1)dx = \left[\frac{x^4}{4} + x\right]_{-2}^2 = \boxed{4}$

(c) $\int_1^2 (4x^3 - 3x^2)dx = \left[x^4 - x^3\right]_1^2 = \boxed{8}$

3.

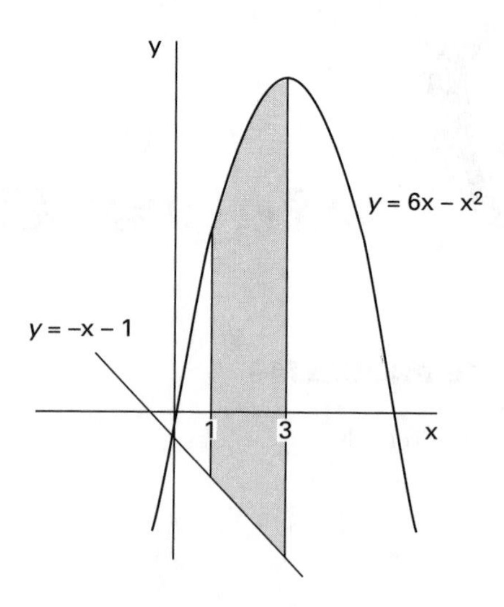

$$\text{area} = \int_1^3 \left((6x - x^2) - (-x-1)\right)dx$$

$$= \int_1^3 (-x^2 + 7x + 1)dx$$

$$= \left[-\frac{x^3}{3} + \frac{7}{2}x^2 + x\right]_1^3 = \boxed{21\frac{1}{3}}$$

4. For waveform (a):

$$a_0 = \frac{1}{2\pi}\int_0^{2\pi} f(t)dt$$

$$= \frac{1}{\pi}\int_0^{\pi} f(t)dt$$

$$= \frac{1}{\pi}\left((r)\left(\frac{\pi}{2}\right) + (-3r)\left(\frac{\pi}{2}\right)\right) = \boxed{-r}$$

For waveform (b):

$$a_0 = \frac{1}{2\pi}\int_0^{2\pi} f(t)dt$$

$$= \frac{1}{\pi}\int_0^{\pi} f(t)dt$$

$$= \left(\frac{1}{\pi}\right)\left(\frac{1}{2}\pi h\right)$$

$$= \boxed{h/2}$$

5. For waveform (a): Since $f(t) = -f(-t)$, $\boxed{\text{it is type B.}}$

For waveform (b): Since $f(t) = f(-t)$, $\boxed{\text{it is type C.}}$

10 Differential Equations

PRACTICE PROBLEMS

1. Solve the following differential equation for y.

$$y'' - 4y' - 12y = 0$$

(A) $A_1e^{6x} + A_2e^{-2x}$
(B) $A_1e^{-6x} + A_2e^{2x}$
(C) $A_1e^{6x} + A_2e^{2x}$
(D) $A_1e^{-6x} + A_2e^{-2x}$

2. Solve the following differential equation for y.

$$y' - y = 2xe^{2x} \qquad y(0) = 1$$

(A) $y = 2e^{-2x}(x-1) + 3e^{-x}$
(B) $y = 2e^{2x}(x-1) + 3e^{x}$
(C) $y = -2e^{-2x}(x-1) + 3e^{-x}$
(D) $y = 2e^{2x}(x-1) + 3e^{-x}$

3. The oscillation exhibited by the top story of a certain building in free motion is given by the following differential equation.

$$x'' + 2x' + 2x = 0 \qquad x(0) = 0 \qquad x'(0) = 1$$

(a) What is x as a function of time?

(A) $e^{-2t}\sin t$
(B) $e^{t}\sin t$
(C) $e^{-t}\sin t$
(D) $e^{-t}\sin t + e^{-t}\cos t$

(b) What is the building's fundamental natural frequency of vibration?

(A) $1/2$
(B) 1
(C) $\sqrt{2}$
(D) 2

(c) What is the amplitude of oscillation?

(A) 0.32
(B) 0.54
(C) 1.7
(D) 6.6

(d) What is x as a function of time if a lateral wind load is applied with a form of $\sin t$?

(A) $\frac{6}{5}e^{-t}\sin t + \frac{2}{5}e^{-t}\cos t$
(B) $\frac{6}{5}e^{t}\sin t + \frac{2}{5}e^{t}\cos t$
(C) $\frac{2}{5}e^{-t}\sin t + \frac{6}{5}e^{-t}\cos t + \frac{2}{5}\sin t - \frac{1}{5}\cos t$
(D) $\frac{6}{5}e^{-t}\sin t + \frac{2}{5}e^{-t}\cos t + \frac{1}{5}\sin t - \frac{2}{5}\cos t$

4. (*Time limit: one hour*) A 90 lbm (40 kg) bag of a chemical is accidentally dropped in an aerating lagoon. The chemical is water soluble and nonreacting. The lagoon is 120 ft (35 m) in diameter and filled to a depth of 10 ft (3 m). The aerators circulate and distribute the chemical evenly throughout the lagoon.

Water enters the lagoon at a rate of 30 gal/min (115 L/min). Fully mixed water is pumped into a reservoir at a rate of 30 gal/min (115 L/min).

The established safe concentration of this chemical is 1 ppb (part per billion). How many days will it take for the concentration of the discharge water to reach this level?

(A) 25 days
(B) 50 days
(C) 100 days
(D) 200 days

5. A tank contains 100 gal (100 L) of brine made by dissolving 60 lbm (60 kg) of salt in pure water. Salt water with a concentration of 1 lbm/gal (1 kg/L) enters the tank at a rate of 2 gal/min (2 L/min). A well-stirred mixture is drawn from the tank at a rate of 3 gal/min (3 L/min). Find the mass of salt in the tank after 1 hr.

(A) 13 lbm (13 kg)
(B) 37 lbm (37 kg)
(C) 43 lbm (43 kg)
(D) 51 lbm (51 kg)

SOLUTIONS

1. Obtain the characteristic equation by replacing each derivative with a polynomial term of equal degree.

$$r^2 - 4r - 12 = 0$$

Factor the characteristic equation.

$$(r - 6)(r + 2) = 0$$

The roots are $r_1 = 6$ and $r_2 = -2$.

Since the roots are real and distinct, the solution is

$$\begin{aligned} y &= A_1 e^{r_1 x} + A_2 e^{r_2 x} \\ &= \boxed{A_1 e^{6x} + A_2 e^{-2x}} \end{aligned}$$

The answer is (A).

2. The equation is a first-order linear differential equation of the form

$$\begin{aligned} y' + p(x)y &= g(x) \\ p(x) &= -1 \\ g(x) &= 2xe^{2x} \end{aligned}$$

The integration factor $u(x)$ is given by

$$\begin{aligned} u(x) &= \exp\left(\int p(x)dx\right) \\ &= \exp\left(\int(-1)dx\right) \\ &= e^{-x} \end{aligned}$$

The closed form of the solution is given by

$$\begin{aligned} y &= \left(\frac{1}{u(x)}\right)\left(\int u(x)g(x)dx + C\right) \\ &= \left(\frac{1}{e^{-x}}\right)\left(\int (e^{-x})(2xe^{2x})dx + C\right) \\ &= e^x\left(2(xe^x - e^x) + C\right) \\ &= e^x\left(2e^x(x-1) + C\right) \\ &= 2e^{2x}(x-1) + Ce^x \end{aligned}$$

Apply the initial condition $y(0) = 1$ to obtain the integration constant C.

$$\begin{aligned} y(0) = 2e^{(2)(0)}(0-1) + Ce^0 &= 1 \\ (2)(1)(-1) + C(1) &= 1 \\ -2 + C &= 1 \\ C &= 3 \end{aligned}$$

Substituting in the value for the integration constant C, the solution is

$$y = \boxed{2e^{2x}(x-1) + 3e^x}$$

The answer is (B).

3. (a) The differential equation is a homogeneous second-order linear differential equation with constant coefficients. Write the characteristic equation.

$$r^2 + 2r + 2 = 0$$

This is a quadratic equation of the form $ar^2 + br + c = 0$ where $a = 1$, $b = 2$, and $c = 2$.

Solve for r.

$$\begin{aligned} r &= \frac{-b \pm \sqrt{b^2 - 4ac}}{2a} \\ &= \frac{-2 \pm \sqrt{(2)^2 - (4)(1)(2)}}{(2)(1)} \\ &= \frac{-2 \pm \sqrt{4-8}}{2} \\ &= \frac{-2 \pm \sqrt{-4}}{2} \\ &= \frac{-2 \pm 2\sqrt{-1}}{2} \\ &= -1 \pm \sqrt{-1} \\ &= -1 \pm i \\ r_1 &= -1 + i \text{and } r_2 = -1 - i \end{aligned}$$

Since the roots are imaginary and of the form $\alpha + i\omega$ and $\alpha - i\omega$ where $\alpha = -1$ and $\omega = 1$, the general form of the solution is given by

$$\begin{aligned} x(t) &= A_1 e^{\alpha t}\cos\omega t + A_2 e^{\alpha t}\sin\omega t \\ &= A_1 e^{-1t}\cos(1t) + A_2 e^{-1t}\sin(1t) \\ &= A_1 e^{-t}\cos t + A_2 e^{-t}\sin t \end{aligned}$$

Apply the initial conditions $x(0) = 0$ and $x'(0) = 1$ to solve for A_1 and A_2.

First, apply the initial condition $x(0) = 0$.

$$\begin{aligned} x(t) = A_1 e^0 \cos 0 + A_2 e^0 \sin 0 &= 0 \\ A_1(1)(1) + A_2(1)(0) &= 0 \\ A_1 &= 0 \end{aligned}$$

Substituting, the solution of the differential equation becomes

$$x(t) = A_2 e^{-t}\sin t$$

To apply the second initial condition, take the first derivative.

$$\begin{aligned} x'(t) &= \frac{d}{dt}(A_2 e^{-t}\sin t) \\ &= A_2\frac{d}{dt}(e^{-t}\sin t) \\ &= A_2\left(\sin t\frac{d}{dt}(e^{-t}) + e^{-t}\frac{d}{dt}\sin t\right) \\ &= A_2\left(\sin t(-e^{-t}) + e^{-t}(\cos t)\right) \\ &= A_2(e^{-t})(-\sin t + \cos t) \end{aligned}$$

Apply the initial condition, $x'(0) = 1$.

$$\begin{aligned} x(0) = A_2(e^0)(-\sin 0 + \cos 0) &= 1 \\ A_2(1)(0+1) &= 1 \\ A_2 &= 1 \end{aligned}$$

The solution is

$$\begin{aligned} x(t) &= A_2 e^{-t}\sin t \\ &= (1)e^{-t}\sin t \\ &= \boxed{e^{-t}\sin t} \end{aligned}$$

The answer is (C).

(b) To determine the natural frequency, set the damping term to zero. The equation has the form

$$x'' + 2x = 0$$

This equation has a general solution of the form

$$x(t) = x_0\cos\omega t + \left(\frac{v_0}{\omega}\right)\sin\omega t$$

ω is the natural frequency. Given the equation $x''+2x = 0$, the characteristic equation is

$$\begin{aligned} r^2 + 2 &= 0 \\ r &= \sqrt{-2} \\ &= \pm\sqrt{2}i \end{aligned}$$

Since the roots are imaginary and of the form $\alpha + i\omega$ and $\alpha - i\omega$ where $\alpha = 0$ and $\omega = \sqrt{2}$, the general form of the solution is given by

$$\begin{aligned} x(t) &= A_1 e^{\alpha t}\cos\omega t + A_2 e^{\alpha t}\sin\omega t \\ &= A_1 e^{0t}\cos\sqrt{2}t + A_2 e^{0t}\sin\sqrt{2}t \\ &= A_1(1)\cos\sqrt{2}t + A_2(1)\sin\sqrt{2}t \\ &= A_1\cos\sqrt{2}t + A_2\sin\sqrt{2}t \end{aligned}$$

Apply the initial conditions, $x(0) = 0$ and $x'(0) = 1$ to solve for A_1 and A_2. Applying the initial condition $x(0) = 0$ gives

$$\begin{aligned} x(0) = A_1\cos\left((\sqrt{2})(0)\right) + A_2\sin(\sqrt{2})(0) &= 0 \\ A_1\cos 0 + A_2\sin 0 &= 0 \\ A_1(1) + A_2(0) &= 0 \\ A_1 &= 0 \end{aligned}$$

Substituting, the solution of the differential equation becomes

$$x(t) = A_2\sin\sqrt{2}t$$

To apply the second initial condition, take the first derivative.

$$\begin{aligned} x'(t) &= \frac{d}{dt}(A_2\sin\sqrt{2}t) \\ &= A_2\sqrt{2}\cos\sqrt{2}t \end{aligned}$$

Apply the second initial condition, $x'(0) = 1$.

$$\begin{aligned} x'(0) = A_2\sqrt{2}\cos(\sqrt{2})(0) &= 1 \\ A_2\sqrt{2}\cos(0) &= 1 \\ A_2(\sqrt{2})(1) &= 1 \\ A_2\sqrt{2} &= 1 \\ A_2 = \frac{1}{\sqrt{2}} &= \frac{\sqrt{2}}{2} \end{aligned}$$

Substituting, the undamped solution becomes

$$x(t) = \left(\frac{\sqrt{2}}{2}\right)\sin\sqrt{2}t$$

Therefore, the undamped natural frequency is $\boxed{\omega = \sqrt{2}.}$

The answer is (C).

(c) The amplitude of the oscillation is the maximum displacement.

Take the derivative of the solution, $x(t) = e^{-t}\sin t$.

$$\begin{aligned} x'(t) &= \frac{d}{dt}(e^{-t}\sin t) \\ &= \sin t\frac{d}{dt}(e^{-t}) + e^{-t}\frac{d}{dt}\sin t \\ &= \sin(t)(-e^{-t}) + e^{-t}\cos t \\ &= e^{-t}(\cos t - \sin t) \end{aligned}$$

The maximum displacement occurs at $x'(t) = 0$.

Since $e^{-t} \neq 0$ except as t approaches infinity,

$$\begin{aligned} \cos t - \sin t &= 0 \\ \tan t &= 1 \\ t &= \tan^{-1}(1) \\ &= 0.785 \text{ rad} \end{aligned}$$

At $t = 0.785$ rad, the displacement is maximum. Substitute into the orginal solution to obtain a value for the maximum displacement.

$$\begin{aligned} x(0.785) &= e^{-0.785}\sin(0.785) \\ &= 0.322 \end{aligned}$$

The amplitude is $\boxed{0.322.}$

The answer is (A).

(d) (An alternative solution using Laplace transforms follows this solution.) The application of a lateral wind load with the form $\sin t$ revises the differential equation to the form

$$x'' + 2x' + 2x = \sin t$$

Express the solution as the sum of the complementary x_c and particular x_p solutions.

$$x(t) = x_c(t) + x_p(t)$$

From part (a),

$$x_c(t) = A_1e^{-t}\cos t + A_2e^{-t}\sin t$$

The general form of the particular solution is given by

$$x_p(t) = x^s(A_3\cos t + A_4\sin t)$$

Determine the value of s; check to see if the terms of the particular solution solve the homogeneous equation.

Examine the term $A_3\cos(t)$.

Take the first derivative.

$$\frac{d}{dx}(A_3\cos t) = -A_3\sin t$$

Take the second derivative.

$$\begin{aligned} \frac{d}{dx}\left(\frac{d}{dx}(A_3\cos t)\right) &= \frac{d}{dx}(-A_3\sin t) \\ &= -A_3\cos t \end{aligned}$$

Substitute the terms into the homogeneous equation.

$$\begin{aligned} x'' + 2x' + 2x &= -A_3\cos t + (2)(-A_3\sin t) \\ &\quad + (2)(-A_3\cos t) \\ &= A_3\cos t - 2A_3\sin t \\ &\neq 0 \end{aligned}$$

Except for the trival solution $A_3 = 0$, the term $A_3\cos t$ does not solve the homogeneous equation.

Examine the second term $A_4\sin t$.

Take the first derivative.

$$\frac{d}{dx}(A_4\sin t) = A_4\cos t$$

Take the second derivative.

$$\begin{aligned} \frac{d}{dx}\left(\frac{d}{dx}(A_4\sin t)\right) &= \frac{d}{dx}(A_4\cos t) \\ &= -A_4\sin t \end{aligned}$$

Substitute the terms into the homogeneous equation.

$$\begin{aligned} x'' + 2x' + 2x &= -A_4\sin t + (2)(A_4\cos t) \\ &\quad + (2)(A_4\sin t) \\ &= A_4\sin t + 2A_4\cos t \\ &\neq 0 \end{aligned}$$

Except for the trival solution $A_4 = 0$, the term $A_4\sin t$ does not solve the homogeneous equation.

Neither of the terms satisfies the homogeneous equation $s = 0$; therefore, the particular solution is of the form

$$x_p(t) = A_3\cos t + A_4\sin t$$

Use the method of undetermined coefficients to solve for A_3 and A_4. Take the first derivative.

$$\begin{aligned} x_p'(t) &= \frac{d}{dx}(A_3\cos t + A_4\sin t) \\ &= -A_3\sin t + A_4\cos t \end{aligned}$$

Take the second derivative.

$$\begin{aligned} x_p''(t) &= \frac{d}{dx}\left(\frac{d}{dx}(A_3\cos t + A_4\sin t)\right) \\ &= \frac{d}{dx}(-A_3\sin t + A_4\cos t) \\ &= -A_3\cos t - A_4\sin t \end{aligned}$$

Substitute the expressions for the derivatives into the differential equation.

$$\begin{aligned} x'' + 2x' + 2x &= (-A_3\cos t - A_4\sin t) \\ &\quad + (2)(-A_3\sin t + A_4\cos t) \\ &\quad + (2)(A_3\cos t + A_4\sin t) \\ &= \sin t \end{aligned}$$

Rearranging terms gives

$$\begin{gathered} (-A_3 + 2A_4 + 2A_3)\cos t \\ +(-A_4 - 2A_3 + 2A_4)\sin t = \sin t \\ (A_3 + 2A_4)\cos t + (-2A_3 + A_4)\sin t = \sin t \end{gathered}$$

Equating coefficients gives

$$\begin{aligned} A_3 + 2A_4 &= 0 \\ -2A_3 + A_4 &= 1 \end{aligned}$$

Multiplying the first equation by 2 and adding equations gives

$$\begin{aligned} A_3 + 2A_4 &= 0 \\ +(-2A_3 + A_4) &= 1 \\ \hline 5A_4 = 1 \ \text{or}\ A_4 &= \tfrac{1}{5} \end{aligned}$$

From the first equation for $A_4 = {}^1/_5$, $A_3 + (2)({}^1/_5) = 0$ and $A_3 = -{}^2/_5$.

Substituting for the coefficients, the particular solution becomes

$$x_p(t) = -\tfrac{2}{5}\cos t + \tfrac{1}{5}\sin t$$

Combining the complementary and particular solutions gives

$$\begin{aligned} x(t) &= x_c(t) + x_p(t) \\ &= A_1e^{-t}\cos t + A_2e^{-t}\sin t - \tfrac{2}{5}\cos t + \tfrac{1}{5}\sin t \end{aligned}$$

Apply the initial conditions to solve for the coefficients A_1 and A_2; then apply the first initial condition, $x(0) = 0$.

$$\begin{aligned} x(t) = A_1e^0\cos 0 + A_2e^0\sin 0 \\ - tfrac25\cos 0 + \tfrac{1}{5}\sin 0 = 0 \\ A_1(1)(1) + A_2(1)(0) + \left(-\tfrac{2}{5}\right)(1) + \left(\tfrac{1}{5}\right)(0) = 0 \\ A_1 - \tfrac{2}{5} = 0 \\ A_1 = \tfrac{2}{5} \end{aligned}$$

Substituting for A_1, the solution becomes

$$x(t) = \tfrac{2}{5}e^{-t}\cos t + A_2e^{-t}\sin t - \tfrac{2}{5}\cos t + \tfrac{1}{5}\sin t$$

Take the first derivative.

$$\begin{aligned} x'(t) &= \frac{d}{dx}\left(\tfrac{2}{5}e^{-t}\cos t + A_2e^{-t}\sin t\right) \\ &\quad + \left(\left(-\tfrac{2}{5}\right)\cos t + \tfrac{1}{5}\sin t\right) \\ &= \left(\tfrac{2}{5}\right)\left(-e^{-t}\cos t - e^{-t}\sin t\right) \\ &\quad + A_2(-e^{-t}\sin t + e^{-t}\cos t) \\ &\quad + \left(-\tfrac{2}{5}\right)(-\sin t) + \tfrac{1}{5}\cos t \end{aligned}$$

Apply the second initial condition, $x'(0) = 1$.

$$\begin{aligned} x'(0) &= \left(\tfrac{2}{5}\right)\left(-e^0\cos 0 - e^0\sin 0\right) \\ &\quad + A_2(-e^0\sin 0 + e^0\cos 0) \\ &\quad + \left(-\tfrac{2}{5}\right)(-\sin 0) \ + \tfrac{1}{5}\cos 0 \\ &= 1 \end{aligned}$$

$$\begin{aligned} \left(\tfrac{2}{5}\right)\big(-(1)(1) - (1)(0)\big) + A_2\big(-(1)(0) \\ +(1)(1)\big) + \left(-\tfrac{2}{5}\right)(0) + \left(\tfrac{1}{5}\right)(1) = 1 \\ \left(\tfrac{2}{5}\right)(-1) + A_2(1) + \left(\tfrac{1}{5}\right) = 1 \\ A_2 = \tfrac{6}{5} \end{aligned}$$

Substituting for A_2, the solution becomes

$$x(t) = \boxed{\begin{aligned} &\tfrac{2}{5}e^{-t}\cos t + \tfrac{6}{5}e^{-t}\sin t \\ &-\tfrac{2}{5}\cos t + \tfrac{1}{5}\sin t \end{aligned}}$$

The answer is (D).

(d) *Alternate solution:*

Use the Laplace transform method.

$$\begin{aligned} x'' + 2x' + 2x &= \sin t \\ \mathcal{L}(x'') + 2\mathcal{L}(x') + 2\mathcal{L}(x) &= \mathcal{L}(\sin t) \\ s^2\mathcal{L}(x) - 1 + 2s\mathcal{L}(x) + 2\mathcal{L}(x) &= \frac{1}{s^2+1} \\ \mathcal{L}(x)(s^2 + 2s + 2) - 1 &= \frac{1}{s^2+1} \end{aligned}$$

$$\begin{aligned} \mathcal{L}(x) &= \frac{1}{s^2+2s+2} + \frac{1}{(s^2+1)(s^2+2s+2)} \\ &= \frac{1}{(s+1)^2+1} + \frac{1}{(s^2+1)(s^2+2s+2)} \end{aligned}$$

Use partial fractions to expand the second term.

$$\frac{1}{(s^2+1)(s^2+2s+2)} = \frac{A_1 + B_1s}{s^2+1} + \frac{A_2 + B_2s}{s^2+2s+2}$$

Cross multiply.

$$\begin{aligned} &= \frac{\begin{array}{c} A_1s^2 + 2A_1s + 2A_1 + B_1s^3 + 2B_1s^2 \\ + A_2s^2 + A_2 + B_2s^3 + B_2s \end{array}}{(s^2+1)(s^2+2s+2)} \\ &= \frac{\begin{array}{c} s^3(B_1 + B_2) + s^2(A_1 + A_2 + 2B_1) \\ + s(2A_1 + 2B_1 + B_2) + 2A_1 + A_2 \end{array}}{(s^2+1)(s^2+2s+2)} \end{aligned}$$

Compare numerators to obtain the following four simultaneous equations.

$$\begin{array}{rrrrl} & & B_1 & + B_2 & = 0 \\ A_1 & + A_2 & + 2B_1 & & = 0 \\ 2A_1 & & + 2B_1 & + B_2 & = 0 \\ 2A_1 & + A_2 & & & = 1 \end{array}$$

Use Cramer's rule to find A_1.

$$A_1 = \frac{\begin{vmatrix} 0 & 0 & 1 & 1 \\ 0 & 1 & 2 & 0 \\ 0 & 0 & 2 & 1 \\ 1 & 1 & 0 & 0 \end{vmatrix}}{\begin{vmatrix} 0 & 0 & 1 & 1 \\ 1 & 1 & 2 & 0 \\ 2 & 0 & 2 & 1 \\ 2 & 1 & 0 & 0 \end{vmatrix}} = \frac{-1}{-5} = \frac{1}{5}$$

The rest of the coefficients are found similarly.

$$\begin{aligned} A_1 &= \tfrac{1}{5} \\ A_2 &= \tfrac{3}{5} \\ B_1 &= -\tfrac{2}{5} \\ B_2 &= \tfrac{2}{5} \end{aligned}$$

Then,

$$\begin{aligned} \mathcal{L}(x) = \frac{1}{(s+1)^2+1} + \frac{\frac{1}{5}}{s^2+1} + \frac{-\frac{2}{5}s}{s^2+1} \\ + \frac{\frac{3}{5}}{s^2+2s+2} + \frac{\frac{2}{5}s}{s^2+2s+2} \end{aligned}$$

Take the inverse transform.

$$\begin{aligned} x(t) &= \mathcal{L}^{-1}\{\mathcal{L}(x)\} \\ &= e^{-t}\sin t + \tfrac{1}{5}\sin t - \tfrac{2}{5}\cos t + \tfrac{3}{5}e^{-t}\sin t \\ &\quad + \tfrac{2}{5}(e^{-t}\cos t - e^{-t}\sin t) \\ &= \boxed{\tfrac{6}{5}e^{-t}\sin t + \tfrac{2}{5}e^{-t}\cos t + \tfrac{1}{5}\sin t - \tfrac{2}{5}\cos t} \end{aligned}$$

The answer is (D).

4. *Customary U.S. Solution*

The differential equation is given as

$$m'(t) = a(t) - \frac{m(t)o(t)}{V(t)}$$

$$\begin{aligned} a(t) &= \text{rate of addition of chemical} \\ m(t) &= \text{mass of chemical at time } t \\ o(t) &= \text{volumetric flow out of the lagoon} \\ &\quad (= 30 \text{ gal/min}) \\ V(t) &= \text{volume in the lagoon at time } t \end{aligned}$$

Water flows into the lagoon at a rate of 30 gal/min, and a water-chemical mix flows out of the lagoon at rate of 30 gal/min. Therefore, the volume of the lagoon at time t is equal to the initial volume.

$$\begin{aligned} V(t) &= \left(\frac{\pi}{4}\right)(\text{diameter of lagoon})^2(\text{depth of lagoon}) \\ &= \left(\frac{\pi}{4}\right)(120 \text{ ft})^2(10 \text{ ft}) \\ &= 113{,}097 \text{ ft}^3 \end{aligned}$$

Use a conversion factor of 7.48 gal/ft^3.

$$\begin{aligned} o(t) &= \frac{30\ \frac{\text{gal}}{\text{min}}}{7.48\ \frac{\text{gal}}{\text{ft}^3}} \\ &= 4.01 \text{ ft}^3/\text{min} \end{aligned}$$

Substituting into the general form of the differential equation gives

$$\begin{aligned} m'(t) &= a(t) - \frac{m(t)o(t)}{V(t)} \\ &= (0) - m(t)\left(\frac{4.01\ \frac{\text{ft}^3}{\text{min}}}{113{,}097 \text{ ft}^3}\right) \\ &= -\left(\frac{3.55\times 10^{-5}}{\text{min}}\right)m(t) \end{aligned}$$

$$m'(t) + \left(\frac{3.55\times 10^{-5}}{\text{min}}\right)m(t) = 0$$

The differential equation of the problem has a characteristic equation.

$$\begin{aligned} r + \frac{3.55\times 10^{-5}}{\text{min}} &= 0 \\ r &= -3.55\times 10^{-5}/\text{min} \end{aligned}$$

The general form of the solution is given by

$$m(t) = Ae^{rt}$$

Substituting for the root, r, gives

$$m(t) = Ae^{(-3.55\times 10^{-5}/\text{min})t}$$

Apply the initial condition $m(0) = 90$ lbm at time $t = 0$.

$$\begin{aligned} m(0) = Ae^{(-3.55\times 10^{-5}/\text{min})(0)} &= 90 \text{ lbm} \\ Ae^0 &= 90 \text{ lbm} \\ A &= 90 \text{ lbm} \end{aligned}$$

Therefore,

$$m(t) = (90 \text{ lbm})\,e^{(-3.55\times 10^{-5}/\text{min})t}$$

Solve for t.

$$\begin{aligned} \frac{m(t)}{90 \text{ lbm}} &= e^{(-3.55\times 10^{-5}/\text{min})t} \\ \ln\left(\frac{m(t)}{90 \text{ lbm}}\right) &= \ln\left(e^{(-3.55\times 10^{-5}/\text{min})t}\right) \\ &= \left(\frac{-3.55\times 10^{-5}}{\text{min}}\right)t \\ t &= \frac{\ln\left(\frac{m(t)}{90 \text{ lbm}}\right)}{\frac{-3.55\times 10^{-5}}{\text{min}}} \end{aligned}$$

The initial mass of the water in the lagoon is given by

$$\begin{aligned} m_i &= V\rho \\ &= (113{,}097\ \text{ft}^3)\left(62.4\ \frac{\text{lbm}}{\text{ft}^3}\right) \\ &= 7.05\times 10^6\ \text{lbm} \end{aligned}$$

The final mass of chemicals is achieved at a concentration of 1 ppb or

$$\begin{aligned} m_f &= \frac{7.06\times 10^6\ \text{lbm}}{1\times 10^9} \\ &= 7.06\times 10^{-3}\ \text{lbm} \end{aligned}$$

Find the time required to achieve a mass of 7.06×10^{-3} lbm.

$$\begin{aligned} t &= \left(\frac{\ln\left(\frac{m(t)}{90\ \text{lbm}}\right)}{\frac{-3.55\times 10^{-5}}{\text{min}}}\right)\left(\frac{1\ \text{hr}}{60\ \text{min}}\right)\left(\frac{1\ \text{day}}{24\ \text{hr}}\right) \\ &= \left(\frac{\ln\left(\frac{7.06\times 10^{-3}\ \text{lbm}}{90\ \text{lbm}}\right)}{\frac{-3.55\times 10^{-5}}{\text{min}}}\right)\left(\frac{1\ \text{hr}}{60\ \text{min}}\right)\left(\frac{1\ \text{day}}{24\ \text{hr}}\right) \\ &= \boxed{185\ \text{days}} \end{aligned}$$

The answer is (D).

SI Solution

The differential equation is given as

$$m'(t) = a(t) - \frac{m(t)o(t)}{V(t)}$$

$$\begin{aligned} a(t) &= \text{rate of addition of chemical} \\ m(t) &= \text{mass of chemical at time } t \\ o(t) &= \text{volumetric flow out of the lagoon} \\ &\quad (= 115\ \text{L/min}) \\ V(t) &= \text{volume in the lagoon at time } t \end{aligned}$$

Water flows into the lagoon at a rate of 115 L/min, and a water-chemical mix flows out of the lagoon at a rate of 115 L/min. Therefore, the volume of the lagoon at time t is equal to the initial volume.

$$\begin{aligned} V(t) &= \left(\frac{\pi}{4}\right)(\text{diameter of lagoon})^2(\text{depth of lagoon}) \\ &= \left(\frac{\pi}{4}\right)(35\ \text{m})^2(3\ \text{m}) \\ &= 2886\ \text{m}^3 \end{aligned}$$

Using a conversion factor of 1 m^3/1000 L gives

$$\begin{aligned} o(t) &= \left(115\ \frac{\text{L}}{\text{min}}\right)\left(\frac{1\ \text{m}^3}{1000\ \text{L}}\right) \\ &= 0.115\ \text{m}^3/\text{min} \end{aligned}$$

Substitute into the general form of the differential equation.

$$\begin{aligned} m'(t) &= a(t) - \frac{m(t)o(t)}{V(t)} \\ &= 0 - m(t)\left(\frac{0.115\ \frac{\text{m}^3}{\text{min}}}{2886\ \text{m}^3}\right) \\ &= -\left(\frac{3.985\times 10^{-5}}{\text{min}}\right)m(t) \end{aligned}$$

$$m'(t) + \left(\frac{3.985\times 10^{-5}}{\text{min}}\right)m(t) = 0$$

The differential equation of the problem has the following characteristic equation.

$$\begin{aligned} r + \frac{3.985\times 10^{-5}}{\text{min}} &= 0 \\ r &= -3.985\times 10^{-5}/\text{min} \end{aligned}$$

The general form of the solution is given by

$$m(t) = Ae^{rt}$$

Substituting in for the root, r, gives

$$m(t) = Ae^{(-3.985\times 10^{-5}/\text{min})t}$$

Apply the initial condition $m(0) = 40$ kg at time $t = 0$.

$$\begin{aligned} m(0) = Ae^{(-3.985\times 10^{-5}/\text{min})(0)} &= 40\ \text{kg} \\ Ae^0 &= 40\ \text{kg} \\ A &= 40\ \text{kg} \end{aligned}$$

Therefore,

$$m(t) = (40\ \text{kg})e^{(-3.985\times 10^{-5}/\text{min})t}$$

Solve for t.

$$\begin{aligned} \frac{m(t)}{40\ \text{kg}} &= e^{(-3.985\times 10^{-5}/\text{min})t} \\ \ln\left(\frac{m(t)}{40\ \text{kg}}\right) &= \ln\left(e^{(-3.985\times 10^{-5}/\text{min})t}\right) \\ &= \left(\frac{-3.985\times 10^{-5}}{\text{min}}\right)t \\ t &= \frac{\ln\left(\frac{m(t)}{40\ \text{kg}}\right)}{\frac{-3.985\times 10^{-5}}{\text{min}}} \end{aligned}$$

The initial mass of water in the lagoon is given by

$$\begin{aligned} m_i &= V\rho \\ &= (2886 \text{ m}^3)\left(1000 \ \frac{\text{kg}}{\text{m}^3}\right) \\ &= 2.886 \times 10^6 \text{ kg} \end{aligned}$$

The final mass of chemicals is achieved at a concentration of 1 ppb or

$$\begin{aligned} m_f &= \frac{2.886 \times 10^6 \text{ kg}}{1 \times 10^9} \\ &= 2.886 \times 10^{-3} \text{ kg} \end{aligned}$$

Find the time required to achieve a mass of 2.886×10^{-3} kg.

$$\begin{aligned} t &= \left(\frac{\ln \frac{m(t)}{40 \text{ kg}}}{\frac{-3.985 \times 10^{-5}}{\text{min}}}\right)\left(\frac{1 \text{ h}}{60 \text{ min}}\right)\left(\frac{1 \text{ day}}{24 \text{ h}}\right) \\ &= \left(\frac{\ln\left(\frac{2.886 \times 10^{-3} \text{ kg}}{40 \text{ kg}}\right)}{\frac{-3.985 \times 10^{-5}}{\text{min}}}\right)\left(\frac{1 \text{ h}}{60 \text{ min}}\right)\left(\frac{1 \text{ day}}{24 \text{ h}}\right) \\ &= \boxed{166 \text{ days}} \end{aligned}$$

The answer is (D).

5. Let

$$\begin{aligned} m(t) &= \text{mass of salt in tank at time } t \\ m_0 &= 60 \text{ mass units} \\ m'(t) &= \text{rate at which salt content is changing} \end{aligned}$$

Two mass units of salt enter each minute, and three volumes leave each minute. The amount of salt leaving each minute is

$$\begin{aligned} &\left(3 \ \frac{\text{vol}}{\text{min}}\right)\left(\text{concentration in } \frac{\text{mass}}{\text{vol}}\right) \\ &\qquad = \left(3 \ \frac{\text{vol}}{\text{min}}\right)\left(\frac{\text{salt content}}{\text{volume}}\right) \\ &\qquad = \left(3 \ \frac{\text{vol}}{\text{min}}\right)\left(\frac{m(t)}{100 - t}\right) \end{aligned}$$

$$\begin{aligned} m'(t) &= 2 - (3)\left(\frac{m(t)}{100 - t}\right) \text{ or } m'(t) + \frac{3m(t)}{100 - t} \\ &= 2 \text{ mass/min} \end{aligned}$$

This is a first-order linear differential equation. The integrating factor is

$$\begin{aligned} m &= \exp\left[3 \int \frac{dt}{100 - t}\right] \\ &= \exp\left[(3)\big(-\ln(100 - t)\big)\right] \\ &= (100 - t)^{-3} \\ m(t) &= (100 - t)^3 \left[2 \int \frac{dt}{(100 - t)^3} + k\right] \\ &= 100 - t + (k)(100 - t)^3 \end{aligned}$$

But $m = 60$ mass units at $t = 0$, so $k = -0.00004$.

$$m(t) = 100 - t - (0.00004)(100 - t)^3$$

At $t = 60$ min,

$$\begin{aligned} m &= 100 - 60 \text{ min} - (0.00004)(100 - 60 \text{ min})^3 \\ &= \boxed{37.44 \text{ mass units}} \end{aligned}$$

The answer is (B).

11 Probability and Statistical Analysis of Data

PRACTICE PROBLEMS

Probability

1. Four military recruits whose respective shoe sizes are 7, 8, 9, and 10 report to the supply clerk to be issued boots. The supply clerk selects one pair of boots in each of the four required sizes and hands them at random to the recruits.

(a) What is the probability that all recruits will receive boots of an incorrect size?

(A) 0.25
(B) 0.38
(C) 0.45
(D) 0.61

(b) What is the probability that exactly three recruits will receive boots of the correct size?

(A) 0
(B) 0.063
(C) 0.17
(D) 0.25

Probability Distributions

2. The time taken by a toll taker to collect the toll from vehicles crossing a bridge is an exponential distribution with a mean of 23 sec. What is the probability that a random vehicle will be processed in 25 sec or more (i.e., will take longer than 25 sec)?

(A) 0.17
(B) 0.25
(C) 0.34
(D) 0.52

3. The number of cars entering a toll plaza on a bridge during the hour after midnight follows a Poisson distribution with a mean of 20.

(a) What is the probability that 17 cars will pass through the toll plaza during that hour on any given night?

(A) 0.076
(B) 0.12
(C) 0.16
(D) 0.23

(b) What is the percent probability that three or fewer cars will pass through the toll plaza at that hour on any given night?

(A) 0.0000032%
(B) 0.0019%
(C) 0.079%
(D) 0.11%

4. A mechanical component exhibits a negative exponential failure distribution with a mean time to failure of 1000 hr. What is the maximum operating time such that the reliability remains above 99%?

(A) 3.3 hr
(B) 5.6 hr
(C) 8.1 hr
(D) 10 hr

5. (*Time limit: one hour*) A survey field crew measures one leg of a traverse four times. The following results are obtained.

repetition	measurement	direction
1	1249.529	forward
2	1249.494	backward
3	1249.384	forward
4	1249.348	backward

The crew chief is under orders to obtain readings with confidence limits of 90%.

(a) Which readings are acceptable?

(A) No readings are acceptable.
(B) Two readings are acceptable.
(C) Three readings are acceptable.
(D) All four readings are acceptable.

(b) Which readings are not acceptable?

(A) No readings are unacceptable.
(B) One reading is unacceptable.
(C) Two readings are unacceptable.
(D) All four readings are unacceptable.

(c) Explain how to determine which readings are not acceptable.

(A) Readings inside the 90% confidence limits are unacceptable.

(B) Readings outside the 90% confidence limits are unacceptable.

(C) Readings outside the upper 90% confidence limit are unacceptable.

(D) Readings outside the lower 90% confidence limit are unacceptable.

(d) What is the most probable value of the distance?

(A) 1249.399

(B) 1249.410

(C) 1249.439

(D) 1249.452

(e) What is the error in the most probable value (at 90% confidence)?

(A) 0.08

(B) 0.11

(C) 0.14

(D) 0.19

(f) If the distance is one side of a square traverse whose sides are all equal, what is the most probable closure error?

(A) 0.14

(B) 0.20

(C) 0.28

(D) 0.35

(g) What is the probable error of part (f) expressed as a fraction?

(A) 1:17,600

(B) 1:14,200

(C) 1:12,500

(D) 1:10,900

(h) What is the order of accuracy of the closure?

(A) first order

(B) second order

(C) third order

(D) fourth order

(i) Define accuracy and distinguish it from precision.

(A) If an experiment can be repeated with identical results, the results are considered accurate.

(B) If an experiment has a small bias, the results are considered precise.

(C) If an experiment is precise, it cannot also be accurate.

(D) If an experiment is unaffected by experimental error, the results are accurate.

(j) Give an example of systematic error.

(A) measuring river depth as a motorized ski boat passes by

(B) using a steel tape that is too short to measure consecutive distances

(C) locating magnetic north near a large iron ore deposit along an overland route

(D) determining local wastewater BOD after a toxic spill

Statistical Analysis

6. *(Time limit: one hour)* California law requires a statistical analysis of the average speed driven by motorists on a road prior to the use of radar speed control. The following speeds (all in mi/hr) were observed in a random sample of 40 cars.

44, 48, 26, 25, 20, 43, 40, 42, 29, 39, 23, 26, 24, 47, 45, 28, 29, 41, 38, 36, 27, 44, 42, 43, 29, 37, 34, 31, 33, 30, 42, 43, 28, 41, 29, 36, 35, 30, 32, 31

(a) Tabulate the frequency distribution of the data.

(b) Draw the frequency histogram.

(c) Draw the frequency polygon.

(d) Tabulate the cumulative frequency distribution.

(e) Draw the cumulative frequency graph.

(f) What is the upper quartile speed?

(A) 30 mph

(B) 35 mph

(C) 40 mph

(D) 45 mph

(g) What is the mean speed?

(A) 31 mph

(B) 33 mph

(C) 35 mph

(D) 37 mph

(h) What is the standard deviation of the sample data?

(A) 2.1 mph

(B) 6.1 mph

(C) 6.8 mph

(D) 7.4 mph

(i) What is the sample standard deviation?

(A) 7.5 mph

(B) 18 mph

(C) 35 mph

(D) 56 mph

(j) What is the sample variance?

(A) 56 mi^2/hr^2

(B) 324 mi^2/hr^2

(C) 1225 mi^2/hr^2

(D) 3136 mi^2/hr^2

7. A spot speed study is conducted for a stretch of roadway. During a normal day, the speeds were found to be normally distributed with a mean of 46 and a standard deviation of 3.

(a) What is the 50th percentile speed?

(A) 39
(B) 43
(C) 46
(D) 49

(b) What is the 85th percentile speed?

(A) 47.1
(B) 48.3
(C) 49.1
(D) 52.7

(c) What is the upper two standard deviation speed?

(A) 47.2
(B) 49.3
(C) 51.1
(D) 52.0

(d) The daily average speeds for the same stretch of roadway on consecutive normal days were determined by sampling 25 vehicles each day. What is the upper two-standard deviation average speed?

(A) 46.6
(B) 47.2
(C) 52.0
(D) 54.7

8. The diameters of bolt holes drilled in structural steel members are normally distributed with a mean of 0.502 in and a standard deviation of 0.005 in. Holes are out of specification if their diameters are less than 0.497 in or more than 0.507 in.

(a) What is the probability that a hole chosen at random will be out of specification?

(A) 0.16
(B) 0.22
(C) 0.32
(D) 0.68

(b) What is the probability that two holes out of a sample of 15 will be out of specification?

(A) 0.074
(B) 0.12
(C) 0.15
(D) 0.32

Hypothesis Testing

9. 100 bearings were tested to failure. The average life was 1520 hr, and the standard deviation was 120 hr. The manufacturer claims a 1600 hr life. Evaluate using confidence limits of 95% and 99%.

(A) The claim is accurate at both 95% and 99% confidence.
(B) The claim is inaccurate only at 95%.
(C) The claim is inaccurate only at 99%.
(D) The claim is inaccurate at both 95% and 99% confidence.

Curve Fitting

10. (a) Find the best equation for a line passing through the points given.

(b) Find the correlation coefficient.

x	y
400	370
800	780
1250	1210
1600	1560
2000	1980
2500	2450
4000	3950

11. Find the best equation for a line passing through the points given.

s	t
20	43
18	141
16	385
14	1099

12. The number of vehicles lining up behind a flashing railroad crossing has been observed for five trains of different lengths, as given. What is the mathematical formula that relates the two variables?

no. of cars in train	no. of vehicles
2	14.8
5	18.0
8	20.4
12	23.0
27	29.9

13. The following yield data are obtained from five identical treatment plants.

(a) Develop a mathematical equation to correlate the yield and average temperature.

(b) What is the correlation coefficient?

treatment plant	average temperature (T)	average yield (Y)
1	207.1	92.30
2	210.3	92.58
3	200.4	91.56
4	201.1	91.63
5	203.4	91.83

14. The following data are obtained from a soil compaction test. What is the mathematical formula that relates the two variables?

x	y
−1	0
0	1
1	1.4
2	1.7
3	2
4	2.2
5	2.4
6	2.6
7	2.8
8	3

15. Two resistances, the meter resistance and a shunt resistor, are connected in parallel in an ammeter. Most of the current passing through the meter goes through the shunt resistor. In order to determine the accuracy of the resistance of shunt resistors being manufactured for a line of ammeters, a manufacturer tests a sample of 100 shunt resistors. The numbers of shunt resistors with the resistance indicated (to the nearest hundredth of an ohm) are as follows.

0.200 Ω, 1; 0.210 Ω, 3; 0.220 Ω, 5; 0.230 Ω, 10; 0.240 Ω, 17; 0.250 Ω, 40; 0.260 Ω, 13; 0.270 Ω, 6; 0.280 Ω, 3; 0.290 Ω, 2

(a) What is the mean resistance?

(A) 0.235 Ω
(B) 0.247 Ω
(C) 0.251 Ω
(D) 0.259 Ω

(b) What is the sample standard deviation?

(A) 0.0003
(B) 0.010
(C) 0.016
(D) 0.24

(c) What is the median resistance?

(A) 0.22 Ω
(B) 0.24 Ω
(C) 0.25 Ω
(D) 0.26 Ω

(d) What is the sample variance?

(A) 0.00027
(B) 0.0083
(C) 0.0114
(D) 0.0163

SOLUTIONS

1. (a) There are 4! = 24 different possible outcomes. By enumeration, there are 9 completely wrong combinations.

$$p\{\text{all wrong}\} = \frac{9}{24} = \boxed{0.375}$$

	sizes				
correct →	7	8	9	10	all wrong
sizes issued	7	8	9	10	
	7	8	10	9	
	7	9	8	10	
	7	9	10	8	
	7	10	8	9	
	7	10	9	8	
	8	9	10	7	X
	8	9	7	10	
	8	10	9	7	
	8	10	7	9	X
	8	7	9	10	
	8	7	10	9	X
	9	10	7	8	X
	9	10	8	7	X
	9	7	10	8	X
	9	7	8	10	
	9	8	7	10	
	9	8	10	7	
	10	7	8	9	X
	10	7	9	8	
	10	8	7	9	
	10	8	9	7	
	10	9	8	7	X
	10	9	7	8	X

The answer is (B).

(b) If three recruits get the correct size, the fourth recruit will also since there will be only one pair remaining.

$$p\{\text{exactly 3}\} = \boxed{0}$$

The answer is (A).

2. For an exponential distribution function, the mean is given as

$$\mu = \frac{1}{\lambda}$$

For a mean of 23,

$$\mu = 23 = \frac{1}{\lambda}$$
$$\lambda = 0.0435$$

For an exponential distribution function,

$$\begin{aligned} p = F(x) &= 1 - e^{-\lambda x} \\ p\{X < x\} &= 1 - p \\ p\{X > x\} &= 1 - F(x) \\ &= 1 - (1 - e^{-\lambda x}) \\ &= e^{-\lambda x} \end{aligned}$$

The probability of a random vehicle being processed in 25 sec or more is given by

$$\begin{aligned} p\{x > 25\} &= e^{-(0.0435)(25)} \\ &= e^{-1.0875} \\ &= \boxed{0.337} \end{aligned}$$

The answer is (C).

3. (a) The distribution is a Poisson distribution with an average of $\lambda = 20$.

The probability for a Poisson distribution is given by

$$p\{x\} = f(x) = \frac{e^{-\lambda}\lambda^x}{x!}$$

The probability of 17 cars is

$$\begin{aligned} p\{x = 17\} = f(17) &= \frac{e^{-20} \times 20^{17}}{17!} \\ &= \boxed{0.076 \quad (7.6\%)} \end{aligned}$$

The answer is (A).

(b) The probability of three or fewer cars is given by

$$\begin{aligned} p\{x \leq 3\} &= p\{x = 0\} + p\{x = 1\} + p\{x = 2\} \\ &\quad + p\{x = 3\} \\ &= f(0) + f(1) + f(2) + f(3) \\ &= \frac{e^{-20} \times 20^0}{0!} + \frac{e^{-20} \times 20^1}{1!} \\ &\quad + \frac{e^{-20} \times 20^2}{2!} + \frac{e^{-20} \times 20^3}{3!} \\ &= 2 \times 10^{-9} + 4.1 \times 10^{-8} \\ &\quad + 4.12 \times 10^{-7} + 2.75 \times 10^{-6} \\ &= 3.2 \times 10^{-6} \\ &= \boxed{0.0000032 \quad (3.2 \times 10^{-4}\%)} \end{aligned}$$

The answer is (A).

4.

$$\begin{aligned} \lambda &= \frac{1}{\text{MTTF}} \\ &= \frac{1}{1000} = 0.001 \end{aligned}$$

The reliability function is

$$R\{t\} = e^{-\lambda t} = e^{-0.001t}$$

Since the reliability is greater than 99%,

$$\begin{aligned} e^{-0.001t} &> 0.99 \\ \ln(e^{-0.001t}) &> \ln(0.99) \\ -0.001t &> \ln(0.99) \\ t &< -1000\ln(0.99) \\ &\boxed{t < 10.05} \end{aligned}$$

The maximum operating time such that the reliability remains above 99% is 10.05 hr.

The answer is (D).

5. Find the average.

$$\begin{aligned} \overline{x} &= \frac{\sum x_i}{n} \\ &= \frac{1249.529 + 1249.494 + 1249.384 + 1249.348}{4} \\ &= 1249.439 \end{aligned}$$

Since the sample population is small, use the sample standard deviation.

$$\begin{aligned} s &= \sqrt{\frac{\sum(x_i - \overline{x})^2}{n-1}} \\ &= \sqrt{\frac{\begin{array}{c}(1249.529-1249.439)^2+(1249.494-1249.439)^2 \\ +\,(1249.384-1249.439)^2+(1249.348-1249.439)^2\end{array}}{4-1}} \\ &= 0.08647 \end{aligned}$$

From the standard deviation table, a two-tail 90% confidence limit falls within $1.645s$ of $\overline{x}$.

$$1249.439 \pm (1.645)(0.08647) = 1249.439 \pm 0.142$$

Therefore, (1249.297, 1249.581) is the 90% confidence range.

(a) By observation, all the readings fall within the 90% confidence range.

The answer is (D).

(b) No readings are unacceptable.

The answer is (A).

(c) Readings outside the 90% confidence limits are unacceptable.

The answer is (B).

(d) The unbiased estimate of the most probable distance is 1249.439.

The answer is (C).

(e) The error for the 90% confidence range is 0.142.

The answer is (C).

(f) If the surveying crew places a marker, measures a distance x, places a second marker, and then measures the same distance x back to the original marker, the ending point should coincide with the original marker. If, due to measurement errors, the ending and starting points do not coincide, the difference is the closure error.

In this example, the survey crew moves around the four sides of a square, so there are two measurements in the x-direction and two measurements in the y-direction. If the errors E_1 and E_2 are known for two measurements, x_1 and x_2, the error associated with the sum or difference $x_1 \pm x_2$ is

$$E\{x_1 \pm x_2\} = \sqrt{E_1^2 + E_2^2}$$

In this case, the error in the x-direction is

$$\begin{aligned} E_x &= \sqrt{(0.1422)^2 + (0.1422)^2} \\ &= 0.2011 \end{aligned}$$

The error in the y-direction is calculated the same way and is also 0.2011. E_x and E_y are combined by the Pythagorean theorem to yield

$$\begin{aligned} E_{\text{closure}} &= \sqrt{(0.2011)^2 + (0.2011)^2} \\ &= \boxed{0.2844} \end{aligned}$$

The answer is (C).

(g) In surveying, error may be expressed as a fraction of one or more legs of the traverse. Assume that the total of all four legs is to be used as the basis.

$$\frac{0.2844}{(4)(1249)} = \boxed{\frac{1}{17{,}567}}$$

The answer is (A).

(h) In surveying, a class 1 third-order error is smaller than 1/10,000. The error of 1/17,567 is smaller than the third-order error; therefore, the error is within the third-order accuracy.

The answer is (C).

(i) An experiment is accurate if it is unchanged by experimental error. Precision is concerned with the repeatability of the experimental results. If an experiment is repeated with identical results, the experiment is said to be precise. However, it is possible to have a highly precise experiment with a large bias.

The answer is (D).

(j) A systematic error is one that is always present and is unchanged from sample to sample. For example, a steel tape that is 0.02 ft short introduces a systematic error.

The answer is (B).

6. (a) and (d) Tabulate the frequency distribution data.

(Note that the lowest speed is 20 mi/hr and the highest speed is 48 mi/hr; therefore, the range is 28 mi/hr. Choose 10 cells with a width of 3 mi/hr.)

midpoint	interval (mi/hr)	frequency	cumulative frequency	cumulative percent
21	20–22	1	1	3
24	23–25	3	4	10
27	26–28	5	9	23
30	29–31	8	17	43
33	32–34	3	20	50
36	35–37	4	24	60
39	38–40	3	27	68
42	41–43	8	35	88
45	44–46	3	38	95
48	47–49	2	40	100

(b)

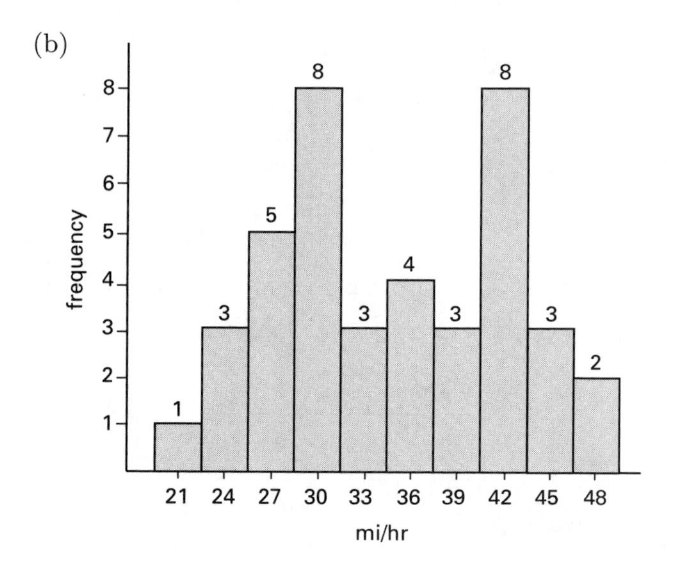

(c)

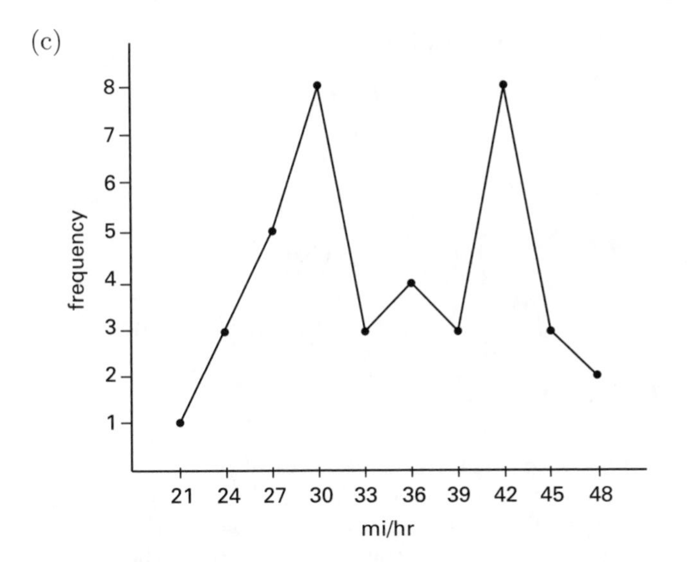

(e)

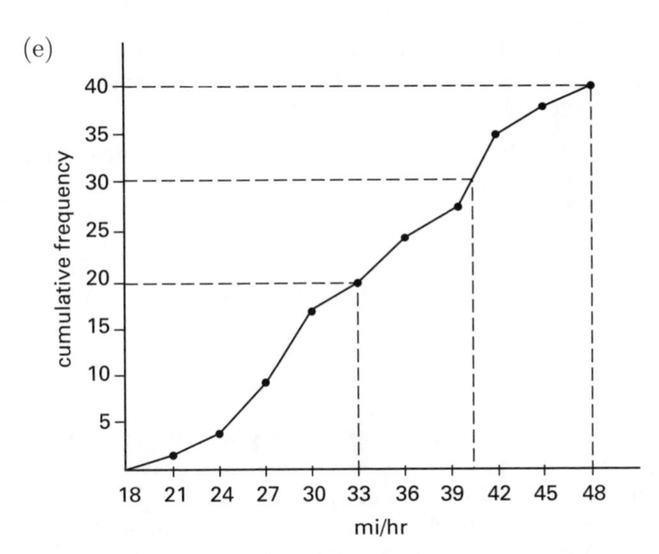

(f) From the cumulative frequency graph in Sol. 6.5, the upper quartile speed occurs at 30 cars or 75%, which corresponds to approximately 40 mi/hr.

The answer is (C).

(g) $\sum x_i = 1390$ mi/hr

$n = 40$

The mean is computed as

$$\overline{x} = \frac{\sum x_i}{n} = \frac{1390\ \frac{\text{mi}}{\text{hr}}}{40} = \boxed{34.75\ \text{mi/hr}}$$

The answer is (C).

(h) The standard deviation of the sample data is given as

$$\sigma = \sqrt{\frac{\sum x^2}{n} - \mu^2}$$

$$\sum x^2 = 50{,}496\ \text{mi}^2/\text{hr}^2$$

Use the sample mean as an unbiased estimator of the population mean, μ.

$$\sigma = \sqrt{\frac{\sum x^2}{n} - \mu^2} = \sqrt{\frac{50{,}496\ \frac{\text{mi}^2}{\text{hr}^2}}{40} - \left(34.75\ \frac{\text{mi}}{\text{hr}}\right)^2} = \boxed{7.405\ \text{mi/hr}}$$

The answer is (D).

(i) The sample standard deviation is given by

$$s = \sqrt{\frac{\sum x^2 - \frac{\left(\sum x\right)^2}{n}}{n-1}} = \sqrt{\frac{50{,}496\ \frac{\text{mi}^2}{\text{hr}^2} - \frac{\left(1390\ \frac{\text{mi}}{\text{hr}}\right)^2}{40}}{40-1}} = \boxed{7.500\ \text{mi/hr}}$$

The answer is (A).

(j) The sample variance is given by the square of the sample standard deviation.

$$s^2 = \left(7.500\ \frac{\text{mi}}{\text{hr}}\right)^2 = \boxed{56.25\ \text{mi}^2/\text{hr}^2}$$

The answer is (A).

7. (a) The 50th percentile speed is the median speed, $\boxed{46,}$ which for a symmetrical normal distribution is the mean speed.

The answer is (C).

(b) The 85th percentile speed is the speed that is exceeded by only 15% of the measurements. Since this is a normal distribution, App. 11.A can be used. 15% in the upper tail corresponds to 35% between the mean and the 85th percentile. This occurs at approximately $= 1.04\sigma$. The 85th percentile speed is

$$\begin{aligned} x_{85\%} &= \mu + 1.04\sigma \\ &= 46 + (1.04)(3) = \boxed{49.12} \end{aligned}$$

The answer is (C).

(c) The upper 2σ speed is

$$\begin{aligned} x_{2\sigma} &= \mu + 2\sigma \\ &= 46 + (2)(3) = \boxed{52} \end{aligned}$$

The answer is (D).

(d) According to the central limit theorem, the mean of the average speeds is the same as the distribution mean, and the standard deviation of sample means (from Eq. 11.66) is

$$\begin{aligned} s_{\overline{x}} &= \frac{\sigma_x}{\sqrt{n}} \\ &= \frac{3}{\sqrt{25}} = 0.6 \\ \overline{x}_{2\sigma} &= \mu + 2\sigma_{\overline{x}} \\ &= 46 + (2)(0.6) = \boxed{47.2} \end{aligned}$$

The answer is (B).

8. (a) From Eq. 11.43,

$$z_{\text{upper}} = \frac{0.507 \text{ in} - 0.502 \text{ in}}{0.005 \text{ in}} = +1$$

From App. 11.A, the area outside $z = +1$ is

$$0.5 - 0.3413 = 0.1587$$

Since these are symmetrical limits, $z_{\text{lower}} = -1$.

$$\text{total fraction defective} = (2)(0.1587) = \boxed{0.3174}$$

The answer is (C).

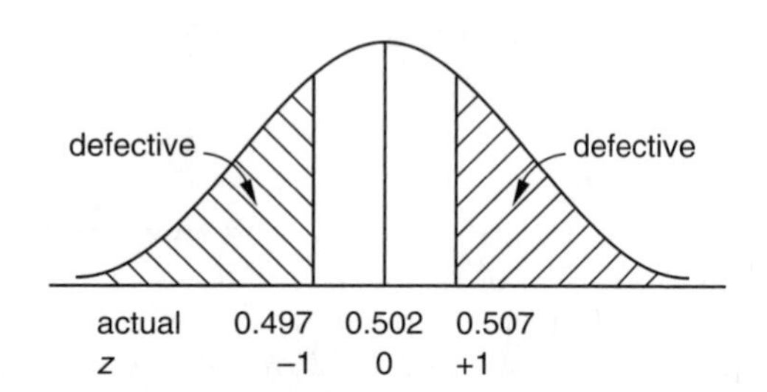

(b) This is a binomial problem.

$$\begin{aligned} p &= p\{\text{defective}\} = 0.3174 \\ q &= 1 - p = 0.6826 \end{aligned}$$

From Eq. 11.28,

$$\begin{aligned} f(2) &= \binom{15}{2}(0.3174)^2(0.6826)^{13} \\ &= \left(\frac{15!}{13!2!}\right)(0.3174)^2(0.6826)^{13} = \boxed{0.0739} \end{aligned}$$

The answer is (A).

9. This is a typical hypothesis test of two sample population means. The two populations are the original population the manufacturer used to determine the 1600 hr average life value and the new population the sample was taken from. The mean ($\overline{x} = 1520$ hr) of the sample and its standard deviation ($s = 120$ hr) are known, but the mean and standard deviation of a population of average lifetimes are unknown.

Assume that the average lifetime population mean and the sample mean are identical.

$$\overline{x} = \mu = 1520 \text{ hr}$$

The standard deviation of the average lifetime population is

$$\sigma_{\overline{x}} = \frac{s}{\sqrt{n}} = \frac{120 \text{ hr}}{\sqrt{100}} = 12 \text{ hr}$$

The manufacturer can be reasonably sure that the claim of a 1600 hr average life is justified if the average test life is near 1600 hr. "Reasonably sure" must be evaluated based on acceptable probability of being incorrect. If the manufacturer is willing to be wrong with a 5% probability, then a 95% confidence level is required.

Since the direction of bias is known, a one-tailed test is required. To determine if the mean has shifted downward, test the hypothesis that 1600 hr is within the 95% limit of a distribution with a mean of 1520 hr and a standard deviation of 12 hr. From a standard normal table, 5% of a standard normal distribution is outside of $z = 1.645$. Therefore, the 95% confidence limit is

$$1520 \text{ hr} + (1.645)(12 \text{ hr}) = 1540 \text{ hr}$$

The manufacturer can be 95% certain that the average lifetime of the bearings is less than 1600 hr.

If the manufacturer is willing to be wrong with a probability of only 1%, then a 99% confidence limit is required. From the normal table, $z = 2.33$ and the 99% confidence limit is

$$1520 \text{ hr} + (2.33)(12 \text{ hr}) = 1548 \text{ hr}$$

The manufacturer can be 99% certain that the average bearing life is less than 1600 hr.

The answer is (D).

10. (a) Plot the data points to determine if the relationship is linear.

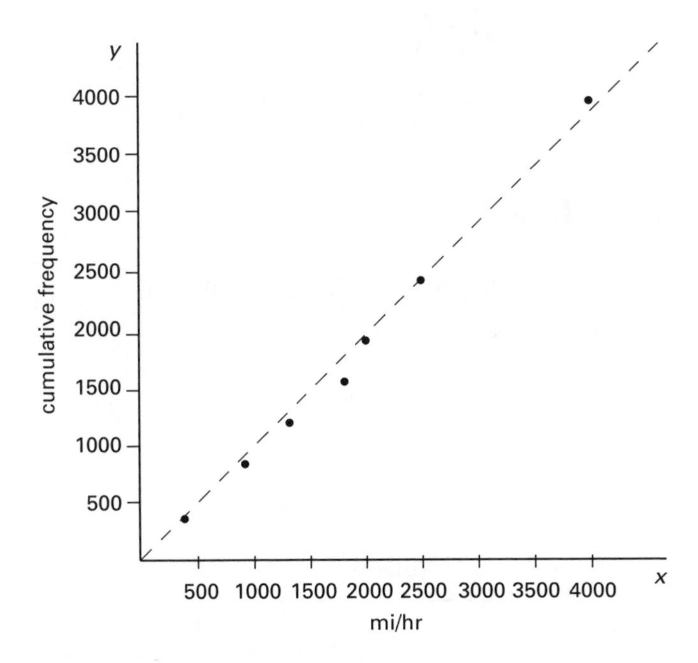

The data appear to be essentially linear. The slope, m, and the y-intercept, b, can be determined using linear regression.

The individual terms are

$$n = 7$$

$$\begin{aligned}\sum x_i &= 400 + 800 + 1250 + 1600 + 2000 + 2500 \\ &\quad + 4000 \\ &= 12{,}550\end{aligned}$$

$$\begin{aligned}\left(\sum x_i\right)^2 &= (12{,}550)^2 \\ &= 1.575 \times 10^8\end{aligned}$$

$$\begin{aligned}\overline{x} &= \frac{\sum x_i}{n} \\ &= \frac{12{,}550}{7} \\ &= 1792.9\end{aligned}$$

$$\begin{aligned}\sum x_i^2 &= (400)^2 + (800)^2 + (1250)^2 + (1600)^2 \\ &\quad + (2000)^2 + (2500)^2 + (4000)^2 \\ &= 3.117 \times 10^7\end{aligned}$$

Similarly,

$$\begin{aligned}\sum y_i &= 370 + 780 + 1210 + 1560 + 1980 \\ &\quad + 2450 + 3950 \\ &= 12{,}300\end{aligned}$$

$$\left(\sum y_i\right)^2 = (12{,}300)^2 = 1.513 \times 10^8$$

$$\overline{y} = \frac{\sum y_i}{n} = \frac{12{,}300}{7} = 1757.1$$

$$\begin{aligned}\sum y_i^2 &= (370)^2 + (780)^2 + (1210)^2 + (1560)^2 \\ &\quad + (1980)^2 + (2450)^2 + (3950)^2 \\ &= 3.017 \times 10^7\end{aligned}$$

Also,

$$\begin{aligned}\sum x_i y_i &= (400)(370) + (800)(780) + (1250)(1210) \\ &\quad + (1600)(1560) + (2000)(1980) \\ &\quad + (2500)(2450) + (4000)(3950) \\ &= 3.067 \times 10^7\end{aligned}$$

The slope is

$$\begin{aligned}m &= \frac{n\sum x_i y_i - \sum x_i \sum y_i}{n\sum x_i^2 - \left(\sum x_i\right)^2} \\ &= \frac{(7)(3.067 \times 10^7) - (12{,}550)(12{,}300)}{(7)(3.117 \times 10^7) - (12{,}550)^2} \\ &= 0.994\end{aligned}$$

The y-intercept is

$$\begin{aligned}b &= \overline{y} - m\overline{x} \\ &= 1757.1 - (0.994)(1792.9) \\ &= -25.0\end{aligned}$$

The least squares equation of the line is

$$\begin{aligned}y &= mx + b \\ &= \boxed{0.994x - 25.0}\end{aligned}$$

(b) The correlation coefficient is

$$\begin{aligned}r &= \frac{n\sum(x_i y_i) - \left(\sum x_i\right)\left(\sum y_i\right)}{\sqrt{\left(n\sum x_i^2 - \left(\sum x_i\right)^2\right)\left(n\sum y_i^2 - \left(\sum y_i\right)^2\right)}} \\ &= \frac{(7)(3.067 \times 10^7) - (12{,}500)(12{,}300)}{\sqrt{\begin{array}{l}\left((7)(3.117 \times 10^7) - (12{,}500)^2\right) \\ \quad \times \left((7)(3.017 \times 10^7) - (12{,}300)^2\right)\end{array}}} \\ &\approx \boxed{1.00}\end{aligned}$$

11. Plotting the data shows that the relationship is nonlinear.

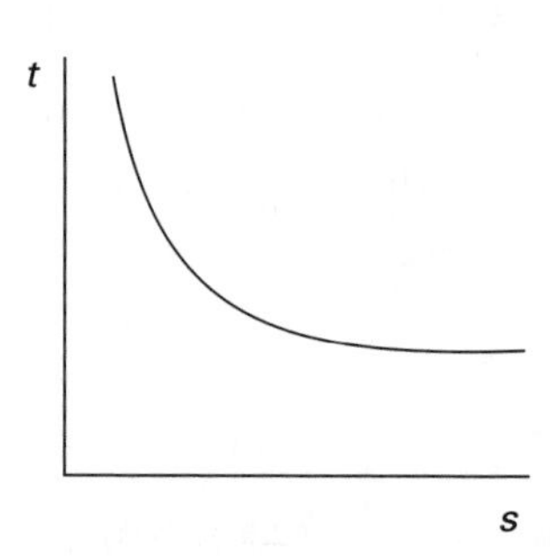

This appears to be an exponential with the form

$$t = ae^{bs}$$

Take the natural log of both sides.

$$\begin{aligned}\ln t &= \ln(ae^{bs})\\ &= \ln a + \ln(e^{bs})\\ &= \ln a + bs\end{aligned}$$

But, $\ln a$ is just a constant, c.

$$\ln t = c + bs$$

Make the transformation $R = \ln t$.

$$R = c + bs$$

s	R
20	3.76
18	4.95
16	5.95
14	7.00

This is linear.

$$\begin{aligned}n &= 4\\ \textstyle\sum s_i &= 20 + 18 + 16 + 14 = 68\\ \bar{s} &= \frac{\sum s}{n} = \frac{68}{4} = 17\\ \textstyle\sum s_i^2 &= (20)^2 + (18)^2 + (16)^2 + (14)^2 = 1176\\ \left(\textstyle\sum s_i\right)^2 &= (68)^2 = 4624\\ \textstyle\sum R_i &= 3.76 + 4.95 + 5.95 + 7.00 = 21.66\\ \overline{R} &= \frac{\sum R_i}{n} = \frac{21.66}{4} = 5.415\\ \textstyle\sum R_i^2 &= (3.76)^2 + (4.95)^2 + (5.95)^2 + (7.00)^2\\ &= 123.04\\ \left(\textstyle\sum R_i\right)^2 &= (21.66)^2 = 469.16\\ \textstyle\sum s_i R_i &= (20)(3.76) + (18)(4.95) + (16)(5.95)\\ &\quad + (14)(7.00)\\ &= 357.5\end{aligned}$$

The slope, b, of the transformed line is

$$\begin{aligned}b &= \frac{n\sum s_i R_i - \sum s_i \sum R_i}{n\sum s_i^2 - \left(\sum s_i\right)^2}\\ &= \frac{(4)(357.5) - (68)(21.66)}{(4)(1176) - (68)^2} = -0.536\end{aligned}$$

The intercept is

$$\begin{aligned}c &= \overline{R} - b\bar{s} = 5.415 - (-0.536)(17)\\ &= 14.527\end{aligned}$$

The transformed equation is

$$\begin{aligned}R &= c + bs\\ &= 14.527 - 0.536s\end{aligned}$$

$$\boxed{\ln t = 14.527 - 0.536s}$$

12. The first step is to graph the data.

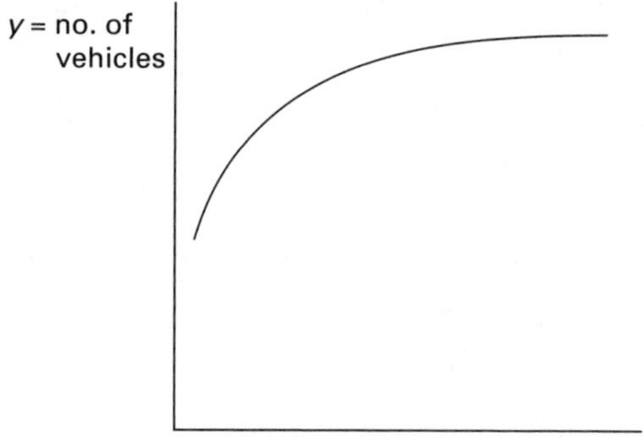

It is assumed that the relationship between the variables has the form $y = a + b\log x$. Therefore, the variable change $z = \log x$ is made, resulting in the following set of data.

z	y
0.301	14.8
0.699	18.0
0.903	20.4
1.079	23.0
1.431	29.9

$$\begin{aligned}\textstyle\sum z_i &= 4.413\\ \textstyle\sum y_i &= 106.1\\ \textstyle\sum z_i^2 &= 4.6082\\ \textstyle\sum y_i^2 &= 2382.2\end{aligned}$$

$$(\sum z_i)^2 = 19.475$$
$$(\sum y_i)^2 = 11{,}257.2$$
$$\overline{z} = 0.8826$$
$$\overline{y} = 21.22$$
$$\sum z_i y_i = 103.06$$
$$n = 5$$

The slope is

$$\begin{aligned} m &= \frac{n\sum z_i y_i - \sum z_i \sum y_i}{n\sum z_i^2 - (\sum z_i)^2} \\ &= \frac{(5)(103.06) - (4.413)(106.1)}{(5)(4.6082) - 19.475} \\ &= 13.20 \end{aligned}$$

The y-intercept is

$$\begin{aligned} b &= \overline{y} - m\overline{z} \\ &= 21.22 - (13.20)(0.8826) \\ &= 9.570 \end{aligned}$$

The resulting equation is

$$y = 9.570 + 13.20z$$

The relationship between x and y is approximately

$$\boxed{y = 9.570 + 13.20 \log x}$$

This is not an optimal correlation, as better correlation coefficients can be obtained if other assumptions about the form of the equation are made. For example, $y = 9.1 + 4\sqrt{x}$ has a better correlation coefficient.

13. (a) Plot the data to verify that they are linear.

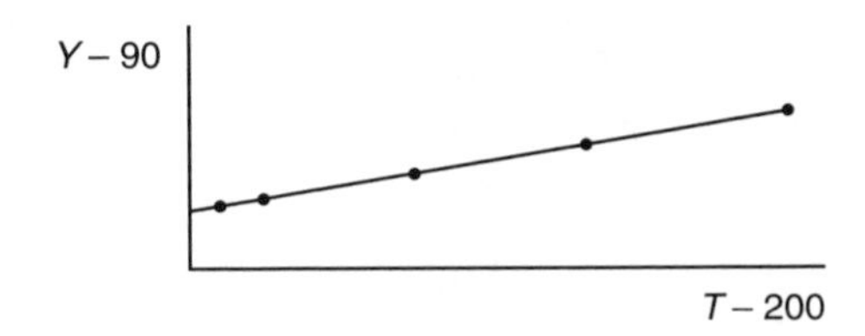

x $T - 200$	y $Y - 90$
7.1	2.30
10.3	2.58
0.4	1.56
1.1	1.63
3.4	1.83

step 1:

$$\sum x_i = 22.3 \qquad \sum y_i = 9.9$$
$$\sum x_i^2 = 169.43 \qquad \sum y_i^2 = 20.39$$
$$(\sum x_i)^2 = 497.29 \qquad (\sum y_i)^2 = 98.01$$
$$\overline{x} = \frac{22.3}{5} = 4.46 \qquad \overline{y} = 1.98$$
$$\sum x_i y_i = 51.54$$

step 2: From Eq. 11.70, the slope is

$$m = \frac{(5)(51.54) - (22.3)(9.9)}{(5)(169.43) - 497.29} = 0.1055$$

step 3: From Eq. 11.71, the y-intercept is

$$b = 1.98 - (0.1055)(4.46) = 1.509$$

The equation of the line is

$$\begin{aligned} y &= 0.1055x + 1.509 \\ Y - 90 &= (0.1055)(T - 200) + 1.509 \\ Y &= \boxed{0.1055T + 70.409} \end{aligned}$$

(b) *step 4:* Use Eq. 11.72 to get the correlation coefficient.

$$\begin{aligned} r &= \frac{(5)(51.54) - (22.3)(9.9)}{\sqrt{\big((5)(169.43) - 497.29\big)\big((5)(20.39) - 98.01\big)}} \\ &= \boxed{0.995} \end{aligned}$$

14. Plot the data to see if they are linear.

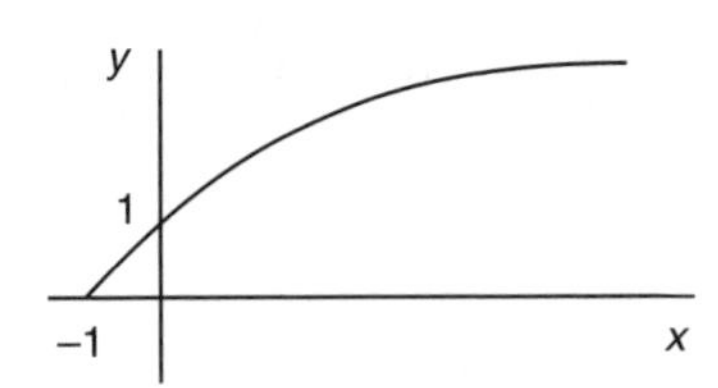

This looks like it could be of the form

$$y = a + b\sqrt{x}$$

However, when x is negative (as in the first point), the function is imaginary. Try shifting the curve to the right, replacing x with $x + 1$.

$$y = a + bz$$
$$z = \sqrt{x+1}$$

z	y
0	0
1	1
1.414	1.4
1.732	1.7
2	2
2.236	2.2
2.45	2.4
2.65	2.6
2.83	2.8
3	3

Since $y \approx z$, the relationship is

$$\boxed{y = \sqrt{x+1}}$$

In this problem, the answer was found accidentally. Usually, regression would be necessary.

15. (a)

R	f	fR	fR^2
0.200	1	0.200	0.0400
0.210	3	0.630	0.1323
0.220	5	1.100	0.2420
0.230	10	2.300	0.5290
0.240	17	4.080	0.9792
0.250	40	10.000	2.5000
0.260	13	3.380	0.8788
0.270	6	1.620	0.4374
0.280	3	0.840	0.2352
0.290	2	0.580	0.1682
	100	24.730	6.1421

$$\overline{R} = \frac{\sum fR}{\sum f} = \frac{24.730\ \Omega}{100} = \boxed{0.2473\ \Omega}$$

The answer is (B).

(b) The sample standard deviation is given by Eq. 11.61.

$$s = \sqrt{\frac{\sum fR^2 - \frac{(\sum fR)^2}{n}}{n-1}} = \sqrt{\frac{6.1421\ \Omega - \frac{(24.73\ \Omega)^2}{100}}{99}} = \boxed{0.0163\ \Omega}$$

The answer is (C).

(c) The 50th and 51st values are both 0.25 Ω. The median is $\boxed{0.25\ \Omega.}$

The answer is (C).

(d) $$s^2 = (0.0163\ \Omega)^2 = \boxed{0.0002656\ \Omega^2}$$

The answer is (A).

12 Numerical Analysis

PRACTICE PROBLEMS

1. A function is given as $y = 3x^{0.93} + 4.2$. What is the percent error if the value of y at $x = 2.7$ is found by using straight-line interpolation between $x = 2$ and $x = 3$?

(A) 0.06%
(B) 0.18%
(C) 2.5%
(D) 5.4%

2. Given the following data points, find y by straight-line interpolation for $x = 2.75$.

x	y
1	4
2	6
3	2
4	−14

(A) 2.1
(B) 2.4
(C) 2.7
(D) 3.0

3. Using the bisection method, find all of the roots of $f(x) = 0$ to the nearest 0.000005.

$$f(x) = x^3 + 2x^2 + 8x - 2$$

SOLUTIONS

1. The actual value at $x = 2.7$ is given by

$$y(x) = 3x^{0.93} + 4.2$$
$$y(2.7) = (3)(2.7)^{0.93} + 4.2 = 11.756$$

At $x = 3$,

$$y(3) = (3)(3)^{0.93} + 4.2 = 12.534$$

At $x = 2$,

$$y(2) = (3)(2)^{0.93} + 4.2 = 9.916$$

Use straight-line interpolation.

$$\frac{x_2 - x}{x_2 - x_1} = \frac{y_2 - y}{y_2 - y_1}$$
$$\frac{3 - 2.7}{3 - 2} = \frac{12.534 - y}{12.534 - 9.916}$$
$$y = 11.749$$

The relative error is given by

$$\frac{\text{actual value} - \text{predicted value}}{\text{actual value}} = \frac{11.756 - 11.749}{11.756} = \boxed{0.0006 \quad (0.06\%)}$$

The answer is (A).

2. Let $x_1 = 2$; therefore, from the table of data points, $y_1 = 6$. Let $x_2 = 3$; therefore, from the table of data points, $y_2 = 2$.

Let $x = 2.75$. By straight-line interpolation,

$$\frac{x_2 - x}{x_2 - x_1} = \frac{y_2 - y}{y_2 - y_1}$$
$$\frac{3 - 2.75}{3 - 2} = \frac{2 - y}{2 - 6}$$
$$\boxed{y = 3}$$

The answer is (D).

3. $f(x) = x^3 + 2x^2 + 8x - 2$

Try to find an interval in which there is a root.

x	$f(x)$
0	−2
1	9

A root exists in the interval [0,1].

Try $x = \left(\frac{1}{2}\right)(0+1) = 0.5$.

$$f(0.5) = (0.5)^3 + (2)(0.5)^2 + (8)(0.5) - 2 = 2.625$$

A root exists in [0,0.5].

Try $x = 0.25$.

$$f(0.25) = (0.25)^3 + (2)(0.25)^2 + (8)(0.25) - 2 = 0.1406$$

A root exists in [0,0.25].

Try $x = 0.125$.

$$\begin{aligned} f(0.125) &= (0.125)^3 + (2)(0.125)^2 + (8)(0.125) - 2 \\ &= -0.967 \end{aligned}$$

A root exists in [0.125,0.25].

Try $x = \left(\frac{1}{2}\right)(0.125 + 0.25) = 0.1875$.

Continuing,

$$\begin{aligned} f(0.1875) &= -0.42 \quad [0.1875, 0.25] \\ f(0.21875) &= -0.144 \quad [0.21875, 0.25] \\ f(0.234375) &= -0.002 \quad [\text{This is close enough.}] \end{aligned}$$

One root is $x_1 \approx \boxed{0.234375.}$

Try to find the other two roots. Use long division to factor the polynomial.

$$\begin{array}{r|l} & x^2 + 2.234375x + 8.52368 \\ \hline x - 0.234375 & x^3 + 2x^2 + 8x - 2 \\ & -(x^3 - 0.234375x^2) \\ \hline & 2.234375x^2 + 8x \\ & -(2.234375x^2 - 0.52368x) \\ \hline & 8.52368x - 2 \\ & -(8.52368x - 1.9977) \\ \hline & \approx 0 \end{array}$$

Use the quadratic equation to find the roots of $x^2 + 2.234375x + 8.52368$.

$$\begin{aligned} x_2, x_3 &= \frac{-2.234375 \pm \sqrt{(2.234375)^2 - (4)(1)(8.52368)}}{(2)(1)} \\ &= \boxed{-1.117189 \pm i2.697327} \quad [\text{both imaginary}] \end{aligned}$$

13 Energy, Work, and Power

PRACTICE PROBLEMS

Energy

1. A solid cast-iron sphere ($\rho = 0.256$ lbm/in^3 (7090 kg/m^3)) of 10 in (25 cm) diameter travels without friction at 30 ft/sec (9 m/s) horizontally. What is its kinetic energy?

(A) 900 ft-lbf (12 kJ)
(B) 1200 ft-lbf (1.6 kJ)
(C) 1600 ft-lbf (2.0 kJ)
(D) 1900 ft-lbf (2.4 kJ)

Work

2. What work is done when a balloon carries a 12 lbm (5.2 kg) load to 40,000 ft (12 000 m) height?

(A) 2.4×10^5 ft-lbf (300 kJ)
(B) 4.8×10^5 ft-lbf (610 kJ)
(C) 7.7×10^5 ft-lbf (980 kJ)
(D) 9.9×10^5 ft-lbf (1.3 MJ)

3. Find the compression of a spring if a 100 lbm (50 kg) weight is dropped from 8 ft (2 m) onto a spring with a constant of 33.33 lbf/in (5.837×10^3 N/m).

(A) 27 in (0.67 m)
(B) 34 in (0.85 m)
(C) 39 in (0.90 m)
(D) 45 in (1.1 m)

4. A punch press flywheel operates at 300 rpm with a moment of inertia of 15 slug-ft^2 (20 kg·m^2). Find the speed in rpm to which the wheel will be reduced after a sudden punching requiring 4500 ft-lbf (6100 J) of work.

(A) 160 rpm
(B) 190 rpm
(C) 220 rpm
(D) 310 rpm

5. A force of 550 lbf (2500 N) making a 40° angle (upward) from the horizontal pushes a box 20 ft (6 m) across the floor. What work is done?

(A) 2200 ft-lbf (3.0 kJ)
(B) 3700 ft-lbf (5.2 kJ)
(C) 4200 ft-lbf (6.0 kJ)
(D) 8400 ft-lbf (12 kJ)

6. A 1000 ft long (300 m long) cable has a mass of 2 lbm per foot (3 kg/m) and is suspended from a winding drum down into a vertical shaft. What work must be done to rewind the cable?

(A) 0.50×10^6 ft-lbf (0.6 MJ)
(B) 0.75×10^6 ft-lbf (0.9 MJ)
(C) 1×10^6 ft-lbf (1.3 MJ)
(D) 2×10^6 ft-lbf (2.6 MJ)

Power

7. What volume in ft^3 (m^3) of water can be pumped to a 130 ft (40 m) height in 1 hr by a 7 hp (5 kW) pump? Assume 85% efficiency.

(A) 1500 ft^3 (40 m^3)
(B) 1800 ft^3 (49 m^3)
(C) 2000 ft^3 (54 m^3)
(D) 2400 ft^3 (65 m^3)

8. What power in horsepower (kW) is required to lift a 3300 lbm (1500 kg) mass 250 ft (80 m) in 14 sec?

(A) 40 hp (30 kW)
(B) 70 hp (53 kW)
(C) 90 hp (68 kW)
(D) 110 hp (84 kW)

SOLUTIONS

1. *Customary U.S. Solution*

Since there is no friction, there is no rotation. The sphere slides.

$$
\begin{aligned}
E_{\text{kinetic}} &= \tfrac{1}{2}\left(\frac{m}{g_c}\right)\mathrm{v}^2 = \tfrac{1}{2}\left(V\left(\frac{\rho}{g_c}\right)\right)\mathrm{v}^2 \\
&= \left(\tfrac{1}{2}\right)\left(\tfrac{4}{3}\pi r^3\right)\left(\frac{\rho}{g_c}\right)\mathrm{v}^2 = \left(\tfrac{2}{3}\pi\right)\left(\frac{10\text{ in}}{2}\right)^3\left(\frac{\rho}{g_c}\right)\mathrm{v}^2 \\
&= \left(\tfrac{2}{3}\pi\right)\left(\left(\frac{10\text{ in}}{2}\right)\left(\frac{1\text{ ft}}{12\text{ in}}\right)\right)^3 \\
&\quad \times \left(\frac{0.256\ \frac{\text{lbm}}{\text{in}^3}}{32.2\ \frac{\text{lbm-ft}}{\text{lbf-sec}^2}}\right)\left(1728\ \frac{\text{in}^3}{\text{ft}^3}\right)\left(30\ \frac{\text{ft}}{\text{sec}}\right)^2 \\
&= \boxed{1873\text{ ft-lbf}}
\end{aligned}
$$

The answer is (D).

SI Solution

Since there is no friction, there is no rotation. The sphere slides.

$$
\begin{aligned}
E_{\text{kinetic}} &= \tfrac{1}{2}m\mathrm{v}^2 = \tfrac{1}{2}(\rho V)\mathrm{v}^2 \\
&= \left(\tfrac{1}{2}\right)\rho\left(\tfrac{4}{3}\pi r^3\right)\mathrm{v}^2 = \left(\tfrac{2}{3}\pi\right)\rho\left(\frac{0.25\text{ m}}{2}\right)^3\mathrm{v}^2 \\
&= \left(\tfrac{2}{3}\pi\right)\left(\frac{0.25\text{ m}}{2}\right)^3\left(7.09\times 10^3\ \frac{\text{kg}}{\text{m}^3}\right)\left(9\ \frac{\text{m}}{\text{s}}\right)^2 \\
&= 2.35\times 10^3\text{ J}\quad \boxed{(2.35\text{ kJ})}
\end{aligned}
$$

The answer is (D).

2. *Customary U.S. Solution*

$$
\begin{aligned}
W &= \Delta E_{\text{potential}} = m\frac{g}{g_c}\Delta h \\
&= (12\text{ lbm})\left(\frac{32.2\ \frac{\text{ft}}{\text{sec}^2}}{32.2\ \frac{\text{lbm-ft}}{\text{sec}^2\text{-lbf}}}\right)(40{,}000\text{ ft}) \\
&= \boxed{4.8\times 10^5\text{ ft-lbf}}
\end{aligned}
$$

The answer is (B).

SI Solution

$$
\begin{aligned}
W &= \Delta E_{\text{potential}} = mg\Delta h \\
&= (5.2\text{ kg})\left(9.81\ \frac{\text{m}}{\text{s}^2}\right)(12\,000\text{ m})\left(\frac{1\text{ kJ}}{1000\text{ J}}\right) \\
&= \boxed{612.1\text{ kJ}}
\end{aligned}
$$

The answer is (B).

3. *Customary U.S. Solution*

$$
\begin{aligned}
\Delta E_{\text{potential}} &= \Delta E_{\text{spring}} \\
W(\Delta h + \Delta x) &= \tfrac{1}{2}k(\Delta x)^2
\end{aligned}
$$

Rearranging,

$$
\tfrac{1}{2}k(\Delta x)^2 - W\Delta x - W\Delta h = 0
$$

$$
\left(\tfrac{1}{2}\right)\left(33.33\ \frac{\text{lbf}}{\text{in}}\right)(\Delta x)^2 - (100\text{ lbf})\Delta x - (100\text{ lbf})(8\text{ ft})\left(12\ \frac{\text{in}}{\text{ft}}\right) = 0
$$

$$
\begin{aligned}
16.665\Delta x^2 - 100\Delta x &= 9600 \\
\Delta x^2 - 6\Delta x &= 576 \\
(\Delta x - 3)^2 &= 576 + 9 \\
\Delta x - 3 &= \sqrt{585} = \pm 24.2 \\
\Delta x &= \boxed{27.2\text{ in}}
\end{aligned}
$$

The answer is (A).

SI Solution

$$
\begin{aligned}
\Delta E_{\text{potential}} &= \Delta E_{\text{spring}} \\
mg(\Delta h + \Delta x) &= \tfrac{1}{2}k(\Delta x)^2
\end{aligned}
$$

Rearranging,

$$
\tfrac{1}{2}k(\Delta x)^2 - mg\Delta x - mg\Delta h = 0
$$

$$
\left(\tfrac{1}{2}\right)\left(5.837\times 10^3\ \frac{\text{N}}{\text{m}}\right)(\Delta x)^2 - (50\text{ kg})\left(9.81\ \frac{\text{m}}{\text{s}^2}\right)\Delta x - (50\text{ kg})\left(9.81\ \frac{\text{m}}{\text{s}^2}\right)(2\text{ m}) = 0
$$

$$
\begin{aligned}
2918.5\Delta x^2 - 490.5\Delta x - 981.0 &= 0 \\
\Delta x^2 - 0.1681\Delta x &= 0.3361 \\
(\Delta x - 0.08403)^2 &= 0.3361 + (0.08403)^2 \\
&= 0.3432 \\
\Delta x - 0.08403 &= \sqrt{0.3432} = \pm 0.5858 \\
\Delta x &= \boxed{0.6699\text{ m}}
\end{aligned}
$$

The answer is (A).

4. *Customary U.S. Solution*

$$W_{\text{done by wheel}} = \Delta E_{\text{rotational}} = \tfrac{1}{2}I\omega_{\text{initial}}^2 - \tfrac{1}{2}I\omega_{\text{final}}^2$$

$$\omega_{\text{final}} = \sqrt{(\omega_{\text{initial}})^2 - \frac{2W}{I}} = 2\pi f$$

$$f_{\text{final}} = \left(\frac{1}{2\pi}\right)\left(\frac{60 \text{ rpm}}{1 \frac{\text{rev}}{\text{sec}}}\right) \times \sqrt{\left(\left(2\pi \frac{\text{rad}}{\text{rev}}\right)(300 \text{ rpm})\left(\frac{1 \frac{\text{rev}}{\text{sec}}}{60 \text{ rpm}}\right)\right)^2 - \frac{(2)(45\times 10^2 \text{ ft-lbf})}{15 \text{ slug-ft}^2}}$$

$$= \boxed{187.8 \text{ rpm}}$$

The answer is (B).

SI Solution

$$W_{\text{done by wheel}} = \Delta E_{\text{rotational}} = \tfrac{1}{2}I\omega_{\text{initial}}^2 - \tfrac{1}{2}I\omega_{\text{final}}^2$$

$$\omega_{\text{final}} = \sqrt{(\omega_{\text{initial}})^2 - \frac{2W}{I}} = 2\pi f$$

$$f_{\text{final}} = \left(\frac{1}{2\pi}\right)\left(\frac{60 \text{ rpm}}{1 \frac{\text{rev}}{\text{s}}}\right) \times \sqrt{\left((2\pi)\left(\frac{300 \frac{\text{rev}}{\text{min}}}{60 \frac{\text{s}}{\text{min}}}\right)\right)^2 - \frac{(2)(6.1\times 10^3 \text{ J})}{20 \text{ kg}\cdot\text{m}^2}}$$

$$= \boxed{185.4 \text{ rpm}}$$

The answer is (B).

5. *Customary U.S. Solution*

$$W_{\text{done on box}} = F_x\Delta x = (F)(\cos\theta)\Delta x = (550 \text{ lbf})(\cos 40^\circ)(20 \text{ ft}) = \boxed{8430 \text{ ft-lbf}}$$

The answer is (D).

SI Solution

$$W_{\text{done on box}} = F_x\Delta x = (F)(\cos\theta)\Delta x = (2500 \text{ N})(\cos 40^\circ)(6 \text{ m})\left(\frac{\text{kJ}}{1000 \text{ J}}\right) = \boxed{11.5 \text{ kJ}}$$

The answer is (D).

6. *Customary U.S. Solution*

$$W_{\text{to retrieve cable}} = \int_0^l F\,dh = \int_0^l ((l-h)w)\,dh = \tfrac{1}{2}wl^2 = \left(\tfrac{1}{2}\right)\left(2 \frac{\text{lbf}}{\text{ft}}\right)(1000 \text{ ft})^2 = \boxed{10^6 \text{ ft-lbf}}$$

The answer is (C).

SI Solution

$$W_{\text{to retrieve cable}} = \int_0^l F\,dh = \int_0^l ((l-h)m_l g)\,dh = \tfrac{1}{2}m_l g l^2 = \left(\tfrac{1}{2}\right)\left(3 \frac{\text{kg}}{\text{m}}\right)\left(9.81 \frac{\text{m}}{\text{s}^2}\right)(300 \text{ m})^2 = 1.32\times 10^6 \text{ J} \quad \boxed{(1.32 \text{ MJ})}$$

The answer is (C).

7. *Customary U.S. Solution*

$$P_{\text{actual}}\Delta t = W_{\text{done by pump}}$$

$$\eta P_{\text{ideal}}\Delta t = \Delta E_{\text{potential}} = m\frac{g}{g_c}\Delta h = (\rho V)\frac{g}{g_c}\Delta h$$

$$V = \frac{\eta P_{\text{ideal}}\Delta t}{\rho \frac{g}{g_c}\Delta h} = \frac{(0.85)(7 \text{ hp})\left(550 \frac{\text{ft-lbf}}{\text{hp-sec}}\right)(3600 \text{ sec})}{\left(62.4 \frac{\text{lbm}}{\text{ft}^3}\right)\left(\frac{32.2 \frac{\text{ft}}{\text{sec}^2}}{32.2 \frac{\text{lbm-ft}}{\text{sec}^2\text{-lbf}}}\right)(130 \text{ ft})} = \boxed{1450 \text{ ft}^3}$$

The answer is (A).

SI Solution

$$P_{\text{actual}}\Delta t = W_{\text{done by pump}}$$

$$\eta P_{\text{ideal}}\Delta t = \Delta E_{\text{potential}} = mg\Delta h = (\rho V)g\Delta h$$

$$V = \frac{\eta P_{\text{ideal}} \Delta t}{\rho g \Delta h}$$

$$= \frac{(0.85)(5 \times 10^3 \text{ W})(3600 \text{ s})}{\left(1000 \ \frac{\text{kg}}{\text{m}^3}\right)\left(9.81 \ \frac{\text{m}}{\text{s}^2}\right)(40 \text{ m})}$$

$$= \boxed{39.0 \text{ m}^3}$$

The answer is (A).

8. *Customary U.S. Solution*

$$P\Delta t = W = m\frac{g}{g_c}\Delta h$$

$$P = \frac{mg\Delta h}{g_c \Delta t}$$

$$= \frac{(3300 \text{ lbm})\left(32.2 \ \frac{\text{ft}}{\text{sec}^2}\right)(250 \text{ ft})}{\left(32.2 \ \frac{\text{lbm-ft}}{\text{sec}^2\text{-lbf}}\right)(14 \text{ sec})\left(550 \ \frac{\text{ft-lbf}}{\text{hp-sec}}\right)}$$

$$= \boxed{107 \text{ hp}}$$

The answer is (D).

SI Solution

$$P\Delta t = W = mg\Delta h$$

$$P = \frac{mg\Delta h}{\Delta t}$$

$$= \left(\frac{(1500 \text{ kg})\left(9.81 \ \frac{\text{m}}{\text{s}^2}\right)(80 \text{ m})}{14 \text{ s}}\right)\left(\frac{1 \text{ kW}}{1000 \text{ W}}\right)$$

$$= \boxed{84.1 \text{ kW}}$$

The answer is (D).

14 Fluid Properties

PRACTICE PROBLEMS

(Use $g = 32.2\ \text{ft/sec}^2$ or $9.81\ \text{m/s}^2$ unless told to do otherwise in the problem.)

Pressure

1. What is the absolute pressure if a gauge reads 8.7 psi (60 kPa) vacuum?

(A) 4 psi (27 kPa)
(B) 6 psi (41 kPa)
(C) 8 psi (55 kPa)
(D) 10 psi (68 kPa)

Viscosity

2. Calculate the kinematic viscosity of air at 80°F (27°C) and 70 psia (480 kPa).

(A) $3.54 \times 10^{-5}\ \text{ft}^2\text{/sec}$ ($3.30 \times 10^{-6}\ \text{m}^2\text{/s}$)
(B) $4.25 \times 10^{-5}\ \text{ft}^2\text{/sec}$ ($3.96 \times 10^{-6}\ \text{m}^2\text{/s}$)
(C) $4.96 \times 10^{-5}\ \text{ft}^2\text{/sec}$ ($4.62 \times 10^{-6}\ \text{m}^2\text{/s}$)
(D) $6.37 \times 10^{-5}\ \text{ft}^2\text{/sec}$ ($5.94 \times 10^{-6}\ \text{m}^2\text{/s}$)

Solutions

3. Volumes of an 8% solution, a 10% solution, and a 20% solution of nitric acid are to be mixed in order to get 100 mL of a 12% solution. If the 8% solution contributes half of the total volume of nitric acid contributed by the 10% and 20% solutions, what volume of 10% acid solution is required?

(A) 20 mL
(B) 30 mL
(C) 50 mL
(D) 80 mL

SOLUTIONS

1. *Customary U.S. Solution*

$$p_{\text{gage}} = -8.7\ \text{lbf/in}^2$$
$$p_{\text{atmospheric}} = 14.7\ \text{lbf/in}^2$$

The relationship between absolute, gage, and atmospheric pressure is given by

$$\begin{aligned} p_{\text{absolute}} &= p_{\text{gage}} + p_{\text{atmospheric}} \\ &= -8.7\ \frac{\text{lbf}}{\text{in}^2} + 14.7\ \frac{\text{lbf}}{\text{in}^2} \\ &= \boxed{6\ \text{lbf/in}^2 \quad (6\ \text{psi})} \end{aligned}$$

The answer is (B).

SI Solution

$$p_{\text{gage}} = -60\ \text{kPa}$$
$$p_{\text{atmospheric}} = 101.3\ \text{kPa}$$

The relationship between absolute, gage, and atmospheric pressure is given by

$$\begin{aligned} p_{\text{absolute}} &= p_{\text{gage}} + p_{\text{atmospheric}} \\ &= -60\ \text{kPa} + 101.3\ \text{kPa} \\ &= \boxed{41.3\ \text{kPa}} \end{aligned}$$

The answer is (B).

2. *Customary U.S. Solution*

From App. 14.D, for air at 14.7 psia and 80°F, the absolute viscosity independent of pressure is $\mu = 3.85 \times 10^{-7}\ \text{lbf-sec/ft}^2$.

Determine the density of air at 70 psia and 80°F. (Assume an ideal gas.)

$$\rho = \frac{p}{RT}$$

For air, $R = 53.3\ \text{lbf-ft/lbm-°R}$.

Substituting gives

$$\rho = \frac{\left(70\ \frac{\text{lbf}}{\text{in}^2}\right)\left(144\ \frac{\text{in}^2}{\text{ft}^2}\right)}{\left(53.3\ \frac{\text{lbf-ft}}{\text{lbm-°R}}\right)(80°\text{F} + 460)} = 0.350\ \text{lbm/ft}^3$$

The kinematic viscosity, ν, is related to the absolute viscosity by

$$\nu = \frac{\mu g_c}{\rho} = \frac{\left(3.85 \times 10^{-7}\ \frac{\text{lbf-sec}}{\text{ft}^2}\right)\left(32.2\ \frac{\text{lbm-ft}}{\text{lbf-sec}^2}\right)}{0.350\ \frac{\text{lbm}}{\text{ft}^3}} = \boxed{3.54 \times 10^{-5}\ \text{ft}^2/\text{sec}}$$

The answer is (A).

SI Solution

From App. 14.E, for air at 480 kPa and 27°C, the absolute viscosity independent of pressure is $\mu = 1.84 \times 10^{-5}$ Pa·s.

Determine the density of air at 480 kPa and 27°C. (Assume an ideal gas.)

$$\rho = \frac{p}{RT}$$

For air, $R = 287$ J/kg·K.

Substituting gives

$$\rho = \frac{(480\ \text{kPa})\left(1000\ \frac{\text{Pa}}{\text{kPa}}\right)}{\left(287\ \frac{\text{J}}{\text{kg·K}}\right)(27°\text{C} + 273)} = 5.575\ \text{kg/m}^3$$

The kinematic viscosity, ν, is related to the absolute viscosity by

$$\nu = \frac{\mu}{\rho} = \frac{1.84 \times 10^{-5}\ \text{Pa·s}}{5.575\ \frac{\text{kg}}{\text{m}^3}} = \boxed{3.30 \times 10^{-6}\ \text{m}^2/\text{s}}$$

The answer is (A).

3. Let

$$\begin{aligned} x &= \text{volume of 8\% solution} \\ y &= \text{volume of 10\% solution} \\ z &= \text{volume of 20\% solution} \end{aligned}$$

The three conditions that must be satisfied are

$$\begin{aligned} x + y + z &= 100\ \text{mL} \\ 0.08x + 0.10y + 0.20z &= (0.12)(100\ \text{mL}) = 12\ \text{mL} \\ 0.08x &= \left(\tfrac{1}{2}\right)(0.10y + 0.20z) \end{aligned}$$

Simplifying these equations,

$$\begin{aligned} x + y + z &= 100 \\ 4x + 5y + 10z &= 600 \\ 8x - 5y - 10z &= 0 \end{aligned}$$

Adding the second and third equations gives

$$\begin{aligned} 12x &= 600 \\ x &= \boxed{50\ \text{mL}} \end{aligned}$$

Work with the first two equations to get

$$\begin{aligned} y + z &= 100 - 50 = 50 \\ 5y + 10z &= 600 - (4)(50) = 400 \end{aligned}$$

Multiplying the top equation by −5 and adding to the bottom equation,

$$\begin{aligned} 5z &= 150 \\ z &= 30\ \text{mL} \end{aligned}$$

From the first equation,

$$y = 20\ \text{mL}$$

The answer is (A).

15 Fluid Statics

PRACTICE PROBLEMS

(Use $g = 32.2$ ft/sec^2 or 9.81 m/s^2 unless told to do otherwise in the problem.)

1. A 4000 lbm blimp contains 10,000 lbm (4500 kg) of hydrogen (specific gas constant = 766.5 ft-lbf/lbm-°R (4124 J/kg·K)) at 56°F (13°C) and 30.2 in Hg (770 mm Hg). What is its lift if the hydrogen and air are in thermal and pressure equilibrium?

(A) 7.6×10^3 lbf (3.4×10^4 N)
(B) 1.2×10^4 lbf (5.3×10^4 N)
(C) 1.3×10^5 lbf (5.9×10^5 N)
(D) 1.7×10^5 lbf (7.7×10^5 N)

2. A hollow 6 ft (1.8 m) diameter sphere floats half-submerged in seawater. What mass of concrete is required as an external anchor to just submerge the sphere completely?

(A) 2700 lbm (1200 kg)
(B) 4200 lbm (1900 kg)
(C) 5500 lbm (2500 kg)
(D) 6300 lbm (2700 kg)

SOLUTIONS

1. *Customary U.S. Solution*

The lift of the hydrogen-filled blimp (F_{lift}) is equal to the difference between the buoyant force (F_b) and the weight of the hydrogen contained in the blimp (W_{H}).

$$F_{\text{lift}} = F_b - W_{\text{H}}$$

The weight of the hydrogen is calculated from the mass of hydrogen by

$$W_{\text{H}} = \frac{mg}{g_c} = \frac{(10{,}000 \text{ lbm})\left(32.2 \ \frac{\text{ft}}{\text{sec}^2}\right)}{32.2 \ \frac{\text{lbm-ft}}{\text{lbf-sec}^2}} = 10{,}000 \text{ lbf}$$

The buoyant force is equal to the weight of the displaced air. The weight of the displaced air is calculated by knowing that the volume of the air displaced is equal to the volume of hydrogen enclosed in the blimp. Compute the volume of the hydrogen contained in the blimp by assuming the hydrogen behaves like an ideal gas.

$$V_{\text{H}} = \frac{mRT}{p}$$

For hydrogen, $R = 766.5$ ft-lbf/lbm-°R.

The temperature of the hydrogen is given as 56°F. Convert to absolute temperature (°R).

$$T = 56°\text{F} + 460 = 516°\text{R}$$

The pressure of the hydrogen is given as 30.2 in Hg. Convert the pressure to units of pounds per square foot.

$$p = (30.2 \text{ in Hg})\left(\frac{1 \ \frac{\text{lbf}}{\text{in}^2}}{2.036 \text{ in Hg}}\right)\left(144 \ \frac{\text{in}^2}{\text{ft}^2}\right) = 2136 \text{ lbf/ft}^2$$

Compute the volume of hydrogen.

$$V_{\text{H}} = \frac{mRT}{p} = \frac{(10{,}000 \text{ lbm})\left(766.5 \ \frac{\text{ft-lbf}}{\text{lbm-°R}}\right)(516°\text{R})}{2136 \ \frac{\text{lbf}}{\text{ft}^2}} = 1.85 \times 10^6 \text{ ft}^3$$

Since the volume of the hydrogen contained in the blimp is equal to the air displaced, the air displaced can be computed from the ideal gas equation by assuming the air behaves like an ideal gas.

$$m = \frac{pV_H}{RT}$$

Since the air and hydrogen are assumed to be in thermal and pressure equilibrium, the temperature and pressure are equal to the value given for the hydrogen.

For air, $R = 53.35$ ft-lbf/lbm-°R.

Substituting gives

$$\begin{aligned} m_{air} &= \frac{pV_H}{RT} \\ &= \frac{\left(2136\ \frac{\text{lbf}}{\text{ft}^2}\right)(1.85 \times 10^6\ \text{ft}^3)}{\left(53.35\ \frac{\text{ft-lbf}}{\text{lbm-°R}}\right)(516\text{°R})} \\ &= 1.435 \times 10^5\ \text{lbm} \end{aligned}$$

Recall that the buoyant force is equal to the weight of the air.

$$\begin{aligned} F_b = W_{air} &= \frac{mg}{g_c} \\ &= \frac{(1.435 \times 10^5\ \text{lbm})\left(32.2\ \frac{\text{ft}}{\text{sec}^2}\right)}{32.2\ \frac{\text{lbm-ft}}{\text{lbf-sec}^2}} \\ &= 1.435 \times 10^5\ \text{lbf} \end{aligned}$$

Therefore, the lift can be calculated as

$$\begin{aligned} F_{lift} &= F_b - W_H - W_{blimp} \\ &= 1.435 \times 10^5\ \text{lbf} - 10{,}000\ \text{lbf} - 4000\ \text{lbf} \\ &= \boxed{1.295 \times 10^5\ \text{lbf}} \end{aligned}$$

The answer is (C).

SI Solution

Assume the mass of the blimp structure is small (negligible) compared with the mass of the hydrogen.

The lift of the hydrogen-filled blimp (F_{lift}) is equal to the difference between the buoyant force (F_b) and the weight of the hydrogen contained in the blimp (W_H).

$$F_{lift} = F_b - W_H$$

The weight of the hydrogen is calculated from the mass of hydrogen by

$$\begin{aligned} W_H &= mg \\ &= (4500\ \text{kg})\left(9.81\ \frac{\text{m}}{\text{s}^2}\right) \\ &= 44\,145\ \text{N} \end{aligned}$$

The buoyant force is equal to the weight of the displaced air. The weight of the displaced air is calculated by knowing that the volume of the air displaced is equal to the volume of hydrogen enclosed in the blimp. Compute the volume of the hydrogen contained in the blimp by assuming the hydrogen behaves like an ideal gas.

$$V_H = \frac{mRT}{p}$$

For hydrogen, $R = 4124$ J/kg·K.

The temperature of the hydrogen is given as 13°C. Convert to absolute temperature (K).

$$T = 13\text{°C} + 273 = 286\text{K}$$

The pressure of the hydrogen is given as 770 mm Hg. Convert the pressure to units of pascals.

$$\begin{aligned} p &= \frac{(770\ \text{mm Hg})\left(133.4\ \frac{\text{kPa}}{\text{m}}\right)}{1000\ \frac{\text{mm}}{\text{m}}} \\ &= 102.7\ \text{kPa} \end{aligned}$$

The volume of hydrogen is

$$\begin{aligned} V_H &= \frac{mRT}{p} \\ &= \frac{(4500\ \text{kg})\left(4124\ \frac{\text{J}}{\text{kg·K}}\right)(286\text{K})}{(102.7\ \text{kPa})\left(1000\ \frac{\text{Pa}}{\text{kPa}}\right)} \\ &= 5.168 \times 10^4\ \text{m}^3 \end{aligned}$$

Since the volume of the hydrogen contained in the blimp is equal to the air displaced, the air displaced can be computed from the ideal gas equation assuming the air behaves like an ideal gas.

$$m = \frac{pV_H}{RT}$$

Since the air and hydrogen are assumed to be in thermal and pressure equilibrium, the temperature and pressure are equal to the value given for the hydrogen.

For air, $R = 287$ J/kg·K.

Substituting gives

$$\begin{aligned} m_{air} &= \frac{pV_H}{RT} \\ &= \frac{(102.7\ \text{kPa})\left(1000\ \frac{\text{Pa}}{\text{kPa}}\right)(5.168 \times 10^4\ \text{m}^3)}{\left(287\ \frac{\text{J}}{\text{kg·K}}\right)(286\text{K})} \\ &= 6.466 \times 10^4\ \text{kg} \end{aligned}$$

The buoyant force is equal to the weight of the air, so

$$\begin{aligned}F_b &= W_{\text{air}} = mg\\ &= (6.466\times 10^4\ \text{kg})\left(9.81\ \frac{\text{m}}{\text{s}^2}\right)\\ &= 6.34\times 10^5\ \text{N}\end{aligned}$$

Therefore, the lift can be calculated as

$$\begin{aligned}F_{\text{lift}} &= F_b - W_{\text{H}}\\ &= 6.34\times 10^5\ \text{N} - 44\,145\ \text{N}\\ &= \boxed{5.90\times 10^5\ \text{N}}\end{aligned}$$

The answer is (C).

2. *Customary U.S. Solution*

The weight of the sphere is equal to the weight of the displaced volume of water when floating.

The buoyant force is given by

$$F_b = \frac{\rho g V_{\text{displaced}}}{g_c}$$

Since the sphere is half submerged,

$$W_{\text{sphere}} = \left(\tfrac{1}{2}\right)\left(\frac{\rho g V_{\text{sphere}}}{g_c}\right)$$

For seawater, $\rho = 64.0\ \text{lbm/ft}^3$.

The volume of the sphere is given by

$$\begin{aligned}V_{\text{sphere}} &= \left(\frac{\pi}{6}\right)d^3\\ &= \left(\frac{\pi}{6}\right)(6\ \text{ft})^3\\ &= 113.1\ \text{ft}^3\end{aligned}$$

The weight of the sphere is

$$\begin{aligned}W_{\text{sphere}} &= \left(\tfrac{1}{2}\right)\left(\frac{\rho g V_{\text{sphere}}}{g_c}\right)\\ &= \left(\tfrac{1}{2}\right)\left(\frac{\left(64.0\ \frac{\text{lbm}}{\text{ft}^3}\right)\left(32.2\ \frac{\text{ft}}{\text{sec}^2}\right)(113.1\ \text{ft}^3)}{32.2\ \frac{\text{lbm-ft}}{\text{lbf-sec}^2}}\right)\\ &= 3619\ \text{lbf}\end{aligned}$$

The buoyant force equation for a fully submerged sphere and anchor can be solved for the concrete volume.

$$W_{\text{sphere}} + W_{\text{concrete}} = (V_{\text{sphere}} + V_{\text{concrete}})\rho_{\text{water}}$$

$$\begin{aligned}W_{\text{sphere}} + \rho_{\text{concrete}}V_{\text{concrete}}\left(\frac{g}{g_c}\right) &= (V_{\text{sphere}} + V_{\text{concrete}})\rho_{\text{water}}\\ &\quad\times\left(\frac{g}{g_c}\right)\end{aligned}$$

$$\begin{aligned}3619\ \text{lbf} + \left(150\ \frac{\text{lbm}}{\text{ft}^3}\right)(V_{\text{concrete}})\left(\frac{32.2\ \frac{\text{ft}}{\text{sec}^2}}{32.2\ \frac{\text{ft-lbm}}{\text{lbf-sec}^2}}\right) &= (113.1\ \text{ft}^3 + V_{\text{concrete}})\\ &\quad\times\left(64.0\ \frac{\text{lbm}}{\text{ft}^3}\right)\\ &\quad\times\left(\frac{32.2\ \frac{\text{ft}}{\text{sec}^2}}{32.2\ \frac{\text{ft-lbm}}{\text{lbf-sec}^2}}\right)\end{aligned}$$

$$\begin{aligned}V_{\text{concrete}} &= 42.09\ \text{ft}^3\\ m_{\text{concrete}} &= \rho_{\text{concrete}}V_{\text{concrete}}\\ &= \left(150\ \frac{\text{lbm}}{\text{ft}^3}\right)(42.09\ \text{ft}^3)\\ &= \boxed{6314\ \text{lbm}}\end{aligned}$$

The answer is (D).

SI Solution

The weight of the sphere is equal to the weight of the displaced volume of water when floating.

The buoyant force is given by

$$F_b = \rho g V_{\text{displaced}}$$

Since the sphere is half submerged,

$$W_{\text{sphere}} = \tfrac{1}{2}\rho g V_{\text{sphere}}$$

For seawater, $\rho = 1025\ \text{kg/m}^3$.

The volume of the sphere is given by

$$\begin{aligned}V_{\text{sphere}} &= \left(\frac{\pi}{6}\right)d^3\\ &= \left(\frac{\pi}{6}\right)(1.8\ \text{m})^3\\ &= 3.054\ \text{m}^3\end{aligned}$$

The weight of the sphere required is

$$\begin{aligned}W_{\text{sphere}} &= \tfrac{1}{2}\rho g V_{\text{sphere}}\\ &= \left(\tfrac{1}{2}\right)\left(1025\ \frac{\text{kg}}{\text{m}^3}\right)\left(9.81\ \frac{\text{m}}{\text{s}^2}\right)(3.054\ \text{m}^3)\\ &= 15\,354\ \text{N}\end{aligned}$$

The buoyant force equation for a fully submerged sphere and anchor can be solved for the concrete volume.

$$W_{\text{sphere}}+W_{\text{concrete}} = (V_{\text{sphere}}+V_{\text{concrete}})\rho_{\text{water}}$$

$$W_{\text{sphere}}+\rho_{\text{concrete}}gV_{\text{concrete}} = g(V_{\text{sphere}}+V_{\text{concrete}})\rho_{\text{water}}$$

$$\begin{aligned}15\,354\ \text{N}+\left(2400\ \frac{\text{kg}}{\text{m}^3}\right)\left(9.81\ \frac{\text{m}}{\text{s}^2}\right)(V_{\text{concrete}}) &= (3.054\ \text{m}^3+V_{\text{concrete}}) \\ &\quad \times\left(1025\ \frac{\text{kg}}{\text{m}^3}\right)\left(9.81\ \frac{\text{m}}{\text{s}^2}\right)\end{aligned}$$

$$V_{\text{concrete}} = 1.138\ \text{m}^3$$

$$\begin{aligned}m_{\text{concrete}} &= \rho_{\text{concrete}}V_{\text{concrete}} \\ &= \left(2400\ \frac{\text{kg}}{\text{m}^3}\right)(1.138\ \text{m}^3) \\ &= \boxed{2731\ \text{kg}}\end{aligned}$$

The answer is (D).

16 Fluid Flow Parameters

PRACTICE PROBLEMS

Use the following values unless told to do otherwise in the problem:

$$g = 32.2 \text{ ft/sec}^2 \ (9.81 \text{ m/s}^2)$$
$$\rho_{\text{water}} = 62.4 \text{ lbm/ft}^3 \ (1000 \text{ kg/m}^3)$$
$$p_{\text{atmospheric}} = 14.7 \text{ psia} \ (101.3 \text{ kPa})$$

Hydraulic Radius

1. A 10 in (25 cm) composition pipe is compressed by a tree root until its inside height is only 7.2 in (18 cm). What is its approximate hydraulic radius when flowing half full? State your assumptions.

(A) 2.2 in (5.5 cm)
(B) 2.7 in (6.9 cm)
(C) 3.2 in (8.1 cm)
(D) 4.5 in (11.4 cm)

2. A pipe with an inside diameter of 18.812 in contains water to a depth of 15.7 in. What is the hydraulic radius? (Work in customary U.S. units only.)

(A) 4.39 in
(B) 5.08 in
(C) 5.72 in
(D) 6.51 in

SOLUTIONS

1. *Customary U.S. Solution*

The perimeter of the pipe is

$$\begin{aligned} p &= \pi d = \pi(10 \text{ in}) \\ &= 31.42 \text{ in} \end{aligned}$$

If the pipe is flowing half-full, the wetted perimeter becomes

$$\begin{aligned} \text{wetted perimeter} &= \tfrac{1}{2}p = \left(\tfrac{1}{2}\right)(31.42 \text{ in}) \\ &= 15.71 \text{ in} \end{aligned}$$

Assume the compressed pipe is an elliptical cross section. The ellipse will have a minor axis, b, equal to one-half the height of the compressed pipe or

$$b = \frac{7.2 \text{ in}}{2} = 3.6 \text{ in}$$

When the pipe is compressed, the perimeter of the pipe will remain constant. The perimeter of an ellipse is given by

$$p \approx 2\pi\sqrt{\tfrac{1}{2}(a^2 + b^2)}$$

Solve for the major axis.

$$\begin{aligned} a &= \sqrt{2\left(\frac{p}{2\pi}\right)^2 - b^2} \\ &= \sqrt{(2)\left(\frac{31.42 \text{ in}}{2\pi}\right)^2 - (3.6 \text{ in})^2} \\ &= 6.09 \text{ in} \end{aligned}$$

The flow area or area of the ellipse is given by

$$\begin{aligned} \text{flow area} &= \tfrac{1}{2}\pi ab \\ &= \tfrac{1}{2}\pi(6.09 \text{ in})(3.6 \text{ in}) \\ &= 34.4 \text{ in}^2 \end{aligned}$$

The hydraulic radius is

$$\begin{aligned} r_h &= \frac{\text{area in flow}}{\text{wetted perimeter}} = \frac{34.4 \text{ in}^2}{15.7 \text{ in}} \\ &= \boxed{2.19 \text{ in}} \end{aligned}$$

The answer is (A).

SI Solution

The perimeter of the pipe is

$$p = \pi d = \pi(25 \text{ cm}) = 78.54 \text{ cm}$$

If the pipe is flowing half-full, the wetted perimeter becomes

$$\text{wetted perimeter} = \tfrac{1}{2}p = \left(\tfrac{1}{2}\right)(78.54 \text{ cm}) = 39.27 \text{ cm}$$

Assume the compressed pipe is an elliptical cross section. The ellipse will have a minor axis, b, equal to one-half the height of the compressed pipe or

$$b = \frac{18 \text{ cm}}{2} = 9 \text{ cm}$$

When the pipe is compressed, the perimeter of the pipe will remain constant. The perimeter of an ellipse is given by

$$p \approx 2\pi\sqrt{\tfrac{1}{2}(a^2 + b^2)}$$

Solve for the major axis.

$$a = \sqrt{2\left(\frac{p}{2\pi}\right)^2 - b^2} = \sqrt{(2)\left(\frac{78.54 \text{ cm}}{2\pi}\right)^2 - (9 \text{ cm})^2} = 15.2 \text{ cm}$$

The flow area or area of the ellipse is given by

$$\text{flow area} = \tfrac{1}{2}\pi ab = \tfrac{1}{2}\pi(15.2 \text{ cm})(9 \text{ cm}) = 214.9 \text{ cm}^2$$

The hydraulic radius is

$$r_h = \frac{\text{area in flow}}{\text{wetted perimeter}} = \frac{214.9 \text{ cm}^2}{39.27 \text{ cm}} = \boxed{5.47 \text{ cm}}$$

The answer is (A).

2. *method 1:* Use App. 7.A for a circular segment.

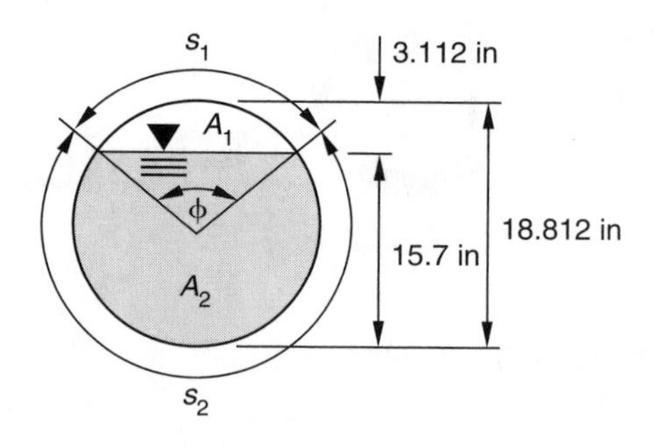

$$r = \frac{d}{2} = \frac{18.812 \text{ in}}{2} = 9.406 \text{ in}$$

$$\phi = (2)\left(\arccos\frac{r-d}{r}\right) = (2)\left(\arccos\frac{9.406 \text{ in} - 3.112 \text{ in}}{9.406 \text{ in}}\right) = 1.675 \text{ rad}$$

$$\sin\phi = 0.9946$$

$$A_1 = \tfrac{1}{2}r^2(\phi - \sin\phi) = \left(\tfrac{1}{2}\right)(9.406 \text{ in})^2(1.675 - 0.9946) = 30.1 \text{ in}^2$$

$$A_{\text{total}} = A_1 + A_2 = \frac{\pi}{4}D^2 = \left(\frac{\pi}{4}\right)(18.812 \text{ in})^2 = 277.95 \text{ in}^2$$

$$A_2 = A_{\text{total}} - A_1 = 277.95 \text{ in}^2 - 30.1 \text{ in}^2 = 247.85 \text{ in}^2$$

$$s_1 = r\phi = (9.406 \text{ in})(1.675) = 15.76 \text{ in}$$

$$s_{\text{total}} = s_1 + s_2 = \pi D = \pi(18.812 \text{ in}) = 59.1 \text{ in}$$

$$s_2 = s_{\text{total}} - s_1 = 59.1 \text{ in} - 15.76 \text{ in} = 43.34 \text{ in}$$

$$r_h = \frac{A_2}{s_2} = \frac{247.85 \text{ in}^2}{43.34 \text{ in}} = \boxed{5.719 \text{ in}}$$

method 2: Use App. 16.A.

$$\frac{d}{D} = \frac{15.7 \text{ in}}{18.812 \text{ in}} = 0.83$$

From App. 16.A, $r_h/D = 0.3041$.

$$r_h = (0.3041)(18.812 \text{ in}) = \boxed{5.72 \text{ in}}$$

The answer is (C).

17 Fluid Dynamics

PRACTICE PROBLEMS

(Use $g = 32.2\ \text{ft/sec}^2$ ($9.81\ \text{m/s}^2$) and 60°F (16°C) water unless told to do otherwise.)

Conservation of Energy

1. 5.0 ft^3/sec (130 L/s) of water flows through a schedule 40 steel pipe that changes gradually in diameter from 6 in at point A to 18 in at point B. Point B is 15 ft (4.6 m) higher than point A. The respective pressures at points A and B are 10 psia (70 kPa) and 7 psia (48.3 kPa). All minor losses are insignificant. What are the direction of flow and velocity at point A?

(A) 3.2 ft/sec (1 m/s); from A to B
(B) 25 ft/sec (7.0 m/s); from A to B
(C) 3.2 ft/sec (1 m/s); from B to A
(D) 25 ft/sec (7.5 m/s); from B to A

2. Points A and B are separated by 3000 ft of new 6 in schedule-40 steel pipe. 750 gal/min of 60°F water flow from point A to point B. Point B is 60 ft above point A. What must be the pressure at point A if the pressure at B must be 50 psig?

(A) 87 psig
(B) 103 psig
(C) 125 psig
(D) 167 psig

3. *(Time limit: one hour)* A pipe network connects junctions A, B, C, and D as shown. All pipe sections have a C-value of 150. Water can be added and removed at any of the junctions to achieve the flows listed. Water flows from point A to point D. No flows are backward. The minimum allowable pressure anywhere in the system is 20 psig. All minor losses are insignificant. For simplification, use the nominal pipe sizes.

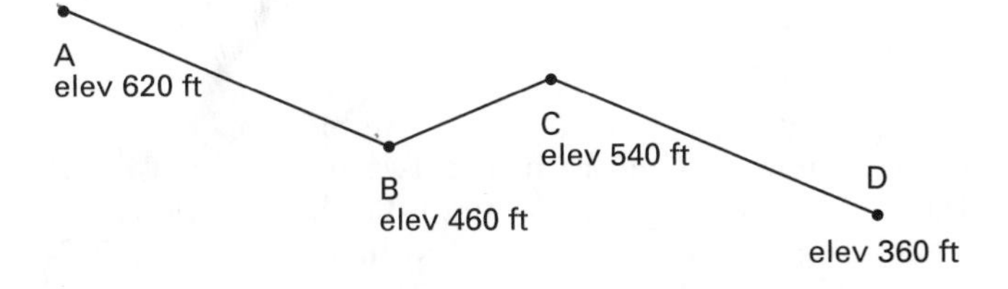

pipe section	length	diameter (in)	flow
A to B	20,000 ft	6	120 gal/min
B to C	10,000 ft	6	160 gal/min
C to D	30,000 ft	4	120 gal/min

(a) What is the pressure at point A?

(A) 14 psig
(B) 23 psig
(C) 31 psig
(D) 47 psig

(b) What is the elevation of the hydraulic grade line at point A referenced to point D?

(A) 290 ft
(B) 330 ft
(C) 470 ft
(D) 610 ft

Friction Loss

4. 1.5 ft^3/sec (40 L/s) of 70°F (20°C) water flows through 1200 ft (355 m) of 6 in (nominal) diameter new schedule-40 steel pipe. What is the friction loss?

(A) 4 ft (1.2 m)
(B) 18 ft (5.2 m)
(C) 36 ft (9.5 m)
(D) 70 ft (2.1 m)

5. 500 gal/min (30 L/s) of 100°F (40°C) water flows through 300 ft (90 m) of 6 in schedule-40 pipe. The pipe contains two 6 in flanged steel elbows, two full-open gate valves, a full-open 90° angle valve, and a swing check valve. The discharge is located 20 ft (6 m) higher than the entrance. What is the pressure difference between the two ends of the pipe?

(A) 12 psi (78 kPa)
(B) 21 psi (140 kPa)
(C) 45 psi (310 kPa)
(D) 87 psi (600 kPa)

6. 70°F (20°C) air is flowing at 60 ft/sec (18 m/s) through 300 ft (90 m) of 6 in schedule-40 pipe. The pipe contains two 6 in flanged steel elbows, two full-open gate valves, a full-open 90° angle valve, and a swing check valve. The discharge is located 20 ft (6 m) higher than the entrance. The average air density is 0.075 lbm/ft^3. The air experiences only small increases in pressure above atmospheric. What is the pressure difference between the two ends of the pipe?

(A) 0.26 psi (1.8 kPa)
(B) 0.49 psi (3.2 kPa)
(C) 1.5 psi (10 kPa)
(D) 13 psi (90 kPa)

Reservoirs

7. Three reservoirs (A, B, and C) are interconnected with a common junction (point D) at elevation 25 ft above an arbitrary reference point. The water levels for reservoirs A, B, and C are at elevations of 50, 40, and 22 ft, respectively. The pipe from reservoir A to the junction is 800 ft of 3 in (nominal) steel pipe. The pipe from reservoir B to the junction is 500 ft of 10 in (nominal) steel pipe. The pipe from reservoir C to the junction is 1000 ft of 4 in (nominal) steel pipe. All pipes are schedule 40 with a friction factor of 0.02. All minor losses and velocity heads can be neglected. What are the direction of flow and pressure at point D?

(A) out of reservoir B; 500 psf
(B) out of reservoir B; 930 psf
(C) into reservoir B; 1100 psf
(D) into reservoir B; 1260 psf

Water Hammer

8. (*Time limit: one hour*) A cast-iron pipe has an inside diameter of 24 in (600 mm) and a wall thickness of 0.75 in (20 mm). The pipe's modulus of elasticity is 20×10^6 psi (140 GPa). The pipeline is 500 ft (150 m) long. 70°F (20°C) water is flowing at 6 ft/sec (2 m/s).

(a) If a valve is closed instantaneously, what will be the pressure increase experienced in the pipe?

(A) 48 psi (330 kPa)
(B) 140 psi (970 kPa)
(C) 320 psi (2.5 MPa)
(D) 470 psi (3.2 MPa)

(b) If the pipe is 500 ft (150 m) long, over what length of time must the valve be closed to create a pressure increase equivalent to instantaneous closure?

(A) 0.25 sec
(B) 0.68 sec
(C) 1.6 sec
(D) 2.1 sec

Parallel Pipe Systems

9. 8 MGD (millions of gallons per day) (350 L/s) of 70°F (20°C) water flows into the new schedule-40 steel pipe network shown. Minor losses are insignificant.

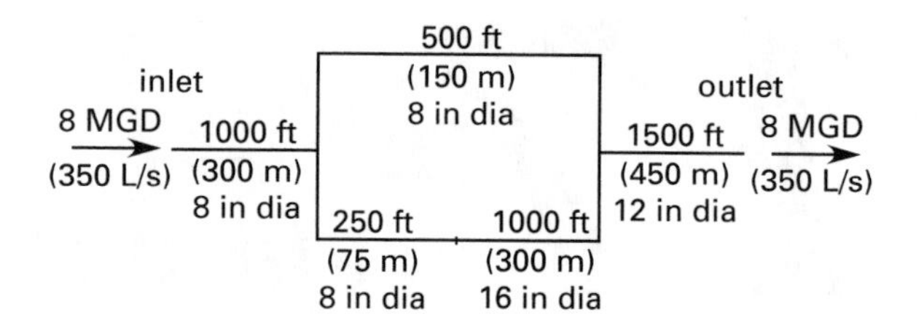

(a) What quantity of water is flowing in the upper branch?

(A) 1.2 ft^3/sec (0.034 m^3/s)
(B) 2.9 ft^3/sec (0.081 m^3/s)
(C) 4.1 ft^3/sec (0.11 m^3/s)
(D) 5.3 ft^3/sec (0.15 m^3/s)

(b) What is the energy loss per unit mass between the inlet and the outlet?

(A) 120 ft-lbf/lbm (0.37 kJ/kg)
(B) 300 ft-lbf/lbm (0.90 kJ/kg)
(C) 480 ft-lbf/lbm (1.4 kJ/kg)
(D) 570 ft-lbf/lbm (1.7 kJ/kg)

Pipe Networks

10. A single-loop pipe network is shown. The distance between each junction is 1000 ft. All junctions are on the same elevation. All pipes have a C-value of 100. The volumetric flow rates are to be determined to within 2 gal/min.

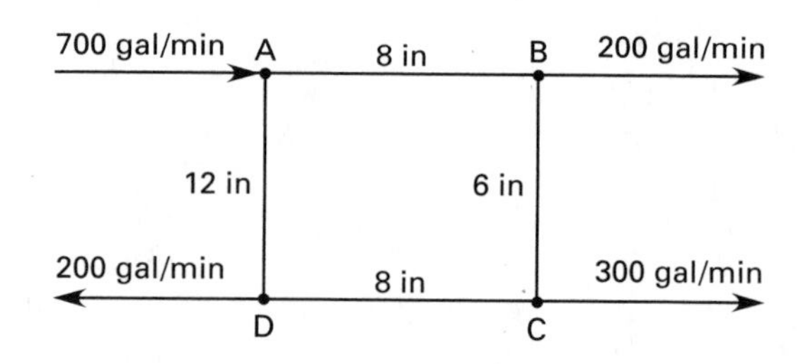

What is the flow rate between junctions B and C?

(A) 36 gal/min from B to C
(B) 58 gal/min from B to C
(C) 84 gal/min from C to B
(D) 112 gal/min from C to B

11. (*Time limit: one hour*) A double-loop pipe network is shown. The distance between each junction is 1000 ft. The water temperature is 60°F. Elevations and some pressure are known for the junctions. All pipes have a C-value of 100.

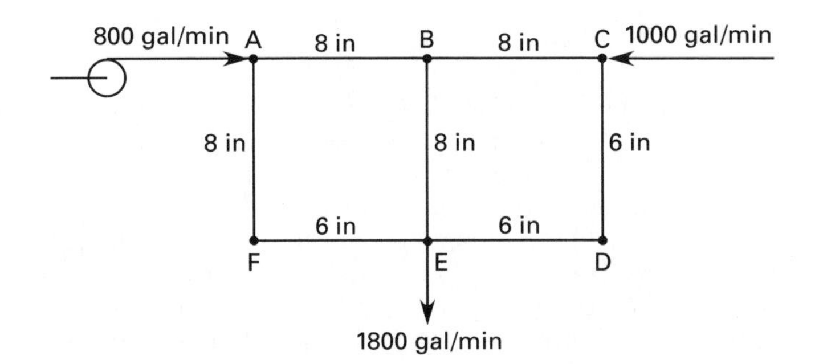

point	pressure	elevation
A		200 ft
B		150 ft
C	40 psig	300 ft
D		150 ft
E		200 ft
F		150 ft

(a) What is the flow rate between junctions B and E?

(A) 540 gal/min
(B) 620 gal/min
(C) 810 gal/min
(D) 980 gal/min

(b) What is the pressure at point D?

(A) 51 psig
(B) 73 psig
(C) 96 psig
(D) 120 psig

(c) If the pump receives water at 20 psig (140 kPa), what hydraulic power is required?

(A) 14 hp
(B) 28 hp
(C) 35 hp
(D) 67 hp

12. *(Time limit: one hour)* The water distribution network shown consists of class A cast-iron pipe installed 1.5 years ago. 1.5 MGD of 50°F water enters at junction A and leaves at junction B. The minimum acceptable pressure at point D is 40 psig (280 kPa).

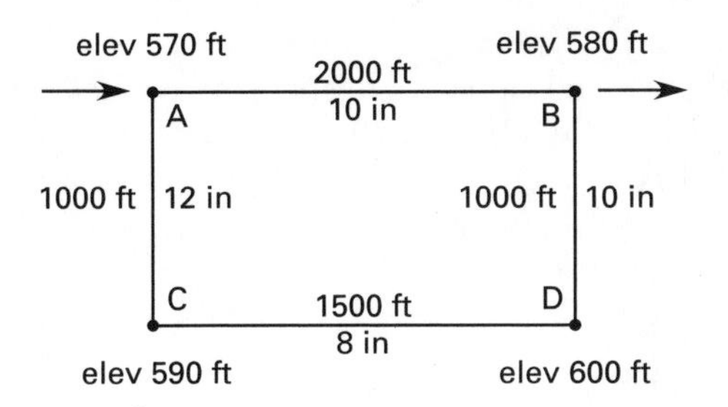

(a) What is the flow rate between junctions A and B?

(A) 320 gal/min
(B) 480 gal/min
(C) 590 gal/min
(D) 660 gal/min

(b) What is the pressure at point B?

(A) 32 psig
(B) 48 psig
(C) 57 psig
(D) 88 psig

Tank Discharge

13. The velocity of discharge from a fire hose is 50 ft/sec (15 m/s). The hose is oriented 45° from the horizontal. Disregarding air friction, what is the maximum range of the discharge?

(A) 45 ft (14 m)
(B) 78 ft (23 m)
(C) 91 ft (24 m)
(D) 110 ft (33 m)

14. A full cylindrical tank 40 ft (12 m) high has a constant diameter of 20 ft. The tank has a 4 in (100 mm) diameter hole in its bottom. The coefficient of discharge for the hole is 0.98. How long will it take for the water level to drop from 40 ft to 20 ft (12 m to 6 m)?

(A) 950 sec
(B) 1200 sec
(C) 1450 sec
(D) 1700 sec

Venturi Meters

15. A venturi meter with an 8 in diameter throat is installed in a 12 in diameter water line. The venturi is perfectly smooth, so that the discharge coefficient is 1.00. An attached mercury manometer registers a 4 in differential. What is the volumetric flow rate?

(A) 1.7 ft^3/sec
(B) 5.2 ft^3/sec
(C) 6.4 ft^3/sec
(D) 18 ft^3/sec

16. 60°F (15°C) benzene (specific gravity at 60°F (15°C) of 0.885) flows through an 8 in/3$^1/_2$ in (200 mm/90 mm) venturi meter whose coefficient of discharge is 0.99. A mercury manometer indicates a 4 in difference in the heights of the mercury columns. What is the volumetric flow rate of the benzene?

(A) 1.2 ft^3/sec (34 L/s)
(B) 9.1 ft^3/sec (250 L/s)
(C) 13 ft^3/sec (360 L/s)
(D) 27 ft^3/sec (760 L/s)

Water Resources

Orifice Meters

17. A sharp-edged orifice meter with a 0.2 ft diameter opening is installed in a 1.0 ft diameter pipe. 70°F water approaches the orifice at 2 ft/sec. What is the indicated pressure drop across the orifice meter?

(A) 5.9 psi
(B) 13 psi
(C) 22 psi
(D) 47 psi

18. A mercury manometer is used to measure a pressure difference across an orifice meter in a water line. The difference in mercury levels is 7 in (17.8 cm). What is the pressure differential?

(A) 1.7 psi (12 kPa)
(B) 3.2 psi (22 kPa)
(C) 7.9 psi (55 kPa)
(D) 23 psi (160 kPa)

19. A sharp-edged ISA orifice is used in a schedule 40 steel 12 in (300 mm) water line. (Figure 17.28 is applicable.) The water temperature is 70°F (20°C), and the flow rate is 10 ft^3/sec (250 L/s). The differential pressure change across the orifice (to the vena contracta) must not exceed 25 ft (7.5 m). What is the smallest orifice that can be used?

(A) 5.5 in (14 cm)
(B) 7.3 in (19 cm)
(C) 8.1 in (20 cm)
(D) 8.9 in (23 cm)

Impulse-Momentum

20. A pipe necks down from 24 in at point A to 12 in at point B. 8 ft^3/sec of 60°F water flow from point A to point B. The pressure head at point A is 20 ft. Friction is insignificant over the distance between points A and B. What are the magnitude and direction of the resultant force on the water?

(A) 2900 lbf; toward A
(B) 3500 lbf; toward A
(C) 2900 lbf; toward B
(D) 3500 lbf; toward B

21. A 2 in (50 mm) diameter horizontal water jet has an absolute velocity (with respect to a stationary point) of 40 ft/sec (12 m/s) as it strikes a curved blade. The blade is moving horizontally away with an absolute velocity of 15 ft/sec (4.5 m/s). Water is deflected 60° from the horizontal. What is the force on the blade?

(A) 18 lbf (80 N)
(B) 26 lbf (110 N)
(C) 35 lbf (160 N)
(D) 47 lbf (210 N)

Pumps

22. 2000 gal/min (125 L/s) of brine with a specific gravity of 1.2 pass through an 85% efficient pump. The centerlines of the pump's 12 in inlet and 8 in outlet are at the same elevation. The inlet suction gauge indicates 6 in (150 mm) of mercury below atmospheric. The discharge pressure gauge is located 4 ft (1.2 m) above the centerline of the pump's outlet and indicates 20 psig (138 kPa). All pipes are schedule 40. What is the input power to the pump?

(A) 12 hp (8.9 kW)
(B) 36 hp (26 kW)
(C) 52 hp (39 kW)
(D) 87 hp (65 kW)

Turbines

23. 100 ft^3/sec (2.6 kL/s) of water pass through a horizontal turbine. The water's pressure is reduced from 30 psig (210 kPa) to 5 psi (35 kPa) vacuum. Disregarding friction, velocity, and other factors, what power is generated?

(A) 110 hp (82 kW)
(B) 380 hp (280 kW)
(C) 730 hp (540 kW)
(D) 920 hp (640 kW)

Drag

24. (*Time limit: one hour*) A car traveling through 70°F (20°C) air has the following characteristics.

frontal area	28 ft^2 (2.6 m^2)
mass	3300 lbm (1500 kg)
drag coefficient	0.42
rolling resistance	1% of weight
engine thermal efficiency	28%
fuel heating value	115,000 Btu/gal (460 MJ/L)

(a) Considering only the drag and rolling resistance, what is the fuel consumption when the car is traveling at 55 mi/hr (90 km/h)?

(A) 0.026 gal/mi (0.0043 L/km)
(B) 0.038 gal/mi (0.0065 L/km)
(C) 0.051 gal/mi (0.0097 L/km)
(D) 0.13 gal/mi (0.022 L/km)

(b) What is the percentage increase in fuel consumption at 65 mi/hr (105 km/h) compared to at 55 mi/hr (90 km/h)?

(A) 10%
(B) 20%
(C) 30%
(D) 40%

SOLUTIONS

1. *Customary U.S. Solution*

Assume schedule-40 pipe.

$$D_A = 0.5054 \text{ ft}$$
$$D_B = 1.4063 \text{ ft}$$

Let point A be at zero elevation.

The total energy at point A from Bernoulli's equation is

$$E_{tA} = E_p + E_v + E_z = \frac{p_A}{\rho} + \frac{v_A^2}{2g_c} + \frac{z_A g}{g_c}$$

At point A, the diameter is 6 in. The velocity at point A is

$$\dot{V} = v_A A_A = v_A\left(\frac{\pi}{4}\right)D_A^2$$

$$v_A = \left(\frac{4}{\pi}\right)\left(\frac{\dot{V}}{D_A^2}\right) = \left(\frac{4}{\pi}\right)\left(\frac{5.0\ \frac{\text{ft}^3}{\text{sec}}}{(0.5054\text{ ft})^2}\right) = \boxed{24.9 \text{ ft/sec}}$$

$$p_A = \left(10\ \frac{\text{lbf}}{\text{in}^2}\right)\left(\frac{144\text{ in}^2}{1\text{ ft}^2}\right) = 1440 \text{ lbf/ft}^2$$

$$z_A = 0$$

For water, $\rho \approx 62.4 \text{ lbm/ft}^3$.

$$E_{tA} = \frac{p_A}{\rho} + \frac{v_A^2}{2g_c} + \frac{z_A g}{g_c} = \frac{1440\ \frac{\text{lbf}}{\text{ft}^2}}{62.4\ \frac{\text{lbm}}{\text{ft}^3}} + \frac{\left(24.9\ \frac{\text{ft}}{\text{sec}}\right)^2}{(2)\left(32.2\ \frac{\text{lbm-ft}}{\text{lbf-sec}^2}\right)} + 0 = 32.7 \text{ ft-lbf/lbm}$$

Similarly, the total energy at point B is

$$v_B = \left(\frac{4}{\pi}\right)\left(\frac{\dot{V}}{D_B^2}\right) = \left(\frac{4}{\pi}\right)\left(\frac{5.0\ \frac{\text{ft}^3}{\text{sec}}}{(1.4063\text{ ft})^2}\right) = 3.22 \text{ ft/sec}$$

$$p_B = \left(7\ \frac{\text{lbf}}{\text{in}^2}\right)\left(\frac{144\text{ in}^2}{1\text{ ft}^2}\right) = 1008 \text{ lbf/ft}^2$$

$$z_B = 15 \text{ ft}$$

$$E_{tB} = \frac{p_B}{\rho} + \frac{v_B^2}{2g_c} + \frac{z_B g}{g_c} = \frac{1008\ \frac{\text{lbf}}{\text{ft}^2}}{62.4\ \frac{\text{lbm}}{\text{ft}^3}} + \frac{\left(3.22\ \frac{\text{ft}}{\text{sec}}\right)^2}{(2)\left(32.2\ \frac{\text{lbm-ft}}{\text{lbf-sec}^2}\right)} + \frac{(15\text{ ft})\left(32.2\ \frac{\text{ft}}{\text{sec}^2}\right)}{32.2\ \frac{\text{lbm-ft}}{\text{lbf-sec}^2}} = 31.3 \text{ ft-lbf/lbm}$$

Since $E_{tA} > E_{tB}$, $\boxed{\text{the flow is from point A to point B.}}$

The answer is (B).

SI Solution

Let point A be at zero elevation.

The total energy at point A from Bernoulli's equation is

$$E_{tA} = E_p + E_v + E_z = \frac{p_A}{\rho} + \frac{v_A^2}{2} + z_A g$$

At point A, from App. 16.C, the diameter is 154 mm (0.154 m). The velocity at point A is

$$\dot{V} = v_A A_A = v_A\left(\frac{\pi}{4}\right)D_A^2$$

$$v_A = \left(\frac{4}{\pi}\right)\left(\frac{\dot{V}}{D_A^2}\right) = \left(\frac{4}{\pi}\right)\left(\frac{\left(130\ \frac{\text{L}}{\text{s}}\right)\left(\frac{1\text{ m}^3}{1000\text{ L}}\right)}{(0.154\text{ m})^2}\right) = \boxed{6.98 \text{ m/s}}$$

$$p_A = 69 \text{ kPa} \quad (69\,000 \text{ Pa})$$

$$z_A = 0$$

For water, $\rho = 1000$ kg/m^3.

$$E_{tA} = \frac{p_A}{\rho} + \frac{v_A^2}{2} + z_A g = \frac{70\,000 \text{ Pa}}{1000 \frac{\text{kg}}{\text{m}^3}} + \frac{\left(6.98 \frac{\text{m}}{\text{s}}\right)^2}{2} + 0 = 94.36 \text{ J/kg}$$

Similarly, at B, the diameter is 0.429 m. The total energy at point B is

$$v_B = \left(\frac{4}{\pi}\right)\left(\frac{\dot{V}}{D_B^2}\right) = \left(\frac{4}{\pi}\right)\left(\frac{\left(130 \frac{\text{L}}{\text{s}}\right)\left(\frac{1 \text{ m}^3}{1000 \text{ L}}\right)}{(0.429 \text{ m})^2}\right) = 0.90 \text{ m/s}$$

$$p_B = 48.3 \text{ kPa} \quad (48\,300 \text{ Pa})$$

$$z_B = 4.6 \text{ m}$$

$$E_{tB} = \frac{p_B}{\rho} + \frac{v_B^2}{2} + z_B g = \frac{48\,300 \text{ Pa}}{1000 \frac{\text{kg}}{\text{m}^3}} + \frac{\left(0.90 \frac{\text{m}}{\text{s}}\right)^2}{2} + (4.6 \text{ m})\left(9.81 \frac{\text{m}}{\text{s}^2}\right) = 93.8 \text{ J/kg}$$

Since $E_{tA} > E_{tB}$, **the flow is from point A to point B.**

The answer is (B).

2.

$$\dot{V} = \left(750 \frac{\text{gal}}{\text{min}}\right)\left(0.002228 \frac{\text{ft}^3\text{-min}}{\text{sec-gal}}\right) = 1.671 \text{ ft}^3/\text{sec}$$

$D = 0.5054$ ft and $A = 0.2006$ ft^2.

$$v = \frac{\dot{V}}{A} = \frac{1.671 \frac{\text{ft}^3}{\text{sec}}}{0.2006 \text{ ft}^2} = 8.33 \text{ ft/sec}$$

For 60°F water,

$$\nu = 1.217 \times 10^{-5} \text{ ft}^2/\text{sec} \quad \text{[App. 14.A]}$$

$$\text{Re} = \frac{vD}{\nu} = \frac{\left(8.33 \frac{\text{ft}}{\text{sec}}\right)(0.5054 \text{ ft})}{1.217 \times 10^{-5} \frac{\text{ft}^2}{\text{sec}}} = 3.46 \times 10^5$$

For steel,

$$\epsilon = 0.0002 \quad \text{[App. 17.A]}$$

$$\frac{\epsilon}{D} = \frac{0.0002 \text{ ft}}{0.5054 \text{ ft}} \approx 0.0004$$

$$f = 0.0175 \quad \text{[App. 17.B]}$$

From Eq. 17.22,

$$h_f = \frac{(0.0175)(3000 \text{ ft})\left(8.33 \frac{\text{ft}}{\text{sec}}\right)^2}{(2)(0.5054 \text{ ft})\left(32.2 \frac{\text{ft}}{\text{sec}^2}\right)} = 111.9 \text{ ft}$$

Use the Bernoulli equation. Since velocity is the same at points A and B, it may be omitted.

$$\frac{\left(144 \frac{\text{in}^2}{\text{ft}^2}\right) p_1}{62.37 \frac{\text{lbf}}{\text{ft}^3}} = \frac{\left(50 \frac{\text{lbf}}{\text{in}^2}\right)\left(144 \frac{\text{in}^2}{\text{ft}^2}\right)}{62.37 \frac{\text{lbf}}{\text{ft}^3}} + 60 \text{ ft} + 111.9 \text{ ft}$$

$$p_1 = \boxed{124.5 \text{ lbf/in}^2 \quad \text{(psig)}}$$

The answer is (C).

3. From Eq. 17.31, the friction loss from A to B is

$$h_{f,A-B} = \frac{(10.44)(20{,}000 \text{ ft})\left(120 \frac{\text{gal}}{\text{min}}\right)^{1.85}}{(150)^{1.85}(6 \text{ in})^{4.8655}} = 22.6 \text{ ft}$$

Calculate the velocity head.

$$v = \frac{\dot{V}}{A} = \frac{\left(120 \frac{\text{gal}}{\text{min}}\right)\left(0.002228 \frac{\text{ft}^3\text{-min}}{\text{sec-gal}}\right)}{\left(\frac{\pi}{4}\right)\left(\frac{6 \text{ in}}{12 \frac{\text{in}}{\text{ft}}}\right)^2} = 1.36 \text{ ft/sec}$$

$$h_v = \frac{v^2}{2g} = \frac{\left(1.36 \frac{\text{ft}}{\text{sec}}\right)^2}{(2)\left(32.2 \frac{\text{ft}}{\text{sec}^2}\right)} = 0.029 \text{ ft}$$

Velocity heads are low and can be disregarded.

$$h_{f,\mathrm{B-C}} = \frac{(10.44)(10{,}000\ \mathrm{ft})\left(160\ \frac{\mathrm{gal}}{\mathrm{min}}\right)^{1.85}}{(150)^{1.85}(6\ \mathrm{in})^{4.8655}} = 19.25\ \mathrm{ft}$$

$$h_{f,\mathrm{C-D}} = \frac{(10.44)(30{,}000\ \mathrm{ft})\left(120\ \frac{\mathrm{gal}}{\mathrm{min}}\right)^{1.85}}{(150)^{1.85}(4\ \mathrm{in})^{4.8655}} = 243.9\ \mathrm{ft}$$

Assume a pressure of 20 psig at point A.

$$h_{p,\mathrm{A}} = \frac{\left(20\ \frac{\mathrm{lbf}}{\mathrm{in}^2}\right)\left(144\ \frac{\mathrm{in}^2}{\mathrm{ft}^2}\right)}{62.4\ \frac{\mathrm{lbf}}{\mathrm{ft}^3}} = 46.2\ \mathrm{ft}$$

From the Bernoulli equation, ignoring velocity head,

$$\begin{aligned} h_{p,\mathrm{A}} + z_{\mathrm{A}} &= h_{p,\mathrm{B}} + z_{\mathrm{B}} + h_{f,\mathrm{A-B}} \\ 46.2\ \mathrm{ft} + 620\ \mathrm{ft} &= h_{p,\mathrm{B}} + 460\ \mathrm{ft} + 22.6\ \mathrm{ft} \\ h_{p,\mathrm{B}} &= 183.6\ \mathrm{ft} \end{aligned}$$

$$p_{\mathrm{B}} = \gamma h = \frac{\left(62.4\ \frac{\mathrm{lbf}}{\mathrm{ft}^3}\right)(183.6\ \mathrm{ft})}{144\ \frac{\mathrm{in}^2}{\mathrm{ft}^2}} = 79.6\ \mathrm{lbf/in}^2\quad(\mathrm{psig})$$

For B to C,

$$\begin{aligned} 183.6\ \mathrm{ft} + 460\ \mathrm{ft} &= h_{p,\mathrm{C}} + 540\ \mathrm{ft} + 19.25\ \mathrm{ft} \\ h_{p,\mathrm{C}} &= 84.35\ \mathrm{ft} \end{aligned}$$

$$p_{\mathrm{C}} = \frac{(84.35\ \mathrm{ft})\left(62.4\ \frac{\mathrm{lbf}}{\mathrm{ft}^3}\right)}{144\ \frac{\mathrm{in}^2}{\mathrm{ft}^2}} = 36.6\ \mathrm{lbf/in}^2\quad(\mathrm{psig})$$

For C to D,

$$\begin{aligned} 84.35\ \mathrm{ft} + 540\ \mathrm{ft} &= h_{p,\mathrm{D}} + 360\ \mathrm{ft} + 243.9\ \mathrm{ft} \\ h_{p,\mathrm{D}} &= 20.45\ \mathrm{ft} \end{aligned}$$

$$p_{\mathrm{D}} = \frac{(20.45\ \mathrm{ft})\left(62.4\ \frac{\mathrm{lbf}}{\mathrm{ft}^3}\right)}{144\ \frac{\mathrm{in}^2}{\mathrm{ft}^2}} = 8.9\ \mathrm{lbf/in}^2\quad(\mathrm{psig})\quad[\text{too low}]$$

Since p_{D} is too low, add 20 – 8.9 = 11.1 lbf/in² (psig) to each point.

(a) $$p_{\mathrm{A}} = 20.0\ \frac{\mathrm{lbf}}{\mathrm{in}^2} + 11.1\ \frac{\mathrm{lbf}}{\mathrm{in}^2} = \boxed{31.1\ \mathrm{lbf/in}^2\quad(\mathrm{psig})}$$

$$p_{\mathrm{B}} = 79.6\ \frac{\mathrm{lbf}}{\mathrm{in}^2} + 11.1\ \frac{\mathrm{lbf}}{\mathrm{in}^2} = 90.7\ \mathrm{lbf/in}^2\quad(\mathrm{psig})$$

$$p_{\mathrm{C}} = 36.6\ \frac{\mathrm{lbf}}{\mathrm{in}^2} + 11.1\ \frac{\mathrm{lbf}}{\mathrm{in}^2} = 47.7\ \mathrm{lbf/in}^2\quad(\mathrm{psig})$$

$$p_{\mathrm{D}} = 8.9\ \frac{\mathrm{lbf}}{\mathrm{in}^2} + 11.1\ \frac{\mathrm{lbf}}{\mathrm{in}^2} = 20.0\ \mathrm{lbf/in}^2\quad(\mathrm{psig})$$

The answer is (C).

(b) The elevation of the hydraulic grade line above point D is the sum of the potential and static heads.

$$\begin{aligned} \Delta h_{\mathrm{A-D}} &= z_{\mathrm{A}} - z_{\mathrm{D}} + \frac{p_{\mathrm{A}} - p_{\mathrm{D}}}{\gamma} \\ &= 620\ \mathrm{ft} - 360\ \mathrm{ft} + \frac{\left(31.1\ \frac{\mathrm{lbf}}{\mathrm{in}^2} - 20\ \frac{\mathrm{lbf}}{\mathrm{in}^2}\right)\left(144\ \frac{\mathrm{in}^2}{\mathrm{ft}^2}\right)}{62.4\ \frac{\mathrm{lbf}}{\mathrm{ft}^3}} \\ &= \boxed{285.6\ \mathrm{ft}} \end{aligned}$$

The answer is (A).

4. *Customary U.S. Solution*

For 6 in schedule-40 pipe, the internal diameter, D_i, is 0.5054 ft. The internal area is 0.2006 ft².

The velocity, v, is calculated from the volumetric flow, $\dot{V}$, and the flow area, A_i, by

$$\mathrm{v} = \frac{\dot{V}}{A_i} = \frac{1.5\ \frac{\mathrm{ft}^3}{\mathrm{sec}}}{0.2006\ \mathrm{ft}^2} = 7.48\ \mathrm{ft/sec}$$

For water at 70°F, the kinematic viscosity, ν, is 1.059×10^{-5} ft²/sec [App. 14.A].

Calculate the Reynolds number.

$$\mathrm{Re} = \frac{D_i \mathrm{v}}{\nu} = \frac{(0.5054\ \mathrm{ft})\left(7.48\ \frac{\mathrm{ft}}{\mathrm{sec}}\right)}{1.059 \times 10^{-5}\ \frac{\mathrm{ft}^2}{\mathrm{sec}}} = 3.57 \times 10^5$$

Since Re > 2100, the flow is turbulent. The friction loss coefficient can be determined from the Moody diagram.

For new steel pipe, the specific roughness, ϵ, is 0.0002 ft.

The relative roughness is

$$\frac{\epsilon}{D_i} = \frac{0.0002 \text{ ft}}{0.5054 \text{ ft}} = 0.0004$$

From the Moody diagram with Re = 3.57×10^5 and $e/D_i = 0.0004$, the friction factor, f, can be determined as 0.0174.

Use Darcy's equation to compute the frictional loss.

$$h_f = \frac{fL\text{v}^2}{2D_i g} = \frac{(0.0174)(1200 \text{ ft})\left(7.48 \ \frac{\text{ft}}{\text{sec}}\right)^2}{(2)(0.5054 \text{ ft})\left(32.2 \ \frac{\text{ft}}{\text{sec}^2}\right)} = \boxed{35.9 \text{ ft}}$$

The answer is (C).

SI Solution

For 6 in pipe, the internal diameter is 154.1 mm and the internal area is 186.5×10^{-4} m^2.

The velocity, v, is calculated from the volumetric flow, $\dot{V}$, and the flow area, A_i, by

$$\text{v} = \frac{\dot{V}}{A_i} = \frac{\left(40 \ \frac{\text{L}}{\text{s}}\right)\left(0.001 \ \frac{\text{m}^3}{\text{L}}\right)}{186.5 \times 10^{-4} \text{ m}^2} = 2.145 \text{ m/s}$$

For water at 20°C, the kinematic viscosity is

$$\nu = \frac{\mu}{\rho} = \frac{1.0050 \times 10^{-3} \text{ Pa·s}}{998.23 \ \frac{\text{kg}}{\text{m}^3}} = 1.007 \times 10^{-6} \text{ m}^2/\text{s}$$

Calculate the Reynolds number.

$$\text{Re} = \frac{D_i \text{v}}{\nu} = \frac{(154.1 \text{ mm})\left(0.001 \ \frac{\text{m}}{\text{mm}}\right)\left(2.145 \ \frac{\text{m}}{\text{s}}\right)}{1.007 \times 10^{-6} \ \frac{\text{m}^2}{\text{s}}} = 3.282 \times 10^5$$

Since Re > 2100, the flow is turbulent. The friction loss coefficient can be determined from the Moody diagram.

For new steel pipe, the specific roughness, ϵ, is 6.0×10^{-5} m.

The relative roughness is

$$\frac{\epsilon}{D_i} = \frac{6.0 \times 10^{-5} \text{ m}}{0.1541 \text{ m}} = 0.0004$$

From the Moody diagram with Re = 3.28×10^5 and $e/D_i = 0.0004$, the friction factor, f, can be determined as 0.0175.

Use Darcy's equation to compute the frictional loss.

$$h_f = \frac{fL\text{v}^2}{2D_i g} = \frac{(0.0175)(355 \text{ m})\left(2.145 \ \frac{\text{m}}{\text{s}}\right)^2}{(2)(0.1541 \text{ m})\left(9.81 \ \frac{\text{m}}{\text{s}^2}\right)} = \boxed{9.45 \text{ m}}$$

The answer is (C).

5. *Customary U.S. Solution*

For 6 in schedule-40 pipe, the internal diameter, D_i, is 0.5054 ft. The internal area is 0.2006 ft^2.

The velocity, v, is calculated from the volumetric flow, $\dot{V}$, and the flow area, A_i.

Convert the volumetric flow rate from gal/min to ft^3/sec.

$$\dot{V} = \left(500 \ \frac{\text{gal}}{\text{min}}\right)\left(\frac{1 \text{ ft}^3}{7.48 \text{ gal}}\right)\left(\frac{1 \text{ min}}{60 \text{ sec}}\right) = 1.114 \text{ ft}^3/\text{sec}$$

The velocity is

$$\text{v} = \frac{\dot{V}}{A_i} = \frac{1.114 \ \frac{\text{ft}^3}{\text{sec}}}{0.2006 \text{ ft}^2} = 5.55 \text{ ft/sec}$$

Use App. 14.A. For water at 100°F, the kinematic viscosity, ν, is 0.739×10^{-5} ft^2/sec and the density is 62.00 lbm/ft^2.

Calculate the Reynolds number.

$$\begin{aligned}\text{Re} &= \frac{D_i \text{v}}{\nu} \\ &= \frac{(0.5054 \text{ ft})\left(5.55 \ \frac{\text{ft}}{\text{sec}}\right)}{0.739 \times 10^{-5} \ \frac{\text{ft}^2}{\text{sec}}} \\ &= 3.80 \times 10^5\end{aligned}$$

Since Re > 2100, the flow is turbulent. The friction loss coefficient can be determined from the Moody diagram.

For new steel pipe, the specific roughness, ϵ, is 0.0002 ft.

The relative roughness is

$$\begin{aligned}\frac{\epsilon}{D_i} &= \frac{0.0002 \text{ ft}}{0.5054 \text{ ft}} \\ &= 0.0004\end{aligned}$$

From the Moody diagram with Re = 3.80×10^5 and $e/D_i = 0.0004$, the friction factor, f, can be determined as 0.0173.

Use App. 17.D. The equivalent lengths of the valves and fittings are

standard radius elbow	2 × 8.9 ft	=	17.8 ft
gate valve (fully open)	2 × 3.2 ft	=	6.4 ft
90° angle valve (fully open)	1 × 63.0 ft	=	63.0 ft
swing check valve	1 × 63.0 ft	=	63.0 ft
			150.2 ft

The equivalent pipe length is the sum of the straight run of pipe and the equivalent length of pipe for the valves and fittings.

$$\begin{aligned}L_e &= L + L_{\text{fittings}} \\ &= 300 \text{ ft} + 150.2 \text{ ft} \\ &= 450.2 \text{ ft}\end{aligned}$$

Use Darcy's equation to compute the frictional loss.

$$\begin{aligned}h_f &= \frac{f L_e \text{v}^2}{2 D_i g} \\ &= \frac{(0.0173)(450.2 \text{ ft})\left(5.55 \ \frac{\text{ft}}{\text{sec}}\right)^2}{(2)(0.5054 \text{ ft})\left(32.2 \ \frac{\text{ft}}{\text{sec}^2}\right)} \\ &= 7.37 \text{ ft}\end{aligned}$$

The total difference in head is the sum of the head loss through the pipe, valves, and fittings and the change in elevation.

$$\begin{aligned}\Delta h &= h_f + \Delta z \\ &= 7.37 \text{ ft} + 20 \text{ ft} \\ &= 27.37 \text{ ft}\end{aligned}$$

The pressure difference between the entrance and discharge can be determined from

$$\begin{aligned}\Delta p &= \gamma \Delta h = \rho h \times \left(\frac{g}{g_c}\right) \\ &= \frac{\left(62.0 \ \frac{\text{lbm}}{\text{ft}^3}\right)(27.37 \text{ ft})\left(32.2 \ \frac{\text{ft}}{\text{sec}^2}\right)\left(\frac{1 \text{ ft}^2}{144 \text{ in}^2}\right)}{32.2 \ \frac{\text{lbm-ft}}{\text{lbf-sec}^2}} \\ &= \boxed{11.8 \text{ lbf/in}^2 \quad (11.8 \text{ psi})}\end{aligned}$$

The answer is (A).

SI Solution

For 6 in pipe, the internal diameter is 154.1 mm (0.1541 m).

The internal area is 186.5×10^{-4} m^2.

The velocity, v, is calculated from the volumetric flow, $\dot{V}$, and the flow area, A_i.

$$\begin{aligned}\text{v} &= \frac{\dot{V}}{A_i} \\ &= \frac{\left(30 \ \frac{\text{L}}{\text{s}}\right)\left(0.001 \ \frac{\text{m}^3}{\text{L}}\right)}{186.5 \times 10^{-4} \text{ m}^2} \\ &= 1.61 \text{ m/s}\end{aligned}$$

Use App. 14.B. For water at 40°C, the kinematic viscosity is 6.611×10^{-7} m^2/s and the density is 992.25 kg/m^3.

Calculate the Reynolds number.

$$\begin{aligned}\text{Re} &= \frac{D_i \text{v}}{\nu} \\ &= \frac{(0.1541 \text{ m})\left(1.61 \ \frac{\text{m}}{\text{s}}\right)}{6.611 \times 10^{-7} \ \frac{\text{m}^2}{\text{s}}} \\ &= 3.75 \times 10^5\end{aligned}$$

Since Re > 2100, the flow is turbulent. The friction loss coefficient can be determined from the Moody diagram.

For new steel pipe, the specific roughness is 6.0×10^{-5} m.

The relative roughness is

$$\begin{aligned}\frac{\epsilon}{D_i} &= \frac{6.0 \times 10^{-5} \text{ m}}{0.1541 \text{ m}} \\ &= 0.0004\end{aligned}$$

From the Moody diagram with $\text{Re} = 3.75 \times 10^5$ and $e/D_i = 0.0004$, the friction factor, f, can be determined as 0.0173.

Use App. 17.D. The equivalent lengths of the valves and fittings are

standard radius elbow	2×2.7 m =	5.4 m
gate valve (fully open)	2×1.0 m =	2.0 m
90° angle valve (fully open)	1×18.9 m =	18.9 m
swing check valve	1×18.9 m =	18.9 m
		45.2 m

The equivalent pipe length is the sum of the straight run of pipe and the equivalent length of pipe for the valves and fittings.

$$\begin{aligned} L_e &= L + L_{\text{fittings}} \\ &= 90 \text{ m} + 45.2 \text{ m} \\ &= 135.2 \text{ m} \end{aligned}$$

Use Darcy's equation to compute the frictional loss.

$$\begin{aligned} h_f &= \frac{fL\text{v}^2}{2D_i g} \\ &= \frac{(0.0173)(135.2 \text{ m})\left(1.61 \ \frac{\text{m}}{\text{s}^2}\right)^2}{(2)(0.1541 \text{ m})\left(9.81 \ \frac{\text{m}}{\text{s}^2}\right)} \\ &= 2.01 \text{ m} \end{aligned}$$

The total difference in head is the sum of the head loss through the pipe, valves, and fittings and the change in elevation.

$$\begin{aligned} \Delta h &= h_f + \Delta z \\ &= 2.01 \text{ m} + 6 \text{ m} \\ &= 8.01 \text{ m} \end{aligned}$$

The pressure difference between the entrance and discharge can be determined from

$$\begin{aligned} \Delta p &= \rho h g \\ &= \left(992.25 \ \frac{\text{kg}}{\text{m}^3}\right)(8.01 \text{ m})\left(9.81 \ \frac{\text{m}}{\text{s}^2}\right) \\ &= \boxed{77\,969 \text{ Pa} \quad (78 \text{ kPa})} \end{aligned}$$

The answer is (A).

6. *Customary U.S. Solution*

For 6 in schedule-40 pipe, the internal diameter, D_i, is 0.5054 ft. The internal area is 0.2006 ft^2.

For air at 70°F and atmospheric pressure, the kinematic viscosity is 16.15×10^{-5} ft^2/sec (App. 14.D).

Calculate the Reynolds number.

$$\begin{aligned} \text{Re} &= \frac{D_i \text{v}}{\nu} \\ &= \frac{(0.5054 \text{ ft})\left(60 \ \frac{\text{ft}}{\text{sec}}\right)}{16.15 \times 10^{-5} \ \frac{\text{ft}^2}{\text{sec}}} \\ &= 1.88 \times 10^5 \end{aligned}$$

Since $\text{Re} > 2100$, the flow is turbulent. The friction loss coefficient can be determined from the Moody diagram.

For new steel pipe, the specific roughness, ϵ, is 0.0002 ft.

The relative roughness is

$$\begin{aligned} \frac{\epsilon}{D_i} &= \frac{0.0002 \text{ ft}}{0.5054 \text{ ft}} \\ &= 0.0004 \end{aligned}$$

From the Moody diagram with $\text{Re} = 1.88 \times 10^5$ and $e/D_i = 0.0004$, the friction factor, f, can be determined as 0.0184.

From App. 17.D, the equivalent length of the valves and fittings is

standard radius elbow	2×8.9 ft =	17.8 ft
gate valve (fully open)	2×3.2 ft =	6.4 ft
90° angle valve (fully open)	1×63.0 ft =	63.0 ft
swing check valve	1×63.0 ft =	63.0 ft
		150.2 ft

The equivalent pipe length is the sum of the straight run of pipe and the equivalent length of pipe for the valves and fittings.

$$\begin{aligned} L_e &= L + L_{\text{fittings}} \\ &= 300 \text{ ft} + 150.2 \text{ ft} \\ &= 450.2 \text{ ft} \end{aligned}$$

Use Darcy's equation to compute the frictional loss.

$$\begin{aligned} h_f &= \frac{fL_e\text{v}^2}{2D_i g} \\ &= \frac{(0.0184)(450.2 \text{ ft})\left(60 \ \frac{\text{ft}}{\text{sec}}\right)^2}{(2)(0.5054 \text{ ft})\left(32.2 \ \frac{\text{ft}}{\text{sec}^2}\right)} \\ &= 916.2 \text{ ft} \end{aligned}$$

The difference in head is the sum of the head loss through the pipe, valves, and fittings and the change in elevation.

$$\begin{aligned} h &= h_f + \Delta z \\ &= 916.2 \text{ ft} + 20 \text{ ft} \\ &= 936.2 \text{ ft} \end{aligned}$$

The pressure difference between the entrance and discharge can be determined from

$$\begin{aligned}\Delta p &= \gamma h = \rho h \times \left(\frac{g}{g_c}\right) \\ &= \left(0.075\ \frac{\text{lbm}}{\text{ft}^3}\right)(936.2\ \text{ft}) \\ &\quad \times \left(\frac{32.2\ \frac{\text{ft}}{\text{sec}^2}}{32.2\ \frac{\text{lbm-ft}}{\text{lbf-sec}^2}}\right)\left(\frac{1\ \text{ft}^2}{144\ \text{in}^2}\right) \\ &= \boxed{0.49\ \text{lbf/in}^2 \quad (0.49\ \text{psi})}\end{aligned}$$

The answer is (B).

SI Solution

For 6 in pipe, the internal diameter, D_i, is 154.1 mm (0.1541 m) and the internal area is 186.5×10^{-4} m^2 [App. 16.C].

For air at 20°C, the kinematic viscosity, ν, is 1.51×10^{-5} m^2/s [App. 14.E].

Calculate the Reynolds number.

$$\begin{aligned}\text{Re} &= \frac{D_i \text{v}}{\nu} \\ &= \frac{(0.1541\ \text{m})\left(18\ \frac{\text{m}}{\text{s}}\right)}{1.51 \times 10^{-5}\ \frac{\text{m}^2}{\text{s}}} \\ &= 1.84 \times 10^5\end{aligned}$$

Since Re > 2100, the flow is turbulent. The friction loss coefficient can be determined from the Moody diagram.

For new steel pipe, the specific roughness, ϵ, is 6.0×10^{-5} m.

The relative roughness is

$$\begin{aligned}\frac{\epsilon}{D_i} &= \frac{6.0 \times 10^{-5}}{0.1541\ \text{m}} \\ &= 0.0004\end{aligned}$$

From the Moody diagram with Re $= 1.84 \times 10^5$ and $e/D_i = 0.0004$, the friction factor, f, can be determined as 0.0185.

Compute the equivalent lengths of the valves and fittings. (Convert from App. 17.D.)

standard radius elbow	2 × 2.7 m	=	5.4 m
gate valve (fully open)	2 × 1.0 m	=	2.0 m
90° angle valve (fully open)	1 × 18.9 m	=	18.9 m
swing check valve	1 × 18.9 m	=	18.9 m
			45.2 m

The equivalent pipe length is the sum of the straight run of pipe and the equivalent length of pipe for the valves and fittings.

$$\begin{aligned}L_e &= L + L_{\text{fittings}} \\ &= 90\ \text{m} + 45.2\ \text{m} \\ &= 135.2\ \text{m}\end{aligned}$$

Use Darcy's equation to compute the frictional loss.

$$\begin{aligned}h_f &= \frac{fL\text{v}^2}{2D_i g} \\ &= \frac{(0.0185)(135.2\ \text{m})\left(18\ \frac{\text{m}}{\text{s}}\right)^2}{(2)(0.1541\ \text{m})\left(9.81\ \frac{\text{m}}{\text{s}^2}\right)} \\ &= 268.0\ \text{m}\end{aligned}$$

The difference in head is the sum of the head loss through the pipe, valves, and fittings and the change in elevation.

$$\begin{aligned}\Delta h &= h_f + \Delta z \\ &= 268.0\ \text{m} + 6\ \text{m} \\ &= 274.0\ \text{m}\end{aligned}$$

Assume the density of the air, ρ, is approximately 1.20 kg/m^3.

The pressure difference between the entrance and discharge can be determined from

$$\begin{aligned}\Delta p &= \rho h g \\ &= \left(1.20\ \frac{\text{kg}}{\text{m}^3}\right)(274.0\ \text{m})\left(9.81\ \frac{\text{m}}{\text{s}^2}\right) \\ &= \boxed{3226\ \text{Pa} \quad (3.23\ \text{kPa})}\end{aligned}$$

The answer is (B).

7. Assume that flows from reservoirs A and B are toward D and then toward C. Then,

$$\dot{V}_{\text{A-D}} + \dot{V}_{\text{B-D}} = \dot{V}_{\text{D-C}}$$

or

$$A_\text{A}\text{v}_{\text{A-D}} + A_\text{B}\text{v}_{\text{B-D}} - A_\text{C}\text{v}_{\text{D-C}} = 0$$

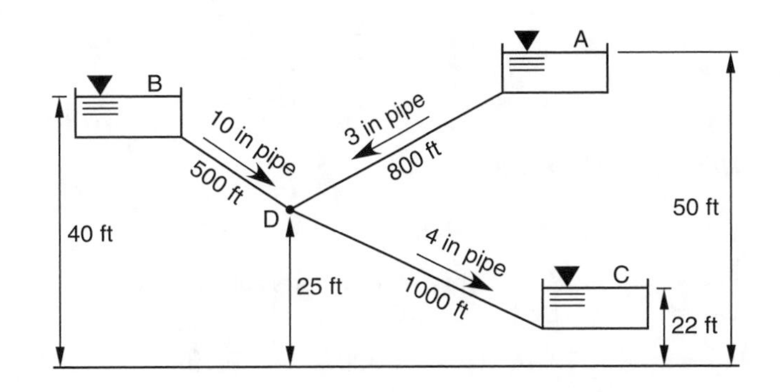

From App. 16.B, for schedule-40 pipe,

$$\begin{aligned} A_A &= 0.05134\ \text{ft}^2 & D_A &= 0.2557\ \text{ft} \\ A_B &= 0.5476\ \text{ft}^2 & D_B &= 0.8350\ \text{ft} \\ A_C &= 0.08841\ \text{ft}^2 & D_C &= 0.3355\ \text{ft} \end{aligned}$$

$$0.05134\text{v}_{A\text{–}D} + 0.5476\text{v}_{B\text{–}D} - 0.08841\text{v}_{D\text{–}C} = 0 \quad \text{[Eq. 1]}$$

Ignoring the velocity heads, the conservation of energy equation between A and D is

$$z_A = \frac{p_D}{\gamma} + z_D + h_{f,A-D}$$

$$50\ \text{ft} = \frac{p_D}{62.4\ \frac{\text{lbf}}{\text{ft}^3}} + 25\ \text{ft} + \frac{(0.02)(800\ \text{ft})(\text{v}_{A\text{–}D})^2}{(2)(0.2557\ \text{ft})\left(32.2\ \frac{\text{ft}}{\text{sec}^2}\right)}$$

or

$$\text{v}_{A\text{–}D} = \sqrt{25.73 - 0.0165p_D} \quad \text{[Eq. 2]}$$

Similarly, for B–D,

$$40\ \text{ft} = \frac{p_D}{62.4\ \frac{\text{lbf}}{\text{ft}^3}} + 25\ \text{ft} + \frac{(0.02)(500\ \text{ft})(\text{v}_{B\text{–}D})^2}{(2)(0.8350\ \text{ft})\left(32.2\ \frac{\text{ft}}{\text{sec}^2}\right)}$$

or

$$\text{v}_{B\text{–}D} = \sqrt{80.66 - 0.0862p_D} \quad \text{[Eq. 3]}$$

For D–C,

$$22\ \text{ft} = \frac{p_D}{62.4\ \frac{\text{lbf}}{\text{ft}^3}} + 25\ \text{ft} - \frac{(0.02)(1000\ \text{ft})(\text{v}_{D\text{–}C})^2}{(2)(0.3355\ \text{ft})\left(32.2\ \frac{\text{ft}}{\text{sec}^2}\right)}$$

or

$$\text{v}_{D\text{–}C} = \sqrt{3.24 + 0.0173p_D} \quad \text{[Eq. 4]}$$

Equations 1, 2, 3, and 4 must be solved simultaneously. To do this, assume a value for p_D. This value then determines all three velocities in Eqs. 2, 3, and 4. These velocities are substituted into Eq. 1. A trial and error solution yields

$$\begin{aligned} \text{v}_{A\text{–}D} &= 3.21\ \text{ft/sec} \\ \text{v}_{B\text{–}D} &= 0.408\ \text{ft/sec} \\ \text{v}_{D\text{–}C} &= 4.40\ \text{ft/sec} \end{aligned}$$

$$\boxed{p_D = 933.8\ \text{lbf/ft}^2\ \ (\text{psf})}$$

$$\boxed{\text{Flow is from B to D.}}$$

The answer is (B).

8. *Customary U.S. Solution*

The composite modulus of elasticity of the pipe and water is given by Eq. 17.205.

For water at 70°F, $E_{\text{water}} = 320 \times 10^3\ \text{lbf/in}^2$.

For cast-iron pipe, $E_{\text{pipe}} = 20 \times 10^6\ \text{lbf/in}^2$.

$$\begin{aligned} E &= \frac{E_{\text{water}} t_{\text{pipe}} E_{\text{pipe}}}{t_{\text{pipe}} E_{\text{pipe}} + d_{\text{pipe}} E_{\text{water}}} \\ &= \frac{\left(320 \times 10^3\ \frac{\text{lbf}}{\text{in}^2}\right)(0.75\ \text{in})\left(20 \times 10^6\ \frac{\text{lbf}}{\text{in}^2}\right)}{(0.75\ \text{in})\left(20 \times 10^6\ \frac{\text{lbf}}{\text{in}^2}\right) + (24\ \text{in})\left(320 \times 10^3\ \frac{\text{lbf}}{\text{in}^2}\right)} \\ &= 2.12 \times 10^5\ \text{lbf/in}^2 \end{aligned}$$

The speed of sound in the pipe is

$$\begin{aligned} a &= \sqrt{\frac{Eg_c}{\rho}} \\ &= \sqrt{\frac{\left(2.12 \times 10^5\ \frac{\text{lbf}}{\text{in}^2}\right)\left(\frac{144\ \text{in}^2}{1\ \text{ft}^2}\right)\left(32.2\ \frac{\text{lbm-ft}}{\text{lbf-sec}^2}\right)}{62.3\ \frac{\text{lbm}}{\text{ft}^3}}} \\ &= 3972\ \text{ft/sec} \end{aligned}$$

(a) The maximum pressure is given by Eq. 17.204.

$$\begin{aligned} \Delta p &= \frac{\rho a \Delta \text{v}}{g_c} \\ &= \left(\frac{\left(62.3\ \frac{\text{lbm}}{\text{ft}^3}\right)\left(3972\ \frac{\text{ft}}{\text{sec}}\right)\left(6\ \frac{\text{ft}}{\text{sec}}\right)}{32.2\ \frac{\text{lbm-ft}}{\text{lbf-sec}^2}}\right) \times \left(\frac{1\ \text{ft}^2}{144\ \text{in}^2}\right) \\ &= \boxed{320.2\ \text{lbf/in}^2\ \ (320.2\ \text{psi})} \end{aligned}$$

The answer is (C).

(b) The length of time the pressure is constant at the valve is given by

$$\begin{aligned} t &= \frac{2L}{a} \\ &= \frac{(2)(500\ \text{ft})}{3972\ \frac{\text{ft}}{\text{sec}}} \\ &= \boxed{0.252\ \text{sec}} \end{aligned}$$

The answer is (A).

SI Solution

The composite modulus of elasticity of the pipe and water is given by Eq. 17.205.

For water at 20°C, $E_{\text{water}} = 2.2 \times 10^9$ Pa.

For cast-iron pipe, $E_{\text{pipe}} = 1.4 \times 10^{11}$ Pa.

$$\begin{aligned} E &= \frac{E_{\text{water}} t_{\text{pipe}} E_{\text{pipe}}}{t_{\text{pipe}} E_{\text{pipe}} + d_{\text{pipe}} E_{\text{water}}} \\ &= \frac{(2.2 \times 10^9 \text{ Pa})(0.02 \text{ m})(1.4 \times 10^{11} \text{ Pa})}{(0.02 \text{ m})(1.4 \times 10^{11} \text{ Pa}) + (0.6 \text{ m})(2.2 \times 10^9 \text{ Pa})} \\ &= 1.50 \times 10^9 \text{ Pa} \end{aligned}$$

The speed of sound in the pipe is

$$\begin{aligned} a &= \sqrt{\frac{E}{\rho}} \\ &= \sqrt{\frac{1.50 \times 10^9 \text{ Pa}}{1000 \ \frac{\text{kg}}{\text{m}^3}}} \\ &= 1225 \text{ m/s} \end{aligned}$$

(a) The maximum pressure is given by

$$\begin{aligned} \Delta p &= \rho a \Delta \text{v} \\ &= \left(1000 \ \frac{\text{kg}}{\text{m}^3}\right)\left(1225 \ \frac{\text{m}}{\text{s}}\right)\left(2 \ \frac{\text{m}}{\text{s}}\right) \\ &= \boxed{2.45 \times 10^6 \text{ Pa} \quad (2450 \text{ kPa})} \end{aligned}$$

The answer is (C).

(b) The length of time the pressure is constant at the valve is given by

$$\begin{aligned} t &= \frac{2L}{a} \\ &= \frac{(2)(150 \text{ m})}{1225 \ \frac{\text{m}}{\text{s}}} \\ &= \boxed{0.245 \text{ s}} \end{aligned}$$

The answer is (A).

9. *Customary U.S. Solution*

First it is necessary to collect data on schedule-40 pipe and water. The fluid viscosity, pipe dimensions, and other parameters can be found in various appendices in Chs. 14 and 16. At 70°F water, $\nu = 1.059 \times 10^{-5}$ ft^2/sec.

From Table 17.2, $\epsilon = 0.0002$ ft.

8 in pipe	$D = 0.6651$ ft $A = 0.3474$ ft^2
12 in pipe	$D = 0.9948$ ft $A = 0.7773$ ft^2
16 in pipe	$D = 1.25$ ft $A = 1.2272$ ft^2

The flow quantity is converted from gallons per minute to cubic feet per second.

$$\begin{aligned} \dot{V} &= \frac{(8 \text{ MGD})\left(10^6 \ \frac{\frac{\text{gal}}{\text{day}}}{\text{MGD}}\right)\left(0.002228 \ \frac{\frac{\text{ft}^3}{\text{sec}}}{\frac{\text{gal}}{\text{min}}}\right)}{\left(24 \ \frac{\text{hr}}{\text{day}}\right)\left(60 \ \frac{\text{min}}{\text{hr}}\right)} \\ &= 12.378 \text{ ft}^3/\text{sec} \end{aligned}$$

For the inlet pipe, the velocity is

$$\text{v} = \frac{\dot{V}}{A} = \frac{12.378 \ \frac{\text{ft}^3}{\text{sec}}}{0.3474 \text{ ft}^2} = 35.63 \text{ ft/sec}$$

The Reynolds number is

$$\begin{aligned} \text{Re} &= \frac{D\text{v}}{\nu} = \frac{(0.6651 \text{ ft})\left(35.63 \ \frac{\text{ft}}{\text{sec}}\right)}{1.059 \times 10^{-5} \ \frac{\text{ft}^2}{\text{sec}}} \\ &= 2.24 \times 10^6 \end{aligned}$$

The relative roughness is

$$\frac{\epsilon}{D} = \frac{0.0002 \text{ ft}}{0.6651 \text{ ft}} = 0.0003$$

From the Moody diagram, $f = 0.015$.

Equation 17.23(b) is used to calculate the frictional energy loss.

$$\begin{aligned} E_{f,1} &= h_f \times \left(\frac{g}{g_c}\right) = \frac{fL\text{v}^2}{2Dg_c} \\ &= \frac{(0.015)(1000 \text{ ft})\left(35.63 \ \frac{\text{ft}}{\text{sec}}\right)^2}{(2)(0.6651 \text{ ft})\left(32.2 \ \frac{\text{lbm-ft}}{\text{lbf-sec}^2}\right)} \\ &= 444.6 \text{ ft-lbf/lbm} \end{aligned}$$

For the outlet pipe, the velocity is

$$\text{v} = \frac{\dot{V}}{A} = \frac{12.378 \ \frac{\text{ft}^3}{\text{sec}}}{0.7773 \text{ ft}^2} = 15.92 \text{ ft/sec}$$

The Reynolds number is

$$\begin{aligned} \text{Re} &= \frac{D\text{v}}{\nu} = \frac{(0.9948 \text{ ft})\left(15.92 \ \frac{\text{ft}}{\text{sec}}\right)}{1.059 \times 10^{-5} \ \frac{\text{ft}^2}{\text{sec}}} \\ &= 1.5 \times 10^6 \end{aligned}$$

The relative roughness is

$$\frac{\epsilon}{D} = \frac{0.0002 \text{ ft}}{0.9948 \text{ ft}} = 0.0002$$

From the Moody diagram, $f = 0.014$.

Equation 17.23(b) is used to calculate the frictional energy loss.

$$\begin{aligned} E_{f,2} &= h_f \times \left(\frac{g}{g_c}\right) = \frac{fL\text{v}^2}{2Dg_c} \\ &= \frac{(0.014)(1500 \text{ ft})\left(15.92 \dfrac{\text{ft}}{\text{sec}}\right)^2}{(2)(0.9948 \text{ ft})\left(32.2 \dfrac{\text{lbm-ft}}{\text{lbf-sec}^2}\right)} \\ &= 83.1 \text{ ft-lbf/lbm} \end{aligned}$$

Assume a 50% split through the two branches. In the upper branch, the velocity is

$$\text{v} = \frac{\dot{V}}{A} = \frac{\left(\frac{1}{2}\right)\left(12.378 \dfrac{\text{ft}^3}{\text{sec}}\right)}{0.3474 \text{ ft}^2} = 17.81 \text{ ft/sec}$$

The Reynolds number is

$$\begin{aligned} \text{Re} &= \frac{D\text{v}}{\nu} = \frac{(0.6651 \text{ ft})\left(17.81 \dfrac{\text{ft}}{\text{sec}}\right)}{1.059 \times 10^{-5} \dfrac{\text{ft}^2}{\text{sec}}} \\ &= 1.1 \times 10^6 \end{aligned}$$

The relative roughness is

$$\frac{\epsilon}{D} = \frac{0.0002 \text{ ft}}{0.6651 \text{ ft}} = 0.0003$$

From the Moody diagram, $f = 0.015$.

For the 16 in pipe in the lower branch, the velocity is

$$\text{v} = \frac{\dot{V}}{A} = \frac{\left(\frac{1}{2}\right)\left(12.378 \dfrac{\text{ft}^3}{\text{sec}}\right)}{1.2272 \text{ ft}^2} = 5.04 \text{ ft/sec}$$

The Reynolds number is

$$\begin{aligned} \text{Re} &= \frac{D\text{v}}{\nu} = \frac{(1.25 \text{ ft})\left(5.04 \dfrac{\text{ft}}{\text{sec}}\right)}{1.059 \times 10^{-5} \dfrac{\text{ft}^2}{\text{sec}}} \\ &= 5.95 \times 10^5 \end{aligned}$$

The relative roughness is

$$\frac{\epsilon}{D} = \frac{0.0002 \text{ ft}}{1.25 \text{ ft}} = 0.00016$$

From the Moody diagram, $f = 0.015$.

These values of f for the two branches are fairly insensitive to changes in $\dot{V}$, so they will be used for the rest of the problem in the upper branch.

Eq. 17.23(b) is used to calculate the frictional energy loss in the upper branch.

$$\begin{aligned} E_{f,\text{upper}} &= h_f \times \left(\frac{g}{g_c}\right) = \frac{fL\text{v}^2}{2Dg_c} \\ &= \frac{(0.015)(500 \text{ ft})\left(17.81 \dfrac{\text{ft}}{\text{sec}}\right)^2}{(2)(0.6651 \text{ ft})\left(32.2 \dfrac{\text{lbm-ft}}{\text{lbf-sec}^2}\right)} \\ &= 55.5 \text{ ft-lbf/lbm} \end{aligned}$$

To calculate a loss for any other flow in the upper branch,

$$\begin{aligned} E_{f,\text{upper 2}} &= E_{f,\text{upper}} \left(\frac{\dot{V}}{\left(\frac{1}{2}\right)\left(12.378 \dfrac{\text{ft}^3}{\text{sec}}\right)}\right)^2 \\ &= \left(55.5 \frac{\text{ft-lbf}}{\text{lbm}}\right)\left(\frac{\dot{V}}{6.189 \dfrac{\text{ft}^3}{\text{sec}}}\right)^2 \\ &= 1.45\dot{V}^2 \end{aligned}$$

Similarly, for the lower branch, in the 8 in section,

$$\begin{aligned} E_{f,\text{lower,8 in}} &= \frac{(0.015)(250 \text{ ft})\left(17.81 \dfrac{\text{ft}}{\text{sec}}\right)^2}{(2)(0.6651 \text{ ft})\left(32.2 \dfrac{\text{lbm-ft}}{\text{lbf-sec}^2}\right)} \\ &= 27.8 \text{ ft-lbf/lbm} \end{aligned}$$

For the lower branch, in the 16 in section,

$$\begin{aligned} E_{f,\text{lower,16 in}} &= \frac{(0.015)(1000 \text{ ft})\left(5.04 \dfrac{\text{ft}}{\text{sec}}\right)^2}{(2)(1.25 \text{ ft})\left(32.2 \dfrac{\text{lbm-ft}}{\text{lbf-sec}^2}\right)} \\ &= 4.7 \text{ ft-lbf/lbm} \end{aligned}$$

The total loss in the lower branch is

$$\begin{aligned} E_{f,\text{lower}} &= E_{f,\text{lower,8 in}} + E_{f,\text{lower,16 in}} \\ &= 27.8 \frac{\text{ft-lbf}}{\text{lbm}} + 4.7 \frac{\text{ft-lbf}}{\text{lbm}} \\ &= 32.5 \text{ ft-lbf/lbm} \end{aligned}$$

To calculate a loss for any other flow in the lower branch,

$$E_{f,\text{lower }2} = E_{f,\text{lower}}\left(\frac{\dot{V}}{\left(\frac{1}{2}\right)\left(12.378\ \frac{\text{ft}^3}{\text{sec}}\right)}\right)^2$$
$$= \left(32.5\ \frac{\text{ft-lbf}}{\text{lbm}}\right)\left(\frac{\dot{V}}{6.189\ \frac{\text{ft}^3}{\text{sec}}}\right)^2$$
$$= 0.85\dot{V}^2$$

Let x be the fraction flowing in the upper branch. Then, because the friction losses are equal,

$$E_{f,\text{upper }2} = E_{f,\text{lower }2}$$
$$1.45x^2 = (0.85)(1-x)^2$$
$$x = 0.432$$

(a)
$$\dot{V}_{\text{upper}} = (0.432)\left(12.378\ \frac{\text{ft}^3}{\text{sec}}\right)$$
$$= \boxed{5.347\ \text{ft}^3/\text{sec}}$$

The answer is (D).

(b)
$$\dot{V}_{\text{lower}} = (1-0.432)\left(12.378\ \frac{\text{ft}^3}{\text{sec}}\right)$$
$$= 7.03\ \text{ft}^3/\text{sec}$$

$$E_{f,\text{total}} = E_{f,1} + E_{f,\text{lower }2} + E_{f,2}$$
$$E_{f,\text{lower }2} = 0.85\dot{V}^2_{\text{lower}}$$
$$= (0.85)\left(7.03\ \frac{\text{ft}^3}{\text{sec}}\right)^2$$
$$= 42.0\ \text{ft}$$
$$E_{f,\text{total}} = 444.6\ \frac{\text{ft-lbf}}{\text{lbm}} + 42.0\ \frac{\text{ft-lbf}}{\text{lbm}} + 83.1\ \frac{\text{ft-lbf}}{\text{lbm}}$$
$$= \boxed{569.7\ \text{ft-lbf/lbm}}$$

The answer is (D).

SI Solution

First it is necessary to collect data on schedule-40 pipe and water. The fluid viscosity, pipe dimensions, and other parameters can be found in various appendices in Chs. 14 and 16. At 20°C water, $\nu = 1.007\times10^{-6}\ \text{m}^2/\text{s}$.

From Table 17.2, $\epsilon = 6\times10^{-5}$ m.

8 in pipe: $D = 202.7$ mm, $A = 322.7\times10^{-4}\ \text{m}^2$

12 in pipe: $D = 303.2$ mm, $A = 721.9\times10^{-4}\ \text{m}^2$

16 in pipe: $D = 381$ mm, $A = 1140\times10^{-4}\ \text{m}^2$

For the inlet pipe, the velocity is

$$\text{v} = \frac{\dot{V}}{A} = \frac{\left(350\ \frac{\text{L}}{\text{s}}\right)\left(\frac{1\ \text{m}^3}{1000\ \text{L}}\right)}{322.7\times10^{-4}\ \text{m}^2} = 10.85\ \text{m/s}$$

The Reynolds number is

$$\text{Re} = \frac{D\text{v}}{\nu} = \frac{(0.2027\ \text{m})\left(10.85\ \frac{\text{m}}{\text{s}}\right)}{1.007\times10^{-6}\ \frac{\text{m}^2}{\text{s}}}$$
$$= 2.18\times10^6$$

The relative roughness is

$$\frac{\epsilon}{D} = \frac{6\times10^{-5}\ \text{m}}{0.2027\ \text{m}} = 0.0003$$

From the Moody diagram, $f = 0.015$.

Equation 17.23(a) is used to calculate the frictional energy loss.

$$E_{f,1} = h_f g = \frac{fL\text{v}^2}{2D}$$
$$= \frac{(0.015)(300\ \text{m})\left(10.85\ \frac{\text{m}}{\text{s}}\right)^2}{(2)(0.2027\ \text{m})}$$
$$= 1307\ \text{J/kg}$$

For the outlet pipe, the velocity is

$$\text{v} = \frac{\dot{V}}{A} = \frac{\left(350\ \frac{\text{L}}{\text{s}}\right)\left(\frac{1\ \text{m}^3}{1000\ \text{L}}\right)}{721.9\times10^{-4}\ \text{m}^2} = 4.848\ \text{m/s}$$

The Reynolds number is

$$\text{Re} = \frac{D\text{v}}{\nu} = \frac{(0.3032\ \text{m})\left(4.848\ \frac{\text{m}}{\text{s}}\right)}{1.007\times10^{-6}\ \frac{\text{m}^2}{\text{s}}}$$
$$= 1.46\times10^6$$

The relative roughness is

$$\frac{\epsilon}{D} = \frac{6\times10^{-5}\ \text{m}}{0.3032\ \text{m}} = 0.0002$$

From the Moody diagram, $f = 0.014$.

Equation 17.23(a) is used to calculate the frictional energy loss.

$$E_{f,2} = h_f g = \frac{fL\text{v}^2}{2D} = \frac{(0.014)(450\ \text{m})\left(4.848\ \frac{\text{m}}{\text{s}}\right)^2}{(2)(0.3032\ \text{m})} = 244.2\ \text{J/kg}$$

Assume a 50% split through the two branches. In the upper branch, the velocity is

$$\text{v} = \frac{\dot{V}}{A} = \frac{\left(\frac{1}{2}\right)\left(350\ \frac{\text{L}}{\text{s}}\right)\left(\frac{1\ \text{m}^3}{1000\ \text{L}}\right)}{322.7\times10^{-4}\ \text{m}^2} = 5.423\ \text{m/s}$$

The Reynolds number is

$$\text{Re} = \frac{D\text{v}}{\nu} = \frac{(0.2027\ \text{m})\left(5.423\ \frac{\text{m}}{\text{s}}\right)}{1.007\times10^{-6}\ \frac{\text{m}^2}{\text{s}}} = 1.1\times10^6$$

The relative roughness is

$$\frac{\epsilon}{D} = \frac{6\times10^{-5}\ \text{m}}{0.2027\ \text{m}} = 0.0003$$

From the Moody diagram, $f = 0.015$.

For the 16 in pipe in the lower branch, the velocity is

$$\text{v} = \frac{\dot{V}}{A} = \frac{\left(\frac{1}{2}\right)\left(350\ \frac{\text{L}}{\text{s}}\right)\left(\frac{1\ \text{m}^3}{1000\ \text{L}}\right)}{1140\times10^{-4}\ \text{m}^2} = 1.535\ \text{m/s}$$

The Reynolds number is

$$\text{Re} = \frac{D\text{v}}{\nu} = \frac{(0.381\ \text{m})\left(1.535\ \frac{\text{m}}{\text{s}}\right)}{1.007\times10^{-6}\ \frac{\text{m}^2}{\text{s}}} = 5.47\times10^5$$

The relative roughness is

$$\frac{\epsilon}{D} = \frac{5.47\times10^{-5}\ \text{m}}{0.381\ \text{m}} = 0.00014$$

From the Moody diagram, $f = 0.015$.

These values of f for the two branches are fairly insensitive to changes in $\dot{V}$, so they will be used for the rest of the problem in the upper branch.

Eq. 17.23(a) is used to calculate the frictional energy loss in the upper branch.

$$E_{f,\text{upper}} = h_f g = \frac{fL\text{v}^2}{2D} = \frac{(0.015)(150\ \text{m})\left(5.423\ \frac{\text{m}}{\text{s}}\right)^2}{(2)(0.2027\ \text{m})} = 163.2\ \text{J/kg}$$

To calculate a loss for any other flow in the upper branch,

$$E_{f,\text{upper 2}} = E_{f,\text{upper}}\left(\frac{\dot{V}}{\left(\frac{1}{2}\right)\left(0.350\ \frac{\text{m}^3}{\text{s}}\right)}\right)^2 = \left(163.2\ \frac{\text{J}}{\text{kg}}\right)\left(\frac{\dot{V}}{0.175\ \frac{\text{m}^3}{\text{s}}}\right)^2 = 5329\dot{V}^2$$

Similarly, for the lower branch, in the 8 in section,

$$E_{f,\text{lower,8 in}} = \frac{(0.015)(75\ \text{m})\left(5.423\ \frac{\text{m}}{\text{s}}\right)^2}{(2)(0.2027\ \text{m})} = 81.61\ \text{J/kg}$$

For the lower branch, in the 16 in section,

$$E_{f,\text{lower,16 in}} = \frac{(0.015)(300\ \text{m})\left(1.585\ \frac{\text{m}}{\text{s}}\right)^2}{(2)(0.381\ \text{m})} = 14.84\ \text{J/kg}$$

The total loss in the lower branch is

$$E_{f,\text{lower}} = E_{f,\text{lower,8 in}} + E_{f,\text{lower,16 in}} = 81.61\ \frac{\text{J}}{\text{kg}} + 14.84\ \frac{\text{J}}{\text{kg}} = 96.45\ \text{J/kg}$$

To calculate a loss for any other flow in the lower branch,

$$E_{f,\text{lower 2}} = E_{f,\text{lower}}\left(\frac{\dot{V}}{\left(\frac{1}{2}\right)\left(0.350\ \frac{\text{m}^3}{\text{s}}\right)}\right)^2 = \left(96.45\ \frac{\text{J}}{\text{kg}}\right)\left(\frac{\dot{V}}{0.175\ \frac{\text{m}^3}{\text{s}}}\right)^2 = 3149\dot{V}^2$$

Let x be the fraction flowing in the upper branch. Then, because the friction losses are equal,

$$E_{f,\text{upper }2} = E_{f,\text{lower }2}$$
$$5329x^2 = (3149)(1-x)^2$$
$$x = 0.435$$

(a)
$$\dot{V}_{\text{upper}} = (0.435)\left(0.350\ \frac{\text{m}^3}{\text{s}}\right) = \boxed{0.152\ \text{m}^3/\text{s}}$$

The answer is (D).

(b)
$$\dot{V}_{\text{lower}} = (1-0.435)\left(0.350\ \frac{\text{m}^3}{\text{s}}\right) = 0.198\ \text{m}^3/\text{s}$$

$$E_{f,\text{total}} = E_{f,1} + E_{f,\text{lower }2} + E_{f,2}$$

$$E_{f,\text{lower }2} = 3149\dot{V}^2_{\text{lower}} = (3149)\left(0.198\ \frac{\text{m}^3}{\text{s}}\right)^2 = 123.5\ \text{J/kg}$$

$$E_{f,\text{total}} = 1307\ \frac{\text{J}}{\text{kg}} + 123.5\ \frac{\text{J}}{\text{kg}} + 244.2\ \frac{\text{J}}{\text{kg}} = \boxed{1675\ \text{J/kg}\ (1.7\ \text{kJ/kg})}$$

The answer is (D).

10. *steps 1, 2, and 3:*

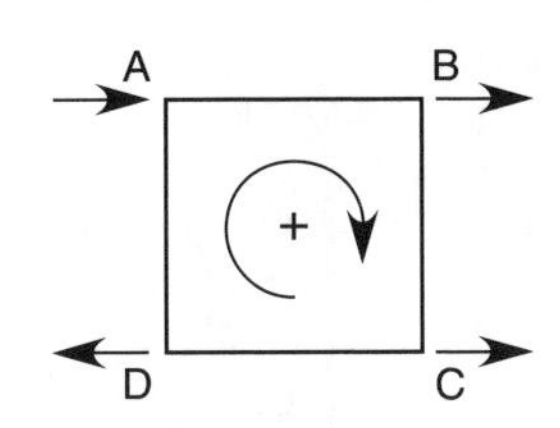

step 4: There is only one loop: ABCD.

step 5:

$$\text{pipe AB}:\quad K' = \frac{(10.44)(1000\ \text{ft})}{(100)^{1.85}(8\ \text{in})^{4.87}} = 8.33\times10^{-5}$$

$$\text{pipe BC}:\quad K' = \frac{(10.44)(1000\ \text{ft})}{(100)^{1.85}(6\ \text{in})^{4.87}} = 3.38\times10^{-4}$$

$$\text{pipe CD}:\quad K' = 8.33\times10^{-5}\quad \text{[same as AB]}$$

$$\text{pipe DA}:\quad K' = \frac{(10.44)(1000\ \text{ft})}{(100)^{1.85}(12\ \text{in})^{4.87}} = 1.16\times10^{-5}$$

step 6: Assume the flows are as shown in the figure.

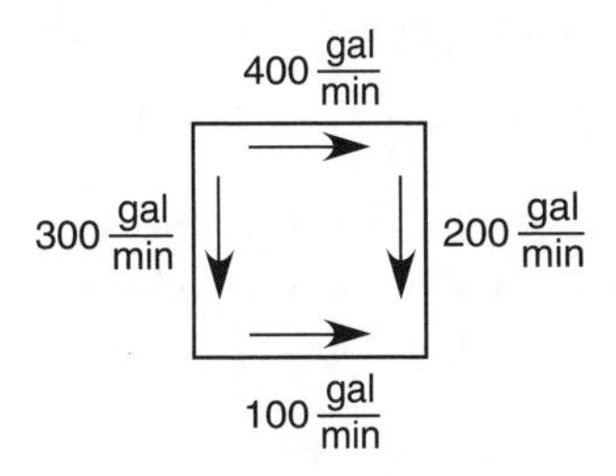

step 7:

$$\delta = \left(\frac{-1}{1.85}\right)\times\left(\frac{(8.33\times10^{-5})(400)^{1.85}+(3.38\times10^{-4})(200)^{1.85} - (8.33\times10^{-5})(100)^{1.85} - (1.16\times10^{-5})(300)^{1.85}}{(8.33\times10^{-5})(400)^{0.85}+(3.38\times10^{-4})(200)^{0.85} + (8.33\times10^{-5})(100)^{0.85} + (1.16\times10^{-5})(300)^{0.85}}\right)$$

$$= \left(\frac{-1}{1.85}\right)\left(\frac{10.67}{4.98\times10^{-2}}\right) = -116\ \text{gal/min}$$

step 8: The adjusted flows are shown.

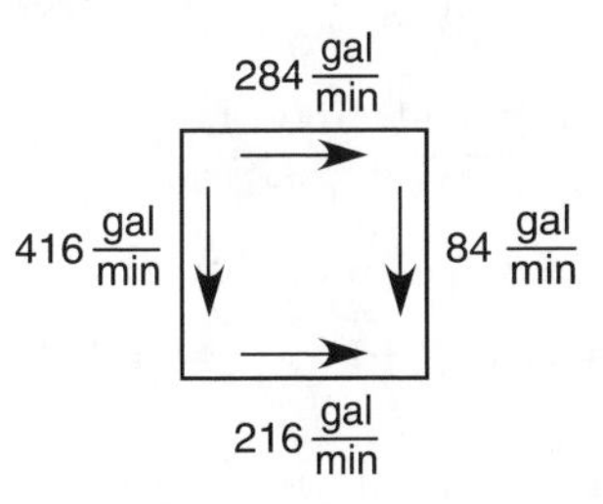

step 7: $\delta = -24$ gal/min

step 8: The adjusted flows are shown.

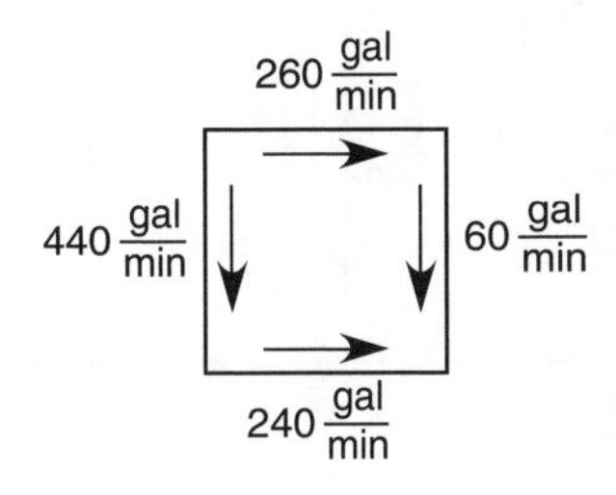

step 7: $\delta = -2$ gal/min [small enough]

step 8: The final adjusted flows are shown.

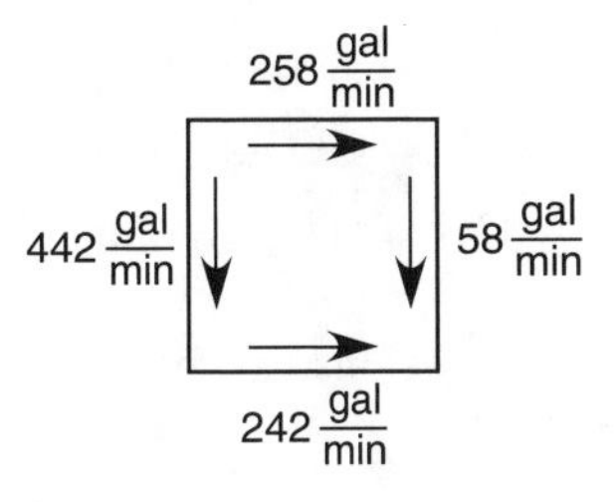

The answer is (B).

11. (a) This is a Hardy Cross problem. The pressure at point C does not change the solution procedure.

step 1: The Hazen-Williams roughness coefficient is given.
step 2: Choose clockwise as positive.
step 3: Nodes are already numbered.
step 4: Choose the loops as shown.

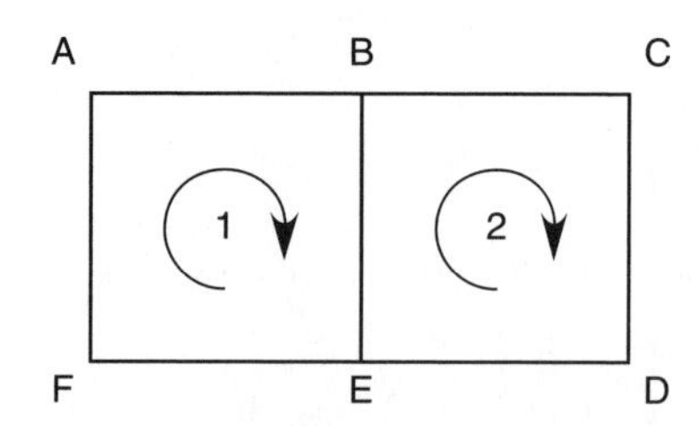

step 5: $\dot{V}$ is in gal/min.
d is in inches.
L is in feet.

Each pipe has the same length.

$$K'_{8\text{ in}} = \frac{(10.44)(1000\text{ ft})}{(100)^{1.85}(8\text{ in})^{4.87}} = 8.33 \times 10^{-5}$$

$$K'_{6\text{ in}} = \frac{(10.44)(1000\text{ ft})}{(100)^{1.85}(6\text{ in})^{4.87}} = 3.38 \times 10^{-4}$$

$$K'_{\text{CE}} = 2K'_{6\text{ in}} = 6.76 \times 10^{-4}$$

$$K'_{\text{EA}} = K'_{6\text{ in}} + K'_{8\text{ in}} = 4.21 \times 10^{-4}$$

step 6: Assume the flows shown.

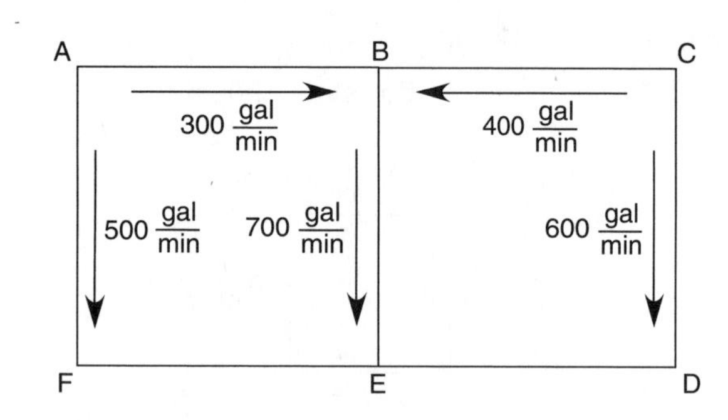

AB = 300 gal/min
BE = 700 gal/min
AE = 500 gal/min
CE = 600 gal/min
CB = 400 gal/min

step 7: If the elevations are included as part of the head loss,

$$\Sigma h = \Sigma K'\dot{V}_a^n + \delta \Sigma nK'\dot{V}_a^{n-1} + z_2 - z_1 = 0$$

However, since the loop closes on itself, $z_2 = z_1$, and the elevations can be omitted.

- First iteration:

Loop 1:

$$\delta_1 = \frac{-\left((8.33 \times 10^{-5})(300)^{1.85} + (8.33 \times 10^{-5})(700)^{1.85} - (4.21 \times 10^{-4})(500)^{1.85}\right)}{(1.85)\left((8.33 \times 10^{-5})(300)^{0.85} + (8.33 \times 10^{-5})(700)^{0.85} + (4.21 \times 10^{-4})(500)^{0.85}\right)}$$

$$= \frac{-(-22.97)}{0.213} = +108 \text{ gal/min}$$

Loop 2:

$$\delta_2 = \frac{-\left((6.76 \times 10^{-4})(600)^{1.85} - (8.33 \times 10^{-5})(700)^{1.85} - (8.33 \times 10^{-5})(400)^{1.85}\right)}{(1.85)\left((6.76 \times 10^{-4})(600)^{0.85} + (8.33 \times 10^{-5})(700)^{0.85} + 8.33 \times 10^{-5})(400)^{0.85}\right)}$$

$$= \frac{-(72.5)}{0.353} = -205 \text{ gal/min}$$

- Second iteration:

AB: 300 + 108 = 408 gal/min
BE: 700 + 108 − (−205) = 1013 gal/min
AE: 500 − 108 = 392 gal/min
CE: 600 + (−205) = 395 gal/min
CB: 400 − (−205) = 605 gal/min

Loop 1:

$$\delta_1 = \frac{-(9.48)}{0.205} = -46 \text{ gal/min}$$

Loop 2:

$$\delta_2 = \frac{-(1.08)}{0.292} = -3.7 \text{ gal/min} \quad [\text{round to } -4]$$

- Third iteration:

AB: 408 + (−46) = 362 gal/min
BE: 1013 + (−46) −(−4) = 971 gal/min
AE: 392 − (−46) = 438 gal/min
CE: 395 + (−4) = 391 gal/min
CB: 605 − (−4) = 609 gal/min

Loop 1:

$$\delta_1 = \frac{-(0.066)}{0.213} = -0.31 \text{ gal/min} \quad [\text{round to } 0]$$

Loop 2:

$$\delta_2 = \frac{-2.4}{0.29} = -8.3 \text{ gal/min} \quad [\text{round to } -8]$$

Use the following flows.

$$\begin{aligned}
\text{AB: } & 362 + 0 &= 362 \text{ gal/min} \\
\text{BE: } & 971 + 0 - (-8) &= \boxed{979 \text{ gal/min}} \\
\text{AE: } & 438 - 0 &= 438 \text{ gal/min} \\
\text{CE: } & 391 + (-8) &= 383 \text{ gal/min} \\
\text{CB: } & 609 - (-8) &= 617 \text{ gal/min}
\end{aligned}$$

The answer is (D).

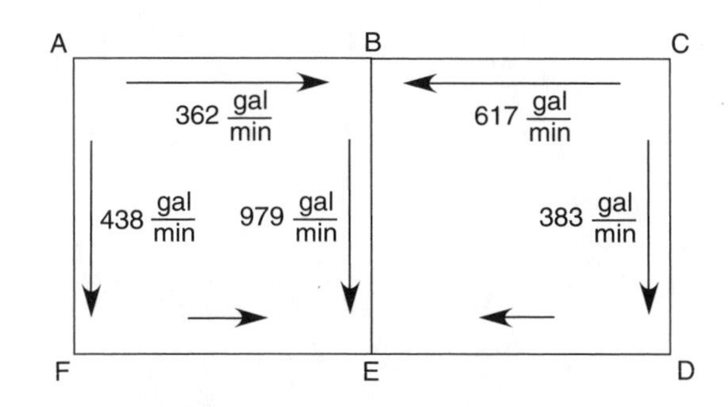

(b) The friction loss in each section is

$$h_{f,\text{AB}} = (8.33 \times 10^{-5})(362)^{1.85} = 4.5 \text{ ft}$$
$$h_{f,\text{BE}} = (8.33 \times 10^{-5})(979)^{1.85} = 28.4 \text{ ft}$$
$$h_{f,\text{AF}} = (8.33 \times 10^{-5})(438)^{1.85} = 6.4 \text{ ft}$$
$$h_{f,\text{FE}} = (3.38 \times 10^{-4})(438)^{1.85} = 26.0 \text{ ft}$$
$$h_{f,\text{CD}} = h_{f,\text{DE}} = (3.38 \times 10^{-4})(383)^{1.85} = 20.3 \text{ ft}$$
$$h_{f,\text{CB}} = (8.33 \times 10^{-5})(617)^{1.85} = 12.1 \text{ ft}$$

$$\gamma = 62.37 \text{ lbf/ft}^3 \quad [\text{at } 60^\circ\text{F}]$$

The pressure at C is 40 psig.

$$h_\text{C} = \frac{\left(40 \ \frac{\text{lbf}}{\text{in}^2}\right)\left(144 \ \frac{\text{in}^2}{\text{ft}^2}\right)}{62.37 \ \frac{\text{lbf}}{\text{ft}^3}} = 92.4 \text{ ft}$$

Next, use the energy continuity equation between adjacent points.

$$h = h_\text{C} + z_\text{C} - z - h_f$$

$$h_\text{D} = 92.4 \text{ ft} + 300 \text{ ft} - 150 \text{ ft} - 20.3 \text{ ft} = 222.1 \text{ ft}$$
$$h_\text{E} = 222.1 \text{ ft} + 150 \text{ ft} - 200 \text{ ft} - 20.3 \text{ ft} = 151.8 \text{ ft}$$
$$h_\text{F} = 151.8 \text{ ft} + 200 \text{ ft} - 150 \text{ ft} + 26 \text{ ft} = 227.8 \text{ ft}$$
$$h_\text{B} = 92.4 \text{ ft} + 300 \text{ ft} - 150 \text{ ft} - 12.1 \text{ ft} = 230.3 \text{ ft}$$
$$h_\text{A} = 230.3 \text{ ft} + 150 \text{ ft} - 200 \text{ ft} + 4.5 \text{ ft} = 184.8 \text{ ft}$$

Using $p = \gamma h$,

$$p_\text{A} = \frac{\left(62.37 \ \frac{\text{lbf}}{\text{ft}^3}\right)(184.8 \text{ ft})}{144 \ \frac{\text{in}^2}{\text{ft}^2}} = 80.0 \text{ lbf/in}^2 \quad (\text{psig})$$

$$p_\text{B} = \frac{\left(62.37 \ \frac{\text{lbf}}{\text{ft}^3}\right)(230.3 \text{ ft})}{144 \ \frac{\text{in}^2}{\text{ft}^2}} = 99.7 \text{ psig}$$

$$p_\text{C} = 40 \text{ psig} \quad [\text{given}]$$

$$p_\text{D} = \frac{\left(62.37 \ \frac{\text{lbf}}{\text{ft}^3}\right)(222.1 \text{ ft})}{144 \ \frac{\text{in}^2}{\text{ft}^2}} = \boxed{96.2 \text{ psig}}$$

$$p_\text{E} = \frac{\left(62.37 \ \frac{\text{lbf}}{\text{ft}^3}\right)(151.8 \text{ ft})}{144 \ \frac{\text{in}^2}{\text{ft}^2}} = 65.7 \text{ psig}$$

$$p_\text{F} = \frac{\left(62.37 \ \frac{\text{lbf}}{\text{ft}^3}\right)(227.8 \text{ ft})}{144 \ \frac{\text{in}^2}{\text{ft}^2}} = 98.7 \text{ psig}$$

The answer is (C).

(c) The pressure increase across the pump is

$$\left(80 \ \frac{\text{lbf}}{\text{in}^2} - 20 \ \frac{\text{lbf}}{\text{in}^2}\right)\left(144 \ \frac{\text{in}^2}{\text{ft}^2}\right) = 8640 \text{ lbf/ft}^2$$

Use Table 18.5 to find the hydraulic horsepower.

$$P = \frac{\left(8640 \ \frac{\text{lbf}}{\text{ft}^2}\right)\left(800 \ \frac{\text{gal}}{\text{min}}\right)}{2.468 \times 10^5 \ \frac{\text{lbf-gal}}{\text{min-ft}^2\text{-hp}}} = \boxed{28 \text{ hp}}$$

The answer is (B).

12. This could be solved as a parallel pipe problem or as a pipe network problem.

(a)• Pipe network solution:

step 1: Use the Darcy equation since Hazen-Williams coefficients are not given (or assume C-values based on the age of the pipe).

step 2: Clockwise is positive.

step 3: All nodes are lettered.

step 4: There is only one loop.

step 5: $\epsilon \approx 0.0008$ ft. Assume full turbulence. For class A cast-iron pipe, $D_i = 10.1$ in.

$$\frac{\epsilon}{D} = \frac{0.0008 \text{ ft}}{\frac{10.1 \text{ in}}{12 \frac{\text{in}}{\text{ft}}}} \approx 0.001$$

$f \approx 0.020$ for full turbulence. (See Fig. 17.4.)

$$K'_{AB} = (1.251 \times 10^{-7})\left(\frac{(0.02)(2000 \text{ ft})}{\left(\frac{10.1 \text{ in}}{12 \frac{\text{in}}{\text{ft}}}\right)^5}\right) = 11.8 \times 10^{-6}$$

$$K'_{BD} = (1.251 \times 10^{-7})\left(\frac{(0.02)(1000 \text{ ft})}{\left(\frac{10.1 \text{ in}}{12 \frac{\text{in}}{\text{ft}}}\right)^5}\right) = 5.92 \times 10^{-6}$$

$$K'_{DC} = (1.251 \times 10^{-7})\left(\frac{(0.02)(1500 \text{ ft})}{\left(\frac{8.13 \text{ in}}{12 \frac{\text{in}}{\text{ft}}}\right)^5}\right) = 26.3 \times 10^{-6}$$

$$K'_{CA} = (1.251 \times 10^{-7})\left(\frac{(0.02)(1000 \text{ ft})}{\left(\frac{12.12 \text{ in}}{12 \frac{\text{in}}{\text{ft}}}\right)^5}\right) = 2.38 \times 10^{-6}$$

step 6: Assume $\dot{V}_{AB} = 1$ MGD.

$$\dot{V}_{ACDB} = 0.5 \text{ MGD}$$

$$\frac{\dot{V}_{AB}}{\dot{V}_{ACDB}} = \frac{1 \text{ MGD}}{0.5 \text{ MGD}} = 2$$

Convert $\dot{V}$ to gal/min.

$$\dot{V}_{AB} = \frac{(1 \text{ MGD})(1 \times 10^6)}{\left(24 \frac{\text{hr}}{\text{day}}\right)\left(60 \frac{\text{min}}{\text{hr}}\right)} = 694 \text{ gal/min}$$

$$\dot{V}_{ACDB} = \frac{(0.5 \text{ MGD})(1 \times 10^6)}{\left(24 \frac{\text{hr}}{\text{day}}\right)\left(60 \frac{\text{min}}{\text{hr}}\right)} = 347 \text{ gal/min}$$

step 7: There is only one loop.

$$694 \frac{\text{gal}}{\text{min}} = (2)\left(347 \frac{\text{gal}}{\text{min}}\right)$$

$$\left(694 \frac{\text{gal}}{\text{min}}\right)^2 = (4)\left(347 \frac{\text{gal}}{\text{min}}\right)^2$$

$$\delta = \frac{(-1)(1 \times 10^{-6})(347)^2\big((11.8)(4) - 5.92 - 26.3 - 2.38\big)}{(2)(1 \times 10^{-6})(347)\big((11.8)(2) + 5.92 + 26.3 + 2.38\big)}$$

$$= -37.6 \text{ gal/min} \quad [\text{use } -38 \text{ gal/min}]$$

step 8:

$$\dot{V}_{AB} = 694 \frac{\text{gal}}{\text{min}} + \left(-38 \frac{\text{gal}}{\text{min}}\right) = 656 \text{ gal/min}$$

$$\dot{V}_{ACDB} = 347 \frac{\text{gal}}{\text{min}} - \left(-38 \frac{\text{gal}}{\text{min}}\right) = 385 \text{ gal/min}$$

Repeat step 7.

$$\frac{\dot{V}_{AB}}{\dot{V}_{ACDB}} = \frac{656 \frac{\text{gal}}{\text{min}}}{385 \frac{\text{gal}}{\text{min}}} = 1.7$$

$$(1.7)^2 = 2.90$$

$$656 \frac{\text{gal}}{\text{min}} = (1.70)\left(385 \frac{\text{gal}}{\text{min}}\right)$$

$$\left(656 \frac{\text{gal}}{\text{min}}\right)^2 = (2.90)\left(385 \frac{\text{gal}}{\text{min}}\right)^2$$

$$\delta = \frac{(-1)(1 \times 10^{-6})(385)^2 \times \big((11.8)(2.90) - 5.92 - 26.3 - 2.38\big)}{(2)(1 \times 10^{-6})(385) \times \big((11.8)(1.70) + 5.92 + 26.3 + 2.38\big)} \approx 0$$

$$\dot{V}_{AB} = \boxed{656 \text{ gal/min}}$$

$$\dot{V}_{ACDB} = 385 \text{ gal/min}$$

Check the Reynolds number in leg AB to verify that $f = 0.02$ was a good choice.

$$A_{10 \text{ in pipe}} = \left(\frac{\pi}{4}\right)\left(\frac{10.1 \text{ in}}{12 \frac{\text{in}}{\text{ft}}}\right)^2 = 0.5564 \text{ [cast-iron pipe]}$$

The flow rate is

$$\left(656 \frac{\text{gal}}{\text{min}}\right)\left(0.002228 \frac{\text{ft}^3\text{-min}}{\text{sec-gal}}\right) = 1.46 \text{ ft}^3/\text{sec}$$

$$\text{v} = \frac{\dot{V}}{A} = \frac{1.46 \frac{\text{ft}^3}{\text{sec}}}{0.5564 \text{ ft}^2} = 2.62 \text{ ft/sec} \quad [\text{reasonable}]$$

For 50°F water,

$$\nu = 1.410 \times 10^{-5}\ \text{ft}^2/\text{sec}$$

$$\text{Re} = \frac{\left(\dfrac{10.1\ \text{in}}{12\ \dfrac{\text{in}}{\text{ft}}}\right)\left(2.62\ \dfrac{\text{ft}}{\text{sec}}\right)}{1.410 \times 10^{-5}\ \dfrac{\text{ft}^2}{\text{sec}}} = 1.56 \times 10^5\ \text{[turbulent]}$$

The assumption of full turbulence made in step 5 is justified.

The answer is (D).

- Alternative closed-form solution:

Use the Darcy equation. Assume $\epsilon = 0.0008$ ft. The relative roughness is

$$\frac{\epsilon}{D} = \frac{0.0008\ \text{ft}}{\dfrac{10.1\ \text{in}}{12\ \dfrac{\text{in}}{\text{ft}}}} = 0.00095\quad \text{[use 0.001]}$$

Assume $\text{v}_{\text{max}} = 5$ ft/sec and 50°F temperature. The Reynolds number is

$$\text{Re} = \frac{\text{v}D}{\nu} = \frac{\left(5\ \dfrac{\text{ft}}{\text{sec}}\right)\left(\dfrac{10.1\ \text{in}}{12\ \dfrac{\text{in}}{\text{ft}}}\right)}{1.41 \times 10^{-5}\ \dfrac{\text{ft}^2}{\text{sec}}} = 2.98 \times 10^5$$

From the Moody diagram, $f \approx 0.0205$.

$$\begin{aligned} h_{f,\text{AB}} &= \frac{fL\text{v}^2}{2Dg} \\ &= \frac{(0.0205)(2000\ \text{ft})\dot{V}_{\text{AB}}^2}{(2)\left(\dfrac{10.1\ \text{in}}{12\ \dfrac{\text{in}}{\text{ft}}}\right)(0.556\ \text{ft}^2)^2\left(32.2\ \dfrac{\text{ft}}{\text{sec}^2}\right)} \\ &= 2.446\dot{V}_{\text{AB}}^2 \end{aligned}$$

$$\begin{aligned} h_{f,\text{ACDB}} &= \frac{(0.0205)(1000)\dot{V}_{\text{ACDB}}^2}{(2)\left(\dfrac{12.12}{12}\right)(0.801)^2(32.2)} \\ &\quad + \frac{(0.0205)(1500)\dot{V}_{\text{ACDB}}^2}{(2)\left(\dfrac{8.13}{12}\right)(0.360)^2(32.2)} \\ &\quad + \frac{(0.0205)(1000)\dot{V}_{\text{ACDB}}^2}{(2)\left(\dfrac{10.1}{12}\right)(0.556\ \text{ft}^2)^2(32.2)} \\ &= 0.4912\dot{V}_{\text{ACDB}}^2 + 5.438\dot{V}_{\text{ACDB}}^2 + 1.223\dot{V}_{\text{ACDB}}^2 \\ &= 7.152\dot{V}_{\text{ACDB}}^2 \end{aligned}$$

$$h_{f,\text{AB}} = h_{f,\text{ACDB}}$$

$$2.446\dot{V}_{\text{AB}}^2 = 7.152\dot{V}_{\text{ACDB}}^2$$

$$\dot{V}_{\text{AB}} = \sqrt{\frac{7.152}{2.446}}\dot{V}_{\text{ACDB}} = 1.71\dot{V}_{\text{ACDB}}\quad \text{[Eq. 1]}$$

The total flow rate is

$$\frac{1.5\ \text{MGD}}{\left(24\ \dfrac{\text{hr}}{\text{day}}\right)\left(60\ \dfrac{\text{min}}{\text{hr}}\right)} = 1041.7\ \text{gal/min}$$

$$\dot{V}_{\text{AB}} + \dot{V}_{\text{ACDB}} = 1041.7\ \text{gal/min}\quad \text{[Eq. 2]}$$

Solving Eqs. 1 and 2 simultaneously,

$$\dot{V}_{\text{AB}} = \boxed{657.3\ \text{gal/min}}$$

$$\dot{V}_{\text{ACDB}} = 384.4\ \text{gal/min}$$

The answer is (D).

This answer is insensitive to the $\text{v}_{\text{max}} = 5$ ft/sec assumption. A second iteration using actual velocities from these flow rates does not change the answer.

- The same technique can be used with the Hazen-Williams equation and an assumed value of C. If $C = 100$ is used, then

$$\dot{V}_{\text{AB}} = \boxed{725.4\ \text{gal/min}}$$

$$\dot{V}_{\text{ACDB}} = 316.3\ \text{gal/min}$$

(b) $f \approx 0.021$.

$$h_{f,\text{AB}} = \frac{(0.021)(2000\ \text{ft})\left(2.62\ \dfrac{\text{ft}}{\text{sec}}\right)^2}{(2)\left(\dfrac{10.1\ \text{in}}{12\ \dfrac{\text{in}}{\text{ft}}}\right)\left(32.2\ \dfrac{\text{ft}}{\text{sec}^2}\right)} = 5.32\ \text{ft}$$

For leg BD, use $f = 0.021$ (assumed).

$$\begin{aligned} \text{v} = \frac{\dot{V}}{A} &= \frac{\left(385\ \dfrac{\text{gal}}{\text{min}}\right)\left(0.002228\ \dfrac{\text{ft}^3\text{-min}}{\text{sec-gal}}\right)}{0.5564\ \text{ft}^2} \\ &= 1.54\ \text{ft/sec} \end{aligned}$$

$$h_{f,\text{DB}} = \frac{(0.021)(1000\ \text{ft})\left(1.54\ \dfrac{\text{ft}}{\text{sec}}\right)^2}{(2)\left(\dfrac{10\ \text{in}}{12\ \dfrac{\text{in}}{\text{ft}}}\right)\left(32.2\ \dfrac{\text{ft}}{\text{sec}^2}\right)} = 0.9\ \text{ft}$$

Water Resources

At 50°F, $\gamma = 62.4$ lbf/ft^3. From the Bernoulli equation (omitting the velocity term),

$$\frac{p_B}{\gamma} + z_B + h_{f,DB} = \frac{p_D}{\gamma} + z_D$$

$$p_B = \left(\frac{62.4 \ \frac{\text{lbf}}{\text{ft}^3}}{144 \ \frac{\text{in}^2}{\text{ft}^2}}\right) \times \left(\frac{\left(40 \ \frac{\text{lbf}}{\text{in}^2}\right)\left(144 \ \frac{\text{in}^2}{\text{ft}^2}\right)}{62.4 \ \frac{\text{lbf}}{\text{ft}^3}} + 600 \text{ ft} - 0.9 \text{ ft} - 580 \text{ ft}\right)$$

$$= \boxed{48.3 \text{ lbf/in}^2 \quad \text{(psi)}}$$

The answer is (B).

$$\frac{p_A}{\gamma} + z_A = \frac{p_B}{\gamma} + z_B + h_{f,AB}$$

$$p_A = \left(\frac{62.4 \ \frac{\text{lbf}}{\text{ft}^3}}{144 \ \frac{\text{in}^2}{\text{ft}^2}}\right) \times \left(\frac{\left(48.3 \ \frac{\text{lbf}}{\text{in}^2}\right)\left(144 \ \frac{\text{in}^2}{\text{ft}^2}\right)}{62.4 \ \frac{\text{lbf}}{\text{ft}^3}} + 580 \text{ ft} + 5.32 \text{ ft} - 540 \text{ ft}\right)$$

$$= 67.9 \text{ lbf/in}^2 \quad \text{(psi)}$$

13. *Customary U.S. Solution*

Use projectile equations.

The maximum range of the discharge is given by

$$R = v_o^2 \left(\frac{\sin 2\phi}{g}\right) \quad \text{[Table 72.2]}$$

$$= \left(50 \ \frac{\text{ft}}{\text{sec}}\right)^2 \left(\frac{\sin((2)(45°))}{32.2 \ \frac{\text{ft}}{\text{sec}^2}}\right)$$

$$= \boxed{77.64 \text{ ft}}$$

The answer is (B).

SI Solution

Use projectile equations.

The maximum range of the discharge is given by

$$R = v_o^2 \left(\frac{\sin 2\phi}{g}\right) \quad \text{[Table 72.2]}$$

$$= \frac{\left(15 \ \frac{\text{m}}{\text{s}}\right)^2 \sin((2)(45°))}{9.81 \ \frac{\text{m}}{\text{s}^2}}$$

$$= \boxed{22.94 \text{ m}}$$

The answer is (B).

14.

$$A_o = \left(\frac{\pi}{4}\right)\left(\frac{4 \text{ in}}{12 \ \frac{\text{in}}{\text{ft}}}\right)^2 = 0.08727 \text{ ft}^2$$

$$A_t = \left(\frac{\pi}{4}\right)(20 \text{ ft})^2 = 314.16 \text{ ft}^2$$

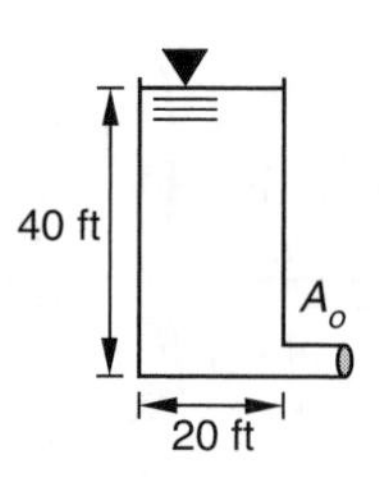

The time to drop from 40 ft to 20 ft is given by Eq. 17.84.

$$t = \frac{(2)(314.16 \text{ ft}^2)\left(\sqrt{40 \text{ ft}} - \sqrt{20 \text{ ft}}\right)}{(0.98)(0.08727 \text{ ft}^2)\sqrt{(2)\left(32.2 \ \frac{\text{ft}}{\text{sec}^2}\right)}}$$

$$= \boxed{1696 \text{ sec}}$$

The answer is (D).

15.

$$C_d = 1.00 \quad \text{[given]}$$

$$F_{va} = \frac{1}{\sqrt{1 - \left(\frac{D_2}{D_1}\right)^4}} = \frac{1}{\sqrt{1 - \left(\frac{8 \text{ in}}{12 \text{ in}}\right)^4}}$$

$$= 1.116$$

$$A_2 = \left(\frac{\pi}{4}\right)\left(\frac{8 \text{ in}}{12 \ \frac{\text{in}}{\text{ft}}}\right)^2 = 0.3491 \text{ ft}^2$$

The specific weight of mercury is 0.491 lbf/in^3; the specific weight of water is 0.0361 lbf/in^3.

$$p_1 - p_2 = \Delta(\gamma h) = \left(\left(0.491\ \frac{\text{lbf}}{\text{in}^3}\right)(4\ \text{in}) - \left(0.0361\ \frac{\text{lbf}}{\text{in}^3}\right)(4\ \text{in}) \right)\left(144\ \frac{\text{in}^2}{\text{ft}^2}\right) = 262.0\ \text{lbf/ft}^2$$

$$Q = F_{\text{va}} C_d A_2 \sqrt{\frac{2g(p_1 - p_2)}{\gamma}} = (1.116)(1)(0.3491\ \text{ft}^2) \times \sqrt{\frac{(2)\left(32.2\ \frac{\text{ft}}{\text{sec}^2}\right)\left(262\ \frac{\text{lbf}}{\text{ft}^2}\right)}{62.4\ \frac{\text{lbf}}{\text{ft}^3}}} = \boxed{6.406\ \text{ft}^3/\text{sec}\quad (\text{cfs})}$$

The answer is (C).

16. *Customary U.S. Solution*

The volumetric flow rate of benzene through the venturi meter is given by

$$\dot{V} = C_f A_2 \sqrt{\frac{2g(\rho_m - \rho)h}{\rho}}$$

The density of mercury, ρ_m, at 60°F is approximately 848 lbm/ft^3.

The density of the benzene at 60°F is

$$\rho = (\text{SG})\rho_{\text{water}} = (0.885)\left(62.4\ \frac{\text{lbm}}{\text{ft}^3}\right) = 55.22\ \text{lbm/ft}^3$$

The throat area is

$$A_2 = \left(\frac{\pi}{4}\right) D_2^2 = \left(\frac{\pi}{4}\right)(3.5\ \text{in})^2\left(\frac{1\ \text{ft}^2}{144\ \text{in}^2}\right) = 0.0668\ \text{ft}^2$$

The flow coefficient is defined as

$$C_f = \frac{C_d}{\sqrt{1 - \beta^4}}$$

β is the ratio of the throat to inlet diameters.

$$\beta = \frac{3.5\ \text{in}}{8\ \text{in}} = 0.4375$$

$$C_f = \frac{C_d}{\sqrt{1 - \beta^4}} = \frac{0.99}{\sqrt{1 - (0.4375)^4}} = 1.00865$$

Find the volumetric flow of benzene.

$$\dot{V} = C_f A_2 \sqrt{\frac{2g(\rho_m - \rho)h}{\rho}} = (1.00865)(0.0668\ \text{ft}^2) \times \sqrt{\frac{(2)\left(32.2\ \frac{\text{ft}}{\text{sec}^2}\right)\left(848\ \frac{\text{lbm}}{\text{ft}^3} - 55.22\ \frac{\text{lbm}}{\text{ft}^3}\right) \times (4\ \text{in})\left(\frac{1\ \text{ft}}{12\ \text{in}}\right)}{55.22\ \frac{\text{lbm}}{\text{ft}^3}}} = \boxed{1.183\ \text{ft}^3/\text{sec}}$$

The answer is (A).

SI Solution

The volumetric flow rate of benzene through the venturi meter is given by

$$\dot{V} = C_f A_2 \sqrt{\frac{2g(\rho_m - \rho)h}{\rho}}$$

ρ_m is the density of mercury at 15°C; ρ_m is approximately 13 600 kg/m^3.

The density of the benzene at 15°C is

$$\rho = (\text{SG})\rho_{\text{water}} = (0.885)\left(1000\ \frac{\text{kg}}{\text{m}^3}\right) = 885\ \text{kg/m}^3$$

The throat area is

$$A_2 = \left(\frac{\pi}{4}\right) D_2^2 = \left(\frac{\pi}{4}\right)(0.09\ \text{m})^2 = 0.0064\ \text{m}^2$$

Water Resources

The flow coefficient is defined as

$$C_f = \frac{C_d}{\sqrt{1-\beta^4}}$$

β is the ratio of the throat to inlet diameters.

$$\begin{aligned}\beta &= \frac{9 \text{ cm}}{20 \text{ cm}} \\ &= 0.45\end{aligned}$$

$$\begin{aligned}C_f &= \frac{C_d}{\sqrt{1-\beta^4}} \\ &= \frac{0.99}{\sqrt{1-(0.45)^4}} \\ &= 1.01094\end{aligned}$$

Find the volumetric flow of benzene.

$$\begin{aligned}\dot{V} &= C_f A_2 \sqrt{\frac{2g(\rho_m-\rho)h}{\rho}} \\ &= (1.01094)(0.0064 \text{ m}^2) \\ &\quad \times \sqrt{\frac{(2)\left(9.81 \ \frac{\text{m}}{\text{s}^2}\right) \times \left(13\,600 \ \frac{\text{kg}}{\text{m}^3} - 885 \ \frac{\text{kg}}{\text{m}^3}\right)(0.1 \text{ m})}{885 \ \frac{\text{kg}}{\text{m}^3}}} \\ &= \boxed{0.0344 \text{ m}^3/\text{s} \quad (34.4 \text{ L/s})}\end{aligned}$$

The answer is (A).

17. For 70°F water,

$$\begin{aligned}D_o &= 0.2 \text{ ft} \\ \text{v}_o &= \text{v}\left(\frac{D}{D_o}\right)^2 = \left(2 \ \frac{\text{ft}}{\text{sec}}\right)\left(\frac{1 \text{ ft}}{0.2 \text{ ft}}\right)^2 \\ &= 50 \text{ ft/sec} \\ \nu &= 1.059 \times 10^{-5} \text{ ft}^2/\text{sec} \\ \gamma &= 62.3 \text{ lbf/ft}^3\end{aligned}$$

$$\text{Re} = \frac{D_o \text{v}_o}{\nu} = \frac{(0.2 \text{ ft})\left(50 \ \frac{\text{ft}}{\text{sec}}\right)}{1.059 \times 10^{-5} \ \frac{\text{ft}^2}{\text{sec}}} = 9.44 \times 10^5$$

$$A_o = \left(\frac{\pi}{4}\right)(0.2 \text{ ft})^2 = 0.0314 \text{ ft}^2$$

$$A_p = \left(\frac{\pi}{4}\right)(1 \text{ ft})^2 = 0.7854 \text{ ft}^2$$

$$\frac{A_o}{A_p} = \frac{0.0314 \text{ ft}^2}{0.7854 \text{ ft}^2} = 0.040$$

From Fig.17.28,

$$C_f \approx 0.60$$

$$\dot{V} = A\text{v} = (0.7854 \text{ ft}^2)\left(2 \ \frac{\text{ft}}{\text{sec}}\right) = 1.571 \text{ ft}^3/\text{sec}$$

From Eq. 17.160,

$$\begin{aligned}\Delta p &= \left(\frac{\gamma}{2g}\right)\left(\frac{\dot{V}}{C_f A_o}\right)^2 \\ &= \left(\frac{62.3 \ \frac{\text{lbf}}{\text{ft}^3}}{(2)\left(32.2 \ \frac{\text{ft}}{\text{sec}^2}\right)}\right)\left(\frac{1.571 \ \frac{\text{ft}^3}{\text{sec}}}{(0.6)(0.0314 \text{ ft}^2)}\right)^2 \\ &= 6727 \text{ lbf/ft}^2 \quad \text{(psf)}\end{aligned}$$

$$\Delta p = \frac{6727 \ \frac{\text{lbf}}{\text{ft}^2}}{144 \ \frac{\text{in}^2}{\text{ft}^2}} = \boxed{46.7 \text{ lbf/in}^2 \quad \text{(psi)}}$$

The answer is (D).

18. *Customary U.S. Solution*

The manometer tube is filled with water above the mercury column. The pressure differential across the orifice meter is given by

$$\Delta p = p_1 - p_2 = (\rho_{\text{mercury}} - \rho_{\text{water}})h \times \left(\frac{g}{g_c}\right)$$

The densities of mercury and water are

$$\begin{aligned}\rho_{\text{mercury}} &= 848 \text{ lbm/ft}^3 \\ \rho_{\text{water}} &= 62.4 \text{ lbm/ft}^3\end{aligned}$$

Substituting gives

$$\begin{aligned}\Delta p &= (\rho_{\text{mercury}} - \rho_{\text{water}})h \times \left(\frac{g}{g_c}\right) \\ &= \frac{\left(848 \ \frac{\text{lbm}}{\text{ft}^3} - 62.4 \ \frac{\text{lbm}}{\text{ft}^3}\right)(7 \text{ in}) \times \left(\frac{1 \text{ ft}}{12 \text{ in}}\right)\left(32.2 \ \frac{\text{ft}}{\text{sec}^2}\right)}{32.2 \ \frac{\text{lbm-ft}}{\text{lbf-sec}^2}} \\ &= \boxed{458.3 \text{ lbf/ft}^2 \quad (3.18 \text{ psi})}\end{aligned}$$

The answer is (B).

SI Solution

The manometer tube is filled with water above the mercury column. The pressure differential across the orifice meter is given by

$$\Delta p = p_1 - p_2 = (\rho_{\text{mercury}} - \rho_{\text{water}})hg$$

The densities of mercury and water are

$$\rho_{\text{mercury}} = 13\,600\ \text{kg/m}^3$$
$$\rho_{\text{water}} = 1000\ \text{kg/m}^3$$

Substituting gives

$$\begin{aligned}\Delta p &= (\rho_{\text{mercury}} - \rho_{\text{water}})gh \\ &= \left(13\,600\ \frac{\text{kg}}{\text{m}^3} - 1000\ \frac{\text{kg}}{\text{m}^3}\right)(0.178\ \text{m})\left(9.81\ \frac{\text{m}}{\text{s}^2}\right) \\ &= \boxed{22\,002\ \text{Pa}\quad (22.0\ \text{kPa})}\end{aligned}$$

The answer is (B).

19. *Customary U.S. Solution*

For 12 in pipe (assuming schedule-40),

$$D_i = 0.99483\ \text{ft}\quad [\text{App. 16.B}]$$
$$A_i = 0.7773\ \text{ft}^2$$

The velocity is

$$\begin{aligned}\text{v} &= \frac{\dot{V}}{A} \\ &= \frac{10\ \frac{\text{ft}^3}{\text{sec}}}{0.7773\ \text{ft}^2} \\ &= 12.87\ \text{ft/sec}\end{aligned}$$

For water at 70°F, $\nu = 1.059 \times 10^{-5}\ \text{ft}^2\text{/sec}$.

The Reynolds number in the pipe is

$$\begin{aligned}\text{Re} &= \frac{\text{v}D_i}{\nu} \\ &= \frac{\left(12.87\ \frac{\text{ft}}{\text{sec}}\right)(0.99483\ \text{ft})}{1.059 \times 10^{-5}\ \frac{\text{ft}^2}{\text{sec}}} \\ &= 1.21 \times 10^6\quad [\text{fully turbulent in the pipe}]\end{aligned}$$

Flow through the orifice will have a higher Reynolds number.

The volumetric flow rate through a sharp-edged orifice is given by

$$\dot{V} = C_f A_o \sqrt{\frac{2g(\rho_m - \rho)h}{\rho}}$$

In terms of pressure,

$$\dot{V} = C_f A_o \sqrt{\frac{2g_c(p_1 - p_2)}{\rho}}$$

Rearranging gives

$$C_f A_o = \frac{\dot{V}}{\sqrt{\frac{2g_c(p_1 - p_2)}{\rho}}}$$

p_1 is the upstream pressure, and p_2 is the downstream pressure.

The maximum head loss must not exceed 25 ft; therefore,

$$\frac{\left(\frac{g_c}{g}\right) \times (p_1 - p_2)}{\rho} = 25\ \text{ft}$$

$$\frac{g_c(p_1 - p_2)}{\rho} = (25\ \text{ft})g$$

Substituting gives

$$\begin{aligned}C_f A_0 &= \frac{10\ \frac{\text{ft}^3}{\text{sec}}}{\sqrt{(2)\left(32.2\ \frac{\text{ft}}{\text{sec}^2}\right)(25\ \text{ft})}} \\ &= 0.249\ \text{ft}^2\end{aligned}$$

Both C_f and A_o depend on the orifice diameter.

For a 7 in diameter orifice,

$$\begin{aligned}A_o &= \frac{\pi D_o^2}{4} = \frac{\pi\left((7\ \text{in})\left(\frac{1\ \text{ft}}{12\ \text{in}}\right)\right)^2}{4} \\ &= 0.267\ \text{ft}^2\end{aligned}$$

$$\frac{A_o}{A_1} = \frac{0.267\ \text{ft}^2}{0.7773\ \text{ft}^2} = 0.343$$

From a chart of flow coefficients (Fig. 17.28), for $A_o/A_1 = 0.343$ and fully turbulent flow,

$$C_f = 0.645$$

$$C_f A_o = (0.645)(0.267\ \text{ft}^2) = 0.172\ \text{ft}^2 < 0.249\ \text{ft}^2$$

Therefore, a 7 in diameter orifice is too small.

Try a 9 in diameter orifice.

$$\begin{aligned}A_o &= \frac{\pi D_o^2}{4} = \frac{\pi\left((9\ \text{in})\left(\frac{1\ \text{ft}}{12\ \text{in}}\right)\right)^2}{4} \\ &= 0.442\ \text{ft}^2\end{aligned}$$

$$\frac{A_o}{A_1} = \frac{0.442\ \text{ft}^2}{0.7773\ \text{ft}^2} = 0.569$$

From Fig. 17.28, for $A_o/A_1 = 0.569$ and fully turbulent flow,

$$C_f = 0.73$$

$$C_f A_o = (0.73)(0.442 \text{ ft}^2) = 0.323 \text{ ft}^2 > 0.249 \text{ ft}^2$$

Therefore, a 9 in orifice is too large.

Interpolating gives

$$\begin{aligned} D_o &= 7 \text{ in} + \frac{(9 \text{ in} - 7 \text{ in})(0.249 \text{ ft}^2 - 0.172 \text{ ft}^2)}{0.323 \text{ ft}^2 - 0.172 \text{ ft}^2} \\ &= 8.0 \text{ in} \end{aligned}$$

Further iterations yield

$$D_o \approx \boxed{8.1 \text{ in}}$$

$$C_f A_o = 0.243 \text{ ft}^2$$

The answer is (C).

SI Solution

For 300 mm pipe (assume the nominal diameter is the inner diameter), $D_i = 0.30$ m.

The velocity is

$$\begin{aligned} \text{v} &= \frac{\dot{V}}{A} = \frac{\dot{V}}{\frac{\pi D_i^2}{4}} \\ &= \frac{\left(250 \ \frac{\text{L}}{\text{s}}\right)\left(\frac{1 \text{ m}^3}{1000 \text{ L}}\right)}{\frac{\pi (0.3 \text{ m})^2}{4}} \\ &= 3.54 \text{ m/s} \end{aligned}$$

From App. 14.B, for water at 20°C,

$$\nu = 1.007 \times 10^{-6} \text{ m}^2/\text{s}$$

The Reynolds number in the pipe is

$$\begin{aligned} \text{Re} &= \frac{\text{v} D_i}{\nu} \\ &= \frac{\left(3.54 \ \frac{\text{m}}{\text{s}}\right)(0.3 \text{ m})}{1.007 \times 10^{-6} \ \frac{\text{m}^2}{\text{s}}} \\ &= 1.05 \times 10^6 \quad \text{[fully turbulent in the pipe]} \end{aligned}$$

Flow through the orifice will have a higher Reynolds number.

The volumetric flow rate through a sharp-edged orifice is given by

$$\dot{V} = C_f A_o \sqrt{\frac{2g(\rho_m - \rho)h}{\rho}}$$

In terms of pressure,

$$\dot{V} = C_f A_o \sqrt{\frac{2(p_1 - p_2)}{\rho}}$$

Rearranging gives

$$C_f A_o = \frac{\dot{V}}{\sqrt{\frac{2(p_1 - p_2)}{\rho}}}$$

p_1 is the upstream pressure, and p_2 is the downstream pressure.

The maximum head loss must not exceed 7.5 m; therefore,

$$\frac{p_1 - p_2}{g\rho} = 7.5 \text{ m}$$

$$\frac{p_1 - p_2}{\rho} = (7.5 \text{ m})g$$

Substituting gives

$$\begin{aligned} C_f A_o &= \frac{0.25 \ \frac{\text{m}^3}{\text{s}}}{\sqrt{(2)\left(9.81 \ \frac{\text{m}}{\text{s}^2}\right)(7.5 \text{ m})}} \\ &= 0.021 \text{ m}^2 \end{aligned}$$

Both C_f and A_o depend on the orifice diameter.

For an 18 cm diameter orifice,

$$A_o = \frac{\pi D_o^2}{4} = \frac{\pi (0.18 \text{ m})^2}{4} = 0.0254 \text{ m}^2$$

$$\frac{A_o}{A_1} = \frac{0.0254 \text{ m}^2}{0.0707 \text{ m}^2} = 0.359$$

From a chart of flow coefficients (Fig. 17.28), for $A_o/A_1 = 0.359$ and fully turbulent flow,

$$C_f = 0.65$$

$$C_f A_o = (0.65)(0.0254 \text{ m}^2) = 0.0165 \text{ m}^2 < 0.021 \text{ m}^2$$

Therefore, an 18 cm diameter orifice is too small.

Try a 23 cm diameter orifice.

$$A_o = \frac{\pi D_o^2}{4} = \frac{\pi (0.23 \text{ m})^2}{4} = 0.0415 \text{ m}^2$$

$$\frac{A_o}{A_1} = \frac{0.0415 \text{ m}^2}{0.0707 \text{ m}^2} = 0.587$$

From Fig. 17.28, for $A_o/A_1 = 0.587$ and fully turbulent flow,

$$C_f = 0.73$$

$$C_f A_o = (0.73)(0.0415 \text{ m}^2) = 0.0303 \text{ m}^2 > 0.021 \text{ m}^2$$

Therefore, a 23 cm orifice is too large.

Interpolating gives

$$\begin{aligned} D_o &= 18 \text{ cm} \\ &\quad + (23 \text{ cm} - 18 \text{ cm})\left(\frac{0.021 \text{ m}^2 - 0.0165 \text{ m}^2}{0.0303 \text{ m}^2 - 0.0165 \text{ m}^2}\right) \\ &= 19.6 \text{ cm} \end{aligned}$$

Further iteration yields

$$D_o = \boxed{20.0 \text{ cm}}$$

$$C_f = 0.675$$

$$C_f A_o = 0.021 \text{ m}^2$$

The answer is (C).

20. $$A_A = \left(\frac{\pi}{4}\right)\left(\frac{24 \text{ in}}{12 \frac{\text{in}}{\text{ft}}}\right)^2 = 3.142 \text{ ft}^2$$

$$A_B = \left(\frac{\pi}{4}\right)\left(\frac{12 \text{ in}}{12 \frac{\text{in}}{\text{ft}}}\right)^2 = 0.7854 \text{ ft}^2$$

$$v_A = \frac{\dot{V}}{A} = \frac{8 \frac{\text{ft}^3}{\text{sec}}}{3.142 \text{ ft}^2} = 2.546 \text{ ft/sec}$$

$$p_A = \gamma h = \left(62.4 \frac{\text{lbf}}{\text{ft}^3}\right)(20 \text{ ft}) = 1248 \text{ lbf/ft}^2$$

Using the Bernoulli equation to solve for p_B,

$$v_B = \frac{\dot{V}}{A} = \frac{8 \frac{\text{ft}^3}{\text{sec}}}{0.7854 \text{ ft}^2} = 10.19 \text{ ft/sec}$$

$$\begin{aligned} p_B &= 1248 \frac{\text{lbf}}{\text{ft}^2} - \left(\frac{\left(10.19 \frac{\text{ft}}{\text{sec}}\right)^2 - \left(2.546 \frac{\text{ft}}{\text{sec}}\right)^2}{(2)\left(32.2 \frac{\text{ft}}{\text{sec}^2}\right)}\right) \\ &\quad \times \left(62.4 \frac{\text{lbf}}{\text{ft}^3}\right) \\ &= 1153.67 \text{ lbf/ft}^2 \end{aligned}$$

With $\theta = 0$, from Eq. 17.200,

$$\begin{aligned} F_x &= \left(1153.67 \frac{\text{lbf}}{\text{ft}^2}\right)(0.7854 \text{ ft}^2) - \left(1248 \frac{\text{lbf}}{\text{ft}^2}\right)(3.142 \text{ ft}^2) \\ &\quad + \left(\frac{\left(8 \frac{\text{ft}^3}{\text{sec}}\right)\left(62.4 \frac{\text{lbf}}{\text{ft}^3}\right)}{32.2 \frac{\text{ft}}{\text{sec}^2}}\right)\left(10.19 \frac{\text{ft}}{\text{sec}} - 2.546 \frac{\text{ft}}{\text{sec}}\right) \\ &= \boxed{-2897 \text{ lbf on the fluid (toward A)}} \end{aligned}$$

$$F_y = 0$$

The answer is (A).

21. *Customary U.S. Solution*

The mass flow rate of the water is

$$\begin{aligned} \dot{m} &= \rho\dot{V} = \rho v A = \frac{\rho v \pi D^2}{4} \\ &= \left(62.4 \frac{\text{lbm}}{\text{ft}^3}\right)\left(40 \frac{\text{ft}}{\text{sec}}\right)\left(\frac{\pi\left((2 \text{ in})\left(\frac{1 \text{ ft}}{12 \text{ in}}\right)\right)^2}{4}\right) \\ &= 54.45 \text{ lbm/sec} \end{aligned}$$

The effective mass flow rate of the water is

$$\begin{aligned} \dot{m}_{\text{eff}} &= \left(\frac{v - v_b}{v}\right)\dot{m} \\ &= \left(\frac{40 \frac{\text{ft}}{\text{sec}} - 15 \frac{\text{ft}}{\text{sec}}}{40 \frac{\text{ft}}{\text{sec}}}\right)\left(54.45 \frac{\text{lbm}}{\text{sec}}\right) \\ &= 34.0 \text{ lbm/sec} \end{aligned}$$

The force in the (horizontal) x-direction is given by

$$\begin{aligned} F_x &= \frac{\dot{m}_{\text{eff}}(v - v_b)(\cos\theta - 1)}{g_c} \\ &= \frac{\left(34.0 \frac{\text{lbm}}{\text{sec}}\right)\left(40 \frac{\text{ft}}{\text{sec}} - 15 \frac{\text{ft}}{\text{sec}}\right)(\cos 60° - 1)}{32.2 \frac{\text{lbm-ft}}{\text{lbf-sec}^2}} \\ &= -13.2 \text{ lbf} \quad [\text{the force is acting to the left}] \end{aligned}$$

The force in the (vertical) y-direction is given by

$$\begin{aligned} F_y &= \frac{\dot{m}_{\text{eff}}(v - v_b)\sin\theta}{g_c} \\ &= \frac{\left(34.0 \frac{\text{lbm}}{\text{sec}}\right)\left(40 \frac{\text{ft}}{\text{sec}} - 15 \frac{\text{ft}}{\text{sec}}\right)(\sin 60°)}{32.2 \frac{\text{lbm-ft}}{\text{lbf-sec}^2}} \\ &= 22.9 \text{ lbf} \quad [\text{the force is acting upward}] \end{aligned}$$

The net resultant force is

$$\begin{aligned} F &= \sqrt{F_x^2 + F_y^2} \\ &= \sqrt{(-13.2\ \text{lbf})^2 + (22.9\ \text{lbf})^2} \\ &= \boxed{26.4\ \text{lbf}} \end{aligned}$$

The answer is (B).

SI Solution

The mass flow rate of the water is

$$\begin{aligned} \dot{m} &= \rho\dot{V} = \rho \text{v} A = \frac{\rho \text{v} \pi D^2}{4} \\ &= \left(1000\ \frac{\text{kg}}{\text{m}^3}\right)\left(12\ \frac{\text{m}}{\text{s}}\right)\left(\frac{\pi(0.05\ \text{m})^2}{4}\right) \\ &= 23.56\ \text{kg/s} \end{aligned}$$

The effective mass flow rate of the water is

$$\begin{aligned} \dot{m}_{\text{eff}} &= \left(\frac{\text{v} - \text{v}_b}{\text{v}}\right)\dot{m} \\ &= \left(\frac{12\ \frac{\text{m}}{\text{s}} - 4.5\ \frac{\text{m}}{\text{s}}}{12\ \frac{\text{m}}{\text{s}}}\right)\left(23.56\ \frac{\text{kg}}{\text{s}}\right) \\ &= 14.73\ \text{kg/s} \end{aligned}$$

The force in the (horizontal) x-direction is given by

$$\begin{aligned} F_x &= \dot{m}_{\text{eff}}(\text{v} - \text{v}_b)(\cos\theta - 1) \\ &= \left(14.73\ \frac{\text{kg}}{\text{s}}\right)\left(12\ \frac{\text{m}}{\text{s}} - 4.5\ \frac{\text{m}}{\text{s}}\right)(\cos 60^\circ - 1) \\ &= -55.2\ \text{N} \quad [\text{the force is acting to the left}] \end{aligned}$$

The force in the (vertical) y-direction is given by

$$\begin{aligned} F_y &= \dot{m}_{\text{eff}}(\text{v} - \text{v}_b)\sin\theta \\ &= \left(14.73\ \frac{\text{kg}}{\text{s}}\right)\left(12\ \frac{\text{m}}{\text{s}} - 4.5\ \frac{\text{m}}{\text{s}}\right)(\sin 60^\circ) \\ &= 95.7\ \text{N} \quad [\text{the force is acting upward}] \end{aligned}$$

The net resultant force is

$$\begin{aligned} F &= \sqrt{F_x^2 + F_y^2} \\ &= \sqrt{(-55.2\ \text{N})^2 + (95.7\ \text{N})^2} \\ &= \boxed{110.5\ \text{N}} \end{aligned}$$

The answer is (B).

22. *Customary U.S. Solution*

The power that must be added to the pump is given by

$$P = \frac{\Delta h \dot{m} \times \left(\frac{g}{g_c}\right)}{\eta}$$

For schedule-40 pipe,

$$\begin{aligned} D_i &= 0.9948\ \text{ft} \quad [\text{App. 16.B}] \\ A_i &= 0.7773\ \text{ft}^2 \\ \text{v} &= \frac{\dot{V}}{A_i} \\ &= \frac{\left(2000\ \frac{\text{gal}}{\text{min}}\right)\left(\frac{0.002228\ \frac{\text{ft}^3}{\text{sec}}}{1\ \frac{\text{gal}}{\text{min}}}\right)}{0.7773\ \text{ft}^2} \\ &= 5.73\ \text{ft/sec} \end{aligned}$$

(Note that the pressures are in terms of gage pressure, and the density of mercury is 0.491 lbm/in^3.)

$$\begin{aligned} p_i &= \left(14.7\ \frac{\text{lbf}}{\text{in}^2} - \frac{(6\ \text{in})\left(0.491\ \frac{\text{lbm}}{\text{in}^3}\right)\left(32.2\ \frac{\text{ft}}{\text{sec}^2}\right)}{32.2\ \frac{\text{lbm-ft}}{\text{lbf-sec}^2}}\right) \\ &\quad \times \left(\frac{144\ \text{in}^2}{1\ \text{ft}^2}\right) \\ &= 1692.6\ \text{lbf/ft}^2 \end{aligned}$$

$$E_{ti} = \frac{p_i}{\rho} + \frac{\text{v}_i^2}{2g_c} + \frac{z_i g}{g_c}$$

Since the pump inlet and outlet are at the same elevation, use $z = 0$ and $\rho = (\text{SG})\rho_{\text{water}}$.

$$\begin{aligned} E_{ti} &= \frac{p_i}{(\text{SG})\rho_{\text{water}}} + \frac{\text{v}_i^2}{2g_c} + 0 \\ &= \frac{1692.6\ \frac{\text{lbf}}{\text{ft}^2}}{(1.2)\left(62.4\ \frac{\text{lbm}}{\text{ft}^3}\right)} + \frac{\left(5.73\ \frac{\text{ft}}{\text{sec}}\right)^2}{(2)\left(32.2\ \frac{\text{lbm-ft}}{\text{lbf-sec}^2}\right)} \\ &= 23.11\ \text{ft-lbf/lbm} \end{aligned}$$

Calculate the total head at the inlet.

$$\begin{aligned} h_{ti} &= E_{ti} \times \left(\frac{g_c}{g}\right) \\ &= \frac{\left(23.11\ \frac{\text{ft-lbf}}{\text{lbm}}\right)\left(32.2\ \frac{\text{lbm-ft}}{\text{lbf-sec}^2}\right)}{32.2\ \frac{\text{ft}}{\text{sec}^2}} \\ &= 23.11\ \text{ft} \end{aligned}$$

At the outlet side of the pump,

$$D_o = 0.6651 \text{ ft}$$
$$A_o = 0.3474 \text{ ft}^2$$
$$Q = \text{v}_o A_o$$
$$\text{v}_o = \frac{Q}{A_o}$$
$$= \frac{\left(2000 \frac{\text{gal}}{\text{min}}\right)\left(\frac{0.002228 \frac{\text{ft}^3}{\text{sec}}}{1 \frac{\text{gal}}{\text{min}}}\right)}{0.3474 \text{ ft}^2}$$
$$= 12.83 \text{ ft/sec}$$

(Note that the pressures are in terms of gage pressure and the gauge is located 4 ft above the pump outlet, which adds 4 ft of pressure head at the pump outlet.)

$$p_o = \left(14.7 \frac{\text{lbf}}{\text{in}^2} + 20 \frac{\text{lbf}}{\text{in}^2}\right)\left(\frac{144 \text{ in}^2}{1 \text{ ft}^2}\right) + 4 \text{ ft}\left(\frac{(1.2)\left(62.4 \frac{\text{lbm}}{\text{ft}^3}\right)\left(32.2 \frac{\text{ft}}{\text{sec}^2}\right)}{32.2 \frac{\text{lbm-ft}}{\text{lbf-sec}^2}}\right)$$
$$= 5296 \text{ lbf/ft}^2$$
$$E_{to} = \frac{p_o}{\rho} + \frac{\text{v}_o^2}{2g_c} + \frac{z_o g}{g_c}$$

Since the pump inlet and outlet are at the same elevation, use $z = 0$ and $\rho = (\text{SG})\rho_{\text{water}}$.

$$E_{to} = \frac{p_o}{(\text{SG})\rho_{\text{water}}} + \frac{\text{v}_o^2}{2g_c} + 0$$
$$= \frac{5296 \frac{\text{lbf}}{\text{ft}^2}}{(1.2)\left(62.4 \frac{\text{lbm}}{\text{ft}^3}\right)} + \frac{\left(12.83 \frac{\text{ft}}{\text{sec}}\right)^2}{(2)\left(32.2 \frac{\text{lbm-ft}}{\text{lbf-sec}^2}\right)}$$
$$= 73.28 \text{ ft-lbf/lbm}$$

Calculate the total head at the outlet.

$$h_{to} = E_{to} \times \left(\frac{g_c}{g}\right)$$
$$= \left(73.28 \frac{\text{ft-lbf}}{\text{lbm}}\right)\left(\frac{32.2 \frac{\text{lbm-ft}}{\text{lbf-sec}^2}}{32.2 \frac{\text{ft}}{\text{sec}^2}}\right)$$
$$= 73.28 \text{ ft}$$

Compute the total head required across the pump.

$$\Delta h = h_{to} - h_{ti}$$
$$= 73.28 \text{ ft} - 23.11 \text{ ft}$$
$$= 50.17 \text{ ft}$$

The mass flow rate is

$$\dot{m} = \rho \dot{V}$$

In terms of the specific gravity, the mass flow rate is

$$\dot{m} = (\text{SG})\rho_{\text{water}}\dot{V}$$
$$= (1.2)\left(62.4 \frac{\text{lbm}}{\text{ft}^3}\right)\left(2000 \frac{\text{gal}}{\text{min}}\right)\left(\frac{0.002228 \frac{\text{ft}^3}{\text{sec}}}{1 \frac{\text{gal}}{\text{min}}}\right)$$
$$= 333.7 \text{ lbm/sec}$$

The power that must be added to the pump is

$$P = \frac{\Delta h \dot{m} \times \left(\frac{g}{g_c}\right)}{\eta}$$
$$= \frac{(50.17 \text{ ft})\left(333.7 \frac{\text{lbm}}{\text{sec}}\right)\left(\frac{32.2 \frac{\text{ft}}{\text{sec}^2}}{32.2 \frac{\text{lbm-ft}}{\text{lbf-sec}^2}}\right)}{(0.85)\left(550 \frac{\text{ft-lbf}}{\text{hp-sec}}\right)}$$
$$= \boxed{35.8 \text{ hp}}$$

(Note that it is not necessary to use absolute pressures as has been done in this solution.)

The answer is (B).

SI Solution

The power that must be added to the pump is given by

$$P = \frac{\Delta h \dot{m} g}{\eta}$$

For schedule-40 pipe,

$$D_i = 303.2 \text{ mm} \quad \text{[App. 16.C]}$$
$$A_i = 0.0722 \text{ m}^2$$
$$\text{v} = \frac{\dot{V}}{A_i}$$
$$= \frac{0.125 \frac{\text{m}^3}{\text{s}}}{0.0722 \text{ m}^2}$$
$$= 1.73 \text{ m/s}$$

(Note that the pressures are in terms of gage pressure, and the density of mercury is 13 600 kg/m³.)

$$p_i = 1.013 \times 10^5 \text{ Pa} - (0.15 \text{ m})\left(13\,600 \frac{\text{kg}}{\text{m}^3}\right)\left(9.81 \frac{\text{m}}{\text{s}^2}\right)$$
$$= 8.13 \times 10^4 \text{ Pa}$$
$$E_{ti} = \frac{p}{\rho} + \frac{\text{v}_i^2}{2} + z_i g$$

Since the pump inlet and outlet are at the same elevation, use $z = 0$ and $\rho = (\text{SG})\rho_{\text{water}}$.

$$\begin{aligned} E_{ti} &= \frac{p}{(\text{SG})\rho_{\text{water}}} + \frac{\text{v}_i^2}{2} + 0 \\ &= \frac{8.13 \times 10^4 \text{ Pa}}{(1.2)\left(1000 \ \frac{\text{kg}}{\text{m}^3}\right)} + \frac{\left(1.73 \ \frac{\text{m}}{\text{s}}\right)^2}{2} \\ &= 69.2 \text{ J/kg} \end{aligned}$$

The total head at the inlet is

$$\begin{aligned} h_{ti} &= \frac{E_{ti}}{g} \\ &= \frac{69.2 \ \frac{\text{J}}{\text{kg}}}{9.81 \ \frac{\text{m}}{\text{s}^2}} \\ &= 7.05 \text{ m} \end{aligned}$$

Assume the pipe nominal diameter is equal to the internal diameter. On the outlet side of the pump,

$$\begin{aligned} D_i &= 202.7 \text{ mm} \\ A_o &= 0.0323 \text{ m}^2 \\ \text{v}_o &= \frac{\dot{V}}{A_o} \\ &= \frac{0.125 \ \frac{\text{m}^3}{\text{s}}}{0.0323 \text{ m}^2} \\ &= 3.87 \text{ m/s} \end{aligned}$$

(Note that the pressures are in terms of gage pressure and the gauge is located 1.2 m above the pump outlet, which adds 1.2 m of pressure head at the pump outlet.)

$$\begin{aligned} p_o &= 1.013 \times 10^5 \text{ Pa} + 138 \times 10^3 \text{ Pa} \\ &\quad + (1.2 \text{ m})\left((1.2)\left(1000 \ \frac{\text{kg}}{\text{m}^3}\right)\left(9.81 \ \frac{\text{m}}{\text{s}^2}\right)\right) \\ &= 2.53 \times 10^5 \text{ Pa} \end{aligned}$$

$$E_{to} = \frac{p_o}{\rho} + \frac{\text{v}_o^2}{2} + z_o g$$

Since the pump inlet and outlet are at the same elevation, use $z = 0$ and $\rho = (\text{SG})\rho_{\text{water}}$.

$$\begin{aligned} E_{to} &= \frac{p_o}{(\text{SG})\rho_{\text{water}}} + \frac{\text{v}_o^2}{2} + 0 \\ &= \frac{2.53 \times 10^5 \text{ Pa}}{(1.2)\left(1000 \ \frac{\text{kg}}{\text{m}^3}\right)} + \frac{\left(3.87 \ \frac{\text{m}}{\text{s}}\right)^2}{2} \\ &= 218.3 \text{ J/kg} \end{aligned}$$

The total head at the outlet is

$$\begin{aligned} h_{to} &= \frac{E_{to}}{g} \\ &= \frac{218.3 \ \frac{\text{J}}{\text{kg}}}{9.81 \ \frac{\text{m}}{\text{s}^2}} \\ &= 22.25 \text{ m} \end{aligned}$$

The total head required across the pump is

$$\begin{aligned} \Delta h &= h_{to} - h_{ti} \\ &= 22.25 \text{ m} - 7.05 \text{ m} \\ &= 15.2 \text{ m} \end{aligned}$$

The mass flow rate is

$$\dot{m} = \rho \dot{V}$$

In terms of the specific gravity, the mass flow rate is

$$\begin{aligned} \dot{m} &= (\text{SG})\rho_{\text{water}} Q \\ &= (1.2)\left(1000 \ \frac{\text{kg}}{\text{m}^3}\right)\left(0.125 \ \frac{\text{m}^3}{\text{s}}\right) \\ &= 150 \text{ kg/s} \end{aligned}$$

The power that must be added to the pump is

$$\begin{aligned} P &= \frac{\Delta h \dot{m} g}{\eta} \\ &= \frac{(15.2 \text{ m})\left(150 \ \frac{\text{kg}}{\text{s}}\right)\left(9.81 \ \frac{\text{m}}{\text{s}^2}\right)}{0.85} \\ &= \boxed{26\,313 \text{ W} \quad (26.3 \text{ kW})} \end{aligned}$$

(Note that it is not necessary to use absolute pressures as has been done in this solution.)

The answer is (B).

23. *Customary U.S. Solution*

The power developed by the horizontal turbine is given by

$$P = \dot{m} h_{\text{loss}} \times \left(\frac{g}{g_c}\right)$$

The mass flow rate is

$$\begin{aligned} \dot{m} &= \dot{V}\rho \\ &= \left(100 \ \frac{\text{ft}^3}{\text{sec}}\right)\left(62.4 \ \frac{\text{lbm}}{\text{ft}^3}\right) \\ &= 6240 \text{ lbm/sec} \end{aligned}$$

The head loss across the horizontal turbine is given by

$$h_{\text{loss}} = \left(\frac{\Delta p}{\rho}\right) \times \left(\frac{g_c}{g}\right)$$

$$= \left(\frac{\left(30\ \frac{\text{lbf}}{\text{in}^2} - \left(-5\ \frac{\text{lbf}}{\text{in}^2}\right)\right)\left(\frac{144\ \text{in}^2}{1\ \text{ft}^2}\right)}{62.4\ \frac{\text{lbm}}{\text{ft}^3}}\right) \times \left(\frac{32.2\ \frac{\text{lbm-ft}}{\text{lbf-sec}^2}}{32.2\ \frac{\text{ft}}{\text{sec}^2}}\right)$$

$$= 80.77\ \text{ft}$$

From Table 18.5, the power developed is

$$P = \dot{m}h_{\text{loss}} \times \left(\frac{g}{g_c}\right)$$

$$= \frac{\left(6240\ \frac{\text{lbm}}{\text{sec}}\right)(80.77\ \text{ft})\left(32.2\ \frac{\text{ft}}{\text{sec}^2}\right)}{\left(32.2\ \frac{\text{lbm-ft}}{\text{lbf-sec}^2}\right)\left(550\ \frac{\text{ft-lbf}}{\text{hp-sec}}\right)}$$

$$= \boxed{916\ \text{hp}}$$

The answer is (D).

SI Solution

The power developed by the horizontal turbine is given by

$$P = \dot{m}h_{\text{loss}}g$$

The mass flow rate is

$$\dot{m} = \dot{V}\rho$$

$$= \left(2.6\ \frac{\text{m}^3}{\text{s}}\right)\left(1000\ \frac{\text{kg}}{\text{m}^3}\right)$$

$$= 2600\ \text{kg/s}$$

The head loss across the horizontal turbine is given by

$$h_{\text{loss}} = \frac{\Delta p}{\rho g}$$

$$= \frac{(210\ \text{kPa} - (-35\ \text{kPa}))\left(1000\ \frac{\text{Pa}}{\text{kPa}}\right)}{\left(1000\ \frac{\text{kg}}{\text{m}^3}\right)\left(9.81\ \frac{\text{m}}{\text{s}^2}\right)}$$

$$= 25.0\ \text{m}$$

From Table 18.5, the power developed is

$$P = \dot{m}h_{\text{loss}}g$$

$$= \left(2600\ \frac{\text{kg}}{\text{s}}\right)(25.0\ \text{m})\left(9.81\ \frac{\text{m}}{\text{s}^2}\right)$$

$$= \boxed{637\,650\ \text{W}\quad (638\ \text{kW})}$$

The answer is (D).

24. *Customary U.S. Solution*

(a) The drag on the car is given by

$$F_D = \frac{C_D A \rho \text{v}^2}{2g_c}$$

For air at 70°F,

$$\rho = \frac{p}{RT} = \frac{\left(14.7\ \frac{\text{lbf}}{\text{in}^2}\right)\left(144\ \frac{\text{in}^2}{\text{ft}^2}\right)}{\left(53.35\ \frac{\text{ft-lbf}}{\text{lbm-°R}}\right)(70°\text{F} + 460)}$$

$$= 0.0749\ \text{lbm/ft}^3$$

$$\text{v} = \left(55\ \frac{\text{mi}}{\text{hr}}\right)\left(5280\ \frac{\text{ft}}{\text{mi}}\right)\left(\frac{1\ \text{hr}}{3600\ \text{sec}}\right)$$

$$= 80.67\ \text{ft/sec}$$

Substituting gives

$$F_D = \frac{C_D A \rho \text{v}^2}{2g_c}$$

$$= \frac{(0.42)(28\ \text{ft}^2)\left(0.0749\ \frac{\text{lbm}}{\text{ft}^3}\right)\left(80.67\ \frac{\text{ft}}{\text{sec}}\right)^2}{(2)\left(32.2\ \frac{\text{lbm-ft}}{\text{lbf-sec}^2}\right)}$$

$$= 89.0\ \text{lbf}$$

The total resisting force is

$$F = F_D + \text{rolling resistance}$$

$$= 89.0\ \text{lbf} + (0.01)(3300\ \text{lbm}) \times \left(\frac{g}{g_c}\right)$$

$$= 89.0\ \text{lbf} + \frac{(0.01)(3300\ \text{lbm})\left(32.2\ \frac{\text{ft}}{\text{sec}^2}\right)}{32.2\ \frac{\text{lbm-ft}}{\text{lbf-sec}^2}}$$

$$= 122.0\ \text{lbf}$$

The power consumed is

$$P = F\text{v}$$

$$= \frac{(122.0\ \text{lbf})\left(80.67\ \frac{\text{ft}}{\text{sec}}\right)}{778\ \frac{\text{ft-lbf}}{\text{Btu}}}$$

$$= 12.65\ \text{Btu/sec}$$

The energy available from the fuel is

$$E_A = (\text{engine thermal efficiency})(\text{fuel heating value})$$

$$= (0.28)\left(115{,}000\ \frac{\text{Btu}}{\text{gal}}\right)$$

$$= 32{,}200\ \text{Btu/gal}$$

The fuel consumption at 55 mi/hr is

$$\frac{P}{E_A} = \frac{12.65 \ \frac{\text{Btu}}{\text{sec}}}{32{,}200 \ \frac{\text{Btu}}{\text{gal}}} = 3.93 \times 10^{-4} \ \text{gal/sec}$$

The fuel consumption is

$$\frac{3.93 \times 10^{-4} \ \frac{\text{gal}}{\text{sec}}}{\left(55 \ \frac{\text{mi}}{\text{hr}}\right)\left(\frac{1 \ \text{hr}}{3600 \ \text{sec}}\right)} = \boxed{0.0257 \ \text{gal/mi}}$$

The answer is (A).

(b) Similarly, the fuel consumption at 65 mi/hr is

$$\begin{aligned} \text{v} &= \left(65 \ \frac{\text{mi}}{\text{hr}}\right)\left(5280 \ \frac{\text{ft}}{\text{mi}}\right)\left(\frac{1 \ \text{hr}}{3600 \ \text{sec}}\right) \\ &= 95.33 \ \text{ft/sec} \end{aligned}$$

$$\begin{aligned} F_D &= \frac{C_D A \rho \text{v}^2}{2g_c} \\ &= \frac{(0.42)(28 \ \text{ft}^2)\left(0.0749 \ \frac{\text{lbm}}{\text{ft}^3}\right)\left(95.33 \ \frac{\text{ft}}{\text{sec}}\right)^2}{(2)\left(32.2 \ \frac{\text{lbm-ft}}{\text{lbf-sec}^2}\right)} \\ &= 124.3 \ \text{lbf} \end{aligned}$$

The total resisting force is

$$\begin{aligned} F &= F_D + \text{rolling resistance} \\ &= 124.3 \ \text{lbf} + (0.01)(3300 \ \text{lbm})\left(\frac{g}{g_c}\right) \\ &= 124.3 \ \text{lbf} + \frac{(0.01)(3300 \ \text{lbm})\left(32.2 \ \frac{\text{ft}}{\text{sec}^2}\right)}{32.2 \ \frac{\text{lbm-ft}}{\text{lbf-sec}^2}} \\ &= 157.3 \ \text{lbf} \end{aligned}$$

The power consumed is

$$\begin{aligned} P &= F\text{v} \\ &= \frac{(157.3 \ \text{lbf})\left(95.33 \ \frac{\text{ft}}{\text{sec}}\right)}{778 \ \frac{\text{ft-lbf}}{\text{Btu}}} \\ &= 19.27 \ \text{Btu/sec} \end{aligned}$$

The fuel consumption at 65 mi/hr is

$$\frac{P}{E_A} = \frac{19.27 \ \frac{\text{Btu}}{\text{sec}}}{32{,}200 \ \frac{\text{Btu}}{\text{gal}}} = 5.98 \times 10^{-4} \ \text{gal/sec}$$

The fuel consumption is

$$\frac{5.98 \times 10^{-4} \ \frac{\text{gal}}{\text{sec}}}{\left(65 \ \frac{\text{mi}}{\text{hr}}\right)\left(\frac{1 \ \text{hr}}{3600 \ \text{sec}}\right)} = 0.0331 \ \text{gal/mi}$$

The relative difference between the fuel consumption at 55 mi/hr and 65 mi/hr is

$$\frac{0.0331 \ \frac{\text{gal}}{\text{mi}} - 0.0257 \ \frac{\text{gal}}{\text{mi}}}{0.0257 \ \frac{\text{gal}}{\text{mi}}} = \boxed{0.288 \quad (28.8\%)}$$

The answer is (C).

SI Solution

(a) The drag on the car is given by

$$F_D = \frac{C_D A \rho \text{v}^2}{2}$$

For air at 20°C,

$$\begin{aligned} \rho &= \frac{p}{RT} = \frac{1.013 \times 10^5 \ \text{Pa}}{\left(287 \ \frac{\text{J}}{\text{kg·K}}\right)(20°\text{C} + 273)} \\ &= 1.205 \ \text{kg/m}^3 \end{aligned}$$

$$\begin{aligned} \text{v} &= \left(90 \ \frac{\text{km}}{\text{h}}\right)\left(1000 \ \frac{\text{m}}{\text{km}}\right)\left(\frac{1 \ \text{h}}{3600 \ \text{s}}\right) \\ &= 25.0 \ \text{m/s} \end{aligned}$$

Substituting gives

$$\begin{aligned} F_D &= \frac{C_D A \rho \text{v}^2}{2} \\ &= \left(\tfrac{1}{2}\right)(0.42)(2.6 \ \text{m}^2)\left(1.205 \ \frac{\text{kg}}{\text{m}^3}\right)\left(25.0 \ \frac{\text{m}}{\text{s}}\right)^2 \\ &= 411.2 \ \text{N} \end{aligned}$$

The total resisting force is

$$\begin{aligned} F &= F_D + \text{rolling resistance} \\ &= 411.2 \ \text{N} + (0.01)(1500 \ \text{kg})g \\ &= 411.2 \ \text{N} + (0.01)(1500 \ \text{kg})\left(9.81 \ \frac{\text{m}}{\text{s}^2}\right) \\ &= 558.4 \ \text{N} \end{aligned}$$

The power consumed is

$$\begin{aligned} P &= Fv \\ &= (558.4\ \text{N})\left(25\ \frac{\text{m}}{\text{s}}\right) \\ &= 13\,960\ \text{W} \end{aligned}$$

The energy available from the fuel is

$$\begin{aligned} E_A &= (\text{engine thermal efficiency})(\text{fuel heating value}) \\ &= (0.28)\left(4.6\times 10^8\ \frac{\text{J}}{\text{L}}\right) \\ &= 1.288\times 10^8\ \text{J/L} \end{aligned}$$

The fuel consumption at 90 km/h is

$$\begin{aligned} \frac{P}{E_A} &= \frac{13\,960\ \text{W}}{1.288\times 10^8\ \frac{\text{J}}{\text{L}}} \\ &= 1.08\times 10^{-4}\ \text{L/s} \end{aligned}$$

The fuel consumption is

$$\frac{1.08\times 10^{-4}\ \frac{\text{L}}{\text{s}}}{\left(90\ \frac{\text{km}}{\text{h}}\right)\left(\frac{1\ \text{h}}{3600\ \text{s}}\right)} = \boxed{0.00434\ \text{L/km}}$$

The answer is (A).

(b) Similarly, the fuel consumption at 105 km/h is

$$\begin{aligned} \text{v} &= \left(105\ \frac{\text{km}}{\text{h}}\right)\left(1000\ \frac{\text{m}}{\text{km}}\right)\left(\frac{1\ \text{h}}{3600\ \text{s}}\right) \\ &= 29.2\ \text{m/s} \\ D &= \frac{C_D A\rho \text{v}^2}{2} \\ &= \left(\tfrac{1}{2}\right)(0.42)(2.6\ \text{m}^2)\left(1.205\ \frac{\text{kg}}{\text{m}^3}\right)\left(29.2\ \frac{\text{m}}{\text{s}}\right)^2 \\ &= 561.0\ \text{N} \end{aligned}$$

The total resisting force is

$$\begin{aligned} F &= F_D + \text{rolling resistance} \\ &= 561.0\ \text{N} + (0.01)(1500\ \text{kg})g \\ &= 561.0\ \text{N} + (0.01)(1500\ \text{kg})\left(9.81\ \frac{\text{m}}{\text{s}^2}\right) \\ &= 708.2\ \text{N} \end{aligned}$$

The power consumed is

$$\begin{aligned} P &= Fv \\ &= (708.2\ \text{N})\left(29.2\ \frac{\text{m}}{\text{s}}\right) \\ &= 20\,679\ \text{W} \end{aligned}$$

The fuel consumption at 105 km/h is

$$\begin{aligned} \frac{P}{E_A} &= \frac{20\,679\ \text{W}}{1.288\times 10^8\ \frac{\text{J}}{\text{L}}} \\ &= 1.61\times 10^{-4}\ \text{L/s} \end{aligned}$$

The fuel consumption is

$$\frac{1.61\times 10^{-4}\ \frac{\text{L}}{\text{s}}}{\left(105\ \frac{\text{km}}{\text{h}}\right)\left(\frac{1\ \text{h}}{3600\ \text{s}}\right)} = 0.00552\ \text{L/km}$$

The relative difference between the fuel consumption at 90 km/h and 105 km/h is

$$\frac{0.00552\ \frac{\text{L}}{\text{km}} - 0.00434\ \frac{\text{L}}{\text{km}}}{0.00434\ \frac{\text{L}}{\text{km}}} = \boxed{0.272\quad (27.2\%)}$$

The answer is (C).

18 Hydraulic Machines

PRACTICE PROBLEMS

Pumping Power

1. 2000 gal/min of 60°F thickened sludge with a specific gravity of 1.2 flow through a pump with an inlet diameter of 12 in and an outlet of 8 in. The centerlines of the inlet and outlet are at the same elevation. The inlet pressure is 8 in of mercury (vacuum). A discharge pressure gauge located 4 ft above the pump discharge centerline reads 20 psig. The pump efficiency is 85%. All pipes are schedule-40. What is the input power of the pump?

(A) 26 hp
(B) 31 hp
(C) 37 hp
(D) 53 hp

2. 1.25 ft^3/sec (35 L/s) of 70°F (21°C) water are pumped from the bottom of a tank through 700 ft (230 m) of 4 in (10.2 cm) schedule-40 steel pipe. The line includes a 50 ft (15 m) rise in elevation, two right-angle elbows, a wide-open gate valve, and a swing check valve. All fittings and valves are regular screwed. The inlet pressure is 50 psig (345 kPa), and a working pressure of 20 psig (140 kPa) is needed at the end of the pipe. What is the hydraulic power for this pumping application?

(A) 16 hp (13 kW)
(B) 23 hp (17 kW)
(C) 49 hp (37 kW)
(D) 66 hp (50 kW)

3. 80 gal/min (5 L/s) of 80°F (27°C) water are lifted 12 ft (4 m) vertically by a pump through a total length of 50 ft (15 m) of a 2 in (5.1 cm) diameter smooth rubber hose. The discharge end of the hose is submerged in 8 ft (2.5 m) of water as shown. What head is added by the pump?

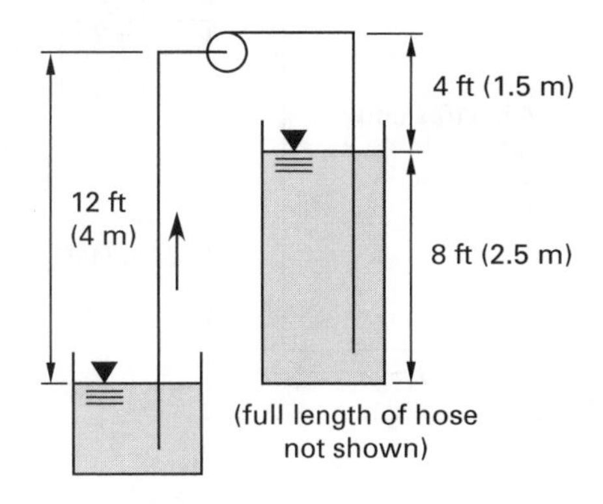

(A) 10 ft (3.0 m)
(B) 13 ft (4.3 m)
(C) 22 ft (6.6 m)
(D) 31 ft (9.3 m)

4. A 20 hp motor drives a centrifugal pump. The pump discharges 60°F (16°C) water at 12 ft/sec (4 m/s) into a 6 in (15.2 cm) steel schedule-40 line. The inlet is 8 in (20.3 cm) schedule-40 steel pipe. The pump suction is 5 psi (35 kPa) below standard atmospheric pressure. The friction head loss in the system is 10 ft (3.3 m). The pump efficiency is 70%. The suction and discharge lines are at the same elevation. What is the maximum height above the pump inlet that water is available at standard atmospheric pressure?

(A) 28 ft (6.9 m)
(B) 37 ft (11 m)
(C) 49 ft (15 m)
(D) 81 ft (25 m)

5. (*Time limit: one hour*) A pump station is used to fill a tank on a hill above from a lake below. The flow rate is 10,000 gal/hr (10.5 L/s) of 60°F (16°C) water. The atmospheric pressure is 14.7 psia (101 kPa). The pump is 12 ft (4 m) above the lake, and the tank surface level is 350 ft (115 m) above the pump. The suction and discharge lines are 4 in (10.2 cm) diameter schedule-40 steel pipe. The equivalent length of the inlet line between the lake and the pump is 300 ft (100 m). The total equivalent length between the lake and the tank is 7000 ft (2300 m), including all fittings, bends, screens, and valves. The cost of electricity is $0.04 per kW-hr. The overall efficiency of the pump and motor set is 70%.

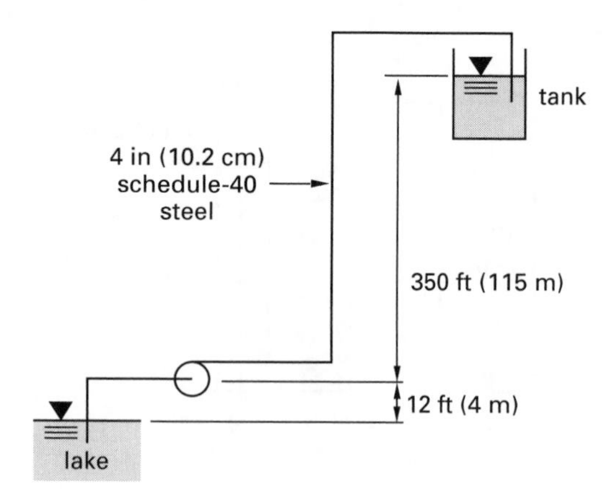

(a) What does it cost to operate the pump for 1 hr?

(A) $0.1
(B) $1
(C) $3
(D) $6

(b) What motor power is required?

(A) 10 hp (7.5 kW)
(B) 30 hp (25 kW)
(C) 50 hp (40 kW)
(D) 75 hp (60 kW)

(c) What is the NPSHA for this application?

(A) 4 ft (1.2 m)
(B) 8 ft (2.4 m)
(C) 12 ft (3.6 m)
(D) 16 ft (4.5 m)

6. (*Time limit: one hour*) A town with a stable, constant population of 10,000 produces sewage at the average rate of 100 gallons per capita day (gpcd), with peak flows of 250 gpcd. The pipe to the pumping station is 5000 ft in length and has a C-value of 130. The elevation drop along the length is 48 ft. Minor losses in infiltration are insignificant. The pump's maximum suction lift is 10 ft.

(a) If all diameters are available and the pipe flows 100% full under gravity flow, what minimum pipe diameter is required?

(A) 8 in
(B) 12 in
(C) 14 in
(D) 18 in

(b) If constant-speed pumps are used, what is the minimum number of pumps (disregarding spares and backups) that should be used?

(A) 2
(B) 3
(C) 4
(D) 5

(c) If variable-speed pumps are used, what is the minimum number of pumps (disregarding spares and backups) that should be used?

(A) 1
(B) 2
(C) 3
(D) 4

(d) If three constant-speed pumps are used, with a fourth as backup, and the pump-motor set efficiency is 60%, what motor power is required?

(A) 2 hp
(B) 3 hp
(C) 5 hp
(D) 8 hp

(e) If two variable-speed pumps are used, with a third as backup, what motor power is required?

(A) 3 hp
(B) 8 hp
(C) 12 hp
(D) 18 hp

(f) Which of the following are ways of controlling sump pump on-off cycles?

I. detecting sump levels
II. detecting pressure in the sump
III. detecting incoming flow rates
IV. using fixed run times
V. detecting outgoing flow rates
VI. operating manually

(A) I, II, and III
(B) I, II, IV, and V
(C) I, III, and V
(D) I, II, IV, V, and VI

(g) With intermittent fan operation, approximately how many air changes should the wet well receive per hour?

(A) 6
(B) 12
(C) 20
(D) 30

(h) With continuous fan operation, approximately how many air changes should the dry well receive per hour?

(A) 6
(B) 12
(C) 20
(D) 30

7. (*Time limit: one hour*) A pump transfers 3.5 MGD of filtered water from the clear well of a 10 ft by 20 ft (plan) rapid sand filter to a higher elevation. The pump efficiency is 85%, and the motor driving pump has an efficiency of 90%. Minor losses are insignificant. Refer to the following illustration for additional information.

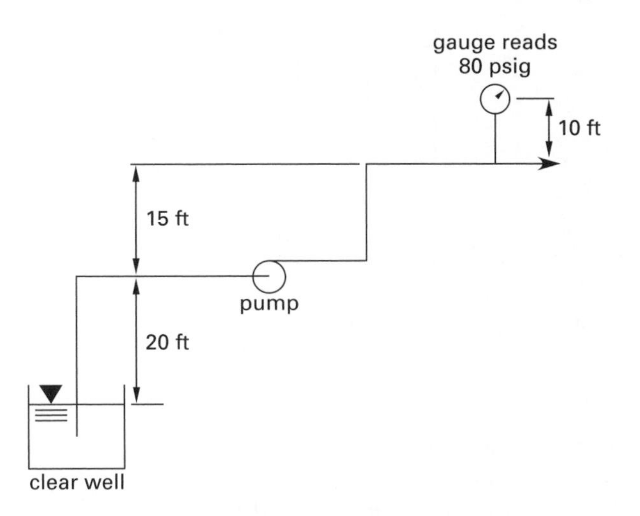

(a) What is the static suction lift?

(A) 15 ft
(B) 20 ft
(C) 35 ft
(D) 40 ft

(b) What is the static discharge head?

(A) 15 ft
(B) 20 ft
(C) 35 ft
(D) 40 ft

(c) Based on the information given, what is the approximate total dynamic head?

(A) 45 ft
(B) 185 ft
(C) 210 ft
(D) 230 ft

(d) What motor power is required?

(A) 50 hp
(B) 100 hp
(C) 150 hp
(D) 200 hp

Pumping Other Fluids

8. (*Time limit: one hour*) Gasoline with a specific gravity of 0.7 and viscosity of 6×10^{-6} ft^2/sec (5.6×10^{-7} m^2/s) is transferred from a tanker to a storage tank. The interior of the storage tank is maintained at atmospheric pressure by a vapor-recovery system. The free surface in the storage tank is 60 ft (20 m) above the tanker's free surface. The pipe consists of 500 ft (170 m) of 3 in (7.62 cm) schedule-40 steel pipe with six flanged elbows and two wide-open gate valves. The pump and motor both have individual efficiencies of 88%. Electricity costs $0.045 per kW-hr. The pump's performance data (based on cold, clear water) are known.

flow rate gpm (L/s)	head ft (m)
0 (0)	127 (42)
100 (6.3)	124 (41)
200 (12)	117 (39)
300 (18)	108 (36)
400 (24)	96 (32)
500 (30)	80 (27)
600 (36)	55 (18)

(a) What is the transfer rate?

(A) 150 gal/min (9.2 L/s)
(B) 180 gal/min (11 L/s)
(C) 200 gal/min (12 L/s)
(D) 230 gal/min (14 L/s)

(b) What is the total cost of operating the pump for 1 hr?

(A) $0.20
(B) $0.80
(C) $1.30
(D) $2.70

Specific Speed

9. A double-suction water pump moving 300 gal/sec (1.1 kL/s) turns at 900 rpm. The pump adds 20 ft (7 m) of head to the water. What is the specific speed?

(A) 3000 rpm (52 rpm)
(B) 6000 rpm (100 rpm)
(C) 9000 rpm (160 rpm)
(D) 12,000 rpm (210 rpm)

Cavitation

10. 100 gal/min (6.3 L/s) of pressurized hot water at 281°F and 80 psia (138°C and 550 kPa) is drawn through 30 ft (10 m) of 1.5 in (3.81 cm) schedule-40 steel pipe into a 2 psig (14 kPa) tank. The inlet and outlet are both 20 ft (6 m) below the surface of the water when the tank is full. The inlet line contains a square mouth inlet, two wide-open gate valves, and two long-radius elbows. All components are regular screwed. The pump's NPSHR is 10 ft (3 m) for this application. The kinematic viscosity of 281°F (138°C) water is 0.239×10^{-5} ft^2/sec (0.222×10^{-6} m^2/s) and the vapor pressure is 50.02 psia (3.431 bar). Will the pump cavitate?

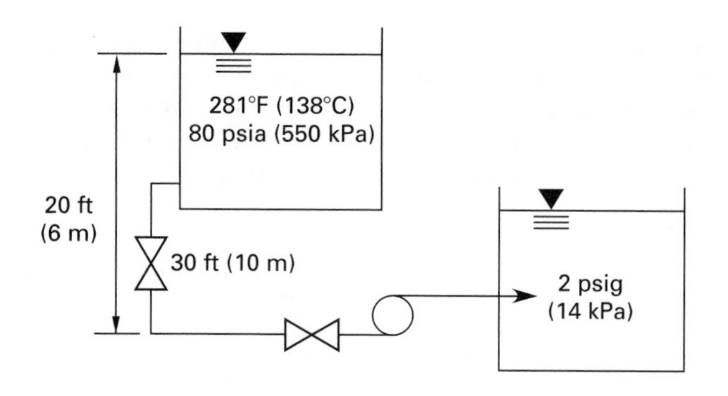

(A) yes; NPSHA = 4 ft (1.2 m)
(B) yes; NPSHA = 9 ft (2.7 m)
(C) no; NPSHA = 24 ft (7.2 m)
(D) no; NPSHA = 68 ft (21 m)

11. The velocity of the tip of a marine propeller is 4.2 times the boat velocity. The propeller is located 8 ft (3 m) below the surface. The temperature of the seawater is 68°F (20°C). The density is approximately 64.0 lbm/ft^3 (1024 kg/m^3), and the salt content is 2.5% by weight. What is the practical maximum boat velocity, as limited strictly by cavitation?

(A) 9.1 ft/sec (2.7 m/s)
(B) 12 ft/sec (3.8 m/s)
(C) 15 ft/sec (4.5 m/s)
(D) 22 ft/sec (6.6 m/s)

Pump and System Curves

12. The inlet of a centrifugal water pump is 7 ft (2.3 m) above the free surface from which it draws. The suction point is a submerged pipe. The supply line consists of 12 ft (4 m) of 2 in (5.08 cm) schedule-40 steel pipe and contains one long-radius elbow and one check valve. The discharge line is 2 in (5.08 cm) schedule-40 steel pipe and includes two long-radius elbows and an 80 ft (27 m) run. The discharge is 20 ft (6.3 m) above the free surface and is to the open atmosphere. All components are regular screwed. The water temperature is 70°F (21°C). The following pump curve data are applicable.

flow rate gpm (L/s)	head ft (m)
0 (0)	110 (37)
10 (0.6)	108 (36)
20 (1.2)	105 (35)
30 (1.8)	102 (34)
40 (2.4)	98 (33)
50 (3.2)	93 (31)
60 (3.6)	87 (29)
70 (4.4)	79 (26)
80 (4.8)	66 (22)
90 (5.7)	50 (17)

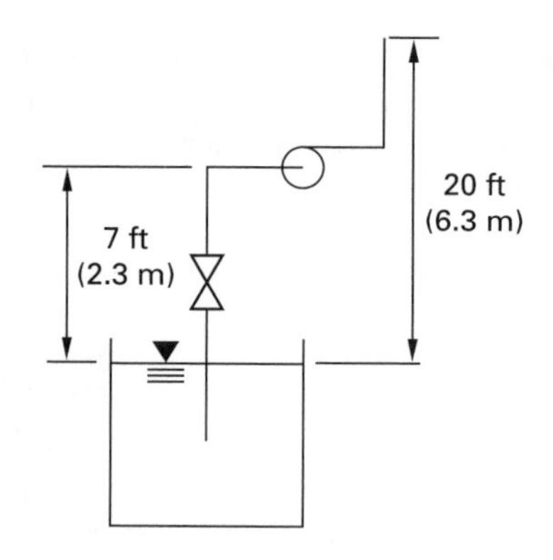

(a) What is the flow rate?

(A) 44 gal/min (2.9 L/s)
(B) 69 gal/min (4.5 L/s)
(C) 82 gal/min (5.5 L/s)
(D) 95 gal/min (6.2 L/s)

(b) What can be said about the use of this pump in this installation?

(A) A different pump should be used.
(B) The pump is operating near its most efficient point.
(C) Pressure fluctuations could result from surging.
(D) Overloading will not be a problem.

Affinity Laws

13. A pump was intended to run at 1750 rpm when driven by a 0.5 hp (0.37 kW) motor. What is the required power rating of a motor that will turn the pump at 2000 rpm?

(A) 0.25 hp (0.19 kW)
(B) 0.45 hp (0.34 kW)
(C) 0.65 hp (0.49 kW)
(D) 0.75 hp (0.55 kW)

14. (*Time limit: one hour*) A centrifugal pump running at 1400 rpm has the curve shown. The pump will be installed in an existing pipeline with known head requirements given by the formula $H = 30 + 2Q^2$. H is the system head in feet of water. Q is the flow rate in ft^3/sec.

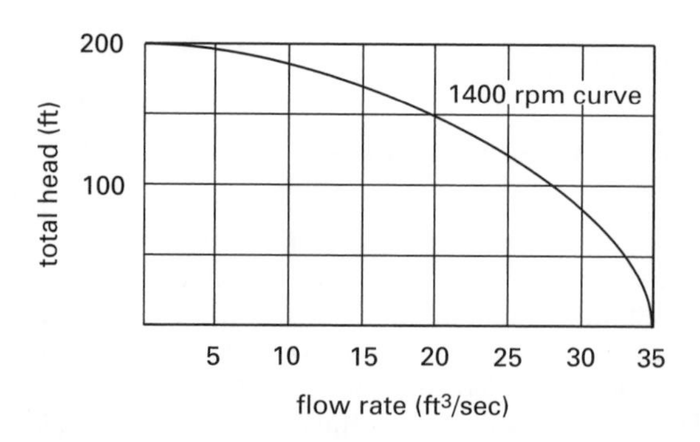

(a) What is the flow rate if the pump is turned at 1400 rpm?

(A) 2000 gal/min
(B) 3500 gal/min
(C) 4000 gal/min
(D) 4500 gal/min

(b) What power is required to drive the pump?

(A) 190 hp
(B) 210 hp
(C) 230 hp
(D) 260 hp

(c) What is the flow rate if the pump is turned at 1200 rpm?

(A) 2000 gal/min
(B) 3500 gal/min
(C) 4000 gal/min
(D) 4500 gal/min

Turbines

15. A horizontal turbine reduces 100 ft^3/sec of water from 30 psia to 5 psia. Friction is negligible. What power is developed?

(A) 350 hp
(B) 500 hp
(C) 650 hp
(D) 800 hp

16. 1000 ft^3/sec of 60°F water flow from a high reservoir through a hydroelectric turbine installation, exiting 625 ft lower. The head loss due to friction is 58 ft. The turbine efficiency is 89%. What power is developed in the turbines?

(A) 40 kW
(B) 18 MW
(C) 43 MW
(D) 71 MW

17. Water at 500 psig and 60°F (3.5 MPa and 16°C) drives a 250 hp (185 kW) turbine at 1750 rpm against a back pressure of 30 psig (210 kPa). The water discharges through a 4 in (100 mm) diameter nozzle at 35 ft/sec (10.5 m/s). The water is deflected 80° by a single blade moving directly away at 10 ft/sec (3 m/s).

(a) What is the specific speed?

(A) 4 (17)
(B) 25 (85)
(C) 75 (260)
(D) 230 (770)

(b) What is the total force acting on the blade?

(A) 100 lbf (450 N)
(B) 140 lbf (570 N)
(C) 160 lbf (720 N)
(D) 280 lbf (1300 N)

18. (*Time limit: one hour*) A Francis-design hydraulic reaction turbine with 22 in (560 mm) diameter blades runs at 610 rpm. The turbine develops 250 hp (185 kW) when 25 ft^3/sec (700 L/s) of water flow through it. The pressure head at the turbine entrance is 92.5 ft (30.8 m). The elevation of the turbine above the tailwater level is 5.26 ft (1.75 m). The inlet and outlet velocity are both 12 ft/sec (3.6 m/s).

(a) What is the effective head?

(A) 90 ft (30 m)
(B) 95 ft (31 m)
(C) 100 ft (33 m)
(D) 105 ft (35 m)

(b) What is the overall turbine efficiency?

(A) 81%
(B) 88%
(C) 93%
(D) 96%

(c) What will be the turbine speed if the effective head is 225 ft (75 m)?

(A) 600 rpm
(B) 920 rpm
(C) 1100 rpm
(D) 1400 rpm

(d) What horsepower is developed if the effective head is 225 ft (75 m)?

(A) 560 hp (420 kW)
(B) 630 hp (470 kW)
(C) 750 hp (560 kW)
(D) 840 hp (630 kW)

(e) What is the flow rate if the effective head is 225 ft (75 m)?

(A) 25 ft^3/sec (700 L/s)
(B) 38 ft^3/sec (1100 L/s)
(C) 56 ft^3/sec (1600 L/s)
(D) 64 ft^3/sec (1800 L/s)

SOLUTIONS

1. $$\left(2000\ \frac{\text{gal}}{\text{min}}\right)\left(0.002228\ \frac{\frac{\text{ft}^3}{\text{sec}}}{\frac{\text{gal}}{\text{min}}}\right) = 4.456\ \text{ft}^3/\text{sec}$$

Assume schedule-40 steel pipe. From App. 16.B,

$$\begin{array}{lll} 12'': & D_1 = 0.9948\ \text{ft} & A_1 = 0.7773\ \text{ft}^2 \\ 8'': & D_2 = 0.6651\ \text{ft} & A_2 = 0.3473\ \text{ft}^2 \end{array}$$

$$\begin{aligned} p_1 &= \left(14.7\ \frac{\text{lbf}}{\text{in}^2} - (8\ \text{in})\left(0.491\ \frac{\text{lbf}}{\text{in}^3}\right)\right)\left(144\ \frac{\text{in}^2}{\text{ft}^2}\right) \\ &= 1551.2\ \text{lbf/ft}^2 \end{aligned}$$

$$\begin{aligned} p_2 &= \left(14.7\ \frac{\text{lbf}}{\text{in}^2} + 20\ \frac{\text{lbf}}{\text{in}^2}\right)\left(144\ \frac{\text{in}^2}{\text{ft}^2}\right) \\ &\quad + (4\ \text{ft})(1.2)\left(62.4\ \frac{\text{lbf}}{\text{ft}^3}\right) \\ &= 5296.3\ \text{lbf/ft}^2 \end{aligned}$$

$$\text{v}_1 = \frac{4.456\ \frac{\text{ft}^3}{\text{sec}}}{0.7773\ \text{ft}^2} = 5.73\ \text{ft/sec}$$

$$\text{v}_2 = \frac{4.456\ \frac{\text{ft}^3}{\text{sec}}}{0.3473\ \text{ft}^2} = 12.83\ \text{ft/sec}$$

From Eq. 18.9, the total heads (in feet of sludge) at points 1 and 2 are

$$\begin{aligned} h_{t,1} = h_{t,s} &= \frac{1551.2\ \frac{\text{lbf}}{\text{ft}^2}}{\left(62.4\ \frac{\text{lbf}}{\text{ft}^3}\right)(1.2)} + \frac{\left(5.73\ \frac{\text{ft}}{\text{sec}}\right)^2}{(2)\left(32.2\ \frac{\text{ft}}{\text{sec}^2}\right)} \\ &= 21.23\ \text{ft} \end{aligned}$$

$$\begin{aligned} h_{t,2} = h_{t,d} &= \frac{5296.3\ \frac{\text{lbf}}{\text{ft}^2}}{\left(62.4\ \frac{\text{lbf}}{\text{ft}^3}\right)(1.2)} + \frac{\left(12.83\ \frac{\text{ft}}{\text{sec}}\right)^2}{(2)\left(32.2\ \frac{\text{ft}}{\text{sec}^2}\right)} \\ &= 73.29\ \text{ft} \end{aligned}$$

The pump must add 73.29 ft − 21.23 ft = 52.06 ft of head (sludge head).

The power required is given in Table 18.5.

$$\begin{aligned} P_{\text{ideal}} &= \frac{(52.06\ \text{ft})(1.2)\left(4.456\ \frac{\text{ft}^3}{\text{sec}}\right)\left(62.4\ \frac{\text{lbf}}{\text{ft}^3}\right)}{550\ \frac{\text{ft-lbf}}{\text{hp-sec}}} \\ &= 31.58\ \text{hp} \end{aligned}$$

The input horsepower is

$$P_{\text{in}} = \frac{P_{\text{ideal}}}{\eta} = \frac{31.58\ \text{hp}}{0.85} = \boxed{37.15\ \text{hp}}$$

The answer is (C).

2. *Customary U.S. Solution*

From App. 16.B, data for 4 in schedule-40 steel pipe are

$$\begin{aligned} D_i &= 0.3355\ \text{ft} \\ A_i &= 0.08841\ \text{ft}^2 \end{aligned}$$

The velocity in the pipe is

$$\text{v} = \frac{\dot{V}}{A} = \frac{1.25\ \frac{\text{ft}^3}{\text{sec}}}{0.08841\ \text{ft}^2} = 14.139\ \text{ft/sec}$$

From Table 17.3, typical equivalent lengths for schedule-40, screwed steel fittings for 4 in pipes are

90° elbow: 13 ft

gate valve: 2.5 ft

check valve: 38.0 ft

The total equivalent length is

$$(2)(13\ \text{ft}) + (1)(2.5\ \text{ft}) + (1)(38\ \text{ft}) = 66.5\ \text{ft}$$

At 70°F, from App. 14.A, the density of water is 62.3 lbm/ft^3 and the kinematic viscosity of water, ν, is 1.059×10^{-5} ft^2/sec. The Reynolds number is

$$\begin{aligned} \text{Re} &= \frac{D\text{v}}{\nu} = \frac{(0.3355\ \text{ft})\left(14.139\ \frac{\text{ft}}{\text{sec}}\right)}{1.059 \times 10^{-5}\ \frac{\text{ft}^2}{\text{sec}}} \\ &= 4.479 \times 10^5 \end{aligned}$$

From App. 17.A, for steel, $\epsilon = 0.0002$ ft.

So,

$$\frac{\epsilon}{D} = \frac{0.0002\ \text{ft}}{0.3355\ \text{ft}} \approx 0.0006$$

From App. 17.B, the friction factor is $f = 0.01835$.

The friction head is given by Eq. 18.6.

$$\begin{aligned} h_f &= \frac{fL\text{v}^2}{2Dg} \\ &= \frac{(0.01835)(700\ \text{ft} + 66.5\ \text{ft})\left(14.139\ \frac{\text{ft}}{\text{sec}}\right)^2}{(2)(0.3355\ \text{ft})\left(32.2\ \frac{\text{ft}}{\text{sec}^2}\right)} \\ &= 130.1\ \text{ft} \end{aligned}$$

The total dynamic head is given by Eq. 18.9. Point s is taken as the bottom of the supply tank.

$$h = \frac{(p_d - p_s)g_c}{\rho g} + \frac{\text{v}_d^2 - \text{v}_s^2}{2g} + z_d - z_s$$

$$\text{v}_s \approx 0$$

$$z_d - z_s = 50\ \text{ft} \quad \text{[given as rise in elevation]}$$

The discharge and suction pressures are

$$p_d = 20 \text{ psig}$$
$$p_s = 50 \text{ psig}$$
$$h = \frac{\left(20\ \frac{\text{lbf}}{\text{in}^2} - 50\ \frac{\text{lbf}}{\text{in}^2}\right)\left(144\ \frac{\text{in}^2}{\text{ft}^2}\right)\left(32.2\ \frac{\text{ft-lbm}}{\text{lbf-sec}^2}\right)}{\left(62.3\ \frac{\text{lbm}}{\text{ft}^3}\right)\left(32.2\ \frac{\text{ft}}{\text{sec}^2}\right)} + \frac{\left(14.139\ \frac{\text{ft}}{\text{sec}}\right)^2}{(2)\left(32.2\ \frac{\text{ft}}{\text{sec}^2}\right)} + 50 \text{ ft}$$
$$= -16.2 \text{ ft}$$

The head added is

$$h_A = h + h_f = -16.2 \text{ ft} + 130.1 \text{ ft} = 113.9 \text{ ft}$$

The mass flow rate is

$$\dot{m} = \rho\dot{V} = \left(62.3\ \frac{\text{lbm}}{\text{ft}^3}\right)\left(1.25\ \frac{\text{ft}^3}{\text{sec}}\right) = 77.875 \text{ lbm/sec}$$

From Table 18.5, the hydraulic horsepower is

$$\text{WHP} = \left(\frac{h_A\dot{m}}{550}\right)\left(\frac{g}{g_c}\right) = \left(\frac{(113.9 \text{ ft})\left(77.875\ \frac{\text{lbm}}{\text{sec}}\right)}{550\ \frac{\text{ft-lbf}}{\text{hp-sec}}}\right) \times \left(\frac{32.2\ \frac{\text{ft}}{\text{sec}^2}}{32.2\ \frac{\text{ft-lbm}}{\text{lbf-sec}^2}}\right) = \boxed{16.13 \text{ hp}}$$

The answer is (A).

SI Solution

From App. 16.C, data for 4 in schedule-40 steel pipe are

$$D_i = 102.3 \text{ mm}$$
$$A_i = 82.19 \times 10^{-4} \text{ m}^2$$

The velocity in the pipe is

$$\text{v} = \frac{\dot{V}}{A} = \frac{\left(35\ \frac{\text{L}}{\text{s}}\right)\left(\frac{1 \text{ m}^3}{1000 \text{ L}}\right)}{82.19 \times 10^{-4} \text{ m}^2} = 4.26 \text{ m/s}$$

From Table 17.3, typical equivalent lengths for schedule-40, screwed steel fittings for 4 in pipes are

90° elbow: 13 ft

gate valve: 2.5 ft

check valve: 38.0 ft

The total equivalent length is

$$(2)(13 \text{ ft}) + (1)(2.5 \text{ ft}) + (1)(38 \text{ ft}) = 66.5 \text{ ft}$$
$$(66.5 \text{ ft})\left(0.3048\ \frac{\text{m}}{\text{ft}}\right) = 20.27 \text{ m}$$

At 21°C, from App. 14.B, the water properties are

$$\rho = 998 \text{ kg/m}^3$$
$$\mu = 0.9827 \times 10^{-3} \text{ Pa·s}$$
$$\nu = \frac{\mu}{\rho} = \frac{0.9827 \times 10^{-3} \text{ Pa·s}}{998\ \frac{\text{kg}}{\text{m}^3}} = 9.85 \times 10^{-7} \text{ m}^2/\text{s}$$

The Reynolds number is

$$\text{Re} = \frac{D\text{v}}{\nu} = \frac{(102.3 \text{ mm})\left(\frac{1 \text{ m}}{1000 \text{ mm}}\right)\left(4.26\ \frac{\text{m}}{\text{s}}\right)}{9.85 \times 10^{-7}\ \frac{\text{m}^2}{\text{s}}} = 4.424 \times 10^5$$

From Table 17.2, for steel, $\epsilon = 6 \times 10^{-5}$ m.

So,

$$\frac{\epsilon}{D} = \frac{6.0 \times 10^{-5} \text{ m}}{(102.3 \text{ mm})\left(\frac{1 \text{ m}}{1000 \text{ mm}}\right)} = 0.0006$$

From App. 17.B, the friction factor is $f = 0.01836$.

From Eq. 18.6, the friction head is

$$h_f = \frac{fL\text{v}^2}{2Dg} = \frac{(0.01836)(230 \text{ m} + 20.27 \text{ m})\left(4.26\ \frac{\text{m}}{\text{s}}\right)^2}{(2)(102.3 \text{ mm})\left(\frac{1 \text{ m}}{1000 \text{ mm}}\right)\left(9.81\ \frac{\text{m}}{\text{s}^2}\right)} = 41.5 \text{ m}$$

The total dynamic head is given by Eq. 18.9. Point s is taken as the bottom of the supply tank.

$$h = \frac{p_d - p_s}{\rho g} + \frac{v_d^2 - v_s^2}{2g} + z_d - z_s$$

$$v_s \approx 0$$

$$z_d - z_s = 15 \text{ m} \quad \text{[given as rise in elevation]}$$

The difference between discharge and suction pressure is

$$p_d - p_s = 140 \text{ kPa} - 345 \text{ kPa} = -205 \text{ kPa}$$

$$\begin{aligned} h &= \frac{(-205 \text{ kPa})\left(1000 \frac{\text{Pa}}{\text{kPa}}\right)}{\left(998 \frac{\text{kg}}{\text{m}^3}\right)\left(9.81 \frac{\text{m}}{\text{s}^2}\right)} + \frac{\left(4.26 \frac{\text{m}}{\text{s}}\right)^2}{(2)\left(9.81 \frac{\text{m}}{\text{s}^2}\right)} + 15 \text{ m} \\ &= -5.0 \text{ m} \end{aligned}$$

The head added by the pump is

$$\begin{aligned} h_A &= h + h_f = -5.0 \text{ m} + 41.5 \text{ m} \\ &= 36.5 \text{ m} \end{aligned}$$

The mass flow rate is

$$\begin{aligned} \dot{m} &= \rho \dot{V} \\ &= \left(998 \frac{\text{kg}}{\text{m}^3}\right)\left(35 \frac{\text{L}}{\text{s}}\right)\left(\frac{1 \text{ m}^3}{1000 \text{ L}}\right) \\ &= 34.93 \text{ kg/s} \end{aligned}$$

From Table 18.6, the hydraulic power is

$$\begin{aligned} \text{WkW} &= \frac{(9.81) h_A \dot{m}}{1000} \\ &= \frac{\left(9.81 \frac{\text{m}}{\text{s}^2}\right)(36.5 \text{ m})\left(34.93 \frac{\text{kg}}{\text{s}}\right)}{1000 \frac{\text{W}}{\text{kW}}} \\ &= \boxed{12.51 \text{ kW}} \end{aligned}$$

The answer is (A).

3. *Customary U.S. Solution*

The area of the rubber hose is

$$A = \left(\frac{\pi}{4}\right) D^2 = \left(\frac{\pi}{4}\right)\left(\frac{2 \text{ in}}{12 \frac{\text{in}}{\text{ft}}}\right)^2 = 0.0218 \text{ ft}^2$$

The velocity of water in the hose is

$$\begin{aligned} \text{v} &= \frac{\dot{V}}{A} = \frac{\left(80 \frac{\text{gal}}{\text{min}}\right)\left(0.002228 \frac{\frac{\text{ft}^3}{\text{sec}}}{\frac{\text{gal}}{\text{min}}}\right)}{0.0218 \text{ ft}^2} \\ &= 8.176 \text{ ft/sec} \end{aligned}$$

At 80°F from App. 14.A, the kinematic viscosity of water is $\nu = 0.93 \times 10^{-5}$ ft^2/sec.

The Reynolds number is

$$\begin{aligned} \text{Re} &= \frac{\text{v}D}{\nu} = \frac{\left(8.176 \frac{\text{ft}}{\text{sec}}\right)(2 \text{ in})\left(\frac{1 \text{ ft}}{12 \text{ in}}\right)}{0.93 \times 10^{-5} \frac{\text{ft}^2}{\text{sec}}} \\ &= 1.47 \times 10^5 \end{aligned}$$

Assume that the rubber hose is smooth. From App. 17.B, the friction factor is $f = 0.0166$.

From Eq. 18.6, the friction head is

$$\begin{aligned} h_f &= \frac{f L \text{v}^2}{2Dg} \\ &= \frac{(0.0166)(50 \text{ ft})\left(8.176 \frac{\text{ft}}{\text{sec}}\right)^2}{(2)(2 \text{ in})\left(\frac{1 \text{ ft}}{12 \text{ in}}\right)\left(32.2 \frac{\text{ft}}{\text{sec}^2}\right)} \\ &= 5.17 \text{ ft} \end{aligned}$$

Neglecting entrance and exit losses, the head added by the pump is

$$\begin{aligned} h_A &= h_f + h_z \\ &= 5.17 \text{ ft} + 12 \text{ ft} - 4 \text{ ft} \\ &= \boxed{13.17 \text{ ft}} \end{aligned}$$

The answer is (B).

SI Solution

The area of the rubber hose is

$$\begin{aligned} A &= \left(\frac{\pi}{4}\right) D^2 = \left(\frac{\pi}{4}\right)(5.1 \text{ cm})^2\left(\frac{1 \text{ m}}{100 \text{ cm}}\right)^2 \\ &= 0.00204 \text{ m}^2 \end{aligned}$$

The velocity of water in the hose is

$$\text{v} = \frac{\dot{V}}{A} = \frac{\left(5 \frac{\text{L}}{\text{s}}\right)\left(\frac{1 \text{ m}^3}{1000 \text{ L}}\right)}{0.00204 \text{ m}^2} = 2.45 \text{ m/s}$$

At 27°C from App. 14.B, the water data are

$$\rho = 996.5\ \text{kg/m}^3$$
$$\mu = 0.8565 \times 10^{-3}\ \text{Pa·s}$$
$$\nu = \frac{\mu}{\rho} = \frac{0.8565 \times 10^{-3}\ \text{Pa·s}}{996.5\ \frac{\text{kg}}{\text{m}^3}} = 8.60 \times 10^{-7}\ \text{m}^2/\text{s}$$

The Reynolds number is

$$\text{Re} = \frac{\text{v}D}{\nu} = \frac{\left(2.45\ \frac{\text{m}}{\text{s}}\right)(5.1\ \text{cm})\left(\frac{1\ \text{m}}{100\ \text{cm}}\right)}{8.60 \times 10^{-7}\ \frac{\text{m}^2}{\text{s}}} = 1.45 \times 10^5$$

Assume that the rubber hose is smooth. From App. 17.B, the friction factor is $f \approx 0.0166$.

From Eq. 18.6, the friction head is

$$h_f = \frac{fL\text{v}^2}{2Dg} = \frac{(0.0166)(15\ \text{m})\left(2.45\ \frac{\text{m}}{\text{s}}\right)^2}{(2)(5.1\ \text{cm})\left(\frac{1\ \text{m}}{100\ \text{cm}}\right)\left(9.81\ \frac{\text{m}}{\text{s}^2}\right)} = 1.49\ \text{m}$$

Neglecting entrance and exit losses, the head added by the pump is

$$h_A = h_f + h_z = 1.49\ \text{m} + 4\ \text{m} - 1.5\ \text{m} = \boxed{3.99\ \text{m}}$$

The answer is (B).

4. *Customary U.S. Solution*

From App. 16.B, the diameters (inside) for 8 in and 6 in schedule-40 steel pipe are

$$D_1 = 7.981\ \text{in}$$
$$D_2 = 6.065\ \text{in}$$

At 60°F from App. 14.A, the density of water is 62.37 lbm/ft^3.

The mass flow rate through 6 in pipe is

$$\dot{m} = A_2\text{v}_2\rho = \left(\frac{\pi}{4}\right)(6.065\ \text{in})^2\left(\frac{1\ \text{ft}^2}{144\ \text{in}^2}\right)\left(12\ \frac{\text{ft}}{\text{sec}}\right)\left(62.37\ \frac{\text{lbm}}{\text{ft}^3}\right) = 150.2\ \text{lbm/sec}$$

The inlet (suction) pressure is

$$14.7\ \text{psia} - 5\ \text{psig} = 9.7\ \text{psia} = \left(9.7\ \frac{\text{lbf}}{\text{in}^2}\right)\left(144\ \frac{\text{in}^2}{\text{ft}^2}\right) = 1397\ \text{lbf/ft}^2$$

From Table 18.5, the head added by the pump is

$$h_A = \left(\frac{(550)(\text{BHP})\eta}{\dot{m}}\right)\left(\frac{g_c}{g}\right) = \left(\frac{\left(550\ \frac{\text{ft-lbf}}{\text{hp-sec}}\right)(20\ \text{hp})(0.70)}{150.2\ \frac{\text{lbm}}{\text{sec}}}\right)\left(\frac{32.2\ \frac{\text{ft-lbm}}{\text{lbf-sec}^2}}{32.2\ \frac{\text{ft}}{\text{sec}^2}}\right) = 51.26\ \text{ft}$$

At 1:

$$p_1 = 1397\ \text{lbf/ft}^2$$
$$z_1 = 0$$
$$\text{v}_1 = \frac{\text{v}_2 A_2}{A_1} = \text{v}_2\left(\frac{D_2}{D_1}\right)^2 = \left(12\ \frac{\text{ft}}{\text{sec}}\right)\left(\frac{6.065\ \text{in}}{7.981\ \text{in}}\right)^2 = 6.93\ \text{ft/sec}$$

At 2:

$$p_2 = \left(14.7\ \frac{\text{lbf}}{\text{in}^2}\right)\left(144\ \frac{\text{in}^2}{\text{ft}^2}\right) = 2117\ \text{lbf/ft}^2$$
$$\text{v}_2 = 12\ \text{ft/sec}\quad [\text{given}]$$

Let z_3 be the additional head above atmospheric. From Eq. 18.9(b), the head added by the pump is

$$h_A = \frac{(p_2 - p_1)g_c}{\rho g} + \frac{\text{v}_2^2 - \text{v}_1^2}{2g} + z_2 - z_1 + h_f + z_3$$

$$51.26\ \text{ft} = \left(\frac{2117\ \frac{\text{lbf}}{\text{ft}^2} - 1397\ \frac{\text{lbf}}{\text{ft}^2}}{62.37\ \frac{\text{lbm}}{\text{ft}^3}}\right)\left(\frac{32.2\ \frac{\text{ft-lbm}}{\text{lbf-sec}^2}}{32.2\ \frac{\text{ft}}{\text{sec}^2}}\right) + \frac{\left(12\ \frac{\text{ft}}{\text{sec}}\right)^2 - \left(6.93\ \frac{\text{ft}}{\text{sec}}\right)^2}{(2)\left(32.2\ \frac{\text{ft}}{\text{sec}^2}\right)} + 0 - 0 + 10\ \text{ft} + z_3$$

$$z_3 = \boxed{28.2\ \text{ft}}$$

The answer is (A).

Water Resources

SI Solution

From App. 16.C, the inside diameters for 8 in and 6 in steel schedule-40 pipe are

$$D_1 = 202.7\ \text{mm}$$
$$D_2 = 154.1\ \text{mm}$$

At 16°C from App. 14.B, the density of water is 998.83 kg/m^3.

The mass flow rate through the 6 in pipe is

$$\begin{aligned}\dot{m} &= A_2 \text{v}_2 \rho \\ &= \left(\frac{\pi}{4}\right)(154.1\ \text{mm})^2\left(\frac{1\ \text{m}}{1000\ \text{mm}}\right)^2\left(4\ \frac{\text{m}}{\text{s}}\right) \\ &\quad \times \left(998.83\ \frac{\text{kg}}{\text{m}^3}\right) \\ &= 74.5\ \text{kg/s}\end{aligned}$$

The inlet (suction) pressure is

$$101.3\ \text{kPa} - 35\ \text{kPa} = 66.3\ \text{kPa}$$

From Table 18.6, the head added by the pump is

$$\begin{aligned}h_A &= \frac{(1000)(\text{BkW})\eta}{(9.81)\dot{m}} \\ &= \frac{\left(1000\ \frac{\text{W}}{\text{kW}}\right)(20\ \text{hp})\left(\frac{0.7457\ \text{kW}}{\text{hp}}\right)(0.70)}{\left(9.81\ \frac{\text{m}}{\text{s}^2}\right)\left(74.5\ \frac{\text{kg}}{\text{s}}\right)} \\ &= 14.28\ \text{m}\end{aligned}$$

At 1:

$$\begin{aligned}p_1 &= 66.3\ \text{kPa} \\ z_1 &= 0 \\ \text{v}_1 &= \text{v}_2\left(\frac{A_2}{A_1}\right) = \text{v}_2\left(\frac{D_2}{D_1}\right)^2 \\ &= \left(4\ \frac{\text{m}}{\text{s}}\right)\left(\frac{154.1\ \text{mm}}{202.7\ \text{mm}}\right)^2 = 2.31\ \text{m/s}\end{aligned}$$

At 2:

$$\begin{aligned}p_2 &= 101.3\ \text{kPa} \\ \text{v}_2 &= 4\ \text{m/s} \quad \text{[given]}\end{aligned}$$

From Eq. 18.9(a), the head added by the pump is

$$h_A = \frac{p_2 - p_1}{\rho g} + \frac{\text{v}_2^2 - \text{v}_1^2}{2g} + z_2 - z_1 + h_f + z_3$$

$$\begin{aligned}14.28\ \text{m} &= \frac{(101.3\ \text{kPa} - 66.3\ \text{kPa})\left(1000\ \frac{\text{Pa}}{\text{kPa}}\right)}{\left(998.83\ \frac{\text{kg}}{\text{m}^3}\right)\left(9.81\ \frac{\text{m}}{\text{s}^2}\right)} \\ &\quad + \frac{\left(4\ \frac{\text{m}}{\text{s}}\right)^2 - \left(2.31\ \frac{\text{m}}{\text{s}}\right)^2}{(2)\left(9.81\ \frac{\text{m}}{\text{s}^2}\right)} \\ &\quad + 0 - 0 + 3.3\ \text{m} + z_3\end{aligned}$$

$$z_3 = \boxed{6.86\ \text{m}}$$

The answer is (A).

5. *Customary U.S. Solution*

The flow rate is

$$\dot{V} = \left(10{,}000\ \frac{\text{gal}}{\text{hr}}\right)\left(0.1337\ \frac{\text{ft}^3}{\text{gal}}\right) = 1337\ \text{ft}^3/\text{hr}$$

From App. 16.B, data for 4 in schedule-40 steel pipe are

$$D_i = 0.3355\ \text{ft}$$
$$A_i = 0.08841\ \text{ft}^2$$

The velocity in the pipe is

$$\text{v} = \frac{\dot{V}}{A} = \frac{\left(1337\ \frac{\text{ft}^3}{\text{hr}}\right)\left(\frac{1\ \text{hr}}{3600\ \text{sec}}\right)}{0.08841\ \text{ft}^2} = 4.20\ \text{ft/sec}$$

At 60°F from App. 14.A, the kinematic viscosity of water is

$$\nu = 1.217 \times 10^{-5}\ \text{ft}^2/\text{sec}$$
$$\rho = 62.37\ \text{lbm/ft}^3$$

The Reynolds number is

$$\begin{aligned}\text{Re} &= \frac{D\text{v}}{\nu} = \frac{(0.3355\ \text{ft})\left(4.20\ \frac{\text{ft}}{\text{sec}}\right)}{1.217 \times 10^{-5}\ \frac{\text{ft}^2}{\text{sec}}} \\ &= 1.16 \times 10^5\end{aligned}$$

From App. 17.A, for welded and seamless steel, $\epsilon = 0.0002$ ft.

$$\frac{\epsilon}{D} = \frac{0.0002\ \text{ft}}{0.3355\ \text{ft}} \approx 0.0006$$

From App. 17.B, the friction factor, f, is 0.0205. The friction head is

$$\begin{aligned} h_f &= \frac{fL\text{v}^2}{2Dg} \\ &= \frac{(0.0205)(7000 \text{ ft})\left(4.2 \ \frac{\text{ft}}{\text{sec}}\right)^2}{(2)(0.3355 \text{ ft})\left(32.2 \ \frac{\text{ft}}{\text{sec}^2}\right)} \\ &= 117.2 \text{ ft} \end{aligned}$$

The head added by the pump is

$$\begin{aligned} h_A &= h_f + h_z \\ &= 117.2 \text{ ft} + 12 \text{ ft} + 350 \text{ ft} \\ &= 479.5 \text{ ft} \end{aligned}$$

From Table 18.5, the hydraulic horsepower is

$$\begin{aligned} \text{WHP} &= \frac{h_A Q(\text{SG})}{3956} \\ &= \frac{(479.2 \text{ ft})\left(10{,}000 \ \frac{\text{gal}}{\text{hr}}\right)\left(\frac{1 \text{ hr}}{60 \text{ min}}\right)(1)}{3956 \ \frac{\text{ft-gal}}{\text{hp-min}}} \\ &= 20.2 \text{ hp} \end{aligned}$$

From Eq. 18.16, the overall efficiency of the pump is

$$\begin{aligned} \eta &= \frac{\text{WHP}}{\text{EHP}} \\ \text{EHP} &= \frac{20.2 \text{ hp}}{0.7} \\ &= 28.9 \text{ hp} \end{aligned}$$

(a) At \$0.04/kW-hr, power costs for 1 hr are

$$(28.9 \text{ hp})\left(\frac{0.7457 \text{ kW}}{\text{hp}}\right)(1 \text{ hr})\left(\frac{\$0.04}{\text{kW-hr}}\right) = \boxed{\$0.86 \text{ per hour}}$$

The answer is (B).

(b) The motor horsepower, EHP, is 28.9 hp. Select a

$\boxed{\text{30 hp motor.}}$

The answer is (B).

(c) From Eq. 18.5(b),

$$\begin{aligned} h_{\text{atm}} &= \left(\frac{p_{\text{atm}}}{\rho}\right)\left(\frac{g_c}{g}\right) \\ &= \left(\frac{\left(14.7 \ \frac{\text{lbf}}{\text{in}^2}\right)\left(144 \ \frac{\text{in}^2}{\text{ft}^2}\right)}{62.37 \ \frac{\text{lbm}}{\text{ft}^3}}\right)\left(\frac{32.2 \ \frac{\text{ft}}{\text{sec}^2}}{32.2 \ \frac{\text{ft-lbf}}{\text{lbm-sec}^2}}\right) \\ &= 33.94 \text{ ft} \end{aligned}$$

The friction losses due to 300 ft is

$$\begin{aligned} h_{f(s)} &= \left(\frac{300 \text{ ft}}{7000 \text{ ft}}\right) h_f \\ &= \left(\frac{300 \text{ ft}}{7000 \text{ ft}}\right)(117.2 \text{ ft}) \\ &= 5.0 \text{ ft} \end{aligned}$$

From App. 14.A, the vapor pressure head at 60°F is 0.59 ft.

The NPSHA from Eq. 18.30(a) is

$$\begin{aligned} \text{NPSHA} &= h_{\text{atm}} + h_{z(s)} - h_{f(s)} - h_{\text{vp}} \\ &= 33.94 \text{ ft} - 12 \text{ ft} - 5.0 \text{ ft} - 0.59 \text{ ft} \\ &= \boxed{16.35 \text{ ft}} \end{aligned}$$

The answer is (D).

SI Solution

From App. 16.C, data for 4 in schedule-40 steel pipe are

$$\begin{aligned} D_i &= 102.3 \text{ mm} \\ A_i &= 82.19 \times 10^{-4} \text{ m}^2 \end{aligned}$$

The velocity in the pipe is

$$\text{v} = \frac{\dot{V}}{A} = \frac{\left(10.5 \ \frac{\text{L}}{\text{s}}\right)\left(\frac{1 \text{ m}^3}{1000 \text{ L}}\right)}{82.19 \times 10^{-4} \text{ m}^2} = 1.28 \text{ m/s}$$

From App. 14.B, at 16°C the water data are

$$\begin{aligned} \rho &= 998.83 \text{ kg/m}^3 \\ \mu &= 1.1261 \times 10^{-3} \text{ Pa·s} \end{aligned}$$

The Reynolds number is

$$\begin{aligned} \text{Re} &= \frac{\rho \text{v} D}{\mu} \\ &= \frac{\left(998.83 \ \frac{\text{kg}}{\text{m}^3}\right)\left(1.28 \ \frac{\text{m}}{\text{s}}\right)(102.3 \text{ mm})\left(\frac{1 \text{ m}}{1000 \text{ mm}}\right)}{1.1261 \times 10^{-3} \text{ Pa·s}} \\ &= 1.16 \times 10^5 \end{aligned}$$

From Table 17.2, for welded and seamless steel, $\epsilon = 6.0 \times 10^{-5}$ m.

$$\frac{\epsilon}{D} = \frac{6.0 \times 10^{-5} \text{ m}}{(102.3 \text{ mm})\left(\frac{1 \text{ m}}{1000 \text{ mm}}\right)} = 0.0006$$

Water Resources

From App. 17.B, the friction factor is $f = 0.0205$.

From Eq. 18.6, the friction head is

$$\begin{aligned} h_f &= \frac{fLv^2}{2Dg} \\ &= \frac{(0.0205)(2300\ \text{m})\left(1.28\ \frac{\text{m}}{\text{s}}\right)^2}{(2)(102.3\ \text{mm})\left(\frac{1\ \text{m}}{1000\ \text{mm}}\right)\left(9.81\ \frac{\text{m}}{\text{s}^2}\right)} \\ &= 38.5\ \text{m} \end{aligned}$$

The head added by the pump is

$$\begin{aligned} h_A &= h_f + h_z \\ &= 38.5\ \text{m} + 4\ \text{m} + 115\ \text{m} = 157.5\ \text{m} \end{aligned}$$

From Table 18.6, the hydraulic power is

$$\begin{aligned} \text{WkW} &= \frac{(9.81)h_A Q(\text{SG})}{1000} \\ &= \frac{\left(9.81\ \frac{\text{m}}{\text{s}^2}\right)(157.5\ \text{m})\left(10.5\ \frac{\text{L}}{\text{s}}\right)(1)}{1000\ \frac{\text{W·L}}{\text{kW·kg}}} \\ &= 16.22\ \text{kW} \end{aligned}$$

From Eq. 18.16,

$$\begin{aligned} \text{EHP} &= \frac{\text{WHP}}{\eta_{\text{overall}}} \\ &= \frac{16.22\ \text{kW}}{0.7} = 23.2\ \text{kW} \end{aligned}$$

(a) At \$0.04/kW·h, power costs for 1 h are

$$(23.2\ \text{kW})(1\ \text{h})\left(\frac{\$0.04}{\text{kW·h}}\right) = \boxed{\$0.93\ \text{per hour}}$$

The answer is (B).

(b) The required motor power is 23.2 kW. Select the next higher standard motor size.

The answer is (B).

(c) From Eq. 18.5(a),

$$\begin{aligned} h_{\text{atm}} &= \frac{p}{\rho g} \\ &= \frac{(101\ \text{kPa})(1000\ \text{Pa})}{\left(998.83\ \frac{\text{kg}}{\text{m}^3}\right)\left(9.81\ \frac{\text{m}}{\text{s}^2}\right)} = 10.31\ \text{m} \end{aligned}$$

The friction loss due to 100 m is

$$\begin{aligned} h_{f(s)} &= \left(\frac{100\ \text{m}}{2300\ \text{m}}\right) h_f \\ &= \left(\frac{100\ \text{m}}{2300\ \text{m}}\right)(38.5\ \text{m}) \\ &= 1.67\ \text{m} \end{aligned}$$

The vapor pressure at 16°C is 0.01818 bar.

From Eq. 18.5(a),

$$\begin{aligned} h_{\text{vp}} &= \frac{p_{\text{vp}}}{g\rho} \\ &= \frac{(0.01818\ \text{bar})\left(1 \times 10^5\ \frac{\text{Pa}}{\text{bar}}\right)}{\left(9.81\ \frac{\text{m}}{\text{s}^2}\right)\left(998.83\ \frac{\text{kg}}{\text{m}^3}\right)} \\ &= 0.19\ \text{m} \end{aligned}$$

The NPSHA from Eq. 18.30(a) is

$$\begin{aligned} \text{NPSHA} &= h_{\text{atm}} + h_{z(s)} - h_{f(s)} - h_{\text{vp}} \\ &= 10.31\ \text{m} - 4\ \text{m} - 1.67\ \text{m} - 0.19\ \text{m} \\ &= \boxed{4.45\ \text{m}} \end{aligned}$$

The answer is (D).

6. (a) Sewers are usually gravity-flow systems. $h_f = \Delta z = 48$ ft since $\Delta p = 0$ and $\Delta v = 0$ for open channel flow.

$$\begin{aligned} Q &= \frac{\left(250\ \frac{\text{gal}}{\text{person·day}}\right)(10{,}000\ \text{people})}{\left(24\ \frac{\text{hr}}{\text{day}}\right)\left(60\ \frac{\text{min}}{\text{hr}}\right)} \\ &= 1736\ \text{gal/min} \end{aligned}$$

Given $C = 130$, solving for d from Eq. 17.31,

$$d_{\text{in}}^{4.8655} = \frac{(10.44)(5000\ \text{ft})\left(1736\ \frac{\text{gal}}{\text{min}}\right)^{1.85}}{(130)^{1.85}(48\ \text{ft})} = 131{,}462$$

$$d = \boxed{11.27\ \text{in} \quad [\text{round to 12 in minimum}]}$$

The answer is (B).

(b) Without having a specific pump curve, the number of pumps can only be specified based on general rules. Use the *Ten States' Standards*, which states:

- No station will have less than two identical pumps.
- Capacity must be met with one pump out of service.
- Provision must be made to alternate pumps automatically.

$$\boxed{\text{Two pumps are required, plus spares.}}$$

The answer is (A).

(c) With a variable speed pump, it will be possible to adjust to the wide variations in flow (100 to 250 gpcd). It may be possible to operate with one pump. However, TSS still requires two.

The answer is (B).

(d) With three constant speed pumps,

$$Q = \frac{1736 \ \frac{\text{gal}}{\text{min}}}{3} = 579 \text{ gal/min at maximum capacity}$$

The pump only has to lift the sewage 10 ft. To get to the pump, the sewage descended 48 ft under the influence of gravity. From Table 18.5, assuming specific gravity ≈ 1.00 and using an average efficiency of $\eta_{\text{pump}} = 0.60$,

$$\begin{aligned} \text{rated motor power} &= \frac{(10 \text{ ft})\left(579 \ \frac{\text{gal}}{\text{min}}\right)}{\left(3956 \ \frac{\text{gal-ft}}{\text{min-hp}}\right)(0.60)} \\ &= \boxed{2.44 \text{ hp} \quad [\text{use } 3.0 \text{ hp}]} \end{aligned}$$

The answer is (B).

(e) With two variable-speed pumps,

$$\begin{aligned} Q &= \frac{1736 \ \frac{\text{gal}}{\text{min}}}{2} = 868 \text{ gal/min} \\ \text{rated motor power} &= \frac{(10 \text{ ft})\left(868 \ \frac{\text{gal}}{\text{min}}\right)}{\left(3956 \ \frac{\text{gal-ft}}{\text{min-hp}}\right)(0.80)} \\ &= \boxed{2.74 \text{ hp} \quad [\text{use } 3.0 \text{ hp}]} \end{aligned}$$

The answer is (A).

(f) Incoming flow rate is independent of sump level.

The answer is (D).

(g) In the wet well, the pump (and perhaps motor) is submerged. Forced ventilation air will prevent a concentration of explosive methane. From the *Ten States' Standards*,

12 air changes per hour if continuous; 30 per hour if intermittent.

The answer is (D).

(h) In the dry well,

6 air changes per hour if continuous; 30 per hour if intermittent.

The answer is (A).

7. (a) $h_{p(s)} = \boxed{20 \text{ ft}}$

The answer is (B).

(b) $h_{p(d)} = \boxed{15 \text{ ft}}$

The answer is (A).

(c) There is no pipe size specified, so h_v cannot be calculated. Even so, v is typically in the 5 to 10 ft/sec range, and $h_\text{v} \approx 0$.

Since pipe lengths are not given, assume $h_f \approx 0$.

$$\begin{aligned} &20 \text{ ft} + 15 \text{ ft} + \frac{\left(80 \ \frac{\text{lbf}}{\text{in}^2}\right)\left(144 \ \frac{\text{in}^2}{\text{ft}^2}\right)}{62.4 \ \frac{\text{lbf}}{\text{ft}^3}} + 10 \text{ ft} \\ &\quad = \boxed{229.6 \text{ ft of water}} \end{aligned}$$

The answer is (D).

(d) The flow rate is

$$\frac{(3.5 \text{ MGD})\left(62.4 \ \frac{\text{lbf}}{\text{ft}^3}\right)}{0.64632 \ \frac{\frac{\text{MGD}}{\text{ft}^3}}{\text{sec}}} = 337.9 \text{ lbf/sec}$$

The rated motor output power does not depend on the motor efficiency.

$$\begin{aligned} P = \frac{h_\text{A}\dot{m}}{550\eta_{\text{pump}}} &= \frac{(229.6 \text{ ft})\left(337.9 \ \frac{\text{lbf}}{\text{sec}}\right)}{\left(550 \ \frac{\text{ft-lbf}}{\text{hp-sec}}\right)(0.85)} \\ &= 166.0 \text{ hp} \quad [\text{Table 18.5}] \end{aligned}$$

$$\boxed{\text{Use a 200 hp motor.}}$$

The answer is (D).

8. *Customary U.S. Solution*

From App. 16.B, the pipe data for 3 in schedule-40 steel pipe are

$$\begin{aligned} D_i &= 0.2557 \text{ ft} \\ A_i &= 0.05134 \text{ ft}^2 \end{aligned}$$

From App. 17.D, the equivalent length for various fittings is

$$\begin{aligned} \text{flanged elbow, } L_e &= 4.4 \text{ ft} \\ \text{wide-open gate valve, } L_e &= 2.8 \text{ ft} \end{aligned}$$

The total equivalent length of pipe and fittings is

$$L_e = 500 \text{ ft} + (6)(4.4 \text{ ft}) + (2)(2.8 \text{ ft}) = 532 \text{ ft}$$

As a first estimate, assume the flow rate is 100 gal/min.

The velocity in the pipe is

$$\text{v} = \frac{\dot{V}}{A} = \frac{\left(100 \frac{\text{gal}}{\text{min}}\right)\left(0.002228 \frac{\frac{\text{ft}^3}{\text{sec}}}{\frac{\text{gal}}{\text{min}}}\right)}{0.05134 \text{ ft}^2} = 4.34 \text{ ft/sec}$$

The Reynolds number is

$$\text{Re} = \frac{\text{v}D}{\nu} = \frac{\left(4.34 \frac{\text{ft}}{\text{sec}}\right)(0.2557 \text{ ft})}{6 \times 10^{-6} \frac{\text{ft}^2}{\text{sec}}} = 1.85 \times 10^5$$

From App. 17.A, $\epsilon = 0.0002$ ft.

So,

$$\frac{\epsilon}{D} = \frac{0.0002 \text{ ft}}{0.2557 \text{ ft}} \approx 0.0008$$

From the friction factor table, $f \approx 0.0204$.

For higher flow rates, f approaches 0.0186. Since the chosen flow rate was almost the lowest, $f = 0.0186$ should be used.

From Eq. 18.6, the friction head loss is

$$h_f = \frac{fL\text{v}^2}{2Dg} = \frac{(0.0186)(532 \text{ ft})\left(4.34 \frac{\text{ft}}{\text{sec}}\right)^2}{(2)(0.2557 \text{ ft})\left(32.2 \frac{\text{ft}}{\text{sec}^2}\right)} = 11.3 \text{ ft of gasoline}$$

This neglects the small velocity head. The other system points can be found using

$$\frac{h_{f_1}}{h_{f_2}} = \left(\frac{Q_1}{Q_2}\right)^2$$

$$h_{f_2} = h_{f_1}\left(\frac{Q_2}{100 \frac{\text{gal}}{\text{min}}}\right)^2 = (11.3 \text{ ft})\left(\frac{Q_2}{100 \frac{\text{gal}}{\text{min}}}\right)^2 = 0.00113 Q_2^2$$

Q (gal/min)	h_f (ft)	$h_f + 60$ (ft)
100	11.3	71.3
200	45.2	105.2
300	101.7	161.7
400	180.8	240.8
500	282.5	342.5
600	406.8	466.8

(a) Plot the system and pump curves.

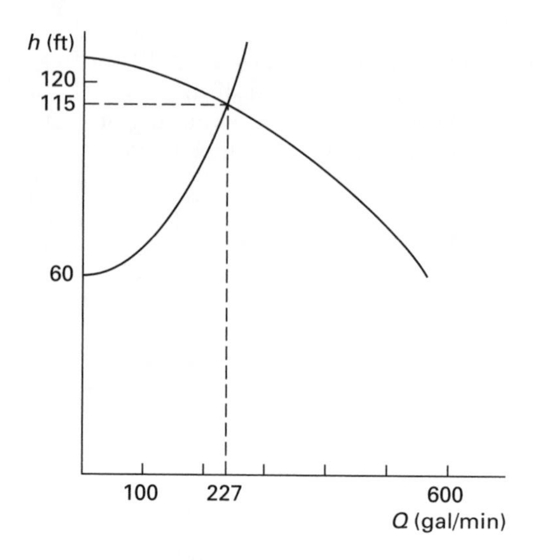

$$h = 115 \text{ ft}$$

$$\boxed{\text{Q} = 227 \text{ gal/min}}$$

(This value could be used to determine a new friction factor.)

The answer is (D).

(b) From Table 18.5, the hydraulic horsepower is

$$\text{WHP} = \frac{h_A Q(\text{SG})}{3956} = \frac{(115 \text{ ft})\left(227 \frac{\text{gal}}{\text{min}}\right)(0.7)}{3956 \frac{\text{ft-gal}}{\text{hp-min}}} = 4.62 \text{ hp}$$

$$\frac{(4.62 \text{ hp})\left(0.7457 \frac{\text{kW}}{\text{hp}}\right) \times (1 \text{ hr})\left(0.045 \frac{\$}{\text{kW-hr}}\right)}{(0.88)(0.88)} = \boxed{\$0.20}$$

The answer is (A).

SI Solution

From App. 16.C, the pipe data for 3 in schedule-40 pipe are

$$D_i = 77.92 \text{ mm}$$

$$A_i = 47.69 \times 10^{-4} \text{ m}^2$$

From App. 17.D, the equivalent length for various fittings is

$$\text{flanged elbow, } L_e = 4.4 \text{ ft}$$

$$\text{wide-open gate valve, } L_e = 2.8 \text{ ft}$$

The total equivalent length of pipe and fittings is

$$\begin{aligned} L_e &= 170 \text{ m} + (6)(4.4 \text{ ft})\left(0.3048 \ \frac{\text{m}}{\text{ft}}\right) \\ &\quad + (2)(2.8 \text{ ft})\left(0.3048 \ \frac{\text{m}}{\text{ft}}\right) \\ &= 180 \text{ m} \end{aligned}$$

As a first estimate, assume flow rate is 6.3 L/s. The velocity in the pipe is

$$\text{v} = \frac{\dot{V}}{A} = \frac{\left(6.3 \ \frac{\text{L}}{\text{s}}\right)\left(\frac{1 \text{ m}^3}{1000 \text{ L}}\right)}{47.69 \times 10^{-4} \text{ m}^2} = 1.32 \text{ m/s}$$

The Reynolds number is

$$\begin{aligned} \text{Re} &= \frac{\text{v}D}{\nu} \\ &= \frac{\left(1.32 \ \frac{\text{m}}{\text{s}}\right)(77.92 \text{ mm})\left(\frac{1 \text{ m}}{1000 \text{ mm}}\right)}{5.6 \times 10^{-7} \ \frac{\text{m}^2}{\text{s}}} \\ &= 1.75 \times 10^5 \end{aligned}$$

From Table 17.2, $\epsilon = 6.0 \times 10^{-5}$ m.

$$\frac{\epsilon}{D} = \frac{6.0 \times 10^{-5} \text{ m}}{(77.92 \text{ mm})\left(\frac{1 \text{ m}}{1000 \text{ mm}}\right)} \approx 0.0008$$

From the friction factor table (App. 17.B), $f = 0.0205$.

For higher flow rates, f approaches 0.0186. Since the chosen flow rate was almost the lowest, $f = 0.0186$ should be used.

From Eq. 17.22, the friction head loss is

$$\begin{aligned} h_f &= \frac{fL\text{v}^2}{2Dg} = \frac{(0.0186)(180 \text{ m})\left(1.32 \ \frac{\text{m}}{\text{s}}\right)^2}{(2)(77.92 \text{ mm})\left(\frac{1 \text{ m}}{1000 \text{ mm}}\right)\left(9.81 \ \frac{\text{m}}{\text{s}^2}\right)} \\ &= 3.82 \text{ m of gasoline} \end{aligned}$$

This neglects the small velocity head. The other system points can be found using

$$\frac{h_{f_1}}{h_{f_2}} = \left(\frac{Q_1}{Q_2}\right)^2$$

$$\begin{aligned} h_{f_2} &= h_{f_1}\left(\frac{Q_2}{Q_1}\right)^2 = (3.82 \text{ m})\left(\frac{Q_2}{6.3 \ \frac{\text{L}}{\text{s}}}\right)^2 \\ &= 0.0962Q_2^2 \end{aligned}$$

Q (L/s)	h_f (m)	$h_f + 20$ (m)
6.3	3.82	23.82
12	13.85	33.85
18	31.2	51.2
24	55.4	75.4
30	86.6	106.6
36	124.7	144.7

(a) Plot the system and pump curves.

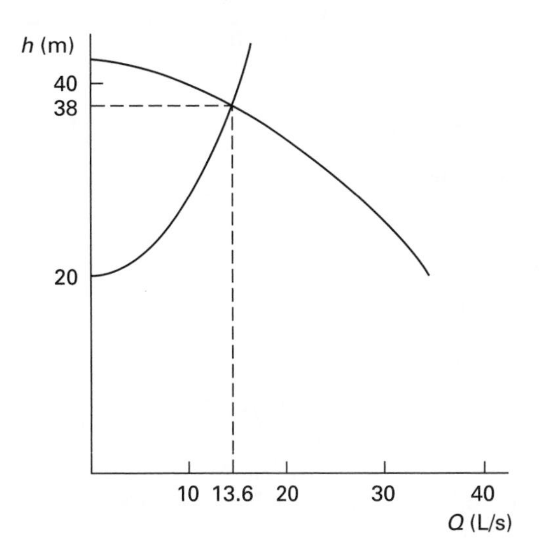

$$h = 38.0 \text{ m}$$

$$\boxed{Q = 13.6 \text{ L/s}}$$

(This value could be used to determine a new friction factor.)

The answer is (D).

(b) From Table 18.6, the hydraulic power is

$$\begin{aligned} \text{WkW} &= \frac{9.81h_AQ(\text{SG})}{1000} \\ &= \frac{\left(9.81 \ \frac{\text{m}}{\text{s}^2}\right)(38.0 \text{ m})\left(13.6 \ \frac{\text{L}}{\text{s}}\right)(0.7)}{1000 \ \frac{\text{W}}{\text{kW}}} \\ &= 3.55 \text{ kW} \end{aligned}$$

The cost per hour is

$$\begin{aligned} &= \left(\frac{3.55 \text{ kW}}{(0.88)(0.88)}\right)(1 \text{ h})\left(0.045 \ \frac{\$}{\text{kW·h}}\right) \\ &= \boxed{\$0.21} \end{aligned}$$

The answer is (A).

9. *Customary U.S. Solution*

From Eq. 18.28(b), the specific speed is

$$n_s = \frac{n\sqrt{Q}}{h_A^{0.75}}$$

For a double-suction pump, Q in the preceding equation is half of the full flow rate.

$$n_s = \frac{(900 \text{ rpm})\sqrt{\left(300 \frac{\text{gal}}{\text{sec}}\right)\left(60 \frac{\text{sec}}{\text{min}}\right)\left(\frac{1}{2}\right)}}{(20 \text{ ft})^{0.75}} = \boxed{9028 \text{ rpm}}$$

The answer is (C).

SI Solution

From Eq. 18.28(a), the specific speed is

$$n_s = \frac{n\sqrt{\dot{V}}}{h_A^{0.75}}$$

For a double-suction pump, $\dot{V}$ in the preceding equation is half of the full flow rate.

$$n_s = \frac{(900 \text{ rpm})\sqrt{\left(1.1 \frac{\text{kL}}{\text{s}}\right)\left(\frac{1 \text{ m}^3}{1 \text{ kL}}\right)\left(\frac{1}{2}\right)}}{(7 \text{ m})^{0.75}} = \boxed{155.1 \text{ rpm}}$$

The answer is (C).

10.

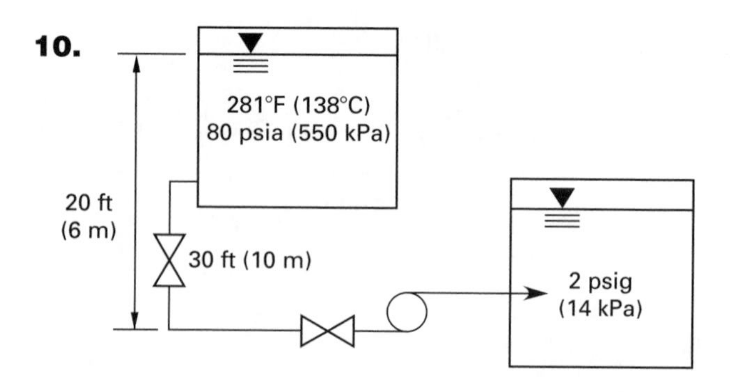

Customary U.S. Solution

From App. 16.B, data for 1.5 in schedule-40 steel pipe are

$$D_i = 0.1342 \text{ ft}$$
$$A_i = 0.01414 \text{ ft}^2$$

The velocity in the pipe is

$$\text{v} = \frac{\dot{V}}{A} = \frac{\left(100 \frac{\text{gal}}{\text{min}}\right)\left(0.002228 \frac{\text{ft}^3\text{-min}}{\text{sec-gal}}\right)}{0.01414 \text{ ft}^2} = 15.76 \text{ ft/sec}$$

From App. 17.D, for screwed steel fittings, the approximate equivalent lengths for fittings are

inlet (square mouth): $L_e = 3.1$ ft

long radius 90° ell: $L_e = 3.4$ ft

wide-open gate valves: $L_e = 1.2$ ft

The total equivalent length is

$$30 \text{ ft} + 3.1 \text{ ft} + (2)(3.4 \text{ ft}) + (2)(1.2 \text{ ft}) = 42.3 \text{ ft}$$

From App. 17.A, for steel, $\epsilon = 0.0002$ ft, so

$$\frac{\epsilon}{D} = \frac{0.0002 \text{ ft}}{0.1342 \text{ ft}} = 0.0015$$

At 281°F, $\nu = 0.239 \times 10^{-5}$ ft^2/sec. The Reynolds number is

$$\text{Re} = \frac{D\text{v}}{\nu} = \frac{(0.1342 \text{ ft})\left(15.76 \frac{\text{ft}}{\text{sec}}\right)}{0.239 \times 10^{-5} \frac{\text{ft}^2}{\text{sec}}} = 8.85 \times 10^5$$

From App. 17.B, the friction factor is $f = 0.022$.

From Eq. 18.6, the friction head is

$$h_f = \frac{fL\text{v}^2}{2Dg} = \frac{(0.022)(42.3 \text{ ft})\left(15.76 \frac{\text{ft}}{\text{sec}}\right)^2}{(2)(0.1342 \text{ ft})\left(32.2 \frac{\text{ft}}{\text{sec}^2}\right)} = 26.74 \text{ ft}$$

At 281°F, the saturated vapor pressure of pressurized water is

$$p_{\text{vapor}} = 50.02 \text{ psia} \quad \text{[from steam tables]}$$

The density of the liquid is the reciprocal of the specific volume, also taken from the steam tables.

$$\rho = \frac{1}{v_f} = \frac{1}{0.01727 \frac{\text{ft}^3}{\text{lbm}}} = 57.9 \text{ lbm/ft}^3$$

From Eq. 18.5(b),

$$h_{vp} = \left(\frac{p_{vapor}}{\rho}\right)\left(\frac{g_c}{g}\right)$$

$$= \left(\frac{\left(50.02\ \frac{lbf}{in^2}\right)\left(144\ \frac{in^2}{ft^2}\right)}{57.9\ \frac{lbm}{ft^3}}\right)\left(\frac{32.2\ \frac{ft\text{-}lbm}{lbf\text{-}sec^2}}{32.2\ \frac{ft}{sec^2}}\right)$$

$$= 124.4\ ft$$

From Eq. 18.5(b), the pressure head is

$$h_p = \left(\frac{p}{\rho}\right)\left(\frac{g_c}{g}\right)$$

$$= \left(\frac{\left(80\ \frac{lbf}{in^2}\right)\left(144\ \frac{in^2}{ft^2}\right)}{57.9\ \frac{lbm}{ft^3}}\right)\left(\frac{32.2\ \frac{ft\text{-}lbm}{lbf\text{-}sec^2}}{32.2\ \frac{ft}{sec^2}}\right)$$

$$= 199.0\ ft$$

From Eq. 18.30(a), the NPSHA is

$$\begin{aligned} \text{NPSHA} &= h_p + h_{z(s)} - h_{f(s)} - h_{vp} \\ &= 199.0\ ft + 20\ ft - 26.74\ ft - 124.4\ ft \\ &= 67.9\ ft \end{aligned}$$

Since NPSHR = 10 ft, the pump will not cavitate.

(Note that a pump may not actually be needed in this configuration.)

The answer is (D).

SI Solution

From App. 16.C, data for 1.5 in schedule-40 steel pipe are

$$D_i = 40.89\ mm$$
$$A_i = 13.13 \times 10^{-4}\ m^2$$

The velocity in the pipe is

$$v = \frac{\dot{V}}{A} = \frac{\left(6.3\ \frac{L}{s}\right)\left(\frac{1\ m^3}{1000\ L}\right)}{13.13 \times 10^{-4}\ m^2} = 4.80\ m/s$$

From App. 17.D, for screwed steel fittings, the approximate equivalent lengths for fittings are

inlet (square mouth): $L_e = 3.1$ ft

long radius 90° ell: $L_e = 3.4$ ft

wide-open gate valves: $L_e = 1.2$ ft

The total equivalent length is

$$30\ ft + 3.1\ ft + (2)(3.4\ ft) + (2)(1.2\ ft) = 42.3\ ft$$

$$(42.3\ ft)\left(0.3048\ \frac{m}{ft}\right) = 12.89\ m$$

From Table 17.2, for steel, $\epsilon = 6.0 \times 10^{-5}$ m.

$$\frac{\epsilon}{D} = \frac{6.0 \times 10^{-5}\ m}{(40.89\ mm)\left(\frac{1\ m}{1000\ mm}\right)} \approx 0.0015$$

At 138°C, $\nu = 0.222 \times 10^{-6}\ m^2/s$. The Reynolds number is

$$\text{Re} = \frac{D\text{v}}{\nu} = \frac{(40.89\ mm)\left(\frac{1\ m}{1000\ mm}\right)\left(4.80\ \frac{m}{s}\right)}{0.222 \times 10^{-6}\ \frac{m^2}{s}} = 8.84 \times 10^5$$

From App. 17.B, the friction factor is $f = 0.022$.

From Eq. 18.6, the friction head is

$$h_f = \frac{fL\text{v}^2}{2Dg} = \frac{(0.022)(12.89\ m)\left(4.8\ \frac{m}{s}\right)^2}{(2)(40.89\ mm)\left(\frac{1\ m}{1000\ mm}\right)\left(9.81\ \frac{m}{s^2}\right)} = 8.14\ m$$

From Eq. 18.5(a),

$$h_{vp} = \frac{p_{vapor}}{\rho g}$$

At 138°C, the saturated vapor pressure of pressurized water is

$$p_{vapor} = 3.431\ bar\quad \text{[from steam tables]}$$

The density of the liquid is the reciprocal of the specific volume, also taken from the steam tables.

$$\rho = \frac{1}{\upsilon_f} = \frac{1}{\left(1.0777\ \frac{cm^3}{g}\right)\left(1000\ \frac{g}{kg}\right)\left(\frac{1\ m^3}{(100\ cm)^3}\right)} = 927.9\ kg/m^3$$

$$h_{vp} = \frac{(3.431\ bar)\left(10^5\ \frac{Pa}{bar}\right)}{\left(927.9\ \frac{kg}{m^3}\right)\left(9.81\ \frac{m}{s^2}\right)} = 37.69\ m$$

From Eq. 18.5(a), the pressure head is

$$h_p = \frac{p}{\rho g} = \frac{(550 \text{ kPa})\left(1000 \frac{\text{Pa}}{\text{kPa}}\right)}{\left(927.9 \frac{\text{kg}}{\text{m}^3}\right)\left(9.81 \frac{\text{m}}{\text{s}^2}\right)} = 60.42 \text{ m}$$

From Eq. 18.30(a), the NPSHA is

$$\begin{aligned} \text{NPSHA} &= h_p + h_{z(s)} - h_{f(s)} - h_{\text{vp}} \\ &= 60.42 \text{ m} + 6 \text{ m} - 8.14 \text{ m} - 37.69 \text{ m} \\ &= 20.6 \text{ m} \end{aligned}$$

Since NPSHR is 3 m, **the pump will not cavitate.**

(Note that a pump may not actually be needed in this configuration.)

The answer is (D).

11. The solvent is the water (fresh), and the solution is the seawater. Since seawater contains approximately $2^1/_2$% salt by weight, 100 lbm of seawater will yield 2.5 lbm salt and 97.5 lbm water. The molecular weight of salt is 23.0 + 35.5 = 58.5 lbm/lbmol. The number of moles of salt in 100 lbm of seawater is

$$n_{\text{salt}} = \frac{2.5 \text{ lbm}}{58.5 \frac{\text{lbm}}{\text{lbmol}}} = 0.043 \text{ lbmol}$$

Similarly, the molecular weight of water is 18.016 lbm/lbmol. The number of moles of water is

$$n_{\text{water}} = \frac{97.5 \text{ lbm}}{18.016 \frac{\text{lbm}}{\text{lbmol}}} = 5.412 \text{ lbmol}$$

The mole fraction of water is

$$\frac{5.412 \text{ lbmol}}{5.412 \text{ lbmol} + 0.043 \text{ lbmol}} = 0.992$$

Customary U.S. Solution

Cavitation will occur when

$$h_{\text{atm}} - h_{\text{v}} < h_{\text{vp}}$$

The density of seawater is 64.0 lbm/ft^3.

From Eq. 18.5(b), the atmospheric head is

$$\begin{aligned} h_{\text{atm}} &= \left(\frac{p}{\rho}\right)\left(\frac{g_c}{g}\right) \\ &= \left(\frac{\left(14.7 \frac{\text{lbf}}{\text{in}^2}\right)\left(144 \frac{\text{in}^2}{\text{ft}^2}\right)}{64.0 \frac{\text{lbm}}{\text{ft}^3}}\right)\left(\frac{32.2 \frac{\text{ft-lbm}}{\text{lbf-sec}^2}}{32.2 \frac{\text{ft}}{\text{sec}^2}}\right) \\ &= 33.075 \text{ ft} \quad \text{[ft of seawater]} \end{aligned}$$

$$h_{\text{depth}} = 8 \text{ ft} \quad \text{[given]}$$

From Eq. 18.7, the velocity head is

$$h_{\text{v}} = \frac{\text{v}_{\text{propeller}}^2}{2g} = \frac{(4.2\text{v}_{\text{boat}})^2}{(2)\left(32.2 \frac{\text{ft}}{\text{sec}^2}\right)} = 0.2739\text{v}_{\text{boat}}^2$$

The vapor pressure of 68°F freshwater is p_{vp} = 0.3391 psia.

From App. 14.A, the density of water at 68°F is 62.32 lbm/ft^3. Raoult's law predicts the actual vapor pressure of the solution.

$$p_{\text{vapor,solution}} = (p_{\text{vapor,solvent}})\left(\begin{array}{c}\text{mole fraction}\\ \text{of the solvent}\end{array}\right)$$

$$\begin{aligned} p_{\text{vapor,seawater}} &= (0.992)(0.3391 \text{ psia}) \\ &= 0.3364 \text{ psia} \end{aligned}$$

From Eq. 18.5(b), the vapor pressure head is

$$\begin{aligned} h_{\text{vapor,seawater}} &= \left(\frac{p}{\rho}\right)\left(\frac{g_c}{g}\right) \\ &= \left(\frac{\left(0.3364 \frac{\text{lbf}}{\text{in}^2}\right)\left(144 \frac{\text{in}^2}{\text{ft}^2}\right)}{64.0 \frac{\text{lbm}}{\text{ft}^3}}\right) \\ &\quad \times \left(\frac{32.2 \frac{\text{ft-lbm}}{\text{lbf-sec}^2}}{32.2 \frac{\text{ft}}{\text{sec}^2}}\right) \\ &= 0.7569 \text{ ft} \end{aligned}$$

Then,

$$8 \text{ ft} + 33.075 \text{ ft} - 0.2739 \text{ v}_{\text{boat}}^2 = 0.7569 \text{ ft}$$

$$\text{v}_{\text{boat}} = \boxed{12.13 \text{ ft/sec}}$$

The answer is (B).

SI Solution

Cavitation will occur when

$$h_{atm} - h_v < h_{vp}$$

The density of seawater is 1024 kg/m^3.

From Eq. 18.5(a), the atmospheric head is

$$\begin{aligned} h_{atm} &= \frac{p}{\rho g} \\ &= \frac{(101.3 \text{ kPa})\left(1000 \ \frac{\text{Pa}}{\text{kPa}}\right)}{\left(1024 \ \frac{\text{kg}}{\text{m}^3}\right)\left(9.81 \ \frac{\text{m}}{\text{s}^2}\right)} \\ &= 10.08 \text{ m} \end{aligned}$$

$$h_{depth} = 3 \text{ m} \quad [\text{given}]$$

From Eq. 18.7, the velocity head is

$$\begin{aligned} h_v &= \frac{v_{propeller}^2}{2g} = \frac{(4.2 v_{boat})^2}{(2)\left(9.81 \ \frac{\text{m}}{\text{s}^2}\right)} \\ &= 0.899 v_{boat}^2 \end{aligned}$$

The vapor pressure of 20°C freshwater is

$$\begin{aligned} p_{vp} &= (0.02339 \text{ bar})\left(100 \ \frac{\text{kPa}}{\text{bar}}\right) \\ &= 2.339 \text{ kPa} \end{aligned}$$

From App. 14.B, the density of water at 20°C is 998.23 kg/m^3. Raoult's law predicts the actual vapor pressure of the solution.

$$p_{vapor,solution} = (p_{vapor,solvent})\left(\begin{array}{c}\text{mole fraction}\\ \text{of the solvent}\end{array}\right)$$

The solvent is the freshwater and the solution is the seawater.

The mole fraction of water is 0.992.

$$p_{vapor,seawater} = (0.992)(2.339 \text{ kPa}) = 2.320 \text{ kPa}$$

From Eq. 18.5(a), the vapor pressure head is

$$\begin{aligned} h_{vapor,seawater} &= \frac{(2.320 \text{ kPa})\left(1000 \ \frac{\text{Pa}}{\text{kPa}}\right)}{\left(9.81 \ \frac{\text{m}}{\text{s}^2}\right)\left(1024 \ \frac{\text{kg}}{\text{m}^3}\right)} \\ &= 0.231 \text{ m} \end{aligned}$$

Then,

$$3 \text{ m} + 10.08 \text{ m} - 0.899 v_{boat}^2 = 0.231 \text{ m}$$

$$v_{boat} = \boxed{3.78 \text{ m/s}}$$

The answer is (B).

12. *Customary U.S. Solution*

From App. 17.D, the approximate equivalent lengths of various screwed steel fittings are

inlet: $L_e = 8.5$ ft [essentially a reentrant inlet]

check valve: $L_e = 19$ ft

long radius elbows: $L_e = 3.6$ ft

The total equivalent length of the 2 in line is

$$\begin{aligned} L_e &= 12 \text{ ft} + 8.5 \text{ ft} + 19.0 \text{ ft} + (3)(3.6 \text{ ft}) + 80 \text{ ft} \\ &= 130.3 \text{ ft} \end{aligned}$$

From App. 16.B, for schedule-40 2 in pipe, the pipe data are

$$D_i = 0.1723 \text{ ft}$$
$$A_i = 0.0233 \text{ ft}^2$$

Since the flow rate is unknown, it must be assumed in order to find velocity. Assume 90 gal/min.

$$\dot{V} = \left(90 \ \frac{\text{gal}}{\text{min}}\right)\left(0.002228 \ \frac{\frac{\text{ft}^3}{\text{sec}}}{\frac{\text{gal}}{\text{min}}}\right) = 0.2005 \text{ ft}^3/\text{sec}$$

The velocity is

$$v = \frac{\dot{V}}{A_i} = \frac{0.2005 \ \frac{\text{ft}^3}{\text{sec}}}{0.0233 \text{ ft}^2} = 8.605 \text{ ft/sec}$$

From App. 14.A, the kinematic viscosity of water at 70°F is $\nu = 1.059 \times 10^{-5}$ ft^2/sec.

The Reynolds number is

$$\begin{aligned} \text{Re} &= \frac{D v}{\nu} = \frac{(0.1723 \text{ ft})\left(8.605 \ \frac{\text{ft}}{\text{sec}}\right)}{1.059 \times 10^{-5} \ \frac{\text{ft}^2}{\text{sec}}} \\ &= 1.4 \times 10^5 \end{aligned}$$

From App. 17.A, the specific roughness of steel pipe is

$$\epsilon = 0.0002 \text{ ft}$$
$$\frac{\epsilon}{D} = \frac{0.0002 \text{ ft}}{0.1723 \text{ ft}} = 0.0012$$

From App. 17.B, $f = 0.022$. At 90 gal/min, the friction loss in the line from Eq. 18.6 is

$$h_f = \frac{fL\text{v}^2}{2Dg} = \frac{(0.022)(130.3\ \text{ft})\left(8.605\ \frac{\text{ft}}{\text{sec}}\right)^2}{(2)(0.1723\ \text{ft})\left(32.2\ \frac{\text{ft}}{\text{sec}^2}\right)} = 19.1\ \text{ft}$$

From Eq. 18.7, the velocity head at 90 gal/min is

$$h_\text{v} = \frac{\text{v}^2}{2g} = \frac{\left(8.605\ \frac{\text{ft}}{\text{sec}}\right)^2}{(2)\left(32.2\ \frac{\text{ft}}{\text{sec}^2}\right)} = 1.1\ \text{ft}$$

In general, the friction head and velocity head are proportional to v^2 and Q^2.

$$h_f = (19.1\ \text{ft})\left(\frac{Q_2}{90\ \frac{\text{gal}}{\text{min}}}\right)^2$$

$$h_\text{v} = (1.1\ \text{ft})\left(\frac{Q_2}{90\ \frac{\text{gal}}{\text{min}}}\right)^2$$

(Note that the 7 ft dimension is included in the 20 ft dimension.)

The total system head is

$$\begin{aligned} h &= h_z + h_\text{v} + h_f \\ &= 20\ \text{ft} + (1.1 + 19.1\ \text{ft})\left(\frac{Q_2}{90\ \frac{\text{gal}}{\text{min}}}\right)^2 \end{aligned}$$

From this equation, the following table for system head can be generated.

Q_2 (gal/min)	system head, h (ft)
0	20.0
10	20.2
20	21.0
30	22.2
40	24.0
50	26.2
60	29.0
70	32.2
80	36.0
90	40.2
100	44.9
110	50.2

(a) The intersection point of the system curve and the pump curve defines the operating flow rate. The flow rate is 95 gal/min.

The answer is (D).

(b) The intersection point is not in an efficient range for the pump because it is so far down on the system curve that the pumping efficiency will be low.

A different pump should be used.

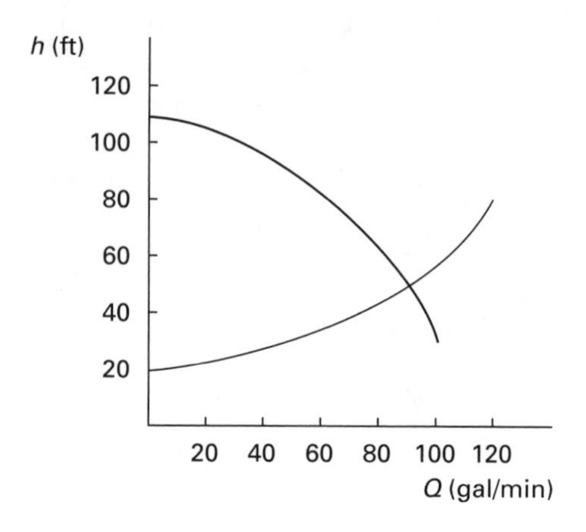

The answer is (A).

SI Solution

Use the approximate equivalent lengths of various screwed steel fittings from the Customary U.S. Solution. The total equivalent length of 5.08 cm schedule-40 pipe is

$$\begin{aligned} L_e &= 4\ \text{m} + \big(8.5\ \text{ft} + 19.0\ \text{ft} + (3)(3.6\ \text{ft})\big)\left(0.3048\ \frac{\text{m}}{\text{ft}}\right) \\ &\quad + 27\ \text{m} \\ &= 42.67\ \text{m} \end{aligned}$$

From App. 16.C, for 2 in schedule-40 pipe, the pipe data are

$$D_i = 52.50\ \text{mm}$$
$$A_i = 21.65 \times 10^{-4}\ \text{m}^2$$

Since the flow rate is unknown, it must be assumed in order to find velocity. Assume 6 L/s.

$$\dot{V} = \left(6\ \frac{\text{L}}{\text{s}}\right)\left(\frac{1\ \text{m}^3}{1000\ \text{L}}\right) = 6 \times 10^{-3}\ \text{m}^3/\text{s}$$

The velocity is

$$\text{v} = \frac{\dot{V}}{A_i} = \frac{6 \times 10^{-3}\ \frac{\text{m}^3}{\text{s}}}{21.65 \times 10^{-4}\ \text{m}^2} = 2.77\ \text{m/s}$$

From App. 14.B, the absolute viscosity of water at 21°C is $\mu = 0.9827 \times 10^{-3}$ Pa·s.

The density of water is $\rho = 998$ kg/m^3.

The Reynolds number is

$$\text{Re} = \frac{\rho \text{v} D_i}{\mu} = \frac{\left(998\ \frac{\text{kg}}{\text{m}^3}\right)\left(2.77\ \frac{\text{m}}{\text{s}}\right) \times (52.50\ \text{mm})\left(\frac{1\ \text{m}}{1000\ \text{mm}}\right)}{0.9827 \times 10^{-3}\ \text{Pa·s}} = 1.5 \times 10^5$$

From Table 17.2, the specific roughness of steel pipe is

$$\epsilon = 6.0 \times 10^{-5}\ \text{m}$$

$$\frac{\epsilon}{D} = \frac{6.0 \times 10^{-5}\ \text{m}}{(52.50\ \text{mm})\left(\frac{1\ \text{m}}{1000\ \text{mm}}\right)} \approx 0.0012$$

From App. 17.B, $f = 0.022$. At 6 L/s, the friction loss in the line from Eq. 18.6 is

$$h_f = \frac{fL\text{v}^2}{2Dg} = \frac{(0.022)(42.67\ \text{m})\left(2.77\ \frac{\text{m}}{\text{s}}\right)^2}{(2)(52.50\ \text{mm})\left(\frac{1\ \text{m}}{1000\ \text{mm}}\right)\left(9.81\ \frac{\text{m}}{\text{s}^2}\right)} = 6.99\ \text{m}$$

At 6 L/s, the velocity head from Eq. 18.7 is

$$h_\text{v} = \frac{\text{v}^2}{2g} = \frac{\left(2.77\ \frac{\text{m}}{\text{s}}\right)^2}{(2)\left(9.81\ \frac{\text{m}}{\text{s}^2}\right)} = 0.39\ \text{m}$$

In general, the friction head and velocity head are proportional to v^2 and Q^2.

$$h_f = (6.99\ \text{m})\left(\frac{Q_2}{6\ \frac{\text{L}}{\text{s}}}\right)^2$$

$$h_\text{v} = (0.39\ \text{m})\left(\frac{Q_2}{6\ \frac{\text{L}}{\text{s}}}\right)^2$$

(Note that the 2.3 m dimension is included in the 6.3 m dimension.)

The total system head is

$$h = h_z + h_\text{v} + h_f = 6.3\ \text{m} + (0.39\ \text{m} + 6.99\ \text{m})\left(\frac{Q_2}{6\ \frac{\text{L}}{\text{s}}}\right)^2$$

From this equation the following table for the system head can be generated.

Q_2 (L/s)	h (m)
0	6.3
0.6	6.37
1.2	6.60
1.8	6.96
2.4	7.48
3.2	8.40
3.6	8.96
4.4	10.27
4.8	11.02
5.7	12.96
6.5	14.96
7.0	16.35
7.5	17.83

The intersection point of the system curve and the pump curve will define the operating flow rate.

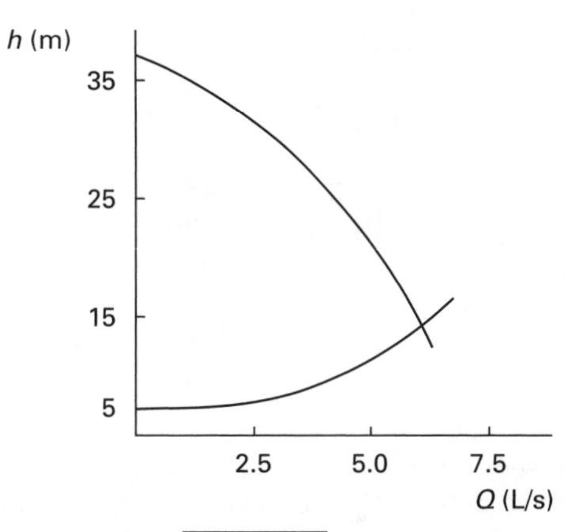

(a) The flow rate is **6.2 L/s.**

The answer is (D).

(b) The intersection point is not in an efficient range of the pump because it is so far down on the system curve that the pumping efficiency will be low.

A different pump should be used.

The answer is (A).

13. From Eq. 18.52,

$$P_2 = P_1\left(\frac{\rho_2 n_2^3 D_2^5}{\rho_1 n_1^3 D_1^5}\right) = P_1\left(\frac{n_2}{n_1}\right)^3 \quad [\rho_2 = \rho_1 \text{ and } D_2 = D_1]$$

Customary U.S. Solution

$$P_2 = (0.5\ \text{hp})\left(\frac{2000\ \text{rpm}}{1750\ \text{rpm}}\right)^3 = \boxed{0.746\ \text{hp}}$$

The answer is (D).

SI Solution

$$P_2 = P_1\left(\frac{n_2}{n_1}\right)^3 = (0.37\ \text{kW})\left(\frac{2000\ \text{rpm}}{1750\ \text{rpm}}\right)^3 = \boxed{0.55\ \text{kW}}$$

The answer is (D).

14. (a) Random values of Q are chosen, and the corresponding values of H are determined by the formula $H = 30 + 2Q^2$.

Q (ft^3/sec)	H (ft)
0	30
2.5	42.5
5	80
7.5	142.5
10	230
15	480
20	830
25	1280
30	1830

The intersection of the system curve and the 1400 rpm pump curve defines the operating point at that rpm.

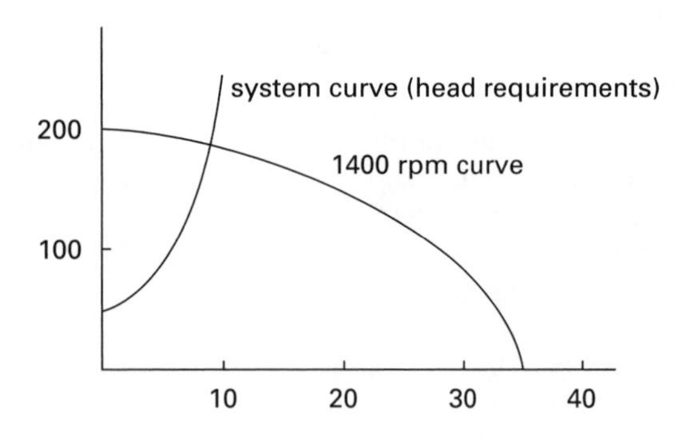

From the intersection of the graphs, at 1400 rpm the flow rate is approximately 9 ft^3/sec and the corresponding head is $30 + (2)(9)^2 \approx 192$ ft.

$$Q = \left(9\ \frac{\text{ft}^3}{\text{sec}}\right)\left(448.8\ \frac{\frac{\text{gal}}{\text{min}}}{\frac{\text{ft}^3}{\text{sec}}}\right) = \boxed{4039\ \text{gal/min}}$$

The answer is (C).

(b) From Table 18.5, the hydraulic horsepower is

$$\text{WHP} = \frac{h_A\dot{V}(\text{SG})}{8.814} = \frac{(192\ \text{ft})\left(9\ \frac{\text{ft}^3}{\text{sec}}\right)(1)}{8.814\ \frac{\text{ft}^4}{\text{hp-sec}}} = 196\ \text{hp}$$

From Eq. 18.28(b), the specific speed is

$$n_s = \frac{n\sqrt{Q}}{h_A^{0.75}} = \frac{(1400\ \text{rpm})\sqrt{4039\ \frac{\text{gal}}{\text{min}}}}{(192\ \text{ft})^{0.75}} = 1725$$

From Fig. 18.8 with curve E, $\eta \approx 86\%$.

The minimum pump motor power should be

$$\frac{196\ \text{hp}}{0.86} = \boxed{228\ \text{hp}}$$

The answer is (C).

(c) From Eq. 18.41,

$$Q_2 = Q_1\left(\frac{n_2}{n_1}\right) = \left(4039\ \frac{\text{gal}}{\text{min}}\right)\left(\frac{1200\ \text{rpm}}{1400\ \text{rpm}}\right) = \boxed{3462\ \text{gal/min}}$$

The answer is (B).

15. Since turbines are essentially pumps running backward, use Table 18.5.

$$\Delta p = \left(30\ \frac{\text{lbf}}{\text{in}^2} - 5\ \frac{\text{lbf}}{\text{in}^2}\right)\left(144\ \frac{\text{in}^2}{\text{ft}^2}\right) = 3600\ \text{lbf/ft}^2$$

$$P = \frac{\left(3600\ \frac{\text{lbf}}{\text{ft}^2}\right)\left(100\ \frac{\text{ft}^3}{\text{sec}}\right)}{550\ \frac{\text{ft-lbf}}{\text{hp-sec}}} = \boxed{654.5\ \text{hp}}$$

The answer is (C).

16. The flow rate is

$$\gamma\dot{V} = \left(62.4\ \frac{\text{lbf}}{\text{ft}^3}\right)\left(1000\ \frac{\text{ft}^3}{\text{sec}}\right) = 6.24 \times 10^4\ \text{lbf/sec}$$

The head available for work is

$$\Delta h = 625\ \text{ft} - 58\ \text{ft} = 567\ \text{ft}$$

Use Table 18.5. The power is

$$\begin{aligned}P &= (0.89)\left(6.24\times 10^4\ \frac{\text{lbf}}{\text{sec}}\right)(567\ \text{ft})\\ &= 3.149\times 10^7\ \text{ft-lbf/sec}\end{aligned}$$

Convert from hp to kW.

$$\frac{\left(3.149\times 10^7\ \frac{\text{ft-lbf}}{\text{sec}}\right)\left(0.7457\ \frac{\text{kW}}{\text{hp}}\right)}{550\ \frac{\text{ft-lbf}}{\text{hp-sec}}} = \boxed{4.27\times 10^4\ \text{kW}\quad (43\ \text{MW})}$$

The answer is (C).

17. *Customary U.S. Solution*

(a) From App. 14.A, the density of water at 60°F is 62.37 lbm/ft^3. From Eq. 18.5(b), the head dropped is

$$\begin{aligned}h &= \left(\frac{\Delta p}{\rho}\right)\left(\frac{g_c}{g}\right)\\ &= \left(\frac{\left(500\ \frac{\text{lbf}}{\text{in}^2} - 30\ \frac{\text{lbf}}{\text{in}^2}\right)\left(144\ \frac{\text{in}^2}{\text{ft}^2}\right)}{62.37\ \frac{\text{lbm}}{\text{ft}^3}}\right)\\ &\quad\times\left(\frac{32.2\ \frac{\text{ft-lbm}}{\text{lbf-sec}^2}}{32.2\ \frac{\text{ft}}{\text{sec}^2}}\right)\\ &= 1085\ \text{ft}\end{aligned}$$

From Eq. 18.62(b), the specific speed of a turbine is

$$\begin{aligned}n_s &= \frac{n\sqrt{P}}{(h_t)^{1.25}} = \frac{(1750\ \text{rpm})\sqrt{250\ \text{hp}}}{(1085\ \text{ft})^{1.25}}\\ &= \boxed{4.443}\end{aligned}$$

The answer is (A).

(b) The flow rate, $\dot{V}$, is

$$\begin{aligned}\dot{V} &= A\text{v} = \left(\frac{\pi}{4}\right)\left(\frac{4\ \text{in}}{12\ \frac{\text{in}}{\text{ft}}}\right)^2\left(35\ \frac{\text{ft}}{\text{sec}}\right)\\ &= 3.054\ \text{ft}^3/\text{sec}\end{aligned}$$

The flow rate, considering the blade's moving away at 10 ft/sec, is

$$\begin{aligned}\dot{V}' &= \frac{\left(35\ \frac{\text{ft}}{\text{sec}} - 10\ \frac{\text{ft}}{\text{sec}}\right)\left(3.054\ \frac{\text{ft}^3}{\text{sec}}\right)}{35\ \frac{\text{ft}}{\text{sec}}}\\ &= 2.181\ \text{ft}^3/\text{sec}\end{aligned}$$

The forces in the x-direction and the y-direction are

$$\begin{aligned}F_x &= \left(\frac{\dot{V}'\rho}{g_c}\right)(\text{v}_j - \text{v}_b)(\cos\theta - 1)\\ &= \left(\frac{\left(2.181\ \frac{\text{ft}^3}{\text{sec}}\right)\left(62.37\ \frac{\text{lbm}}{\text{ft}^3}\right)}{32.2\ \frac{\text{lbm-ft}}{\text{lbf-sec}^2}}\right)\\ &\quad\times\left(35\ \frac{\text{ft}}{\text{sec}} - 10\ \frac{\text{ft}}{\text{sec}}\right)(\cos 80^\circ - 1)\\ &= -87.27\ \text{lbf}\end{aligned}$$

$$\begin{aligned}F_y &= \left(\frac{\dot{V}'\rho}{g_c}\right)(\text{v}_j - \text{v}_b)\sin\theta\\ &= \left(\frac{\left(2.181\ \frac{\text{ft}^3}{\text{sec}}\right)\left(62.37\ \frac{\text{lbm}}{\text{ft}^3}\right)}{32.2\ \frac{\text{lbm-ft}}{\text{lbf-sec}^2}}\right)\\ &\quad\times\left(35\ \frac{\text{ft}}{\text{sec}} - 10\ \frac{\text{ft}}{\text{sec}}\right)\sin 80^\circ\\ &= 104.0\ \text{lbf}\end{aligned}$$

$$\begin{aligned}R &= \sqrt{F_x^2 + F_y^2}\\ &= \sqrt{(-87.27\ \text{lbf})^2 + (104.0\ \text{lbf})^2} = \boxed{135.8\ \text{lbf}}\end{aligned}$$

The answer is (B).

SI Solution

(a) From App. 14.B, the density of water at 16°C is 998.83 kg/m^3. From Eq. 18.5(a), the head dropped is

$$\begin{aligned}h &= \frac{\Delta p}{\rho g}\\ &= \frac{(3.5\ \text{MPa})\left(10^6\ \frac{\text{Pa}}{\text{MPa}}\right) - (210\ \text{kPa})\left(1000\ \frac{\text{Pa}}{\text{kPa}}\right)}{\left(998.83\ \frac{\text{kg}}{\text{m}^3}\right)\left(9.81\ \frac{\text{m}}{\text{s}^2}\right)}\\ &= 335.8\ \text{m}\end{aligned}$$

Water Resources

From Eq. 18.62(a), the specific speed of a turbine is

$$\begin{aligned} n_s &= \frac{n\sqrt{P}}{(h_t)^{1.25}} \\ &= \frac{(1750\text{ rpm})\sqrt{185\text{ kW}}}{(335.8\text{ m})^{1.25}} \\ &= \boxed{16.56} \end{aligned}$$

The answer is (A).

(b) The flow rate, $\dot{V}$, is

$$\begin{aligned} \dot{V} = A\text{v} &= \left(\frac{\pi}{4}\right)\left((100\text{ mm})\left(\frac{1\text{ m}}{1000\text{ mm}}\right)\right)^2\left(10.5\ \frac{\text{m}}{\text{s}}\right) \\ &= 0.08247\text{ m}^3/\text{s} \end{aligned}$$

The flow rate, considering the blade's moving away at 3 m/s, is

$$\begin{aligned} \dot{V}' &= \frac{\left(10.5\ \frac{\text{m}}{\text{s}} - 3\ \frac{\text{m}}{\text{s}}\right)\left(0.08247\ \frac{\text{m}^3}{\text{s}}\right)}{10.5\ \frac{\text{m}}{\text{s}}} \\ &= 0.05891\text{ m}^3/\text{s} \end{aligned}$$

The forces in the x-direction and the y-direction are

$$\begin{aligned} F_x &= \dot{V}'\rho(\text{v}_j - \text{v}_b)(\cos\theta - 1) \\ &= \left(0.05891\ \frac{\text{m}^3}{\text{s}}\right)\left(998.83\ \frac{\text{kg}}{\text{m}^3}\right) \\ &\quad \times\left(10.5\ \frac{\text{m}}{\text{s}} - 3\ \frac{\text{m}}{\text{s}}\right)(\cos 80^\circ - 1) \\ &= -364.7\text{ N} \end{aligned}$$

$$\begin{aligned} F_y &= \dot{V}'\rho(\text{v}_j - \text{v}_b)\sin\theta \\ &= \left(0.05891\ \frac{\text{m}^3}{\text{s}}\right)\left(998.83\ \frac{\text{kg}}{\text{m}^3}\right) \\ &\quad \times\left(10.5\ \frac{\text{m}}{\text{s}} - 3\ \frac{\text{m}}{\text{s}}\right)\sin 80^\circ \\ &= 434.6\text{ N} \end{aligned}$$

$$\begin{aligned} R &= \sqrt{F_x^2 + F_y^2} \\ &= \sqrt{(-364.7\text{ N})^2 + (434.6\text{ N})^2} = \boxed{567.3\text{ N}} \end{aligned}$$

The answer is (B).

18. *Customary U.S. Solution*

(a) The total effective head is due to the pressure head, velocity head, and tailwater head.

$$\begin{aligned} h_{\text{eff}} &= h_p + h_\text{v} - h_{z,\text{(tailwater)}} \\ &= 92.5\text{ ft} + \frac{\left(12\ \frac{\text{ft}}{\text{sec}}\right)^2}{(2)\left(32.2\ \frac{\text{ft}}{\text{sec}^2}\right)} - (-5.26\text{ ft}) \\ &= \boxed{100\text{ ft}} \end{aligned}$$

The answer is (C).

(b) From Table 18.5, the theoretical hydraulic horsepower is

$$\begin{aligned} P_{\text{th}} &= \frac{h_A\dot{V}(\text{SG})}{8.814} = \frac{(100\text{ ft})\left(25\ \frac{\text{ft}^3}{\text{sec}}\right)(1)}{8.814\ \frac{\text{ft}^4}{\text{hp-sec}}} \\ &= 283.6\text{ hp} \end{aligned}$$

The overall turbine efficiency is

$$\eta = \frac{P_{\text{brake}}}{P_{\text{th}}} = \frac{250\text{ hp}}{283.6\text{ hp}} = \boxed{0.882\quad(88.2\%)}$$

The answer is (B).

(c) From Eq. 18.42 (the affinity laws),

$$\begin{aligned} n_2 &= n_1\sqrt{\frac{h_2}{h_1}} = (610\text{ rpm})\sqrt{\frac{225\text{ ft}}{100\text{ ft}}} \\ &= \boxed{915\text{ rpm}} \end{aligned}$$

The answer is (B).

(d) Combine Eqs. 18.42 and 18.43.

$$\begin{aligned} P_2 &= P_1\left(\frac{h_2}{h_1}\right)^{1.5} = (250\text{ hp})\left(\frac{225\text{ ft}}{100\text{ ft}}\right)^{1.5} \\ &= \boxed{843.8\text{ hp}} \end{aligned}$$

The answer is (D).

(e) Combine Eqs. 18.41 and 18.42.

$$\begin{aligned} Q_2 &= Q_1\sqrt{\frac{h_2}{h_1}} = \left(25\ \frac{\text{ft}^3}{\text{sec}}\right)\sqrt{\frac{225\text{ ft}}{100\text{ ft}}} \\ &= \boxed{37.5\text{ ft}^3/\text{sec}} \end{aligned}$$

The answer is (B).

SI Solution

(a) The total effective head is due to the pressure head, velocity head, and tailwater head.

$$\begin{aligned} h_{\text{eff}} &= h_p + h_{\text{v}} - h_{z,(\text{tailwater})} \\ &= 30.8\ \text{m} + \frac{\left(3.6\ \frac{\text{m}}{\text{s}}\right)^2}{(2)\left(9.81\ \frac{\text{m}}{\text{s}^2}\right)} - (-1.75\ \text{m}) \\ &= \boxed{33.21\ \text{m}} \end{aligned}$$

The answer is (C).

(b) From Table 18.6, the theoretical hydraulic kilowatt is

$$\begin{aligned} P_{\text{th}} &= \frac{9.81 h_A Q(\text{SG})}{1000} \\ &= \frac{\left(9.81\ \frac{\text{m}}{\text{s}^2}\right)(33.21\ \text{m})\left(700\ \frac{\text{L}}{\text{s}}\right)(1)}{1000\ \frac{\text{W}}{\text{kW}}} \\ &= 228.1\ \text{kW} \end{aligned}$$

The overall turbine efficiency is

$$\begin{aligned} \eta &= \frac{P_{\text{brake}}}{P_{\text{th}}} = \frac{185\ \text{kW}}{228.1\ \text{kW}} \\ &= \boxed{0.811\quad(81.1\%)} \end{aligned}$$

The answer is (A).

(c) From Eq. 18.42 (the affinity laws),

$$\begin{aligned} n_2 &= n_1\sqrt{\frac{h_2}{h_1}} = (610\ \text{rpm})\sqrt{\frac{75\ \text{m}}{33.21\ \text{m}}} \\ &= \boxed{917\ \text{rpm}} \end{aligned}$$

The answer is (B).

(d) Combine Eqs. 18.42 and 18.43.

$$\begin{aligned} P_2 &= P_1\left(\frac{h_2}{h_1}\right)^{1.5} = (185\ \text{kW})\left(\frac{75\ \text{m}}{33.21\ \text{m}}\right)^{1.5} \\ &= \boxed{627.3\ \text{kW}} \end{aligned}$$

The answer is (D).

(e) Combine Eqs. 18.41 and 18.42.

$$\begin{aligned} Q_2 &= Q_1\sqrt{\frac{h_2}{h_1}} = \left(700\ \frac{\text{L}}{\text{s}}\right)\sqrt{\frac{75\ \text{m}}{33.21\ \text{m}}} \\ &= \boxed{1051.9\ \text{L/s}} \end{aligned}$$

The answer is (B).

19 Open Channel Flow

PRACTICE PROBLEMS

Rectangular Channels

1. A wooden flume ($n = 0.012$) with a rectangular cross section is 2 ft wide. The flume carries 3 ft^3/sec of water down a 1% slope. What is the depth of flow?

(A) 0.3 ft
(B) 0.4 ft
(C) 0.5 ft
(D) 0.6 ft

2. A rectangular open channel is to be constructed with smooth concrete on a slope of 0.08. The design flow rate is 17 m^3/s. The channel design must be optimum. What are the channel dimensions?

(A) 0.6 m deep; 1.2 m wide
(B) 0.8 m deep; 1.6 m wide
(C) 1.0 m deep; 2.0 m wide
(D) 1.2 m deep; 2.4 m wide

3. (*Time limit: one hour*) A dam's outfall structure has three rectangular inlets sized 1 ft high by 2 ft wide as shown. The orifice coefficient for each inlet is 0.7. The vertical riser joins with a 100 ft long square box culvert ($n = 0.013$) placed on a 0.05 slope. The box culvert drains into a tailwater basin whose water level remains constant at the top of the culvert opening. During steady-state operation, the water level in the vertical barrel of the structure is 4 ft above the top of the box culvert. The orifice coefficient of the box culvert entrance is $^2/_3$. What are the dimensions of the box culvert?

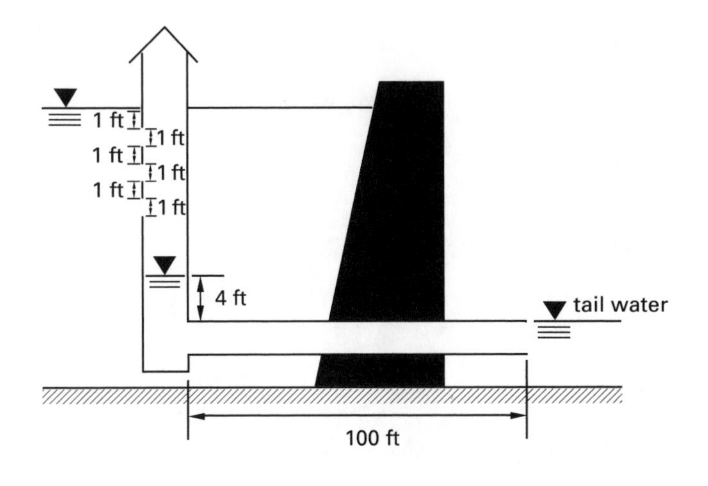

(A) 1.5 ft by 1.5 ft
(B) 2.0 ft by 2.0 ft
(C) 2.5 ft by 2.5 ft
(D) 4.0 ft by 4.0 ft

Circular Channels

4. A 24 in diameter pipe ($n = 0.013$) was installed 30 years ago on a 0.001 slope. Recent tests indicate that the full-flow capacity of the pipe is 6.0 ft^3/sec.

(a) What was the original full-flow capacity?

(A) 4.1 ft^3/sec
(B) 6.3 ft^3/sec
(C) 6.7 ft^3/sec
(D) 7.2 ft^3/sec

(b) What was the original full-flow velocity?

(A) 1.8 ft/sec
(B) 2.3 ft/sec
(C) 2.7 ft/sec
(D) 3.7 ft/sec

(c) What is the current full-flow velocity?

(A) 1.9 ft/sec
(B) 2.4 ft/sec
(C) 3.1 ft/sec
(D) 5.8 ft/sec

(d) What is the current Manning coefficient?

(A) 0.011
(B) 0.013
(C) 0.016
(D) 0.024

5. A circular sewer is to be installed on a 1% grade. Its Manning coefficient is $n = 0.013$. The maximum full-flow capacity is to be 3.5 ft^3/sec.

(a) What size pipe would you recommend?

(A) 8 in
(B) 12 in
(C) 16 in
(D) 18 in

(b) Assuming that the diameter chosen is 12 in, what is the full-flow capacity?

(A) 2.7 ft^3/sec
(B) 3.3 ft^3/sec
(C) 3.6 ft^3/sec
(D) 4.5 ft^3/sec

(c) What is the full-flow velocity?

(A) 2.9 ft/sec
(B) 3.3 ft/sec
(C) 4.1 ft/sec
(D) 4.6 ft/sec

(d) What is the depth of flow when the flow is 0.7 ft^3/sec?

(A) 4 in
(B) 7 in
(C) 9 in
(D) 12 in

(e) What minimum velocity will prevent solids from settling out in the sewer?

(A) 1 to 1.5 ft/sec
(B) 1.5 to 2.0 ft/sec
(C) 2.0 to 2.5 ft/sec
(D) 2.5 to 3.0 ft/sec

6. A 4 ft diameter concrete storm drain on a 0.02 slope carries water at a depth of 1.5 ft. Manning's roughness coefficient for a full storm drain is $n_{\text{full}} = 0.013$, but n varies with depth.

(a) What is the velocity of the water in the storm drain?

(A) 3 ft/sec
(B) 6 ft/sec
(C) 8 ft/sec
(D) 11 ft/sec

(b) What is the maximum velocity that can be achieved in the storm drain?

(A) 8 ft/sec
(B) 13 ft/sec
(C) 17 ft/sec
(D) 21 ft/sec

(c) What is the maximum capacity of the storm drain?

(A) 130 ft^3/sec
(B) 160 ft^3/sec
(C) 190 ft^3/sec
(D) 210 ft^3/sec

7. A circular storm sewer ($n = 0.012$) is being designed to carry a peak flow of 5 ft^3/sec. To allow for excess capacity, the depth at peak flow is to be 75% of the sewer diameter. The sewer is to be installed in a bed with a slope of 2%. What is the required sewer diameter?

(A) 12 in
(B) 16 in
(C) 20 in
(D) 24 in

8. (*Time limit: one hour*) Flow in a 6 ft diameter, newly formed concrete culvert is 150 ft^3/sec at a point where the depth of flow is 3 ft.

(a) What is the hydraulic radius?

(A) 1.1 ft
(B) 1.5 ft
(C) 2.3 ft
(D) 3.0 ft

(b) What is the slope of the pipe?

(A) 0.0043
(B) 0.0080
(C) 0.010
(D) 0.030

(c) What type of flow occurs?

(A) subcritical
(B) critical
(C) supercritical
(D) choked

(d) What is the hydraulic radius if the pipe flows full?

(A) 1.5 ft
(B) 3.0 ft
(C) 4.0 ft
(D) 6.0 ft

(e) What is the flow quantity if the culvert flows full?

(A) 100 ft^3/sec
(B) 120 ft^3/sec
(C) 200 ft^3/sec
(D) 300 ft^3/sec

(f) If the flow is 150 ft^3/sec, what is the critical depth?

(A) 1.1 ft
(B) 1.5 ft
(C) 3.3 ft
(D) 4.3 ft

(g) What is the critical slope?

(A) 0.002
(B) 0.003
(C) 0.004
(D) 0.05

(h) If the exit depth is less than critical, which culvert flow types are most probable?

(A) 1, 5
(B) 1, 3, 5
(C) 1, 2, 3
(D) 5 only

(i) What is the Froude number for a depth of flow of 3 ft?

(A) 0.86
(B) 0.99
(C) 1.22
(D) 2.84

(j) After the water exits the circular culvert, it flows at a velocity of 15 ft/sec and a depth of 2 ft in a rectangular channel. Assuming a hydraulic jump is possible, what will be the depth after the jump?

(A) 2.5 ft
(B) 3.0 ft
(C) 3.3 ft
(D) 4.4 ft

Trapezoidal Channels

9. The trapezoidal channel shown has a Manning coefficient of $n = 0.013$ and is laid at a slope of 0.002. The depth of flow is 2.0 ft. What is the flow rate?

(A) 120 ft^3/sec
(B) 150 ft^3/sec
(C) 180 ft^3/sec
(D) 210 ft^3/sec

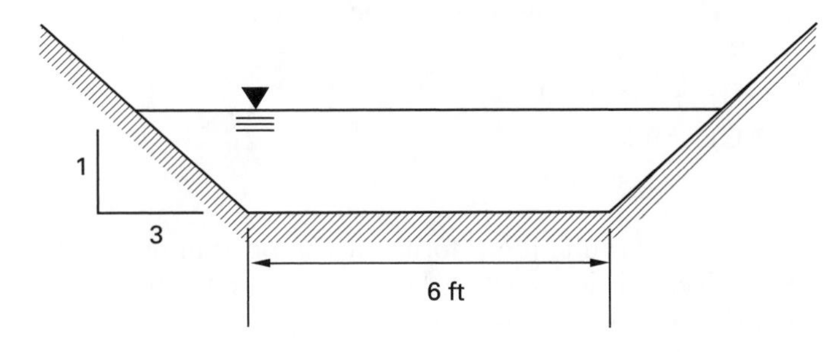

10. A trapezoidal open channel is to be constructed with smooth concrete on a slope of 0.08. The design flow rate is 17 m^3/s. The channel design must be optimum. What are most nearly the channel dimensions?

(A) depth 0.8 m; base 1.0 m
(B) depth 1.0 m; base 1.2 m
(C) depth 1.2 m; base 1.4 m
(D) depth 1.4 m; base 1.8 m

Weirs

11. A sharp-crested rectangular weir with two end contractions is 5 ft wide. The weir height is 6 ft. The head over the weir is 0.43 ft. What is the flow rate?

(A) 3.9 ft^3/sec
(B) 4.3 ft^3/sec
(C) 4.6 ft^3/sec
(D) 6.4 ft^3/sec

12. A weir is constructed with a trapezoidal opening. The base of the opening is 18 in wide, and the sides have a slope of 4:1 (vertical:horizontal). The depth of flow over the weir is 9 in. What is the rate of discharge?

(A) 1.8 ft^3/sec
(B) 3.3 ft^3/sec
(C) 5.1 ft^3/sec
(D) 8.9 ft^3/sec

13. (*Time limit: one hour*) A 17 ft by 20 ft rectangular tank is fed from the bottom and drains through a double-constricted rectangular weir into a trough. The weir opening is to be designed using the following parameters.

discharge rate:	2 MGD minimum
	4 MGD average
	8 MGD maximum
minimum tank freeboard:	4 in
minimum head over weir:	10 in
maximum weir elevation:	590 ft
tank bottom elevation:	580 ft

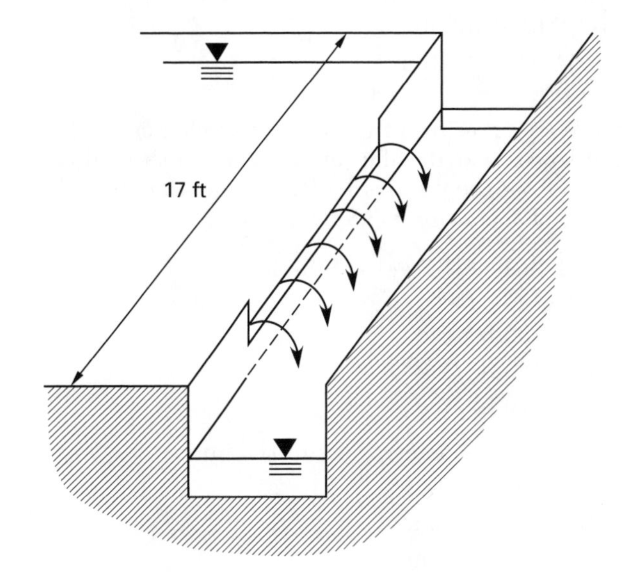

(a) What is the required weir opening length?

(A) 1.2 ft
(B) 1.4 ft
(C) 2.1 ft
(D) 2.7 ft

Water Resources

(b) What is the surface elevation in the tank at maximum flow?

(A) 581.9 ft
(B) 589.8 ft
(C) 592.6 ft
(D) 593.1 ft

(c) What is the surface elevation in the tank at minimum flow?

(A) 590.8 ft
(B) 591.2 ft
(C) 591.7 ft
(D) 592.0 ft

Hydraulic Jumps and Drops

14. Water flowing in a rectangular channel 5 ft wide experiences a hydraulic jump. The depth of flow upstream from the jump is 1 ft. The depth of flow downstream of the jump is 2.4 ft. What quantity is flowing?

(A) 39 ft^3/sec
(B) 45 ft^3/sec
(C) 52 ft^3/sec
(D) 57 ft^3/sec

15. A hydraulic jump forms at the toe of a spillway. The water surface levels are 0.2 ft and 6 ft above the apron before and after the jump, respectively. The velocity before the jump is 54.7 ft/sec. What is the energy loss in the jump?

(A) 41 ft-lbf/lbm
(B) 58 ft-lbf/lbm
(C) 65 ft-lbf/lbm
(D) 92 ft-lbf/lbm

16. Water flowing in a rectangular channel 6 ft wide experiences a hydraulic jump. The depth of flow upstream from the jump is 1.2 ft. The depth of flow downstream of the jump is 3.8 ft.

(a) What quantity is flowing?

(A) 75 ft^3/sec
(B) 120 ft^3/sec
(C) 180 ft^3/sec
(D) 210 ft^3/sec

(b) What is the energy loss in the jump?

(A) 1 ft-lbf/lbm
(B) 3 ft-lbf/lbm
(C) 5 ft-lbf/lbm
(D) 12 ft-lbf/lbm

Spillways

17. A spillway operates with 2 ft of head. The toe of the spillway is 40 ft below the crest of the spillway. The spillway discharge coefficient is 3.5.

(a) What is the discharge per foot of crest?

(A) 9.9 ft^3/sec
(B) 11 ft^3/sec
(C) 15 ft^3/sec
(D) 27 ft^3/sec

(b) What is the depth of flow at the toe?

(A) 0.19 ft
(B) 0.35 ft
(C) 0.78 ft
(D) 1.4 ft

18. 10,000 ft^3/sec of water flow over a 100 ft wide spillway and continue down a constant-width discharge chute on a 5% grade. The discharge chute has a Manning coefficient of $n = 0.012$.

(a) What is the normal depth of the water flowing down the chute?

(A) 1.9 ft
(B) 2.2 ft
(C) 2.7 ft
(D) 3.5 ft

(b) What is the critical depth of the water flowing down the chute?

(A) 3.8 ft
(B) 5.1 ft
(C) 6.8 ft
(D) 7.7 ft

(c) What type of flow occurs on the chute?

(A) tranquil
(B) subcritical
(C) rapid
(D) nonsteady

(d) If a hydraulic jump forms at the bottom of the 5% slope where it joins with a horizontal apron, what is the depth after the jump?

(A) 9.8 ft
(B) 12 ft
(C) 16 ft
(D) 21 ft

19. A spillway has a crest length of 60 ft. The stilling basin below the crest has a width of 60 ft and a level bottom. The spillway discharge coefficient is 3.7. The head loss in the chute is 20% of the difference in level between the reservoir surface and the stilling basin bottom.

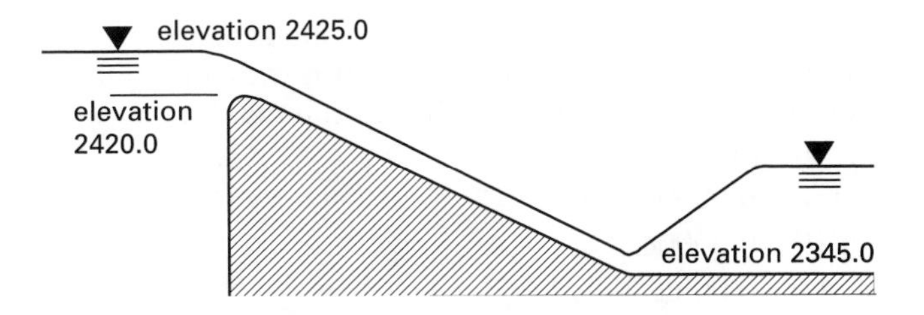

(a) What is the flow rate over the spillway?
- (A) 1800 ft^3/sec
- (B) 2100 ft^3/sec
- (C) 2500 ft^3/sec
- (D) 2900 ft^3/sec

(b) What is the depth of flow at the toe of the spillway?
- (A) 0.65 ft
- (B) 1.2 ft
- (C) 1.6 ft
- (D) 2.0 ft

(c) What tailwater depth is required to cause a hydraulic jump to form at the toe?
- (A) 4 ft
- (B) 7 ft
- (C) 10 ft
- (D) 13 ft

(d) What energy is lost in the hydraulic jump?
- (A) 30 ft-lbf/lbm
- (B) 50 ft-lbf/lbm
- (C) 75 ft-lbf/lbm
- (D) 100 ft-lbf/lbm

Parshall Flumes

20. A Parshall flume has a throat width of 6 ft. The upstream head measured from the throat floor is 18 in.

(a) What is the flow rate?
- (A) 41 ft^3/sec
- (B) 46 ft^3/sec
- (C) 54 ft^3/sec
- (D) 63 ft^3/sec

(b) Which of the following is not normally considered to be a feature of Parshall flumes?
- (A) They are self-cleansing.
- (B) They can be operated by unskilled personnel.
- (C) Head loss is small.
- (D) They can be placed in temporary installations.

Backwater Curves

21. A rectangular channel 8 ft wide carries a flow of 150 ft^3/sec. The channel slope is 0.0015, and the Manning coefficient is $n = 0.015$. A weir installed across the channel raises the depth at the weir to 6 ft.

(a) What is the normal depth of flow?
- (A) 2.9 ft
- (B) 3.3 ft
- (C) 3.6 ft
- (D) 4.5 ft

(b) enspace What is the distance between the weir and the point where the depth is 5 ft?
- (A) 610 ft
- (B) 840 ft
- (C) 1100 ft
- (D) 1700 ft

(c) What is the distance between the points where the depth is 4 ft and 5 ft?
- (A) 700 ft
- (B) 800 ft
- (C) 950 ft
- (D) 1100 ft

Culverts

22. A 42 in diameter concrete culvert is 250 ft long and laid at a slope of 0.006. The culvert entrance is flush and square-edged. The tailwater level at the outlet is just above the crown of the barrel, and the headwater is 5.0 ft above the crown of the culvert's inlet. What is the capacity?
- (A) 60 ft^3/sec
- (B) 80 ft^3/sec
- (C) 100 ft^3/sec
- (D) 120 ft^3/sec

23. (*Time limit: one hour*) A dam spills 2000 m^3/s of water over its crest as shown. The crest height is 30 m. The width of the dam is 150 m. The depth at point B is 0.6 m. The stilling basin, constructed of rough concrete, is rectangular in shape, 150 m in width, and has a slope of 0.002.

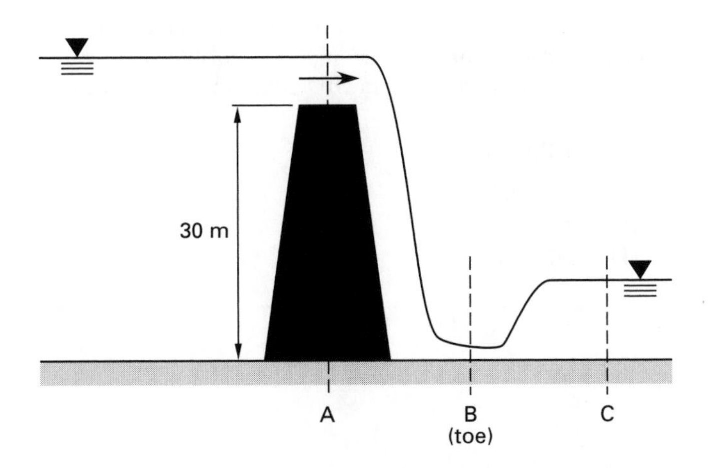

Water Resources

(a) What is the head over the crest?

(A) 0.8 m
(B) 1.2 m
(C) 2.2 m
(D) 3.3 m

(b) What is the total upstream specific energy head?

(A) 32.2 m
(B) 34.1 m
(C) 36.1 m
(D) 38.0 m

(c) What is the total specific energy head at point B?

(A) 25.8 m
(B) 32.2 m
(C) 34.1 m
(D) 43.3 m

(d) What is the friction head loss between points A and B?

(A) 2.2 m
(B) 4.2 m
(C) 8.3 m
(D) 9.8 m

(e) What is the normal downstream depth?

(A) 2.0 m
(B) 2.6 m
(C) 5.6 m
(D) 10.0 m

(f) What is the critical depth?

(A) 2.0 m
(B) 2.6 m
(C) 7.0 m
(D) 8.3 m

(g) At what point will a hydraulic jump occur?

(A) on the spillway slope
(B) at point B
(C) downstream of point B
(D) nowhere (a hydraulic jump will not occur)

(h) The stilling basin feeds into a natural channel composed of graded material from silt to cobble size. What is the maximum velocity in this channel to prevent erosion?

(A) 1.2 m/s
(B) 6.7 m/s
(C) 15.0 m/s
(D) 26.0 m/s

(i) If the natural channel has the same geometry as the stilling basin (i.e., rectangular with a width of 150 m), what slope is required to maintain the maximum velocity and prevent erosion?

(A) 0.00001
(B) 0.00004
(C) 0.0002
(D) 0.002

(j) Assuming that the slope in the natural channel remains at 0.002, what is the friction loss over a 1000 m reach?

(A) 0.1 m
(B) 0.6 m
(C) 1.5 m
(D) 2.0 m

SOLUTIONS

1.

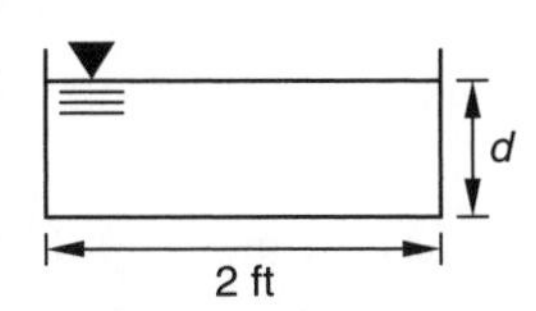

$$R = \frac{2d}{2+2d} = \frac{d}{1+d}$$

From Eq. 19.13,

$$Q = \left(\frac{1.49}{n}\right) AR^{2/3}\sqrt{S}$$

$$3\ \frac{\text{ft}^3}{\text{sec}} = \left(\frac{1.49}{0.012}\right)(2d)\left(\frac{d}{d+1}\right)^{2/3}\sqrt{0.01}$$

$$0.120805 = \left(\frac{d^{5/2}}{d+1}\right)^{2/3}$$

$$0.042 = \frac{d^{5/2}}{d+1}$$

By trial and error, $d = \boxed{0.314\text{ ft.}}$

The answer is (A).

2. From App. 19.A, $n = 0.011$. $d = w/2$ for an optimum channel.

$$A = wd = \frac{w^2}{2}$$

$$P = w + 2\left(\frac{w}{2}\right) = 2w$$

$$R = \frac{A}{P} = \frac{\frac{w^2}{2}}{2w} = \frac{w}{4}$$

(Note that the units conversion factor of 1.49 is not needed in problems with SI units.)

$$17\ \frac{\text{m}^3}{\text{s}} = \left(\frac{w^2}{2}\right)\left(\frac{1.0}{0.011}\right)\left(\frac{w}{4}\right)^{2/3}\sqrt{0.08}$$

$$d = \boxed{0.79\text{ m}}$$

$$w = \boxed{1.57\text{ m}}$$

The answer is (B).

3.

$$Q_o = C_d A_o \sqrt{2gh}$$

$$Q_1 = (0.7)(2\text{ ft}^2)\sqrt{(2)\left(32.2\ \frac{\text{ft}}{\text{sec}^2}\right)(1.5\text{ ft})}$$

$$= (11.235)\sqrt{1.5}\quad\text{[top inlet]}$$

$$Q_1 + Q_2 + Q_3 = (11.235)\left(\sqrt{1.5} + \sqrt{3.5} + \sqrt{5.5}\right)$$

$$= 61.13\ \text{ft}^3/\text{sec}$$

Culvert size (and hence R) is unknown. To get a trial value, initially disregard entrance loss and barrel friction.

$$H = 4\text{ ft} + (100\text{ ft})(0.05) = 9\text{ ft}$$

$$\text{v} = \sqrt{2gh} = \sqrt{(2)\left(32.2\ \frac{\text{ft}}{\text{sec}^2}\right)(9\text{ ft})}$$

$$= 24.07\ \text{ft/sec}$$

$$A = \frac{Q}{C_d\text{v}} = \frac{61.13\ \frac{\text{ft}^3}{\text{sec}}}{\left(\frac{2}{3}\right)\left(24.07\ \frac{\text{ft}}{\text{sec}}\right)} = 3.81\text{ ft}^2$$

$$\text{box size} = \sqrt{3.81\text{ ft}^2} = 1.95\text{ ft}$$

(Round to $2^1/_4$ ft^2 to account for friction.)

Due to the hydrostatic pressure, assume the culvert box flows full.

Use Eq. 19.100. $k_e = 0.50$ for a square-edged entrance.

$$R = \frac{(2.25\text{ ft})^2}{(4)(2.25\text{ ft})} = 0.5625\text{ ft}$$

$$\text{v} = \sqrt{\frac{9\text{ ft}}{\frac{1+0.5}{(2)\left(32.2\ \frac{\text{ft}}{\text{sec}^2}\right)} + \frac{(0.013)^2(100\text{ ft})}{(2.21)(0.5625\text{ ft})^{4/3}}}}$$

$$= 15.05\ \text{ft/sec}$$

$$\text{box size} = \sqrt{\frac{61.13\ \frac{\text{ft}^3}{\text{sec}}}{\left(\frac{2}{3}\right)\left(15.05\ \frac{\text{ft}}{\text{sec}}\right)}}$$

$$= \boxed{2.47\text{ ft}\quad\text{[2.5 ft or more]}}$$

The answer is (C).

4. From Eq. 19.13,

$$\text{(a)}\ Q_o = \left(\frac{1.49}{0.013}\right)\left(\frac{\pi}{4}\right)\left(\frac{24\text{ in}}{12\ \frac{\text{in}}{\text{ft}}}\right)^2\left(\frac{24\text{ in}}{\left(12\ \frac{\text{in}}{\text{ft}}\right)(4)}\right)^{2/3}$$

$$\times \sqrt{0.001}$$

$$= \boxed{7.173\ \text{ft}^3/\text{sec}}$$

The answer is (D).

(b)

$$\text{v}_o = \frac{Q_o}{A} = \frac{7.173\ \frac{\text{ft}^3}{\text{sec}}}{\left(\frac{\pi}{4}\right)(2\text{ ft})^2}$$

$$= \boxed{2.28\text{ ft/sec}}$$

The answer is (B).

(c)

$$\mathrm{v}_{\text{current}} = \frac{6.00\ \frac{\text{ft}^3}{\text{sec}}}{\left(\frac{\pi}{4}\right)(2\ \text{ft})^2} = \boxed{1.91\ \text{ft/sec}}$$

The answer is (A).

(d) Since Q is inversely proportional to n,

$$n_{\text{current}} = \left(\frac{Q_o}{Q_{\text{current}}}\right) n_o = \left(\frac{7.173\ \frac{\text{ft}^3}{\text{sec}}}{6\ \frac{\text{ft}^3}{\text{sec}}}\right)(0.013) = \boxed{0.01554}$$

The answer is (C).

5. (a) From Eq. 19.16,

$$D = (1.33)\left(\frac{\left(3.5\ \frac{\text{ft}^3}{\text{sec}}\right)(0.013)}{\sqrt{0.01}}\right)^{3/8} = 0.99\ \text{ft}$$

$$\boxed{\text{Use 12 in pipe.}}$$

The answer is (B).

(b) From Eq. 19.13,

$$Q = \left(\frac{1.49}{0.013}\right)\left(\frac{\pi}{4}\right)(1\ \text{ft})^2\left(\frac{1\ \text{ft}}{4}\right)^{2/3}\sqrt{0.01} = \boxed{3.57\ \text{ft}^3/\text{sec}}$$

The answer is (C).

(c)

$$\mathrm{v} = \frac{Q}{A} = \frac{3.57\ \frac{\text{ft}^3}{\text{sec}}}{\left(\frac{\pi}{4}\right)(1\ \text{ft})^2} = \boxed{4.55\ \text{ft/sec}}$$

The answer is (D).

(d)

$$\frac{Q}{Q_{\text{full}}} = \frac{0.7\ \frac{\text{ft}^3}{\text{sec}}}{3.57\ \frac{\text{ft}^3}{\text{sec}}} \approx 0.2$$

From App. 19.C, letting n vary with depth,

$$\frac{d}{D} \approx 0.35$$

$$d = (0.35)(12\ \text{in}) = \boxed{4.2\ \text{in}}$$

The answer is (A).

(e) To be self-cleansing, $\mathrm{v} = \boxed{2\ \text{to}\ 2.5\ \text{ft/sec.}}$

The answer is (C).

6. *Full flowing:*

$$R = \frac{D}{4} = \frac{4\ \text{ft}}{4} = 1\ \text{ft}$$

From Eqs. 19.12 and 19.13,

$$\mathrm{v} = \left(\frac{1.49}{0.013}\right)(1\ \text{ft})^{2/3}\sqrt{0.02} = 16.21\ \text{ft/sec}$$

$$Q = A\mathrm{v} = \left(\frac{\pi}{4}\right)(4\ \text{ft})^2\left(16.21\ \frac{\text{ft}}{\text{sec}}\right) = 203.7\ \text{ft}^3/\text{sec}$$

Actual:

$$\frac{d}{D} = \frac{1.5\ \text{ft}}{4\ \text{ft}} = 0.375$$

From App. 19.C, assuming that n varies with depth,

$$\frac{\mathrm{v}}{\mathrm{v}_{\text{full}}} = 0.68$$

(a) $\mathrm{v} = (0.68)\left(16.21\ \frac{\text{ft}}{\text{sec}}\right) = \boxed{11.0\ \text{ft/sec.}}$

The answer is (D).

(b) From App. 19.C, the maximum value of $\mathrm{v}/\mathrm{v}_{\text{full}}$ is 1.04 (at $d/D \approx 0.9$).

$$\mathrm{v}_{\max} = (1.04)\left(16.21\ \frac{\text{ft}}{\text{sec}}\right) = \boxed{16.86\ \text{ft/sec}}$$

The answer is (C).

(c) Similarly, $Q_{\max}/Q_{\text{full}} = 1.02$ (at $d/D \approx 0.96$).

$$Q_{\max} = (1.02)\left(203.7\ \frac{\text{ft}^3}{\text{sec}}\right) = \boxed{207.8\ \text{ft}^3/\text{sec}}$$

The answer is (D).

7. From App. 16.A,

$$A = 0.6318D^2$$

$$R = 0.3017D$$

From Eq. 19.13,

$$Q = \left(\frac{1.49}{n}\right) AR^{2/3}\sqrt{S}$$

$$5\ \frac{\text{ft}^3}{\text{sec}} = \left(\frac{1.49}{0.012}\right)(0.6318D^2)(0.3017D)^{2/3}\sqrt{0.02}$$

$$1.0 = D^{8/3}$$

$$D = \boxed{1.0 \text{ ft} \quad (12 \text{ in})}$$

The answer is (A).

(Appendix 19.C can also be used to solve this problem.)

8. (a) For a circular channel flowing half-full, the hydraulic radius is

$$R = \frac{D}{4} = \frac{6 \text{ ft}}{4} = \boxed{1.5 \text{ ft}}$$

The answer is (B).

(b) From App. 19.A, $n = 0.012$. From Eq. 19.13,

$$Q = \left(\frac{1.49}{n}\right) AR^{2/3}\sqrt{S}$$

Rearranging and recognizing that the culvert flows half full,

$$S = \left(\frac{Qn}{1.49AR^{2/3}}\right)^2$$

$$= \left(\frac{\left(150\ \frac{\text{ft}^3}{\text{sec}}\right)(0.012)}{(1.49)(0.5)\left(\frac{\pi(6 \text{ ft})^2}{4}\right)(1.5 \text{ ft})^{2/3}}\right)^2$$

$$= \boxed{0.0043}$$

The answer is (A).

(c) Equation 19.77 could be used, but App. 19.D is more convenient for use with circular channels. Entering with $Q = 150$ ft^3/sec and $D = 6$ ft, $d_c = 3.3$ ft. The actual flow depth is less than the critical depth; therefore, the flow is supercritical.

The answer is (C).

(d) For a circular channel flowing full, the hydraulic radius is the same as for half-full flow.

$$R = \frac{D}{4} = \frac{6 \text{ ft}}{4} = \boxed{1.5 \text{ ft}}$$

The answer is (A).

(e) From Eq. 19.13,

$$Q = \left(\frac{1.49}{n}\right) AR^{2/3}\sqrt{S}$$

$$= \left(\frac{1.49}{0.012}\right)\left(\frac{\pi(6 \text{ ft})^2}{4}\right)(1.5 \text{ ft})^{2/3}\sqrt{0.0043}$$

$$= \boxed{300 \text{ ft}^3/\text{sec}}$$

The answer is (D).

(f) This part was solved in part (c): $d_c = \boxed{3.3 \text{ ft.}}$

The answer is (C).

(g) From Eq. 19.13, use $d = d_c$.

$$Q = \left(\frac{1.49}{n}\right) AR^{2/3}\sqrt{S}$$

R is calculated using the formula in Table 19.2.

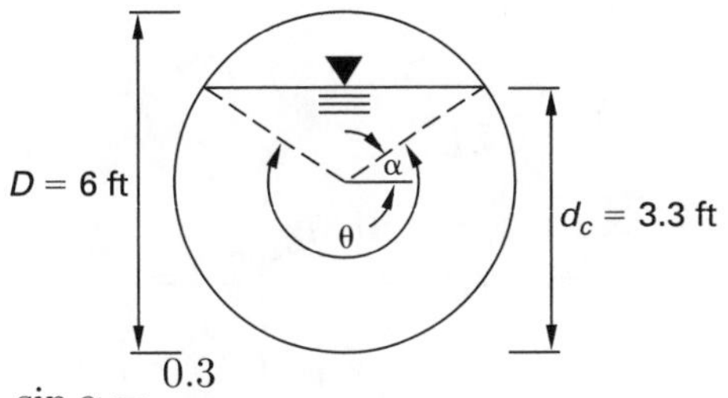

D = 6 ft
α
θ
d_c = 3.3 ft

$$\sin\alpha = \frac{0.3}{3}$$

$$\alpha = 5.74^\circ$$

$$\theta = 180^\circ + 2\alpha = 180^\circ + (2)(5.74^\circ)$$

$$= 191.48^\circ$$

Convert to radians.

$$\theta = (191.48^\circ)\left(\frac{2\pi \text{ rad}}{360^\circ}\right) = 3.34 \text{ rad}$$

$$R = \tfrac{1}{4}\left(1 - \frac{\sin\theta}{\theta}\right)D$$

$$= \left(\tfrac{1}{4}\right)\left(1 - \frac{\sin(3.34 \text{ rad})}{3.34 \text{ rad}}\right)(6 \text{ ft}) = 1.59 \text{ ft}$$

$$A = \tfrac{1}{8}(\theta - \sin\theta)D^2$$

$$= \left(\tfrac{1}{8}\right)(3.34 \text{ rad} - \sin(3.34 \text{ rad}))(6 \text{ ft})^2 = 15.9 \text{ ft}^2$$

$$S = \frac{Qn}{(1.49AR^{2/3})}^2$$

$$= \left(\frac{\left(150\ \frac{\text{ft}^3}{\text{sec}}\right)(0.012)}{(1.49)(15.9 \text{ ft}^2)(1.59 \text{ ft})^{2/3}}\right)^2$$

$$= \boxed{0.003}$$

The answer is (B).

(h) The pipe flows partially full, and the exit is not submerged. This eliminates types 4 and 6 and leaves types 1, 2, 3, and 5. Type 2 is eliminated because the flow is not at critical depth at the outlet. Type 3 is tranquil flow throughout, so it is also eliminated. Distinguishing between types 1 and 5 cannot be done without further information about the entrance conditions.

The answer is (A).

(i)

$$\begin{aligned} \text{v} &= \left(\frac{1.49}{n}\right) R^{2/3}\sqrt{S} \\ &= \left(\frac{1.49}{0.012}\right)(1.5\ \text{ft})^{2/3}\sqrt{0.0043} \\ &= 10.67\ \text{ft/sec} \end{aligned}$$

$$\begin{aligned} D_h &= \frac{\pi D}{8} = \frac{\pi(6\ \text{ft})}{8} \\ &= 2.36\ \text{ft} \end{aligned}$$

From Eq. 19.78,

$$\begin{aligned} \text{Fr} &= \frac{\text{v}}{\sqrt{gD_h}} \\ &= \frac{10.67\ \frac{\text{ft}}{\text{sec}}}{\sqrt{\left(32.2\ \frac{\text{ft}}{\text{sec}^2}\right)(2.36\ \text{ft})}} \\ &= \boxed{1.22} \end{aligned}$$

Since Fr > 1, the flow is supercritical.

The answer is (C).

(j) From Eq. 19.91,

$$\begin{aligned} d_2 &= -\tfrac{1}{2}d_1 + \sqrt{\frac{2\text{v}_1^2 d_1}{g} + \frac{d_1^2}{4}} \\ &= -\left(\tfrac{1}{2}\right)(2\ \text{ft}) + \sqrt{\frac{(2)\left(15\ \frac{\text{ft}}{\text{sec}}\right)^2(2\ \text{ft})}{32.2\ \frac{\text{ft}}{\text{sec}^2}} + \frac{(2\ \text{ft})^2}{4}} \\ &= \boxed{4.4\ \text{ft}} \end{aligned}$$

The answer is (D).

9.

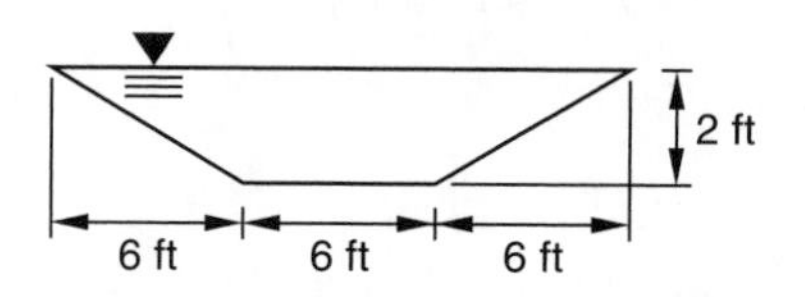

$$A = (6\ \text{ft})(2\ \text{ft}) + \frac{(2)(6\ \text{ft})(2\ \text{ft})}{2} = 24\ \text{ft}^2$$

$$P = 6\ \text{ft} + 2\sqrt{(6\ \text{ft})^2 + (2\ \text{ft})^2} = 18.65\ \text{ft}$$

$$R = \frac{24\ \text{ft}^2}{18.65\ \text{ft}} = 1.287\ \text{ft}$$

$$\begin{aligned} Q &= (24\ \text{ft}^2)\left(\frac{1.49}{0.013}\right)(1.287\ \text{ft})^{2/3}\sqrt{0.002} \\ &= \boxed{145.6\ \text{ft}^3/\text{sec}} \end{aligned}$$

The answer is (B).

Alternate solution:

Solve using App. 19.E.

$$m = 3.0\ (18.4^\circ)\ \text{column}$$

$$\frac{D}{b} = \frac{2.0\ \text{ft}}{6.0\ \text{ft}} = 0.33$$

Interpolating, $K \approx 6.69$.

$$\begin{aligned} Q &= K\left(\frac{1}{n}\right)D^{8/3}\sqrt{S} \\ &= (6.69)\left(\frac{1}{0.013}\right)(2.0\ \text{ft})^{8/3}\sqrt{0.002} \\ &= 146.1\ \text{ft}^3/\text{sec} \end{aligned}$$

The answer is (B).

10. The optimum trapezoidal channel has side slopes of 60° and a depth that is twice the hydraulic radius.

From App. 19.A, $n = 0.011$ for smooth concrete.

Rearranging Eq. 19.13,

$$\begin{aligned} AR^{2/3} &= \frac{Qn}{\sqrt{S}} \\ &= \frac{\left(17\ \frac{\text{m}^3}{\text{s}}\right)(0.011)}{\sqrt{0.08}} = 0.661 \end{aligned}$$

From Table 19.2,

$$A = \left(b + \frac{d}{\tan\theta}\right)d$$

$$R = \frac{bd\sin\theta + d^2\cos\theta}{b\sin\theta + 2d}$$

Since $R = d/2$ and $b = 2d/\sqrt{3}$,

$$AR^{2/3} = \left(\frac{2d}{\sqrt{3}} + \frac{d}{\tan\theta}\right)d\left(\frac{d}{2}\right)^{2/3} = 0.661$$

Simplifying, $d^{8/3} = 0.606$ with $\theta = 60°$, the optimal channel has a depth, d, of 0.828 m and a base, b, of 0.957 m.

The answer is (A).

Alternate solution:

Use App. 19.F.

"Optimum" means $\alpha = 60°$ and $D/b = \sqrt{3}/2 = 0.866$ [see p. 19.9].

Double interpolate $K' \approx 1.12$.

This is metric, so from footnote (c),

$$K' = \frac{1.12}{1.486} \approx 0.754$$

$$Q = K'\left(\frac{1}{n}\right) b^{8/3}\sqrt{S}$$

$$17\ \frac{\text{m}^2}{\text{s}} = (0.754)\left(\frac{1}{0.011}\right) b^{8/3}\sqrt{0.8}$$

$$b = 0.877\ \text{m}$$

$$d = \frac{\sqrt{3}}{2}b = \left(\frac{\sqrt{3}}{2}\right)(0.877\ \text{m}) = 0.76\ \text{m}$$

The answer is (A).

11. From Eq. 19.52,

$$b = 5\ \text{ft} - (0.1)(2)(0.43\ \text{ft}) = 4.914\ \text{ft}$$

From Eq. 19.50,

$$C_1 = \left(0.6035 + (0.0813)\left(\frac{0.43\ \text{ft}}{6\ \text{ft}}\right) + \frac{0.000295}{6}\right) \times \left(1 + \frac{0.00361}{0.43\ \text{ft}}\right)^{3/2}$$
$$= (0.60938)(1.0126) = 0.617$$

From Eq. 19.49,

$$Q = \left(\tfrac{2}{3}\right)(0.617)(4.914\ \text{ft})\sqrt{(2)\left(32.2\ \frac{\text{ft}}{\text{sec}^2}\right)}(0.43\ \text{ft})^{3/2}$$
$$= \boxed{4.574\ \text{ft}^3/\text{sec}}$$

The answer is (C).

12. This is a Cipoletti weir. From Eq. 19.57,

$$Q = (3.367)\left(\frac{18\ \text{in}}{12\ \frac{\text{in}}{\text{ft}}}\right)\left(\frac{9\ \text{in}}{12\ \frac{\text{in}}{\text{ft}}}\right)^{3/2} = \boxed{3.28\ \text{ft}^3/\text{sec}}$$

The answer is (B).

13.

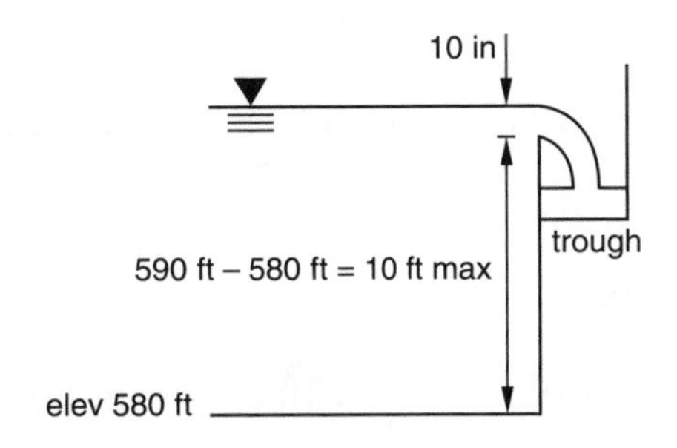

(a) Use Eq. 19.50.

$$C_1 \approx 0.602 + (0.083)\left(\frac{10\ \text{in}}{(590\ \text{ft} - 580\ \text{ft})\left(12\ \frac{\text{in}}{\text{ft}}\right)}\right)$$
$$= 0.609$$

Use Eq. 19.49 with $C_1 = 0.609$.

$$Q_{\text{MGD}} = \left(0.6463\ \frac{\text{MGD-sec}}{\text{ft}^3}\right) Q_{\text{ft}^3/\text{sec}}$$
$$= \left(0.6463\ \frac{\text{MGD-sec}}{\text{ft}^3}\right)\left(\tfrac{2}{3}\right)(0.609)bH^{3/2} \times \sqrt{(2)\left(32.2\ \frac{\text{ft}}{\text{sec}^2}\right)}$$
$$= 2.11bH^{3/2}$$

To satisfy the minimum head requirement, the minimum flow must be used. The minimum head requirement will then be automatically satisfied at maximum flow.

At minimum flow, $Q = 2$ MGD.

$$H = \frac{10\ \text{in}}{12\ \frac{\text{in}}{\text{ft}}} = 0.833\ \text{ft}$$

$$b_{\text{effective}} = \frac{2\ \text{MGD}}{(2.11)(0.833\ \text{ft})^{3/2}} = 1.25\ \text{ft}$$

From Eq. 19.52,

$$b_{\text{actual}} = 1.25\ \text{ft} + (0.1)(2)(0.833\ \text{ft}) = \boxed{1.42\ \text{ft}}$$

The answer is (B).

(b) At maximum flow,

$$8\ \text{MGD} = (2.11)(1.25)H^{3/2}$$
$$H \approx 2.1\ \text{ft}$$

Recalculate C_1 from Eq. 19.50.

$$C_1 \approx 0.602 + (0.083)\left(\frac{2.1 \text{ ft}}{590 \text{ ft} - 580 \text{ ft}}\right) = 0.619$$

From Eq. 19.49,

$$Q = \left(0.6463 \frac{\text{MGD-sec}}{\text{ft}^3}\right)\left(\tfrac{2}{3}\right)(0.619)bH^{3/2} \times \sqrt{(2)\left(32.2 \frac{\text{ft}}{\text{sec}^2}\right)} = 2.14bH^{3/2}$$

Use Eq. 19.52 to calculate $b_{\text{effective}}$. Continue iterating until H stabilizes.

$$b_{\text{effective}} = 1.42 \text{ ft} - (0.1)(2)(2.1 \text{ ft}) = 1.0 \text{ ft}$$
$$8 \text{ MGD} = (2.14)(1)H^{3/2}$$
$$H = 2.41 \text{ ft}$$
$$b_{\text{effective}} = 1.42 \text{ ft} - (0.1)(2)(2.41 \text{ ft}) = 0.94 \text{ ft}$$
$$8 \text{ MGD} = (2.14)(0.94)H^{3/2}$$
$$H = 2.51 \text{ ft}$$

$$b_{\text{effective}} = 1.42 \text{ ft} - (0.1)(2)(2.51 \text{ ft}) = 0.918 \text{ ft}$$
$$8 \text{ MGD} = (2.14)(0.918)H^{3/2}$$
$$H = 2.55 \text{ ft}$$
$$b_{\text{effective}} = 1.42 \text{ ft} - (0.1)(2)(2.55 \text{ ft}) = 0.920 \text{ ft}$$
$$8 \text{ MGD} = (2.14)(0.920)H^{3/2}$$
$$H = 2.55 \text{ ft} \quad \text{[acceptable]}$$

Use a freeboard elevation of

$$580 \text{ ft} + 10 \text{ ft} + 2.55 \text{ ft} + \frac{4 \text{ in}}{12 \frac{\text{in}}{\text{ft}}} = 592.88 \text{ ft}$$

The elevation at maximum flow is

$$592.88 \text{ ft} - \text{freeboard height} = 592.88 \text{ ft} - \frac{4 \text{ in}}{12 \frac{\text{in}}{\text{ft}}} = \boxed{592.55 \text{ ft} \quad (592.6 \text{ ft})}$$

The answer is (C).

(c) The channel elevation is

$$580 \text{ ft} + 10 \text{ ft} = 590 \text{ ft}$$

The elevation at minimum flow is

$$590 \text{ ft} + 0.833 \text{ ft} = \boxed{590.833 \text{ ft}}$$

The answer is (A).

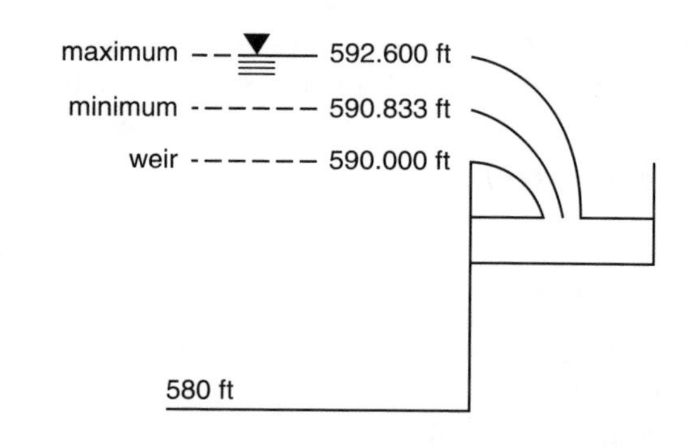

14. From Eq. 19.93,

$$v_1 = \sqrt{\left(\frac{32.2 \frac{\text{ft}}{\text{sec}^2}}{2}\right)\left(\frac{2.4 \text{ ft}}{1 \text{ ft}}\right)(1 \text{ ft} + 2.4 \text{ ft})} = 11.46 \text{ ft/sec}$$

$$Q = A\text{v} = (5 \text{ ft})(1 \text{ ft})\left(11.46 \frac{\text{ft}}{\text{sec}}\right) = \boxed{57.30 \text{ ft}^3/\text{sec}}$$

The answer is (D).

15. From Eq. 19.8,

$$(0.2 \text{ ft})(\text{width})\left(54.7 \frac{\text{ft}}{\text{sec}}\right) = (6 \text{ ft})(\text{width})\text{v}_2$$
$$\text{v}_2 = 1.82 \text{ ft/sec}$$

From Eq. 19.94,

$$\Delta E = 0.2 \text{ ft} + \frac{\left(54.7 \frac{\text{ft}}{\text{sec}}\right)^2}{(2)\left(32.2 \frac{\text{ft}}{\text{sec}^2}\right)} - \left(6 \text{ ft} + \frac{\left(1.82 \frac{\text{ft}}{\text{sec}}\right)^2}{(2)\left(32.2 \frac{\text{ft}}{\text{sec}^2}\right)}\right) = \boxed{40.61 \text{ ft (ft-lbf/lbm)}}$$

The answer is (A).

16. (a) From Eq. 19.93,

$$v_1 = \sqrt{\left(\frac{\left(32.2 \frac{\text{ft}}{\text{sec}^2}\right)(3.8 \text{ ft})}{(2)(1.2 \text{ ft})}\right)(1.2 \text{ ft} + 3.8 \text{ ft})} = 15.97 \text{ ft/sec}$$

$$Q = A\text{v} = (1.2 \text{ ft})(6 \text{ ft})\left(15.97 \frac{\text{ft}}{\text{sec}}\right) = \boxed{115 \text{ ft}^3/\text{sec}}$$

The answer is (B).

(b) $$\mathrm{v}_2 = \frac{Q}{A} = \frac{115\ \frac{\text{ft}^3}{\text{sec}}}{(3.8\ \text{ft})(6\ \text{ft})} = 5.04\ \text{ft/sec}$$

From Eq. 19.94,

$$\Delta E = \left(1.2\ \text{ft} + \frac{\left(15.97\ \frac{\text{ft}}{\text{sec}}\right)^2}{(2)\left(32.2\ \frac{\text{ft}}{\text{sec}^2}\right)}\right) - \left(3.8\ \text{ft} + \frac{\left(5.04\ \frac{\text{ft}}{\text{sec}}\right)^2}{(2)\left(32.2\ \frac{\text{ft}}{\text{sec}^2}\right)}\right)$$

$$= \boxed{0.97\ \text{ft}\quad(\text{ft-lbf/lbm})}$$

The answer is (A).

17. Disregard the velocity of approach. Use $C_s = 3.5$ in Eq. 19.60.

(a) $$Q = (3.5)(1\ \text{ft})(2\ \text{ft})^{3/2} = \boxed{9.9\ \text{ft}^3/\text{sec per ft of width}}$$

The answer is (A).

(b) The upstream energy is

$$E_1 = d = 2\ \text{ft} \quad [\mathrm{v}_1 \approx 0]$$

At the toe, from Eq. 19.72,

$$E_2 = E_1 + z_1 - z_2 = 2\ \text{ft} + 40\ \text{ft} - 0 = 42\ \text{ft}$$

$$E_2 = d_2 + \frac{\mathrm{v}_2^2}{2g} = d_2 + \frac{Q^2}{2gA_2^2}$$

$$d_2 + \frac{\left(9.9\ \frac{\text{ft}^3}{\text{sec}}\right)^2}{(2)\left(32.2\ \frac{\text{ft}}{\text{sec}^2}\right)(1\ \text{ft}^2)d_2^2} = 42\ \text{ft}$$

By trial and error, $d_2 = \boxed{0.19\ \text{ft.}}$

The answer is (A).

18.

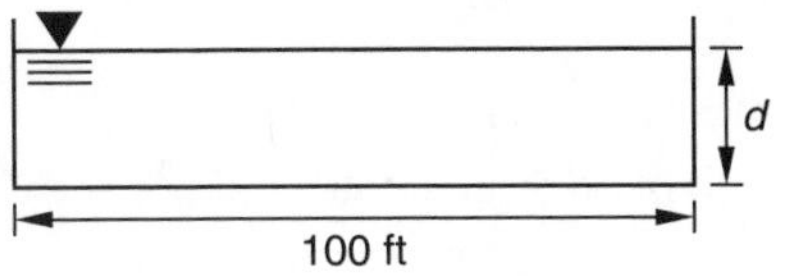

[figure: rectangular channel, width 100 ft, depth d]

(a) $$R = \frac{(100\ \text{ft})d}{100\ \text{ft} + 2d} = \frac{50d}{50+d}$$

From Eq. 19.13,

$$10{,}000\ \frac{\text{ft}^3}{\text{sec}} = \left(\frac{1.49}{0.012}\right)(100d)\left(\frac{50d}{50+d}\right)^{2/3}\sqrt{0.05}$$

$$0.26538 = \left(\frac{d^{5/2}}{50+d}\right)^{2/3}$$

$$0.1367 = \frac{d^{5/2}}{50+d}$$

By trial and error, $d = \boxed{2.2\ \text{ft.}}$

The answer is (B).

(b) From Eq. 19.74,

$$d_c = \sqrt[3]{\frac{\left(10{,}000\ \frac{\text{ft}^3}{\text{sec}}\right)^2}{\left(32.2\ \frac{\text{ft}}{\text{sec}^2}\right)(100\ \text{ft})^2}} = \boxed{6.77\ \text{ft}}$$

The answer is (C).

(c) Since $d < d_c$, $\boxed{\text{flow is rapid (shooting).}}$

The answer is (C).

(d) $$\mathrm{v} = \frac{Q}{A} = \frac{10{,}000\ \frac{\text{ft}^3}{\text{sec}}}{(100\ \text{ft})(2.2\ \text{ft})} = 45.45\ \text{ft/sec}$$

From Eq. 19.91,

$$d_2 = -\left(\tfrac{1}{2}\right)(2.2\ \text{ft}) + \sqrt{\frac{(2)\left(45.45\ \frac{\text{ft}}{\text{sec}}\right)^2(2.2\ \text{ft})}{32.2\ \frac{\text{ft}}{\text{sec}^2}} + \frac{(2.2\ \text{ft})^2}{4}}$$

$$= \boxed{15.74\ \text{ft}}$$

The answer is (C).

19. (a) From Eq. 19.60 using $C_s = 3.7$,

$$H = 2425\ \text{ft} - 2420\ \text{ft} = 5\ \text{ft}$$

$$Q = (3.7)(60\ \text{ft})(5\ \text{ft})^{3/2} = \boxed{2482\ \text{ft}^3/\text{sec}}$$

The answer is (C).

(b) Disregarding the velocity of approach, the initial energy is

$$E_1 = (\text{elev})_1 = 2425 \text{ ft}$$

At the toe (point 2),

$$E_1 = E_2 + h_f$$

$$2425 = (\text{elev})_2 + d_2 + \frac{Q^2}{2gw^2d_2^2} + (0.20)(2425 \text{ ft} - 2345 \text{ ft})$$

$$(1 - 0.20) \times (2425 \text{ ft} - 2345 \text{ ft}) = d_2 + \frac{\left(2482 \ \frac{\text{ft}^3}{\text{sec}}\right)^2}{(2)\left(32.2 \ \frac{\text{ft}}{\text{sec}^2}\right)(60 \text{ ft})^2 d_2^2}$$

$$64 \text{ ft} = d_2 + \frac{26.57}{d_2^2}$$

By trial and error, $d_2 = \boxed{0.647 \text{ ft.}}$

The answer is (A).

(c) $d_{\text{toe}} = d_1$ is implicitly a conjugate depth.

If there was a hydraulic jump, it would be between the conjugate depths, d_1 to d_2.

$$d_1 = 0.647 \text{ ft}$$

$$\text{v}_1 = \frac{Q}{A_1} = \frac{Q}{d_1 w} = \frac{2482 \ \frac{\text{ft}^3}{\text{sec}}}{(0.647 \text{ ft})(60 \text{ ft})} = 63.94 \text{ ft/sec}$$

From Eq. 19.91,

$$d_2 = -\tfrac{1}{2}d_1 + \sqrt{\frac{2\text{v}_1^2 d_1}{g} + \frac{d_1^2}{4}}$$
$$= \left(-\tfrac{1}{2}\right)(0.647 \text{ ft}) + \sqrt{\frac{(2)\left(63.9 \ \frac{\text{ft}}{\text{sec}}\right)^2 (0.647 \text{ ft})}{32.2 \ \frac{\text{ft}}{\text{sec}^2}} + \frac{(0.647 \text{ ft})^2}{4}}$$
$$= 12.49 \text{ ft}$$

The answer is (D).

(d) $$\text{v}_2 = \frac{Q}{A_2} = \frac{Q}{d_2 w} = \frac{2482 \ \frac{\text{ft}^3}{\text{sec}}}{(12.49 \text{ ft})(60 \text{ ft})} = 3.31 \text{ ft/sec}$$

Use d_2 and v_2 in Eq. 19.94.

$$\Delta E = \left(d_1 + \frac{\text{v}_1^2}{2g}\right) - \left(d_2 + \frac{\text{v}_2^2}{2g}\right)$$
$$= \left(0.647 \text{ ft} + \frac{\left(63.9 \ \frac{\text{ft}}{\text{sec}}\right)^2}{(2)\left(32.2 \ \frac{\text{ft}}{\text{sec}^2}\right)}\right) - \left(12.49 \text{ ft} + \frac{\left(3.31 \ \frac{\text{ft}}{\text{sec}}\right)^2}{(2)\left(32.2 \ \frac{\text{ft}}{\text{sec}^2}\right)}\right)$$
$$= \boxed{51.4 \text{ ft} \quad (\text{ft-lbf/lbm})}$$

The answer is (B).

20. (a) From Eq. 19.64,

$$n = (1.522)(6 \text{ ft})^{0.026} = 1.595$$

From Eq. 19.63,

$$Q = (4)(6 \text{ ft})\left(\frac{18 \text{ in}}{12 \ \frac{\text{in}}{\text{ft}}}\right)^{1.595} = \boxed{45.8 \text{ ft}^3/\text{sec}}$$

The answer is (B).

(b) The flume is self-cleansing, the head loss is small, and operation by unskilled personnel is possible. Parshall flumes are usually permanent structures.

The answer is (D).

21. (a) The normal depth is d.

$$R = \frac{8d}{8 + 2d}$$

From Eq. 19.13,

$$150 \ \frac{\text{ft}^3}{\text{sec}} = (8d)\left(\frac{1.49}{0.015}\right)\left(\frac{8d}{8 + 2d}\right)^{2/3}\sqrt{0.0015}$$

$$4.875 = d\left(\frac{8d}{8 + 2d}\right)^{2/3}$$

By trial and error, $d = \boxed{3.28 \text{ ft.}}$

The answer is (B).

(b) To get the backwater curve, choose depths and compute distances to those depths.

Preliminary parameters:

d	$A=8d$	$v=\frac{Q}{A}$	$\frac{v^2}{2g}$	$E=d+\frac{v^2}{2g}$	$P=2d+8$	R
6	48	3.12	0.151	6.151	20	2.40
5	40	3.75	0.218	5.218	18	2.22
4	32	4.69	0.342	4.342	16	2.00
3.28	26.24	5.72	0.502	3.782	14.56	1.80

The average velocities and average hydraulic radii are

d	v	v_{ave}	R	R_{ave}
6.00	3.12		2.40	
		3.435		2.31
5.00	3.75		2.22	
		4.22		2.11
4.00	4.69		2.00	
		5.21		1.90
3.28	5.72		1.80	

From Eq. 19.87, the average energy gradient is

$$S_{ave} = \left(\frac{n v_{ave}}{(1.486)(R_{ave})^{2/3}}\right)^2$$

d	S_{ave}	ΔE	$S_o - S_{ave}$
6.00			
	0.000394	0.933	0.001106
5.00			
	0.000671	0.876	0.000829
4.00			
	0.001175	0.552	0.000325
3.28			

From Eq. 19.89,

$$L_{6\text{-}5}: \quad \frac{\Delta E}{S_o - S_{ave}} = \frac{0.933 \text{ ft}}{0.001106} = \boxed{844 \text{ ft}}$$

The answer is (B).

(c) $$L_{5\text{-}4}: \quad \frac{0.876 \text{ ft}}{0.000829} = \boxed{1057 \text{ ft}}$$

The answer is (D).

22. For concrete, $n = 0.012$.

The pipe diameter is

$$D = \frac{42 \text{ in}}{12 \frac{\text{in}}{\text{ft}}} = 3.5 \text{ ft}$$

The hydraulic radius is

$$R = \frac{D}{4} = \frac{3.5 \text{ ft}}{4} = 0.875 \text{ ft}$$

From a table of orifice coefficients, Table 17.5, $C_d = 0.82$.

$$h_1 = 5 \text{ ft} + 3.5 \text{ ft} + (250 \text{ ft})(0.006) = 10 \text{ ft}$$

Determine the culvert flow type.

$$\frac{h_1 - z}{D} = \frac{10 \text{ ft} - (250 \text{ ft})(0.006)}{3.5 \text{ ft}} = 2.4$$

The tailwater submerges the culvert outlet, so this is type-4 flow (outlet control as per Table 19.9).

$h_1 = 10$ ft

$h_4 = D = 3.5$ ft

$$Q = C_d A_0 \sqrt{2g\left(\frac{h_1 - h_4}{1 + \frac{29 C_d^2 n^2 L}{R^{4/3}}}\right)}$$

$$= \left(\frac{(0.82\pi)(3.5 \text{ ft})^2}{4}\right) \times \sqrt{(2)\left(32.2 \frac{\text{ft}}{\text{sec}^2}\right) \times \left(\frac{10 \text{ ft} - 3.5 \text{ ft}}{1 + \frac{(29)(0.82)^2(0.012)^2(250 \text{ ft})}{(0.875 \text{ ft})^{4/3}}}\right)}$$

$$= 119 \text{ ft}^3/\text{sec}$$

Check to see if the flow rate is limited by inlet geometry. Evaluate type-6 flow. From Eq. 19.107 with $h_3 = d$ and $h_f = 0$,

$$Q = C_d A_0 \sqrt{2g(h_1 - h_3 - h_f)}$$

$$= (0.82)\left(\frac{\pi(3.5 \text{ ft})^2}{4}\right) \times \sqrt{(2)\left(32.2 \frac{\text{ft}}{\text{sec}^2}\right)(10 \text{ ft} - 3.5 \text{ ft} - 0)}$$

$$= 161.4 \text{ ft}^3/\text{sec}$$

The culvert appears to have the capacity.

$$Q = \boxed{119 \text{ ft}^3/\text{sec}}$$

The answer is (D).

23. (a) Rearrange Eq. 19.60.

$$H = \left(\frac{Q}{C_s b}\right)^{2/3} = \left(\frac{2000 \frac{\text{m}^3}{\text{s}}}{(2.2)(150 \text{ m})}\right)^{2/3}$$

$$= \boxed{3.32 \text{ m}}$$

The answer is (D).

(b)

$$\mathrm{v}_1 = \frac{Q}{A_1} = \frac{2000\ \frac{\mathrm{m}^3}{\mathrm{s}}}{(150\ \mathrm{m})(3.32\ \mathrm{m})} = 4.02\ \mathrm{m/s}$$

From Eq. 19.96,

$$E_1 = y_{\mathrm{crest}} + H + \frac{\mathrm{v}_1^2}{2g} = 30\ \mathrm{m} + 3.32\ \mathrm{m} + \frac{\left(4.02\ \frac{\mathrm{m}}{\mathrm{s}}\right)^2}{(2)\left(9.81\ \frac{\mathrm{m}}{\mathrm{s}^2}\right)} = \boxed{34.1\ \mathrm{m}}$$

The answer is (B).

(c)

$$\mathrm{v}_2 = \frac{Q}{A_2} = \frac{2000\ \frac{\mathrm{m}^3}{\mathrm{s}}}{(150\ \mathrm{m})(0.6\ \mathrm{m})} = 22.2\ \mathrm{m/s}$$

$$E_2 = d_2 + \frac{\mathrm{v}_2^2}{2g} = 0.6\ \mathrm{m} + \frac{\left(22.2\ \frac{\mathrm{m}}{\mathrm{s}}\right)^2}{(2)\left(9.81\ \frac{\mathrm{m}}{\mathrm{s}^2}\right)} = \boxed{25.8\ \mathrm{m}}$$

The answer is (A).

(d) The friction loss is

$$E_1 - E_2 = 34.1\ \mathrm{m} - 25.8\ \mathrm{m} = \boxed{8.3\ \mathrm{m}}$$

The answer is (C).

(e) From Eq. 19.13,

$$Q = \left(\frac{1.00}{n}\right) A R^{2/3} \sqrt{S}$$

$$2000\ \frac{\mathrm{m}^3}{\mathrm{s}} = \left(\frac{1.00}{0.016}\right)(150\ \mathrm{m})d_n\left(\frac{(150\ \mathrm{m})d_n}{2d_n + 150\ \mathrm{m}}\right)^{2/3} \times \sqrt{0.002}$$

By trial and error, $d_n = \boxed{2.6\ \mathrm{m.}}$

The answer is (B).

(f) From Eq. 19.74,

$$d_c = \left(\frac{Q^2}{g w^2}\right)^{1/3} = \left(\frac{\left(2000\ \frac{\mathrm{m}^3}{\mathrm{s}}\right)^2}{\left(9.81\ \frac{\mathrm{m}}{\mathrm{s}^2}\right)(150\ \mathrm{m})^2}\right)^{1/3} = \boxed{2.6\ \mathrm{m}}$$

The answer is (B).

(g) The normal depth of flow coincides with the critical depth. A hydraulic jump is an abrupt rise from below the critical depth to above the critical depth. In this case the water will rise gradually and no jump occurs.

The answer is (D).

(h) From Table 19.7, the maximum velocity (assuming the water is clear) is

$$\mathrm{v}_{\mathrm{max}} = \left(4.0\ \frac{\mathrm{ft}}{\mathrm{sec}}\right)\left(0.3\ \frac{\mathrm{m}}{\mathrm{ft}}\right) = \boxed{1.2\ \mathrm{m/s}}$$

The answer is (A).

(i) From App. 19.A, for natural channels in good condition, $n = 0.025$.

The depth of flow is

$$d = \frac{Q}{\mathrm{v}b} = \frac{2000\ \frac{\mathrm{m}^3}{\mathrm{sec}}}{\left(1.2\ \frac{\mathrm{m}}{\mathrm{s}}\right)(150\ \mathrm{m})} = 11.11\ \mathrm{m}$$

$$R = \frac{bd}{2d + b} = \frac{(150\ \mathrm{m})(11.1\ \mathrm{m})}{(2)(11.1\ \mathrm{m}) + 150\ \mathrm{m}} = 9.67\ \mathrm{m}$$

Rearranging Eq. 19.12,

$$S = \left(\frac{\mathrm{v}n}{R^{2/3}}\right)^2 = \left(\frac{\left(1.2\ \frac{\mathrm{m}}{\mathrm{s}}\right)(0.025)}{(9.67\ \mathrm{m})^{2/3}}\right)^2 = \boxed{0.000044}$$

The answer is (B).

(j) From Eq. 19.29,

$$h_f = LS = (1000\ \mathrm{m})(0.002) = \boxed{2\ \mathrm{m}}$$

The answer is (D).

20 Meteorology, Climatology, and Hydrology

PRACTICE PROBLEMS

Hydrographs

1. A 2 h storm over a 111 km^2 area produces a total runoff volume of 4×10^6 m^3 with a peak discharge of 260 m^3/s.

(a) What is the total excess precipitation?

(A) 1.4 cm
(B) 2.6 cm
(C) 3.6 cm
(D) 4.0 cm

(b) What is the unit hydrograph peak discharge?

(A) 72 $m^3/s{\cdot}cm$
(B) 120 $m^3/s{\cdot}cm$
(C) 210 $m^3/s{\cdot}cm$
(D) 260 $m^3/s{\cdot}cm$

(c) If a 2 h storm producing 6.5 cm of runoff is to be used to design a culvert, what is the design flood hydrograph volume?

(A) 4.0×10^6 m^3
(B) 7.2×10^6 m^3
(C) 2.6×10^7 m^3
(D) 3.6×10^7 m^3

(d) What is the design discharge?

(A) 89 m^3/s
(B) 130 m^3/s
(C) 260 m^3/s
(D) 470 m^3/s

(e) The recurrence interval of the 6.5 cm storm is 50 yr, and the culvert is to be designed for a 30 yr life. What is the probability that the capacity will be exceeded during the design life?

(A) 0%
(B) 33%
(C) 45%
(D) 92%

(f) The unit hydrograph represents water flowing into a stream from

I. base flow
II. evapotranspiration
III. overland flow
IV. surface flow
V. interflow

(A) I only
(B) I and III
(C) II, III, and IV
(D) III, IV, and V

Use the following information for parts (g) through (j).

Stream discharges recorded during a 3 h storm on a 40 km^2 watershed are as follows.

t (h)	$Q\left(\frac{m^3}{s}\right)$
0	17
1	16
2	48
3	90
4	108
5	85
6	58
7	38
8	26
9	20
10	18
11	17
12	16

(g) What is the unit hydrograph peak discharge from surface runoff for this 3 h storm?

(A) 5.2 $m^3/s{\cdot}cm$
(B) 29 $m^3/s{\cdot}cm$
(C) 65 $m^3/s{\cdot}cm$
(B) 120 $m^3/s{\cdot}cm$

(h) What is the peak discharge for a 6 h storm producing 5 cm of runoff in this watershed?

(A) 92 m^3/s
(B) 140 m^3/s
(C) 170 m^3/s
(D) 300 m^3/s

(i) What is the peak discharge for a 5 h storm producing 5 cm of runoff in this watershed?

(A) 63 m^3/s
(B) 90 m^3/s
(C) 110 m^3/s
(D) 260 m^3/s

(j) What is the design flood hydrograph volume for a 5 h storm producing 5 cm of runoff in this watershed?

(A) 1.3×10^4 m^3
(B) 4.0×10^5 m^3
(C) 8.3×10^5 m^3
(D) 2.0×10^6 m^3

2. (*Time limit: one hour*) A stream gauging station recorded the following discharges from a 1.2 mi^2 drainage area after a storm.

t (hr)	$Q\left(\frac{\text{ft}^3}{\text{sec}}\right)$
0	102
1	99
2	101
3	215
4	507
5	625
6	455
7	325
8	205
9	145
10	100
11	70
12	55
13	49
14	43
15	38

(a) Draw the actual hydrograph. (b) Separate the groundwater and surface water. (c) Draw the unit hydrograph.

(d) What is the length of the direct runoff recession limb?

(A) 5 hr
(B) 6 hr
(C) 10 hr
(D) 13 hr

Rational Equation

3. A 0.5 mi^2 drainage area has a suggested runoff coefficient of 0.6 and a time of concentration of 60 min. The drainage area is in Steel region no. 3, and a 10 yr storm is to be used for design purposes. What is the peak runoff?

(A) 310 ft^3/sec
(B) 390 ft^3/sec
(C) 460 ft^3/sec
(D) 730 ft^3/sec

4. Four contiguous 5 ac watersheds are served by an adjacent 1200 ft storm drain (n = 0.013 and slope = 0.005). Inlets to the storm drain are placed every 300 ft along the storm drain. The inlet time for each area served by an inlet is 15 min, and the area's runoff coefficient is 0.55. A storm to be used for design purposes has the following characteristics (I is in in/hr, t is in min).

$$I = \frac{100}{t_c + 10}$$

All flows are maximum, and all pipe sizes are available. n is constant. What is the diameter of the last section of storm drain?

(A) 28 in
(B) 32 in
(C) 36 in
(D) 42 in

5. (*Time limit: one hour*) A 75 ac urbanized section of land drains into a rectangular 5 ft by 7 ft channel that directs runoff through a round culvert under a roadway. The culvert is concrete and is 60 in in diameter. It is 36 ft long and placed on a 1% slope.

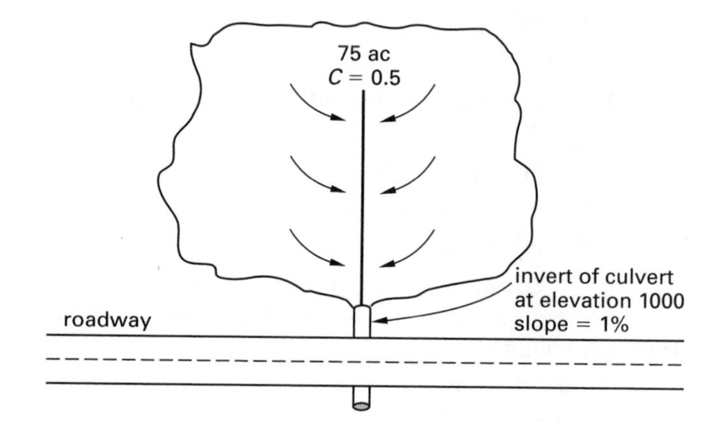

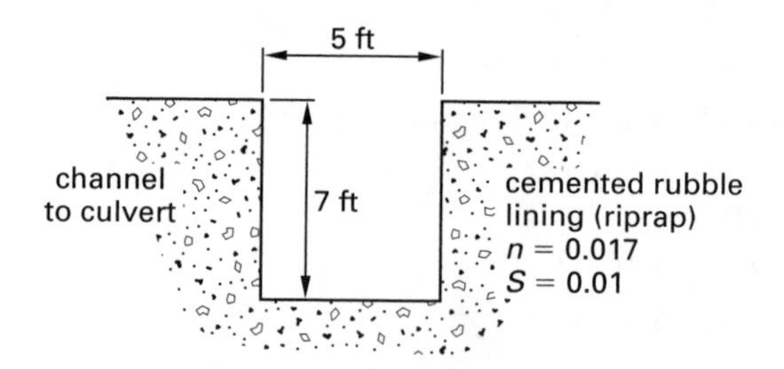

It is desired to evaluate the culvert design based on a 50 yr storm. It is known that the time of concentration to the entrance of the culvert is 30 min. The corresponding rainfall intensity is 5.3 in/hr.

(a) If the minimum road surface elevation is 1010 ft, will the road surface be flooded?

(A) yes; the culvert needs 10 ft^3/sec additional capacity
(B) yes; the culvert needs 70 ft^3/sec additional capacity
(C) no; the culvert has 20 ft^3/sec excess capacity
(D) no; the culvert has 70 ft^3/sec excess capacity

(b) What is the depth of flow elevation upstream at the culvert entrance after 30 min?

(A) 2.7 ft
(B) 3.1 ft
(C) 3.6 ft
(D) 4.1 ft

(c) Assuming n varies with depth of flow, what is the depth of flow at the culvert exit after 30 min?

(A) 2.7 ft
(B) 3.1 ft
(C) 3.6 ft
(D) 4.1 ft

Hydrograph Synthesis

6. (*Time limit: one hour*) A standard 4 hr storm produces 2 in net of runoff. A stream gauging report is produced from successive sampling every few hours. Two weeks later, the first 4 hr of an 8 hr storm over the same watershed produces 1 in of runoff. The second 4 hr produces 2 in of runoff. Neglecting ground water, draw a hydrograph of the 8 hr storm.

t (hr)	$Q\left(\frac{\text{ft}^3}{\text{sec}}\right)$
0	0
2	100
4	350
6	600
8	420
10	300
12	250
14	150
16	100
18	–
20	50
22	–
24	0

Reservoir Sizing

7. A class A evaporation pan located near a reservoir shows a 1 day evaporation loss of 0.8 in. If the pan coefficient is 0.7, what is the approximate evaporation loss in the reservoir?

(A) 0.56 in
(B) 0.63 in
(C) 0.98 in
(D) 1.1 in

8. A reservoir has a total capacity of 7 volume units. At the beginning of a study, the reservoir contains 5.5 units. The monthly demand on the reservoir from a nearby city is 0.7 units. The monthly inflow to the reservoir is normally distributed with a mean of 0.9 units and a standard deviation of 0.2 units. Simulate one year of reservoir operation with a 99.97% confidence level.

9. Repeat Prob. 8 assuming that the monthly demand on the reservoir is normally distributed with a mean of 0.7 units and a standard deviation of 0.2 units.

10. (*Time limit: one hour*) A reservoir is needed to provide 20 ac-ft of water each month. The inflow for each of 13 representative months is given. The reservoir starts full. Size the reservoir.

month	inflow (ac-ft)
February	30
March	60
April	20
May	10
June	5
July	10
August	5
September	10
October	20
November	90
December	85
January	75
February	50

(A) 40 ac-ft
(B) 60 ac-ft
(C) 90 ac-ft
(D) 120 ac-ft

11. (*Time limit: one hour*) A reservoir with a constant draft of 240 MG/mi^2-yr is being designed. The inflow distribution is known.

month	inflow (MG/mi^2)	inflow (m^3/m^2)
January	20	0.029
February	30	0.044
March	45	0.066
April	30	0.044
May	40	0.058
June	30	0.044
July	15	0.022
August	5	0.007
September	15	0.022
October	60	0.088
November	90	0.132
December	40	0.058

(a) What should be the minimum reservoir size?

(A) 25 MG/mi^2
(B) 35 MG/mi^2
(C) 45 MG/mi^2
(D) 65 MG/mi^2

(b) When would the reservoir start to spill?

(A) at the end of March

(B) at the end of June

(C) at the end of September

(D) at the end of March and the end of October

SOLUTIONS

1. (a) From Eq. 20.21,

$$P_{\text{ave}} = \frac{V}{A_d} = \frac{4\times 10^6 \text{ m}^3}{(111 \text{ km}^2)\left(1000 \ \frac{\text{m}}{\text{km}}\right)^2} = \boxed{0.036 \text{ m } (3.6 \text{ cm})}$$

The answer is (C).

(b) The unit hydrograph discharge is the peak discharge divided by the average precipitation.

$$Q_{\text{hydrograph,unit}} = \frac{Q_p}{P} = \frac{260 \ \frac{\text{m}^3}{\text{s}}}{3.6 \text{ cm}} = \boxed{72.2 \text{ m}^3/\text{s·cm}}$$

The answer is (A).

(c) The design flood hydrograph volume for a 6.5 cm storm is determined by multiplying the unit hydrograph volume by 6.5. For the unit hydrograph,

$$V_{\text{hydrograph}} = \frac{V}{P} = \frac{4\times 10^6 \text{ m}^3}{3.6 \text{ cm}} = 1.11\times 10^6 \text{ m}^3/\text{cm}$$

For the 6.5 cm storm,

$$V = \left(1.11\times 10^6 \ \frac{\text{m}^3}{\text{cm}}\right)(6.5 \text{ cm}) = \boxed{7.2\times 10^6 \text{ m}^3}$$

The answer is (B).

(d) The design discharge is determined by multiplying the unit hydrograph discharge by 6.5 cm.

$$Q_p = (Q_{\text{hydrograph}})(6.5) = \left(72.2 \ \frac{\text{m}^3}{\text{s·cm}}\right)(6.5 \text{ cm}) = \boxed{469.3 \text{ m}^3/\text{s}}$$

The answer is (D).

(e) From Eq. 20.20,

$$p\{F \text{ event in } n \text{ years}\} = 1 - \left(1 - \frac{1}{F}\right)^n$$

$$p\{50 \text{ yr flood in 30 yr}\} = 1 - \left(1 - \frac{1}{50}\right)^{30} = \boxed{0.45 \ \ (45\%)}$$

The answer is (C).

(f) The unit hydrograph represents all discharge into a stream except for groundwater or base flow, which are separated out. Evapotranspiration refers to water that is returned to the atmosphere and, therefore, is not measured in the stream discharge. The unit hydrograph includes overland flow, surface flow, and interflow.

The answer is (D).

(g) The actual runoff is plotted. To separate base flow from overland flow, use the straight line method. Draw a horizontal line from the start of the rising limb to the falling limb. In the table, subtract the base flow from the overland flow.

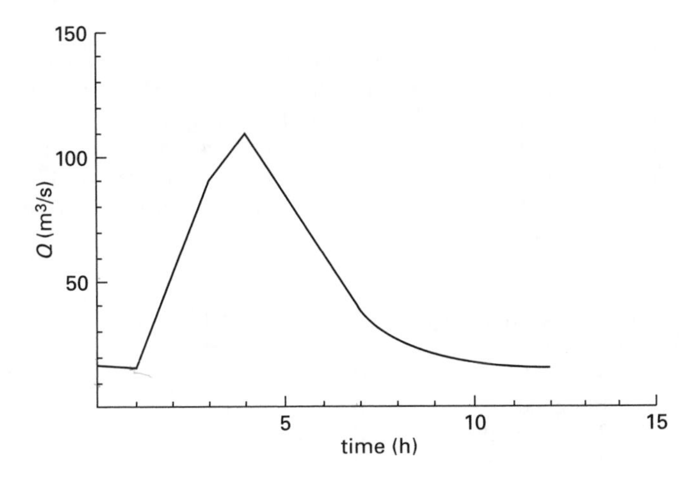

hour	runoff	ground water	surface water	surface water / 3.14
0	17	16	1	0.32
1	16	16	0	0
2	48	16	32	10.19
3	90	16	74	23.56
4	108	16	92	29.29
5	85	16	69	21.97
6	58	16	42	13.37
7	38	16	22	7.00
8	26	16	10	3.18
9	20	16	4	1.27
10	18	16	2	0.64
11	17	16	1	0.32
12	16	16	0	0
		total	349 m^3/s	

The average precipitation is given by Eq. 20.21.

$$P = \frac{V}{A_d}$$

The volume of runoff is the area under the separated hydrograph curve. Since data are given for each hour, the "width" of each histogram cell is 1 h.

$$V = \left(349\ \frac{\text{m}^3}{\text{s}}\right)(1\ \text{h})\left(3600\ \frac{\text{s}}{\text{h}}\right)$$
$$= 1\,256\,400\ \text{m}^3$$

$$P = \frac{1\,256\,400\ \text{m}^3}{(40\ \text{km}^2)\left(1000\ \frac{\text{m}}{\text{km}}\right)^2}$$
$$= 0.0314\ \text{m} \quad (3.14\ \text{cm})$$

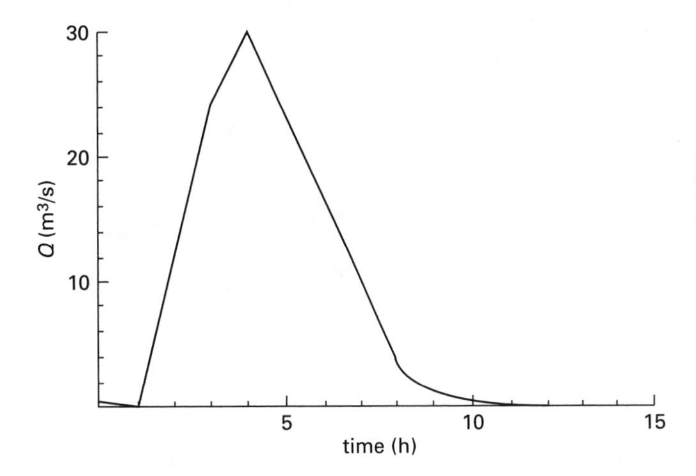

The unit hydrograph peak discharge is

$$\frac{108\ \frac{\text{m}^3}{\text{s}} - 16\ \frac{\text{m}^3}{\text{s}}}{3.14\ \text{cm}} = \boxed{29.3\ \text{m}^3/\text{s}\cdot\text{cm}}$$

The answer is (B).

(h) There are two methods for constructing hydrographs for a longer storm than that of the unit hydrograph. For storms that are whole multiples of the unit hydrograph duration, the lagging storm method is simplest.

In a table, add another unit hydrograph separated by $t_r = 3$ h. Then, add the ordinates for a hydrograph of $Nt_r = (2)(3\ \text{h}) = 6$ h duration. The ordinates must then be divided by $N = 2$ to get the unit hydrograph ordinates.

hour	unit hydrograph 3 h storm	second storm	total	unit hydrograph 6 h storm
0	0.32	0	0.32	0.16
1	0	0	0	0
2	10.19	0	10.19	5.09
3	23.56	0.32	23.88	11.94
4	29.29	0	29.29	14.65
5	21.97	10.19	32.16	16.08
6	13.37	23.56	36.93	18.47
7	7.00	29.29	36.29	18.15
8	3.18	21.97	25.15	12.58
9	1.27	13.37	14.65	7.32
10	0.64	7.00	7.64	3.82
11	0.32	3.18	3.50	1.75
12	0	1.27	1.27	0.64
13	0	0.64	0.64	0.32
14	0	0.32	0.32	0.16
15	0	0	0	0

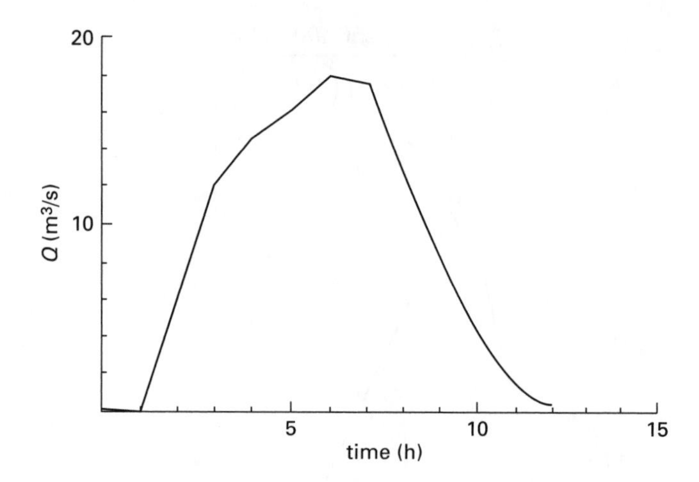

The peak discharge for a 5 cm storm is calculated from the peak unit hydrograph discharge.

$$Q_{p,\text{unit hydrograph}} = 18.47\ \text{m}^3/\text{s}\cdot\text{cm}$$

$$\begin{aligned} Q_{p,5\ \text{cm storm}} &= (5\ \text{cm})\left(18.47\ \frac{\text{m}^3}{\text{s}\cdot\text{cm}}\right) \\ &= \boxed{92.35\ \text{m}^3/\text{s}} \end{aligned}$$

The answer is (A).

(i) If a storm is not a multiple of the unit hydrograph duration, the S-curve method must be used to construct a hydrograph for that storm. In the table, add the ordinates of five 3 h unit hydrographs, each offset by 3 h, to produce an S-curve.

hour	3 h unit hydro- graph	second storm	third storm	fourth storm	fifth storm	total (S-curve)
0	0.32	0	0	0	0	0.32
1	0	0	0	0	0	0
2	10.19	0	0	0	0	10.19
3	23.56	0.32	0	0	0	23.88
4	29.29	0	0	0	0	29.29
5	21.97	10.19	0	0	0	32.16
6	13.37	23.56	0.32	0	0	37.25
7	7.00	29.29	0	0	0	36.29
8	3.18	21.97	10.19	0	0	35.33
9	1.27	13.37	23.56	0.32	0	38.52
10	0.64	7.00	29.29	0	0	36.93
11	0.32	3.18	21.97	10.19	0	35.66
12	0	1.27	13.37	23.56	0.32	38.52
13	0	0.64	7.00	29.29	0	36.93
14	0	0.32	3.18	21.97	10.19	35.66
15	0	0	1.27	13.37	23.56	38.52

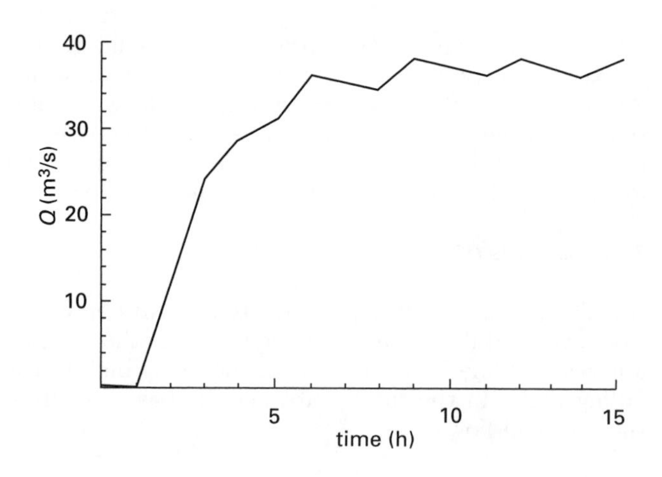

After approximately 5 h, the S-curve levels off to a steady value. (It actually alternates slightly between 35 and 39 m^3/s.) The 5 h unit hydrograph is calculated by tabulating another S-curve lagging the first by 5 h, finding the difference, and scaling to a ratio of 3:5. (The negative values obtained reflect the approximate nature of the method.)

hour	S-curve	lagging S-curve	5 h hydrograph
0	0.32	0	0.19
1	0	0	0
2	10.19	0	6.11
3	23.88	0	14.32
4	29.29	0	17.57
5	32.16	0.32	19.10
6	37.25	0	22.35
7	36.29	10.19	15.66
8	35.33	23.88	6.88
9	38.52	29.29	5.54
10	36.93	32.16	2.87
11	35.66	37.25	−0.96
12	38.52	36.29	1.34
13	36.93	35.33	0.96
14	35.66	38.52	−1.72
15	38.52	36.93	0.96
		total	111.17

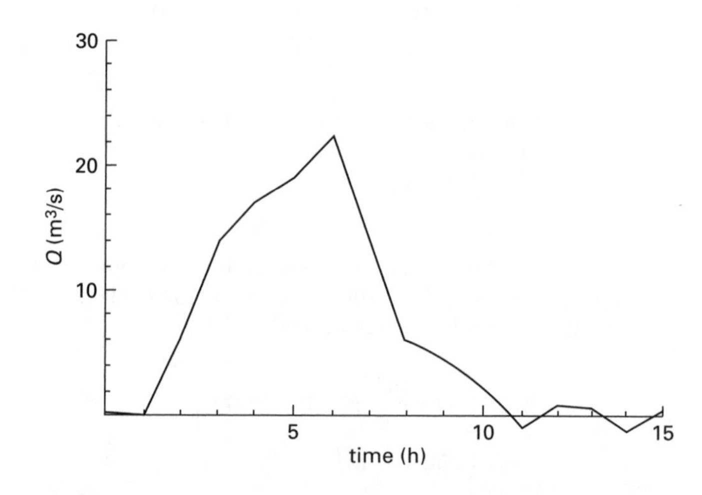

The peak discharge for a 5 cm storm is calculated from the peak unit hydrograph discharge.

$$Q_{p,\text{unit hydrograph}} = 22.35 \text{ m}^3/\text{s·cm} \quad \text{[per cm of rainfall]}$$

$$Q_{p,5 \text{ cm storm}} = (5 \text{ cm})\left(22.35 \ \frac{\text{m}^3}{\text{s·cm}}\right) = \boxed{111.75 \text{ m}^3/\text{s}}$$

The answer is (C).

(j) The design flood volume is derived from the 5 h unit hydrograph.

$$V_{\text{unit hydrograph}} = \left(111.75 \ \frac{\text{m}^3}{\text{s·cm}}\right)(1 \text{ h})(3600 \text{ s/h}) = 400\,300 \text{ m}^3/\text{cm}$$

For a 5 cm storm,

$$V_{5 \text{ cm storm}} = \left(400\,300 \ \frac{\text{m}^3}{\text{cm}}\right)(5 \text{ cm}) = \boxed{2.12 \times 10^6 \text{ m}^3}$$

The answer is (D).

2. (a) The actual hydrograph is a plot of time versus flow quantity, both of which are given in the problem statement.

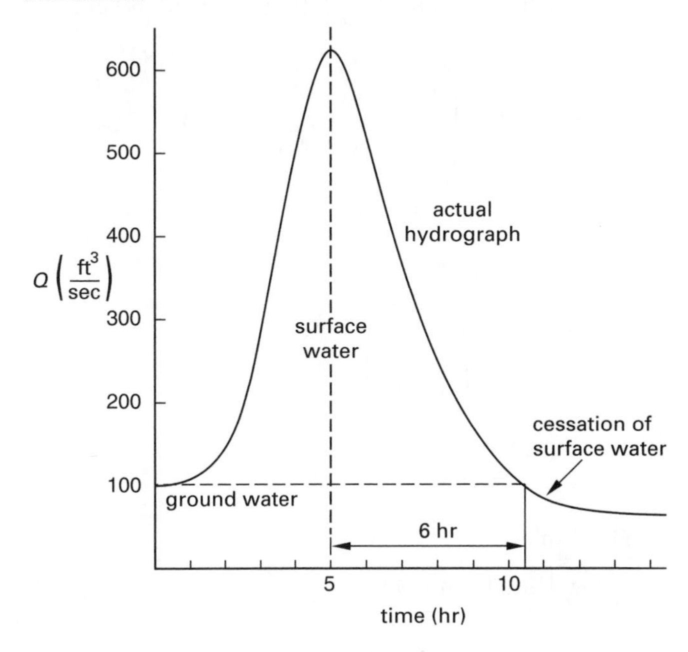

(b) Use the fixed-base method to separate the base flow from the overland flow. The base flow before the storm is projected to a point directly under the peak (in this case a more or less horizontal line), and then a straight line connects the projection to the falling limb. The connection point is 5 to 7 hr later, so use 6 hr.

(c) The ordinates of the unit hydrograph are found by separating the base flow and then dividing the ordinates of the actual hydrograph by the average precipitation. The average precipitation is determined by tabulating the surface water flow and adding to find the total volume.

hour	runoff (ft^3/sec)	ground water (ft^3/sec)	surface water (ft^3/sec)	surface water/2.43 (ft^3/sec)
0	102	100	0	0
1	99	100	0	0
2	101	100	0	0
3	215	100	115	47.3
4	507	100	407	167.5
5	625	100	525	215.0
6	455	93	362	149.0
7	325	87	238	97.9
8	205	80	125	51.4
9	145	74	71	29.2
10	100	67	33	13.6
11	70	60	10	4.1
12	55	55	0	0
13	49	49	0	0
14	43	43	0	0
15	38	38	0	0
		total	1886 ft^3/sec	

The total volume of surface water is the total flow multiplied by the time interval (1 hr).

$$V = \left(1886 \ \frac{\text{ft}^3}{\text{sec}}\right)(1 \text{ hr})\left(3600 \ \frac{\text{sec}}{\text{hr}}\right) = 6.79 \times 10^6 \text{ ft}^3$$

The watershed area is

$$A_d = (1.2 \text{ mi}^2)\left(5280 \ \frac{\text{ft}}{\text{mi}}\right)^2 = 3.35 \times 10^7 \text{ ft}^2$$

The average precipitation is the total volume divided by the drainage area.

$$P = \frac{(6.79 \times 10^6 \text{ ft}^3)\left(12 \ \frac{\text{in}}{\text{ft}}\right)}{3.35 \times 10^7 \text{ ft}^2} = 2.43 \text{ in}$$

The ordinates for the unit hydrograph are tabulated in the last column of the table.

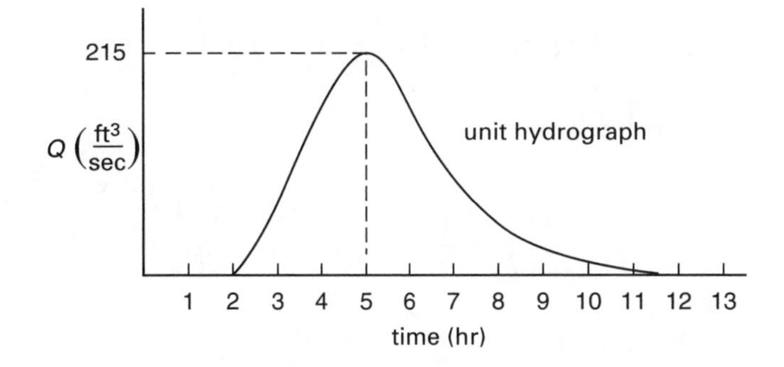

(d) The time base for direct runoff is about 11 hr − 2 hr = 9 hr. The time base includes all the time in which direct runoff is observed, that is, both the rising and falling limbs.

The actual hydrograph is as shown. From the graph,

$$t_b \approx \boxed{6 \text{ hr}}$$

The answer is (B).

3. Use Table 20.2.

$$K = 170$$
$$b = 23$$

From Eq. 20.14, the intensity is

$$I = \frac{170}{60 \text{ min} + 23} = 2.05 \text{ in/hr}$$
$$A_d = (0.5)(640 \text{ ac}) = 320 \text{ ac}$$

From Eq. 20.36,

$$Q_p = CIA_d = (0.6)\left(2.05 \ \frac{\text{in}}{\text{hr}}\right)(320 \text{ ac})$$
$$= \boxed{393.6 \text{ ft}^3/\text{sec}}$$

The answer is (B).

4.

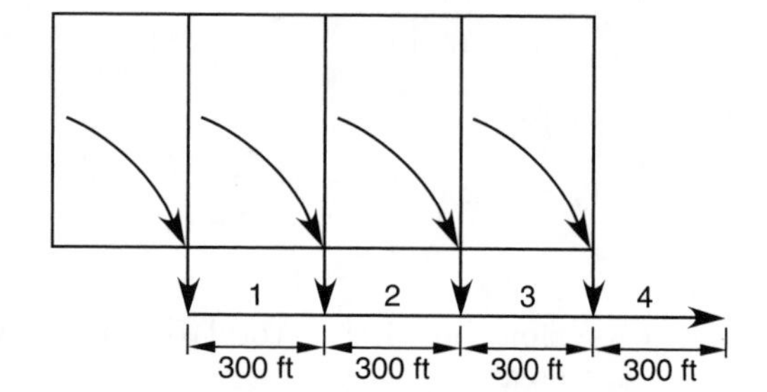

- First inlet:

Peak flow for the first inlet begins at $t = 15$ min.

$$I = \frac{100}{15 \text{ min} + 10} = 4 \text{ in/hr}$$

From Eq. 20.36,

$$Q_p = CIA_d = (0.55)\left(4 \ \frac{\text{in}}{\text{hr}}\right)(5 \text{ ac}) = 11 \text{ ft}^3/\text{sec}$$

Use Eq. 19.16.

$$D = (1.33)\left(\frac{nQ}{\sqrt{S}}\right)^{3/8} = (1.33)\left(\frac{(0.013)\left(11 \ \frac{\text{ft}^3}{\text{sec}}\right)}{\sqrt{0.005}}\right)^{3/8}$$
$$= 1.73 \text{ ft} \quad (21 \text{ in})$$

$$\text{v}_{\text{full}} = \frac{Q}{A} = \frac{11 \ \frac{\text{ft}^3}{\text{sec}}}{\left(\frac{\pi}{4}\right)\left(\frac{21 \text{ in}}{12 \ \frac{\text{in}}{\text{ft}}}\right)^2}$$
$$= 4.6 \text{ ft/sec}$$

(Note that an integer-inch pipe diameter must be used to calculate velocity, because 1.73 ft diameter pipes are not manufactured.)

- Second inlet:

The flow time from inlet 1 to inlet 2 is

$$\frac{300 \text{ ft}}{4.6 \ \frac{\text{ft}}{\text{sec}}} = 65.2 \text{ sec} \quad (1.09 \text{ min})$$

The intensity at $t = 15.0 \text{ min} + 1.09 \text{ min} = 16.09 \text{ min}$ is

$$I = \frac{100}{16.09 \text{ min} + 10} = 3.83 \text{ in/hr}$$

The runoff is

$$Q = (0.55)\left(3.83 \ \frac{\text{in}}{\text{hr}}\right)(10 \text{ ac}) = 21.07 \text{ ft}^3/\text{sec}$$

$$D = (1.33)\left(\frac{(0.013)\left(21.07 \ \frac{\text{ft}^3}{\text{sec}}\right)}{\sqrt{0.005}}\right)^{3/8}$$
$$= 2.21 \text{ ft} \quad (27 \text{ in})$$

$$\text{v}_{\text{full}} = \frac{21.07 \ \frac{\text{ft}^3}{\text{sec}}}{\left(\frac{\pi}{4}\right)\left(\frac{27 \text{ in}}{12 \ \frac{\text{in}}{\text{ft}}}\right)^2}$$
$$\approx 5.3 \text{ ft/sec}$$

- Third inlet:

The flow time from inlet 2 is

$$t = \frac{\text{L}}{\text{v}} = \frac{300 \text{ ft}}{\left(5.3 \ \frac{\text{ft}}{\text{sec}}\right)\left(60 \ \frac{\text{sec}}{\text{min}}\right)} = 0.94 \text{ min}$$
$$t_c = 16.09 \text{ min} + 0.94 \text{ min} = 17.03 \text{ min}$$
$$I = \frac{100}{17.03 \text{ min} + 10} = 3.70 \text{ in/hr}$$

$$Q = (0.55)\left(3.70\ \frac{\text{in}}{\text{hr}}\right)(15) = 30.53\ \text{ft}^3/\text{sec}$$

$$D = (1.33)\left(\frac{(0.013)\left(30.53\ \frac{\text{ft}^3}{\text{sec}}\right)}{\sqrt{0.005}}\right)^{3/8}$$

$$= 2.54\ \text{ft}\quad (31\ \text{in})$$

$$\text{v} = \frac{Q}{A} = \frac{30.53\ \frac{\text{ft}^3}{\text{sec}}}{\left(\frac{\pi}{4}\right)\left(\frac{31\ \text{in}}{12\ \frac{\text{in}}{\text{ft}}}\right)^2}$$

$$= 5.82\ \text{ft/sec}$$

- Fourth inlet:

The flow time from inlet 3 is

$$\frac{300\ \text{ft}}{\left(5.82\ \frac{\text{ft}}{\text{sec}}\right)\left(60\ \frac{\text{sec}}{\text{min}}\right)} = 0.86\ \text{min}$$

$$t_c = 17.03\ \text{min} + 0.86\ \text{min} = 17.89\ \text{min}$$

$$I = \frac{100}{17.89\ \text{min} + 10} = 3.59\ \text{in/hr}$$

$$Q = (0.55)\left(3.59\ \frac{\text{in}}{\text{hr}}\right)(20\ \text{ac}) = 39.49\ \text{ft}^3/\text{sec}$$

$$D = (1.33)\left(\frac{(0.013)\left(39.49\ \frac{\text{ft}^3}{\text{sec}}\right)}{\sqrt{0.005}}\right)^{3/8}$$

$$= 2.80\ \text{ft}\quad (33.6\ \text{in})$$

Use 36 in pipe.

The answer is (C).

5.

$$I = 5.3\ \text{in/hr}$$

$$C = 0.5\quad [\text{given}]$$

$$Q = CIA$$

$$= (0.5)\left(5.3\ \frac{\text{in}}{\text{hr}}\right)(75\ \text{ac})$$

$$= 198.8\ \text{ac-in/hr}\quad (200\ \text{ft}^3/\text{sec})$$

(a) Take a worst-case approach. See if the culvert has a capacity equal to or greater than 200 ft^3/sec when water is at the elevation of the roadway.

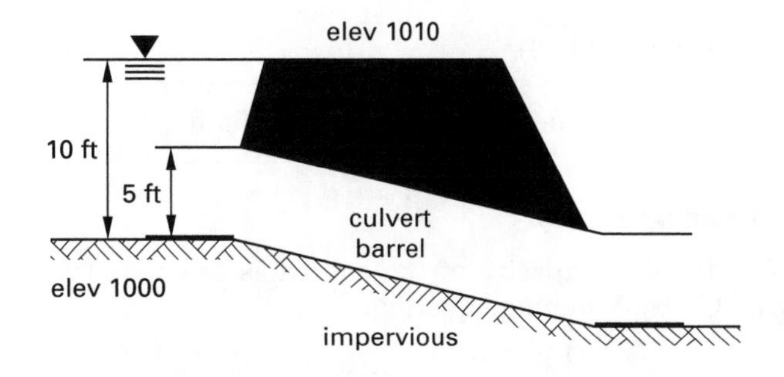

Disregard barrel friction. $K_e = 0.90$ for the projecting, square end. Assume type 6 flow.

$$H = 10\ \text{ft} + (36\ \text{ft})(0.01) - 5\ \text{ft} = 5.36\ \text{ft}$$

From Eq. 19.100,

$$\text{v} = \sqrt{\frac{\frac{5.36\ \text{ft}}{1+0.9}}{(2)\left(32.2\ \frac{\text{ft}}{\text{sec}^2}\right)}} = 13.48\ \text{ft/sec}$$

$$Q = A\text{v} = \left(\frac{\pi}{4}\right)(5\ \text{ft})^2\left(13.48\ \frac{\text{ft}}{\text{sec}}\right) = 265\ \text{ft}^3/\text{sec}$$

The excess capacity is

$$265\ \frac{\text{ft}^3}{\text{sec}} - 200\ \frac{\text{ft}^3}{\text{sec}} = 65\ \text{ft}^3/\text{sec}$$

There will be no flooding, as the culvert has excess capacity. The water cannot be maintained at elevation 1010.

The answer is (D).

- Alternate method:

$$C_d = 0.72\quad [\text{Table 17.5}]$$

$$H = 5.36\ \text{ft}$$

$$Q = C_d A\sqrt{2gh} = (0.72)\left(\frac{\pi}{4}\right)(5\ \text{ft})^2 \times \sqrt{(2)\left(32.2\ \frac{\text{ft}}{\text{sec}^2}\right)(5.36\ \text{ft})}$$

$$= 263\ \text{ft}^3/\text{sec}$$

(b) For the rubble channel,

$$n = 0.017$$

$$S = 0.01$$

$$R = \frac{5d}{2d+5}$$

$$C = \left(\frac{1.49}{0.017}\right)\left(\frac{5d}{2d+5}\right)^{1/6}$$

$$Q = A\text{v} = (5d)\left(\frac{1.49}{0.017}\right)\left(\frac{5d}{2d+5}\right)^{0.667}\sqrt{0.01}$$

$$200\ \frac{\text{ft}^3}{\text{sec}} = (43.8d)\left(\frac{5d}{2d+5}\right)^{0.667}$$

Water Resources

By trial and error,

$$d = \text{depth in channel} = \boxed{3.55 \text{ ft}}$$

The answer is (C).

(c) The full capacity of the culvert is calculated from the Chezy-Manning equation.

$$R = \frac{D}{4} = \frac{5 \text{ ft}}{4} = 1.25 \text{ ft}$$

$$A = \left(\frac{\pi}{4}\right) D^2 = \left(\frac{\pi}{4}\right) (5 \text{ ft})^2 = 19.63 \text{ ft}^2$$

$$n = 0.013 \quad [\text{assumed}]$$

From Eq. 19.13,

$$Q_{\text{full}} = \left(\frac{1.49}{n}\right) AR^{2/3}\sqrt{S}$$

$$= \left(\frac{1.49}{0.013}\right) (19.63 \text{ ft}^2)(1.25 \text{ ft})^{2/3}\sqrt{0.01}$$

$$= 261 \text{ ft}^3/\text{sec}$$

This is close to (but not the same as) the 265 ft^3/sec calculated for the pressure flow.

$$\frac{Q}{Q_{\text{full}}} = \frac{200 \frac{\text{ft}^3}{\text{sec}}}{261 \frac{\text{ft}^3}{\text{sec}}} = 0.77$$

Use the table of circular channel ratios. From App. 19.C,

$$\frac{d}{D} \approx 0.72$$

$$d_{\text{barrel}} = (0.72)(5 \text{ ft}) = \boxed{3.6 \text{ ft}}$$

The answer is (C).

6. Since the first storm produces 1 in net and the second storm produces 2 in, divide all runoffs by two. Offset the second storm by 4 hr.

hour	first storm	second storm	total
0	0		0
2	50		50
4	175	0	175
6	300	100	400
8	210	350	560
10	150	600	750
12	125	420	545
14	75	300	375
16	50	250	300
18	≈ 38	150	188
20	25	100	125
22	≈ 12	≈ 75	87
24	0	50	50
26		≈ 25	25
28		0	0

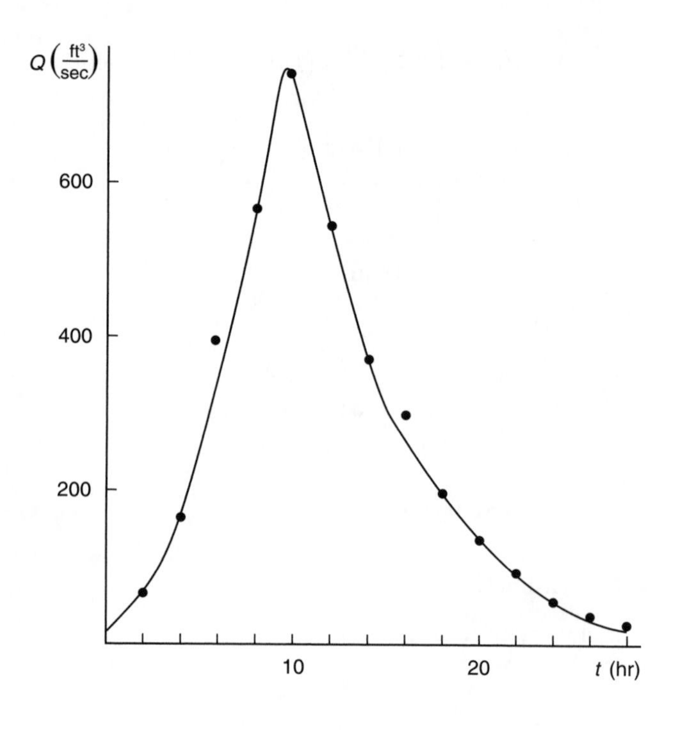

7. From Eq. 20.48,

$$E_{\text{reservoir}} = C_p E_p = (0.7)(0.8 \text{ in}) = \boxed{0.56 \text{ in}}$$

The answer is (A).

8. For 99.97% confidence, $z = \pm 3$. The inflow distribution is a normal distribution that extends $\pm 3\sigma$, or from 0.3 to 1.5. Take the cell width as $\frac{1}{2}\sigma$ or 0.1. Then, for the first cell,

actual endpoints: 0.3 to 0.4

midpoint: 0.35

z limits: -3.0 to -2.5

area under curve: $0.5 - 0.49 = 0.01$

The following inflow distribution is produced similarly.

endpoints	midpoint	z limits	area under curve	cum. × 100
0.3 to 0.4	0.35	−3.0 to −2.5	0.01	1
0.4 to 0.5	0.45	−2.5 to −2.0	0.02	3
0.5 to 0.6	0.55	−2.0 to −1.5	0.04	7
0.6 to 0.7	0.65	−1.5 to −1.0	0.09	16
0.7 to 0.8	0.75	−1.0 to −0.5	0.15	31
0.8 to 0.9	0.85	−0.5 to 0	0.19	50
0.9 to 1.0	0.95	0 to 0.5	0.19	69
1.0 to 1.1	1.05	0.5 to 1.0	0.15	84
1.1 to 1.2	1.15	1.0 to 1.5	0.09	93
1.2 to 1.3	1.25	1.5 to 2.0	0.04	97
1.3 to 1.4	1.35	2.0 to 2.5	0.02	99
1.4 to 1.5	1.45	2.5 to 3.0	0.01	100

Choose 12 random numbers less than 100 from App. 20.B. Use the third row, reading to the right.

- Inflow distribution:

month	random number	corresponding midpoint
1	06	0.55
2	40	0.85
3	18	0.75
4	73	1.05
5	97	1.25
6	72	1.05
7	89	1.15
8	83	1.05
9	24	0.75
10	41	0.85
11	88	1.15
12	86	1.15

Simulate the reservoir operation.

month	starting volume	+ inflow	− constant use	= ending volume	+ spill
1	5.5	0.55	0.7	5.35	
2	5.35	0.85	0.7	5.5	
3	5.5	0.75	0.7	5.55	
4	5.55	1.05	0.7	5.9	
5	5.9	1.25	0.7	6.45	
6	6.45	1.05	0.7	6.8	
7	6.8	1.15	0.7	7.0	0.25
8	7.0	1.05	0.7	7.0	0.35
9	7.0	0.75	0.7	7.0	0.05
10	7.0	0.85	0.7	7.0	0.15
11	7.0	1.15	0.7	7.0	0.45
12	7.0	1.15	0.7	7.0	0.45

9. Use the same simulation procedure for inflow as in Prob. 8. Proceed similarly.

- Demand distribution:

endpoints	midpoint	z limits	cum. × 100
0.1 to 0.2	0.15	−3.0 to −2.5	1
0.2 to 0.3	0.25	−2.5 to −2.0	3
0.3 to 0.4	0.35	−2.0 to −1.5	7
0.4 to 0.5	0.45	−1.5 to −1.0	16
0.5 to 0.6	0.55	−1.0 to −0.5	31
0.6 to 0.7	0.65	−0.5 to 0	50
0.7 to 0.8	0.75	0 to 0.5	69
0.8 to 0.9	0.85	0.5 to 1.0	84
0.9 to 1.0	0.95	1.0 to 1.5	93
1.0 to 1.1	1.05	1.5 to 2.0	97
1.1 to 1.2	1.15	2.0 to 2.5	99
1.2 to 1.3	1.25	2.5 to 3.0	100

Choose 12 random numbers. Use the fourth row, reading to the right.

- Demand distribution:

month	random number	use
1	04	0.35
2	75	0.85
3	41	0.65
4	44	0.65
5	89	0.95
6	39	0.65
7	42	0.65
8	09	0.45
9	42	0.65
10	11	0.45
11	58	0.75
12	04	0.35

Simulate the reservoir operation.

month	starting volume	+ inflow	− monthly demand	= ending volume	+ spill
1	5.5	0.55	0.35	5.7	
2	5.7	0.85	0.85	5.7	
3	5.7	0.75	0.65	5.8	
4	5.8	1.05	0.65	6.2	
5	6.2	1.25	0.95	6.5	
6	6.5	1.05	0.65	6.9	
7	6.9	1.15	0.65	7.0	0.4
8	7.0	1.05	0.45	7.0	0.6
9	7.0	0.75	0.65	7.0	0.1
10	7.0	0.85	0.45	7.0	0.4
11	7.0	1.15	0.75	7.0	0.4
12	7.0	1.15	0.35	7.0	0.8

10. Note that all dates correspond to the end of the month. Draw the mass diagram.

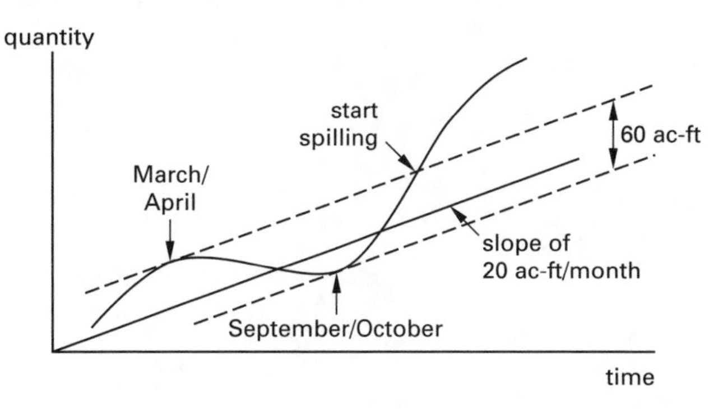

If the reservoir had been full in March or April, the maximum shortfall would have been 60 ac-ft, and the reservoir would be essentially empty in September and October.

60 ac-ft

The answer is (B).

11. (a) Assume the reservoir is initially empty. Draw the mass diagram.

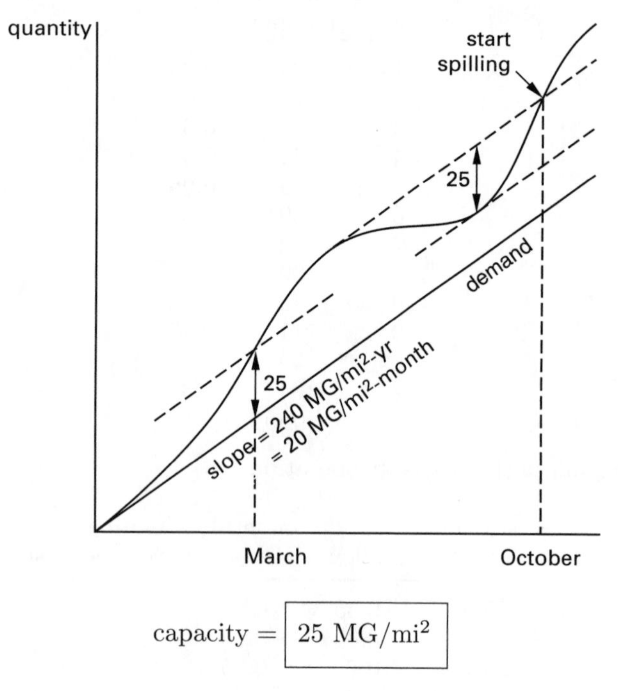

$$\text{capacity} = \boxed{25\ \text{MG/mi}^2}$$

The answer is (A).

(b) The reservoir would start spilling around the end of March. It would also begin spilling around the end of October.

The answer is (D).

21 Groundwater

PRACTICE PROBLEMS

Wells

1. A well extends from the ground surface at elevation 383 ft through a gravel bed to a layer of bedrock at elevation 289 ft. The screened well is 1500 ft from a river whose surface level is 363 ft. The well is pumped by a 10 in diameter schedule-40 steel pipe which draws 120,000 gal/day. The hydraulic conductivity of the well is 1600 gal/day-ft^2. The pump discharges into a piping network whose friction head is 100 ft. What net power is required for steady flow?

(A) 1.1 hp
(B) 2.6 hp
(C) 7.8 hp
(D) 15 hp

2. (*Time limit: one hour*) An aquifer consists of a homogeneous material 300 ft thick. The surface of the water table in this aquifer is 100 ft below ground surface. An 18 in diameter well extends through the top 100 ft and then 200 ft below the water table, for a total depth of 300 ft. The aquifer transmissivity is 10,000 gal/day-ft. The well's radius of influence is 900 ft with a 20 ft drawdown at the well.

(a) The geologic formation in which the well is installed is called

(A) an aquifuge.
(B) an artesian well.
(C) a connate aquifer.
(D) an unconfined aquifer.

(b) What is the hydraulic conductivity of the aquifer?

(A) 50 gal/day-ft^2
(B) 500 gal/day-ft^2
(C) 1300 gal/day-ft^2
(D) 10,000 gal/day-ft^2

(c) What steady discharge is possible?

(A) 0.26 ft^3/sec
(B) 0.52 ft^3/sec
(C) 1.2 ft^3/sec
(D) 5.2 ft^3/sec

(d) What is the drawdown 100 ft from the well?

(A) 1 ft
(B) 6 ft
(C) 13 ft
(D) 18 ft

(e) Assuming a reasonable pump efficiency, what horsepower motor should be selected to achieve the steady discharge?

(A) 3.6 hp
(B) 5.5 hp
(C) 7.5 hp
(D) 12.2 hp

The same well conditions are found at an adjacent site where the aquifer extends 100 ft below a layer of low permeability that is 200 ft thick. The piezometric surface is 100 ft below the ground surface.

(f) The geologic formation in which the well is installed is called

(A) an aquifuge.
(B) a confined aquifer.
(C) a spring.
(D) a vadose well.

(g) The 18 in diameter well extends 300 ft below the ground surface to the bottom of the aquifer. The aquifer transmissivity is 10,000 gal/day-ft. What is the hydraulic conductivity of the aquifer?

(A) 50 gal/day-ft^2
(B) 100 gal/day-ft^2
(C) 500 gal/day-ft^2
(D) 800 gal/day-ft^2

(h) What is the steady discharge in the well if the radius of influence is 900 ft with a 20 ft drawdown at the well?

(A) 0.27 ft^3/sec
(B) 0.62 ft^3/sec
(C) 1.2 ft^3/sec
(D) 3.5 ft^3/sec

(i) At what distance from the well is the drawdown equal to 10 ft?

(A) 3 ft
(B) 26 ft
(C) 130 ft
(D) 450 ft

(j) After pumping is stopped, groundwater flows in one direction through the aquifer. If the aquifer is only 100 ft wide and the porosity is 0.4, what is the area in clear flow?

(A) 2000 ft^2
(B) 4000 ft^2
(C) 8000 ft^2
(D) 20,000 ft^2

Flow Nets

3. (*Time limit: one hour*) An impervious concrete dam is shown. The dam reduces seepage by using two impervious sheets extending 5 m below the dam bottom.

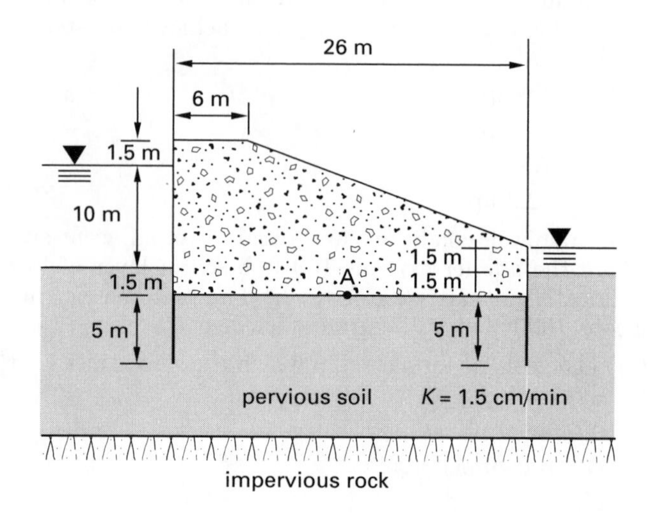

(a) Sketch the flow net.

(b) Determine the seepage per meter of width.

(A) 0.03 m^3/min
(B) 0.1 m^3/min
(C) 0.3 m^3/min
(D) 0.9 m^3/min

(c) What is the uplift pressure on the dam at point A, midway between the left and right edges?

(A) 17 kPa
(B) 33 kPa
(C) 71 kPa
(D) 91 kPa

4. (*Time limit: one hour*) A cofferdam in a river is shown. Sheet piles extend below the mud line, and the cofferdam floor is unlined. (The figure is drawn to scale.) Draw the initial flow net.

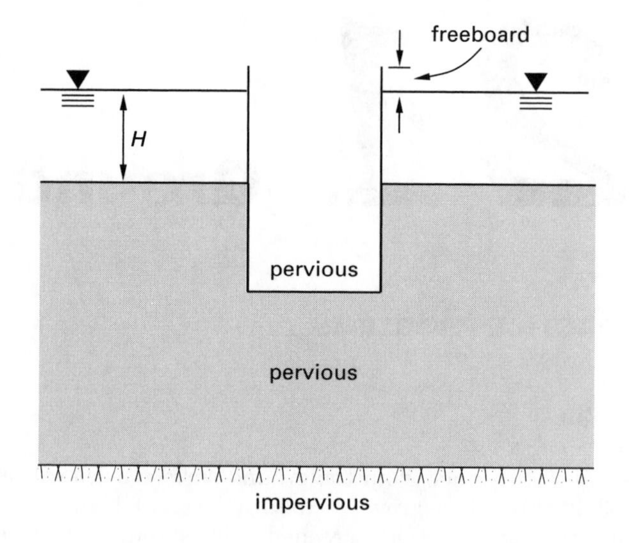

5. (*Time limit: one hour*) The concrete dam shown is drawn to scale. The silt has a hydraulic conductivity of 0.15 ft/hr.

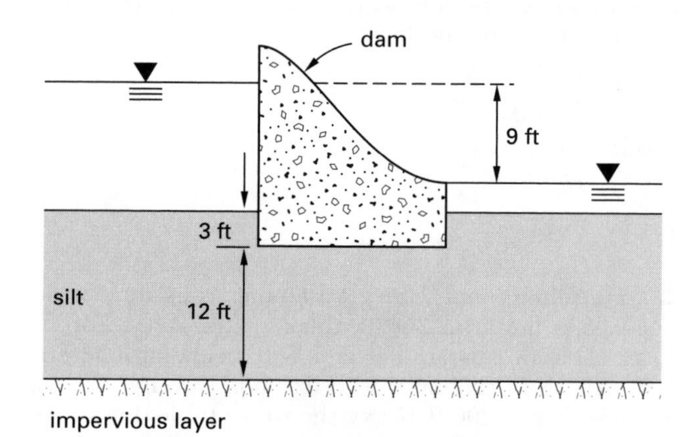

(a) Sketch the flow net for the dam.

(b) What is the approximate seepage rate per foot of width?

(A) 0.05 ft^3/hr-ft width
(B) 0.20 ft^3/hr-ft width
(C) 0.40 ft^3/hr-ft width
(D) 0.90 ft^3/hr-ft width

SOLUTIONS

1.

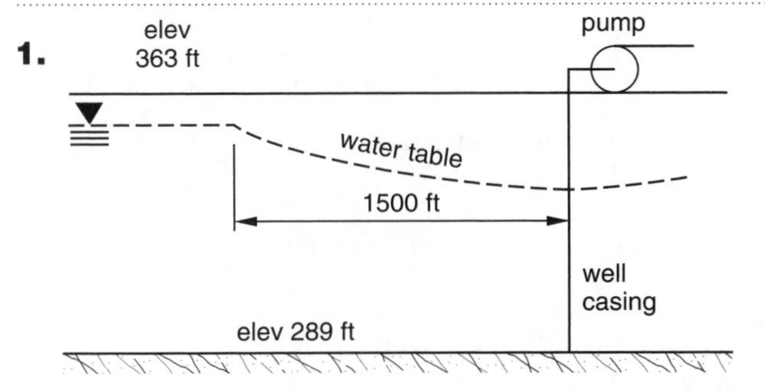

From App. 16.B, for schedule-40 pipe,

$$D_i = 0.835 \text{ ft}$$
$$A_i = 0.5476 \text{ ft}^2$$
$$Q = \left(120{,}000 \ \frac{\text{gal}}{\text{day}}\right)\left(1.547\times10^{-6} \ \frac{\text{ft}^3\text{-day}}{\text{sec-gal}}\right) = 0.1856 \text{ ft}^3/\text{sec}$$
$$K_p = \left(1600 \ \frac{\text{gal}}{\text{day-ft}^2}\right)\left(0.1337 \ \frac{\text{ft}^3}{\text{gal}}\right) = 213.9 \text{ ft}^3/\text{day-ft}^2$$
$$y_1 = 363 \text{ ft} - 289 \text{ ft} = 74 \text{ ft}$$
$$r_1 = 1500 \text{ ft}$$
$$r_2 = \frac{0.835 \text{ ft}}{2} = 0.4175 \text{ ft}$$

From Eq. 21.25,

$$0.1856 \ \frac{\text{ft}^3}{\text{sec}} = \frac{\pi\left(213.9 \ \frac{\text{ft}^3}{\text{day-ft}^2}\right)\left((74 \text{ ft})^2 - y_2^2\right)}{\left(86{,}400 \ \frac{\text{sec}}{\text{day}}\right)\ln\left(\frac{1500 \text{ ft}}{0.4175 \text{ ft}}\right)}$$
$$195.36 \text{ ft}^2 = (74 \text{ ft})^2 - y_2^2$$
$$y_2 = 72.67 \text{ ft}$$

The drawdown is

$$d = 74 \text{ ft} - 72.67 \text{ ft} = 1.33 \text{ ft}$$

d is small compared to y.

The velocity of the water in the pipe is

$$\text{v} = \frac{Q}{A} = \frac{0.1856 \ \frac{\text{ft}^3}{\text{sec}}}{0.5476 \text{ ft}^2} = 0.34 \text{ ft/sec}$$

This velocity is too small to include the velocity head.

The suction lift is

$$383 \text{ ft} - 363 \text{ ft} + 1.33 \text{ ft} = 21.33 \text{ ft}$$

$$H = 21.33 \text{ ft} + 100 \text{ ft} = 121.33 \text{ ft}$$

The mass flow of the water is

$$\dot{m} = \left(0.1856 \ \frac{\text{ft}^3}{\text{sec}}\right)\left(62.4 \ \frac{\text{lbm}}{\text{ft}^3}\right) = 11.58 \text{ lbm/sec}$$

From Table 18.5, the net water horsepower is

$$P = \left(\frac{h_A\dot{m}}{550}\right)\left(\frac{g}{g_c}\right)$$
$$= \left(\frac{(121.33 \text{ ft})\left(11.58 \ \frac{\text{lbm}}{\text{sec}}\right)}{550 \ \frac{\text{ft-lbf}}{\text{hp-sec}}}\right)\left(\frac{32.2 \ \frac{\text{ft}}{\text{sec}^2}}{32.2 \ \frac{\text{ft-lbm}}{\text{lbf-sec}^2}}\right)$$
$$= \boxed{2.55 \text{ hp}}$$

The answer is (B).

2. (a) The geologic formation is an unconfined aquifer. If the aquifer were overlain by an impermeable layer, it would be a confined, or artesian, aquifer.

The answer is (D).

(b) From Eq. 21.13,

$$K = \frac{T}{Y} = \frac{10{,}000 \ \frac{\text{gal}}{\text{day-ft}}}{200 \text{ ft}} = \boxed{50 \text{ gal/day-ft}^2}$$

The answer is (A).

(c)

$$y_1 = 200 \text{ ft at } r_1 = 900 \text{ ft}$$
$$y_2 = 200 \text{ ft} - 20 \text{ ft} = 180 \text{ ft at } r_2$$
$$r_2 = \frac{\frac{18 \text{ in}}{2}}{12 \ \frac{\text{in}}{\text{ft}}} = 0.75 \text{ ft}$$

From Eq. 21.25,

$$Q = \frac{\pi K(y_1^2 - y_2^2)}{\ln\left(\frac{r_1}{r_2}\right)}$$
$$= \frac{\pi\left(50 \ \frac{\text{gal}}{\text{day-ft}^2}\right)\left(0.13368 \ \frac{\text{ft}^3}{\text{gal}}\right) \times \left((200 \text{ ft})^2 - (180 \text{ ft})^2\right)}{\ln\left(\frac{900 \text{ ft}}{0.75 \text{ ft}}\right)\left(86{,}400 \ \frac{\text{sec}}{\text{day}}\right)}$$
$$= \boxed{0.261 \text{ ft}^3/\text{sec}}$$

The answer is (A).

(d)

$$y_1 = 200 \text{ ft at } r_1$$
$$r_1 = 900 \text{ ft}$$
$$r_2 = 100 \text{ ft}$$
$$Q = 0.261 \text{ ft}^3/\text{sec}$$

Rearranging Eq. 21.25,

$$\begin{aligned} y_2^2 &= y_1^2 - \frac{Q \ln\left(\frac{r_1}{r_2}\right)}{\pi K} \\ &= (200 \text{ ft})^2 \\ &\quad - \frac{\left(0.261 \frac{\text{ft}^3}{\text{sec}}\right) \ln\left(\frac{900 \text{ ft}}{100 \text{ ft}}\right) \left(86{,}400 \frac{\text{sec}}{\text{day}}\right)}{\pi \left(50 \frac{\text{gal}}{\text{day-ft}^2}\right) \left(0.13368 \frac{\text{ft}^3}{\text{gal}}\right)} \\ y_2 &= 194 \text{ ft} \end{aligned}$$

The drawdown is

$$200 \text{ ft} - 194 \text{ ft} = \boxed{6 \text{ ft}}$$

The answer is (B).

(e) This is a low-capacity pump, so the efficiency will be lower. Assume a pump efficiency of 0.65. From Table 18.5, the hydraulic horsepower is

$$\begin{aligned} P &= \frac{h_A \dot{V}(\text{SG})}{8.814} \\ &= \frac{(100 \text{ ft} + 20 \text{ ft}) \left(0.261 \frac{\text{ft}^3}{\text{sec}}\right)(1)}{\left(8.814 \frac{\text{sec}}{\text{ft}^4\text{-hp}}\right)(0.65)} \\ &= \boxed{5.47 \text{ hp}} \end{aligned}$$

Choose a standard motor size of 7.5 hp.

The answer is (C).

(f) The geologic formation is a confined aquifer. The well is an artesian well.

The answer is (B).

(g) The aquifer depth, Y, is 100 ft. Y is the thickness of the aquifer, not the height of the water table or piezometric surface.

From Eq. 21.13,

$$\begin{aligned} K &= \frac{T}{Y} = \frac{10{,}000 \frac{\text{gal}}{\text{day-ft}}}{100 \text{ ft}} \\ &= \boxed{100 \text{ gal/day-ft}^2} \end{aligned}$$

The answer is (B).

(h) The well radius of influence is

$$r_1 = 900 \text{ ft}$$
$$y_1 = 200 \text{ ft at } r_1$$

Calculate the well casing radius.

$$r_2 = \frac{\frac{18 \text{ in}}{2}}{12 \frac{\text{in}}{\text{ft}}} = 0.75 \text{ ft}$$

$$y_2 = 200 \text{ ft} - 20 \text{ ft} = 180 \text{ ft at } r_2$$

From Eq. 21.27,

$$\begin{aligned} Q &= \frac{2\pi K(y_1 - y_2)Y}{\ln\left(\frac{r_1}{r_2}\right)} \\ &= \frac{2\pi \left(100 \frac{\text{gal}}{\text{day-ft}^2}\right)(200 \text{ ft} - 180 \text{ ft})(100 \text{ ft})}{\ln\left(\frac{900 \text{ ft}}{0.75 \text{ ft}}\right)} \\ &= \frac{\left(1.772 \times 10^5 \frac{\text{gal}}{\text{day}}\right)\left(0.002228 \frac{\text{ft}^3\text{-min}}{\text{gal-sec}}\right)}{\left(24 \frac{\text{hr}}{\text{day}}\right)\left(60 \frac{\text{min}}{\text{hr}}\right)} \\ &= \boxed{0.274 \text{ ft}^3/\text{sec}} \end{aligned}$$

The answer is (A).

(i) Rearrange Eq. 21.27 to find the drawdown distance, r_2.

$$y_2 = 200 \text{ ft} - 10 \text{ ft} = 190 \text{ ft}$$

$$\begin{aligned} \ln\left(\frac{r_1}{r_2}\right) &= 2\pi K(y_1 - y_2)\left(\frac{Y}{Q}\right) \\ &= \frac{2\pi \left(100 \frac{\text{gal}}{\text{day-ft}^2}\right)(200 \text{ ft} - 190 \text{ ft})(100 \text{ ft})}{177{,}239 \frac{\text{gal}}{\text{day}}} \\ &= 3.545 \\ r_2 &= \frac{r_1}{e^{3.545}} = \frac{900 \text{ ft}}{e^{3.545}} \\ &= \boxed{26 \text{ ft}} \end{aligned}$$

The answer is (B).

(j) Darcy's law assumes the total cross-sectional area. To obtain the cross-sectional area of the pores (voids), multiply by the porosity.

$$\begin{aligned} A_{\text{clear flow}} &= nA = nbY \\ &= (0.4)(100 \text{ ft})(100 \text{ ft}) \\ &= \boxed{4000 \text{ ft}^2} \end{aligned}$$

The answer is (B).

3. (a) One possible flow net is shown.

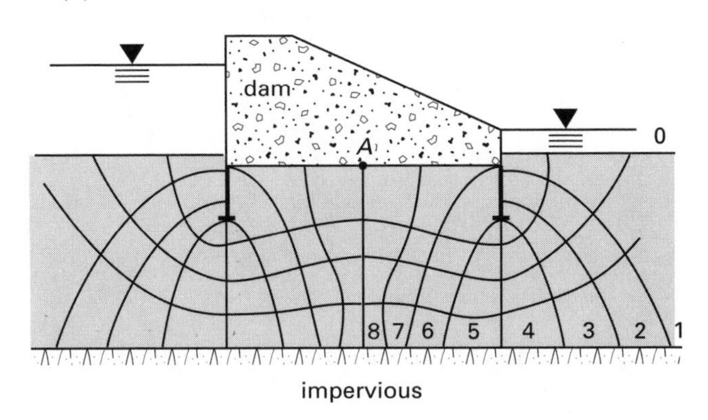

(b) Use Eq. 21.33 to solve for the flow rate.

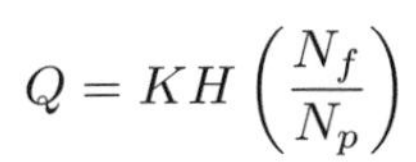

$$Q = KH\left(\frac{N_f}{N_p}\right)$$

From the flow net as drawn, $N_f = 4$ and $N_p = 16$.

$$H = 10\ \text{m} - 1.5\ \text{m} = 8.5\ \text{m}$$

$$Q = \left(1.5\ \frac{\text{cm}}{\text{min}}\right)\left(\frac{1\ \text{m}}{100\ \text{cm}}\right)(8.5\ \text{m})\left(\tfrac{4}{16}\right)\left(1\ \frac{\text{m}}{\text{m}}\right)$$

$$= \boxed{0.031875\ \text{m}^3/\text{min}\quad [\text{per meter of width}]}$$

The answer is (A).

(c) Point A is located at an elevation 3 ft lower than the downstream water level. From Eq. 21.36,

$$p_u = \left(\left(\frac{j}{N_p}\right)H + z\right)\rho_w g$$

$$= \left(\left(\tfrac{8}{16}\right)(8.5\ \text{m}) + 3.0\ \text{m}\right)\left(1000\ \frac{\text{kg}}{\text{m}^3}\right)\left(9.81\ \frac{\text{m}}{\text{s}^2}\right)$$

$$= \boxed{71\,122\ \text{Pa}\quad (71\ \text{kPa})}$$

The answer is (C).

4.

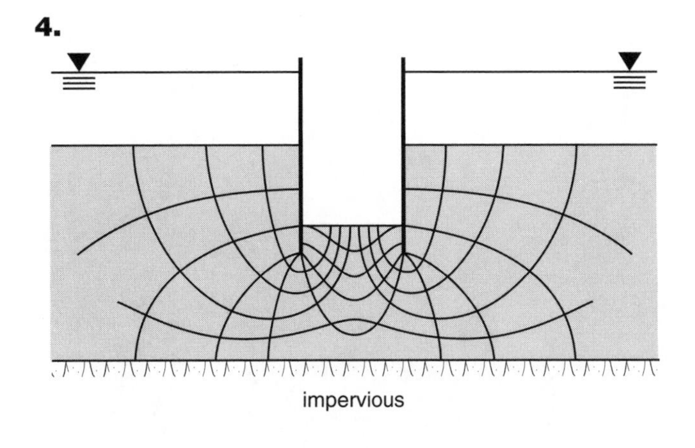

5. (a)

(b) From Eq. 21.33, with 4 flow paths and 13 equipotential drops,

$$Q = \left(0.15\ \frac{\text{ft}}{\text{hr}}\right)(9.0\ \text{ft})\left(\tfrac{4}{13}\right)$$

$$= \boxed{0.42\ \text{ft}^3/\text{hr-ft width}}$$

The answer is (C).

22 Inorganic Chemistry

PRACTICE PROBLEMS

Empirical Formula Development

1. The gravimetric analysis of a compound is 40% carbon, 6.7% hydrogen, and 53.3% oxygen. What is the simplest formula for the compound?

(A) HCO
(B) HCO_2
(C) CH_2O
(D) CHO_2

SOLUTIONS

1. Calculate the mole ratios of the atoms by assuming there are 100 g of sample.

For 100 g of sample,

substance	mass	$\frac{m}{AW}$ = no. moles	mole ratio
C	40 g	$\frac{40}{12} = 3.33$	1
H	6.7 g	$\frac{6.7}{1} = 6.7$	2
O	53.3 g	$\frac{53.3}{16} = 3.33$	1

The empirical formula is $\boxed{CH_2O.}$

The answer is (C).

23 Organic Chemistry

PRACTICE PROBLEMS

There are no problems in this book corresponding to Ch. 23 of the *Civil Engineering Reference Manual*.

24 Combustion and Incineration

PRACTICE PROBLEMS

1. Methane (MW = 16.043) with a heating value of 24,000 Btu/lbm (55.8 MJ/kg) is burned in a furnace with a 50% efficiency. How much water can be heated from 60 to 200°F (15 to 95°C) when 7 ft^3 (200 L) at 60°F and 14.73 psia are burned?

(A) 25 lbm (11 kg)
(B) 35 lbm (16 kg)
(C) 50 lbm (23 kg)
(D) 95 lbm (43 kg)

2. 15 lbm/hr (6.8 kg/h) of propane (C_3H_8, MW = 44.097) are burned stoichiometrically in air. What volume of dry carbon dioxide (CO_2) is formed after cooling to 70°F (21°C) and 14.7 psia (101 kPa)?

(A) 180 ft^3/hr (5.0 m^3/h)
(B) 270 ft^3/hr (7.6 m^3/h)
(C) 390 ft^3/hr (11 m^3/h)
(D) 450 ft^3/hr (13 m^3/h)

3. In a particular installation, 30% excess air at 15 psia (103 kPa) and 100°F (40°C) is needed for the combustion of methane. How much nitrogen (MW = 28.016) passes through the furnace if methane is burned at the rate of 4000 ft^3/hr (31 L/s)?

(A) 270 lbm/hr (0.033 kg/s)
(B) 930 lbm/hr (0.11 kg/s)
(C) 1800 lbm/hr (0.22 kg/s)
(D) 2700 lbm/hr (0.34 kg/s)

4. Propane (C_3H_8) is burned with 20% excess air. What is the gravimetric percentage of carbon dioxide in the flue gas?

(A) 8%
(B) 12%
(C) 15%
(D) 22%

5. (*Time limit: one hour*) A natural gas is 93% methane, 3.73% nitrogen, 1.82% hydrogen, 0.45% carbon monoxide, 0.35% oxygen, 0.25% ethylene, 0.22% carbon dioxide, and 0.18% hydrogen sulfide by volume. The gas is burned with 40% excess air. Atmospheric air is 60°F and at standard atmospheric pressure.

(a) What is the gas density?

(A) 0.017 lbm/ft^3
(B) 0.043 lbm/ft^3
(C) 0.069 lbm/ft^3
(D) 0.110 lbm/ft^3

(b) What are the theoretical air requirements?

(A) 5 ft^3 air/ft^3 fuel
(B) 7 ft^3 air/ft^3 fuel
(C) 9 ft^3 air/ft^3 fuel
(D) 13 ft^3 air/ft^3 fuel

(c) What is the percentage of CO_2 in the flue gas (wet basis)?

(A) 6.9%
(B) 7.7%
(C) 8.1%
(D) 11%

(d) What is the percentage of CO_2 in the flue gas (dry basis)?

(A) 6.9%
(B) 7.7%
(C) 8.1%
(D) 11%

6. (*Time limit: one hour*) A utility boiler burns coal with an ultimate analysis of 76.56% carbon, 7.7% oxygen, 6.1% silicon, 5.5% hydrogen, 2.44% sulfur, and 1.7% nitrogen. 410 lbm/hr of refuse are removed with a composition of 30% carbon and 0% sulfur. All the sulfur and the remaining carbon is burned. The power plant has the following characteristics.

- coal feed rate: 15,300 lbm/hr
- electric power rating: 17 MW
- generator efficiency: 95%
- steam generator efficiency: 86%
- cooling water rate: 225 ft^3/sec

(a) What is the emission rate of solid particulates in lbm/hr?

(A) 23 lbm/hr
(B) 150 lbm/hr
(C) 810 lbm/hr
(C) 1700 lbm/hr

(b) How much sulfur dioxide is produced per hour?

(A) 220 lbm/hr
(B) 340 lbm/hr
(C) 750 lbm/hr
(D) 1100 lbm/hr

(c) What is the temperature rise of the cooling water?

(A) 2.4°F
(B) 6.5°C
(C) 9.8°F
(D) 13°F

(d) What efficiency must the flue gas particulate collectors have in order to meet a limit of 0.1 lbm of particulates per million Btu per hour (0.155 kg/MW)?

(A) 93.1%
(B) 97.4%
(C) 98.8%
(D) 99.1%

SOLUTIONS

1. *Customary U.S. Solution*

$$T = 60°\text{F} + 460 = 520°\text{R}$$
$$p = 14.73 \text{ psia}$$

The specific gas constant, R, is calculated from the universal gas constant, R^*, and the molecular weight.

$$R = \frac{R^*}{\text{MW}} = \frac{1545.33 \ \frac{\text{ft-lbf}}{\text{lbmol-°R}}}{16.043 \ \frac{\text{lbm}}{\text{lbmol}}}$$
$$= 96.32 \text{ ft-lbf/lbm-°R}$$
$$m = \frac{pV}{RT}$$
$$= \frac{\left(14.73 \ \frac{\text{lbf}}{\text{in}^2}\right)\left(144 \ \frac{\text{in}^2}{\text{ft}^2}\right)(7 \text{ ft}^3)}{\left(96.32 \ \frac{\text{ft-lbf}}{\text{lbm-°R}}\right)(520°\text{R})}$$
$$= 0.296 \text{ lbm}$$

The energy available from methane is

$$Q = \eta m(\text{HHV})$$
$$= (0.5)(0.296 \text{ lbm})\left(24{,}000 \ \frac{\text{Btu}}{\text{lbm}}\right)$$
$$= 3552 \text{ Btu}$$

This energy is used by water to heat from 60°F to 200°F.

$$Q = m_{\text{water}} c_p (T_2 - T_1)$$
$$m_{\text{water}} = \frac{3552 \text{ Btu}}{\left(1 \ \frac{\text{Btu}}{\text{lbm-°F}}\right)(200°\text{F} - 60°\text{F})} = \boxed{25.37 \text{ lbm}}$$

The answer is (A).

SI Solution

$$T = (60°\text{F} + 460)\left(\frac{1 \text{ K}}{1.8°\text{R}}\right) = 288.89\text{K}$$
$$p = (14.73 \text{ psia})\left(\frac{101.325 \text{ kPa}}{14.696 \text{ psia}}\right) = 101.56 \text{ kPa}$$

The specific gas constant, R, is calculated from the universal gas constant, R^*, and the molecular weight.

$$R = \frac{R^*}{\text{MW}} = \frac{8314.3 \ \frac{\text{J}}{\text{kmol·K}}}{16.043 \ \frac{\text{kg}}{\text{kmol}}}$$

$$= 518.25 \ \text{J/kg·K}$$

$$m = \frac{pV}{RT}$$

$$= \frac{(101.56 \ \text{kPa})\left(1000 \ \frac{\text{Pa}}{\text{kPa}}\right)(200 \ \text{L})\left(\frac{1 \ \text{m}^3}{1000 \ \text{L}}\right)}{\left(518.25 \ \frac{\text{J}}{\text{kg·K}}\right)(288.89\text{K})}$$

$$= 0.136 \ \text{kg}$$

The energy available from methane is

$$Q = \eta m(\text{HHV})$$

$$= (0.5)(0.136 \ \text{kg})\left(55.8 \ \frac{\text{MJ}}{\text{kg}}\right)\left(1000 \ \frac{\text{kJ}}{\text{MJ}}\right)$$

$$= 3794 \ \text{kJ}$$

This energy is used by water to heat it from 15°C to 95°C.

$$Q = m_{\text{water}} c_p (T_2 - T_1)$$

$$m_{\text{water}} = \frac{3794 \ \text{kJ}}{\left(4.1868 \ \frac{\text{kJ}}{\text{kg·C}}\right)(95°\text{C} - 15°\text{C})} = \boxed{11.33 \ \text{kg}}$$

The answer is (A).

2. *Customary U.S. Solution*

From Table 24.7,

	C_3H_8	+	$5O_2$	$\longrightarrow$	$3CO_2$	+ $4H_2O$
MW	44.097		(5)(32)		(3)(44.011)	
	44.097		160		132.033	

The amount of carbon dioxide produced is 132.033 lbm/44.097 lbm propane. For 15 lbm/hr of propane, the amount of carbon dioxide produced is

$$\left(\frac{132.033 \ \text{lbm}}{44.097 \ \text{lbm}}\right)\left(15 \ \frac{\text{lbm}}{\text{hr}}\right) = 44.91 \ \text{lbm/hr}$$

$$R = \frac{R^*}{\text{MW}} = \frac{1545.33 \ \frac{\text{ft-lbf}}{\text{lbmol-°R}}}{44.011 \ \frac{\text{lbm}}{\text{lbmol}}}$$

$$= 35.11 \ \text{ft-lbf/lbm-°R}$$

$$T = 70°\text{F} + 460 = 530°\text{R}$$

$$\dot{V} = \frac{\dot{m}RT}{p}$$

$$= \frac{\left(44.91 \ \frac{\text{lbm}}{\text{hr}}\right)\left(35.11 \ \frac{\text{ft-lbf}}{\text{lbm-°R}}\right)(530°\text{R})}{\left(14.7 \ \frac{\text{lbf}}{\text{in}^2}\right)\left(144 \ \frac{\text{in}^2}{\text{ft}^2}\right)}$$

$$= \boxed{394.8 \ \text{ft}^3/\text{hr} \quad (395 \ \text{ft}^3/\text{hr})}$$

The answer is (C).

SI Solution

From Table 24.7,

	C_3H_8	+	$5O_2$	$\longrightarrow$	$3CO_2$	+ $4H_2O$
MW	44.097		(5)(32)		(3)(44.011)	
	44.097		160		132.033	

The amount of carbon dioxide produced is 132.033 kg/44.097 kg propane. For 6.8 kg/h of propane, the amount of carbon dioxide produced is

$$\left(\frac{132.033 \ \text{kg}}{44.097 \ \text{kg}}\right)\left(6.8 \ \frac{\text{kg}}{\text{h}}\right) = 20.36 \ \text{kg/h}$$

$$R = \frac{R^*}{\text{MW}} = \frac{8314.3 \ \frac{\text{J}}{\text{kmol·K}}}{44.01 \ \frac{\text{kg}}{\text{kmol}}}$$

$$= 188.92 \ \text{J/kg·K}$$

$$T = 21°\text{C} + 273 = 294\text{K}$$

$$V = \frac{mRT}{p}$$

$$= \frac{\left(20.36 \ \frac{\text{kg}}{\text{h}}\right)\left(188.92 \ \frac{\text{J}}{\text{kg·K}}\right)(294\text{K})}{(101 \ \text{kPa})\left(1000 \ \frac{\text{Pa}}{\text{kPa}}\right)}$$

$$= \boxed{11.20 \ \text{m}^3/\text{h} \quad (11 \ \text{m}^3/\text{h})}$$

The answer is (C).

3. Use the balanced chemical reaction equation from Table 24.7.

$$CH_4 + 2O_2 \longrightarrow CO_2 + 2H_2O$$

With 30% excess air and considering there are 3.773 volumes of nitrogen for every volume of oxygen (Table 24.6), the reaction equation is

$$CH_4 + (1.3)(2)O_2 + (1.3)(2)(3.773)N_2 \longrightarrow CO_2 + 2H_2O + (1.3)(2)(3.773)N_2 + 0.6O_2$$

$$CH_4 + 2.6O_2 + 9.81N_2 \longrightarrow CO_2 + 2H_2O + 9.81N_2 + 0.6O_2$$

Water Resources

Customary U.S. Solution

The volume of nitrogen that accompanies 4000 ft^3/hr of entering methane is

$$V_{N_2} = \left(\frac{9.81\ ft^3\ N_2}{1\ ft^3\ CH_4}\right)\left(4000\ \frac{ft^3}{hr}\ CH_4\right)$$
$$= 39{,}240\ ft^3\ N_2/hr$$

This is the "partial volume" of nitrogen in the input stream.

$$R = \frac{R^*}{MW} = \frac{1545.33\ \frac{ft\text{-}lbf}{lbmol\text{-}°R}}{28.016\ \frac{lbm}{lbmol}}$$
$$= 55.16\ ft\text{-}lbf/lbm\text{-}°R$$

The absolute temperature is

$$T = 100°F + 460 = 560°R$$

$$m_{N_2} = \frac{p_{N_2}V_{N_2}}{RT}$$
$$= \frac{\left(15\ \frac{lbf}{in^2}\right)\left(144\ \frac{in^2}{ft^2}\right)\left(39{,}240\ \frac{ft^3}{hr}\right)}{\left(55.16\ \frac{ft\text{-}lbf}{lbm\text{-}°R}\right)(560°R)}$$
$$= \boxed{2744\ lbm/hr}$$

The answer is (D).

SI Solution

The volume of nitrogen that accompanies 31 L/s of entering methane is

$$\left(\frac{9.81\ m^3\ N_2}{1\ m^3\ CH_4}\right)\left(31\ \frac{L}{s}\right)\left(\frac{1\ m^3}{1000\ L}\right) = 0.3041\ m^3/s$$

This is the "partial volume" of nitrogen in the input stream.

$$R = \frac{R^*}{MW} = \frac{8314.3\ \frac{J}{kmol\cdot K}}{28.016\ \frac{kg}{kmol}}$$
$$= 296.8\ J/kg\cdot K$$

The absolute temperature is

$$T = 40°C + 273 = 313K$$

$$m_{N_2} = \frac{p_{N_2}V_{N_2}}{RT}$$
$$= \frac{(103\ kPa)\left(1000\ \frac{Pa}{kPa}\right)\left(0.3041\ \frac{m^3}{s}\right)}{\left(296.8\ \frac{J}{kg\cdot K}\right)(313K)}$$
$$= \boxed{0.337\ kg/s}$$

The answer is (D).

4. The balanced chemical reaction equation is

$$C_3H_8 + 5O_2 \longrightarrow 3CO_2 + 4H_2O$$

With 20% excess air, the oxygen volume is $(1.2)(5) = 6$.

$$C_3H_8 + 6O_2 \longrightarrow 3CO_2 + 4H_2O + O_2$$

From Table 24.6, there are 3.773 volumes of nitrogen for every volume of oxygen.

$$(6)(3.773) = 22.6$$

$$C_3H_8 + 6O_2 + 22.6\ N_2 \longrightarrow 3CO_2 + 4H_2O + O_2 + 22.6\ N_2$$

The percentage of carbon dioxide by weight in flue gas is

$$G_{CO_2} = \frac{(3)(44.011)}{(3)(44.011) + (4)(18.016) + 32 + (22.6)(28.016)}$$
$$= \boxed{0.152\quad (15.2\%)}$$

The answer is (C).

5. (a) For methane, $B = 0.93$.

From ideal gas laws (R for methane = 96.32 ft-lbf/lbm-°R),

$$\rho = \frac{p}{RT}$$
$$= \frac{\left(14.7\ \frac{lbf}{in^2}\right)\left(144\ \frac{in^2}{ft^2}\right)}{\left(96.32\ \frac{ft\text{-}lbf}{lbm\text{-}°R}\right)(60°F + 460)}$$
$$= 0.0422\ lbm/ft^3$$

From Table 24.9, $K = 9.55\ ft^3$ air/ft^3 fuel.

From Table 24.8,

products: $1ft^3\ CO_2$, $2\ ft^3\ H_2O$

From App. 24.A, HHV = 1013 Btu/ft^3.

Similar results for all the other fuel components are tabulated in the table on the next page.

The composite density is

$$\rho = \sum B_i\rho_i$$
$$= (0.93)\left(0.0422\ \frac{lbm}{ft^3}\right) + (0.0373)\left(0.0738\ \frac{lbm}{ft^3}\right)$$
$$+ (0.0045)\left(0.0738\ \frac{lbm}{ft^3}\right) + (0.0182)\left(0.0053\ \frac{lbm}{ft^3}\right)$$
$$+ (0.0025)\left(0.0739\ \frac{lbm}{ft^3}\right) + (0.0018)\left(0.0900\ \frac{lbm}{ft^3}\right)$$
$$+ (0.0035)\left(0.0843\ \frac{lbm}{ft^3}\right) + (0.0022)\left(0.1160\ \frac{lbm}{ft^3}\right)$$
$$= \boxed{0.0433\ lbm/ft^3}$$

The answer is (B).

Reaction Products for Prob. 5

gas	B	ρ $\left(\frac{\text{lbm}}{\text{ft}^3}\right)$	ft^3 air	HHV $\left(\frac{\text{Btu}}{\text{ft}^3}\right)$	volumes of products CO_2	H_2O	other
CH_4	0.93	0.0422	9.556	1013	1	2	–
N_2	0.0373	0.0738	–	–	–	–	1 N_2
CO	0.0045	0.0738	2.389	322	1	–	–
H_2	0.0182	0.0053	2.389	325	–	1	–
C_2H_4	0.0025	0.0739	14.33	1614	2	2	–
H_2S	0.0018	0.0900	7.167	647	–	1	1 SO_2
O_2	0.0035	0.0843	–	–	–	–	–
CO_2	0.0022	0.1160	–	–	1	–	–

(b) The air is 20.9% oxygen by volume. The theoretical air requirements are

$$\begin{aligned}\sum B_i V_{\text{air},i} - \frac{O_2 \text{ in fuel}}{0.209} &= (0.93)(9.556 \text{ ft}^3) + (0.0373)(0) \\ &\quad + (0.0045)(2.389 \text{ ft}^3) + (0.0182)(2.389 \text{ ft}^3) \\ &\quad + (0.0025)(14.33 \text{ ft}^3) + (0.0018)(7.167 \text{ ft}^3) \\ &\quad + (0.0035)(0) + (0.0022)(0) - \frac{0.0035}{0.209} \\ &= 8.990 \text{ ft}^3 - 0.01675 \text{ ft}^3 \\ &= \boxed{8.9733 \text{ ft}^3 \text{ air/ft}^3 \text{ fuel}}\end{aligned}$$

The answer is (C).

(c) and (d) The theoretical oxygen will be

$$(8.9733 \text{ ft}^3)(0.209) = 1.875 \text{ ft}^3/\text{ft}^3$$

The excess oxygen will be

$$(0.4)(1.875 \text{ ft}^3) = 0.75 \text{ ft}^3/\text{ft}^3$$

Similarly, the total nitrogen in the stack gases is

$$(1.4)(0.791)(8.9733 \text{ ft}^3) + 0.034 = 9.971 \text{ ft}^3/\text{ft}^3 \text{ fuel}$$

The stack gases per ft^3 of fuel are

$$\begin{aligned}&\text{excess } O_2: && = 0.7500 \text{ ft}^3 \\ &\text{excess } N_2: && = 9.971 \text{ ft}^3 \\ &\text{excess } SO_2: && = 0.0018 \text{ ft}^3 \\ &\text{excess } CO_2: (0.93)(1) + (0.0045)(1) \\ &\qquad + (0.0025)(2) + (0.0022)(1) && = 0.942 \text{ ft}^3 \\ &\text{excess } H_2O: (0.93)(2) + (0.0182)(1) \\ &\qquad + (0.0025)(2) + (0.0018)(1) && = 1.885 \text{ ft}^3 \\ &\text{total} && = 13.55 \text{ ft}^3\end{aligned}$$

The total wet volume is 13.55 ft^3/ft^3 fuel.

The total dry volume is 11.66 ft^3/ft^3 fuel.

The volumetric analyses are

	O_2	N_2	SO_2	CO_2	H_2O
wet:	$\frac{0.7500 \text{ ft}^3}{13.55 \text{ ft}^3}$	$\frac{9.971 \text{ ft}^3}{13.55 \text{ ft}^3}$	$\frac{0.0018 \text{ ft}^3}{13.55 \text{ ft}^3}$	$\frac{0.942 \text{ ft}^3}{13.55 \text{ ft}^3}$	$\frac{1.885}{13.55}$
	= 0.0554	0.736	–	0.070	0.139
dry:	$\frac{0.7500 \text{ ft}^3}{11.66 \text{ ft}^3}$	$\frac{9.971 \text{ ft}^3}{11.66 \text{ ft}^3}$	$\frac{0.0018 \text{ ft}^3}{11.66 \text{ ft}^3}$	$\frac{0.942 \text{ ft}^3}{11.66 \text{ ft}^3}$	–
	= 0.0643	0.855	–	0.081	–

(c) $B_{CO_2,\text{wet}} = \boxed{6.9\%}$

The answer is (A).

(d) $B_{CO_2,\text{dry}} = \boxed{8.1\%}$

The answer is (C).

6. (a) Silicon in ash is SiO_2 with a molecular weight of

$$28.09 \frac{\text{lbm}}{\text{lbmol}} + (2)\left(16 \frac{\text{lbm}}{\text{lbmol}}\right) = 60.09 \text{ lbm/lbmol}$$

The oxygen used with 6.1% by mass silicon is

$$\left(\frac{(2)(16 \text{ lbm})}{28.09 \text{ lbm}}\right)\left(0.061 \frac{\text{lbm}}{\text{lbm coal}}\right) = 0.0695 \text{ lbm/lbm coal}$$

Silicon ash produced per lbm of coal is

$$0.061 \frac{\text{lbm}}{\text{lbm coal}} + 0.0695 \frac{\text{lbm}}{\text{lbm coal}} = 0.1305 \text{ lbm/lbm coal}$$

Silicon ash produced per hour is

$$\left(0.1305\ \frac{\text{lbm}}{\text{lbm coal}}\right)\left(15{,}300\ \frac{\text{lbm coal}}{\text{hr}}\right) = 1996.7\ \text{lbm/hr}$$

The silicon in 410 lbm/hr refuse is

$$\left(410\ \frac{\text{lbm}}{\text{hr}}\right)(1-0.3) = 287\ \text{lbm/hr}$$

The emission rate is

$$1996.7\ \text{lbm/hr} - 287\ \text{lbm/hr} = \boxed{1709.7\ \text{lbm/hr}}$$

The answer is (D).

(b) From Table 24.7, the stoichiometric reaction for sulfur is

$$\begin{array}{lcccc} & \text{S} & + \ \text{O}_2 & \longrightarrow & \text{SO}_2 \\ \text{MW} & 32 & 32 & & 64 \end{array}$$

Sulfur dioxide produced for 15,300 lbm/hr of coal feed is

$$\left(15{,}300\ \frac{\text{lbm}}{\text{hr}}\right)(0.0244\ \text{lbm S})\left(\frac{64\ \text{lbm SO}_2}{32\ \text{lbm S}}\right) = \boxed{746.6\ \text{lbm/hr}}$$

The answer is (C).

(c) From Eq. 24.16(b), the heating value of the fuel is

$$\begin{aligned}\text{HHV} &= 14{,}093G_\text{C} + (60{,}958)\left(G_\text{H} - \frac{G_\text{O}}{8}\right) + 3983G_\text{S} \\ &= \left(14{,}093\ \frac{\text{Btu}}{\text{lbm}}\right)(0.7656) \\ &\quad + \left(60{,}958\ \frac{\text{Btu}}{\text{lbm}}\right)\left(0.055 - \frac{0.077}{8}\right) \\ &\quad + \left(3983\ \frac{\text{Btu}}{\text{lbm}}\right)(0.0244) \\ &= 13{,}653\ \text{Btu/lbm}\end{aligned}$$

The gross available combustion power is

$$\dot{m}_f(\text{HV}) = \left(15{,}300\ \frac{\text{lbm}}{\text{hr}}\right)\left(13{,}653\ \frac{\text{Btu}}{\text{lbm}}\right) = 2.089 \times 10^8\ \text{Btu/hr}$$

The carbon in 410 lbm/hr refuse is

$$\left(410\ \frac{\text{lbm}}{\text{hr}}\right)(0.3) = 123\ \text{lbm/hr}$$

Power lost in unburned carbon in refuse is $\dot{m}_\text{C}(\text{HV})$.

From App. 24.A, the gross heat of combustion for carbon is 14,093 Btu/lbm.

$$\left(123\ \frac{\text{lbm}}{\text{hr}}\right)\left(14{,}093\ \frac{\text{Btu}}{\text{lbm}}\right) = 1.733 \times 10^6\ \text{Btu/hr}$$

The remaining combustion power is

$$2.089 \times 10^8\ \frac{\text{Btu}}{\text{hr}} - 1.733 \times 10^6\ \frac{\text{Btu}}{\text{hr}} = 2.072 \times 10^8\ \text{Btu/hr}$$

Losses in the steam generator and electrical generator will further reduce this to

$$(0.86)\left(2.072 \times 10^8\ \frac{\text{Btu}}{\text{hr}}\right) = 1.782 \times 10^8\ \text{Btu/hr}$$

With an electrical output of 17 MW, thermal energy removed by cooling water is

$$\begin{aligned}Q &= 1.782 \times 10^8\ \frac{\text{Btu}}{\text{hr}} - (17\ \text{MW})\left(1000\ \frac{\text{kW}}{\text{MW}}\right) \\ &\quad \times \left(3413\ \frac{\frac{\text{Btu}}{\text{hr}}}{\text{kW}}\right) \\ &= 1.202 \times 10^8\ \text{Btu/hr}\end{aligned}$$

The temperature rise of the cooling water is

$$\Delta T = \frac{Q}{\dot{m}c_p}$$

At 60°F, the specific heat of water is $c_p = 1$ Btu/lbm-°F.

$$\Delta T = \frac{1.202 \times 10^8\ \frac{\text{Btu}}{\text{hr}}}{\left(225\ \frac{\text{ft}^3}{\text{sec}}\right)\left(62.4\ \frac{\text{lbm}}{\text{ft}^3}\right)\left(3600\ \frac{\text{sec}}{\text{hr}}\right)\left(1\ \frac{\text{Btu}}{\text{lbm-°F}}\right)} = \boxed{2.38°\text{F}}$$

The answer is (A).

The electrical generation is not cooled by the cooling water. Therefore, it is not correct to include the generation efficiency in the calculation of losses.

(d) Limiting 0.1 lbm of particulates per million Btu per hour, the allowable emission rate is

$$\begin{aligned}&\left(0.1\ \frac{\text{lbm}}{\text{MBtu}}\right)\left(15{,}300\ \frac{\text{lbm}}{\text{hr}}\right) \\ &\quad \times \left(13{,}653\ \frac{\text{Btu}}{\text{lbm}}\right)\left(\frac{1\ \text{MBtu}}{10^6\ \text{Btu}}\right) = 20.89\ \text{lbm/hr}\end{aligned}$$

The efficiency of the flue gas particulate collectors is

$$\eta = \frac{\text{actual emission rate} - \text{allowable emission rate}}{\text{actual emission rate}}$$

$$= \frac{1709.7 \ \frac{\text{lbm}}{\text{hr}} - 20.89 \ \frac{\text{lbm}}{\text{hr}}}{1709.7 \ \frac{\text{lbm}}{\text{hr}}}$$

$$= \boxed{0.988 \ \ (98.8\%)}$$

The answer is (C).

Water Resources

25 Water Supply Quality and Testing

PRACTICE PROBLEMS

Phosphorus

1. (*Time limit: one hour*) In a study of a small pond to determine phosphorus impact, the following information was collected.

pond size: 12 ac-ft (14.8×10^6 L)
watershed area: 4 ac (grassland)
average annual rainfall: 5 in
bioavailable P in runoff: 0.01 mg/L
runoff coefficient: 0.1

Biological processes in the pond biota convert phosphorus to a nonbioavailable form at the rate of 22% per year. Recycling of sediment phosphorus by rooted plants and by anaerobic conditions in the hypolimnion converts 12% per year of the nonbioavailable phosphorus back to bioavailable forms.

Runoff into the pond evaporates during the year, so no change in pond volume occurs.

(a) Starting from an initial condition of 0.1 mg/L bioavailable P, what is the P concentration after five annual cycles?

(A) 0.03 mg/L
(B) 0.11 mg/L
(C) 0.13 mg/L
(D) 0.18 mg/L

(b) What can be done to reduce the P accumulation?

(A) Add chemicals to precipitate phosphorus in the pond.
(B) Add chemicals to combine with phosphorus in the pond.
(C) Reduce use of phosphorus-based fertilizers in the surrounding fields.
(D) Use natural-based soaps and detergents in the home.

Alkalinity

2. (*Time limit: one hour*) Groundwater is used for a water supply. It is taken from the ground at 25°C. The initial properties of the water are as follows.

CO_2	60 mg/L (as $CaCO_3$)
alkalinity	200 mg/L (as $CaCO_3$)
pH	7.1

The water is treated for CO_2 removal by spraying into the atmosphere through a nozzle. The final CO_2 concentration is 5.6 mg/L (as $CaCO_3$) at 25°C. The first ionization constant of carbonic acid is 4.45×10^{-7}. What is the final pH of the water after spraying and recovery, assuming the alkalinity is unchanged in the process?

(A) 7.3
(B) 7.9
(C) 8.4
(D) 8.8

Solids

3. (*Time limit: one hour*) The solids concentration of a stream water sample is to be determined. The total solids concentration is determined by placing a portion of the sample into a porcelain evaporating dish, drying the sample at 105°C, and igniting the residue by placing the dried sample in a muffle furnace at 550°C. The following masses are recorded.

mass of empty dish: 50.326 g
mass of dish and sample: 118.400 g
mass of dish and dry solids: 50.437 g
mass of dish and ignited solids: 50.383 g

(a) The total solids concentration is

(A) 900 mg/L
(B) 1000 mg/L
(C) 1100 mg/L
(D) 1600 mg/L

(b) The total volatile solids concentration is

(A) 630 mg/L
(B) 710 mg/L
(C) 790 mg/L
(D) 830 mg/L

(c) The total fixed solids concentration is

(A) 270 mg/L
(B) 300 mg/L
(C) 420 mg/L
(D) 840 mg/L

The suspended solids concentration is determined by filtering a portion of the sample through a glass-fiber filter disk, drying the disk at 105°C, and igniting the residue by placing the dried sample in a muffle furnace at 550°C. The follow masses are recorded.

volume of sample: 30 mL
mass of filter disk: 0.1170 g
mass of disk and dry solids: 0.1278 g
mass of disk and ignited solids: 0.1248 g

(d) The total suspended solids concentration is

(A) 240 mg/L
(B) 360 mg/L
(C) 370 mg/L
(D) 820 mg/L

(e) The volatile suspended solids concentration is

(A) 100 mg/L
(B) 120 mg/L
(C) 230 mg/L
(D) 640 mg/L

(f) The fixed suspended solids concentration is

(A) 120 mg/L
(B) 180 mg/L
(C) 240 mg/L
(D) 260 mg/L

The dissolved solids concentration is determined by filtering a portion of the sample through a glass-fiber filter disk into a porcelain evaporating dish, drying the sample at 105°C, and igniting the residue by placing the dried sample in a muffle furnace at 550°C. The following masses are recorded.

volume of sample: 25 mL
mass of empty dish: 51.494 g
mass of dish and dry solids: 51.524 g
mass of dish and ignited solids: 51.506 g

(g) The total dissolved solids concentration is

(A) 1000 mg/L
(B) 1100 mg/L
(C) 1200 mg/L
(D) 1400 mg/L

(h) The volatile dissolved solids concentration is

(A) 680 mg/L
(B) 720 mg/L
(C) 810 mg/L
(D) 900 mg/L

(i) The fixed dissolved solids concentration is

(A) 230 mg/L
(B) 300 mg/L
(C) 410 mg/L
(D) 480 mg/L

(j) We would expect the total solids determined in part (a) to be equal to the sum of the total suspended solids determined in (d) plus the total dissolved solids determined in (g). Does the total solids concentration equal the sum of the suspended and dissolved solids?

(A) yes
(B) no, probably due to poor laboratory technique
(C) no, probably due to rounding of measured values
(D) no, probably because the samples were representative but not identical

Hardness

4. (*Time limit: one hour*) The laboratory analysis of a water sample is as follows. All concentrations are "as substance."

Ca^{++}	74.0 mg/L
Mg^{++}	18.3 mg/L
Na^{+}	27.6 mg/L
K^{+}	39.1 mg/L
pH	7.8
HCO_3^{-}	274.5 mg/L
SO_4^{--}	72.0 mg/L
Cl^{-}	49.7 mg/L

(a) The hardness of the water in terms of mg/L of calcium carbonate equivalent is

(A) 1.3 mg/L as $CaCO_3$
(B) 4.7 mg/L as $CaCO_3$
(C) 66 mg/L as $CaCO_3$
(D) 260 mg/L as $CaCO_3$

(b) Based on the laboratory analysis, which of the following can be said?

(A) It is surprising that no carbonates were found in the sample.
(B) The cations in solution, when converted to milliequivalents, will equal the anions, when converted to milliequivalents.
(C) The large amount of bicarbonate in the solution tends to make the water acidic.
(D) None of the above can be said.

(c) Assuming that hypothetical compounds are formed proportionally to the relative concentrations of ions, the hypothetical concentration of calcium bicarbonate in the sample is

(A) 3.7 meq/L
(B) 4.5 meq/L
(C) 4.7 meq/L
(D) 74 meq/L

(d) The hypothetical concentration of magnesium bicarbonate is

(A) 0.8 meq/L
(B) 1.5 meq/L
(C) 1.6 meq/L
(D) 5.2 meq/L

(e) The hypothetical concentration of magnesium sulfate is

(A) 0.7 meq/L
(B) 1.5 meq/L
(C) 4.5 meq/L
(D) 5.2 meq/L

(f) The hypothetical concentration of sodium sulfate is

(A) 0.8 meq/L
(B) 1.2 meq/L
(C) 1.5 meq/L
(D) 6.4 meq/L

(g) The hypothetical concentration of sodium chloride is

(A) 0.2 meq/L
(B) 0.4 meq/L
(C) 1.2 meq/L
(D) 1.4 meq/L

(h) The amount of lime (CaO) necessary in a lime softening process to remove the hardness caused by calcium bicarbonate is

(A) 17 mg/L
(B) 28 mg/L
(C) 95 mg/L
(D) 100 mg/L

(i) To remove hardness caused by magnesium bicarbonate, it is necessary to raise the pH by adding 35 mg/L of CaO in excess of the stoichiometric requirements. The amount of lime (CaO) necessary to remove the carbonate hardness caused by magnesium bicarbonate is

(A) 12 mg/L
(B) 45 mg/L
(C) 56 mg/L
(D) 80 mg/L

(j) The water is subsequently recarbonated to reduce the pH. Assume that by this softening process calcium hardness can be reduced to 30 mg/L and magnesium hardness can be reduced to 10 mg/L, both measured in terms of equivalent calcium carbonate. The amount of hardness that remains in the water is

(A) 10 mg/L as $CaCO_3$
(B) 30 mg/L as $CaCO_3$
(C) 45 mg/L as $CaCO_3$
(D) 75 mg/L as $CaCO_3$

SOLUTIONS

1. (a)

$$\text{annual rainfall} = (4\text{ ac})(5\text{ in}) = 20\text{ ac-in}$$

$$\begin{aligned}\text{runoff to pond} &= (0.1)(20\text{ ac-in}) \times \left(102{,}790\ \frac{\text{L}}{\text{ac-in}}\right) \\ &= 205{,}600\text{ L}\end{aligned}$$

$$\begin{aligned}\begin{array}{r}\text{bioavailable P}\\ \text{reaching pond}\end{array} &= (205{,}600\text{ L})\left(0.01\ \frac{\text{mg}}{\text{L}}\right) \\ &= 2056\text{ mg}\end{aligned}$$

$$\begin{aligned}\begin{array}{r}\text{initial bioavailable}\\ \text{P in pond}\end{array} &= (14.8\times 10^6\text{ L})\left(0.1\ \frac{\text{mg}}{\text{L}}\right) \\ &= 14.8\times 10^5\text{ mg}\end{aligned}$$

The total bioavailable P in the pond is

$$2056\text{ mg} + 14.8\times 10^5\text{ mg} = 14.82058\times 10^5\text{ mg}$$

Of this, a net percentage of $100\% - 22\% + ((12\%)(22\%)/100\%) = 80.64\%$ remains after biological processes and recycling.

$$(0.8064)(14.82058\times 10^5\text{ mg}) = 11.95132\times 10^5\text{ mg}$$

The following table is prepared in a similar manner.

year	P at start of year	P from rainfall	total P available for activity	P at end of year
1	14.8×10^5	2056	14.82058×10^5	11.95132×10^5
2	11.95132×10^5	2056	11.97188×10^5	9.65412×10^5
3	9.65412×10^5	2056	9.67468×10^5	7.80166×10^5
4	7.80166×10^5	2056	7.82222×10^5	6.30784×10^5
5	6.30784×10^5	2056	6.32840×10^5	5.10322×10^5

The concentration after five years will be

$$\frac{5.10322\times 10^5\text{ mg}}{14.8\times 10^6\text{ L}} = \boxed{0.0345\text{ mg/L}}$$

The answer is (A).

(b) To reduce the phosphorus accumulation in the pond, the arrival of additional bioavailable phosphorous must be reduced. This entails watershed management to reduce the amount of phosphorus applied as fertilizer and released through other sources. As eutrophication is a natural process accelerated by the availability of plant nutrients (nitrogen and phosphorus especially), reducing phosphorus can slow the process.

While there are chemical means to alter the forms of phosphorus to nonbioavailable states, this usually is not practical on a large scale and is potentially harmful in itself.

It is unlikely that the pond receives untreated discharge from local homes. Use of phosphate-rich detergents in the home will not affect the pond.

The answer is (C).

2. With CO_2 present, all alkalinity is in the bicarbonate (HCO_3^-) form. Therefore, the equilibrium expression for the first ionization of carbonic acid may be used.

$$CO_2 + H_2O \rightarrow H_2CO_3 \leftrightarrow H^+ + HCO_3^-$$

$$K_1 = \frac{[H^+][HCO_3^-]}{[H_2CO_3]} = 4.45 \times 10^{-7}$$

The coefficients of CO_2 and H_2CO_3 are both 1. The number of moles of each compound is the same. One mole of CO_2 produces one mole of H_2CO_3.

$$[H_2CO_3] = [CO_2] = \frac{5.6\ \frac{\text{mg}}{\text{L}}}{\left(100\ \frac{\text{g}}{\text{mol}}\right)\left(1000\ \frac{\text{mg}}{\text{g}}\right)} = 5.6 \times 10^{-5}$$

$$[HCO_3^-] = \frac{200\ \frac{\text{mg}}{\text{L}}}{\left(100\ \frac{\text{g}}{\text{mol}}\right)\left(1000\ \frac{\text{mg}}{\text{g}}\right)} = 2 \times 10^{-3}$$

Solve for the hydrogen ion concentration.

$$[H^+] = \frac{K_1[H_2CO_3]}{[HCO_3^-]} = \frac{K_1[H_2CO_3]}{[CO_2]} = \frac{(4.45 \times 10^{-7})(5.6 \times 10^{-5})}{2 \times 10^{-3}} = 1.246 \times 10^{-8}$$

$$\text{pH} = -\log[H^+] = -\log(1.246 \times 10^{-8}) = \boxed{7.9}$$

The answer is (B).

3. (a) The density of water is 1 g/mL. The volume of the tested sample is

$$\frac{118.4\ \text{g} - 50.326\ \text{g}}{1\ \frac{\text{g}}{\text{mL}}} = 68.1\ \text{mL}$$

$$\text{TS} = \frac{(50.437\ \text{g} - 50.326\ \text{g})\left(1000\ \frac{\text{mg}}{\text{g}}\right)\left(1000\ \frac{\text{mL}}{\text{L}}\right)}{68.1\ \text{mL}} = \boxed{1630\ \text{mg/L}}$$

The answer is (D).

(b)
$$\text{TVS} = \frac{(50.437\ \text{g} - 50.383\ \text{g}) \times \left(1000\ \frac{\text{mg}}{\text{g}}\right)\left(1000\ \frac{\text{mL}}{\text{L}}\right)}{68.1\ \text{mL}} = \boxed{793\ \text{mg/L}}$$

The answer is (C).

(c)
$$\text{TFS} = 1630\ \frac{\text{mg}}{\text{L}} - 793\ \frac{\text{mg}}{\text{L}} = \boxed{837\ \text{mg/L}}$$

The answer is (D).

(d)
$$\text{TSS} = \frac{(0.1278\ \text{g} - 0.1170\ \text{g}) \times \left(1000\ \frac{\text{mg}}{\text{g}}\right)\left(1000\ \frac{\text{mL}}{\text{L}}\right)}{30\ \text{mL}} = \boxed{360\ \text{mg/L}}$$

The answer is (B).

(e)
$$\text{VSS} = \frac{(0.1278\ \text{g} - 0.1248\ \text{g}) \times \left(1000\ \frac{\text{mg}}{\text{g}}\right)\left(1000\ \frac{\text{mL}}{\text{L}}\right)}{30\ \text{mL}} = \boxed{100\ \text{mg/L}}$$

The answer is (A).

(f)
$$\text{FSS} = 360\ \frac{\text{mg}}{\text{L}} - 100\ \frac{\text{mg}}{\text{L}} = \boxed{260\ \text{mg/L}}$$

The answer is (D).

(g)
$$\text{TDS} = \frac{(51.524\ \text{g} - 51.494\ \text{g}) \times \left(1000\ \frac{\text{mg}}{\text{g}}\right)\left(1000\ \frac{\text{mL}}{\text{L}}\right)}{25\ \text{mL}} = \boxed{1200\ \text{mg/L}}$$

The answer is (C).

(h)
$$\text{VDS} = \frac{(51.524\ \text{g} - 51.506\ \text{g}) \times \left(1000\ \frac{\text{mg}}{\text{g}}\right)\left(1000\ \frac{\text{mL}}{\text{L}}\right)}{25\ \text{mL}} = \boxed{720\ \text{mg/L}}$$

The answer is (B).

(i)
$$\text{FDS} = 1200\ \frac{\text{mg}}{\text{L}} - 720\ \frac{\text{mg}}{\text{L}} = \boxed{480\ \text{mg/L}}$$

The answer is (D).

(j) If only the first two procedures were performed, the dissolved solids would be calculated as

$$\begin{aligned}\text{TDS} &= \text{TS} - \text{TSS} = 1630\ \frac{\text{mg}}{\text{L}} - 360\ \frac{\text{mg}}{\text{L}} \\ &= 1270\ \text{mg/L (vs. 1200mg/L)} \\ \text{VDS} &= \text{TVS} - \text{VSS} = 793\ \frac{\text{mg}}{\text{L}} - 100\ \frac{\text{mg}}{\text{L}} \\ &= 693\ \text{mg/L (vs. 720mg/L)} \\ \text{FDS} &= \text{TFS} - \text{FSS} = 837\ \frac{\text{mg}}{\text{L}} - 260\ \frac{\text{mg}}{\text{L}} \\ &= 577\ \text{mg/L (vs. 480mg/L)}\end{aligned}$$

The results of the three sets of tests are not the same. These differences are too great to be caused by improper rounding or faulty laboratory technique. This is probably the result of using samples that are not truly identical. For this reason, the suspended solids values are mathematically determined from the total solids and dissolved solids tests.

The answer is (D).

4. (a) Hardness is caused by multivalent cations: Ca^{++} and Mg^{++} in this example. Calculate the milliequivalents by dividing the measured concentration by the milliequivalent weight.

$$Ca^{++}\text{:}\quad \frac{74.0\ \frac{\text{mg}}{\text{L}}}{20\ \frac{\text{mg}}{\text{meq}}} = 3.7\ \text{meq/L}$$

$$Mg^{++}\text{:}\quad \frac{18.3\ \frac{\text{mg}}{\text{L}}}{12.2\ \frac{\text{mg}}{\text{meq}}} = 1.5\ \text{meq/L}$$

$$\text{total hardness} = 3.7\ \frac{\text{meq}}{\text{L}} + 1.5\ \frac{\text{meq}}{\text{L}} = 5.2\ \text{meq/L}$$

Determine the calcium carbonate equivalent by multiplying by the equivalent weight of calcium carbonate.

$$\begin{aligned}\text{hardness} &= \left(5.2\ \frac{\text{meq}}{\text{L}}\right)\left(50\ \frac{\text{mg}}{\text{meq}}\right) \\ &= \boxed{260\ \text{mg/L as}\ CaCO_3}\end{aligned}$$

The answer is (D).

(b) Alkalinity in the form of carbonate radical does not exist at a pH below 8.3. Bicarbonate is a form of alkalinity, and it neutralizes, not creates, acidity. To determine the cation/anion relationship, convert all of the concentrations to milliequivalents by dividing the measured concentrations by the milliequivalent weights.

$$Ca^{++}\text{:}\quad \frac{74.0\ \frac{\text{mg}}{\text{L}}}{20\ \frac{\text{mg}}{\text{meq}}} = 3.7\ \text{meq/L}$$

$$Mg^{++}\text{:}\quad \frac{18.3\ \frac{\text{mg}}{\text{L}}}{12.2\ \frac{\text{mg}}{\text{meq}}} = 1.5\ \text{meq/L}$$

$$Na^{+}\text{:}\quad \frac{27.6\ \frac{\text{mg}}{\text{L}}}{23\ \frac{\text{mg}}{\text{meq}}} = 1.2\ \text{meq/L}$$

$$K^{+}\text{:}\quad \frac{39.1\ \frac{\text{mg}}{\text{L}}}{39.1\ \frac{\text{mg}}{\text{meq}}} = 1.0\ \text{meq/L}$$

$$\begin{aligned}\text{total cations} &= 3.7\ \frac{\text{meq}}{\text{L}} + 1.5\ \frac{\text{meq}}{\text{L}} + 1.2\ \frac{\text{meq}}{\text{L}} \\ &\quad + 1.0\ \frac{\text{meq}}{\text{L}} \\ &= \boxed{7.4\ \text{meq/L}}\end{aligned}$$

$$HCO_3^{-}\text{:}\quad \frac{274.5\ \frac{\text{mg}}{\text{L}}}{61\ \frac{\text{mg}}{\text{meq}}} = 4.5\ \text{meq/L}$$

$$SO_4^{--}\text{:}\quad \frac{72.0\ \frac{\text{mg}}{\text{L}}}{48\ \frac{\text{mg}}{\text{meq}}} = 1.5\ \text{meq/L}$$

$$Cl^{-}\text{:}\quad \frac{49.7\ \frac{\text{mg}}{\text{L}}}{35.5\ \frac{\text{mg}}{\text{meq}}} = 1.4\ \text{meq/L}$$

$$\begin{aligned}\text{total anions} &= 4.5\ \frac{\text{meq}}{\text{L}} + 1.5\ \frac{\text{meq}}{\text{L}} + 1.4\ \frac{\text{meq}}{\text{L}} \\ &= \boxed{7.4\ \text{meq/L}}\end{aligned}$$

The answer is (B).

(c) To determine the hypothetical compounds, construct a milliequivalent per liter bar graph for cations and anions, as shown.

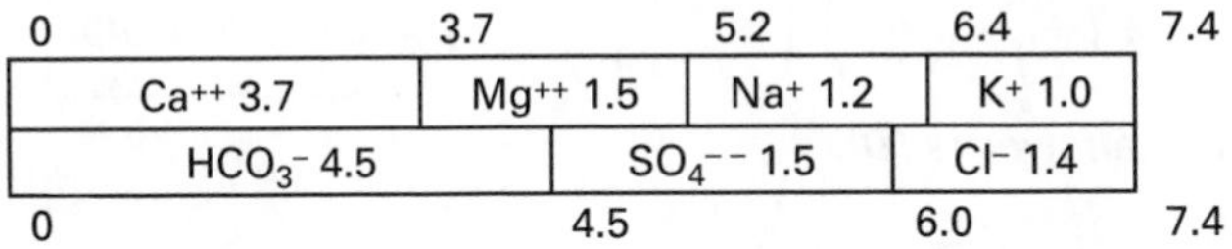

0 | 3.7 | 5.2 | 6.4 | 7.4

Ca^{++} 3.7	Mg^{++} 1.5	Na^{+} 1.2	K^{+} 1.0
HCO_3^{-} 4.5	SO_4^{--} 1.5	Cl^{-} 1.4	

0 | 4.5 | 6.0 | 7.4

The hypothetical compounds are determined by moving from left to right.

hypothetical	compound concentration (meq/L)	remarks
$Ca(HCO_3)_2$	3.7	Ca^{++} exhausted
$Mg(HCO_3)_2$	0.8	HCO_3^- exhausted
$MgSO_4$	0.7	Mg^{++} exhausted
Na_2SO_4	0.8	SO_4^{--} exhausted
NaCl	0.4	Na^+ exhausted
KCl	1.0	K^+ and Cl^- exhausted

The answer is (A).

(d) From part (c), the hypothetical concentration of magnesium bicarbonate is **0.8 meq/L.**

The answer is (A).

(e) From part (c), the hypothetical concentration of magnesium sulfate is **0.7 meq/L.**

The answer is (A).

(f) From part (c), the hypothetical concentration of sodium sulfate is **0.8 meq/L.**

The answer is (A).

(g) From part (c), the hypothetical concentration of sodium chloride is **0.4 meq/L.**

The answer is (B).

(h) The reactions are

$$CaO + H_2O \rightarrow Ca(OH)_2$$
$$Ca(HCO_3)_2 + Ca(OH)_2 \rightarrow CaCO_3 \downarrow + 2H_2O$$

One molecule of CaO forms one molecule of $Ca(OH)_2$, which in turn reacts with one molecule of $Ca(HCO_3)_2$. There are 3.7 meq/L of $Ca(HCO_3)_2$ in solution. The equivalent weight of CaO is $(40+16)/2 = 28$. Therefore, 3.7 meq/L of CaO are needed.

$$\left(3.7\ \frac{\text{meq}}{\text{L}}\ \text{CaO}\right)\left(28\ \frac{\text{mg}}{\text{meq}}\right) = \boxed{103.6\ \text{mg/L CaO}}$$

The answer is (D).

(i) The reactions are

$$CaO + H_2O \rightarrow Ca(OH)_2$$
$$Mg(HCO_3)_2 + 2Ca(OH)_2 \rightarrow$$
$$2CaCO_3 \downarrow + Mg(OH)_2 \downarrow + 2H_2O$$

Two molecules of $Ca(OH)_2$ are required for each molecule of $Mg(HCO_3)_2$.

$$\left(0.8\ \frac{\text{meq}}{\text{L}}\ Mg(HCO_3)_2\right)(2)\left(28\ \frac{\text{mg}}{\text{meq}}\right) + 35\ \frac{\text{mg}}{\text{L}}\ \text{excess} = \boxed{79.8\ \text{mg/L}}$$

The answer is (D).

(j) The original hardness was in the hypothetical forms of calcium bicarbonate, magnesium bicarbonate, and magnesium sulfate. The bicarbonates have been removed in parts (h) and (i), leaving the residuals of 30 mg/L calcium hardness and 10 mg/L magnesium hardness. However, no attempt was made to remove the noncarbonate magnesium hardness represented by magnesium sulfate (this would have required the addition of soda ash and additional lime).

The residual hardness is as follows.

$$\text{calcium hardness:}\ \frac{30\ \frac{\text{mg}}{\text{L}}}{50\ \frac{\text{mg}}{\text{meq}}} = 0.6\ \text{meq/L}$$

$$\text{magnesium carbonate hardness:}\ \frac{10\ \frac{\text{mg}}{\text{L}}}{50\ \frac{\text{mg}}{\text{meq}}} = 0.2\ \text{meq/L}$$

From part (c), the magnesium noncarbonate hardness is 0.7 meq/L.

$$\begin{aligned}\text{residual hardness} &= 0.6\ \frac{\text{meq}}{\text{L}} + 0.2\ \frac{\text{meq}}{\text{L}} + 0.7\ \frac{\text{meq}}{\text{L}}\\ &= 1.5\ \text{meq/L}\end{aligned}$$

$$\left(1.5\ \frac{\text{meq}}{\text{L}}\right)\left(50\ \frac{\text{mg}}{\text{meq}}\right) = \boxed{75\ \text{mg/L as}\ CaCO_3}$$

The answer is (D).

26 Water Supply Treatment and Distribution

PRACTICE PROBLEMS

Plain Sedimentation

1. A settling tank has an overflow rate of 100,000 gal/ft^2-day. Water carrying sediment of various sizes enters the tank. The sediment has the following distribution of settling velocities.

settling velocity (ft/min)	mass fraction remaining
10.0	0.54
5.0	0.45
2.0	0.35
1.0	0.20
0.75	0.10
0.50	0.03

(a) What is the gravimetric percentage of sediment particles completely removed?

(A) 32%
(B) 39%
(C) 47%
(D) 55%

(b) What is the total gravimetric percentage of all sediment particles removed?

(A) 35%
(B) 40%
(C) 50%
(D) 60%

2. A spherical sand particle has a specific gravity of 2.6 and a diameter of 1 mm. What is the settling velocity?

(A) 0.2 ft/sec
(B) 0.7 ft/sec
(C) 1.1 ft/sec
(D) 1.6 ft/sec

3. A mechanically cleaned circular clarifier is to be designed with the following characteristics.

flow rate	2.8 MGD
detention period	2 hr
surface loading	700 gal/ft^2-day

(a) What is the approximate diameter?

(A) 45 ft
(B) 60 ft
(C) 70 ft
(D) 90 ft

(b) What is the approximate depth?

(A) 6 ft
(B) 8 ft
(C) 12 ft
(D) 15 ft

(c) If the initial flow rate is reduced to 1.1 MGD, what is the surface loading?

(A) 190 gal/day-ft^2
(B) 230 gal/day-ft^2
(C) 250 gal/day-ft^2
(D) 280 gal/day-ft^2

(d) If the initial flow rate is reduced to 1.1 MGD, what is the average detention period?

(A) 4 hr
(B) 5 hr
(C) 6 hr
(D) 8 hr

4. (*Time limit: one hour*) A water treatment plant is designed to handle a total flow rate of 1.5 MGD. The current design includes two identical sedimentation basins running in parallel with the following characteristics.

plan area	90 ft × 16 ft
depth	12 ft
total weir length	48 ft per basin
three-month sustained average low	70% of the design average daily flow
three-month sustained average high	200% of the design average daily flow

The basins must meet the following government standards.

minimum retention time	4.0 hrs
maximum weir load	20,000 gpd/ft
maximum velocity	0.5 ft/min

Do the basins meet the standards?

(A) Yes, specifications are met at both peak and low flows.
(B) No, specifications are not met at low flow.
(C) No, specifications are not met at high flow.
(D) No, specifications are not met at either high or low flow.

Mixing Physics

5. (*Time limit: one hour*) A flocculator tank with a volume of 200,000 ft^3 uses a paddle wheel to mix the coagulant in 60°F water. The operating characteristics are as follows.

mean velocity gradient	45 sec^{-1}
paddle drag coefficient	1.75
paddle tip velocity	2 ft/sec
relative water/paddle velocity	1.5 ft/sec

(a) What is the theoretical power required to drive the paddle?

(A) 12 hp
(B) 17 hp
(C) 60 hp
(D) 120 hp

(b) What is the drag force on the paddle?

(A) 450 lbf
(B) 910 lbf
(C) 5500 lbf
(D) 6400 lbf

(c) What is the required paddle area?

(A) 900 ft^2
(B) 1200 ft^2
(C) 1500 ft^2
(D) 1700 ft^2

Filtration

6. A water treatment plant has four square rapid sand filters. The flow rate is 4 gal/min-ft^2. Each filter has a treatment capacity of 600,000 gal/day. Each filter is backwashed once a day for 8 min. The rate of rise during washing is 24 in/min.

(a) What are the inside dimensions of each filter?

(A) 6 ft × 6 ft
(B) 8 ft × 8 ft
(C) 10 ft × 10 ft
(D) 12 ft × 12 ft

(b) What percentage of the filtered water is used for backwashing?

(A) 2%
(B) 4%
(C) 8%
(D) 10%

7. A water treatment plant has five identical rapid sand filters. Each is square in plan and has a capacity of 1.0 MGD. The application rate is 4 gal/min-ft^2, and the total wetted depth is 10 ft.

(a) What should be the cross-section dimensions of each filter?

(A) 7 ft × 7 ft
(B) 9 ft × 9 ft
(C) 10 ft × 10 ft
(D) 13 ft × 13 ft

(b) Each filter is backwashed every day for 5 min. The rate of rise of the backwash water is 2 ft/min. What percentage of the plant's filtered water will be used for backwashing?

(A) 1.3%
(B) 1.9%
(C) 2.4%
(D) 3.1%

Precipitation Softening

8. A town's water supply has the following ionic concentrations.

Al^{+++}	0.5 mg/L
Ca^{++}	80.2 mg/L
Cl^-	85.9 mg/L
CO_2	19 mg/L
CO_3^{--}	0
Fe^{++}	1.0 mg/L
Fl^-	0
HCO_3^-	185 mg/L
Mg^{++}	24.3 mg/L
Na^+	46.0 mg/L
NO_3^-	0
SO_4^{--}	125 mg/L

(a) What is the total hardness?

(A) 160 mg/L as $CaCO_3$
(B) 200 mg/L as $CaCO_3$
(C) 260 mg/L as $CaCO_3$
(D) 300 mg/L as $CaCO_3$

(b) How much slaked lime is required to combine with the carbonate hardness?

(A) 45 mg/L as substance
(B) 90 mg/L as substance
(C) 130 mg/L as substance
(D) 150 mg/L as substance

(c) How much soda ash is required to react with the carbonate hardness?

(A) none
(B) 15 mg/L as substance
(C) 35 mg/L as substance
(D) 60 mg/L as substance

9. A city's water supply contains the following ionic concentrations.

$Ca(HCO_3)_2$	137 mg/L as $CaCO_3$
$MgSO_4$	72 mg/L as $CaCO_3$
CO_2	0

(a) How much slaked lime is required to soften 1,000,000 gal of this water to a hardness of 100 mg/L (as $CaCO_3$) if 30 mg/L (as $CaCO_3$) of excess lime is used?

(A) 930 lbm
(B) 1200 lbm
(C) 1300 lbm
(D) 1700 lbm

(b) How much soda ash is required to soften 1,000,000 gal of this water to a hardness of 100 mg/L (as $CaCO_3$) if 30 mg/L (as $CaCO_3$) of excess lime is used?

(A) none
(B) 300 lbm
(C) 500 lbm
(D) 700 lbm

(c) How much soda ash is required to soften 1,000,000 gal of this water to a hardness of 50 mg/L (as $CaCO_3$) if 30 mg/L (as $CaCO_3$) of excess lime is used?

(A) 10 mg/L as $CaCO_3$
(B) 20 mg/L as $CaCO_3$
(C) 40 mg/L as $CaCO_3$
(D) 50 mg/L as $CaCO_3$

Zeolite Softening

10. The water described in Prob. 9 is to be softened using a zeolite process with the following characteristics: exchange capacity, 10,000 grains/ft^3; and salt requirement, 0.5 lbm per 1000 grains hardness removed. How much salt is required to soften the water to 100 mg/L hardness?

(A) 700 lbm/MG
(B) 1800 lbm/MG
(C) 2500 lbm/MG
(D) 3200 lbm/MG

11. The hardness of water from an underground aquifer is to be reduced from 245 mg/L to 80 mg/L using a zeolite process. The volumetric flow rate is 20,000 gal/day. The process has the following characteristics: resin exchange capacity, 20,000 grains/ft^3; and zeolite volume, 2 ft^3.

(a) What fraction of the water is bypassed around the process?

(A) 0.15
(B) 0.33
(C) 0.67
(D) 0.85

(b) What is the time between regenerations of the softener?

(A) 5 hr
(B) 16 hr
(C) 24 hr
(D) 30 hr

Demand

12. (*Time limit: one hour*) An area is expected to attain a population of 40,000 in 20 years. The cumulative per capita water demand for a peak day in the area is shown in the following diagram.

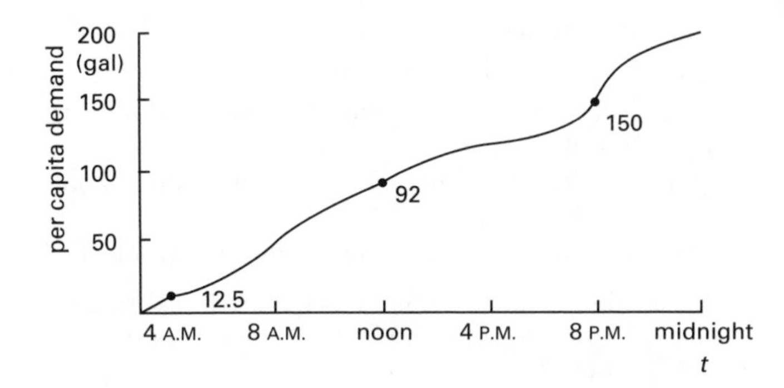

(a) What is the total per capita demand for a peak day?

(A) 160 gal
(B) 200 gal
(C) 240 gal
(D) 290 gal

(b) Assuming uniform operation over 24 hours, what storage volume is required in the treatment plant for all uses, including fire fighting demand?

(A) 1.1 MG
(B) 2.5 MG
(C) 3.2 MG
(D) 5.5 MG

(c) Assume that the pumping station runs uniformly from 4 a.m. until 8 a.m. to fill the storage tanks for the day. What storage is required to meet all uses, including fire fighting?

(A) 3.2 MG
(B) 3.8 MG
(C) 5.5 MG
(D) 8.0 MG

13. The water supply for a town of 15,000 people is taken from a river. The average consumption is 110 gpcd. The water has the following characteristics.

turbidity	20 to 100 NTU (varies)
total hardness	less than 60 mg/L as $CaCO_3$
coliform count	200 to 1000 per 100 mL (varies)

(a) What capacity should the distribution and treatment system have?

(A) 6000 gpm
(B) 7500 gpm
(C) 9500 gpm
(D) 12,000 gpm

(b) If the application rate is 4 gal/min-ft^2, what total filter area is required?

(A) 290 ft^2
(B) 580 ft^2
(C) 910 ft^2
(D) 1100 ft^2

(c) Is softening required?

(A) No, 60 mg/L is soft water.
(B) No, softening would interfere with turbidity removal.
(C) Yes, the turbidity and coliform counts would also benefit.
(D) Yes, all municipal water should be softened.

(d) If a chlorine dose of 2 mg/L is required to obtain the desired chlorine residual, how much chlorine is required every 24 hours?

(A) 15 lbm
(B) 22 lbm
(C) 28 lbm
(D) 32 lbm

SOLUTIONS

1. (a)
$$v^* = \frac{\left(100{,}000\ \frac{\text{gal}}{\text{ft}^2\text{-day}}\right)\left(0.1337\ \frac{\text{ft}^3}{\text{gal}}\right)}{\left(24\ \frac{\text{hr}}{\text{day}}\right)\left(60\ \frac{\text{min}}{\text{hr}}\right)} = 9.28 \text{ ft/min}$$

Using interpolation, the mass fraction remaining is

$$0.45 + \left(\frac{9.28\ \frac{\text{ft}}{\text{min}} - 5.0\ \frac{\text{ft}}{\text{min}}}{10.0\ \frac{\text{ft}}{\text{min}} - 5.0\ \frac{\text{ft}}{\text{min}}}\right)(0.54 - 0.45) = 0.527 \text{ remains in flow}$$

$$1 - 0.527 = 0.473$$

$$\boxed{47.3\% \text{ completely removed}}$$

The answer is (C).

(b) Plot the mass fraction remaining versus the settling velocity.

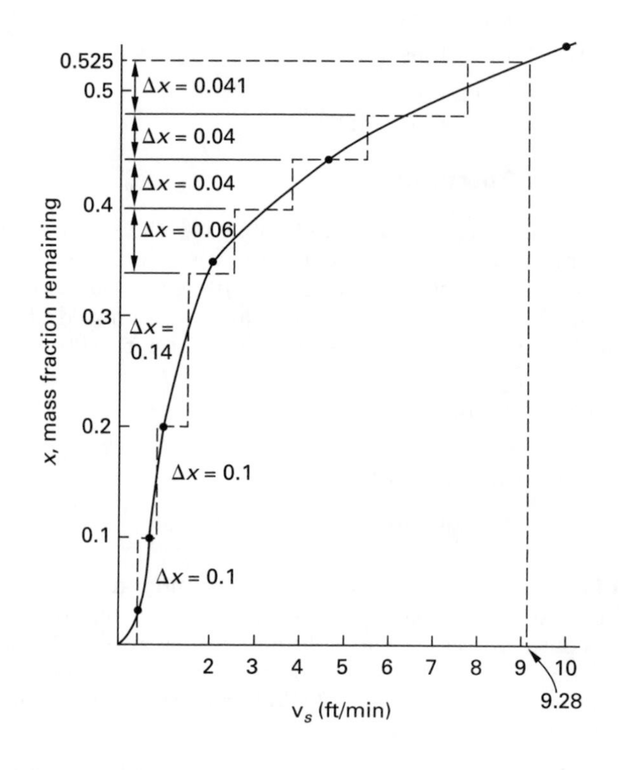

$$v_o = \left(\frac{100{,}000 \text{ gal}}{\text{ft}^2\text{-day}}\right)\left(\frac{1 \text{ ft}^3}{7.48 \text{ gal}}\right)\left(\frac{1 \text{ day}}{24 \text{ hr}}\right)\left(\frac{1 \text{ hr}}{60 \text{ min}}\right) = 9.28 \text{ ft/min}$$

$X_o \approx 0.525 \quad (52.5\%)$

Determine $\Delta x \text{v}_t$ by graphical integration.

Δx	v_t	$(\Delta x)(\text{v}_t)$
0.1	0.5	0.05
0.1	0.8	0.08
0.14	1.4	0.196
0.06	2.5	0.15
0.04	3.75	0.15
0.04	5.5	0.22
0.041	7.75	0.318
	total	1.164

The overall removal efficiency is

$$\begin{aligned} X &= 1 - X_o + \sum\left(\frac{\Delta x \text{v}_t}{\text{v}_o}\right) \\ &= 1 - 0.525 + \frac{1.164}{9.28} \\ &= 0.60 \end{aligned}$$

The answer is (D).

2. From Fig. 26.2, $\text{v}_s = \boxed{0.7 \text{ ft/sec.}}$

The answer is (B).

3. (a) The surface area is

$$A_{\text{surface}} = \frac{2.8 \times 10^6 \ \frac{\text{gal}}{\text{day}}}{700 \ \frac{\text{gal}}{\text{ft}^2\text{-day}}} = 4000 \text{ ft}^2$$

Since $A = \left(\frac{\pi}{4}\right) D^2$,

$$D = \sqrt{\frac{(4)(4000 \text{ ft}^2)}{\pi}} = \boxed{71.4 \text{ ft}}$$

The answer is (C).

(b) The volume is

$$V = \frac{\left(2.8 \times 10^6 \ \frac{\text{gal}}{\text{day}}\right)(2 \text{ hr})}{24 \ \frac{\text{hr}}{\text{day}}} = 2.333 \times 10^5 \text{ gal}$$

The depth is

$$\begin{aligned} d &= \frac{V}{A_{\text{surface}}} = \frac{(2.333 \times 10^5 \text{ gal})\left(0.1337 \ \frac{\text{ft}^3}{\text{gal}}\right)}{4000 \text{ ft}^2} \\ &= \boxed{7.8 \text{ ft}} \end{aligned}$$

The answer is (B).

(c) $$v^* = \frac{1.1 \times 10^6 \ \frac{\text{gal}}{\text{day}}}{4000 \text{ ft}^2} = \boxed{275 \text{ gal/day-ft}^2}$$

The answer is (D).

(d) Use Eq. 26.6.

$$t = \frac{(2.333 \times 10^5 \text{ gal})\left(24 \ \frac{\text{hr}}{\text{day}}\right)}{1.1 \times 10^6 \ \frac{\text{gal}}{\text{day}}} = \boxed{5.09 \text{ hr}}$$

The answer is (B).

4.

$$\begin{aligned} \text{volume per basin} &= (90 \text{ ft})(16 \text{ ft})(12 \text{ ft}) \\ &= 17{,}280 \text{ ft}^3 \end{aligned}$$

(The freeboard is not given.)

The area per basin is

$$A = (90 \text{ ft})(16 \text{ ft}) = 1440 \text{ ft}^2$$

The three-month peak flow per basin is

$$(2)\left(\frac{1.5 \text{ MGD}}{2}\right) = 1.5 \text{ MGD}$$

$$(1.5 \text{ MGD})\left(1.547 \ \frac{\frac{\text{ft}^3}{\text{sec}}}{\text{MGD}}\right) = 2.32 \text{ ft}^3\text{/sec} \quad (\text{cfs})$$

The detention time at peak flow is given by Eq. 26.6.

$$t = \frac{V}{Q} = \frac{17{,}280 \text{ ft}^3}{\left(2.32 \ \frac{\text{ft}^3}{\text{sec}}\right)\left(60 \ \frac{\text{min}}{\text{hr}}\right)\left(60 \ \frac{\text{ft}^3}{\text{hr}}\right)} = 2.07 \text{ hr}$$

Since 2.07 hr < 4 hr, this is not acceptable.

The weir loading is

$$\frac{1.5 \times 10^6 \ \frac{\text{gal}}{\text{day}}}{48 \text{ ft}} = 31{,}250 \text{ gal/day-ft}$$

Since 31,250 > 20,000, this is not acceptable either.

The overflow rate is

$$v^* = \frac{Q}{A} = \frac{\left(2.32 \ \frac{\text{ft}^3}{\text{sec}}\right)\left(60 \ \frac{\text{sec}}{\text{min}}\right)}{1440 \text{ ft}^2} = 0.0967 \text{ ft/min}$$

Since 0.0967 < 0.5, this is acceptable.

At low flow,

$$\text{flow} = \frac{0.7}{2.0} = 0.35 \quad [35\% \text{ of high flow}]$$

$$t = \frac{2.07 \text{ hr}}{0.35} = 5.91 \text{ hr} \quad [\text{acceptable}]$$

$$\text{weir loading} = (0.35)\left(31{,}250 \ \frac{\text{gal}}{\text{day-ft}}\right) = 10{,}938 \ \frac{\text{gal}}{\text{day-ft}} \quad [\text{acceptable}]$$

$$v^* = (0.35)\left(0.1 \ \frac{\text{ft}}{\text{min}}\right) = 0.035 \text{ ft/min} \quad [\text{acceptable}]$$

The basins have been correctly designed for low flow but not for peak flow. One or more basins should be used.

The answer is (C).

5. For 60°F water, $\mu = 2.359 \times 10^{-5}$ lbf-sec/ft^2.

From Eq. 26.21,

$$P = \mu G^2 V_{\text{tank}} = \left(2.359 \times 10^{-5} \ \frac{\text{lbf-sec}}{\text{ft}^2}\right)\left(45 \ \frac{1}{\text{sec}}\right)^2 \times (200{,}000 \text{ ft}^3) = 9554 \text{ ft-lbf/sec}$$

(a)

$$\text{water horsepower} = \frac{9554 \ \frac{\text{ft-lbf}}{\text{sec}}}{550 \ \frac{\text{ft-lbf}}{\text{hp-sec}}} = \boxed{17.4 \text{ hp}}$$

The answer is (B).

(b) Since work = force × distance, then power = force× velocity.

$$D = \frac{P}{\text{v}} = \frac{9554 \ \frac{\text{ft-lbf}}{\text{sec}}}{1.5 \ \frac{\text{ft}}{\text{sec}}} = \boxed{6369 \text{ lbf}}$$

The answer is (D).

(c) Use Eq. 26.15.

$$A = \frac{2gF_D}{C_D \gamma \text{v}^2} = \frac{(2)\left(32.2 \ \frac{\text{ft}}{\text{sec}^2}\right)(6369 \text{ lbf})}{(1.75)\left(62.4 \ \frac{\text{lbf}}{\text{ft}^3}\right)\left(1.5 \ \frac{\text{ft}}{\text{sec}}\right)^2} = \boxed{1669 \text{ ft}^2}$$

The answer is (D).

6. (a) The required area is

$$A = \frac{600{,}000 \ \frac{\text{gal}}{\text{day}}}{\left(4 \ \frac{\text{gal}}{\text{min-ft}^2}\right)\left(24 \ \frac{\text{hr}}{\text{day}}\right)\left(60 \ \frac{\text{min}}{\text{hr}}\right)} = 104.2 \text{ ft}^2$$

$$\boxed{\text{use } 10 \text{ ft} \times 10 \text{ ft}}$$

The answer is (C).

(b) The required volume is

$$V = \left(8 \ \frac{\text{min}}{\text{day}}\right)(100 \text{ ft}^2)\left(2 \ \frac{\text{ft}}{\text{min}}\right) = 1600 \text{ ft}^3/\text{day}$$

$$\left(1600 \ \frac{\text{ft}^3}{\text{day}}\right)\left(7.481 \ \frac{\text{gal}}{\text{ft}^3}\right) = 11{,}970 \text{ gal/day}$$

$$\frac{11{,}970 \ \frac{\text{gal}}{\text{day}}}{600{,}000 \ \frac{\text{gal}}{\text{day}}} = \boxed{0.01 \ (2\%)}$$

The answer is (A).

7. (a)

$$A = \frac{1 \times 10^6 \ \frac{\text{gal}}{\text{day}}}{\left(24 \ \frac{\text{hr}}{\text{day}}\right)\left(60 \ \frac{\text{min}}{\text{hr}}\right)\left(4 \ \frac{\text{gal}}{\text{min-ft}^2}\right)} = 173.6 \text{ ft}^2$$

$$\text{width} = \sqrt{173.6 \text{ ft}^2} = 13.2 \text{ ft}$$

$$\boxed{13.2 \text{ ft} \times 13.2 \text{ ft}}$$

The answer is (D).

(b) The water volume is

$$V = \left(5 \ \frac{\text{min}}{\text{day}}\right)(173.6 \text{ ft}^2)\left(2 \ \frac{\text{ft}}{\text{min}}\right)(5 \text{ filters}) = 8680 \text{ ft}^3/\text{day}$$

Convert to gallons.

$$\left(8680 \ \frac{\text{ft}^3}{\text{day}}\right)\left(7.481 \ \frac{\text{gal}}{\text{ft}^3}\right) = 64{,}935 \text{ gal/day}$$

$$\frac{64{,}935 \ \frac{\text{gal}}{\text{day}}}{5 \times 10^6 \ \frac{\text{gal}}{\text{day}}} = \boxed{0.013 \ (1.3\%)}$$

The answer is (A).

8. (a)

	mg/L as substance		factor from App. 22.C		
Ca^{++}:	80.2	×	2.5	=	200.5 mg/L
Mg^{++}:	24.3	×	4.1	=	99.63 mg/L
Fe^{++}:	1	×	1.79	=	1.79 mg/L
Al^{+++}:	0.5	×	5.56	=	2.78 mg/L

$$\text{hardness} = \boxed{304.7 \text{ mg/L}}$$

The answer is (D).

(b) To remove the carbonate hardness,

$$CO_2{:}19 \ \frac{\text{mg}}{\text{L}} \times 2.27 = 43.13 \text{ mg/L as } CaCO_3$$

Add lime to remove the carbonate hardness. It does not matter whether the HCO_3^- comes from Mg^{++}, Ca^{++}, or Fe^{++}; adding lime will remove it.

There may be Mg^{++}, Ca^{++}, or Fe^{++} ions left over in the form of noncarbonate hardness, but the problem asked for carbonate hardness. Converting from mg/L of substance to mg/L as $CaCO_3$,

$$HCO_3^-{:}\ 185 \ \frac{\text{mg}}{\text{L}} \times 0.82 = 151.7 \text{ mg/L}$$

The total equivalents to be neutralized are

$$43.13 \ \frac{\text{mg}}{\text{L}} + 151.7 \ \frac{\text{mg}}{\text{L}} = 194.83 \text{ mg/L}$$

Convert $Ca(OH)_2$ using App. 22.C.

$$\frac{\text{mg}}{\text{L}} \text{ of } Ca(OH)_2 = \frac{194.83 \ \frac{\text{mg}}{\text{L}}}{1.35} = \boxed{144.3 \text{ mg/L as substance}}$$

The answer is (D).

(c) No soda ash is required since it is used to remove noncarbonate hardness.

The answer is (A).

9. (a) $Ca(HCO_3)_2$ and $MgSO_4$ both contribute to hardness. Since 100 mg/L of hardness is the goal, leave all $MgSO_4$ in the water. Take out 137 mg/L + 72 mg/L − 100 mg/L = 109 mg/L of $Ca(HCO_3)_2$. From App. 22.C (including the excess even though the reaction is not complete),

$$\text{pure } Ca(OH)_2 = 30 \ \frac{\text{mg}}{\text{L}} + \frac{109 \ \frac{\text{mg}}{\text{L}}}{1.35} = 110.74 \text{ mg/L}$$

$$\left(110.74 \ \frac{\text{mg}}{\text{L}}\right)\left(8.345 \ \frac{\text{lbm-L}}{\text{mg-MG}}\right) = \boxed{924 \text{ lbm/MG}}$$

The answer is (A).

(b) Since the goal is a residual hardness of 100 mg/L and the magnesium sulfate only contributes 72 mg/L, it does not have to be removed. No soda ash is required.

The answer is (A).

(c) In order to reduce hardness to 50 mg/L, 72 mg/L − 50 mg/L = 22 mg/L of $MgSO_4$ must be removed. (This assumes that all of the carbonate hardness has already been removed using lime softening.) This will require 22 mg/L (as $CaCO_3$) of soda ash.

The answer is (B).

10. There are 7000 grains in a pound. The hardness removed is

$$137 \ \frac{\text{mg}}{\text{L}} + 72 \ \frac{\text{mg}}{\text{L}} - 100 \ \frac{\text{mg}}{\text{L}} = 109 \text{ mg/L}$$

$$\left(109 \ \frac{\text{mg}}{\text{L}}\right)\left(8.345 \ \frac{\text{lbm-L}}{\text{mg-MG}}\right) = 909.6 \text{ lbm hardness/MG}$$

$$\left(\frac{0.5 \text{ lbm}}{1000 \text{ gr}}\right)\left(909.6 \ \frac{\text{lbm}}{\text{MG}}\right)\left(7000 \ \frac{\text{gr}}{\text{lbm}}\right) = \boxed{3.18 \times 10^3 \text{ lbm/MG} \quad (3200 \text{ lbm/MG})}$$

The answer is (D).

11. (a) A bypass process is required.

$$\text{fraction bypassed:} \frac{80 \ \frac{\text{mg}}{\text{L}}}{245 \ \frac{\text{mg}}{\text{L}}} = \boxed{0.326}$$

$$\text{fraction processed:} 1 - 0.326 = 0.674$$

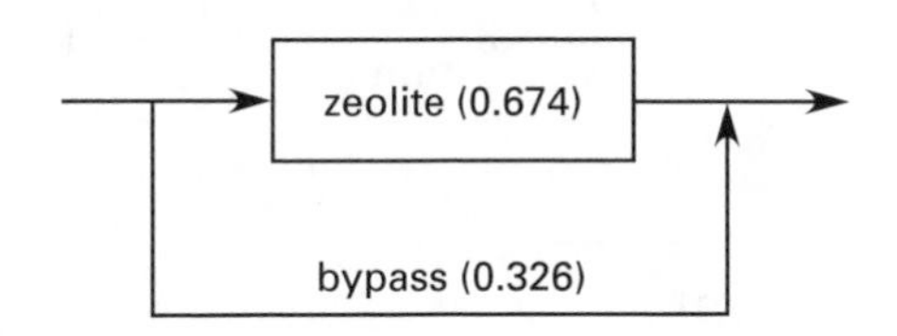

The answer is (B).

(b) The maximum hardness reduction is

$$\frac{(2 \text{ ft}^3)\left(20{,}000 \ \frac{\text{gr}}{\text{ft}^3}\right)}{7000 \ \frac{\text{gr}}{\text{lbm}}} = 5.71 \text{ lbm}$$

The hardness removal rate is

$$\frac{(0.674)\left(245 \ \frac{\text{mg}}{\text{L}}\right)\left(8.345 \times 10^{-6} \ \frac{\text{lbm}}{\text{gal}}\right)\left(20{,}000 \ \frac{\text{gal}}{\text{day}}\right)}{24 \ \frac{\text{hr}}{\text{day}}} = 1.15 \text{ lbm/hr}$$

$$t = \frac{5.71 \text{ lbm}}{1.15 \ \frac{\text{lbm}}{\text{hr}}} = \boxed{4.97 \text{ hr}}$$

The answer is (A).

12. (a) The given graph shows that by the end of the day, the cumulative demand has risen to 200 gal per person. The daily demand is $\boxed{\text{200 gal.}}$

The answer is (B).

(b)

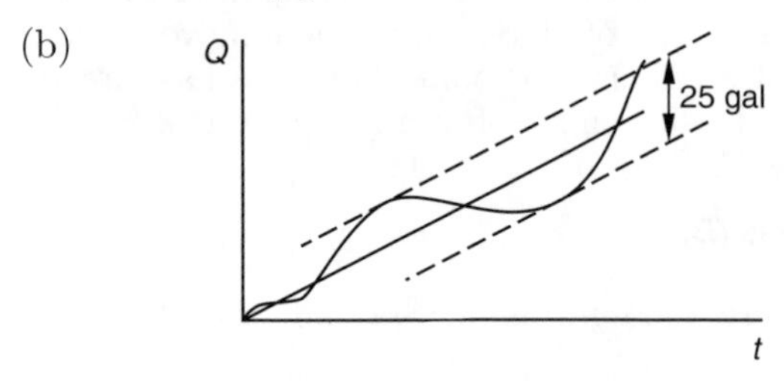

From the cumulative flow quantity, the storage requirement per capita-day is 25 gal.

The population use is

$$\left(25\ \frac{\text{gal}}{\text{day-person}}\right)(40{,}000\ \text{people}) = 1{,}000{,}000\ \text{gal/day}$$

From Eq. 26.53, the fire fighting requirement is

$$Q = 1020\sqrt{40}\left(1 - 0.01\sqrt{40}\right) = 6043\ \text{gal/min}$$

The flow rate is 6043/1000 = 6 thousands of gallons, so maintain the flow for 4 hr (approximate ISO specifications).

$$\begin{aligned}\text{capacity} &= 1{,}000{,}000\ \text{gal} \\ &\quad + \left(6043\ \frac{\text{gal}}{\text{min}}\right)\left(60\ \frac{\text{min}}{\text{hr}}\right)(4\ \text{hr}) \\ &= \boxed{2{,}450{,}000\ \text{gal}\quad (2.5\ \text{MG})}\end{aligned}$$

The answer is (B).

(c) The pump supplies demand from 4 a.m. until 8 a.m. The storage supplies demand from 8 a.m. until 4 a.m.

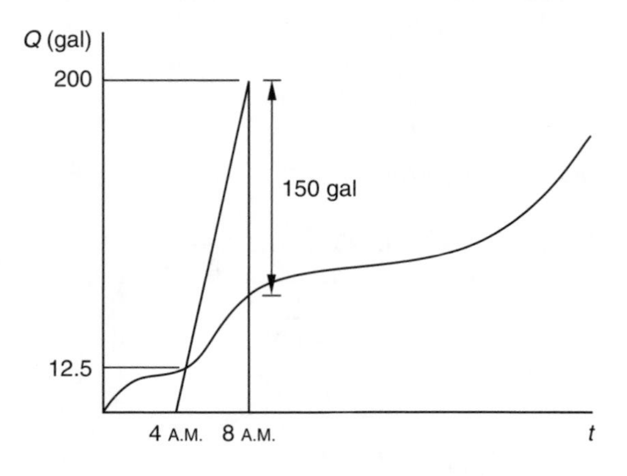

The population use is

$$(150\ \text{gal} + 12.5\ \text{gal})(40{,}000) = 6{,}500{,}000\ \text{gal}$$

Now add fire fighting.

$$\begin{aligned}6{,}500{,}000\ \text{gal} &+ \left(6043\ \frac{\text{gal}}{\text{min}}\right)\left(60\ \frac{\text{min}}{\text{hr}}\right)(4\ \text{hr}) \\ &= \boxed{7{,}950{,}000\ \text{gal}\quad (8.0\ \text{MG})}\end{aligned}$$

The answer is (D).

13. (a) From Table 26.5, use 3 as the peak multiplier.

$$\frac{\left(110\ \frac{\text{gal}}{\text{day}}\right)(15{,}000)(3)}{\left(24\ \frac{\text{hr}}{\text{day}}\right)\left(60\ \frac{\text{min}}{\text{hr}}\right)} = 3437\ \text{gal/min}$$

The fire fighting requirements are given by Eq. 26.53.

$$Q = 1020\sqrt{15}\left(1 - 0.01\sqrt{15}\right) = 3797\ \text{gal/min}$$

The total maximum demand for which the distribution system should be designed is

$$3437\ \frac{\text{gal}}{\text{min}} + 3797\ \frac{\text{gal}}{\text{min}} = \boxed{7234\ \text{gal/min}}$$

The answer is (B).

(b) The filter area should not be based on the maximum hourly rate since some of the demand during the peak hours can come from storage (clearwell or tanks). Also, fire requirements can bypass the filters if necessary.

$$\left(110\ \frac{\text{gal}}{\text{day-person}}\right)(15{,}000\ \text{people}) = 1.65\times 10^6\ \text{gal/day}$$

Using a flow rate of 4 gpm/ft^2, the required filter area is

$$A = \frac{1.65\times 10^6\ \frac{\text{gal}}{\text{day}}}{\left(4\ \frac{\text{gal}}{\text{min-ft}^2}\right)\left(24\ \frac{\text{hr}}{\text{day}}\right)\left(60\ \frac{\text{min}}{\text{hr}}\right)} = \boxed{286.5\ \text{ft}^2}$$

The answer is (A).

(c) $\boxed{\text{No, 60 mg/L is soft water.}}$

The answer is (A).

(d) Disregarding the fire flow, the required average (not peak) daily chlorine mass is given by Eq. 26.11.

$$\begin{aligned}&\frac{\left(110\ \frac{\text{gal}}{\text{day-person}}\right)(15{,}000\ \text{people}) \times \left(2\ \frac{\text{mg}}{\text{L}}\right)\left(8.345\ \frac{\text{lbm-L}}{\text{MG-mg}}\right)}{10^6\ \frac{\text{gal}}{\text{MG}}} \\ &= \boxed{27.54\ \text{lbm/day}}\end{aligned}$$

The answer is (C).

27 Biochemistry, Biology, and Bacteriology

PRACTICE PROBLEMS

1. (*Time limit: one hour*) A fresh wastewater sample is taken containing nitrate ions, sulfate ions, and dissolved oxygen, and it is placed in a sealed jar absent of air.

(a) What is the sequence of oxidation of the compounds?

(A) nitrate, dissolved oxygen, and then sulfate
(B) sulfate, nitrate, and then dissolved oxygen
(C) dissolved oxygen, nitrate, and then sulfate
(D) none of the above

(b) Obnoxious odors will

(A) appear in the sample when the dissolved oxygen is exhausted.
(B) appear in the sample when the dissolved oxygen and nitrates are exhausted.
(C) appear in the sample when the dissolved oxygen, nitrate, and sulfate are exhausted.
(D) not appear.

(c) Bacteria will convert ammonia to nitrate if the bacteria are

(A) phototropic.
(B) autotrophic.
(C) thermophilic.
(D) obligate anaerobes.

(d) Bacteria will generally reduce nitrate to nitrogen gas only if the bacteria are

(A) photosynthetic.
(B) obligate aerobic.
(C) facultative heterotrophic.
(D) aerobic phototropic.

(e) A bacteriophage is a

(A) bacterial enzyme.
(B) virus that infects bacteria.
(C) mesophilic organism.
(D) virus that stimulates bacterial growth.

(f) Algal growth in the wastewater

(A) will be inhibited when the dissolved oxygen is exhausted.
(B) will be inhibited when toxins are also found in the solution.
(C) will be inhibited when toxins are found in the solution or when the dissolved oxygen is exhausted.
(D) cannot be inhibited by chemical means.

(g) In the presence of nitrifying bacteria, nontoxic inorganic compounds, and sunlight, algal growth in the wastewater sample

(A) will be prevented.
(B) will be inhibited.
(C) will be unaffected.
(D) will be enhanced.

(h) The addition of protozoa to the wastewater will

(A) not change the wastewater's biochemical composition.
(B) increase the growth of algae in wastewater.
(C) increase the growth of bacteria in wastewater.
(D) decrease the growth of algae and bacteria in the wastewater.

(i) Coliform bacteria in wastewater from human, animal, or soil sources

(A) can be distinguished in the multiple-tube fermentation test.
(B) can be categorized into only two groups: human/animal and soil.
(C) can be distinguished if multiple-tube fermentation and Eschericheiae coli (EC) tests are both used.
(D) cannot be distinguished.

(j) Under what specific condition would a presence-absence (P-A) coliform test be used on this sample?

(A) if the sample contains known pathogens
(B) when multiple-tube fermentation indicates positive dilution in a presumptive test
(C) when multiple-tube fermentation indicates negative dilution in a presumptive test
(D) if the sample will be used as drinking water

2. A waste stabilization pond will be used in a municipal wastewater treatment system.

(a) Explain the function of bacteria and algae in the stabilization pond's application.

(b) Explain why algae would be a problem if it were present in the discharge from the pond.

(c) Define mechanisms that could be utilized to control algae and that would prove helpful in the development and use of this wastewater treatment system.

(d) Explain what is meant by *facultative pond.*

SOLUTIONS

1. (a) The sequence of oxygen usage reduced by bacteria is dissolved oxygen, nitrate, and then sulfate.

The answer is (C).

(b) Following the sequence of oxidation, obnoxious odors will occur when dissolved oxygen and nitrate are exhausted.

The answer is (B).

(c) Nitrification is performed by autotrophic bacteria to gain energy for growth by synthesis of carbon dioxide in an aerobic environment.

The answer is (B).

(d) Facultative heterotropic bacteria can decompose organic matter to gain energy under anaerobic conditions by removing the oxygen from nitrate, releasing nitrogen gas.

The answer is (C).

(e) A bacteriophage is a virus that infects bacteria.

The answer is (B).

(f) Algae are photosynthetic (gaining energy from light), releasing oxygen during metabolism. They thrive in aerobic environments. The presence of nitrifying bacteria and toxins and the depletion of dissolved oxygen would inhibit the growth process.

The answer is (C).

(g) Nitrifying bacteria are nonphotosynthetic, gaining energy by taking in oxygen to oxidize reduced inorganic nitrogen. The presence of inorganic nutrients would maintain the nitrification process, thereby limiting algal growth.

The answer is (B).

(h) Protozoa consume bacteria and algae in wastewater treatment and in the aquatic food chain.

The answer is (D).

(i) Fecal coliforms from humans and other warm-blooded animals are the same bacterial species. Coliforms originating from the soil can be separated by a confirmatory procedure using EC medium broth incubated at the elevated temperature of 44.5°C (112°F).

The answer is (B).

(j) The presence-absence technique is used for the testing of drinking water.

The answer is (D).

2. (a) A waste stabilization pond's operation is dependent on the reaction of bacteria and algae. Organic matter is metabolized by bacteria to produce the principle products of carbon dioxide, water, and a small amount of ammonia nitrogen. Algae convert sunlight into energy through photosynthesis. They utilize the end products of cell synthesis and other nutrients to synthesize new cells and produce oxygen. The most important role of the algae is in the production of oxygen in the pond for use by aerobic bacteria. In the absence of sunlight, the algae will consume oxygen in the same manner as bacteria. Algae removal is important in producing a high-quality effluent from the pond.

(b) The discharge of algae increases suspended solids in the discharge and may present a problem in meeting water quality criteria. The algae exert an oxygen demand when they settle to the bottom of the stream and undergo respiration.

(c) The following methods have been suggested for control of algae: (a) multiple ponds in series, (2) drawing off of effluent from below the surface by use of a good baffling arrangement to avoid algae concentrations, (3) sand filter or rock filter for algae removal, (4) alum addition and flocculation, (5) microscreening, and (6) chlorination to kill algae. Chlorination may increase BOD loading due to dead algae cells releasing stored organic material.

(d) Facultative ponds have two zones of treatment: an aerobic surface layer in which oxygen is used by aerobic bacteria for waste stabilization and an anaerobic bottom zone in which sludge decomposition occurs. No artificially induced aeration is used.

28 Wastewater Quantity and Quality

PRACTICE PROBLEMS

Sewer Velocities and Sizing

1. (*Time limit: one hour*) A town of 10,000 people (125 gpcd) has its own primary treatment plant.

(a) What mass of total solids should the treatment plant expect?

(A) 57 lbm/day
(B) 740 lbm/day
(C) 3800 lbm/day
(D) 8300 lbm/day

(b) If the town is 4 mi from the treatment plant and 400 ft above it in elevation, what minimum size pipe should be used between the town and the plant, assuming that the pipe flows full? Disregard infiltration.

(A) 8 in
(B) 12 in
(C) 14 in
(D) 18 in

2. (*Time limit: one hour*) Your client has just completed a subdivision, and his sewage lines hook up to a collector that goes into a trunk. A problem has developed in the first manhole up from the trunk, and the collector pipe overflows periodically. An industrial plant is hooked directly into the trunk, and it is this plant's flow that is making your client's line back up. The subdivision is in a flood plain, and the sewer lines are very flat and cannot be steepened. (a) Describe two possible solutions to this problem. (b) Sketch plan views of your solutions.

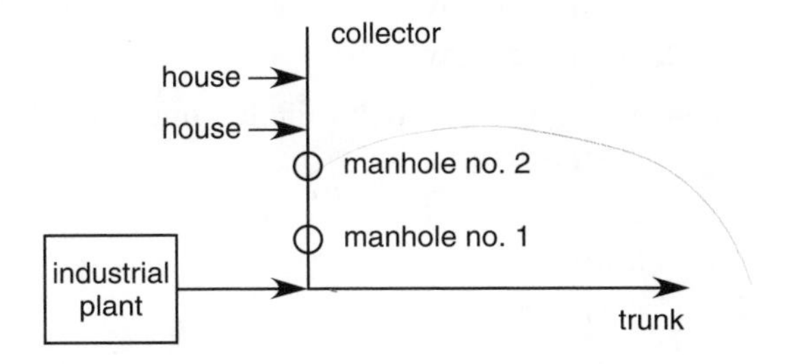

Dilution

3. Town A wants to discharge wastewater into a river 5.79 km upstream of town B. The state standard level for dissolved oxygen is 5 mg/L. Other data pertinent to this problem are given in the following table.

parameter	town A	river
flow (m^3/s)	0.280	0.877
ultimate BOD at 28°C (mg/L)	6.44	7.00
DO (mg/L)	1.00	6.00
K_d at 28°C, day^{-1} (base-10)	N/A	0.199
K_r at 28°C, day^{-1} (base-10)	N/A	0.370
velocity (m/s)	N/A	0.650
temperature (°C)	28	28

(a) Will the dissolved oxygen level at town A be reduced below the state standard level?

(A) yes; DO_{min} = 4.1 mg/L
(B) yes; DO_{min} = 4.8 mg/L
(C) no; DO_{min} = 5.3 mg/L
(D) no; DO_{min} = 5.9 mg/L

(b) Will the dissolved oxygen level be below the state standard level at town B?

(A) yes; DO_{min} = 4.1 mg/L
(B) yes; DO_{min} = 4.8 mg/L
(C) no; DO_{min} = 5.3 mg/L
(D) no; DO_{min} = 5.9 mg/L

(c) At any point downstream, will the dissolved oxygen level be below the state standard level?

(A) yes; DO_{min} = 4.1 mg/L
(B) yes; DO_{min} = 4.7 mg/L
(C) no; DO_{min} = 5.1 mg/L
(D) no; DO_{min} = 5.5 mg/L

(d) Determine if the discharge will reduce the dissolved oxygen in the river at town B below the state standard level if the winter river temperature drops to 12°C and all other characteristics remain the same.

(A) yes; DO_{min} = 4.5 mg/L
(B) yes; DO_{min} = 4.9 mg/L
(C) no; DO_{min} = 5.1 mg/L
(D) no; DO_{min} = 5.9 mg/L

4. To protect aquatic life, the limit on increase in temperature in a certain stream is 2°C at seasonal low flow. In addition, the stream must not contain more than 0.002 mg/L of un-ionized ammonia.

A manufacturing facility proposes to draw cooling water from the stream and return it to the stream at a higher temperature. They will also be discharging some un-ionized ammonia in the cooling water.

The critical low flow in the stream is 1200 m^3/s at a temperature of 20°C. The stream has no measurable un-ionized ammonia in its natural state. The facility intends to withdraw 60 m^3/s of cooling water and return it to the stream without loss of mass.

(a) What is the maximum temperature for the discharge water if the stream is not to be damaged?

(A) 40°C
(B) 60°C
(C) 80°C
(D) 90°C

(b) What is the maximum ammonia concentration for the discharge water if the stream is not to be damaged?

(A) 0.04 mg/L
(B) 0.06 mg/L
(C) 0.09 mg/L
(D) 0.13 mg/L

5. A rural town of 10,000 people discharges partially treated wastewater into a stream. The stream has the following characteristics.

minimum flow rate	120 ft^3/sec
velocity	3 mi/hr
minimum dissolved oxygen	7.5 mg/L at 15°C
temperature	15°C
BOD	0

The reoxygenation and deoxygenation coefficients (base-10) are

reoxygenation coefficient of stream and effluent mixture	0.2 day^{-1} at 20°C
deoxygenation coefficient	0.15 day^{-1} at 20°C

The town's effluent has the following characteristics.

source	volume	BOD at 20°C	temp.
domestic	122 gpcd	0.191 lbm/cd	64°F
infiltration	116,000 gpd		51°F
industrial no. 1	180,000 gpd	800 mg/L	95°F
industrial no. 2	76,000 gpd	1700 mg/L	84°F

(a) What is the domestic waste BOD in mg/L at 20°C?

(A) 140 mg/L
(B) 160 mg/L
(C) 190 mg/L
(D) 220 mg/L

(b) What is the approximate total effluent 20°C BOD in mg/L just before discharge into the stream?

(A) 245 mg/L
(B) 270 mg/L
(C) 295 mg/L
(D) 315 mg/L

(c) What is the average temperature of the effluent just before discharge into the stream?

(A) 20°C
(B) 30°C
(C) 50°C
(D) 70°C

(d) If the wastewater does not contribute to dissolved oxygen at all, how far downstream is the theoretical point of minimum dissolved oxygen concentration?

(A) 40 mi
(B) 70 mi
(C) 80 mi
(D) 120 mi

(e) What is the theoretical minimum dissolved oxygen concentration in the stream?

(A) 4.7 mg/L
(B) 5.5 mg/L
(C) 7.3 mg/L
(D) 8.3 mg/L

6. A sewage treatment plant is being designed to handle both domestic and industrial wastewaters. The city population is 20,000, and an excess capacity factor of 15% is to be used for future domestic expansion. The wastewaters have the following characteristics.

source	volume	BOD
domestic	100 gpcd	0.18 lbm/cd
industrial no. 1	1.3 MGD	1100 mg/L
industrial no. 2	1.0 MGD	500 mg/L

(a) What is the design population equivalent for the plant?

(A) 60,000
(B) 75,000
(C) 90,000
(D) 110,000

(b) What is the plant's organic loading?

(A) 2300 lbm/day
(B) 8100 lbm/day
(C) 14,000 lbm/day
(D) 20,000 lbm/day

(c) What is the plant's hydraulic loading?

(A) 4.6 MGD
(B) 5.8 MGD
(C) 6.3 MGD
(D) 7.5 MGD

SOLUTIONS

1. Assume: 125 gpcd for average flow
800 mg/L total solids

(a)
$$\frac{(10{,}000)\left(125\ \frac{\text{gal}}{\text{day}}\right)\left(800\ \frac{\text{mg}}{\text{L}}\right)\left(8.345\ \frac{\text{lbm-L}}{\text{MG-mg}}\right)}{1\times10^{6}\ \frac{\text{gal}}{\text{MG}}}$$

$$=\boxed{8345\ \text{lbm/day}}$$

(Minor variations in the assumptions should not affect the answer choice.)

The answer is (D).

(b)
$$S=\frac{400\ \text{ft}}{(4\ \text{mi})\left(5280\ \frac{\text{ft}}{\text{mi}}\right)}=0.01894$$

Use Eq. 28.1.

$$\frac{Q_{\text{peak}}}{Q_{\text{ave}}}=\frac{18+\sqrt{\text{P}}}{4+\sqrt{\text{P}}}=\frac{18+\sqrt{10}}{4+\sqrt{10}}$$
$$\approx 3.0$$

$$Q_{\text{peak}}=\frac{\left(125\ \frac{\text{gal}}{\text{day}}\right)(3)(10{,}000)\left(0.1337\ \frac{\text{ft}^3}{\text{gal}}\right)}{\left(24\ \frac{\text{hr}}{\text{day}}\right)\left(60\ \frac{\text{min}}{\text{hr}}\right)\left(60\ \frac{\text{sec}}{\text{min}}\right)}$$
$$=5.80\ \text{ft}^3/\text{sec}\quad(\text{cfs})$$

Assume $n=0.013$. From Eq. 19.16,

$$D=1.33\left(\frac{nQ}{\sqrt{S}}\right)^{3/8}$$
$$=(1.33)\left(\frac{(0.013)\left(5.80\ \frac{\text{ft}^3}{\text{sec}}\right)}{\sqrt{0.01894}}\right)^{3/8}$$
$$=1.06\ \text{ft}\quad(12.7\ \text{in})$$

$$\boxed{14\ \text{in}}$$

(A 12 in pipe does not have the capacity of a 12.7 in pipe.)

The answer is (C).

2. (a) If the first manhole overflows, the piezometric head at the rim must be less than the head in the trunk. Further up at manhole no. 2, the increase in elevation is sufficient to raise rim no. 2, so the head in the trunk must be lowered.

Alternate solutions:

- relief storage (surge chambers) between trunk and manhole no. 1
- storage at plant and gradual release
- private trunk for plant
- private treatment and discharge for plant
- larger trunk capacity using larger or parallel pipes
- backflow preventors

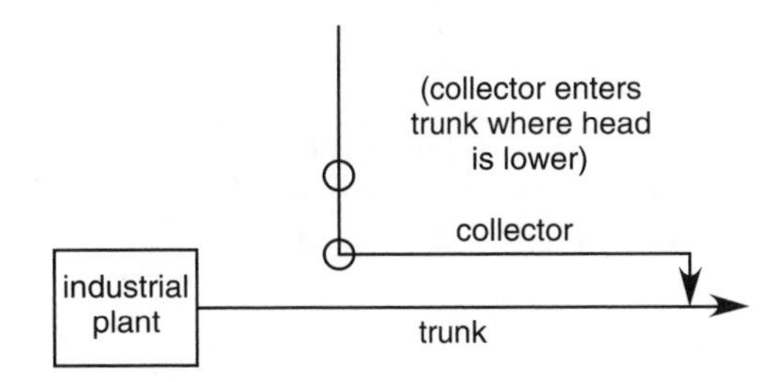

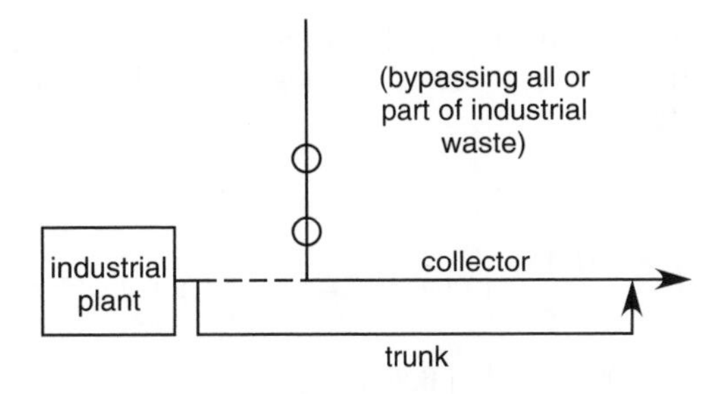

3. (a) Find the dissolved oxygen at town A from Eq. 28.33.

$$\text{DO}=\frac{Q_w\text{DO}_w+Q_r\text{DO}_r}{Q_w+Q_r}$$
$$=\frac{\left(0.280\ \frac{\text{m}^3}{\text{s}}\right)\left(1\ \frac{\text{mg}}{\text{L}}\right)+\left(0.877\ \frac{\text{m}^3}{\text{s}}\right)\left(6\ \frac{\text{mg}}{\text{L}}\right)}{0.280\ \frac{\text{m}^3}{\text{s}}+0.877\ \frac{\text{m}^3}{\text{s}}}$$
$$=\boxed{4.79\ \text{mg/L}}$$

So, immediately after mixing, the water will be reduced below the state standard of 5 mg/L.

The answer is (B).

(b) The composite ultimate BOD is

$$\text{BOD}_{u,28^\circ\text{C}}=\frac{Q_w\text{BOD}_w+Q_r\text{BOD}_r}{Q_w+Q_r}$$
$$=\frac{\left(0.280\ \frac{\text{m}^3}{\text{s}}\right)\left(6.44\ \frac{\text{mg}}{\text{L}}\right)+\left(0.877\ \frac{\text{m}^3}{\text{s}}\right)\left(7\ \frac{\text{mg}}{\text{L}}\right)}{0.280\ \frac{\text{m}^3}{\text{s}}+0.877\ \frac{\text{m}^3}{\text{s}}}$$
$$=6.8645\ \text{mg/L}$$

From App. 22.D, $DO_{sat} = 7.92$ mg/L at 28°C. The initial deficit is

$$\begin{aligned} D &= DO_{sat} - DO = 7.92\ \frac{mg}{L} - 4.79\ \frac{mg}{L} \\ &= 3.13\ mg/L \end{aligned}$$

Calculate travel time from town A to town B.

$$\begin{aligned} t &= \frac{(5.79\ km)\left(1000\ \frac{m}{km}\right)}{\left(0.650\ \frac{m}{s}\right)\left(86{,}400\ \frac{s}{day}\right)} \\ &= 0.1031\ day \end{aligned}$$

Calculate the deficit 5.79 km downstream. Use base-10 exponents because the K values are in base-10. Use Eq. 28.34.

$$\begin{aligned} D_t &= \left(\frac{K_d BOD_u}{K_r - K_d}\right)(10^{-K_d t} - 10^{-K_r t}) + D_0(10^{-K_r t}) \\ &= \frac{(0.199\ day^{-1})\left(6.8645\ \frac{mg}{L}\right)}{0.370\ \frac{mg}{L} - 0.199\ \frac{mg}{L}} \\ &\quad \times \left(10^{-(0.199\ day^{-1})(0.1031\ day)}\right. \\ &\quad \left. - 10^{-(0.370\ day^{-1})(0.1031\ day)}\right) \\ &\quad + \left(3.13\ \frac{mg}{L}\right)\left(10^{-(0.370\ day^{-1})(0.1031\ day)}\right) \\ &= 3.17\ mg/L \end{aligned}$$

Calculate the dissolved oxygen downstream.

$$\begin{aligned} DO_{town\ B} &= DO_{sat} - D \\ &= 7.92\ \frac{mg}{L} - 3.17\ \frac{mg}{L} \\ &= \boxed{4.75\ mg/L} \end{aligned}$$

This is still below the state standard of 5 mg/L. More distance is required for reoxygenation to increase the oxygen concentration.

The answer is (B).

(c) Determine the critical time. Use Eq. 28.36.

$$\begin{aligned} t_c &= \left(\frac{1}{K_r - K_d}\right) \\ &\quad \times \log_{10}\left(\left(\frac{K_d BOD_u - K_r D_0 + K_d D_0}{K_d BOD_u}\right)\left(\frac{K_r}{K_d}\right)\right) \\ &= \left(\frac{1}{0.370\ day^{-1} - 0.199\ day^{-1}}\right) \\ &\quad \times \log_{10}\left(\left(\frac{\begin{array}{r}(0.199\ day^{-1})\left(6.8645\ \frac{mg}{L}\right) \\ - (0.370\ day^{-1})\left(3.13\ \frac{mg}{L}\right) \\ + (0.199\ day^{-1})\left(3.13\ \frac{mg}{L}\right)\end{array}}{(0.199\ day^{-1})\left(6.8645\ \frac{mg}{L}\right)}\right) \right. \\ &\quad \left. \times \left(\frac{0.370\ day^{-1}}{0.199\ day^{01}}\right)\right) \\ &= 0.3122\ day \end{aligned}$$

Calculate critical deficit and dissolved oxygen. Use Eq. 28.37.

$$\begin{aligned} D_c &= \left(\frac{K_d BOD_u}{K_r}\right)10^{-K_d t} \\ &= \left(\frac{(0.199\ day^{-1})\left(6.8645\ \frac{mg}{L}\right)}{0.370\ day^{-1}}\right) \\ &\quad \times \left(10^{-(0.199\ day^{-1})(0.3122\ day)}\right) \\ &= 3.20\ mg/L \\ DO &= DO_{sat} - D_c \\ &= 7.92\ \frac{mg}{L} - 3.20\ \frac{mg}{L} \\ &= \boxed{4.72\ mg/L} \end{aligned}$$

This is below the state standard of 5 mg/L.

The answer is (B).

(d) From Eq. 28.33, the composite dissolved oxygen is

$$\begin{aligned}\text{DO} &= \frac{Q_w\text{DO}_w + Q_r\text{DO}_r}{Q_w + Q_r} \\ &= \frac{\left(0.280 \frac{\text{m}^3}{\text{s}}\right)\left(1 \frac{\text{mg}}{\text{L}}\right) + \left(0.877 \frac{\text{m}^3}{\text{s}}\right)\left(6 \frac{\text{mg}}{\text{L}}\right)}{0.280 \frac{\text{m}^3}{\text{s}} + 0.877 \frac{\text{m}^3}{\text{s}}} \\ &= 4.79 \text{ mg/L} \quad \text{[no change]}\end{aligned}$$

Calculate the temperature of the river water and wastewater mixture.

$$\begin{aligned}T_1 &= \frac{Q_wT_w + Q_rT_r}{Q_w + Q_r} \\ &= \frac{\left(0.280 \frac{\text{m}^3}{\text{s}}\right)(28^\circ\text{C}) + \left(0.877 \frac{\text{m}^3}{\text{s}}\right)(12^\circ\text{C})}{0.280 \frac{\text{m}^3}{\text{s}} + 0.877 \frac{\text{m}^3}{\text{s}}} \\ &= 15.87^\circ\text{C}\end{aligned}$$

Calculate K_d at this temperature. Two different values of θ_d are used for the two temperature ranges. Use Eq. 28.27.

$$\begin{aligned}K_{d,T_1} &= K_{d,T_2}\theta^{T_1-T_2} \\ K_{d,20^\circ C} &= (0.199 \text{ day}^{-1})(1.056)^{20^\circ C-28^\circ C} \\ &= 0.1287 \text{ day}^{-1} \\ K_{d,T_1} &= K_{d,T_2}\theta^{T_1-T_2} \\ K_{d,15.87^\circ C} &= (0.1287 \text{ day}^{-1})(1.135)^{15.87^\circ C-20^\circ C} \\ &= 0.07628 \text{ day}^{-1}\end{aligned}$$

Similarly, use Eq. 28.24 to correct K_r.

$$\begin{aligned}K_{r,T_1} &= K_{r,T_2}(1.024)^{T_1-T_2} \\ K_{r,15.87^\circ C} &= K_{r,28^\circ C}(1.024)^{15.87^\circ C-28^\circ C} \\ &= (0.370 \text{ day}^{-1})(1.024)^{15.87^\circ C-28^\circ C} \\ &= 0.2775 \text{ day}^{-1}\end{aligned}$$

Correct BOD_u to the new temperature. Use Eq. 28.31 twice.

$$\begin{aligned}\text{BOD}_{u,20^\circ C} &= \frac{\text{BOD}_{u,28^\circ C}}{(0.02)(28^\circ\text{C}) + 0.6} \\ \text{BOD}_{u,15.87^\circ C} &= (\text{BOD}_{u,20^\circ C})\big((0.02)(15.87^\circ\text{C}) + 0.6\big) \\ &= \frac{\left(6.8645 \frac{\text{mg}}{\text{L}}\right)\big((0.02)(15.87^\circ\text{C}) + 0.06\big)}{(0.02)(28^\circ\text{C}) + 0.6} \\ &= 5.43 \text{ mg/L}\end{aligned}$$

At 15.87°C, the saturation dissolved oxygen is found from App. 22.D. $\text{DO}_{\text{sat},15.87^\circ\text{C}} = 9.98$ mg/L.

Calculate the initial deficit from Eq. 28.33.

$$\begin{aligned}D_0 &= 9.98 \frac{\text{mg}}{\text{L}} - 4.79 \frac{\text{mg}}{\text{L}} \\ &= 5.19 \text{ mg/L}\end{aligned}$$

Calculate the deficit at town B. Use Eq. 28.34.

$$\begin{aligned}D_\text{B} &= \left(\frac{K_d\text{BOD}_u}{K_r - K_d}\right)(10^{-K_dt} - 10^{-K_rt}) + D_0(10^{-K_rt}) \\ &= \left(\frac{(0.07628 \text{ day}^{-1})\left(5.43 \frac{\text{mg}}{\text{L}}\right)}{0.2775 \text{ day}^{-1} - 0.07628 \text{ day}^{-1}}\right) \\ &\quad \times \Big(10^{-(0.07628 \text{ day}^{-1})(0.1031 \text{ day})} \\ &\quad -10^{-(0.2775 \text{ day}^{-1})(0.1031 \text{ day})}\Big) \\ &\quad + \left(5.19 \frac{\text{mg}}{\text{L}}\right)\left(10^{-(0.2775 \text{ day}^{-1})(0.1031 \text{ day})}\right) \\ &= 4.95 \text{ mg/L}\end{aligned}$$

The dissolved oxygen at town B is

$$\begin{aligned}\text{DO}_\text{B} &= \text{DO}_\text{sat} - D_\text{B} \\ &= 10.83 \frac{\text{mg}}{\text{L}} - 4.95 \frac{\text{mg}}{\text{L}} \\ &= \boxed{5.88 \text{ mg/L}}\end{aligned}$$

This meets the state standard of 5 mg/L.

The answer is (D).

4. (a)

$$\begin{aligned}T &= 20^\circ\text{C} + 2^\circ\text{C} = 22^\circ\text{C} \\ T_1 &= 20^\circ\text{C} \\ Q_1 + Q_2 &= 1200 \text{ m}^3/\text{s} \\ Q_2 &= 60 \text{ m}^3/\text{s} \\ Q_1 &= 1200 \frac{\text{m}^3}{\text{s}} - 60 \frac{\text{m}^3}{\text{s}} = 1140 \text{ m}^3/\text{s} \\ T &= \frac{T_1Q_1 + T_2Q_2}{Q_1 + Q_2} \\ 22^\circ\text{C} &= \frac{(20^\circ\text{C})\left(1140 \frac{\text{m}^3}{\text{s}}\right) + C_2\left(60 \frac{\text{m}^3}{\text{s}}\right)}{1200 \frac{\text{m}^3}{\text{s}}} \\ &= \boxed{60^\circ\text{C}}\end{aligned}$$

The answer is (B).

(b)

$$\begin{aligned}C &= 0 + 0.002 \frac{\text{mg}}{\text{L}} \\ C_1 &= 0 \\ 0.002 \frac{\text{mg}}{\text{L}} &= \frac{(0)\left(1140 \frac{\text{m}^3}{\text{s}}\right) + C_2\left(60 \frac{\text{m}^3}{\text{s}}\right)}{1200 \frac{\text{m}^3}{\text{s}}} \\ &= \boxed{0.04 \text{ mg/L}}\end{aligned}$$

The answer is (A).

5. (a) The domestic BOD concentration is

$$\frac{\left(0.191\ \frac{\text{lbm}}{\text{capita-day}}\right)\left(10^6\ \frac{\text{gal}}{\text{MG}}\right) \times \left(0.1198\ \frac{\text{mg-MG}}{\text{L-lbm}}\right)}{122\ \frac{\text{gal}}{\text{capita-day}}} = \boxed{187.6\ \text{mg/L}}$$

The answer is (C).

(b) Use Eq. 28.33.

$$\frac{\left(122\ \frac{\text{gal}}{\text{day}}\right)(10{,}000)\left(187.6\ \frac{\text{mg}}{\text{L}}\right) + \left(116{,}000\ \frac{\text{gal}}{\text{day}}\right)(0) + \left(180{,}000\ \frac{\text{gal}}{\text{day}}\right)\left(800\ \frac{\text{mg}}{\text{L}}\right) + \left(76{,}000\ \frac{\text{gal}}{\text{day}}\right)\left(1700\ \frac{\text{mg}}{\text{L}}\right)}{\left(122\ \frac{\text{gal}}{\text{day}}\right)(10{,}000) + 116{,}000\ \frac{\text{gal}}{\text{day}} 0 + 180{,}000\ \frac{\text{gal}}{\text{day}} + 76{,}000\ \frac{\text{gal}}{\text{day}}} = \boxed{315.4\ \text{mg/L}}$$

The answer is (D).

(c) Use Eq. 28.33.

$$\frac{\left(122\ \frac{\text{gal}}{\text{day}}\right)(10{,}000)(64^\circ\text{F}) + \left(116{,}000\ \frac{\text{gal}}{\text{day}}\right)(51^\circ\text{F}) + \left(180{,}000\ \frac{\text{gal}}{\text{day}}\right)(95^\circ\text{F}) + \left(76{,}000\ \frac{\text{gal}}{\text{day}}\right)(84^\circ\text{F})}{\left(122\ \frac{\text{gal}}{\text{day}}\right)(10{,}000) + 116{,}000\ \frac{\text{gal}}{\text{day}} + 180{,}000\ \frac{\text{gal}}{\text{day}} + 76{,}000\ \frac{\text{gal}}{\text{day}}} = \boxed{67.5^\circ\text{F}\quad(19.7^\circ\text{C})}$$

The answer is (A).

(d) The total discharge into the river is

$$\left(\left(122\ \frac{\text{gal}}{\text{day}}\right)(10{,}000) + 116{,}000\ \frac{\text{gal}}{\text{day}} + 180{,}000\ \frac{\text{gal}}{\text{day}} + 76{,}000\ \frac{\text{gal}}{\text{day}}\right)\left(1.547\times10^{-6}\ \frac{\text{ft}^3\text{-day}}{\text{sec-gal}}\right) = 2.46\ \text{ft}^3/\text{sec}\quad(\text{cfs})$$

step 1: Find the stream conditions immediately after mixing.

$$\text{BOD}_{5,20^\circ\text{C}} = \frac{\left(2.46\ \frac{\text{ft}^3}{\text{sec}}\right)\left(315.4\ \frac{\text{mg}}{\text{L}}\right) + \left(120\ \frac{\text{ft}^3}{\text{sec}}\right)(0)}{2.46\ \frac{\text{ft}^3}{\text{sec}} + 120\ \frac{\text{ft}^3}{\text{sec}}} = 6.34\ \text{mg/L}$$

$$\text{DO} = \frac{\left(2.46\ \frac{\text{ft}^3}{\text{sec}}\right)(0) + \left(120\ \frac{\text{ft}^3}{\text{sec}}\right)\left(7.5\ \frac{\text{mg}}{\text{L}}\right)}{2.46\ \frac{\text{ft}^3}{\text{sec}} + 120\ \frac{\text{ft}^3}{\text{sec}}} = 7.35\ \text{mg/L}$$

$$T = \frac{\left(2.46\ \frac{\text{ft}^3}{\text{sec}}\right)(19.7^\circ\text{C}) + \left(120\ \frac{\text{ft}^3}{\text{sec}}\right)(15^\circ\text{C})}{2.46\ \frac{\text{ft}^3}{\text{sec}} + 120\ \frac{\text{ft}^3}{\text{sec}}} = 15.1^\circ\text{C}$$

step 2: Calculate the rate constants at 15.1°C. Use Eq. 28.26.

$$K_{d,15.1^\circ C} = (0.15\ \text{day}^{-1})(1.135)^{15.1^\circ C - 20^\circ C} = 0.0807\ \text{day}^{-1}$$

$$K_{r,15.1^\circ C} = (0.2\ \text{day}^{-1})(1.024)^{15.1^\circ C - 20^\circ C} = 0.178\ \text{day}^{-1}$$

step 3: Estimate BOD_u. Use Eq. 28.29.

$$\text{BOD}_{u,20^\circ C} = \frac{6.34\ \frac{\text{mg}}{\text{L}}}{1 - 10^{-(0.15\ \text{day}^{-1})(5\ \text{days})}} = 7.71\ \text{mg/L}$$

Use Eq. 28.31 to convert BOD_u to 15.1°C.

$$\begin{aligned} BOD_{u,15.1^\circ C} &= BOD_{u,20^\circ C}(0.02T_{^\circ C} + 0.6) \\ &= \left(7.71 \ \frac{mg}{L}\right)\left((0.02)(15.1^\circ C)\right. \\ &\quad \left. + 0.6\right) \\ &= 6.95 \text{ mg/L} \end{aligned}$$

step 4: From App. 22.D at 15°C, saturated DO = 10.15 mg/L. Since the actual is 7.35 mg/L, the deficit is

$$D_0 = 10.15 \ \frac{mg}{L} - 7.35 \ \frac{mg}{L} = 2.8 \text{ mg/L}$$

step 5: Calculate t_c. Use Eq. 28.36.

$$t_c = \left(\frac{1}{0.178 \text{ day}^{-1} - 0.0807 \text{ day}^{-1}}\right) \times \log_{10}\left(\frac{(0.0807 \text{ day}^{-1})\left(6.95 \ \frac{mg}{L}\right) - (0.178 \text{ day}^{-1})\left(2.8 \ \frac{mg}{L}\right) + (0.0807 \text{ day}^{-1})\left(2.8 \ \frac{mg}{L}\right)}{(0.0807 \text{ day}^{-1})\left(6.95 \ \frac{mg}{L}\right)} \times \left(\frac{0.178 \text{ day}^{-1}}{0.0807 \text{ day}^{-1}}\right)\right) = 0.562 \text{ days}$$

step 6: The distance downstream is

$$(0.562 \text{ days})\left(3 \ \frac{mi}{hr}\right)\left(24 \ \frac{hr}{day}\right) = \boxed{40.5 \text{ mi}}$$

The answer is (A).

(e) *step 7:* Use Eq. 28.37.

$$\begin{aligned} D_c &= \left(\frac{(0.0807 \text{ day}^{-1})\left(6.95 \ \frac{mg}{L}\right)}{0.178 \text{ day}^{-1}}\right) \\ &\quad \times 10^{-(0.0807 \text{ day}^{-1})(0.566 \text{ day})} \\ &= 2.84 \text{ mg/L} \end{aligned}$$

step 8:

$$\begin{aligned} DO_{min} &= 10.15 \ \frac{mg}{L} - 2.84 \ \frac{mg}{L} \\ &= \boxed{7.31 \text{ mg/L}} \end{aligned}$$

The answer is (C).

6. (a) Do not apply the population expansion factor to the industrial effluents. Use Eq. 28.2, modified for the given population equivalent for the domestic flow contribution.

$$\begin{aligned} P_e &= P_{\text{domestic flow}} + P_{\text{industrial source 1}} \\ &\quad + P_{\text{industrial source 2}} \\ &= (20)(1.15) \\ &\quad + \frac{\left(1100 \ \frac{mg}{L}\right)\left(1.3 \times 10^6 \ \frac{gal}{day}\right) \times \left(8.345 \times 10^{-9} \ \frac{lbm\text{-}L}{MG\text{-}mg}\right)}{0.18 \ \frac{lbm}{day}} \\ &\quad + \frac{(500)\left(1.0 \times 10^6 \ \frac{gal}{day}\right) \times \left(8.345 \times 10^{-9} \ \frac{lbm\text{-}L}{MG\text{-}mg}\right)}{0.18 \ \frac{lbm}{day}} \\ &= \boxed{112.5 \quad [\text{thousands of people}]} \end{aligned}$$

The answer is (D).

(b) Since the plant loading is requested, the organic loading can be given in lbm/day. The population from part (a) is 112,500.

$$L_{BOD} = (112{,}500)\left(0.18 \ \frac{lbm}{day}\right) = \boxed{20{,}250 \text{ lbm/day}}$$

The answer is (D).

$$\begin{aligned} \text{(c)}L_H &= (20{,}000)(1.15)\left(100 \ \frac{gal}{day}\right) \\ &\quad + 1.3 \times 10^6 \ \frac{gal}{day} + 1.0 \times 10^6 \ rfracgalday \\ &= \boxed{4.6 \times 10^6 \text{ gal/day (4.6 MGD)}} \end{aligned}$$

The answer is (A).

Environmental

29 Wastewater Treatment: Equipment and Processes

PRACTICE PROBLEMS

Lagoons

1. A cheese factory located in a normally warm state has liquid waste with the following characteristics.

waste no. 1	
volume	10,000 gal/day
BOD	1000 mg/L
waste no. 2	
volume	25,000 gal/day
BOD	250 mg/L

The factory will use a 4 ft deep on-site nonaerated lagoon to stabilize the waste.

(a) What is the total BOD loading?

(A) 60 lbm/day
(B) 85 lbm/day
(C) 110 lbm/day
(D) 140 lbm/day

(b) What lagoon size is required if the BOD loading is 20 lbm BOD/ac-day?

(A) 2.7 ac
(B) 6.8 ac
(C) 8.3 ac
(D) 10.9 ac

(c) What is the detention time?

(A) 5 wk
(B) 14 wk
(C) 36 wk
(D) 52 wk

Trickling Filters

2. It is estimated that the BOD of raw sewage received at a treatment plant serving a population of 20,000 will be 300 mg/L. It is estimated that the per capita BOD loading is 0.17 lbm/day. 30% of the influent BOD is removed by settling. One single-stage high-rate trickling filter is to be used to reduce the plant effluent to 50 mg/L. Recirculation is from the filter effluent to the primary settling influent. *Ten States' Standards* is in effect.

(a) What is the design flow rate?

(A) 0.9 MGD
(B) 1.1 MGD
(C) 1.4 MGD
(D) 2.0 MGD

(b) Assuming a design flow rate of 1.35 MGD, what is the total organic load on the filter?

(A) 1700 lbm/day
(B) 2100 lbm/day
(C) 2400 lbm/day
(D) 3400 lbm/day

(c) Assuming an incoming volume of 1.35 MGD and an organic loading of 60 lbm/day-1000 ft^3, what flow should be recirculated?

(A) 0.6 MGD
(B) 0.9 MGD
(C) 1.2 MGD
(D) 2.2 MGD

(d) What is the overall plant efficiency?

(A) 45%
(B) 76%
(C) 83%
(D) 91%

3. The average wastewater flow from a community of 20,000 is 125 gpcd. The 5 day, 20°C BOD is 250 mg/L. The suspended solids content is 300 mg/L. A final plant effluent of 50 mg/L of BOD is to be achieved through the use of two sets of identical settling tanks and trickling filters operating in parallel. The settling tanks are to be designed to a standard of 1000 gpd/ft^2. The trickling filters are to be 6 ft deep. There is no recirculation.

(a) What settling tank surface area is required?

(A) 2500 ft^2
(B) 3000 ft^2
(C) 3500 ft^2
(D) 4500 ft^2

(b) What settling tank diameter is required?

(A) 40 ft
(B) 50 ft
(C) 65 ft
(D) 80 ft

(c) Estimate the BOD removal in the settling tanks.

(A) 15%
(B) 30%
(C) 45%
(D) 60%

(d) What is the trickling filter diameter?

(A) 65 ft
(B) 75 ft
(C) 85 ft
(D) 95 ft

4. Wastewater from a city with a population of 40,000 has an average daily flow of 4.4 MGD. The sewage has the following characteristics.

BOD_5 at 20°C	160 mg/L
COD	800 mg/L
total solids	900 mg/L
suspended solids	180 mg/L
volatile solids	320 mg/L
settleable solids	8 mg/L
pH	7.8

The wastewater is to be treated with primary settling and secondary trickling filtration. The settling basins are to be circular, 8 ft deep, and designed to a standard overflow rate of 1000 gal/day-ft^2.

(a) Assuming two identical basins operating in parallel, what should be the diameter of each sedimentation basin in order to remove about 30% of the BOD?

(A) 45 ft
(B) 53 ft
(C) 77 ft
(D) 86 ft

(b) What is the detention time?

(A) 1.4 hr
(B) 1.8 hr
(C) 2.3 hr
(D) 3.1 hr

(c) What is the weir loading?

(A) 8200 gal/day-ft
(B) 11,000 gal/day-ft
(C) 13,000 gal/day-ft
(D) 15,000 gal/day-ft

5. (*Time limit: one hour*) A small community has a projected average flow of 1 MGD. Incoming wastewater has the following properties: BOD, 250 mg/L; grit specific gravity, 2.65; and total suspended solids, 400 mg/L. The community wants to have a wastewater treatment plant consisting of a single aerated grit chamber, a single primary clarifier, two identical circular trickling filters in parallel, and a single secondary clarifier. There will be no equalization basin. Recirculation from the second clarifier to the entrance of the trickling filters will be 100% of the average flow. The final effluent is to have a BOD of 30 mg/L.

(a) What is the peak design flow at the grit chamber?

(A) 1.0 MGD
(B) 1.5 MGD
(C) 2.0 MGD
(D) 2.5 MGD

(b) Determine the aerated grit chamber width assuming a 3 min detention time, 20 ft length, and a width:depth ratio of 1.25.

(A) 4 ft
(B) 6 ft
(C) 8 ft
(D) 10 ft

(c) Determine the approximate air requirements for the grit chamber in order to capture approximately 95% of the grit.

(A) 60 cfm
(B) 140 cfm
(C) 280 cfm
(D) 420 cfm

(d) If the clarifier is 12 ft deep, what should be its diameter?

(A) 50 ft
(B) 56 ft
(C) 63 ft
(D) 81 ft

(e) What is the peak flow entering the trickling filters?

(A) 1.0 MGD
(B) 2.0 MGD
(C) 3.0 MGD
(D) 4.0 MGD

(f) Assume the primary clarifier removes 30% of the incoming BOD, and the filters see the average flow only. Determine the diameter of the trickling filters assuming a 6 ft deep rock bed.

(A) 45 ft
(B) 55 ft
(C) 65 ft
(D) 85 ft

(g) Determine the diameter of the final clarifier.

(A) 35 ft
(B) 50 ft
(C) 65 ft
(D) 80 ft

Recirculating Biological Contactors

6. (*Time limit: one hour*) 1.5 MGD of wastewater with a BOD of 250 mg/L is processed by a high-rate rock trickling filter followed by recirculating biological contactor (RBC) processing. The trickling filter is 75 ft in diameter and 6 ft deep and was designed to NRC standards. The BOD removal efficiency of the RBC process is given by the following equation. (k has a value of 2.45 gal/day-ft^2, and Q (in units of gal/day) does not include recirculation. A is the immersed area of the RBC in ft^2.)

$$\eta_{\text{BOD}} = \frac{1}{\left(1+\frac{kA}{Q}\right)^3}$$

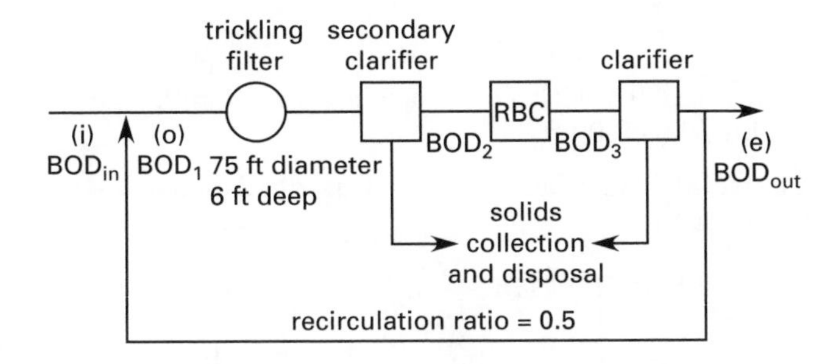

(a) What total exposed RBC surface area is required to achieve an effluent BOD$_{out}$ of 30 mg/L?

(A) 4.2×10^4 ft^2
(B) 8.4×10^4 ft^2
(C) 1.3×10^5 ft^2
(D) 2.0×10^5 ft^2

(b) The recirculation pick-up point is relocated from after the final clarifier to after the trickling filter. The efficiency of the RBC process is 65%. Determine the recirculation ratio such that BOD$_{out}$ is 30 mg/L.

(A) 25%
(B) 50%
(C) 75%
(D) 100%

(c) If BOD$_2$ = 85 mg/L, BOD$_{out}$ = 30 mg/L, the yield is 0.4 lbm/lbm BOD removed, and the sludge specific gravity is essentially 1.0, what is the approximate sludge volume produced from the clarifiers?

(A) 4 ft^3/day
(B) 12 ft^3/day
(C) 21 ft^3/day
(D) 35 ft^3/day

SOLUTIONS

1. (a) The total BOD is

$$\left(1000\ \frac{\text{mg}}{\text{L}}\right)\left(8.345\ \frac{\text{lbm-L}}{\text{mg-MG}}\right)\left(\frac{10{,}000\ \frac{\text{gal}}{\text{day}}}{1{,}000{,}000\ \frac{\text{gal}}{\text{MG}}}\right)$$
$$+\left(250\ \frac{\text{mg}}{\text{L}}\right)\left(8.345\ \frac{\text{lbm-L}}{\text{mg-MG}}\right)\left(\frac{25{,}000\ \frac{\text{gal}}{\text{day}}}{1{,}000{,}000\ \frac{\text{gal}}{\text{MG}}}\right)$$
$$= \boxed{135.6\ \text{lbm/day}\ (1.4\ \text{MGD})}$$

The answer is (D).

(b) The warm weather and depth contribute to a decrease in pond effectiveness. Assume 20 lbm BOD/ac-day for a nonaerated stabilization pond. From Eq. 29.4, the required area is

$$A = \frac{Q}{\text{v}^*} = \frac{135.6\ \frac{\text{lbm}}{\text{day}}}{20\ \frac{\text{lbm}}{\text{ac-day}}} = \boxed{6.8\ \text{ac}}$$

The answer is (B).

(c) Use Eq. 29.5.

$$t_d = \frac{\text{V}}{Q}$$
$$= \frac{(6.8\ \text{ac})\left(43{,}560\ \frac{\text{ft}^2}{\text{ac}}\right)(4\ \text{ft})}{\left(35{,}000\ \frac{\text{gal}}{\text{day}}\right)\left(1.547\times 10^{-6}\ \frac{\text{ft}^3\text{-day}}{\text{sec-gal}}\right)\left(3600\ \frac{\text{sec}}{\text{hr}}\right)}$$
$$= \boxed{6078\ \text{hr}\quad (36.2\ \text{weeks})}$$

The answer is (C).

2. (a)
$$Q = \frac{\dot{m}}{C} = \frac{\left(0.17\ \frac{\text{lbm}}{\text{capita-day}}\right)(20{,}000\ \text{people})}{\left(8.345\times 10^{-6}\ \frac{\text{lbm-L}}{\text{gal-mg}}\right)\left(300\ \frac{\text{mg}}{\text{L}}\right)}$$
$$= \boxed{1.358\times 10^6\ \text{gal/day}}$$

The answer is (C).

(*Ten States' Standards* specifies 100 gpcd in the absence of other information. In that case, Q = (100 gal/capita-day)(20,000 people) = 2×10^6 gal/day.)

(b) The total BOD load leaving the primary clarifier and entering the filter is

$$\text{BOD}_i = (1 - 0.30)\left(300\ \frac{\text{mg}}{\text{L}}\right) = 210\ \text{mg/L}$$

$$L_{\text{BOD}} = (1.35\ \text{MGD})\left(210\ \frac{\text{mg}}{\text{L}}\right)\left(8.345\ \frac{\text{lbm-L}}{\text{mg-MG}}\right) = \boxed{2366\ \text{lbm/day}}$$

The answer is (C).

(c) The efficiency of the filter and secondary clarifier is found from Eq. 29.9.

$$\eta = \frac{210\ \frac{\text{mg}}{\text{L}} - 50\ \frac{\text{mg}}{\text{L}}}{210\ \frac{\text{mg}}{\text{L}}} = 0.762\ (76.2\%)$$

Use Eq. 29.13.

$$0.762 = \frac{1}{1 + 0.0561\sqrt{\frac{60\ \frac{\text{lbm}}{\text{day-1000 ft}^3}}{F}}}$$

$$F = 1.936$$

Use Eq. 29.15.

$$1.936 = \frac{1 + R}{\left(1 + (0.1)(R)\right)^2}$$

$$R = 1.6$$

Use Eq. 29.10.

$$Q_r = (1.6)(1.35\ \text{MGD}) = \boxed{2.16\ \text{MGD}}$$

The answer is (D).

(d) Use Eq. 29.9.

$$\eta = \frac{300\ \frac{\text{mg}}{\text{L}} - 50\ \frac{\text{mg}}{\text{L}}}{300\ \frac{\text{mg}}{\text{L}}} = \boxed{0.833\quad(83.3\%)}$$

The answer is (C).

3. Use Eq. 29.4. Disregarding variations in peak flow, the average design volume is

$$\frac{\left(125\ \frac{\text{gal}}{\text{capita-day}}\right)(20{,}000\ \text{people})}{1{,}000{,}000\ \frac{\text{gal}}{\text{MG}}} = 2.5\ \text{MGD}$$

(a) The settling tank surface area is

$$\frac{2.5 \times 10^6\ \frac{\text{gal}}{\text{day}}}{1000\ \frac{\text{gal}}{\text{day-ft}^2}} = \boxed{2500\ \text{ft}^2}$$

The answer is (A).

(b) The required diameter when using two tanks in parallel is

$$D = \sqrt{\frac{(4)(2500\ \text{ft}^2)}{2\pi}} = \boxed{39.9\ \text{ft (use 40 ft) each}}$$

The answer is (A).

(c) A 30% removal is typical.

The answer is (B).

(d) BOD entering the filter is

$$(1 - 0.30)\left(250\ \frac{\text{mg}}{\text{L}}\right) = 175\ \text{mg/L}$$

The filter efficiency is

$$\eta = \frac{175\ \frac{\text{mg}}{\text{L}} - 50\ \frac{\text{mg}}{\text{L}}}{175\ \frac{\text{mg}}{\text{L}}} = 0.71$$

From Fig. 29.4 with 71% efficiency and $R = 0$, $L_{\text{BOD}} = 55\ \text{lbm/day-1000 ft}^3$.

The total load is found from Eq. 29.12 rearranged in terms of V_1.

$$\frac{(2.5\ \text{MGD})\left(175\ \frac{\text{mg}}{\text{L}}\right) \times \left(8.345\ \frac{\text{lbm-L}}{\text{mg-MG}}\right)\left(1000\ \frac{\text{ft}^3}{1000\ \text{ft}^3}\right)}{\left(55\ \frac{\text{lbm}}{\text{day-1000 ft}^3}\right)(2\ \text{filters})}$$

$$= 33{,}190\ \text{ft}^3/\text{filter}$$

With a depth of 6 ft, the total required surface area is

$$A = \frac{V}{\text{depth}} = \frac{33{,}190\ \text{ft}^3}{6\ \text{ft}} = 5532\ \text{ft}^2$$

The required diameter per filter is

$$D = \sqrt{\frac{4A}{\pi}} = \sqrt{\frac{(4)(5532\ \text{ft}^2)}{\pi}} = \boxed{83.9\ \text{ft}}$$

The answer is (C).

4. (a) There is nothing particularly special about a basin that removes 30% BOD. Choose two basins in parallel, each working with half of the total flow. Choose an overflow rate of 1000 gpd/ft². The area per basin is given by Eq. 29.4.

$$A = \frac{4.4 \times 10^6\ \frac{\text{gal}}{\text{day}}}{(2)\left(1000\ \frac{\text{gal}}{\text{day-ft}^2}\right)} = 2200\ \text{ft}^2$$

$$d = \sqrt{\left(\frac{4}{\pi}\right)(2200\ \text{ft}^2)} = \boxed{52.9\ \text{ft}}$$

The answer is (B).

(b) The detention time is given by Eq. 29.5.

$$t = \frac{(2200\ \text{ft}^2)(8\ \text{ft})}{\left(2.2\ \frac{\text{MGD}}{\text{tank}}\right)\left(1.547\ \frac{\frac{\text{ft}^3}{\text{sec}}}{\text{MGD}}\right)\left(3600\ \frac{\text{sec}}{\text{hr}}\right)} = \boxed{1.436\ \text{hr}}$$

The answer is (A).

(c)
$$\begin{aligned}\text{circumference} &= \pi(52.9\ \text{ft}) \\ &= 166.2\ \text{ft} \\ \text{weir loading} &= \frac{2.2\times 10^6\ \frac{\text{gal}}{\text{day}}}{166.2\ \text{ft}} \\ &= \boxed{13{,}240\ \text{gal/day-ft (gpd/ft)}}\end{aligned}$$

The answer is (C).

5.

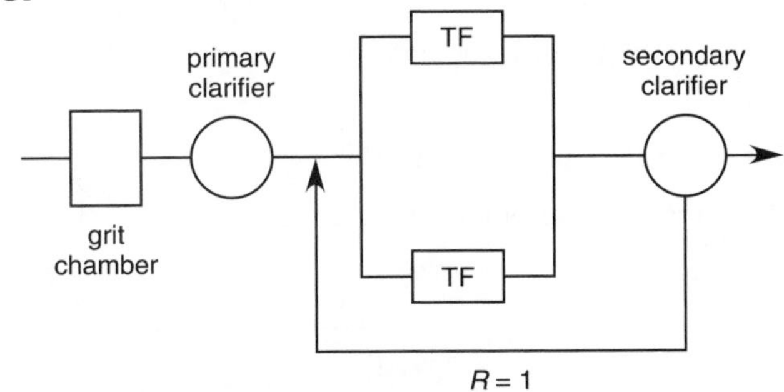

(a) From Table 28.1, assume a peak flow multiplier of 2 since the population size is unknown. (Alternatively, a population "equivalent" can be used to estimate the population, and then Eq. 28.1 can be used. This will produce a different answer.) The peak flow is

$$(2)(1\ \text{MGD}) = \boxed{2\ \text{MGD}}$$

The answer is (C).

(b) The peak flow rate per second is

$$Q_{\text{peak}} = \frac{(2)(1\ \text{MGD})\left(10^6\ \frac{\text{gal}}{\text{MG}}\right)\left(0.1337\ \frac{\text{ft}^3}{\text{gal}}\right)}{\left(24\ \frac{\text{hr}}{\text{day}}\right)\left(60\ \frac{\text{min}}{\text{hr}}\right)\left(60\ \frac{\text{sec}}{\text{min}}\right)} = 3.095\ \text{ft}^3/\text{sec}$$

With a detention time of 3 min, the volume of the grit chamber would be

$$V = Qt = \left(3.095\ \frac{\text{ft}^3}{\text{sec}}\right)\left(60\ \frac{\text{sec}}{\text{min}}\right)(3\ \text{min}) = 557.1\ \text{ft}^3$$

$$\begin{aligned}557.1\ \text{ft}^3 &= \text{length}\times\text{width}\times\text{depth} \\ &= (20\ \text{ft})(1.25)(\text{depth})^2 \\ \text{water depth} &= 4.72\ \text{ft}\quad [\text{round to } 4\tfrac{3}{4}\ \text{ft}] \\ \text{chamber width} &= (1.25)(4.72\ \text{ft}) \\ &= \boxed{5.9\ \text{ft}\quad [\text{round to 6 ft}]}\end{aligned}$$

The answer is (B).

(c) This is a shallow chamber. Use 3 cfm/ft.

$$\left(3\ \frac{\text{ft}^3}{\text{min-ft}}\right)(20\ \text{ft}) = \boxed{60\ \text{ft}^3/\text{min (cfm) of air}}$$

The answer is (A).

(d) (Note that the *Ten States' Standards* requires two basins.) Surface loading is the primary design parameter. Choose a surface loading of 1000 gal/day-ft^2. The volume is

$$A = \frac{Q}{v^*} = \frac{(2\ \text{MGD})\left(10^6\ \frac{\text{gal}}{\text{day-MGD}}\right)}{1000\ \frac{\text{gal}}{\text{day-ft}^2}} = 2000\ \text{ft}^2$$

$$D = \sqrt{\frac{4A}{\pi}} = \sqrt{\frac{(4)(2000\ \text{ft}^2)}{\pi}} = \boxed{50.5\ \text{ft}}$$

The answer is (A).

(e) The volume of the primary clarifier is fixed by the weir height. Therefore, the clarifier does not provide any storage (i.e., no damping of the flow rates). The fluctuations in flow will be passed on to the trickling filters.

$$Q_p + Q_r = 2\ \text{MGD} + 1\ \text{MGD} = 3\ \text{MGD}$$

The answer is (C).

(f) Assume the primary sedimentation basin removes 30% of the BOD. BOD incoming to the trickle filter process is given by Eq. 29.12.

$$(1-0.30)\left(250\ \frac{\text{mg}}{\text{L}}\right) = 175\ \text{mg/L}$$

$$\left(175\ \frac{\text{mg}}{\text{L}}\right)\left(8.345\ \frac{\text{lbm-L}}{\text{mg-MG}}\right)\times(1\ \text{MGD}) = 1460\ \text{lbm/day}$$

The required trickling filter-clarifier process efficiency is

$$\eta = \frac{\text{BOD}_{\text{in}} - \text{BOD}_{\text{out}}}{\text{BOD}_{\text{in}}} = \frac{175\ \frac{\text{mg}}{\text{L}} - 30\ \frac{\text{mg}}{\text{L}}}{175\ \frac{\text{mg}}{\text{L}}}$$

$$= 0.829\ (82.9\%)$$

From Eq. 29.15 with $w = 0.1$ and $R = 1$,

$$F = \frac{1+R}{(1+wR)^2} = \frac{1+1}{\left(1+(0.1)(1)\right)^2} = 1.65$$

From Eq. 29.13,

$$\eta = \frac{1}{1 + 0.0561\sqrt{\frac{L_{\text{BOD}}}{F}}}$$

$$0.83 = \frac{1}{1 + 0.0561\sqrt{\frac{L_{\text{BOD}}}{1.65}}}$$

$$L_{\text{BOD}} \approx 22\ \text{lbm/day-1000 ft}^3$$

$$\text{filter volume} = \frac{\left(1460\ \frac{\text{lbm}}{\text{day}}\right)\left(1000\ \frac{\text{ft}^3}{1000\ \text{ft}^3}\right)}{\left(22\ \frac{\text{lbm}}{\text{day-1000 ft}^3}\right)(2\ \text{filters})}$$

$$= 33{,}180\ \text{ft}^3$$

$$D = \sqrt{\frac{(4)(33{,}180\ \text{ft}^3)}{\pi(6\ \text{ft})}} = \boxed{83.9\ \text{ft}}$$

[round to 85 ft]

Use two 85 ft diameter, 6 ft deep filters.

The answer is (D).

(g) For the final clarifier, the maximum overflow rate is 1100 gpd/ft^2, and the minimum depth is 10 ft. Assume volumetric flow fluctuations will be damped out by previous processes.

$$A = \frac{1 \times 10^6\ \frac{\text{gal}}{\text{day}}}{1100\ \frac{\text{gal}}{\text{day-ft}^2}} = 909\ \text{ft}^2$$

$$D = \sqrt{\frac{(4)(909\ \text{ft}^2)}{\pi}} = \boxed{34.0\ \text{ft}} \quad \text{[round to 35 ft]}$$

Use a 35 ft diameter, 10 ft deep basin.

The answer is (A).

6. (a) Assume the last clarifier removes no BOD.

$$\text{BOD}_3 = 30\ \text{mg/L}$$

There is no recirculation that matches the NRC model. The NRC model "recirculation" is from the trickling filter discharge directly back to the entrance to the filter. This problem's recirculation is from several processes beyond the trickling filter. The recirculation increases the BOD loading and dilutes the influent.

The BOD loading to the trickling filter must include recirculation and is given by Eq. 29.12.

$$L_{\text{BOD}} = \frac{(1.5\ \text{MGD})\left(250\ \frac{\text{mg}}{\text{L}}\right) \times \left(8.345\ \frac{\text{lbm-L}}{\text{MG-mg}}\right)\left(1000\ \frac{\text{ft}^3}{1000\ \text{ft}^3}\right)}{\pi\left(\frac{75\ \text{ft}}{2}\right)^2(6\ \text{ft})} + \frac{(0.5)(1.5\ \text{MGD})\left(30\ \frac{\text{mg}}{\text{L}}\right) \times \left(8.345\ \frac{\text{lbm-L}}{\text{MG-mg}}\right)\left(1000\ \frac{\text{ft}^3}{1000\ \text{ft}^3}\right)}{\pi\left(\frac{75\ \text{ft}}{2}\right)^2(6\ \text{ft})}$$

$$= 118.1\ \frac{\text{lbm}}{\text{day-1000 ft}^3} + 7.1\ \frac{\text{lbm}}{\text{day-1000 ft}^3}$$

$$= 125.2\ \text{lbm/day-1000 ft}^3$$

Since $R = 0$, $F = 1$ [Eq. 29.15].

The filter/clarifier efficiency is given by Eq. 29.13.

$$\eta = \frac{1}{1 + 0.0561\sqrt{\frac{125.2\ \frac{\text{lbm}}{\text{day-1000 ft}^3}}{1.00}}}$$

$$= 0.61\quad(61\%)$$

$$\text{BOD}_2 = (1 - 0.61)\left(250\ \frac{\text{mg}}{\text{L}}\right) = 97.5\ \text{mg/L}$$

The removal fraction in the RBC must be

$$\frac{97.5\ \frac{\text{mg}}{\text{L}} - 30\ \frac{\text{mg}}{\text{L}}}{97.5\ \frac{\text{mg}}{\text{L}}} = 0.69$$

Solving the given performance equation, the immersed area is

$$0.69 = \frac{1}{\left(1 + \frac{2.45A}{1.5 \times 10^6}\right)^3}$$

$$A = 80{,}609\ \text{ft}^2$$

From Table 29.11, only 40% of the total RBC area is immersed at one time. The total RBC area is

$$A_{\text{total}} = \frac{80{,}609\ \text{ft}^2}{0.4} = \boxed{201{,}523\ \text{ft}^2 \quad (200{,}000\ \text{ft}^2)}$$

The answer is (D).

(b) This is one of the recirculation modes to which the NRC model applies. Solve the problem backward to get $\text{BOD}_{\text{out}} = 30$ mg/L.

$$\text{BOD}_3 = 30\ \text{mg/L}$$

$$\text{BOD}_2 = \frac{30\ \frac{\text{mg}}{\text{L}}}{1 - 0.65} = 85.7\ \text{mg/L}$$

The efficiency in the NRC model includes the effect of the clarifier, even though the recirculation occurs before the clarifier.

$$\eta_{\text{trickling filter and clarifier}} = \frac{250\ \frac{\text{mg}}{\text{L}} - 85.7\ \frac{\text{mg}}{\text{L}}}{250\ \frac{\text{mg}}{\text{L}}} = 0.657$$

In this configuration, the BOD loading does not include the effects of recirculation, as the NRC model places a higher emphasis on organic loading than on hydraulic loading. From part (a), $L_{\text{BOD}} = 118.1$ lbm/day-1000 ft^3.

Equation 29.13 could be solved for F, and then that value used in Eq. 29.15 to find R. It is easier to use Fig. 29.4 with $L_{\text{BOD}} = 118.1$ and $\eta = 65.7\%$.

$$R \approx \boxed{0.5 \quad (50\%)}$$

The answer is (B).

(c) Approximate the BOD removal of the secondary (first in-line) clarifier.

$$\text{BOD}_{\text{entering}} = \frac{\text{BOD}_2}{1 - \eta} = \frac{85\ \frac{\text{mg}}{\text{L}}}{1 - 0.30} = 121.4\ \text{mg/L}$$

[assuming 30% removal]

The sludge production is

$$\left(0.4\ \frac{\text{lbm}}{\text{lbm}}\right)\left(\left(121.4\ \frac{\text{mg}}{\text{L}} - 85\ \frac{\text{mg}}{\text{L}}\right) + \left(43\ \frac{\text{mg}}{\text{L}} - 30\ \frac{\text{mg}}{\text{L}}\right)\right)\left(8.345\ \frac{\text{lbm-L}}{\text{MG-mg}}\right)(1.5\,\text{MGD})$$

$$= 247.3\ \text{lbm/day}$$

The sludge volume is

$$V = \frac{247.3\ \frac{\text{lbm}}{\text{day}}}{62.4\ \frac{\text{lbm}}{\text{ft}^3}} = \boxed{3.96\ \text{ft}^3/\text{day}}$$

The answer is (A).

Environmental

30 Activated Sludge and Sludge Processing

PRACTICE PROBLEMS

Sludge Quantities

1. 33 m^3/day of thickened sludge with a suspended solids content of 3.8% and 13 m^3/day of anaerobic digester sludge with a suspended solids content of 7.8% are produced in a wastewater treatment plant.

(a) What would be the effect of using a filter press to increase the solids content of the thickened sludge to 24%?

(A) 4000 m^3/yr
(B) 8000 m^3/yr
(C) 10,000 m^3/yr
(D) 15,000 m^3/yr

(b) What volume of digester sludge must be disposed of from sand drying beds that increase the solids concentration to 35%?

(A) 500 m^3/yr
(B) 1000 m^3/yr
(C) 2000 m^3/yr
(D) 4000 m^3/yr

2. An activated sludge plant processes 10 MGD of wastewater with 240 mg/L BOD and 225 mg/L suspended solids. 70% of the suspended solids are inorganic. The discharge from the final clarifier contains 15 mg/L BOD (all organic) and 20 mg/L suspended solids (all inorganic). Primary clarification removes 60% of the suspended solids and 35% of the BOD. The BOD reduction in the primary clarifier does not contribute to sludge production. The sludge produced has a specific gravity of 1.02 and a solids content of 6%. The cell yield (conversion of BOD reduction to biological solids) is 60%. The final clarifier does not reduce BOD.

(a) What is the daily mass of dry sludge solids produced?

(A) 2500 lbm/day
(B) 5000 lbm/day
(C) 10,000 lbm/day
(D) 20,000 lbm/day

(b) Assuming that the sludge is completely dried and compressed solid, what is the daily sludge volume?

(A) 300 ft^3/day
(B) 500 ft^3/day
(C) 1000 ft^3/day
(D) 1600 ft^3/day

(c) Assuming that the final gravimetric fraction of water in the sludge is 70% and air voids increase the volume by 10%, what is the daily sludge volume?

(A) 200 ft^3/day
(B) 500 ft^3/day
(C) 1000 ft^3/day
(D) 1600 ft^3/day

The actual application rate depends on sludge strength (primarily nitrogen content), the condition of the receiving soil, the crop or plants using the applied nutrients, and the spreading technology.

SOLUTIONS

1. (a) Sludge volume is inversely proportional to the solids content. The sludge volume from the filter press is

$$V_2 = V_1\left(\frac{SS_1}{SS_2}\right) = \left(33\ \frac{m^3}{day}\right)\left(\frac{0.038}{0.24}\right) = 5.23\ m^3/day$$

The yearly decrease in volume is

$$\left(33\ \frac{m^3}{day} - 5.23\ \frac{m^3}{day}\right)\left(365\ \frac{day}{yr}\right) = \boxed{10{,}136\ m^3/yr}$$

The answer is (C).

(b) The sludge volume from the sand drying bed is

$$V_2 = V_1\left(\frac{SS_1}{SS_2}\right) = \left(13\ \frac{m^3}{day}\right)\left(\frac{0.078}{0.35}\right) = 2.90\ m^3/day$$

The yearly disposal volume is

$$\left(2.9\ \frac{m^3}{day}\right)\left(365\ \frac{day}{yr}\right) = \boxed{1059\ m^3/yr}$$

The answer is (B).

2. The BOD and SS are not mutually exclusive. Some of the SS is organic in nature, and this shows up as BOD.

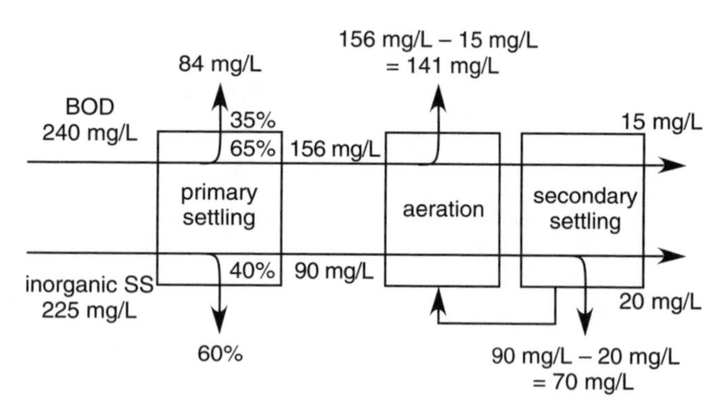

(a) The total dry weight of SS removed in all processes is

$$\left(225\ \frac{mg}{L} - 20\ \frac{MG}{L}\right)\left(8.345\ \frac{lbm\text{-}L}{MG\text{-}MG}\right)\left(10\ \frac{MG}{day}\right) = 17{,}107\ lbm/day$$

The BOD entering the secondary process is

$$(1 - 0.35)\left(240\ \frac{MG}{L}\right) = 156\ MG/L$$

Solids from BOD reduction in the secondary process are

$$\begin{aligned} Y(\Delta BOD)Q &= (0.60)\left(156\ \frac{mg}{L} - 15\ \frac{mg}{L}\right) \\ &\quad\times\left(10\ \frac{MG}{day}\right)\left(8.345\ \frac{lbm\text{-}L}{mg\text{-}MG}\right) \\ &= 7060\ lbm/day \end{aligned}$$

The total dry sludge mass is

$$17{,}107\ \frac{lbm}{day} + 7060\ \frac{lbm}{day} = \boxed{24{,}167\ lbm/day}$$

The answer is (D).

(b) The specific gravity of the sludge solids can be found from Eq. 30.47.

$$\begin{aligned} \frac{1}{SG} &= \frac{1-s}{1} + \frac{s}{SG_{solids}} \\ \frac{1}{1.02} &= \frac{1-0.06}{1} + \frac{0.06}{SG_{solids}} \\ SG_{solids} &= 1.485 \end{aligned}$$

The density of the solids is

$$\rho_{solids} = (1.485)\left(62.4\ \frac{lbm}{ft^3}\right) = 92.66\ lbm/ft^3$$

The solid dry sludge volume is

$$V_{solid} = \frac{m}{\rho} = \frac{24{,}167\ \frac{lbm}{day}}{92.66\ \frac{lbm}{ft^3}} = \boxed{260.8\ ft^3/day}$$

The answer is (A).

(c) $1 - s$ (the moisture content) is given as 0.70. The disposal volume is

$$\begin{aligned} V_t &= V_{solid} + V_{water} \\ V_{solid} &= \frac{m_{solid}}{\rho} \\ V_{water} &= \frac{m_{water}}{62.4\ \frac{lbm}{ft^3}} = \frac{0.70m_t}{62.4\ \frac{lbm}{ft^3}} \\ &= \frac{0.70m_{solids}}{\left(62.4\ \frac{lbm}{ft^3}\right)(1-0.70)} \\ V &\approx (1.10)\left(\frac{24{,}167\ \frac{lbm}{day}}{92.66\ \frac{lbm}{ft^3}} + \frac{\left(24{,}167\ \frac{lbm}{day}\right)(0.70)}{\left(62.4\ \frac{lbm}{ft^3}\right)(1-0.70)}\right) \\ &= \boxed{1280.9\ ft^3/day} \end{aligned}$$

Notice that the basis of moisture content, s, is the total sludge mass, whereas the traditional "water content" basis used in geotechnical calculations is the solids mass.

The answer is (C).

31 Municipal Solid Waste

PRACTICE PROBLEMS

Landfills

1. A town has a current population of 10,000, which is expected to double in 15 yr. The town intends to dispose of its municipal solid waste in a 30 ac landfill that will be converted to a park in 20 yr. Solid waste is generated at the rate of 5 lbm/capita-day. The average compacted density in the landfill will be 1000 lbm/yd^3. Disregarding any soil addition for cover and cell construction, how long will it take to fill the landfill to a uniform height of 6 ft?

(A) 2100 days
(B) 4200 days
(C) 6300 days
(D) 9500 days

2. (*Time limit: one hour*) A town of 10,000 people has selected a 50 ac square landfill site to deposit its solid waste. The minimum unused side borders are 50 ft. The landfill currently consists of a square depression with an average depth of 20 ft below the surrounding grade. When the landfill is at final capacity, it will be covered with 10 ft of earth cover. The maximum height of the covered landfill is 20 ft above the surrounding grade. Solid waste is generated at the rate of 5 lbm/capita-day. The average compacted density in the landfill will be 1000 lbm/yd^3.

(a) What is the volumetric capacity of the landfill site?

(A) 1.1×10^6 ft^3
(B) 6.5×10^6 ft^3
(C) 3.1×10^7 ft^3
(D) 5.7×10^7 ft^3

(b) Using a loading factor of 1.25, what is the volume of landfill used each day?

(A) 60 yd^3/day
(B) 120 yd^3/day
(C) 180 yd^3/day
(D) 240 yd^3/day

(c) What is the service life of the landfill site?

(A) 30 yr
(B) 45 yr
(C) 60 yr
(D) 90 yr

SOLUTIONS

1. Find the rate of increase of waste production.

The mass of waste deposited in the first day will be

$$(10{,}000 \text{ people})\left(5\ \frac{\text{lbm}}{\text{person-day}}\right) = 50{,}000 \text{ lbm/day}$$

The mass of waste deposited on the last day will be

$$(20{,}000 \text{ people})\left(5\ \frac{\text{lbm}}{\text{person-day}}\right) = 100{,}000 \text{ lbm/day}$$

The increase in rate is

$$\frac{\Delta m}{\Delta t} = \frac{100{,}000\ \frac{\text{lbm}}{\text{day}} - 50{,}000\ \frac{\text{lbm}}{\text{day}}}{(15 \text{ yr})\left(365\ \frac{\text{days}}{\text{yr}}\right)} = 9.132 \text{ lbm/day}^2$$

The mass deposited on day D is

$$\begin{aligned} m_D &= 50{,}000 + (9.132)(D - 1) \\ &\approx 50{,}000 + 9.132D \end{aligned}$$

The cumulative mass deposited is

$$m_t = \int_0^t m_D\, dt = 50{,}000t + \frac{9.132t^2}{2}$$

With a compacted density of 1000 lbm/yd^3 and a loading factor of 1.00 (no soil cover), the capacity of the site with a 6 ft lift is

$$\begin{aligned} m_{\text{max}} &= \frac{(30 \text{ ac})\left(43{,}560\ \frac{\text{ft}^2}{\text{ac}}\right)(6 \text{ ft})\left(1000\ \frac{\text{lbm}}{\text{yd}^3}\right)}{27\ \frac{\text{ft}^3}{\text{yd}^3}} \\ &= 2.9 \times 10^8 \text{ lbm} \end{aligned}$$

The time to fill is found by solving the quadratic equation.

$$\begin{aligned} 2.9 \times 10^8 &= 50{,}000t + \frac{9.132t^2}{2} \\ t^2 + 10{,}951t &= 6.351 \times 10^7 \\ t &= \boxed{4193 \text{ days} \quad (11.5 \text{ yr})} \end{aligned}$$

The answer is (B).

2. The side length of the square disposal site is

(a)
$$\text{side} = \sqrt{(50\ \text{ac})\left(43{,}560\ \frac{\text{ft}^2}{\text{ac}}\right)} = 1476\ \text{ft}$$

With 50 ft borders, the usable area is

$$A = \left(1476\ \text{ft} - (2)(50\ \text{ft})\right)^2 = 1.893 \times 10^6\ \text{ft}^2$$

If the site is excavated 20 ft, 10 ft of soil is used as cover, and the maximum above-ground height is 20 ft, the service capacity of compacted waste is

$$(1.893 \times 10^6\ \text{ft}^2)(30\ \text{ft}) = \boxed{5.68 \times 10^7\ \text{ft}^3}$$

The answer is (D).

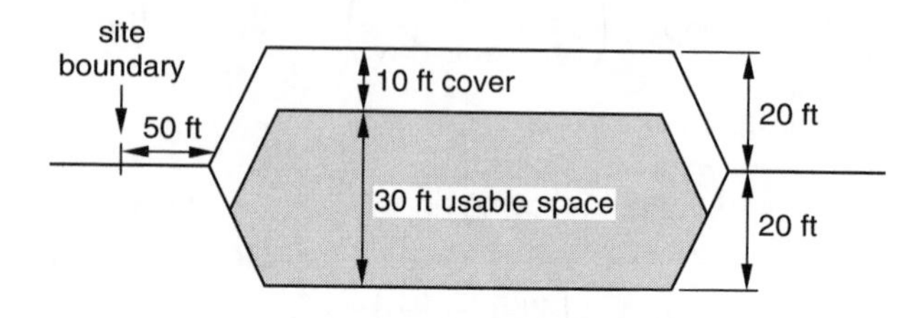

(not to scale or representative of actual construction)

(b) The volume of landfill per day is

$$\frac{(10{,}000\ \text{people})\left(5\ \frac{\text{lbm}}{\text{day-person}}\right)(1.25)}{1000\ \frac{\text{lbm}}{\text{yd}^3}} = \boxed{62.5\ \text{yd}^3/\text{day}}$$

The answer is (A).

(c) The service life is

$$\frac{5.68 \times 10^7\ \text{ft}^3}{\left(27\ \frac{\text{ft}^3}{\text{yd}^3}\right)\left(62.5\ \frac{\text{yd}^3}{\text{day}}\right)\left(365\ \frac{\text{days}}{\text{yr}}\right)} = \boxed{92.2\ \text{yr}}$$

The answer is (D).

32 Pollutants in the Environment

PRACTICE PROBLEMS

There are no problems in this book corresponding to Ch. 32 of the *Civil Engineering Reference Manual*.

33 Disposition of Hazardous Materials

PRACTICE PROBLEMS

There are no problems in this book corresponding to Ch. 33 of the *Civil Engineering Reference Manual*.

34 Environmental Remediation

PRACTICE PROBLEMS

There are no problems in this book corresponding to Ch. 34 of the *Civil Engineering Reference Manual*.

35 Soil Properties and Testing

PRACTICE PROBLEMS

Properties and Classification

1. A clay has the following Atterberg limits: liquid limit = 60, plastic limit = 40, and shrinkage limit = 25. The clay shrinks from 15 cm^3 to 9.57 cm^3 when the moisture content is decreased from the liquid limit to the shrinkage limit in the Atterberg tests. What is the clay's specific gravity (dry)?

(A) 2.0
(B) 2.3
(C) 2.5
(D) 2.7

2. A sample of soil has the following characteristics.

% passing no. 40 screen	95
% passing no. 200 screen	57
liquid limit	37
plastic limit	18

What are the AASHTO classification and group index number?

(A) A-5(6)
(B) A-5(8)
(C) A-6(8)
(D) A-7(5)

3. (*Time limit: one hour*) A sample of moist soil was found to have the following characteristics:

volume	0.514 ft^3 (as sampled)
mass	56.74 lbm (as sampled)
	48.72 lbm (after oven drying)
specific gravity of solids	2.69

After drying, the soil was run through a set of sieves with the following percentages passing through each sieve.

sieve	% finer by mass
1/2 in	52
no. 4	37
no. 10	32
no. 20	23
no. 40	11
no. 60	7
no. 100	4

(a) What is the density of the in situ soil?

(A) 92 lbm/ft^3
(B) 98 lbm/ft^3
(C) 110 lbm/ft^3
(D) 120 lbm/ft^3

(b) What is the unit weight of the in situ soil?

(A) 92 lbf/ft^3
(B) 98 lbf/ft^3
(C) 110 lbf/ft^3
(D) 120 lbf/ft^3

(c) What is the void ratio of the in situ soil?

(A) 0.77
(B) 0.80
(C) 1.00
(D) 1.60

(d) What is the porosity of the in situ soil?

(A) 0.44
(B) 0.57
(C) 0.77
(D) 0.98

(e) What is the degree of saturation of the in situ soil?

(A) 44%
(B) 57%
(C) 78%
(D) 95%

(f) Which of the figures represents the particle size distribution of the soil?

(A)

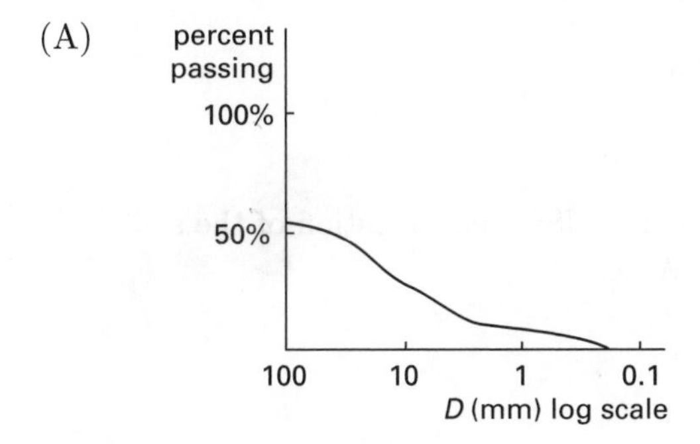

(B)

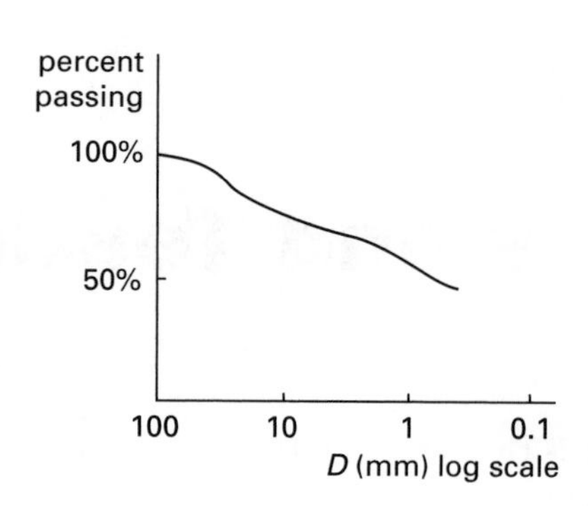

(C)

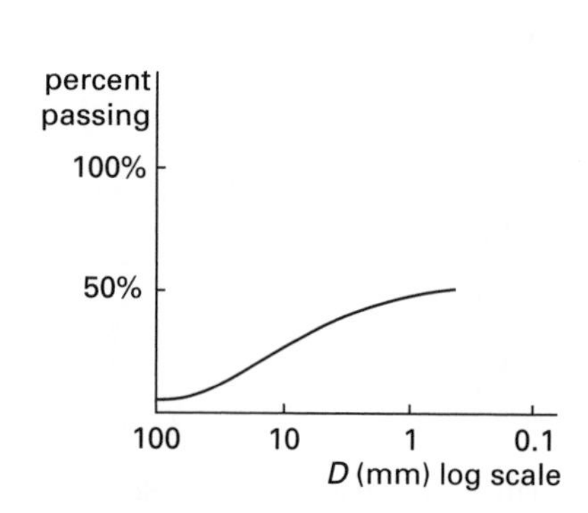

(D)

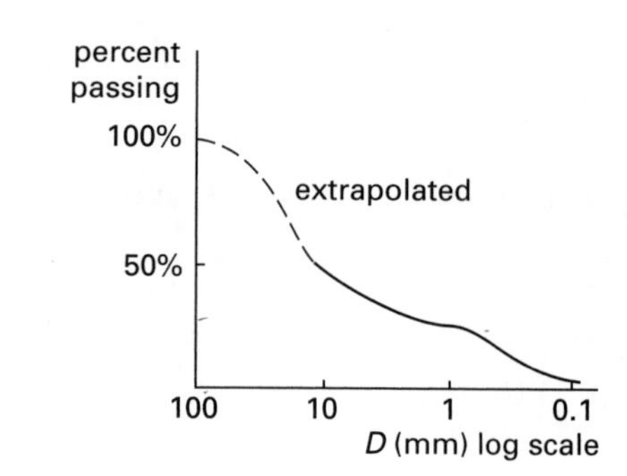

(g) What is the uniformity coefficient?

(A) 35
(B) 40
(C) 43
(D) 57

(h) What is the coefficient of curvature?

(A) 0.28
(B) 0.35
(C) 0.51
(D) 0.68

(i) What is the USCS classification of the soil?

(A) GW
(B) GP
(C) SW
(D) SM

(j) What is the AASHTO classification of the soil?

(A) A-1-a
(B) A-1-b
(C) A-2-4
(D) A-2-7

Permeability Test

4. A permeability test is conducted with a sample of soil that is 60 mm in diameter and 120 mm long. The head is kept constant at 225 mm. The flow is 1.5 mL in 6.5 min. What is the coefficient of permeability?

(A) 4.4 m/yr
(B) 12 m/yr
(C) 17 m/yr
(D) 23 m/yr

Proctor Test

5. A Proctor test was performed on a soil that has a specific gravity of solids of 2.71. The following results were obtained.

water content (%)	total unit weight (lbf/ft^3)
10	98
13	106
16	119
18	125
20	129
22	128
25	123

(a) Plot the moisture versus dry density curve.

(b) What is the maximum dry unit weight?

(A) 97 lbf/ft^3
(B) 103 lbf/ft^3
(C) 108 lbf/ft^3
(D) 112 lbf/ft^3

(c) What is the optimum moisture content?

(A) 12%
(B) 17%
(C) 20%
(D) 24%

(d) What range of moisture is permitted if a contractor must achieve 90% compaction?

(A) 12–23%
(B) 14–26%
(C) 18–28%
(D) 20–24%

6. For the soil in Prob. 5, what volume of water must be added to obtain 1 yd^3 of soil at the maximum density if the soil is originally at 10% water content (dry basis)?

(A) 6 gal
(B) 14 gal
(C) 18 gal
(D) 31 gal

Borrow Soil

7. Specifications on a job require a fill using borrow soil to be compacted to 95% of its standard Proctor maximum dry density. Tests indicate that this maximum is 124.0 lbm/ft^3 when dry. The soil now has 12% moisture. The borrow material has a void ratio of 0.60 and a solid specific gravity of 2.65. What is the minimum volume of borrow soil required per 1.0 ft^3 of fill volume?

(A) 0.90 ft^3
(B) 1.1 ft^3
(C) 1.3 ft^3
(D) 1.5 ft^3

8. (*Time limit: one hour*) Two choices for borrow soil are available.

	borrow A	borrow B
density in place	115 lbm/ft^3	120 lbm/ft^3
density in transport		95 lbm/ft^3
void ratio in transport	0.92	
water content in place	25%	20%
cost to excavate	$0.20/yd^3$	$0.10/yd^3$
cost to haul	$0.30/yd^3$	$0.40/yd^3$
SG of solids	2.7	2.7
maximum Proctor dry density	112 lbm/ft^3	110 lbm/ft^3

It will be necessary to fill a 200,000 yd^3 depression, and the fill material must be compacted to 95% of the standard Proctor (maximum) density. A 10% (dry basis) final moisture content is desired in either case.

(a) What would be the volume of borrow from site A?

(A) 4.8×10^6 ft^3
(B) 5.4×10^6 ft^3
(C) 5.6×10^6 ft^3
(D) 6.3×10^6 ft^3

(b) What would be the volume of borrow from site B?

(A) 4.8×10^6 ft^3
(B) 5.4×10^6 ft^3
(C) 5.6×10^6 ft^3
(D) 6.3×10^6 ft^3

(c) What soil would be cheaper to use?

(A) borrow A by $6200
(B) borrow A by $7400
(C) borrow B by $4500
(D) borrow B by $19,000

Triaxial Shear Test

9. A triaxial shear test is performed on a well-drained sand sample. At failure, the normal stress on the failure plane was 6260 lbf/ft^2 and the shear stress on the failure plane was 4175 lbf/ft^2.

(a) What is the angle of internal friction?

(A) 16°
(B) 28°
(C) 34°
(D) 42°

(b) What is the maximum principal stress?

(A) 9000 lbf/ft^2
(B) 14,000 lbf/ft^2
(C) 16,000 lbf/ft^2
(D) 22,000 lbf/ft^2

10. (*Time limit: one hour*) Two series of triaxial shear tests on a soil were performed with the following results. (All values are in lbf/in^2.)

Undrained series:

confining pressure	total axial stress
0	60
50	110
100	160

Drained series:

confining pressure	total axial stress
50	250
100	400
150	550

(a) What is the angle of internal friction for the undrained series?

(A) 0°
(B) 5°
(C) 12°
(D) 23°

(b) What is the angle of internal friction for the drained series?

(A) 10°
(B) 20°
(C) 30°
(D) 40°

Geotechnical

(c) What is the cohesion for the undrained series?

(A) 30 psi
(B) 90 psi
(C) 180 psi
(D) 280 psi

(d) What is the cohesion for the drained series?

(A) 17 psi
(B) 23 psi
(C) 29 psi
(D) 45 psi

(e) What is the angle of the failure plane (with respect to the horizontal axis) for the undrained series?

(A) 0
(B) 15°
(C) 30°
(D) 45°

(f) What is the angle of the failure plane (with respect to the horizontal axis) for the drained series?

(A) 15°
(B) 39°
(C) 45°
(D) 60°

(g) Given a fourth test of the drained sample with a radial confining pressure of 300 lbf/in^2, what is the expected axial load at failure?

(A) 550 lbf/in^2
(B) 1000 lbf/in^2
(C) 1200 lbf/in^2
(D) 1600 lbf/in^2

11. (*Time limit: one hour*) A consolidated, undrained (CU) triaxial test was performed on a normally consolidated, saturated clay. The minor principal stress, σ_3, at the start of the test was 100 kPa. The deviator stress, σ_D, at failure was 60 kPa. The excess pore pressure, ΔU, was measured during the test, and at failure it was 45 kPa.

(a) What is the undrained strength at failure, S_u?

(A) 15 kPa
(B) 30 kPa
(C) 40 kPa
(D) 60 kPa

(b) What is the total minor principal stress at failure, σ_{3f}?

(A) 40 kPa
(B) 55 kPa
(C) 60 kPa
(D) 100 kPa

(c) What is the total major principal stress at failure, σ_{1f}?

(A) 30 kPa
(B) 40 kPa
(C) 130 kPa
(D) 160 kPa

(d) What is the effective minor principal stress at failure, σ'_{3f}?

(A) 30 kPa
(B) 55 kPa
(C) 60 kPa
(D) 100 kPa

(e) What is the effective major principal stress at failure, σ'_{1f}?

(A) 60 kPa
(B) 100 kPa
(C) 120 kPa
(D) 160 kPa

(f) What is the effective friction angle, ϕ'?

(A) 21°
(B) 22°
(C) 26°
(D) 31°

(g) What is the angle of the effective failure plane, α?

(A) 37°
(B) 45°
(C) 55°
(D) 61°

(h) What is the effective normal stress on the failure plane at failure, σ_{nf}?

(A) 66 kPa
(B) 74 kPa
(C) 85 kPa
(D) 130 kPa

(i) What is the effective shear stress on the failure plane at failure, τ_{nf}?

(A) 28 kPa
(B) 30 kPa
(C) 37 kPa
(D) 60 kPa

(j) What is another name for this type of test?

(A) S-test
(B) R-test
(C) unconfined compressive test
(D) CBR test

Direct Shear Test

12. (*Time limit: one hour*) A sandy clay soil was tested in a direct shear apparatus, with the following results.

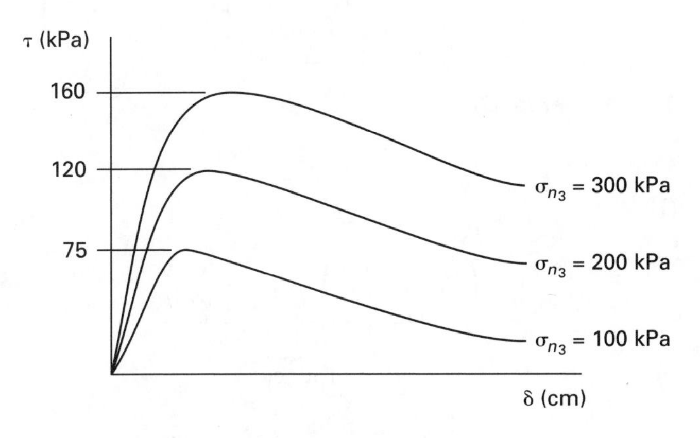

(a) What is the friction angle?

(A) 7°
(B) 14°
(C) 23°
(D) 29°

(b) What is the cohesion intercept?

(A) 24 kpa
(B) 33 kPa
(C) 57 kPa
(D) 86 kPa

Consolidation Test

13. A sample of sand has a relative density of 40% with a solids specific gravity of 2.65. The minimum void ratio is 0.45, and the maximum void ratio is 0.97.

(a) What is the specific weight of the sand in a saturated condition?

(A) 94 lbf/ft^3
(B) 116 lbf/ft^3
(C) 121 lbf/ft^3
(D) 137 lbf/ft^3

(b) If the sand is compacted to a relative density of 65%, what will be the decrease in thickness of a layer 4 ft thick?

(A) 0.1 ft
(B) 0.2 ft
(C) 0.3 ft
(D) 0.4 ft

14. A consolidation test is performed on a soil with the following results.

pressure (lbf/ft^2)	void ratio, e
250	0.755
520	0.754
1040	0.753
2090	0.750
4180	0.740
8350	0.724
16,700	0.704
33,400	0.684
8350	0.691
250	0.710

(a) Graph the curve of stress versus void ratio on log or semilog paper.

(b) What is the preconsolidation pressure?

(A) 4400 lbf/ft^2
(B) 6200 lbf/ft^2
(C) 8900 lbf/ft^2
(D) 12,000 lbf/ft^2

(c) What is the compression index?

(A) 0.0053
(B) 0.090
(C) 0.17
(D) 0.25

SOLUTIONS

1. The water reduction is

$$15 \text{ cm}^3 - 9.57 \text{ cm}^3 = 5.43 \text{ cm}^3$$

Since 1 cm^3 of water has a mass of 1 g, the mass loss is 5.43 g. The percentage mass loss (dry basis) is

$$60\% - 25\% = 35\%$$

Therefore, from Eq. 35.6, and using Δw to indicate the change in the water content, the solid mass is

$$m_s = \frac{\Delta m_w}{\Delta w} = \frac{5.43 \text{ g}}{0.35} = 15.5 \text{ g}$$

The water volume at the shrinkage limit (where there are no air voids) is

$$\begin{aligned} V_w &= \frac{m_w}{\rho_w} = \frac{w m_s}{\rho_w} \\ &= \frac{(0.25)(15.5 \text{ g})}{1 \ \frac{\text{g}}{\text{cm}^3}} \\ &= 3.875 \text{ cm}^3 \end{aligned}$$

Since at and above the shrinkage limit there are no air voids, the volume of solid at the shrinkage limit is

$$9.57 \text{ cm}^3 - 3.875 \text{ cm}^3 = 5.695 \text{ cm}^3$$

The density of the solid is

$$\begin{aligned} \rho &= \frac{15.5 \text{ g}}{5.695 \text{ cm}^3} = 2.72 \text{ g/cm}^3 \\ \text{SG} &= \frac{\rho_s}{\rho_w} \\ &= \frac{2.72 \ \frac{\text{g}}{\text{cm}^3}}{1 \ \frac{\text{g}}{\text{cm}^3}} \\ &= \boxed{2.72 \quad (2.7)} \end{aligned}$$

The answer is (D).

2. The plasticity index is given by Eq. 35.23.

$$\text{PI} = \text{LL} - \text{PL} = 37 - 18 = 19$$

The group index is given by Eq. 35.3.

$$\begin{aligned} I_g &= (F_{200} - 35)\big(0.2 + 0.005(\text{LL} - 40)\big) \\ &\quad + 0.01(F_{200} - 15)(\text{PI} - 10) \\ &= (57 - 35)\big(0.2 + (0.005)(37 - 40)\big) \\ &\quad + (0.01)(57 - 15)(19 - 10) \\ &= 7.85 \quad \text{[round to 8]} \end{aligned}$$

$$\text{classification} = \boxed{\text{A-6(8)}}$$

The answer is (C).

3. (a)

$$\begin{aligned} \rho &= \frac{m_t}{V_t} \\ &= \frac{56.74 \text{ lbm}}{0.514 \text{ ft}^3} \\ &= \boxed{110.4 \text{ lbm/ft}^3} \end{aligned}$$

The answer is (C).

(b)

$$\begin{aligned} \gamma &= \frac{\rho g}{g_c} \\ &= \frac{\left(110.4 \ \frac{\text{lbm}}{\text{ft}^3}\right)\left(32.2 \ \frac{\text{ft}}{\text{sec}^2}\right)}{32.2 \ \frac{\text{ft-lbm}}{\text{lbf-sec}^2}} \\ &= \boxed{110.4 \text{ lbf/ft}^3} \end{aligned}$$

The answer is (C).

(c)

$$\begin{aligned} m_w &= m_t - m_s = 56.74 \text{ lbm} - 48.72 \text{ lbm} \\ &= 8.02 \text{ lbm} \\ V_s &= \frac{m_s}{\rho_s} = \frac{48.72 \text{ lbm}}{(2.69)\left(62.4 \ \frac{\text{lbm}}{\text{ft}^3}\right)} \\ &= 0.29025 \text{ ft}^3 \\ V_w &= \frac{8.02 \text{ lbm}}{62.4 \ \frac{\text{lbm}}{\text{ft}^3}} = 0.12853 \text{ ft}^3 \\ V_g &= 0.514 \text{ ft}^3 - 0.12853 \text{ ft}^3 - 0.29025 \text{ ft}^3 \\ &= 0.09522 \text{ ft}^3 \end{aligned}$$

From Eq. 35.5,

$$\begin{aligned} e &= \frac{V_g + V_w}{V_s} \\ &= \frac{0.09522 \text{ ft}^3 + 0.12853 \text{ ft}^3}{0.29025 \text{ ft}^3} \\ &= \boxed{0.771} \end{aligned}$$

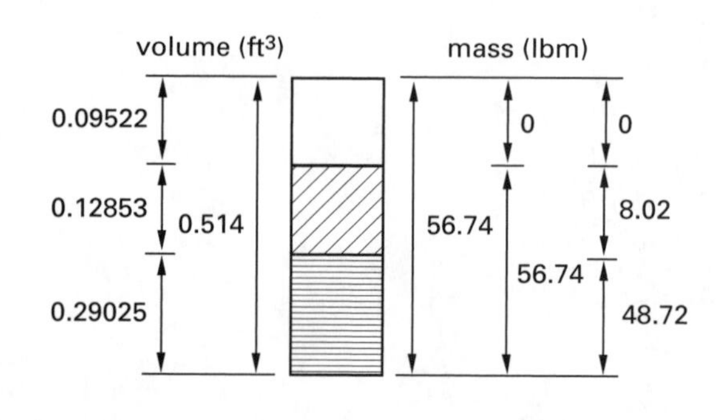

The answer is (A).

(d) Use Eq. 35.4.

$$n = \frac{V_g + V_w}{V_t} = \frac{0.09522 \text{ ft}^3 + 0.12853 \text{ ft}^3}{0.514 \text{ ft}^3} = \boxed{0.435}$$

The answer is (A).

(e) Use Eq. 35.7.

$$S = \frac{V_w}{V_g + V_w} \times 100\% = \frac{0.12853 \text{ ft}^3}{0.09522 \text{ ft}^3 + 0.12853 \text{ ft}^3} \times 100\% = \boxed{57\%}$$

The answer is (B).

(f) Use Table 35.2 to determine the opening size for each sieve. Then graph the values.

sieve	opening size (mm)	% finer by weight
1/2 in	12.70	52
no. 4	4.76	37
no. 10	2.00	32
no. 20	0.85	23
no. 40	0.42	11
no. 60	0.25	7
no. 100	0.15	4

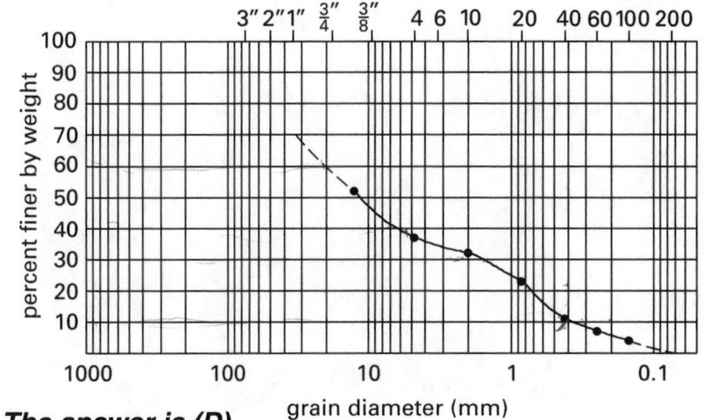

The answer is (D).

(g) Picking off values from the graph,

$$D_{10} = 0.4 \text{ mm} \quad \text{[interpolated]}$$

$$D_{60} \approx 23 \text{ mm} \quad \text{[extrapolated]}$$

From Eq. 35.1,

$$C_u = \frac{D_{60}}{D_{10}} = \frac{23 \text{ mm}}{0.4 \text{ mm}} = \boxed{57.5}$$

The answer is (D).

(h) From the graph, $D_{30} = 1.6$ mm. Use Eq. 35.2.

$$C_z = \frac{(D_{30})^2}{D_{10}D_{60}} = \frac{(1.6 \text{ mm})^2}{(0.4 \text{ mm})(23 \text{ mm})} = \boxed{0.278}$$

The answer is (A).

(i) Use Table 35.5 and the particle size indicators from parts (f)–(g).

The soil is coarse because over 50% is coarser than the no. 200 sieve. Over half is larger than the no. 4 sieve, so the soil is gravelly. The amount finer than the no. 200 sieve is between 0 and 5%, so the soil is either in the GW or GP group. From part (g), $C_u = 42.5$ and $C_z = 0.376$. Therefore, the first requirement for GW is met but not the second, and the soil is classified as $\boxed{\text{GP.}}$ The uniformity coefficient is too high for the soil to be uniform, so the soil is a gap-graded gravel.

The answer is (B).

(j) Use Table 35.4 and the particle size indicators previously calculated.

The first group from the left consistent with the test data is the correct classification. The soil does not exceed 50% passing the no. 10, 30% passing the no. 40, or 15% passing the no. 200. Therefore the soil is classified as $\boxed{\text{A-1-a.}}$

The answer is (A).

4. From Eq. 35.29,

$$k = \frac{VL}{hAt} = \frac{(1.5 \text{ mL})(120 \text{ mm})}{(225 \text{ mm})\left(\frac{\text{PI}}{4}\right)(60 \text{ mm})^2(6.5 \text{ min})}$$

$$= 4.35 \times 10^{-5} \text{ mL/mm}^2\text{·min}$$

$$= \frac{\left(4.35 \times 10^{-5} \frac{\text{mL}}{\text{mm}^2\text{·mm}}\right) \times \left(1000 \frac{\text{mm}^3}{\text{mL}}\right)\left(525{,}600 \frac{\text{min}}{\text{yr}}\right)}{1000 \frac{\text{mm}}{\text{m}}}$$

$$= \boxed{22.86 \text{ m/yr}}$$

The answer is (D).

5. (a) For the first sample, from Eq. 35.11,

$$\gamma_d = \frac{\gamma}{1+w} = \frac{98\ \frac{\text{lbf}}{\text{ft}^3}}{1+0.10} = 89.1\ \text{lbf/ft}^3$$

Similarly, for the rest of the samples,

w	γ_d
10%	89.1
13%	93.8
16%	102.6
18%	105.9
20%	107.5
22%	104.9
25%	98.4

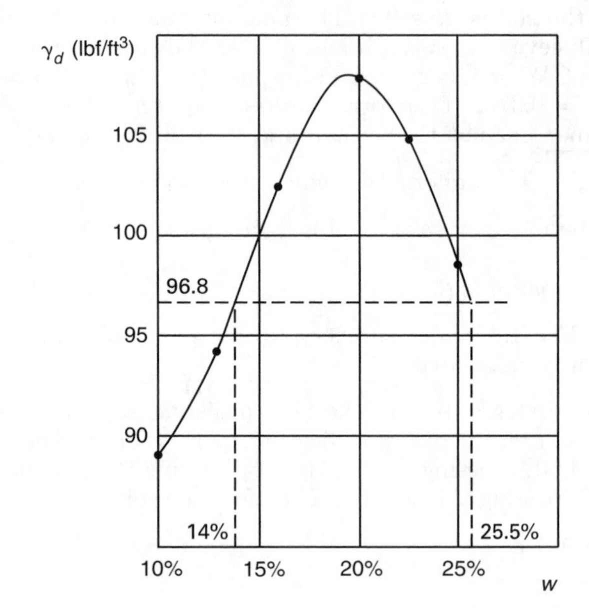

(b) $\gamma_{d,\text{max}} = \boxed{107.5\ \text{lbf/ft}^3}$

The answer is (C).

(c) Depending on how the curve is drawn through the data points, $w = \boxed{19\text{–}20\%}$

The answer is (C).

(d) $(0.90)(107.5\ \text{lbf/ft}^3) = 96.8\ \text{lbf/ft}^3$

$$\boxed{14\% \leq w \leq 25.5\%}$$

The answer is (B).

6. At maximum density, the masses of water and solids in 1 ft^3 of soil are

$$m_s = 107.5\ \text{lbm}$$
$$m_w = (0.19)(107.5\ \text{lbm}) = 20.43\ \text{lbm}$$

At 10% water content, $\gamma_d \approx 89.1\ \text{lbf/ft}^3$. For 1 ft^3,

$$m_s = 89.1\ \text{lbm}$$
$$m_w = (0.10)(89.1\ \text{lbm}) = 8.91\ \text{lbm}$$

To get 1 yd^3 (27 ft^3) of maximum density soil, the requirement is

$$\frac{\left(27\ \frac{\text{ft}^3}{\text{yd}^3}\right)(107.5\ \text{lbm})}{89.1\ \text{lbm}} = \begin{array}{c}32.58\ \text{ft}^3/\text{yd}^3\ \text{of}\\ 10\%\ \text{moisture soil}\end{array}$$

The required water for 1 yd^3 of soil is

$$\frac{(1\ \text{yd}^3)\left(\left(27\ \frac{\text{ft}^3}{\text{yd}^3}\right)\left(20.43\ \frac{\text{lbm}}{\text{ft}^3}\right) - (32.58\ \text{ft}^3)(8.91\ \text{lbm})\right)\left(7.48\ \frac{\text{gal}}{\text{ft}^3}\right)}{62.4\ \frac{\text{lbm}}{\text{ft}^3}}$$

$$= \boxed{31.33\ \text{gal}}$$

The answer is (D).

7. Start with 1 ft^3 of fill in final form.

$$\gamma_{d,\text{required}} = (0.95)\left(124.0\ \frac{\text{lbf}}{\text{ft}^3}\right) = 117.8\ \text{lbf/ft}^3$$

The weight of solids in 1 ft^3 of fill is

$$W_s = \left(117.8\ \frac{\text{lbf}}{\text{ft}^3}\right)(1\ \text{ft}^3) = 117.8\ \text{lbf}$$

Determine the borrow required.

$$\gamma_s = (2.65)\left(62.4\ \frac{\text{lbf}}{\text{ft}^3}\right) = 165.36\ \text{lbf/ft}^3$$
$$W_s = 117.8\ \text{lbf} \quad [\text{same as fill}]$$
$$V_s = \frac{W_s}{\gamma_s} = \frac{117.8\ \text{lbf}}{165.36\ \frac{\text{lbf}}{\text{ft}^3}} = 0.712\ \text{ft}^3$$
$$V_t = V_s(1+e) = (0.712\ \text{ft}^3)(1+0.6)$$
$$= \boxed{1.139\ \text{ft}^3}$$

The answer is (B).

The moisture content is irrelevant.

8. (a) Soil A:

Calculate the required fill soil density.

$$\gamma_{F,\text{dry}} = (0.95)\left(112\ \frac{\text{lbf}}{\text{ft}^3}\right) = 106.4\ \text{lbf/ft}^3$$

This is also the weight of the solid in the fill.

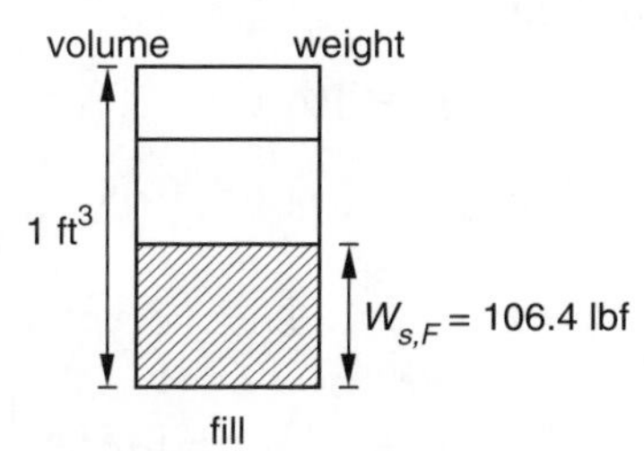

The weight of solid in the fill is the same as the weight of solid in the borrow.

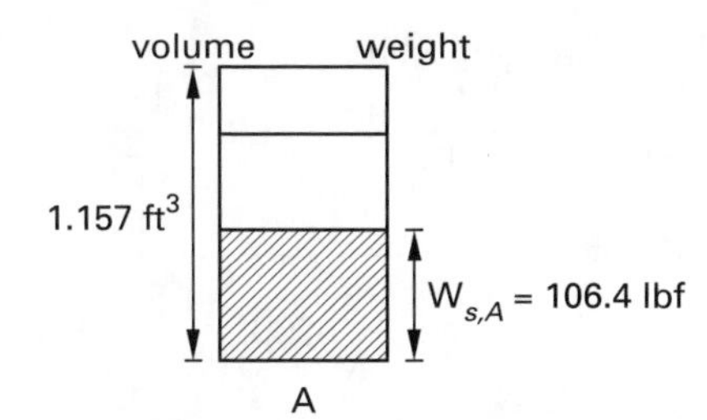

The weight of water in the fill is

$$W_{w,F} = wW_{s,F} = (0.10)(106.4\ \text{lbf}) = 10.64\ \text{lbf}$$

Air has essentially no weight, so the total fill density is

$$W_{t,F} = 106.4\ \text{lbf} + 10.64\ \text{lbf} = 117.04\ \text{lbf}$$

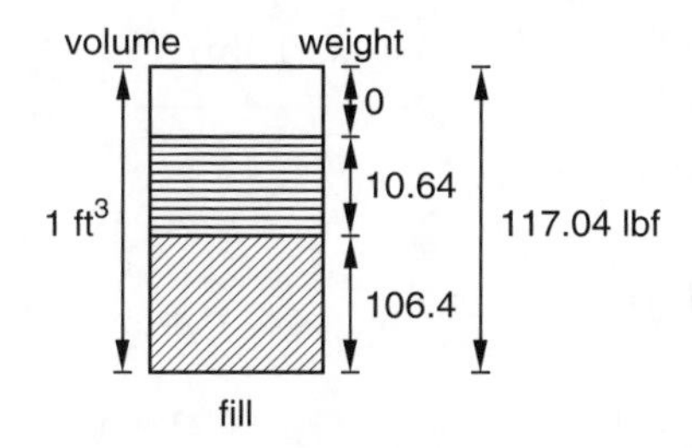

The volume is 200,000 yd^3. The weight of borrow soil solids is

$$(200{,}000\ \text{yd}^3)\left(27\ \frac{\text{ft}^3}{\text{yd}^3}\right)\left(106.4\ \frac{\text{lbf}}{\text{ft}^3}\right) = 5.75\times 10^8\ \text{lbf}$$

Determine the borrow soil weight from the requirements.

$$W_{s,A} = W_{s,F} = 106.4\ \text{lbf}$$
$$W_{w,A} = wW_{s,A} = (0.25)(106.4\ \text{lbf}) = 26.6\ \text{lbf}$$
$$W_{a,A} = 0$$

The weight to be excavated to get 1 ft^3 of fill is

$$W_{t,A} = 26.6\ \text{lbf} + 106.4\ \text{lbf} = 133\ \text{lbf}$$

The volume of borrow A per ft^3 of fill is

$$\frac{133}{\gamma_A} = \frac{133\ \text{lbf}}{115\ \frac{\text{lbf}}{\text{ft}^3}} = 1.157\ \text{ft}^3$$

The volume of soil A is

$$\left(\frac{1.157\ \text{ft}^3}{1\ \text{ft}^3}\right)(200{,}000\ \text{yd}^3)\left(27\ \frac{\text{ft}^3}{\text{yd}^3}\right) = \boxed{6.25\times 10^6\ \text{ft}^3}$$

The answer is (D).

Compute the volume of solids in transport.

$$V_{s,\text{transport}} = \frac{5.75\times 10^8\ \text{lbf}}{(2.70)\left(62.4\ \frac{\text{lbf}}{\text{ft}^3}\right)} = 3.41\times 10^6\ \text{ft}^3$$

The volume of voids in transport is

$$V_\text{void} = eV_s = (0.92)(3.41\times 10^6\ \text{ft}^3) = 3.14\times 10^6\ \text{ft}^3$$

The total transport volume is

$$V_t = V_v + V_s = 3.41\times 10^6\ \text{ft}^3 + 3.14\times 10^6\ \text{ft}^3 = 6.55\times 10^6\ \text{ft}^3$$

The cost to excavate and transport A is

$$C_\text{A} = \left(0.20\ \frac{\$}{\text{yd}^3}\right)\left(\frac{6.25\times 10^6\ \text{ft}^3}{27\ \frac{\text{ft}^3}{\text{yd}^3}}\right) + \left(0.30\ \frac{\$}{\text{yd}^3}\right)\left(\frac{6.55\times 10^6\ \text{ft}^3}{27\ \frac{\text{ft}^3}{\text{yd}^3}}\right) = \$119{,}074$$

(b) Soil B:

$$\gamma_\text{dry} = (0.95)\left(110\ \frac{\text{lbf}}{\text{ft}^3}\right) = 104.5\ \text{lbf/ft}^3$$
$$W_{s,F} = 104.5\ \text{lbf}$$
$$W_{s,F} = W_{s,B} = 104.5\ \text{lbf}$$
$$W_{w,F} = wW_{s,F} = (0.10)(104.5\ \text{lbf}) = 10.45\ \text{lbf}$$
$$W_{t,F} = 104.5\ \text{lbf} + 10.45\ \text{lbf} = 114.95\ \text{lbf}$$

The weight of borrow soil solids is

$$(200{,}000\ \text{yd}^3)\left(27\ \frac{\text{ft}^3}{\text{yd}^3}\right)\left(104.5\ \frac{\text{lbf}}{\text{ft}^3}\right) = 5.64\times 10^8\ \text{lbf}$$
$$W_{s,B} = 104.5\ \text{lbf}$$
$$W_{w,B} = (0.20)(104.5\ \text{lbf}) = 20.9\ \text{lbf}$$
$$W_{t,B} = 20.9\ \text{lbf} + 104.5\ \text{lbf} = 125.4\ \text{lbf}$$

The volume of borrow B per ft^3 of fill is

$$\frac{125.4 \text{ lbf}}{120 \ \frac{\text{lbf}}{\text{ft}^3}} = 1.045 \text{ ft}^3$$

The volume of soil B is

$$\left(\frac{1.045}{1}\right)(200{,}000 \text{ yd}^3)\left(27 \ \frac{\text{ft}^3}{\text{yd}^3}\right) = \boxed{5.64 \times 10^6 \text{ ft}^3}$$

The answer is (C).

The volume in transport is

$$\begin{aligned} V_{s,\text{transport}} &= \frac{\text{weight}}{\gamma_{\text{transport}}} \\ &= \frac{(5.64 \times 10^6 \text{ ft}^3)\left(120 \ \frac{\text{lbf}}{\text{ft}^3}\right)}{95 \ \frac{\text{lbf}}{\text{ft}^3}} \\ &= 7.13 \times 10^6 \text{ ft}^3 \end{aligned}$$

Soil A has the lesser transport volume.

The transport cost is

$$\begin{aligned} C_{\text{B}} &= \left(0.10 \ \frac{\$}{\text{yd}^3}\right)\left(\frac{5.64 \times 10^6 \text{ ft}^3}{27 \ \frac{\text{ft}^3}{\text{yd}^3}}\right) \\ &\quad + \left(0.40 \ \frac{\$}{\text{yd}^3}\right)\left(\frac{7.13 \times 10^6 \text{ ft}^3}{27 \ \frac{\text{ft}^3}{\text{yd}^3}}\right) \\ &= \$126{,}518 \end{aligned}$$

(c) C_{A} is the lesser cost by

$$\$126{,}518 - \$119{,}074 = \boxed{\$7444}$$

The answer is (B).

9. (a) For a dry or drained sample, $c = 0$. Use Mohr's circle.

$$\phi = \arctan\left(\frac{4175 \ \frac{\text{lbf}}{\text{ft}^2}}{6260 \ \frac{\text{lbf}}{\text{ft}^2}}\right) = \arctan(0.667) = \boxed{33.7°}$$

The answer is (C).

(b)

$$\begin{aligned} \text{slope of line PO} &= \frac{-1}{0.667} = -1.5 \\ y &= mx + b \\ 4175 \ \frac{\text{lbf}}{\text{ft}^2} &= (-1.5)\left(6260 \ \frac{\text{lbf}}{\text{ft}^2}\right) + b \\ b &= 13{,}565 \ \frac{\text{lbf}}{\text{ft}^2} \\ y &= -1.5x + 13{,}565 \ \frac{\text{lbf}}{\text{ft}^2} \end{aligned}$$

When $y = 0$,

$$x = \frac{13{,}565 \ \frac{\text{lbf}}{\text{ft}^2}}{1.5} = 9043 \ \frac{\text{lbf}}{\text{ft}^2}$$

$$\begin{aligned} \text{length PO} &= \sqrt{\left(6260 \ \frac{\text{lbf}}{\text{ft}^2} - 9043 \ \frac{\text{lbf}}{\text{ft}^2}\right)^2 + \left(4175 \ \frac{\text{lbf}}{\text{ft}^2}\right)^2} \\ &= 5018 \ \frac{\text{lbf}}{\text{ft}^2} \end{aligned}$$

$$\begin{aligned} \text{maximum principal stresses} &= 9043 \ \frac{\text{lbf}}{\text{ft}^2} + 5018 \ \frac{\text{lbf}}{\text{ft}^2} \\ &= \boxed{14{,}061 \text{ lbf/ft}^2} \end{aligned}$$

The answer is (B).

10. Graph the results.

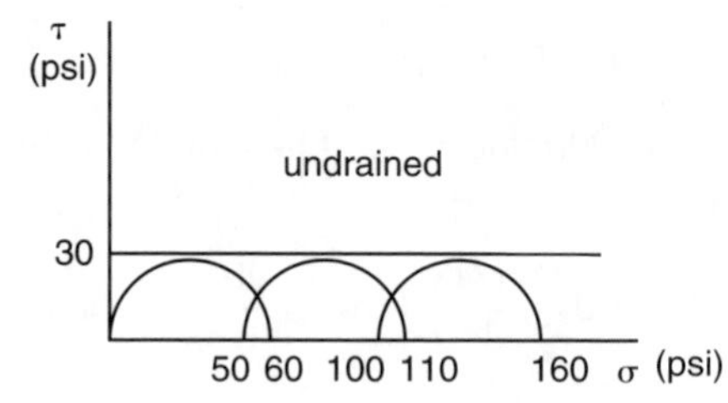

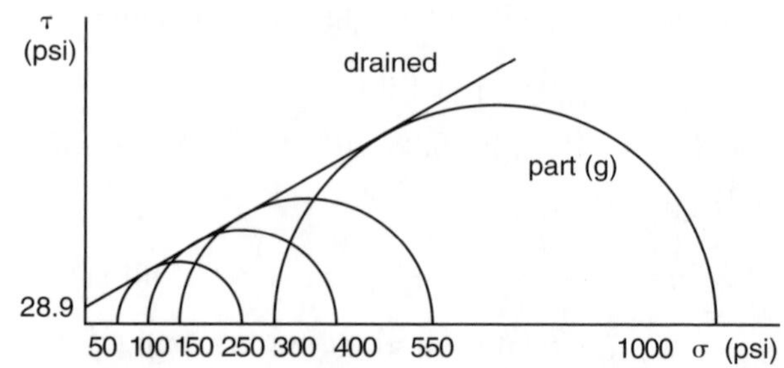

(a) undrained: $\phi = \boxed{0°}$

The answer is (A).

(b) drained: $\phi = \boxed{30°}$ [found using a protractor]

The answer is (C).

(c) undrained: $c = \boxed{30 \text{ psi}}$

The answer is (A).

(d) drained: $c = \boxed{28.9 \text{ psi}}$ [determined graphically]

The answer is (C).

(e) Use Eq. 35.41.

$$\text{undrained}: \theta = 45° + \tfrac{1}{2}(0) = \boxed{45°}$$

The answer is (D).

(f) Use Eq. 35.41.

$$\text{drained}: \theta = 45° + \tfrac{1}{2}(30°) = \boxed{60°}$$

The answer is (D).

(g) By drawing circles to touch the rupture line and 300 psi by trial and error,

$$\sigma_A = \boxed{1000 \text{ lbf/in}^2}$$

The answer is (B).

11.

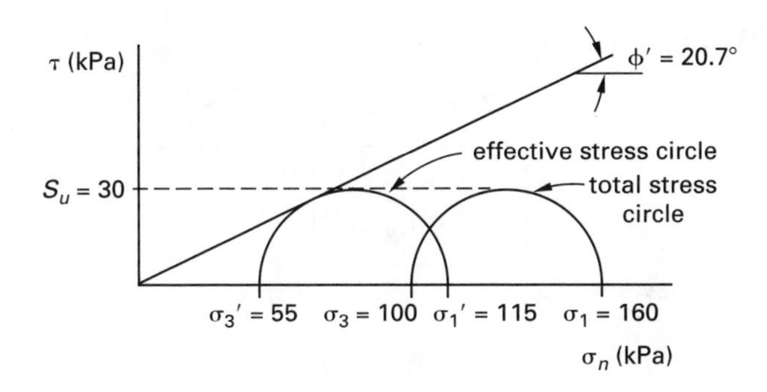

(a) For clays in the undrained condition, $\phi = 0°$ is assumed. Use Eq. 35.42.

$$\begin{aligned} s_u = c &= \frac{\sigma_D}{2} \\ &= \frac{60 \text{ kPa}}{2} = \boxed{30 \text{ kPa}} \end{aligned}$$

The answer is (B).

(b) The chamber pressure remains constant during the test, so in terms of total stress, the minor principal stress is the same as at the beginning of the test.

$$\sigma_{3f} = \boxed{100 \text{ kPa}}$$

The answer is (D).

(c) The major principal stress is equal to the minor principal stress plus the deviator stress.

$$\begin{aligned} \sigma_{1f} &= \sigma_{3f} + \sigma_D \\ &= 100 \text{ kPa} + 60 \text{ kPa} \\ &= \boxed{160 \text{ kPa}} \end{aligned}$$

The total stress conditions can be represented by the Mohr's circle. The diameter of the circle is equal to the deviator stress.

The answer is (D).

(d) The sample starts with the minor principal stress equal to 100 kPa, then develops an excess pore pressure of 45 kPa. Use Eq. 35.17.

$$\begin{aligned} \sigma_{3'} &= \sigma_3 - u \\ &= 100 \text{ kPa} - 45 \text{ kPa} \\ &= \boxed{55 \text{ kPa}} \end{aligned}$$

The answer is (B).

(e) The total major principal stress at failure is 160 kPa. Use Eq. 35.17.

$$\begin{aligned} \sigma_{1'} &= \sigma_1 - u \\ &= 160 \text{ kPa} - 45 \text{ kPa} \\ &= \boxed{115 \text{ kPa}} \end{aligned}$$

The answer is (C).

(f) The Mohr's circle for the effective stress conditions in the test can be drawn from the known conditions. The size of the effective and total stress circles is the same; therefore, the diameter of the circle is 60 kPa.

The friction angle can be computed from Eq. 35.40.

$$\begin{aligned} \frac{\sigma_{1f'}}{\sigma_{3f'}} &= \frac{1 + \sin\phi'}{1 - \sin\phi'} \\ \frac{115 \text{ kPa}}{55 \text{ kPa}} &= \frac{1 + \sin\phi'}{1 - \sin\phi'} \end{aligned}$$

By trial and error, $\phi' = \boxed{20.7°}$.

The answer is (A).

Geotechnical

(g) The failure angle is the theoretical angle of the plane of failure from the horizontal. Use Eq. 35.41.

$$\begin{aligned}\alpha &= 45° + \frac{\phi'}{2}\\ &= 45° + \frac{20.7°}{2}\\ &= \boxed{55.3°}\end{aligned}$$

The answer is (C).

(h) The major and minor principal stresses are known. Use Eq. 35.38.

$$\begin{aligned}\sigma_\theta &= \tfrac{1}{2}(\sigma_1 + \sigma_3) + \tfrac{1}{2}(\sigma_1 - \sigma_3)\cos 2\theta\\ &= \tfrac{1}{2}(115 \text{ kPa} + 55 \text{ kPa})\\ &\quad + \left(\tfrac{1}{2}\right)(115 \text{ kPa} - 55 \text{ kPa})\cos\left((2)(55.3°)\right)\\ &= \boxed{74.4 \text{ kPa}}\end{aligned}$$

The answer is (B).

(i) Use Eq. 35.39.

$$\begin{aligned}\tau_\theta &= \tfrac{1}{2}(\sigma_A - \sigma_R)\sin 2\theta\\ &= \left(\tfrac{1}{2}\right)(115 \text{ kPa} - 55 \text{ kPa})\sin\left((2)(55.3°)\right)\\ &= \boxed{28.1 \text{ kPa}}\end{aligned}$$

The answer is (A).

(j) Another name for the consolidated, undrained test is the $\boxed{\text{R-test.}}$

The answer is (B).

12. Plot the maximum shear stress from the stress-displacement curve versus the normal stress from each individual test.

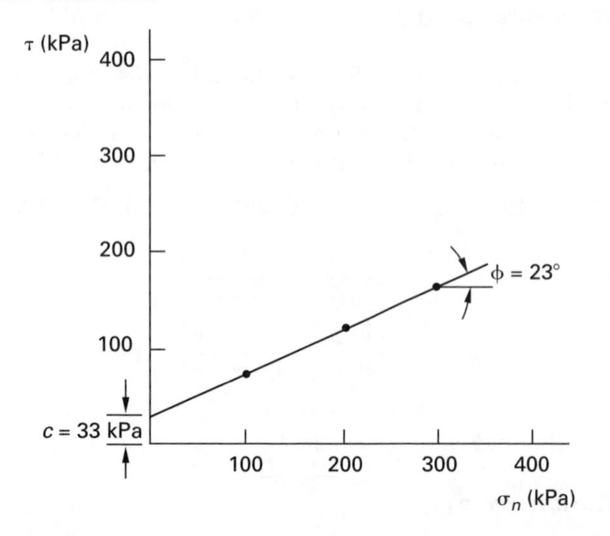

The best fit line through the three data points represents the failure envelope for the soil. Measuring from the plot,

(a) $\phi = \boxed{23°}$

The answer is (C).

(b) From the graph, $c = \boxed{33 \text{ kPa}}$

The answer is (B).

13. (a)

$$\begin{aligned}D_r &= \frac{e_{\text{max}} - e}{e_{\text{max}} - e_{\text{min}}}\\ 0.4 &= \frac{0.97 - e}{0.97 - 0.45}\\ e &= 0.76\end{aligned}$$

Since $e = V_v/V_s$,

$$V_{\text{total}} = V_s + V_v = V_s + eV_s = (1+e)V_s$$

Consider 1 ft^3 of saturated sand.

$$\text{weight of solids} = \frac{(2.65)\left(62.4\ \frac{\text{lbf}}{\text{ft}^3}\right)}{1 + 0.76} = 93.95 \text{ lbf/ft}^3$$

$$\text{weight of water} = \frac{(0.76)\left(62.4\ \frac{\text{lbf}}{\text{ft}^3}\right)}{1 + 0.76} = 26.95 \text{ lbf/ft}^3$$

The total weight is

$$93.95\ \frac{\text{lbf}}{\text{ft}^3} + 26.95\ \frac{\text{lbf}}{\text{ft}^3} = \boxed{120.9 \text{ lbf/ft}^3}$$

Alternate solution:

From Table 35.7,

$$\rho_{\text{sat}} = \frac{(SG + e)\rho_w}{1 + e}$$

Writing this in terms of specific weight,

$$\begin{aligned}\gamma_{\text{sat}} &= \frac{(2.65 + 0.76)\left(62.4\ \frac{\text{lbf}}{\text{ft}^2}\right)}{1 + 0.76}\\ &= \boxed{120.9 \text{ lbf/ft}^3}\end{aligned}$$

The answer is (C).

(b)

$$\begin{aligned}\frac{\Delta V}{V_{t,40}} &= \frac{V_{t,40} - V_{t,65}}{V_{t,40}}\\ &= \frac{V_s(1 + e_{40}) - V_s(1 + e_{65})}{V_s(1 + e_{40})}\\ &= \frac{e_{40} - e_{65}}{1 + e_{40}} = \frac{0.76 - 0.63}{1.76} = 0.0739\end{aligned}$$

$$\Delta t = (0.0739)(4 \text{ ft}) = \boxed{0.295 \text{ ft}}$$

The answer is (C).

14. (a) Refer to the graph.

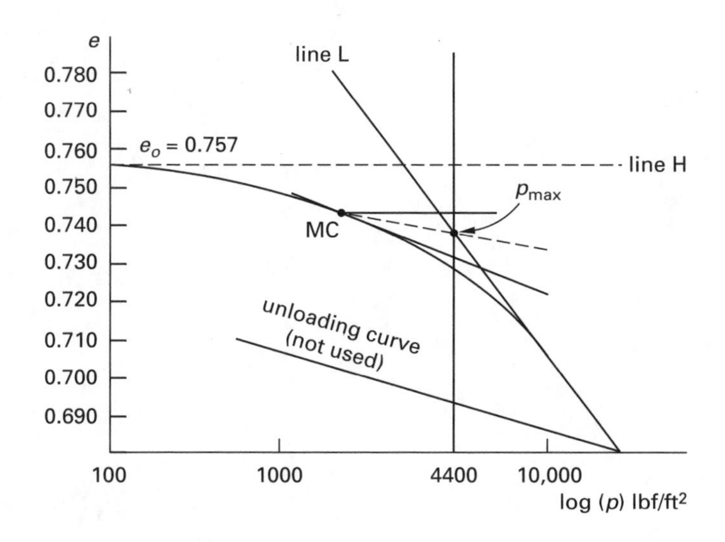

(b) Use the Casagrande procedure below to locate p'_{max}.

step 1: Extend the curve to the left and estimate $e_o = 0.757$.

step 2: Draw a horizontal line from e_o (line H).

step 3: Extend the tail tangent upward (line L).

step 4: Find the point of maximum curvature by inspection (point MC). At point MC, $e = 0.746$ and $p = 3000 \text{ lbf/ft}^2$.

step 5: Draw horizontal and tangent lines from point MC.

step 6: Bisect the angle (dotted line).

step 7: The intersection of the dotted line and line L gives $p'_{max} = 4400 \text{ lbf/ft}^2$.

The answer is (A).

(c) Choose two points on line L, which is the virgin compression line. For example, $(4400 \text{ lbf/ft}^2, 0.739)$ and $(10{,}000 \text{ lbf/ft}^2, 0.707)$.

Use Eq. 35.34.

$$C_c = -\frac{e_1 - e_2}{\log\left(\frac{p_1}{p_2}\right)}$$

$$= -\frac{0.739 - 0.707}{\log\left(\frac{4400 \frac{\text{lbf}}{\text{ft}^2}}{10{,}000 \frac{\text{lbf}}{\text{ft}^2}}\right)}$$

$$= \boxed{0.090}$$

The answer is (B).

Geotechnical

36 Shallow Foundations

PRACTICE PROBLEMS

1. A mat foundation is to be used to support a building with dimensions of 80 ft × 40 ft and a total load of 5200 tons. The mat is located at a depth of $D_f = 8$ ft below the ground surface. The soil beneath the mat is a sand with a density of 120 lbf/ft^3 and an average SPT N-value of 18.

(a) What is the allowable bearing capacity of the mat?

(A) 2.2 tons/ft^2
(B) 2.5 tons/ft^2
(C) 2.9 tons/ft^2
(D) 5.4 tons/ft^2

(b) What is the factor of safety against bearing capacity failure?

(A) 1.3
(B) 1.9
(C) 2.2
(D) 4.4

2. A total force of 1600 kN is to be supported by a square footing. The footing is to rest directly on sand that has a density of 1900 kg/m^3 and an angle of internal friction of 38°. Use the Terzaghi bearing capacity factors.

(a) Size the footing.

(A) 1.8 m square
(B) 2.4 m square
(C) 2.7 m square
(D) 3.4 m square

(b) Size the footing if it is to be placed 1.5 m below the surface.

(A) 1.0 m square
(B) 1.1 m square
(C) 1.3 m square
(D) 1.8 m square

3. (*Time limit: one hour*) The three foundations shown are situated in a sandy clay with the following characteristics: specific weight = 108 lbf/ft^3, angle of internal friction = 25°, and cohesion = 400 lbf/ft^2. The water table is 35 ft below the ground surface. Use a factor of safety of 2.5 where required.

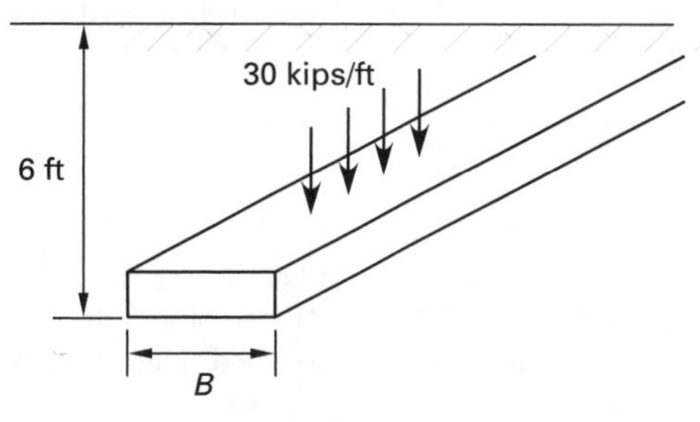

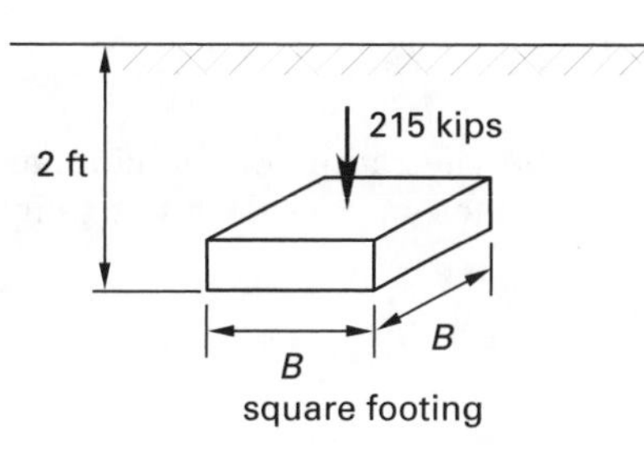

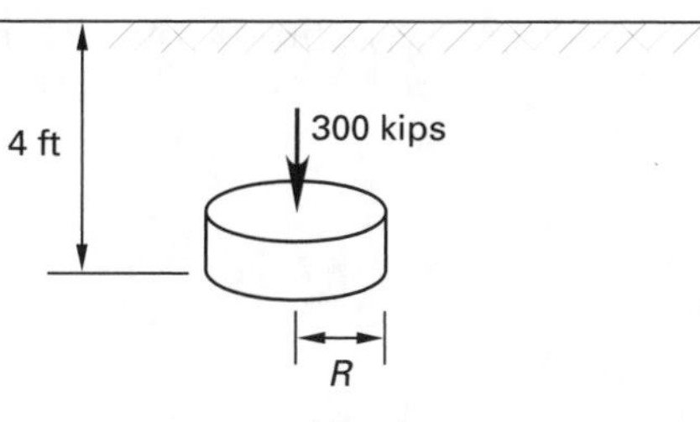

(a) What is the bearing capacity factor N_c according to the Terzaghi and Meyerhof/Vesic theories?

(A) 5.7, 5.1
(B) 11, 9.7
(C) 25, 21
(D) 350, 270

(b) What is the bearing capacity factor N_q according to the Terzaghi and Meyerhof/Vesic theories?

(A) 0, 0
(B) 1.0, 1.0
(C) 3.6, 3.2
(D) 13, 11

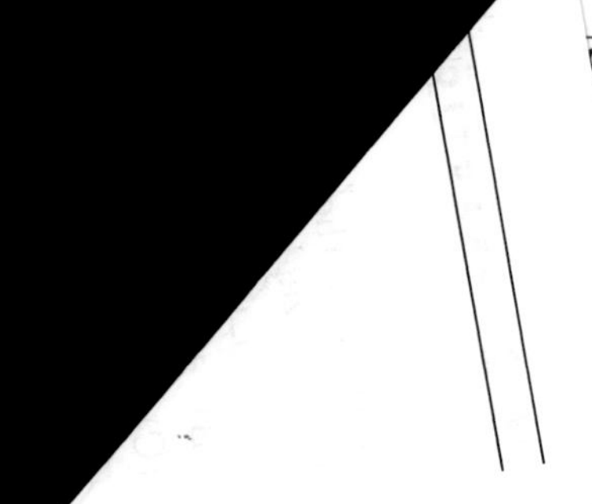

ty factor N_γ according
/esic theories?

ne wall footing using
eight of the footing.

(C) 3.8 ft
(D) 4.6 ft

(e) What should be the width of the square footing using the Terzaghi factors? Neglect the weight of the footing.

(A) 2.5 ft
(B) 3.2 ft
(C) 4.8 ft
(D) 5.5 ft

(f) What should be the radius of the circular footing, using the Terzaghi factors? Neglect the weight of the footing.

(A) 3.5 ft
(B) 3.8 ft
(C) 4.2 ft
(D) 5.5 ft

(g) What should be the radius of the circular footing using the Meyerhof/Vesic factors? Neglect the weight of the footing.

(A) 3.5 ft
(B) 3.7 ft
(C) 4.7 ft
(D) 5.8 ft

(h) What is the allowable bearing capacity of the square footing assuming a width of 4 ft and assuming that the water table is at 2.0 ft? Use the Meyerhof/Vesic factors.

(A) 5300 lbf/ft^2
(B) 6300 lbf/ft^2
(C) 7600 lbf/ft^2
(D) 11,000 lbf/ft^2

(i) What is the allowable bearing capacity of the square footing assuming a width of 4 ft and assuming that the water table is at 1 ft? Use the Meyerhof/Vesic factors.

(A) 4900 lbf/ft^2
(B) 5100 lbf/ft^2
(C) 6200 lbf/ft^2
(D) 8900 lbf/ft^2

(j) What is the allowable bearing capacity of the circular footing assuming a radius of 4 ft and assuming that the water table is at the ground surface? Use the Meyerhof/Vesic factors.

(A) 5200 lbf/ft^2
(B) 5700 lbf/ft^2
(C) 7600 lbf/ft^2
(A) 9300 lbf/ft^2

4. (*Time limit: one hour*) A 2 m × 2 m square spread footing is designed to support an axial column load of 600 kN and a moment of 150 kN·m, as shown. The footing is placed 1 m into a sandy soil with a density of 2000 kg/m^3 and an angle of internal friction of 30°.

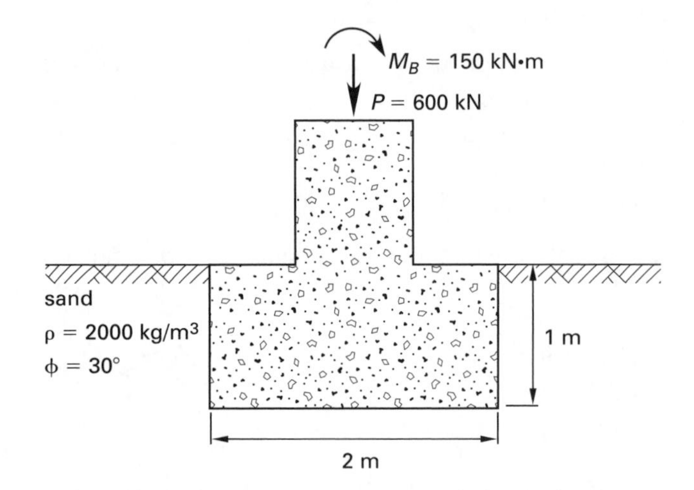

(a) What is the eccentricity of the resultant load?

(A) 0.10 m
(B) 0.25 m
(C) 0.50 m
(D) 1.3 m

(b) What equivalent width should be used for the footing design?

(A) 1.3 m
(B) 1.5 m
(C) 1.8 m
(D) 2.0 m

(c) Using the Meyerhof theory, what is the value of the bearing capacity factor, N_γ, to be used in calculating the bearing capacity?

(A) 16
(B) 22
(C) 27
(D) 35

(d) Using the Meyerhof theory, what is the value of the bearing capacity factor, N_q, to be used in calculating the bearing capacity?

(A) 11
(B) 16
(C) 18
(D) 23

(e) What is the value of the shape factor for N_γ to be used in calculating the bearing capacity?

(A) 0.85
(B) 0.88
(C) 0.90
(D) 1.00

(f) What is the value of the net bearing capacity?

(A) 100 kPa
(B) 300 kPa
(C) 550 kPa
(D) 1300 kPa

(g) What is the value of the allowable bearing capacity?

(A) 50 kPa
(B) 150 kPa
(C) 270 kPa
(D) 600 kPa

(h) What is the total allowable load on the shallow foundation?

(A) 800 kN
(B) 1100 kN
(C) 1300 kN
(D) 2100 kN

(i) What is the maximum soil pressure on the bottom of the footing?

(A) 55 kPa
(B) 90 kPa
(C) 200 kPa
(D) 260 kPa

(j) What is the factor of safety against bearing capacity failure for this footing?

(A) 1.6
(B) 2.1
(C) 2.7
(D) 3.0

SOLUTIONS

1. (a) The SPT N-value should be corrected for the overburden pressure. At the base of the mat foundation, the overburden pressure is

$$\begin{aligned} p_{\text{overburden}} &= \gamma D_f = \frac{\left(120\ \frac{\text{lbf}}{\text{ft}^3}\right)(8\ \text{ft})}{2000\ \frac{\text{lbf}}{\text{ton}}} \\ &= \frac{960\ \frac{\text{lbf}}{\text{ft}^2}}{2000\ \frac{\text{lbf}}{\text{ton}}} \\ &= 0.48\ \text{tons/ft}^2 \end{aligned}$$

From Table 36.6, the correction factor is $C_n \approx 1.21$.

At a depth of $D_f + B$ below the ground surface, the overburden pressure is

$$\begin{aligned} p_{\text{overburden}} &= \gamma(D_f + B) \\ &= \frac{\left(120\ \frac{\text{lbf}}{\text{ft}^3}\right)(8\ \text{ft} + 40\ \text{ft})}{2000\ \frac{\text{lbf}}{\text{ton}}} \\ &= \frac{5760\ \frac{\text{lbf}}{\text{ft}^2}}{2000\ \frac{\text{lbf}}{\text{ton}}} \\ &= 2.88\ \text{tons/ft}^2 \end{aligned}$$

From Table 36.6, the correction factor is $C_n \approx 0.63$. This is the factor that should be used for design, since it will result in the lowest N-value.

The net allowable bearing capacity is given by Eq. 36.22.

$$\begin{aligned} q_{\text{net,allowable}} &= 0.22C_nN \\ &= (0.22)(0.63)(18) \\ &= \boxed{2.49\ \text{tons/ft}^2} \end{aligned}$$

The answer is (B).

(b) By Eq. 36.23,

$$\begin{aligned} p_{\text{net,actual}} &= \frac{\text{total load}}{\text{raft area}} - \gamma D_f \\ &= \frac{5200\ \text{tons}}{(80\ \text{ft})(40\ \text{ft})} - \frac{\left(120\ \frac{\text{lbf}}{\text{ft}^3}\right)(8\ \text{ft})}{2000\ \frac{\text{lbf}}{\text{ton}}} \\ &= 1.15\ \text{tons/ft}^2 \end{aligned}$$

Geotechnical

The factor of safety against bearing capacity failure based on allowable stress is given by Eq. 36.4.

$$\frac{q_{\text{net,allowable}}}{p_{\text{net,actual}}} = \frac{2.49 \ \frac{\text{tons}}{\text{ft}^2}}{1.15 \ \frac{\text{tons}}{\text{ft}^2}} = 2.2$$

Since Eq. 36.22 already has a factor of safety of 2 based on the ultimate bearing capacity,

$$F = (2)(2.2) = \boxed{4.4}$$

The answer is (D).

2. (a) If the footing rests on the sand, then $D_f = 0$. From Table 36.2 or Fig. 36.2, for $\phi = 38°$, $N_\gamma \approx 77$.

The shape factor from Table 36.5 is 0.85.

The net bearing capacity equation for sand is given by Eq. 36.10.

$$\begin{aligned} q_{\text{net}} &= \tfrac{1}{2}B\rho g N_\gamma + \rho g D_f(N_q - 1) \\ &= \frac{\tfrac{1}{2}B\left(1900 \ \frac{\text{kg}}{\text{m}^3}\right)\left(9.81 \ \frac{\text{m}}{\text{s}^2}\right)(77)(0.85) + 0}{1000 \ \frac{\text{Pa}}{\text{kPa}}} \\ &= 609.96B \text{ kPa} \end{aligned}$$

From Eq. 36.4, using $F = 2$,

$$q_a = \frac{q_{\text{net}}}{F}$$

$$\begin{aligned} q_{\text{net}} = Fq_a &= (2)\left(\frac{1600 \text{ kN}}{B^2}\right) \\ &= \frac{3200}{B^2} \quad \text{[in kPa]} \end{aligned}$$

Equating these two expressions for q_{net},

$$\frac{3200}{B^2} = 609.96B$$

$$B = \boxed{1.75 \text{ m}}$$

The answer is (A).

(b) From Fig. 36.2 and Table 36.2, $N_q \approx 65$.

The net bearing capacity equation for sand is given by Eq. 36.10.

$$\begin{aligned} q_{\text{net}} &= \tfrac{1}{2}B\rho g N_\gamma + \rho g D_f(N_q - 1) \\ &= \tfrac{1}{2}B\left(1900 \ \frac{\text{kg}}{\text{m}^3}\right)\left(9.81 \ \frac{\text{m}}{\text{s}^2}\right)(77)(0.85) \\ &\quad + \left(1900 \ \frac{\text{kg}}{\text{m}^3}\right)\left(9.81 \ \frac{\text{m}}{\text{s}^2}\right)(1.5 \text{ m})(65 - 1) \\ &= 609.96\text{B kPa} + 1789.34 \text{ kPa} \end{aligned}$$

The design capacity is the same as for part (a). Equating the two expressions for q_{net},

$$\begin{aligned} \frac{3200}{B^2} &= 609.96\text{B kPa} + 1789.34 \text{ kPa} \\ 3200 &= 609.96B^3 + 1789.34B^2 \end{aligned}$$

By trial and error,

$$B = \boxed{1.14 \text{ m}}$$

The answer is (B).

3. (a) From Tables 36.2 and 36.3, for $\phi = 25°$, $N_c = 25.1$ (Terzaghi) and $N_c = 20.7$ (Meyerhof).

The answer is (C).

(b) From Tables 36.2 and 36.3, for $\phi = 25°$, $N_q = 12.7$ (Terzaghi) and $N_q = 10.7$ (Meyerhof).

The answer is (D).

(c) From Tables 36.2 and 36.3, for $\phi = 25°$, $N_\gamma = 9.7$ (Terzaghi) and $N_\gamma = 10.8$ (Meyerhof).

The answer is (D).

(d) The water table is too deep to affect the bearing capacity. The net bearing pressure on the footing is

$$\begin{aligned} q_{\text{net}} &= q_{\text{ult}} - \gamma D_f \\ &= \tfrac{1}{2}\gamma B N_\gamma + cN_c + \gamma D_f(N_q - 1) \end{aligned}$$

For a wall footing, the shape factor for N_c from Table 36.4 is 1.0; the shape factor for N_γ from Table 36.5 is also 1.0.

$$\begin{aligned} q_{\text{net}} &= \left(\tfrac{1}{2}\right)\left(108 \ \frac{\text{lbf}}{\text{ft}^3}\right)B(9.7)(1.0) \\ &\quad + \left(400 \ \frac{\text{lbf}}{\text{ft}^2}\right)(25.1)(1.0) \\ &\quad + \left(108 \ \frac{\text{lbf}}{\text{ft}^3}\right)(6 \text{ ft})(12.7 - 1) \\ &= 523.8B + 17{,}622 \text{ lbf/ft}^2 \end{aligned}$$

By Eq. 36.4, per foot of wall length,

$$\begin{aligned} q_{\text{net}} = Fq_a &= \frac{(2.5)(30{,}000 \text{ lbf})}{\text{B}} \\ &= \frac{75{,}000}{B} \quad [\text{in lbf/ft}^2] \end{aligned}$$

Therefore,

$$\begin{aligned} q_{\text{net}} = \frac{75{,}000}{B} &= 523.8B + 17{,}622 \\ 523.8B^2 + 17{,}622B &= 75{,}000 \\ B^2 + 33.64B &= 143.18 \end{aligned}$$

$$B = \boxed{3.8 \text{ ft}}$$

The answer is (C).

(e) The water table is too deep to affect the bearing capacity. The net bearing pressure on the footing is

$$\begin{aligned} q_{\text{net}} &= q_{\text{ult}} - \gamma D_f \\ &= \tfrac{1}{2}\gamma B N_\gamma + c N_c + \gamma D_f (N_q - 1) \end{aligned}$$

For a square footing, the shape factor for N_c from Table 36.4 is 1.25; the shape factor for N_γ from Table 36.5 is 0.85.

$$\begin{aligned} q_{\text{net}} &= \left(\tfrac{1}{2}\right)\left(108\ \frac{\text{lbf}}{\text{ft}^3}\right) B(9.7)(0.85) \\ &\quad + \left(400\ \frac{\text{lbf}}{\text{ft}^2}\right)(25.1)(1.25) \\ &\quad + \left(108\ \frac{\text{lbf}}{\text{ft}^3}\right)(2\ \text{ft})(12.7 - 1) \\ &= 445.2B + 12{,}550 + 2527.2 \quad [\text{in lbf/ft}^2] \\ &= 445.2B + 15{,}077.2\ \text{lbf/ft}^2 \end{aligned}$$

By Eq. 36.4,

$$\begin{aligned} q_{\text{net}} &= F q_a \\ &= \frac{(2.5)(215{,}000\ \text{lbf})}{\text{B}^2} \\ &= \frac{537{,}500}{B^2} \quad [\text{in lbf/ft}^2] \end{aligned}$$

Therefore,

$$\begin{aligned} q_{\text{net}} &= \frac{537{,}500}{B^2} \quad [\text{in lbf/ft}^2] \\ &= 445.2B + 15{,}077.2\ \text{lbf/ft}^2 \end{aligned}$$

$$445.2B^3 + 15{,}077B^2 = 537{,}500$$

By trial and error,

$$B = \boxed{5.5\ \text{ft}}$$

The answer is (D).

(f) The water table is too deep to affect the bearing capacity. The net bearing pressure on the footing is

$$\begin{aligned} q_{\text{net}} &= q_{\text{ult}} - \gamma D_f \\ &= \tfrac{1}{2}\gamma B N_\gamma + c N_c + \gamma D_f (N_q - 1) \end{aligned}$$

For a circular footing, the shape factor for N_c from Table 36.4 is 1.20; the shape factor for N_γ from Table 36.5 is 0.70.

$$\begin{aligned} q_{\text{net}} &= \left(\tfrac{1}{2}\right)\left(108\ \frac{\text{lbf}}{\text{ft}^3}\right)(2R)(9.7)(0.70) \\ &\quad + \left(400\ \frac{\text{lbf}}{\text{ft}^2}\right)(25.1)(1.20) \\ &\quad + \left(108\ \frac{\text{lbf}}{\text{ft}^3}\right)(4\ \text{ft})(12.7 - 1) \\ &= 733.32R + 12{,}048 + 5054.4 \quad [\text{in lbf/ft}^2] \end{aligned}$$

By Eq. 36.4,

$$\begin{aligned} q_{\text{net}} &= F q_a \\ &= \frac{(2.5)(300{,}000\ \text{lbf})}{\pi R^2} \\ &= \frac{238{,}732}{R^2} \quad [\text{in lbf/ft}^2] \end{aligned}$$

Therefore,

$$\begin{aligned} q_{\text{net}} &= \frac{238{,}732}{R^2} \quad [\text{in lbf/ft}^2] \\ &= 733.32R + 17{,}102.4 \\ &\quad [\text{in lbf/ft}^2] \end{aligned}$$

$$733.32R^3 + 17{,}102.4R^2 = 238{,}732$$

By trial and error,

$$R = \boxed{3.5\ \text{ft}}$$

The answer is (A).

(g) Solve the problem as in part (f), substituting the Meyerhof/Vesic factors for the Terzaghi factors in the bearing capacity equation. The shape factors are the same.

$$\begin{aligned} q_{\text{net}} &= \tfrac{1}{2}\gamma B N_\gamma + c N_c + \gamma D_f (N_q - 1) \\ &= \left(\tfrac{1}{2}\right)\left(108\ \frac{\text{lbf}}{\text{ft}^3}\right)(2R)(10.8)(0.70) \\ &\quad + \left(400\ \frac{\text{lbf}}{\text{ft}^2}\right)(20.7)(1.20) \\ &\quad + \left(108\ \frac{\text{lbf}}{\text{ft}^3}\right)(4\ \text{ft})(10.7 - 1) \\ &= 816.48R + 9936 + 4190.4 \quad [\text{in lbf/ft}^2] \end{aligned}$$

By Eq. 36.4,

$$\begin{aligned} q_{\text{net}} &= F q_a \\ &= \frac{(2.5)(300{,}000\ \text{lbf})}{\pi R^2} \\ &= \frac{238{,}732}{R^2}\ \text{lbf/ft}^2 \end{aligned}$$

Therefore,

$$\begin{aligned} q_{\text{net}} &= \frac{238{,}732}{R^2} \quad [\text{in lbf/ft}^2] \\ &= 816.48R + 14{,}126.4\ \text{lbf/ft}^2 \end{aligned}$$

$$816.48R^3 + 14{,}126.4R^2 = 238{,}732$$

By trial and error,

$$R = \boxed{3.7\ \text{ft}}$$

The answer is (B).

(h) The water table is at the base of the footing. By Eqs. 36.1 and 36.3, the net bearing pressure on the footing is

$$\begin{aligned} q_{\text{net}} &= q_{\text{ult}} - \gamma D_f \\ &= \tfrac{1}{2}\gamma_b B N_\gamma + c N_c + \gamma_d D_f (N_q - 1) \end{aligned}$$

Geotechnical

For a square footing, the shape factor for N_c from Table 36.4 is 1.25; the shape factor for N_γ from Table 36.5 is 0.85.

$$\begin{aligned} q_{\text{net}} &= \left(\tfrac{1}{2}\right)\left(108\ \frac{\text{lbf}}{\text{ft}^3} - 62.4\ \frac{\text{lbf}}{\text{ft}^3}\right)(4\text{ ft})(10.8)(0.85) \\ &\quad + \left(400\ \frac{\text{lbf}}{\text{ft}^2}\right)(20.7)(1.25) \\ &\quad + \left(108\ \frac{\text{lbf}}{\text{ft}^3}\right)(2\text{ ft})(10.7-1) \\ &= 837.2\ \frac{\text{lbf}}{\text{ft}^2} + 10{,}350\ \frac{\text{lbf}}{\text{ft}^2} + 2095.2\ \frac{\text{lbf}}{\text{ft}^2} \\ &= 13{,}282.4\ \text{lbf/ft}^2 \end{aligned}$$

By Eq. 36.4,

$$\begin{aligned} q_a &= \frac{q_{\text{net}}}{F} \\ &= \frac{13{,}282.4\ \frac{\text{lbf}}{\text{ft}^2}}{2.5} \\ &= \boxed{5313\ \text{lbf/ft}^2} \end{aligned}$$

The answer is (A).

(i) The water table is between the base of the footing and the surface. By Eqs. 36.1 and 36.15, the net bearing pressure on the footing is

$$\begin{aligned} q_{\text{net}} &= q_{\text{ult}} - \gamma D_f \\ &= \tfrac{1}{2}\gamma_b B N_\gamma + cN_c \\ &\quad + \left(\gamma_d D_f + \left(62.4\ \frac{\text{lbf}}{\text{ft}^3}\right)(D_w - D_f)\right)(N_q - 1) \end{aligned}$$

For a square footing, the shape factor for N_c from Table 36.4 is 1.25; the shape factor for N_γ from Table 36.5 is 0.85.

$$\begin{aligned} q_{\text{net}} &= \left(\tfrac{1}{2}\right)\left(108\ \frac{\text{lbf}}{\text{ft}^3} - 62.4\ \frac{\text{lbf}}{\text{ft}^3}\right)(4\text{ ft})(10.8)(0.85) \\ &\quad + \left(400\ \frac{\text{lbf}}{\text{ft}^2}\right)(20.7)(1.25) \\ &\quad + \left(\left(108\ \frac{\text{lbf}}{\text{ft}^3}\right)(2\text{ ft}) \right. \\ &\quad \left. + \left(62.4\ \frac{\text{lbf}}{\text{ft}^3}\right)(1\text{ ft} - 2\text{ ft})\right)(10.7-1) \\ &= 837.2\ \frac{\text{lbf}}{\text{ft}^2} + 10{,}350\ \frac{\text{lbf}}{\text{ft}^2} + 1489.9\ \frac{\text{lbf}}{\text{ft}^2} \\ &= 12{,}677\ \text{lbf/ft}^2 \end{aligned}$$

By Eq. 36.4,

$$\begin{aligned} q_a &= \frac{q_{\text{net}}}{F} \\ &= \frac{12{,}677\ \frac{\text{lbf}}{\text{ft}^2}}{2.5} \\ &= \boxed{5070\ \text{lbf/ft}^2} \end{aligned}$$

The answer is (B).

(j) The water table is at the surface. By Eqs. 36.1 and 36.14, the net bearing pressure on the footing is

$$\begin{aligned} q_{\text{net}} &= q_{\text{ult}} - \gamma D_f \\ &= \tfrac{1}{2}\gamma_b B N_\gamma + cN_c + \gamma_b D_f (N_q - 1) \end{aligned}$$

The cN_c term is included because the soil has a non-zero cohesion, unlike pure sand.

For a round footing, the shape factor for N_c from Table 36.4 is 1.20; the shape factor for N_γ from Table 36.5 is 0.70.

$$\begin{aligned} q_{\text{net}} &= \left(\tfrac{1}{2}\right)\left(108\ \frac{\text{lbf}}{\text{ft}^3} - 62.4\ \frac{\text{lbf}}{\text{ft}^3}\right)(2)(4\text{ ft})(10.8)(0.70) \\ &\quad + \left(400\ \frac{\text{lbf}}{\text{ft}^2}\right)(20.7)(1.20) \\ &\quad + \left(\left(108\ \frac{\text{lbf}}{\text{ft}^3} - 62.4\ \frac{\text{lbf}}{\text{ft}^3}\right)(4\text{ ft})\right)(10.7-1) \\ &= 13{,}084\ \text{lbf/ft}^2 \end{aligned}$$

By Eq. 36.4,

$$\begin{aligned} q_a &= \frac{q_{\text{net}}}{F} \\ &= \frac{13{,}084\ \frac{\text{lbf}}{\text{ft}^2}}{2.5} \\ &= \boxed{5234\ \text{lbf/ft}^2} \end{aligned}$$

The answer is (A).

4. (a) ($M_L = 0$.) The eccentricity is given by Eq. 36.17.

$$\begin{aligned} \epsilon_B &= \frac{M_B}{P} = \frac{150\ \text{kN·m}}{600\ \text{kN}} \\ &= \boxed{0.25\ \text{m}} \end{aligned}$$

The answer is (B).

(b) The footing width to be used for analysis of bearing capacity is reduced by twice the eccentricity. Using Eq. 36.18, the effective width is

$$\begin{aligned} B' &= B - 2\epsilon_B = 2\ \text{m} - (2)(0.25\ \text{m}) \\ &= \boxed{1.5\ \text{m}} \end{aligned}$$

The answer is (B).

(c) From Table 36.3, for $\phi = 30°$, $N_\gamma = \boxed{15.7.}$

The answer is (A).

(d) From Table 36.3, for $\phi = 30°$, $N_q = \boxed{18.4.}$

The answer is (C).

(e) The shape factor for N_γ is found by interpolation from Table 36.5 and using the effective width. For $B'/L' = 1.5\ \text{m}/2.0\ \text{m} = 0.75$, the shape factor is $\boxed{0.875.}$

(Note that $L' = L$ since $M_L = 0$.)

The answer is (B).

(f) The net bearing capacity is given by Eq. 36.10.

(If available, a shape factor for the N_q term can be used.)

$$\begin{aligned} q_{\text{net}} &= q_{\text{ult}} - \rho g D_f \\ &= \tfrac{1}{2} B \rho g N_\gamma + \rho g D_f (N_q - 1) \\ &= \left(\tfrac{1}{2}\right)(1.5\ \text{m})\left(2000\ \frac{\text{kg}}{\text{m}^3}\right)\left(9.81\ \frac{\text{m}}{\text{s}^2}\right)(15.7)(0.875) \\ &\quad + \left(2000\ \frac{\text{kg}}{\text{m}^3}\right)\left(9.81\ \frac{\text{m}}{\text{s}^2}\right)(1\ \text{m})(18.4 - 1) \\ &= \boxed{543{,}535\ \text{Pa}\quad (550\ \text{kPa})} \end{aligned}$$

The answer is (C).

(g) The allowable bearing capacity is given by Eq. 36.11. The factor of safety is typically taken as 2 for sand.

$$\begin{aligned} q_a &= \frac{q_{\text{net}}}{F} = \frac{544\ \text{kPa}}{2} \\ &= \boxed{272\ \text{kPa}} \end{aligned}$$

The answer is (C).

(h) The allowable load is calculated using the reduced area.

$$\begin{aligned} p_a &= q_a B' L' = (272\ \text{kPa})(1.5\ \text{m})(2.0\ \text{m}) \\ &= \boxed{816\ \text{kN}} \end{aligned}$$

The answer is (A).

(i) The maximum pressure beneath the footing is given by Eq. 36.20. The actual footing dimensions are used to calculate the pressure distribution.

$$\begin{aligned} p_{\text{max}} &= \left(\frac{P}{BL}\right)\left(1 + \frac{6\epsilon}{B}\right) \\ &= \left(\frac{600\ \text{kN}}{(2\ \text{m})(2\ \text{m})}\right)\left(1 + \frac{(6)(0.25\ \text{m})}{2\ \text{m}}\right) \\ &= \boxed{262.5\ \text{kPa}} \end{aligned}$$

(This pressure is just within the allowable limit.)

The answer is (D).

(j) The factor of safety against bearing capacity failure is determined by dividing the net bearing capacity by the maximum contact pressure.

$$\begin{aligned} F &= \frac{q_{\text{net}}}{p_{\text{max}}} = \frac{544\ \text{kPa}}{262.5\ \text{kPa}} \\ &= \boxed{2.07} \end{aligned}$$

The answer is (B).

37 Rigid Retaining Walls

PRACTICE PROBLEMS

Retaining Wall Analysis

1. A retaining wall is designed for free-draining granular backfill. After several years of operation, the weepholes become plugged and the water table rises to within 10 ft of the top of the wall. (a) What is the resultant force for the drained case? (b) What is the point of application for the drained case? (c) What is the resultant force for the plugged case? (d) What is the point of application of the resultant force for the plugged case?

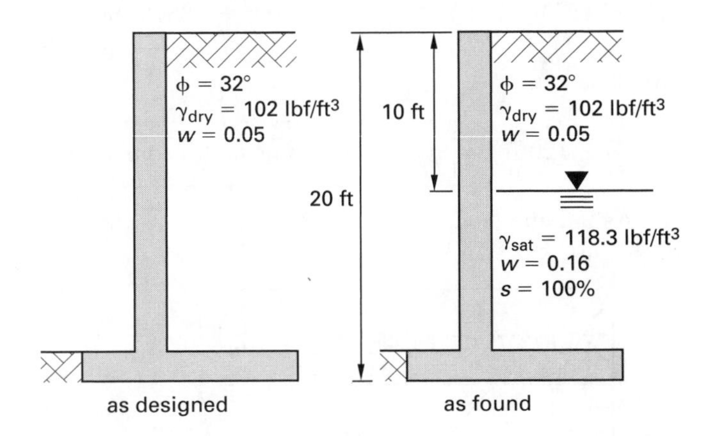

2. *(Time limit: one hour)* A 26 ft high retaining wall holds back sand with a 96 lbf/ft^3 drained specific weight. The water table is 10 ft below the top of the wall. The saturated specific weight is 121 lbf/ft^3. The angle of internal friction is 36°. (a) What is the active earth resultant? (b) What is the location of the active earth pressure resultant? (c) If the water table elevation could be reduced 16 ft to the bottom of the wall, what would be the reduction in overturning moment?

Retaining Wall Design

3. *(Time limit: one hour)* A reinforced concrete retaining wall is used to support a 14 ft cut in sandy soil. The backfill is level, but a surcharge of 500 lbf/ft^2 is present for a considerable distance behind the wall. Factors of safety of 1.5 against sliding and overturning are required. Customary and reasonable assumptions regarding the proportions can be made. Passive pressure is to be disregarded. The need for a key must be established.

soil drained specific weight	130 lbf/ft^3
angle of internal friction	35°
coefficient of friction against concrete	0.5
allowable soil pressure	4500 lbf/ft^2
frost line	4 ft below grade

(a) What is the approximate minimum stem height?

(A) 14 ft
(B) 16 ft
(C) 18 ft
(D) 20 ft

For parts (b) through (j), assume a base length, B, of 11.5 ft; a base thickness, d, of 1.75 ft; a stem thickness at the base of 1.75 ft; a stem thickness at the top of 1 ft; a stem height (above base) of 18 ft; and a heel extension (past the back of the stem) of 6.5 ft.

(b) The surcharge is equivalent to what thickness of backfill soil?

(A) 2 ft
(B) 3 ft
(C) 4 ft
(D) 5 ft

(c) What is the horizontal reaction due to the surcharge?

(A) 2400 lbf/ft
(B) 2700 lbf/ft
(C) 3500 lbf/ft
(D) 3900 lbf/ft

(d) What is the active soil resultant?

(A) 5700 lbf/ft
(B) 6800 lbf/ft
(C) 7500 lbf/ft
(D) 8300 lbf/ft

(e) What is the total overturning moment, taken about the toe, per foot of wall?

(A) 26,000 ft-lbf
(B) 45,000 ft-lbf
(C) 70,000 ft-lbf
(D) 190,000 ft-lbf

(f) What is the factor of safety against overturning?
(A) 1.5
(B) 1.8
(C) 2.3
(D) 2.6

(g) What is the maximum vertical pressure at the toe?
(A) 4000 lbf/ft^2
(B) 4500 lbf/ft^2
(C) 5000 lbf/ft^2
(D) 5500 lbf/ft^2

(h) What is the minimum vertical pressure at the heel?
(A) 600 lbf/ft^2
(B) 1300 lbf/ft^2
(C) 2400 lbf/ft^2
(D) 2700 lbf/ft^2

(i) What is the factor of safety against sliding without a key?
(A) 1.1
(B) 1.4
(C) 1.6
(D) 1.8

(j) What is the factor of safety against sliding if a key, 1.75 ft wide and 1.0 ft deep, is used?
(A) 1.2
(B) 1.5
(C) 1.7
(D) 2.1

SOLUTIONS

1. (a) As designed,

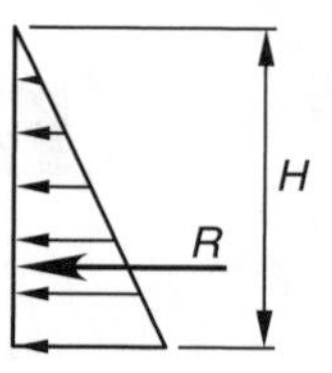

Use Eq. 37.7.

$$k_a = \frac{1 - \sin 32°}{1 + \sin 32°} = 0.307$$

The soil is drained but not dry.

$$\gamma = (1 + 0.05)\left(102\ \frac{\text{lbf}}{\text{ft}^3}\right) = 107.1\ \text{lbf/ft}^3$$

Use Eq. 37.10.

$$R = \left(\tfrac{1}{2}\right)(0.307)\left(107.1\ \frac{\text{lbf}}{\text{ft}^3}\right)(20\ \text{ft})^2 = \boxed{6576\ \text{lbf/ft}}$$

(b) This acts

$$\left(\tfrac{1}{3}\right)(20\ \text{ft}) = \boxed{6.67\ \text{ft up from the bottom}}$$

(c) As it happened, the pressure distributions (see Eq. 37.43) are

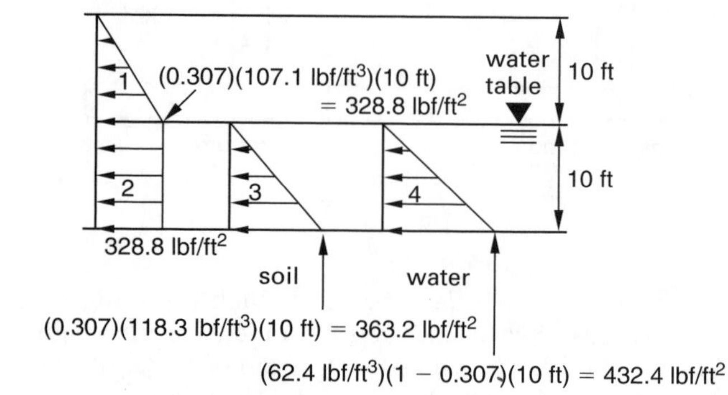

$$R_1 = \left(\tfrac{1}{2}\right)(10\ \text{ft})\left(328.8\ \frac{\text{lbf}}{\text{ft}^2}\right) = 1644\ \text{lbf/ft}$$
[13.33 ft up]

$$R_2 = (10\ \text{ft})\left(328.8\ \frac{\text{lbf}}{\text{ft}^2}\right) = 3288\ \text{lbf/ft}\quad [5\ \text{ft up}]$$

$$R_3 + R_4 = \left(\tfrac{1}{2}\right)(10\ \text{ft})\left(363.2\ \frac{\text{lbf}}{\text{ft}^2} + 432.4\ \frac{\text{lbf}}{\text{ft}^2}\right) = 3978\ \text{lbf/ft}\quad [10/3\ \text{ft up}]$$

$$R_{\text{total}} = 1644\ \frac{\text{lbf}}{\text{ft}} + 3288\ \frac{\text{lbf}}{\text{ft}} + 3978\ \frac{\text{lbf}}{\text{ft}} = \boxed{8910\ \text{lbf/ft}}$$

(d) This acts

$$\frac{\left(1644\ \frac{\text{lbf}}{\text{ft}}\right)(13.33\text{ ft}) + \left(3288\ \frac{\text{lbf}}{\text{ft}}\right)(5\text{ ft}) + \left(3978\ \frac{\text{lbf}}{\text{ft}}\right)\left(\frac{10\text{ ft}}{3}\right)}{8910\ \frac{\text{lbf}}{\text{ft}}}$$

$$= \boxed{5.79\text{ ft up from the bottom}}$$

2. (a)

$$k_a = \frac{1 - \sin 36^\circ}{1 + \sin 36^\circ} = 0.26$$

At $H = 10$ ft,

$$\begin{aligned} p_h &= k_a p_v = k_a \gamma H \\ &= (0.26)\left(96\ \frac{\text{lbf}}{\text{ft}^3}\right)(10\text{ ft}) = 249.6\text{ lbf/ft}^2 \end{aligned}$$

$$\begin{aligned} p &= 249.6\ \frac{\text{lbf}}{\text{ft}^2} \\ &\quad + (0.26)\left(121\ \frac{\text{lbf}}{\text{ft}^3} - 62.4\ \frac{\text{lbf}}{\text{ft}^3}\right)(H - 10\text{ ft}) \\ &\quad + \left(62.4\ \frac{\text{lbf}}{\text{ft}^3}\right)(H - 10\text{ ft}) \\ &= 249.6\ \frac{\text{lbf}}{\text{ft}^2} + \left(77.6\ \frac{\text{lbf}}{\text{ft}^3}\right)(H - 10\text{ ft}) \end{aligned}$$

The pressure distributions on the wall are as shown.

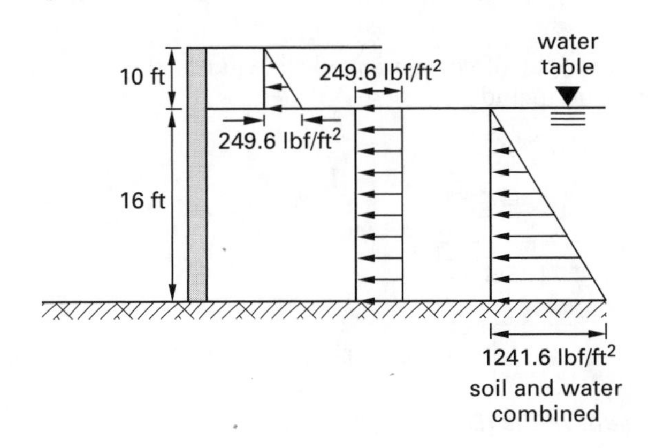

The resultants of each of the preceding three distributions are

$\left(\frac{1}{2}\right)(10\text{ ft})\left(249.6\text{ lbf/ft}^2\right) = 1248$ lbf/ft, located $\left(\frac{2}{3}\right)(10\text{ ft}) = 6.67$ ft from the top (19.33 ft from the bottom)

$(16\text{ ft})\left(249.6\ \frac{\text{lbf}}{\text{ft}^2}\right) = 3994$ lbf/ft, located 8 ft from the bottom

$\left(\frac{1}{2}\right)(16\text{ ft})\left(1241.66\ \frac{\text{lbf}}{\text{ft}^2}\right) = 9933$ lbf/ft, located $\left(\frac{1}{3}\right)(16\text{ ft})$ from the bottom

The active resultant is

$$1248\ \frac{\text{lbf}}{\text{ft}} + 3994\ \frac{\text{lbf}}{\text{ft}} + 9933\ \frac{\text{lbf}}{\text{ft}} = \boxed{15{,}175\text{ lbf/ft}}$$

(b) Taking moments about the base,

$$\begin{aligned} \text{moment arm} &= \frac{\left(1248\ \frac{\text{lbf}}{\text{ft}}\right)(19.33\text{ ft}) + \left(3994\ \frac{\text{lbf}}{\text{ft}}\right)(8\text{ ft}) + \left(9933\ \frac{\text{lbf}}{\text{ft}}\right)(5.33\text{ ft})}{15{,}175\ \frac{\text{lbf}}{\text{ft}}} \\ &= \frac{109{,}019\ \frac{\text{ft-lbf}}{\text{ft}}}{15{,}175\ \frac{\text{lbf}}{\text{ft}}} \\ &= \boxed{7.18\text{ ft up from the bottom}} \end{aligned}$$

(c) The original overturning moment is

$$M_{\text{original}} = 109{,}019\text{ ft-lbf/ft}$$

If the water table were lowered, the resultant would be

$$\begin{aligned} R_a &= \left(\tfrac{1}{2}\right)(0.26)\left(96\ \frac{\text{lbf}}{\text{ft}^3}\right)(26\text{ ft})^2 \\ &= 8436\text{ lbf/ft located }\left(\tfrac{1}{3}\right)(26\text{ ft}) \\ &= 8.67\text{ ft from the bottom} \end{aligned}$$

$$M_{\text{drained}} = \left(8436\ \frac{\text{lbf}}{\text{ft}}\right)(8.67\text{ ft}) = 73{,}123\text{ ft-lbf/ft}$$

$$\begin{aligned} M_{\text{reduction}} &= 109{,}019\ \frac{\text{ft-lbf}}{\text{ft}} - 73{,}123\ \frac{\text{ft-lbf}}{\text{ft}} \\ &= \boxed{35{,}896\text{ ft-lbf/ft}} \end{aligned}$$

3. (a) To prevent horizontal frost heave that could move the retaining wall, the base must be at least 4 ft below the surface to be below the frost line. Therefore, the overall height of the wall is at least $\boxed{18\text{ ft.}}$

The answer is (C).

(b)

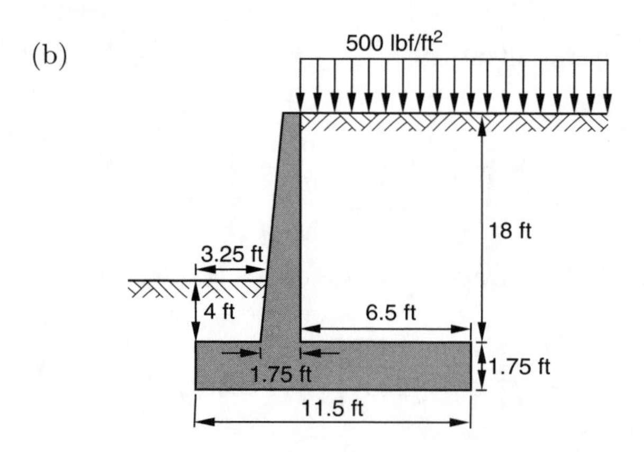

Convert the surcharge to an equivalent weight (depth) of soil.

$$\frac{500\ \frac{\text{lbf}}{\text{ft}^2}}{130\ \frac{\text{lbf}}{\text{ft}^3}} = \boxed{3.85\ \text{ft}}$$

The answer is (C).

(c)

$$k_a = \frac{1 - \sin 35^\circ}{1 + \sin 35^\circ} = 0.27$$

The surcharge pressure is

$$p_{\text{surcharge}} = (0.27)\left(500\ \frac{\text{lbf}}{\text{ft}^2}\right) = 135\ \text{lbf/ft}^2$$

The surcharge reaction is

$$R_{\text{surcharge}} = p_{\text{surcharge}}H = \left(135\ \frac{\text{lbf}}{\text{ft}^2}\right)(19.75\ \text{ft})$$

$$= \boxed{2666\ \text{lbf/ft}}$$

The answer is (B).

(d) The active soil resultant is

$$R_a = \left(\tfrac{1}{2}\right)(0.27)\left(130\ \frac{\text{lbf}}{\text{ft}^3}\right)(19.75\ \text{ft})^2 = \boxed{6846\ \text{lbf/ft}}$$

Since the soil above the heel is horizontal, R_a is horizontal and there is no vertical component of R_a.

The answer is (B).

(e) The surcharge resultant acts halfway up from the base, and the active soil resultant acts one-third up from the base.

$$M_{\text{OT}} = \left(2666\ \frac{\text{lbf}}{\text{ft}}\right)\left(\frac{19.75\ \text{ft}}{2}\right) + \left(6846\ \frac{\text{lbf}}{\text{ft}}\right)\left(\frac{19.75\ \text{ft}}{3}\right)$$

$$= 71{,}396\ \text{ft-lbf} \quad \text{[per foot of wall]}$$

The answer is (C).

(f) To calculate the factor of safety against overturning, it is necessary to take resisting moments about the pivot point (which, in this case, is the toe).

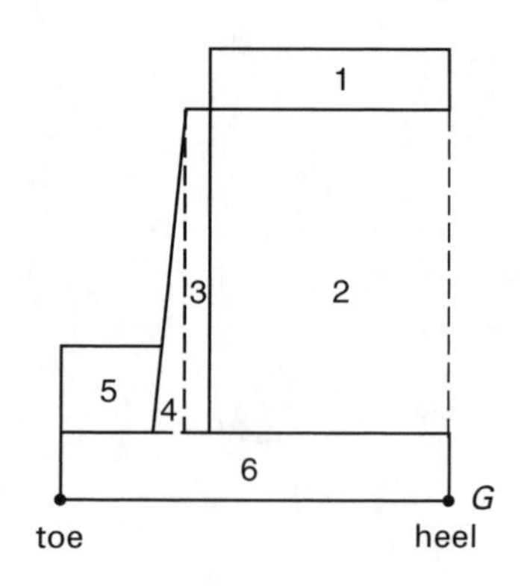

i	area	γ	W_i	x_i from toe	$M_{\text{resisting}}$
1	(6.5)(3.85) = 25.03	130	3254	8.25	26,846
2	(6.5)(18) = 117	130	15,210	8.25	125,483
3	(1)(18) = 18	150	2700	4.50	12,150
4	$\left(\frac{1}{2}\right)$(18)(0.75) = 6.75	150	1013	3.75	3799
5	(3.25)(4)=13	130	1690	1.63	2755
6	(11.5)(1.75) = 20.13	150	3020	5.75	17,365
			26,887		188,398

(The weight of component 6 disregards the key, which could be included.)

Use Eq. 37.49.

$$F_{\text{OT}} = \frac{188{,}398\ \frac{\text{ft-lbf}}{\text{ft}}}{71{,}396\ \frac{\text{ft-lbf}}{\text{ft}}}$$

$$= \boxed{2.64} \quad \text{[acceptable]}$$

The answer is (D).

(g) Use Eq. 37.47.

$$x_R = \frac{M_{\text{resisting}} - M_{\text{OT}}}{\sum W_i}$$

$$= \frac{188{,}398\ \frac{\text{ft-lbf}}{\text{ft}} - 71{,}396\ \frac{\text{ft-lbf}}{\text{ft}}}{26{,}887\ \frac{\text{lbf}}{\text{ft}}}$$

$$= 4.35\ \text{ft}$$

Use Eq. 37.48.

$$\epsilon^* = \left|\left(\tfrac{1}{2}\right)(11.5\ \text{ft}) - 4.35\ \text{ft}\right| = 1.4\ \text{ft}$$

This ϵ^* is less than 11.5 ft/6 = 1.92 ft, so it is acceptable. Use Eq. 37.50.

$$p_{\text{max,toe}} = \left(\frac{26{,}887\ \frac{\text{lbf}}{\text{ft}}}{11.5\ \text{ft}}\right)\left(1 + \frac{(6)(1.4\ \text{ft})}{11.5\ \text{ft}}\right) = \boxed{4045\ \text{lbf/ft}^2}$$

The answer is (A).

(h) $$p_{\text{min,heel}} = \left(\frac{26{,}887\ \frac{\text{lbf}}{\text{ft}}}{11.5\ \text{ft}}\right)\left(1 - \frac{(6)(1.4\ \text{ft})}{11.5\ \text{ft}}\right) = \boxed{630\ \text{lbf/ft}^2}$$

The answer is (A).

(i) Since there is no key, the friction between concrete and soil resists sliding.

$$R_s = (\Sigma W)(0.5) = \left(26{,}887\ \frac{\text{lbf}}{\text{ft}}\right)(0.5) = 13{,}444\ \text{lbf/ft}$$

$$F_{\text{SL}} = \frac{R_s}{R_{\text{surcharge}} + R_a} = \frac{13{,}444\ \frac{\text{lbf}}{\text{ft}}}{2666\ \frac{\text{lbf}}{\text{ft}} + 6846\ \frac{\text{lbf}}{\text{ft}}} = \boxed{1.41}$$

(This is less than 1.5, so it is not acceptable.)

The answer is (B).

(j) A key is needed for this design, so extend the stem down about 1 ft.

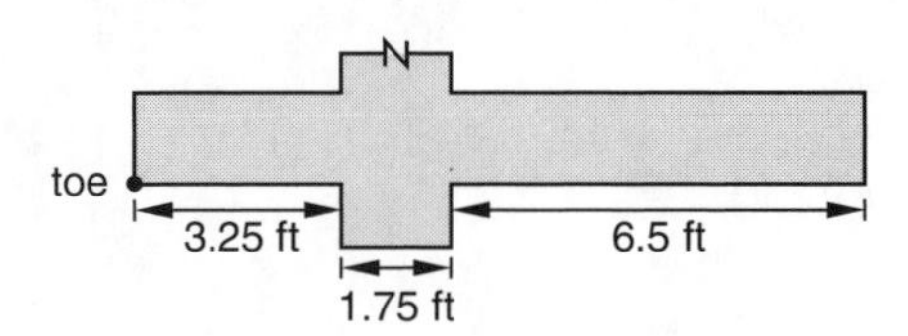

The upward earth pressure distribution on the base is

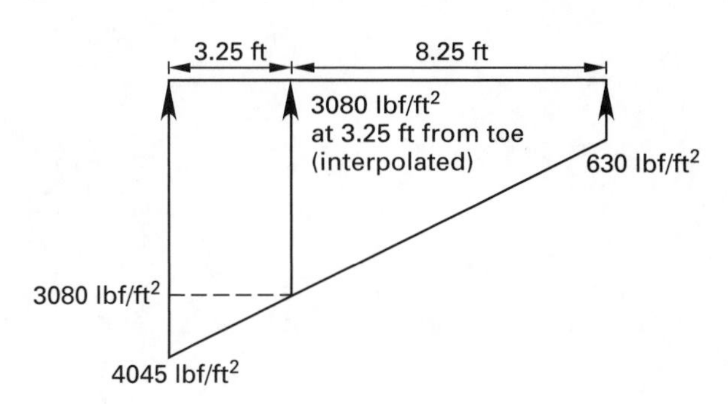

$$p_{\text{key face}} = 630\ \frac{\text{lbf}}{\text{ft}^2} + \left(\frac{8.25\ \text{ft}}{8.25\ \text{ft} + 3.25\ \text{ft}}\right) \times \left(4045\ \frac{\text{lbf}}{\text{ft}^2} - 630\ \frac{\text{lbf}}{\text{ft}^2}\right) = 3080\ \text{lbf/ft}^2$$

Check the resistance to sliding again. From the toe to the key, the soil must shear. From the toe to the key, the frictional resisting force is found as the product of the total upward normal force and the coefficient of shearing (internal) friction.

$$\begin{aligned} R_{s_1} &= \Bigg((3.25\ \text{ft})\left(3080\ \frac{\text{lbf}}{\text{ft}^2}\right) \\ &\quad + \left(\tfrac{1}{2}\right)(3.25\ \text{ft})\left(4045\ \frac{\text{lbf}}{\text{ft}^2} - 3080\ \frac{\text{lbf}}{\text{ft}^2}\right)\Bigg)(\tan 35^\circ) \\ &= (11{,}578\ \text{lbf})(\tan 35^\circ) \\ &= 8107\ \text{lbf/ft} \end{aligned}$$

From the key to the heel, the base slides on the soil. From the key to the heel, the frictional resisting force is found as the product of the total upward normal force and the coefficient of sliding friction.

$$R_{s_2} = \left(26{,}887\ \frac{\text{lbf}}{\text{ft}} - 11{,}578\ \frac{\text{lbf}}{\text{ft}}\right)(0.5) = 7655\ \text{lbf/ft}$$

$$F_{\text{SL}} = \frac{8107\ \frac{\text{lbf}}{\text{ft}} + 7655\ \frac{\text{lbf}}{\text{ft}}}{2666\ \frac{\text{lbf}}{\text{ft}} + 6846\ \frac{\text{lbf}}{\text{ft}}} = \boxed{1.66} \quad \text{[acceptable]}$$

The answer is (C).

38 Piles and Deep Foundations

PRACTICE PROBLEMS

Single Piles

1. A 10.75 in diameter steel pile is driven 65 ft into stiff, insensitive clay. The clay has an undrained shear strength of 1300 lbf/ft^2. The clay's specific weight is 115 lbf/ft^3. The water table is at the ground surface. The entire pile length is effective. (a) What is the end-bearing capacity? (b) What is the friction capacity? (c) What is the allowable bearing capacity?

Pile Groups

2. *(Time limit: one hour)* Tests have shown that a single pile would have, by itself, an uplift capacity of 150 tons and a compressive capacity of 500 tons. 36 piles are installed on a grid with a spacing of 3.5 ft. The pile group is capped and connected at the top by a thick steel-reinforced concrete slab adding its own axial load of 600 tons. Find the maximum moment that the pile group can take in the x- and y-directions assuming axial load is applied to permit the moment.

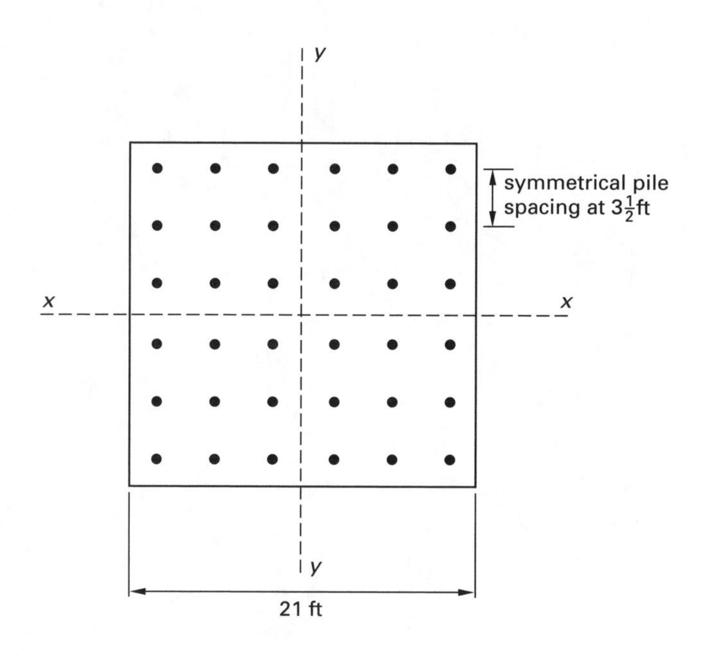

SOLUTIONS

1. (a) $B = \dfrac{10.75 \text{ in}}{12 \dfrac{\text{in}}{\text{ft}}} = 0.896 \text{ ft}$

For saturated clay, $\phi = 0°$.

$$N_c = 9.0$$

Use Eq. 38.10.

$$Q_p = \frac{\left(\dfrac{\pi(0.896 \text{ ft})^2}{4}\right)\left(1300 \dfrac{\text{lbf}}{\text{ft}^2}\right)(9)}{1000 \dfrac{\text{lbf}}{\text{kip}}} = \boxed{7.38 \text{ kips}}$$

(b) Assume

$$\begin{aligned} c_A &= 0.5c \\ &= (0.5)\left(1300 \frac{\text{lbf}}{\text{ft}^2}\right) \\ &= 650 \text{ lbf/ft}^2 \end{aligned}$$

Use Eq. 38.18.

$$\begin{aligned} Q_s &= pc_AL \\ &= \frac{\pi\left(\dfrac{10.75 \text{ in}}{12 \dfrac{\text{in}}{\text{ft}}}\right)\left(650 \dfrac{\text{lbf}}{\text{ft}^2}\right)(65 \text{ ft})}{1000 \dfrac{\text{lbf}}{\text{kip}}} \\ &= \boxed{119 \text{ kips}} \end{aligned}$$

(c) The allowable load is

$$\begin{aligned} Q_a &= \frac{Q_{\text{ult}}}{F} \\ &= \frac{7377 \text{ lbf} + 118{,}906 \text{ lbf}}{(3)\left(1000 \dfrac{\text{lbf}}{\text{kip}}\right)} \\ &= \boxed{42.1 \text{ kips}} \end{aligned}$$

2. The moment capacity is limited by the tension capacity of the piles. Assume rotation about the y-y axis. Piles to the left of the y-y will be in tension. Work on the basis of one row of six piles.

$$x_A = (0.5)(3.5\text{ ft}) = 1.75\text{ ft} = x_D$$
$$x_B = (1.5)(3.5\text{ ft}) = 5.25\text{ ft} = x_E$$
$$x_C = (2.5)(3.5\text{ ft}) = 8.75\text{ ft} = x_F$$

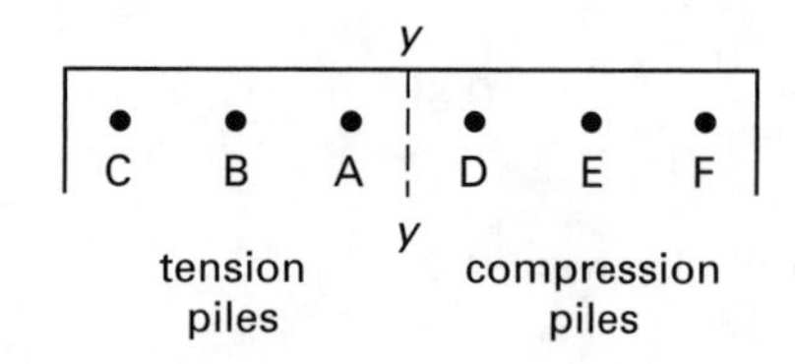

The force in outermost tension pile C is

$$F_C = \frac{W}{n} + \frac{Mx_C}{X_A^2 + X_B^2 + X_C^2 + X_D^2 + X_E^2 + X_F^2}$$

Use a convention that tension is positive and compression is negative.

$$\begin{aligned} 150\text{ tons} &= -\frac{600\text{ tons}}{36\text{ piles}} \\ &\quad + M\left(\frac{8.75\text{ ft}}{(2)\left((1.75\text{ ft})^2 + (5.25\text{ ft})^2 + (8.75\text{ ft})^2\right)}\right) \\ &= -16.67\text{ tons} + M\left(0.0408\ \frac{1}{\text{ft}}\right) \\ M &= 4085\text{ ft-tons}\quad [\text{per row}] \end{aligned}$$

For six rows,

$$\begin{aligned} M_{max} &= (6)(4085\text{ ft-tons}) \\ &= \boxed{24{,}510\text{ ft-tons}} \end{aligned}$$

39 Temporary Excavations

PRACTICE PROBLEMS

Braced Cuts

1. A 30 ft deep, 40 ft square excavation in sand is being designed. The sand will be dewatered before excavation. The angle of internal friction is 40°, and the specific weight is 121 lbf/ft^3. Bracing consists of horizontal lagging supported by 8 in soldier piles separated horizontally by 8 ft. (a) What is the approximate shape of the pressure diagram on the lagging? (b) What is the bending moment on the lagging?

Flexible Bulkheads

2. *(Time limit: one hour)* A 35 ft long sheet pile is driven through 10 ft of clay to bedrock below. The sheet pile supports a 25 ft vertical cut through drained sand. A tie rod is located 8 ft below the surface, terminating at a deadman behind the failure plane. There is no significant water table. Soil parameters are shown in the accompanying illustration. What is the tensile force in the tie rod?

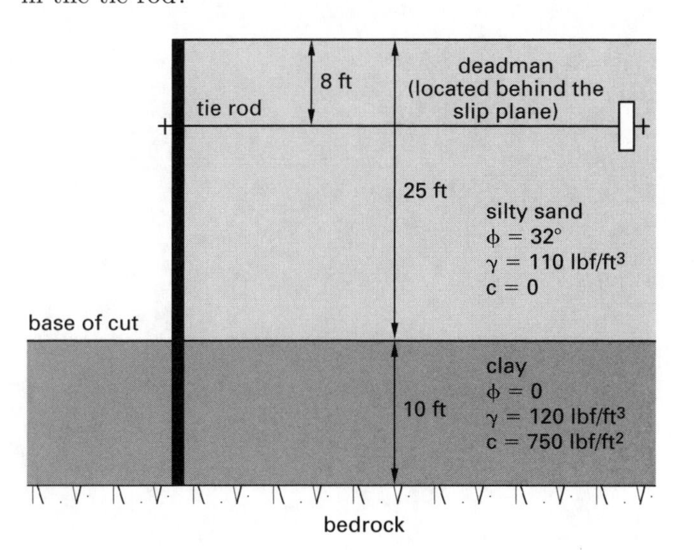

SOLUTIONS

1. (a) Use Eqs. 39.1 and 37.7.

$$p_{\text{max}} = (0.65)\left(121\ \frac{\text{lbf}}{\text{ft}^3}\right)(30\ \text{ft})\tan^2\left(45° - \frac{40°}{2}\right) = 513\ \text{lbf/ft}^2$$

The distribution is rectangular.

(b) Assume simple supports of each horizontal lagging timber.

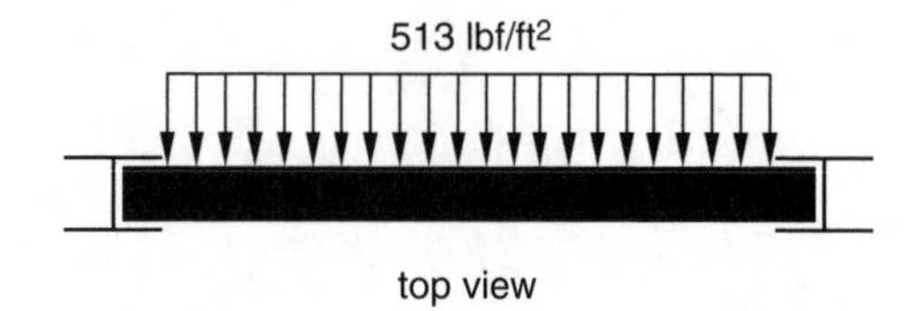

From beam equations, the maximum moment per foot of timber width is

$$M = \frac{\left(513\ \frac{\text{lbf}}{\text{ft}^2}\right)(8\ \text{ft})^2}{8} = 4104\ \text{ft-lbf}$$

2. (a)
$$k_{a,\text{sand}} = \frac{1 - \sin 32°}{1 + \sin 32°} = 0.307$$

$$p_v = \left(110\ \frac{\text{lbf}}{\text{ft}^3}\right)(25\ \text{ft}) = 2750\ \text{lbf/ft}^2$$

$$p_{a,25} = (0.307)\left(2750\ \frac{\text{lbf}}{\text{ft}^2}\right) = 844\ \text{lbf/ft}^2$$

The reaction per foot of wall is

$$R = \left(\tfrac{1}{2}\right)(25\ \text{ft})\left(844\ \frac{\text{lbf}}{\text{ft}^2}\right)(1\ \text{ft}) = 10{,}550\ \text{lbf}$$

R acts 10 ft + (25 ft/3) = 18.33 ft up from bedrock.

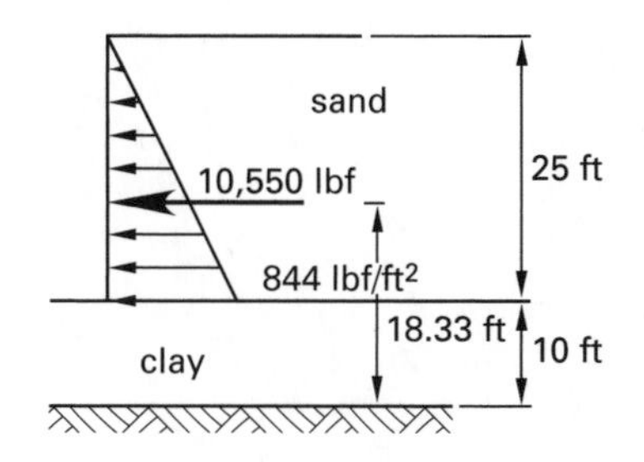

Determine the horizontal clay pressure.

$$\phi = 0°$$

$$k_a = 1 \quad [\text{since } \phi = 0°]$$

From Eq. 37.8,

$$p_{a,\text{clay}} = p_v - 2c$$

p_v is the surcharge at the top of the clay layer plus the self-load (γH), which is duplicated on the passive side.

$$p_a = 2750 \ \frac{\text{lbf}}{\text{ft}^2} + (\text{clay depth})\left(120 \ \frac{\text{lbf}}{\text{ft}^3}\right) - 2c$$

$$\begin{aligned} p_p &= p_v + 2c \\ &= (\text{clay depth})\left(120 \ \frac{\text{lbf}}{\text{ft}^3}\right) + 2c \end{aligned}$$

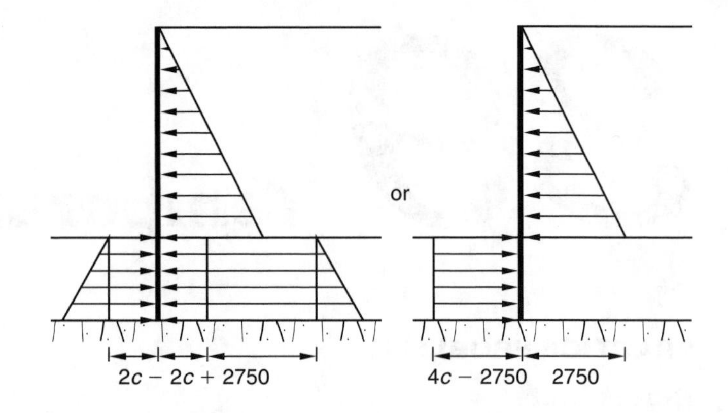

$$\begin{aligned} 4c - 2750 \ \frac{\text{lbf}}{\text{ft}^2} &= (4)\left(750 \ \frac{\text{lbf}}{\text{ft}^2}\right) - 2750 \ \frac{\text{lbf}}{\text{ft}^2} \\ &= 250 \ \text{lbf/ft}^2 \end{aligned}$$

$$R_p = \left(250 \ \frac{\text{lbf}}{\text{ft}^2}\right)(10 \ \text{ft})(1 \ \text{ft}) = 2500 \ rmlbf$$

$$F = 10{,}550 \ \text{lbf} - 2500 \ \text{lbf} = \boxed{8050 \ \text{lbf}}$$

40 Special Soil Topics

PRACTICE PROBLEMS

Pressure at Depths

1. A concentrated vertical load of 6000 lbf is applied at the ground surface. What is the increase in vertical pressure 3.5 ft below the surface and 4 ft from the line of action of the force?

2. A 10 ft × 10 ft footing exerts a pressure of 3000 lbf/ft^2 on the soil below it. What is the increase in vertical soil stress 10 ft below and 8 ft from the center of the footing?

3. A building weighing 20 tons is placed on the L-shaped slab shown. The soil below the basement is a 100 ft thick layer of dense sand. (a) What is the increase in vertical soil pressure at a depth of 30 ft below point A? (b) What is the increase in vertical soil pressure at a depth of 45 ft below point B?

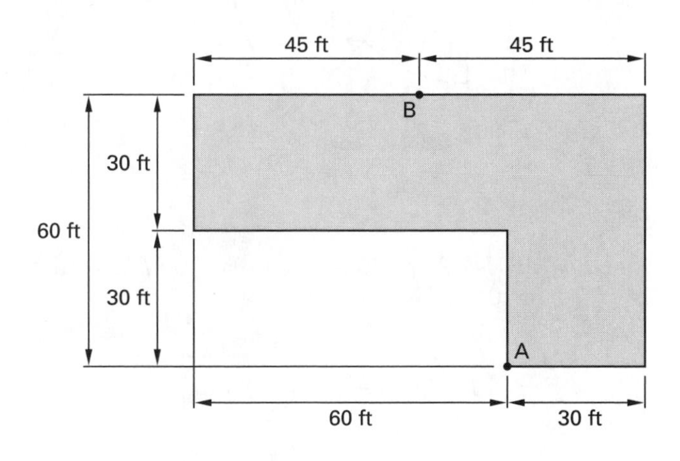

Consolidation

4. A large manufacturing plant with a uniform floor loading of 500 lbf/ft^2 is supported by a large mat foundation. A 38 ft thick layer of silty soil is underlain by sand and gravel. The plant was constructed on the silty soil 10 ft above the water table. After a number of years and after all settlement of the building had stopped, a series of wells were drilled, which dropped the water table from 10 ft below the surface to 18 ft below the surface. The unit weight of the soil above the water table is 100 lbf/ft^3. It is 120 lbf/ft^3 below the water table. The silt compression index is 0.02. The void ratio before the water table was lowered was 0.6. How much more will the building settle?

5. The compression index of a normally loaded 16 ft thick clay soil is 0.31. The clay is bounded above and below by layers of sand. When the effective stress on the soil is 2600 lbf/ft^2, the void ratio is 1.04, and the permeability is 4×10^{-7} mm/s. The stress is increased gradually to 3900 lbf/ft^2. (a) What is the change in the void ratio? (b) What is the settlement in the clay layer? (c) How much time is required for settlement to reach 75%?

6. At a building site, a rock layer is covered by 8 ft of soft clay. The clay is covered with 15 ft of silty sand. The water table is 18 ft above the rock layer. Properties of the layers are shown on the illustration. A construction firm wants to consolidate the clay layer by surcharging the site with 10 more ft of sandy fill (with a specific weight of 110 lbf/ft^3) and by lowering the water table 5 ft. What consolidation will be achieved by this process?

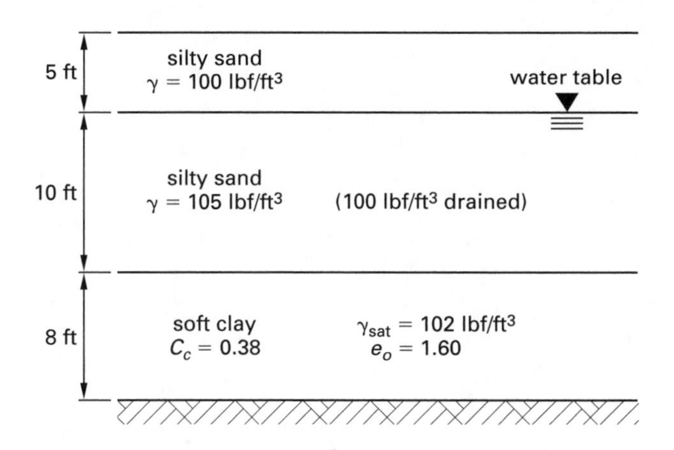

Stresses on Buried Pipes

7. A 12 in concrete sewer pipe is to be installed under a backfill of 11 ft of saturated topsoil with a specific weight of 120 lbf/ft^3. The trench width is 2 ft. The pipe strength is 1500 lbf/ft. There are no live loads. A factor of safety of 1.5 is required. Specify the bedding design.

Slope Stability

8. A 2:1 (horizontal:vertical) sloped cut is made in homogeneous saturated clay. The clay's specific weight is 112 lbf/ft^3, and its ultimate shear strength is 1100 lbf/ft^2. The cut is 43 ft deep. The clay extends 15 ft below the toe of the cut to a rock layer. What is the cohesive factor of safety of this cut?

9. *(Time limit: one hour)* A 2 in PVC pipe is to be buried in a trench 18 in wide and 3½ ft deep. The angle of internal friction of the cohesive soil is 18°. The cohesion is 200 lbf/ft^2. The moisture content is 20%. The dry specific weight of the soil is 115 lbf/ft^3. (a) Will the soil stand or slide during trenching if the spoils are placed directly alongside the trench sides? (b) How far away from the trench must the spoils be placed to satisfy OSHA? (c) When the spoils are spread evenly for a distance of 2 ft from the trench, will the soil remain stable? (d) Recommend initial bedding and backfill materials. (e) What additional precautions would you recommend to protect any workers who enter the trench?

SOLUTIONS

1. Use Eq. 40.1.

$$p_v = \left(\frac{(3)(6000 \text{ lbf})}{(2\pi)(3.5 \text{ ft})^2}\right)\left(\frac{1}{1+\left(\frac{4 \text{ ft}}{3.5 \text{ ft}}\right)^2}\right)^{5/2}$$

$$= \boxed{28.96 \text{ lbf/ft}^2}$$

2. *method 1:* Use Fig. 40.3. About 25 squares are covered. Use Eq. 40.5.

$$p = (25)(0.005)\left(3000 \frac{\text{lbf}}{\text{ft}^2}\right) = \boxed{375 \text{ lbf/ft}^2}$$

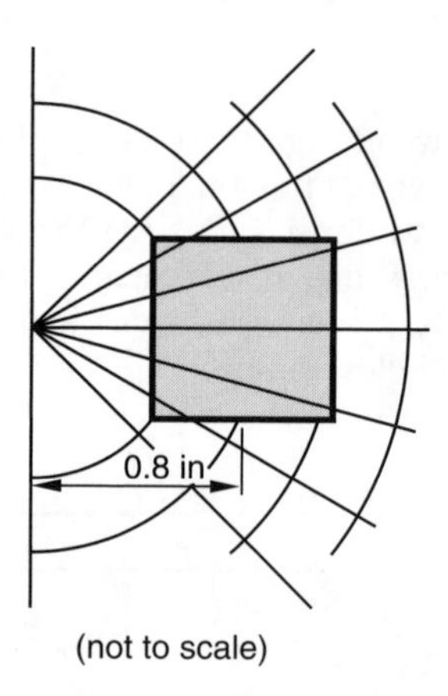

(not to scale)

method 2: Use App. 40.A. At $(-B, +0.8B)$, read $\approx 0.1p$. Use Eq. 40.5.

$$p = (0.1)\left(3000 \frac{\text{lbf}}{\text{ft}^2}\right) = \boxed{300 \text{ lbf/ft}^2}$$

3. (a) $A = (90 \text{ ft})(30 \text{ ft}) + (30 \text{ ft})(30 \text{ ft}) = 3600 \text{ ft}^2$

$$p_{\text{applied}} = \frac{(20 \text{ tons})\left(2000 \frac{\text{lbf}}{\text{ton}}\right)}{3600 \text{ ft}^2} = 11.11 \text{ lbf/ft}^2$$

Use Fig. 40.3.

Draw the slab such that 1 in = 30 ft. Approximately 46 squares are covered. Use Eq. 40.5.

$$p_{30 \text{ ft}} = (46)(0.005)\left(11.11 \frac{\text{lbf}}{\text{ft}^2}\right) = \boxed{2.56 \text{ lbf/ft}^2}$$

(b) Draw the slab such that 1 in = 45 ft. Approximately 63 squares are covered.

$$p_{45\text{ ft}} = (63)(0.005)\left(11.11\ \frac{\text{lbf}}{\text{ft}^2}\right) = \boxed{3.5\ \text{lbf/ft}^2}$$

4. *steps 1 and 2:* The top 10 ft of soil will have settled under the 500 psf load as much as it is going to. The next 8 ft will settle due to the drop in the water table. The soil under the water table will also experience a reconsolidation. The midpoints of these two settling layers are at depths of 14 ft and 28 ft, respectively. The settlement will be due to the change in effective pressure at these midpoints. Because the raft is so large, its pressure load is assumed not to decrease with depth. Besides, its size was not given.

Before the drop in the water table,

$$\begin{aligned} p_{o,14} &= 500\ \frac{\text{lbf}}{\text{ft}^2} + (10\ \text{ft})\left(100\ \frac{\text{lbf}}{\text{ft}^3}\right) \\ &\quad + (4\ \text{ft})\left(120\ \frac{\text{lbf}}{\text{ft}^3} - 62.4\ \frac{\text{lbf}}{\text{ft}^3}\right) \\ &= 1730\ \text{lbf/ft}^2 \end{aligned}$$

$$\begin{aligned} p_{o,28} &= 500\ \frac{\text{lbf}}{\text{ft}^2} + (10\ \text{ft})\left(100\ \frac{\text{lbf}}{\text{ft}^3}\right) \\ &\quad + (18\ \text{ft})\left(120\ \frac{\text{lbf}}{\text{ft}^3} - 62.4\ \frac{\text{lbf}}{\text{ft}^3}\right) \\ &= 2537\ \text{lbf/ft}^2 \end{aligned}$$

After the drop in the water table,

$$\begin{aligned} p_{14} &= 500\ \frac{\text{lbf}}{\text{ft}^2} + (14\ \text{ft})\left(100\ \frac{\text{lbf}}{\text{ft}^3}\right) \\ &= 1900\ \text{lbf/ft}^2 \end{aligned}$$

$$\begin{aligned} p_{28} &= 500\ \frac{\text{lbf}}{\text{ft}^2} + (18\ \text{ft})\left(100\ \frac{\text{lbf}}{\text{ft}^3}\right) \\ &\quad + (10\ \text{ft})\left(120\ \frac{\text{lbf}}{\text{ft}^3} - 62.4\ \frac{\text{lbf}}{\text{ft}^3}\right) \\ &= 2876\ \text{lbf/ft}^2 \end{aligned}$$

step 3: $e_o = 0.6$ [given]

step 4: $C_c = 0.02$ [given]

step 5: Use Eq. 40.16. Because of the 8 ft layer,

$$\begin{aligned} S_1 &= \left(\frac{0.02}{1+0.6}\right)(8\ \text{ft})\log_{10}\left(\frac{1900\ \frac{\text{lbf}}{\text{ft}^2}}{1730\ \frac{\text{lbf}}{\text{ft}^2}}\right) \\ &= 0.00407\ \text{ft} \end{aligned}$$

Because of the submerged layer,

$$\begin{aligned} S_2 &= \left(\frac{0.02}{1+0.6}\right)(20\ \text{ft})\log_{10}\left(\frac{2876\ \frac{\text{lbf}}{\text{ft}^2}}{2537\ \frac{\text{lbf}}{\text{ft}^2}}\right) \\ &= 0.01362\ \text{ft} \end{aligned}$$

The total settlement is

$$\begin{aligned} S_1 + S_2 &= 0.00407\ \text{ft} + 0.01362\ \text{ft} \\ &= \boxed{0.0177\ \text{ft}} \end{aligned}$$

5. (a) Though reported as a positive number, the compression index represents a negative slope. Use Eq. 40.7.

$$-0.31 = \frac{1.04 - e_2}{\log_{10}\left(\frac{2600\ \frac{\text{lbf}}{\text{ft}^2}}{3900\ \frac{\text{lbf}}{\text{ft}^2}}\right)}$$

$$e_2 = 0.985$$

$$\Delta e = 0.985 - 1.04 = \boxed{-0.055}$$

(b) Use Eq. 40.15.

$$S = (16\ \text{ft})\left(\frac{-0.055}{1+1.04}\right) = \boxed{-0.43\ \text{ft}}$$

(c) Use Eq. 40.22.

$$a_v = \frac{1.04 - 0.985}{3900\ \frac{\text{lbf}}{\text{ft}^2} - 2600\ \frac{\text{lbf}}{\text{ft}^2}} = 4.23\times 10^{-5}\ \text{ft}^2/\text{lbf}$$

The permeability is

$$\begin{aligned} &\frac{4\times 10^{-7}\ \frac{\text{mm}}{\text{sec}}}{\left(10\ \frac{\text{mm}}{\text{cm}}\right)\left(2.54\ \frac{\text{cm}}{\text{in}}\right)\left(12\ \frac{\text{in}}{\text{ft}}\right)} \\ &\quad = 1.31\times 10^{-9}\ \text{ft/sec} \end{aligned}$$

$$\begin{aligned} C_v &= \frac{\left(1.31\times 10^{-9}\ \frac{\text{ft}}{\text{sec}}\right)(1+1.04)}{\left(62.4\ \frac{\text{lbf}}{\text{ft}^3}\right)\left(4.23\times 10^{-5}\ \frac{\text{ft}^2}{\text{lbf}}\right)} \\ &= 1.01\times 10^{-6}\ \text{ft}^2/\text{sec} \end{aligned}$$

For $U_z = 75\%$, $T_v = 0.48$.

$$\begin{aligned} t &= \frac{(0.48)\left(\frac{16\ \text{ft}}{2}\right)^2}{\left(1.01\times 10^{-6}\ \frac{\text{ft}^2}{\text{sec}}\right)\left(24\ \frac{\text{hr}}{\text{day}}\right)\left(3600\ \frac{\text{sec}}{\text{hr}}\right)} \\ &= \boxed{352\ \text{days}} \end{aligned}$$

Geotechnical

6. The effective pressure at the midpoint of the clay layer is

$$\begin{aligned} p_o &= (5 \text{ ft})\left(100 \ \frac{\text{lbf}}{\text{ft}^3}\right) + (10 \text{ ft})\left(105 \ \frac{\text{lbf}}{\text{ft}^3} - 62.4 \ \frac{\text{lbf}}{\text{ft}^3}\right) \\ &\quad + \left(\tfrac{1}{2}\right)\left(102 \ \frac{\text{lbf}}{\text{ft}^3} - 62.4 \ \frac{\text{lbf}}{\text{ft}^3}\right)(8 \text{ ft}) \\ &= 1084.4 \text{ lbf/ft}^2 \end{aligned}$$

The final effective pressure is

$$\begin{aligned} p &= (10 \text{ ft})\left(110 \ \frac{\text{lbf}}{\text{ft}^3}\right) + (10 \text{ ft})\left(100 \ \frac{\text{lbf}}{\text{ft}^3}\right) \\ &\quad + (5 \text{ ft})\left(105 \ \frac{\text{lbf}}{\text{ft}^3} - 62.4 \ \frac{\text{lbf}}{\text{ft}^3}\right) \\ &\quad + \left(\tfrac{1}{2}\right)\left(102 \ \frac{\text{lbf}}{\text{ft}^3} - 62.4 \ \frac{\text{lbf}}{\text{ft}^3}\right)(8 \text{ ft}) \\ &= 2471.4 \text{ lbf/ft}^2 \end{aligned}$$

Use Eq. 40.16.

$$\begin{aligned} S &= \left(\frac{0.38}{1+1.60}\right)(8 \text{ ft})\log_{10}\left(\frac{2471.4 \ \frac{\text{lbf}}{\text{ft}^2}}{1084.4 \ \frac{\text{lbf}}{\text{ft}^2}}\right) \\ &= \boxed{0.42 \text{ ft}} \end{aligned}$$

Note that C_c can also be estimated from e_o, but the data provided in the problem takes priority.

7.

$$\frac{h}{B} = \frac{11 \text{ ft}}{2 \text{ ft}} = 5.5 \quad [\text{round to } 6.0]$$

Use Table 40.2. For $\gamma = 120 \text{ lbf/ft}^3$, $C = 3.04$. (The load on the pipe is more important than the soil type.)

The actual load is given by Eq. 40.30.

$$w = (3.04)\left(120 \ \frac{\text{lbf}}{\text{ft}^3}\right)(2 \text{ ft})^2 = 1459.2 \text{ lbf/ft}$$

Solve for the load factor. Use Eq. 40.36.

$$\text{LF} = \frac{\left(1459.2 \ \frac{\text{lbf}}{\text{ft}}\right)(1.5)}{1500 \ \frac{\text{lbf}}{\text{ft}}} = 1.46$$

$$\boxed{\text{Select Class C or better.}}$$

8. For clay and $\phi = 0°$, the cohesion is the same as the shear strength. ("Saturated clay" is not the same as "submerged clay.")

$$\begin{aligned} c &= 1100 \text{ lbf/in}^2 \\ \beta &= \arctan\left(\tfrac{1}{2}\right) = 26.6° \\ d &= \frac{15 \text{ ft}}{43 \text{ ft}} = 0.35 \end{aligned}$$

From the Taylor chart, the stability number is $N_o \approx 6.5$. From Eq. 40.28, the cohesion safety factor is

$$F = \frac{N_o c}{\gamma H} = \frac{(6.5)\left(1100 \ \frac{\text{lbf}}{\text{ft}^2}\right)}{\left(112 \ \frac{\text{lbf}}{\text{ft}^3}\right)(43 \text{ ft})} = \boxed{1.48}$$

$$\left[\begin{array}{c}\text{within 1.3 to 1.5 range;} \\ \text{acceptable}\end{array}\right]$$

9.

$$\gamma = (1+0.20)\left(115 \ \frac{\text{lbf}}{\text{ft}^3}\right) = 138 \text{ lbf/ft}^3$$

The weight of the soil excavated per foot of trench is

$$\left(\frac{18 \text{ in}}{12 \ \frac{\text{in}}{\text{ft}}}\right)(3.5 \text{ ft})\left(138 \ \frac{\text{lbf}}{\text{ft}^3}\right) = 725 \text{ lbf/ft}$$

Half of the spoils will appear on each side of the trench.

$$\frac{725 \ \frac{\text{lbf}}{\text{ft}}}{2} = 362 \text{ lbf/ft per side}$$

(a) Since $\phi > 5°$, failure will be toe slope. Investigate the stability with the trial wedge method. Since the soil is somewhat cohesive ($c > 0$), the failure plane would make an angle of 18° or greater. The horizontal distance BC is

$$\text{BC} = (3.5 \text{ ft})\tan(90° - 18°) = 10.77 \text{ ft}$$

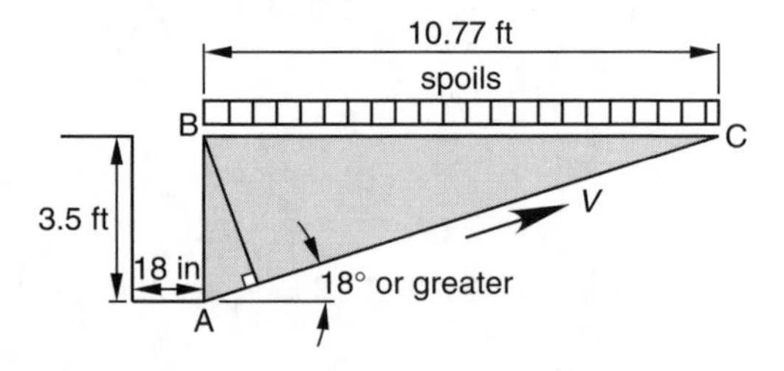

The length of line AC is

$$\sqrt{(3.5 \text{ ft})^2 + (10.77 \text{ ft})^2} = 11.3 \text{ ft}$$

The soil weight in ABC is

$$\left(\tfrac{1}{2}\right)(3.5 \text{ ft})(10.77 \text{ ft})\left(138 \ \frac{\text{lbf}}{\text{ft}^3}\right) = 2600 \text{ lbf}$$

The average pressure occurs at the average depth.

$$p_v = \left(\frac{3.5 \text{ ft}}{2}\right)\left(138 \ \frac{\text{lbf}}{\text{ft}^3}\right) = 241.5 \text{ lbf/ft}^2$$

The surcharge from the spoils is

$$p_q = \frac{362\ \frac{\text{lbf}}{\text{ft}}}{10.77\ \text{ft}} = 33.6\ \text{lbf/ft}^2$$

The normal stress on an 18° failure plane would be

$$\sigma_n = (p_v + p_q)(\cos 18°)$$

The shear strength of the soil is

$$\begin{aligned} S_{us} &= c + \sigma_n \tan\phi \\ &= 200\ \frac{\text{lbf}}{\text{ft}^2} + \left(241.5\ \frac{\text{lbf}}{\text{ft}^2} + 33.6\ \frac{\text{lbf}}{\text{ft}^2}\right) \\ &\quad \times (\cos 18°)(\tan 18°) \\ &= 285\ \text{lbf/ft}^2 \end{aligned}$$

The ultimate resisting shear force is

$$V = \left(285\ \frac{\text{lbf}}{\text{ft}^2}\right)(11.3\ \text{ft}) = 3221\ \text{lbf/ft}$$

The total vertical weight is

$$W_{\text{wedge}} + W_{\text{spoils}} = 2600\ \frac{\text{lbf}}{\text{ft}} + 362\ \frac{\text{lbf}}{\text{ft}} = 2962\ \text{lbf/ft}$$

Resolve the soil weight into a force parallel to the assumed shear plane.

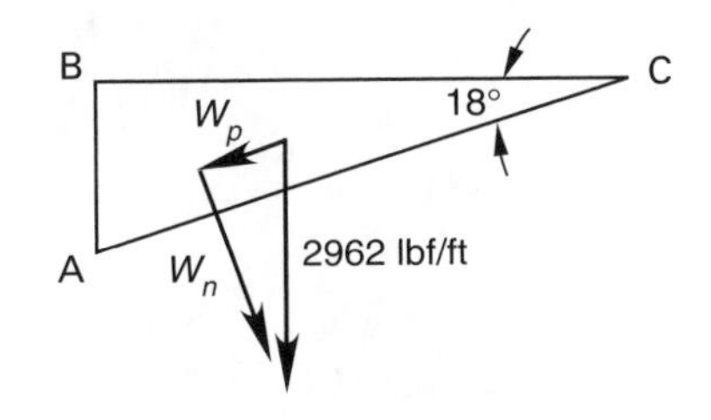

$$W_p = \left(2962\ \frac{\text{lbf}}{\text{ft}}\right)(\sin 18°) = 915\ \text{lbf/ft}$$

Since 915 lbf/ft < 3221 lbf/ft, **the soil will not slide.**

The factor of safety against sliding is

$$F_{\text{sliding}} = \frac{3221\ \frac{\text{lbf}}{\text{ft}}}{915\ \frac{\text{lbf}}{\text{ft}}} = 3.52$$

The failure plane angle is unknown and may not be 18°. This process can be duplicated for other angles.

(b) **2 ft**

(c) **yes** [see table]

(d) **For a flexible pipe of this small size and buried as deep as this, there is no special bedding. Backfill with granular soil to eliminate settling of fill.**

(e)
- Place barricades along open trench.
- Join pipes above ground, not in trench.
- Have someone outside of trench spotting (less than 5 ft deep does not normally get shoring).
- Place exit ladders every 50 ft.

ϕ	BC (ft)	AC (ft)	soil mass $\left(\frac{\text{lbf}}{\text{ft}^2}\right)$	p_q $\left(\frac{\text{lbf}}{\text{ft}^2}\right)$	S_{us} $\left(\frac{\text{lbf}}{\text{ft}^2}\right)$	V $\left(\frac{\text{lbf}}{\text{ft}}\right)$	W_p $\left(\frac{\text{lbf}}{\text{ft}}\right)$	F	failure?
18°	10.77	11.3	2600	33.6	285	3221	915	3.52	no
30°	6.06	7.0	1463	59.7	285	1995	913	2.19	no
40°	4.17	5.44	1007	86.8	282	1534	880	1.74	no
60°	2.02	4.04	488	179.2	269	1087	736	1.48	no
80°	0.62	3.55	150	584.0	247	877	504	1.74	no

Geotechnical

41 Determinate Statics

PRACTICE PROBLEMS

1. Two towers are located on level ground 100 ft (30 m) apart. They support a transmission line with a mass of 2 lbm/ft (3 kg/m). The midpoint sag is 10 ft (3 m).

(a) What is the midpoint tension?

(A) 125 lbf (0.55 kN)
(B) 250 lbf (1.1 kN)
(C) 375 lbf (1.6 kN)
(D) 500 lbf (2.2 kN)

(b) What is the maximum tension in the transmission line?

(A) 170 lbf (0.70 kN)
(B) 210 lbf (0.86 kN)
(C) 270 lbf (1.2 kN)
(D) 330 lbf (1.4 kN)

(c) If the maximum tension is 500 lbf (2200 N), what is the sag in the cable?

(A) 1.3 ft (0.4 m)
(B) 3.2 ft (1.0 m)
(C) 4.0 ft (1.2 m)
(D) 5.1 ft (1.5 m)

2. Two legs of a tripod are mounted on a vertical wall. Both legs are horizontal. The apex is 12 distance units from the wall. The right leg is 13.4 units long. The wall mounting points are 10 units apart. A third leg is mounted on the wall 6 units to the left of the right upper leg and and 9 units below the two top legs. A vertical downward load of 200 is supported at the apex. What is the reaction at the lowest mounting point?

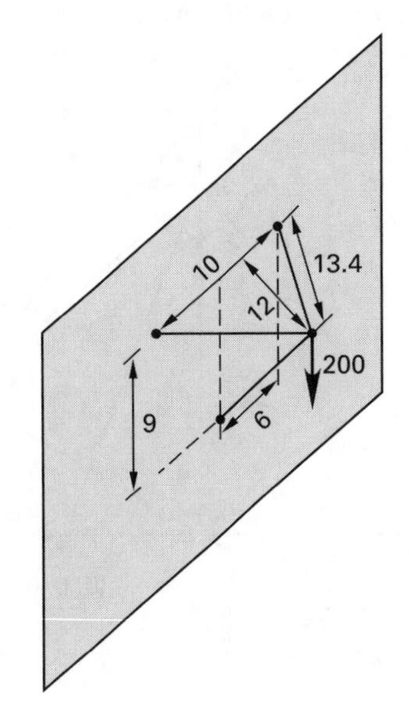

(A) 120
(B) 170
(C) 250
(D) 330

3. The ideal truss shown is supported by a pinned connection at point D and a roller connection at point C. Loads are applied at points A and F. What is the force in member DE?

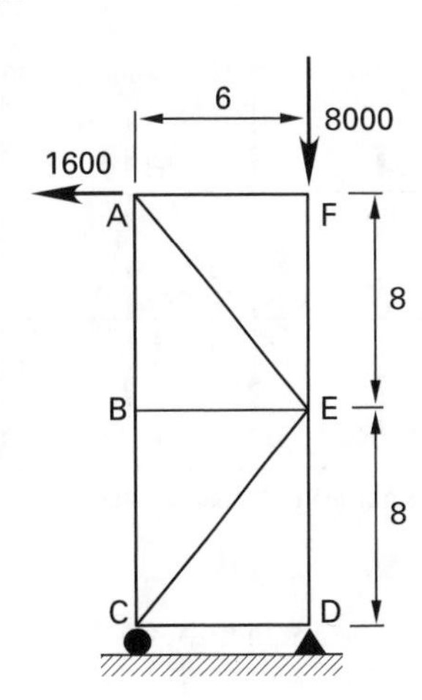

(A) 1200
(B) 2700
(C) 3300
(D) 3700

4. A pin-connected tripod is loaded at the apex by a horizontal force of 1200 as shown. What is the magnitude of the force in member AD?

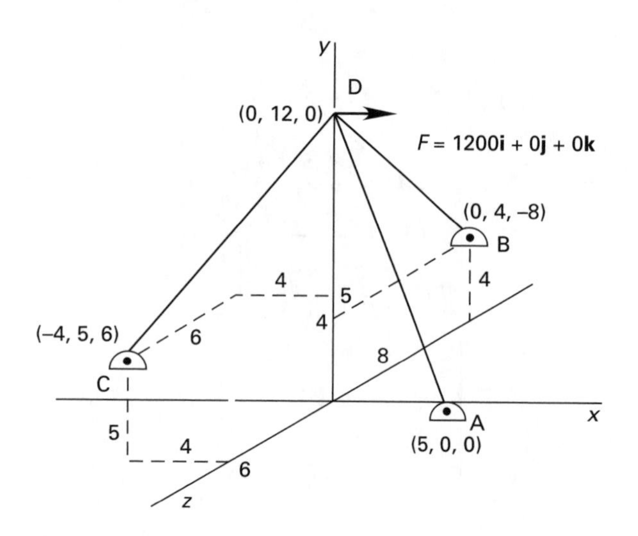

(A) 1100
(B) 1300
(C) 1800
(D) 2500

5. A truss is loaded by forces of 4000 at each upper connection point and forces of 60,000 at each lower connection point as shown.

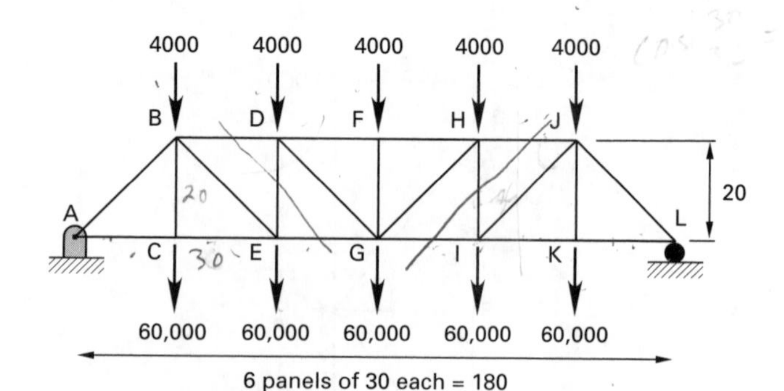

(a) What is the force in member DE?
(A) 36,000
(B) 45,000
(C) 60,000
(D) 160,000

(b) What is the force in member HJ?
(A) 24,000
(B) 60,000
(C) 160,000
(D) 380,000

6. The rigid rod AO is supported by guy wires BO and CO, as shown. (Points A, B, and C are all in the same vertical plane. Points A, O, and C are all in the same horizontal plane.) Vertical and horizontal forces are 12,000 and 6000, respectively, as carried at the end of the rod. What are the x-, y-, and z-components of the reactions at point C?

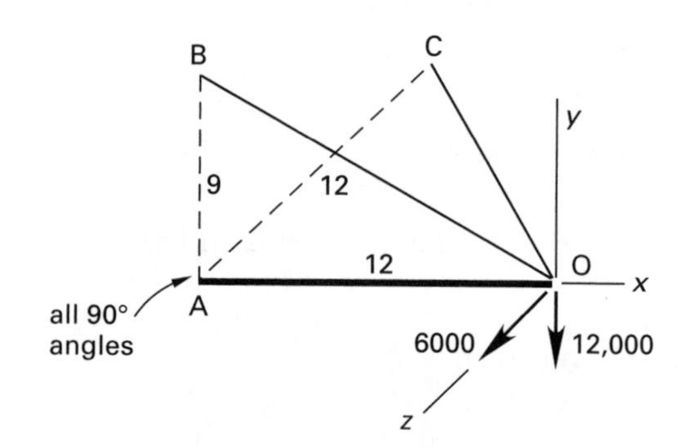

(A) $(C_x, C_y, C_z) = (0, 6000, 0)$
(B) $(C_x, C_y, C_z) = (6000, 0, 6000)$
(C) $(C_x, C_y, C_z) = (6000, 0, 4200)$
(D) $(C_x, C_y, C_z) = (4200, 0, 8500)$

7. When the temperature is 70°F (21.11°C), sections of steel railroad rail are welded end to end to form a continuous, horizontal track exactly 1 mi long (1.6 km). Both ends of the track are constrained by preexisting installed sections of rail. Before the 1 mi section of track can be nailed to the ties, however, the sun warms it to a uniform temperature of 99.14°F (37.30°C). Laborers watch in amazement as the rail pops up in the middle and takes on a parabolic shape. The laborers prop the rail up (so that it does not buckle over) while they take souvenir pictures. How high off the ground is the midpoint of the hot rail?

(A) 0.8 ft (2.4 m)
(B) 2.1 ft (3.6 m)
(C) 17 ft (5.1 m)
(D) 45 ft (14 m)

8. (a) Find the force in member FC.

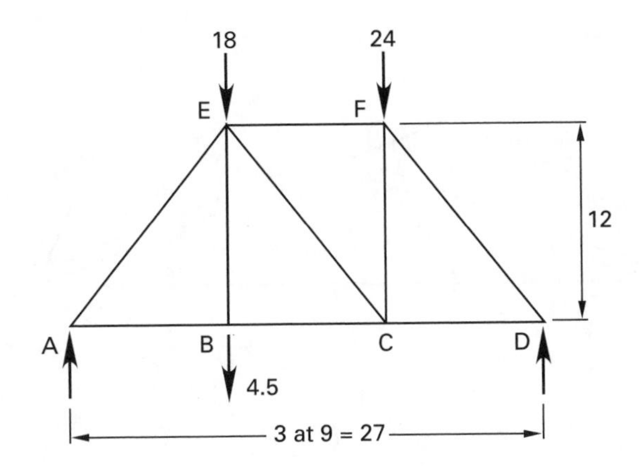

(A) 0.5
(B) 17
(C) 23
(D) 29

(b) What is the force in member CE?

(A) 0.50
(B) 0.63
(C) 11
(D) 17

SOLUTIONS

1. *Customary U.S. Solution*

Use Eq. 41.67 to relate the midpoint sag S to the constant c.

$$S = c\left(\cosh\left(\frac{a}{c}\right) - 1\right)$$

$$10 \text{ ft} = c\left(\cosh\left(\frac{50 \text{ ft}}{c}\right) - 1\right)$$

Solve by trial and error.

$$c = 126.6 \text{ ft}$$

(a) Use Eq. 41.69 to find the midpoint tension.

$$\begin{aligned} H = wc &= m\left(\frac{g}{g_c}\right)c \\ &= \left(2 \ \frac{\text{lbm}}{\text{ft}}\right)\left(\frac{32.2 \ \frac{\text{ft}}{\text{sec}^2}}{32.2 \ \frac{\text{ft-lbm}}{\text{lbf-sec}^2}}\right)(126.6 \text{ ft}) \\ &= \boxed{253.2 \text{ lbf}} \end{aligned}$$

The answer is (B).

(b) Use Eq. 41.71 to find the maximum tension.

$$\begin{aligned} T = wy = w(c+S) &= m\left(\frac{g}{g_c}\right)(c+S) \\ &= \left(2 \ \frac{\text{lbm}}{\text{ft}}\right)\left(\frac{32.2 \ \frac{\text{ft}}{\text{sec}^2}}{32.2 \ \frac{\text{ft-lbm}}{\text{lbf-sec}^2}}\right)(126.6 \text{ ft} + 10 \text{ ft}) \\ &= \boxed{273.2 \text{ lbf}} \end{aligned}$$

The answer is (C).

(c) From $T = wy$,

$$y = \frac{T}{w} = \frac{500 \text{ lbf}}{2 \ \frac{\text{lbf}}{\text{ft}}} = 250 \text{ ft} \quad [\text{at right support}]$$

$$250 \text{ ft} = c\left(\cosh\left(\frac{50 \text{ ft}}{c}\right)\right)$$

By trial and error, $c = 245$ ft.

Substitute into Eq. 41.67.

$$\begin{aligned} S &= c\left(\cosh\left(\frac{a}{c}\right) - 1\right) \\ &= (245 \text{ ft})\left(\cosh\left(\frac{50 \text{ ft}}{245 \text{ ft}}\right) - 1\right) \\ &= \boxed{5.12 \text{ ft}} \end{aligned}$$

The answer is (D).

SI Solution

Use Eq. 41.67 to relate the midpoint sag S to the constant c.

$$S = c\left(\cosh\left(\frac{a}{c}\right) - 1\right)$$

$$3\text{ m} = c\left(\cosh\left(\frac{15\text{ m}}{c}\right) - 1\right)$$

Solve by trial and error.

$$c = 38.0\text{ m}$$

(a) Use Eq. 41.69 to find the midpoint tension.

$$H = wc = mgc = \left(3\ \frac{\text{kg}}{\text{m}}\right)\left(9.81\ \frac{\text{m}}{\text{s}^2}\right)(38.0\text{ m})$$

$$= \boxed{1118.3\text{ N}}$$

The answer is (B).

(b) Use Eq. 41.71 to find the maximum tension.

$$T = wy = w(c+S) = mg(c+S)$$

$$= \left(3\ \frac{\text{kg}}{\text{m}}\right)\left(9.81\ \frac{\text{m}}{\text{s}^2}\right)(38.0\text{ m} + 3.0\text{ m})$$

$$= \boxed{1206.6\text{ N}}$$

The answer is (C).

(c) From $T = wy$,

$$y = \frac{T}{w} = \frac{2200\text{ N}}{\left(3\ \frac{\text{kg}}{\text{m}}\right)\left(9.81\ \frac{\text{m}}{\text{s}^2}\right)}$$

$$= 74.75\text{ m}\quad\text{[at right support]}$$

$$74.75\text{ m} = c\cosh\left(\frac{15\text{ m}}{c}\right)$$

By trial and error, $c = 73.2$ m.

Substitute into Eq. 41.67.

$$S = c\left(\cosh\left(\frac{a}{c}\right) - 1\right)$$

$$= (73.72\text{ m})\left(\cosh\left(\frac{15\text{ m}}{73.2\text{ m}}\right) - 1\right)$$

$$= \boxed{1.54\text{ m}}$$

The answer is (D).

2. *step 1:* Draw the tripod with the origin at the apex.

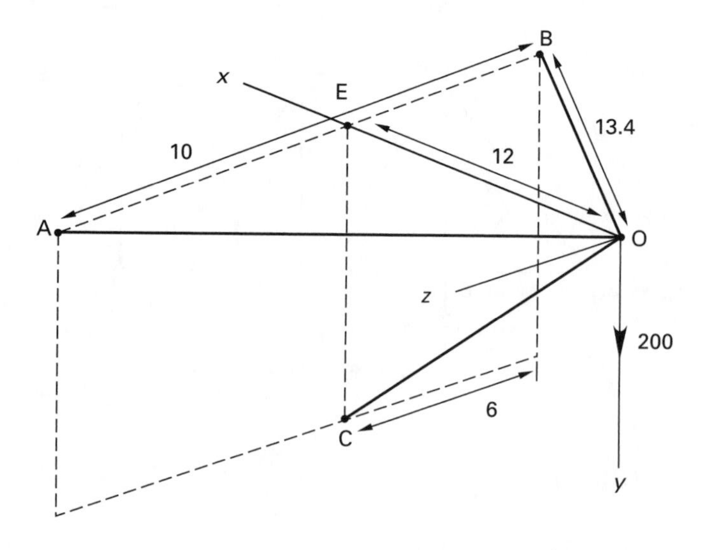

step 2: By inspection, the force components are $F_x = 0$, $F_y = 200$, and $F_z = 0$.

step 3: First, from triangle EBO, length BE is

$$\text{BE} = \sqrt{(13.4\text{ units})^2 - (12\text{ units})^2}$$

$$= 5.96\text{units}\quad\text{[use 6 units]}$$

The (x, y, z) coordinates of the three support points are

point A: $(12, 0, 4)$
point B: $(12, 0, -6)$
point C: $(12, 9, 0)$

step 4: Find the lengths of the legs.

$$\text{AO} = \sqrt{x^2 + y^2 + z^2}$$

$$= \sqrt{(12\text{ units})^2 + (0\text{ units})^2 + (4\text{ units})^2}$$

$$= 12.65\text{ units}$$

$$\text{BO} = \sqrt{x^2 + y^2 + z^2}$$

$$= \sqrt{(12\text{ units})^2 + (0\text{ units})^2 + (-6\text{ units})^2}$$

$$= 13.4\text{ units}$$

$$\text{CO} = \sqrt{x^2 + y^2 + z^2}$$

$$= \sqrt{(12\text{ units})^2 + (9\text{ units})^2 + (0\text{ units})^2}$$

$$= 15.0\text{ units}$$

step 5: Use Eqs. 41.75, 41.76, and 41.77 to find the direction cosines for each leg.

For leg AO,

$$\cos\theta_{\text{A},x} = \frac{x_\text{A}}{\text{AO}} = \frac{12\text{ units}}{12.65\text{ units}} = 0.949$$

$$\cos\theta_{\text{A},y} = \frac{y_\text{A}}{\text{AO}} = \frac{0\text{ units}}{12.65\text{ units}} = 0$$

$$\cos\theta_{\text{A},z} = \frac{z_\text{A}}{\text{AO}} = \frac{4\text{ units}}{12.65\text{ units}} = 0.316$$

For leg BO,

$$\cos\theta_{B,x} = \frac{x_B}{BO} = \frac{12 \text{ units}}{13.4 \text{ units}} = 0.896$$

$$\cos\theta_{B,y} = \frac{y_B}{BO} = \frac{0 \text{ units}}{13.4 \text{ units}} = 0$$

$$\cos\theta_{B,z} = \frac{z_B}{BO} = \frac{-6 \text{ units}}{13.4 \text{ units}} = -0.448$$

For leg CO,

$$\cos\theta_{C,x} = \frac{x_C}{CO} = \frac{12 \text{ units}}{15.0 \text{ units}} = 0.80$$

$$\cos\theta_{C,y} = \frac{y_C}{CO} = \frac{9 \text{ units}}{15.0 \text{ units}} = 0.60$$

$$\cos\theta_{C,z} = \frac{z_C}{CO} = \frac{0 \text{ units}}{15.0 \text{ units}} = 0$$

steps 6 and 7: Substitute Eqs. 41.78, 41.79, and 41.80 into equilibrium Eqs. 41.81, 41.82, and 41.83.

$$F_A\cos\theta_{A,x} + F_B\cos\theta_{B,x} + F_C\cos\theta_{C,x} + F_x = 0$$
$$F_A\cos\theta_{A,y} + F_B\cos\theta_{B,y} + F_C\cos\theta_{C,y} + F_y = 0$$
$$F_A\cos\theta_{A,z} + F_B\cos\theta_{B,z} + F_C\cos\theta_{C,z} + F_z = 0$$
$$0.949F_A + 0.896F_B + 0.80F_C + 0 = 0$$
$$0F_A + 0F_B + 0.60F_C + 200 = 0$$
$$0.316F_A - 0.448F_B + 0F_C + 0 = 0$$

Solve the three equations simultaneously.

$$F_A = 168.6 \quad \text{(T)}$$
$$F_B = 118.9 \quad \text{(T)}$$
$$F_C = \boxed{-333.3 \text{ (C)}}$$

The answer is (D).

3. First, find the vertical reaction at point D.

$$\sum M_C = (CD)D_y - (AF)(8000) + (AC)(1600) = 0$$
$$6D_y - (6)(8000) + (16)(1600) = 0$$

Solve for $D_y = 3733.3$.

The free-body diagram of pin D is as follows.

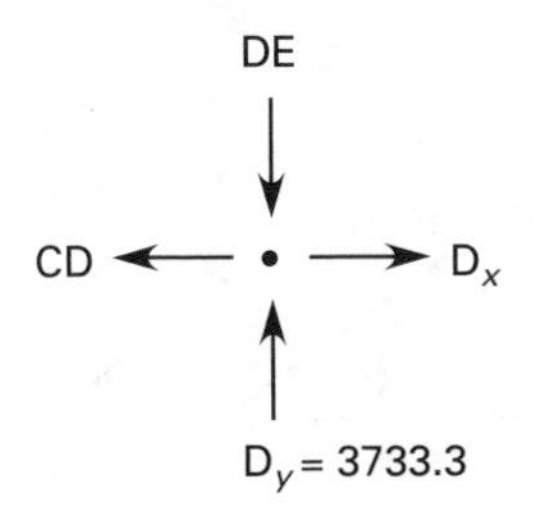

$$\sum F_y = D_y - DE = 0$$

Therefore,

$$DE = D_y = \boxed{3733.3 \quad \text{(C)}}$$

The answer is (D).

4. *step 1:* Move the origin to the apex of the tripod. Call this point O.

step 2: By inspection, the force components are $F_x = 1200$, $F_y = 0$, and $F_z = 0$.

step 3: The (x, y, z) coordinates of the three support points are

point A: $(5, -12, 0)$
point B: $(0, -8, -8)$
point C: $(-4, -7, 6)$

step 4: Find the lengths of the legs.

$$AO = \sqrt{x^2 + y^2 + z^2} = \sqrt{(5)^2 + (-12)^2 + (0)^2} = 13.0$$

$$BO = \sqrt{x^2 + y^2 + z^2} = \sqrt{(0)^2 + (-8)^2 + (-8)^2} = 11.31$$

$$CO = \sqrt{x^2 + y^2 + z^2} = \sqrt{(-4)^2 + (-7)^2 + (6)^2} = 10.05$$

step 5: Use Eqs. 41.75, 41.76, and 41.77 to find the direction cosines for each leg.

For leg AO,

$$\cos\theta_{A,x} = \frac{x_A}{AO} = \frac{5}{13.0} = 0.385$$

$$\cos\theta_{A,y} = \frac{y_A}{AO} = \frac{-12}{13.0} = -0.923$$

$$\cos\theta_{A,z} = \frac{z_A}{AO} = \frac{0}{13.0} = 0$$

For leg BO,

$$\cos\theta_{B,x} = \frac{x_B}{BO} = \frac{0}{11.31} = 0$$

$$\cos\theta_{B,y} = \frac{y_B}{BO} = \frac{-8}{11.31} = -0.707$$

$$\cos\theta_{B,z} = \frac{z_B}{BO} = \frac{-8}{11.31} = -0.707$$

For leg CO,

$$\cos\theta_{C,x} = \frac{x_C}{CO} = \frac{-4}{10.05} = -0.398$$

$$\cos\theta_{C,y} = \frac{y_C}{CO} = \frac{-7}{10.05} = -0.697$$

$$\cos\theta_{C,z} = \frac{z_C}{CO} = \frac{6}{10.05} = 0.597$$

steps 6 and 7: Substitute Eqs. 41.78, 41.79, and 41.80 into equilibrium Eqs. 41.81, 41.82, and 41.83.

$$F_A\cos\theta_{A,x} + F_B\cos\theta_{B,x} + F_C\cos\theta_{C,x} + F_x = 0$$
$$F_A\cos\theta_{A,y} + F_B\cos\theta_{B,y} + F_C\cos\theta_{C,y} + F_y = 0$$
$$F_A\cos\theta_{A,z} + F_B\cos\theta_{B,z} + F_C\cos\theta_{C,z} + F_z = 0$$
$$0.385F_A + 0F_B - 0.398F_C + 1200 = 0$$
$$-0.923F_A - 0.707F_B - 0.697F_C + 0 = 0$$
$$0F_A - 0.707F_B + 0.597F_C + 0 = 0$$

Solve the three equations simultaneously.

$$\boxed{F_A = -1793 \quad (C)}$$
$$F_B = 1080 \quad (T)$$
$$F_C = 1279 \quad (T)$$

If needed, the resultant force can be calculated as

$$F = \sqrt{F_A^2 + F_B^2 + F_C^2} = \sqrt{(-1793)^2 + (1080)^2 + (1279)^2} = 2453$$

The answer is (C).

5. First, find the vertical reactions.

$$\sum F_y = A_y + L_y - (5)(4000) - (5)(60{,}000) = 0$$

By symmetry, $A_y = L_y$.

$$2A_y = (5)(4000) + (5)(60{,}000)$$
$$A_y = 160{,}000$$
$$L_y = 160{,}000$$

(a) For DE, make a cut in members BD, DE, and EG.

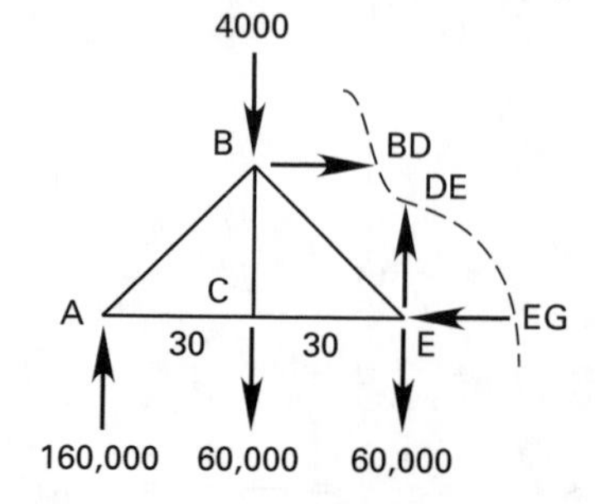

$$\sum F_y = 160{,}000 - 60{,}000 - 60{,}000 + DE - 4000 = 0$$

$$DE = \boxed{-36{,}000 \quad (C)}$$

The answer is (A).

(b) For HJ, make a cut in members HJ, HI, and GI.

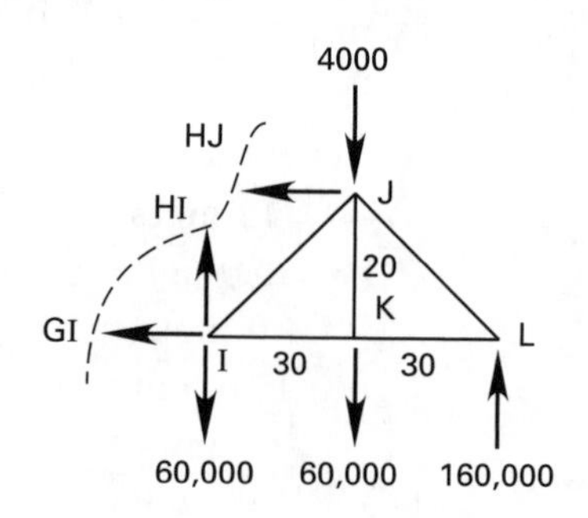

$$\sum M_I = (160{,}000)(60) - (60{,}000)(30) - (4000)(30) + HJ(20) = 0$$

$$HJ = \boxed{-384{,}000 \quad (C)}$$

The answer is (D).

6. First, consider a free-body diagram at point O in the x-y plane.

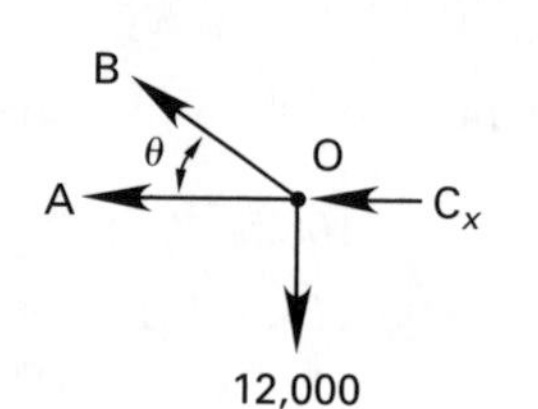

θ is obtained from triangle AOB.

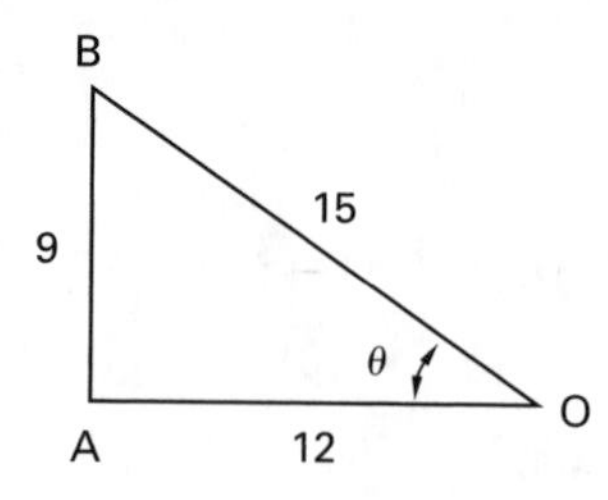

C_x is obtained from triangle AOC.

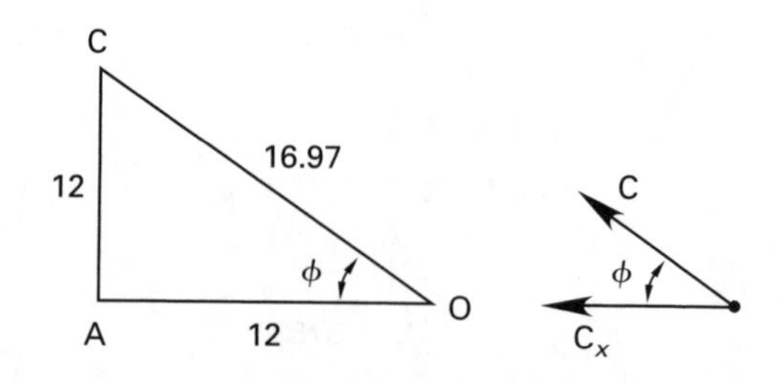

Equilibrium in the x-y plane at point O is $\sum F_y = 0$.

$$\mathrm{B}\sin\theta = 12{,}000$$

$$\mathrm{B} = \frac{12{,}000}{\sin\theta} = \frac{12{,}000}{\frac{9}{15}} = 20{,}000$$

$$\mathrm{B}_x = \mathrm{B}\cos\theta = (20{,}000)\left(\frac{12}{15}\right) = 16{,}000$$

$$\mathrm{B}_y = \mathrm{B}\sin\theta = (20{,}000)\left(\frac{9}{15}\right) = 12{,}000$$

Since BO is in the x-y plane, $\mathrm{B}_z = 0$.

Next, consider a free-body diagram at point O in the x-z plane.

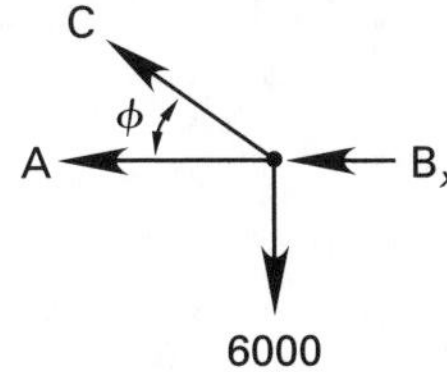

$\sum F_x = 0$:

$$\mathrm{A} + \mathrm{B}_x + \mathrm{C}\cos\phi = 0$$

$$\begin{aligned}\mathrm{A} + \mathrm{C}\cos\phi &= -\mathrm{B}_x = -\mathrm{B}\cos\theta \\ &= (-20{,}000)\left(\frac{12}{15}\right) \\ &= -16{,}000\end{aligned}$$

$\sum F_y = 0$:

$$\begin{aligned}\mathrm{C}\sin\phi &= 6000 \\ \mathrm{C} &= \frac{6000}{\sin\phi} \\ &= \frac{6000}{\frac{12}{16.97}} = 8485\end{aligned}$$

Thus,

$$\begin{aligned}\mathrm{A} &= -\mathrm{C}\cos\phi - 16{,}000 \\ &= (-8485)\left(\frac{12}{16.97}\right) - 16{,}000 \\ &= -22{,}000\end{aligned}$$

Since AO is on the x-axis,

$$\mathrm{A}_x = -22{,}000$$

$$\mathrm{A}_y = 0$$

$$\mathrm{A}_z = 0$$

Solve for C reactions.

$$\mathrm{C}_x = \mathrm{C}\cos\phi = (8485)\left(\frac{12}{16.97}\right) = 6000$$

$$\mathrm{C}_z = \mathrm{C}\sin\phi = (8485)\left(\frac{12}{16.97}\right) = 6000$$

Since CO is in the x-z plane, $\mathrm{C}_y = 0$.

The answer is (B).

7. *Customary U.S. Solution*

First, find the amount of thermal expansion. From Table 44.2, the coefficient of thermal expansion for steel is 6.5×10^{-6} 1/°F. Use Eq. 44.9.

$$\begin{aligned}\delta &= \alpha L_o(T_2 - T_1) \\ &= \left(6.5 \times 10^{-6}\ \frac{1}{^\circ\mathrm{F}}\right)(1\ \mathrm{mi})\left(5280\ \frac{\mathrm{ft}}{\mathrm{mi}}\right) \\ &\quad \times (99.14^\circ\mathrm{F} - 70^\circ\mathrm{F}) \\ &= 1.000085\ \mathrm{ft} \approx 1\ \mathrm{ft}\end{aligned}$$

For thermal expansion, assume the distributed load is uniform along the length of the rail. This resembles the case of a cable under its own weight. From the parabolic cable figure (Fig. 41.18), when distance S is small relative to distance a, the problem can be solved by using the parabolic equation, Eq. 41.57.

$$L \approx a\left(1 + \left(\tfrac{2}{3}\right)\left(\frac{S}{a}\right)^2 - \left(\tfrac{2}{5}\right)\left(\frac{S}{a}\right)^4\right)$$

$$\frac{5280\ \mathrm{ft} + 1\ \mathrm{ft}}{2} \approx (2640\ \mathrm{ft})\left(1 + \left(\tfrac{2}{3}\right)\left(\frac{S}{2640\ \mathrm{ft}}\right)^2 - \left(\tfrac{2}{5}\right)\left(\frac{S}{2640\ \mathrm{ft}}\right)^4\right)$$

Using trial and error, $S \approx$ **44.5 ft.**

The answer is (D).

SI Solution

First, find the amount of thermal expansion. From Table 44.2, the coefficient of thermal expansion for steel is 11.7×10^{-6} 1/°C. Use 44.9.

$$\begin{aligned}\delta &= \alpha L_o(T_2 - T_1) \\ &= \left(11.7 \times 10^{-6}\ \frac{1}{^\circ\mathrm{C}}\right)(1.6\ \mathrm{km})\left(1000\ \frac{\mathrm{m}}{\mathrm{km}}\right) \\ &\quad \times (37.30^\circ\mathrm{C} - 21.11^\circ\mathrm{C}) \\ &= 0.30308\ \mathrm{m} \approx 0.30\ \mathrm{m}\end{aligned}$$

For thermal expansion, assume the distributed load is uniform along the length of the rail. This resembles the case of a cable under its own weight. From the parabolic cable figure (Fig. 41.18), when distance S is small relative to distance a, the problem can be solved by using the parabolic equation, Eq. 41.57.

$$L \approx a\left(1 + \left(\tfrac{2}{3}\right)\left(\frac{S}{a}\right)^2 - \left(\tfrac{2}{5}\right)\left(\frac{S}{a}\right)^4\right)$$

$$\frac{1600\ \mathrm{m} + 0.3\ \mathrm{m}}{2} \approx (800\ \mathrm{m})\left(1 + \left(\tfrac{2}{3}\right)\left(\frac{S}{800\ \mathrm{m}}\right)^2 - \left(\tfrac{2}{5}\right)\left(\frac{S}{800\ \mathrm{m}}\right)^4\right)$$

Using trial and error, $S \approx \boxed{13.6 \text{ m.}}$

The answer is (D).

8. (a) First, find the reactions. Take clockwise moments about A as positive.

$$\sum M_A = (27)(-D_y) + (18)(24) + (9)(22.5) = 0$$
$$D_y = 23.5$$
$$A_y = 18 + 24 + 4.5 - 23.5 = 23$$

Either the method of sections (easiest) or a member-by-member analysis can be used.

The general force triangle is

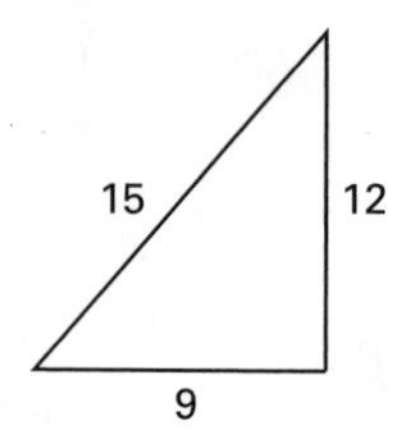

At pin A,

$$AE_y = 23$$
$$AE_x = \left(\frac{9}{12}\right)(23) = 17.25$$
$$AE = \left(\frac{15}{12}\right)(23) = 28.75 \text{ (C)}$$
$$AB = AE_x = 17.25 \text{ (T)}$$

At pin B,

$$BE = 4.5 \text{ (T)}$$
$$BC = AB = 17.25 \text{ (T)}$$

At pin D,

$$DF_y = 23.5$$
$$DF_x = \left(\frac{9}{12}\right)(23.5) = 17.63$$
$$DF = \left(\frac{15}{12}\right)(23.5) = 29.38 \text{ (C)}$$
$$DC = DF_x = 17.63 \text{ (T)}$$

At pin F,

$$FE = DF_x = 17.63 \text{ (C)}$$
$$FC = 24 - DF_y = \boxed{0.5 \text{ (C)}}$$

The answer is (A).

(b) At pin C,

$$CE_y = FC = 0.5$$
$$CE = \left(\frac{15}{12}\right)(0.5) = \boxed{0.625 \text{ (T)}}$$

The answer is (B).

42 Properties of Areas

PRACTICE PROBLEMS

1. Locate the centroid of the area.

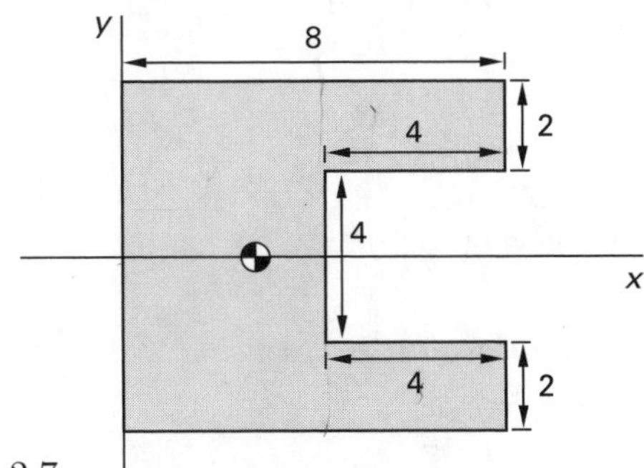

(A) 2.7
(B) 2.9
(C) 3.1
(D) 3.3

2. Replace the distributed load with three concentrated loads, and indicate the points of application.

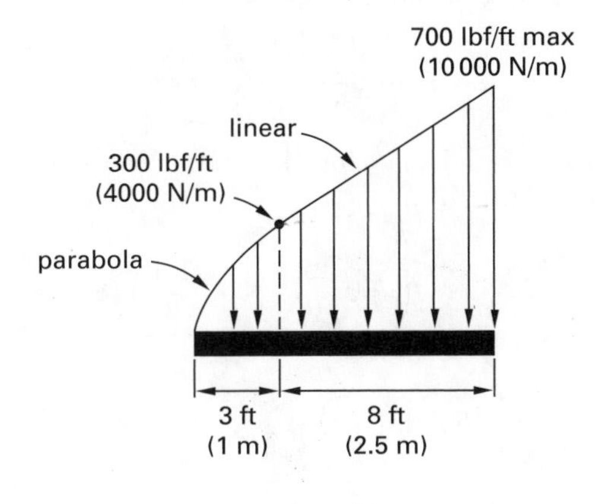

3. Find the centroidal moment of inertia about an axis parallel to the x-axis.

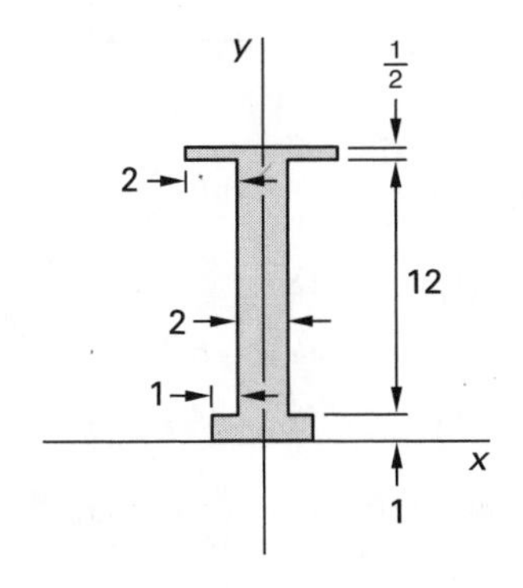

(A) 160 units^4
(B) 290 units^4
(C) 570 units^4
(D) 740 units^4

Structural

SOLUTIONS

1. The area is divided into three basic shapes.

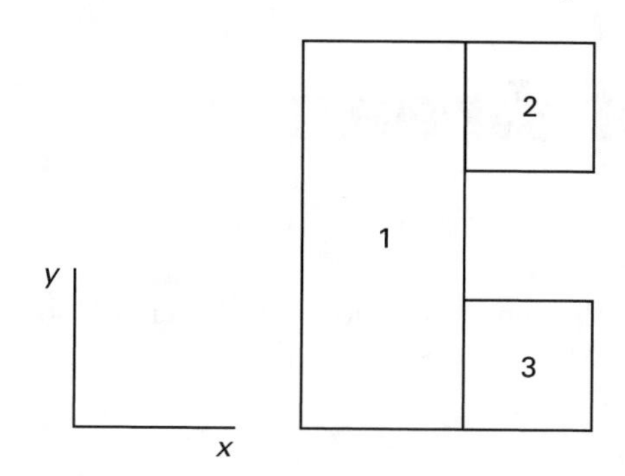

First, calculate the areas of the basic shapes.

$$A_1 = (4)(8) = 32 \text{ units}^2$$
$$A_2 = (4)(2) = 8 \text{ units}^2$$
$$A_3 = (4)(2) = 8 \text{ units}^2$$

Next, find the x-components of the centroids of the basic shapes.

$$x_{c1} = 2 \text{ units}$$
$$x_{c2} = 6 \text{ units}$$
$$x_{c3} = 6 \text{ units}$$

Finally, use Eq. 42.5.

$$x_c = \frac{\sum A_i x_{ci}}{\sum A_i} = \frac{(32)(2) + (8)(6) + (8)(6)}{32 + 8 + 8} = \boxed{3.33 \text{ units}}$$

The answer is (D).

2. *Customary U.S. Solution*

The parabolic shape is

$$f(x) = \left(300 \, \frac{\text{lbf}}{\text{ft}}\right)\sqrt{\frac{x}{3}}$$

First, use Eq. 42.3 to find the concentrated load given by the area.

$$A = \int f(x)dx = \int_0^{3 \text{ ft}} 300\sqrt{\frac{x}{3}}dx = \left(\frac{300 \, \frac{\text{lbf}}{\text{ft}}}{\sqrt{3}}\right)\left[\frac{x^{3/2}}{\frac{3}{2}}\right]_0^{3 \text{ ft}}$$
$$= \left(\frac{300 \, \frac{\text{lbf}}{\text{ft}}}{\left(\frac{3}{2}\right)\sqrt{3}}\right)\left(3^{3/2} - 0^{3/2}\right) \text{ ft} = \boxed{600 \text{ lbf}}$$

From Eq. 42.4,

$$dA = f(x)dx = 300\sqrt{\frac{x}{3}}$$

Finally, use Eq. 42.1 to find the location x_c of the concentrated load from the left end.

$$x_c = \frac{\int x dx}{A} = \frac{1}{600 \text{ lbf}}\int_0^3 300x\sqrt{\frac{x}{3}}dx$$
$$= \frac{300}{600\sqrt{3}}\int_0^3 x^{3/2}dx = \left(\frac{300}{600\sqrt{3}}\right)\left[\frac{x^{5/2}}{\frac{5}{2}}\right]_0^3$$
$$= \left(\frac{300}{600\sqrt{3}\left(\frac{5}{2}\right)}\right)\left(3^{5/2} - 0^{5/2}\right) = \boxed{1.8 \text{ ft}}$$

Alternative solution for the parabola:

Use App. 42.A.

$$\text{A} = \frac{2bh}{3} = \frac{(2)\left(300 \, \frac{\text{lbf}}{\text{ft}}\right)(3 \text{ ft})}{3} = \boxed{600 \text{ lbf}}$$

The centroid is located at a distance from the left end of

$$\frac{3h}{5} = \frac{(3)(3 \text{ ft})}{5} = \boxed{1.8 \text{ ft}}$$

The concentrated load for the triangular shape is the area from App. 42.A.

$$A = \frac{bh}{2} = \frac{\left(700 \, \frac{\text{lbf}}{\text{ft}} - 300 \, \frac{\text{lbf}}{\text{ft}}\right)(8 \text{ ft})}{2} = \boxed{1600 \text{ lbf}}$$

From App. 42.A, the location of the concentrated load from the right end is

$$\frac{h}{3} = \frac{8 \text{ ft}}{3} = \boxed{2.67 \text{ ft}}$$

The concentrated load for the rectangular shape is the area from App. 42.A.

$$A = bh = \left(300 \, \frac{\text{lbf}}{\text{ft}}\right)(8 \text{ ft}) = \boxed{2400 \text{ lbf}}$$

From App. 42.A, the location of the concentrated load from the right end is

$$\frac{h}{2} = \frac{8 \text{ ft}}{2} = \boxed{4 \text{ ft}}$$

SI Solution

The parabolic shape is

$$f(x) = 4000 \frac{\text{N}}{\text{m}} \sqrt{x}$$

First, use Eq. 42.3 to find the concentrated load given by the area.

$$A = \int f(x)dx = \int_0^1 4000\sqrt{x}\,dx = \left[\frac{4000\ x^{3/2}}{\frac{3}{2}}\right]_0^1$$

$$= \left(\frac{4000 \frac{\text{N}}{\text{m}}}{\frac{3}{2}}\right)\left(1^{3/2} - 0^{3/2}\right) \text{ m} = \boxed{2666.7 \text{ N}}$$

[first concentrated load]

From Eq. 42.4,

$$dA = f(x)dx = 4000 \frac{\text{N}}{\text{m}}\sqrt{x}\,dx$$

Finally, use Eq. 42.1 to find the location, x_c, of the concentrated load from the left end.

$$x_c = \frac{\int x dA}{A} = \frac{1}{2666.7 \text{ N}} \int_0^{1 \text{ m}} \left(4000 \frac{\text{N}}{\text{m}}\right) x\sqrt{x}\,dx$$

$$= \frac{4000}{2666.7} \int_0^{1 \text{ m}} x^{3/2} dx = \left(\frac{4000}{2666.7}\right)\left[\frac{x^{5/2}}{\frac{5}{2}}\right]_0^{1 \text{ m}}$$

$$= \left(\frac{4000 \frac{\text{N}}{\text{m}}}{2666.7}\right)\left(\frac{1^{5/2} - 0^{5/2}}{\frac{5}{2}}\right) = \boxed{0.60 \text{ m}} \quad \text{[location]}$$

Alternative solution for the parabola:

Use App. 42.A.

$$\text{A} = \frac{2bh}{3} = \frac{(2)\left(4000 \frac{\text{N}}{\text{m}}\right)(1 \text{ m})}{3}$$

$$= \boxed{2666.7 \text{ N}}$$

The centroid is located at

$$\frac{3h}{5} = \frac{(3)(1 \text{ m})}{5} = \boxed{0.6 \text{ m}}$$

The concentrated load for the triangular shape is the area from App. 42.A.

$$A = \frac{bh}{2} = \frac{\left(10\,000 \frac{\text{N}}{\text{m}} - 4000 \frac{\text{N}}{\text{m}}\right)(2.5 \text{ m})}{2}$$

$$= \boxed{7500 \text{ N}} \quad \text{[second concentrated load]}$$

From App. 42.A, the location of the concentrated load from the right end is

$$\frac{h}{3} = \frac{2.5 \text{ m}}{3} = \boxed{0.83 \text{ m}}$$

The concentrated load for the rectangular shape is the area from App. 42.A.

$$A = bh = \left(4000 \frac{\text{N}}{\text{m}}\right)(2.5 \text{ m})$$

$$= \boxed{10\,000 \text{ N} \quad \text{[third concentrated load]}}$$

From App. 42.A, the location of the concentrated load from the right end is

$$\frac{h}{2} = \frac{2.5 \text{ m}}{2} = \boxed{1.25 \text{ m}}$$

3. The area is divided into three basic shapes.

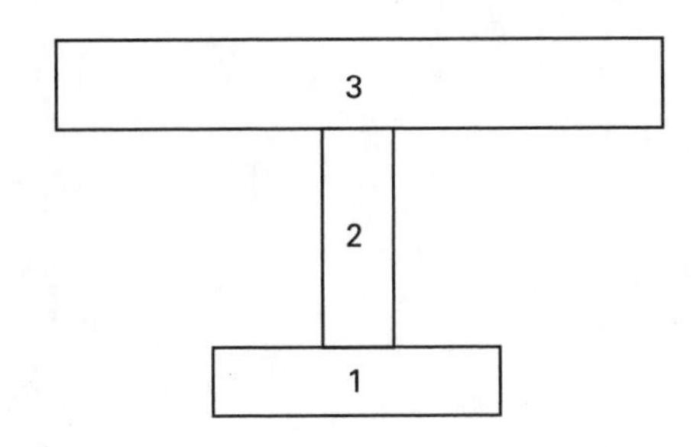

First, calculate the areas of the basic shapes.

$$A_1 = (4)(1) = 4 \text{ units}^2$$
$$A_2 = (2)(12) = 24 \text{ units}^2$$
$$A_3 = (6)(0.5) = 3 \text{ units}^2$$

Next, find the y-components of the centroids of the basic shapes.

$$y_{c1} = 0.5 \text{ units}$$
$$y_{c2} = 7 \text{ units}$$
$$y_{c3} = 13.25 \text{ units}$$

From Eq. 42.6, the centroid of the area is

$$\begin{aligned} y_c &= \frac{\sum A_i y_{ci}}{\sum A_i} = \frac{(4)(0.5) + (24)(7) + (3)(13.25)}{4 + 24 + 3} \\ &= 6.77 \text{ units} \end{aligned}$$

From App. 42.A, the moment of inertia of basic shape 1 about its own centroid is

$$I_{cx1} = \frac{bh^3}{12} = \frac{(4)(1)^3}{12} = 0.33 \text{ units}^4$$

The moment of inertia of basic shape 2 about its own centroid is

$$I_{cx2} = \frac{bh^3}{12} = \frac{(2)(12)^3}{12} = 288 \text{ units}^4$$

The moment of inertia of basic shape 3 about its own centroid is

$$I_{cx3} = \frac{bh^3}{12} = \frac{(6)(0.5)^3}{12} = 0.063 \text{ units}^4$$

From the parallel axis theorem, Eq. 42.20, the moment of inertia of basic shape 1 about the centroidal axis of the section is

$$\begin{aligned} I_{x1} &= I_{cx1} + A_1 d_1^2 = 0.33 + (4)(6.77 - 0.5)^2 \\ &= 157.6 \text{ units}^4 \end{aligned}$$

The moment of inertia of basic shape 2 about the centroidal axis of the section is

$$\begin{aligned} I_{x2} &= I_{cx2} + A_2 d_2^2 = 288 + (24)(7.0 - 6.77)^2 \\ &= 289.3 \text{ units}^4 \end{aligned}$$

The moment of inertia of basic shape 3 about the centroidal axis of the section is

$$\begin{aligned} I_{x3} &= I_{cx3} + A_3 d_3^2 = 0.063 + (3)(13.25 - 6.77)^2 \\ &= 126.0 \text{ units}^4 \end{aligned}$$

The total moment of inertia about the centroidal axis of the section is

$$\begin{aligned} I_x &= I_{x1} + I_{x2} + I_{x3} \\ &= 157.6 \text{ units}^4 + 289.3 \text{ units}^4 + 126.0 \text{ units}^4 \\ &= \boxed{572.9 \text{ units}^4} \end{aligned}$$

The answer is (C).

43 Material Properties and Testing

PRACTICE PROBLEMS

1. The engineering stress and engineering strain for a copper specimen are 20,000 lbf/in^2 (140 MPa) and 0.0200 in/in (0.0200 mm/mm), respectively. Poisson's ratio for the specimen is 0.3.

(a) What is the true stress?

(A) 14,000 lbf/in^2 (98 MPa)
(B) 18,000 lbf/in^2 (130 MPa)
(C) 20,000 lbf/in^2 (140 MPa)
(D) 22,000 lbf/in^2 (160 MPa)

(b) What is the true strain?

(A) 0.0182 in/in (0.0182 mm/mm)
(B) 0.0189 in/in (0.0189 mm/mm)
(C) 0.0194 in/in (0.0194 mm/mm)
(D) 0.0198 in/in (0.0198 mm/mm)

2. A graph of engineering stress-strain is shown. Poisson's ratio for the material is 0.3.

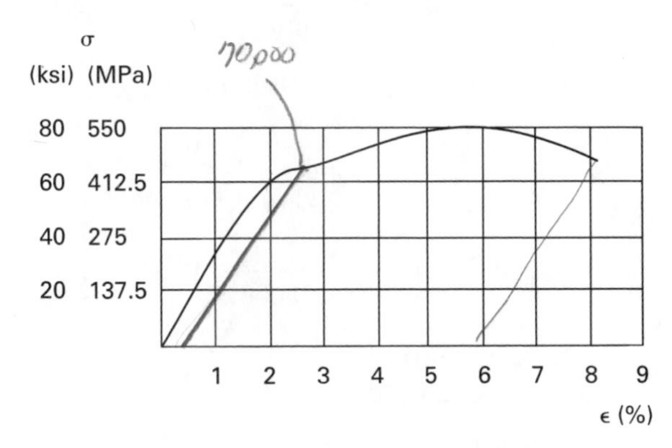

(a) Find the 0.5% yield strength.

(A) 70,000 lbf/in^2 (480 MPa)
(B) 76,000 lbf/in^2 (530 MPa)
(C) 84,000 lbf/in^2 (590 MPa)
(D) 98,000 lbf/in^2 (690 MPa)

(b) Find the elastic modulus.

(A) 2.4×10^6 lbf/in^2 (17 GPa)
(B) 2.7×10^6 lbf/in^2 (19 GPa)
(C) 2.9×10^6 lbf/in^2 (20 GPa)
(D) 3.0×10^6 lbf/in^2 (21 GPa)

(c) Find the ultimate strength.

(A) 72,000 lbf/in^2 (500 MPa)
(B) 76,000 lbf/in^2 (530 MPa)
(C) 80,000 lbf/in^2 (550 MPa)
(D) 84,000 lbf/in^2 (590 MPa)

(d) Find the fracture strength.

(A) 62,000 lbf/in^2 (430 MPa)
(B) 70,000 lbf/in^2 (480 MPa)
(C) 76,000 lbf/in^2 (530 MPa)
(D) 84,000 lbf/in^2 (590 MPa)

(e) Find the percentage of elongation at fracture.

(A) 6%
(B) 8%
(C) 10%
(D) 12%

(f) Find the shear modulus.

(A) 0.9×10^6 lbf/in^2 (6.3 GPa)
(B) 1.1×10^6 lbf/in^2 (7.7 GPa)
(C) 1.2×10^6 lbf/in^2 (7.9 GPa)
(D) 1.5×10^6 lbf/in^2 (11 GPa)

(g) Find the toughness.

(A) 5100 in-lbf/in^3 (35 MJ/m^3)
(B) 5700 in-lbf/in^3 (38 MJ/m^3)
(C) 6300 in-lbf/in^3 (42 MJ/m^3)
(D) 8900 in-lbf/in^3 (60 MJ/m^3)

3. A specimen with an unstressed cross-sectional area of 4 in^2 (25 cm^2) necks down to 3.42 in^2 (22 cm^2) before breaking in a standard tensile test. What is the reduction in area of the material?

(A) 0.094 (0.89)
(B) 0.10 (0.94)
(C) 0.13 (0.10)
(D) 0.15 (0.12)

4. A constant 15,000 lbf/in^2 (100 MPa) tensile stress is applied to a specimen. The stress is known to be less than the material's yield strength. The strain is measured at various times. What is the steady-state creep rate for the material?

time (hr)	strain (in/in)
5	0.018
10	0.022
20	0.026
30	0.031
40	0.035
50	0.040
60	0.046
70	0.058

(A) $0.00037\ \text{hr}^{-1}$
(B) $0.0041\ \text{hr}^{-1}$
(C) $0.00046\ \text{hr}^{-1}$
(D) $0.00049\ \text{hr}^{-1}$

SOLUTIONS

1. *Customary U.S. Solution*

The fractional reduction in diameter is

$$\nu e = (0.3)\left(0.020\ \frac{\text{in}}{\text{in}}\right) = 0.006$$

(a) The true stress is given by Eq. 43.5.

$$\begin{aligned}\sigma &= \frac{F}{A_o(1-\nu e)^2} = \left(\frac{F_o}{A_o}\right)\left(\frac{1}{(1-\nu e)^2}\right)\\ &= \left(20{,}000\ \frac{\text{lbf}}{\text{in}^2}\right)\left(\frac{1}{(1-0.006)^2}\right)\\ &= \boxed{20{,}242\ \text{lbf/in}^2}\end{aligned}$$

The answer is (C).

(b) The true strain is given by Eq. 43.6.

$$\epsilon = \ln(1+e) = \ln(1+0.020) = \boxed{0.0198\ \text{in/in}}$$

The answer is (D).

SI Solution

The fractional reduction in diameter is

$$\nu e = (0.3)\left(0.020\ \frac{\text{mm}}{\text{mm}}\right) = 0.006$$

(a) The true stress is given by Eq. 43.5.

$$\begin{aligned}\sigma &= \frac{F}{A_o(1-\nu e)^2} = \left(\frac{F}{A_o}\right)\left(\frac{1}{(1-\nu e)^2}\right)\\ &= (140\ \text{MPa})\left(\frac{1}{(1-0.006)^2}\right)\\ &= \boxed{141.7\ \text{MPa}}\end{aligned}$$

The answer is (C).

(b) The true strain is

$$\epsilon = \ln(1+e) = \ln(1+0.020) = \boxed{0.0198\ \text{mm/mm}}$$

The answer is (D).

2. *Customary U.S. Solution*

(a) Extend a line from the 0.5% offset strain value parallel to the linear portion of the curve.

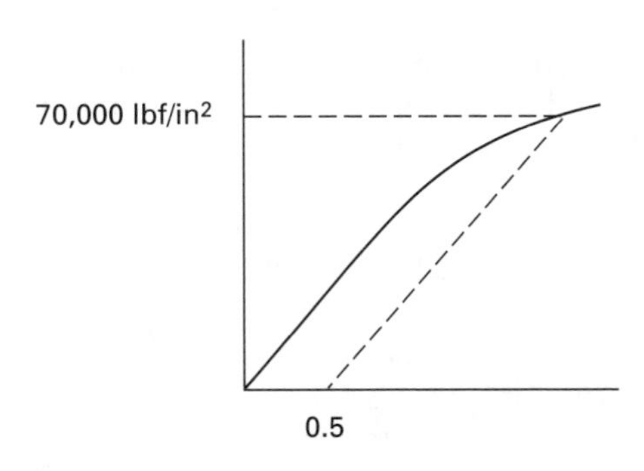

The 0.5% yield strength is $\boxed{70{,}000 \text{ lbf/in}^2.}$

The answer is (A).

(b) At the elastic limit, the stress is 60,000 lbf/in^2 and the percent strain is 2. The elastic modulus is

$$E = \frac{\text{stress}}{\text{strain}} = \frac{60{,}000 \ \frac{\text{lbf}}{\text{in}^2}}{0.02}$$

$$= \boxed{3 \times 10^6 \text{ lbf/in}^2}$$

The answer is (D).

(c) The ultimate strength is the highest point of the curve. This value is $\boxed{80{,}000 \text{ lbf/in}^2.}$

The answer is (C).

(d) The fracture strength is at the end of the curve. This value is $\boxed{70{,}000 \text{ lbf/in}^2.}$

The answer is (B).

(e) The percent elongation at fracture is determined by extending a straight line parallel to the initial strain line from the fracture point. This gives an approximate value of $\boxed{6\%.}$

The answer is (A).

(f) The shear modulus is given by Eq. 43.21.

$$G = \frac{E}{2(1+\nu)} = \frac{3 \times 10^6 \ \frac{\text{lbf}}{\text{in}^2}}{(2)(1+0.3)}$$

$$= \boxed{1.15 \times 10^6 \text{ lbf/in}^2}$$

The answer is (C).

(g) The toughness is the area under the stress-strain curve. Divide the area into squares of 20 ksi × 1%. There are about 25.5 squares covered.

$$(25.5)\left(20{,}000 \ \frac{\text{lbf}}{\text{in}^2}\right)\left(0.01 \ \frac{\text{in}}{\text{in}}\right) = \boxed{5100 \text{ in-lbf/in}^3}$$

The answer is (A).

SI Solution

(a) Extend a line from the 0.5% offset strain value parallel to the linear portion of the curve.

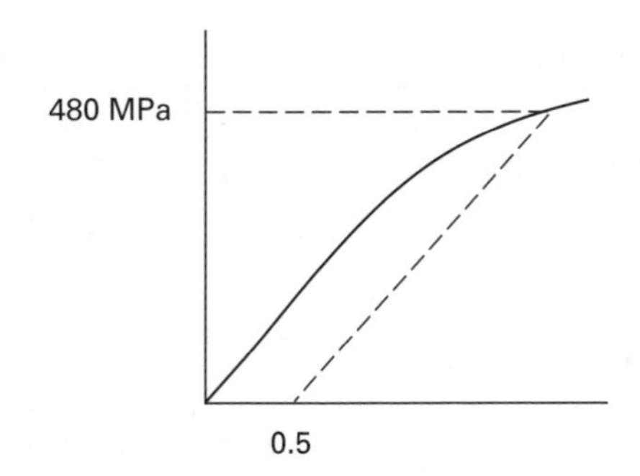

The 0.5% yield strength is $\boxed{480 \text{ MPa.}}$

The answer is (A).

(b) At the elastic limit, the stress is 410 MPa and the percent strain is 2. The elastic modulus is

$$E = \frac{\text{stress}}{\text{strain}} = \frac{410 \text{ MPa}}{(0.02)\left(1000 \ \frac{\text{MPa}}{\text{GPa}}\right)}$$

$$= \boxed{20.5 \text{ GPa}}$$

The answer is (D).

(c) The ultimate strength is the highest point of the curve. This value is $\boxed{550 \text{ MPa.}}$

The answer is (C).

(d) The fracture strength is at the end of the curve. This value is $\boxed{480 \text{ MPa.}}$

The answer is (B).

(e) The percent elongation at fracture is determined by extending a straight line parallel to the initial strain line from the fracture point. This gives an approximate value of $\boxed{6\%.}$

The answer is (A).

(f) The shear modulus is given by Eq. 43.21.

$$G = \frac{E}{2(1+\nu)} = \frac{20.5 \text{ GPa}}{(2)(1+0.3)}$$

$$= \boxed{7.88 \text{ GPa}}$$

The answer is (C).

(g) The toughness is the area under the stress-strain curve. Divide the area into squares of 137.5 MPa × 1%. There are about 25.5 squares covered.

$$(25.5)(137.5 \text{ MPa})\left(0.01 \ \frac{\text{m}}{\text{m}}\right) = \boxed{35 \text{ MJ/m}^3}$$

The answer is (A).

3. *Customary U.S. Solution*

The reduction in area is found from Eq. 43.12.

$$\begin{aligned}\text{reduction in area} &= \frac{A_o - A_f}{A_o} \\ &= \frac{4.0 \text{ in}^2 - 3.42 \text{ in}^2}{4.0 \text{ in}^2} = \boxed{0.145}\end{aligned}$$

The answer is (D).

SI Solution

Use the reduction in area as the measure of ductility. Use Eq. 43.12.

$$\begin{aligned}\text{ductility} = \text{reduction in area} &= \frac{A_o - A_f}{A_o} \\ &= \frac{25 \text{ cm}^2 - 22 \text{ cm}^2}{25 \text{ cm}^2} = \boxed{0.12}\end{aligned}$$

The answer is (D).

4. Plot the data and draw a straight line. Disregard the first and last data points, as these represent primary and tertiary creep, respectively.

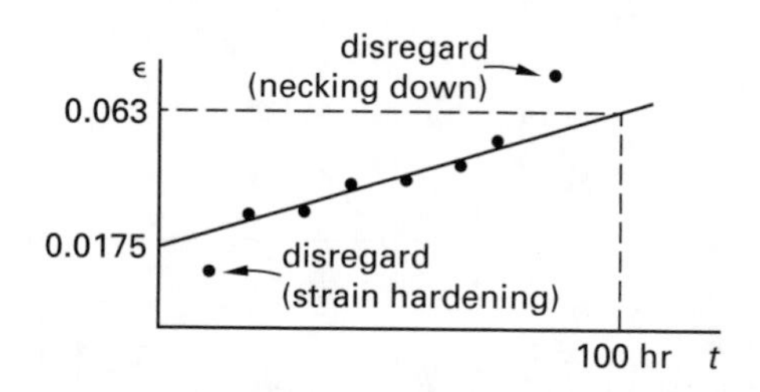

The creep rate is the slope of the line.

$$\begin{aligned}\text{creep rate} &= \frac{\Delta\epsilon}{\Delta t} = \frac{0.063 \ \frac{\text{in}}{\text{in}} - 0.0175 \ \frac{\text{in}}{\text{in}}}{100 \text{ hr}} \\ &= \boxed{0.000455 \ \frac{1}{\text{hr}}}\end{aligned}$$

The answer is (C).

44 Strength of Materials

PRACTICE PROBLEMS

Elastic Deformation

1. A 1 in (25 mm) diameter soft steel rod carries a tensile load of 15,000 lbf (67 kN). The elongation is 0.158 in (4 mm). The modulus of elasticity is 2.9×10^7 lbf/in^2 (200 GPa). What is the total length of the rod?

(A) 239.46 in (5.853 m)
(B) 239.93 in (5.857 m)
(C) 240.03 in (5.861 m)
(D) 240.07 in (5.865 m)

Thermal Deformation

2. A straight steel beam 200 ft (60 m) long is installed when the temperature is 40°F (4°C). It is supported in such a manner as to allow only 0.5 in (12 mm) longitudinal expansion. Lateral support is provided to prevent buckling. If the temperature increases to 110°F (43°C), what will be the compressive stress in the member?

(A) 3900 lbf/in^2 (28 MPa)
(B) 7400 lbf/in^2 (51 MPa)
(C) 9200 lbf/in^2 (67 MPa)
(D) 12,000 lbf/in^2 (88 MPa)

Shear and Moment Diagrams

3. A beam 25 ft long is simply supported at two points: at its left and 5 ft from its right end. A uniform load of 2 kips/ft extends over a 10 ft length starting from the left end. There is a concentrated 10 kip load at the right end. Draw the shear and moment diagrams.

4. A beam 14 ft (3.6 m) long is supported at the left end and 2 ft (0.6 m) from the right end. The beam has a mass of 20 lbm/ft (30 kg/m). A 100 lbf (450 N) load is applied 2 ft (0.6 m) from the left end. An 80 lbf (350 N) load is applied at the right end.

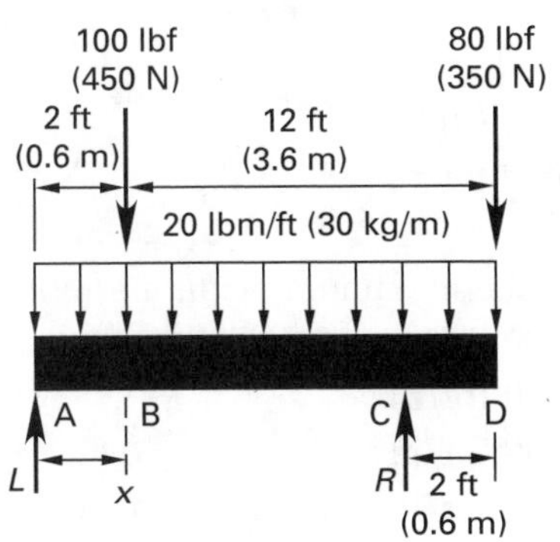

(a) What is the maximum moment?

(A) 150 ft-lbf (200 N·m)
(B) 250 ft-lbf (340 N·m)
(C) 390 ft-lbf (520 N·m)
(D) 830 ft-lbf (1100 N·m)

(b) What is the maximum shear?

(A) 80 lbf (360 N)
(B) 120 lbf (530 N)
(C) 150 lbf (650 N)
(D) 190 lbf (830 N)

Stresses in Beams

5. The allowable stress in a steel beam is 20 ksi. The maximum moment carried by the beam is 1.5×10^5 ft-lbf. What is the required section modulus?

(A) 90 in^3
(B) 110 in^3
(C) 120 in^3
(D) 350 in^3

6. A simply supported beam 14 ft long carries a uniform load of 200 lbf/ft over its entire length. The beam has a width of 3.625 in and a depth of 7.625 in.

(a) What is the maximum bending stress?

(A) 1700 lbf/in^2
(B) 2500 lbf/in^2
(C) 3700 lbf/in^2
(D) 9200 lbf/in^2

(b) What is the maximum shear stress?

(A) 38 lbf/in^2
(B) 76 lbf/in^2
(C) 150 lbf/in^2
(D) 300 lbf/in^2

Beam Deflections

7. A 40 ft long simply supported steel beam with a moment of inertia of I is reinforced along the central 20 ft, leaving the two 10 ft ends unreinforced. The moment of inertia of the reinforced sections is $2I$. A 20,000 lbf load is applied at midspan, 20 ft from each end. What is the deflection?

(A) $1.1 \times 10^6/EI$ ft
(B) $2.7 \times 10^6/EI$ ft
(C) $1.5 \times 10^7/EI$ ft
(D) $8.2 \times 10^7/EI$ ft

8. A 17 ft steel beam carries a uniform load of 500 lbf/ft over its entire length and a 2000 lbf concentrated load 5 ft from the right end, as shown. The beam is simply supported at each end. The centroidal moment of inertia of the cross section is 200 in^4. What is the midspan deflection?

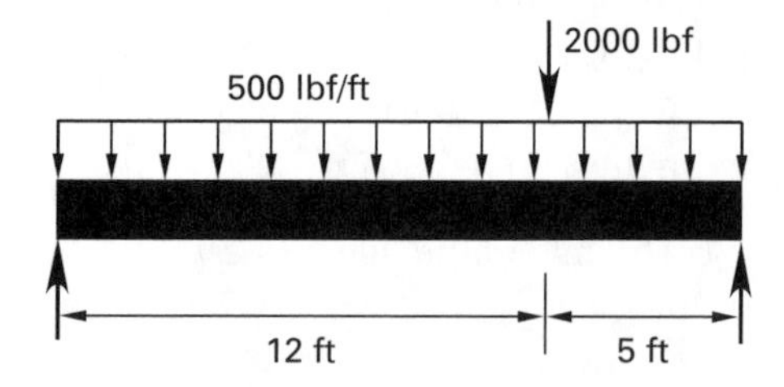

(A) 0.05 in
(B) 0.20 in
(C) 0.40 in
(D) 0.70 in

9. A cantilever beam is 6 ft (1.8 m) in length. The cross section is 6 in wide by 4 in high (150 mm wide by 100 mm high). The modulus of elasticity is 1.5×10^6 psi (10 GPa). The beam is loaded by two concentrated forces: 200 lbf (900 N) located 1 ft (0.3 m) from the free end and 120 lbf (530 N) located 2 ft (0.6 m) from the free end. What is the tip deflection?

(A) 0.29 in (0.0084 m)
(B) 0.31 in (0.0090 m)
(C) 0.47 in (0.014 m)
(D) 0.55 in (0.015 m)

Truss Deflections

10. A steel truss with pinned joints is constructed as shown. Support R_1 is pinned. Support R_2 is a roller. The length:area ratio for each member is 50 in^{-1}. The modulus of elasticity of the steel is 2.9×10^7 lbf/in^2. What is the vertical deflection at the point where the 10,000 lbf load is applied?

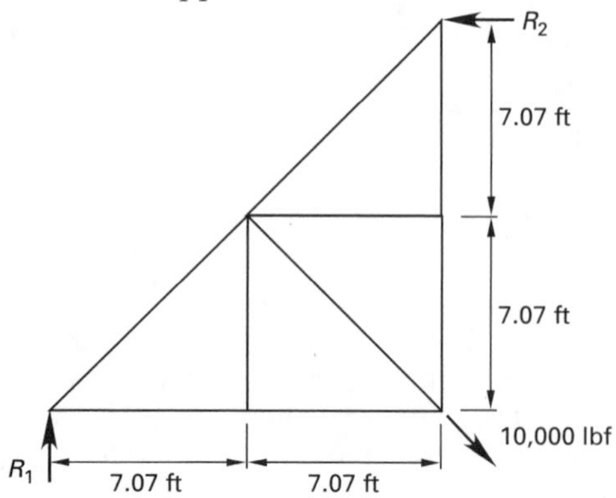

(A) 0.07 in
(B) 0.13 in
(C) 0.24 in
(D) 0.35 in

Composite Structures

11. A bimetallic spring is constructed of an aluminum strip with a $^1/_8$ in by $1^1/_2$ in (3.2 mm by 38 mm) cross section bonded on top of a $^1/_{16}$ in by $1^1/_2$ in (1.6 mm by 38 mm) steel strip. The modulus of elasticity of the aluminum is 10×10^6 lbf/in^2 (70 GPa). The modulus of elasticity of the steel is 30×10^6 lbf/in^2 (200 GPa). What is the equivalent, all-aluminum centroidal area moment of inertia of the cross section?

(A) 2.4×10^{-4} in^4 (190 mm^4)
(B) 5.2×10^{-4} in^4 (390 mm^4)
(C) 1.3×10^{-3} in^4 (550 mm^4)
(D) 2.6×10^{-3} in^4 (1100 mm^4)

12. The reinforced concrete beam illustrated is subjected to a maximum moment of 8125 ft-lbf. The total steel cross-sectional area is 1 in^2, assumed to be concentrated 10 in from the top surface. The modulus of elasticity of the concrete is 2×10^6 lbf/in^2, and the modulus of elasticity of the steel is 2.9×10^7 lbf/in^2. The bond between the steel and the concrete is perfect. The beam is balanced such that the concrete and steel fail simultaneously. The load-carrying contribution of concrete in tension is to be disregarded. Analyze the beam using the transformed area method.

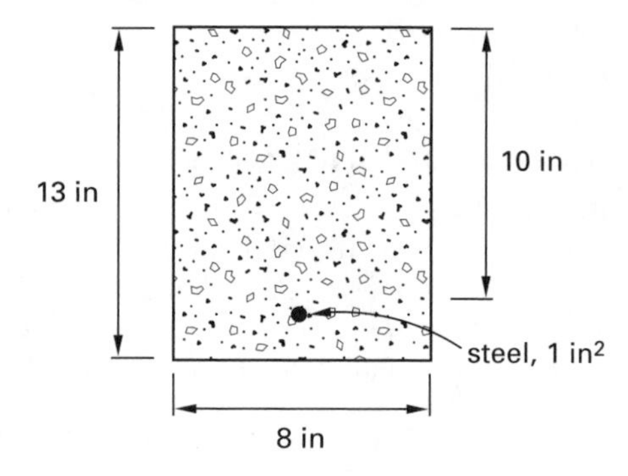

(a) What is the maximum flexural stress in the concrete?

(A) 320 lbf/in^2
(B) 640 lbf/in^2
(C) 910 lbf/in^2
(D) 15,300 lbf/in^2

(b) What is the maximum flexural stress in the steel?

(A) 11,500 lbf/in^2
(B) 16,300 lbf/in^2
(C) 29,500 lbf/in^2
(D) 36,000 lbf/in^2

SOLUTIONS

1. *Customary U.S. Solution*

First, the cross-sectional area of the rod is

$$A = \frac{\pi d^2}{4}$$

From Eq. 44.4, the unstretched length of the rod is

$$L_o = \frac{\delta EA}{F} = \frac{(0.158\text{ in})\left(2.9\times 10^7\ \frac{\text{lbf}}{\text{in}^2}\right)\left(\frac{\pi}{4}\right)(1\text{ in})^2}{15{,}000\text{ lbf}}$$
$$= 239.913\text{ in}$$

The total length of the rod is given by Eq. 44.5.

$$L = L_o + \delta = 239.913\text{ in} + 0.158\text{ in} = \boxed{240.071\text{ in}}$$

The answer is (D).

SI Solution

First, the cross-sectional area of the rod is

$$A = \frac{\pi d^2}{4}$$

From Eq. 44.4, the unstretched length of the rod is

$$L_o = \frac{\delta EA}{F}$$
$$= \frac{(0.004\text{ m})\left(20\times 10^{10}\ \frac{\text{N}}{\text{m}^2}\right)\left(\frac{\pi}{4}\right)(0.025\text{ m})^2}{67\times 10^3\text{ N}}$$
$$= 5.861\text{ m}$$

The total length of the rod is

$$L = L_o + \delta = 5.861\text{ m} + 0.004\text{ m} = \boxed{5.865\text{ m}}$$

The answer is (D).

2. *Customary U.S. Solution*

Use Eq. 44.9 to find the amount of expansion for an unconstrained beam. Use Table 44.2.

$$\Delta L = \alpha L_o(T_2 - T_1)$$
$$= \left(6.5\times 10^{-6}\ \frac{1}{^\circ\text{F}}\right)(200\text{ ft})\left(12\ \frac{\text{in}}{\text{ft}}\right) \times (110^\circ\text{F} - 40^\circ\text{F})$$
$$= 1.092\text{ in}$$

The constrained length is

$$\Delta L_c = 1.092\text{ in} - 0.5\text{ in} = 0.592\text{ in}$$

From Eq. 44.14, the thermal strain is

$$\epsilon_{\text{th}} = \frac{\Delta L_c}{L_o + 0.5\text{ in}} = \frac{0.592\text{ in}}{2400\text{ in} + 0.5\text{ in}}$$
$$= 2.466\times 10^{-4}\text{ in/in}$$

From Eq. 44.15, the compressive thermal stress is

$$\sigma_{\text{th}} = E\epsilon_{\text{th}} = \left(30\times 10^6\ \frac{\text{lbf}}{\text{in}^2}\right)(2.466\times 10^{-4})$$
$$= \boxed{7398\text{ lbf/in}^2}$$

The answer is (B).

SI Solution

Use Eq. 44.9 to find the amount of expansion for an unconstrained beam. Use Table 44.2.

$$\Delta L = \alpha L_o(T_2 - T_1)$$
$$= \left(11.7\times 10^{-6}\ \frac{1}{^\circ\text{C}}\right)(60\text{ m})(43^\circ\text{C} - 4^\circ\text{C})$$
$$= 0.02738\text{ m}$$

The constrained length is

$$\Delta L_c = 0.02738\text{ m} - 0.012\text{ m} = 0.01538\text{ m}$$

From Eq. 44.14, the thermal strain is

$$\epsilon_{\text{th}} = \frac{\Delta L_c}{L_o + 0.012\text{ m}} = \frac{0.01538\text{ m}}{60\text{ m} + 0.012\text{ m}}$$
$$= 0.00256\text{ m/m}$$

From Eq. 44.15, the compressive thermal stress is

$$\sigma_{\text{th}} = E\epsilon_{\text{th}} = (20\times 10^{10}\text{ Pa})\left(0.000256\ \frac{\text{m}}{\text{m}}\right)$$
$$= \boxed{5.12\times 10^7\text{ Pa}\quad(51.2\text{ MPa})}$$

The answer is (B).

3.

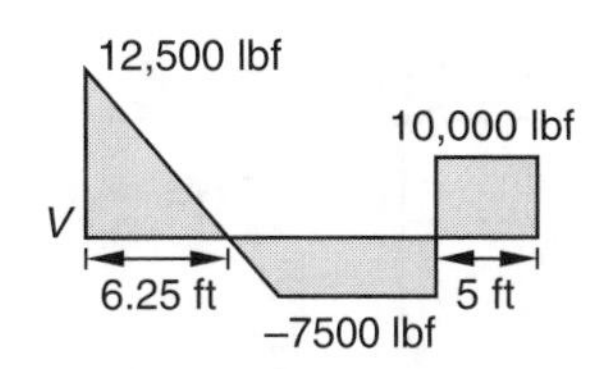

Structural

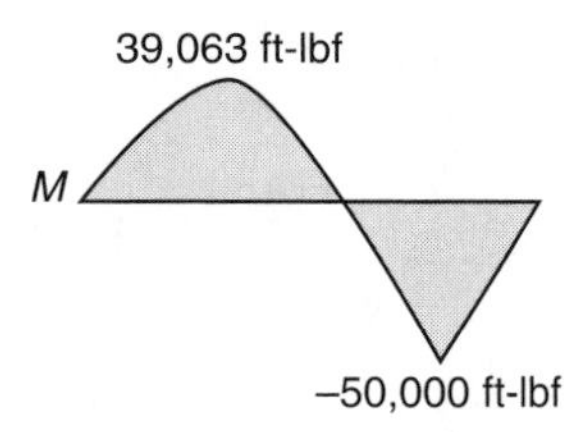

4. *Customary U.S. Solution*

First, determine the reactions. The uniform load can be assumed to be concentrated at the center of the beam.

Sum the moments about A.

$$\begin{aligned}&(100\text{ lbf})(2\text{ ft}) + (80\text{ lbf})(14\text{ ft})\\&\quad + \left(20\ \frac{\text{lbm}}{\text{ft}}\right)\left(\frac{32.2\ \frac{\text{ft}}{\text{sec}^2}}{32.2\ \frac{\text{lbm-ft}}{\text{lbf-sec}^2}}\right)\\&\quad \times(14\text{ ft})(7\text{ ft}) - R(12\text{ ft}) = 0\\&\qquad R = 273.3\text{ lbf}\end{aligned}$$

Sum the forces in the vertical direction.

$$\begin{aligned}L + 273.3\text{ lbf} &= 100\text{ lbf} + 80\text{ lbf}\\&\quad + \left(20\ \frac{\text{lbm}}{\text{ft}}\right)\left(\frac{32.2\ \frac{\text{ft}}{\text{sec}^2}}{32.2\ \frac{\text{lbm-ft}}{\text{lbf-sec}^2}}\right)(14\text{ ft})\\L &= 186.7\text{ lbf}\end{aligned}$$

The shear diagram starts at +186.7 lbf at the left reaction and decreases linearly at a rate of 20 lbf/ft to 146.7 lbf at point B. The concentrated load reduces the shear to 46.7 lbf. The shear then decreases linearly at a rate of 20 lbf/ft to point C. Measuring x from the left, the shear line goes through zero at

$$x = 2\text{ ft} + \frac{46.7\text{ lbf}}{20\ \frac{\text{lbf}}{\text{ft}}} = 4.3\text{ ft}$$

The shear at the right of the beam at point D is 80 lbf and increases linearly at a rate of 20 lbf/ft to 120 lbf at point C. The reaction R at point C decreases the shear to −153.3 lbf. This is sufficient to draw the shear diagram.

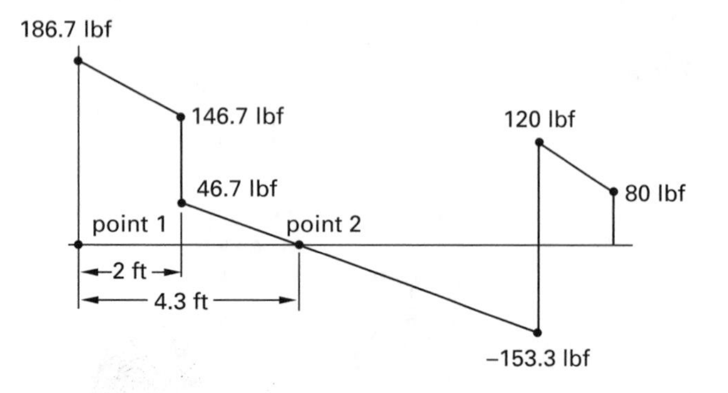

(a) From the shear diagram, the maximum moment occurs when the shear is zero. Call this point 2. The moment at the left reaction is zero. Call this point 1. Use Eq. 44.27.

$$M_2 = M_1 + \int_{x_1}^{x_2} V\,dx$$

The integral is the area under the curve from $x_1 = 0$ to $x_2 = 4.3$ ft.

$$\begin{aligned}M_2 &= 0 + (146.7\text{ lbf})(2\text{ ft})\\&\quad + \left(\tfrac{1}{2}\right)(186.7\text{ lbf} - 146.7\text{ lbf})(2\text{ ft})\\&\quad + \left(\tfrac{1}{2}\right)(46.7\text{ lbf})(4.3\text{ ft} - 2\text{ ft})\\&= \boxed{387.1\text{ ft-lbf}}\end{aligned}$$

The answer is (C).

(b) From the shear diagram, the maximum shear is $\boxed{186.7\text{ lbf.}}$

The answer is (D).

SI Solution

First, determine the reactions. The uniform load can be assumed to be concentrated at the center of the beam.

Sum the moments about A:

$$\begin{aligned}&(450\text{ N})(0.6\text{ m}) + (350\text{ N})(4.2\text{ m})\\&\quad + \left(30\ \frac{\text{kg}}{\text{m}}\right)(4.2\text{ m})\left(9.81\ \frac{\text{N}}{\text{kg}}\right)(2.1\text{ m})\\&\quad - R(3.6\text{ m}) = 0\\&\qquad R = 1204.4\text{ N}\end{aligned}$$

Sum the forces in the vertical direction.

$$\begin{aligned}L + 1204.4\text{ N} &= 450\text{ N} + 350\text{ N}\\&\quad + \left(30\ \frac{\text{kg}}{\text{m}}\right)(4.2\text{ m})\left(9.81\ \frac{\text{N}}{\text{kg}}\right)\\L &= 831.7\text{ N}\end{aligned}$$

The shear diagram starts at +831.7 N at the left reaction and decreases linearly at a rate of 294.3 N/m to 655.1 N at point B. The concentrated load reduces the shear to 205.1 N. The shear then decreases linearly at a rate of 294.3 N/m to point C. Measuring x from the left, the shear line goes through zero at

$$x = 0.6\text{ m} + \frac{205.1\text{ N}}{294.3\ \frac{\text{N}}{\text{m}}} = 1.3\text{ m}$$

The shear at the right of the beam at point D is 350 N and increases linearly at a rate of 294.3 N/m to 526.3 N at point C. The reaction R at point C decreases the shear to −677.8 N. This is sufficient to draw the shear diagram.

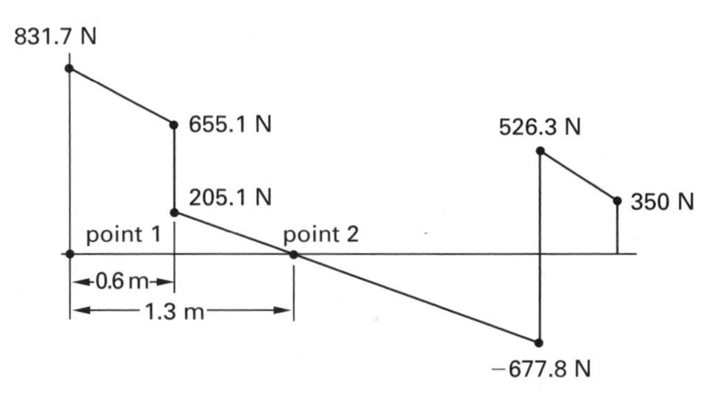

(a) From the shear diagram, the maximum moment occurs when the shear = 0. Call this point 2. The moment at the left reaction = 0. Call this point 1.

Use Eq. 44.27.

$$M_2 = M_1 + \int_{x_1}^{x_2} V\,dx$$

The integral is the area under the curve from $x_1 = 0$ to $x_2 = 1.3$ m.

$$\begin{aligned} M_2 &= 0 + (655.1\text{ N})(0.6\text{ m}) \\ &\quad + \left(\tfrac{1}{2}\right)(831.7\text{ N} - 655.1\text{ N})\,(0.6\text{ m}) \\ &\quad + \left(\tfrac{1}{2}\right)(205.1\text{ N})(1.3\text{ m} - 0.6\text{ m}) \\ &= \boxed{517.8\text{ N·m}} \end{aligned}$$

The answer is (C).

(b) From the shear diagram, the maximum shear is $\boxed{831.7\text{ N.}}$

The answer is (D).

5. Use Eq. 44.38.

$$S = \frac{M}{\sigma} = \frac{(1.5\times 10^5\text{ ft-lbf})\left(12\ \frac{\text{in}}{\text{ft}}\right)}{20{,}000\ \frac{\text{lbf}}{\text{in}^2}} = \boxed{90\text{ in}^3}$$

The answer is (A).

6. (a)

$$= \frac{bh^3}{12} = \frac{(3.625\text{ in})(7.625\text{ in})^3}{12} = 133.92\text{ in}^4$$

$$M_{\text{max}} = (7\text{ ft})\left(\frac{7\text{ ft}}{2}\right)\left(200\ \frac{\text{lbf}}{\text{ft}}\right)\left(12\ \frac{\text{in}}{\text{ft}}\right) = 58{,}800\text{ in-lbf}$$

$$\sigma_{b,\text{max}} = \frac{Mc}{I} = \frac{(58{,}800\text{in-lbf})\left(\frac{7.625\text{ in}}{2}\right)}{133.92\text{ in}^4} = \boxed{1674\text{ lbf/in}^2}$$

The answer is (A).

(b)

$$V_{\text{max}} = \frac{(14\text{ ft})\left(200\ \frac{\text{lbf}}{\text{ft}}\right)}{2} = 1400\text{ lbf}$$

$$\tau_{\text{max}} = \frac{3V}{2A} = \frac{(3)(1400\text{ lbf})}{(2)(3.625\text{ in})(7.625\text{ in})} = \boxed{76.0\text{ lbf/in}^2}$$

The answer is (B).

7.

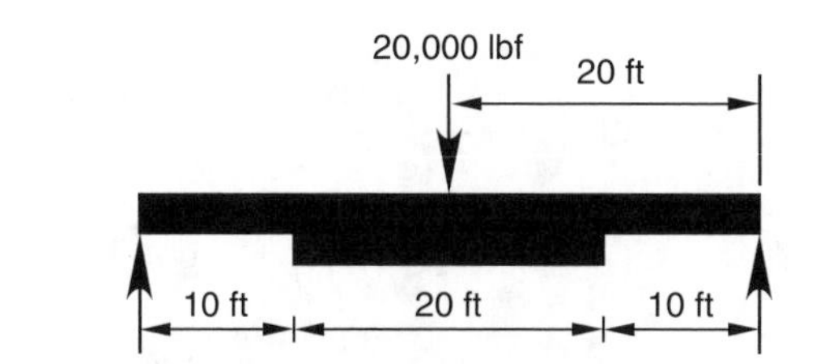

Assume simple supports. Use the conjugate beam method.

step 1:

$$M_{\text{max}} = (20\text{ ft})\left(\frac{20{,}000\text{ lbf}}{2}\right) = 200{,}000\text{ ft-lbf}$$

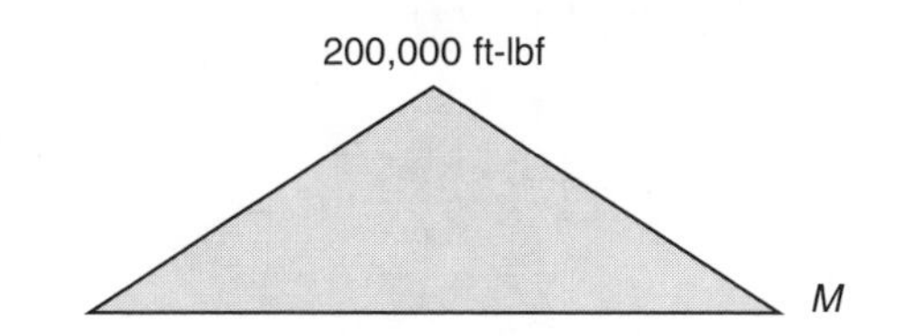

step 2:

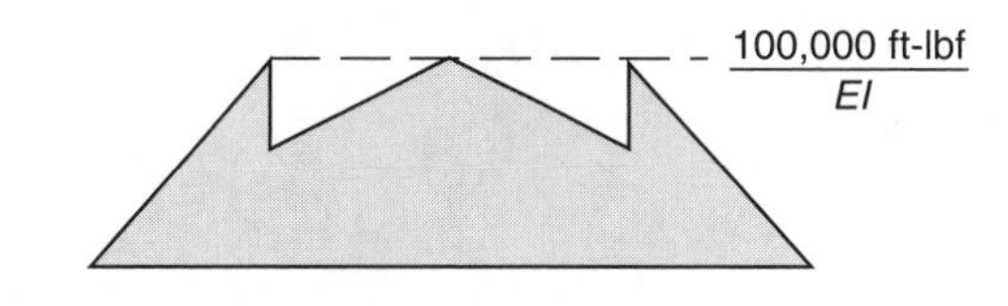

steps 3 and 4:

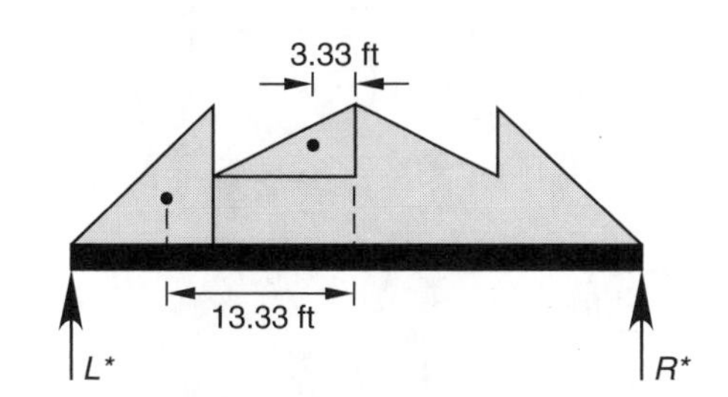

The total load on the conjugate beam is

$$(2)\left(\left(\tfrac{1}{2}\right)(10)\left(\frac{100{,}000}{EI}\right)+\left(\tfrac{1}{2}\right)(10)\left(\frac{50{,}000}{EI}\right)+(10)\left(\frac{50{,}000}{EI}\right)\right)$$

$$=\frac{2.5\times 10^6\ \text{ft}^2\text{-lbf}}{EI}$$

$$L^*=R^*=\frac{2.5\times 10^6\ \text{ft}^2\text{-lbf}}{2EI}=\frac{1.25\times 10^6\ \text{ft}^2\text{-lbf}}{EI}$$

step 5: Taking clockwise moments as positive, the conjugate moment at the midpoint is

$$\begin{aligned}M_{\text{mid}}&=(20)\left(\frac{1.25\times 10^6}{EI}\right)-\left(\tfrac{1}{2}\right)(10)\left(\frac{100{,}000}{EI}\right)(13.33)\\&\quad-\left(\tfrac{1}{2}\right)(10)\left(\frac{50{,}000}{EI}\right)(3.33)-(10)\left(\frac{50{,}000}{EI}\right)(5)\\&=\frac{1.5\times 10^7\ \text{ft}^3\text{-lbf}}{EI}\end{aligned}$$

If E is in lbf/ft^2 and I is in ft^4, then

$$\delta=\boxed{\frac{1.5\times 10^7}{EI}\ \text{ft}}$$

The answer is (C).

8. Use superposition.

Uniform load:

$$y_{\text{center}}=y_{\text{max}}=\frac{(5)\left(\dfrac{500\ \frac{\text{lbf}}{\text{ft}}}{12\ \frac{\text{in}}{\text{ft}}}\right)\left((17\ \text{ft})\left(12\ \frac{\text{in}}{\text{ft}}\right)\right)^4}{(384)\left(3\times 10^7\ \frac{\text{lbf}}{\text{in}^2}\right)(200\ \text{in}^4)}$$

$$=0.1566\ \text{in}$$

Concentrated load:

Use case 7 from App. 44.A.

$$L=(17\ \text{ft})\left(12\ \frac{\text{in}}{\text{ft}}\right)=204\ \text{in}$$

$$b=(5\ \text{ft})\left(12\ \frac{\text{in}}{\text{ft}}\right)=60\ \text{in}$$

At midspan,

$$x_a=\left(\frac{17\ \text{ft}}{2}\right)\left(12\ \frac{\text{in}}{\text{ft}}\right)=102\ \text{in}$$

$$\begin{aligned}y_{\text{center}}&=y_a\\&=\left(\frac{(2000\ \text{lbf})(60\ \text{in})(102\ \text{in})}{(6)\left(3\times 10^7\ \frac{\text{lbf}}{\text{in}^2}\right)(200\ \text{in}^4)(204\ \text{in})}\right)\\&\quad\times\left((204\ \text{in})^2-(60\ \text{in})^2-(102\ \text{in})^2\right)\\&=0.0460\ \text{in}\end{aligned}$$

$$y_{\text{total}}=0.1566\ \text{in}+0.0460\ \text{in}=\boxed{0.2026\ \text{in}}$$

The answer is (B).

9. *Customary U.S. Solution*

First, the moment of inertia of the beam cross section is

$$I=\frac{bh^3}{12}=\frac{(6\ \text{in})(4\ \text{in})^3}{12}=32\ \text{in}^4$$

From case 1 in App. 44.A, the deflection of the 200 lbf load is

$$y_1=\frac{PL^3}{3EI}=\frac{(200\ \text{lbf})(72\ \text{in}-12\ \text{in})^3}{(3)\left(1.5\times 10^6\ \frac{\text{lbf}}{\text{in}^2}\right)(32\ \text{in}^4)}$$

$$=0.30\ \text{in}$$

The slope at the 200 lbf load is

$$\theta=\frac{PL^2}{2EI}=\frac{(200\ \text{lbf})(72\ \text{in}-12\ \text{in})^2}{(2)\left(1.5\times 10^6\ \frac{\text{lbf}}{\text{in}^2}\right)(32\ \text{in}^4)}$$

$$=0.0075\ \text{rad}$$

The additional deflection at the tip of the beam is

$$y_{1'}=(0.0075\ \text{rad})(12\ \text{in})=0.09\ \text{in}$$

From case 1 in App. 44.A, the deflection at the 120 lbf load is

$$y_2=\frac{PL^3}{3EI}=\frac{(120\ \text{lbf})(72\ \text{in}-24\ \text{in})^3}{(3)\left(1.5\times 10^6\ \frac{\text{lbf}}{\text{in}^2}\right)(32\ \text{in}^4)}$$

$$=0.0922\ \text{in}$$

The slope at the 120 lbf load is

$$\theta_2 = \frac{PL^2}{2EI} = \frac{(120\ \text{lbf})(72\ \text{in} - 24\ \text{in})^2}{(2)\left(1.5\times 10^6\ \frac{\text{lbf}}{\text{in}^2}\right)(32\ \text{in}^4)}$$
$$= 0.00288\ \text{rad}$$

The additional deflection at the tip of the beam is

$$y_{2'} = (0.00288\ \text{rad})(24\ \text{in}) = 0.0691\ \text{in}$$

The total deflection is the sum of the preceding four parts.

$$y_{\text{tip}} = y_1 + y_{1'} + y_2 + y_{2'}$$
$$= 0.30\ \text{in} + 0.09\ \text{in} + 0.0922\ \text{in} + 0.0691\ \text{in}$$
$$= \boxed{0.5513\ \text{in}}$$

The answer is (D).

SI Solution

First, the moment of inertia of the beam cross section is

$$I = \frac{bh^3}{12} = \frac{(0.15\ \text{m})(0.10\ \text{m})^3}{12} = 1.25\times 10^{-5}\ \text{m}^4$$

From case 1 in App. 44.A, the deflection at the 900 N load is

$$y_1 = \frac{PL^3}{3EI} = \frac{(900\ \text{N})(1.8\ \text{m} - 0.3\ \text{m})^3}{(3)(10\times 10^9\ \text{Pa})(1.25\times 10^{-5}\ \text{m}^4)}$$
$$= 0.0081\ \text{m}$$

The slope at the 900 N load is

$$\theta_1 = \frac{PL^2}{2EI} = \frac{(900\ \text{N})(1.8\ \text{m} - 0.3\ \text{m})^2}{(2)(10\times 10^9\ \text{Pa})(1.25\times 10^{-5}\ \text{m}^4)}$$
$$= 0.0081\ \text{rad}$$

The additional deflection at the tip of the beam is

$$y_{1'} = (0.0081\ \text{rad})(0.3\ \text{m}) = 0.00243\ \text{m}$$

From case 1 in App. 44.A, the deflection at the 530 N load is

$$y_2 = \frac{PL^3}{3EI} = \frac{(530\ \text{N})(1.8\ \text{m} - 0.6\ \text{m})^3}{(3)(10\times 10^9\ \text{Pa})(1.25\times 10^{-5}\ \text{m}^4)}$$
$$= 0.00244\ \text{m}$$

The slope at the 530 N load is

$$\theta_2 = \frac{PL^2}{2EI} = \frac{(530\ \text{N})(1.8\ \text{m} - 0.6\ \text{m})^2}{(2)(10\times 10^9\ \text{Pa})(1.25\times 10^{-5}\ \text{m}^4)}$$
$$= 0.00305\ \text{rad}$$

The additional deflection at the tip of the beam is

$$y_{2'} = (0.00305\ \text{rad})(0.6\ \text{m}) = 0.00183\ \text{m}$$

The total deflection is the sum of the preceding four parts.

$$y_{\text{tip}} = y_1 + y_{1'} + y_2 + y_{2'}$$
$$= 0.0081\ \text{m} + 0.00243\ \text{m} + 0.00244\ \text{m} + 0.00183\ \text{m}$$
$$= \boxed{0.0148\ \text{m}}$$

The answer is (D).

10. Use the virtual work method. The reactions are

$$R_1 = 7071\ \text{lbf}$$
$$R_2 = 7071\ \text{lbf}$$

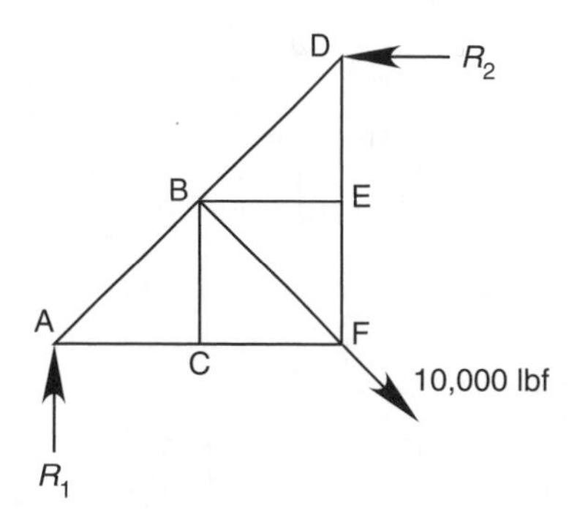

The member forces, S, tabulated below can be found from principles of static equilibrium.

member	force, S (lbf)	u	Su (lbf)
AB	−10,000	−1.414	14,140
AC	7071	0	0
CB	0	0	0
CF	7071	0	0
BF	0	0	0
FE	7071	1	7071
EB	0	0	0
ED	7071	1	7071
BD	−10,000	−1.414	14,140
total			42,420

Apply a vertical load of 1 at point F. Calculate the forces u in each member.

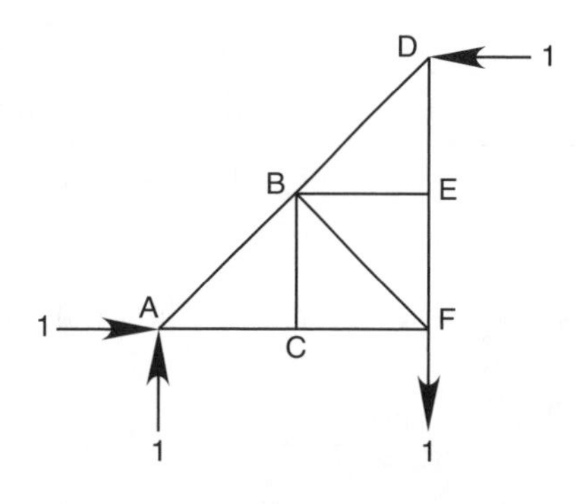

Structural

$$\delta = \sum \frac{SuL}{AE} = \frac{(42{,}420 \text{ lbf})\left(50 \ \frac{1}{\text{in}}\right)}{2.9 \times 10^7 \ \frac{\text{lbf}}{\text{in}^2}} = \boxed{0.0731 \text{ in}}$$

The answer is (A).

($\delta = 0.0693$ in if the strain energy method is used. However, that method is less accurate.)

11. *Customary U.S. Solution*

First, determine an equivalent aluminum area for the steel. The ratio of equivalent aluminum area to steel area is the modular ratio.

$$n = \frac{30 \times 10^6 \ \frac{\text{lbf}}{\text{in}^2}}{10 \times 10^6 \ \frac{\text{lbf}}{\text{in}^2}} = 3$$

The equivalent aluminum width to replace the steel is

$$(1.5 \text{ in})(3) = 4.5 \text{ in}$$

The equivalent all-aluminum cross section is

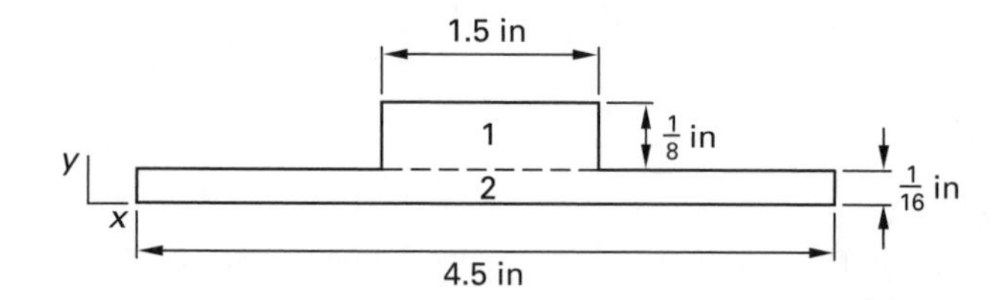

To find the centroid of the section, first calculate the areas of the basic shapes.

$$A_1 = (1.5 \text{ in})\left(\tfrac{1}{8} \text{ in}\right) = 0.1875 \text{ in}^2$$
$$A_2 = (4.5 \text{ in})\left(\tfrac{1}{16} \text{ in}\right) = 0.28125 \text{ in}^2$$

Next, find the y-components of the centroids of the basic shapes.

$$y_{c1} = \tfrac{1}{16} \text{ in} + \left(\tfrac{1}{2}\right)\left(\tfrac{1}{8} \text{ in}\right) = 0.125 \text{ in}$$

$$y_{c2} = \frac{\frac{1}{16} \text{ in}}{2} = 0.03125 \text{ in}$$

The centroid of the section is

$$y_c = \frac{\sum_i A_i y_{ci}}{\sum A_i} = \frac{(0.1875 \text{ in}^2)(0.125 \text{ in}) + (0.28125 \text{ in}^2)(0.03125 \text{ in})}{0.1875 \text{ in}^2 + 0.28125 \text{ in}^2} = 0.06875 \text{ in}$$

The moment of inertia of basic shape 1 about its own centroid is

$$I_{cy1} = \frac{bh^3}{12} = \frac{(1.5 \text{ in})(0.125 \text{ in})^3}{12} = 2.441 \times 10^{-4} \text{ in}^4$$

The moment of inertia of basic shape 2 about its own centroid is

$$I_{cy2} = \frac{bh^3}{12} = \frac{(4.5 \text{ in})(0.0625 \text{ in})^3}{12} = 9.155 \times 10^{-5} \text{ in}^4$$

The distance from the centroid of basic shape 1 to the section centroid is

$$d_1 = y_{c1} - y_c = 0.125 \text{ in} - 0.06875 \text{ in} = 0.05625 \text{ in}$$

The distance from the centroid of basic shape 2 to the section centroid is

$$d_2 = y_c - y_{c2} = 0.06875 \text{ in} - 0.03125 \text{ in} = 0.0375 \text{ in}$$

From the parallel axis theorem (Eq. 42.20), the moment of inertia of basic shape 1 about the centroid of the section is

$$\begin{aligned} I_{y1} &= I_{yc1} + A_1 d_1^2 \\ &= 2.441 \times 10^{-4} \text{ in}^4 + (0.1875 \text{ in}^2)(0.05625 \text{ in})^2 \\ &= 8.374 \times 10^{-4} \text{ in}^4 \end{aligned}$$

The moment of inertia of basic shape 2 about the centroid of the section is

$$\begin{aligned} I_{y2} &= I_{yc2} + A_2 d_2^2 \\ &= 9.155 \times 10^{-5} \text{ in}^4 + (0.28125 \text{ in}^2)(0.0375 \text{ in})^2 \\ &= 4.871 \times 10^{-4} \text{ in}^4 \end{aligned}$$

The total moment of inertia of the section about the centroid of the section is

$$\begin{aligned} I_y &= I_{y1} + I_{y2} = 8.374 \times 10^{-4} \text{ in}^4 + 4.871 \times 10^{-4} \text{ in}^4 \\ &= \boxed{1.325 \times 10^{-3} \text{ in}^4} \end{aligned}$$

The answer is (C).

SI Solution

First, determine an equivalent aluminum area for the steel. The ratio of equivalent aluminum area to steel area is the modular ratio.

$$n = \frac{20 \times 10^4 \text{ MPa}}{70 \times 10^3 \text{ MPa}} = 2.86$$

The equivalent aluminum width to replace the steel is

$$(38 \text{ mm})(2.86) = 108.7 \text{ mm}$$

The equivalent all-aluminum cross section is

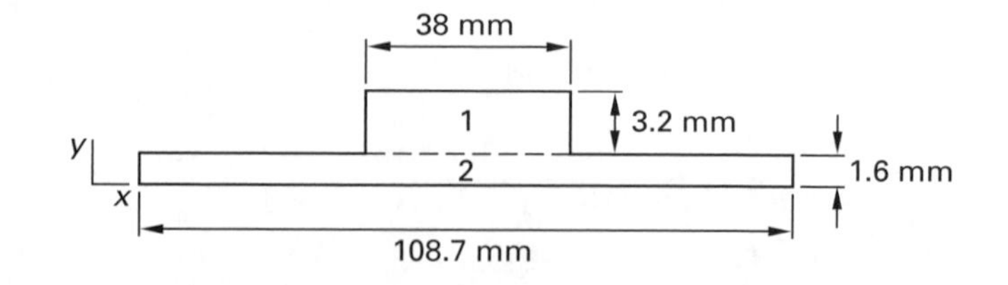

To find the centroid of the section, first calculate the areas of the basic shapes.

$$A_1 = (38\ \text{mm})(3.2\ \text{mm}) = 121.6\ \text{mm}^2$$
$$A_2 = (108.7\ \text{mm})(1.6\ \text{mm}) = 173.9\ \text{mm}^2$$

Next, find the y-components of the centroids of the basic shapes.

$$y_{c1} = 1.6\ \text{mm} + \left(\tfrac{1}{2}\right)(3.2\ \text{mm}) = 3.2\ \text{mm}$$
$$y_{c2} = \frac{1.6\ \text{mm}}{2} = 0.8\ \text{mm}$$

The centroid of the section is

$$\begin{aligned} y_c &= \frac{\sum_i A_i y_{ci}}{\sum A_i} \\ &= \frac{(121.6\ \text{mm}^2)(3.2\ \text{mm}) + (173.9\ \text{mm}^2)(0.8\ \text{mm})}{121.6\ \text{mm}^2 + 173.9\ \text{mm}^2} \\ &= 1.79\ \text{mm} \end{aligned}$$

The moment of inertia of basic shape 1 about its own centroid is

$$I_{cy1} = \frac{bh^3}{12} = \frac{(38\ \text{mm})(3.2\ \text{mm})^3}{12} = 103.77\ \text{mm}^4$$

The moment of inertia of basic shape 2 about its own centroid is

$$I_{cy2} = \frac{bh^3}{12} = \frac{(108.7\ \text{mm})(1.6\ \text{mm})^3}{12} = 37.10\ \text{mm}^4$$

The distance from the centroid of basic shape 1 to the section centroid is

$$d_1 = y_{c1} - y_c = 3.2\ \text{mm} - 1.79\ \text{mm} = 1.41\ \text{mm}$$

The distance from the centroid of basic shape 2 to the section centroid is

$$d_2 = y_c - y_{c2} = 1.79\ \text{mm} - 0.8\ \text{mm} = 0.99\ \text{mm}$$

From the parallel axis theorem, the moment of inertia of basic shape 1 about the centroid of the section is

$$\begin{aligned} I_{y1} &= I_{yc1} + A_1 d_1^2 \\ &= 103.77\ \text{mm}^4 + (121.6\ \text{mm}^2)(1.41\ \text{mm})^2 \\ &= 345.52\ \text{mm}^4 \end{aligned}$$

The moment of inertia of basic shape 2 about the centroid of the section is

$$\begin{aligned} I_{y2} &= I_{yc2} + A_2 d_2^2 \\ &= 37.10\ \text{mm}^4 + (173.9\ \text{mm}^2)(0.99\ \text{mm})^2 \\ &= 207.54\ \text{mm}^4 \end{aligned}$$

The total moment of inertia of the section about the centroid of the section is

$$\begin{aligned} I_y &= I_{y1} + I_{y2} = 345.52\ \text{mm}^4 + 207.54\ \text{mm}^4 \\ &= \boxed{553.06\ \text{mm}^4} \end{aligned}$$

The answer is (C).

12. Use Eq. 44.54.

$$n = \frac{2.9\times 10^7\ \frac{\text{lbf}}{\text{in}^2}}{2\times 10^6\ \frac{\text{lbf}}{\text{in}^2}} = 14.5$$
$$A_s = 1\ \text{in}^2$$
$$nA_s = (14.5)(1\ \text{in}^2) = 14.5\ \text{in}^2$$

The two relevant sections are

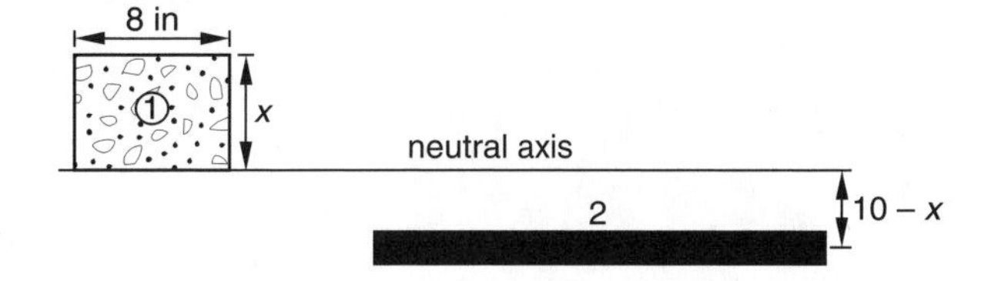

$$A_1 = 8x\ \text{in}^2$$
$$A_2 = 14.5\ \text{in}^2$$

Since the neutral axis is the centroid axis of the transformed section, the moment of area above the neutral axis must equal the moment of area below the neutral axis.

$$\begin{aligned} (8x)\left(\tfrac{1}{2}x\right) &= (10 - x)(14.5) \\ 4x^2 + 14.5x - 145 &= 0 \\ x &= 4.4752\ \text{in} \\ 10 - x &= 5.5248\ \text{in} \end{aligned}$$

The steel's contribution to I is found from the parallel axis theorem, with the steel treated as a line. Disregarding the concrete below the neutral axis, the centroidal moment of inertia is

$$\begin{aligned} I_c &= \frac{(8\ \text{in})(4.4752\ \text{in})^3}{3} + (14.5\ \text{in}^2)(5.5248\ \text{in})^2 \\ &= 681.6\ \text{in}^4 \end{aligned}$$

Structural

(a) The maximum concrete bending stress is

$$\sigma_c = \frac{(8125 \text{ ft-lbf})\left(12 \frac{\text{in}}{\text{ft}}\right)(4.4752 \text{ in})}{681.6 \text{ in}^4} = \boxed{640.2 \text{ lbf/in}^2}$$

The answer is (B).

(b) The maximum steel bending stress is

$$\sigma_s = \frac{(14.5)(8125 \text{ ft-lbf})\left(12 \frac{\text{in}}{\text{ft}}\right)(5.5248 \text{ in})}{681.6 \text{ in}^4} = \boxed{11{,}459 \text{ lbf/in}^2}$$

The answer is (A).

45 Basic Elements of Design

PRACTICE PROBLEMS

Note: Unless instructed otherwise in a problem, use the following properties:

steel: $E = 29 \times 10^6$ lbf/in^2 (20×10^4 MPa)
$G = 11.5 \times 10^6$ lbf/in^2 (8.0×10^4 MPa)
$\alpha = 6.5 \times 10^{-6}$ 1/°F (1.2×10^{-5} 1/°C)
$\nu = 0.3$

Columns

1. A structural steel member 50 ft (15 m) long is used as a long column to support 75,000 lbf (330 kN). Both ends are built-in, and there are no intermediate supports. A factor of safety of 2.5 is used. What is the required moment of inertia?

(A) 67 in^4 (2.8×10^{-5} m^4)
(B) 96 in^4 (4.0×10^{-5} m^4)
(C) 130 in^4 (5.6×10^{-5} m^4)
(D) 190 in^4 (8.0×10^{-5} m^4)

2. A long steel column with a yield strength of 36,000 lbf/in^2 (250 MPa) has pinned ends. The column is 25 ft (7.5 m) long and has a cross-sectional area of 25.6 in^2 (165 cm^2), a centroidal moment of inertia of 350 in^4 (14,600 cm^2), and a distance from the neutral axis to the extreme fiber of 7 in (180 mm). It carries an axial concentric load of 100,000 lbf (440 kN) and an eccentric load of 150,000 lbf (660 kN) located 3.33 in (80 mm) from the longitudinal axial axis. Use the secant formula to determine the stress factor of safety.

(A) 1.6
(B) 2.1
(C) 2.4
(D) 3.0

3. A 4 in × 4 in (nominal size) timber post is used to support a sign. The post's modulus of elasticity is 1.5×10^6 lbf/in^2. One end of the post is embedded in a deep concrete base. The other end supports a sign 9 ft above the ground. Neglect torsion and wind effects. What is the Euler load sign weight that will cause failure by buckling?

(A) 2700 lbf
(B) 3700 lbf
(C) 4000 lbf
(D) 5500 lbf

Bolts

4. A $^3/_4$-16 UNF steel bolt is used without washers to clamp two rigid steel plates, each 2 in (50 mm) thick. 0.75 in (19 mm) of the threaded section of the bolt remains under the nut. The nut has six threads. The nut is tightened until the stress in the bolt body is 40,000 lbf/in^2 (280 MPa). The bolt's modulus of elasticity is 2.9×10^7 lbf/in^2 (200 GPa). How much does the bolt stretch?

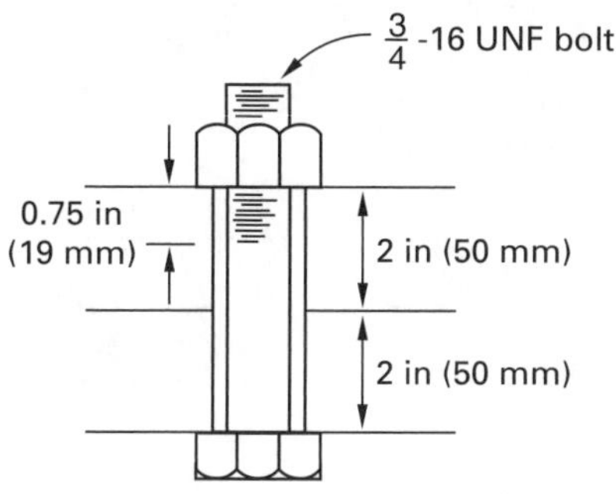

(A) 0.006 in (0.15 mm)
(B) 0.012 in (0.30 mm)
(C) 0.024 in (0.60 mm)
(D) 0.048 in (1.2 mm)

Eccentrically Loaded Bolted Connections

5. The bracket shown is attached to a column with three 0.75 in (19 mm) bolts arranged in an equilateral triangular layout. A force is applied with a moment arm of 20 in (500 mm) measured to the centroid of the bolt group. The maximum shear stress in the bolts is limited to 15,000 lbf/in^2 (100 MPa). What is the maximum force that the connection can support?

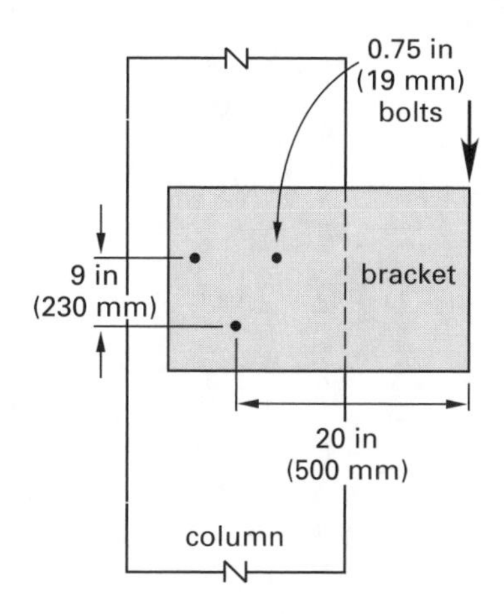

(A) 700 lbf (3.1 kN)
(B) 1200 lbf (5.3 kN)
(C) 2400 lbf (11 kN)
(D) 4700 lbf (20 kN)

Eccentrically Loaded Welded Connections

6. A fillet weld is used to secure a steel bracket to a column. The bracket supports a 10,000 lbf (44 kN) force applied 12 in (300 mm) from the face of the column, as shown. The maximum shear stress in the weld material is 8000 lbf/in^2 (55 MPa). What size fillet weld is required?

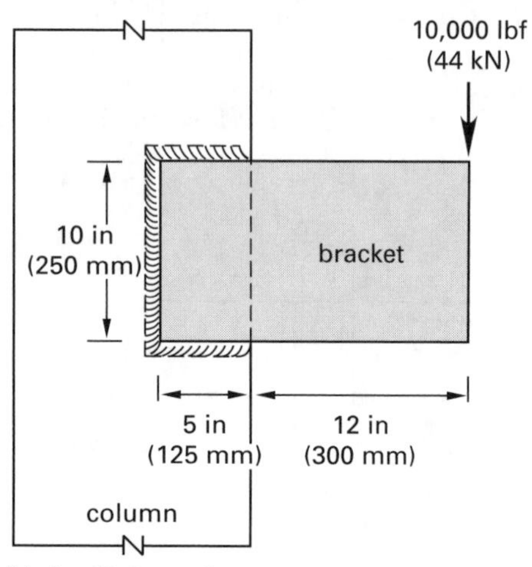

(A) $^3/_8$ in (9.5 mm)
(B) $^1/_2$ in (13 mm)
(C) $^5/_8$ in (16 mm)
(D) $^3/_4$ in (19 mm)

Wire Rope

7. (*Time limit: one hour*) Wet concrete is mixed at a point that is separated from the pour location by a deep, 200 ft wide ravine. It is decided to use a $^1/_2$ in steel cable and a series of $^1/_2$ in steel cable hangers to support a 6 in schedule-40 steel pipe. The cables are 6×19 standard hoisting rope with a mass of 0.4 lbm/ft and a breaking strength of 11 tons. The pipe can be assumed to be completely filled with concrete. The cabling geometry is as illustrated. Horizontal restraint is infinite, and the impulse effects of periodic pump stroke pulsations are to be neglected. The vertical wire rope hangers remain vertical. The two pipe ends are supported. Use a tensile factor of safety of 5, and consider a 50% loss of strength at cable connections. What is the factor of safety in the cable?

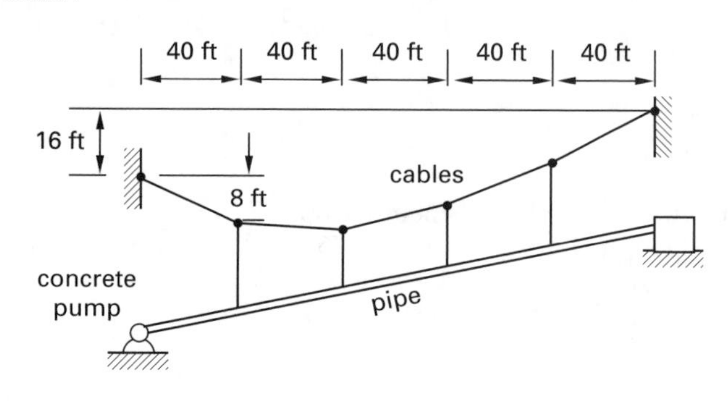

(A) 0.15
(B) 1.5
(C) 1.9
(D) 2.4

SOLUTIONS

1. *Customary U.S. Solution*

The design load for a factor of safety of 2.5 is

$$F = (2.5)(75{,}000 \text{ lbf}) = 187{,}500 \text{ lbf}$$

From Table 45.1, the theoretical end restraint coefficient for built-in ends is $C = 0.5$, and the minimum design value is 0.65.

From Eq. 45.3, the effective length of the column is

$$L' = CL = (0.65)(50 \text{ ft})\left(12 \ \frac{\text{in}}{\text{ft}}\right) = 390 \text{ in}$$

Set the design load equal to the Euler load F_e and use Eq. 45.1 to find the required moment of inertia.

$$\begin{aligned} I &= \frac{F_e(L')^2}{\pi^2 E} \\ &= \frac{(187{,}500 \text{ lbf})(390 \text{ in})^2}{\pi^2\left(29\times 10^6 \ \frac{\text{lbf}}{\text{in}^2}\right)} = \boxed{99.64 \text{ in}^4} \end{aligned}$$

The answer is (B).

SI Solution

The design load for a factor of safety of 2.5 is

$$F = (2.5)(330\times 10^3 \text{ N}) = 825\times 10^3 \text{ N}$$

From Table 45.1, the theoretical end restraint coefficient for built-in ends is $C = 0.5$, and the minimum design value is 0.65.

From Eq. 45.3, the effective length of the column is

$$L' = CL = (0.65)(15 \text{ m}) = 9.75 \text{ m}$$

Set the design load equal to the Euler load F_e and use Eq. 45.1 to find the required moment of inertia.

$$\begin{aligned} I &= \frac{F_e(L')^2}{\pi^2 E} \\ &= \frac{(825\times 10^3 \text{ N})(9.75 \text{ m})^2}{\pi^2\left(20\times 10^{10} \ \frac{\text{N}}{\text{m}^2}\right)} = \boxed{3.973\times 10^{-5} \text{ m}^4} \end{aligned}$$

The answer is (B).

2. *Customary U.S. Solution*

The radius of gyration is

$$k = \sqrt{\frac{I}{A}} = \sqrt{\frac{350 \text{ in}^4}{25.6 \text{ in}^2}} = 3.70 \text{ in}$$

The slenderness ratio is

$$\frac{L}{k} = \frac{(25 \text{ ft})\left(12 \ \frac{\text{in}}{\text{ft}}\right)}{3.70 \text{ in}} = 81.08$$

The total buckling load is

$$F = 100{,}000 \text{ lbf} + 150{,}000 \text{ lbf} = 250{,}000 \text{ lbf}$$

From Eq. 45.9,

$$\begin{aligned} \phi &= \tfrac{1}{2}\left(\frac{L}{k}\right)\sqrt{\frac{F}{AE}} \\ &= \left(\tfrac{1}{2}\right)(81.08)\sqrt{\frac{250{,}000 \text{ lbf}}{(25.6 \text{ in}^2)\left(29\times 10^6 \ \frac{\text{lbf}}{\text{in}^2}\right)}} \\ &= 0.744 \text{ rad} \end{aligned}$$

The eccentricity is

$$e = \frac{M}{F} = \frac{(150{,}000 \text{ lbf})(3.33 \text{ in})}{250{,}000 \text{ lbf}} = 2.0 \text{ in}$$

From Eq. 45.8, the critical column stress is

$$\begin{aligned} \sigma_{\text{max}} &= \left(\frac{F}{A}\right)\left(1 + \left(\frac{ec}{k^2}\right)(\sec\phi)\right) \\ &= \left(\frac{250{,}000 \text{ lbf}}{25.6 \text{ in}^2}\right) \\ &\quad \times\left(1 + \left(\frac{(2.0 \text{ in})(7 \text{ in})}{(3.70 \text{ in})^2}\right)(\sec(0.744 \text{ rad}))\right) \\ &= 23{,}339 \text{ lbf/in}^2 \end{aligned}$$

The stress factor of safety for the column is

$$\text{FS} = \frac{S_y}{\sigma_{\text{max}}} = \frac{36{,}000 \ \frac{\text{lbf}}{\text{in}^2}}{23{,}339 \ \frac{\text{lbf}}{\text{in}^2}} = \boxed{1.54}$$

The answer is (A).

SI Solution

The radius of gyration is

$$k = \sqrt{\frac{I}{A}} = \sqrt{\frac{14\,600 \text{ cm}^4}{165 \text{ cm}^2}} = 9.41 \text{ cm}$$

The slenderness ratio is

$$\frac{L}{k} = \frac{(7.5 \text{ m})\left(100 \ \frac{\text{cm}}{\text{m}}\right)}{9.41 \text{ cm}} = 79.70$$

The total buckling load is

$$\begin{aligned} F &= 440 \text{ kN} + 660 \text{ kN} = (1100 \text{ kN})\left(1000 \ \frac{\text{N}}{\text{kN}}\right) \\ &= 1.1\times 10^6 \text{ N} \end{aligned}$$

Structural

From Eq. 45.9,

$$\phi = \tfrac{1}{2}\left(\frac{L}{k}\right)\sqrt{\frac{F}{AE}}$$

$$= \left(\tfrac{1}{2}\right)(79.70)\sqrt{\frac{1.1\times10^6\ \text{N}}{\left(\frac{165\ \text{cm}^2}{\left(100\ \frac{\text{cm}}{\text{m}}\right)^2}\right)\left(20\times10^{10}\ \frac{\text{N}}{\text{m}^2}\right)}}$$

$$= 0.728\ \text{rad}$$

The eccentricity is

$$e = \frac{M}{F} = \frac{(660\ \text{kN})(80\ \text{mm})}{1100\ \text{kN}} = 48\ \text{mm}$$

From Eq. 45.8, the critical column stress is

$$\sigma_{\text{max}} = \left(\frac{F}{A}\right)\left(1+\left(\frac{ec}{k^2}\right)(\sec\phi)\right)$$

$$= \left(\frac{1.1\times10^6\ \text{N}}{\frac{165\ \text{cm}^2}{\left(100\ \frac{\text{cm}}{\text{m}}\right)^2}}\right) \times\left(1+\left(\frac{(48\ \text{mm})(180\ \text{mm})}{(9.41\ \text{cm})^2\left(100\ \frac{\text{mm}^2}{\text{cm}^2}\right)} \times\left(\sec(0.728\ \text{rad})\right)\right)\right) \times\left(\frac{1\ \text{MPa}}{10^6\ \text{Pa}}\right)$$

$$= 153.8\ \text{MPa}$$

The stress factor of safety for the column is

$$\text{FS} = \frac{S_y}{\sigma_{\text{max}}} = \frac{250\ \text{MPa}}{153.8\ \text{MPa}} = \boxed{1.63}$$

The answer is (A).

3. $I = \frac{(3.5\ \text{in})^4}{12} = 12.51\ \text{in}^4$ [finished lumber size]

$$k = \sqrt{\frac{I}{A}} = \sqrt{\frac{12.51\ \text{in}^4}{(3.5\ \text{in})^2}} = 1.01\ \text{in}$$

The end-restraint coefficient is $C = 2.1$ (NDS). The slenderness ratio is

$$\frac{L}{k} = \frac{(2.1)(9\ \text{ft})\left(12\ \frac{\text{in}}{\text{ft}}\right)}{1.01\ \text{in}} = 224.6$$

L/k is well above 100. (Note that most timber codes limit L/k to 50, so this would not be a permitted application.)

$$F_e = \frac{\pi^2\left(1.5\times10^6\ \frac{\text{lbf}}{\text{in}^2}\right)(12.51\ \text{in}^4)}{\left((2.1)(9\ \text{ft})\left(12\ \frac{\text{in}}{\text{ft}}\right)\right)^2} = \boxed{3673\ \text{lbf}}$$

The answer is (B).

4. *Customary U.S. Solution*

The elongation in the unthreaded part of the bolt is

$$\delta_1 = \frac{\sigma_1 L_1}{E} = \frac{\left(40{,}000\ \frac{\text{lbf}}{\text{in}^2}\right)(4.0\ \text{in} - 0.75\ \text{in})}{2.9\times10^7\ \frac{\text{lbf}}{\text{in}^2}}$$

$$= 0.00448\ \text{in}$$

From Table 45.5, the stress area for a $^3/_4$-16 UNF bolt is $A_1 = 0.373\ \text{in}^2$.

The stress area for the threaded part of the bolt is

$$A_2 = \frac{\pi d^2}{4} = \frac{\pi(0.75\ \text{in})^2}{4} = 0.4418\ \text{in}^2$$

The stress in the threaded part of the bolt is

$$\sigma_2 = \sigma_1\left(\frac{A_2}{A_1}\right) = \left(40{,}000\ \frac{\text{lbf}}{\text{in}^2}\right)\left(\frac{0.4418\ \text{in}^2}{0.373\ \text{in}^2}\right)$$

$$= 47{,}378\ \text{lbf/in}^2$$

Assume half of the bolt in the nut contributes to elongation. The elongation in the threaded part of the bolt, including three threads in the nut, is

$$\delta_2 = \frac{\sigma_2 L_2}{E}$$

$$= \frac{\left(47{,}378\ \frac{\text{lbf}}{\text{in}^2}\right)\times\left(0.75\ \text{in} + (3\ \text{threads})\left(\frac{1}{16\ \frac{\text{threads}}{\text{in}}}\right)\right)}{2.9\times10^7\ \frac{\text{lbf}}{\text{in}^2}}$$

$$= 0.00153\ \text{in}$$

The total stretch of the bolt is

$$\delta = \delta_1 + \delta_2 = 0.00448\ \text{in} + 0.00153\ \text{in} = \boxed{0.00601\ \text{in}}$$

The answer is (A).

SI Solution

The elongation in the unthreaded part of the bolt is

$$\delta_1 = \frac{\sigma_1 L_1}{E} = \frac{(280\text{ MPa})(100\text{ mm} - 19\text{ mm})}{(200\text{ GPa})\left(1000\,\frac{\text{MPa}}{\text{GPa}}\right)}$$
$$= 0.1134\text{ mm}$$

From Table 45.5, the stress area for a $^3/_4$-16 UNF bolt is

$$A_1 = (0.373\text{ in}^2)\left(25.4\,\frac{\text{mm}}{\text{in}}\right)^2 = 240.64\text{ mm}^2$$

The stress area for the threaded part of the bolt is

$$A_2 = \frac{\pi d^2}{4} = \frac{\pi(19\text{ mm})^2}{4} = 283.53\text{ mm}^2$$

The stress in the threaded part of the bolt is

$$\sigma_2 = \sigma_1\left(\frac{A_2}{A_1}\right) = (280\text{ MPa})\left(\frac{283.53\text{ mm}^2}{240.64\text{ mm}^2}\right)$$
$$= 330\text{ MPa}$$

Assume half of the bolt in the nut contributes to elongation.

$$\delta_2 = \frac{\sigma_2 L_2}{E}$$
$$= \frac{(330\text{ MPa})\left(\begin{array}{l}19\text{ mm} + (3\text{ threads})\\ \times\left(\dfrac{1}{16\,\dfrac{\text{threads}}{\text{in}}}\right)\left(25.4\dfrac{\text{mm}}{\text{in}}\right)\end{array}\right)}{(200\text{ GPa}\left(1000\,\frac{\text{MPa}}{\text{GPa}}\right)}$$
$$= 0.0392\text{ mm}$$

The total stretch of the bolt is

$$\delta = \delta_1 + \delta_2 = 0.1134\text{ mm} + 0.0392\text{ mm} = \boxed{0.1526\text{ mm}}$$

The answer is (A).

5. *Customary U.S. Solution*

First, find the properties of the bolt area. For an equilateral triangular layout, the distance from the centroid of the bolt group to the center of each bolt is

$$r = \left(\tfrac{2}{3}\right)(9\text{ in}) = 6\text{ in}$$

The area of each bolt is

$$A = \frac{\pi d^2}{4} = \frac{\pi(0.75\text{ in})^2}{4} = 0.442\text{ in}^2$$

The polar moment of inertia of a bolt about the centroid is

$$J = Ar^2 = (0.442\text{ in}^2)(6\text{ in})^2 = 15.91\text{ in}^4$$

The vertical shear load at each bolt is

$$F_v = \frac{F}{3} = 0.333F \quad [\text{in lbf}]$$

The moment applied to each bolt is

$$M = \frac{(20\text{ in})F}{3} = 6.67F \quad [\text{in in-lbf}]$$

The vertical shear stress in each bolt is

$$\tau_v = \frac{F_v}{A} = \frac{0.333F}{0.442\text{ in}^2} = 0.753F \quad [\text{in lbf/in}^2]$$

The torsional shear stress in each bolt is

$$\tau = \frac{Mr}{J} = \frac{(6.67F)(6\text{ in})}{15.91\text{ in}^4} = 2.52F \quad [\text{in lbf/in}^2]$$

The most highly stressed bolt is the right-most bolt. The shear stress configuration is

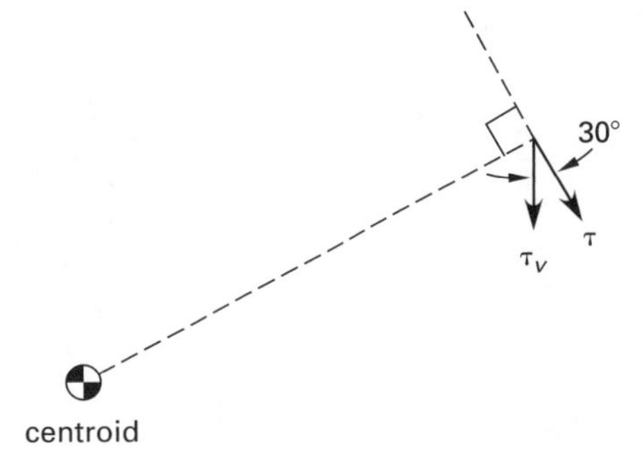

The stresses are combined to find the maximum stress.

$$\tau_{\text{max}} = \sqrt{(\tau\sin 30°)^2 + (\tau_v + \tau\cos 30°)^2}$$

$$15{,}000\,\frac{\text{lbf}}{\text{in}^2} = \sqrt{\begin{array}{l}\left((2.52F)(\sin 30°)\right)^2\\ \quad + \left(0.753F + (2.52F)(\cos 30°)\right)^2\end{array}}$$
$$= 3.19F$$

$$F = \boxed{4702\text{ lbf}}$$

The answer is (D).

Structural

SI Solution

First, find the properties of the bolt area. For an equilateral triangular layout, the distance from the centroid of the bolt group to each bolt is

$$r = \left(\tfrac{2}{3}\right)(230\ \text{mm}) = 153\ \text{mm}$$

The area of each bolt is

$$A = \frac{\pi d^2}{4} = \frac{\pi (19\ \text{mm})^2 \left(\dfrac{1\ \text{m}}{10^3\ \text{mm}}\right)^2}{4} = 2.835{\times}10^{-4}\ \text{m}^2$$

The polar moment of inertia of a bolt about the centroid is

$$\begin{aligned} J &= Ar^2 = (2.835 \times 10^{-4}\ \text{m}^2)(0.153\ \text{m})^2 \\ &= 6.636 \times 10^{-6}\ \text{m}^4 \end{aligned}$$

The vertical shear load at each bolt is

$$F_v = \frac{F}{3} = 0.333F \quad [\text{in N}]$$

The moment applied to each bolt is

$$M = \frac{(0.5\ \text{m})F}{3} = 0.167F\ \text{N·m}$$

The vertical shear stress in each bolt is

$$\tau_v = \frac{F_v}{A} = \frac{0.333F}{2.835 \times 10^{-4}\ \text{m}^2} = 1.1746 \times 10^3 F\ [\text{in N/m}^2]$$

The torsional shear stress in each bolt is

$$\begin{aligned} \tau &= \frac{Mr}{J} = \frac{(0.167F)(0.153\ \text{m})}{6.636 \times 10^{-6}\ \text{m}^4} \\ &= 3.850 \times 10^3 F \quad [\text{in N/m}^2] \end{aligned}$$

The most highly stressed bolt is the right-most bolt. The shear stress configuration is

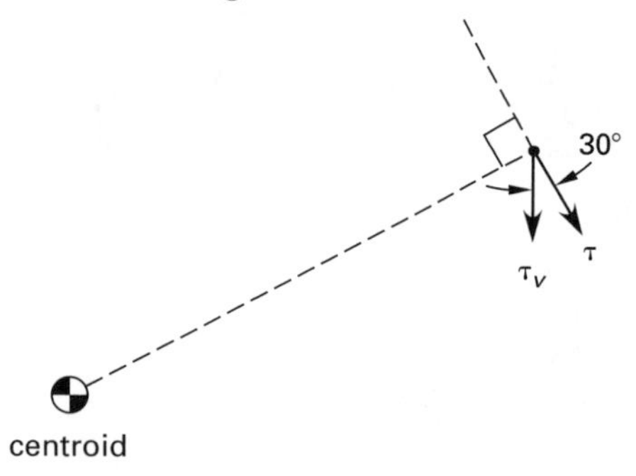

The stresses are combined to find the maximum stress.

$$\tau_{\max} = \sqrt{(\tau \sin 30°)^2 + (\tau_v + \tau \cos 30°)^2}$$

$$(100\ \text{MPa}) \times \left(1000\ \frac{\text{kPa}}{\text{MPa}}\right) = \sqrt{\begin{aligned} &\left((3.850 \times 10^3 F)(\sin 30°)\right)^2 \\ &+ \left(1.1746 \times 10^3 F \right. \\ &\left. + (3.850 \times 10^3 F)(\cos 30°)\right)^2 \end{aligned}}$$

$$= 4.90 \times 10^3 F \quad [\text{in N/m}^2]$$

$$F = \boxed{20.4\ \text{kN}}$$

The answer is (D).

6. *Customary U.S. Solution*

The fillet weld consists of three basic shapes, each with a throat size of t.

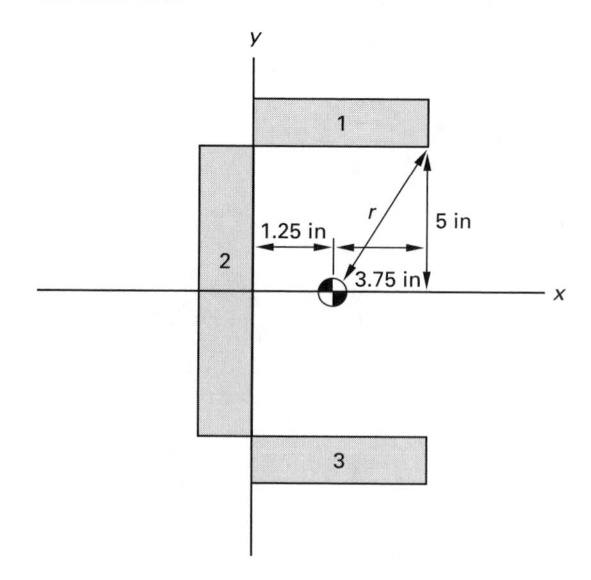

First, find the centroid.

By inspection, $y_c = 0$.

The areas of the basic shapes are

$$\begin{aligned} A_1 &= 5t \quad [\text{in in}^2] \\ A_2 &= 10t \quad [\text{in in}^2] \\ A_3 &= 5t \quad [\text{in in}^2] \end{aligned}$$

The x-components of the centroids of the basic shapes are

$$\begin{aligned} x_{c1} &= \frac{5\ \text{in}}{2} = 2.5\ \text{in} \\ x_{c2} &= -\frac{t}{2} = 0 \quad [\text{for small } t] \\ x_{c3} &= \frac{5\ \text{in}}{2} = 2.5\ \text{in} \\ x_c &= \frac{\sum_i A_i x_{ci}}{A_i} \\ &= \frac{(5t)(2.5\ \text{in}) + (10t)(0) + (5t)(2.5\ \text{in})}{5t + 10t + 5t} \\ &= 1.25\ \text{in} \end{aligned}$$

Next, find the centroidal moments of inertia.

The moments of inertia of basic shape 1 about its own centroid are

$$I_{cx1} = \frac{bh^3}{12} = \frac{(5\ \text{in})t^3}{12} = 0.417t^3 \quad [\text{in in}^4]$$

$$I_{cy1} = \frac{bh^3}{12} = \frac{t(5\ \text{in})^3}{12} = 10.417t \quad [\text{in in}^4]$$

The moments of inertia of basic shape 2 about its own centroid are

$$I_{cx2} = \frac{bh^3}{12} = \frac{t(10 \text{ in})^3}{12} = 83.333t \quad [\text{in in}^4]$$

$$I_{cy2} = \frac{bh^3}{12} = \frac{(10 \text{ in})t^3}{12} = 0.833t^3 \quad [\text{in in}^4]$$

The moments of inertia of basic shape 3 about its own centroid are the same as for basic shape 1.

$$I_{cx3} = I_{cx1} = 0.417t^3 \quad [\text{in in}^4]$$
$$I_{cy3} = I_{cy1} = 10.417t \quad [\text{in in}^4]$$

From the parallel axis theorem, the moments of inertia of basic shape 1 about the centroidal axis of the section are

$$\begin{aligned} I_{x1} &= I_{cx1} + A_1 d_1^2 = 0.417t^3 + (5t)(5 \text{ in})^2 \\ &= 0.417t^3 + 125t \\ I_{y1} &= I_{cy1} + A_1 d_1^2 \\ &= 10.417t + 5t\left(\frac{5 \text{ in}}{2} - 1.25 \text{ in}\right)^2 \\ &= 18.23t \quad [\text{in in}^4] \end{aligned}$$

The moments of inertia of basic shape 2 about the centroidal axis of the section are

$$\begin{aligned} I_{x2} &= I_{cx2} + A_2 d_2^2 = 83.333t + (10t)(0) \\ &= 83.333t \quad [\text{in in}^4] \\ I_{y2} &= I_{cy2} + A_2 d_2^2 = 0.833t^3 + (10t)(1.25 \text{ in})^2 \\ &= 0.833t^3 + 15.625t \end{aligned}$$

The moment of inertia of basic shape 3 about the centroidal axis of the section are the same as for basic shape 1.

$$I_{x3} = I_{x1} = 0.417t^3 + 125t$$
$$I_{y3} = I_{y1} = 18.23t \quad [\text{in in}^4]$$

The total moments of inertia about the centroidal axis of the section are

$$\begin{aligned} I_x &= I_{x1} + I_{x2} + I_{x3} \\ &= 0.417t^3 + 125t + 83.333t + 0.417t^3 + 125t \\ &= 0.834t^3 + 333.33t \\ I_y &= I_{y1} + I_{y2} + I_{y3} \\ &= 18.23t + 0.833t^3 + 15.625t + 18.23t \\ &= 0.833t^3 + 52.09t \end{aligned}$$

Since t will be small and since the coefficient of the t^3 term is smaller than the coefficient of the t term, the higher-order t^3 term may be neglected.

$$I_x = 333.33t \quad [\text{in in}^4]$$
$$I_y = 52.09t \quad [\text{in in}^4]$$

The polar moment of inertia for the section is

$$\begin{aligned} J &= I_x + I_y = 333.33t + 52.09t \\ &= 385.42t \quad [\text{in in}^4] \end{aligned}$$

The maximum shear stress will occur at the right-most point of basic shape 1 since this point is farthest from the centroid of the section.

This distance is

$$\begin{aligned} r &= \sqrt{(5 \text{ in} - 1.25 \text{ in})^2 + \left(\frac{10 \text{ in}}{2}\right)^2} \\ &= 6.25 \text{ in} \end{aligned}$$

The applied moment is

$$\begin{aligned} M &= Fe = (10{,}000 \text{ lbf})(12 \text{ in} + 5 \text{ in} - 1.25 \text{ in}) \\ &= 157{,}500 \text{ in-lbf} \end{aligned}$$

The torsional shear stress is

$$\tau = \frac{Mr}{J} = \frac{(157{,}500 \text{ in-lbf})(6.25 \text{ in})}{385.42t} = \frac{2554.0}{t} \text{ lbf/in}^2$$

The shear stress is perpendicular to the line from the point to the centroid. The x- and y-components are shown on the figure.

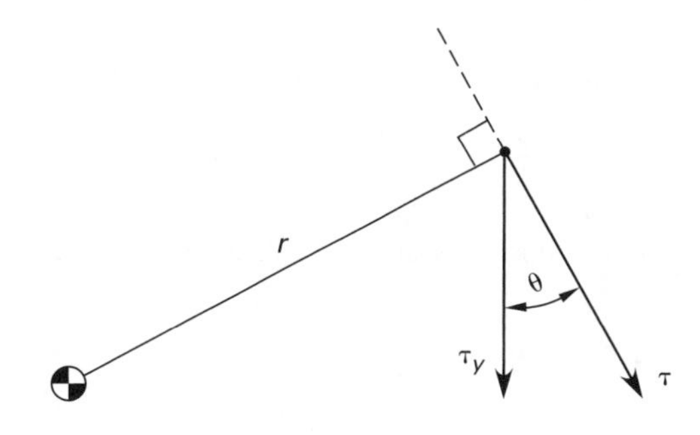

The angle θ is

$$\theta = 90^\circ - \tan^{-1}\left(\frac{5 \text{ in} - 1.25 \text{ in}}{\frac{10 \text{ in}}{2}}\right) = 53.1^\circ$$

$$\begin{aligned} \tau_x &= \tau \sin\theta = \left(\frac{2554.0 \ \frac{\text{lbf}}{\text{in}^2}}{t}\right)(\sin 53.1^\circ) \\ &= \frac{2042.4}{t} \text{ lbf/in}^2 \end{aligned}$$

$$\begin{aligned} \tau_y &= \tau \cos\theta = \left(\frac{2554.0 \ \frac{\text{lbf}}{\text{in}^2}}{t}\right)(\cos 53.1^\circ) \\ &= \frac{1533.5}{t} \text{ lbf/in}^2 \end{aligned}$$

Structural

The vertical shear stress due to the load is

$$\tau_{vy} = \frac{F}{\sum A_i} = \frac{10{,}000 \text{ lbf}}{5t + 10t + 5t} = \frac{500}{t} \text{ lbf/in}^2$$

The resultant shear stress is

$$\begin{aligned}\tau &= \sqrt{\tau_x^2 + (\tau_y + \tau_{vy})^2} \\ &= \sqrt{\left(\frac{2042.4}{t}\right)^2 + \left(\frac{1533.5}{t} + \frac{500}{t}\right)^2} \\ &= \frac{2882.1}{t} \text{ lbf/in}^2\end{aligned}$$

For a maximum shear stress of 8000 lbf/in^2,

$$\begin{aligned}8000 \frac{\text{lbf}}{\text{in}^2} &= \frac{2882.1}{t} \text{ lbf/in}^2 \\ t &= 0.360 \text{ in}\end{aligned}$$

From Eq. 45.42, the required weld size is

$$y = \frac{t}{0.707} = \frac{0.360 \text{ in}}{0.707} = \boxed{0.509 \text{ in} \quad (1/2 \text{ in})}$$

The answer is (B).

SI Solution

The fillet weld consists of three basic shapes, each with a throat size of t.

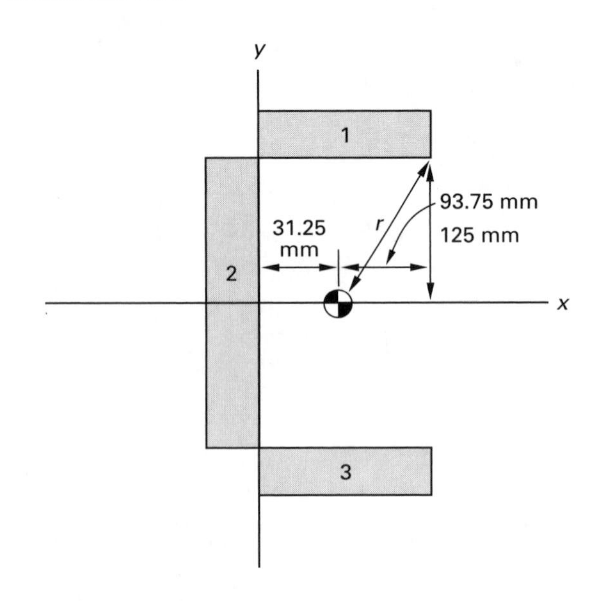

First, find the centroid.

By inspection, $y_c = 0$.

The areas of the basic shapes are

$$\begin{aligned}A_1 &= 125t \quad [\text{in mm}^2] \\ A_2 &= 250t \quad [\text{in mm}^2] \\ A_3 &= 125t \quad [\text{in mm}^2]\end{aligned}$$

The x-components of the centroids of the basic shapes are

$$\begin{aligned}x_{c1} &= \frac{125 \text{ mm}}{2} = 62.5 \text{ mm} \\ x_{c2} &= -\frac{t}{2} = 0 \quad [\text{for small } t] \\ x_{c3} &= \frac{125 \text{ mm}}{2} = 62.5 \text{ mm} \\ x_c &= \frac{\sum_i A_i x_{ci}}{A_i} \\ &= \frac{(125t)(62.5 \text{ mm}) + (250t)(0) + (125t)(62.5 \text{ mm})}{125t + 250t + 125t} \\ &= 31.25 \text{ mm}\end{aligned}$$

Next, find the centroidal moments of inertia.

The moments of inertia of basic shape 1 about its own centroid are

$$\begin{aligned}I_{cx1} &= \frac{bh^3}{12} = \frac{(125 \text{ mm})t^3}{12} = 10.42t^3 \quad [\text{in mm}^4] \\ I_{cy1} &= \frac{bh^3}{12} = \frac{t(125 \text{ mm})^3}{12} = 162\,760t \quad [\text{in mm}^4]\end{aligned}$$

The moments of inertia of basic shape 2 about its own centroid are

$$\begin{aligned}I_{cx2} &= \frac{bh^3}{12} = \frac{t(250 \text{ mm})^3}{12} = 1\,302\,083t \quad [\text{in mm}^4] \\ I_{cy2} &= \frac{bh^3}{12} = \frac{(250 \text{ mm})t^3}{12} = 20.83t^3 \quad [\text{in mm}^4]\end{aligned}$$

The moments of inertia of basic shape 3 about its own centroid are the same as for basic shape 1.

$$\begin{aligned}I_{cx3} &= I_{cx1} = 10.42t^3 \quad [\text{in mm}^4] \\ I_{cy3} &= I_{cy1} = 162\,760t \quad [\text{in mm}^4]\end{aligned}$$

From the parallel axis theorem, the moments of inertia of basic shape 1 about the centroidal axis of the section are

$$\begin{aligned}I_{x1} &= I_{cx1} + A_1 d_1^2 \\ &= 10.42t^3 + (125t)(125 \text{ mm})^2 \\ &= 10.42t^3 + 1\,953\,125t \\ I_{y1} &= I_{cy1} + A_1 d_1^2 \\ &= 162\,760t + (125t)\left(\frac{125 \text{ mm}}{2} - 31.25 \text{ mm}\right)^2 \\ &= 284\,830t \quad [\text{in mm}^4]\end{aligned}$$

The moments of inertia of basic shape 2 about the centroidal axis of the section are

$$\begin{aligned} I_{x2} &= I_{cx2} + A_2 d_2^2 \\ &= 1\,302\,083t + (250t)(0)^2 \\ &= 1\,302\,083t \quad [\text{in mm}^4] \\ I_{y2} &= I_{cy2} + A_2 d_2^2 \\ &= 20.83t^3 + (250t)(31.25 \text{ mm})^2 \\ &= 20.83t^3 + 244\,140t \end{aligned}$$

The moments of inertia of basic shape 3 about the centroidal axis of the section are the same as for basic shape 1.

$$\begin{aligned} I_{x3} &= I_{x1} = 10.42t^3 + 1\,953\,125t \\ I_{y3} &= I_{y1} = 284\,830t \quad [\text{in mm}^4] \end{aligned}$$

The total moments of inertia about the centroidal axis of the section are

$$\begin{aligned} I_x &= I_{x1} + I_{x2} + I_{x3} \\ &= 10.42t^3 + 1\,953\,125t + 1\,302\,083t + 10.42t^3 \\ &\quad + 1\,953\,125t \\ &= 20.84t^3 + 5\,208\,333t \\ I_y &= I_{y1} + I_{y2} + I_{y3} \\ &= 284\,830t + 20.83t^3 + 244\,140t \\ &\quad + 284\,830t \\ &= 20.83t^3 + 813\,800t \end{aligned}$$

Since t will be small and since the coefficient of the t^3 term is smaller than the coefficient of the t term, the t^3 term may be neglected.

$$\begin{aligned} I_x &= 5\,208\,333t \quad [\text{in mm}^4] \\ I_y &= 813\,800t \quad [\text{in mm}^4] \end{aligned}$$

The maximum shear stress will occur at the right-most point of basic shape 1 since this point is farthest from the centroid of the section.

The distance is

$$\begin{aligned} r &= \sqrt{(125 \text{ mm} - 31.25 \text{ mm})^2 + \left(\frac{250 \text{ mm}}{2}\right)^2} \\ &= 156.25 \text{ mm} \end{aligned}$$

The polar moment of inertia for the section is

$$\begin{aligned} J &= I_x + I_y = 5\,208\,333t + 813\,800t \\ &= 6\,022\,133t \quad [\text{in mm}^4] \end{aligned}$$

The applied moment is

$$\begin{aligned} M &= Fe = (44 \text{ kN})(300 \text{ mm} + 125 \text{ mm} - 31.25 \text{ mm}) \\ &= 17\,325 \text{ kN·mm} \end{aligned}$$

The torsional shear stress is

$$\begin{aligned} \tau &= \frac{Mr}{J} \\ &= \left(\frac{(17\,325 \text{ kN·mm})(156.25 \text{ mm})}{6\,022\,133t}\right)\left(10^3 \frac{\text{mm}}{\text{m}}\right)^2 \\ &\quad \times \left(\frac{1 \text{ MPa}}{1000 \text{ kPa}}\right) \\ &= \frac{449.5}{t} \text{ MPa} \end{aligned}$$

The shear stress is perpendicular to the line from the point to the centroid. The x- and y-components are shown on the figure.

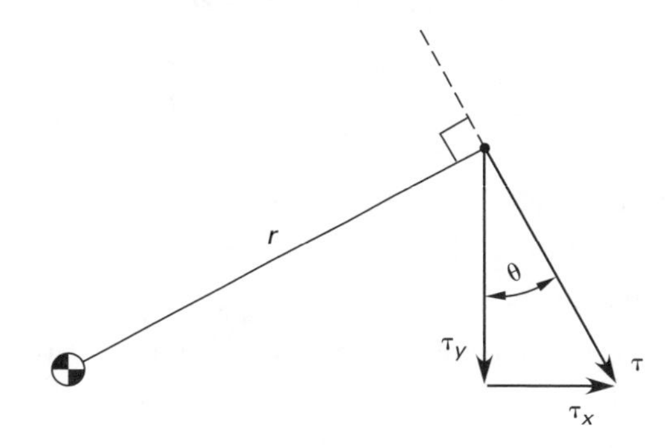

The angle θ is

$$\theta = 90° - \tan^{-1}\left(\frac{125 \text{ mm} - 31.25 \text{ mm}}{\frac{250 \text{ mm}}{2}}\right) = 53.1°$$

$$\begin{aligned} \tau_x &= \tau \sin\theta = \left(\frac{449.5 \text{ MPa}}{t}\right)(\sin 53.1°) \\ &= \frac{359.5 \text{ MPa}}{t} \\ \tau_y &= \tau \cos\theta = \left(\frac{449.5 \text{ MPa}}{t}\right)(\cos 53.1°) \\ &= \frac{269.9 \text{ MPa}}{t} \end{aligned}$$

The vertical shear stress due to the load is

$$\begin{aligned} \tau_{vy} &= \frac{F}{\sum A_i} \\ &= \frac{(44 \text{ kN})\left(\frac{1 \text{ MPa}}{1000 \text{ kPa}}\right)}{(125t + 250t + 125t)\left(\frac{1 \text{ m}}{10^3 \text{ mm}}\right)^2} \\ &= \frac{88.0}{t} \text{ MPa} \end{aligned}$$

The resultant shear stress is

$$\begin{aligned}\tau &= \sqrt{\tau_x^2 + (\tau_y + \tau_{vy})^2} \\ &= \sqrt{\left(\frac{359.5}{t}\right)^2 + \left(\frac{269.9}{t} + \frac{88.0}{t}\right)^2} \\ &= \frac{507.3}{t}\ \text{MPa}\end{aligned}$$

For a maximum shear stress of 55 MPa,

$$\begin{aligned}55\ \text{MPa} &= \frac{507.3}{t}\ \text{MPa} \\ t &= 9.22\ \text{mm}\end{aligned}$$

From Eq. 45.42, the required weld size is

$$y = \frac{t}{0.707} = \frac{9.22\ \text{mm}}{0.707} = \boxed{13.0\ \text{mm}}$$

The answer is (B).

7. Assumptions:

- concrete specific weight = 150 lbf/ft^3 (wet and dry)
- steel specific weight = 490 lbf/ft^3
- ignore cable self-weight
- cable sections act as tension links
- For the pipe,

$$\begin{aligned}D_o &= 6.625\ \text{in} \\ D_i &= 6.065\ \text{in} \\ A_i &= 0.2006\ \text{ft}^2\end{aligned}$$

The volume and weight of pipe per 40 ft section is

$$\begin{aligned}V &= \left(\frac{\pi}{4}\right)\left(\left(\frac{6.625\ \text{in}}{12\ \frac{\text{in}}{\text{ft}}}\right)^2 - \left(\frac{6.065\ \text{in}}{12\ \frac{\text{in}}{\text{ft}}}\right)^2\right)(40\ \text{ft}) \\ &= 1.55\ \text{ft}^3\end{aligned}$$

$$\text{weight} = \left(490\ \frac{\text{lbf}}{\text{ft}^3}\right)(1.55\ \text{ft}^3) = 759.7\ \text{lbf} \quad [\text{use } 760\ \text{lbf}]$$

The weight of concrete per 40 ft section is

$$\begin{aligned}V &= (0.2006\ \text{ft}^2)(40\ \text{ft}) = 8.0\ \text{ft}^3 \\ \text{weight} &= \left(150\ \frac{\text{lbf}}{\text{ft}^3}\right)(8.0\ \text{ft}^3) = 1200\ \text{lbf}\end{aligned}$$

The load per vertical hanger is

$$760\ \text{lbf} + 1200\ \text{lbf} = 1960\ \text{lbf}$$

Note: Since the horizontal force component is constant, the cable tension will be greatest where the cable is steepest. This appears to be the right-hand support.

At the left cable support,

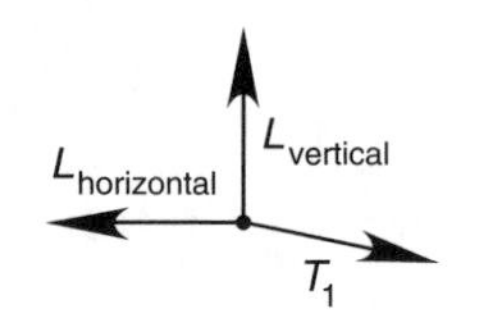

8 ft

40.8 ft

40 ft

$$\begin{aligned}L_{\text{vertical}} &= \left(\frac{8\ \text{ft}}{40.8\ \text{ft}}\right)T_1 = 0.196T_1 \quad [\text{use } 0.20\text{T}_1] \\ L_{\text{horizontal}} &= \left(\frac{40\ \text{ft}}{40.8\ \text{ft}}\right)T_1 = 0.98T_1\end{aligned}$$

Sum the moments about the right cable support to find T_1. Take counterclockwise moments to be positive.

$$\begin{aligned}\sum M &= (1960\ \text{lbf})(40\ \text{ft} + 80\ \text{ft} + 120\ \text{ft} + 160\ \text{ft}) \\ &\quad - 16L_{\text{horizontal}} - 200L_{\text{vertical}} \\ &= 784{,}000 - (16)(0.98T_1) - (200)(0.20T_1) \\ T_1 &= 14{,}080\ \text{lbf} \\ L_{\text{horizontal}} &= 0.98T_1 = (0.98)(14{,}080\ \text{lbf}) \\ &= 13{,}798\ \text{lbf} \\ L_{\text{vertical}} &= 0.20T_1 = (0.20)(14{,}080\ \text{lbf}) \\ &= 2816\ \text{lbf} \\ R_{\text{horizontal}} &= L_{\text{horizontal}} = 13{,}798\ \text{lbf} \\ R_{\text{vertical}} &= \sum \text{loads} - L_{\text{vertical}} \\ &= (4)(1960\ \text{lbf}) - 2816\ \text{lbf} \\ &= 5024\ \text{lbf} \\ T_5 &= \sqrt{(13{,}798\ \text{lbf})^2 + (5024\ \text{lbf})^2} \\ &\approx 14{,}680\ \text{lbf}\end{aligned}$$

The factor of safety includes the stress concentration factor.

$$\begin{aligned}\text{FS} &= \frac{\text{allowable tension}}{\text{actual force}} = \frac{\dfrac{(11\ \text{tons})\left(2000\ \dfrac{\text{lbf}}{\text{ton}}\right)}{5}}{(2)(14{,}680\ \text{lbf})} \\ &= \boxed{0.15}\end{aligned}$$

The answer is (A).

46 Structural Analysis I

PRACTICE PROBLEMS

Degree of Indeterminacy

1. What is the degree of indeterminacy of the following structure?

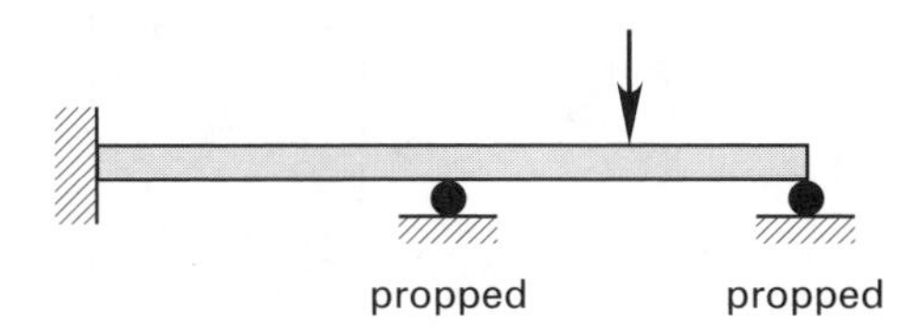

(A) 1
(B) 2
(C) 3
(D) 4

2. What is the degree of indeterminacy of the following truss?

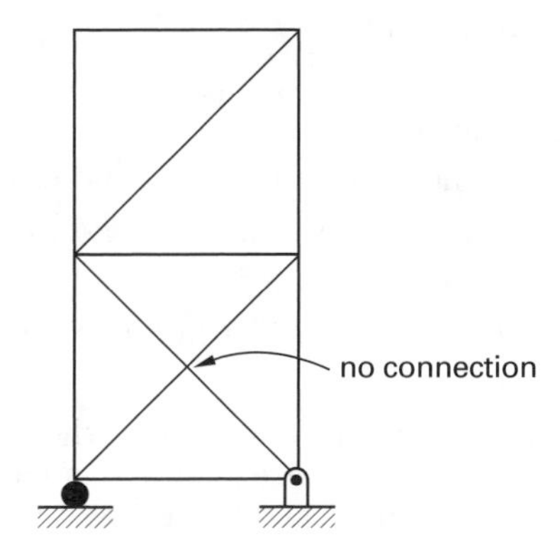

(A) 1
(B) 2
(C) 3
(D) 4

Elastic Deformation

3. What is the compressive load if a 1 in × 2 in × 10 in (2 cm × 5 cm × 30 cm) copper tie rod ($E = 18 \times 10^6$ lbf/in^2 (12×10^4 MPa)) experiences a 55°F (30°C) increase from the no-stress temperature?

(A) 8000 lbf (26 kN)
(B) 12,000 lbf (38 kN)
(C) 15,000 lbf (48 kN)
(D) 18,000 lbf (58 kN)

4. A 2 in diameter (5 cm diameter) and 15 in long (40 cm long) steel rod ($E = 30 \times 10^6$ lbf/in^2 (20×10^4 MPa)) supports a 2250 lbf (10 000 N) compressive load.

(a) What is the stress?

(A) 570 lbf/in^2 (4.0 MPa)
(B) 720 lbf/in^2 (5.1 MPa)
(C) 1100 lbf/in^2 (7.8 MPa)
(D) 1400 lbf/in^2 (9.9 MPa)

(b) What is the decrease in length?

(A) 8.7×10^{-5} in (2.3×10^{-6} m)
(B) 1.1×10^{-4} in (3.0×10^{-6} m)
(C) 3.6×10^{-4} in (1.0×10^{-5} m)
(D) 9.3×10^{-4} in (2.6×10^{-5} m)

Consistent Deformation Method

5.

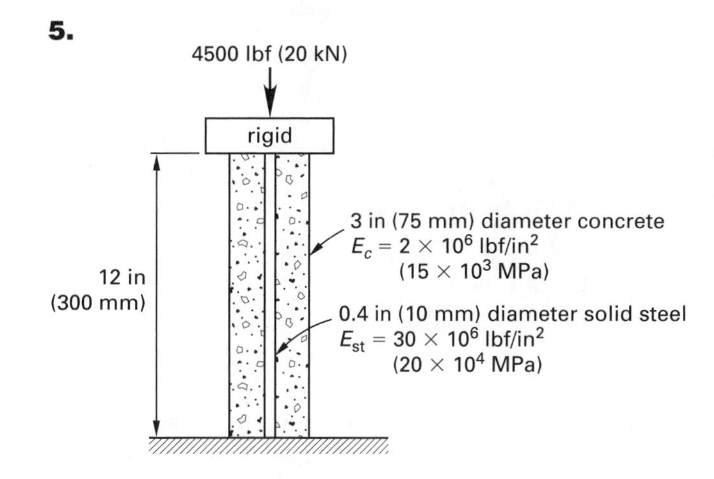

(a) Find the force and stress in each member for the structure shown.

(b) Find the total deflection for the structure shown.

6. The concentric pipe assembly shown is rigidly attached at both ends. The rigid ends are unconstrained. What is the stress generated in each pipe if the temperature of the assembly rises 200°F (110°C)?

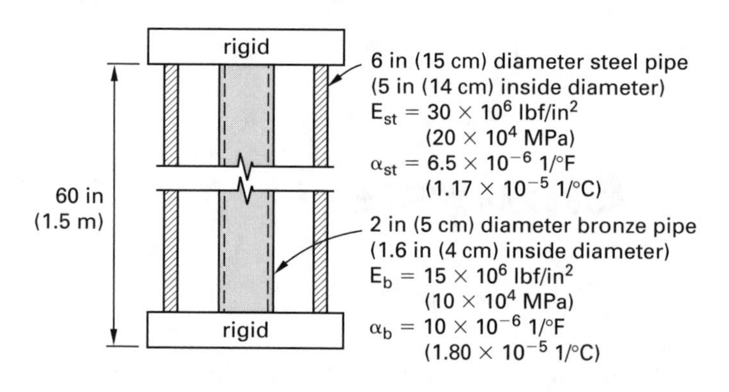

7. For the structure shown, what is the load carried by the steel cable? The beam is made of steel, $I = 10.0 \text{ in}^4$ ($4.17 \times 10^6 \text{ mm}^4$). The cable cross section is 0.0124 in^2 (8 mm^2). Before the 270 lbf load is applied, the cable is taut but carries no load.

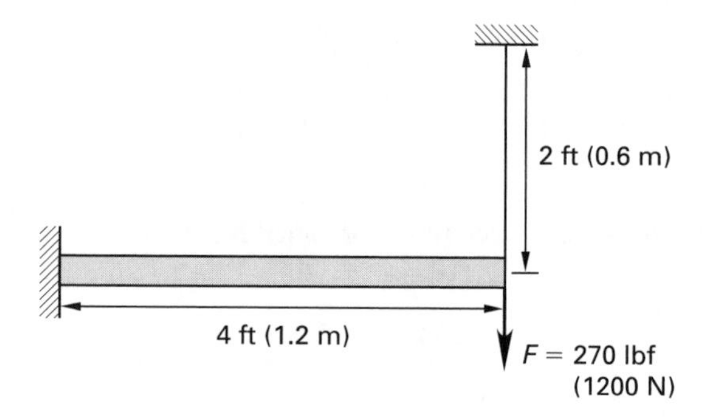

(A) 180 lbf (780 N)
(B) 200 lbf (870 N)
(C) 220 lbf (950 N)
(D) 240 lbf (1.0 kN)

8. A beam is simply supported at its ends and by a column of area A at its center. The beam and column are the same material. If the beam is subject to uniform load, w, what are the reactions and the force in the column?

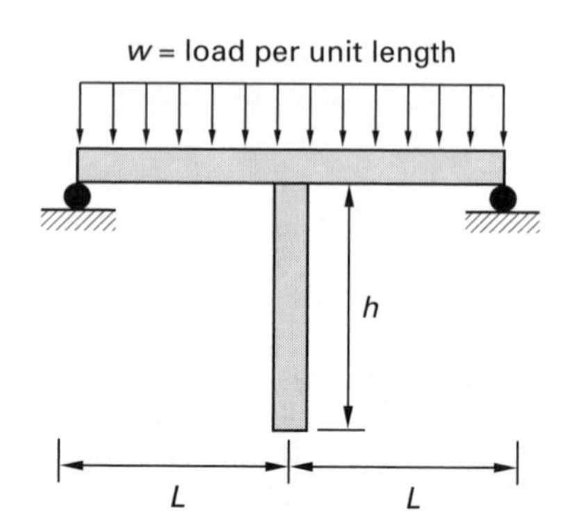

9.

$$a = 6 \text{ ft } (2 \text{ m})$$
$$b = 3 \text{ ft } (1 \text{ m})$$
$$c = 3 \text{ ft } (1 \text{ m})$$
$$d = 12 \text{ ft } (4 \text{ m})$$
$$P = 4500 \text{ lbf } (20 \text{ kN})$$
$$\text{rod area} = 0.124 \text{ in}^2 \ (80 \text{ mm}^2)$$
$$E = 10 \times 10^6 \text{ lbf/in}^2 \ (70 \times 10^3 \text{ MPa})$$

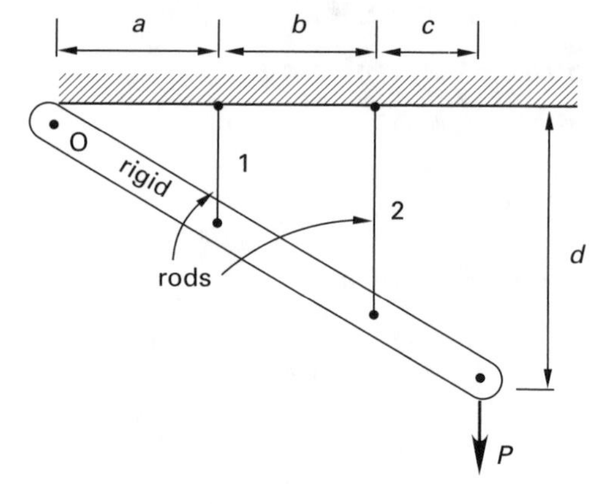

(a) What is the load of each of the aluminum rods for the structure shown?

(A) 3200 lbf (14 kN)
(B) 3600 lbf (16 kN)
(C) 4000 lbf (18 kN)
(D) 4300 lbf (19 kN)

(b) What is the elongation of each of the aluminum rods for the structure shown?

(A) 0.13 in and 0.26 in (3.3 mm and 6.6 mm)
(B) 0.21 in and 0.47 in (5.7 mm and 12 mm)
(C) 0.26 in and 0.62 in (6.6 mm and 16 mm)
(D) 0.21 in and 0.31 in (5.7 mm and 8.6 mm)

10. What are the reactions at each of the supports for the structure? The beams are identical in length, cross-sectional shape and area, and material.

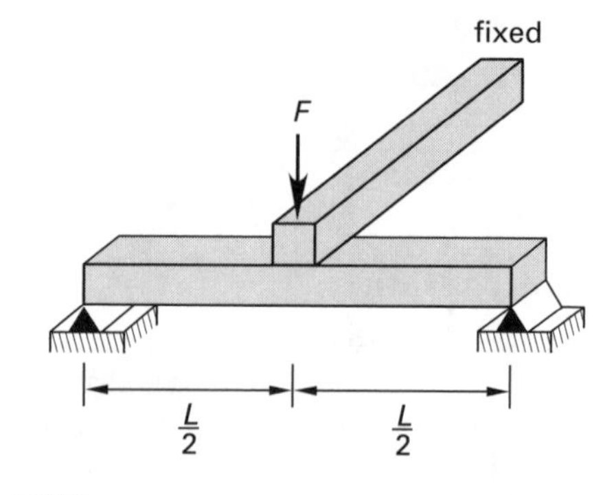

(A) $F/13$
(B) $4F/21$
(C) $8F/17$
(D) $2F/3$

Superposition Method

11. Using the superposition method, find the intermediate support reaction, R_2, for the structure.

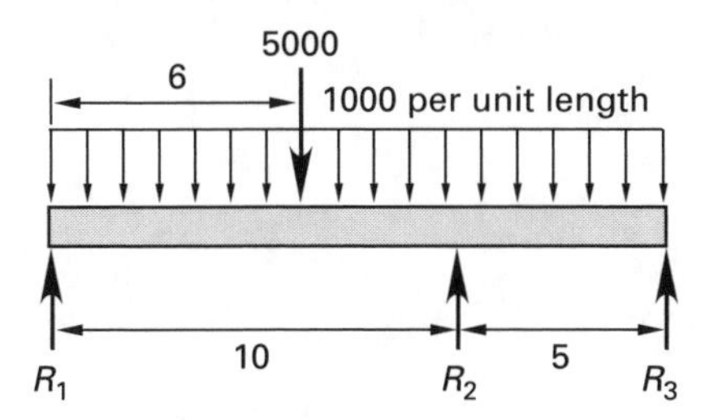

(A) 9000
(B) 11,000
(C) 15,000
(D) 18,000

12. Using the superposition method, determine the reaction, R, at the prop.

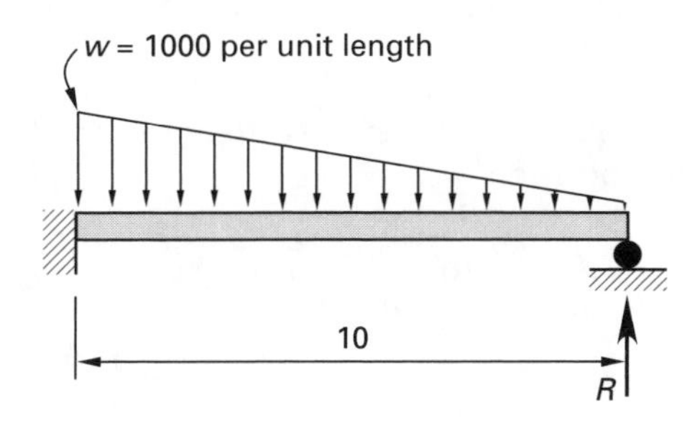

(A) 800
(B) 1000
(C) 1200
(D) 1400

Three-Moment Equation

13. What are the reactions R_1, R_2, and R_3?

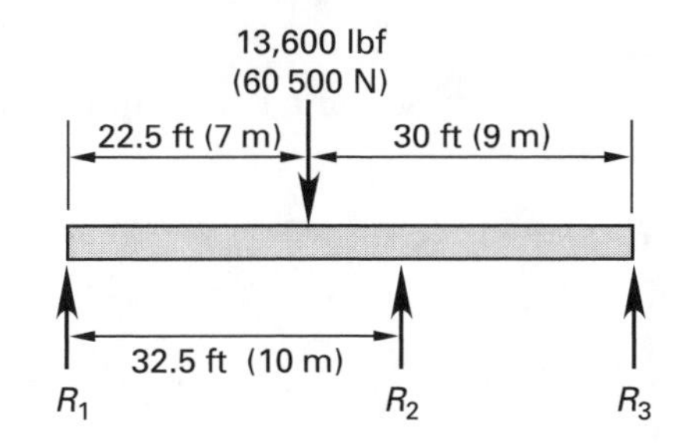

14. What are the reactions R_1, R_2, and R_3?

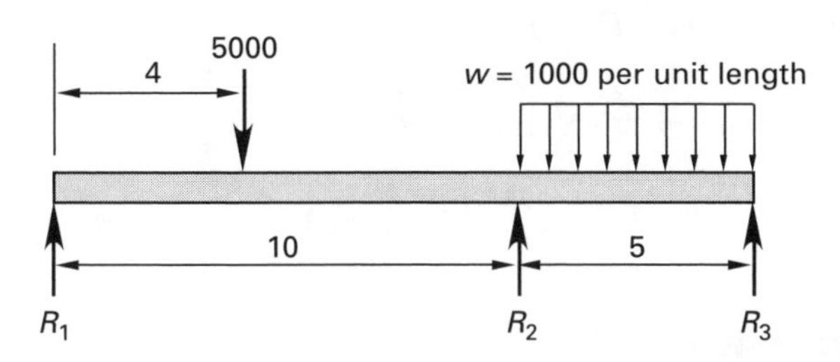

Fixed-End Moments

15. What are the vertical reactions at both ends of the structure?

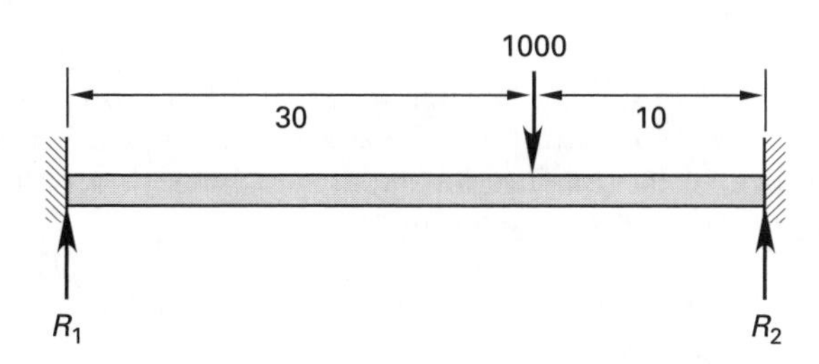

16. The truss shown carries a moving uniform live load of 2 kips/ft and a moving concentrated live load of 15 kips.

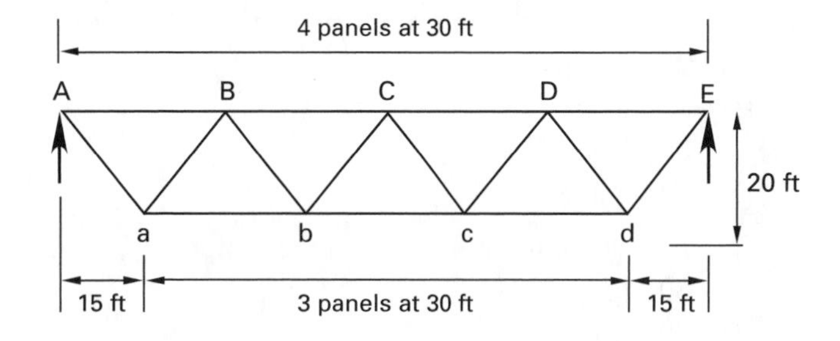

(a) What is the maximum force in member Bb?
(A) 40 kips
(B) 51 kips
(C) 59 kips
(D) 73 kips

(b) What is the maximum force in member BC?
(A) 15 kips
(B) 80 kips
(C) 95 kips
(D) 170 kips

17. The truss shown carried a group of four live loads along its bottom chord. What is the maximum force in member CD?

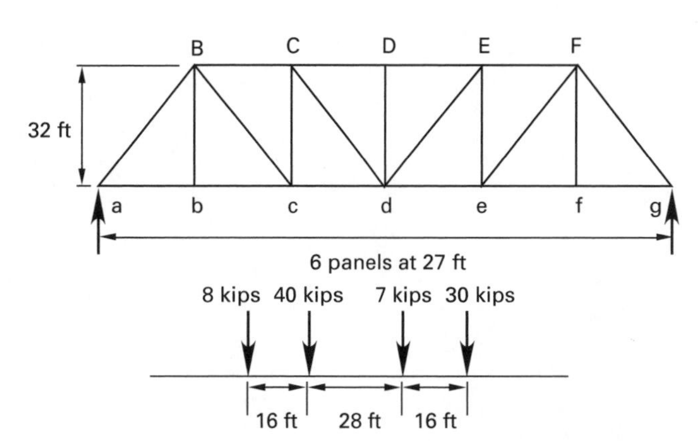

Structural

(A) 34 kips
(B) 82 kips
(C) 180 kips
(D) 220 kips

18. A moving load consisting of two 30 kip forces separated by a constant 6 ft travels over a two-span bridge. The bridge has an interior expansion joint that can be considered to be an ideal hinge.

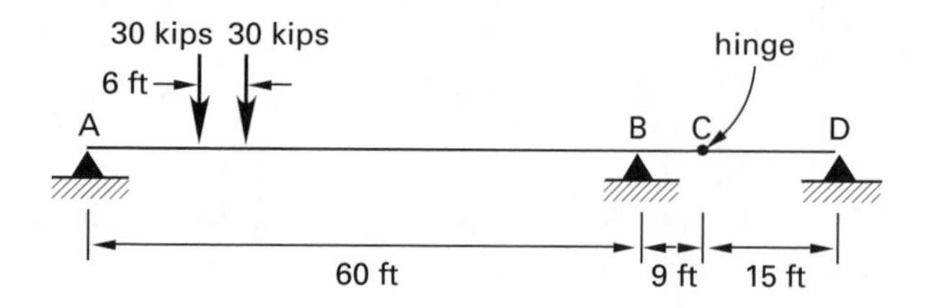

(a) What is the maximum moment at point B?
(A) 250 ft-kips
(B) 310 ft-kips
(C) 380 ft-kips
(D) 430 ft-kips

(b) What is the maximum reaction at point B?
(A) 30 kips
(B) 60 kips
(C) 66 kips
(D) 90 kips

SOLUTIONS

1. The degree of indeterminacy is 2. Remove the two props (vertical reactions) in order to make the structure statically determinate.

The answer is (B).

2.

$$\begin{aligned}\text{degree of indeterminacy} &= 3 + \text{no. members} - (2)\left(\text{no. joints}\right)\\ &= 3 + 10 - (2)(6) = \boxed{1}\end{aligned}$$

The answer is (A).

3. *Customary U.S. Solution*

The thermal strain is

$$\varepsilon_{\text{th}} = \alpha\Delta T = \left(8.9\times10^{-6}\ \frac{1}{^\circ\text{F}}\right)(55^\circ\text{F}) = 0.000490$$

The strain must be counteracted to maintain the rod in its original position.

$$\sigma = E\varepsilon_{\text{th}} = \left(18\times10^{6}\ \frac{\text{lbf}}{\text{in}^2}\right)(0.000490) = 8820\ \text{lbf/in}^2$$

$$F = \sigma A = \left(8820\ \frac{\text{lbf}}{\text{in}^2}\right)(1\ \text{in})(2\ \text{in}) = \boxed{17{,}640\ \text{lbf}}$$

The answer is (D).

SI Solution

The thermal strain is

$$\varepsilon_{\text{th}} = \alpha\Delta T = \left(16.0\times10^{-6}\ \frac{1}{^\circ\text{C}}\right)(30^\circ\text{C}) = 0.00048$$

The strain must be counteracted to maintain the rod in its original position.

$$\begin{aligned}\sigma = E\varepsilon_{\text{th}} &= \left(12\times10^{4}\ \text{MPa}\right)(0.00048)\\ &= 57.6\ \text{MPa}\end{aligned}$$

$$\begin{aligned}F = \sigma A &= \left(57.6\times10^{6}\ \text{Pa}\right)(0.02\ \text{m})(0.05\ \text{m})\\ &= 57\,600\ \text{N} = \boxed{57.6\ \text{kN}}\end{aligned}$$

The answer is (D).

4. *Customary U.S. Solution*

(a)

$$\sigma = \frac{F}{A} = \frac{2250\ \text{lbf}}{\frac{(\pi)(2\ \text{in})^2}{4}} = \boxed{716.2\ \text{lbf/in}^2}$$

The answer is (B).

(b) $$\varepsilon = \frac{\sigma}{E} = \frac{\delta}{L_o}$$

$$\delta = L_o \frac{\sigma}{E} = (15 \text{ in}) \left(\frac{716.2 \ \frac{\text{lbf}}{\text{in}^2}}{30 \times 10^6 \ \frac{\text{lbf}}{\text{in}^2}} \right)$$

$$= \boxed{3.58 \times 10^{-4} \text{ in}}$$

The answer is (C).

SI Solution

(a) $$\sigma = \frac{F}{A} = \frac{10\,000 \text{ N}}{\frac{\pi(0.05 \text{ m})^2}{4}}$$

$$= 5.093 \times 10^6 \text{ Pa} = \boxed{5.093 \text{ MPa}}$$

The answer is (B).

(b) $$\varepsilon = \frac{\sigma}{E} = \frac{\delta}{L_o}$$

$$\delta = L_o \left(\frac{\sigma}{E} \right) = (0.4 \text{ m}) \left(\frac{5.093 \text{ MPa}}{20 \times 10^4 \text{ MPa}} \right)$$

$$= \boxed{1.02 \times 10^{-5} \text{ m}}$$

The answer is (C).

5. *Customary U.S. Solution*

(a) Let F_c and F_{st} be the loads carried by the concrete and steel, respectively.

$$F = F_c + F_{st} = 4500 \text{ lbf} \quad \text{[Eq. 1]}$$

The deformation of the steel is

$$\delta_{st} = \frac{F_{st} L}{A_{st} E_{st}}$$

The deformation of the concrete is

$$\delta_c = \frac{F_c L}{A_c E_c}$$

The geometric constraint is $\delta_{st} = \delta_c$, or

$$\frac{F_{st} L}{A_{st} E_{st}} = \frac{F_c L}{A_c E_c} \quad \text{[Eq. 2]}$$

Solving Eqs. 1 and 2,

$$F_c = \frac{F}{1 + \frac{A_{st} E_{st}}{A_c E_c}}$$

$$= \frac{4500 \text{ lbf}}{1 + \frac{\frac{(\pi)(0.4 \text{ in})^2}{4} \left(30 \times 10^6 \ \frac{\text{lbf}}{\text{in}^2} \right)}{\frac{(\pi)\left((3.0 \text{ in})^2 - (0.4 \text{ in})^2\right)}{4} \left(2 \times 10^6 \ \frac{\text{lbf}}{\text{in}^2} \right)}}$$

$$= \boxed{3539 \text{ lbf}}$$

$$F_{st} = 4500 \text{ lbf} - F_c = 4500 \text{ lbf} - 3539 \text{ lbf} = \boxed{961 \text{ lbf}}$$

$$\sigma_{st} = \frac{F_{st}}{A_{st}} = \frac{961 \text{ lbf}}{\frac{(\pi)(0.4 \text{ in})^2}{4}} = \boxed{7647 \text{ lbf/in}^2}$$

$$\sigma_c = \frac{F_c}{A_c} = \frac{3539 \text{ lbf}}{\frac{(\pi)\left((3 \text{ in})^2 - (0.4 \text{ in})^2\right)}{4}}$$

$$= \boxed{510 \text{ lbf/in}^2}$$

(b) $$\delta = \frac{F_{st} L}{A_{st} E_{st}} = \frac{(961 \text{ lbf})(12 \text{ in})}{\frac{(\pi)(0.4 \text{ in})^2}{4} \left(30 \times 10^6 \ \frac{\text{lbf}}{\text{in}^2} \right)}$$

$$= \boxed{3.06 \times 10^{-3} \text{ in}}$$

SI Solution

(a) Let F_c and F_{st} be the loads carried by the concrete and steel, respectively.

$$F = F_c + F_{st} = 20 \text{ kN} \quad \text{[Eq. 1]}$$

The deformation of the steel is

$$\delta_{st} = \frac{F_{st} L}{A_{st} E_{st}}$$

The deformation of the concrete is

$$\delta_c = \frac{F_c L}{A_c E_c}$$

The geometric constraint is $\delta_{st} = \delta_c$, or

$$\frac{F_{st} L}{A_{st} E_{st}} = \frac{F_c L}{A_c E_c} \quad \text{[Eq. 2]}$$

Solving Eqs. 1 and 2,

$$F_c = \frac{F}{1 + \frac{A_{st}E_{st}}{A_cE_c}}$$

$$= \frac{20\text{ kN}}{1 + \frac{\left(\frac{\pi}{4}\right)(0.01\text{ m})^2(20 \times 10^4\text{ MPa})}{\left(\frac{\pi}{4}\right)\left((0.075\text{ m})^2 - (0.01\text{ m})^2\right)(15 \times 10^3\text{ MPa})}}$$

$$= \boxed{16.11\text{ kN}}$$

$$F_{st} = 20\text{ kN} - F_c = 20\text{ kN} - 16.11\text{ kN} = \boxed{3.89\text{ kN}}$$

$$\sigma_{st} = \frac{F_{st}}{A_{st}} = \frac{3.89 \times 10^{-3}\text{ MN}}{\left(\frac{\pi}{4}\right)(0.01\text{ m})^2} = \boxed{49.5\text{ MPa}}$$

$$\sigma_c = \frac{F_c}{A_c} = \frac{16.11 \times 10^{-3}\text{ MN}}{\left(\frac{\pi}{4}\right)\left((0.075\text{ m})^2 - (0.01\text{ m})^2\right)}$$

$$= \boxed{3.71\text{ MPa}}$$

(b)

$$\delta = \frac{F_{st}L}{A_{st}E_{st}}$$

$$= \frac{(3.89 \times 10^3\text{ N})(0.3\text{ m})}{\left(\frac{\pi}{4}\right)(0.01\text{ m})^2(20 \times 10^{10}\text{ Pa})}$$

$$= 7.43 \times 10^{-5}\text{ m} = \boxed{74.3\ \mu\text{m}}$$

6. *Customary U.S. Solution*

$$A_{st} = \left(\frac{\pi}{4}\right)\left((6\text{ in})^2 - (5\text{ in})^2\right) = 8.639\text{ in}^2$$

$$A_b = \left(\frac{\pi}{4}\right)\left((2\text{ in})^2 - (1.6\text{ in})^2\right) = 1.131\text{ in}^2$$

The thermal deformation that would occur if the pipes were free is

$$\delta_b = \alpha_b L\Delta T = \left(10 \times 10^{-6}\frac{1}{°\text{F}}\right)(60\text{ in})(200°\text{F})$$

$$= 0.120\text{ in}$$

$$\delta_{st} = \alpha_{st} L\Delta T = \left(6.5 \times 10^{-6}\frac{1}{°\text{F}}\right)(60\text{ in})(200°\text{F})$$

$$= 0.078\text{ in}$$

The two pipes expand by the same amount, δ, so that $\delta_{st} < \delta < \delta_b$.

$$\delta - \delta_{st} = \frac{F_{st}L}{A_{st}E_{st}} \quad \text{[Eq. 1]}$$

$$\delta_b - \delta = \frac{F_bL}{A_bE_b} \quad \text{[Eq. 2]}$$

F_{st} is a tensile force, F_b is a compressive force. Since there is no external force,

$$F_{st} = F_b = F$$

Adding Eq. 1 to Eq. 2,

$$\delta_b - \delta_{st} = \frac{FL}{A_{st}E_{st}} + \frac{FL}{A_bE_b}$$

$$= FL\left(\frac{1}{A_{st}E_{st}} + \frac{1}{A_bE_b}\right)$$

$$F = \frac{\frac{\delta_b - \delta_{st}}{L}}{\frac{1}{A_{st}E_{st}} + \frac{1}{A_bE_b}}$$

$$= \frac{\frac{0.120\text{ in} - 0.078\text{ in}}{60\text{ in}}}{\frac{1}{(8.639\text{ in}^2)\left(30 \times 10^6\ \frac{\text{lbf}}{\text{in}^2}\right)} + \frac{1}{(1.131\text{ in}^2)\left(15 \times 10^6\ \frac{\text{lbf}}{\text{in}^2}\right)}}$$

$$= 11{,}146\text{ lbf}$$

$$\sigma_{st} = \frac{F}{A_{st}} = \frac{11{,}146\text{ lbf}}{8.639\text{ in}^2} = \boxed{1290\text{ psi}}$$

$$\sigma_b = \frac{F}{A_b} = \frac{11{,}146\text{ lbf}}{1.131\text{ in}^2} = \boxed{9855\text{ psi}}$$

SI Solution

$$A_{st} = \left(\frac{\pi}{4}\right)\left((15\text{ cm})^2 - (14\text{ cm})^2\right) = 22.78\text{ cm}^2$$

$$= 0.002\,278\text{ m}^2$$

$$A_b = \left(\frac{\pi}{4}\right)\left((5\text{ cm})^2 - (4\text{ cm})^2\right) = 7.069\text{ cm}^2$$

$$= 0.000\,706\,9\text{ m}^2$$

The thermal deformation that would occur if the pipes were free is

$$\delta_b = \alpha_b L \Delta T = \left(1.8 \times 10^{-5} \frac{1}{^\circ\text{C}}\right)(1.5\ \text{m})(110^\circ\text{C})$$
$$= 0.00297\ \text{m}$$

$$\delta_{\text{st}} = \alpha_{\text{st}} L \Delta T = \left(1.17 \times 10^{-5} \frac{1}{^\circ\text{C}}\right)(1.5\ \text{m})(110^\circ\text{C})$$
$$= 0.00193\ \text{m}$$

The two pipes expand by the same amount, δ, so that $\delta_{\text{st}} < \delta < \delta_b$.

$$\delta - \delta_{\text{st}} = \frac{F_{\text{st}} L}{A_{\text{st}} E_{\text{st}}} \qquad \text{[Eq. 1]}$$

$$\delta_b - \delta = \frac{F_b L}{A_b E_b} \qquad \text{[Eq. 2]}$$

F_{st} is a tensile force, F_b is a compressive force. Since there is no external force,

$$F_{\text{st}} = F_b = F$$

Adding Eq. 1 to Eq. 2,

$$\delta_b - \delta_{\text{st}} = \frac{FL}{A_{\text{st}} E_{\text{st}}} + \frac{FL}{A_b E_b}$$
$$= FL\left(\frac{1}{A_{\text{st}} E_{\text{st}}} + \frac{1}{A_b E_b}\right)$$

$$F = \frac{\frac{\delta_b - \delta_{\text{st}}}{L}}{\frac{1}{A_{\text{st}} E_{\text{st}}} + \frac{1}{A_b E_b}}$$

$$= \frac{\frac{0.00297\ \text{m} - 0.00193\ \text{m}}{1.5\ \text{m}}}{\frac{1}{(0.002278\ \text{m}^2)(20 \times 10^{10}\ \text{Pa})} + \frac{1}{(0.0007069\ \text{m}^2)(10 \times 10^{10}\ \text{Pa})}}$$

$$= 4.24 \times 10^4\ \text{N}$$

$$\sigma_{\text{st}} = \frac{F}{A_{\text{st}}} = \frac{4.24 \times 10^4\ \text{N}}{0.002278\ \text{m}^2} = 1.86 \times 10^7\ \text{Pa}$$
$$= \boxed{18.6\ \text{MPa}}$$

$$\sigma_b = \frac{F}{A_b} = \frac{4.24 \times 10^4\ \text{N}}{0.0007069\ \text{m}^2} = 6.00 \times 10^7\ \text{Pa}$$
$$= \boxed{60.0\ \text{MPa}}$$

7. *Customary U.S. Solution*

The deflection of the beam is $\delta_b = (P_b L^3)/(3EI)$ where P_b is the net load at the beam tip. If P_c is the tension in the cable,

$$\delta_c = \frac{P_c L_c}{AE}$$

$\delta_c = \delta_b$ is the constraint on the deformation. Therefore,

$$\frac{P_b L_b^3}{3EI} = \frac{P_c L_c}{AE} \qquad \text{[Eq. 1]}$$

Another equation is the equilibrium equation.

$$F - P_c = P_b \qquad \text{[Eq. 2]}$$

Solving Eqs. 1 and 2 simultaneously,

$$P_c = \frac{\frac{F L_b^3}{3I}}{\frac{L_c}{A} + \frac{L_b^3}{3I}}$$

$$= \frac{\frac{(270\ \text{lbf})\left((4\ \text{ft})\left(12\ \frac{\text{in}}{\text{ft}}\right)\right)^3}{(3)(10.0\ \text{in}^4)}}{\frac{(2\ \text{ft})\left(12\ \frac{\text{in}}{\text{ft}}\right)}{0.0124\ \text{in}^2} + \frac{\left((4\ \text{ft})\left(12\ \frac{\text{in}}{\text{ft}}\right)\right)^3}{(3)(10.0\ \text{in}^4)}}$$

$$= \boxed{177\ \text{lbf}}$$

The answer is (A).

SI Solution

The deflection of the beam is $\delta_b = (P_b L^3)/(3EI)$ where P_b is the net load at the beam tip. If P_c is the tension in the cable,

$$\delta_c = \frac{P_c L_c}{AE}$$

$\delta_c = \delta_b$ is the constraint on the deformation. Therefore,

$$\frac{P_b L_b^3}{3EI} = \frac{P_c L_c}{AE} \qquad \text{[Eq. 1]}$$

Another equation is the equilibrium equation.

$$F - P_c = P_b \qquad \text{[Eq. 2]}$$

Structural

Solving Eqs. 1 and 2 simultaneously,

$$P_C = \frac{\frac{FL_b^3}{3I}}{\frac{L_C}{A} + \frac{L_b^3}{3I}}$$

$$= \frac{\frac{(1200\text{ N})(1.2\text{ m})^3}{(3)(4.17\times 10^6\text{ mm}^4)\left(\frac{1\text{ m}^4}{10^{12}\text{ mm}^4}\right)}}{\frac{0.6\text{ m}}{(8\text{ mm}^2)\left(\frac{1\text{ m}^2}{10^6\text{ mm}^2}\right)} + \frac{(1.2\text{ m})^3}{(3)(4.17\times 10^6\text{ mm}^4)\left(\frac{1\text{ m}^4}{10^{12}\text{ mm}^4}\right)}}$$

$$= \boxed{778\text{ N}}$$

The answer is (A).

8. *Customary U.S. Solution*

Let deflection down be positive.

The deflection at the center of the beam is

$$\delta_b = \frac{5w(2L)^4}{384EI} - \frac{F(2L)^3}{48EI}$$

F is the force applied by the column at the beam center. The beam deflection must be equal to the shortening of the column.

$$\delta_c = \frac{Fh}{EA}$$

Since $\delta_b = \delta_c$,

$$\frac{Fh}{EA} = \frac{5w(2L)^4}{384EI} - \frac{F(2L)^3}{48EI}$$

$$F = \boxed{\frac{5AwL^4}{24hI + 4AL^3}} \quad \text{[Eq. 1]}$$

Another equation is the equilibrium equation. Let R_1 and R_2 be the left and right support reactions, respectively, on the beam. By symmetry,

$$R_1 = R_2 = R$$
$$2R + F - 2wL = 0 \quad \text{[Eq. 2]}$$

Solving Eqs. 1 and 2 simultaneously,

$$R = \frac{2wL - F}{2}$$

$$= \boxed{wL\left(\frac{48hI + 3AL^3}{48hI + 8AL^3}\right)}$$

SI Solution

Let deflection down be positive.

The deflection at the center of the beam is

$$\delta_b = \frac{5w(2L)^4}{384EI} - \frac{F(2L)^3}{48EI}$$

F is the force applied by the column at the beam center. The beam deflection must be equal to the shortening of the column.

$$\delta_c = \frac{Fh}{EA}$$

Since $\delta_b = \delta_c$,

$$\frac{Fh}{EA} = \frac{5w(2L)^4}{384EI} - \frac{F(2L)^3}{48EI}$$

$$F = \boxed{\frac{5AwL^4}{24hI + 4AL^3}} \quad \text{[Eq. 1]}$$

Another equation is the equilibrium equation. Let R_1 and R_2 be the left and right support reactions, respectively, on the beam. By symmetry,

$$R_1 = R_2 = R$$
$$2R + F - 2wL = 0 \quad \text{[Eq. 2]}$$

Solving Eqs. 1 and 2 simultaneously,

$$R = \frac{2wL - F}{2}$$

$$= \boxed{wL\left(\frac{48hI + 3AL^3}{48hI + 8AL^3}\right)}$$

9. *Customary U.S. Solution*

(a) Let F_1 and F_2 and δ_1 and δ_2 be the tensions and the deformations in the cables, respectively. The moment equilibrium equation taken at the hinge of the rigid bar is

$$\sum M_o = aF_1 + (a+b)F_2 - (a+b+c)P = 0 \quad \text{[Eq. 1]}$$

The relationship between the elongations is

$$\frac{\delta_1}{a} = \frac{\delta_2}{a+b}$$

This can be rewritten as

$$\frac{F_1L_1}{AEa} = \frac{F_2L_2}{AE(a+b)}$$

Since $L_1/a = L_2/(a+b)$,

$$F_1 = F_2 \quad \text{[Eq. 2]}$$

Solving Eqs. 1 and 2,

$$aF + (a+b)F - (a+b+c)P = 0$$

$$\begin{aligned} F = F_1 = F_2 &= \left(\frac{a+b+c}{2a+b}\right)P \\ &= \left(\frac{6\text{ ft} + 3\text{ ft} + 3\text{ ft}}{(2)(6\text{ ft}) + 3\text{ ft}}\right)(4500\text{ lbf}) = \boxed{3600\text{ lbf}} \end{aligned}$$

The answer is (B).

(b) Take downward as a positive deflection. The slope of the rigid member is

$$\frac{d}{a+b+c}$$

$$\begin{aligned} \delta_1 &= \left(\frac{F_1}{AE}\right)L_1 = \left(\frac{F_1}{AE}\right)\left(\frac{d}{a+b+c}\right)a \\ &= \left(\frac{3600\text{ lbf}}{(0.124\text{ in}^2)\left(10\times10^6\ \frac{\text{lbf}}{\text{in}^2}\right)}\right) \\ &\quad \times\left(\frac{12\text{ ft}}{6\text{ ft}+3\text{ ft}+3\text{ ft}}\right)(6\text{ ft}) \\ &= 1.74\times10^{-2}\text{ ft} = \boxed{0.209\text{ in}} \end{aligned}$$

$$\begin{aligned} \delta_2 &= \left(\frac{F_2}{AE}\right)L_2 = \left(\frac{F_2}{AE}\right)\left(\frac{d}{a+b+c}\right)(a+b) \\ &= \left(\frac{3600\text{ lbf}}{(0.124\text{ in}^2)\left(10\times10^6\ \frac{\text{lbf}}{\text{in}^2}\right)}\right) \\ &\quad \times\left(\frac{12\text{ ft}}{6\text{ ft}+3\text{ ft}+3\text{ ft}}\right)(6\text{ ft}+3\text{ ft}) \\ &= 2.61\times10^{-2}\text{ ft} = \boxed{0.314\text{ in}} \end{aligned}$$

The answer is (D).

SI Solution

(a) Let F_1 and F_2 and δ_1 and δ_2 be the tensions and the deformations in the cables, respectively. The moment equilibrium equation taken at the hinge of the rigid bar is

$$\sum M_O = aF_1 + (a+b)F_2 - (a+b+c)P = 0 \quad \text{[Eq. 1]}$$

The relationship between the elongations is

$$\frac{\delta_1}{a} = \frac{\delta_2}{a+b}$$

This can be rewritten as

$$\frac{F_1L_1}{AEa} = \frac{F_2L_2}{AE(a+b)}$$

Since $\dfrac{L_1}{a} = \dfrac{L_2}{a+b}$,

$$F_1 = F_2 \quad \text{[Eq. 2]}$$

Solving Eqs. 1 and 2,

$$aF + (a+b)F - (a+b+c)P = 0$$

$$\begin{aligned} F = F_1 = F_2 &= \left(\frac{a+b+c}{2a+b}\right)P \\ &= \left(\frac{2\text{ m}+1\text{ m}+1\text{ m}}{4\text{ m}+1\text{ m}}\right)(20\text{ kN}) = \boxed{16\text{ kN}} \end{aligned}$$

The answer is (B).

(b) Take downward as a positive deflection. The slope of the rigid member is

$$\frac{d}{a+b+c}$$

$$\begin{aligned} \delta_1 &= \left(\frac{F_1}{AE}\right)L_1 = \left(\frac{F_1}{AE}\right)\left(\frac{d}{a+b+c}\right)a \\ &= \left(\frac{16\,000\text{ N}}{(80\text{ mm}^2)\left(\frac{1\text{ m}^2}{10^6\text{ mm}^2}\right)(70\times10^9\text{ Pa})}\right) \\ &\quad \times\left(\frac{4\text{ m}}{2\text{ m}+1\text{ m}+1\text{ m}}\right)(2\text{ m}) \\ &= 5.71\times10^{-3}\text{ m} = \boxed{5.71\text{ mm}} \end{aligned}$$

$$\begin{aligned} \delta_2 &= \left(\frac{F_2}{AE}\right)L_2 = \left(\frac{F_2}{AE}\right)\left(\frac{d}{a+b+c}\right)(a+b) \\ &= \left(\frac{16\,000\text{ N}}{(80\text{ mm}^2)\left(\frac{1\text{ m}^2}{10^6\text{ mm}^2}\right)(70\times10^9\text{ Pa})}\right) \\ &\quad \times\left(\frac{4\text{ m}}{2\text{ m}+1\text{ m}+1\text{ m}}\right)(2\text{ m}+1\text{ m}) \\ &= 8.57\times10^{-3}\text{ m} = \boxed{8.57\text{ mm}} \end{aligned}$$

The answer is (D).

10. *Customary U.S. Solution*

Let subscript s refer to the supported beam and subscript c refer to the cantilever beam. The deflections are equal.

$$\begin{aligned} \delta_s &= \delta_c \\ \frac{P_sL^3}{48EI} &= \frac{P_cL^3}{3EI} \\ P_s &= 16P_c \end{aligned} \quad \text{[Eq. 1]}$$

Structural

P_S and P_C are the net loads exerted on the supported beam center and the cantilever beam tip, respectively. The equilibrium equation for the supported beam is

$$\sum F_{S,y} = 2R - P_S = 0 \quad \text{[Eq. 2]}$$

The equilibrium equation for the cantilever beam is

$$\sum F_{C,y} = P_C = F - P_S \quad \text{[Eq. 3]}$$

Solving Eqs. 1 and 3 simultaneously,

$$P_C = \frac{F}{17}$$

$$P_S = \frac{16}{17}F$$

From Eq. 2,

$$R = \boxed{\frac{8}{17}F}$$

The answer is (C).

SI Solution

Let subscript s refer to the supported beam and subscript c refer to the cantilever beam. The deflections are equal.

$$\delta_S = \delta_C$$

$$\frac{P_S L^3}{48EI} = \frac{P_C L^3}{3EI}$$

$$P_S = 16P_C \quad \text{[Eq. 1]}$$

P_S and P_C are the net loads exerted on the supported beam center and the cantilever beam tip, respectively. The equilibrium equation for the supported beam is

$$\sum F_{S,y} = 2R - P_S = 0 \quad \text{[Eq. 2]}$$

The equilibrium equation for the cantilever beam is

$$\sum F_{C,y} = P_C = F - P_S \quad \text{[Eq. 3]}$$

Solving Eqs. 1 and 3 simultaneously,

$$P_C = \frac{F}{17}$$

$$P_S = \frac{16}{17}F$$

From Eq. 2,

$$R = \boxed{\frac{8}{17}F}$$

The answer is (C).

11. *Customary U.S. Solution*

Assume deflection downward is positive.

step 1: Remove support 2 to make the structure statically determinate.

step 2: The deflection at the location of (removed) support 2 is the sum of the deflections induced by the discrete load and the distributed load.

From a beam deflection table,

$$\delta_{\text{discrete}} = \frac{Pb}{6EIL}\left(\left(\frac{L}{b}\right)(x-a)^3 + (L^2 - b^2)x - x^3\right)$$

$$(P = 5000,\ L = 15,\ a = 6,\ b = 9,\ x = 10)$$

From a beam deflection table,

$$\delta_{\text{distributed}} = -\frac{w}{24EI}(L^3x - 2Lx^3 + x^4)$$

$$(w = 1000)$$

step 3: The deflection induced by R_2 alone considered as a load is

$$\delta_{R_2} = \frac{-R_2a^2b^2}{3EIL} \quad (a = 10,\ b = 5)$$

step 4: The total deflection at the location of support 2 is zero.

$$\delta_{\text{discrete}} + \delta_{\text{distributed}} + \delta_{R_2} = 0$$

$$\begin{aligned}&\left(\frac{(5000)(9)}{(6)(15)}\right)\\ &\times\left(\left(\frac{15}{9}\right)(10-6)^3 + \left((15)^2 - (9)^2\right)(10) - (10)^3\right)\\ &+\left(\frac{1000}{24}\right)\left((15)^3(10) - (2)(15)(10)^3 + (10)^4\right)\\ &-\frac{R_2(10)^2(5)^2}{(3)(15)} = 0\end{aligned}$$

$$R_2 = \boxed{15{,}233}$$

The answer is (C).

SI Solution

Assume deflection downward is positive.

step 1: Remove support 2 to make the structure statically determinate.

step 2: The deflection at the location of (removed) support 2 is the sum of the deflections induced by the discrete load and the distributed load.

From a beam deflection table,

$$\delta_{\text{discrete}} = \frac{Pb}{6EIL}\left(\left(\frac{L}{b}\right)(x-a)^3 + (L^2-b^2)x - x^3\right)$$

$$(P = 5000,\ L = 15,\ a = 6,\ b = 9,\ x = 10)$$

From a beam deflection table,

$$\delta_{\text{distributed}} = -\frac{w}{24EI}(L^3x - 2Lx^3 + x^4)$$

$$(w = 1000)$$

step 3: The deflection induced by R_2 alone considered as a load is

$$\delta_{R_2} = \frac{-R_2a^2b^2}{3EIL} \quad (a = 10,\ b = 5)$$

step 4: The total deflection at the location of support 2 is zero.

$$\begin{aligned}
&\delta_{\text{discrete}} + \delta_{\text{distributed}} + \delta_{R_2} = 0\\
&\left(\frac{(5000)(9)}{(6)(15)}\right)\\
&\quad\times\left(\left(\frac{15}{9}\right)(10-6)^3 + \left((15)^2-(9)^2\right)(10) - (10)^3\right)\\
&\quad+\left(\frac{1000}{24}\right)\left((15)^3(10) - (2)(15)(10)^3 + (10)^4\right)\\
&\quad-\frac{R_2(10)^2(5)^2}{(3)(15)} = 0\\
&R_2 = \boxed{15\,233}
\end{aligned}$$

The answer is (C).

12. *Customary U.S. Solution*

First, remove the prop to make the structure statically determinate. The deflection induced by the distributed load at the tip is

$$\delta_{\text{distributed}} = \frac{-wL^4}{30EI} \quad \text{(down)}$$

The deflection caused by load R_1 is

$$\delta_R = \frac{RL^3}{3EI} \quad \text{(up)}$$

Since the deflection is actually zero at the tip,

$$\begin{aligned}
\delta_{\text{distributed}} + \delta_R &= 0\\
\frac{-wL^4}{30EI} + \frac{RL^3}{3EI} &= 0\\
R = \frac{wL}{10} = \frac{(1000)(10)}{10} &= \boxed{1000}
\end{aligned}$$

The answer is (B).

SI Solution

First, remove the prop to make the structure statically determinate. The deflection induced by the distributed load at the tip is

$$\delta_{\text{distributed}} = \frac{-wL^4}{30EI} \quad \text{(down)}$$

The deflection caused by load R_1 is

$$\delta_R = \frac{RL^3}{3EI} \quad \text{(up)}$$

Since the deflection is actually zero at the tip,

$$\begin{aligned}
\delta_{\text{distributed}} + \delta_R &= 0\\
\frac{-wL^4}{30EI} + \frac{RL^3}{3EI} &= 0\\
R = \frac{wL}{10} = \frac{(1000)(10)}{10} &= \boxed{1000}
\end{aligned}$$

The answer is (B).

13. *Customary U.S. Solution*

Use the three-moment method. The first moment of the area is

$$\begin{aligned}
A_1a &= \tfrac{1}{6}Fc(L^2 - c^2)\\
&= \left(\tfrac{1}{6}\right)(13{,}600)(22.5)\left((32.5)^2 - (22.5)^2\right)\\
&= 28{,}050{,}000
\end{aligned}$$

Since there is no force between R_2 and R_3, $A_2b = 0$.

The left and right ends of the beam are simply supported; M_1 and M_3 are zero. Therefore, the three-moment equation becomes

$$\begin{aligned}
2M_2(32.5 + 20) &= (-6)\left(\frac{28{,}050{,}000}{32.5}\right)\\
M_2 &= -49{,}318.7
\end{aligned}$$

M_2 can be written in terms of the load and reactions to the left of support 2.

$$\begin{aligned}
M_2 &= (-13{,}600)(10) + (32.5)(R_1) = -49{,}318.7\\
R_1 &= \boxed{2667.1 \text{ lbf}}
\end{aligned}$$

Structural

Now that R_1 is known, moments can be taken about support 3 to the left.

$$\sum M_3 = (2667.1)(52.5) - (13,600)(30) + (R_2)(20) = 0$$

$$R_2 = \boxed{13,398.9 \text{ lbf}}$$

R_3 can be obtained by taking moments about support 1.

$$\sum M_1 = (22.5)(13,600) - (32.5)(13,398.9) - (52.5)(R_3) = 0$$

$$R_3 = \boxed{-2466.0 \text{ lbf}}$$

SI Solution

Use the three-moment method. The first moment of the area is

$$\begin{aligned} A_1 a &= \frac{1}{6} Fc(L^2 - c^2) \\ &= \left(\frac{1}{6}\right)(60\,500 \text{ N})(7 \text{ m})\big((10 \text{ m})^2 - (7 \text{ m})^2\big) \\ &= 3.60 \times 10^6 \text{ N·m}^3 \end{aligned}$$

Since there is no force between R_2 and R_3, $A_2 b = 0$.

The left and right ends of the beam are simply supported; M_1 and M_3 are zero. Therefore, the three-moment equation becomes

$$2M_2(10 \text{ m} + 6 \text{ m}) = (-6)\left(\frac{3.6 \times 10^6 \text{ N·m}^3}{10 \text{ m}}\right)$$

$$M_2 = -67\,500 \text{ N·m}$$

M_2 can be written in terms of the load and reactions to the left of support 2.

$$\begin{aligned} M_2 &= (-60\,500 \text{ N})(10 \text{ m} - 7 \text{ m}) + (10 \text{ m})(R_1) \\ &= -67\,500 \text{ N·m} \end{aligned}$$

$$R_1 = \boxed{11\,400 \text{ N}}$$

Now that R_1 is known, moments can be taken about support 3 to the left.

$$\sum M_3 = (11\,400 \text{ N})(16 \text{ m}) - (60\,500 \text{ N})(9 \text{ m}) + (R_2)(6 \text{ m}) = 0$$

$$R_2 = \boxed{60\,350 \text{ N}}$$

R_3 can be obtained by taking moments about support 1.

$$\sum M_1 = (7 \text{ m})(60\,500 \text{ N}) - (10 \text{ m})(60\,350 \text{ N}) - (16 \text{ m})(R_3) = 0$$

$$R_3 = \boxed{-11\,250 \text{ N}}$$

14. *Customary U.S. Solution*

Use the three-moment method. The first moments of the areas are

$$\begin{aligned} A_1 a &= \frac{1}{6} Fc(L_1^2 - c^2) = \left(\frac{1}{6}\right)(5000)(4)\big((10)^2 - (4)^2\big) \\ &= 280{,}000 \end{aligned}$$

$$A_2 b = \frac{wL_2^4}{24} = \frac{(1000)(5)^4}{24} = 26{,}042$$

The two ends of the beam are simply supported; M_1 and M_3 are zero. Therefore, the three moment equation becomes

$$2M_2(10 + 5) = (-6)\left(\frac{280{,}000}{10} + \frac{26{,}042}{5}\right)$$

$$M_2 = -6642$$

M_2 can also be written in terms of the load and reactions to the left of support 2.

$$M_2 = (-5000)(6) + 10R_1 = -6642$$

$$R_1 = \boxed{2336}$$

Sum the moments around support 3.

$$\sum M_3 = (2336)(15) - (5000)(11) + 5R_2 - (5000)(2.5) = 0$$

$$R_2 = \boxed{6492}$$

Sum the moments around support 1.

$$\sum M_1 = (5000)(4) - (6492)(10) + (5000)(12.5) - 15R_3 = 0$$

$$R_3 = \boxed{1172}$$

SI Solution

Use the three-moment method. The first moments of the areas are

$$\begin{aligned} A_1 a &= \frac{1}{6} Fc(L_1^2 - c^2) = \left(\frac{1}{6}\right)(5000)(4)\big((10)^2 - (4)^2\big) \\ &= 280\,000 \end{aligned}$$

$$A_2 b = \frac{wL_2^4}{24} = \frac{(1000)(5)^4}{24} = 26\,042$$

The two ends of the beam are simply supported; M_1 and M_3 are zero. Therefore, the three moment equation becomes

$$2M_2(10+5) = (-6)\left(\frac{280\,000}{10} + \frac{26\,042}{5}\right)$$
$$M_2 = -6642$$

M_2 can also be written in terms of the load and reactions to the left of support 2.

$$M_2 = (-5000)(6) + 10R_1 = -6642$$
$$R_1 = \boxed{2336}$$

Sum the moments around support 3.

$$\sum M_3 = (2336)(15) - (5000)(11) + 5R_2 - (5000)(2.5) = 0$$
$$R_2 = \boxed{6492}$$

Sum the moments around support 1.

$$\sum M_1 = (5000)(4) - (6492)(10) + (5000)(12.5) - 15R_3 = 0$$
$$R_3 = \boxed{1172}$$

15. *Customary U.S. Solution*

The equilibrium requirement is

$$R_1 + R_2 = 1000$$

The moment equation at support 1 is

$$\sum M_{R_1} = M_1 + M_2 + (1000)(30) - 40R_2 = 0$$

From a table of fixed-end moments,

$$M_1 = \frac{-Fb^2a}{L^2} = \frac{-(1000)(10)^2(30)}{(40)^2} = -1875$$
$$M_2 = \frac{Fa^2b}{L^2} = \frac{(1000)(30)^2(10)}{(40)^2} = 5625$$
$$R_2 = \boxed{843.75}$$

The moment equation at support 2 is

$$\sum M_{R_2} = M_1 + M_2 - (1000)(10) + 40R_1 = 0$$
$$R_1 = \boxed{156.25}$$

SI Solution

From a table of fixed-end moments,

$$M_1 = \frac{-Fb^2a}{L^2} = \frac{-(1000)(10)^2(30)}{(40)^2} = -1875$$
$$M_2 = \frac{Fa^2b}{L^2} = \frac{(1000)(30)^2(10)}{(40)^2} = 5625$$

The moment equation at support 1 is

$$\sum M_{R_1} = M_1 + M_2 + (1000)(30) - 40R_2 = 0$$
$$-1875 + 5625 + (1000)(30) - 40R_2 = 0$$
$$R_2 = \boxed{843.75}$$

The moment equation at support 2 is

$$\sum M_{R_2} = M_1 + M_2 - (1000)(10) + 40R_1 = 0$$
$$-1875 + 5625 - (1000)(10) + 40R_1 = 0$$
$$R_1 = \boxed{156.25}$$

The equilibrium requirement is

$$R_1 + R_2 = 1000$$
$$156.25 + 843.75 = 1000 \quad \text{(check)}$$

16. The force in member Bb depends on the shear, V, across the cut shown.

(a)

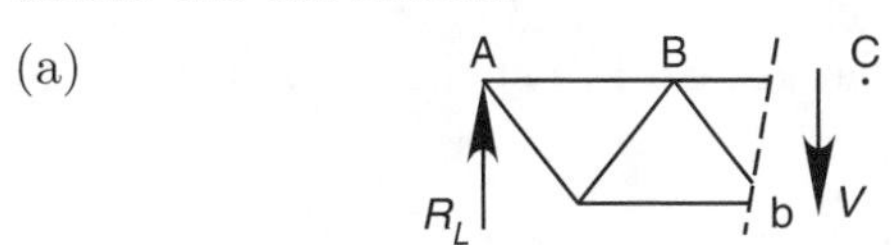

- Influence diagram for shear across panel Bb:

If the unit load is to the right of point C, the reaction R_L is

$$R_L = \frac{x}{120 \text{ ft}}$$

(x is the distance from the right reaction to the unit load.)

$$V = R_L = \frac{x}{120 \text{ ft}}$$

If the unit load is to the left of point B,

$$R_L = \frac{x}{120 \text{ ft}}$$
$$V = R_L - 1 = \frac{x}{120 \text{ ft}} - 1$$

At points B and C,

$$V_B = \frac{90 \text{ ft}}{120 \text{ ft}} - 1 = -0.25$$
$$V_C = \frac{60 \text{ ft}}{120 \text{ ft}} = 0.5$$

The influence diagram is

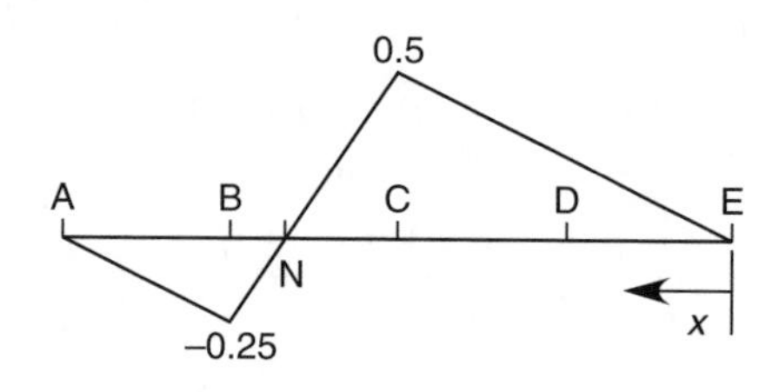

The neutral point, N, is located at

$$x = (2)(30\ \text{ft}) + \left(\frac{0.5}{0.5 + 0.25}\right)(30\ \text{ft}) = 80\ \text{ft}$$

• Maximum shear due to moving uniform load:

The moving load, (perhaps) representing a stream of cars, is allowed to be over any part or all of the bridge deck. The shear will be maximum in member Bb if the load is distributed from N to E.

The area under the influence line from N to E is

$$\left(\tfrac{1}{2}\right)\big((20\ \text{ft})(0.5) + (2)(30\ \text{ft})(0.5)\big) = 20\ \text{ft}$$

The maximum shear, V, is

$$(20\ \text{ft})\left(2\ \frac{\text{kips}}{\text{ft}}\right) = 40\ \text{kips}$$

• Maximum shear due to moving concentrated load:

From the influence diagram, maximum shear will occur when the concentrated load is at point C. The shear in panel Bb is

$$(0.5)(15\ \text{kips}) = 7.5\ \text{kips}$$

• Tension in member Bb:

The force triangle is

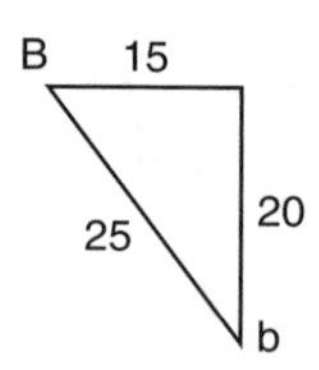

The total maximum shear across panel Bb is

$$40\ \text{kips} + 7.5\ \text{kips} = 47.5\ \text{kips}$$

The total maximum tension in member Bb is

$$(47.5\ \text{ft})\left(\frac{25\ \text{ft}}{20\ \text{ft}}\right) = \boxed{59.375\ \text{kips}}$$

The answer is (C).

(b) • Influence diagram for moment at point b:

Since the horizontal member BC cannot resist vertical shear, the shear influence diagram previously used will not work.

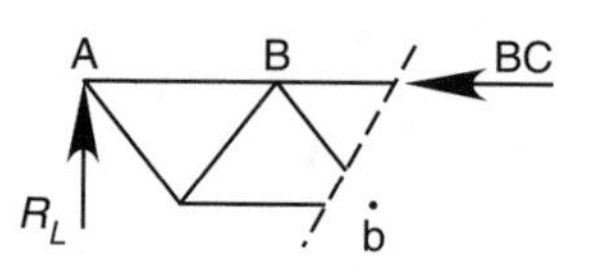

With no loads between A and C, the force in member BC can be found by summing the moments about point b. Taking clockwise moments as positive,

$$\sum M_\text{b} = (45\ \text{ft})R_L - (20\ \text{ft})(\text{BC}) = 0$$
$$\text{BC} = \frac{45R_L}{20\ \text{ft}}$$

$45R_L$ is the moment that the moment from force BC opposes. In general,

$$\text{BC} = \frac{M_\text{b}}{20\ \text{ft}}$$

If the load is between C and E,

$$R_L = \frac{x}{120\ \text{ft}} \quad [x \text{ is measured from E}]$$

The moment caused by R_L is

$$M_\text{b} = (45\ \text{ft})\left(\frac{x}{120\ \text{ft}}\right) = 0.375x$$

If the load is between A and B, the reaction is

$$R_L = \frac{x}{120\ \text{ft}}$$

The moment at b is also affected by the load between A and B.

$$M_\text{b} = 0.375x - (1)(x - 75\ \text{ft})$$
$$= 75\ \text{ft} - 0.625x$$

Plotting these values versus x,

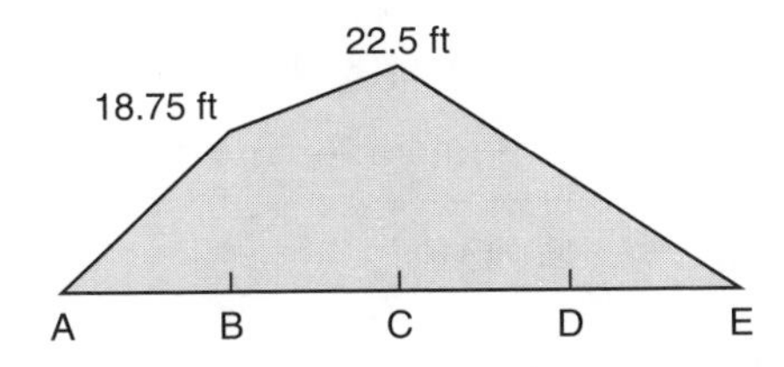

• Maximum moment due to uniform load:

The moment at b is maximum when the entire truss is loaded from A to E. The area under the curve is

$$\begin{aligned}&\left(\tfrac{1}{2}\right)(30\text{ ft})(18.75\text{ ft}) + \left(\tfrac{1}{2}\right)(60\text{ ft})(22.5\text{ ft})\\&\quad + (30\text{ ft})(18.75\text{ ft})\\&\quad + \left(\tfrac{1}{2}\right)(30\text{ ft})(22.5\text{ ft} - 18.75\text{ ft})\\&= 1575\text{ ft}^2\end{aligned}$$

The maximum moment is

$$M_\text{b} = \left(2\ \frac{\text{kips}}{\text{ft}}\right)(1575\text{ ft}^2) = 3150\text{ ft-kips}$$

• Maximum moment due to concentrated load:

Maximum moment will occur when the load is at C.

$$M_\text{b} = (15\text{ ft})(22.5\text{ kips}) = 337.5\text{ ft-kips}$$

• Total maximum moment:

$$M_\text{b} = 337.5\text{ ft-kips} + 3150\text{ ft-kips} = 3487.5\text{ ft-kips}$$

• Compression in BC:

$$\text{BC} = \frac{M_\text{b}}{20\text{ ft}} = \frac{3487.5\text{ ft-kips}}{20\text{ ft}} = \boxed{174.4\text{ kips}}$$

The answer is (D).

17. The force in member CD is a function of the moment at point d.

If the unit load is between d and g, the left reaction is

$$R_L = \frac{x}{162\text{ ft}}$$

The moment at point d is

$$M_\text{d} = (3)(27\text{ ft})\left(\frac{x}{162\text{ ft}}\right) = 0.5x \quad [\text{d to g}]$$

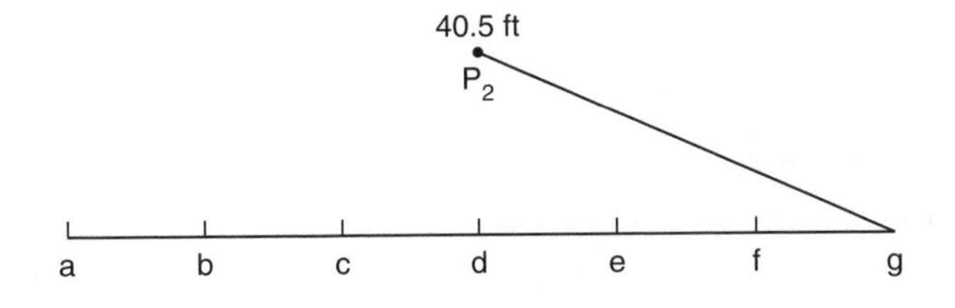

If the unit load is between a and c, the moment at point d is

$$\begin{aligned}M_\text{d} &= 0.5x - (1)\big(x - (3)(27\text{ ft})\big)\\&= 81\text{ ft} - 0.5x \quad [\text{a to c}]\end{aligned}$$

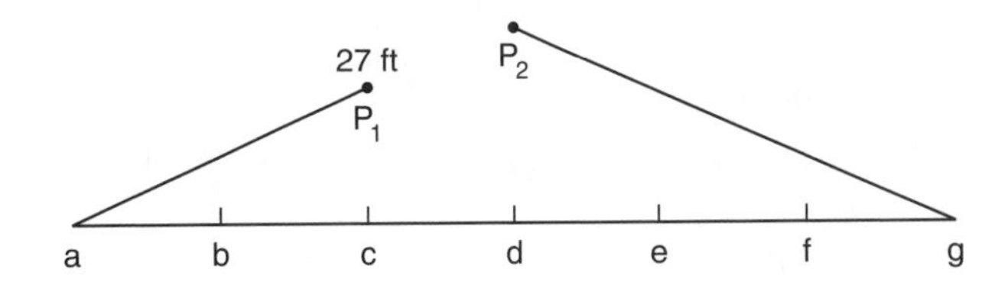

Complete the influence diagram by joining points P_1 and P_2. Observe that the slope of P_1 is the same as that of P_2. This is because point d is at the center of the truss.

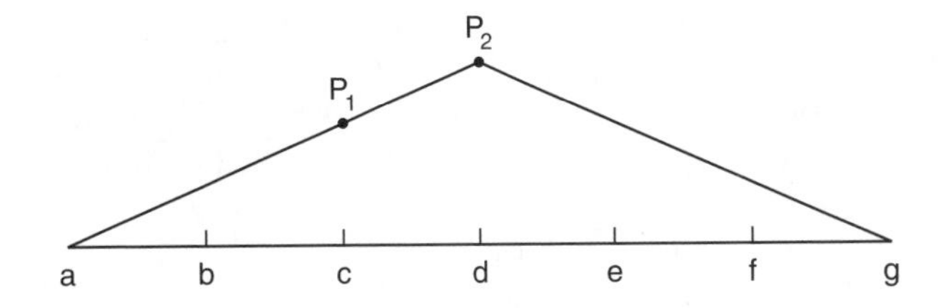

The resultant of the load group and its location are

$$8\text{ kips} + 40\text{ kips} + 7\text{ kips} + 30\text{ kips} = 85\text{ kips}$$

$$\frac{\begin{array}{c}(40\text{ kips})(16\text{ ft}) + (7\text{ kips})(16\text{ ft} + 28\text{ ft}\\ + (30\text{ kips})(16\text{ ft} + 28\text{ ft} + 16\text{ ft})\end{array}}{85\text{ kips}} = 32.33\text{ ft}$$

The resultant is located 32.33 ft to the right of the 8 kip load.

Assume the load group moves from right to left.

• Case 1:

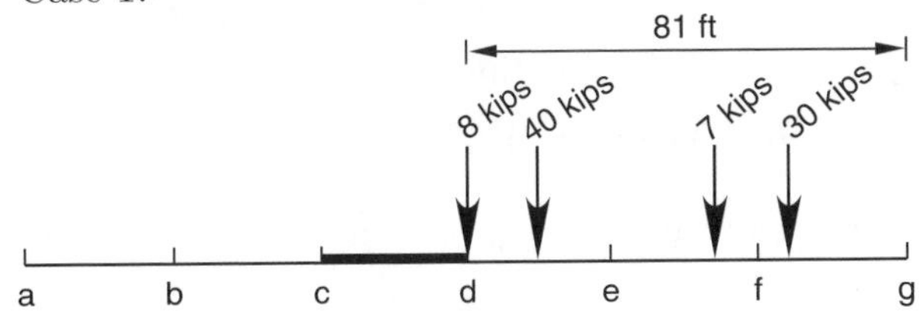

Taking counterclockwise moments as positive,

$$\begin{aligned}\sum M_\text{g} &= (85\text{ kips})(81\text{ ft} - 32.33\text{ ft}) - (162\text{ ft})R_L = 0\\R_L &= 25.54\text{ kips}\end{aligned}$$

The shear does not change sign under panel cd.

• Case 2:

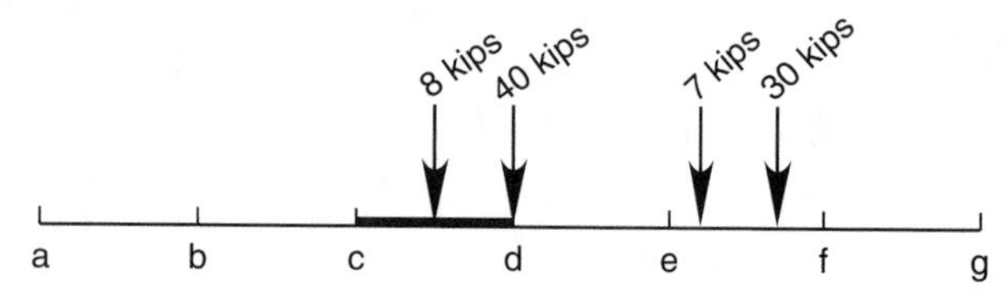

$$R_L = 33.9 \text{ kips}$$
$$V_{CD} = 33.9 \text{ kips} - 8 \text{ kips} - 40 \text{ kips} = -14.1 \text{ kips}$$

Since the shear changes sign (goes through zero), the moment is maximum when the 40 kip load is at point d.

load	x	influence diagram height	moment
8 kips	97 ft	32.5	$(8)(32.5) = 260$ ft-kips
40 kips	81 ft	40.5	$(40)(40.5) = 1620$ ft-kips
7 kips	53 ft	26.5	$(7)(26.5) = 185.5$ ft-kips
30 kips	37 ft	18.5	$(30)(18.5) = 555$ ft-kips
total			2620.5 ft-kips

The compression in CD is

$$\text{CD} = \frac{2620.5 \text{ ft-kips}}{32 \text{ ft}} = \boxed{81.9 \text{ kips}}$$

The answer is (B).

Since $7 + 30 < 8 + 40$, if the load moves from left to right it must reach the same position as case 2 for the moment to be maximum. Therefore, the left-to-right analysis is not needed.

• Alternate solution:

Having determined that the 40 kip load should be at point d, find the left reaction.

$$\begin{aligned} \sum M_{R_R} &= (162 \text{ ft})R_L - (8 \text{ kips})(97 \text{ ft}) \\ &\quad - (40 \text{ kips})(81 \text{ ft}) - (7 \text{ kips})(53 \text{ ft}) \\ &\quad - (30 \text{ kips})(37 \text{ ft}) = 0 \\ R_L &= 33.93 \text{ kips} \end{aligned}$$

Sum the moments about point d and use the method of sections.

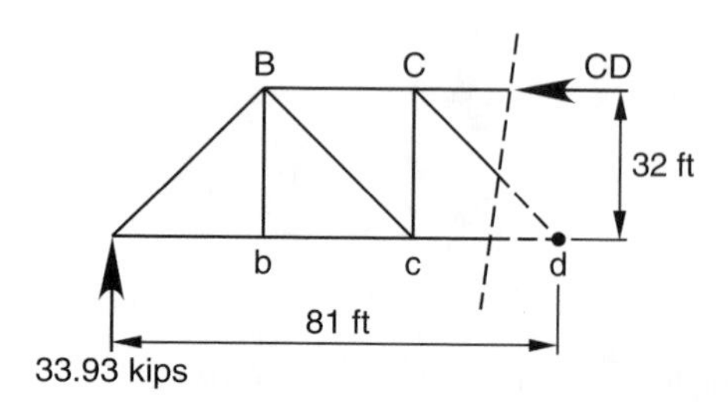

Taking clockwise moments as positive,

$$\begin{aligned} \sum M_{\text{d}} &= (33.93 \text{ kips})(81 \text{ ft}) - (32 \text{ ft})(\text{CD}) \\ &\quad - (8 \text{ kips})(16 \text{ ft}) = 0 \\ \text{CD} &= \boxed{81.9 \text{ kips}} \end{aligned}$$

The answer is (B).

18. (a) • Moment:

Put a hinge at point B and rotate.

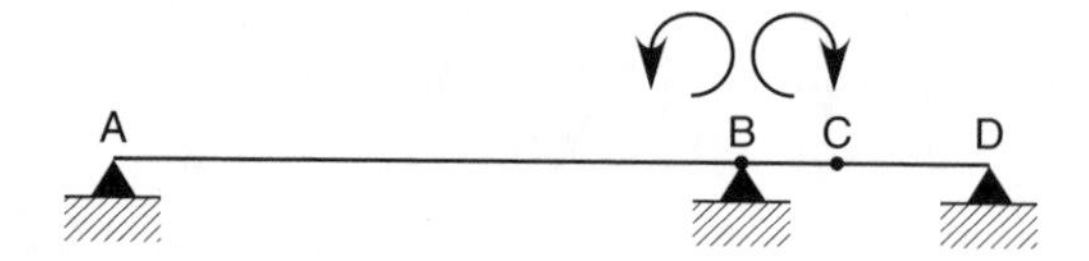

The moment influence diagram is

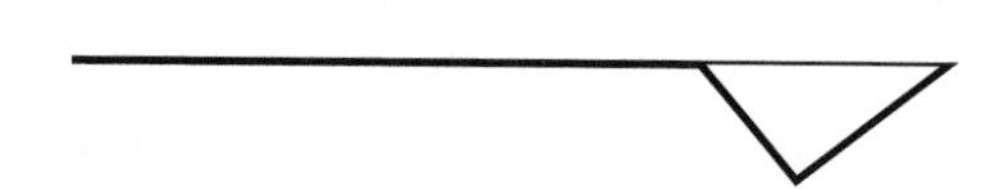

One of the loads should be at point C. Since the slope of the influence line is less between C and D, the ordinate 6 ft to the right of point C will be larger than the ordinate 6 ft to the left of point C. The reaction at the hinge due to the 30 kip load 6 ft to the right of point C is

$$R = (30 \text{ kips})\left(\frac{15 \text{ ft} - 6 \text{ ft}}{15 \text{ ft}}\right) = 18 \text{ kips}$$

The moment at point B is

$$\begin{aligned} M_{\text{B}} &= (9 \text{ ft})(30 \text{ kips} + 18 \text{ kips}) \\ &= \boxed{432 \text{ ft-kips}} \end{aligned}$$

The answer is (D).

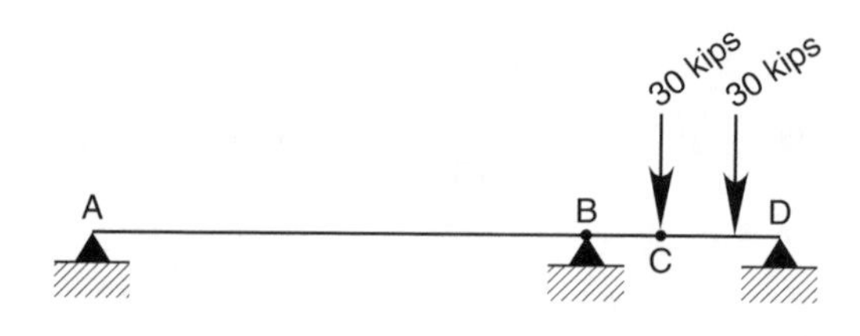

(b) • Shear:

Use the method of virtual displacement to draw the shear influence diagram. Since the point is a reaction point, lift the point a distance of 1. The shear diagram is

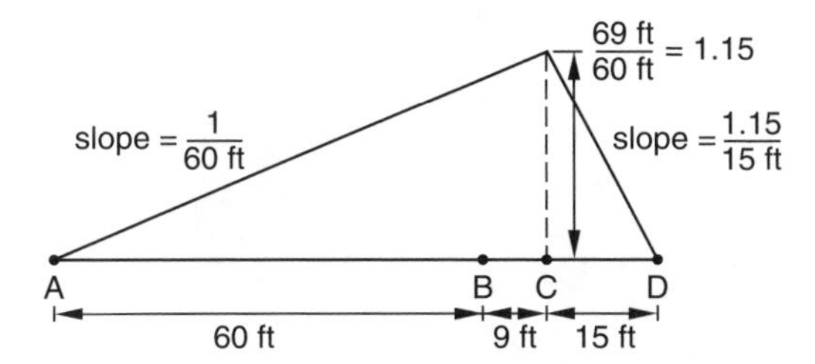

For maximum shear, one load or the other must be at point C.

By inspection, the effect of having both loads to the left of C is greater than having them to the right.

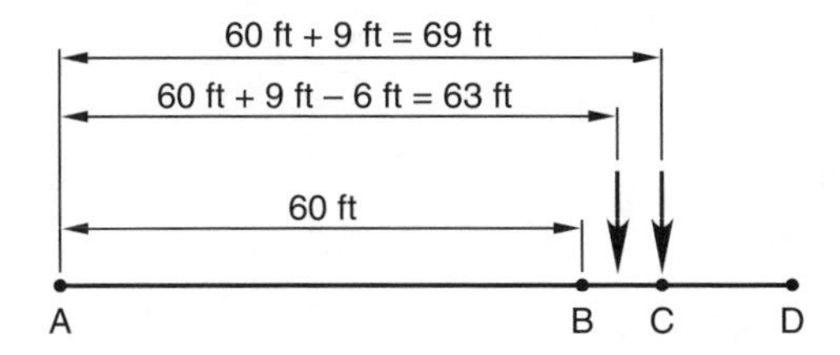

The maximum shear is

$$V_{\text{max}} = (30 \text{ kips})\left(\frac{63 \text{ ft}}{60 \text{ ft}} + \frac{69 \text{ ft}}{60 \text{ ft}}\right) = \boxed{66 \text{ kips}}$$

The answer is (C).

Structural

47 Structural Analysis II

PRACTICE PROBLEMS

1. (*Time limit: one hour*) A single-story rigid timber frame is subjected to a uniform load acting over the beam plus a concentrated load at midheight of the left side column. The two columns are solid square, 14 in by 14 in. The beam is 12 in wide and 24 in deep. Determine the moment distribution.

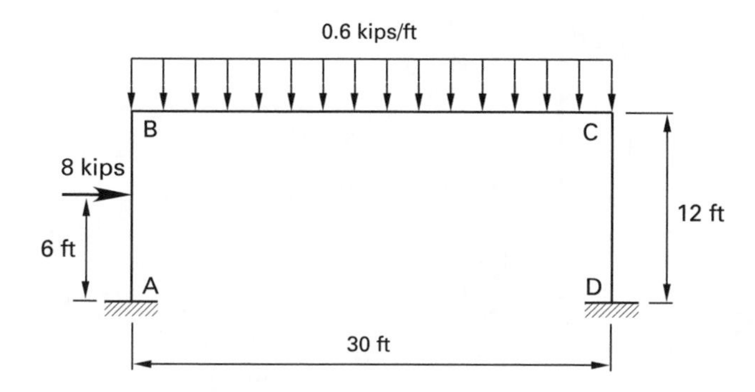

2. A cantilever beam with a cable attached to its free end is loaded uniformly as shown. For the dimensions, properties, and loading indicated, determine (a) the force in the cable and (b) the maximum moment in the beam.

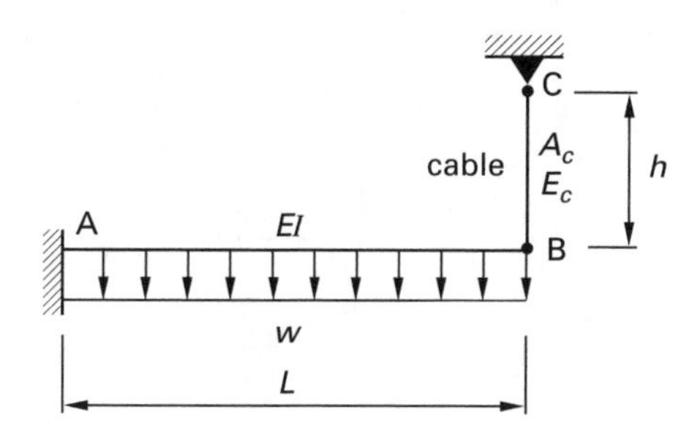

w = 1 kip/ft
L = 10 ft
h = 5 ft
E = 29,000 kips/in^2
E_c = 20,000 kips/in^2
A_c = 0.4 in^2
I = 200 in^4

3. The column is loaded as shown. The product EI has a value of 10^4 kip-ft^2.

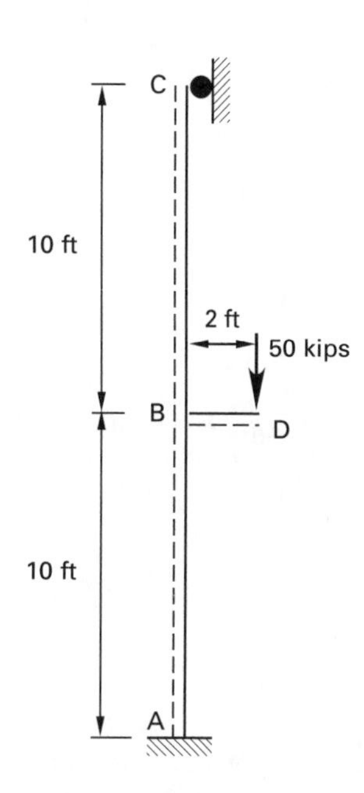

(a) What is the degree of static indeterminacy?

(A) 0 (determinate)
(B) 1
(C) 2
(D) 3

(b) Using the convention that compression on the side of the dashed line is positive, the moment at point A is most nearly

(A) −25 ft-kips
(B) 13 ft-kips
(C) 25 ft-kips
(D) 50 ft-kips

(c) The shear at point A is most nearly

(A) 0
(B) 5.6 kips
(C) 10 kips
(D) 50 kips

(d) The axial force in bar CB is most nearly

(A) 0
(B) 25 kips (tension)
(C) 50 kips (compression)
(D) 50 kips (tension)

(e) The shear in bar CB is most nearly

(A) 0
(B) 5.6 kips
(C) 10 kips
(D) 50 kips

(f) The point of inflection in bar AB is most nearly

(A) at A (there is no inflection point)
(B) 2.2 ft from A
(C) 5 ft from A
(D) 7.5 ft from A

(g) The maximum moment in either AB or BC is most nearly

(A) −56 ft-kips
(B) 50 ft-kips
(C) 56 ft-kips
(D) 100 ft-kips

(h) Using the convention that deflection to the right is positive, the horizontal displacement of point B is most nearly

(A) −0.03 ft
(B) 0
(C) 0.03 ft
(D) 0.06 ft

(i) If the support A is replaced by a pin, the moment at B in member BA will be most nearly

(A) −50 ft-kips
(B) −25 ft-kips
(C) 25 ft-kips
(D) 100 ft-kips

(j) If the support at A is replaced by a pin, the maximum shear in member CB will be most nearly

(A) 0
(B) 5 kips
(C) 7.5 kips
(D) 10 kips

4. The three structures shown are designated as S1, S2, and S3.

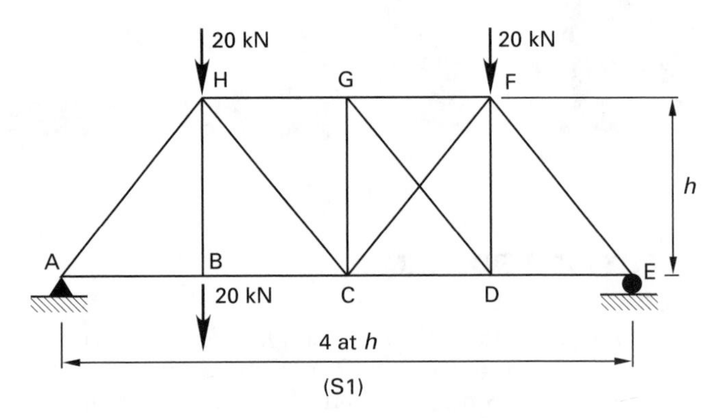

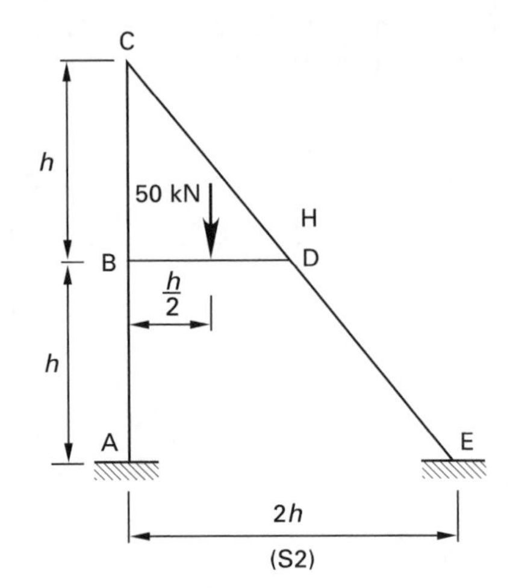

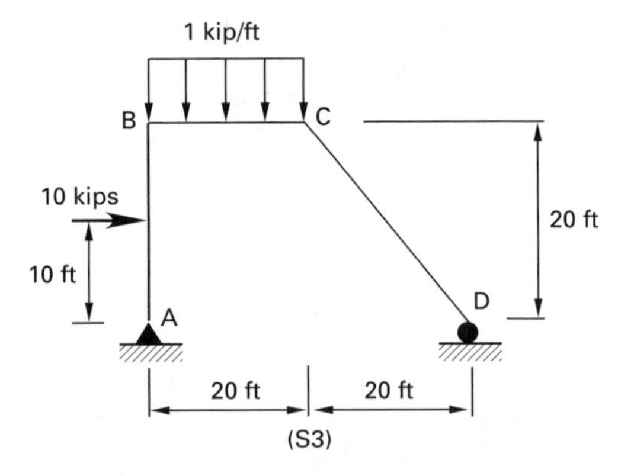

(a) The truss S1 is

(A) unstable
(B) statically indeterminate to the first degree
(C) statically indeterminate to the second degree
(D) determinate

(b) The frame S2 is

(A) statically indeterminate to the third degree
(B) statically indeterminate to the sixth degree
(C) statically indeterminate to the ninth degree
(D) determinate

(c) For truss S1, the force in bar HB is most nearly

(A) 0
(B) 20 kN (compression)
(C) 20 kN (tension)
(D) 40 kN (compression)

(d) For the truss S1, the force in bar HC is more nearly

(A) 0
(B) 7.1 kN (compression)
(C) 7.1 kN (tension)
(D) 40 kN (compression)

(e) The degree of kinematic indeterminacy in frame S2 is

(A) 0 (determinate)
(B) 3
(C) 6
(D) 9

(f) The sum of the horizontal reactions in frame S2 is most nearly

(A) 0
(B) 13 kN
(C) 25 kN
(D) 50 kN

(g) For frame S3, the axial force in bar CD is most nearly

(A) 0
(B) 5.3 kips (compression)
(C) 11 kips (compression)
(D) 21 kips (compression)

(h) For frame S3, the maximum shear force on member BC is most nearly

(A) 10 kips
(B) 13 kips
(C) 15 kips
(D) 20 kips

(i) For frame S3, the maximum positive moment on bar BC is most nearly

(A) 0
(B) 50 ft-kips
(C) 75 ft-kips
(D) 180 ft-kips

(j) For the frame S2, if the axial force on bar BC is assumed negligible and the axial force on AB is 5 kN, the maximum shear on beam BD is most nearly

(A) 0
(B) 10 kN
(C) 45 kN
(D) 50 kN

5. All supports on the continuous beam shown are simple. What are the reactions?

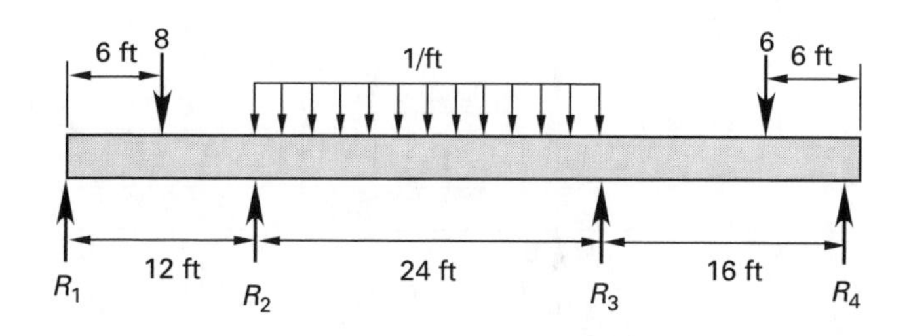

6. In the structure shown, the two support points are pinned and the two joints are rigid. (a) What are the magnitudes of the moments at points B and C? (b) What are the reactions?

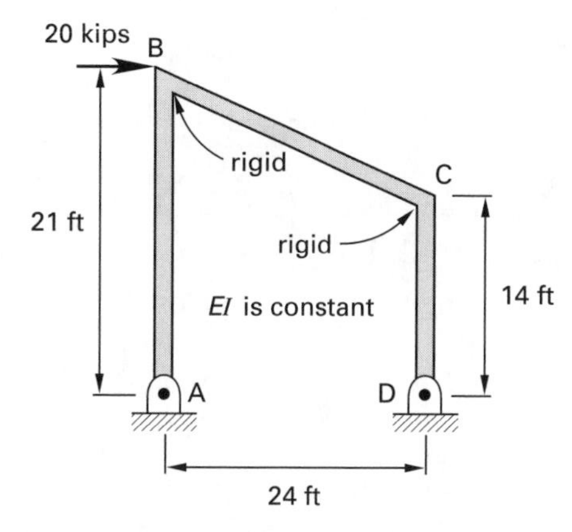

7. A frame with rigid joints is shown. (a) What are the horizontal reactions? (b) What are the joint moments at points B, C, and D?

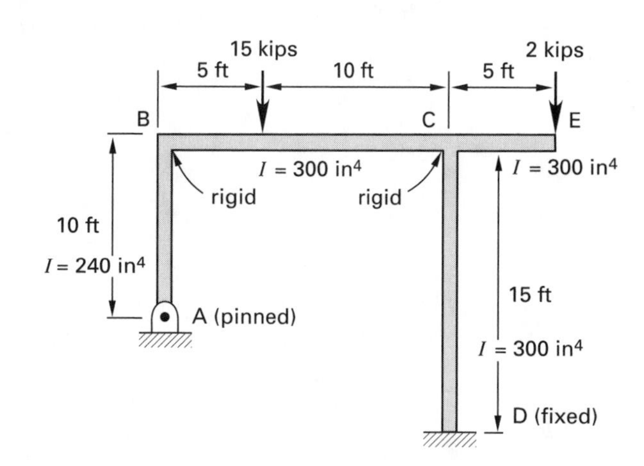

Structural

8. What are the magnitudes of the maximum moments on spans AB, BC, and CD?

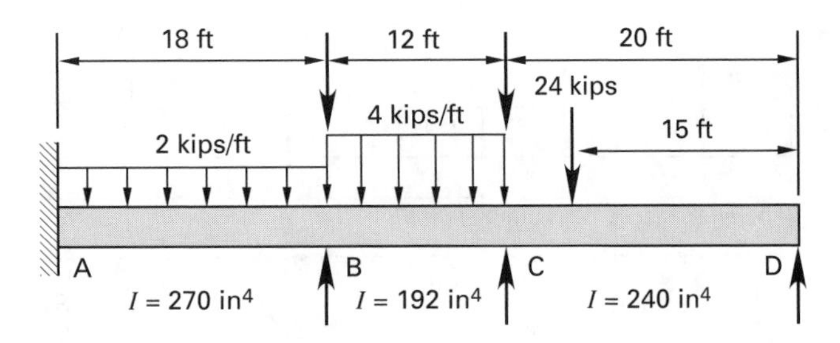

9. Determine the deflection of the frame shown assuming that the two connections are (a) pinned and (b) rigid.

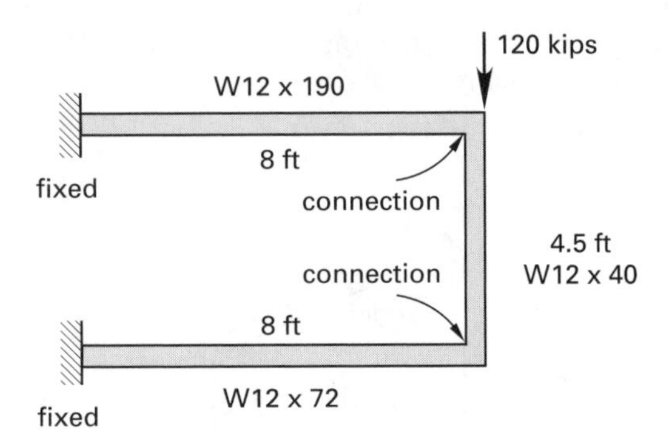

10. (*Time limit: one hour*) A continuous beam (W24 by 76, A36 steel) is loaded with a series of 20 kip point loads as shown. The beam contains two expansion joints that can be assumed to act as frictionless hinges. Determine the (a) maximum shear stress, (b) maximum bending stress, and (c) midpoint deflection.

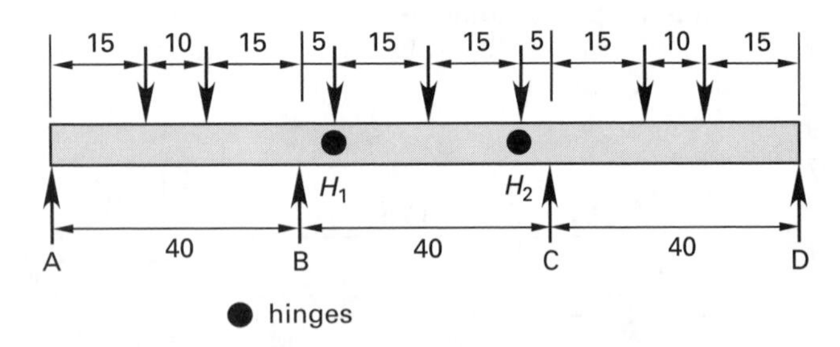

11. (*Time limit: one hour*) All joints are rigid in the structure shown. The beam self-weights are to be disregarded. (a) Determine the horizontal reactions at A, C, and D. (b) Draw the moment diagram on the tension side of the frame.

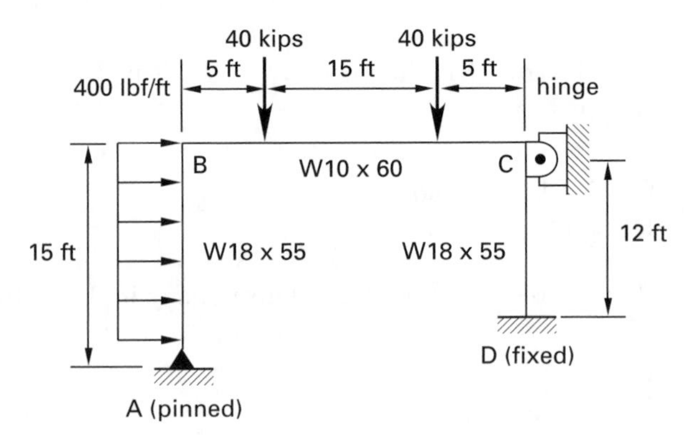

SOLUTIONS

1. The structure is indeterminate to the third degree, so a flexibility analysis requires the computation of nine independent deflection terms. For a stiffness solution, the degree of kinematic indeterminacy is 6 if axial deformations are considered and is still 3 if axial deformations are neglected. The moment distribution method requires only a single sway restraint, and thus requires the least amount of effort.

The centroidal moments of inertia are

$$I_{\text{column}} = \frac{1}{12}bh^3 = \left(\frac{1}{12}\right)(14 \text{ in})^4$$
$$= 3201.3 \text{ in}^4$$
$$I_{\text{beam}} = \left(\frac{1}{12}\right)(12 \text{ in})(24 \text{ in})^3$$
$$= 13{,}824 \text{ in}^4$$

Stiffness is $4EI/L$. However, the $4E$ term appears in each stiffness, so that quantity can be omitted. The final units are irrelevant as long as they are consistent.

$$K_{\text{column}} = \frac{I}{L} = \frac{3201.3 \text{ in}^4}{(12 \text{ ft})\left(12 \ \frac{\text{in}}{\text{ft}}\right)}$$
$$= 22.23$$
$$K_{\text{beams}} = \frac{13{,}824 \text{ in}^4}{(30 \text{ ft})\left(12 \ \frac{\text{in}}{\text{ft}}\right)}$$
$$= 38.4$$

The distribution factors are

$$DF_{\text{AB}} = DF_{\text{DC}} = 0 \quad \text{[fixed ends]}$$
$$DF_{\text{BA}} = DF_{\text{CD}} = \frac{22.23}{22.33 + 38.4} = 0.37$$
$$DF_{\text{BC}} = DF_{\text{CB}} = \frac{38.4}{22.33 + 38.4} = 0.63$$

The fixed-end moments due to loading are found from App. 47.A.

$$\text{FEM}_{\text{AB}} = \text{FEM}_{\text{BA}} = \frac{PL}{8}$$
$$= \frac{(8 \text{ kips})(12 \text{ ft})}{8} = 12 \text{ ft-kips}$$
$$\text{FEM}_{\text{BC}} = \text{FEM}_{\text{CB}} = \frac{wL^2}{12}$$
$$= \frac{\left(0.6 \ \frac{\text{kips}}{\text{ft}}\right)(30 \text{ ft})^2}{12} = 45 \text{ ft-kips}$$

The moment distribution procedure is recorded on the line drawing of the structure.

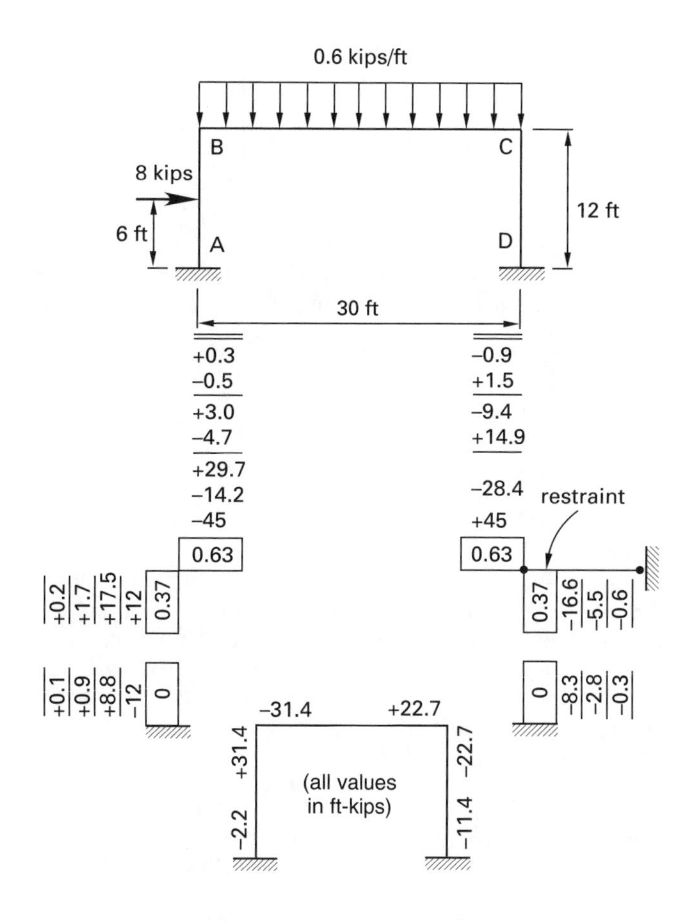

The horizontal reactions at A and D are found from the free-body diagrams of the columns. Summing moments about points B and C, respectively,

$$V_{\text{A}} = \frac{31.4 \text{ ft-kips} - 2.2 \text{ ft-kips} - (8 \text{ kips})(6 \text{ ft})}{12 \text{ ft}}$$
$$= 1.57 \text{ kips}$$
$$V_{\text{D}} = \frac{22.7 \text{ ft-kips} + 11.4 \text{ ft-kips}}{12 \text{ ft}}$$
$$= 2.84 \text{ kips}$$

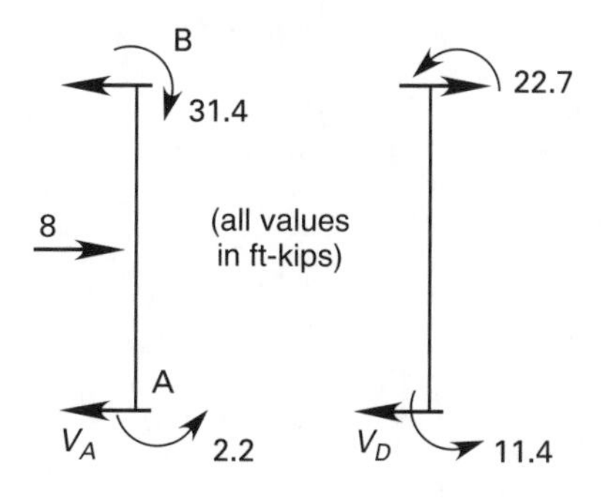

The force in the restraint is found from horizontal equilibrium considerations of the complete structure.

$$A = 8 \text{ kips} - 1.57 \text{ kips} - 2.84 \text{ kips} = 3.59 \text{ kips}$$

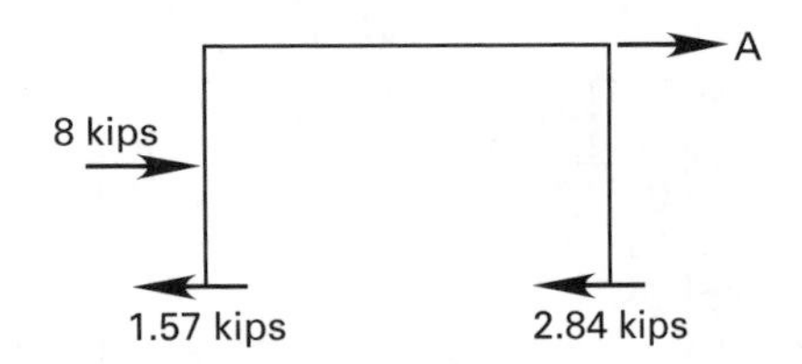

The deformed shape shown results when the restraint is dragged an arbitrary amount.

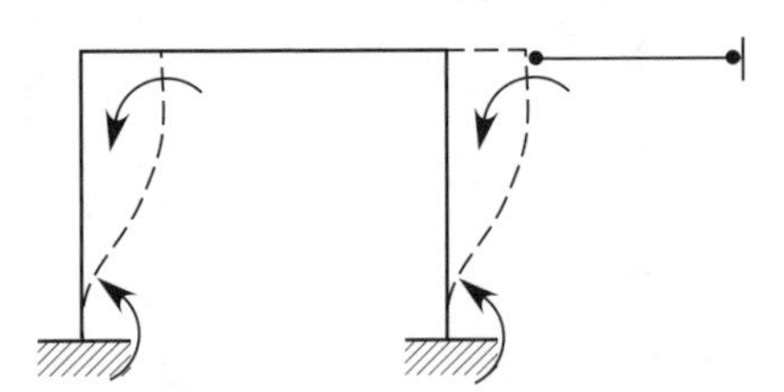

The fixed-end moments that appear on the columns are given by Eq. 47.36. The moments of inertia and lengths are the same for both columns. To simplify the calculations, use an arbitrary value of 100 ft-kips as the fixed-end moment. The fixed-end moments are distributed, and the final results are as shown.

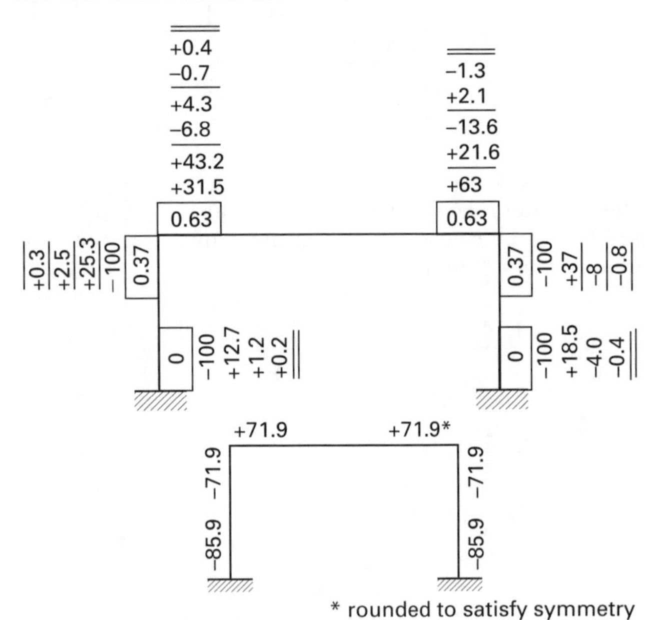

The horizontal reactions in the columns can be found from the free-body diagrams of the column.

$$V_A = V_B = \frac{85.9 \text{ ft-kips} + 71.9 \text{ ft-kips}}{12 \text{ ft}}$$
$$= 13.15 \text{ kips}$$

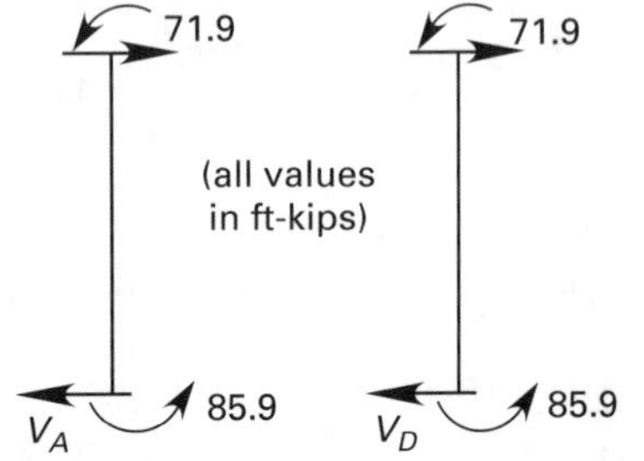

The force in the restraint is found from a free-body diagram of the entire structure.

$$B = 13.15 \text{ kips} + 13.15 \text{ kips} = 26.30 \text{ kips}$$

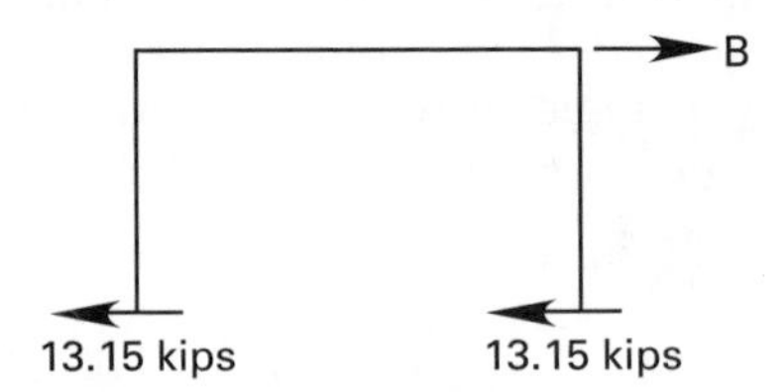

Finally, the effects that derive from force A need to be canceled. The scaling factor is

$$\alpha = \frac{\text{A}}{\text{B}} = \frac{3.59 \text{ kips}}{26.30 \text{ kips}} = 0.14$$

At the fixed end on the left side column, for example, the final moment is

$$M_{\text{A,final}} = -2.2 \text{ ft-kips} + (0.14)(-85.9 \text{ ft-kips})$$
$$= -14.1 \text{ ft-kips}$$

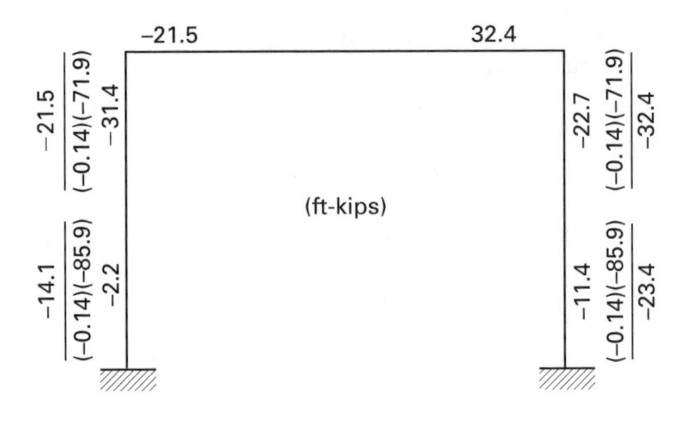

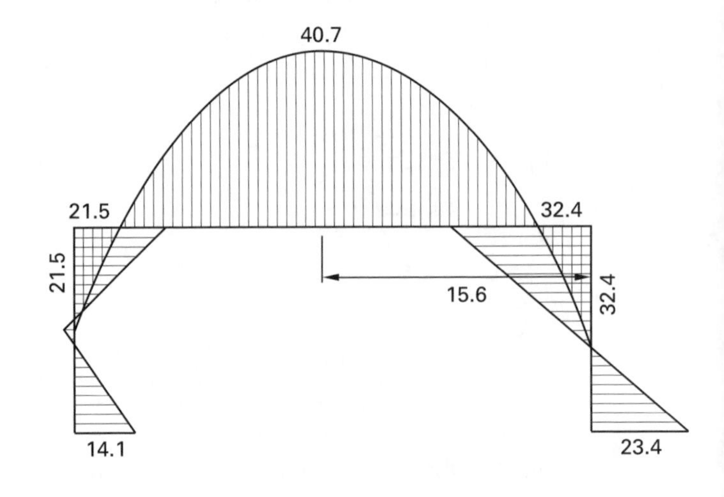

2. (a) Since the cable properties were given, the cable cannot be treated as an unyielding prop. This structure is indeterminate to the first degree, so the flexibility method can be used. First, obtain a general solution for the force in the cable as a function of the parameters involved. Then, the moments in the beam can be calculated.

The illustration shows the selection of the redundant and the internal forces for the $X = 0$ case (the load case) and for the $X_1 = 1$ case.

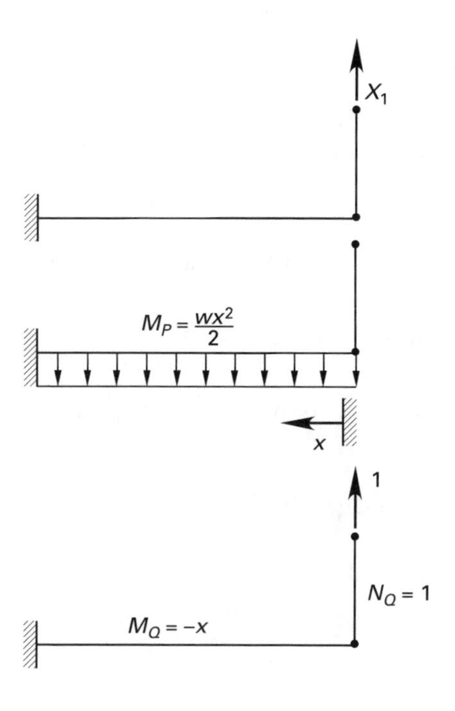

Follow the procedure outlined in the chapter.

$$\delta_1 = \int_0^L \frac{wx^2(-x)}{2EI}dx = -\frac{wL^4}{8EI}$$

$$f_{1,1} = \int_0^L \frac{(-x)^2}{EI}dx + \int_0^{L_c} \frac{1^2}{A_cE_c}dy = \frac{L^3}{3EI} + \frac{L_c}{E_cA_c}$$

Substitute into Eq. 47.24.

$$\left(\frac{L^3}{3EI} + \frac{L_c}{E_cA_c}\right)X_1 = \frac{wL^4}{8EI}$$

Solve for X_1, the force in the cable.

$$\begin{aligned} X_1 &= \frac{wL}{8\left(\frac{1}{3} + \frac{EIL_c}{E_cA_cL^3}\right)} \\ &= \frac{\left(1\ \frac{\text{kip}}{\text{ft}}\right)(10\ \text{ft})}{(8)\left(\frac{1}{3} + \frac{\left(29{,}000\ \frac{\text{kips}}{\text{in}^2}\right)(200\ \text{in}^4)(5\ \text{ft})\left(12\ \frac{\text{in}}{\text{ft}}\right)}{\left(20{,}000\ \frac{\text{kips}}{\text{in}^2}\right)(0.4\ \text{in}^2)(120\ \text{in})^3}\right)} \\ &= \boxed{3.49\ \text{kips}} \end{aligned}$$

(b) Once the force in the cable is known, the moment diagram in the beam can be determined. The maximum moment occurs at the fixed end and has a value of $\boxed{15.10\ \text{ft-kips.}}$

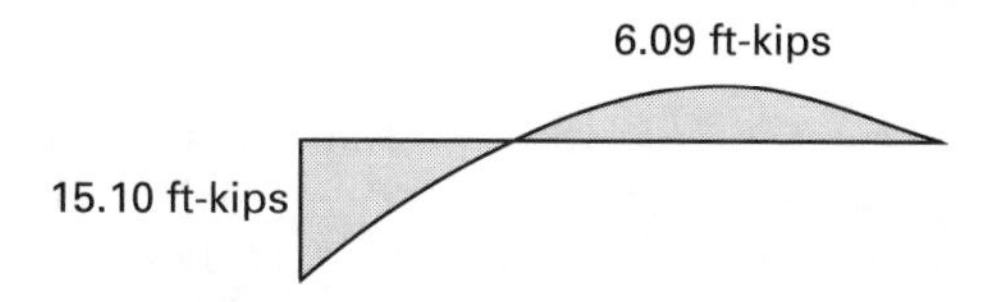

3. (a) When the roller support at C is removed, the structure becomes a statically determinate cantilever.

The degree of indeterminacy is $\boxed{1.}$

The answer is (B).

(b) The 50 kip load introduces a moment of 100 ft-kips at point B. From App. 47.A, case 20, the fixed-end moments due to this applied external moment at each end are

$$a = b = \frac{L}{2}$$

$$\begin{aligned} \text{FEM}_{\text{AB}} = \text{FEM}_{\text{BA}} &= \left(\frac{Ma}{L}\right)\left(\frac{3b}{L} - 1\right) \\ &= (100\ \text{ft-kips})\left(\frac{1}{2}\right)\left(\frac{3}{2} - 1\right) \\ &= 25\ \text{ft-kips} \end{aligned}$$

Distribute moments as follows: The unbalance at C is 25 ft-kips, so change the sign and multiply by a distribution factor of 1 (since there is a pin at C); carry half (−12.5 ft-kips) of the distributed moment to A; the final moment at A is

$$25\ \text{ft-kips} - 12.5\ \text{ft-kips} = \boxed{12.5\ \text{ft-kips}} \quad \text{[clockwise]}$$

This moment induces compression on the side of the dotted line.

The answer is (B).

(c) Using the fact that the moment at A is 12.5 ft-kips, sum the moments about C.

$$V_A = \frac{12.5 \text{ ft-kips} + 100 \text{ ft-kips}}{20 \text{ ft}} = \boxed{5.63 \text{ kips}}$$

The answer is (B).

(d) Since the vertical reaction at C is zero, the axial force in CB is $\boxed{\text{zero.}}$

The answer is (A).

(e) The shear in bar CB is the same as in bar AB: $\boxed{5.63 \text{ kips.}}$

The answer is (B).

(f) A free-body diagram shows that the moment (in ft-kips) in bar AB is given by $12.5 - 5.625y$, where y is the distance measured from point A. Solving, the point of inflection is located at $y = \boxed{2.22 \text{ ft.}}$

The answer is (B).

(g) The maximum moment occurs at point B on bar BC. This moment equals the product of the reaction at C times the distance to B, or $(5.63 \text{ kips})(10 \text{ ft}) = \boxed{56.3 \text{ ft-kips.}}$

The answer is (C).

(h) Use the dummy load method to compute this deflection. Treating the reaction at C as an applied load, the unit load case consists of a horizontal force acting (to the right) at B on a cantilever fixed at A. For this condition, the moments are

$$m_Q = -10 + y \quad \text{[bar AB]}$$
$$m_Q = 0 \quad \text{[bar BC and bar BD]}$$

Only the moment m_P in bar AB (since the others are multiplied by zero) is needed. From part (f), $m_P = 12.5 - 5.625y$. Therefore, the deflection is given by Eq. 47.9.

$$\Delta_B = \int_0^{10} \frac{(12.5 - 5.625y)(-10 + y)}{EI} dy$$

This is integrated to give the deflection.

$$\Delta_B = \frac{-125y + 34.375y^2 - 1.875y^3}{EI}$$

At $y = 10$,

$$\Delta_B = \frac{312.5}{EI} = \frac{312.5 \text{ ft}^3\text{-kip}}{10^4 \text{ ft}^2\text{-kip}} = \boxed{0.0325 \text{ ft}}$$

The answer is (C).

(i) If A is replaced by a pin support, the structure becomes statically determinate. The horizontal reactions at A and C are

$$\frac{100 \text{ ft-kips}}{20 \text{ ft}} = 5 \text{ kips} \quad \left[\begin{array}{c}\text{to the right at A} \\ \text{and to the left at C}\end{array}\right]$$

The moment at B in member AB is

$$(-5 \text{ kips})(10 \text{ ft}) = \boxed{-50 \text{ ft-kips}}$$

The negative sign indicates tension on the side of the dotted line.

The answer is (A).

(j) As determined in part (i), the horizontal reactions at both ends are $\boxed{5 \text{ kips.}}$

The answer is (B).

4. (a) There are 14 bars, 3 reactions, and 8 joints.

$$\text{bars} + \text{reactions} = 14 + 3 = 17$$
$$2(\text{joints}) = (2)(8) = 16$$

> The truss is statically indeterminate to the first degree.

The answer is (B).

(b) The structure becomes statically determinate if the three reactions at E are eliminated and bar BD is cut.

> The degree of indeterminacy is 6.

The answer is (B).

(c) From an equilibrium of vertical forces at joint B, the force in HB is $\boxed{20 \text{ kN (tension).}}$

The answer is (C).

(d) The reaction at support A is

$$\frac{(20\text{ kN} + 20\text{ kN})(3h) + (20\text{ kN})(\text{h})}{4\text{h}} = 35\text{ kN}$$

Taking a vertical section that cuts through HG, HC, and BC and looking to the left side, the vertical component of the force in HC is seen to be 20 kN + 20 kN − 35 kN = 5 kN (upward—which indicates compression). The total force in HC is

$$(5\text{ kN})\sqrt{2} = \boxed{7.07\text{ kN}}$$

The answer is (B).

(e) The structure has three free joints, so the degree of kinematic indeterminacy is $3^2 = \boxed{9.}$

The answer is (D).

(f) Since there are no horizontal loads, the horizontal reactions must add to $\boxed{\text{zero.}}$

The answer is (A).

(g) Summing moments about A, the vertical reaction at D is

$$\frac{(10\text{ ft})(20\text{ kips}) + (10\text{ ft})(10\text{ kips})}{40\text{ ft}} = 7.5\text{ kips}$$

Since CD is inclined at 45°, the axial force in CD is

$$\frac{7.5\text{ kips}}{\sin 45°} = \boxed{10.6\text{ kips}\quad[\text{compression}]}$$

The answer is (C).

(h) From part (g), the vertical reaction at A is 20 kips − 7.5 kips = 12.5 kips (upward). The shear in bar BC is 12.5 kips at point B and decreases 1 kip/ft linearly to −7.5 kips at point C. The maximum shear is $\boxed{12.5\text{ kips.}}$

The answer is (B).

(i) The maximum positive moment takes place where the shear is zero. From parts (g) and (h), this point is located 7.5 ft to the left of point C. Summing moments about this location to the right (on a free-body diagram),

$$M = (7.5\text{ kips})(20\text{ ft} + 7.5\text{ ft}) - \frac{(7.5\text{ kips})(7.5\text{ ft})}{2}$$

$$= \boxed{178.12\text{ ft-kips}}$$

The answer is (D).

(j) Cut a free-body through BC and BD. There are two cases to consider, depending on where the cutting plane intersects along BD. If the axial force on BC is assumed to be zero, then the shear on BD at B is 5 kN, changing to $\boxed{45\text{ kN}}$ to the right of the load.

The answer is (C).

5.

	A	AB	BC	BC	CD	CD
L	12		24		16	
EI	1		1		1	
R	$\frac{1}{12}$		$\frac{1}{24}$		$\frac{1}{16}$	
F	0	1	1	1	1	0
K		$\frac{1}{4}$	$\frac{1}{6}$	$\frac{1}{6}$	$\frac{3}{16}$	
D	1	0.6	0.4	0.471	0.529	1
C	$\frac{1}{2}$	0	$\frac{1}{2}$	$\frac{1}{2}$	0	$\frac{1}{2}$
FEM	−12	12	−48	48	−8.44	14.1
BAL	12.0	21.6	14.4	−18.63	−20.93	−14.1
COM	0	6.0	−9.32	7.2	−7.05	0
BAL	0	1.99	1.33	−0.07	−0.08	0
COM	0	0	−0.04	0.67	0	0
BAL	0	0.02	0.02	−0.32	−0.35	0
COM	0	0	X	X	0	0
total	0	41.61	−41.61	36.85	−36.85	0

Find reactions.

(b) • Member AB:

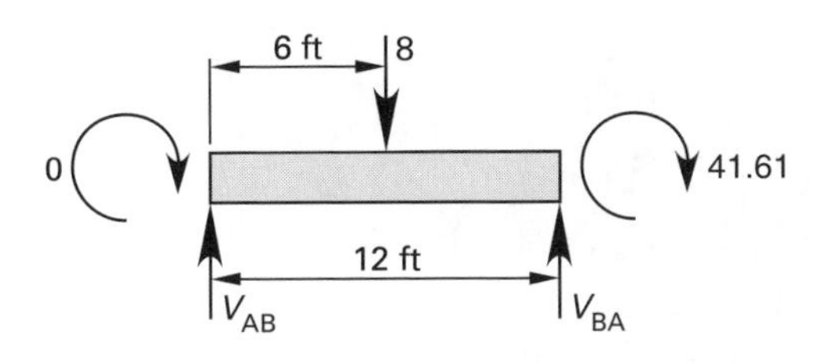

Taking clockwise moments and upward forces as positive,

$$\sum M_A = 0 + 41.61 + (6)(8) - 12V_{BA} = 0$$
$$V_{BA} = 7.47$$
$$\sum F_y = V_{AB} + 7.47 - 8 = 0$$
$$V_{AB} = 0.53$$

- Member BC:

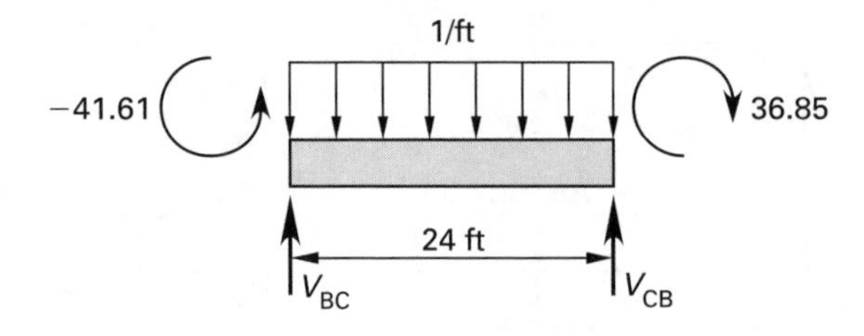

$$\sum M_B = -41.61 + 36.85 + (\tfrac{1}{2})(24)^2(1) - 24V_{CB} = 0$$
$$V_{CB} = 11.8$$
$$\sum F_y = V_{BC} + 11.8 - (24)(1) = 0$$
$$V_{BC} = 12.2$$

- Member CD:

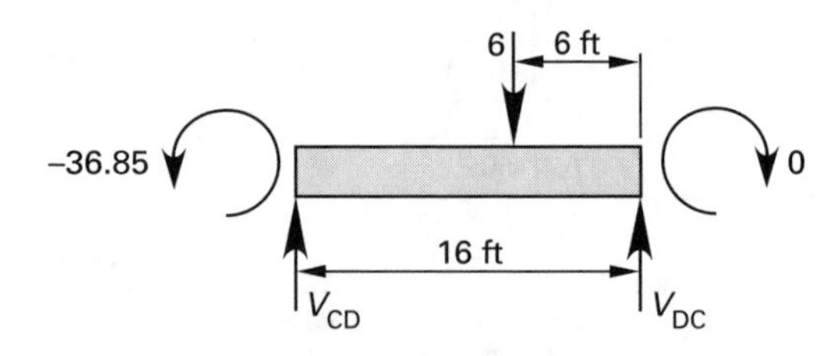

$$\sum M_C = -36.85 + (10)(6) - 16V_{DC} = 0$$
$$V_{DC} = 1.45$$
$$\sum F_y = V_{CD} + 1.45 - 6 = 0$$
$$V_{CD} = 4.55$$

The reactions are

$$R_1 = V_{AB} = 0.53$$
$$R_2 = V_{BA} + V_{BC} = 7.47 + 12.2 = 19.67$$
$$R_3 = V_{CB} + V_{CD} = 11.80 + 4.55 = 16.35$$
$$R_4 = V_{DC} = 1.45$$

6. Use the moment distribution worksheet to calculate the end moments produced when a total moment of 100 is distributed to the two vertical members. Draw freebodies.

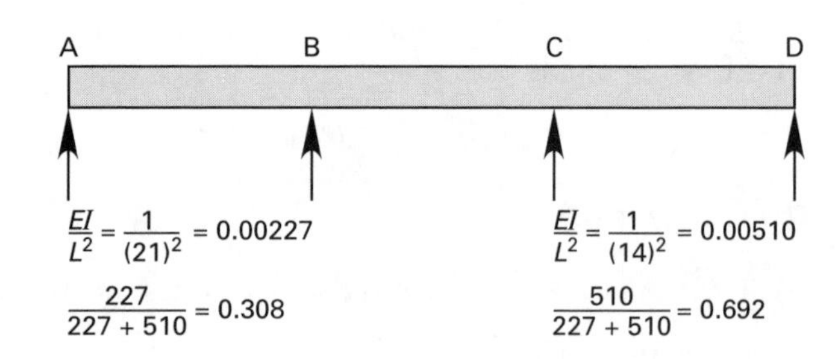

	A	AB	BC	BC	CD	CD
L	21		25		14	
EI	1		1		1	
R	$\frac{1}{21}$		$\frac{1}{25}$		$\frac{1}{14}$	
F	0	1	1	1	1	0
K		$\frac{1}{7}$	$\frac{4}{25}$	$\frac{4}{25}$	$\frac{3}{14}$	
D		0.472	0.528	0.427	0.573	
C	$\frac{1}{2}$	0	$\frac{1}{2}$	$\frac{1}{2}$	0	$\frac{1}{2}$
FEM	30.8	30.8	0	0	69.2	69.2
BAL	−30.8	−14.54	−16.26	−29.55	−39.65	−69.2
COM	0	−15.4	−14.78	−8.13	−34.6	0
BAL	0	14.24	15.94	18.25	24.48	0
COM	0	0	9.13	7.97	0	0
BAL	0	−4.31	−4.82	−3.40	−4.57	0
COM	0	0	−1.70	−2.41	0	0
BAL	0	0.80	0.90	1.03	1.38	0
COM	0	0	0.51	0.45	0	0
BAL	0	−0.24	−0.27	−0.19	−0.26	0
COM	0	0	X	X	0	0
total	0	11.35	−11.35	−15.98	15.98	0

(a) • Member AB:

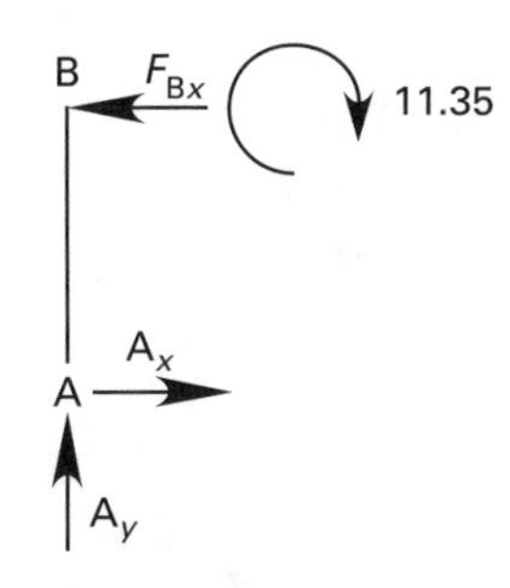

Take clockwise moments as positive.

$$\sum M_\text{A} = 11.35 - 21F_{\text{B}x} = 0$$
$$F_{\text{B}x} = 0.54 \text{ kips}$$

• Member DC:

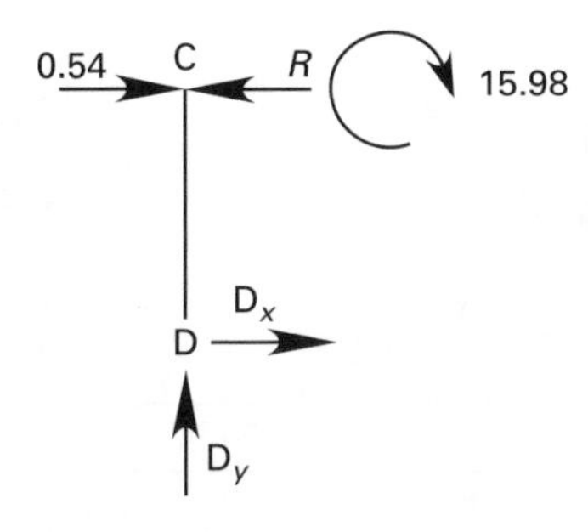

$$\sum M_\text{D} = 15.98 + (0.54)(14) - 14R = 0$$
$$R = 1.68 \text{ kips} \quad \text{[to the left]}$$

P_I was 20 to the left to keep the original frame from swaying.

$$\frac{20}{1.68} = 11.90$$

$$M_\text{AB} = -(11.90)(11.35) = \boxed{-135.07 \text{ ft-kips}}$$

$$M_\text{CD} = -(11.90)(15.98) = \boxed{-190.16 \text{ ft-kips}}$$

The actual conditions are developed as follows.

(b) • Member AB:

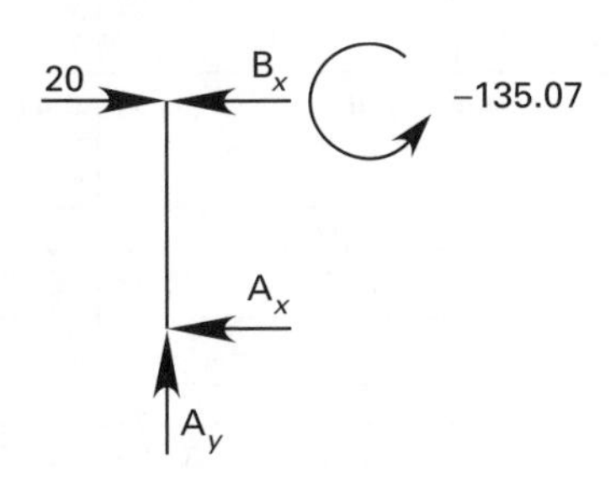

Taking clockwise moments as positive,

$$\sum M_\text{A} = -135.07 + (20)(21) - 21\text{B}_x = 0$$
$$\text{B}_x = 13.57 \text{ kips} \quad \text{[to the left]}$$
$$\sum M_\text{B} = 21\text{A}_x - 135.07 = 0$$
$$\text{A}_x = \boxed{6.43 \text{ kips} \quad \text{[to the left]}}$$

Taking the frame as a whole unit with forces to the right and clockwise moments as positive,

$$\sum F_x = 20 - 6.43 - \text{D}_x = 0$$
$$\text{D}_x = \boxed{13.57 \text{ kips} \quad \text{[to the left]}}$$
$$\sum M_\text{A} = (21)(20) - 24\text{D}_y = 0$$
$$\text{D}_y = \boxed{17.5 \text{ kips} \quad \text{[up]}}$$
$$\text{A}_y = \boxed{-17.5 \text{ kips} \quad \text{[down]}}$$

7. (a) *step 1:* Complete the moment distribution worksheet to find the end moments (assuming sidesway is prevented).

Structural

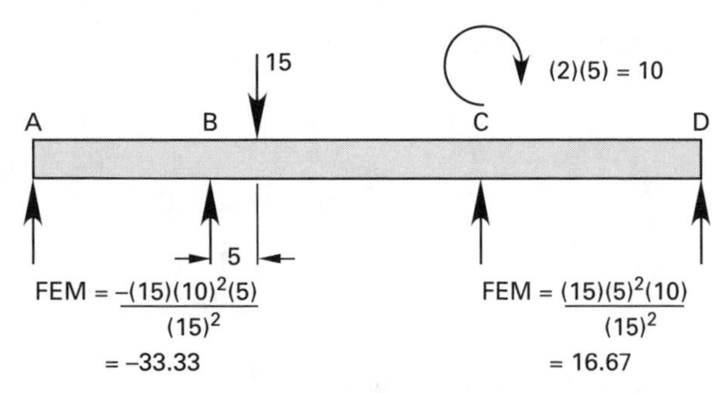

Note: The applied joint moment is clockwise, hence it is positive. The resisting FEM is negative, since the distribution factors are 0.5 for spans BD–DC; –5 goes to each.

	A	AB	BC	BC	CD	CD
L	10		15		15	
EI	240		300		300	
R	24		20		20	
F	0	1	1	1	1	1
K		72	80	80	80	
D		0.474	0.526	0.5	0.5	
C	$\frac{1}{2}$	0	$\frac{1}{2}$	$\frac{1}{2}$	$\frac{1}{2}$	$\frac{1}{2}$
FEM	0	0	–33.33	16.67		0
				–5	–5 (do not add to total)	
BAL	0	15.80	17.53	–3.34	–3.34	0
COM	0	0	–1.67	8.77	0	–1.67
BAL	0	0.79	0.88	–4.39	–4.39	0
COM	0	0	–2.19	0.44	0	–2.19
BAL	0	1.04	1.15	–0.22	–0.22	0
COM	0	0	–0.11	0.58	0	–0.11
BAL	0	0.05	0.06	–0.29	–0.29	0
COM	0	0	X	X	0	X
total	0	17.68	–17.68	18.22	–8.24	–3.97

step 2: Now find the reaction needed to prevent sideway.

- Member AB:

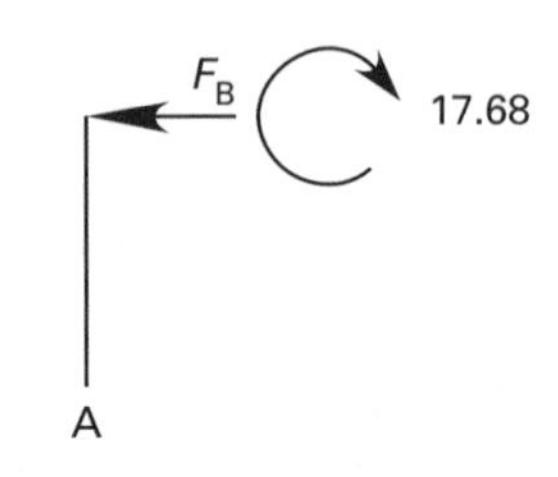

Taking clockwise moments as positive,

$$\sum M_A = 17.68 - 10F_B = 0$$

$$F_B = 1.768 \text{ kips} \quad \text{[to the left]}$$

- Member DC:

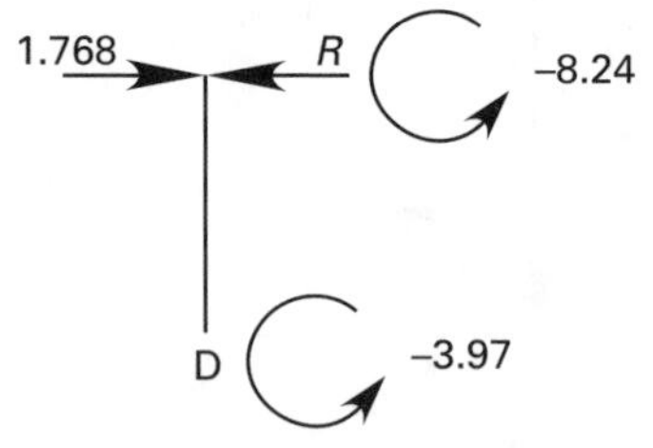

$$\sum M_D = -3.97 - 8.24 + (1.768)(15) - 15R = 0$$

$$R = 0.95 \text{ kips} \quad \text{[to the left]}$$

steps 3 and 4:

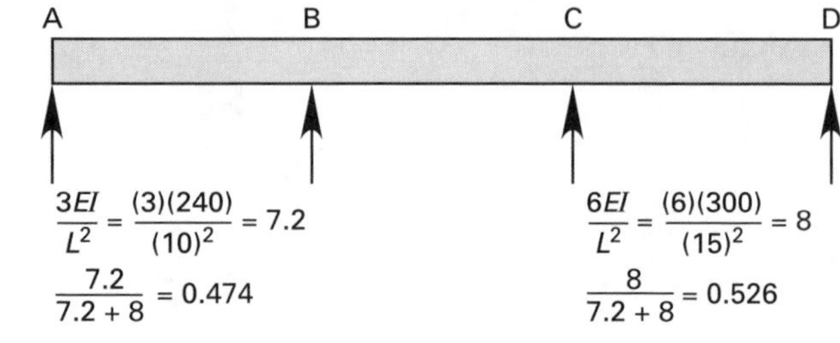

	A	AB	BC	BC	CD	CD
L	10		15		15	
EI	240		300		300	
R	24		20		20	
F	0	1	1	1	1	1
K		72	80	80	80	
D		0.474	0.526	0.5	0.5	
C	$\frac{1}{2}$	0	$\frac{1}{2}$	$\frac{1}{2}$	$\frac{1}{2}$	$\frac{1}{2}$
	total distributed moment = 47.4 + 52.6 = 100					
FEM	0	47.4	0	0	52.6	52.6
	no moment develops when deflected					
BAL	0	–22.47	–24.93	–26.3	–26.3	0
COM	0	0	–13.15	–12.47	0	–13.15
BAL	0	6.23	6.92	6.24	6.24	0
COM	0	0	3.12	3.46	0	3.12
BAL	0	–1.48	–1.64	–1.73	–1.73	0
COM	0	0	–0.87	–0.82	0	–0.87
BAL	0	0.41	0.46	0.41	0.41	0
COM	0	0	0.20	0.23	0	0.20
BAL	0	–0.09	–0.11	–0.11	–0.11	0
COM	X	0	X	X	0	X
total	0	30.0	–30.0	–31.09	31.11	41.9

step 5: Calculate the force required to prevent sidesway.

- Member AB:

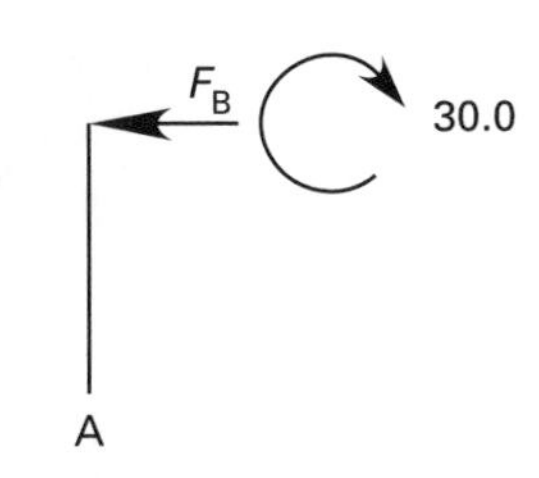

$$\sum M_A = 30.0 - 10F_B = 0$$
$$F_B = 3.0 \text{ kips}$$

- Member CD:

$$\sum M_D = 31.09 + 41.9 + (3)(15) - 15R = 0$$
$$R' = 7.87 \text{ kips} \quad \text{[to the left]}$$

steps 6 and 7:

correction ratio: $\dfrac{0.95}{7.87} = 0.12$

moment at A: $0 + (0.12)(-1)(0) = 0$
moment at B: $17.68 + (0.12)(-1)(30)$
$= \boxed{14.08 \text{ ft-kips}}$
moment at C: $18.22 + (0.12)(-1)(-31.09)$
$= \boxed{21.95 \text{ ft-kips}}$
moment at D: $-3.97 + (0.12)(-1)(41.9)$
$= \boxed{-9.0 \text{ ft-kips}}$

Since R and R' are in the same directions, the derived moments must be reversed in sign.

(b) • Member AB:

$$\sum M_B = 14.08 \text{ ft-kips} - (10 \text{ ft})(A_x) = 0$$
$$\boxed{A_x = 1.408 \text{ kips}}$$

- Frame as a whole:

$$\sum F_x = 1.408 \text{ kips} + D_x = 0$$
$$\boxed{D_x = -1.408 \text{ kips}}$$

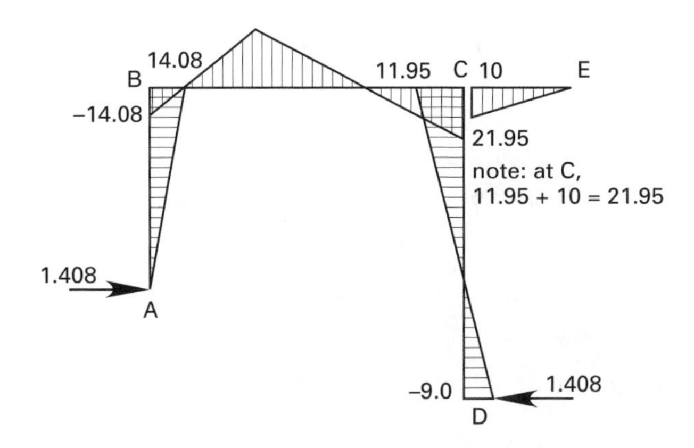

8.

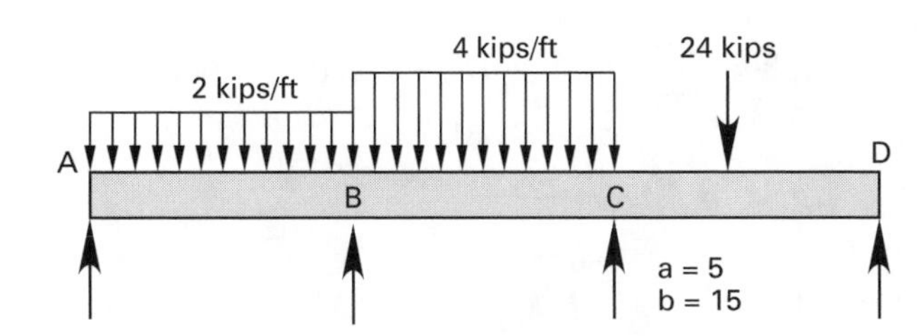

	A	AB	BC	BC	CD	CD
L	18		12		20	
EI	270		192		240	
R	15		16		12	
F	1	1	1	1	1	0
K		60	64	64	36	
D		0.484	0.516	0.64	0.36	
C	$\frac{1}{2}$	$\frac{1}{2}$	$\frac{1}{2}$	$\frac{1}{2}$	0	$\frac{1}{2}$
FEM	−54	54	−48	48	−67.5	22.5
BAL	0	−2.9	−3.1	12.48	7.02	−22.5
COM	−1.45	0	6.24	−1.55	−11.25	0
BAL	0	−3.02	−3.22	8.19	4.61	0
COM	−1.51	0	4.10	−1.61	0	0
BAL	0	−1.98	−2.12	1.03	0.58	0
COM	−0.99	0	0.52	−1.06	0	0
BAL	0	−0.25	−0.27	0.68	0.38	0
COM	−0.13	0	0.34	−0.14	0	0
BAL	0	−0.16	−0.18	0.09	0.05	0
COM	X	0	X	X	X	X
total	−58.08	45.69	−45.69	66.11	−66.11	0

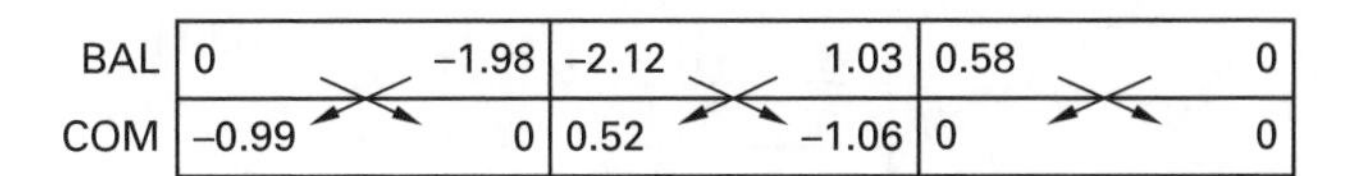
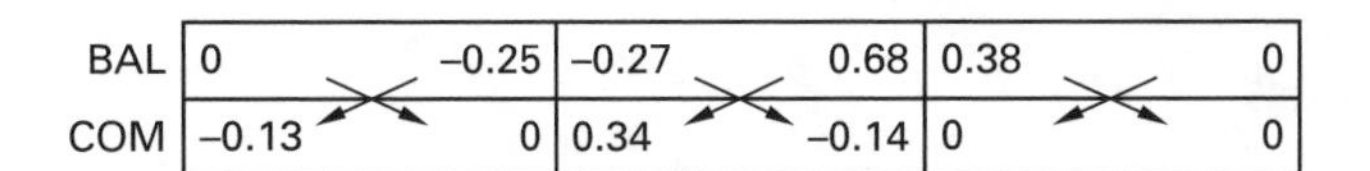
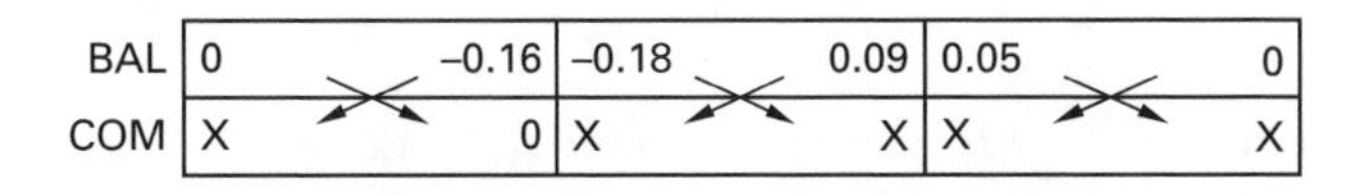

Start at span CD and work to the left.

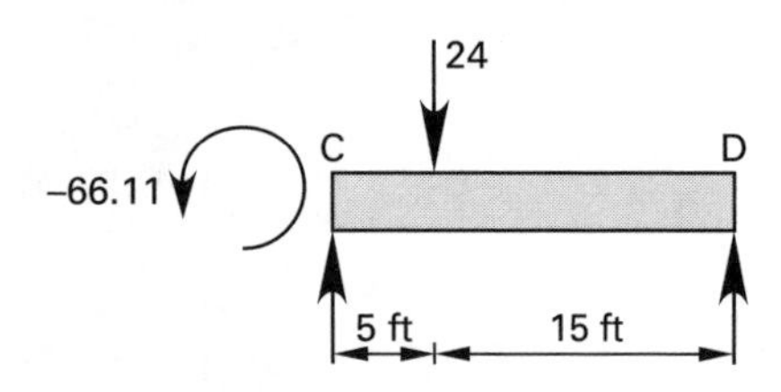

Taking clockwise moments as positive,

$$\text{applied} \sum M_C = -20R_D + (5)(24) - 66.11 = 0$$
$$R_D = 2.69 \text{ kips}$$

$$\text{applied} \sum M_D = 20R_C - (15)(24) - 66.11 = 0$$
$$R_C = 21.31 \text{ kips}$$

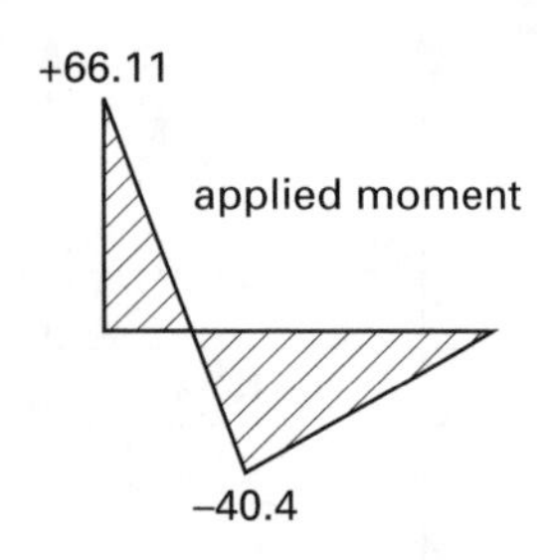

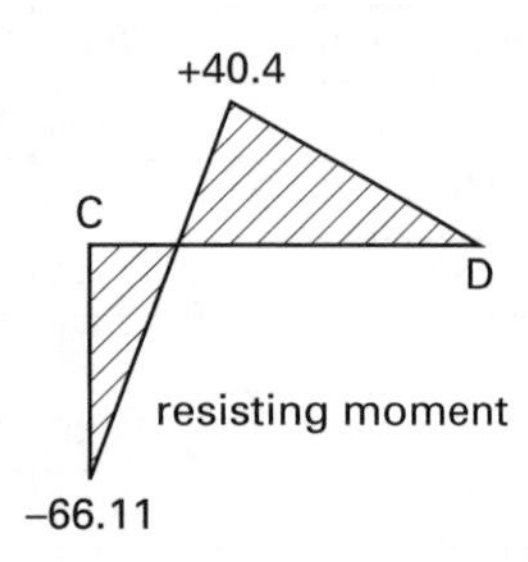

For span BC,

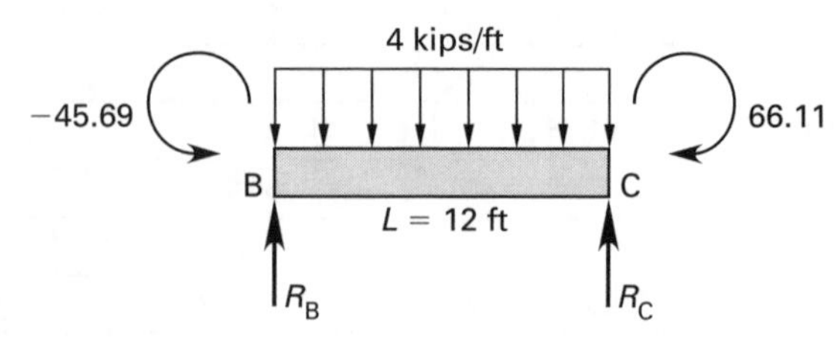

$$\text{applied} \sum M_B = 66.11 - 45.69 + (4)(12)(6) - 12R_C = 0$$
$$R_C = 25.7 \text{ kips}$$

$$\text{applied} \sum M_C = 66.11 - 45.69 - (4)(12)(6) + 12R_B = 0$$
$$R_B = 22.3 \text{ kips}$$

Taking counterclockwise moments as positive,

$$\text{resisting} \sum M = -66.11 + 25.7x - \left(\tfrac{1}{2}\right)(4)x^2$$

At $x = 12$ ft,

$$M = -66.11 + (25.7)(12) - \left(\tfrac{1}{2}\right)(4)(12)^2 = -45.71 \text{ ft-kips}$$

The first derivative is

$$\frac{dM}{dx} = 25.7 - 4x$$

M is maximum at

$$x = \frac{25.7}{4} = 6.4 \text{ ft}$$

At $x = 6.4$ ft,

$$M_B = -66.11 + (25.7)(6.4) - \left(\tfrac{1}{2}\right)(4)(6.4)^2 = 16.45 \text{ ft-kips}$$

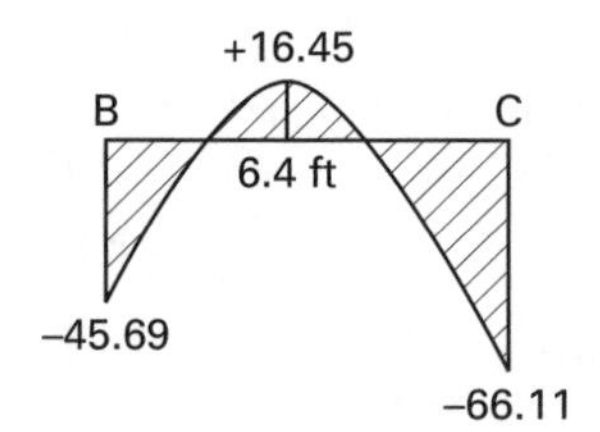

For span AB,

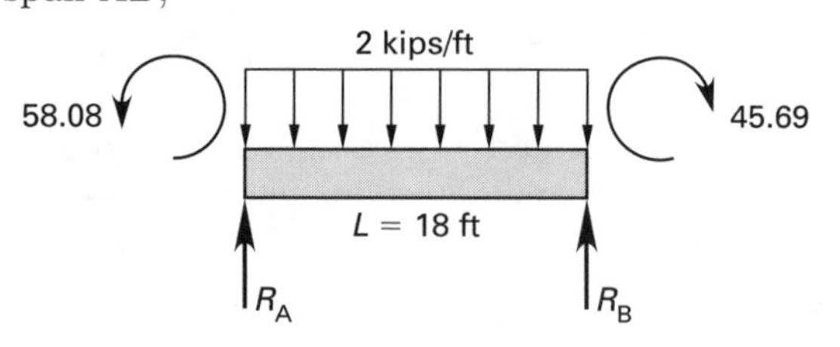

Taking clockwise moments as positive,

$$\text{applied} \sum M_A = 45.69 - 58.08 + (2)(18)(9) - 18R_B = 0$$
$$R_B = 17.31 \text{ kips}$$

$$\text{applied} \sum M_B = 45.69 - 58.08 - (2)(18)(9) + 18R_A = 0$$
$$R_A = 18.69 \text{ kips}$$

Taking counterclockwise moments as positive,

$$\text{resisting} \sum M = -45.69 + 17.31x - \left(\tfrac{1}{2}\right)(2)x^2$$

At $x = 18$ ft,

$$\begin{aligned} M &= -45.69 + (17.31)(18) - \left(\tfrac{1}{2}\right)(2)(18)^2 \\ &= 58.11 \text{ ft-kips} \end{aligned}$$

The first derivative is

$$\frac{dM}{dx} = 17.31 - 2x$$

M is maximum at

$$x = \frac{17.31}{2} = 8.66 \text{ ft}$$

At $x = 8.66$ ft,

$$\begin{aligned} M &= -45.69 + (17.31)(8.66) - \left(\tfrac{1}{2}\right)(2)(8.66)^2 \\ &= 29.22 \text{ ft-kips} \end{aligned}$$

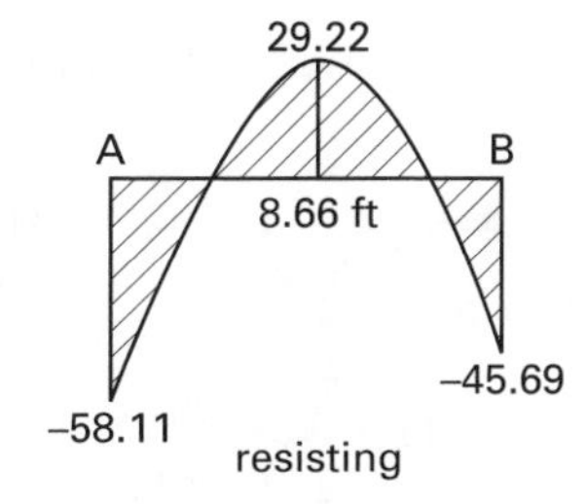

Based on the previous analysis,

$$\begin{aligned} R_A &= 18.69 \text{ kips} \\ R_B &= 17.31 + 22.3 = 39.61 \text{ kips} \\ R_C &= 25.70 + 21.31 = 47.01 \text{ kips} \\ R_D &= 2.69 \text{ kips} \end{aligned}$$

The maximum moments are

on span AB: −58.11 ft-kips on span BC: −66.11 ft-kips on span CD: −66.11 ft-kips

9. (a) Assume the shapes bend against their largest moment of inertia. Assume the vertical bar remains vertical so that both horizontal members deflect the same amount.

$$\begin{aligned} \text{W } 12 \times 190\text{: } I &= 1890 \text{ in}^4 \\ \text{W } 12 \times 72\text{: } I &= \underline{\ 597 \text{ in}^4} \\ \text{total} &\quad 2487 \text{ in}^4 \end{aligned}$$

The two horizontal beams act together like a combined cantilever.

$$L = \left(12 \ \frac{\text{in}}{\text{ft}}\right)(8 \text{ ft}) = 96 \text{ in}$$

$$y_{max} = \frac{(120{,}000 \text{ lbf})(96 \text{ in})^3}{(3)\left(2.9 \times 10^7 \ \frac{\text{lbf}}{\text{in}^2}\right)(2487 \text{ in}^4)} = \boxed{0.49 \text{ in}}$$

(b) *step 1:* Since there are no fixed-end moments, all total moments are zero.

step 2: In order to prevent all of the sidesway, all of the applied lateral load must be canceled. Therefore, $P_I = 120$ to the left.

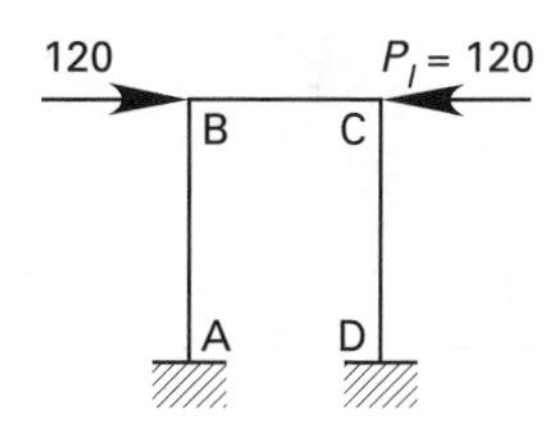

step 3: Apply a total FEM of 100.

For AB,

$$\frac{I}{L^2} = \frac{1890}{(8)^2} = 29.53$$

For DC,

$$\frac{I}{L^2} = \frac{597}{(8)^2} = 9.33$$

Give 29.53/(29.53 + 9.33) = 0.76 (76%) to AB.

Give 1 − 0.76 = 0.24 (24%) to DC.

Structural

step 4: The moment distribution worksheet for the application of these FEMs is

	A	AB	BC	BC	CD	CD
L	8		4.5		8	
EI	1890		310		597	
R	236.25		68.89		74.625	
F	1	1	1	1	1	1
K	945	945	275.56	275.56	298.5	298.5
D		0.774	0.226	0.48	0.52	
C	$\frac{1}{2}$	$\frac{1}{2}$	$\frac{1}{2}$	$\frac{1}{2}$	$\frac{1}{2}$	$\frac{1}{2}$
FEM	–76	–76	0	0	–24	–24
BAL	0	58.82	17.18	11.52	12.48	0
COM	29.41	0	5.76	8.59	0	6.24
BAL	0	–4.46	–1.30	–4.12	–4.47	0
COM	–2.23	0	–2.06	–0.65	0	–2.23
BAL	0	1.59	0.47	0.31	0.34	0
COM	0.79	0	0.15	0.24	0	0.17
BAL	0	–0.12	–0.03	–0.11	–0.13	0
COM	X	0	X	X	0	X
total	–48.03	–20.17	20.17	15.78	–15.78	–19.95

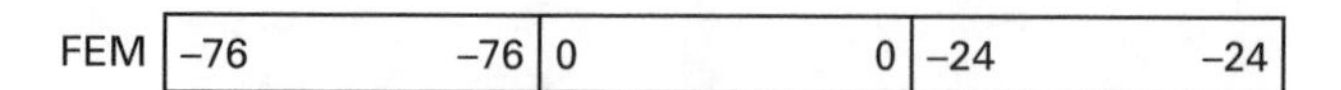

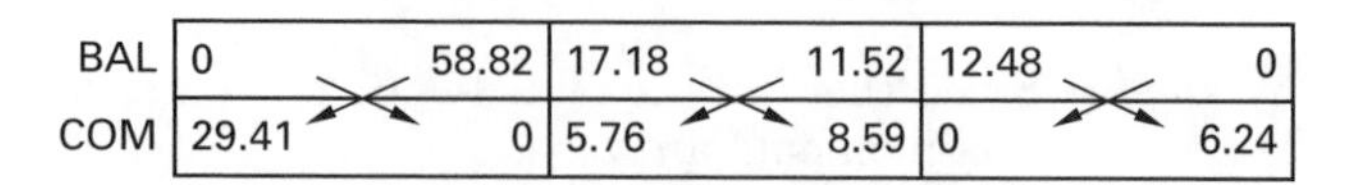

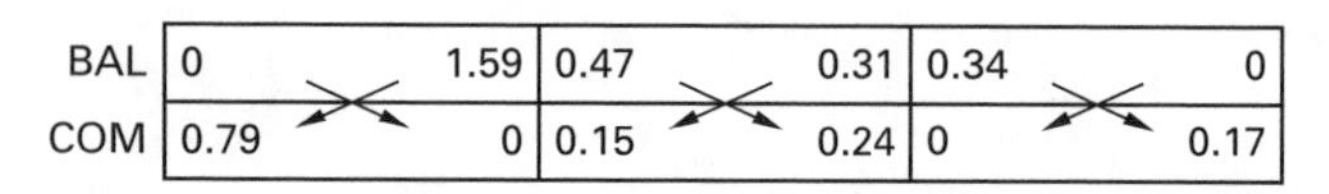

step 5: Take member AB as a free body.

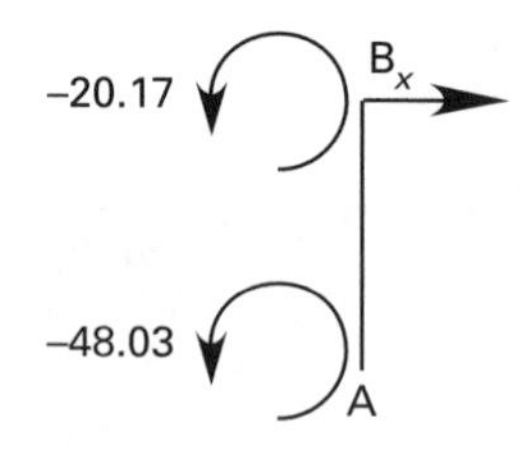

Taking clockwise moments as positive,

$$\sum M_A = 8B_x - 20.17 - 48.03 = 0$$
$$B_x = 8.53 \text{ kips} \quad \text{[to the right]}$$

Take member DC as a free body.

$$\sum M_D = 8R - (8)(8.53) - 15.78 - 19.95 = 0$$
$$R = 13.0 \text{ kips} \quad \text{[to the right]}$$

step 6: The ratio is

$$-\left(\frac{-120}{13.0}\right) = 9.23$$

$$M_A = (9.23)(-48.03) = -443.3 \text{ ft-kips}$$
$$M_B = (9.23)(\pm 20.17) = \pm 186.17 \text{ ft-kips}$$
$$M_C = (9.23)(\pm 15.78) = \pm 145.6 \text{ ft-kips}$$
$$M_D = (9.23)(-19.95) = -184.1 \text{ ft-kips}$$

step 7: Draw the free body of member AB.

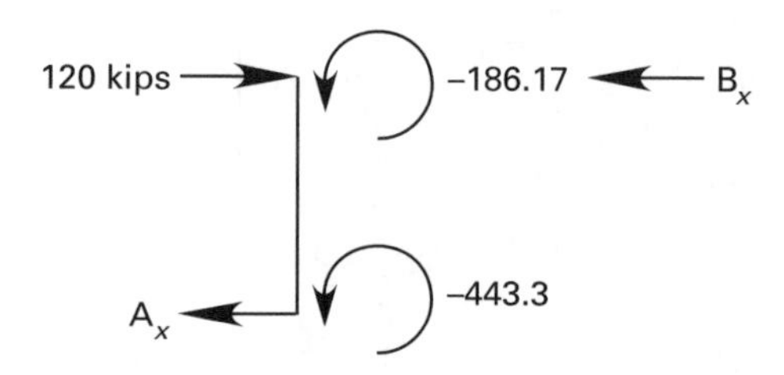

B_x can be found by summing moments about A.

$$\sum M_A = -443.3 \text{ ft-kips} - 186.17 \text{ ft-kips} + (8 \text{ ft})(120 \text{ kips}) - (8 \text{ ft})B_x = 0$$
$$B_x = 41.32 \text{ kips} \quad \text{[to the left]}$$

Break the loads into two parts.

- Concentrated loads:

The net load is $120 - 41.32 = 78.68$ kips.

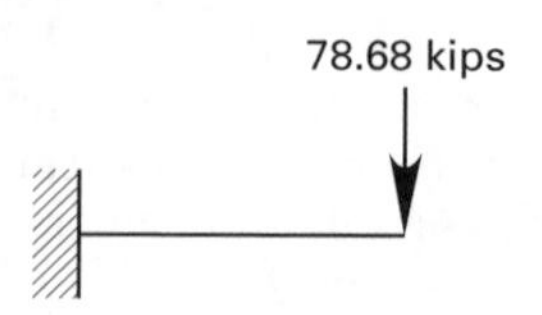

The deflection is

$$y_1 = \frac{FL^3}{3EI} = \frac{(78{,}680 \text{ lbf})(96 \text{ in})^3}{(3)\left(2.9 \times 10^7 \ \frac{\text{lbf}}{\text{in}^2}\right)(1890 \text{ in}^4)}$$
$$= 0.423 \text{ in} \quad [\text{down}]$$

- Applied moments:

The moment diagram for the pure moments on the span is

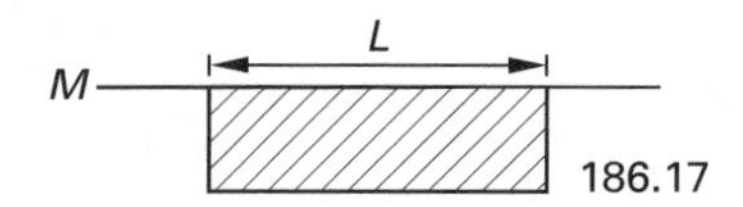

From the area-moment method,

$$EIy = ML\left(\frac{L}{2}\right)$$
$$y_2 = \frac{ML^2}{2EI}$$

The moment at the built-in end does not cause a deflection, so $M = 186.17$ ft-kips.

$$y_2 = \frac{ML^2}{2EI} = \frac{(-186{,}170 \text{ ft-lbf})\left(12 \ \frac{\text{in}}{\text{ft}}\right)(96 \text{ in})^2}{(2)\left(2.9 \times 10^7 \ \frac{\text{lbf}}{\text{in}^2}\right)(1890 \text{ in}^4)}$$
$$= -0.188 \text{ in} \quad [\text{up}]$$

$$y_{\text{total}} = 0.423 \text{ in} - 0.188 \text{ in} = \boxed{0.235 \text{ in} \quad [\text{down}]}$$

10. (a) Although the beam has four supports, it is determinate. Since moment = 0 at H_1, taking clockwise moments to the left of H_1 as positive,

$$\sum M_{H_1} = (5 \text{ ft})R_{\text{B}} + (45 \text{ ft})R_{\text{A}} - (20 \text{ kips})(20 \text{ ft} + 30 \text{ ft}) = 0$$
$$5R_{\text{B}} + 45R_{\text{A}} = 1000 \text{ kips}$$

Since the beam and loading are symmetrical,

$$R_{\text{A}} + R_{\text{B}} = \left(\tfrac{1}{2}\right)(7)(20 \text{ kips}) = 70 \text{ kips}$$

Solving for R_{A} and R_{B} simultaneously,

$$R_{\text{A}} = 16.25 \text{ kips} = R_{\text{D}}$$
$$R_{\text{B}} = 53.75 \text{ kips} = R_{\text{C}}$$

For a W24 × 76 beam,

$$A = 22.4 \text{ in}^2$$
$$d = 23.92 \text{ in}^2$$
$$S = 176 \text{ in}^3$$
$$t_w = 0.440 \text{ in}$$
$$I = 2100 \text{ in}^4$$

The shear diagram is

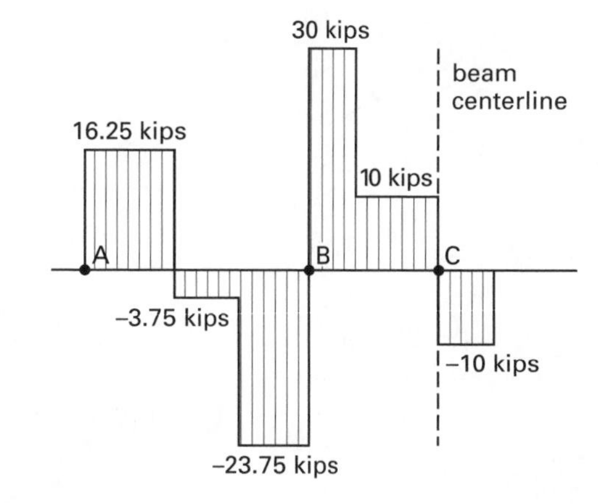

The web takes all the shear. The maximum shear stress is

$$\tau_{\max} = \frac{V}{dt_w} = \frac{30 \times 10^3 \text{ lbf}}{(23.92 \text{ in})(0.440 \text{ in})}$$
$$= \boxed{2850 \text{ lbf/in}^2}$$

(b) The moment is maximum where $V = 0$. Under the first inboard load,

$$M = (15 \text{ ft})(16.25 \text{ kips}) = 243.75 \text{ ft-kips}$$

At R_{B},

$$M = (40 \text{ ft})(16.25 \text{ kips}) - (15 \text{ ft} + 25 \text{ ft})(20 \text{ kips})$$
$$= -150 \text{ ft-kips}$$

At the center,

$$M = (15 \text{ ft})\left(\frac{20 \text{ kips}}{2}\right) = 150 \text{ ft-kips}$$
$$M_{\max} = 243.75 \text{ ft-kips}$$

$$\sigma_{\max} = \frac{M}{S} = \frac{(243.75 \times 10^3 \text{ ft-lbf})\left(12 \ \frac{\text{in}}{\text{ft}}\right)}{176 \text{ in}^3}$$
$$= \boxed{16{,}619 \text{ lbf/in}^2}$$

(c) The deflection between hinges is

$$y_{max} = \frac{FL^3}{48EI} = \frac{(20 \times 10^3 \text{ lbf})\left((30 \text{ ft})\left(12 \frac{\text{in}}{\text{ft}}\right)\right)^3}{(48)\left(2.9 \times 10^7 \frac{\text{lbf}}{\text{in}^2}\right)(2100 \text{ in}^4)}$$
$$= 0.319 \text{ in}$$

The deflection at the midpoint is composed of four terms.

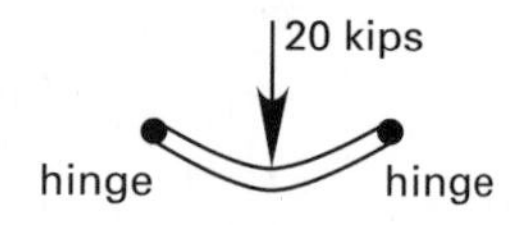

$$y_1 = \frac{FL^3}{48EI} = \frac{-(20 \times 10^3 \text{ lbf})\left((30 \text{ ft})\left(12 \frac{\text{in}}{\text{ft}}\right)\right)^3}{(48)\left(2.9 \times 10^7 \frac{\text{lbf}}{\text{in}^2}\right)(2100 \text{ in}^4)}$$
$$= -0.319 \text{ in}$$

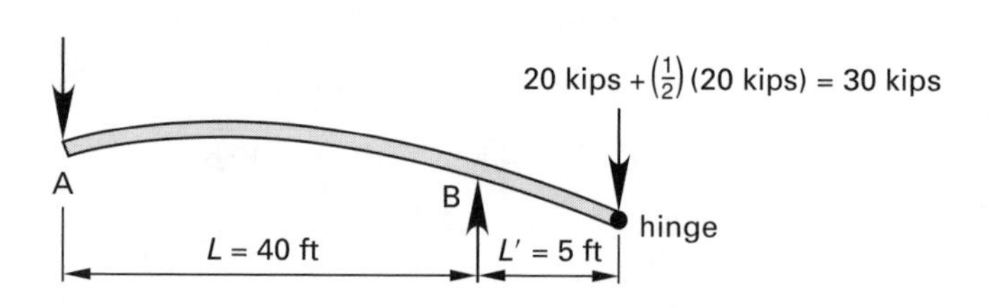

$$y_2 = \left(\frac{FL^2}{3EI}\right)(L + L')$$
$$= \frac{-(30 \times 10^3 \text{ lbf})\left((5 \text{ ft})\left(12 \frac{\text{in}}{\text{ft}}\right)\right)^2 \times \left((45 \text{ ft})\left(12 \frac{\text{in}}{\text{ft}}\right)\right)}{(3)\left(2.9 \times 10^7 \frac{\text{lbf}}{\text{In}^2}\right)(2100 \text{ in}^4)}$$
$$= -0.319 \text{ in}$$

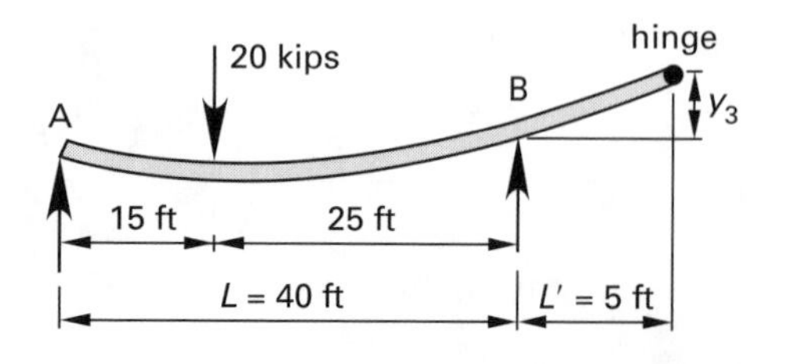

The slope at B (App. 44.A, case 7) is

$$m_B = \left(\frac{Pab}{6EI}\right)\left(1 + \frac{a}{L}\right)$$
$$= \left(\frac{(20 \times 10^3 \text{ lbf})(15 \text{ ft})\left(12 \frac{\text{in}}{\text{ft}}\right) \times (25 \text{ ft})\left(12 \frac{\text{in}}{\text{ft}}\right)}{(6)\left(2.9 \times 10^7 \frac{\text{lbf}}{\text{in}^2}\right)(2100 \text{ in}^4)}\right)\left(1 + \frac{15 \text{ ft}}{40 \text{ ft}}\right)$$
$$= 0.00406 \text{ rad}$$
$$y_3 = m_B L'$$
$$= (0.00406 \text{ rad})(5 \text{ ft})\left(12 \frac{\text{in}}{\text{ft}}\right)$$
$$= 0.244 \text{ in}$$

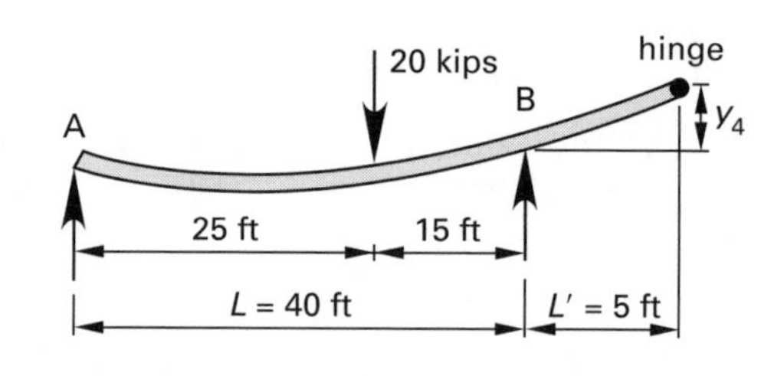

$$m_B = \left(\frac{Pab}{6EI}\right)\left(1 + \frac{a}{L}\right)$$
$$= \left(\frac{(20 \times 10^3 \text{ lbf})(25 \text{ ft})\left(12 \frac{\text{in}}{\text{ft}}\right) \times (15 \text{ ft})\left(12 \frac{\text{in}}{\text{ft}}\right)}{(6)\left(2.9 \times 10^7 \frac{\text{lbf}}{\text{in}^2}\right)(2100 \text{ in}^4)}\right)\left(1 + \frac{25 \text{ ft}}{40 \text{ ft}}\right)$$
$$= 0.00480 \text{ rad}$$
$$y_4 = M_B L' = (0.00480 \text{ rad})(5 \text{ ft})\left(12 \frac{\text{in}}{\text{ft}}\right)$$
$$= 0.288 \text{ in}$$

The total midpoint deflection is

$$y_t = y_1 + y_2 + y_3 + y_4$$
$$= -0.319 - 0.319 + 0.244 + 0.288$$
$$= \boxed{-0.106 \text{ in}}$$

11. The moment distribution worksheet is

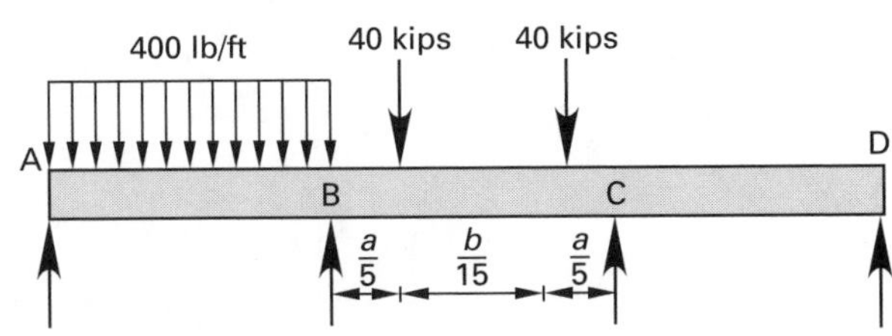

$$\frac{wL^2}{12} = \frac{(0.4)(15)^2}{12} = 7.5 \text{ ft-kips}$$

$$\left(\frac{Pa}{L^2}\right)(2a^2 + 3ab + b^2) = \left(\frac{(40)(5)}{(25)^2}\right)\left((2)(5)^2 + (3)(5)(15) + (15)^2\right)$$

$$= 160 \text{ ft-kips}$$

	AB	BA	BC	CB	CD	DC
L	15		25		12	
EI	890		341		890	
R	59.33		13.64		74.17	
F	0	1	1	1	1	1
K	237.3	178	54.6	54.6	296.7	296.7
D	1	0.765	0.235	0.155	0.845	1
C	$\frac{1}{2}$	0	$\frac{1}{2}$	$\frac{1}{2}$	$\frac{1}{2}$	$\frac{1}{2}$
FEM	−7.5	7.5	−160	160	0	0
BAL	7.5	116.66	35.84	−24.8	−135.2	0
COM	0	3.75	−12.4	17.92	0	−67.6
BAL	0	6.62	2.03	−2.78	−15.14	0
COM	0	0	−1.39	1.01	0	−7.57
BAL	0	1.06	0.33	−0.16	−0.85	0
COM	X	0	X	X	0	X
total	0	136.71	−136.71	151.19	−151.19	−75.17

Taking clockwise moments as positive,

$$\sum M_\mathrm{A} = 136.71 \text{ ft-kips} + \left(\tfrac{1}{2}\right)\left(0.4\ \frac{\text{kips}}{\text{ft}}\right)(15 \text{ ft})^2 - (15 \text{ ft})\mathrm{B}_x = 0$$

$$\mathrm{B}_x = 12.11 \text{ kips} \quad [\text{to the left}]$$

$$\left(0.4\ \frac{\text{kips}}{\text{ft}}\right)(15 \text{ ft}) - 12.11 \text{ kips} + \mathrm{A}_x = 0$$

$$\boxed{\mathrm{A}_x = 6.11 \text{ kips} \quad [\text{to the right}]}$$

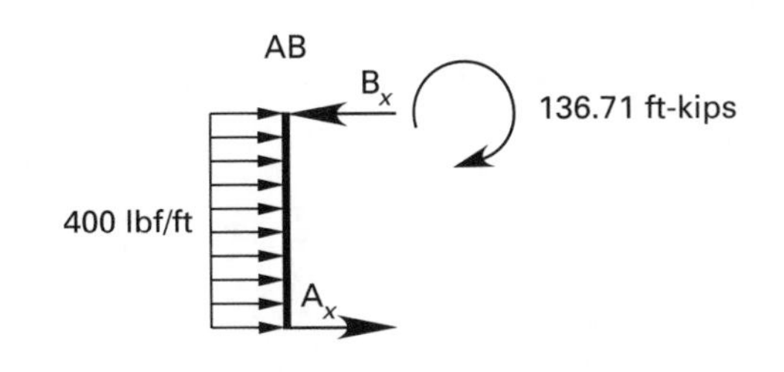

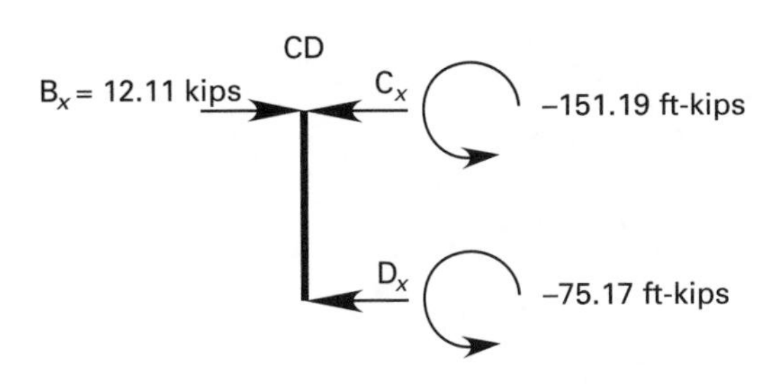

$$\sum M_\mathrm{D} = (12 \text{ ft})(12.11 \text{ kips}) - (12 \text{ ft})R - 151.19 \text{ ft-kips} - 75.17 \text{ ft-kips} = 0$$

$$\boxed{C_x = -6.75 \text{ kips} \quad [\text{to the right}]}$$

$$12.11 \text{ kips} + 6.75 \text{ kips} - \mathrm{D}_x = 0$$

$$\boxed{\mathrm{D}_x = 18.86 \text{ kips} \quad [\text{to the left}]}$$

(b)

136.71 ft-kips
(40)(5) = 200 ft-kips
63.29 ft-kips
151.19 ft-kips
(40)(5) = 200 ft-kips
48.81 ft-kips
75.17 ft-kips

48 Properties of Concrete and Reinforcing Steel

PRACTICE PROBLEMS

1. A batch of fine aggregate was sieve graded. The percentages retained on each sieve are shown. What is the sand's fineness modulus?

sieve	percentage retained
4	4
8	11
16	21
30	22
50	24
100	17
dust (pan)	1

(A) 0.99
(B) 1.8
(C) 2.9
(D) 99

2. (*Time limit: one hour*) The 6 in by 6 in concrete beam section shown was tested in a third-point loading apparatus. The failure occurred outside the middle third as shown under a maximum total load of 5000 lbf. The beam was normal-weight concrete with a compressive strength of 5000 psi. Neglect the beam's self-weight.

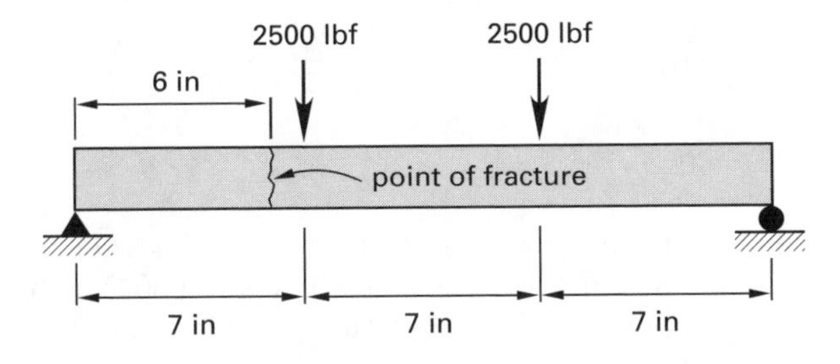

(a) The maximum shear force is most nearly

(A) 625 lbf
(B) 1250 lbf
(C) 2500 lbf
(D) 5000 lbf

(b) The maximum bending moment is most nearly

(A) 5000 in-lbf
(B) 10,000 in-lbf
(C) 15,000 in-lbf
(D) 17,500 in-lbf

(c) The shear force within the middle third of the beam is most nearly

(A) zero
(B) 625 lbf
(C) 1250 lbf
(D) 2500 lbf

(d) The bending moment within the middle third of the beam

(A) is zero.
(B) increases linearly from right to left.
(C) is parabolic in shape.
(D) is a straight line with zero slope.

(e) The modulus of rupture is most nearly

(A) 320 psi
(B) 420 psi
(C) 520 psi
(D) 620 psi

(f) If the fracture occurred within the middle third, the modulus of rupture is most nearly

(A) 420 psi
(B) 450 psi
(C) 470 psi
(D) 490 psi

(g) The modulus of rupture using the ACI empirical equation is most nearly

(A) 530 psi
(B) 550 psi
(C) 570 psi
(D) 590 psi

(h) If the beam is made of lightweight concrete, the modulus of rupture is most nearly

(A) 380 psi
(B) 400 psi
(C) 420 psi
(D) 440 psi

(i) The center-point loading test is no longer standard in determining the modulus of rupture for concrete because

(A) it yields higher values of modulus of rupture than the true values.
(B) it is difficult to perform accurately in the laboratory.
(C) it costs more than the third-point loading test.
(D) of all of the above.

(j) The modulus of rupture test yields a higher value of strength than a direct tensile test and splitting tensile test made on the same specimen because

(A) the assumed stress block shape does not match the real shape.
(B) direct tensile tests are sensitive to any accidental eccentricity.
(C) the concrete is assumed to be perfectly elastic.
(D) of all of the above.

3. If a concrete has a density of 2300 kg/m^3 and a compressive strength of 15 MPa, what modulus of elasticity is predicted by ACI 318?

(A) 14 MPa
(B) 18 MPa
(C) 28 MPa
(D) 18 GPa

4. (*Time limit: one hour*) A normal-weight concrete specimen tested at 28 days yielded the following results.

(a) A 6 in by 12 in concrete cylinder failed at an axial compressive force of 105,000 lbf. The ultimate compressive strength is most nearly

(A) 1200 psi
(B) 1900 psi
(C) 3700 psi
(D) 17,500 psi

(b) A 6 in by 12 in concrete cylinder resisted a transverse force of 45,000 lbf in a split tensile cylinder test. The concrete tensile strength is most nearly

(A) 400 psi
(B) 450 psi
(C) 625 psi
(D) 666 psi

(c) A 6 in square unreinforced beam section resisted a force of 5400 lbf on a 21 in span length. A third-point loading test was used, and the fracture occurred within the middle third. The modulus of rupture is most nearly

(A) 150 psi
(B) 400 psi
(C) 525 psi
(D) 900 psi

(d) The ratio of the modulus of rupture to the compressive strength is most nearly

(A) 4%
(B) 5%
(C) 12%
(D) 14%

(e) The ratio of the split tensile strength to the compressive strength is most nearly

(A) 8%
(B) 11%
(C) 17%
(D) 20%

(f) The conclusion that could be stated based on the answers of parts (d) and (e) is

(A) concrete is weak in tension and strong in compression.
(B) concrete is weak in tension and strong in shear.
(C) concrete is strong in tension and weak in compression.
(D) none of the above.

(g) Given a compressive strength of 3700 lbf/in^2, the approximate value of the modulus of rupture is most nearly

(A) 460 psi
(B) 500 psi
(C) 550 psi
(D) 600 psi

(h) Given a compressive strength of 3700 psi, the approximate value of the split tensile strength is most nearly

(A) 300 psi
(B) 410 psi
(C) 500 psi
(D) 600 psi

(i) From the compression test, the axial strain was 0.0015 in/in and the lateral strain was 0.00027 in/in. The Poission's ratio is most nearly

(A) 0.00027
(B) 0.0015
(C) 0.10
(D) 0.18

(j) For the beam in part (c), if the fracture occurred 1 in away from the left support, the value of the modulus of rupture obtained from this test

(A) is less than the correct value.
(B) equals the correct value.
(C) is higher than the correct value.
(D) becomes indeterminate.

5. Stress-strain curves are shown for grade-60 steel rebar and 4000 psi normal-weight concrete.

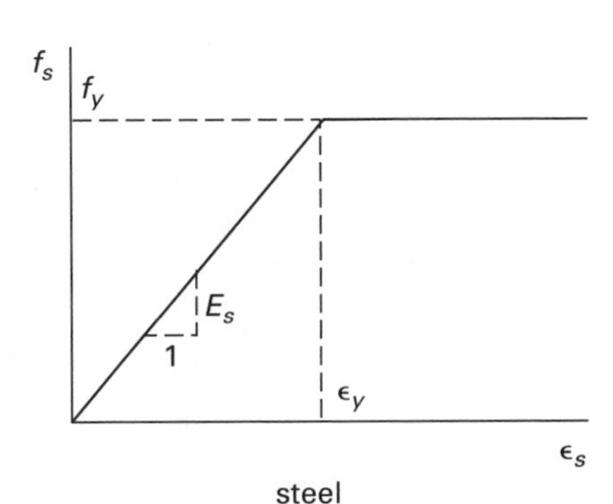

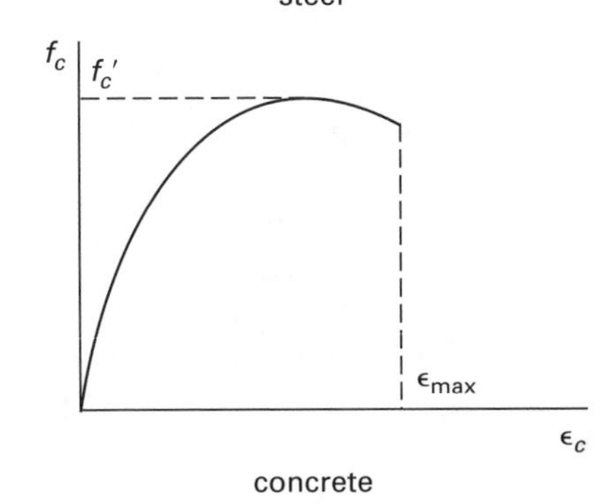

(a) What is the yield strength of the steel?

(A) 30 ksi
(B) 36 ksi
(C) 50 ksi
(D) 60 ksi

(b) What is the modulus of elasticity of the steel?

(A) 10×10^6 psi
(B) 29×10^6 psi
(C) 30×10^6 psi
(D) 36×10^6 psi

(c) What is the steel strain at the yield point?

(A) 0.002 in/in
(B) 0.003 in/in
(C) 0.06 in/in
(D) 0.09 in/in

(d) What is the concrete's compressive strength?

(A) 2000 psi
(B) 4000 psi
(C) 6000 psi
(D) 8000 psi

(e) What is the concrete's modulus of elasticity?

(A) 1.1×10^6 psi
(B) 1.7×10^6 psi
(C) 2.9×10^6 psi
(D) 3.6×10^6 psi

(f) What is the concrete's ultimate compressive strain?

(A) 0.001 in/in
(B) 0.003 in/in
(C) 0.005 in/in
(D) 0.009 in/in

SOLUTIONS

1. Consider the sieves in sequence from finest to coarsest. If the no. 100 sieve had been used first, only the pan (dust) would have passed through, and the retained percentage would have been $100\% - 1\% = 99\%$.

For the no. 50 sieve, the retained amount would have been $100\% - 1\% - 17\% = 82\%$.

The following table is prepared similarly.

sieve	cumulative retained
100	$100\% - 1\% = 99\%$
50	$99\% - 17\% = 82\%$
30	$82\% - 24\% = 58\%$
16	$58\% - 22\% = 36\%$
8	$36\% - 21\% = 15\%$
4	$15\% - 11\% = 4\%$
	TOTAL: 294

The fineness modulus is $294/100 = \boxed{2.94.}$

The answer is (C).

2. (a) From the shear diagram, the maximum shear force is $\boxed{2500 \text{ lbf.}}$

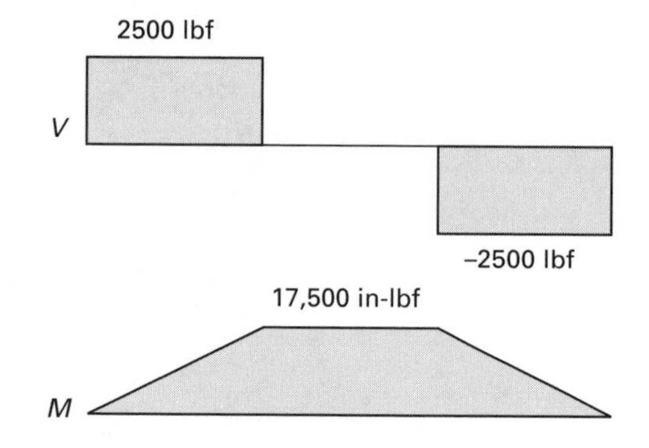

The answer is (C).

(b) From the moment diagram, the maximum moment is $\boxed{17{,}500 \text{ in-lbf.}}$

The answer is (D).

(c) From the shear diagram, the shear force in the middle third is $\boxed{\text{zero.}}$

The answer is (A).

(d) From the moment diagram, the moment is constant in the middle third.

The answer is (D).

(e) Use Eq. 48.7.

$$f_r = \frac{Mc}{I} = \frac{(2500 \text{ lbf})(6 \textit{ in})\left(\frac{6 \text{ in}}{2}\right)}{\frac{(6 \text{ in})(6 \text{ in})^3}{12}} = \boxed{417 \text{ lbf/in}^2}$$

The answer is (B).

(f) Use Eq. 48.7.

$$f_r = \frac{Mc}{I} = \frac{(17{,}500 \text{ in-lbf})\left(\frac{6 \text{ in}}{2}\right)}{\frac{(6 \text{ in})(6 \text{ in})^3}{12}} = \boxed{486 \text{ lbf/in}^2}$$

The answer is (D).

(g) Use the empirical ACI equation, Eq. 48.8.

$$f_r = 7.5\sqrt{f'_c} = 7.5\sqrt{5000 \frac{\text{lbf}}{\text{in}^2}} = \boxed{530 \text{ lbf/in}^2}$$

The answer is (A).

(h) Use the empirical ACI equation, Eq. 48.8.

$$f_r = (0.75)(7.5)\sqrt{f'_c} = (0.75)(7.5)\sqrt{5000 \frac{\text{lbf}}{\text{in}^2}} = \boxed{397 \text{ lbf/in}^2}$$

The answer is (B).

(i) ***The answer is (B).***

(j) ***The answer is (D).***

3. Use Eq. 48.1. (Equation 48.2 could also be used.)

$$E_c = w^{1.5}(0.043)\sqrt{f'_c} = \left(2300 \frac{\text{kg}}{\text{m}^3}\right)^{1.5}(0.043)\sqrt{15 \text{ MPa}} = \boxed{18{,}369 \text{ MPa} \quad (18.4 \text{ GPa})}$$

The answer is (D).

4. (a) Use Eq. 48.3.

$$f'_c = \frac{P}{A} = \frac{105{,}000 \text{ lbf}}{\left(\frac{\pi}{4}\right)(6 \text{ in})^2} = \boxed{3714 \text{ lbf/in}^2}$$

The answer is (C).

(b) Use Eq. 48.7.

$$f_{ct} = \frac{2P}{\pi DL} = \frac{(2)(45{,}000 \text{ lbf})}{\pi(6 \text{ in})(12 \text{ in})} = \boxed{398 \text{ lbf/in}^2}$$

The answer is (A).

(c)
$$f_r = \frac{Mc}{I} = \frac{\left(\frac{5400 \text{ lbf}}{2}\right)\left(\frac{21 \text{ in}}{3}\right)\left(\frac{6 \text{ in}}{2}\right)}{\frac{(6 \text{ in})(6 \text{ in})^3}{12}} = \boxed{525 \text{ lbf/in}^2}$$

The answer is (C).

(d)
$$\frac{f_r}{f'_c} = \frac{525 \frac{\text{lbf}}{\text{in}^2}}{3714 \frac{\text{lbf}}{\text{in}^2}} = \boxed{0.141 \quad (14\%)}$$

The answer is (D).

(e)
$$\frac{f_{ct}}{f'_c} = \frac{398 \frac{\text{lbf}}{\text{in}^2}}{3714 \frac{\text{lbf}}{\text{in}^2}} = \boxed{0.107 \ (11\%)}$$

The answer is (B).

(f) ***The answer is (A).***

(g) Use Eq. 48.8.

$$f_r = 7.5\sqrt{f'_c} = 7.5\sqrt{3700 \frac{\text{lbf}}{\text{in}^2}} = \boxed{456 \text{ lbf/in}^2}$$

The answer is (A).

(h) Use Eq. 48.4.

$$f_{ct} = 6.7\sqrt{f'_c} = 6.7\sqrt{3700 \frac{\text{lbf}}{\text{in}^2}} = \boxed{408 \text{ lbf/in}^2}$$

The answer is (B).

(i)
$$\nu = \frac{\text{lateral strain}}{\text{axial strain}} = \frac{0.00027 \frac{\text{in}}{\text{in}}}{0.0015 \frac{\text{in}}{\text{in}}} = \boxed{0.18}$$

The answer is (D).

(j) Refer to ASTM C78.

The answer is (D).

5. (a) The yield strength is 60 ksi for grade-60 steel.

The answer is (D).

(b) The standard modulus of elasticity for reinforcing steel is $\boxed{29 \times 10^6 \text{ psi.}}$

The answer is (B).

(c) The strain at yield is

$$\epsilon_y = \frac{f_y}{E} = \frac{60{,}000 \frac{\text{lbf}}{\text{in}^2}}{29 \times 10^6 \frac{\text{lbf}}{\text{in}^2}} = \boxed{0.00207 \text{ in/in}}$$

The answer is (A).

(d) The compressive strength is the concrete's designation.

The answer is (B).

(e) For normal-weight concrete,

$$E_c = 57{,}000\sqrt{f'_c} = 57{,}000\sqrt{4000 \frac{\text{lbf}}{\text{in}^2}} = \boxed{3.6 \times 10^6 \text{ lbf/in}^2}$$

The answer is (D).

(f) Concrete is assumed to crack when it reaches a standard strain value of $\boxed{0.003 \text{ in/in.}}$

The answer is (B).

Structural

49 Concrete Proportioning, Mixing, and Placing

PRACTICE PROBLEMS

1. (*Time limit: one hour*) The following information is submitted for a proposed concrete mix.

cement	
specific gravity	3.15
fine aggregate	
fineness modulus	2.65
specific gravity	2.48
absorption	3.0%
coarse aggregate	
specific gravity	2.68
dry bulk density	105 lbf/ft^3
absorption	0.7%
concrete	
slump	5 in
water	5.0 gal/sack
air content	4%
cement content	6.5 sack mix
coarse aggregate	0.57 ft^3/ft^3 bulk

(a) What is the dry weight of the coarse aggregate in 1 yd^3 of concrete?

(A) 1120 lbf/yd^3
(B) 1300 lbf/yd^3
(C) 1500 lbf/yd^3
(D) 1600 lbf/yd^3

(b) What is the absolute volume of the coarse aggregate in 1 yd^3 of concrete?

(A) 8.2 ft^3/yd^3
(B) 8.3 ft^3/yd^3
(C) 8.7 ft^3/yd^3
(D) 9.7 ft^3/yd^3

(c) What is the water-cement ratio?

(A) 0.42
(B) 0.44
(C) 0.46
(D) 0.48

(d) What is the weight of cement in 1 yd^3 of fresh concrete?

(A) 610 lbf/yd^3
(B) 720 lbf/yd^3
(C) 820 lbf/yd^3
(D) 910 lbf/yd^3

(e) What is the absolute volume of fine aggregate in 1 yd^3 of fresh concrete?

(A) 7.8 ft^3
(B) 8.8 ft^3
(C) 9.8 ft^3
(D) 11 ft^3

(f) What is the SSD weight of the sand in 1 yd^3 of fresh concrete?

(A) 1250 lbf
(B) 1270 lbf
(C) 1300 lbf
(D) 1370 lbf

(g) What is the absolute volume of the cement in 1 yd^3 of this fresh concrete?

(A) 3.1 ft^3
(B) 3.2 ft^3
(C) 3.3 ft^3
(D) 3.4 ft^3

(h) What is the volume of water designed for use in 1 yd^3 of fresh concrete?

(A) 3.3 ft^3
(B) 3.8 ft^3
(C) 4.3 ft^3
(D) 4.8 ft^3

(i) If the oven-dry weight of the coarse aggregate were 1600 lbf, how much water would it need to absorb to reach SSD conditions?

(A) 5.3 lbf
(B) 11 lbf
(C) 32 lbf
(D) 1100 lbf

(j) The unit weight of this concrete mix is most nearly

(A) 140 lbf/ft^3
(B) 143 lbf/ft^3
(C) 145 lbf/ft^3
(D) 150 lbf/ft^3

2. (*Time limit: one hour*) A trial batch of concrete is needed according to the following requirements and characteristics.

no. of sacks of cement per yd^3 of concrete	7
entrained air	6%
total water-cement ratio	0.54
cement specific gravity	3.15

For economical reasons, two different sizes of coarse aggregate were used. The aggregates have the following characteristics.

	specific gravity	moisture	percentage of aggregate by volume
fine aggregate	2.60	5% excess	36%
$\frac{3}{4}$ in coarse aggregate	2.65	2% excess	18%
1.0 in coarse aggregate	2.62	3% deficit	46%

(a) The weight of cement needed for 2 yd^3 of concrete is most nearly

(A) 660 lbf
(B) 1300 lbf
(C) 1500 lbf
(D) none of the above

(b) The volume of cement required for 2 yd^3 is

(A) 3.4 ft^3
(B) 5.4 ft^3
(C) 6.7 ft^3
(D) 7.0 ft^3

(c) The weight of water needed for 2 yd^3 is

(A) 670 lbf
(B) 710 lbf
(C) 870 lbf
(D) 940 lbf

(d) The volume of water needed for 2 yd^3 is

(A) 8.4 ft^3
(B) 9.7 ft^3
(C) 11 ft^3
(D) 12 ft^3

(e) The absolute volume of fine aggregate in 2 yd^3 is

(A) 11.7 ft^3
(B) 12.5 ft^3
(C) 13.4 ft^3
(D) 14.2 ft^3

(f) The SSD weight of fine aggregate in 2 yd^3 is

(A) 1800 lbf
(B) 1850 lbf
(C) 1910 lbf
(D) 1990 lbf

(g) The absolute volume of the $^3/_4$ in coarse aggregate in 2 yd^3 is

(A) 5.9 ft^3
(B) 6.2 ft^3
(C) 6.8 ft^3
(D) 7.1 ft^3

(h) The absolute volume of the 1 in coarse aggregate in 2 yd^3 is

(A) 15.0 ft^3
(B) 16.8 ft^3
(C) 17.3 ft^3
(D) 18.1 ft^3

(i) The SSD weight of the $^3/_4$ in coarse aggregate in 2 yd^3 is

(A) 860 lbf
(B) 970 lbf
(C) 1050 lbf
(D) 1120 lbf

(j) The SSD weight of the 1 in coarse aggregate in 2 yd^3 is

(A) 2460 lbf
(B) 2480 lbf
(C) 2510 lbf
(D) 2630 lbf

3. 1.5 yd^3 of portland cement concrete with the following specifications are needed.

cement content	6.5 sacks/yd^3
water-cement ratio	5.75 gal/sack
cement specific gravity	3.10
fine-aggregate specific gravity	2.65
coarse-aggregate specific gravity	2.00
aggregate grading	30% fine; 70% coarse (by volume)
free moisture (SSD basis)	1.5% excess in fine aggregate 3.0% deficient in coarse aggregate
entrained air	5%

(a) What is the as-delivered weight of the cement?

(A) 710 lbf
(B) 850 lbf
(C) 920 lbf
(D) 1100 lbf

(b) What is the as-delivered weight of the fine aggregate?

(A) 1300 lbf
(B) 1500 lbf
(C) 1600 lbf
(D) 1900 lbf

(c) What is the as-delivered weight of the coarse aggregate?

(A) 1850 lbf
(B) 1970 lbf
(C) 2140 lbf
(D) 2220 lbf

(d) What is the as-delivered weight of the water?

(A) 440 lbf
(B) 520 lbf
(C) 690 lbf
(D) 820 lbf

SOLUTIONS

1. (a) The weight of the coarse dry aggregate is

$$\left(0.57\ \frac{\text{ft}^3}{\text{ft}^3}\right)\left(105\ \frac{\text{lbf}}{\text{ft}^3}\right)\left(27\ \frac{\text{ft}^3}{\text{yd}^3}\right) = \boxed{1616\ \text{lbf/yd}^3}$$

The answer is (D).

(b) The volume of coarse aggregate is

$$\frac{1616\ \dfrac{\text{lbf}}{\text{yd}^3}}{(2.68)\left(62.4\ \dfrac{\text{lbf}}{\text{ft}^3}\right)} = \boxed{9.66\ \text{ft}^3/\text{yd}^3}$$

The answer is (D).

(c) The water-cement ratio is

$$\frac{\left(5\ \dfrac{\text{gal}}{\text{sack}}\right)\left(8.34\ \dfrac{\text{lbf}}{\text{gal}}\right)}{94\ \text{lbf/sack}} = \boxed{0.44}$$

The answer is (B).

(d) The weight of cement in 1 yd^3 is

$$\left(6.5\ \frac{\text{sacks}}{\text{yd}^3}\right)\left(94\ \frac{\text{lbf}}{\text{sack}}\right) = \boxed{611\ \text{lbf/yd}^3}$$

The answer is (A).

(e) To find the volume of fine aggregate, the volume of all other components must be found.

ingredient	weight (lbf)	volume (ft^3)
cement	611	$\dfrac{611\ \text{lbf}}{(3.15)\left(62.4\ \dfrac{\text{lbf}}{\text{ft}^3}\right)} = 3.11$
coarse aggregate	1616	9.66 (from part (b))
water	(0.44)(611 lbf) = 269	$\dfrac{269\ \text{lbf}}{62.4\ \dfrac{\text{lbf}}{\text{ft}^3}} = 4.31$
air	0	$(0.04)(1\ \text{yd}^3)\left(27\ \dfrac{\text{ft}^3}{\text{yd}^3}\right) = 1.08$
		$18.16\ \text{ft}^3$

The volume of fine aggregate in 1 yd^3 of the mix is

$$(1\ \text{yd}^3)\left(27\ \frac{\text{ft}^3}{\text{yd}^3}\right) - 18.16\ \text{ft}^3 = \boxed{8.84\ \text{ft}^3}$$

The answer is (B).

Structural

(f) The weight of fine aggregate is

$$(8.84\ \text{ft}^3)\left(62.4\ \frac{\text{lbf}}{\text{ft}^3}\right)(2.48) = \boxed{1368\ \text{lbf}}$$

The answer is (D).

(g) Refer to part (e).

The answer is (A).

(h) Refer to part (e).

The answer is (C).

(i) The weight of the water is

$$(0.007)(1600\ \text{lbf}) = \boxed{11.2\ \text{lbf}}$$

The answer is (B).

(j) The unit weight of the concrete is

$$\frac{611\ \frac{\text{lbf}}{\text{yd}^3} + 1616\ \frac{\text{lbf}}{\text{yd}^3} + 269\ \frac{\text{lbf}}{\text{yd}^3} + 1368\ \frac{\text{lbf}}{\text{yd}^3}}{27\ \frac{\text{ft}^3}{\text{yd}^3}}$$

$$= \boxed{143\ \text{lbf/ft}^3}$$

The answer is (B).

2. (a) $\left(7\ \frac{\text{sacks}}{\text{yd}^3}\right)\left(94\ \frac{\text{lbf}}{\text{sack}}\right)(2\ \text{yd}^3) = \boxed{1316\ \text{lbf}}$

The answer is (B).

(b) $$\frac{1316\ \text{lbf}}{(3.15)\left(62.4\ \frac{\text{lbf}}{\text{ft}^3}\right)} = \boxed{6.70\ \text{ft}^3}$$

The answer is (C).

(c) Determine the weights and volumes for 2 yd^3.

ingredient	weight (lbf)	volume (ft^3)
cement	1316	6.70
water	710.6	11.39
air	0	$(0.06)(2\ \text{yd}^3)\left(27\ \frac{\text{ft}^3}{\text{yd}^3}\right) = 3.24$
		21.33

The volume of fine aggregate is

$$\left((2\ \text{yd}^3)\left(27\ \frac{\text{ft}^3}{\text{yd}^3}\right) - 21.33\ \text{ft}^3\right)(0.36) = 11.76\ \text{ft}^3$$

The weight of fine aggregate is

$$(11.76\ \text{ft}^3)(2.6)\left(62.4\ \frac{\text{lbf}}{\text{yd}^3}\right) = 1907.9\ \text{lbf}$$

The volume of $^3/_4$ in coarse aggregate is

$$\left((2\ \text{yd}^3)\left(27\ \frac{\text{ft}^3}{\text{yd}^3}\right) - 21.33\ \text{ft}^3\right)(0.18) = 5.88\ \text{ft}^3$$

The weight of $^3/_4$ in coarse aggregate is

$$(5.88\ \text{ft}^3)(2.65)\left(62.4\ \frac{\text{lbf}}{\text{ft}^3}\right) = 972.3\ \text{lbf}$$

The volume of 1 in coarse aggregate is

$$\left((2\ \text{yd}^3)\left(27\ \frac{\text{ft}^3}{\text{yd}^3}\right) - 21.33\ \text{ft}^3\right)(0.46) = 15.03\ \text{ft}^3$$

The weight of 1 in coarse aggregate is

$$(15.03\ \text{ft}^3)(2.62)\left(62.4\ \frac{\text{lbf}}{\text{ft}^3}\right) = 2457\ \text{lbf}$$

The total volume as calculated is

$$21.33\ \text{ft}^3 + 11.76\ \text{ft}^3 + 5.88\ \text{ft}^3 + 15.03\ \text{ft}^3 = 54.00\ \text{ft}^3$$

The total weight is

$$\begin{aligned} &1316\ \text{lbf} + 710.6\ \text{lbf} + 1907.9\ \text{lbf} \\ &\quad + 972.3\ \text{lbf} + 2457.2\ \text{lbf} \\ &= 7364\ \text{lbf} \end{aligned}$$

The uncorrected water required is calculated from the water-cement ratio.

$$(0.54)(1316\ \text{lbf}) = 710.6\ \text{lbf} \quad [\text{before correction}]$$

The corrected water depends on the excesses and deficits in the aggregates.

fine aggregate: $(1907.9\ \text{lbf})(1 + 0.05) = 2003.3\ \text{lbf}$
$^3/_4$ in coarse aggregate: $(972.3\ \text{lbf})(1 + 0.02) = 991.7\ \text{lbf}$
1 in coarse aggregate: $(2457.2\ \text{lbf})(1 - 0.03) = 2383.5\ \text{lbf}$

The correction for surface water is

$$\begin{aligned} &7364\ \text{lbf} \\ &\quad - (1316\ \text{lbf} + 2003.2\ \text{lbf} + 991.7\ \text{lbf} + 2383.5\ \text{lbf}) \\ &= \boxed{669.5\ \text{lbf}} \end{aligned}$$

The answer is (A).

(d) The volume of water is

$$\frac{669.5 \text{ lbf}}{62.4 \frac{\text{lbf}}{\text{ft}^3}} = \boxed{10.73 \text{ ft}^3}$$

The answer is (C).

(e) Refer to part (c).

The answer is (A).

(f) Refer to part (c).

The answer is (C).

(g) Refer to part (c).

The answer is (A).

(h) Refer to part (c).

The answer is (A).

(i) Refer to part (c).

The answer is (B).

(j) Refer to part (c).

The answer is (A).

3. On a per cubic yard basis,

$$\begin{aligned}\text{volume of cement used} &= \frac{\left(6.5 \frac{\text{sacks}}{\text{yd}^3}\right)\left(94 \frac{\text{lbf}}{\text{sack}}\right)}{\left(62.4 \frac{\text{lbf}}{\text{ft}^3}\right)(3.10)} \\ &= 3.16 \text{ ft}^3/\text{yd}^3\end{aligned}$$

$$\begin{aligned}\text{volume of water used} &= \frac{\left(6.5 \frac{\text{sacks}}{\text{yd}^3}\right)\left(5.75 \frac{\text{gal}}{\text{sack}}\right)}{7.48 \frac{\text{gal}}{\text{ft}^3}} \\ &= 5.00 \text{ ft}^3/\text{yd}^3\end{aligned}$$

$$\text{volume of air} = (0.05)\left(27 \frac{\text{ft}^3}{\text{yd}^3}\right) = 1.35 \text{ ft}^3/\text{yd}^3$$

$$\text{total: } 3.16 \frac{\text{ft}^3}{\text{yd}^3} + 5.00 \frac{\text{ft}^3}{\text{yd}^3} + 1.35 \frac{\text{ft}^3}{\text{yd}^3} = 9.51 \text{ ft}^3/\text{yd}^3$$

The remainder of the cubic yard ($27 \text{ ft}^3/\text{yd}^3 - 9.51 \text{ ft}^3/\text{yd}^3 = 17.49 \text{ ft}^3/\text{yd}^3$) must be aggregate. The volumes of fine and coarse aggregates are

$$\text{fine:} \quad \left(17.49 \frac{\text{ft}^3}{\text{yd}^3}\right)(0.30) = 5.25 \text{ ft}^3/\text{yd}^3$$

$$\text{coarse:} \quad \left(17.49 \frac{\text{ft}^3}{\text{yd}^3}\right)(0.70) = 12.24 \text{ ft}^3/\text{yd}^3$$

The weights are

$$\begin{aligned}\text{cement:} \quad &\left(3.16 \frac{\text{ft}^3}{\text{yd}^3}\right)(3.10)\left(62.4 \frac{\text{lbf}}{\text{ft}^3}\right) \\ &= 611.3 \text{ lbf/yd}^3\end{aligned}$$

$$\begin{aligned}\text{water:} \quad &\left(5.0 \frac{\text{ft}^3}{\text{yd}^3}\right)\left(62.4 \frac{\text{lbf}}{\text{ft}^3}\right) \\ &= 312.0 \text{ lbf/yd}^3\end{aligned}$$

$$\begin{aligned}\text{fine aggregate:} \quad &\left(5.25 \frac{\text{ft}^3}{\text{yd}^3}\right)(2.65)\left(62.4 \frac{\text{lbf}}{\text{ft}^3}\right) \\ &= 868.1 \text{ lbf/yd}^3\end{aligned}$$

$$\begin{aligned}\text{coarse aggregate:} \quad &\left(12.24 \frac{\text{ft}^3}{\text{yd}^3}\right)(2.00)\left(62.4 \frac{\text{lbf}}{\text{ft}^3}\right) \\ &= 1527.6 \text{ lbf/yd}^3\end{aligned}$$

The mix ratio is (611.3/611.3, 868.1/611.3, 1527.6/611.3) or (1:1.42:2.5) by weight.

constituent	ratio	weight per sack cement (lbf)	weight density (lbf/ft^3)	absolute volume (ft^3)
cement	1	94	193.4	0.486
fine	1.42	133.5	165.4	0.807
coarse	2.5	235.0	124.8	1.883
water			$\left(\frac{5.75}{7.48}\right)$ =	0.769
				3.945 ft^3

The solid yield is 3.945 ft^3/sack. The yield with 5% air is

$$\frac{3.945 \frac{\text{ft}^3}{\text{sack}}}{1 - 0.05} = 4.153 \text{ ft}^3/\text{sack}$$

The number of one-sack batches is

$$\frac{(1.5 \text{ yd}^3)\left(27 \frac{\text{ft}^3}{\text{yd}^3}\right)}{4.153 \frac{\text{ft}^3}{\text{sack}}} = 9.752 \text{ sacks}$$

(a) The required cement weight is

$$(9.752 \text{ sacks})\left(94 \frac{\text{lbf}}{\text{sack}}\right) = \boxed{916.7 \text{ lbf}}$$

The answer is (C).

Structural

(b) The required fine aggregate weight is

$$(9.752 \text{ sacks})(1.015)\left(94 \ \frac{\text{lbf}}{\text{sack}}\right)(1.42)$$

$$= \boxed{1321.2 \text{ lbf as delivered} \qquad [1301.7 \text{ lbf SSD}]}$$

The answer is (A).

Note: Dividing by 1.015 converts to SSD conditions.

(c) The required coarse aggregate is

$$(9.752 \text{ sacks})(0.97)\left(94 \ \frac{\text{lbf}}{\text{sack}}\right)(2.5)$$

$$= \boxed{2223.0 \text{ lbf as delivered} \qquad [2291.7 \text{ lbf SSD}]}$$

The answer is (D).

(d) The weight of the excess water in the fine aggregate is

$$(1321.2 \text{ lbf})\left(\frac{0.015}{1.015}\right) = 19.53 \text{ lbf}$$

The water needed to bring the coarse aggregate to SSD condition is

$$(2223.0 \text{ lbf})\left(\frac{0.03}{0.97}\right) = 68.75 \text{ lbf}$$

The total weight of water required is

$$\left(\frac{\left(5.75 \ \frac{\text{gal}}{\text{sack}}\right)(9.752 \text{ sacks})}{7.48 \ \frac{\text{gal}}{\text{ft}^3}}\right)\left(62.4 \ \frac{\text{lbf}}{\text{ft}^3}\right)$$
$$+ 68.75 \text{ lbf} - 19.53 \text{ lbf}$$
$$= \boxed{517.0 \text{ lbf}}$$

The answer is (B).

50 Reinforced Concrete: Beams

PRACTICE PROBLEMS

(Use the ultimate strength method to solve all of these problems.)

1. The simply supported beam shown spans 20 ft and, in addition to its own weight, carries a uniformly distributed service dead load of 1.75 kips/ft and a uniformly distributed live load of 3.0 kips/ft. Five no. 9 bars running the full length of the beam are used as flexural reinforcement. $f'_c = 4000$ lbf/in^2, and $f_y = 60{,}000$ lbf/in^2. No. 3 stirrups are used. (a) Determine the theoretical spacing of stirrups at the critical section using the refined expression for the shear strength of the concrete. (b) Determine the theoretical spacing of the stirrups at the critical section using the simplified expression for the concrete shear strength. (c) Identify the portion of the beam where stirrups are not required.

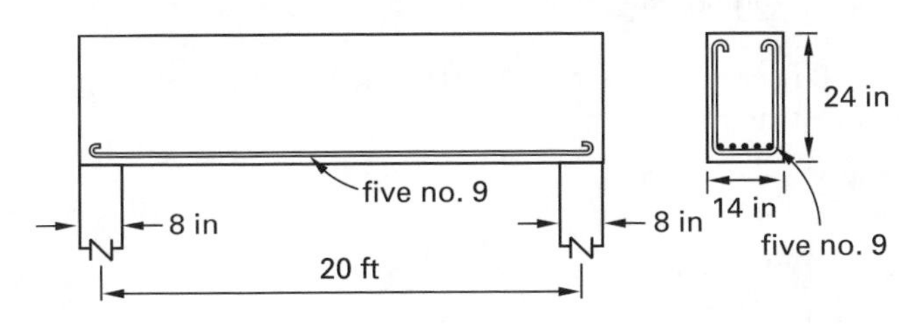

2. (*Time limit: one hour*) A simply supported reinforced concrete beam spans 20 ft as shown. The beam is subjected to a uniform service dead load equal to 2.0 kips/ft (exclusive of beam weight) and to a uniform service live load of 2.4 kips/ft. $f'_c = 3000$ lbf/in^2, and $f_y = 40{,}000$ lbf/in^2. The depth to the reinforcing is 25.5 in.

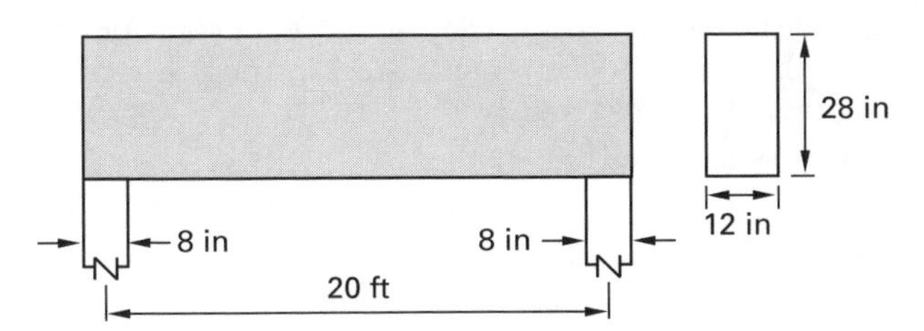

(a) The factored uniform load is most nearly

(A) 6.9 kips/ft
(B) 7.4 kips/ft
(C) 8.0 kips/ft
(D) 9.2 kips/ft

(b) The tension steel required at the section of maximum moment is most nearly

(A) 4.5 in^2
(B) 5.0 in^2
(C) 5.6 in^2
(D) 6.1 in^2

(c) The maximum area of tension steel permitted (based on flexure requirements) is most nearly

(A) 5.6 in^2
(B) 7.7 in^2
(C) 8.5 in^2
(D) 9.2 in^2

(d) The minimum area of tension steel permitted is most nearly

(A) 1.0 in^2
(B) 1.3 in^2
(C) 1.5 in^2
(D) 1.8 in^2

(e) The maximum uniform factored load that the beam can sustain if no compression steel is used is most nearly

(A) 8.0 kips/ft
(B) 9.0 kips/ft
(C) 10 kips/ft
(D) 13 kips/ft

(f) The spacing of no. 3 U-shaped stirrups at the location where the shear is maximum is most nearly

(A) 5.0 in
(B) 6.0 in
(C) 6.4 in
(D) 8.3 in

(g) The total length of the beam where no shear reinforcement is required is most nearly

(A) 10 in
(B) 12 in
(C) 15 in
(D) 20 in

(h) Based on the maximum total shear permitted in the beam, the maximum uniform factored load that the beam can be designed for, with the given dimensions, is most nearly

(A) 16 kips/ft
(B) 17 kips/ft
(C) 18 kips/ft
(D) 20 kips/ft

(i) If the live load is subsequently reduced to 50% of its maximum value, the service moment that should be used to compute the effective moment of inertia is most nearly

(A) 200 ft-kips
(B) 220 ft-kips
(C) 240 ft-kips
(D) 250 ft-kips

(j) The cracking moment is most nearly

(A) 40 ft-kips
(B) 45 ft-kips
(C) 51 ft-kips
(D) 54 ft-kips

3. (*Time limit: one hour*) A reinforced concrete T-beam with a 30 ft span and a 30 in effective width in a floor slab system is fixed at both ends and is reinforced as shown. $f'_c = 3000$ lbf/in^2, and $f_y = 60{,}000$ lbf/in^2.

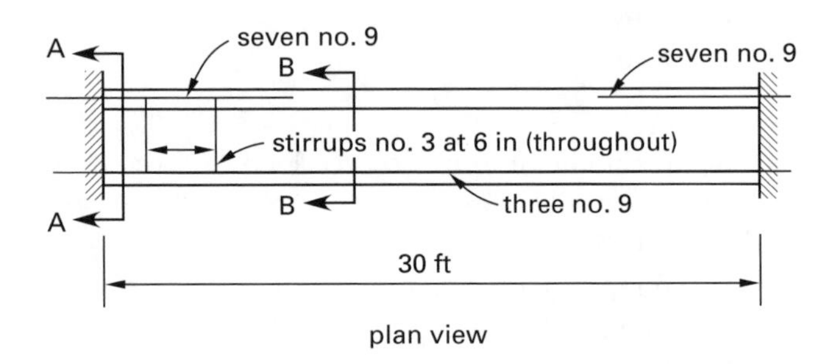

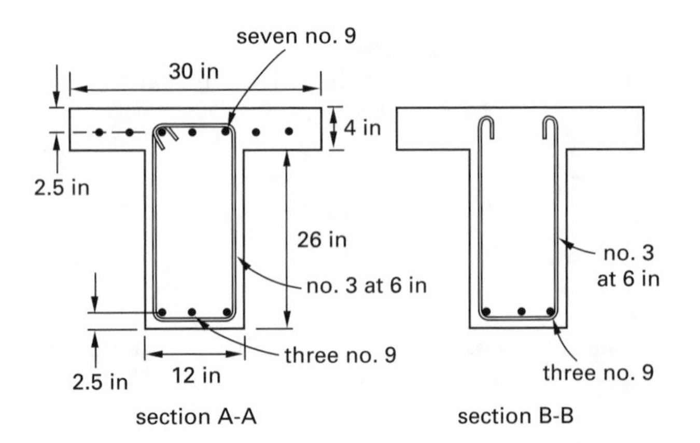

(a) The design moment capacity in the positive moment region is most nearly

(A) 300 ft-kips
(B) 360 ft-kips
(C) 400 ft-kips
(D) 410 ft-kips

(b) At capacity, the stress in the compression steel in the negative moment region is most nearly

(A) 35 kips/in^2
(B) 47 kips/in^2
(C) 52 kips/in^2
(D) 60 kips/in^2

(c) The gross moment of inertia is most nearly

(A) 27,000 in^4
(B) 37,000 in^4
(C) 41,000 in^4
(D) 68,000 in^4

(d) The cracked moment of inertia in the positive moment region is most nearly

(A) 10,000 in^4
(B) 13,000 in^4
(C) 15,000 in^4
(D) 20,000 in^4

(e) The cracking moment in the positive moment region is most nearly

(A) 22 ft-kips
(B) 47 ft-kips
(C) 60 ft-kips
(D) 74 ft-kips

(f) The cracking moment in the negative moment region is most nearly

(A) 60 ft-kips
(B) 74 ft-kips
(C) 99 ft-kips
(D) 120 ft-kips

(g) If a uniformly distributed service load of 4 kips/ft acts on the beam, using the refined equation for steel stress, the stress in the tension steel in the positive moment region is most nearly

(A) 18 kips/in^2
(B) 23 kips/in^2
(C) 30 kips/in^2
(D) 35 kips/in^2

(h) The maximum spacing of the tension steel bars based on cracking is most nearly

(A) 4 in
(B) 8 in
(C) 13 in
(D) 18 in

(i) The maximum uniform factored load that the beam can sustain as controlled by the shear reinforcement is most nearly

(A) 4.0 kips/ft
(B) 5.9 kips/ft
(C) 6.5 kips/ft
(D) 7.5 kips/ft

(j) The maximum uniform factored load that the beam can sustain as governed by the flexural capacity is most nearly

(A) 7.1 kips/ft
(B) 8.2 kips/ft
(C) 9.5 kips/ft
(D) 14 kips/ft

4. A singly reinforced rectangular beam carries a moment from dead loads of 50,000 ft-lbf and a moment from live loads of 200,000 ft-lbf. $f'_c = 3000$ lbf/in^2 and $f_y = 50{,}000$ lbf/in^2. No. 4 stirrups are to be used. Design and detail the cross section.

5. A beam must withstand an ultimate factored moment of 400,000 ft-lbf. $f'_c = 4000$ lbf/in^2, and $f_y = 40{,}000$ lbf/in^2. No. 3 stirrups will be used. (Do not design the shear reinforcement, check for cracking, or check deflection.) (a) Determine the beam width and depth. (b) Determine the required steel area. (c) How many layers of steel are needed? (d) What is the overall beam depth?

6. The slab shown supports a moment of 20,000 ft-lbf per foot of width. The concrete's unit weight is 145 lbf/ft^3, and $f'_c = 3000$ lbf/in^2. The response is elastic. Use the transformed area method of evaluating the stresses in the (a) concrete and (b) steel.

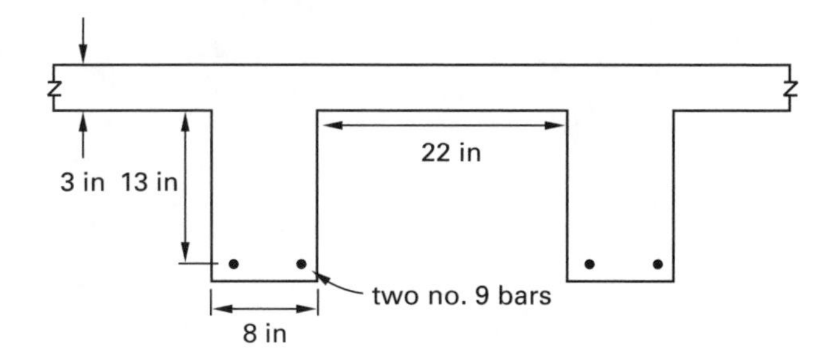

7. (*Time limit: one hour*) A rectangular, singly reinforced beam is shown. The beam supports two concentrated dead loads and a self-weight dead load of 2000 lbf/ft. $f'_c = 4000$ lbf/in^2, and $f_y = 60{,}000$ lbf/in^2. No. 11 steel bars must be used. The modular ratio is 8. (Do not design shear reinforcement. Do not consider live loading.) (a) Determine the beam width and depth. (b) Determine the required steel area using half of the maximum steel permitted. (c) How many layers of steel are needed?

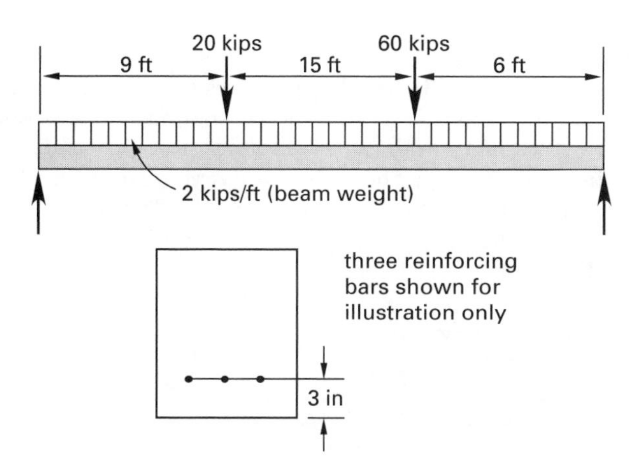

8. (*Time limit: one hour*) A simply supported, singly reinforced rectangular concrete beam supporting a roof is needed to span 27 ft. It carries a dead load of 1 kip/ft (which includes the beam weight) and a live load of 2 kips/ft. $f'_c = 4000$ lbf/in^2, $f_y = 60{,}000$ lbf/in^2, and $E_c = 3.6 \times 10^6$ lbf/in^2. The maximum permitted steel reinforcing is to be used. (a) Use no. 11 bars and a beam width of 14 in to design the beam. (b) If the beam has an interior exposure, will cracking be within limits? (c) If 30% of the live load is sustained, calculate the instantaneous and long-term centerline deflections.

9. (*Time limit: one hour*) The beam shown carries a dead load of 1300 lbf/ft (including its own weight) and a live load of 1900 lbf/ft. $f'_c = 3000$ lbf/in^2, and $f_y = 60{,}000$ lbf/in^2. The clear distance between the two beam supports is 28 ft. The beam's depth is 21 in. (a) What additional shear reinforcement strength is required to be provided by the steel? (b) Over what length of the beam is shear reinforcement not required? (c) Assuming no. 3 bars are used, what stirrup spacing should be used between the face of the support out to a distance of approximately 19 in from the face of the support?

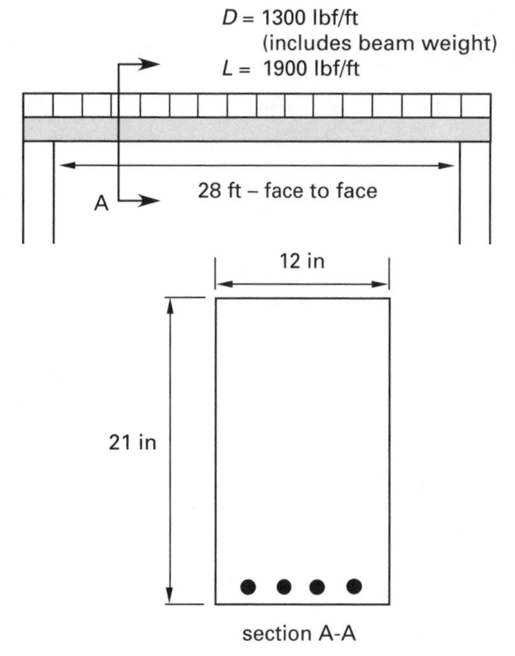

10. (*Time limit: one hour*) The simply supported tension-reinforced beam shown carries two identical sets of point loads and a set of uniform loadings. The beam has a width of 16 in. $f'_c = 3000$ lbf/in^2, and $f_y = 60{,}000$ lbf/in^2. (a) What is the ultimate moment? (b) Specify the minimum reinforcement depth. (c) Detail the reinforcing steel. (d) Use no. 3 bars to design the shear reinforcing.

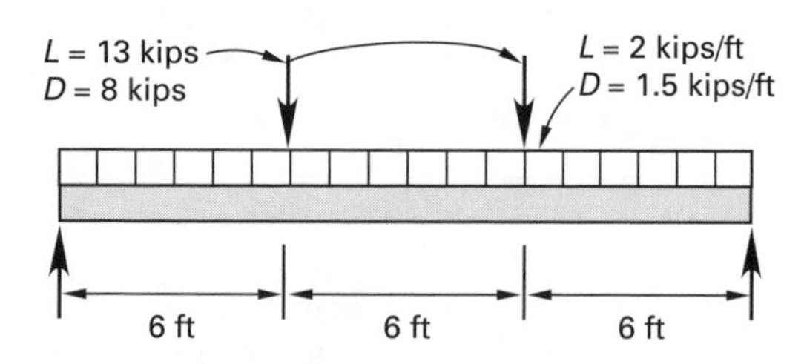

11. (*Time limit: one hour*) The reinforced concrete beam shown in the first illustration supports two 322 kip concentrated loads. (The loads have already been factored and include beam self-weights.) The size of the bearing plates is 18 in by 20 in. Compressive strength of concrete $f'_c = 4$ ksi, and yield strength of reinforcement $f_y = 60$ ksi. A strut-and-tie model for this beam is shown in the second illustration. The strut-and-tie model consists of three struts (AB, BC, and CD), one tie (AD), and four nodes (A, B, C, and D). The effective width of strut BC, w_c, is equal to 7.5 in and the effective width of tie AD, w_t, is equal to 9 in. The vertical position of node A is 9 in/2 = 4.5 in from the bottom of the beam, and for node B it is 7.5 in/2 = 3.75 in from the top of the beam.

(a) Determine if this beam is a deep beam according to code provisions.

(b) Check bearing capacity at all nodal zones, assuming that no confining reinforcement is provided in the nodal zones.

(c) Check design capacity of strut BC.

(d) Determine reinforcement required in tie AD.

(e) Check design capacity of strut AB.

(f) Design the anchorage of the tie bars from step (d).

(g) Determine the horizontal and vertical reinforcement required to resist splitting of diagonal struts.

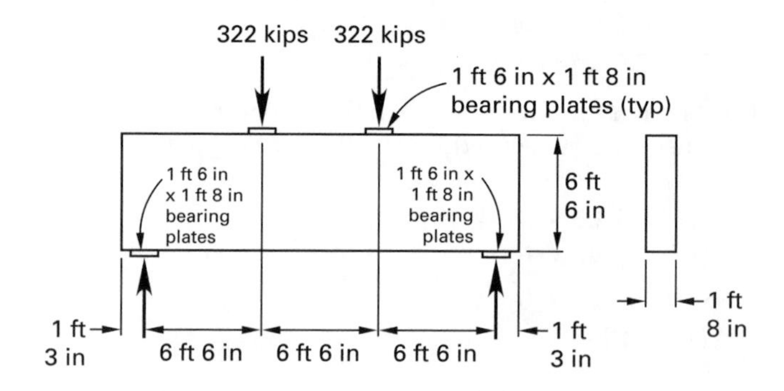

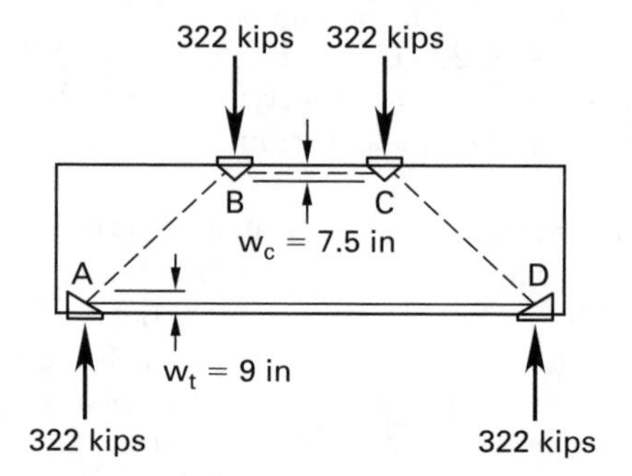

SOLUTIONS

1. Use either **ACI 318 App. C** or **ACI 318 Ch. 9** (see p. 50-6) to solve this problem.

The weight of the beam is

$$\frac{(14\text{ in})(24\text{ in})\left(0.15\ \frac{\text{kip}}{\text{ft}^3}\right)}{\left(12\ \frac{\text{in}}{\text{ft}}\right)^2} = 0.35\text{ kip/ft}$$

Solution Using ACI 318 App. C

The factored uniform load is

$$\begin{aligned} w_u &= 1.4D + 1.7L \\ &= (1.4)\left(1.75\ \frac{\text{kips}}{\text{ft}} + 0.35\ \frac{\text{kip}}{\text{ft}}\right) + (1.7)\left(3\ \frac{\text{kips}}{\text{ft}}\right) \\ &= 8.04\text{ kips/ft} \end{aligned}$$

Obtain the approximate shear envelope.

$$\begin{aligned} \text{maximum reaction} &= w_u\left(\frac{L}{2}\right) \\ &= \frac{\left(8.04\ \frac{\text{kips}}{\text{ft}}\right)(20\text{ ft})}{2} \\ &= 80.4\text{ kips} \end{aligned}$$

The centerline shear when the factored live load is positioned over half of the span is

$$\begin{aligned} V &= 1.7\left(\frac{w\left(\frac{L}{2}\right)\left(\frac{L}{4}\right)}{L}\right) = \frac{1.7wL}{8} \\ &= \frac{(1.7)\left(3.0\ \frac{\text{kips}}{\text{ft}}\right)(20\text{ ft})}{8} \\ &= 12.75\text{ kips} \end{aligned}$$

Since the reaction induces compression, the critical section is located at a distance d from the face of the support. The shear envelope, with the shear at the critical section identified, is illustrated. (The effective depth has been taken as $d = h - 2.5\text{ in} = 21.5\text{ in}$. The critical shear of 66.02 kips is found by interpolation.)

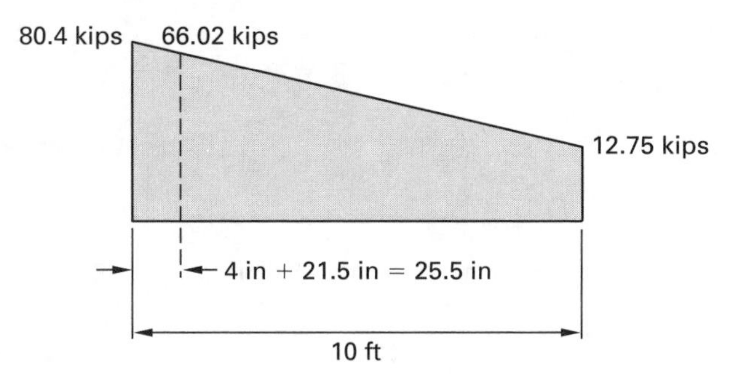

(a) The critical section is located (measured from the center of the column) at

$$\frac{4\text{ in} + 21.5\text{ in}}{12\ \frac{\text{in}}{\text{ft}}} = 2.125\text{ ft}$$

The factored moment at the critical section is

$$\begin{aligned} M_u &= (80.4\text{ kips})(2.125\text{ ft}) - \frac{\left(8.04\ \frac{\text{kips}}{\text{ft}}\right)(2.125\text{ ft})^2}{2} \\ &= 152.70\text{ ft-kips} \end{aligned}$$

The flexural reinforcement steel ratio at the critical section is given by Eq. 50.31.

$$\begin{aligned} \rho_w &= \frac{A_s}{b_w d} \\ &= \frac{5\text{ in}^2}{(14\text{ in})(21.5\text{ in})} \\ &= 0.0166 \end{aligned}$$

From the shear envelope, $V_u = 66.02$ kips. Use Eq. 50.52 to compute the shear strength of the concrete.

$$\begin{aligned} V_c &= \left(1.9\sqrt{f'_c} + 2500\rho_w\left(\frac{V_u d}{M_u}\right)\right)b_w d \\ &= \frac{\left(\begin{array}{c} 1.9\sqrt{4000\ \frac{\text{lbf}}{\text{in}^2}} + (2500)(0.0166) \\ \times\left(\frac{(66.02\text{ kips})(21.5\text{ in})}{(152.7\text{ ft-kips})\left(12\ \frac{\text{in}}{\text{ft}}\right)}\right)\end{array}\right)(14\text{ in})(21.5\text{ in})}{1000\ \frac{\text{lbf}}{\text{kip}}} \\ &= 45.85\text{ kips} \end{aligned}$$

The shear that must be resisted by the shear reinforcement is obtained from Eq. 50.63.

$$\begin{aligned} V_{s,\text{req}} &= \frac{V_u}{\phi} - V_c \\ &= \frac{66.02\text{ kips}}{0.85} - 45.85\text{ kips} \\ &= 31.83\text{ kips} \end{aligned}$$

The spacing of the no. 3 stirrups is found from Eq. 50.64.

$$\begin{aligned} s &= \frac{A_v f_y d}{V_{s,\text{req}}} \\ &= \frac{(0.22\text{ in}^2)\left(60\ \frac{\text{kips}}{\text{in}^2}\right)(21.5\text{ in})}{31.83\text{ kips}} \\ &= \boxed{8.92\text{ in}} \end{aligned}$$

Check the maximum spacing.

$$4\sqrt{f'_c}b_w d = \frac{4\sqrt{4000\ \frac{\text{lbf}}{\text{in}^2}}(14\ \text{in})(21.5\ \text{in})}{1000\ \frac{\text{lbf}}{\text{kip}}}$$
$$= 76.15\ \text{kips}$$

Since $V_{s,\text{req}} < 4\sqrt{f'_c}b_w d$ (Eq. 50.59), the maximum spacing permitted is

$$\frac{d}{2} = \frac{21.5\ \text{in}}{2} = 10.75\ \text{in}\quad [\text{does not control}]$$

Check the minimum A_v from Eq. 50.61.

$$\frac{50 b_w s}{f_y} = \frac{(50)(14\ \text{in})(8.92\ \text{in})}{60{,}000\ \frac{\text{lbf}}{\text{in}^2}}$$
$$= 0.1041\ \text{in}^2$$

$0.1041\ \text{in}^2$ is less than the $A_v = 0.22\ \text{in}^2$ provided by the U-shaped no. 3 stirrup, so the minimum requirement is satisfied.

(b) From Eq. 50.53,

$$V_c = 2\sqrt{f'_c}b_w d$$
$$= \frac{2\sqrt{4000\ \frac{\text{lbf}}{\text{in}^2}}(14\ \text{in})(21.5\ \text{in})}{1000\ \frac{\text{lbf}}{\text{kip}}}$$
$$= 38.07\ \text{kips}$$

The same procedure as in part (a) is used. From Eq. 50.63,

$$V_{s,\text{req}} = \frac{V_u}{\phi} - V_c$$
$$= \frac{66.02\ \text{kips}}{0.85} - 38.07\ \text{kips}$$
$$= 39.60\ \text{kips}$$
$$s = \frac{A_v f_y d}{V_{s,\text{req}}}$$
$$= \frac{(0.22\ \text{in}^2)\left(60\ \frac{\text{kips}}{\text{in}^2}\right)(21.5\ \text{in})}{39.60\ \text{kips}}$$
$$= \boxed{7.17\ \text{in}}$$

(c) Stirrups are unnecessary when $V_u \le \phi V_c/2$. Use the simplified expression for V_c. From Eq. 50.62,

$$\frac{\phi V_c}{2} = \frac{(0.85)(38.07\ \text{kips})}{2}$$
$$= 16.18\ \text{kips}$$

Using simple geometric calculations, this value of the shear is found in the shear envelope at 6.08 in from the centerline of the beam. The section where stirrups can be omitted is, therefore, $(2)(6.08\ \text{in}) = \boxed{12.16\ \text{in}}$ long in the center of the beam. This is negligible for practical purposes and would not be considered.

Solution Using ACI 318 Ch. 9

The factored uniform load is

$$w_u = 1.2D + 1.6L$$
$$= (1.2)\left(1.75\ \frac{\text{kips}}{\text{ft}} + 0.35\ \frac{\text{kip}}{\text{ft}}\right) + (1.6)\left(3\ \frac{\text{kips}}{\text{ft}}\right)$$
$$= 7.32\ \text{kips/ft}$$

$1.4D$ does not control.

Obtain the approximate shear envelope.

$$\text{maximum reaction} = w_u\left(\frac{L}{2}\right)$$
$$= \frac{\left(7.32\ \frac{\text{kips}}{\text{ft}}\right)(20\ \text{ft})}{2}$$
$$= 73.2\ \text{kips}$$

The centerline shear when the factored live load is positioned over half of the span is

$$V = 1.6\left(\frac{w\left(\frac{L}{2}\right)\left(\frac{L}{4}\right)}{L}\right) = \frac{1.6wL}{8}$$
$$= \frac{(1.6)\left(3.0\ \frac{\text{kips}}{\text{ft}}\right)(20\ \text{ft})}{8}$$
$$= 12.0\ \text{kips}$$

Since the reaction induces compression, the critical section is located at a distance d from the face of the support. The shear envelope, with the shear at the critical section identified, is illustrated. (The effective depth has been taken as $d = h - 2.5\ \text{in} = 21.5\ \text{in}$. The critical shear of 66.02 kips is found by interpolation.)

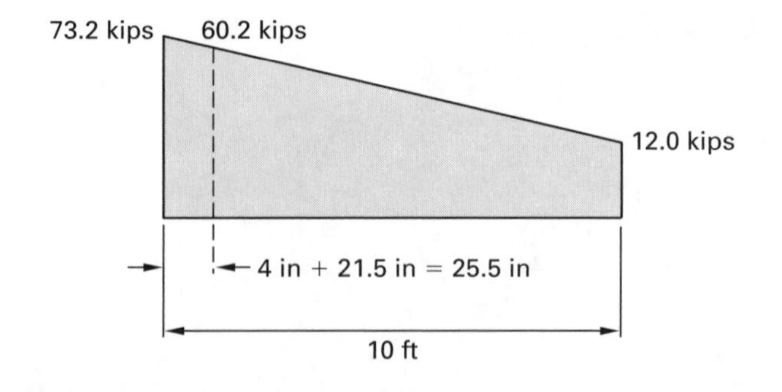

(a) The critical section is located (measured from the center of the column) at

$$\frac{4\text{ in} + 21.5\text{ in}}{12\,\frac{\text{in}}{\text{ft}}} = 2.125\text{ ft}$$

The factored moment at the critical section is

$$\begin{aligned}M_u &= (73.2\text{ kips})(2.125\text{ ft}) - \frac{\left(7.32\,\frac{\text{kips}}{\text{ft}}\right)(2.125\text{ ft})^2}{2}\\ &= 139.02\text{ ft-kips}\end{aligned}$$

The flexural reinforcement steel ratio at the critical section is given by Eq. 50.31.

$$\begin{aligned}\rho_w &= \frac{A_s}{b_w d}\\ &= \frac{5\text{ in}^2}{(14\text{ in})(21.5\text{ in})}\\ &= 0.0166\end{aligned}$$

From the shear envelope, $V_u = 60.20$ kips. Use Eq. 50.52 to compute the shear strength of the concrete.

$$\begin{aligned}V_c &= \left(1.9\sqrt{f'_c} + 2500\rho_w\left(\frac{V_u d}{M_u}\right)\right)b_w d\\ &= \frac{\left(\begin{array}{l}1.9\sqrt{4000\,\frac{\text{lbf}}{\text{in}^2}} + (2500)(0.0166)\\ \times\left(\frac{(60.2\text{ kips})(21.5\text{ in})}{(139.02\text{ ft-kips})\left(12\,\frac{\text{in}}{\text{ft}}\right)}\right)\end{array}\right)(14\text{ in})(21.5\text{ in})}{1000\,\frac{\text{lbf}}{\text{kip}}}\\ &= 45.86\text{ kips}\end{aligned}$$

The shear that must be resisted by the shear reinforcement is obtained from Eq. 50.63.

$$\begin{aligned}V_{s,\text{req}} &= \frac{V_u}{\phi} - V_c\\ &= \frac{60.2\text{ kips}}{0.75} - 45.86\text{ kips}\\ &= 34.41\text{ kips}\end{aligned}$$

The spacing of the no. 3 stirrups is found from Eq. 50.64.

$$\begin{aligned}s &= \frac{A_v f_y d}{V_{s,\text{req}}}\\ &= \frac{(0.22\text{ in}^2)\left(60\,\frac{\text{kips}}{\text{in}^2}\right)(21.5\text{ in})}{34.41\text{ kips}}\\ &= \boxed{8.25\text{ in}}\end{aligned}$$

Check the maximum spacing.

$$\begin{aligned}4\sqrt{f'_c}b_w d &= \frac{4\sqrt{4000\,\frac{\text{lbf}}{\text{in}^2}}(14\text{ in})(21.5\text{ in})}{1000\,\frac{\text{lbf}}{\text{kip}}}\\ &= 76.15\text{ kips}\end{aligned}$$

Since $V_{s,\text{req}} < 4\sqrt{f'_c}b_w d$ (Eq. 50.59), the maximum spacing permitted is

$$\frac{d}{2} = \frac{21.5\text{ in}}{2} = 10.75\text{ in}\quad\text{[does not control]}$$

Check the minimum A_v from Eq. 50.61.

$$\begin{aligned}A_v &= 0.75\sqrt{f'_c}\frac{b_w s}{f_y}\\ &= 0.75\sqrt{4000}\frac{(14)(8.25)}{60{,}000}\\ &= 0.0913\\ \frac{50 b_w s}{f_y} &= \frac{(50)(14\text{ in})(8.25\text{ in})}{60{,}000\,\frac{\text{lbf}}{\text{in}^2}}\\ &= 0.09625\text{ in}^2\quad\text{[controls]}\end{aligned}$$

0.09625 in² is less than the $A_v = 0.22$ in² provided by the U-shaped no. 3 stirrup, so the minimum requirement is satisfied.

(b) From Eq. 50.53,

$$\begin{aligned}V_c &= 2\sqrt{f'_c}b_w d\\ &= \frac{2\sqrt{4000\,\frac{\text{lbf}}{\text{in}^2}}(14\text{ in})(21.5\text{ in})}{1000\,\frac{\text{lbf}}{\text{kip}}}\\ &= 38.07\text{ kips}\end{aligned}$$

The same procedure as in part (a) is used. From Eq. 50.63,

$$\begin{aligned}V_{s,\text{req}} &= \frac{V_u}{\phi} - V_c\\ &= \frac{60.20\text{ kips}}{0.75} - 38.07\text{ kips}\\ &= 42.20\text{ kips}\\ s &= \frac{A_v f_y d}{V_{s,\text{req}}}\\ &= \frac{(0.22\text{ in}^2)\left(60\,\frac{\text{kips}}{\text{in}^2}\right)(21.5\text{ in})}{42.20\text{ kips}}\\ &= \boxed{6.73\text{ in}}\end{aligned}$$

(c) Stirrups are unnecessary when $V_u \leq \phi V_c/2$. Use the simplified expression for V_c. From Eq. 50.62,

$$\begin{aligned}\frac{\phi V_c}{2} &= \frac{(0.75)(38.07 \text{ kips})}{2}\\ &= 14.28 \text{ kips}\end{aligned}$$

Using simple geometric calculations, this value of the shear is found in the shear envelope at 4.47 in from the centerline of the beam. The section where stirrups can be omitted is, therefore, (2)(4.47 in) = $\boxed{8.94 \text{ in}}$ long in the center of the beam. This is negligible for practical purposes and would not be considered.

2. Use either **ACI 318 App. C** or **ACI 318 Ch. 9** (see p. 50-11) to solve this problem.

Solution Using ACI 318 App. C

(a) The weight of the beam is

$$\frac{(12 \text{ in})(28 \text{ in})\left(0.15 \frac{\text{kip}}{\text{ft}^3}\right)}{\left(12 \frac{\text{in}}{\text{ft}}\right)^2} = 0.35 \text{ kip/ft}$$

The factored uniform load is

$$\begin{aligned}w_u &= 1.4D + 1.7L\\ &= (1.4)\left(2.0 \frac{\text{kips}}{\text{ft}} + 0.35 \frac{\text{kip}}{\text{ft}}\right) + (1.7)\left(2.4 \frac{\text{kips}}{\text{ft}}\right)\\ &= \boxed{7.37 \text{ kips/ft}}\end{aligned}$$

The answer is (B).

(b) The maximum moment is

$$\begin{aligned}M_u &= \frac{wL^2}{8} = \frac{\left(7.37 \frac{\text{kips}}{\text{ft}}\right)(20 \text{ ft})^2}{8}\\ &= 368.50 \text{ ft-kips}\end{aligned}$$

Estimate $\lambda = 0.1d = (0.1)(25.5 \text{ in}) = 2.55$ in. From Eq. 50.28,

$$\begin{aligned}A_s &= \frac{M_u}{\phi f_y(d-\lambda)}\\ &= \frac{(368.50 \text{ ft-kips})\left(12 \frac{\text{in}}{\text{ft}}\right)}{(0.90)\left(40 \frac{\text{kips}}{\text{in}^2}\right)(25.5 \text{ in} - 2.55 \text{ in})}\\ &= 5.35 \text{ in}^2\end{aligned}$$

Compute M_n using the standard approach. Use Eq. 50.26.

$$\begin{aligned}A_c &= \frac{f_y A_s}{0.85 f'_c}\\ &= \frac{(5.35 \text{ in}^2)\left(40 \frac{\text{kips}}{\text{in}^2}\right)}{(0.85)\left(3 \frac{\text{kips}}{\text{in}^2}\right)}\\ &= 83.92 \text{ in}^2\end{aligned}$$

Since the compressed zone is rectangular,

$$\begin{aligned}a &= \frac{A_c}{b} = \frac{83.92 \text{ in}^2}{12 \text{ in}}\\ &= 6.99 \text{ in}\\ \lambda &= \frac{a}{2} = \frac{6.99 \text{ in}}{2} = 3.50 \text{ in}\end{aligned}$$

The nominal moment capacity is obtained from Eq. 50.27.

$$\begin{aligned}M_n &= A_s f_y(d-\lambda)\\ &= \frac{(5.35 \text{ in}^2)\left(40 \frac{\text{kips}}{\text{in}^2}\right)(25.5 \text{ in} - 3.50 \text{ in})}{12 \frac{\text{in}}{\text{ft}}}\\ &= 392.33 \text{ ft-kips}\end{aligned}$$

The design moment capacity is

$$\begin{aligned}\phi M_n &= (0.9)(392.33 \text{ ft-kips})\\ &= 353.10 \text{ ft-kips}\end{aligned}$$

The adjusted steel area is

$$\begin{aligned}A_s &= \left(\frac{368.50 \text{ ft-kips}}{353.10 \text{ ft-kips}}\right)(5.35 \text{ in}^2)\\ &= \boxed{5.58 \text{ in}^2}\end{aligned}$$

The answer is (C).

(c) The depth of the equivalent rectangular stress block at balance is obtained from Eq. 50.20 with $\beta_1 = 0.85$ from Eq. 50.21.

$$\begin{aligned}a_b &= \beta_1\left(\frac{87{,}000}{87{,}000 + f_y}\right)d\\ &= (0.85)\left(\frac{87{,}000}{87{,}000 + 40{,}000}\right)(25.5 \text{ in})\\ &= 14.84 \text{ in}\end{aligned}$$

The area A_{cb} included in the depth a_b is

$$A_{cb} = (12 \text{ in})(14.84 \text{ in}) = 178.08 \text{ in}^2$$

Substituting this into Eq. 50.19,

$$A_{sb} = (0.85)\left(\frac{f'_c A_{cb}}{f_y}\right) = \frac{(0.85)\left(3\ \frac{\text{kips}}{\text{in}^2}\right)(178.08\ \text{in}^2)}{40\ \frac{\text{kips}}{\text{in}^2}} = 11.35\ \text{in}^2$$

Without the use of compression steel, the maximum steel permitted is given by Eq. 50.18.

$$A_{s,\text{max}} = 0.75A_{sb} = (0.75)(11.35\ \text{in}^2) = \boxed{8.51\ \text{in}^2}$$

The answer is (C).

(d) Use Eq. 50.16.

$$\frac{3\sqrt{f'_c}b_w d}{f_y} = \left(\frac{3\sqrt{3000\ \frac{\text{lbf}}{\text{in}^2}}}{40{,}000\ \frac{\text{lbf}}{\text{in}^2}}\right)(12\ \text{in})(25.5\ \text{in}) = 1.26\ \text{in}^2$$

$$\frac{200b_w d}{f_y} = \frac{(200)(12\ \text{in})(25.5\ \text{in})}{40{,}000\ \frac{\text{lbf}}{\text{in}^2}} = \boxed{1.53\ \text{in}^2 \quad \text{[controls]}}$$

The answer is (C).

(e) Since $A_{s,\text{max}} = 8.51\ \text{in}^2$, from Eq. 50.26,

$$A_c = \frac{f_y A_s}{0.85f'_c} = \frac{\left(40\ \frac{\text{kips}}{\text{in}^2}\right)(8.51\ \text{in}^2)}{(0.85)\left(3\ \frac{\text{kips}}{\text{in}^2}\right)} = 133.49\ \text{in}^2$$

Since the compressed zone is rectangular,

$$a = \frac{A_c}{b} = \frac{133.49\ \text{in}^2}{12\ \text{in}} = 11.12\ \text{in}$$

$$\lambda = \frac{a}{2} = \frac{11.12\ \text{in}}{2} = 5.56\ \text{in}$$

The nominal moment capacity is obtained from Eq. 50.27.

$$M_n = A_s f_y (d - \lambda) = \frac{(8.51\ \text{in}^2)\left(40\ \frac{\text{kips}}{\text{in}^2}\right)(25.5\ \text{in} - 5.56\ \text{in})}{12\ \frac{\text{in}}{\text{ft}}} = 565.63\ \text{ft-kips}$$

Determine the corresponding ϕ factor.

$$c = \frac{a}{\beta_1} = \frac{11.12\ \text{in}}{0.85} = 13.08\ \text{in}$$

$$\epsilon_t = \left(\frac{d-c}{c}\right)\epsilon_{\text{max}} = \left(\frac{25.5\ \text{in} - 13.08\ \text{in}}{13.08}\right)(0.003) = 0.0028$$

Interpolate between the following.

$$\epsilon_t = \frac{f_y}{E} = \frac{40\ \frac{\text{kips}}{\text{in}^2}}{29{,}000\ \frac{\text{kips}}{\text{in}^2}} = 0.00138 \quad (\phi = 0.7)$$

$$\epsilon_t = 0.005 \quad (\phi = 0.09)$$

$$\phi = 0.778$$

The design moment capacity is

$$\phi M_n = (0.778)(565.63\ \text{ft-kips}) = 440.06\ \text{ft-kips}$$

The uniform factored load is, therefore,

$$w_u = \frac{8\phi M_n}{L^2} = \frac{(8)(440.06\ \text{ft-kips})}{(20\ \text{ft})^2} = \boxed{8.80\ \text{kips/ft}}$$

The answer is (A).

(f) Obtain the approximate shear envelope.

$$\text{maximum reaction} = w_u\left(\frac{L}{2}\right) = \frac{\left(7.37\ \frac{\text{kips}}{\text{ft}}\right)(20\ \text{ft})}{2} = 73.70\ \text{kips}$$

The shear at centerline when the factored live load is positioned over half of the span is

$$V = 1.7\left(\frac{w\left(\frac{L}{2}\right)\left(\frac{L}{4}\right)}{L}\right) = \frac{1.7wL}{8} = \frac{(1.7)\left(2.4\ \frac{\text{kips}}{\text{ft}}\right)(20\ \text{ft})}{8} = 10.20\ \text{kips}$$

Since the reaction induces compression, the critical section is located at a distance d from the face of the support. The shear envelope, with the shear at the critical section identified, is illustrated. The effective depth has been taken as $d = h - 2.5$ in.

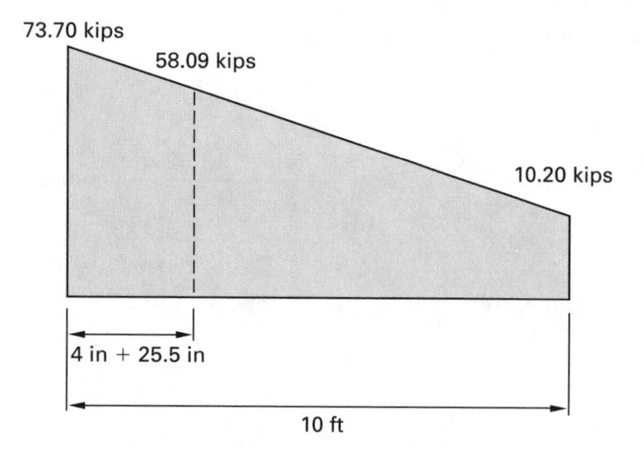

From Eq. 50.53,

$$V_c = 2\sqrt{f'_c}b_w d = \frac{2\sqrt{3000\ \frac{\text{lbf}}{\text{in}^2}}(12\text{ in})(25.5\text{ in})}{1000\ \frac{\text{lbf}}{\text{kip}}} = 33.52\text{ kips}$$

Use Eq. 50.63.

$$V_{s,\text{req}} = \frac{V_u}{\phi} - V_c = \frac{58.09\text{ kips}}{0.85} - 33.52\text{ kips} = 34.82\text{ kips}$$

$$s = \frac{A_v f_y d}{V_{s,\text{req}}} = \frac{(0.22\text{ in}^2)\left(40\ \frac{\text{kips}}{\text{in}^2}\right)(25.5\text{ in})}{34.82\text{ kips}} = \boxed{6.44\text{ in}}$$

The answer is (C).

(g) Stirrups can be omitted when $V_u \le \phi V_c/2$. Use the simplified expression for V_c.

$$\frac{\phi V_c}{2} = \frac{(0.85)(33.52\text{ kips})}{2} = 14.25\text{ kips}$$

Using simple geometric calculations, this value of the shear is reached in the shear envelope at 7.65 in from the centerline of the beam. The section where stirrups can be omitted is, therefore, $(2)(7.65\text{ in}) = \boxed{15.30\text{ in}}$ long in the center of the beam.

The answer is (C).

(h) Use Eq. 50.58.

$$\phi 10\sqrt{f'_c}b_w d = \frac{(0.85)(10)\sqrt{3000\ \frac{\text{lbf}}{\text{in}^2}}(12\text{ in})(25.5\text{ in})}{1000\ \frac{\text{lbf}}{\text{kip}}} = 142.46\text{ kips}$$

Assuming a uniform scaling of the shear envelope,

$$w_u = \left(\frac{142.46\text{ kips}}{58.09\text{ kips}}\right)\left(7.37\ \frac{\text{kips}}{\text{ft}}\right) = \boxed{18.07\text{ kips/ft}}$$

The answer is (C).

(i) Since cracking is not reversible, the moment due to the full service load should be used.

$$\text{service load} = 2\ \frac{\text{kips}}{\text{ft}} + 0.35\ \frac{\text{kip}}{\text{ft}} + 2.4\ \frac{\text{kips}}{\text{ft}} = 4.75\text{ kips/ft}$$

$$M_s = \frac{wL^2}{8} = \frac{\left(4.75\ \frac{\text{kips}}{\text{ft}}\right)(20\text{ ft})^2}{8} = \boxed{237.5\text{ ft-kips}}$$

The answer is (C).

(j)

$$I_g = \frac{(12\text{ in})(28\text{ in})^3}{12} = 21{,}952\text{ in}^4$$

$$f_r = 7.5\sqrt{f'_c} = 7.5\sqrt{3000\ \frac{\text{lbf}}{\text{in}^2}} = 410\text{ lbf/in}^2$$

$$y_t = \frac{h}{2} = \frac{28\text{ in}}{2} = 14\text{ in}$$

Use Eq. 50.44.

$$M_{cr} = \frac{f_r I_g}{y_t} = \frac{\left(0.410\ \frac{\text{kip}}{\text{in}^2}\right)(21{,}952\text{ in}^4)}{(14\text{ in})\left(12\ \frac{\text{in}}{\text{ft}}\right)} = \boxed{53.57\text{ ft-kips}}$$

The answer is (D).

Solution Using ACI 318 Ch. 9

(a) The factored uniform load is

$$\begin{aligned} w_u &= 1.2D + 1.6L \\ &= (1.2)\left(2.0\ \frac{\text{kips}}{\text{ft}} + 0.35\ \frac{\text{kip}}{\text{ft}}\right) + (1.6)\left(2.4\ \frac{\text{kips}}{\text{ft}}\right) \\ &= \boxed{6.66\ \text{kips/ft}} \end{aligned}$$

$w_u = 1.4D$ does not control.

The answer is (A).

(b) The maximum moment is

$$\begin{aligned} M_u &= \frac{wL^2}{8} = \frac{\left(6.66\ \frac{\text{kips}}{\text{ft}}\right)(20\ \text{ft})^2}{8} \\ &= 333.0\ \text{ft-kips} \end{aligned}$$

Estimate $\lambda = 0.1d = (0.1)(25.5\ \text{in}) = 2.55\ \text{in}$, in which case the section is tension controlled because $c/d < 0.375$. From Eq. 50.28,

$$\begin{aligned} A_s &= \frac{M_u}{\phi f_y(d-\lambda)} \\ &= \frac{(333.0\ \text{ft-kips})\left(12\ \frac{\text{in}}{\text{ft}}\right)}{(0.90)\left(40\ \frac{\text{kips}}{\text{in}^2}\right)(25.5\ \text{in} - 2.55\ \text{in})} \\ &= 4.84\ \text{in}^2 \end{aligned}$$

Compute M_n using the standard approach. Use Eq. 50.26.

$$\begin{aligned} A_c &= \frac{f_y A_s}{0.85 f'_c} \\ &= \frac{(4.84\ \text{in}^2)\left(40\ \frac{\text{kips}}{\text{in}^2}\right)}{(0.85)\left(3\ \frac{\text{kips}}{\text{in}^2}\right)} \\ &= 75.92\ \text{in}^2 \end{aligned}$$

Since the compressed zone is rectangular,

$$\begin{aligned} a &= \frac{A_c}{b} = \frac{75.92\ \text{in}^2}{12\ \text{in}} \\ &= 6.33\ \text{in} \\ \lambda &= \frac{a}{2} = \frac{6.33\ \text{in}}{2} = 3.17\ \text{in} \end{aligned}$$

Check to see if the section is tension controlled.

$$c = \frac{a}{\beta_1} = \frac{6.33\ \text{in}}{0.85} = 7.45\ \text{in}$$

$$\epsilon_t = \left(\frac{d-c}{c}\right)(0.003) = 0.007 > 0.005$$

The section is tension controlled.

The nominal moment capacity is obtained from Eq. 50.27.

$$\begin{aligned} M_n &= A_s f_y (d-\lambda) \\ &= \frac{(4.84\ \text{in}^2)\left(40\ \frac{\text{kips}}{\text{in}^2}\right)(25.5\ \text{in} - 3.17\ \text{in})}{12\ \frac{\text{in}}{\text{ft}}} \\ &= 360.26\ \text{ft-kips} \end{aligned}$$

The design moment capacity is

$$\begin{aligned} \phi M_n &= (0.9)(360.26\ \text{ft-kips}) \\ &= 324.23\ \text{ft-kips} \end{aligned}$$

The adjusted steel area is

$$\begin{aligned} A_s &= \left(\frac{333.0\ \text{ft-kips}}{324.23\ \text{ft-kips}}\right)(4.84\ \text{in}^2) \\ &= \boxed{4.97\ \text{in}^2} \end{aligned}$$

The answer is (B).

(c) ACI 318 10.3.5 requires that, for nonprestressed flexural members with axial load less than $0.10 f'_c A_g$, the net tensile strain, ϵ_t, at nominal strength must not be less than 0.004.

$$\frac{0.003}{0.004} = \frac{c}{d-c}$$

$$a = \beta_1 c$$

$$\begin{aligned} A_{s,\text{max}} &= \frac{0.85 f'_c b \beta_1 c}{f_y} = \frac{0.85 f'_c b \beta_1 \left(\frac{0.003}{0.007}\right) d}{f_y} \\ &= \frac{(0.85)(3\ \text{ksi})(12\ \text{in})(0.85)\left(\frac{0.003}{0.007}\right)(25.5\ \text{in})}{40\ \text{ksi}} \\ &= 7.1\ \text{in}^2 \end{aligned}$$

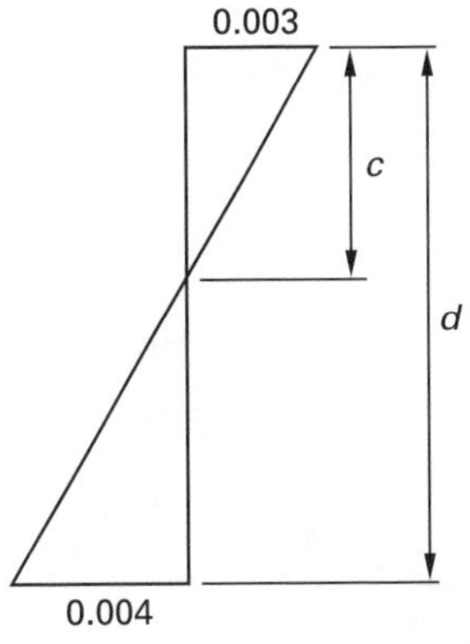

The answer is (B).

(d) Use Eq. 50.16.

$$\frac{3\sqrt{f'_c}b_w d}{f_y} = \left(\frac{3\sqrt{3000\ \frac{\text{lbf}}{\text{in}^2}}}{40{,}000\ \frac{\text{lbf}}{\text{in}^2}}\right)(12\ \text{in})(25.5\ \text{in}) = 1.26\ \text{in}^2$$

$$\frac{200b_w d}{f_y} = \frac{(200)(12\ \text{in})(25.5\ \text{in})}{40{,}000\ \frac{\text{lbf}}{\text{in}^2}} = \boxed{1.53\ \text{in}^2 \quad [\text{controls}]}$$

The answer is (C).

(e) Since $A_{s,\text{max}} = 7.10\ \text{in}^2$, from Eq. 50.26,

$$A_c = \frac{f_y A_s}{0.85f'_c} = \frac{\left(40\ \frac{\text{kips}}{\text{in}^2}\right)(7.10\ \text{in}^2)}{(0.85)\left(3\ \frac{\text{kips}}{\text{in}^2}\right)} = 111.37\ \text{in}^2$$

Since the compressed zone is rectangular,

$$a = \frac{A_c}{b} = \frac{111.37\ \text{in}^2}{12\ \text{in}} = 9.28\ \text{in}$$

$$\lambda = \frac{a}{2} = \frac{9.28\ \text{in}}{2} = 4.64\ \text{in}$$

The nominal moment capacity is obtained from Eq. 50.27.

$$M_n = A_s f_y(d - \lambda) = \frac{(7.10\ \text{in}^2)\left(40\ \frac{\text{kips}}{\text{in}^2}\right)(25.5\ \text{in} - 4.64\ \text{in})}{12\ \frac{\text{in}}{\text{ft}}} = 493.69\ \text{ft-kips}$$

The ϕ factor corresponding to $\epsilon_t = 0.004$ is

$$\phi = 0.812$$

The design moment capacity is

$$\phi M_n = (0.812)(493.69\ \text{ft-kips}) = 400.88\ \text{ft-kips}$$

The uniform factored load is, therefore,

$$w_u = \frac{8\phi M_n}{L^2} = \frac{(8)(400.88\ \text{ft-kips})}{(20\ \text{ft})^2} = \boxed{8.02\ \text{kips/ft}}$$

The answer is (A).

(f) Obtain the approximate shear envelope.

$$\text{maximum reaction} = w_u\left(\frac{L}{2}\right) = \frac{\left(6.66\ \frac{\text{kips}}{\text{ft}}\right)(20\ \text{ft})}{2} = 66.60\ \text{kips}$$

The shear at centerline when the factored live load is positioned over half of the span is

$$V = 1.6\left(\frac{w\left(\frac{L}{2}\right)\left(\frac{L}{4}\right)}{L}\right) = \frac{1.6wL}{8} = \frac{(1.6)\left(2.4\ \frac{\text{kips}}{\text{ft}}\right)(20\ \text{ft})}{8} = 9.60\ \text{kips}$$

Since the reaction induces compression, the critical section is located at a distance d from the face of the support. The shear envelope, with the shear at the critical section identified, is illustrated. The effective depth has been taken as $d = h - 2.5$ in.

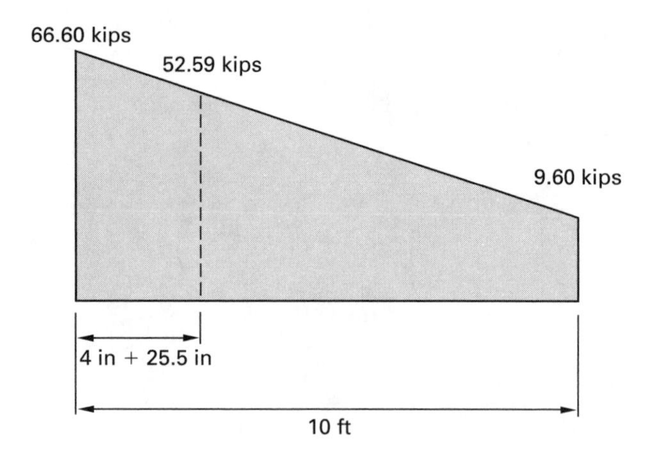

From Eq. 50.53,

$$V_c = 2\sqrt{f'_c}b_w d = \frac{2\sqrt{3000\ \frac{\text{lbf}}{\text{in}^2}}(12\ \text{in})(25.5\ \text{in})}{1000\ \frac{\text{lbf}}{\text{kip}}} = 33.52\ \text{kips}$$

Use Eq. 50.63.

$$V_{s,\text{req}} = \frac{V_u}{\phi} - V_c = \frac{52.59 \text{ kips}}{0.75} - 33.52 \text{ kips} = 36.60 \text{ kips}$$

$$s = \frac{A_v f_y d}{V_{s,\text{req}}} = \frac{(0.22 \text{ in}^2)\left(40 \ \frac{\text{kips}}{\text{in}^2}\right)(25.5 \text{ in})}{36.60 \text{ kips}} = \boxed{6.13 \text{ in}}$$

The answer is (B).

(g) Stirrups can be omitted when $V_u \leq \phi V_c/2$. Use the simplified expression for V_c.

$$\frac{\phi V_c}{2} = \frac{(0.75)(33.52 \text{ kips})}{2} = 12.57 \text{ kips}$$

Using simple geometric calculations, this value of the shear is reached in the shear envelope at 6.25 in from the centerline of the beam. The section where stirrups can be omitted is, therefore, (2)(6.25 in) = $\boxed{12.50 \text{ in}}$ long in the center of the beam.

The answer is (B).

(h) Use Eq. 50.58.

$$\phi 10\sqrt{f'_c}b_w d = \frac{(0.75)(10)\sqrt{3000 \ \frac{\text{lbf}}{\text{in}^2}}(12 \text{ in})(25.5 \text{ in})}{1000 \ \frac{\text{lbf}}{\text{kip}}} = 125.70 \text{ kips}$$

Assuming a uniform scaling of the shear envelope,

$$w_u = \left(\frac{125.70 \text{ kips}}{52.59 \text{ kips}}\right)\left(6.66 \ \frac{\text{kips}}{\text{ft}}\right) = \boxed{15.92 \text{ kips/ft}}$$

The answer is (A).

(i) Since cracking is not reversible, the moment due to the full service load should be used.

$$\text{service load} = 2 \ \frac{\text{kips}}{\text{ft}} + 0.35 \ \frac{\text{kip}}{\text{ft}} + 2.4 \ \frac{\text{kips}}{\text{ft}} = 4.75 \text{ kips/ft}$$

$$M_s = \frac{wL^2}{8} = \frac{\left(4.75 \ \frac{\text{kips}}{\text{ft}}\right)(20 \text{ ft})^2}{8} = \boxed{237.5 \text{ ft-kips}}$$

The answer is (C).

(j)

$$I_g = \frac{(12 \text{ in})(28 \text{ in})^3}{12} = 21{,}952 \text{ in}^4$$

$$f_r = 7.5\sqrt{f'_c} = 7.5\sqrt{3000 \ \frac{\text{lbf}}{\text{in}^2}} = 410 \text{ lbf/in}^2$$

$$y_t = \frac{h}{2} = \frac{28 \text{ in}}{2} = 14 \text{ in}$$

Use Eq. 50.44.

$$M_{cr} = \frac{f_r I_g}{y_t} = \frac{\left(0.410 \ \frac{\text{kip}}{\text{in}^2}\right)(21{,}952 \text{ in}^4)}{(14 \text{ in})\left(12 \ \frac{\text{in}}{\text{ft}}\right)} = \boxed{53.57 \text{ ft-kips}}$$

The answer is (D).

3. Use either **ACI 318 Apps. B and C** or **ACI 318 Ch. 9 and Unified Design Method** (see p. 50-16) to solve this problem.

Solution Using ACI 318 Apps. B and C

(a) In the positive moment region, $d = 27.5$ in and $A_s = 3.0 \text{ in}^2$. Therefore,

$$A_c = \frac{f_y A_s}{0.85 f'_c} = \frac{\left(60 \ \frac{\text{kips}}{\text{in}^2}\right)(3 \text{ in}^2)}{(0.85)\left(3 \ \frac{\text{kips}}{\text{in}^2}\right)} = 70.59 \text{ in}^2$$

Structural

Assume the stress block is confined to the flange. Then, from Eq. 50.29,

$$a = \frac{A_c}{b} = \frac{70.79 \text{ in}^2}{30 \text{ in}} = 2.35 \text{ in}$$

(The assumption proves correct.)

$$\lambda = \frac{a}{2} = \frac{2.35 \text{ in}}{2} = 1.18 \text{ in}$$

The nominal moment capacity is obtained from Eq. 50.27.

$$\begin{aligned} M_n &= A_s f_y (d - \lambda) \\ &= \frac{(3 \text{ in}^2)\left(60 \frac{\text{kips}}{\text{in}^2}\right)(27.5 \text{ in} - 1.18 \text{ in})}{12 \frac{\text{in}}{\text{ft}}} \\ &= 394.85 \text{ ft-kips} \end{aligned}$$

Determine the ϕ factor.

$$\begin{aligned} c &= \frac{a}{0.85} = \frac{2.35 \text{ in}}{0.85} \\ &= 2.76 \text{ in} \\ \epsilon_t &= \left(\frac{d-c}{c}\right)(0.003) \\ &= 0.027 > 0.005 \\ \phi &= 0.9 \end{aligned}$$

The section is tension controlled.

The design moment capacity (maximum positive moment) is

$$\begin{aligned} \phi M_n &= (0.9)(394.85 \text{ ft-kips}) \\ &= \boxed{355.37 \text{ ft-kips}} \end{aligned}$$

The answer is (B).

(b) In the negative moment region, $d = 27.5$ in, $A_s = 7.0$ in^2, and $A_s' = 3.0$ in^2.

Assuming the compression steel yields, from Eq. 50.70,

$$\begin{aligned} A_c &= \frac{f_y A_s - f_s' A_s'}{0.85 f_c'} \\ &= \frac{\left(60 \frac{\text{kips}}{\text{in}^2}\right)(7 \text{ in}^2) - \left(60 \frac{\text{kips}}{\text{in}^2}\right)(3 \text{ in}^2)}{(0.85)\left(3 \frac{\text{kips}}{\text{in}^2}\right)} \\ &= 94.12 \text{ in}^2 \end{aligned}$$

Using Eq. 50.29,

$$\begin{aligned} a &= \frac{A_c}{b_w} = \frac{94.12 \text{ in}^2}{12 \text{ in}} = 7.84 \text{ in} \\ \lambda &= \frac{a}{2} = \frac{7.84 \text{ in}}{2} = 3.92 \text{ in} \\ c &= \frac{a}{\beta_1} = \frac{7.84 \text{ in}}{0.85} = 9.23 \text{ in} \end{aligned}$$

From Eq. 50.67,

$$\begin{aligned} \varepsilon_s' &= \left(\frac{0.003}{c}\right)(c - d') \\ &= \left(\frac{0.003}{9.23 \text{ in}}\right)(9.23 \text{ in} - 2.5 \text{ in}) \\ &= 0.0022 \end{aligned}$$

Use Eq. 50.68.

$$\begin{aligned} f_s' &= E_s \varepsilon_s' = \left(29{,}000 \frac{\text{kips}}{\text{in}^2}\right)(0.0022) \\ &= 63.43 \text{ kips/in}^2 \\ f_s' &= f_y = \boxed{60 \text{ kips/in}^2} \quad \text{[confirms assumption]} \end{aligned}$$

The answer is (D).

(c) Divide the section into two elementary areas as shown. The computations are summarized in the following table.

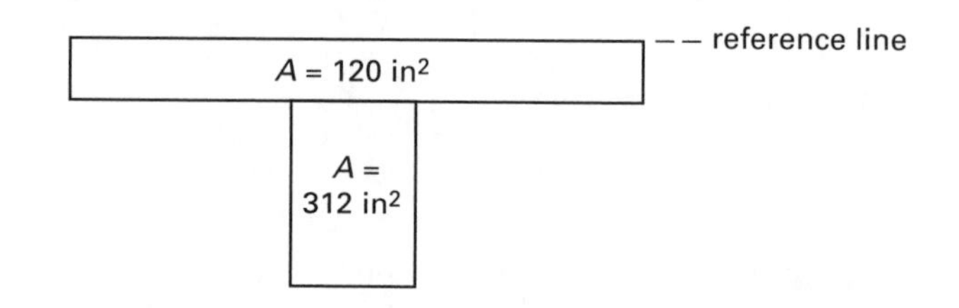

	A (in^2)	y (in)	Ay (in^3)	$A(\overline{y} - y)^2$ (in^4)	I_0 (in^4)
	120	2	240	14,083	160
	312	17	5304	5417	17,576
totals	432		5544	19,500	17,736

$$\overline{y} = \frac{5544 \text{ in}^3}{432 \text{ in}^2} = 12.83 \text{ in}$$

$$I_g = 19{,}500 \text{ in}^4 + 17{,}736 \text{ in}^4 = \boxed{37{,}236 \text{ in}^4}$$

The answer is (B).

(d) From Eq. 50.38,

$$\begin{aligned} E_c &= 57{,}000\sqrt{f_c'} \\ &= \frac{57{,}000\sqrt{3000 \frac{\text{lbf}}{\text{in}^2}}}{1000 \frac{\text{lbf}}{\text{kip}}} \\ &= 3122 \text{ kips/in}^2 \\ n &= \frac{E_s}{E_c} = \frac{29{,}000 \frac{\text{kips}}{\text{in}^2}}{3122 \frac{\text{kips}}{\text{in}^2}} \\ &= 9.29 \end{aligned}$$

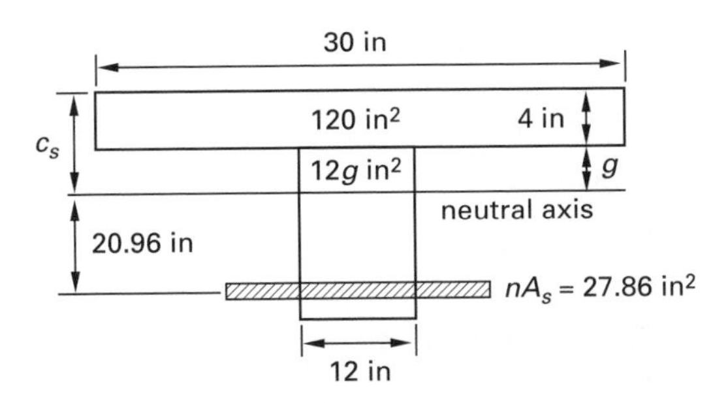

From the sketch of the transformed section,

$$(120 \text{ in}^2)(g + 2 \text{ in}) + \frac{(12 \text{ in})g^2}{2} = (27.86 \text{ in}^2)(27.5 \text{ in} - 4 \text{ in} - g)$$
$$g = 2.54 \text{ in}$$

The neutral axis is located at $c_s = 6.54$ in.

The transformed cracked moment of inertia is calculated using the parallel axis theorem.

$$\begin{aligned} I_{cr} &= \left(\tfrac{1}{12}\right)(30 \text{ in})(4 \text{ in})^3 + \left(\tfrac{1}{3}\right)(12 \text{ in})(2.54 \text{ in})^3 \\ &\quad + (120 \text{ in}^2)(4.54 \text{ in})^2 \\ &\quad + (27.86 \text{ in}^2)(20.96 \text{ in})^2 \\ &= \boxed{14{,}938 \text{ in}^4} \end{aligned}$$

The answer is (C).

(e) From Eq. 50.45,

$$f_r = 7.5\sqrt{f'_c} = \frac{7.5\sqrt{3000 \ \frac{\text{lbf}}{\text{in}^2}}}{1000 \ \frac{\text{lbf}}{\text{kip}}} = 0.41 \text{ kip/in}^2$$

From Eq. 50.44,

$$\begin{aligned} M_{cr} &= \frac{f_r I_g}{y_t} \\ &= \frac{\left(0.41 \ \frac{\text{kip}}{\text{in}^2}\right)(37{,}236 \text{ in}^4)}{(30 \text{ in} - 12.83 \text{ in})\left(12 \ \frac{\text{in}}{\text{ft}}\right)} \\ &= \boxed{74.25 \text{ ft-kips}} \end{aligned}$$

The answer is (D).

(f) Use Eq. 50.44.

$$\begin{aligned} M_{cr} &= \frac{f_r I_g}{y_t} \\ &= \frac{\left(0.41 \ \frac{\text{kip}}{\text{in}^2}\right)(37{,}236 \text{ in}^4)}{(12.83 \text{ in})\left(12 \ \frac{\text{in}}{\text{ft}}\right)} \\ &= \boxed{99.33 \text{ ft-kips}} \end{aligned}$$

The answer is (C).

(g) The maximum positive service moment is

$$M_s = \left(\frac{1}{24}\right)wL^2 = \left(\frac{1}{24}\right)\left(4 \ \frac{\text{kips}}{\text{ft}}\right)(30 \text{ ft})^2 = 150 \text{ ft-kips}$$

Therefore,

$$\begin{aligned} f_s &= \frac{nM_s(d - c_s)}{I_{cr}} \\ &= \frac{(9.29)(150 \text{ ft-kips})\left(12 \ \frac{\text{in}}{\text{ft}}\right)(27.5 \text{ in} - 6.54 \text{ in})}{14{,}938 \text{ in}^4} \\ &= \boxed{23.46 \text{ kips/in}^2} \end{aligned}$$

The answer is (B).

(h)

$$\begin{aligned} c_c &= d_c - \frac{d_b}{2} = 2.5 \text{ in} - \frac{1.125 \text{ in}}{2} \\ &= 1.9375 \text{ in} \end{aligned}$$
$$f_s = 60 \text{ kips/in}^2$$

From Eq. 50.36,

$$\begin{aligned} s &\le \frac{540}{f_s} - 2.5c_c \\ &\le \frac{540}{23.46 \ \frac{\text{kips}}{\text{in}^2}} - (2.5)(1.9375 \text{ in}) \\ &\le 18.2 \text{ in} \quad \text{[controls]} \end{aligned}$$
$$s \le \frac{432}{f_s} = \frac{432}{23.46 \ \frac{\text{kips}}{\text{in}^2}} = 18.4 \text{ in}$$

The answer is (D).

(i) From Eq. 50.53,

$$V_c = 2\sqrt{f'_c}b_w d = \frac{2\sqrt{3000\ \frac{\text{lbf}}{\text{in}^2}}(12\text{ in})(27.5\text{ in})}{1000\ \frac{\text{lbf}}{\text{kip}}} = 36.15\text{ kips}$$

From Eq. 50.54,

$$V_s = \frac{A_v f_y d}{s} = \frac{(0.22\text{ in}^2)\left(60\ \frac{\text{kips}}{\text{in}^2}\right)(27.5\text{ in})}{6\text{ in}} = 60.50\text{ kips}$$

From Eq. 50.51,

$$\phi V_n = \phi(V_c + V_s) = (0.85)(36.15\text{ kips} + 60.50\text{ kips}) = 82.15\text{ kips}$$

The critical section is at a distance d from the face of the support. The shear at the critical section is

$$V_u = \tfrac{1}{2}w_u\left(30\text{ ft} - \frac{(2)(27.5\text{ in})}{12\ \frac{\text{in}}{\text{ft}}}\right) = 12.71w_u$$

Equating the maximum shear to the shear capacity,

$$w_u = \frac{82.15\text{ kips}}{12.71\text{ ft}} = \boxed{6.50\text{ kips/ft}}$$

The answer is (C).

(j) Use Eq. 50.69 with top and bottom steel yielding. (λ was calculated in part (b).)

$$M_n = (A_s - A'_s)f_y(d-\lambda) + A'_s f'_s(d-d')$$

$$= \frac{(7\text{ in}^2 - 3\text{ in}^2)\left(60\ \frac{\text{kips}}{\text{in}^2}\right)(27.5\text{ in} - 3.92\text{ in}) + (3\text{ in}^2)\left(60\ \frac{\text{kips}}{\text{in}^2}\right)(27.5\text{ in} - 2.5\text{ in})}{12\ \frac{\text{in}}{\text{ft}}}$$

$$= 846.6\text{ ft-kips}\quad\text{[global maximum]}$$

The maximum positive moment is

$$M_u = \frac{1}{24}w_u L^2 = \frac{1}{24}w_u(30\text{ ft})^2 = 37.5w_u$$

The maximum positive moment was calculated in part (a). Equating this value to the capacity ϕM_n for the positive moment region,

$$w_u = \frac{355.37\text{ ft-kips}}{37.5\text{ ft}^2} = \boxed{9.48\text{ kips/ft}}$$

The maximum negative moment from this fixed-end beam is

$$M_u = \frac{1}{12}w_u L^2 = \frac{1}{12}w_u(30\text{ ft})^2 = 75w_u$$

Equate this value to the capacity ϕM_n for the negative moment region, using $\phi = 0.9$. (ϵ_t can be calculated as 0.0059.)

$$\phi M_n = M_u$$
$$(0.9)(846.6\text{ ft-kips}) = 75w_u$$
$$w_u = 10.2\text{ kips/ft}$$

The positive region controls.

The answer is (C).

Solution Using ACI 318 Ch. 9 and Unified Design Method

(a) Determine the ϕ factor.

$$c = \frac{a}{\beta_1} = \frac{2.35\text{ in}}{0.85} = 2.76\text{ in}$$

$$\epsilon_t = \left(\frac{d-c}{c}\right)(0.003) = \left(\frac{27.5\text{ in} - 2.76\text{ in}}{2.76\text{ in}}\right)(0.003) = 0.026 > 0.005$$

Therefore, the section is tension controlled, and $\phi = 0.9$.

Determine the ϕ factor.

$$c = \frac{a}{0.85} = \frac{2.35\text{ in}}{0.85} = 2.76\text{ in}$$

$$\epsilon_t = \left(\frac{d-c}{c}\right)(0.003) = 0.027 > 0.005$$

$$\phi = 0.9$$

The section is tension controlled.

The design moment capacity (maximum positive moment) is

$$\phi M_n = (0.9)(394.85\text{ ft-kips}) = \boxed{355.37\text{ ft-kips}}$$

The answer is (B).

(b) In the negative moment region, $d = 27.5$ in, $A_s = 7.0\ \text{in}^2$, and $A'_s = 3.0\ \text{in}^2$.

Assuming the compression steel yields, from Eq. 50.70,

$$A_c = \frac{f_y A_s - f'_s A'_s}{0.85 f'_c} = \frac{\left(60\ \frac{\text{kips}}{\text{in}^2}\right)(7\ \text{in}^2) - \left(60\ \frac{\text{kips}}{\text{in}^2}\right)(3\ \text{in}^2)}{(0.85)\left(3\ \frac{\text{kips}}{\text{in}^2}\right)} = 94.12\ \text{in}^2$$

Using Eq. 50.29,

$$a = \frac{A_c}{b_w} = \frac{94.12\ \text{in}^2}{12\ \text{in}} = 7.84\ \text{in}$$

$$\lambda = \frac{a}{2} = \frac{7.84\ \text{in}}{2} = 3.92\ \text{in}$$

$$c = \frac{a}{\beta_1} = \frac{7.84\ \text{in}}{0.85} = 9.23\ \text{in}$$

From Eq. 50.67,

$$\varepsilon'_s = \left(\frac{0.003}{c}\right)(c - d') = \left(\frac{0.003}{9.23\ \text{in}}\right)(9.23\ \text{in} - 2.5\ \text{in}) = 0.0022$$

Use Eq. 50.68.

$$f'_s = E_s\varepsilon'_s = \left(29{,}000\ \frac{\text{kips}}{\text{in}^2}\right)(0.0022) = 63.43\ \text{kips/in}^2$$

$$f'_s = f_y = \boxed{60\ \text{kips/in}^2} \quad \text{[confirms assumption]}$$

The answer is (D).

(c) Divide the section into two elementary areas as shown. The computations are summarized in the following table.

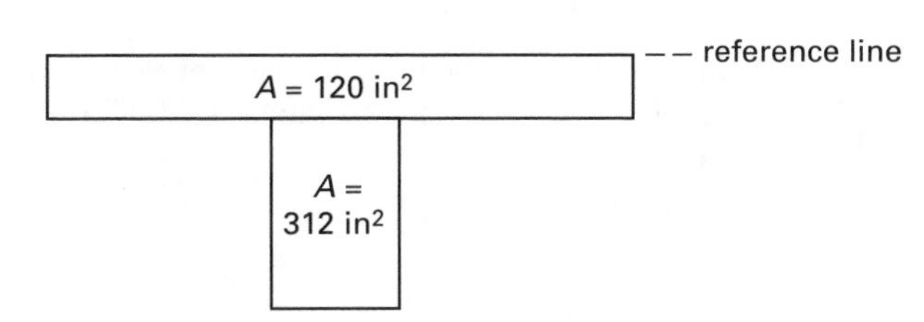

	A (in²)	y (in)	Ay (in³)	$A(\overline{y}-y)^2$ (in⁴)	I_0 (in⁴)
	120	2	240	14,083	160
	312	17	5304	5417	17,576
totals	432		5544	19,500	17,736

$$\overline{y} = \frac{5544\ \text{in}^3}{432\ \text{in}^2} = 12.83\ \text{in}$$

$$I_g = 19{,}500\ \text{in}^4 + 17{,}736\ \text{in}^4 = \boxed{37{,}236\ \text{in}^4}$$

The answer is (B).

(d) From Eq. 50.38,

$$E_c = 57{,}000\sqrt{f'_c} = \frac{57{,}000\sqrt{3000\ \frac{\text{lbf}}{\text{in}^2}}}{1000\ \frac{\text{lbf}}{\text{kip}}} = 3122\ \text{kips/in}^2$$

$$n = \frac{E_s}{E_c} = \frac{29{,}000\ \frac{\text{kips}}{\text{in}^2}}{3122\ \frac{\text{kips}}{\text{in}^2}} = 9.29$$

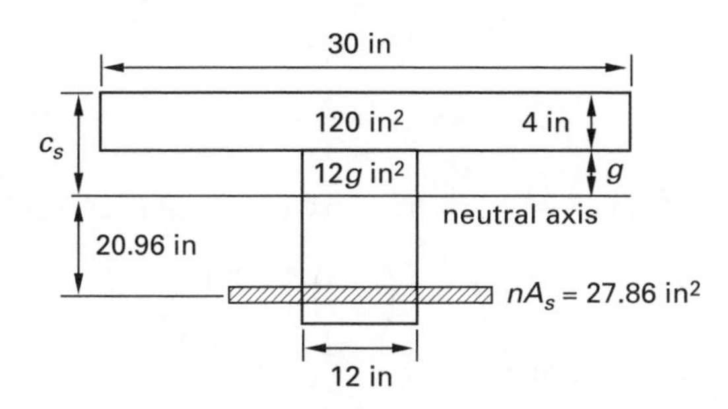

From the sketch of the transformed section,

$$(120\ \text{in}^2)(g + 2\ \text{in}) + \frac{(12\ \text{in})g^2}{2} = (27.86\ \text{in}^2)(27.5\ \text{in} - 4\ \text{in} - g)$$

$$g = 2.54\ \text{in}$$

The neutral axis is located at $c_s = 6.54$ in.

The transformed cracked moment of inertia is calculated using the parallel axis theorem.

$$I_{cr} = \left(\tfrac{1}{12}\right)(30\ \text{in})(4\ \text{in})^3 + \left(\tfrac{1}{3}\right)(12\ \text{in})(2.54\ \text{in})^3 + (120\ \text{in}^2)(4.54\ \text{in})^2 + (27.86\ \text{in}^2)(20.96\ \text{in})^2 = \boxed{14{,}938\ \text{in}^4}$$

The answer is (C).

(e) From Eq. 50.45,

$$f_r = 7.5\sqrt{f'_c} = \frac{7.5\sqrt{3000\ \frac{\text{lbf}}{\text{in}^2}}}{1000\ \frac{\text{lbf}}{\text{kip}}} = 0.41\ \text{kip/in}^2$$

Structural

From Eq. 50.44,

$$M_{cr} = \frac{f_r I_g}{y_t} = \frac{\left(0.41\ \frac{\text{kip}}{\text{in}^2}\right)(37{,}236\ \text{in}^4)}{(30\ \text{in} - 12.83\ \text{in})\left(12\ \frac{\text{in}}{\text{ft}}\right)} = \boxed{74.25\ \text{ft-kips}}$$

The answer is (D).

(f) Use Eq. 50.44.

$$M_{cr} = \frac{f_r I_g}{y_t} = \frac{\left(0.41\ \frac{\text{kip}}{\text{in}^2}\right)(37{,}236\ \text{in}^4)}{(12.83\ \text{in})\left(12\ \frac{\text{in}}{\text{ft}}\right)} = \boxed{99.33\ \text{ft-kips}}$$

The answer is (C).

(g) The maximum positive service moment is

$$M_s = \left(\frac{1}{24}\right)wL^2 = \left(\frac{1}{24}\right)\left(4\ \frac{\text{kips}}{\text{ft}}\right)(30\ \text{ft})^2 = 150\ \text{ft-kips}$$

Therefore,

$$f_s = \frac{nM_s(d - c_s)}{I_{cr}} = \frac{(9.29)(150\ \text{ft-kips})\left(12\ \frac{\text{in}}{\text{ft}}\right)(27.5\ \text{in} - 6.54\ \text{in})}{14{,}938\ \text{in}^4} = \boxed{23.46\ \text{kips/in}^2}$$

The answer is (B).

(h)

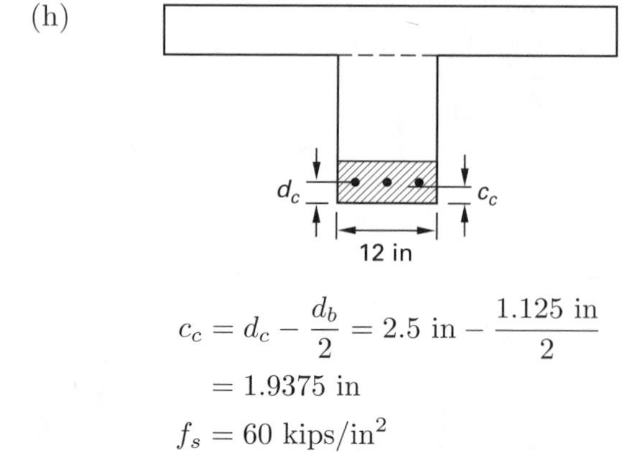

$$c_c = d_c - \frac{d_b}{2} = 2.5\ \text{in} - \frac{1.125\ \text{in}}{2} = 1.9375\ \text{in}$$

$$f_s = 60\ \text{kips/in}^2$$

From Eq. 50.36,

$$s \le \frac{540}{f_s} - 2.5c_c \le \frac{540}{23.46\ \frac{\text{kips}}{\text{in}^2}} - (2.5)(1.9375\ \text{in}) \le 18.2\ \text{in}\quad \text{[controls]}$$

$$s \le \frac{432}{f_s} = \frac{432}{23.46\ \frac{\text{kips}}{\text{in}^2}} = 18.4\ \text{in}$$

The answer is (D).

(i) The ϕ factor for shear should be 0.75. Therefore,

$$\phi V_n = V_c + V_s = (0.75)(36.15\ \text{kips} + 60.50\ \text{kips}) = 72.49\ \text{kips}$$

$$w_u = \frac{72.49\ \text{kips}}{12.71\ \text{ft}} = \boxed{5.70\ \text{kips/ft}}$$

The answer is (B).

(j) Use Eq. 50.69 with top and bottom steel yielding. (λ was calculated in part (b).)

$$M_n = (A_s - A'_s)f_y(d - \lambda) + A'_s f'_s(d - d')$$

$$= \frac{(7\ \text{in}^2 - 3\ \text{in}^2)\left(60\ \frac{\text{kips}}{\text{in}^2}\right)(27.5\ \text{in} - 3.92\ \text{in}) + (3\ \text{in}^2)\left(60\ \frac{\text{kips}}{\text{in}^2}\right)(27.5\ \text{in} - 2.5\ \text{in})}{12\ \frac{\text{in}}{\text{ft}}}$$

$$= 846.6\ \text{ft-kips}\quad \text{[global maximum]}$$

The maximum positive moment is

$$M_u = \frac{1}{24}w_u L^2 = \frac{1}{24}w_u(30\ \text{ft})^2 = 37.5w_u$$

The maximum positive moment was calculated in part (a). Equating this value to the capacity ϕM_n for the positive moment region,

$$w_u = \frac{355.37\ \text{ft-kips}}{37.5\ \text{ft}^2} = \boxed{9.48\ \text{kips/ft}}$$

The maximum negative moment from this fixed-end beam is

$$M_u = \frac{1}{12}w_u L^2 = \frac{1}{12}w_u(30\ \text{ft})^2 = 75w_u$$

Equate this value to the capacity ϕM_n for the negative moment region, using $\phi = 0.9$. (ϵ_t can be calculated as 0.0059.)

$$\phi M_n = M_u$$
$$(0.9)(846.6 \text{ ft-kips}) = 75w_u$$
$$w_u = 10.2 \text{ kips/ft}$$

The positive region controls.

The answer is (C).

4. Use either **ACI 318 Apps. B and C** or **ACI 318 Ch. 9 and Unified Design Method** to solve this problem.

Solution Using ACI 318 Apps. B and C

step 1: Calculate ρ_{max}. Use Eq. 50.33.

$$\rho_{max} = 0.75\rho_{balanced}$$
$$= (0.75)\left(\frac{(0.85)(0.85)\left(3000 \ \frac{\text{lbf}}{\text{in}^2}\right)}{50{,}000 \ \frac{\text{lbf}}{\text{in}^2}}\right) \times \left(\frac{87{,}000}{87{,}000 + 50{,}000}\right)$$
$$= 0.0206$$

$$\text{Try} \quad \rho = \frac{(0.18)\left(3000 \ \frac{\text{lbf}}{\text{in}^2}\right)}{50{,}000 \ \frac{\text{lbf}}{\text{in}^2}} = 0.0108$$

[< 0.0206, so acceptable]

step 2: Solve for the required nominal moment strength.

$$M_u = (1.4)(50{,}000 \text{ ft-lbf}) + (1.7)(200{,}000 \text{ ft-lbf})$$
$$= 410{,}000 \text{ ft-lbf}$$
$$M_n = \frac{M_u}{\phi} = \frac{410{,}000 \text{ ft-lbf}}{0.9} = 455{,}560 \text{ ft-lbf}$$

step 3: Substitute into Eq. 50.32 and solve for bd^2.

$$bd^2 = \frac{M_n}{\rho f_y\left(1 - \frac{\rho f_y}{1.7f'_c}\right)}$$
$$= \frac{(455{,}560 \text{ ft-lbf})\left(12 \ \frac{\text{in}}{\text{ft}}\right)}{(0.0108)\left(50{,}000 \ \frac{\text{lbf}}{\text{in}^2}\right) \times \left(1 - \frac{(0.0108)\left(50{,}000 \ \frac{\text{lbf}}{\text{in}^2}\right)}{(1.7)\left(3000 \ \frac{\text{lbf}}{\text{in}^2}\right)}\right)}$$
$$= 11{,}323 \text{ in}^3$$

step 4: Choose $d = 1.75b$.

$$(1.75)^2b^3 = 11{,}323 \text{ in}^3$$
$$b = \boxed{15.46 \text{ in} \quad (15.5 \text{ in})}$$
$$d = (1.75)(15.46 \text{ in})$$
$$= \boxed{27.06 \text{ in} \quad (27 \text{ in})}$$

step 5:

$$\text{actual } bd^2 = (15.5 \text{ in})(27 \text{ in})^2$$
$$= 11{,}300 \text{ in}^3 \quad \text{[close]}$$

No change in ρ is needed.

step 6:

$$A_{st} = \rho bd = (0.0108)(15.5 \text{ in})(27 \text{ in})$$
$$= 4.52 \text{ in}^2$$

step 7: Use six no. 8 bars (4.74 in^2). These can all fit in a single layer.

step 8: Assume no. 4 stirrups.

$$d_c = 1.5 \text{ in} + \tfrac{1}{2} \text{ in} + 0.5 \text{ in} = \boxed{2.5 \text{ in}}$$

Solution Using ACI 318 Ch. 9 and Unified Design Method

step 1: Try

$$\rho = \frac{(0.18)\left(3000 \ \frac{\text{lbf}}{\text{in}^2}\right)}{50{,}000 \ \frac{\text{lbf}}{\text{in}^2}} = 0.0108$$

step 2: Solve for the required nominal moment strength.

$$M_u = (1.2)(50{,}000 \text{ ft-lbf}) + (1.6)(200{,}000 \text{ ft-lbf})$$
$$= 380{,}000 \text{ ft-lbf}$$

Assume the section is tension controlled.

$$M_n = \frac{M_u}{\phi} = \frac{380{,}000 \text{ ft-lbf}}{0.9} = 422{,}222 \text{ ft-lbf}$$

step 3: Substitute into Eq. 50.32 and solve for bd^2.

$$bd^2 = \frac{M_n}{\rho f_y\left(1 - \frac{\rho f_y}{1.7f'_c}\right)}$$
$$= \frac{(422{,}222 \text{ ft-lbf})\left(12 \ \frac{\text{in}}{\text{ft}}\right)}{(0.0108)\left(50{,}000 \ \frac{\text{lbf}}{\text{in}^2}\right) \times \left(1 - \frac{(0.0108)\left(50{,}000 \ \frac{\text{lbf}}{\text{in}^2}\right)}{(1.7)\left(3000 \ \frac{\text{lbf}}{\text{in}^2}\right)}\right)}$$
$$= 10{,}494 \text{ in}^3$$

Structural

step 4: Choose $d = 1.75b$.

$$(1.75)^2b^3 = 10{,}494 \text{ in}^3$$
$$b = \boxed{15.08 \text{ in} \quad (15.5 \text{ in})}$$
$$d = (1.75)(15.08 \text{ in}) = \boxed{26.39 \text{ in} \quad (27 \text{ in})}$$

step 5:

$$\text{actual } bd^2 = (15.5 \text{ in})(27 \text{ in})^2 = 11{,}300 \text{ in}^3 \quad \text{[close]}$$

No change in ρ is needed.

For this low reinforcement ratio, the section is going to be tension controlled. No formal check is necessary.

step 6:

$$A_{st} = \rho bd = (0.0108)(15.5 \text{ in})(27 \text{ in}) = 4.52 \text{ in}^2$$

step 7: Use six no. 8 bars (4.74 in^2). These can all fit in a single layer.

step 8: Assume no. 4 stirrups.

$$d_c = 1.5 \text{ in} + \tfrac{1}{2} \text{ in} + 0.5 \text{ in} = \boxed{2.5 \text{ in}}$$

5. Use either **ACI 318 Apps. B and C** or **ACI 318 Ch. 9 and Unified Design Method** to solve this problem. There is no difference between the two solutions.

(a) *step 1:*

$$\rho \approx \frac{(0.18)\left(4000 \ \frac{\text{lbf}}{\text{in}^2}\right)}{40{,}000 \ \frac{\text{lbf}}{\text{in}^2}} = 0.018$$

step 2: Solve for the required nominal moment strength. Assume the section is tension controlled.

$$M_n = \frac{M_u}{\phi} = \frac{400{,}000 \text{ ft-lbf}}{0.9} = 444{,}444 \text{ ft-lbf}$$

step 3: Substitute into Eq. 50.32 and solve for bd^2.

$$bd^2 = \frac{M_n}{\rho f_y\left(1 - \frac{\rho f_y}{1.7f'_c}\right)}$$
$$= \frac{(444{,}444 \text{ ft-lbf})\left(12 \ \frac{\text{in}}{\text{ft}}\right)}{(0.018)\left(40{,}000 \ \frac{\text{lbf}}{\text{in}^2}\right) \times \left(1 - \frac{(0.018)\left(40{,}000 \ \frac{\text{lbf}}{\text{in}^2}\right)}{(1.7)\left(4000 \ \frac{\text{lbf}}{\text{in}^2}\right)}\right)}$$
$$= 8284 \text{ in}^3$$

step 4: Choose $d = 1.75b$.

$$8284 \text{ in}^3 = b(1.75b)^2$$
$$b = \boxed{13.93 \text{ in} \quad (14 \text{ in})}$$
$$d = (1.75)(14) = \boxed{24.5 \text{ in}}$$

(b) and (c) *step 5:* ρ is not substantially different.

step 6:

$$A_{st} = \rho bd = (0.018)(14 \text{ in})(24.5 \text{ in}) = 6.17 \text{ in}^2$$

Check to see if the section is tension controlled.

$$c = \frac{A_s f_y}{0.85 f'_c b \beta_1} = 9.16 \text{ in}$$
$$\epsilon_t = \left(\frac{d - c}{c}\right)(0.003) = \left(\frac{24.5 \text{ in} - 9.16 \text{ in}}{9.16 \text{ in}}\right)(0.003) = 0.00502 > 0.005$$

The section is tension controlled.

step 7: Use four no. 11 bars (6.24 in^2) in one layer. Use no. 3 stirrups.

(d) Including room for reinforcing and stirrups, the overall beam depth will be

$$24.5 \text{ in} + \frac{1.410 \text{ in}}{2} + 0.375 \text{ in} + 1.5 \text{ in} = \boxed{27.08 \text{ in} \quad (27.5 \text{ in})}$$

6. Use either **ACI 318 Apps. B and C** or **ACI 318 Ch. 9 and Unified Design Method** to solve this problem. There is no difference between the two solutions.

The effective beam width is

$$b_e = \text{minimum}\left\{\begin{array}{l} \frac{1}{4}l_{\text{beam}} \\ 8 \text{ in} + (16)(3 \text{ in}) = 56 \text{ in} \\ 8 \text{ in} + 22 \text{ in} = 30 \text{ in} \end{array}\right\}$$
$$= 30 \text{ in}$$

Work with the following beam.

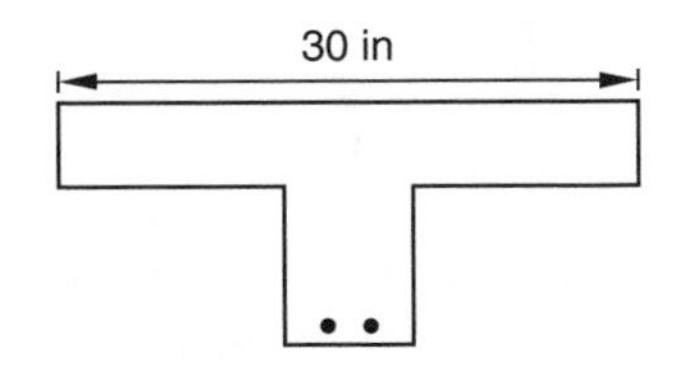

step 1: $E_{\text{steel}} = 2.9 \times 10^7 \text{ lbf/in}^2$

From Eq. 48.1,

$$E_{\text{concrete}} = \left(145 \frac{\text{lbf}}{\text{in}^3}\right)^{1.5}(33)\sqrt{3000 \frac{\text{lbf}}{\text{in}^2}}$$
$$= 3.16 \times 10^6 \text{ psi}$$

step 2: $$n = \frac{2.9 \times 10^7 \frac{\text{lbf}}{\text{in}^2}}{3.16 \times 10^6 \frac{\text{lbf}}{\text{in}^2}} = 9.18$$

Note: The ACI code allows the use of the rounded value of n, but does not require it.

step 3: The expanded steel area is

$$nA_b = (9.18)(2)(1 \text{ in}^2) = 18.36 \text{ in}^2$$

step 4: Assume the neutral axis is below the flange.

step 5: (skip)

step 6: The area of concrete is

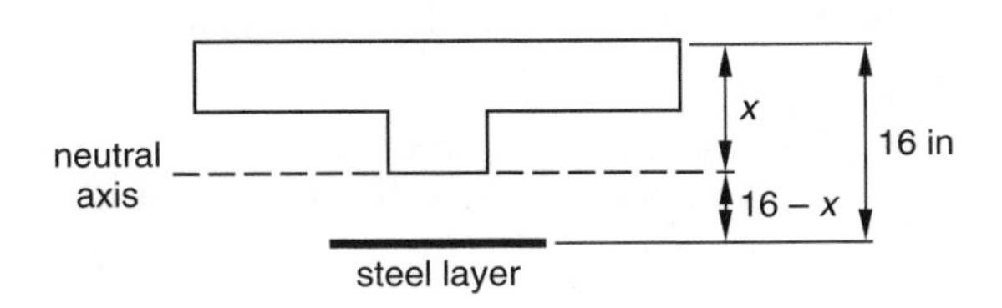

- Stem to flange bottom:

$$A = (8 \text{ in})(x - 3 \text{ in}) = 8x - 24 \text{ in}^2$$

- All of flange:

$$A = (30 \text{ in})(3 \text{ in}) = 90 \text{ in}^2$$

Balance the moments of the transformed areas.

$$(8x - 24)\left(\frac{x-3}{2}\right) + (90)(x - 1.5) = (16 - x)(18.36)$$
$$= 4x^2 + 84.36x - 392.76$$
$$x = 3.92 \text{ in}$$
$$16 - x = 16 - 3.92 = 12.08 \text{ in}$$

step 7: The moment of inertia of the beam is

$$I_{\text{steel}} = (18.36 \text{ in}^2)(12.08)^2 = 2679.2 \text{ in}^4$$
$$I_{\text{concrete}} = \frac{(8 \text{ in})(0.92 \text{ in})^3}{3} + \frac{(30 \text{ in})(3 \text{ in})^3}{12} + (3 \text{ in})(30 \text{ in})(2.42 \text{ in})^2$$
$$= 594.8 \text{ in}^4$$
$$I_{\text{total}} = 2679.2 \text{ in}^4 + 594.8 \text{ in}^4$$
$$= 3274.0 \text{ in}^4$$

step 8: $$M = \left(20{,}000 \frac{\text{ft-lbf}}{\text{ft}}\right)\left(\frac{30 \text{ in}}{12 \frac{\text{in}}{\text{ft}}}\right)$$
$$= 50{,}000 \text{ ft-lbf}$$

step 9: (a) $$\sigma_{\text{concrete}} = \frac{(50{,}000 \text{ ft-lbf})\left(12 \frac{\text{in}}{\text{ft}}\right) \times (3.92 \text{ in})}{3274.0 \text{ in}^4}$$
$$= \boxed{718.4 \text{ lbf/in}^2}$$

(b) $$\sigma_{\text{steel}} = \frac{(9.18)(50{,}000 \text{ ft-lbf}) \times \left(12 \frac{\text{in}}{\text{ft}}\right)(12.08 \text{ in})}{3274.0 \text{ in}^4}$$
$$= \boxed{20{,}323 \text{ lbf/in}^2}$$

Structural

7. Use either **ACI 318 Apps. B and C** or **ACI 318 Ch. 9 and Unified Design Method** (see p. 50-23) to solve this problem.

Solution Using ACI 318 Apps. B and C

(a) All loads are dead loads.

The total distributed load is (30 ft)(2 kips/ft) = 60 kips.

Taking clockwise moments as positive,

$$\begin{aligned}\sum M_L &= -(30\text{ ft})R + (9\text{ ft})(20\text{ kips}) \\ &\quad + (24\text{ ft})(60\text{ kips}) + (60\text{ kips})\left(\frac{30\text{ ft}}{2}\right) \\ &= 0 \\ R &= 84\text{ kips}\end{aligned}$$

$$\begin{aligned}\sum M_R &= (30\text{ ft})L - (60\text{ kips})(6\text{ ft}) \\ &\quad - (20\text{ kips})(21\text{ ft}) - (60\text{ kips})\left(\frac{30\text{ ft}}{2}\right) \\ &= 0 \\ L &= 56\text{ kips}\end{aligned}$$

Check: 84 kips + 56 kips = 20 kips + 60 kips + 60 kips = 140 kips [correct]

Taking clockwise moments as positive, the moment between the two center loads is

$$\begin{aligned}M_L &= 56x - x^2 - (20)(x-9) \\ \frac{dM}{dx} &= 56 - 2x - 20 = 36 - 2x = 0 \\ x &= 18\text{ ft} \quad [\text{at maximum moment}]\end{aligned}$$

$$\begin{aligned}M_{18} &= (56\text{ kips})(18\text{ ft}) - \frac{\left(2\ \frac{\text{kips}}{\text{ft}}\right)(18\text{ ft})^2}{2} \\ &\quad - (20\text{ kips})(9\text{ ft}) \\ &= 504\text{ ft-kips} = M_w\end{aligned}$$

When ACI 318 Apps. B and C are used,

$$M_u = 1.4M_w = (1.4)(504\text{ ft-kips}) = 705.6\text{ ft-kips}$$

step 1: Find ρ_{balanced}. $\beta_1 = 0.85$ since $f'_c = 4000$ psi. From Eq. 50.33,

$$\begin{aligned}\rho_{\text{balanced}} &= \left(\frac{(0.85)(0.85)\left(4000\ \frac{\text{lbf}}{\text{in}^2}\right)}{60{,}000\ \frac{\text{lbf}}{\text{in}^2}}\right) \\ &\quad \times \left(\frac{87{,}000}{87{,}000 + 60{,}000}\right) \\ &= 0.0285\end{aligned}$$

Choose $\rho = 0.375\rho_{\text{balanced}}$.

$$\rho = (0.375)(0.0285) = 0.0107$$

step 2: Solve for the required nominal moment strength. Assume the section is tension controlled.

$$\begin{aligned}M_n &= \frac{M_u}{\phi} = \frac{705{,}600\text{ ft-lbf}}{0.9} \\ &= 784{,}000\text{ ft-lbf}\end{aligned}$$

step 3: Substitute into Eq. 50.32 and solve for bd^2.

$$\begin{aligned}bd^2 &= \frac{M_n}{\rho f_y\left(1 - \frac{\rho f_y}{1.7f'_c}\right)} \\ &= \frac{(784{,}000\text{ ft-lbf})\left(12\ \frac{\text{in}}{\text{ft}}\right)}{(0.0107)\left(60{,}000\ \frac{\text{lbf}}{\text{in}^2}\right) \times \left(1 - \frac{(0.0107)\left(60{,}000\ \frac{\text{lbf}}{\text{in}^2}\right)}{(1.7)\left(4000\ \frac{\text{lbf}}{\text{in}^2}\right)}\right)} \\ &= 16{,}182\text{ in}^3\end{aligned}$$

step 4: Assume $d/b = 2.0$.

$$\begin{aligned}bd^2 &= b4b^2 = 16{,}182\text{ in}^3 \\ b &= \boxed{15.93\text{ in} \quad (16\text{ in})} \\ d &= \boxed{(2)(16\text{ in}) = 32\text{ in}}\end{aligned}$$

(b) *step 5* :
$$\begin{aligned}\rho_{\text{revised}} &= \rho_{\text{old}}\left(\frac{15.93\text{ in}}{16.00\text{ in}}\right) \\ &= (0.0107\text{ in}^2)\left(\frac{15.93\text{ in}}{16.00\text{ in}}\right) \\ &= 0.0107\end{aligned}$$

Check the limits.

$$0.0107 < (0.75)(0.0285) = 0.0214 \quad [\text{acceptable}]$$

At this low reinforcement ratio, the section is tension controlled. There is no need for a formal check.

step 6: $A_{st} = \rho bd = (0.0107)(16 \text{ in})(32 \text{ in})$
$= 5.48 \text{ in}^2$

(c) *step 7:* Use four no. 11 bars (6.24 in^2) in one layer.

Refer to Table 50.2.

beam width: $16 \text{ in} > 13.7 \text{ in}$ [acceptable]

$$\text{actual } \rho = \frac{6.24 \text{ in}^2}{(16 \text{ in})(32 \text{ in})}$$
$$= 0.0122 < \rho_{\max} \quad \text{[acceptable]}$$

Solution Using ACI 318 Ch. 9 and Unified Design Method

(a) All loads are dead loads.

The total distributed load is (30 ft)(2 kips/ft) = 60 kips.

Taking clockwise moments as positive,

$$\sum M_L = -(30 \text{ ft})R + (9 \text{ ft})(20 \text{ kips}) + (24 \text{ ft})(60 \text{ kips}) + (60 \text{ kips})\left(\frac{30 \text{ ft}}{2}\right) = 0$$

$$R = 84 \text{ kips}$$

$$\sum M_R = (30 \text{ ft})L - (60 \text{ kips})(6 \text{ ft}) - (20 \text{ kips})(21 \text{ ft}) - (60 \text{ kips})\left(\frac{30 \text{ ft}}{2}\right) = 0$$

$$L = 56 \text{ kips}$$

Check: 84 kips + 56 kips = 20 kips + 60 kips + 60 kips = 140 kips [correct]

Taking clockwise moments as positive, the moment between the two center loads is

$$M_L = 56x - x^2 - (20)(x - 9)$$
$$\frac{dM}{dx} = 56 - 2x - 20 = 36 - 2x = 0$$
$$x = 18 \text{ ft} \quad \text{[at maximum moment]}$$

$$M_{18} = (56 \text{ kips})(18 \text{ ft}) - \frac{\left(2\ \frac{\text{kips}}{\text{ft}}\right)(18 \text{ ft})^2}{2} - (20 \text{ kips})(9 \text{ ft})$$
$$= 504 \text{ ft-kips} = M_w$$

Because only dead load is considered when ACI 318 Unified Design Method and Ch. 9 are used, the controlling load combination is $U = 1.4D$.

$$M_u = 1.4M_w = (1.4)(504 \text{ ft-kips}) = 705.6 \text{ ft-kips}$$

step 1: The maximum permitted reinforcement ratio corresponds to a net tensile strain of $\epsilon_t = 0.004$.

$$\rho_{\max} = \frac{0.85 f'_c \beta_1}{f_f}\left(\frac{0.003}{0.003 + 0.004}\right) = 0.0206$$

Use $\rho = 0.5\rho_{\max} = 0.0103$.

step 2: Solve for the required nominal moment strength, assuming the section is tension controlled.

$$M_n = \frac{M_u}{\phi} = \frac{705{,}600 \text{ ft-lbf}}{0.9} = 784{,}000 \text{ ft-lbf}$$

step 3: Substitute into Eq. 50.32 and solve for bd^2.

$$bd^2 = \frac{M_n}{\rho f_y\left(1 - \frac{\rho f_y}{1.7 f'_c}\right)}$$
$$= \frac{(784{,}000 \text{ ft-lbf})\left(12\ \frac{\text{in}}{\text{ft}}\right)}{(0.0103)\left(60{,}000\ \frac{\text{lbf}}{\text{in}^2}\right) \times \left(1 - \frac{(0.0103)\left(60{,}000\ \frac{\text{lbf}}{\text{in}^2}\right)}{(1.7)\left(4000\ \frac{\text{lbf}}{\text{in}^2}\right)}\right)}$$
$$= 16{,}745 \text{ in}^3$$

Structural

step 4: Assume $d/b = 2.0$.

$$bd^2 = b4b^2 = 16{,}745 \text{ in}^3$$

$$b = \boxed{16.0 \text{ in}}$$

$$d = \boxed{(2)(16 \text{ in}) = 32 \text{ in}}$$

(b) *step 5:* $A_{st} = \rho bd = (0.0107)(16 \text{ in})(32 \text{ in}) = 5.48 \text{ in}^2$

(c) *step 6:* $\boxed{\text{Use four no. 11 bars } (6.24 \text{ in}^2) \text{ in one layer.}}$

Refer to Table 50.2.

beam width: $16 \text{ in} > 13.7 \text{ in}$ [acceptable]

$$\text{actual } \rho = \frac{6.24 \text{ in}^2}{(16 \text{ in})(32 \text{ in})} = 0.0122 < \rho_{max} \quad \text{[acceptable]}$$

With this low reinforcement ratio, the section is tension controlled.

8. Use either **ACI 318 Apps. B and C** or **ACI 318 Ch. 9** (see p. 50-25) to solve this problem.

Solution Using ACI 318 Apps. B and C

(a) *step 1:* (See Prob. 7.)

$$\rho_{balanced} = 0.0285$$

$$\rho_{max} = (0.75)(0.0285) = 0.0214$$

Use $\rho = 0.0214$ as directed.

The net tensile strain corresponding to this reinforcement ratio is $\epsilon_t = 0.00376$, and $\phi = 0.80$.

step 2: Solve for the required nominal moment strength.

$$w_u = (1.4)\left(1 \ \frac{\text{kip}}{\text{ft}}\right) + (1.7)\left(2 \ \frac{\text{kips}}{\text{ft}}\right) = 4.8 \text{ kips/ft}$$

$$M_u = \frac{w_u L^2}{8} = \frac{\left(480{,}000 \ \frac{\text{lbf}}{\text{ft}}\right)(27 \text{ ft})^2}{8} = 437{,}400 \text{ ft-lbf}$$

$$M_n = \frac{M_u}{\phi} = \frac{437{,}400 \text{ ft-lbf}}{0.8} = 546{,}750 \text{ ft-lbf}$$

step 3: Substitute into Eq. 50.32 and solve for bd^2.

$$bd^2 = \frac{M_n}{\rho f_y\left(1 - \frac{\rho f_y}{1.7 f'_c}\right)}$$

$$= \frac{(546{,}750 \text{ ft-lbf})\left(12 \ \frac{\text{in}}{\text{ft}}\right)}{(0.0214)\left(60{,}000 \ \frac{\text{lbf}}{\text{in}^2}\right) \times \left(1 - \frac{(0.0214)\left(60{,}000 \ \frac{\text{lbf}}{\text{in}^2}\right)}{(1.7)\left(4000 \ \frac{\text{lbf}}{\text{in}^2}\right)}\right)}$$

$$= 6299.0 \text{ in}^3$$

step 4: $b = \boxed{14 \text{ in} \quad \text{[given]}}$

$$d = \sqrt{\frac{6299.0 \text{ in}^3}{14 \text{ in}}} = 21.2 \text{ in} \quad \boxed{(21 \text{ in})}$$

step 5:

$$A_{st,max} = (0.0214)(14 \text{ in})(21 \text{ in}) = 6.29 \text{ in}^2 \quad \text{[maximum]}$$

This is a maximum value, and the steel chosen should be slightly less than this.

step 6: $\boxed{\text{Try four no. 11 bars.}}$

$$A_{st} = (4)\left(1.56 \text{ in}^2\right) = 6.24 \text{ in}^2 < 6.29 \text{ in}^2$$

$$\rho = \frac{A_{st}}{bd} = \frac{6.24 \text{ in}^2}{(14 \text{ in})(21 \text{ in})} = 0.02122 < 0.0214 \quad \text{[acceptable]}$$

The actual ρ value is very close to the assumed value. Therefore, no revision is necessary.

step 7:

$$h = \text{beam depth} = 21 \text{ in} + \frac{1.410 \text{ in}}{2} + 0.375 \text{ in} + 1.5 \text{ in}$$

$$= \boxed{23.58 \text{ in} \quad (24 \text{ in})}$$

(b) *step 8:*

$$f_s = (0.60)\left(60{,}000\ \frac{\text{lbf}}{\text{in}^2}\right) = 36{,}000\ \text{lbf/in}^2$$

With a no. 3 stirrup and 3/4 in side cover (interior service), the spacing of the four no. 11 bars is

$$s = \frac{14\ \text{in} - (2)(0.75\ \text{in}) - (2)(0.375\ \text{in}) - 1.41\ \text{in}}{3\ \text{spaces between 4 bars}} = 3.45\ \text{in}$$

From Eq. 50.36 with $f_s = 36$ ksi,

$$\begin{aligned} s_{\text{max}} &= \frac{540}{f_s} - 2.5c_c \le \frac{432}{f_s} \\ &= \frac{540}{36\ \text{ksi}} - (2.5)(1.5\ \text{in} + 0.375\ \text{in}) \le \frac{432}{36\ \text{ksi}} \\ &= 10.3\ \text{in} \le 12\ \text{in} \end{aligned}$$

Since 3.45 in < 10.3 in, cracking will be acceptable for interior service.

(c) *step 9:*

$$I_g = \frac{(14\ \text{in})(24\ \text{in})^3}{12} = 16{,}128\ \text{in}^4$$

$$\begin{aligned} M_{\text{max}} &= \frac{wL^2}{8} \\ &= \frac{\left(2\ \frac{\text{kips}}{\text{ft}} + 1\ \frac{\text{kip}}{\text{ft}}\right)(27\ \text{ft})^2}{8} \\ &= 273.4\ \text{ft-kips}\quad [\text{unfactored}] \\ &= (273.4\ \text{ft-kips})\left(12\ \frac{\text{lbf}}{\text{kip}}\right)\left(1000\ \frac{\text{lbf}}{\text{kip}}\right) \\ &= 3.28 \times 10^6\ \text{in-lbf} \end{aligned}$$

$$f_r = 7.5\sqrt{4000\ \frac{\text{lbf}}{\text{in}^2}} = 474.3\ \text{lbf/in}^2$$

$$y_t = \frac{24\ \text{in}}{2} = 12\ \text{in}$$

$$M_{\text{cr}} = \frac{\left(474.3\ \frac{\text{lbf}}{\text{in}^2}\right)(16{,}128\ \text{in}^4)}{12\ \frac{\text{in}}{\text{ft}}} = 6.37 \times 10^5\ \text{in-lbf}$$

$$\left(\frac{M_{\text{cr}}}{M_{\text{max}}}\right)^3 = \left(\frac{0.637 \times 10^6\ \text{in-lbf}}{3.28 \times 10^6\ \text{in-lbf}}\right)^3 = 0.0073$$

To calculate I_{cr}, find the neutral axis.

$$n = \frac{2.9 \times 10^7\ \frac{\text{lbf}}{\text{in}^2}}{3.6 \times 10^6\ \frac{\text{lbf}}{\text{in}^2}} = 8.06$$

$$\begin{aligned} c_s &= \left(\frac{(8.06)(6.24\ \text{in}^2)}{14\ \text{in}}\right) \times \left(\sqrt{1 + \frac{(2)(14\ \text{in})(21\ \text{in})}{(8.06)(6.24\ \text{in}^2)}} - 1\right) \\ &= 9.21\ \text{in} \end{aligned}$$

$$\begin{aligned} I_{\text{cr}} &= \frac{(14\ \text{in})(9.21\ \text{in})^3}{3} + (8.06)(6.24\ \text{in}^2)(21\ \text{in} - 9.21\ \text{in})^2 \\ &= 10{,}637\ \text{in}^3 \end{aligned}$$

$$\begin{aligned} I_e &= (0.0073)(16{,}128\ \text{in}^4) + (1 - 0.0073)(10{,}637\ \text{in}^4) \\ &= 10{,}677\ \text{in}^4 \end{aligned}$$

Find the deflection for a uniform load.

$$\begin{aligned} \delta_{\text{inst}} &= \frac{5wL^4}{384EI} \\ &= \frac{(5)\left(3\ \frac{\text{kips}}{\text{ft}}\right)\left(1000\ \frac{\text{lbf}}{\text{kip}}\right)\left((27\ \text{ft})\left(12\ \frac{\text{in}}{\text{ft}}\right)\right)^4}{(384)\left(3.6 \times 10^6\ \frac{\text{lbf}}{\text{in}^2}\right)(10{,}677\ \text{in}^4)\left(12\ \frac{\text{in}}{\text{ft}}\right)} \\ &= \boxed{0.933\ \text{in}\quad [\text{with 100\% live load}]} \end{aligned}$$

For a long-term loading, the percentage of the load to be sustained is

$$\frac{1 + (0.30)(2)}{1 + 2} = 0.53\quad (53\%)$$

The factored long-term deflection is

$$\delta = (0.53)(0.933\ \text{in}) = 0.494\ \text{in}$$

$$\rho' = 0\quad [\text{no compression steel}]$$

$$\lambda = 2$$

$$\begin{aligned} \delta_{\text{long term}} &= (2)(0.494\ \text{in}) = 0.988\ \text{in} \\ \delta_{\text{total}} &= 0.933\ \text{in} + 0.988\ \text{in} \\ &= \boxed{1.92\ \text{in}\quad \left[\begin{array}{c}\text{with 30\% of} \\ \text{live load sustained}\end{array}\right]} \end{aligned}$$

Solution Using ACI 318 Ch. 9

(a) *step 1:* (See Prob. 7.)

The reinforcement ratio corresponding to $\epsilon_t = 0.004$ is

$$\rho = 0.0206$$

Use $\rho = 0.0206$ as directed. The corresponding ϕ factor is

$$\phi = 0.48 + 83\epsilon_t = 0.812$$

Structural

step 2: Solve for the required nominal strength. The controlling load combination is

$$w_u = (1.2)\left(1\ \frac{\text{kip}}{\text{ft}}\right) + (1.6)\left(2\ \frac{\text{kips}}{\text{ft}}\right) = 4.4\ \text{kips/ft}$$

$$M_u = \frac{w_u L^2}{8} = \frac{\left(4400\ \frac{\text{lbf}}{\text{ft}}\right)(27\ \text{ft})^2}{8} = 400{,}950\ \text{ft-lbf}$$

$$M_n = \frac{M_u}{\phi} = \frac{400{,}950\ \text{ft-lbf}}{0.812} = 493{,}781\ \text{ft-lbf}$$

step 3: Substitute into Eq. 50.32 and solve for bd^2.

$$bd^2 = \frac{M_n}{\rho f_y\left(1 - \frac{\rho f_y}{1.7f'_c}\right)} = \frac{(493{,}781\ \text{ft-lbf})\left(12\ \frac{\text{in}}{\text{ft}}\right)}{(0.0206)\left(60{,}000\ \frac{\text{lbf}}{\text{in}^2}\right) \times \left(1 - \frac{(0.0206)\left(60{,}000\ \frac{\text{lbf}}{\text{in}^2}\right)}{(1.7)\left(4000\ \frac{\text{lbf}}{\text{in}^2}\right)}\right)} = 5859\ \text{in}^3$$

step 4:

$$\boxed{b = 14\ \text{in}\quad [\text{given}]}$$

$$d = \sqrt{\frac{5859\ \text{in}^3}{14\ \text{in}}} = 20.5\ \text{in}$$

step 5:

$$A_{\text{st,max}} = (0.0206)(14\ \text{in})(20.5\ \text{in}) = 5.91\ \text{in}^2\quad [\text{maximum}]$$

This is a maximum value, and the steel chosen should be slightly less than this.

step 6: Try four no. 11 bars.

$$A_{\text{st}} = (4)\left(1.56\ \text{in}^2\right) = 6.24\ \text{in}^2$$

Since 6.24 in² > $A_{\text{st,max}}$, increase beam depth to 22 in.

$$\rho = \frac{A_{\text{st}}}{bd} = \frac{6.24\ \text{in}^2}{(14\ \text{in})(22\ \text{in})} = 0.02026 < 0.0206\quad [\text{acceptable}]$$

step 7:

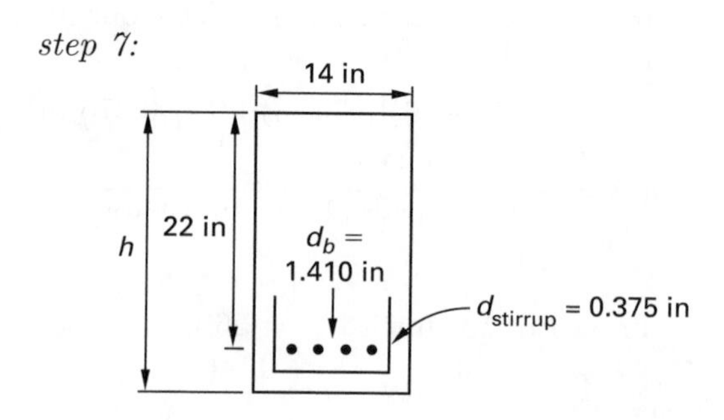

$$h = \begin{array}{c}\text{beam}\\ \text{depth}\end{array} = 22\ \text{in} + \frac{1.410\ \text{in}}{2} + 0.375\ \text{in} + 1.5\ \text{in} = \boxed{24.58\ \text{in}\quad (25\ \text{in})}$$

(b) *step 8:*

$$f_s = (0.60)\left(60{,}000\ \frac{\text{lbf}}{\text{in}^2}\right) = 36{,}000\ \text{lbf/in}^2$$

With a no. 3 stirrup and 3/4 in side cover (interior service), the spacing of the four no. 11 bars is

$$s = \frac{14\ \text{in} - (2)(0.75\ \text{in}) - (2)(0.375\ \text{in}) - 1.41\ \text{in}}{3\ \text{spaces between 4 bars}} = 3.45\ \text{in}$$

From Eq. 50.36 with $f_s = 36$ ksi,

$$s_{\text{max}} = \frac{540}{f_s} - 2.5c_c \le \frac{432}{f_s} = \frac{540}{36\ \text{ksi}} - (2.5)(1.5\ \text{in} + 0.375\ \text{in}) \le \frac{432}{36\ \text{ksi}} = 10.3\ \text{in} \le 12\ \text{in}$$

Since 3.45 in < 10.3 in, cracking will be acceptable for interior service.

(c) *step 9*: $I_g = \frac{(14 \text{ in})(25 \text{ in})^3}{12} = 18{,}229 \text{ in}^4$

$$M_{\max} = \frac{wL^2}{8}$$

$$= \frac{\left(2 \frac{\text{kips}}{\text{ft}} + 1 \frac{\text{kip}}{\text{ft}}\right)(27 \text{ ft})^2}{8}$$

$$= 273.4 \text{ ft-kips} \quad \text{[unfactored]}$$

$$= (273.4 \text{ ft-kips})\left(12 \frac{\text{lbf}}{\text{kip}}\right)\left(1000 \frac{\text{lbf}}{\text{kip}}\right)$$

$$= 3.28 \times 10^6 \text{ in-lbf}$$

$$f_r = 7.5\sqrt{4000 \frac{\text{lbf}}{\text{in}^2}} = 474.3 \text{ lbf/in}^2$$

$$y_t = \frac{25 \text{ in}}{2} = 12.5 \text{ in}$$

$$M_{\text{cr}} = \frac{\left(474.3 \frac{\text{lbf}}{\text{in}^2}\right)(18{,}229 \text{ in}^4)}{12 \frac{\text{in}}{\text{ft}}}$$

$$= 7.21 \times 10^5 \text{ in-lbf}$$

$$\left(\frac{M_{\text{cr}}}{M_{\max}}\right)^3 = \left(\frac{0.721 \times 10^6 \text{ in-lbf}}{3.28 \times 10^6 \text{ in-lbf}}\right)^3$$

$$= 0.01062$$

To calculate I_{cr}, find the neutral axis.

$$n = \frac{2.9 \times 10^7 \frac{\text{lbf}}{\text{in}^2}}{3.6 \times 10^6 \frac{\text{lbf}}{\text{in}^2}} = 8.06$$

$$c_s = \left(\frac{(8.06)(6.24 \text{ in}^2)}{14 \text{ in}}\right) \times \left(\sqrt{1 + \frac{(2)(14 \text{ in})(22 \text{ in})}{(8.06)(6.24 \text{ in}^2)}} - 1\right)$$

$$= 9.48 \text{ in}$$

$$I_{\text{cr}} = \frac{(14 \text{ in})(9.48 \text{ in})^3}{3} + (8.06)(6.24 \text{ in}^2)(22 \text{ in} - 9.48 \text{ in})^2$$

$$= 11{,}860 \text{ in}^3$$

$$I_e = (0.01062)(18{,}229 \text{ in}^4) + (1 - 0.01062)(11{,}860 \text{ in}^4)$$

$$= 11{,}928 \text{ in}^4$$

Find the deflection for a uniform load.

$$\delta_{\text{inst}} = \frac{5wL^4}{384EI}$$

$$= \frac{(5)\left(3 \frac{\text{kips}}{\text{ft}}\right)\left(1000 \frac{\text{lbf}}{\text{kip}}\right)\left((27 \text{ ft})\left(12 \frac{\text{in}}{\text{ft}}\right)\right)^4}{(384)\left(3.6 \times 10^6 \frac{\text{lbf}}{\text{in}^2}\right)(11{,}928 \text{ in}^4)\left(12 \frac{\text{in}}{\text{ft}}\right)}$$

$$= \boxed{0.835 \text{ in} \quad \text{[with 100\% live load]}}$$

For a long-term loading, the percentage of the load to be sustained is

$$\frac{1 + (0.30)(2)}{1 + 2} = 0.53 \; (53\%)$$

The factored long-term deflection is

$$\delta = (0.53)(0.835 \text{ in}) = 0.443 \text{ in}$$

$$\rho' = 0 \quad \text{[no compression steel]}$$

$$\lambda = 2$$

$$\delta_{\text{long term}} = (2)(0.443 \text{ in}) = 0.886 \text{ in}$$

$$\delta_{\text{total}} = 0.835 \text{ in} + 0.886 \text{ in}$$

$$= \boxed{1.72 \text{ in} \quad \left[\begin{array}{c}\text{with 30\% of} \\ \text{live load sustained}\end{array}\right]}$$

9. Use either **ACI 318 App. C** or **ACI 318 Ch. 9** (see p. 50-28) to solve this problem.

Solution Using ACI 318 App. C

(a) The factored load is

$$(1.4)\left(1300 \frac{\text{lbf}}{\text{ft}}\right) + (1.7)\left(1900 \frac{\text{lbf}}{\text{ft}}\right) = 5050 \text{ lbf/ft}$$

The total factored load is (28 ft)(5050 lbf/ft) = 141,400 lbf. Each reaction is

$$\frac{141{,}400 \text{ lbf}}{2} = 70{,}700 \text{ lbf} \quad \text{[factored]}$$

The column dimensions are unknown, so the reaction is placed at the face.

The centerline shear when the factored live load is positioned over half of the span is

$$V = \frac{(1.7)\left(1.9 \frac{\text{kips}}{\text{ft}}\right)(28 \text{ ft})}{8} = 11.31 \text{ kips}$$

Since the reaction induces compression, the critical section is located at a distance d from the face of the support. The shear envelope, with the shear at the critical

Structural

section identified, is illustrated. (The effective depth has been taken as $d = 21 \text{ in} - 2.5 \text{ in} = 18.5 \text{ in}$. The critical shear, V_u, is found by interpolation.)

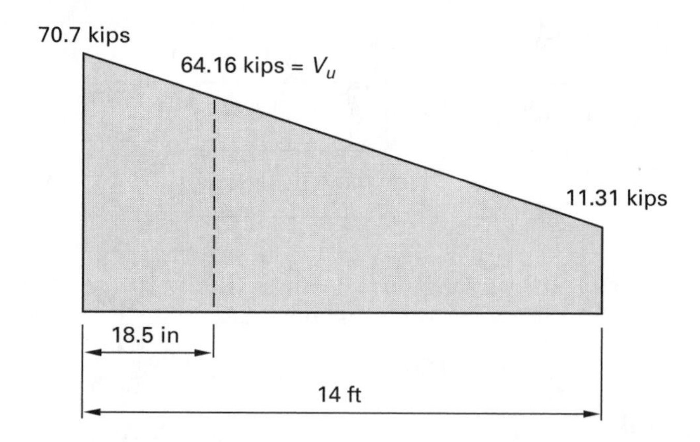

The critical section is located (from the face of the column) at

$$\frac{18.5 \text{ in}}{12 \frac{\text{in}}{\text{ft}}} = 1.54 \text{ ft}$$

The nominal concrete shear strength is given by Eq. 50.53.

$$V_c = \frac{2\sqrt{3000 \frac{\text{lbf}}{\text{in}^2}}(12 \text{ in})(18.5 \text{ in})}{1000 \frac{\text{lbf}}{\text{kip}}} = 24.31 \text{ kips}$$

Since $V_u > \phi V_c$, the beam needs extra shear reinforcement until V_u drops to ϕV_c. The steel shear contribution is given by Eq. 50.63.

$$V_{\text{st}} = \frac{64.16 \text{ kips}}{0.85} - 27.61 \text{ kips} = \boxed{47.87 \text{ kips}}$$

Check to see if the beam should be redesigned. Use Eq. 50.57.

$$\begin{aligned} V_{\text{st,max}} &= 8\sqrt{3000 \frac{\text{lbf}}{\text{in}^2}}(12 \text{ in})(18.5 \text{ in}) \\ &= 97{,}276 \text{ lbf} \quad \text{[acceptable]} \end{aligned}$$

(b) No reinforcement is needed after V_u drops below $^1/_2\phi V_c$.

$$\tfrac{1}{2}\phi V_c = (0.5)(0.85)(24.31 \text{ kips}) = 10.33 \text{ kips}$$

$$70{,}700 \text{ lbf} + \frac{x_{\text{feet}}(11{,}310 \text{ lbf} - 70{,}700 \text{ lbf})}{14 \text{ ft}} = 10{,}330 \text{ lbf}$$

$x = 14.2$ ft from the face of the support. Since $x > L/2$, stirrups are required over the entire span.

(c) $4\sqrt{3000 \frac{\text{lbf}}{\text{in}^2}}(12 \text{ in})(18.5 \text{ in}) = 48{,}638 \text{ lbf}$

Since $V_{\text{st}} < 48{,}638$ lbf,

$$\text{maximum spacing} = \frac{d}{2} = \frac{18.5 \text{ in}}{2} = 9.25 \text{ in}$$

However, even closer spacing may be needed. For no. 3 bars,

$$A_{\text{stirrup}} = (2)(0.11 \text{ in}^2) = 0.22 \text{ in}^2$$

The spacing is

$$s = \frac{(2)(0.11 \text{ in}^2)\left(60 \frac{\text{kips}}{\text{in}^2}\right)(18.5 \text{ in})}{47.87 \text{ kips}} = \boxed{5.1 \text{ in}}$$

Use 5 in spacing starting $s/2 \approx 3$ in from the face of support.

Solution Using ACI 318 Ch. 9

(a) The controlling factored load is

$$(1.2)\left(1300 \frac{\text{lbf}}{\text{ft}}\right) + (1.6)\left(1900 \frac{\text{lbf}}{\text{ft}}\right) = 4600 \text{ lbf/ft}$$

The total factored load is $(28 \text{ ft})(4600 \text{ lbf/ft}) = 128{,}800$ lbf. Each reaction is

$$\frac{128{,}800 \text{ lbf}}{2} = 64{,}400 \text{ lbf} \quad \text{[factored]}$$

The column dimensions are unknown, so the reaction is placed at the face.

The centerline shear when the factored live load is positioned over half of the span is

$$V = \frac{(1.6)\left(1.9 \frac{\text{kips}}{\text{ft}}\right)(28 \text{ ft})}{8} = 10.64 \text{ kips}$$

Since the reaction induces compression, the critical section is located at a distance d from the face of the support. The shear envelope, with the shear at the critical section identified, is illustrated. (The effective depth has been taken as $d = 21 \text{ in} - 2.5 \text{ in} = 18.5 \text{ in}$. The critical shear, V_u, is found by interpolation.)

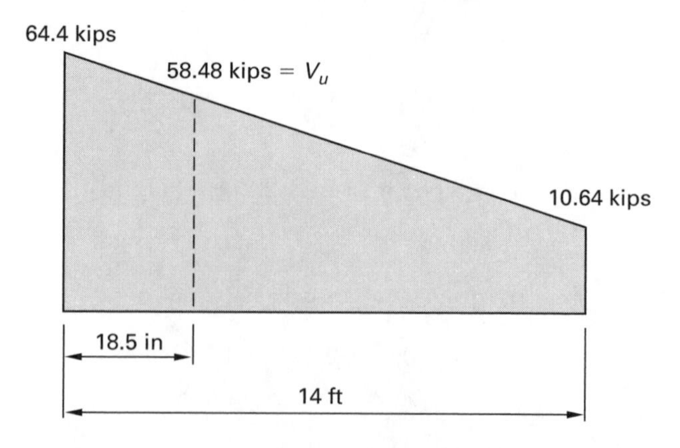

The critical section is located (from the face of the column) at

$$\frac{18.5 \text{ in}}{12 \frac{\text{in}}{\text{ft}}} = 1.54 \text{ ft}$$

The nominal concrete shear strength is given by Eq. 50.53.

$$V_c = \frac{2\sqrt{3000 \frac{\text{lbf}}{\text{in}^2}}(12 \text{ in})(18.5 \text{ in})}{1000 \frac{\text{lbf}}{\text{kip}}} = 24.31 \text{ kips}$$

Since $V_u > \phi V_c$, the beam needs extra shear reinforcement until V_u drops to ϕV_c. The steel shear contribution is given by Eq. 50.63.

$$V_{\text{st}} = \frac{58.48 \text{ kips}}{0.75} - 24.31 \text{ kips} = \boxed{53.66 \text{ kips}}$$

Check to see if the beam should be redesigned. Use Eq. 50.57.

$$\begin{aligned} V_{\text{st,max}} &= 8\sqrt{3000 \frac{\text{lbf}}{\text{in}^2}}(12 \text{ in})(18.5 \text{ in}) \\ &= 97{,}276 \text{ lbf} \quad \text{[acceptable]} \end{aligned}$$

(b) No reinforcement is needed after V_u drops below $1/2\phi V_c$.

$$\tfrac{1}{2}\phi V_c = (0.5)(0.75)(24.31 \text{ kips}) = 9.12 \text{ kips}$$

Since 9.12 kips < 10.64 kips, stirrups are required over the entire span.

(c) $4\sqrt{3000 \frac{\text{lbf}}{\text{in}^2}}(12 \text{ in})(18.5 \text{ in}) = 48{,}638 \text{ lbf}$

Since $V_{\text{st}} < 48{,}638$ lbf,

$$\text{maximum spacing} = \frac{d}{2} = \frac{18.5 \text{ in}}{2} = 9.25 \text{ in}$$

However, even closer spacing may be needed. For no. 3 bars,

$$A_{\text{stirrup}} = (2)(0.11 \text{ in}^2) = 0.22 \text{ in}^2$$

The spacing is

$$s = \frac{(2)(0.11 \text{ in}^2)\left(60 \frac{\text{kips}}{\text{in}^2}\right)(18.5 \text{ in})}{53.66 \text{ kips}} = \boxed{4.6 \text{ in}}$$

Use 4.5 in spacing starting $s/2 \approx 2.5$ in from the face of support.

10. Use either **ACI 318 Apps. B and C** or **ACI 318 Ch. 9 and Unified Design Method** (see p. 50-31) to solve this problem.

Solution Using ACI 318 Apps. B and C

(a)

$$\begin{aligned} M_{DL} &= \frac{\left(1.5 \frac{\text{kips}}{\text{ft}}\right)(18 \text{ ft})^2}{8} + (8 \text{ kips})(6 \text{ ft}) \\ &= 108.75 \text{ ft-kips} \\ M_{LL} &= \frac{\left(2 \frac{\text{kips}}{\text{ft}}\right)(18 \text{ ft})^2}{8} + (13 \text{ kips})(6 \text{ ft}) \\ &= 159 \text{ ft-kips} \\ M_u &= (1.4)(108.75 \text{ ft-kips}) \\ &\quad + (1.7)(159 \text{ ft-kips}) \\ &= \boxed{422.6 \text{ ft-kips}} \end{aligned}$$

(b) *step 1:* To achieve minimum beam depth as required by the problem, choose the maximum allowable reinforcement.

With $\beta = 0.85$,

$$\begin{aligned} \rho_{\text{balanced}} &= \left(\frac{(0.85)(0.85)\left(3000 \frac{\text{lbf}}{\text{in}^2}\right)}{60{,}000 \frac{\text{lbf}}{\text{in}^2}}\right) \\ &\quad \times \left(\frac{87{,}000}{87{,}000 + 60{,}000 \frac{\text{lbf}}{\text{in}^2}}\right) \\ &= 0.0214 \end{aligned}$$

$$\rho_{\text{max}} = (0.75)(0.0214) = 0.016 \quad \text{[acceptable]}$$

The corresponding strain and strength reduction factor are

$$\epsilon_t = 0.00376$$

$$\phi = 0.8$$

step 2: Solve for the required nominal moment strength.

$$\begin{aligned} M_n &= \frac{M_u}{\phi} = \frac{422{,}600 \text{ ft-lbf}}{0.8} \\ &= 528{,}250 \text{ ft-lbf} \end{aligned}$$

Structural

step 3: Substitute into Eq. 50.32 and solve for bd^2.

$$bd^2 = \frac{M_n}{\rho f_y\left(1 - \frac{\rho f_y}{1.7f'_c}\right)}$$

$$= \frac{(528{,}250\ \text{ft-lbf})\left(12\ \frac{\text{in}}{\text{ft}}\right)}{(0.016)\left(60{,}000\ \frac{\text{lbf}}{\text{in}^2}\right)} \times \left(1 - \frac{(0.016)\left(60{,}000\ \frac{\text{lbf}}{\text{in}^2}\right)}{(1.7)\left(3000\ \frac{\text{lbf}}{\text{in}^2}\right)}\right)$$

$$= 8143\ \text{in}^3$$

step 4: b was given as 16 in.

$$d = \sqrt{\frac{8143\ \text{in}^3}{16\ \text{in}}}$$

$$= \boxed{22.56\ \text{in}\quad (22.5\ \text{in})}$$

(c) *step 5:* $A_{st} = (0.016)(16\ \text{in})(22.5\ \text{in}) = 5.76\ \text{in}^2$

step 6: Select reinforcement.

> Use two no. 10 bars and two no. 11 bars (5.66 in^2).

(At this point, it may be desirable to recalculate ρ and resize the beam.)

step 7: $d_c = \frac{1.270\ \text{in}}{2} + 0.375\ \text{in} + 1.5\ \text{in}$

$= 2.51\ \text{in}$ [round to 2.75 in]

(d) The factored uniform load is

$$(1.4)\left(1.5\ \frac{\text{kips}}{\text{ft}}\right) + (1.7)\left(2\ \frac{\text{kips}}{\text{ft}}\right) = 5.5\ \text{kips/ft}$$

The factored concentrated load is

$$(1.4)(8\ \text{kips}) + (1.7)(13\ \text{kips}) = 33.3\ \text{kips}$$

ACI 318 12.13.2.1 requires a hook.

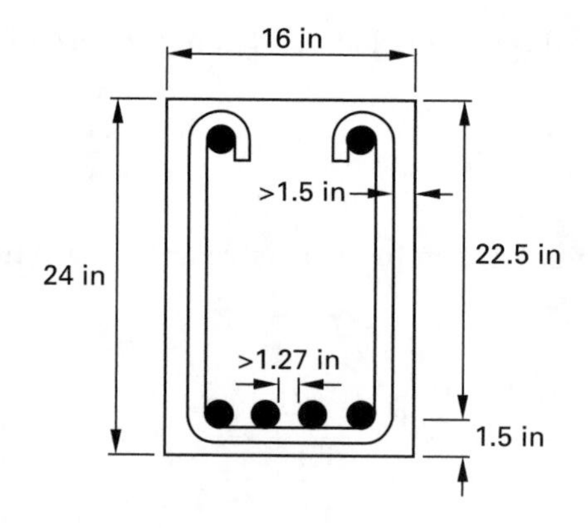

The factored reactions are

$$R = \frac{\left(5.5\ \frac{\text{kips}}{\text{ft}}\right)(18\ \text{ft}) + (2)(33.3\ \text{kips})}{2} = 82.8\ \text{kips}$$

The shear diagram is

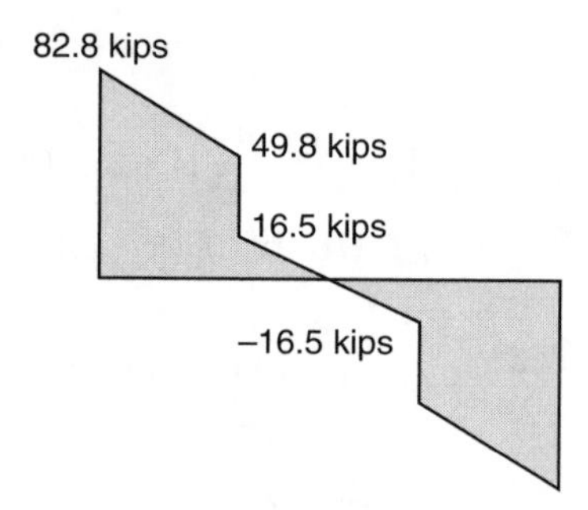

At a distance of 21.25 in from the support, the shear is

$$V_u = 82.8\ \text{kips} - \left(\frac{21.25\ \text{in}}{12\ \frac{\text{in}}{\text{ft}}}\right)\left(5.5\ \frac{\text{kips}}{\text{ft}}\right) = 73.1\ \text{kips}$$

Use Eq. 50.53.

$$V_c = 2\sqrt{3000\ \frac{\text{lbf}}{\text{in}^2}}(16\ \text{in})(21.25\ \text{in})$$

$$= 37{,}245\ \text{lbf}\quad [37.2\ \text{kips}]$$

The required steel strength is given by Eq. 50.63.

$$V_{st} = \frac{73.1\ \text{kips}}{0.85} - 37.2\ \text{kips} = 48.8\ \text{kips}$$

Check $V_{st,max}$.

$$V_{st,max} = 8\sqrt{3000\ \frac{\text{lbf}}{\text{in}^2}}(16\ \text{in})(21.25\ \text{in})$$

$$= 148{,}981\ \text{lbf}\quad (149\ \text{kips})\quad [\text{acceptable}]$$

Using $A_v = (2)(0.11 \text{ in}^2)$ for no. 3 bars, the spacing is

$$s = \frac{(2)(0.11 \text{ in}^2)\left(60{,}000 \ \frac{\text{lbf}}{\text{in}^2}\right)(21.25 \text{ in})}{48{,}800 \text{ lbf}} = \boxed{5.75 \text{ in}}$$

Start the first stirrup $d/2 = 10$ in from support face.

(It may be desirable to recalculate spacing required at $x = 3$ ft and continue that new spacing to $x = 6$ rather than using $s = 5.75$ in over the entire 6 ft.)

At $x = 6$ ft, the shear reinforcement required is

$$V_{st} = \frac{16.5 \text{ kips}}{0.85} - 37.2 \text{ kips} = -17.8 \text{ kips} \quad [\text{zero}]$$

Between $x = 6$ ft and $x = 12$ ft, only minimum reinforcement is required (ACI 11.5.5.1).

Solution Using ACI 318 Ch. 9 and Unified Design Method

(a) $M_u = (1.2)(108.75 \text{ ft-kips}) + (1.6)(159 \text{ ft-kips})$

$$= \boxed{384.9 \text{ ft-kips}}$$

1.4D does not control.

(b) *step 1:* To achieve minimum beam depth as required by the problem, choose the maximum allowable reinforcement.

The maximum reinforcement ratio corresponding to $\epsilon_t = 0.004$ is

$$\rho_{\max} = \left(\frac{0.85 f'_c \beta_1}{f_y}\right)\left(\frac{0.003}{0.003 + \epsilon_t}\right) = 0.0155$$

The corresponding ϕ factor is

$$\phi = 0.48 + 83\epsilon_t = 0.812$$

step 2: Solve for the required nominal moment strength.

$$M_n = \frac{M_u}{\phi} = \frac{384{,}900 \text{ ft-lbf}}{0.812} = 474{,}015 \text{ ft-lbf}$$

step 3: Substitute into Eq. 50.32 and solve for bd^2.

$$bd^2 = \frac{M_n}{\rho f_y \left(1 - \frac{\rho f_y}{1.7 f'_c}\right)}$$

$$= \frac{(474{,}015 \text{ ft-lbf})\left(12 \ \frac{\text{in}}{\text{ft}}\right)}{(0.0155)\left(60{,}000 \ \frac{\text{lbf}}{\text{in}^2}\right) \times \left(1 - \frac{(0.0155)\left(60{,}000 \ \frac{\text{lbf}}{\text{in}^2}\right)}{(1.7)\left(3000 \ \frac{\text{lbf}}{\text{in}^2}\right)}\right)}$$

$$= 7480 \text{ in}^3$$

step 4: b was given as 16 in.

$$d = \sqrt{\frac{7480 \text{ in}^3}{16 \text{ in}}} = \boxed{21.62 \text{ in} \quad (21.5 \text{ in})}$$

(c) *step 5:* $A_{st} = (0.0155)(16 \text{ in})(21.5 \text{ in}) = 5.33 \text{ in}^2$

step 6: Select reinforcement.

Use two no. 10 bars and two no. 11 bars (5.66 in^2).

(At this point, it may be desirable to recalculate ρ and resize the beam.)

step 7: $d_c = \frac{1.270 \text{ in}}{2} + 0.375 \text{ in} + 1.5 \text{ in}$

$$= 2.51 \text{ in} \quad [\text{round to 2.5 in}]$$

(d) The controlling factored uniform load is

$$(1.2)\left(1.5 \ \frac{\text{kips}}{\text{ft}}\right) + (1.6)\left(2 \ \frac{\text{kips}}{\text{ft}}\right) = 5.0 \text{ kips/ft}$$

The controlling factored concentrated load is

$$(1.2)(8 \text{ kips}) + (1.6)(13 \text{ kips}) = 30.4 \text{ kips}$$

ACI 318 12.13.2.1 requires a hook.

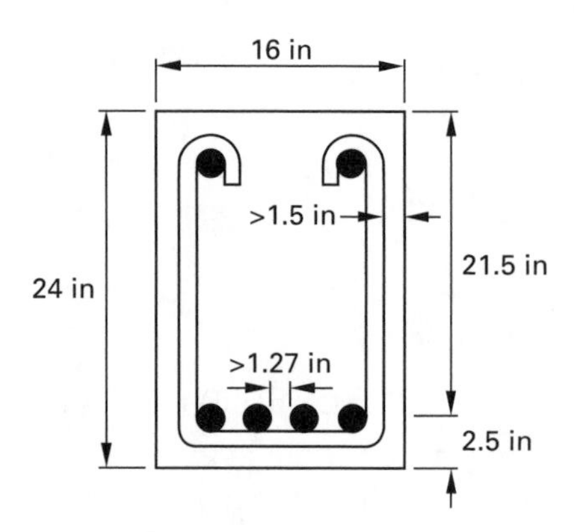

The factored reactions are

$$R = \frac{\left(5.0\ \frac{\text{kips}}{\text{ft}}\right)(18\text{ ft}) + (2)(30.4\text{ kips})}{2} = 75.4\text{ kips}$$

The shear diagram is

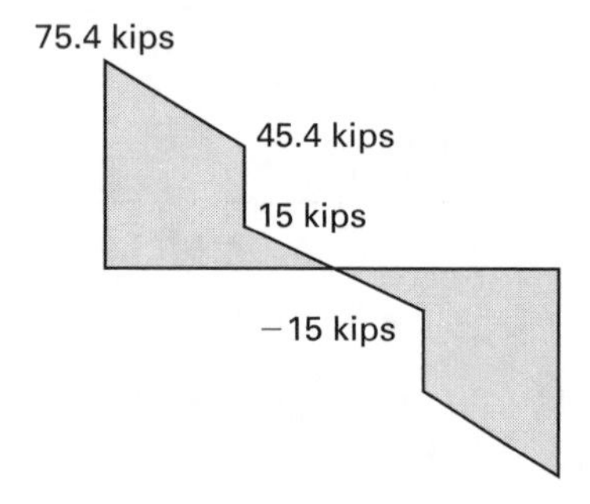

At a distance of 20.5 in from the support, the shear is

$$V_u = 75.4\text{ kips} - \left(\frac{20.5\text{ in}}{12\ \frac{\text{in}}{\text{ft}}}\right)\left(5.0\ \frac{\text{kips}}{\text{ft}}\right) = 66.9\text{ kips}$$

Use Eq. 50.53.

$$\begin{aligned} V_c &= 2\sqrt{3000\ \frac{\text{lbf}}{\text{in}^2}}(16\text{ in})(20.5\text{ in}) \\ &= 35{,}930\text{ lbf}\quad [35.9\text{ kips}] \end{aligned}$$

The required steel strength is given by Eq. 50.63.

$$V_{\text{st}} = \frac{66.9\text{ kips}}{0.75} - 35.9\text{ kips} = 53.3\text{ kips}$$

Check $V_{\text{st,max}}$.

$$\begin{aligned} V_{\text{st,max}} &= 8\sqrt{3000\ \frac{\text{lbf}}{\text{in}^2}}(16\text{ in})(20.5\text{ in}) \\ &= 143{,}723\text{ lbf}\quad (144\text{ kips})\quad [\text{acceptable}] \end{aligned}$$

Using $A_v = (2)(0.11\text{ in}^2)$ for no. 3 bars, the spacing is

$$s = \frac{(2)(0.11\text{ in}^2)\left(60{,}000\ \frac{\text{lbf}}{\text{in}^2}\right)(20.5\text{ in})}{53{,}300\text{ lbf}} = \boxed{5.08\text{ in}}$$

Use 5 in spacing.

Start the first stirrup $s/2 = 2.5$ in from support face.

(It may be desirable to recalculate spacing required at $x = 3$ ft and continue that new spacing to $x = 6$ rather than using $s = 5$ in over the entire 6 ft.)

At $x = 6$ ft, the shear reinforcement required is

$$V_{\text{st}} = \frac{15\text{ kips}}{0.75} - 35.9\text{ kips} = -15.9\text{ kips}\quad [\text{zero}]$$

Between $x = 6$ ft and $x = 12$ ft, only minimum reinforcement is required (ACI 11.5.5.1).

11. (a) According to ACI 318 Sec. 10.7.1(a), a deep beam is one in which $l_n/h \le 4$. In this case,

$$\frac{l_n}{h} = \frac{(3)(6.5\text{ ft})}{6.5\text{ ft}} = 3 < 4$$

Therefore, this is a deep beam.

(b) The entire deep beam is a D-region. It can be seen that the nodal zones at B and C are C-C-C nodal zones and the ones at A and D are C-C-T nodal zones. The bearing strength is given by ACI 318 Eq. A-7.

$$\phi F_{nn} = \phi f_{cu} A_n$$

The ϕ factor for a strut and tie model is 0.75, and A_n can be taken as the area of the bearing pads. The effective compressive strength of the concrete in the nodal zone, f_{cu}, can be determined from ACI 318 Eq. A-8.

$$f_{cu} = 0.85\beta_n f'_c$$

$\beta_n = 1.0$ in nodal zones bounded by struts or bearing areas or both (Sec. A.5.2.1, ACI 318). Therefore, for nodal zones B and C (C-C-C nodal zones),

$$\begin{aligned} \phi F_{nn} &= \phi f_{cu} A_n \\ &= (0.75)(0.85)(1.0)\left(4\ \frac{\text{kips}}{\text{in}^2}\right)(18\text{ in})(20\text{ in}) \\ &= 918\text{ kips} > 322\text{ kips}\quad [\text{OK}] \end{aligned}$$

$\beta_n = 0.80$ in nodal zones anchoring one tie. For nodal zones A and D (C-C-T nodal zones),

$$\begin{aligned} \phi F_{nn} &= \phi f_{cu} A_n \\ &= (0.75)(0.85)(0.8)\left(4\ \frac{\text{kips}}{\text{in}^2}\right)(18\text{ in})(20\text{ in}) \\ &= 734\text{ kips} > 322\text{ kips}\quad [\text{OK}] \end{aligned}$$

(c) From the free-body diagram, sum the moments about point A.

$$(322\text{ kips})(78\text{ in}) = F_{BC}\left(78\text{ in} - \frac{w_c}{2} - \frac{w_t}{2}\right)$$

$$F_{BC} = (322\text{ kips})\frac{78\text{ in}}{78\text{ in} - \frac{7.5\text{ in}}{2} - \frac{9\text{ in}}{2}}$$

$$= 360\text{ kips}$$

The design strength of strut BC is

$$\begin{aligned}\phi F_{ns} &= \phi f_{cu}A_c \geq F_{BC}\\ &= \phi 0.85\beta_s f'_c w_c b \quad [\text{strut}] \quad [\text{ACI Eq. A-2}]\\ &= \phi 0.85\beta_n f'_c w_c b \quad [\text{nodal zone}] \quad [\text{ACI Eq. A-8}]\end{aligned}$$

Since strut BC has a uniform cross section over its entire length, $\beta_s = 1.0$ per Sec. A.3.2.1 of ACI 318. Also, $\beta_n = 1.0$, since the nodal zone at B is C-C-C (Sec. A.5.2.1, ACI 318). Check the capacity of strut BC.

$$\phi F_{ns} = (0.75)(0.85)(1)\left(4\ \frac{\text{kips}}{\text{in}^2}\right)(7.5\text{ in})(20\text{ in})$$

$$= 383\text{ kips} > 360\text{ kips} \quad [\text{OK}]$$

(d) From the free-body diagram of half of the beam, it can be seen that $F_{AD} = F_{BC}$.

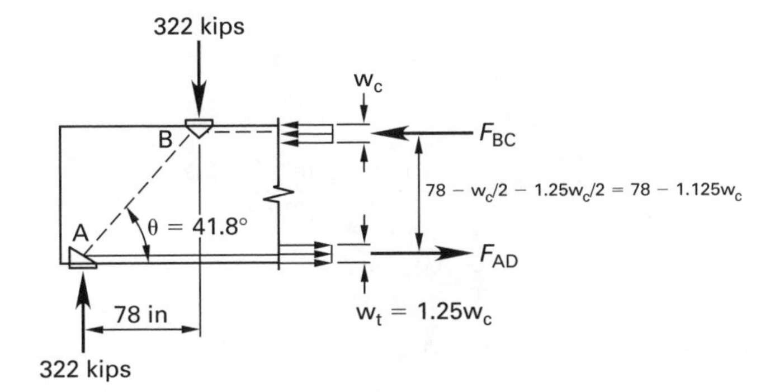

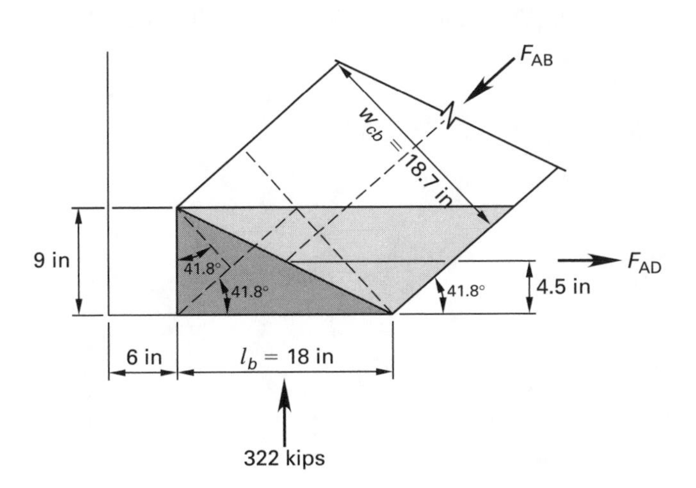

$$F_{AD} = 360\text{ kips} \leq \phi F_{nt} = \phi A_{st} f_y$$

$$A_{st} = \frac{360\text{ kips}}{(0.75)\left(60\ \frac{\text{kips}}{\text{in}^2}\right)} = 8.0\text{ in}^2 > \frac{200bd}{f_y}$$

$$= 4.9\text{ in}^2 \quad \left[\begin{array}{c}\text{minimum reinforcement per}\\ \text{Sec. 10.7.3, ACI 318}\end{array}\right]$$

Use two layers of four no. 9 bars ($A_s = 8.0$ in^2).

(e) Find the angle of strut AB with respect to tie AD.

$$\tan\theta = \frac{78\text{ in} - \frac{7.5\text{ in}}{2} - \frac{9\text{ in}}{2}}{78\text{ in}}$$

$$\theta = 41.8° > 25° \quad [\text{OK per Sec. A.2.5, ACI 318}]$$

$$F_{AB} = \frac{360\text{ kips}}{\cos 41.8°} = 483\text{ kips}$$

As shown in the problem illustrations, the width at the top of the strut is

$$\begin{aligned}w_{ct} &= l_b \sin\theta + w_c\cos\theta\\ &= (18\text{ in})\sin 41.8° + (7.5\text{ in})\cos 41.8°\\ &= 17.6\text{ in}\end{aligned}$$

The width at the bottom of the strut is

$$\begin{aligned}w_{cb} &= l_b \sin\theta + w_t\cos\theta\\ &= (18\text{ in})\sin 41.8° + (9\text{ in})\cos 41.8°\\ &= 18.7\text{ in}\end{aligned}$$

$\beta_s = 0.75$, assuming that reinforcement satisfying Sec. A.3.3 of ACI 318 is provided (Sec. A.3.1, ACI 318). The design strength of the strut is based on the smaller of the two widths at the ends of the strut.

$$\begin{aligned}\phi F_{ns} &= \phi 0.85\beta_s f'_c w_{ct} b\\ &= (0.75)(0.85)(0.75)\left(4\ \frac{\text{kips}}{\text{in}^2}\right)(17.6\text{ in})(20\text{ in})\\ &= 673\text{ kips} > 483\text{ kips} \quad [\text{OK}]\end{aligned}$$

(f) The two layers of four no. 9 tie bars must be properly anchored at both ends of the beam.

The critical section for development of the tie reinforcement is determined in accordance with Sec. A.4.3.2 of ACI 318. The distance, x, from the edge of the bearing plate (nodal zone) to the critical section is determined from geometry.

$$\tan 41.8° = \frac{4.5\text{ in}}{x}$$

$$x = 5\text{ in}$$

Structural

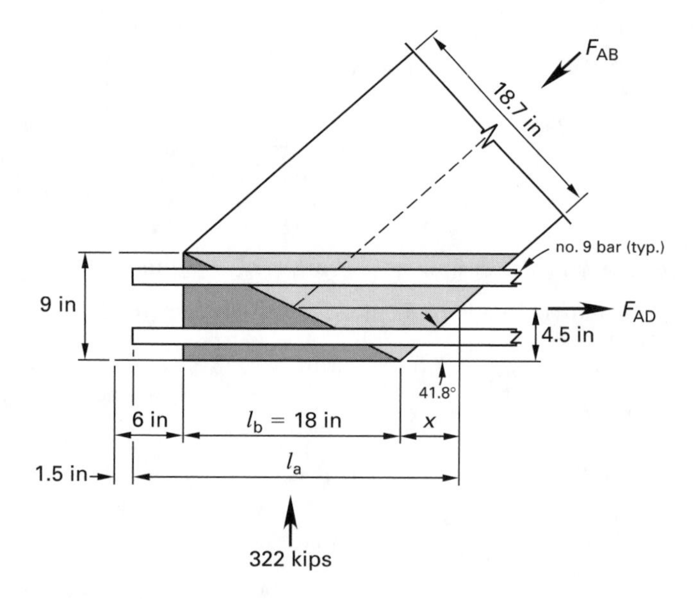

The available length for straight bar development is

$$24 \text{ in} - 1.5 \text{ in} + 5 \text{ in} = 27.5 \text{ in}$$

To develop a straight no. 9 bar, the required development length is determined from

$$l_d = \left(\frac{3}{40}\right)\left(\frac{f_y}{\sqrt{f'_c}}\right)\left(\frac{\alpha\beta\gamma\lambda}{\frac{c+K_{tr}}{d_b}}\right)d_b \quad \text{[ACI Eq. 12-1]}$$

$$= \left(\frac{3}{40}\right)\left(\frac{60{,}000\ \frac{\text{lbf}}{\text{in}^2}}{\sqrt{4000\ \frac{\text{lbf}}{\text{in}^2}}}\right)\left(\frac{(1.0)(1.0)(1.0)(1.0)}{\frac{2.0 \text{ in}}{1.128 \text{ in}}}\right)$$

$$= 40.1 \text{ in} > 27.5 \text{ in}$$

Thus, provide a standard hook at the ends of the no. 9 bars.

The development length of a no. 9 bar with a standard 90° hook is determined from Sec. 12.5.2 of ACI 318.

$$l_{dh} = \frac{0.02\beta\gamma f_y d_b}{\sqrt{f'_c}}$$

$$= \frac{(0.02)(1.0)(1.0)\left(60{,}000\ \frac{\text{lbf}}{\text{in}^2}\right)(1.128 \text{ in})}{\sqrt{4000\ \frac{\text{lbf}}{\text{in}^2}}}$$

$$= 21.4 \text{ in} < 27.5 \text{ in} \quad \text{[OK]}$$

(g) Provide horizontal and vertical reinforcement in accordance with Sec. A.3.3 of ACI 318.

For horizontal reinforcement, try two no. 5 bars (one bar on each face) on 12 in spacing.

$$\gamma_2 = \theta = 41.8°$$

$$\frac{A_{s2}\sin\gamma_2}{bs_2} = \frac{(2 \text{ in})(0.31 \text{ in})\sin 41.8°}{(20 \text{ in})(12 \text{ in})} = 0.0017$$

For vertical reinforcement, try two sets of two no. 4 overlapping ties on 12 in spacing.

$$\gamma_1 = 90° - \gamma_2 = 90° - 41.8° = 48.2°$$

$$\frac{A_{s1}\sin\gamma_1}{bs_1} = \frac{(4 \text{ in})(0.2 \text{ in})\sin 48.2°}{(20 \text{ in})(12 \text{ in})} = 0.0025$$

Check Eq. A-4 in ACI 318.

$$0.0017 + 0.0025 = 0.0042 > 0.0030 \quad \text{[OK]}$$

51 Reinforced Concrete: Slabs

PRACTICE PROBLEMS

Two-Way Slabs

1. A plan of a floor system supported on beams is shown. In addition to its own weight, the floor is subjected to uniformly distributed dead and live loads of 30 lbf/ft^2 and 50 lbf/ft^2, respectively. Material properties are f'_c = 4000 lbf/in^2 and f_y = 60,000 lbf/in^2. (a) Calculate the relative beam stiffness for beam A2-B2. (b) Determine if the thickness of the slab is adequate for deflection control. (c) Design the steel reinforcement for positive moment in beam A2-B2.

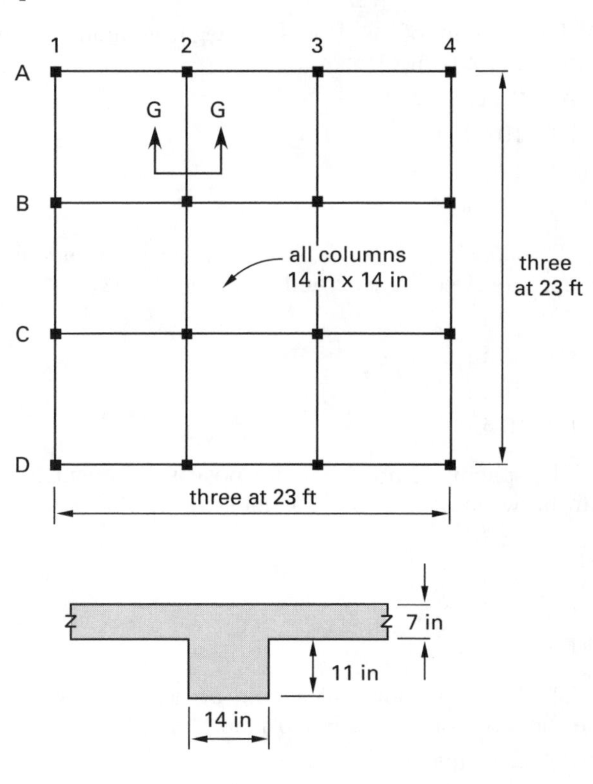

2. A flat plate with an edge beam is shown. In addition to its own weight, the floor slab is subjected to uniformly distributed dead and live loads of 20 lbf/ft^2 and 50 lbf/ft^2, respectively. Material properties are f'_c = 4000 lbf/in^2 and f_y = 40,000 lbf/in^2. (a) Determine the torsional constant, C, for the edge beam in span A2-A3. (b) Determine the adequacy of the slab thickness for deflection control. (c) Design the positive reinforcement needed in the column strip whose center lies on the axis A2-B2.

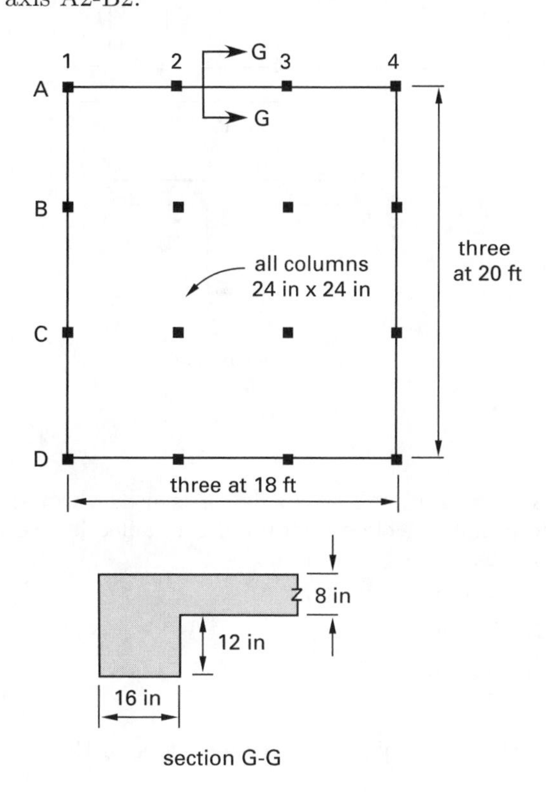

3. (*Time limit: one hour*) The floor system shown is subjected, in addition to its own weight, to uniformly distributed dead and live loads of 20 lbf/ft^2 and 80 lbf/ft^2, respectively. Material properties are $f'_c = 3000$ lbf/in^2 and $f_y = 40{,}000$ lbf/in^2.

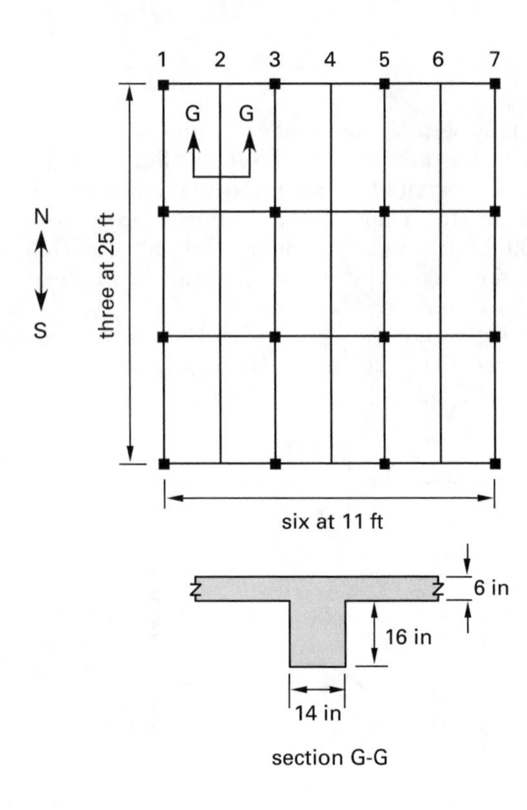

section G-G

(a) Assuming a single slab depth is used throughout, the minimum thickness required for deflection control is most nearly

(A) 5.0 in
(B) 5.5 in
(C) 6.0 in
(D) 7.5 in

For parts (b) through (j), assume the slab thickness is 6 in.

(b) The factored uniform load is most nearly

(A) 130 lbf/ft^2
(B) 160 lbf/ft^2
(C) 270 lbf/ft^2
(D) 300 lbf/ft^2

(c) The maximum live load that the slab can carry, as limited by shear, is most nearly

(A) 400 lbf/ft^2
(B) 470 lbf/ft^2
(C) 500 lbf/ft^2
(D) 600 lbf/ft^2

(d) The required spacing of no. 3 bars parallel to the N-S direction is most nearly

(A) 7 in
(B) 8 in
(C) 9 in
(D) 10 in

(e) The maximum positive moment in this floor system is most nearly

(A) 1.5 ft-kip/ft
(B) 1.6 ft-kip/ft
(C) 1.9 ft-kip/ft
(D) 2.1 ft-kip/ft

(f) The maximum negative moment in this floor system is most nearly

(A) 2.5 ft-kip/ft
(B) 2.6 ft-kip/ft
(C) 2.9 ft-kip/ft
(D) 3.1 ft-kip/ft

(g) The spacing of no. 4 bars for negative moment normal to axis 2 is most nearly

(A) 9 in
(B) 10 in
(C) 12 in
(D) 14 in

(h) The spacing of no. 6 bars for negative moment in the span between axes 2 and 3 is most nearly

(A) 12 in
(B) 16 in
(C) 18 in
(D) 27 in

(i) The spacing of no. 4 bars for positive moment in the span between axes 2 and 3 is most nearly

(A) 14 in
(B) 17 in
(C) 19 in
(D) 20 in

(j) The load per foot that should be used to design the beam located on axis 2 is most nearly

(A) 2.8 kips/ft
(B) 3.0 kips/ft
(C) 3.2 kips/ft
(D) 3.5 kips/ft

4. (*Time limit: one hour*) A flat plate is supported on 24 in by 24 in columns that are spaced 18 ft in the E-W and 22 ft in the N-S direction. In addition to its own weight, the floor is subjected to uniformly distributed dead and live loads of 20 lbf/ft^2 and 50 lbf/ft^2, respectively. Material properties are $f'_c = 4000$ lbf/in^2 and $f_y = 60{,}000$ lbf/in^2.

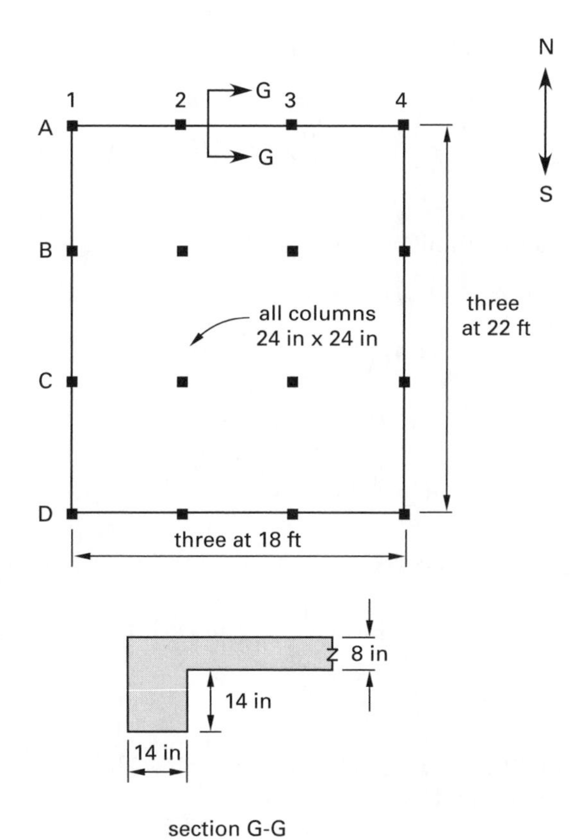

(a) The torsional constant C for the edge beam is most nearly

(A) 9000 in^4
(B) 12,000 in^4
(C) 14,000 in^4
(D) 34,000 in^4

(b) The minimum thickness for deflection control is most nearly

(A) 6.0 in
(B) 7.3 in
(C) 8.0 in
(D) 8.5 in

(c) The minimum thickness if the edge beam is eliminated is most nearly

(A) 6.0 in
(B) 7.5 in
(C) 8.0 in
(D) 8.5 in

(d) The relative torsional stiffness, β_t, for the edge beam spanning in the E-W direction is most nearly

(A) 0.50
(B) 0.75
(C) 1.00
(D) 1.3

(e) The maximum positive moment in the column strip that runs from B2 to C2 is most nearly

(A) 40 ft-kips
(B) 48 ft-kips
(C) 52 ft-kips
(D) 62 ft-kips

(f) The maximum positive moment in the column strip that runs from B1 to C1 is most nearly

(A) 21 ft-kips
(B) 24 ft-kips
(C) 28 ft-kips
(D) 31 ft-kips

(g) The maximum positive moment in the N-S middle strips in panel B2-B3-C2-C3 is most nearly

(A) 28 ft-kips
(B) 32 ft-kips
(C) 38 ft-kips
(D) 40 ft-kips

(h) If drop panels that meet ACI 318 requirements are provided, the minimum thickness for deflection control (outside the drop panel) is most nearly

(A) 5.5 in
(B) 6.7 in
(C) 7.5 in
(D) 8.0 in

(i) The required total thickness for the drop panels is most nearly

(A) 7.0 in
(B) 7.5 in
(C) 8.5 in
(D) 9.0 in

(j) The minimum dimension in the E-W direction for interior drop panels is most nearly

(A) 3 ft
(B) 5 ft
(C) 6 ft
(D) 7 ft

SOLUTIONS

1. Use either **ACI 318 Apps. B and C** or **ACI 318 Ch. 9 and Unified Design Method** (see p. 51-5) to solve this problem.

Solution Using ACI 318 Apps. B and C

(a) The beam section is

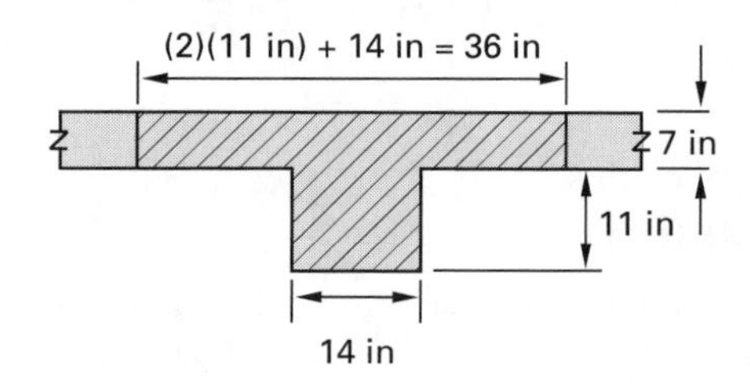

Use a table to calculate the moment of inertia of the beam.

	A (in^2)	y (in)	Ay (in^3)	$A(y-\overline{y})^2$ (in^4)	I_o (in^4)
	252	3.5	882	2936.8	1029.0
	154	12.5	1925	4805.7	1551.8
totals	406		2807	7742.5	2581.8

$$\overline{y} = \frac{2807 \text{ in}^3}{406 \text{ in}^2} = 6.91 \text{ in}$$

$$I_b = 7742.5 \text{ in}^4 + 2581.8 \text{ in}^4 = 10{,}324 \text{ in}^4$$

The moment of inertia of the slab is

$$I_s = \left(\tfrac{1}{12}\right)(23 \text{ ft})\left(12 \frac{\text{in}}{\text{ft}}\right)(7 \text{ in})^3 = 7889 \text{ in}^4$$

$$\alpha = \frac{10{,}324 \text{ in}^4}{7889 \text{ in}^4} = \boxed{1.31}$$

(b)

$$\alpha_m = 1.31$$

$$l_n = (23 \text{ ft})\left(12 \frac{\text{in}}{\text{ft}}\right) - 14 \text{ in} = 262 \text{ in}$$

$$\beta = 1$$

Use Eq. 51.7.

$$t = \frac{l_n\left(0.8 + \frac{f_y}{200{,}000}\right)}{36 + 5\beta(\alpha_m - 0.2)} \geq 5 \text{ in}$$

$$= \frac{(262 \text{ in})\left(0.8 + \frac{60{,}000 \frac{\text{lbf}}{\text{in}^2}}{200{,}000 \frac{\text{lbf}}{\text{in}^2}}\right)}{36 + (5)(1)(1.31 - 0.2)}$$

$$= 6.94 \text{ in} < 7 \text{ in} \quad \boxed{[\text{adequate}]}$$

(c) The weight of the slab is

$$\left(\frac{7 \text{ in}}{12 \frac{\text{in}}{\text{ft}}}\right)\left(0.15 \frac{\text{kip}}{\text{ft}^3}\right) = 0.0875 \text{ kip/ft}^2$$

The factored uniform load is

$$\begin{aligned} w_u &= 1.4D + 1.7L \\ &= (1.4)\left(0.03 \frac{\text{kip}}{\text{ft}^2} + 0.0875 \frac{\text{kip}}{\text{ft}^2}\right) + (1.7)\left(0.05 \frac{\text{kip}}{\text{ft}^2}\right) \\ &= 0.25 \frac{\text{kip}}{\text{ft}^2} \end{aligned}$$

The distances needed for the computation of M_o are

$$l_n = 23 \text{ ft} - \frac{14 \text{ in}}{12 \frac{\text{in}}{\text{ft}}} = 21.83 \text{ ft}$$

$$l_2 = 23 \text{ ft}$$

The additional weight of concrete below the slab must be included in the design load.

$$w_u = \left(0.25 \frac{\text{kip}}{\text{ft}^2}\right)(23 \text{ ft}) + \frac{(1.4)(11 \text{ in})(14 \text{ in})\left(0.15 \frac{\text{kip}}{\text{ft}^3}\right)}{\left(12 \frac{\text{in}}{\text{ft}}\right)^2} = 6.0 \text{ kips/ft}$$

From Table 51.3, the statical moment for a typical bay is

$$M_o = \left(\tfrac{1}{8}\right)\left(6.0 \frac{\text{kips}}{\text{ft}}\right)(21.83 \text{ ft})^2 = 357.4 \text{ ft-kips}$$

The positive design moment is

$$0.57M_o = (0.57)(357.4 \text{ ft-kips}) = 203.7 \text{ ft-kips}$$

From Table 51.4(c), the portion of this moment assigned to the column strip is

$$\begin{aligned} M_{\text{column strip}} &= (0.75)(0.57M_o) \\ &= (0.75)(203.7 \text{ ft-kips}) = 152.8 \text{ ft-kips} \end{aligned}$$

$$\alpha\left(\frac{l_2}{l_1}\right) = (1.31)\left(\frac{23 \text{ ft}}{23 \text{ ft}}\right) = 1.31 > 1$$

Therefore, the beam takes 85% of the column strip moment. (See ACI 318 Sec. 13.6.5.1.)

$$M_u = (0.85)(152.8 \text{ ft-kips}) = 130 \text{ ft-kips}$$

Estimate $d = t - 2.5 \text{ in} = 15.5 \text{ in}$.

Assume the section is tension controlled and $\lambda = (0.1)(15.5 \text{ in}) = 1.55$ in. Use Eq. 50.28.

$$A_s = \frac{M_u}{\phi f_y (d-\lambda)} = \frac{(130 \text{ ft-kips})\left(12 \ \frac{\text{in}}{\text{ft}}\right)}{(0.9)\left(60 \ \frac{\text{kips}}{\text{in}^2}\right)(15.5 \text{ in} - 1.55 \text{ in})} = 2.07 \text{ in}^2$$

Use Eq. 50.26.

$$A_c = \frac{f_y A_s}{0.85 f_c'} = \frac{(2.07 \text{ in}^2)\left(60 \ \frac{\text{kips}}{\text{in}^2}\right)}{(0.85)\left(4 \ \frac{\text{kips}}{\text{in}^2}\right)} = 36.5 \text{ in}^2$$

$$a = \frac{36.5 \text{ in}^2}{36 \text{ in}} = 1.01 \text{ in}$$

$$\lambda = \frac{a}{2} = \frac{1.01 \text{ in}}{2} = 0.51 \text{ in}$$

Check to see if the section is tension controlled.

$$c = \frac{a}{\beta_1} = \frac{1.01 \text{ in}}{0.85} = 1.19 \text{ in}$$

$$\epsilon_t = \frac{d-c}{c} 0.003 = \left(\frac{15.5 \text{ in} - 1.19 \text{ in}}{1.19 \text{ in}}\right)(0.003) = 0.036 > 0.005$$

Therefore, the section is tension controlled.

Use Eq. 50.27.

$$\phi M_n = \frac{(0.9)(2.07)\left(60 \ \frac{\text{kips}}{\text{in}^2}\right)(15.5 \text{ in} - 0.51 \text{ in})}{12 \ \frac{\text{in}}{\text{ft}}} = 139.6 \text{ ft-kips}$$

Adjust the area of steel.

$$A_s = \frac{(130 \text{ ft-kips})(2.07 \text{ in}^2)}{139.6 \text{ ft-kips}} = 1.92 \text{ in}^2$$

Use two no. 8 bars plus one no. 7 bar ($A_s = 2.18 \text{ in}^2$).

Solution Using ACI 318 Ch. 9 and Unified Design Method

(a) The beam section is

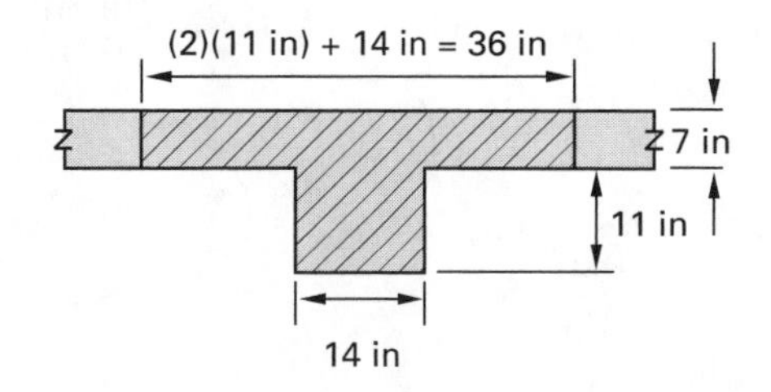

Use a table to calculate the moment of inertia of the beam.

	A (in²)	y (in)	Ay (in³)	$A(y-\bar{y})^2$ (in⁴)	I_o (in⁴)
	252	3.5	882	2936.8	1029.0
	154	12.5	1925	4805.7	1551.8
totals	406		2807	7742.5	2581.8

$$\bar{y} = \frac{2807 \text{ in}^3}{406 \text{ in}^2} = 6.91 \text{ in}$$

$$I_b = 7742.5 \text{ in}^4 + 2581.8 \text{ in}^4 = 10{,}324 \text{ in}^4$$

The moment of inertia of the slab is

$$I_s = \left(\tfrac{1}{12}\right)(23 \text{ ft})\left(12 \ \frac{\text{in}}{\text{ft}}\right)(7 \text{ in})^3 = 7889 \text{ in}^4$$

$$\alpha = \frac{10{,}324 \text{ in}^4}{7889 \text{ in}^4} = \boxed{1.31}$$

(b)

$$\alpha_m = 1.31$$

$$l_n = (23 \text{ ft})\left(12 \ \frac{\text{in}}{\text{ft}}\right) - 14 \text{ in} = 262 \text{ in}$$

$$\beta = 1$$

Use Eq. 51.7.

$$t = \frac{l_n\left(0.8 + \frac{f_y}{200{,}000}\right)}{36 + 5\beta(\alpha_m - 0.2)} \geq 5 \text{ in}$$

$$= \frac{(262 \text{ in})\left(0.8 + \frac{60{,}000 \ \frac{\text{lbf}}{\text{in}^2}}{200{,}000 \ \frac{\text{lbf}}{\text{in}^2}}\right)}{36 + (5)(1)(1.31 - 0.2)}$$

$$= 6.94 \text{ in} < 7 \text{ in} \quad \boxed{\text{[adequate]}}$$

(c) The weight of the slab is

$$\left(\frac{7 \text{ in}}{12 \ \frac{\text{in}}{\text{ft}}}\right)\left(0.15 \ \frac{\text{kip}}{\text{ft}^3}\right) = 0.0875 \text{ kip/ft}^2$$

Structural

The factored uniform load is

$$\begin{aligned} w_u &= 1.2D + 1.6L \\ &= (1.2)\left(0.03\ \frac{\text{kip}}{\text{ft}^2} + 0.0875\ \frac{\text{kip}}{\text{ft}^2}\right) \\ &\quad + (1.6)\left(0.05\ \frac{\text{kip}}{\text{ft}^2}\right) \\ &= 0.22\ \frac{\text{kip}}{\text{ft}^2} \end{aligned}$$

The distances needed for the computation of M_o are

$$l_n = 23\text{ ft} - \frac{14\text{ in}}{12\ \frac{\text{in}}{\text{ft}}} = 21.83\text{ ft}$$

$$l_2 = 23\text{ ft}$$

The additional weight of concrete below the slab must be included in the design load.

$$\begin{aligned} w_u &= \left(0.22\ \frac{\text{kip}}{\text{ft}^2}\right)(23\text{ ft}) \\ &\quad + \frac{(1.2)(11\text{ in})(14\text{ in})\left(0.15\ \frac{\text{kip}}{\text{ft}^3}\right)}{\left(12\ \frac{\text{in}}{\text{ft}}\right)^2} \\ &= 5.25\text{ kips/ft} \end{aligned}$$

From Table 51.3, the statical moment for a typical bay is

$$M_o = \left(\tfrac{1}{8}\right)\left(5.25\ \frac{\text{kips}}{\text{ft}}\right)(21.83\text{ ft})^2 = 312.7\text{ ft-kips}$$

The positive design moment is

$$0.57M_o = (0.57)(312.7\text{ ft-kips}) = 178.2\text{ ft-kips}$$

From Table 51.4(c), the portion of this moment assigned to the column strip is

$$\begin{aligned} M_{\text{column strip}} &= (0.75)(0.57M_o) \\ &= (0.75)(178.2\text{ ft-kips}) = 133.7\text{ ft-kips} \end{aligned}$$

$$\alpha\left(\frac{l_2}{l_1}\right) = (1.31)\left(\frac{23\text{ ft}}{23\text{ ft}}\right) = 1.31 > 1$$

Therefore, the beam takes 85% of the column strip moment. (See ACI 318 Sec. 13.6.5.1.)

$$M_u = (0.85)(133.7\text{ ft-kips}) = 113.6\text{ ft-kips}$$

Estimate $d = t - 2.5\text{ in} = 15.5\text{ in}$.

Assume $\lambda = (0.1)(15.5\text{ in}) = 1.55\text{ in}$ and the section is tension controlled. Use Eq. 50.28.

$$\begin{aligned} A_s &= \frac{M_u}{\phi f_y(d-\lambda)} \\ &= \frac{(113.6\text{ ft-kips})\left(12\ \frac{\text{in}}{\text{ft}}\right)}{(0.9)\left(60\ \frac{\text{kips}}{\text{in}^2}\right)(15.5\text{ in} - 1.55\text{ in})} \\ &= 1.81\text{ in}^2 \end{aligned}$$

Use Eq. 50.26.

$$\begin{aligned} A_c &= \frac{f_yA_s}{0.85f_c'} = \frac{(1.81\text{ in}^2)\left(60\ \frac{\text{kips}}{\text{in}^2}\right)}{(0.85)\left(4\ \frac{\text{kips}}{\text{in}^2}\right)} \\ &= 31.9\text{ in}^2 \end{aligned}$$

$$a = \frac{31.9\text{ in}^2}{36\text{ in}} = 0.89\text{ in}$$

$$\lambda = \frac{a}{2} = \frac{0.89\text{ in}}{2} = 0.445\text{ in}$$

With a value of a this small, the section is going to be tension controlled.

Use Eq. 50.27.

$$\begin{aligned} \phi M_n &= \frac{(0.9)(1.81)\left(60\ \frac{\text{kips}}{\text{in}^2}\right)(15.5\text{ in} - 0.445\text{ in})}{12\ \frac{\text{in}}{\text{ft}}} \\ &= 122.6\text{ ft-kips} \end{aligned}$$

Adjust the area of steel.

$$\begin{aligned} A_s &= \frac{(113.6\text{ ft-kips})(1.81\text{ in}^2)}{122.6\text{ ft-kips}} \\ &= 1.68\text{ in}^2 \end{aligned}$$

Use three no. 7 bars ($A_s = 1.8\text{ in}^2$).

2. Use either **ACI 318 Apps. B and C** or **ACI 318 Ch. 9 and Unified Design Method** (see p. 51-8) to solve this problem.

Solution Using ACI 318 Apps. B and C

(a) Refer to Fig. 51.4.

$$A = 0;\ B = 12\text{ in};\ 4t = (4)(8\text{ in}) = 32\text{ in}$$
$$12\text{ in} < 32\text{ in}$$

For the computation of C, the beam section is treated as the sum of two rectangles. The division that leads to the largest C is shown.

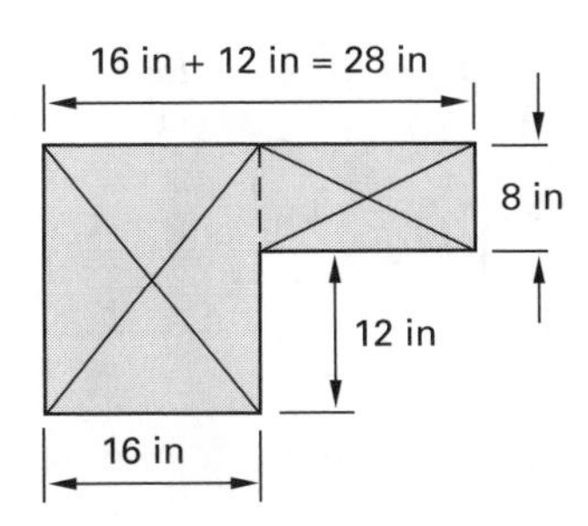

Use Eq. 51.6.

$$C = \sum \left(1 - 0.63\left(\frac{x}{y}\right)\right)\left(\frac{x^3 y}{3}\right)$$
$$= \left(1 - (0.63)\left(\frac{16 \text{ in}}{20 \text{ in}}\right)\right)\left(\frac{(16 \text{ in})^3(20 \text{ in})}{3}\right) + \left(1 - (0.63)\left(\frac{8 \text{ in}}{12 \text{ in}}\right)\right)\left(\frac{(8 \text{ in})^3(12 \text{ in})}{3}\right)$$
$$= \boxed{14{,}732 \text{ in}^4}$$

(b) From Table 51.5, without drop panels or exterior panels, with an edge beam,

$$h \geq \frac{l_n}{36} = \frac{(20 \text{ ft})\left(12 \ \frac{\text{in}}{\text{ft}}\right) - (2 \text{ ft})\left(12 \ \frac{\text{in}}{\text{ft}}\right)}{36}$$
$$= 6.0 \text{ in} \quad \left[< 8 \text{ in, so } \boxed{\text{adequate}}\right]$$

(c) The weight of the slab is

$$\left(\frac{8 \text{ in}}{12 \ \frac{\text{in}}{\text{ft}}}\right)\left(0.15 \ \frac{\text{kip}}{\text{ft}^3}\right) = 0.1 \text{ kip/ft}^2$$

The factored uniform load is

$$w_u = 1.4D + 1.7L$$
$$= (1.4)\left(0.1 \ \frac{\text{kip}}{\text{ft}^2} + 0.02 \ \frac{\text{kip}}{\text{ft}^2}\right) + (1.7)\left(0.05 \ \frac{\text{kip}}{\text{ft}^2}\right)$$
$$= 0.253 \text{ kip/ft}^2$$

The distances needed for the computation of M_o are

$$l_n = 20 \text{ ft} - 2 \text{ ft} = 18 \text{ ft} \quad \left[\begin{array}{c}\text{column face-to-column}\\ \text{face free distances}\end{array}\right]$$
$$l_2 = 18 \text{ ft} \quad [\text{square}]$$
$$M_o = \left(\frac{1}{8}\right)\left(0.253 \ \frac{\text{kip}}{\text{ft}^2}\right)(18 \text{ ft})(18 \text{ ft})^2$$
$$= 184.43 \text{ ft-kips}$$

From Table 51.3, the positive moment is

$$0.50M_o = (0.50)(184.43 \text{ ft-kips}) = 92.22 \text{ ft-kips}$$

From Table 51.4(c),

$$M_{\text{column strip}} = (0.60)(\text{positive moment}) = (0.60)(92.22 \text{ ft-kips}) = 55.33 \text{ ft-kips}$$

Estimate $d = t - \lambda = 8 \text{ in} - 1 \text{ in} = 7 \text{ in}$.

Assume the section is tension controlled and $\lambda = (0.1)(7 \text{ in}) = 0.7$ in. Use Eq. 50.28.

$$A_s = \frac{M_u}{\phi f_y (d - \lambda)}$$
$$= \frac{(55.33 \text{ ft-kips})\left(12 \ \frac{\text{in}}{\text{ft}}\right)}{(0.9)\left(40 \ \frac{\text{kips}}{\text{in}^2}\right)(7.0 \text{ in} - 0.7 \text{ in})}$$
$$= 2.93 \text{ in}^2$$

Use Eq. 50.26.

$$A_c = \frac{f_y A_s}{0.85 f_c'} = \frac{(2.93 \text{ in}^2)\left(40 \ \frac{\text{kips}}{\text{in}^2}\right)}{(0.85)\left(4 \ \frac{\text{kips}}{\text{in}^2}\right)}$$
$$= 34.44 \text{ in}^2$$

The width of a column strip is

$$(0.25)(18 \text{ ft})(2) = 9 \text{ ft}$$
$$a = \frac{34.44 \text{ in}^2}{(9 \text{ ft})\left(12 \ \frac{\text{in}}{\text{ft}}\right)} = 0.31 \text{ in}$$
$$\lambda = \frac{a}{2} = \frac{0.31 \text{ in}}{2} = 0.16 \text{ in}$$

Check to see if the section is tension controlled.

$$c = \frac{a}{\beta_1} = \frac{0.31 \text{ in}}{0.85} = 0.36 \text{ in}$$
$$\epsilon_t = \left(\frac{d - c}{c}\right)(0.003)$$
$$= \frac{7 \text{ in} - 0.36 \text{ in}}{0.36 \text{ in}}$$
$$= 0.055 > 0.005$$

The section is tension controlled.

Use Eq. 50.27.

$$\phi M_n = \frac{(0.9)(2.93)\left(40 \ \frac{\text{kips}}{\text{in}^2}\right)(7 \text{ in} - 0.16 \text{ in})}{12 \ \frac{\text{in}}{\text{ft}}}$$
$$= 60.13 \text{ ft-kips}$$

Structural

Adjust the area of steel.

$$\begin{aligned} A_s &= \frac{(55.33 \text{ ft-kips})(2.93 \text{ in}^2)}{60.13 \text{ ft-kips}} \\ &= 2.70 \text{ in}^2 \end{aligned}$$

$$\begin{aligned} A_{s,\min} &= (0.002)(9 \text{ ft})\left(12 \frac{\text{in}}{\text{ft}}\right)(8 \text{ in}) \\ &= 1.73 \text{ in}^2 \quad \text{[does not control]} \end{aligned}$$

Use fourteen no. 4 bars ($A_s = 2.80 \text{ in}^2$).

Solution Using ACI 318 Ch. 9 and Unified Design Method

(a) Refer to Fig. 51.4.

$$A = 0;\; B = 12 \text{ in};\; 4t = (4)(8 \text{ in}) = 32 \text{ in}$$
$$12 \text{ in} < 32 \text{ in}$$

For the computation of C, the beam section is treated as the sum of two rectangles. The division that leads to the largest C is shown.

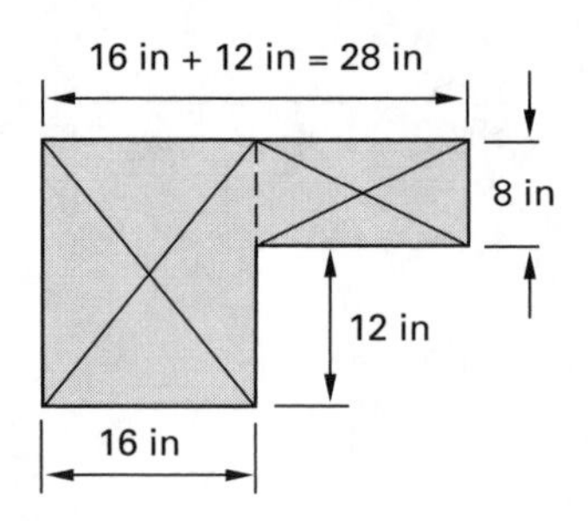

Use Eq. 51.6.

$$\begin{aligned} C &= \sum\left(1 - 0.63\left(\frac{x}{y}\right)\right)\left(\frac{x^3y}{3}\right) \\ &= \left(1 - (0.63)\left(\frac{16 \text{ in}}{20 \text{ in}}\right)\right)\left(\frac{(16 \text{ in})^3(20 \text{ in})}{3}\right) \\ &\quad + \left(1 - (0.63)\left(\frac{8 \text{ in}}{12 \text{ in}}\right)\right)\left(\frac{(8 \text{ in})^3(12 \text{ in})}{3}\right) \\ &= \boxed{14{,}732 \text{ in}^4} \end{aligned}$$

(b) From Table 51.5, without drop panels or exterior panels, with an edge beam,

$$\begin{aligned} h &\geq \frac{l_n}{36} \\ &= \frac{(20 \text{ ft})\left(12 \frac{\text{in}}{\text{ft}}\right) - (2 \text{ ft})\left(12 \frac{\text{in}}{\text{ft}}\right)}{36} \\ &= 6.0 \text{ in} \quad \left[< 8 \text{ in, so } \boxed{\text{adequate}}\right] \end{aligned}$$

(c) The weight of the slab is

$$\left(\frac{8 \text{ in}}{12 \frac{\text{in}}{\text{ft}}}\right)\left(0.15 \frac{\text{kip}}{\text{ft}^3}\right) = 0.1 \text{ kip/ft}^2$$

The controlling factored uniform load is

$$\begin{aligned} w_u &= 1.2D + 1.6L \\ &= (1.2)\left(0.1 \frac{\text{kip}}{\text{ft}^2} + 0.02 \frac{\text{kip}}{\text{ft}^2}\right) \\ &\quad + (1.6)\left(0.05 \frac{\text{kip}}{\text{ft}^2}\right) \\ &= 0.224 \text{ kip/ft}^2 \end{aligned}$$

The distances needed for the computation of M_o are

$$l_n = 20 \text{ ft} - 2 \text{ ft} = 18 \text{ ft} \quad \left[\begin{array}{c}\text{column face-to-column} \\ \text{face free distances}\end{array}\right]$$

$$l_2 = 18 \text{ ft} \quad \text{[square]}$$

$$\begin{aligned} M_o &= \left(\frac{1}{8}\right)\left(0.224 \frac{\text{kip}}{\text{ft}^2}\right)(18 \text{ ft})(18 \text{ ft})^2 \\ &= 163.29 \text{ ft-kips} \end{aligned}$$

From Table 51.3, the positive moment is

$$0.50M_o = (0.50)(163.29 \text{ ft-kips}) = 81.65 \text{ ft-kips}$$

From Table 51.4(c),

$$\begin{aligned} M_{\text{column strip}} &= (0.60)(\text{positive moment}) \\ &= (0.60)(81.65 \text{ ft-kips}) = 48.99 \text{ ft-kips} \end{aligned}$$

Estimate $d = 8 \text{ in} - 1 \text{ in} = 7 \text{ in}$.

Assume the section is tension controlled and $\lambda = (0.1)(7 \text{ in}) = 0.7 \text{ in}$. Use Eq. 50.28.

$$\begin{aligned} A_s &= \frac{M_u}{\phi f_y(d - \lambda)} \\ &= \frac{(48.99 \text{ ft-kips})\left(12 \frac{\text{in}}{\text{ft}}\right)}{(0.9)\left(40 \frac{\text{kips}}{\text{in}^2}\right)(7.0 \text{ in} - 0.7 \text{ in})} \\ &= 2.59 \text{ in}^2 \end{aligned}$$

Use Eq. 50.26.

$$\begin{aligned} A_c &= \frac{f_y A_s}{0.85 f_c'} = \frac{(2.59 \text{ in}^2)\left(40 \frac{\text{kips}}{\text{in}^2}\right)}{(0.85)\left(4 \frac{\text{kips}}{\text{in}^2}\right)} \\ &= 30.44 \text{ in}^2 \end{aligned}$$

The width of a column strip is

$$(0.25)(18\text{ ft})(2) = 9\text{ ft}$$

$$a = \frac{30.44\text{ in}^2}{(9\text{ ft})\left(12\ \frac{\text{in}}{\text{ft}}\right)} = 0.27\text{ in}$$

$$\lambda = \frac{a}{2} = \frac{0.27\text{ in}}{2} = 0.135\text{ in}$$

With the small value of a, the section is going to be tension controlled.

Use Eq. 50.27.

$$\phi M_n = \frac{(0.9)(2.59)\left(40\ \frac{\text{kips}}{\text{in}^2}\right)(7\text{ in} - 0.135\text{ in})}{12\ \frac{\text{in}}{\text{ft}}} = 53.34\text{ ft-kips}$$

Adjust the area of steel.

$$A_s = \frac{(48.99\text{ ft-kips})(2.59\text{ in}^2)}{53.34\text{ ft-kips}} = 2.38\text{ in}^2$$

$$A_{s,\min} = (0.002)(9\text{ ft})\left(12\ \frac{\text{in}}{\text{ft}}\right)(8\text{ in}) = 1.73\text{ in}^2 \quad [\text{does not control}]$$

Use 12 no. 4 bars ($A_s = 2.4\text{ in}^2$).

3. Use either **ACI 318 Apps. B and C** or **ACI 318 Ch. 9 and Unified Design Method** (see p. 51-11) to solve this problem.

Solution Using ACI 318 Apps. B and C

(a) From Table 51.1, since this is a one-way slab system (long-to-short span ratio greater than 2),

$$l_n = (11\text{ ft})\left(12\ \frac{\text{in}}{\text{ft}}\right) - 14\text{ in} = 118\text{ in}$$

$$t \geq \frac{l_n}{24} = \frac{118\text{ in}}{24} = 4.92\text{ in}$$

The answer is (A).

(b) The weight of the slab is

$$\frac{(6\text{ in})\left(150\ \frac{\text{lbf}}{\text{ft}^3}\right)}{12\ \frac{\text{in}}{\text{ft}}} = 75\text{ lbf/ft}^2$$

The factored uniform load is

$$\begin{aligned} w_u &= 1.4D + 1.7L \\ &= (1.4)\left(75\ \frac{\text{lbf}}{\text{ft}^2} + 20\ \frac{\text{lbf}}{\text{ft}^2}\right) + (1.7)\left(80\ \frac{\text{lbf}}{\text{ft}^2}\right) \\ &= 269\text{ lbf/ft}^2 \quad (0.269\text{ kip/ft}^2) \end{aligned}$$

The answer is (C).

(c) $l_n = (11\text{ ft})\left(12\ \frac{\text{in}}{\text{ft}}\right) - 14\text{ in} = 118\text{ in}$

Calculate the shear capacity per foot ($b_w = 12$ in) of slab. Approximate the slab depth as $d = h - 1$ in $= 6$ in $- 1$ in $= 5$ in. From Eq. 50.57,

$$\begin{aligned} V_c &= 2\sqrt{f'_c}b_w d \\ &= \frac{2\sqrt{3000\ \frac{\text{lbf}}{\text{in}^2}}(12\text{ in})(5\text{ in})}{1000\ \frac{\text{lbf}}{\text{kip}}} \\ &= 6.57\text{ kips} \end{aligned}$$

$$\phi V_c = (0.85)(6.57\text{ kips}) = 5.58\text{ kips}$$

Use Eq. 47.43 and equate the maximum shear from a load, w, to the capacity.

$$\frac{1.15 w l_n}{2} = 5.58\text{ kips}$$

$$w = 0.99\text{ kip/ft}^2$$

Subtract the total dead load weight and divide by the load factor of 1.7.

$$w_l = \frac{990\ \frac{\text{lbf}}{\text{ft}^2} - (1.4)\left(75 + 20\ \frac{\text{lbf}}{\text{ft}^2}\right)}{1.7} = 504\text{ lbf/ft}^2$$

The answer is (C).

(d) The temperature steel required for grade-40 reinforcement is

$$A_s = (0.002)(6\text{ in})(12\text{ in}) = 0.144\text{ in}^2/\text{ft}$$

The number of bars per ft is

$$\frac{0.144\ \frac{\text{in}^2}{\text{ft}}}{0.11\ \frac{\text{in}^2}{\text{bar}}} = 1.309$$

The spacing is

$$s = \frac{12\text{ in}}{1.309} = 9.16\text{ in}$$

The answer is (C).

Structural

(e) $l_n = \dfrac{118 \text{ in}}{12 \dfrac{\text{in}}{\text{ft}}} = 9.83 \text{ ft}$

From Eq. 47.42,

$$\frac{w_u l_n^2}{14} = \frac{\left(0.269 \dfrac{\text{kip}}{\text{ft}^2}\right)(9.83 \text{ ft})^2}{14} = \boxed{1.86 \text{ ft-kips/ft}}$$

The answer is (C).

(f) From Eq. 47.42,

$$\frac{w_u l_n^2}{10} = \frac{\left(0.269 \dfrac{\text{kip}}{\text{ft}^2}\right)(9.83 \text{ ft})^2}{10} = \boxed{2.60 \text{ ft-kips/ft}}$$

The answer is (B).

(g) $M_u = 2.60 \dfrac{\text{ft-kips}}{\text{ft}}$

Estimate $d = 6 \text{ in} - 1 \text{ in} = 5 \text{ in}$.

Assume $\lambda = (0.1)(5 \text{ in}) = 0.5 \text{ in}$ and that the section is tension controlled.

Use Eq. 50.28.

$$\begin{aligned} A_s &= \frac{M_u}{\phi f_y (d - \lambda)} \\ &= \frac{(2.60 \text{ ft-kips})\left(12 \dfrac{\text{in}}{\text{ft}}\right)}{(0.9)\left(40 \dfrac{\text{kips}}{\text{in}^2}\right)(5 \text{ in} - 0.5 \text{ in})} \\ &= 0.193 \text{ in}^2/\text{ft} \end{aligned}$$

(Assume λ is close enough.)

The number of bars per ft is

$$\frac{0.193 \dfrac{\text{in}^2}{\text{ft}}}{0.2 \text{ in}^2} = 0.963$$

$$s = \frac{12 \text{ in}}{0.963} = \boxed{12.46 \text{ in}}$$

The answer is (C).

(h) Scale the answer from the previous question.

$$s = \left(\frac{0.444 \text{ in}^2}{0.2 \text{ in}^2}\right)(12.46 \text{ in}) = 27.68 \text{ in}$$

The maximum spacing is

$$3t = (3)(6 \text{ in}) = \boxed{18 \text{ in} \quad [\text{controls}]}$$

The answer is (C).

(i) From Eq. 47.42,

$$\frac{w_u l_n^2}{16} = \frac{\left(0.269 \dfrac{\text{kips}}{\text{ft}^2}\right)(9.83 \text{ ft})^2}{16} = 1.63 \frac{\text{ft-kips}}{\text{ft}}$$

Use Eq. 50.28.

$$\begin{aligned} A_s &= \frac{M_u}{\phi f_y (d - \lambda)} \\ &= \frac{(1.63 \text{ ft-kips})\left(12 \dfrac{\text{in}}{\text{ft}}\right)}{(0.9)\left(40 \dfrac{\text{kips}}{\text{in}^2}\right)(5 \text{ in} - 0.5 \text{ in})} \\ &= 0.12 \text{ in}^2/\text{ft} \quad [\text{less than } A_{s,\text{min}}] \\ A_{s,\text{min}} &= (0.002)(6 \text{ in})\left(12 \frac{\text{in}}{\text{ft}}\right) = 0.144 \text{ in}^2/\text{ft} \end{aligned}$$

The number of bars per ft is

$$\frac{0.144 \dfrac{\text{in}^2}{\text{ft}}}{0.2 \dfrac{\text{in}^2}{\text{ft}}} = 0.72$$

$$s = \frac{12 \text{ in}}{0.72} = \boxed{16.67 \text{ in}} \quad [\text{less than } 3t\text{, so OK}]$$

The answer is (B).

(j) From Eq. 47.43,

$$\begin{aligned} (1.15)\left(\frac{w_u l}{2}\right) + \frac{w_u l}{2} &= \frac{(2.15)\left(0.269 \dfrac{\text{kip}}{\text{ft}^2}\right)(11 \text{ ft})}{2} \\ &= 3.18 \text{ kip/ft} \end{aligned}$$

The part of the weight of the beam not considered in the previous computation (including the 1.4 load factor) is

$$\frac{(1.4)(14 \text{ in})(16 \text{ in})\left(0.15 \dfrac{\text{kip}}{\text{ft}^3}\right)}{\left(12 \dfrac{\text{in}}{\text{ft}}\right)^2} = 0.33 \text{ kip/ft}$$

The total factored load is

$$3.18 \frac{\text{kips}}{\text{ft}} + 0.33 \frac{\text{kip}}{\text{ft}} = \boxed{3.51 \text{ kips/ft}}$$

The answer is (D).

Solution Using ACI 318 Ch. 9 and Unified Design Method

(a) From Table 51.1, since this is a one-way slab system (long-to-short span ratio greater than 2),

$$l_n = (11 \text{ ft})\left(12 \ \frac{\text{in}}{\text{ft}}\right) - 14 \text{ in} = 118 \text{ in}$$

$$t \geq \frac{l_n}{24} = \frac{118 \text{ in}}{24} = \boxed{4.92 \text{ in}}$$

The answer is (A).

(b) The weight of the slab is

$$\frac{(6 \text{ in})\left(150 \ \frac{\text{lbf}}{\text{ft}^3}\right)}{12 \ \frac{\text{in}}{\text{ft}}} = 75 \text{ lbf/ft}^2$$

The controlling factored uniform load is

$$\begin{aligned} w_u &= 1.2D + 1.6L \\ &= (1.2)\left(75 \ \frac{\text{lbf}}{\text{ft}^2} + 20 \ \frac{\text{lbf}}{\text{ft}^2}\right) + (1.6)\left(80 \ \frac{\text{lbf}}{\text{ft}^2}\right) \\ &= \boxed{242 \text{ lbf/ft}^2 \quad (0.242 \text{ kip/ft}^2)} \end{aligned}$$

The answer is (C).

(c) $l_n = (11 \text{ ft})\left(12 \ \frac{\text{in}}{\text{ft}}\right) - 14 \text{ in} = 118 \text{ in}$

Calculate the shear capacity per foot ($b_w = 12$ in) of slab. Approximate the slab depth as $d = h - 1$ in $= 6$ in $-$ 1 in $=$ 5 in. From Eq. 50.57,

$$\begin{aligned} V_c &= 2\sqrt{f'_c}b_w d \\ &= \frac{2\sqrt{3000 \ \frac{\text{lbf}}{\text{in}^2}}(12 \text{ in})(5 \text{ in})}{1000 \ \frac{\text{lbf}}{\text{kip}}} \\ &= 6.57 \text{ kips} \end{aligned}$$

$$\phi V_c = (0.75)(6.57 \text{ kips}) = 4.93 \text{ kips}$$

Use Eq. 47.43 and equate the maximum shear from a load, w, to the capacity.

$$\frac{1.15 w l_n}{2} = 4.93 \text{ kips}$$

$$w = 0.87 \text{ kip/ft}^2$$

Subtract the total dead load weight and divide by the load factor of 1.6.

$$w_l = \frac{870 \ \frac{\text{lbf}}{\text{ft}^2} - (1.2)\left(75 + 20 \ \frac{\text{lbf}}{\text{ft}^2}\right)}{1.6} = \boxed{473 \text{ lbf/ft}^2}$$

The answer is (B).

(d) The temperature steel required for grade-40 reinforcement is

$$A_s = (0.002)(6 \text{ in})(12 \text{ in}) = 0.144 \text{ in}^2/\text{ft}$$

The number of bars per ft is

$$\frac{0.144 \ \frac{\text{in}^2}{\text{ft}}}{0.11 \ \frac{\text{in}^2}{\text{bar}}} = 1.309$$

The spacing is

$$s = \frac{12 \text{ in}}{1.309} = \boxed{9.16 \text{ in}}$$

The answer is (C).

(e) $l_n = \dfrac{118 \text{ in}}{12 \ \frac{\text{in}}{\text{ft}}} = 9.83 \text{ ft}$

From Eq. 47.42,

$$\frac{w_u l_n^2}{14} = \frac{\left(0.242 \ \frac{\text{kip}}{\text{ft}^2}\right)(9.83 \text{ ft})^2}{14} = \boxed{1.67 \text{ ft-kips/ft}}$$

The answer is (B).

(f) From Eq. 47.42,

$$\frac{w_u l_n^2}{10} = \frac{\left(0.242 \ \frac{\text{kip}}{\text{ft}^2}\right)(9.83 \text{ ft})^2}{10} = \boxed{2.34 \text{ ft-kips/ft}}$$

The answer is (A).

(g) $M_u = 2.34 \ \dfrac{\text{ft-kips}}{\text{ft}}$

Estimate $d = 6$ in $-$ 1 in $=$ 5 in.

Assume $\lambda = (0.1)(5 \text{ in}) = 0.5$ in and that the section is tension controlled.

Use Eq. 50.28.

$$\begin{aligned} A_s &= \frac{M_u}{\phi f_y (d - \lambda)} \\ &= \frac{(2.34 \text{ ft-kips})\left(12 \ \frac{\text{in}}{\text{ft}}\right)}{(0.9)\left(40 \ \frac{\text{kips}}{\text{in}^2}\right)(5 \text{ in} - 0.5 \text{ in})} \\ &= 0.174 \text{ in}^2/\text{ft} \end{aligned}$$

(Assume λ is close enough, in which case the section will be tension controlled.)

Structural

The number of bars per ft is

$$\frac{0.174\ \frac{\text{in}^2}{\text{ft}}}{0.2\ \text{in}^2} = 0.87$$

$$s = \frac{12\ \text{in}}{0.87} = \boxed{13.79\ \text{in}}$$

The answer is (D).

(h) Scale the answer from the previous question.

$$s = \left(\frac{0.444\ \text{in}^2}{0.2\ \text{in}^2}\right)(13.79\ \text{in}) = 30.61\ \text{in}$$

The maximum spacing is

$$3t = (3)(6\ \text{in}) = \boxed{18\ \text{in}\quad [\text{controls}]}$$

The answer is (C).

(i) From Eq. 47.42,

$$\frac{w_u l_n^2}{16} = \frac{\left(0.242\ \frac{\text{kips}}{\text{ft}^2}\right)(9.83\ \text{ft})^2}{16} = 1.47\ \frac{\text{ft-kips}}{\text{ft}}$$

Use Eq. 50.28.

$$\begin{aligned} A_s &= \frac{M_u}{\phi f_y (d-\lambda)} \\ &= \frac{(1.47\ \text{ft-kips})\left(12\ \frac{\text{in}}{\text{ft}}\right)}{(0.9)\left(40\ \frac{\text{kips}}{\text{in}^2}\right)(5\ \text{in} - 0.5\ \text{in})} \\ &= 0.108\ \text{in}^2/\text{ft}\quad [\text{less than } A_{s,\text{min}}] \end{aligned}$$

$$A_{s,\text{min}} = (0.002)(6\ \text{in})\left(12\ \frac{\text{in}}{\text{ft}}\right) = 0.144\ \text{in}^2/\text{ft}$$

The number of bars per ft is

$$\frac{0.144\ \frac{\text{in}^2}{\text{ft}}}{0.2\ \frac{\text{in}^2}{\text{ft}}} = 0.72$$

$$s = \frac{12\ \text{in}}{0.72} = \boxed{16.67\ \text{in}}\quad [\text{less than } 3t, \text{ so OK}]$$

The answer is (B).

(j) From Eq. 47.43,

$$(1.15)\left(\frac{w_u l}{2}\right) + \frac{w_u l}{2} = \frac{(2.15)\left(0.242\ \frac{\text{kip}}{\text{ft}^2}\right)(11\ \text{ft})}{2} = 2.86\ \text{kip/ft}$$

The part of the weight of the beam not considered in the previous computation (including the 1.4 load factor) is

$$\frac{(1.2)(14\ \text{in})(16\ \text{in})\left(0.15\ \frac{\text{kip}}{\text{ft}^3}\right)}{\left(12\ \frac{\text{in}}{\text{ft}}\right)^2} = 0.28\ \text{kip/ft}$$

The total factored load is

$$2.86\ \frac{\text{kips}}{\text{ft}} + 0.28\ \frac{\text{kip}}{\text{ft}} = \boxed{3.14\ \text{kips/ft}}$$

The answer is (C).

4. Use either **ACI 318 Apps. B and C** or **ACI 318 Ch. 9 and Unified Design Method** (see p. 51-13) to solve this problem.

Solution Using ACI 318 Apps. B and C

(a) For the computation of C, the beam section is treated as the sum of two rectangles. The division that leads to the largest C is shown.

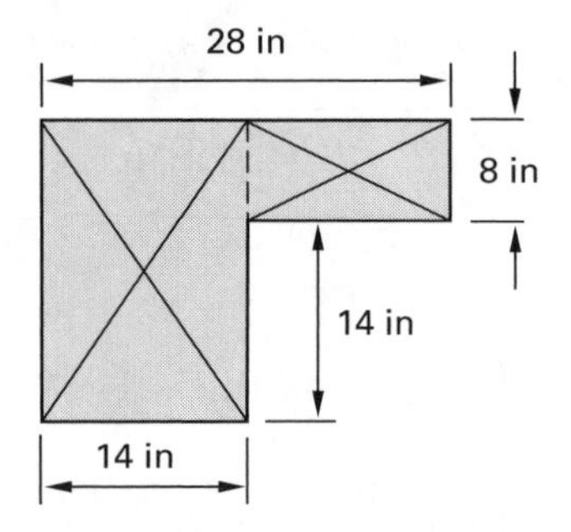

Use Eq. 51.6.

$$\begin{aligned} C &= \sum\left(1 - 0.63\left(\frac{x}{y}\right)\right)\left(\frac{x^3 y}{3}\right) \\ &= \left(1 - (0.63)\left(\frac{8\ \text{in}}{14\ \text{in}}\right)\right)\left(\frac{(8\ \text{in})^3(14\ \text{in})}{3}\right) \\ &\quad + \left(1 - (0.63)\left(\frac{14\ \text{in}}{22\ \text{in}}\right)\right)\left(\frac{(14\ \text{in})^3(22\ \text{in})}{3}\right) \\ &= \boxed{13{,}584\ \text{in}^4} \end{aligned}$$

The answer is (C).

(b) $l_n = 22\ \text{ft} - 2\ \text{ft} = 20\ \text{ft}$

From Table 51.5,

$$t \geq \frac{l_n}{33} = \frac{(20\ \text{ft})\left(12\ \frac{\text{in}}{\text{ft}}\right)}{33} = \boxed{7.27\ \text{in}}$$

The answer is (B).

(c) From Table 51.5,

$$t \geq \frac{l_n}{30} = \frac{(20\ \text{ft})\left(12\ \frac{\text{in}}{\text{ft}}\right)}{30} = \boxed{8\ \text{in}}$$

The answer is (C).

(d)

$$I_s = \left(\tfrac{1}{12}\right)(18\ \text{ft})\left(12\ \frac{\text{in}}{\text{ft}}\right)(8\ \text{in})^3$$
$$= 9216\ \text{in}^4$$

Use Eq. 51.5.

$$\beta_t = \frac{E_{cb}C}{2E_{cs}I_s}$$
$$= \frac{13{,}584\ \text{in}^4}{(2)(9216\ \text{in}^4)} = \boxed{0.74}$$

The answer is (B).

(e) The slab weight is

$$\frac{(8\ \text{in})\left(0.15\ \frac{\text{kip}}{\text{ft}^3}\right)}{12\ \frac{\text{in}}{\text{ft}}} = 0.1\ \text{kip/ft}^2$$

The factored uniform load is

$$w_u = 1.4D + 1.7L$$
$$= (1.4)\left(0.1\ \frac{\text{kip}}{\text{ft}^2} + 0.02\ \frac{\text{kip}}{\text{ft}^2}\right) + (1.7)\left(0.05\ \frac{\text{kip}}{\text{ft}^2}\right)$$
$$= 0.253\ \text{kip/ft}^2$$

The factors needed for the computation of M_o are

$$l_n = 22\ \text{ft} - 2\ \text{ft} = 20\ \text{ft}$$
$$l_2 = 18\ \text{ft}$$

The positive moment is

$$M_o = \frac{\left(0.253\ \frac{\text{kips}}{\text{ft}}\right)(18\ \text{ft})(20\ \text{ft})^2}{8}$$
$$= 227.70\ \text{ft-kips}$$
$$0.35M_o = (0.35)(227.70\ \text{ft-kips})$$
$$= 79.70\ \text{ft-kips}$$

From Table 51.4(c),

$$\text{column strip} = (0.60)(\text{positive moment})$$
$$= (0.60)(79.70\ \text{ft-kips}) = \boxed{47.82\ \text{ft-kips}}$$

The answer is (B).

(f) This is an edge, so

$$l_2 = \frac{18\ \text{ft}}{2} = 9\ \text{ft}$$

From Table 51.4, since 60% of the moment is required, as in part (e), the answer is 50% of the result in part (e).

$$(0.50)(47.82\ \text{ft-kips}) = \boxed{23.91\ \text{ft-kips}}$$

The answer is (B).

(g) From part (e), the total positive moment is 79.70 ft-kips. Since the column strip takes 60%, the middle strip takes 40%. The middle strip's positive moment is

$$(0.4)(79.70\ \text{ft-kips}) = \boxed{31.88\ \text{ft-kips}}$$

The answer is (B).

(h) From Table 51.5,

$$t \geq \frac{l_n}{36} = \frac{(20\ \text{ft})\left(12\ \frac{\text{in}}{\text{ft}}\right)}{36}$$
$$= \boxed{6.67\ \text{in}}$$

The answer is (B).

(i) At the drop panel location,

$$\text{thickness} \geq 1.25t = (1.25)(6.67\ \text{in}) = \boxed{8.33\ \text{in}}$$

The answer is (C).

(j) The minimum dimension is

$$(2)\left(\frac{18\ \text{ft}}{6}\right) = \boxed{6\ \text{ft}}$$

The answer is (C).

Solution Using ACI 318 Ch. 9 and Unified Design Method

(a) For the computation of C, the beam section is treated as the sum of two rectangles. The division that leads to the largest C is shown.

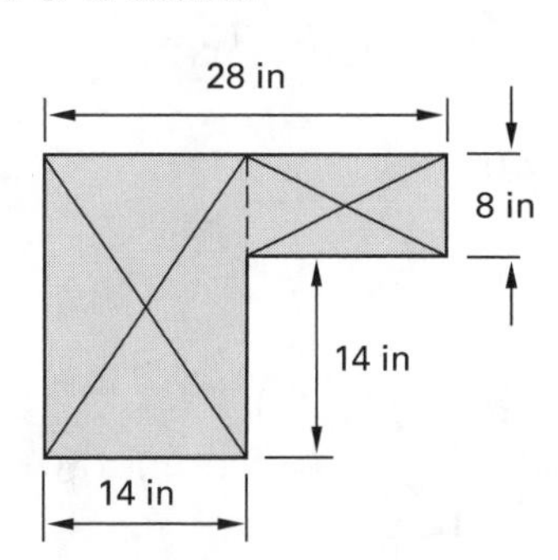

Structural

Use Eq. 51.6.

$$C = \sum \left(1 - 0.63\left(\frac{x}{y}\right)\right)\left(\frac{x^3y}{3}\right)$$
$$= \left(1 - (0.63)\left(\frac{8 \text{ in}}{14 \text{ in}}\right)\right)\left(\frac{(8 \text{ in})^3(14 \text{ in})}{3}\right) + \left(1 - (0.63)\left(\frac{14 \text{ in}}{22 \text{ in}}\right)\right)\left(\frac{(14 \text{ in})^3(22 \text{ in})}{3}\right)$$
$$= \boxed{13{,}584 \text{ in}^4}$$

The answer is (C).

(b) $l_n = 22 \text{ ft} - 2 \text{ ft} = 20 \text{ ft}$

From Table 51.5,

$$t \geq \frac{l_n}{33} = \frac{(20 \text{ ft})\left(12 \frac{\text{in}}{\text{ft}}\right)}{33} = \boxed{7.27 \text{ in}}$$

The answer is (B).

(c) From Table 51.5,

$$t \geq \frac{l_n}{30} = \frac{(20 \text{ ft})\left(12 \frac{\text{in}}{\text{ft}}\right)}{30} = \boxed{8 \text{ in}}$$

The answer is (C).

(d)

$$I_s = \left(\tfrac{1}{12}\right)(18 \text{ ft})\left(12 \frac{\text{in}}{\text{ft}}\right)(8 \text{ in})^3 = 9216 \text{ in}^4$$

Use Eq. 51.5.

$$\beta_t = \frac{E_{cb}C}{2E_{cs}I_s} = \frac{13{,}584 \text{ in}^4}{(2)(9216 \text{ in}^4)} = \boxed{0.74}$$

The answer is (B).

(e) The slab weight is

$$\frac{(8 \text{ in})\left(0.15 \frac{\text{kip}}{\text{ft}^3}\right)}{12 \frac{\text{in}}{\text{ft}}} = 0.1 \text{ kip/ft}^2$$

The factored uniform load is

$$w_u = 1.2D + 1.6L$$
$$= (1.2)\left(0.1 \frac{\text{kip}}{\text{ft}^2} + 0.02 \frac{\text{kip}}{\text{ft}^2}\right) + (1.6)\left(0.05 \frac{\text{kip}}{\text{ft}^2}\right)$$
$$= 0.224 \text{ kip/ft}^2$$

The factors needed for the computation of M_o are

$$l_n = 22 \text{ ft} - 2 \text{ ft} = 20 \text{ ft}$$
$$l_2 = 18 \text{ ft}$$

The positive moment is

$$M_o = \frac{\left(0.224 \frac{\text{kips}}{\text{ft}}\right)(18 \text{ ft})(20 \text{ ft})^2}{8} = 201.6 \text{ ft-kips}$$
$$0.35M_o = (0.35)(201.6 \text{ ft-kips}) = 70.56 \text{ ft-kips}$$

From Table 51.4(c),

$$\text{column strip} = (0.60)(\text{positive moment}) = (0.60)(70.56 \text{ ft-kips}) = \boxed{42.34 \text{ ft-kips}}$$

The answer is (A).

(f) This is an edge, so

$$l_2 = \frac{18 \text{ ft}}{2} = 9 \text{ ft}$$

From Table 51.4, since 60% of the moment is required, as in part (e), the answer is 50% of the result in part (e).

$$(0.50)(42.34 \text{ ft-kips}) = \boxed{21.17 \text{ ft-kips}}$$

The answer is (A).

(g) From part (e), the total positive moment is 70.56 ft-kips. Since the column strip takes 60%, the middle strip takes 40%. The middle strip's positive moment is

$$(0.4)(70.56 \text{ ft-kips}) = \boxed{28.22 \text{ ft-kips}}$$

The answer is (A).

(h) From Table 51.5,

$$t \geq \frac{l_n}{36} = \frac{(20 \text{ ft})\left(12 \frac{\text{in}}{\text{ft}}\right)}{36} = \boxed{6.67 \text{ in}}$$

The answer is (B).

(i) At the drop panel location,

$$\text{thickness} \geq 1.25t = (1.25)(6.67\text{ in}) = \boxed{8.33\text{ in}}$$

The answer is (C).

(j) The minimum dimension is

$$(2)\left(\frac{18\text{ ft}}{6}\right) = \boxed{6\text{ ft}}$$

The answer is (C).

Structural

52 Reinforced Concrete: Short Columns

PRACTICE PROBLEMS

1. (*Time limit: one hour*) A short column in a building is subjected to a factored axial load of 600 kips. Moments are negligible. Material properties are $f'_c = 4000$ lbf/in^2 and $f_y = 60{,}000$ lbf/in^2. Design the column as a square tied column using a reinforcement ratio of 4%.

2. A short column is subjected to an axial load of $P_u = 650$ kips and a moment $M_u = 250$ ft-kips. Material properties are $f'_c = 4000$ lbf/in^2 and $f_y = 60{,}000$ lbf/in^2. Design a tied square column with approximately 3% steel. Use steel distributed in all four faces.

3. (*Time limit: one hour*) A short column is subjected to a factored axial load of $P_u = 850$ kips. Material properties are $f'_c = 4000$ lbf/in^2 and $f_y = 60{,}000$ lbf/in^2.

(a) Assume moments are negligible. If the column is designed as a tied column, the minimum gross cross-sectional area is most nearly

(A) 170 in^2
(B) 190 in^2
(C) 210 in^2
(D) 220 in^2

(b) Assume moments are negligible. If the column is designed as a spiral column, the minimum gross cross-sectional area is most nearly

(A) 150 in^2
(B) 170 in^2
(C) 180 in^2
(D) 190 in^2

(c) If the column is an 18 in by 18 in square tied column, the required area of steel is most nearly

(A) 6.0 in^2
(B) 7.5 in^2
(C) 8.5 in^2
(D) 9.0 in^2

(d) If the column is an 18 in by 18 in square tied column, the maximum moment that can act without affecting the design (for pure axial loading) is most nearly

(A) 50 ft-kips
(B) 130 ft-kips
(C) 200 ft-kips
(D) 250 ft-kips

(e) If the column is an 18 in by 18 in square tied column and no. 9 bars are used for the longitudinal reinforcement, the maximum spacing of no. 3 ties is most nearly

(A) 14 in
(B) 16 in
(C) 18 in
(D) 20 in

(f) If the column is a spiral circular column with an outside diameter of 20 in, the required area of steel is most nearly

(A) 4.4 in^2
(B) 4.7 in^2
(C) 5.0 in^2
(D) 5.2 in^2

(g) If the column is a circular spiral column with an outside diameter of 20 in, the maximum moment that can act without affecting the design (for pure axial loading) is most nearly

(A) 50 ft-kips
(B) 60 ft-kips
(C) 70 ft-kips
(D) 140 ft-kips

(h) If the column is a spiral circular column with an outside diameter of 20 in and the diameter of the spiral wire is $^3/_8$ in, the maximum spiral pitch is most nearly

(A) 1.8 in
(B) 2.3 in
(C) 2.5 in
(D) 3.0 in

(i) When minimum reinforcement is used, the gross cross-sectional area for a tied column is most nearly

(A) 330 in^2
(B) 350 in^2
(C) 380 in^2
(D) 410 in^2

(j) When minimum reinforcement is used, the gross cross-sectional area for a spiral column is most nearly

(A) 310 in^2
(B) 340 in^2
(C) 360 in^2
(D) 410 in^2

4. (*Time limit: one hour*) A 20 in by 20 in tied column is built with $f'_c = 4000$ lbf/in^2 and $f_y = 60{,}000$ lbf/in^2.

(a) The maximum axial load that can be designed for is most nearly

(A) 1600 kips
(B) 1800 kips
(C) 2000 kips
(D) 2100 kips

(b) Assume moments are negligible. For $P_u = 1000$ kips, the required steel area is most nearly

(A) 5.0 in^2
(B) 6.5 in^2
(C) 7.5 in^2
(D) 8.5 in^2

(c) For $P_u = 800$ kips, the largest moment that does not affect the design is most nearly

(A) 65 ft-kips
(B) 130 ft-kips
(C) 200 ft-kips
(D) 260 ft-kips

(d) For $P_u = 800$ kips and $M_u = 400$ ft-kips, the required steel area is most nearly

(A) 10 in^2
(B) 14 in^2
(C) 18 in^2
(D) 22 in^2

(e) When the steel ratio equals 4%, the maximum moment that can be resisted when $P_u = 800$ kips is most nearly

(A) 300 ft-kips
(B) 380 ft-kips
(C) 480 ft-kips
(D) 500 ft-kips

(f) When the steel ratio is 4%, the nonzero axial load at which the moment capacity is largest is most nearly

(A) 300 kips
(B) 400 kips
(C) 500 kips
(D) 600 kips

(g) Assuming the column is unbraced and the effective length factor is 1.5, the maximum clear height for which the column is short is most nearly

(A) 7.5 ft
(B) 8 ft
(C) 9 ft
(D) 10 ft

(h) If the column is reinforced with no. 11 bars, the appropriate transverse reinforcement is most nearly

(A) no. 3 ties at 20 in
(B) no. 4 ties at 20 in
(C) no. 3 ties at 18 in
(D) no. 4 ties at 24 in

(i) If the column is reinforced with eight no. 8 bars placed in four bundles of two bars at each corner, the appropriate transverse reinforcement is most nearly

(A) no. 3 ties at 20 in
(B) no. 4 ties at 16 in
(C) no. 3 ties at 18 in
(D) no. 4 ties at 24 in

(j) If the steel ratio is 3%, the maximum eccentricity for a load $P_u = 800$ kips is most nearly

(A) 3.0 in
(B) 4.6 in
(C) 5.5 in
(D) 6.5 in

5. Design a short spiral column to carry an axial dead load of 175 kips and an axial live load of 300 kips. Use $f'_c = 3000$ lbf/in^2 and $f_y = 40{,}000$ lbf/in^2.

6. Design a short square tied column to carry an axial dead load of 100 kips and an axial live load of 125 kips. Use $f'_c = 3500$ lbf/in^2 and $f_y = 40{,}000$ lbf/in^2. Use no. 5 bars or larger for the longitudinal reinforcement. Specify the ties.

SOLUTIONS

1. Use either **ACI 318 App. C** or **ACI 318 Ch. 9** to solve this problem.

Solution Using ACI 318 App. C

Use Eq. 52.11.

$$A_g = \frac{P_u}{\phi\beta(0.85f'_c(1-\rho_g)+\rho_g f_y)}$$

$$= \frac{600 \text{ kips}}{(0.70)(0.80)\left((0.85)\left(4 \ \frac{\text{kips}}{\text{in}^2}\right) \times (1-0.04) + (0.04)\left(60 \ \frac{\text{kips}}{\text{in}^2}\right)\right)}$$

$$= 189.2 \text{ in}^2$$

The cross-sectional dimension is

$$h = \sqrt{189.2 \text{ in}^2} = 13.75 \text{ in} \quad \boxed{(14 \text{ in})}$$

The steel area is

$$A_{st} = \rho_g A_g = (0.04)(189.2 \text{ in}^2)$$

$$= 7.57 \text{ in}^2 \quad \boxed{\text{(use eight no. 9 bars)}}$$

The ties can be no. 3 since the bars are not larger than no. 10 and there are no bundles. The spacing of the ties is the smallest of (a) 16 diameters of the long bars, (16)(9/8 in) = 18 in; (b) 48 diameters of the tie, (48)(3/8 in) = 18 in; and (c) the minimum dimension of the cross section, **14 in, which controls.**

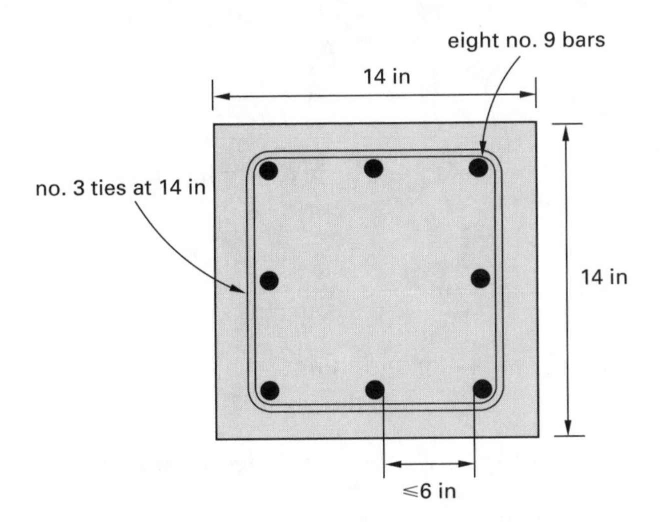

Solution Using ACI 318 Ch. 9

Use Eq. 52.11.

$$A_g = \frac{P_u}{\phi\beta(0.85f'_c(1-\rho_g)+\rho_g f_y)}$$

$$= \frac{600 \text{ kips}}{(0.65)(0.80)\left((0.85)\left(4 \ \frac{\text{kips}}{\text{in}^2}\right) \times (1-0.04) + (0.04)\left(60 \ \frac{\text{kips}}{\text{in}^2}\right)\right)}$$

$$= 203.7 \text{ in}^2$$

The cross-sectional dimension is

$$h = \sqrt{203.7 \text{ in}^2} = 14.27 \text{ in} \quad \boxed{(15 \text{ in})}$$

The steel area is

$$A_{st} = \rho_g A_g = (0.04)(203.8 \text{ in}^2)$$

$$= 8.15 \text{ in}^2$$ **(use eight no. 9 bars, which provide an area slightly smaller than required)**

The ties can be no. 3 since the bars are not larger than no. 10 and there are no bundles. The spacing of the ties is the smallest of (a) 16 diameters of the long bars, (16)(9/8 in) = 18 in; (b) 48 diameters of the tie, (48)(3/8 in) = 18 in; and (c) the minimum dimension of the cross section, **15 in, which controls.**

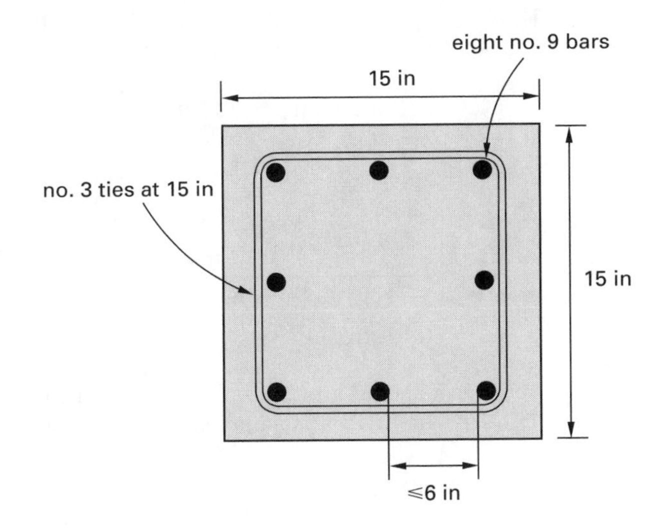

Structural

2. Use either **ACI 318 App. C** or **ACI 318 Ch. 9** to solve this problem.

Solution Using ACI 318 App. C

Select a trial value of $\rho_g = 0.02$, which is smaller than the target of 0.03. Use Eq. 52.12 to obtain an initial solution.

$$\begin{aligned} A_g &= \frac{P_u}{\phi\beta\big(0.85f'_c(1-\rho_g)+\rho_g f_y\big)} \\ &= \frac{650 \text{ kips}}{(0.70)(0.80)\left((0.85)\left(4\ \frac{\text{kips}}{\text{in}^2}\right) \times (1-0.02)+(0.02)\left(60\ \frac{\text{kips}}{\text{in}^2}\right)\right)} \\ &= 256.1 \text{ in}^2 \end{aligned}$$

The cross-sectional dimension is

$$h = \sqrt{256.1 \text{ in}^2} = 16 \text{ in}$$

Use an interaction diagram. First, calculate

$$\frac{P_u}{A_g} = \frac{650 \text{ kips}}{256 \text{ in}^2} = 2.53 \text{ kips/in}^2$$

$$\frac{M_u}{A_g h} = \frac{(250 \text{ ft-kips})\left(12\ \frac{\text{in}}{\text{ft}}\right)}{(256 \text{ in}^2)(16 \text{ in})} = 0.73 \text{ kip/in}^2$$

The value of γ can be estimated as

$$\frac{16 \text{ in} - 5 \text{ in}}{16 \text{ in}} = 0.68$$

Assume the steel will be distributed on all four faces. From the appropriate diagram in App. 52.A, per ACI 318 App. B, ρ_g is approximately 0.07. The column cross section must be increased. Try an 18 in by 18 in section.

$$A_g = (18 \text{ in})(18 \text{ in}) = 324 \text{ in}^2$$

$$\frac{P_u}{A_g} = \frac{650 \text{ kips}}{324 \text{ in}^2} = 2.01 \text{ kips/in}^2$$

$$\frac{M_u}{A_g h} = \frac{(250 \text{ ft-kips})\left(12\ \frac{\text{in}}{\text{ft}}\right)}{(324 \text{ in}^2)(18 \text{ in})} = 0.51 \text{ kip/in}^2$$

From the interaction diagram, a steel ratio of approximately 3.4% is needed. This is sufficiently close to the target of 3%.

Check that γ is sufficiently close to 0.75.

$$\gamma = \frac{18 \text{ in} - 5 \text{ in}}{18 \text{ in}} = 0.72 \quad \text{[OK]}$$

The area of steel is calculated from Eq. 52.12.

$$\begin{aligned} A_{st} &= \rho_g A_g = (0.034)(324 \text{ in}^2) \\ &= 11.01 \text{ in}^2 \quad \boxed{\text{(use twelve no. 9 bars)}} \end{aligned}$$

The ties can be no. 3 since the bars are not larger than no. 10 and there are no bundles. The spacing of the ties is the smallest of (a) 16 diameters of the long bars, (16)(9/8 in) = 18 in; (b) 48 diameters of the tie, (48)(3/8 in) = 18 in; and (c) the minimum dimension of the cross section, $\boxed{18 \text{ in.}}$

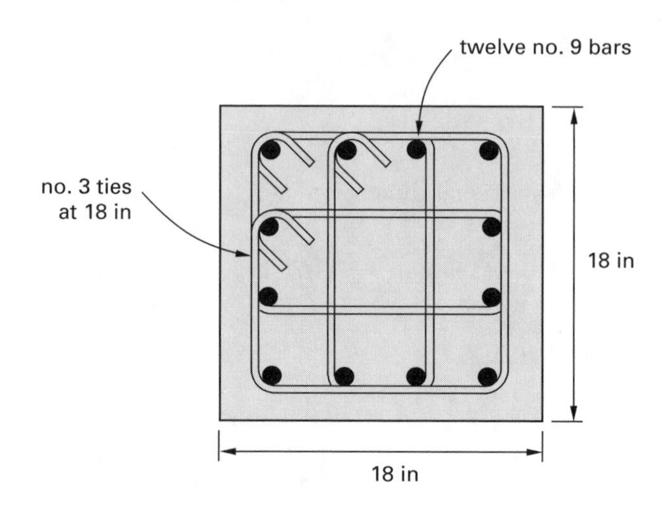

Solution Using ACI 318 Ch. 9

Start with a 16 in by 16 in section.

Use an interaction diagram. First, assuming $\phi = 0.65$, calculate

$$\frac{P_u}{\phi f'_c A_g} = \frac{650 \text{ kips}}{(0.65)\left(4\ \frac{\text{kips}}{\text{in}^2}\right)(256 \text{ in}^2)} = 0.98$$

$$\frac{M_u}{\phi f'_c A_g h} = \frac{250 \text{ ft-kips}}{(0.65)\left(4\ \frac{\text{kips}}{\text{in}^2}\right)(256 \text{ in}^2)(16 \text{ in})\left(\frac{1 \text{ ft}}{12 \text{ in}}\right)} = 0.28$$

The value of γ can be estimated as

$$\frac{16 \text{ in} - 5 \text{ in}}{16 \text{ in}} = 0.68$$

Assume the steel will be distributed on all four faces. From the appropriate interaction diagram in App. 52.A, ρ_g is approximately 0.08. The column cross section must be increased. Try a 20 in by 20 in section.

$$A_g = (20 \text{ in})(20 \text{ in}) = 400 \text{ in}^2$$

$$\frac{P_u}{\phi f'_c A_g} = \frac{650 \text{ kips}}{(0.65)\left(4 \frac{\text{kips}}{\text{in}^2}\right)(400 \text{ in}^2)} = 0.625$$

$$\frac{M_u}{\phi f'_c A_g h} = \frac{250 \text{ ft-kips}}{(0.65)\left(4 \frac{\text{kips}}{\text{in}^2}\right)(400 \text{ in}^2)(20 \text{ in})\left(\frac{1 \text{ ft}}{12 \text{ in}}\right)} = 0.144$$

$$\gamma = \frac{20 \text{ in} - 5 \text{ in}}{20 \text{ in}} = 0.75$$

From interaction diagrams in App. 52.A, a steel ratio of approximately 2.1% is needed. This ratio is sufficiently below the target of 3%, and therefore acceptable.

The area of steel is calculated from Eq. 52.12.

$$A_{st} = \rho_g A_g = (0.021)(400 \text{ in}^2) = 8.4 \text{ in}^2$$

$$\boxed{\text{(use twelve no. 8 bars)}}$$

With the chosen reinforcement,

$$\rho_g = \frac{nA_{\text{bar}}}{A_g} = \frac{(12)(0.79 \text{ in}^2)}{400 \text{ in}^2} = 0.237$$

From interaction diagrams in App. 52.A, the steel tensile strain, ϵ_t, corresponding to this column design is less than 0.002. Therefore, the section is compression controlled and $\phi = 0.65$, as assumed.

The ties can be no. 3 since the bars are not larger than no. 10 and there are no bundles. The spacing of the ties is the smallest of (a) 16 diameters of the long bars, which is $(16)(^8/_8 \text{ in}) = 16$ in; (b) 48 diameters of the tie, which is $(48)(^3/_8 \text{ in}) = 18$ in; and (c) the minimum dimension of the cross section, which is 20 in. Tie spacing of 16 in controls.

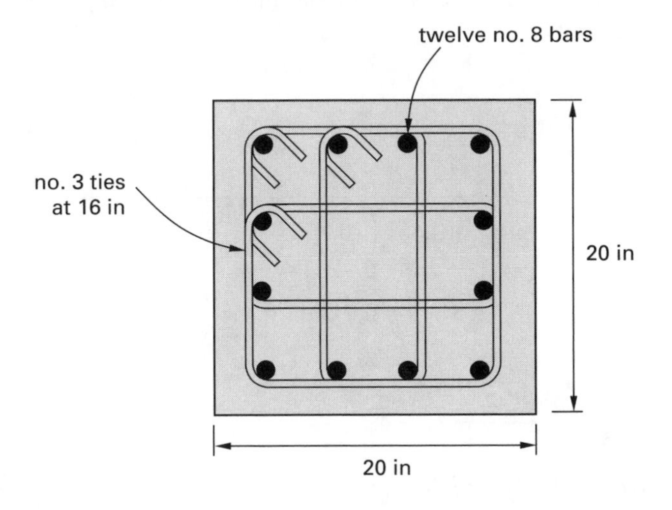

3. Use either **ACI 318 App. C** or **ACI 318 Ch. 9** (see p. 52-7) to solve this problem.

Solution Using ACI 318 App. C

(a) $\phi = 0.70$ for tied column. Taking ρ_g as the maximum allowed, 0.08, and using Eq. 52.11,

$$A_g = \frac{P_u}{\phi\beta(0.85 f'_c(1-\rho_g) + \rho_g f_y)}$$

$$= \frac{850 \text{ kips}}{(0.70)(0.80)\left((0.85)\left(4 \frac{\text{kips}}{\text{in}^2}\right) \times (1-0.08) + (0.08)\left(60 \frac{\text{kips}}{\text{in}^2}\right)\right)}$$

$$= \boxed{191.5 \text{ in}^2}$$

The answer is (B).

(b) Taking ρ_g as the maximum allowed, 0.08, and using Eq. 52.11,

$$A_g = \frac{P_u}{\phi\beta(0.85 f'_c(1-\rho_g) + \rho_g f_y)}$$

$$= \frac{850 \text{ kips}}{(0.75)(0.85)\left((0.85)\left(4 \frac{\text{kips}}{\text{in}^2}\right) \times (1-0.08) + (0.08)\left(60 \frac{\text{kips}}{\text{in}^2}\right)\right)}$$

$$= \boxed{168.2 \text{ in}^2}$$

The answer is (B).

(c) $A_g = (18 \text{ in})^2 = 324 \text{ in}^2$

Using Eq. 52.11,

$$0.85 f'_c(1-\rho_g) + \rho_g f_y = \frac{P_u}{\phi\beta A_g}$$

$$(0.85)\left(4 \frac{\text{kips}}{\text{in}^2}\right)(1-\rho_g) + \rho_g\left(60 \frac{\text{kips}}{\text{in}^2}\right) = \frac{850 \text{ kips}}{(0.70)(0.80)(324 \text{ in}^2)}$$

$$\rho_g = 0.0226$$

$$A_{st} = \rho_g A_g = (0.0226)(324 \text{ in}^2) = \boxed{7.32 \text{ in}^2}$$

The answer is (B).

Structural

(d)

$$\frac{e}{h} = 0.1$$
$$e = (0.1)(18\text{ in}) = 1.8\text{ in}$$
$$M = P_u e = \frac{(850\text{ kips})(1.8\text{ in})}{12\ \frac{\text{in}}{\text{ft}}}$$
$$= \boxed{127.5\text{ ft-kips}}$$

The answer is (B).

(e) Use the minimum of (a) 16 diameters of the long bars, (16)(9/8 in) = 18 in; (b) 48 diameters of the tie, (48)(3/8 in) = 18 in; and (c) the minimum dimension of the cross section, $\boxed{18\text{ in.}}$

The answer is (C).

(f)

$$A_g = \pi\left(\frac{D_g}{2}\right)^2 = \pi\left(\frac{20\text{ in}}{2}\right)^2$$
$$= 314\text{ in}^2$$

From Eq. 52.11,

$$0.85f'_c(1-\rho_g) + \rho_g f_y = \frac{P_u}{\phi\beta A_g}$$
$$(0.85)\left(4\ \frac{\text{kips}}{\text{in}^2}\right)(1-\rho_g) + \rho_g\left(60\ \frac{\text{kips}}{\text{in}^2}\right) = \frac{850\text{ kips}}{(0.75)(0.85)(314\text{ in}^2)}$$
$$\rho_g = 0.0149$$
$$A_{st} = \rho_g A_g$$
$$= (0.0149)(314\text{ in}^2)$$
$$= \boxed{4.68\text{ in}^2}$$

The answer is (B).

(g)

$$\frac{e}{h} = 0.05$$
$$e = (0.05)(20\text{ in}) = 1.0\text{ in}$$
$$M = P_u e = \frac{(850\text{ kips})(1.0\text{ in})}{12\ \frac{\text{in}}{\text{ft}}}$$
$$= \boxed{70.8\text{ ft-kips}}$$

The answer is (C).

(h)

$$D_c = 20\text{ in} - 3\text{ in} = 17\text{ in}$$
$$\frac{A_g}{A_c} = \left(\frac{D_g}{D_c}\right)^2 = \left(\frac{20\text{ in}}{17\text{ in}}\right)^2$$
$$= 1.384$$

Use Eq. 52.4.

$$\rho_s = 0.45\left(\frac{A_g}{A_c} - 1\right)\left(\frac{f'_c}{f_y}\right)$$
$$= (0.45)(1.384 - 1)\left(\frac{4\ \frac{\text{kips}}{\text{in}^2}}{60\ \frac{\text{kips}}{\text{in}^2}}\right)$$
$$= 0.0115$$

Use Eq. 52.5.

$$s = \frac{4A_{sp}}{\rho_s D_c} = \frac{(4)(0.11\text{ in}^2)}{(0.0115)(17\text{ in})}$$
$$= \boxed{2.25\text{ in}}$$

The answer is (B).

(i) Taking ρ_g as the minimum allowed, 0.01, and using Eq. 52.11,

$$A_g = \frac{P_u}{\phi\beta(0.85f'_c(1-\rho_g) + \rho_g f_y)}$$
$$= \frac{850\text{ kips}}{(0.70)(0.80)\left((0.85)\left(4\ \frac{\text{kips}}{\text{in}^2}\right) \times (1-0.01) + (0.01)\left(60\ \frac{\text{kips}}{\text{in}^2}\right)\right)}$$
$$= \boxed{382.7\text{ in}^2}$$

The answer is (C).

(j) Taking ρ_g as the minimum allowed, 0.01, and using Eq. 52.11,

$$A_g = \frac{P_u}{\phi\beta(0.85f'_c(1-\rho_g) + \rho_g f_y)}$$
$$= \frac{850\text{ kips}}{(0.75)(0.85)\left((0.85)\left(4\ \frac{\text{kips}}{\text{in}^2}\right) \times (1-0.01) + (0.01)\left(60\ \frac{\text{kips}}{\text{in}^2}\right)\right)}$$
$$= \boxed{336.2\text{ in}^2}$$

The answer is (B).

Solution Using ACI 318 Ch. 9

(a) $\phi = 0.65$ for tied column. Taking ρ_g as the maximum allowed, 0.08, and using Eq. 52.11,

$$\begin{aligned} A_g &= \frac{P_u}{\phi\beta\left(0.85f'_c(1-\rho_g)+\rho_g f_y\right)} \\ &= \frac{850 \text{ kips}}{(0.65)(0.80)\left((0.85)\left(4 \frac{\text{kips}}{\text{in}^2}\right) \times (1-0.08)+(0.08)\left(60 \frac{\text{kips}}{\text{in}^2}\right)\right)} \\ &= \boxed{206.2 \text{ in}^2} \end{aligned}$$

The answer is (C).

(b) Taking ρ_g as the maximum allowed, 0.08, and using Eq. 52.11,

$$\begin{aligned} A_g &= \frac{P_u}{\phi\beta\left(0.85f'_c(1-\rho_g)+\rho_g f_y\right)} \\ &= \frac{850 \text{ kips}}{(0.70)(0.85)\left((0.85)\left(4 \frac{\text{kips}}{\text{in}^2}\right) \times (1-0.08)+(0.08)\left(60 \frac{\text{kips}}{\text{in}^2}\right)\right)} \\ &= \boxed{180.2 \text{ in}^2} \end{aligned}$$

The answer is (C).

(c)

$$A_g = (18 \text{ in})^2 = 324 \text{ in}^2$$

Using Eq. 52.11,

$$0.85f'_c(1-\rho_g)+\rho_g f_y = \frac{P_u}{\phi\beta A_g}$$

$$(0.85)\left(4 \frac{\text{kips}}{\text{in}^2}\right)(1-\rho_g) + \rho_g\left(60 \frac{\text{kips}}{\text{in}^2}\right) = \frac{850 \text{ kips}}{(0.65)(0.80)(324 \text{ in}^2)}$$

$$\begin{aligned} \rho_g &= 0.0292 \\ A_{st} &= \rho_g A_g \\ &= (0.0292)(324 \text{ in}^2) \\ &= \boxed{9.46 \text{ in}^2} \end{aligned}$$

The answer is (D).

(d)

$$\begin{aligned} \frac{e}{h} &= 0.1 \\ e &= (0.1)(18 \text{ in}) = 1.8 \text{ in} \\ M = P_u e &= \frac{(850 \text{ kips})(1.8 \text{ in})}{12 \frac{\text{in}}{\text{ft}}} \\ &= \boxed{127.5 \text{ ft-kips}} \end{aligned}$$

The answer is (B).

(e) Use the minimum of (a) 16 diameters of the long bars, $(16)(^9/_8 \text{ in}) = 18$ in; (b) 48 diameters of the tie, $(48)(^3/_8 \text{ in}) = 18$ in; and (c) the minimum dimension of the cross section, $\boxed{18 \text{ in.}}$

The answer is (C).

(f) From Eq. 52.11,

$$0.85f'_c(1-\rho_g)+\rho_g f_y = \frac{P_u}{\phi\beta A_g}$$

$$(0.85)\left(4 \frac{\text{kips}}{\text{in}^2}\right)(1-\rho_g) + \rho_g\left(60 \frac{\text{kips}}{\text{in}^2}\right) = \frac{850 \text{ kips}}{(0.70)(0.85)(314 \text{ in}^2)}$$

$$\begin{aligned} \rho &= 0.0203 \\ A_{st} &= \rho_g A_g \\ &= (0.0203)(314 \text{ in}^2) \\ &= \boxed{6.37 \text{ in}^2} \end{aligned}$$

The answer is (D).

(g)

$$\begin{aligned} \frac{e}{h} &= 0.05 \\ e &= (0.05)(20 \text{ in}) = 1.0 \text{ in} \\ M = P_u e &= \frac{(850 \text{ kips})(1.0 \text{ in})}{12 \frac{\text{in}}{\text{ft}}} \\ &= \boxed{70.8 \text{ ft-kips}} \end{aligned}$$

The answer is (C).

(h)

$$\begin{aligned} D_c &= 20 \text{ in} - 3 \text{ in} = 17 \text{ in} \\ \frac{A_g}{A_c} &= \left(\frac{D_g}{D_c}\right)^2 = \left(\frac{20 \text{ in}}{17 \text{ in}}\right)^2 \\ &= 1.384 \end{aligned}$$

Use Eq. 52.4.

$$\begin{aligned} \rho_s &= 0.45\left(\frac{A_g}{A_c}-1\right)\left(\frac{f'_c}{f_y}\right) \\ &= (0.45)(1.384-1)\left(\frac{4 \frac{\text{kips}}{\text{in}^2}}{60 \frac{\text{kips}}{\text{in}^2}}\right) \\ &= 0.0115 \end{aligned}$$

Structural

Use Eq. 52.5.

$$s = \frac{4A_{sp}}{\rho_s D_c} = \frac{(4)(0.11 \text{ in}^2)}{(0.0115)(17 \text{ in})} = \boxed{2.25 \text{ in}}$$

The answer is (B).

(i) Taking ρ_g as the minimum allowed, 0.01, and using Eq. 52.11,

$$A_g = \frac{P_u}{\phi\beta(0.85f_c'(1-\rho_g)+\rho_g f_y)} = \frac{850 \text{ kips}}{(0.65)(0.80)\left((0.85)\left(4 \frac{\text{kips}}{\text{in}^2}\right) \times (1-0.01)+(0.01)\left(60 \frac{\text{kips}}{\text{in}^2}\right)\right)} = \boxed{412.1 \text{ in}^2}$$

The answer is (D).

(j) Taking ρ_g as the minimum allowed, 0.01, and using Eq. 52.11,

$$A_g = \frac{P_u}{\phi\beta(0.85f_c'(1-\rho_g)+\rho_g f_y)} = \frac{850 \text{ kips}}{(0.70)(0.85)\left((0.85)\left(4 \frac{\text{kips}}{\text{in}^2}\right) \times (1-0.01)+(0.01)\left(60 \frac{\text{kips}}{\text{in}^2}\right)\right)} = \boxed{360.2 \text{ in}^2}$$

The answer is (C).

4. Use either **ACI 318 App. C** or **ACI 318 Ch. 9** (see p. 52-9) to solve this problem.

Solution Using ACI 318 App. C

(a)

$$\begin{aligned} P_u &= \phi\beta A_g(0.85f_c'(1-\rho_g)+\rho_g f_y) \\ &= (0.7)(0.8)(400 \text{ in}^2)\left((0.85)\left(4 \frac{\text{kips}}{\text{in}^2}\right) \times (1-0.08)+\left(60 \frac{\text{kips}}{\text{in}^2}\right)(0.08)\right) \\ &= \boxed{1775.9 \text{ kips}} \end{aligned}$$

The answer is (B).

(b) Use Eq. 52.11.

$$0.85f_c'(1-\rho_g)+\rho_g f_y = \frac{P_u}{\phi\beta A_g}$$

$$(0.85)\left(4 \frac{\text{kips}}{\text{in}^2}\right)(1-\rho_g) + \rho_g\left(60 \frac{\text{kips}}{\text{in}^2}\right) = \frac{1000 \text{ kips}}{(0.70)(0.80)(400 \text{ in}^2)}$$

$$\rho = 0.0188$$

$$A_{st} = (0.0188)(400 \text{ in}^2) = \boxed{7.52 \text{ in}^2}$$

The answer is (C).

(c)

$$\frac{e}{h} = 0.1$$

$$e = (0.1)(20 \text{ in}) = 2.0 \text{ in}$$

$$M = P_u e = \frac{(800 \text{ kips})(2.0 \text{ in})}{12 \frac{\text{in}}{\text{ft}}} = \boxed{133.3 \text{ ft-kips}}$$

The answer is (B).

(d)

$$\frac{P_u}{A_g} = \frac{800 \text{ kips}}{400 \text{ in}^2} = 2.0 \frac{\text{kips}}{\text{in}^2}$$

$$\frac{M_u}{A_g h} = \frac{(400 \text{ ft-kips})\left(12 \frac{\text{in}}{\text{ft}}\right)}{(400 \text{ in}^2)(20 \text{ in})} = 0.60 \text{ kip/in}^2$$

$$\gamma = \frac{20 \text{ in} - 5 \text{ in}}{20 \text{ in}} = 0.75$$

Use the appropriate interaction diagram in App. 52.A. ρ_g is approximately 0.045.

The area of steel is given by Eq. 52.12.

$$A_{st} = \rho_g A_g = (0.045)(400 \text{ in}^2) = \boxed{18 \text{ in}^2}$$

The answer is (C).

(e) From the same interaction diagram,

$$\frac{M_u}{A_g h} = 0.57 \text{ kip/in}^2$$

$$M_u = \frac{\left(0.57 \frac{\text{kip}}{\text{in}^2}\right)(400 \text{ in}^2)(20 \text{ in})}{12 \frac{\text{in}}{\text{ft}}} = \boxed{380 \text{ ft-kips}}$$

The answer is (B).

(f) $P_u = \phi P_n$

Maximum moment occurs at the balanced point, where the steel fails when the concrete crushes. At this point, from the interaction diagram where $\rho_g = 0.04$,

$$\frac{P_u}{A_g} = 1.0 \text{ kip/in}^2$$

$$P_u = \left(1.0 \ \frac{\text{kip}}{\text{in}^2}\right)(20 \text{ in})(20 \text{ in}) = \boxed{400 \text{ kips}}$$

The answer is (B).

(g) Use Eq. 52.2.

$$\frac{k_u l_u}{r} = 22$$

$$r = 0.3h = (0.3)(20 \text{ in}) = 6 \text{ in} \quad \begin{bmatrix}\text{square tied}\\ \text{column}\end{bmatrix}$$

$$l_u = \frac{22r}{k_u} = \frac{(22)(6 \text{ in})}{(1.5)\left(12 \ \frac{\text{in}}{\text{ft}}\right)} = \boxed{7.3 \text{ ft}}$$

The answer is (A).

(h) Since longitudinal bars are larger than no. 10,

$\boxed{\text{no. 4 ties are required.}}$

The spacing is the minimum of (a) 16 diameters of the long bars, (16)(1.41 in) = 22.56 in; (b) 48 diameters of the tie, (48)(0.5 in) = 24 in; and (c) the minimum dimension of the cross section, $\boxed{\text{20 in, which controls.}}$

The answer is (B).

(i) Since the longitudinal bars are bundled,

$\boxed{\text{no. 4 ties are required.}}$

The spacing is the minimum of (a) 16 diameters of the long bars, (16)(1.0 in) = 16 in; (b) 48 diameters of the tie, (48)(0.5) = 24 in; and (c) the minimum dimension of the cross section, 20 in.

$$\boxed{16 \text{ in}}$$

The answer is (B).

(j) $\frac{P_u}{A_g} = 2.0 \text{ kips/in}^2$

From R4.60-75 at $\rho_g = 0.03$,

$$\frac{M_u}{A_g h} = 0.46 \text{ kip/in}^2$$

Therefore,

$$M_u = \left(0.46 \ \frac{\text{kip}}{\text{in}^2}\right)(400 \text{ in}^2)(20 \text{ in}) = 3680 \text{ in-kips}$$

$$\varepsilon = \frac{M_u}{P_u} = \frac{3680 \text{ in-kips}}{800 \text{ kips}} = \boxed{4.6 \text{ in}}$$

The answer is (B).

Solution Using ACI 318 Ch. 9

(a)

$$\begin{aligned} P_u &= \phi\beta A_g\left(0.85 f'_c(1-\rho_g) + \rho_g f_y\right) \\ &= (0.65)(0.8)(400 \text{ in}^2)\left((0.85)\left(4 \ \frac{\text{kips}}{\text{in}^2}\right) \times (1-0.08) + \left(60 \ \frac{\text{kips}}{\text{in}^2}\right)(0.08)\right) \\ &= \boxed{1649.0 \text{ kips}} \end{aligned}$$

The answer is (A).

(b) Use Eq. 52.11.

$$0.85 f'_c(1-\rho_g) + \rho_g f_y = \frac{P_u}{\phi\beta A_g}$$

$$(0.85)\left(4 \ \frac{\text{kips}}{\text{in}^2}\right)(1-\rho_g) + \rho_g\left(60 \ \frac{\text{kips}}{\text{in}^2}\right) = \frac{1000 \text{ kips}}{(0.65)(0.80)(400 \text{ in}^2)}$$

$$\rho = 0.0248$$

$$A_{st} = (0.0248)(400 \text{ in}^2) = \boxed{9.92 \text{ in}^2}$$

The answer is (D).

(c)

$$\frac{e}{h} = 0.1$$

$$e = (0.1)(20 \text{ in}) = 2.0 \text{ in}$$

$$M = P_u e = \frac{(800 \text{ kips})(2.0 \text{ in})}{12 \ \frac{\text{in}}{\text{ft}}} = \boxed{133.3 \text{ ft-kips}}$$

The answer is (B).

(d) Assume a compression-controlled section, $\phi = 0.65$.

$$\frac{P_u}{\phi f'_c A_g} = \frac{800 \text{ kips}}{(0.65)\left(4 \ \frac{\text{kips}}{\text{in}^2}\right)(400 \text{ in}^2)} = 0.77$$

$$\frac{M_u}{\phi f'_c A_g h} = \frac{(400 \text{ ft-kips})(12)}{(0.65)\left(4 \frac{\text{kips}}{\text{in}^2}\right)(400 \text{ in}^2)(20 \text{ in})\left(\frac{1 \text{ ft}}{12 \text{ in}}\right)} = 0.23$$

The value of γ can be estimated as

$$\frac{20 \text{ in} - 5 \text{ in}}{20 \text{ in}} = 0.75$$

From interaction diagrams in App. 52.A, the steel ratio is approximately 0.0051. The area of steel is calculated from Eq. 52.12.

$$A_{st} = \rho_g A_g = (0.051)(400 \text{ in}^2) = \boxed{20.4 \text{ in}^2}$$

With this high amount of reinforcement, the section is compression controlled.

The answer is (D).

(e) From interaction diagrams in App. 52.A, when $\rho_g = 4\%$ and $P_u = 800$ kips,

$$\frac{M_u}{\phi f'_c A_g h} = 0.192$$

$$M_u = \frac{(0.65)\left(4 \frac{\text{kips}}{\text{in}^2}\right)(400 \text{ in}^2)(20 \text{ in})(0.192)}{12 \frac{\text{in}}{\text{ft}}} = \boxed{333 \text{ ft-kips}}$$

From the interaction diagram, the section is clearly compression controlled.

The answer is (A).

(f) Maximum moment occurs at the balanced point. At this point where $\rho_g = 0.004$, from interaction diagrams in App. 52.A,

$$\frac{P_u}{\phi f'_c A_g} = 0.373$$

$$P_u = (0.65)\left(4 \frac{\text{kips}}{\text{in}^2}\right)(400 \text{ in}^2)(0.373) = \boxed{388 \text{ kips}}$$

Note that ϕ for a balanced section is very close to that of a compression-controlled section, so $\phi = 0.65$ is used.

The answer is (B).

(g) Use Eq. 52.2.

$$\frac{k_u l_u}{r} = 22$$

$$r = 0.3h = (0.3)(20 \text{ in}) = 6 \text{ in} \quad \left[\begin{array}{c}\text{square tied} \\ \text{column}\end{array}\right]$$

$$l_u = \frac{22r}{k_u} = \frac{(22)(6 \text{ in})}{(1.5)\left(12 \frac{\text{in}}{\text{ft}}\right)} = \boxed{7.3 \text{ ft}}$$

The answer is (A).

(h) Since longitudinal bars are larger than no. 10, no. 4 ties are required.

The spacing is the minimum of (a) 16 diameters of the long bars, (16)(1.41 in) = 22.56 in; (b) 48 diameters of the tie, (48)(0.5 in) = 24 in; and (c) the minimum dimension of the cross section, 20 in, which controls.

The answer is (B).

(i) Since the longitudinal bars are bundled, no. 4 ties are required.

The spacing is the minimum of (a) 16 diameters of the long bars, (16)(1.0 in) = 16 in; (b) 48 diameters of the tie, (48)(0.5) = 24 in; and (c) the minimum dimension of the cross section, 20 in.

$$\boxed{16 \text{ in}}$$

The answer is (B).

(j) Assume the section is compression controlled.

$$\frac{P_u}{\phi f'_c A_g} = \frac{800 \text{ kips}}{(0.65)\left(4 \frac{\text{kips}}{\text{in}^2}\right)(400 \text{ in}^2)} = 0.77$$

From interaction diagrams in App. 52.A, at $\rho_g = 0.03$,

$$\frac{M_u}{\phi f'_c A_g h} = 0.15$$

$$M_u = (0.15)(0.65)\left(4 \frac{\text{kips}}{\text{in}^2}\right)(400 \text{ in}^2)(20 \text{ in}) = 3120 \text{ in-kips}$$

$$e = \frac{M_u}{P_u} = \frac{3120 \text{ in-kips}}{800 \text{ kips}} = \boxed{3.9 \text{ in}}$$

The answer is (B).

5. Use either **ACI 318 App. C** or **ACI 318 Ch. 9** to solve this problem.

Solution Using ACI 318 App. C

$$\begin{aligned} P_u &= (1.4)(175{,}000 \text{ lbf}) + (1.7)(300{,}000 \text{ lbf}) \\ &= 755{,}000 \text{ lbf} \end{aligned}$$

Assume $\rho_g = 0.02$.

$$\begin{aligned} 755{,}000 \text{ lbf} &= A_g(0.85)(0.75) \\ &\quad \times \left((0.85)(1-0.02)\left(3000 \ \frac{\text{lbf}}{\text{in}^2}\right) \right. \\ &\quad \left. + \left(40{,}000 \ \frac{\text{lbf}}{\text{in}^2}\right)(0.02) \right) \\ A_g &= 359 \text{ in}^2 \end{aligned}$$

$$A_g = \frac{\pi}{4} D_g^2$$

$$D_g = \sqrt{\frac{(4)(359 \text{ in}^2)}{\pi}} = 21.38 \text{ in} \quad \boxed{(21.5 \text{ in})}$$

Use $1^1/_2$ in cover. Assuming no. 8 bars and no. 3 spiral wire, the steel diameter is

$$21.5 \text{ in} - (2)\left(1\tfrac{1}{2} \text{ in}\right) - (2)\left(\tfrac{3}{8} \text{ in}\right) - 1 \text{ in} = 16.75 \text{ in}$$

The core diameter is

$$21.5 \text{ in} - (2)\left(1\tfrac{1}{2} \text{ in}\right) = 18.5 \text{ in}$$

The required steel area is given by Eq. 52.12.

$$(0.02)(359 \text{ in}^2) = 7.18 \text{ in}^2$$

Some possibilities are

16 no. 6 bars
12 no. 7 bars
9 no. 8 bars
8 no. 9 bars
6 no. 10 bars

Try the 12 no. 7 bars. The steel circumference is

$$\pi(16.75 \text{ in}) = 52.62 \text{ in}$$

The clear spacing between bars is

$$\frac{52.62 \text{ in} - (12)(0.875 \text{ in})}{12 \text{ spaces}} = 3.51 \text{ in/space}$$

Since 3.51 in > (1.5)(0.875 in) = 1.31 in, this is acceptable.

Use Eq. 52.4.

$$\begin{aligned} \rho_s &= (0.45)\left(\left(\frac{21.5 \text{ in}}{18.5 \text{ in}}\right)^2 - 1 \right) \left(\frac{3000 \ \frac{\text{lbf}}{\text{in}^2}}{40{,}000 \ \frac{\text{lbf}}{\text{in}^2}} \right) \\ &= 0.0118 \end{aligned}$$

With a spiral pitch of 2 in,

$$\rho_s = \frac{(4)(0.11 \text{ in}^2)}{(2 \text{ in})(18.5 \text{ in})} = 0.0119 \text{ in} > 0.0118 \quad \text{[acceptable]}$$

The clear spacing between spirals is

$$2 \text{ in} - \tfrac{3}{8} \text{ in} = 1.625 \text{ in}$$

Since 1.0 in < 1.625 in < 3 in, this is acceptable.

$$\boxed{\text{Use twelve no. 7 bars.}}$$

Solution Using ACI 318 Ch. 9

$$\begin{aligned} P_u &= (1.2)(175{,}000 \text{ lbf}) + (1.6)(300{,}000 \text{ lbf}) \\ &= 690{,}000 \text{ lbf} \end{aligned}$$

Assume $\rho_g = 0.02$.

$$\begin{aligned} 690{,}000 \text{ lbf} &= A_g(0.85)(0.70) \\ &\quad \times \left((0.85)(1-0.02)\left(3000 \ \frac{\text{lbf}}{\text{in}^2}\right) \right. \\ &\quad \left. + \left(40{,}000 \ \frac{\text{lbf}}{\text{in}^2}\right)(0.02) \right) \\ A_g &= 352 \text{ in}^2 \end{aligned}$$

$$A_g = \frac{\pi}{4} D_g^2$$

$$D_g = \sqrt{\frac{(4)(352 \text{ in}^2)}{\pi}} = 21.17 \text{ in} \quad \boxed{(21.5 \text{ in})}$$

Use $1^1/_2$ in cover. Assuming no. 8 bars and no. 3 spiral wire, the steel diameter is

$$21.5 \text{ in} - (2)\left(1\tfrac{1}{2} \text{ in}\right) - (2)\left(\tfrac{3}{8} \text{ in}\right) - 1 \text{ in} = 16.75 \text{ in}$$

The core diameter is

$$21.5 \text{ in} - (2)\left(1\tfrac{1}{2} \text{ in}\right) = 18.5 \text{ in}$$

The required steel area is given by Eq. 52.12.

$$(0.02)(359 \text{ in}^2) = 7.18 \text{ in}^2$$

Some possibilities are

16 no. 6 bars
12 no. 7 bars
9 no. 8 bars
8 no. 9 bars
6 no. 10 bars

Try the 12 no. 7 bars. The steel circumference is

$$\pi(16.75 \text{ in}) = 52.62 \text{ in}$$

Structural

The clear spacing between bars is

$$\frac{52.62 \text{ in} - (12)(0.875 \text{ in})}{12 \text{ spaces}} = 3.51 \text{ in/space}$$

Since 3.51 in > (1.5)(0.875 in) = 1.31 in, this is acceptable.

Use Eq. 52.4.

$$\rho_s = (0.45)\left(\left(\frac{21.5 \text{ in}}{18.5 \text{ in}}\right)^2 - 1\right)\left(\frac{3000 \frac{\text{lbf}}{\text{in}^2}}{40{,}000 \frac{\text{lbf}}{\text{in}^2}}\right)$$
$$= 0.0118$$

With a spiral pitch of 2 in,

$$\rho_s = \frac{(4)(0.11 \text{ in}^2)}{(2 \text{ in})(18.5 \text{ in})} = 0.0119 \text{ in} > 0.0118 \quad \text{[acceptable]}$$

The clear spacing between spirals is

$$2 \text{ in} - \tfrac{3}{8} \text{ in} = 1.625 \text{ in}$$

Since 1.0 in < 1.625 in < 3 in, this is acceptable.

Use twelve no. 7 bars.

6. Use either **ACI 318 App. C** or **ACI 318 Ch. 9** to solve this problem.

Solution Using ACI 318 App. C

$$P_u = (1.4)(100{,}000 \text{ lbf}) + (1.7)(125{,}000 \text{ lbf})$$
$$= 352{,}500 \text{ lbf}$$

With $\phi = 0.70$ and $\rho_g = 0.02$,

$$352{,}500 \text{ lbf} = (0.80)(0.70)A_g \times \left((0.85)\left(3500 \frac{\text{lbf}}{\text{in}^2}\right)(1 - 0.02) + \left(40{,}000 \frac{\text{lbf}}{\text{in}^2}\right)(0.02)\right)$$
$$A_g = 169.4 \text{ in}^2$$

The column width is

$$\sqrt{169.4 \text{ in}^2} \approx \boxed{13 \text{ in square}}$$

The actual gross area is $A_g = (13 \text{ in})^2 = 169 \text{ in}^2$.

Assume no. 6 bars and no. 3 ties. With 1½ in cover, the steel core size is

$$13 \text{ in} - (2)\left(1\tfrac{1}{2} \text{ in}\right) - (2)\left(\tfrac{3}{8} \text{ in}\right) - (2)\left(\frac{0.75 \text{ in}}{2}\right) = 8.5 \text{ in}$$

The required steel area is

$$(0.02)(169.4 \text{ in}^2) = 3.39 \text{ in}^2$$

To distribute bars evenly around the column, the number of bars must be in multiples of four. Try 12 bars.

$$A_{\text{bar}} = \frac{3.39 \text{ in}^2}{12} = 0.28 \text{ in}^2$$

0.28 in² is too small since no. 5 bars or larger need to be used. Try eight bars.

$$A_{\text{bar}} = \frac{3.39 \text{ in}^2}{8} = 0.42 \text{ in}^2$$

Use eight no. 6 (0.44 in² each) bars.

The clear spacing between long bars is

$$\frac{(4)(8.5 \text{ in}) - (8)(0.750 \text{ in})}{8} = 3.5 \text{ in} \quad \text{[acceptable]}$$

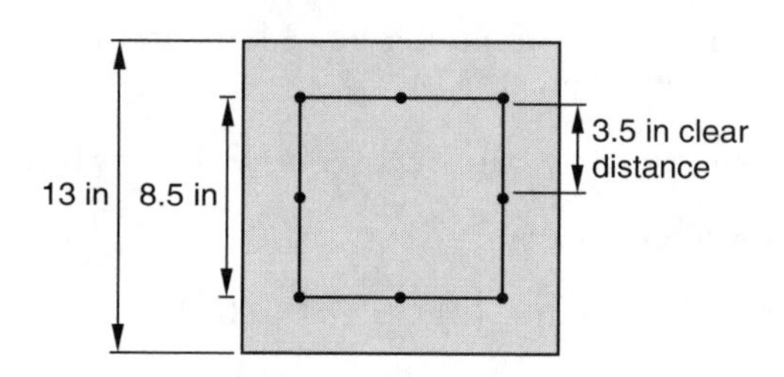

Use no. 3 ties. The tie spacing should be less than the minimum of

$$\left\{\begin{array}{l} (48)\left(\tfrac{3}{8} \text{ in}\right) = 18 \text{ in} \\ (16)\left(\tfrac{6}{8} \text{ in}\right) = 12 \text{ in} \\ \text{gross width} = 13 \text{ in} \end{array}\right\} = \boxed{12 \text{ in}}$$

With only eight bars, every alternate bar is supported. The spacing between bars is less than 6 in, so no additional cross ties are needed.

Solution Using ACI 318 Ch. 9

$$P_u = (1.2)(100{,}000 \text{ lbf}) + (1.6)(125{,}000 \text{ lbf})$$
$$= 320{,}000 \text{ lbf}$$

With $\phi = 0.65$ and $\rho_g = 0.02$,

$$320{,}000 \text{ lbf} = (0.80)(0.65)A_g \times \left((0.85)\left(3500 \frac{\text{lbf}}{\text{in}^2}\right)(1 - 0.02) + \left(40{,}000 \frac{\text{lbf}}{\text{in}^2}\right)(0.02)\right)$$
$$A_g = 165.6 \text{ in}^2$$

The column width is

$$\sqrt{165.6 \text{ in}^2} \approx \boxed{13 \text{ in square}}$$

The actual gross area is $A_g = (13 \text{ in})^2 = 169 \text{ in}^2$.

Assume no. 6 bars and no. 3 ties. With $1^1/_2$ in cover, the steel core size is

$$13 \text{ in} - (2)\left(1\tfrac{1}{2} \text{ in}\right) - (2)\left(\tfrac{3}{8} \text{ in}\right) - (2)\left(\frac{0.75 \text{ in}}{2}\right) = 8.5 \text{ in}$$

The required steel area is

$$(0.02)(169.4 \text{ in}^2) = 3.39 \text{ in}^2$$

To distribute bars evenly around the column, the number of bars must be in multiples of four. Try 12 bars.

$$A_{\text{bar}} = \frac{3.39 \text{ in}^2}{12} = 0.28 \text{ in}^2$$

0.28 in^2 is too small since no. 5 bars or larger need to be used. Try eight bars.

$$A_{\text{bar}} = \frac{3.39 \text{ in}^2}{8} = 0.42 \text{ in}^2$$

$\boxed{\text{Use eight no. 6 (0.44 in}^2 \text{ each) bars.}}$

The clear spacing between long bars is

$$\frac{(4)(8.5 \text{ in}) - (8)(0.750 \text{ in})}{8} = 3.5 \text{ in} \quad \text{[acceptable]}$$

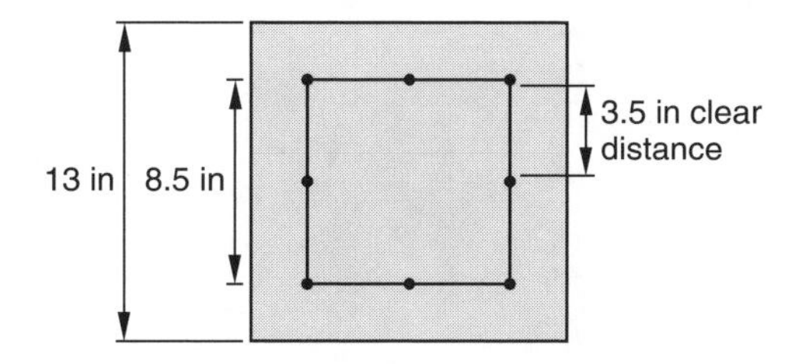

$\boxed{\text{Use no. 3 ties.}}$ The tie spacing should be less than the minimum of

$$\left\{\begin{array}{l} (48)\left(\tfrac{3}{8} \text{ in}\right) = 18 \text{ in} \\ (16)\left(\tfrac{6}{8} \text{ in}\right) = 12 \text{ in} \\ \text{gross width} = 13 \text{ in} \end{array}\right\} = \boxed{12 \text{ in}}$$

With only eight bars, every alternate bar is supported. The spacing between bars is less than 6 in, so no additional cross ties are needed.

Structural

53 Reinforced Concrete: Long Columns

PRACTICE PROBLEMS

1. The frame shown is one of several parallel frames in an unbraced building. The frames are spaced on 20 ft centers and are constructed using 4000 lbf/in^2 concrete. All columns are 18 in by 18 in. Assume a pin connection between columns and footings. All slabs are two-way. Compute the effective length factor k_u for the columns on the second story.

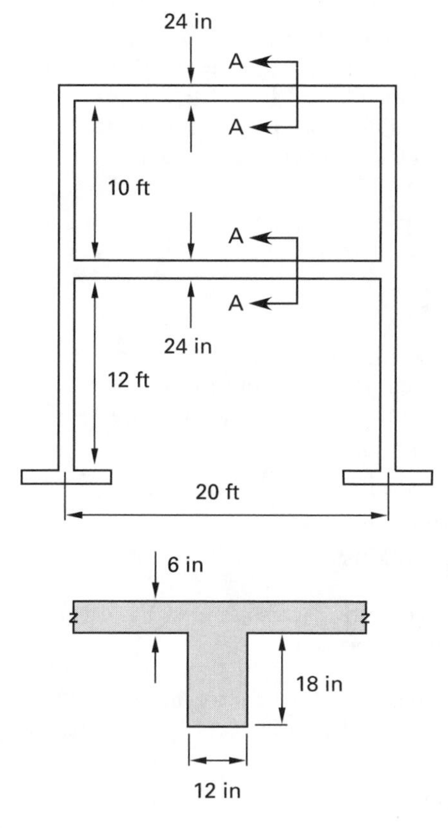

section A–A

(A) 1.1
(B) 1.2
(C) 1.5
(D) 1.8

2. The frame shown is part of a building that can be considered braced by the presence of stiff concrete walls surrounding the elevator shafts. The soil under the footings is soft, and a relative stiffness of $\Psi = 5$ is considered appropriate at the base. Other values of Ψ have been computed and are given in the illustration. Structural loadings for column B between points 0 and 1 are 400 kips dead load and 100 kips live load. The concrete modulus of elasticity, E_c, is 4000 kips/in^2. All columns are 16 in by 16 in square.

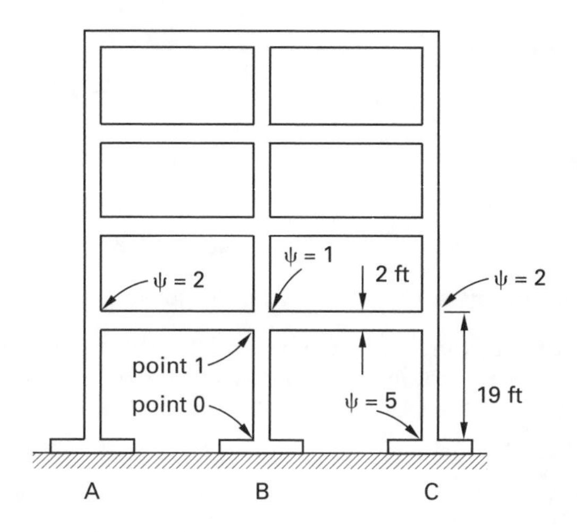

(a) The buckling load is most nearly
(A) 1100 kips
(B) 1600 kips
(C) 2200 kips
(D) 2800 kips

(b) The moment that should be used to design column B in the first story is most nearly
(A) 120 ft-kips
(B) 140 ft-kips
(C) 160 ft-kips
(D) 200 ft-kips

3. (*Time limit: one hour*) A six-floor (six-story) reinforced concrete building is supported on columns placed on a grid spaced 30 ft in the E-W direction and 20 ft in the N-S direction. The story height, measured from the top of one slab to the top of the slab above, is 14 ft. At the first level, the distance from the top of the footings to the top of the first story slab is 18 ft. The columns are 20 in by 20 in. Columns are connected in both directions by beams with an overall depth of 2 ft and a web width of 1 ft. The slab thickness is 6 in. All slabs are two way. 4000 lbf/in^2 concrete ($E = 4 \times 10^6$ psi) is used throughout.

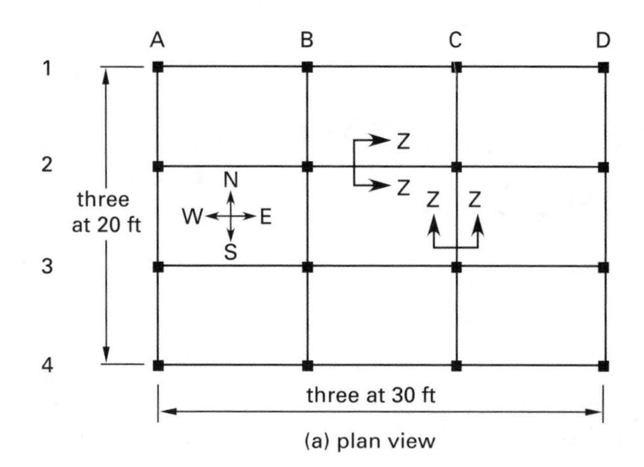

(a) plan view

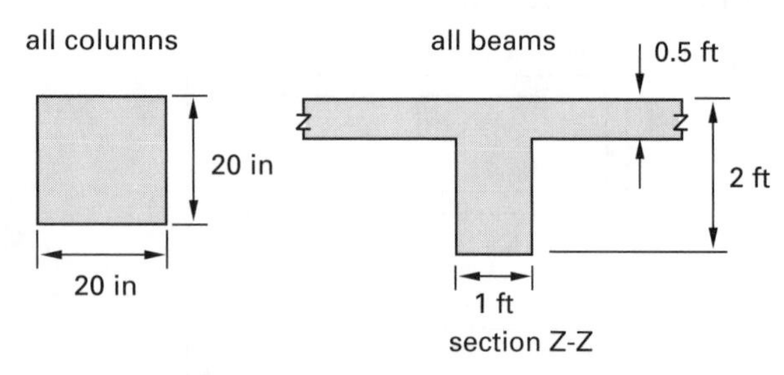

(b) column construction (c) beam construction

(a) For the computation of the relative stiffness parameter, Ψ, the moment of inertia of the beams is most nearly

(A) 4800 in^4
(B) 6100 in^4
(C) 8600 in^4
(D) 14,000 in^4

In parts (b) through (h), assume the building is unbraced.

(b) The effective length factor for bending about the W-E axis in column B2 at the third level is most nearly

(A) 0.85
(B) 1.0
(C) 1.5
(D) 2.0

(c) The effective length factor for bending about the W-E axis in column B1 at the third level is most nearly

(A) 0.9
(B) 1.0
(C) 1.5
(D) 1.9

(d) Assuming the footing provides absolute ideal fixity at the base and using $\beta_d = 0$, the effective length for bending about the W-E axis in column B2 at the first level is most nearly

(A) 14 ft
(B) 16 ft
(C) 19 ft
(D) 21 ft

(e) The buckling load about the W-E axis for column B2 in the first level is most nearly

(A) 2800 kips
(B) 4000 kips
(C) 4900 kips
(D) 5600 kips

(f) Assume the dead load is 0.1 kip/ft^2 and the live load is 0.06 kips/ft^2 on floors and the roof. Consider the load combinations $U = 0.75(1.4D + 1.7L + 1.7W)$ [ACI 318 App. C] and $U = 1.2D + 1.0L + 1.6W$ [ACI 318 Ch. 9]. Neglecting reductions in the live load, the $\sum P_u$ term used for the calculation of δ_s in the first level is most nearly

(A) 1100 kips
(B) 4500 kips
(C) 4900 kips
(D) 5900 kips

(g) Assume the unfactored wind load is 30 lbf/ft^2 and the load combinations are $U = 0.75(1.4D+1.7L+1.7W)$ [ACI 318 App. C] and $U = 1.2D + 1.0L + 1.6W$) [ACI 318 Ch. 9]. The first-story column drift is 0.3 in. The value of Q at this level is most nearly

(A) 0.024
(B) 0.032
(C) 0.15
(D) 0.20

(h) The maximum axial load in the columns of the first level for which the maximum moment can be safely assumed to occur at one of the two ends is most nearly

(A) 990 kips
(B) 1600 kips
(C) 1900 kips
(D) 2400 kips

For parts (i) and (j), assume walls are added so that the building is braced.

(i) The effective length factor for bending about W-E in column B2 at the first level is most nearly

(A) 0.64
(B) 0.87
(C) 1.0
(D) 1.2

(j) Assume the unfactored axial forces from dead and live loads on a particular column are 200 kips and 50 kips, respectively. The value of β_d for the load combinations $U = 1.4D + 1.7L$ [ACI 318 App. C] and $U = 1.2D + 1.6L$ [ACI 318 Ch. 9] is most nearly

(A) 0.50
(B) 0.77
(C) 0.80
(D) 1.0

4. (*Time limit: one hour*) The single-story frame shown is built with concrete having a modulus of elasticity, E_c, of 3605 kips/in^2. The actions in the columns due to dead and live load are as shown. Column ends A, F, and E are fixed, but a realistic finite flexibility is to be considered.

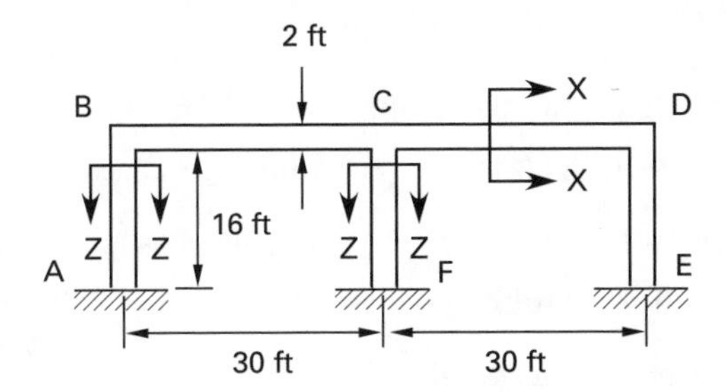

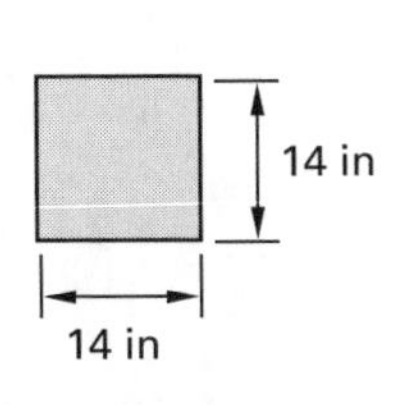

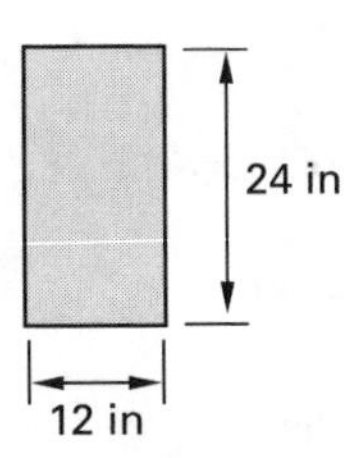

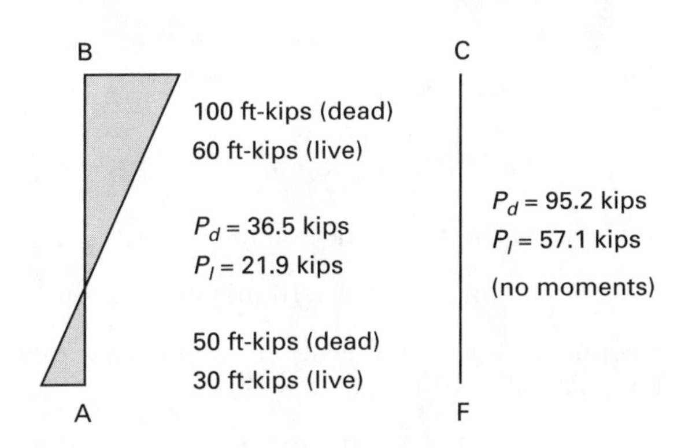

Assume the frame is part of a building that is braced. Assume the load combination is $U = 1.4D + 1.7L$ (ACI 318 App. C) and $U = 1.2D + 1.6L$ (ACI 318 Ch. 9).

(a) The moment $M_{2,ns}$ in column AB is most nearly

(A) 120 ft-kips
(B) 140 ft-kips
(C) 160 ft-kips
(D) 240 ft-kips

(b) The effective length for column AB is most nearly

(A) 9.0 ft
(B) 11 ft
(C) 12 ft
(D) 13 ft

(c) The effective length of column CF is most nearly

(A) 10 ft
(B) 11 ft
(C) 13 ft
(D) 15 ft

(d) The factor β_d that should be used to compute the buckling load for column AB is most nearly

(A) 0.00
(B) 0.45
(C) 0.58
(D) 0.65

(e) The buckling load for column AB is most nearly

(A) 1200 kips
(B) 1400 kips
(C) 1500 kips
(D) 1800 kips

(f) The parameter C_m for column AB (used to compute the amplification factor δ_{ns}) is most nearly

(A) 0.3
(B) 0.4
(C) 0.5
(D) 0.6

(g) The parameter C_m for column CF (used to compute the amplification factor δ_{ns}) is most nearly

(A) 0.4
(B) 0.5
(C) 0.6
(D) 1.0

(h) The buckling load for column CF is most nearly

(A) 1000 kips
(B) 1300 kips
(C) 1600 kips
(D) 2200 kips

(i) The amplification factor δ_{ns} in column CF is most nearly

(A) 0.9
(B) 1.0
(C) 1.2
(D) 1.5

(j) The amplification factor δ_{ns} in column AB is most nearly

(A) 0.44
(B) 1.0
(C) 1.1
(D) 1.3

SOLUTIONS

1. The solution and answer remain the same whether ACI 318 Apps. B and C are used or ACI 318 Unified Design Method and Ch. 9 are used.

Since 18 in < (4)(6 in), the effective beam section is as follows.

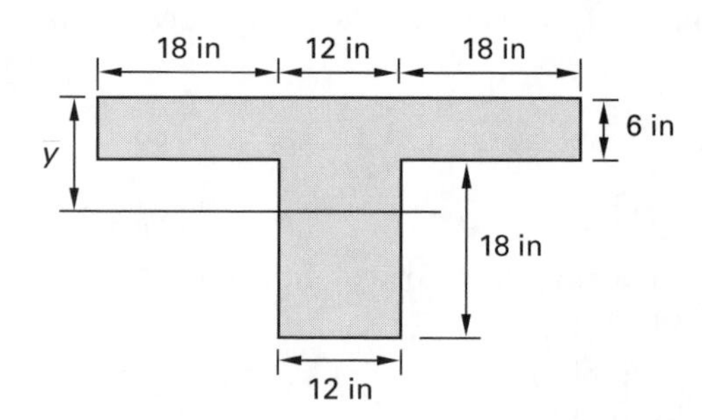

Compute I_b.

	A (in^2)	y (in)	Ay (in^3)	$A(y-\overline{y})^2$ (in^4)	I_0 (in^4)
	288	3	864	7617.3	864
	216	15	3240	10,156.4	5832
totals	504		4104	17,773.7	6696

$$\overline{y} = \frac{4104 \text{ in}^3}{504 \text{ in}^2} = 8.14 \text{ in}$$

$$I_b = 17{,}773.7 \text{ in}^4 + 6696 \text{ in}^4 = 24{,}470 \text{ in}^4$$

The column moment of inertia is

$$I_c = \left(\tfrac{1}{12}\right)(18 \text{ in})^4 = 8748 \text{ in}^4$$

Adjust for cracking and creep.

$$I_{be} = 0.35I_b = (0.35)(24{,}470 \text{ in}^4) = 8564.5 \text{ in}^4$$

$$I_{ce} = 0.70I_c = (0.70)(8748 \text{ in}^4) = 6123.6 \text{ in}^4$$

$$L_{c,1} = (12 \text{ ft}) + \frac{24 \text{ in}}{(2)\left(12 \ \frac{\text{in}}{\text{ft}}\right)} = 13 \text{ ft}$$

$$L_{c,2} = 10 \text{ ft} + (2)(1 \text{ ft}) = 12 \text{ ft}$$

Use Eq. 53.3.

$$\Psi_c = \frac{\dfrac{6123.6 \text{ in}^4}{13 \text{ ft}}}{\dfrac{8564.5 \text{ in}^4}{20 \text{ ft}}} = 1.19$$

$$\Psi_b = \frac{\dfrac{6123.6 \text{ in}^4}{12 \text{ ft}} + \dfrac{6123.6 \text{ in}^4}{13 \text{ ft}}}{\dfrac{8564.5 \text{ in}^4}{20 \text{ ft}}} = 2.29$$

From Fig. 53.3,

$$k = \boxed{1.5}$$

The answer is (C).

2. Use either **ACI 318 App. C** or **ACI 318 Ch. 9** (see p. 53-5) to solve this problem.

Solution Using ACI 318 App. C

(a) From Fig. 53.3 for $\Psi_1 = 1$ and $\Psi_2 = 5$, $k_b \approx 0.85$.

$$l_u = 19 \text{ ft} - 2 \text{ ft} = 17 \text{ ft} \quad [\text{clear distance}]$$

$$P_u = (1.4)(400 \text{ kips}) + (1.7)(100 \text{ kips}) = 730 \text{ kips}$$

There is no lateral loading. So,

$$M_1 = M_2 = 0$$

$$C_m = 0.6 + (0.4)\left(\frac{M_1}{M_2}\right) \quad [\text{Eq. 53.9}]$$

For $M_1 = M_2 = 0$, $C_m = 1$. Use Eq. 53.10.

$$M_{2,\text{min}} = P_u(0.6 + 0.03h) = \frac{(730 \text{ kips})\big(0.6 \text{ in} + (0.03)(16 \text{ in})\big)}{12 \ \frac{\text{in}}{\text{ft}}} = 65.7 \text{ ft-kips}$$

$$\beta_d = \frac{(1.4)(400 \text{ kips})}{730 \text{ kips}} = 0.767$$

The moment of inertia of the column is

$$I_g = \tfrac{1}{12}bh^3 = \left(\tfrac{1}{12}\right)(16 \text{ in})(16 \text{ in})^3 = 5461.3 \text{ in}^4$$

To account for the magnification of moments due to cracking [ACI 318 10.11.1],

$$I'_g = 0.70I_g = (0.70)(5461.3 \text{ in}^4) = 3822.9 \text{ in}^4$$

Use Eq. 53.5.

$$EI = \frac{0.4E_cI_g}{1+\beta_d} = \frac{(0.4)\left(4000 \ \frac{\text{kips}}{\text{in}^2}\right)(3822.9 \text{ in}^4)}{1 + 0.767} = 3.46 \times 10^6 \text{ kip-in}^2$$

The buckling load is given by Eq. 53.6.

$$P_c = \frac{\pi^2 EI}{(kl_u)^2} = \frac{\pi^2(3.46 \times 10^6 \text{ kip-in}^2)}{\left((0.85)(17 \text{ ft})\left(12 \ \frac{\text{in}}{\text{ft}}\right)\right)^2} = \boxed{1136 \text{ kips}}$$

The answer is (A).

(b) The amplifications factor is given by Eq. 53.8. Since $M_1 = M_2 = 0$, use $M_1/M_2 = 1$, and $C_m = 1$.

$$\delta_{ns} = \frac{C_m}{1 - \frac{P_u}{0.75P_c}} = \frac{1}{1 - \frac{730 \text{ kips}}{(0.75)(1622 \text{ kips})}} = 2.50$$

The design loads for column B are

$$P_u = 730 \text{ kips}$$
$$M_c = \delta_{ns}M_2 = (2.50)(65.70 \text{ ft-kips}) = \boxed{164.3 \text{ ft-kips}}$$

(The value of 2.50 for δ_{ns} is large and, as such, is indicative of a very flexible column. An increase in the section size to reduce δ_{ns} would be good practice in this case.)

The answer is (C).

Solution Using ACI 318 Ch. 9

(a) From Fig. 53.3 for $\Psi_1 = 1$ and $\Psi_2 = 5$, $k_b \approx 0.85$.

$$l_u = 19 \text{ ft} - 2 \text{ ft} = 17 \text{ ft} \quad \text{[clear distance]}$$
$$P_u = (1.2)(400 \text{ kips}) + (1.6)(100 \text{ kips}) = 640 \text{ kips}$$
$$P_u = 1.4D = 560 \text{ kips} \quad \text{[does not count]}$$

There is no lateral loading. So,

$$M_1 = M_2 = 0$$
$$C_m = 0.6 + (0.4)\left(\frac{M_1}{M_2}\right) \quad \text{[Eq. 53.9]}$$

For $M_1 = M_2 = 0$, $C_m = 1$. Use Eq. 53.10.

$$M_{2,\text{min}} = P_u(0.6 + 0.03h) = \frac{(640 \text{ kips})\big(0.6 \text{ in} + (0.03)(16 \text{ in})\big)}{12 \frac{\text{in}}{\text{ft}}} = 57.6 \text{ ft-kips}$$

$$\beta_d = \frac{(1.2)(400 \text{ kips})}{640 \text{ kips}} = 0.75$$

The moment of inertia of the column is

$$I_g = \tfrac{1}{12}bh^3 = \left(\tfrac{1}{12}\right)(16 \text{ in})(16 \text{ in})^3 = 5461.3 \text{ in}^4$$

Use Eq. 53.5.

$$EI = \frac{0.4E_cI_g}{1 + \beta_d} = \frac{(0.4)\left(4000 \frac{\text{kips}}{\text{in}^2}\right)(5461.3 \text{ in}^4)}{1 + 0.75} = 4.99 \times 10^6 \text{ kip-in}^2$$

The buckling load is given by Eq. 53.6.

$$P_c = \frac{\pi^2 EI}{(kl_u)^2} = \frac{\pi^2(4.99 \times 10^6 \text{ kip-in}^2)}{\left((0.85)(17 \text{ ft})\left(12 \frac{\text{in}}{\text{ft}}\right)\right)^2} = \boxed{1638 \text{ kips}}$$

The answer is (B).

(b) The amplifications factor is given by Eq. 53.8. Since $M_1 = M_2 = 0$, use $M_1/M_2 = 1$, and $C_m = 1$.

$$\delta_{ns} = \frac{C_m}{1 - \frac{P_u}{0.75P_c}} = \frac{1}{1 - \frac{640 \text{ kips}}{(0.75)(1638 \text{ kips})}} = 2.09$$

The design loads for column B are

$$P_u = 640 \text{ kips}$$
$$M_c = \delta_{ns}M_2 = (2.09)(57.6 \text{ ft-kips}) = \boxed{120.4 \text{ ft-kips}}$$

The answer is (A).

3. Use either **ACI 318 App. C** or **ACI 318 Ch. 9** (see p. 53-7) to solve this problem.

Solution Using ACI 318 App. C

(a) The effective beam section is

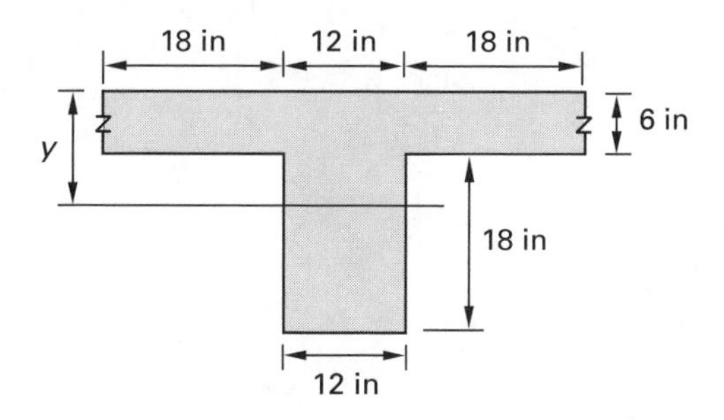

Compute I_b. $I_b = 24{,}469.7 \text{ in}^4$. For the computation of Ψ, the beam inertia is reduced by multiplying by 0.35.

$$I_{b,\text{effective}} = 0.35I_b = (0.35)(24{,}469.7 \text{ in}^4) = \boxed{8564.4 \text{ in}^4}$$

The answer is (C).

(b)

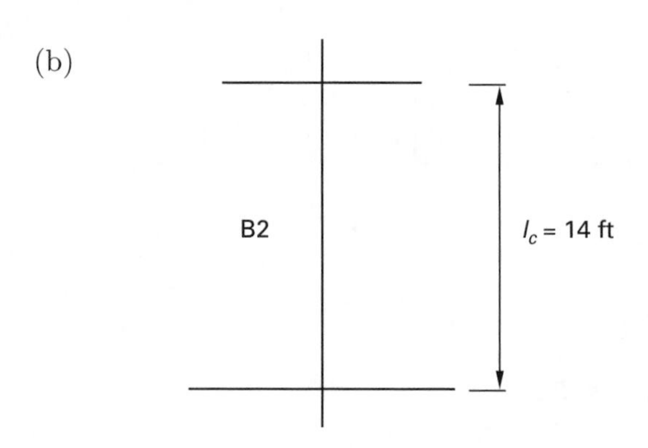

The effective moments of inertia are

$$\text{beam} = 8564.4 \text{ in}^4$$

$$I_{g,\text{column}} = \left(\frac{1}{12}\right)(20 \text{ in})^4 = 13{,}333.3 \text{ in}^4$$

$$\text{column} = 0.70 I_{g,\text{column}} = (0.70)(13{,}333.3 \text{ in}^4) = 9333.3 \text{ in}^4$$

Since the conditions are identical at the top and the bottom, the values of Ψ_1 and Ψ_2 are the same. Use Eq. 53.3.

$$\Psi = \frac{(2)\left(\frac{9333.3 \text{ in}^4}{14 \text{ ft}}\right)}{(2)\left(\frac{8564.4 \text{ in}^4}{20 \text{ ft}}\right)} = 1.56$$

$$k_u \approx \boxed{1.47}$$

The answer is (C).

(c) In column B1, there is only one beam framing in the N-S direction (bending about W-E), so from Eq. 53.3,

$$\Psi = \frac{(2)\left(\frac{9333.3 \text{ in}^4}{14 \text{ ft}}\right)}{\frac{8564.4 \text{ in}^4}{20 \text{ ft}}} = 3.11$$

$$k_u \approx \boxed{1.85}$$

The answer is (D).

(d)

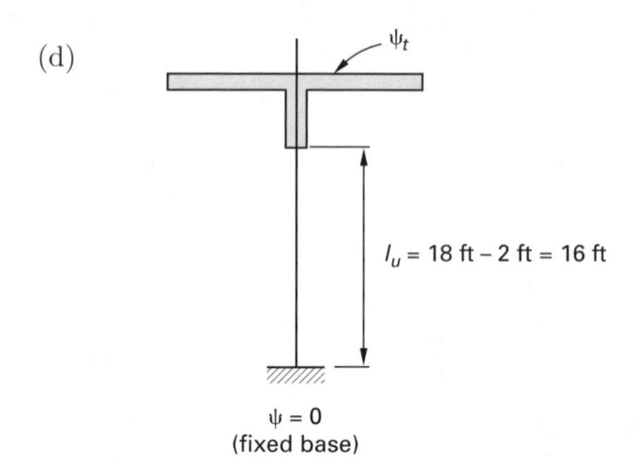

$$\Psi_t = \frac{(9333.3 \text{ in}^4)\left(\frac{1}{14 \text{ ft}} + \frac{1}{18 \text{ ft}}\right)}{(2)\left(\frac{8564.4 \text{ in}^4}{20 \text{ ft}}\right)} = 1.38$$

$$k_u = 1.2$$

The effective length is

$$k_u l_u = (1.2)(16 \text{ ft}) = \boxed{19.2 \text{ ft}}$$

The answer is (C).

(e) $\beta_d = 0$

Use Eq. 53.5.

$$EI = \frac{0.4 E_c I_g}{1 + \beta_d} = \frac{(0.4)\left(4000 \ \frac{\text{kips}}{\text{in}^2}\right)(13{,}333.3 \text{ in}^4)}{1 + 0} = 2.13 \times 10^7 \text{ kip-in}^2$$

Use Eq. 53.6.

$$P_c = \frac{\pi^2 EI}{(k_u l_u)^2} = \frac{\pi^2 (2.13 \times 10^7 \text{ kip-in}^2)}{\left((1.2)(16 \text{ ft})\left(12 \ \frac{\text{in}}{\text{ft}}\right)\right)^2} = \boxed{3960 \text{ kips}}$$

The answer is (B).

(f)

$$\begin{aligned} w_u &= 0.75(1.4D + 1.7L) \\ &= (0.75)\left((1.4)\left(0.1 \ \frac{\text{kip}}{\text{ft}^2}\right) + (1.7)\left(0.06 \ \frac{\text{kip}}{\text{ft}^2}\right)\right) \\ &= 0.1815 \text{ kip/ft}^2 \end{aligned}$$

The area per floor is

$$(60 \text{ ft})(90 \text{ ft}) = 5400 \text{ ft}^2$$

Since the first level carries five stories and a roof,

$$\sum P_u = (6)\left(0.1815 \ \frac{\text{kip}}{\text{ft}^2}\right)(5400 \text{ ft}^2) = \boxed{5881 \text{ kips} \quad (5880 \text{ kips})}$$

The answer is (D).

(g) Assume the foundation carries half of the first story's wind load.

$$V_u = (0.75)(1.7)\left(30\ \frac{\text{lbf}}{\text{ft}^2}\right)\left(\frac{1\ \text{kip}}{1000\ \text{lbf}}\right) \times (90\ \text{ft})\left((5)(14\ \text{ft}) + \frac{18\ \text{ft}}{2}\right) = 272.0\ \text{kips}$$

Use Eq. 53.1.

$$Q = \frac{\sum P_u \Delta_o}{V_u l_c} = \frac{(4900\ \text{kips})(0.3\ \text{in})}{(272.0\ \text{kips})(18\ \text{ft})\left(12\ \frac{\text{in}}{\text{ft}}\right)} = \boxed{0.025}$$

The answer is (A).

(h) The criterion to be satisfied is given by Eq. 53.15.

$$\frac{l_u}{r} \le \frac{35}{\sqrt{\frac{P_u}{f'_c A_g}}}$$

$$r = (0.3)(20\ \text{in}) = 6\ \text{in}$$

$$l_u = 16\ \text{ft}$$

$$\frac{l_u}{r} = \frac{(16\ \text{ft})\left(12\ \frac{\text{in}}{\text{ft}}\right)}{6\ \text{in}} = 32$$

$$\frac{P_u}{f'_c A_g} = \left(\frac{35}{32}\right)^2 = 1.196$$

$$P_u = \left(4000\ \frac{\text{lbf}}{\text{in}^2}\right)(400\ \text{in}^2)(1.196)\left(\frac{1\ \text{kip}}{1000\ \text{lbf}}\right) = \boxed{1914\ \text{kips}}$$

The answer is (C).

(i) From part (d),

$$\Psi_t = 1.38$$

$$\Psi_b = 0$$

From the alignment chart,

$$k_b = \boxed{0.64}$$

The answer is (A).

(j)

$$\beta_d = \frac{\text{permanent factored load}}{\text{total factored load}} = \frac{(1.4)(200\ \text{kips})}{(1.4)(200\ \text{kips}) + (1.7)(50\ \text{kips})} = \boxed{0.767}$$

The answer is (B).

Solution Using ACI 318 Ch. 9

(a) The effective beam section is

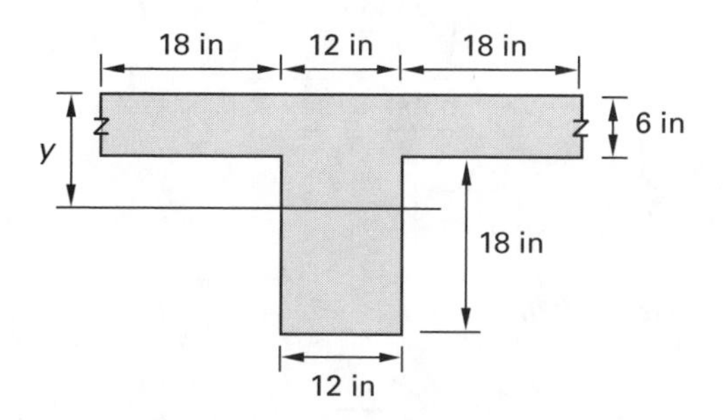

Compute I_b. $I_b = 24{,}469.7\ \text{in}^4$. For the computation of Ψ, the beam inertia is reduced by multiplying by 0.35.

$$I_{b,\text{effective}} = 0.35 I_b = (0.35)(24{,}469.7\ \text{in}^4) = \boxed{8564.4\ \text{in}^4}$$

The answer is (C).

(b)

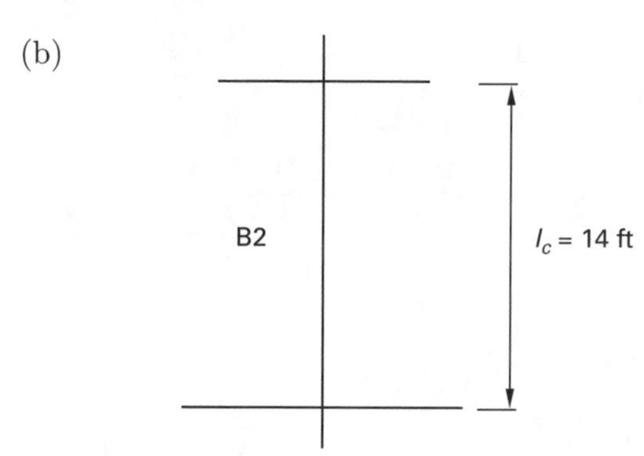

The effective moments of inertia are

$$\text{beam} = 8564.4\ \text{in}^4$$

$$I_{g,\text{column}} = \left(\frac{1}{12}\right)(20\ \text{in})^4 = 13{,}333.3\ \text{in}^4$$

$$\text{column} = 0.70 I_{g,\text{column}} = (0.70)(13{,}333.3\ \text{in}^4) = 9333.3\ \text{in}^4$$

Since the conditions are identical at the top and the bottom, the values of Ψ_1 and Ψ_2 are the same. Use Eq. 53.3.

$$\Psi = \frac{(2)\left(\frac{9333.3\ \text{in}^4}{14\ \text{ft}}\right)}{(2)\left(\frac{8564.4\ \text{in}^4}{20\ \text{ft}}\right)} = 1.56$$

$$k_u \approx \boxed{1.47}$$

The answer is (C).

Structural

(c) In column B1, there is only one beam framing in the N-S direction (bending about W-E), so from Eq. 53.3,

$$\Psi = \frac{(2)\left(\dfrac{9333.3 \text{ in}^4}{14 \text{ ft}}\right)}{\dfrac{8564.4 \text{ in}^4}{20 \text{ ft}}} = 3.11$$

$$k_u \approx \boxed{1.85}$$

The answer is (D).

(d)

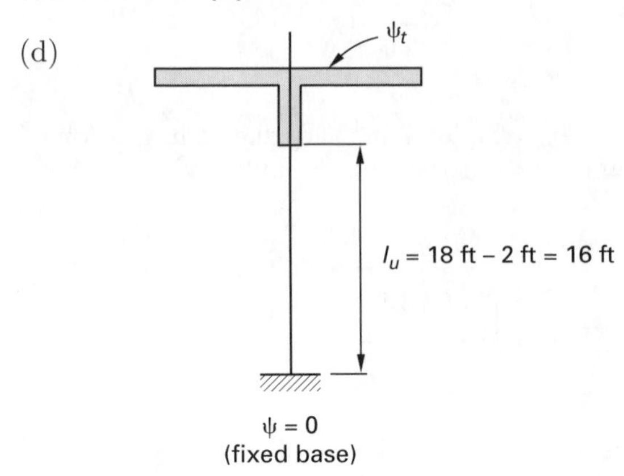

$$\Psi_t = \frac{(9333.3 \text{ in}^4)\left(\dfrac{1}{14 \text{ ft}} + \dfrac{1}{18 \text{ ft}}\right)}{(2)\left(\dfrac{8564.4 \text{ in}^4}{20 \text{ ft}}\right)} = 1.38$$

$$k_u = 1.2$$

The effective length is

$$k_u l_u = (1.2)(16 \text{ ft}) = \boxed{19.2 \text{ ft}}$$

The answer is (C).

(e) $\beta_d = 0$

Use Eq. 53.5.

$$\begin{aligned} EI &= \frac{0.4E_c I_g}{1+\beta_d} \\ &= \frac{(0.4)\left(4000 \ \dfrac{\text{kips}}{\text{in}^2}\right)(13{,}333.3 \text{ in}^4)}{1+0} \\ &= 2.13 \times 10^7 \text{ kip-in}^2 \end{aligned}$$

Use Eq. 53.6.

$$\begin{aligned} P_c &= \frac{\pi^2 EI}{(k_u l_u)^2} = \frac{\pi^2 (2.13 \times 10^7 \text{ kip-in}^2)}{\left((1.2)(16 \text{ ft})\left(12 \ \dfrac{\text{in}}{\text{ft}}\right)\right)^2} \\ &= \boxed{3960 \text{ kips}} \end{aligned}$$

The answer is (B).

(f) $$w_u = 1.2D + 1.0L + 1.6W$$

Because the live load is less than 100 lb/ft^2, the live load factor can be reduced to 0.5. Therefore,

$$\begin{aligned} w_u &= 1.2D + 0.5L + 1.6W \\ &= (1.2)\left(0.1 \ \frac{\text{kips}}{\text{ft}^2}\right) + (0.5)\left(0.06 \ \frac{\text{kips}}{\text{ft}^2}\right) \\ &= 0.15 \text{ kips/ft}^2 \end{aligned}$$

The area per floor is

$$(60 \text{ ft})(90 \text{ ft}) = 5400 \text{ ft}^2$$

Since the first level carries five stories and a roof,

$$\begin{aligned} \sum P_u &= (6)\left(0.15 \ \frac{\text{kips}}{\text{ft}^2}\right)(5400 \text{ ft}^2) \\ &= \boxed{4860 \text{ kips}} \end{aligned}$$

The answer is (C).

(g) $1.3W$ can be used in place of $1.6W$ to keep the App. C and Ch. 9 load combinations comparable in view of ACI 9.2.1(b).

$$\begin{aligned} V_u &= (1.3)\left(30 \ \frac{\text{lbf}}{\text{ft}^2}\right)\left(\frac{1 \text{ kip}}{1000 \text{ lbf}}\right)(90 \text{ ft}) \\ &\quad \times \left((5)(14 \text{ ft}) + \frac{18 \text{ ft}}{2}\right) \\ &= 277.3 \text{ kips} \end{aligned}$$

Use Eq. 53.1.

$$\begin{aligned} Q &= \frac{\sum P_u \Delta_o}{V_u l_c} \\ &= \frac{(4860 \text{ kips})(0.3 \text{ in})}{(277.3 \text{ kips})(18 \text{ ft})\left(12 \ \dfrac{\text{in}}{\text{ft}}\right)} \\ &= \boxed{0.024} \end{aligned}$$

The answer is (A).

(h) The criterion to be satisfied is given by Eq. 53.15.

$$\frac{l_u}{r} \le \frac{35}{\sqrt{\dfrac{P_u}{f'_c A_g}}}$$

$$r = (0.3)(20 \text{ in}) = 6 \text{ in}$$

$$l_u = 16 \text{ ft}$$

$$\frac{l_u}{r} = \frac{(16\ \text{ft})\left(12\ \frac{\text{in}}{\text{ft}}\right)}{6\ \text{in}} = 32$$

$$\frac{P_u}{f_c' A_g} = \left(\frac{35}{32}\right)^2 = 1.196$$

$$P_u = \left(4000\ \frac{\text{lbf}}{\text{in}^2}\right)(400\ \text{in}^2)(1.196)\left(\frac{1\ \text{kip}}{1000\ \text{lbf}}\right) = \boxed{1914\ \text{kips}}$$

The answer is (C).

(i) From part (d),

$$\Psi_t = 1.38$$
$$\Psi_b = 0$$

From the alignment chart,

$$k_b = \boxed{0.64}$$

The answer is (A).

(j)
$$\beta_d = \frac{\text{permanent factored load}}{\text{total factored load}} = \frac{(1.2)(200\ \text{kips})}{(1.2)(200\ \text{kips}) + (1.6)(50\ \text{kips})} = \boxed{0.75}$$

The answer is (B).

4. Use either **ACI 318 App. C** or **ACI 318 Ch. 9** (see p. 53-10) to solve this problem.

Solution Using ACI 318 App. C

(a)
$$M_{2,ns} = (1.4)(100\ \text{ft-kips}) + (1.7)(60\ \text{ft-kips}) = \boxed{242\ \text{ft-kips}}$$

The answer is (D).

(b) Compute Ψ at the top.

For the beams,

$$I_g = \left(\tfrac{1}{12}\right)(12\ \text{in})(24\ \text{in})^3 = 13{,}824\ \text{in}^4$$

To account for cracking,

$$I_b = 0.35 I_g = (0.35)(13{,}824\ \text{in}^4) = 4838.4\ \text{in}^4$$

For the columns,

$$I_g = \left(\tfrac{1}{12}\right)(14\ \text{in})(14\ \text{in})^3 = 3201.3\ \text{in}^4$$

To account for cracking,

$$I_c = 0.70 I_g = (0.70)(3201.3\ \text{in}^4) = 2240.9\ \text{in}^4$$

Use Eq. 53.3.

$$\Psi_t = \frac{\frac{2240.9\ \text{in}^4}{17\ \text{ft}}}{\frac{4838.4\ \text{in}^4}{30\ \text{ft}}} = 0.81$$

At the base, the column is assumed fixed. In theory, this implies $\Psi_b = 0$. In practice, it is customary to substitute 1.0 for the theoretical value of zero to account for the finite flexibility at the "fixed end." Taking $\Psi_b = 1$, $K_b = 0.75$.

The effective length is

$$K_b l_u = (0.75)(16\ \text{ft}) = \boxed{12\ \text{ft}}$$

The answer is (C).

(c) Use Eq. 53.3.

$$\Psi_t = \frac{\frac{2240.9\ \text{in}^4}{17\ \text{in}}}{(2)\left(\frac{4838.4\ \text{in}^4}{30\ \text{in}}\right)} = 0.41$$

$$\Psi_b = 1$$
$$K_b = 0.72$$

$$K_b l_u = (0.72)(16\ \text{ft}) = \boxed{11.52\ \text{ft}}$$

The answer is (B).

(d)
$$\beta_d = \frac{\text{permanent factored axial load}}{\text{total factored axial load}} = \frac{(1.4)(36.5\ \text{kips})}{(1.4)(36.5\ \text{kips}) + (1.7)(21.9\ \text{kips})} = \boxed{0.58}$$

The answer is (C).

Structural

(e) Use Eq. 53.3.

$$P_c = \frac{\pi^2 EI}{(K_b l_u)^2}$$

From Eq. 53.5,

$$EI = \frac{0.4E_cI_g}{1+\beta_d} = \frac{(0.4)\left(3605\ \frac{\text{kips}}{\text{in}^2}\right)(3201.3\ \text{in}^4)}{1+0.58}$$
$$= 2.92 \times 10^6\ \text{kip-in}^2$$

$$P_c = \frac{\pi^2(2.92 \times 10^6\ \text{kip-in}^2)}{\left((0.75)(16\ \text{ft})\left(12\ \frac{\text{in}}{\text{ft}}\right)\right)^2} = \boxed{1390.6\ \text{kips}}$$

The answer is (B).

(f) From Eq. 53.9,

$$C_m = 0.6 + (0.4)\left(\frac{M_1}{M_2}\right) \ge 0.4$$
$$\frac{M_1}{M_2} = -\frac{(1.4)(50\ \text{ft-kips}) + (1.7)(30\ \text{ft-kips})}{(1.4)(100\ \text{ft-kips}) + (1.7)(60\ \text{ft-kips})}$$
$$= -0.5$$
$$C_m = \boxed{0.4}$$

The answer is (B).

(g) In this column, $M_1 = M_2 = 0$. Therefore,

$$C_m = \boxed{1}$$

The answer is (D).

(h) $$\beta_d = \frac{\text{permanent factored axial load}}{\text{total factored axial load}}$$
$$= \frac{(1.4)(95.2\ \text{kips})}{(1.4)(95.2\ \text{kips}) + (1.7)(57.1\ \text{kips})} = 0.58$$

From part (e), $EI = 2.92 \times 10^6\ \text{kip-in}^2$.

From part (c), $K_b = 0.72$.

Use Eq. 53.6.

$$P_c = \frac{\pi^2 EI}{(K_b l_u)^2} = \frac{\pi^2(2.92 \times 10^6\ \text{kip-in}^2)}{\left((0.72)(16\ \text{ft})\left(12\ \frac{\text{in}}{\text{ft}}\right)\right)^2}$$
$$= \boxed{1508.0\ \text{kips}}$$

The answer is (C).

(i) $$P_u = (1.4)(95.2\ \text{kips}) + (1.7)(57.1\ \text{kips})$$
$$= 230.4\ \text{kips}$$

From Eq. 53.8,

$$\delta_{ns} = \frac{C_m}{1 - \frac{P_u}{0.75P_c}} = \frac{1}{1 - \frac{230.4\ \text{kips}}{(0.75)(1508.0\ \text{kips})}}$$
$$= \boxed{1.256}$$

The answer is (C).

(j) $$P_u = (1.4)(36.5\ \text{kips}) + (1.7)(21.9\ \text{kips})$$
$$= 88.3\ \text{kips}$$

From part (e), $P_c = 1390.6$ kips.

From part (f), $C_m = 0.4$.

$$\delta_{ns} = \frac{C_m}{1 - \frac{P_u}{0.75P_c}} = \frac{0.4}{1 - \frac{88.3\ \text{kips}}{(0.75)(1390.6\ \text{kips})}}$$
$$= 0.437 < 1.0$$
$$\delta_{ns} = \boxed{1.0}$$

The answer is (B).

Solution Using ACI 318 Ch. 9

(a) $$M_{2,ns} = (1.2)(100\ \text{ft-kips}) + (1.6)(60\ \text{ft-kips})$$
$$= \boxed{216\ \text{ft-kips}}$$

The answer is (D).

(b) Compute Ψ at the top.

For the beams,

$$I_g = \left(\tfrac{1}{12}\right)(12\ \text{in})(24\ \text{in})^3 = 13{,}824\ \text{in}^4$$

To account for cracking,

$$I_b = 0.35I_g = (0.35)(13{,}824\ \text{in}^4) = 4838.4\ \text{in}^4$$

For the columns,

$$I_g = \left(\tfrac{1}{12}\right)(14\ \text{in})(14\ \text{in})^3 = 3201.3\ \text{in}^4$$

To account for cracking,

$$I_c = 0.70I_g = (0.70)(3201.3\ \text{in}^4) = 2240.9\ \text{in}^4$$

Use Eq. 53.3.

$$\Psi_t = \frac{\frac{2240.9 \text{ in}^4}{17 \text{ ft}}}{\frac{4838.4 \text{ in}^4}{30 \text{ ft}}} = 0.81$$

At the base, the column is assumed fixed. In theory, this implies $\Psi_b = 0$. In practice, it is customary to substitute 1.0 for the theoretical value of zero to account for the finite flexibility at the "fixed end." Taking $\Psi_b = 1$, $K_b = 0.75$.

The effective length is

$$K_b l_u = (0.75)(16 \text{ ft}) = \boxed{12 \text{ ft}}$$

The answer is (C).

(c) Use Eq. 53.3.

$$\Psi_t = \frac{\frac{2240.9 \text{ in}^4}{17 \text{ in}}}{(2)\left(\frac{4838.4 \text{ in}^4}{30 \text{ in}}\right)} = 0.41$$

$$\Psi_b = 1$$

$$K_b = 0.72$$

$$K_b l_u = (0.72)(16 \text{ ft}) = \boxed{11.52 \text{ ft}}$$

The answer is (B).

(d)
$$\begin{aligned} \beta_d &= \frac{\text{permanent factored axial load}}{\text{total factored axial load}} \\ &= \frac{(1.2)(36.5 \text{ kips})}{(1.2)(36.5 \text{ kips}) + (1.6)(21.9 \text{ kips})} \\ &= \boxed{0.56} \end{aligned}$$

The answer is (C).

(e) Use Eq. 53.3.

$$P_c = \frac{\pi^2 EI}{(K_b l_u)^2}$$

From Eq. 53.5,

$$\begin{aligned} EI &= \frac{0.4E_cI_g}{1+\beta_d} = \frac{(0.4)\left(3605 \frac{\text{kips}}{\text{in}^2}\right)(3201.3 \text{ in}^4)}{1+0.56} \\ &= 2.96 \times 10^6 \text{ kip-in}^2 \end{aligned}$$

$$P_c = \frac{\pi^2 (2.96 \times 10^6 \text{ kip-in}^2)}{\left((0.75)(16 \text{ ft})\left(12 \frac{\text{in}}{\text{ft}}\right)\right)^2} = \boxed{1409.6 \text{ kips}}$$

The answer is (B).

(f) From Eq. 53.9,

$$\begin{aligned} C_m &= 0.6 + (0.4)\left(\frac{M_1}{M_2}\right) \geq 0.4 \\ \frac{M_1}{M_2} &= -\frac{(1.2)(50 \text{ ft-kips}) + (1.6)(30 \text{ ft-kips})}{(1.2)(100 \text{ ft-kips}) + (1.6)(60 \text{ ft-kips})} \\ &= -0.5 \\ C_m &= \boxed{0.4} \end{aligned}$$

The answer is (B).

(g) In this column, $M_1 = M_2 = 0$. Therefore,

$$C_m = \boxed{1}$$

The answer is (D).

(h)
$$\begin{aligned} \beta_d &= \frac{\text{permanent factored axial load}}{\text{total factored axial load}} \\ &= \frac{(1.2)(95.2 \text{ kips})}{(1.2)(95.2 \text{ kips}) + (1.6)(57.1 \text{ kips})} = 0.56 \end{aligned}$$

From part (e), $EI = 2.96 \times 10^6$ kip-in^2.

From part (c), $K_b = 0.72$.

Use Eq. 53.6.

$$\begin{aligned} P_c &= \frac{\pi^2 EI}{(K_b l_u)^2} = \frac{\pi^2 (2.96 \times 10^6 \text{ kip-in}^2)}{\left((0.72)(16 \text{ ft})\left(12 \frac{\text{in}}{\text{ft}}\right)\right)^2} \\ &= \boxed{1528.7 \text{ kips}} \end{aligned}$$

The answer is (C).

(i)
$$\begin{aligned} P_u &= (1.2)(95.2 \text{ kips}) + (1.6)(57.1 \text{ kips}) \\ &= 205.6 \text{ kips} \end{aligned}$$

From Eq. 53.8,

$$\begin{aligned} \delta_{ns} &= \frac{C_m}{1 - \frac{P_u}{0.75P_c}} = \frac{1}{1 - \frac{205.6 \text{ kips}}{(0.75)(1528.7 \text{ kips})}} \\ &= \boxed{1.219} \end{aligned}$$

The answer is (C).

(j)
$$\begin{aligned} P_u &= (1.2)(36.5 \text{ kips}) + (1.6)(21.9 \text{ kips}) \\ &= 78.8 \text{ kips} \end{aligned}$$

From part (e), $P_c = 1409.6$ kips.

From part (f), $C_m = 0.4$.

$$\begin{aligned} \delta_{ns} &= \frac{C_m}{1 - \frac{P_u}{0.75P_c}} = \frac{0.4}{1 - \frac{78.8 \text{ kips}}{(0.75)(1409.6 \text{ kips})}} \\ &= 0.432 < 1.0 \end{aligned}$$

$$\delta_{ns} = \boxed{1.0}$$

The answer is (B).

54 Reinforced Concrete: Walls and Retaining Walls

PRACTICE PROBLEMS

1. A 14 ft high concrete bearing wall supports concrete roof beams spaced on 7 ft centers. Using the tributary area method, each end reaction is found to carry a dead load of 25 kips and a live load of 20 kips. The bearing width of the beam is 12 in. Both the top and bottom of the wall are restrained against translation and rotation. The concrete strength and the yield strength of steel are 4000 psi and 60,000 psi, respectively. The wall weight can be neglected. The ACI empirical formula is to be used. (a) What is the required wall thickness? (b) Is the bearing strength adequate if the wall is 7 in thick? (c) What is the vertical load capacity of a 7 in wall? (d) Design the vertical reinforcing steel. (e) Design the horizontal reinforcing steel.

2. (a) What is the vertical capacity of the wall described in Prob. 1 if the wall is not restrained against rotation at the top or bottom? Assume the wall thickness is 7 in. (b) Is the wall capacity still adequate?

3. *(Time limit: one hour)* The retaining wall shown supports a 400 lbf/ft^2 surcharge in addition to the active backfill loading. There is no batter on the visible face. 3000 psi concrete and 60,000 psi steel are used. The specific weights of the backfill and the concrete are 100 lbf/ft^3 and 150 lbf/ft^3, respectively. The active earth pressure coefficient is 0.5. The passive earth pressure coefficient is 2.0. (a) What is the factor of safety against overturning? (b) What is the minimum theoretical heel thickness? (c) Using a heel depth of 30 in and no. 8 reinforcing bars, what bar spacing is required in the heel?

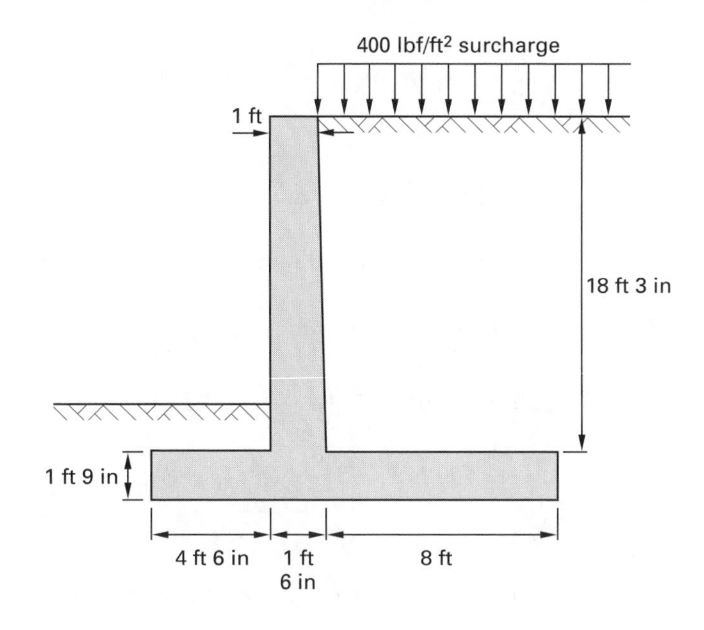

Structural

SOLUTIONS

1. Use either **ACI 318 App. C** or **ACI 318 Ch. 9** to solve this problem.

Solution Using ACI 318 App. C

(a) According to ACI 318 14.5.3.1, the wall thickness, h, is the larger of 4 in or 1/25 of the supported height or length.

$$\left(\frac{1}{25}\right)(14\text{ ft})\left(12\ \frac{\text{in}}{\text{ft}}\right) = \boxed{6.72\text{ in}}$$

Use 7 in.

(b) The factored end reactions are

$$(1.4)(25\text{ kips}) + (1.7)(20\text{ kips}) = 69\text{ kips}$$

The design bearing strength is

$$\phi(0.85f'_cA) = (0.70)(0.85)\left(4\ \frac{\text{kips}}{\text{in}^2}\right)(12\text{ in})(7\text{ in})$$

$$= 199.9\text{ kips}\ \boxed{[> 69\text{ kips, so OK}]}$$

(c) $A_g = (\text{horizontal wall length})(\text{wall thickness})$

According to ACI 318 14.2.4, for the horizontal length of the wall to be considered effective for each concentrated load, the length cannot exceed the center-to-center distance between loads or the width of the bearing plus four times the wall thickness.

The center-to-center spacing is

$$(7\text{ ft})\left(12\ \frac{\text{in}}{\text{ft}}\right) = 84\text{ in}$$

The width of bearing plus four times the wall thickness, h, is

$$12\text{ in} + (4)(7\text{ in}) = 40\text{ in}$$

The 40 in dimension controls.

$$\phi P_{n,w} \le 0.55\phi f'_cA_g\left(1 - \left(\frac{kl_c}{32h}\right)^2\right)$$

$$(0.55)(0.70)\left(4\ \frac{\text{kips}}{\text{in}^2}\right)(40\text{ in})(7\text{ in}) \times \left(1 - \left(\frac{(0.8)(14\text{ ft})\left(12\ \frac{\text{in}}{\text{ft}}\right)}{(32)(7\text{ in})}\right)^2\right)$$

$$= \boxed{276\text{ kips}}\quad [> 69\text{ kips, so OK}]$$

(d) This is a bearing wall. Per ACI 318 14.3.2 and 14.3.3, the steel will be as follows.

The maximum spacing of vertical and horizontal bars is

$$(3)(7\text{ in}) = 21\text{ in or }18\text{ in}\quad [18\text{ in controls}]$$

The area of vertical steel is

$$(0.0015)(12\text{ in})(7\text{ in}) = 0.126\text{ in}^2/\text{ft}$$

$$\boxed{\text{Use no. 4 bars at 18 in (provides 0.133 in}^2/\text{ft).}}$$

(e) The area of horizontal steel is

$$(0.0025)(12\text{ in})(7\text{ in}) = 0.21\text{ in}^2/\text{ft}$$

$$\boxed{\text{Use no. 4 bars at 10 in (provides 0.236 in}^2/\text{ft).}}$$

Solution Using ACI 318 Ch. 9

(a) According to ACI 318 14.5.3.1, the wall thickness, h, is the larger of 4 in or 1/25 of the supported height or length.

$$\left(\frac{1}{25}\right)(14\text{ ft})\left(12\ \frac{\text{in}}{\text{ft}}\right) = \boxed{6.72\text{ in}}$$

Use 7 in.

(b) The factored end reactions are

$$(1.2)(25\text{ kips}) + (1.6)(20\text{ kips}) = 62\text{ kips}$$

The design bearing strength is

$$\phi(0.85f'_cA) = (0.65)(0.85)\left(4\ \frac{\text{kips}}{\text{in}^2}\right)(12\text{ in})(7\text{ in})$$

$$= 185.6\text{ kips}\ \boxed{[> 62\text{ kips, so OK}]}$$

(c) $A_g = (\text{horizontal wall length})(\text{wall thickness})$

According to ACI 318 14.2.4, for the horizontal length of the wall to be considered effective for each concentrated load, the length cannot exceed the center-to-center distance between loads or the width of the bearing plus four times the wall thickness.

The center-to-center spacing is

$$(7\text{ ft})\left(12\ \frac{\text{in}}{\text{ft}}\right) = 84\text{ in}$$

The width of bearing plus four times the wall thickness, h, is

$$12\text{ in} + (4)(7\text{ in}) = 40\text{ in}$$

The 40 in dimension controls.

$$\phi P_{n,w} \le 0.55\phi f'_c A_g \left(1 - \left(\frac{kl_c}{32h}\right)^2\right)$$

$$(0.55)(0.70)\left(4\ \frac{\text{kips}}{\text{in}^2}\right)(40\ \text{in})(7\ \text{in}) \times \left(1 - \left(\frac{(0.8)(14\ \text{ft})\left(12\ \frac{\text{in}}{\text{ft}}\right)}{(32)(7\ \text{in})}\right)^2\right)$$

$$= \boxed{276\ \text{kips}} \quad [> 69\ \text{kips, so OK}]$$

(d) This is a bearing wall. Per ACI 318 14.3.2 and 14.3.3, the steel will be as follows.

The maximum spacing of vertical and horizontal bars is

$$(3)(7\ \text{in}) = 21\ \text{in or 18 in} \quad [18\ \text{in controls}]$$

The area of vertical steel is

$$(0.0015)(12\ \text{in})(7\ \text{in}) = 0.126\ \text{in}^2/\text{ft}$$

Use no. 4 bars at 18 in (provides 0.133 in^2/ft).

(e) The area of horizontal steel is

$$(0.0025)(12\ \text{in})(7\ \text{in}) = 0.21\ \text{in}^2/\text{ft}$$

Use no. 4 bars at 10 in (provides 0.236 in^2/ft).

2. Use either **ACI 318 App. C** or **ACI 318 Ch. 9** to solve this problem.

Solution Using ACI 318 App. C

(a) The effective length factor, k, is 1.0 when the wall is not restrained against rotation at both ends. Use the ACI empirical formula.

$$\phi P_{n,w} \le 0.55\phi f'_c A_g \left(1 - \left(\frac{kl_c}{32h}\right)^2\right)$$

$$(0.55)(0.70)\left(4\ \frac{\text{kips}}{\text{in}^2}\right)(40\ \text{in})(7\ \text{in}) \times \left(1 - \left(\frac{(1.0)(14\ \text{ft})\left(12\ \frac{\text{in}}{\text{ft}}\right)}{(32)(7\ \text{in})}\right)^2\right)$$

$$= \boxed{188.7\ \text{kips}}$$

(b) $\boxed{> 69\ \text{kips, so OK}}$

Solution Using ACI 318 Ch. 9

(a) The effective length factor, k, is 1.0 when the wall is not restrained against rotation at both ends. Use the ACI empirical formula.

$$\phi P_{n,w} \le 0.55\phi f'_c A_g \left(1 - \left(\frac{kl_c}{32h}\right)^2\right)$$

$$(0.55)(0.70)\left(4\ \frac{\text{kips}}{\text{in}^2}\right)(40\ \text{in})(7\ \text{in}) \times \left(1 - \left(\frac{(1.0)(14\ \text{ft})\left(12\ \frac{\text{in}}{\text{ft}}\right)}{(32)(7\ \text{in})}\right)^2\right)$$

$$= \boxed{188.7\ \text{kips}}$$

(b) $\boxed{> 62\ \text{kips, so OK}}$

3. Use either **ACI 318 App. C** or **ACI 318 Ch. 9** (see p. 54-5) to solve this problem.

Solution Using ACI 318 App. C

(a) $H = 18.25\ \text{ft} + 1.75\ \text{ft} = 20\ \text{ft}$

The soil pressure resultant is

$$R_a = \left(\tfrac{1}{2}\right)(0.5)\left(100\ \frac{\text{lbf}}{\text{ft}^3}\right)(20\ \text{ft})^2 = 10{,}000\ \text{lbf/ft}$$

R_a acts at 20 ft/3 = 6.67 ft up from the base.

The surcharge loading is

$$R_{q,h} = (0.5)\left(400\ \frac{\text{lbf}}{\text{ft}^2}\right)(20\ \text{ft}) = 4000\ \text{lbf}$$

$R_{q,h}$ acts at 20 ft/2 = 10 ft up from the base.

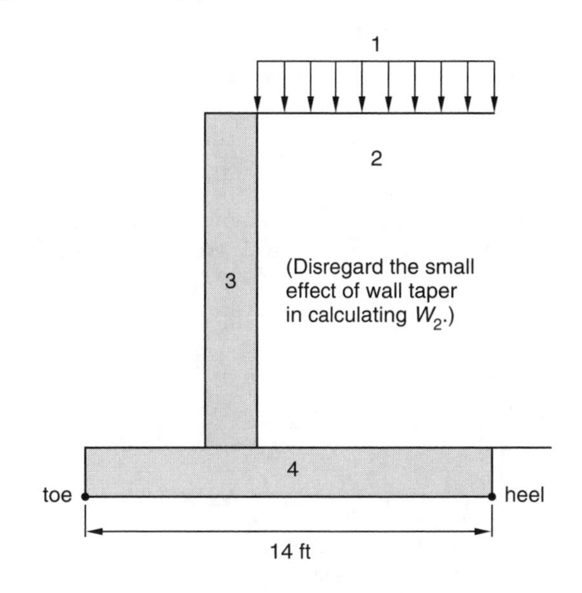

Structural

element	length or area (ft/ft^2)	q or γ (lbf/ft^2)	F or W (lbf)	$r_{i,heel}$ (ft)
1	8	400	3200	
2	$8 \times 18.25 = 146$	100	14,600	
3	$1.25 \times 18.25 = 22.8$	150	3420	$\approx$
4	$14 \times 1.75 = 24.5$	150	3675	7
total			24,895 lbf	

element	F (lbf)	$14 - r_i = x_{i,toe}$ (ft)	moment about toe (ft-lbf)
1	3200	10	32,000
2	14,600	10	146,000
3	3420	5.5	18,810
4	3675	7	25,725
total			222,535 ft-lbf

$$\begin{aligned}\sum R_a y_a &= (10{,}000 \text{ lbf})(6.67 \text{ ft}) + (4000 \text{ lbf})(10 \text{ ft})\\ &= 106{,}700 \text{ ft-lbf}\end{aligned}$$

$$\begin{aligned}F_{\text{overturning}} &= \frac{222{,}535 \text{ ft-lbf}}{106{,}700 \text{ ft-lbf}}\\ &= \boxed{2.09 \quad [\text{acceptable}]}\end{aligned}$$

(Normally, a check of the factor of safety against sliding would also be made. F_{sliding} is inadequate in this problem, but a check has not been requested.)

(b)

$$\begin{aligned}\text{heel weight} &= (8 \text{ ft})(1.75 \text{ ft})\left(150 \ \frac{\text{lbf}}{\text{ft}^3}\right)\\ &= 2100 \text{ lbf/ft} \quad [\text{per foot of width}]\end{aligned}$$

$$\begin{aligned}V_u &= (1.4)\left(2100 \ \frac{\text{lbf}}{\text{ft}} + 14{,}600 \ \frac{\text{lbf}}{\text{ft}}\right)\\ &\quad + (1.7)\left(3200 \ \frac{\text{lbf}}{\text{ft}}\right)\\ &= 28{,}820 \text{ lbf/ft}\end{aligned}$$

The nominal shear strength of concrete is

$$\phi v_n = \phi v_c = (2)(0.85)\sqrt{3000 \ \frac{\text{lbf}}{\text{in}^2}} = 93.1 \text{ lbf/in}^2$$

The required heel thickness, without requiring shear reinforcement and neglecting soil pressure beneath the footing, is

$$\frac{28{,}820 \ \frac{\text{lbf}}{\text{ft}}}{\left(93.1 \ \frac{\text{lbf}}{\text{in}^2}\right)\left(12 \ \frac{\text{in}}{\text{ft}}\right)} = 25.8 \text{ in} \quad [\text{round to 26 in}]$$

To include at least a 2 in cover, use a heel thickness of

$$t_b = \boxed{30 \text{ in}}$$

(c) The heel weight becomes

$$(8 \text{ ft})\left(\frac{30 \text{ in}}{12 \ \frac{\text{in}}{\text{ft}}}\right)(1)\left(150 \ \frac{\text{lbf}}{\text{ft}^3}\right) = 3000 \text{ lbf/ft}$$

Take moments about the stem face.

element	F (lbf)	distance (ft)	M (ft-lbf)
1	3200	4	12,800
2	14,600	4	58,400
heel	3000	4	12,000

The ultimate moment on the heel at the stem face is given by Eq. 54.10.

$$\begin{aligned}M_u &= (1.4)\left(58{,}400 \ \frac{\text{ft-lbf}}{\text{ft}} + 12{,}000 \ \frac{\text{ft-lbf}}{\text{ft}}\right)\\ &\quad + (1.7)\left(12{,}800 \ \frac{\text{ft-lbf}}{\text{ft}}\right)\\ &= 120{,}320 \text{ ft-lbf/ft}\end{aligned}$$

Use Eq. 54.13, assuming the section is tension controlled.

$$\begin{aligned}M_n &= \frac{M_u}{\phi} = \frac{120{,}320 \ \frac{\text{ft-lbf}}{\text{ft}}}{0.9}\\ &= 133{,}689 \text{ ft-lbf/ft}\end{aligned}$$

$$\begin{aligned}&\left(133{,}689 \ \frac{\text{ft-lbf}}{\text{ft}}\right)\\ &\quad \times \left(12 \ \frac{\text{in}}{\text{ft}}\right) = \rho(12 \text{ in})(26 \text{ in})^2\\ &\quad \times \left(60{,}000 \ \frac{\text{lbf}}{\text{in}^2}\right)\\ &\quad \times \left(1 - \frac{\rho\left(60{,}000 \ \frac{\text{lbf}}{\text{in}^2}\right)}{(1.7)\left(3000 \ \frac{\text{lbf}}{\text{in}^2}\right)}\right)\\ &\rho = 0.0034\end{aligned}$$

With this low reinforcement ratio, the section is tension controlled.

Check.

$$\rho_{\min} = \frac{200}{60{,}000 \ \frac{\text{lbf}}{\text{in}^2}} = 0.0033 \quad [\text{acceptable}]$$

No. 8 steel has an area of 0.79 in^2.

$$\text{spacing} = \frac{0.79 \text{ in}^2}{(0.0034)(26 \text{ in})} = 8.9 \text{ in}$$

Use one no. 8 bar every 9 in.

Solution Using ACI 318 Ch. 9

(a) $H = 18.25 \text{ ft} + 1.75 \text{ ft} = 20 \text{ ft}$

The soil pressure resultant is

$$R_a = \left(\tfrac{1}{2}\right)(0.5)\left(100 \frac{\text{lbf}}{\text{ft}^3}\right)(20 \text{ ft})^2 = 10{,}000 \text{ lbf/ft}$$

R_a acts at 20 ft/3 = 6.67 ft up from the base.

The surcharge loading is

$$R_{q,h} = (0.5)\left(400 \frac{\text{lbf}}{\text{ft}^2}\right)(20 \text{ ft}) = 4000 \text{ lbf}$$

$R_{q,h}$ acts at 20 ft/2 = 10 ft up from the base.

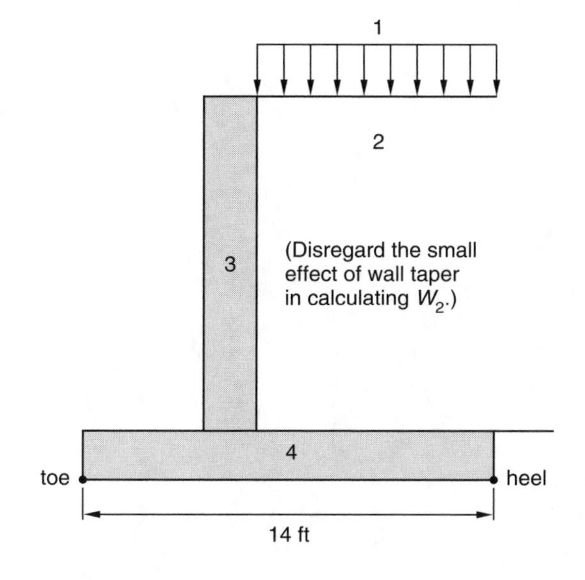

element	length or area (ft/ft^2)	q or γ (lbf/ft^2)	F or W (lbf)	$r_{i,heel}$ (ft)
1	8	400	3200	
2	8 × 18.25 = 146	100	14,600	
3	1.25 × 18.25 = 22.8	150	3420	≈
4	14 × 1.75 = 24.5	150	3675	7
total			24,895 lbf	

element	F (lbf)	$14 - r_i = x_{i,toe}$ (ft)	moment about toe (ft-lbf)
1	3200	10	32,000
2	14,600	10	146,000
3	3420	5.5	18,810
4	3675	7	25,725
total			222,535 ft-lbf

$$\sum R_a y_a = (10{,}000 \text{ lbf})(6.67 \text{ ft}) + (4000 \text{ lbf})(10 \text{ ft}) = 106{,}700 \text{ ft-lbf}$$

$$F_{\text{overturning}} = \frac{222{,}535 \text{ ft-lbf}}{106{,}700 \text{ ft-lbf}} = 2.09 \quad [\text{acceptable}]$$

(Normally, a check of the factor of safety against sliding would also be made. F_{sliding} is inadequate in this problem, but a check has not been requested.)

(b) $$\text{heel weight} = (8 \text{ ft})(1.75 \text{ ft})\left(150 \frac{\text{lbf}}{\text{ft}^3}\right) = 2100 \text{ lbf/ft} \quad [\text{per foot of width}]$$

$$V_u = (1.2)\left(2100 \frac{\text{lbf}}{\text{ft}} + 14{,}600 \frac{\text{lbf}}{\text{ft}}\right) + (1.6)\left(3200 \frac{\text{lbf}}{\text{ft}}\right) = 25{,}160 \text{ lbf/ft}$$

The nominal shear strength of concrete is

$$\phi v_n = \phi v_c = (2)(0.75)\sqrt{3000 \frac{\text{lbf}}{\text{in}^2}} = 82.1 \text{ lbf/in}^2$$

The required heel thickness, without requiring shear reinforcement and neglecting soil pressure beneath the footing, is

$$\frac{25{,}160 \frac{\text{lbf}}{\text{ft}}}{\left(82.1 \frac{\text{lbf}}{\text{in}^2}\right)\left(12 \frac{\text{in}}{\text{ft}}\right)} = 25.5 \text{ in} \quad [\text{round to 26 in}]$$

To include at least a 2 in cover, use a heel thickness of

$$t_b = 30 \text{ in}$$

(c) The heel weight becomes

$$(8 \text{ ft})\left(\frac{30 \text{ in}}{12 \frac{\text{in}}{\text{ft}}}\right)(1)\left(150 \frac{\text{lbf}}{\text{ft}^3}\right) = 3000 \text{ lbf/ft}$$

Take moments about the stem face.

element	F (lbf)	distance (ft)	M (ft-lbf)
1	3200	4	12,800
2	14,600	4	58,400
heel	3000	4	12,000

Structural

The ultimate moment on the heel at the stem face is given by Eq. 54.10.

$$\begin{aligned} M_u &= (1.2)\left(58{,}400\ \frac{\text{ft-lbf}}{\text{ft}} + 12{,}000\ \frac{\text{ft-lbf}}{\text{ft}}\right) \\ &\quad + (1.6)\left(12{,}800\ \frac{\text{ft-lbf}}{\text{ft}}\right) \\ &= 104{,}960\ \text{ft-lbf/ft} \end{aligned}$$

Use Eq. 54.13, assuming the section is tension controlled.

$$\begin{aligned} M_n &= \frac{M_u}{\phi} = \frac{104{,}960\ \frac{\text{ft-lbf}}{\text{ft}}}{0.9} \\ &= 116{,}622\ \text{ft-lbf/ft} \end{aligned}$$

$$\begin{aligned} &\left(116{,}622\ \frac{\text{ft-lbf}}{\text{ft}}\right) \\ &\quad \times \left(12\ \frac{\text{in}}{\text{ft}}\right) = \rho(12\ \text{in})(26\ \text{in})^2 \\ &\qquad \times \left(60{,}000\ \frac{\text{lbf}}{\text{in}^2}\right) \\ &\qquad \times \left(1 - \frac{\rho\left(60{,}000\ \frac{\text{lbf}}{\text{in}^2}\right)}{(1.7)\left(3000\ \frac{\text{lbf}}{\text{in}^2}\right)}\right) \\ &\rho = 0.0030 \end{aligned}$$

Check.

$$\rho_{\text{min}} = \frac{200}{60{,}000\ \frac{\text{lbf}}{\text{in}^2}} = 0.0033 \quad \text{[controls]}$$

Since ρ_{min} controls, the section is tension controlled.

No. 8 steel has an area of 0.79 in^2.

$$\text{spacing} = \frac{0.79\ \text{in}^2}{(0.0033)(26\ \text{in})} = 9.2\ \text{in}$$

Use one no. 8 bar every 9 in.

55 Reinforced Concrete: Footings

PRACTICE PROBLEMS

(In the following problems, none of the loads is the result of wind or earthquake action. Unless specified in the problem, assume concrete has a specific weight of 150 lbf/ft^3. All bars are uncoated.)

1. A reinforced concrete column is supported at the center of a square footing. The footing is in contact with the bare ground 4 ft below grade. The allowable gross soil pressure is 2000 lbf/ft^2. The specific weight of the soil is 110 lbf/ft^3. The footing thickness is 18 in. The axial loads, moments, and shears (directed along the top of the footing surface) acting on the footing through the column are

$$P_d = 100 \text{ kips}$$
$$P_l = 58 \text{ kips}$$
$$M_d = 20 \text{ ft-kips}$$
$$M_l = 40 \text{ ft-kips}$$
$$V_d = 2 \text{ kips} \quad \text{[to the right]}$$
$$V_l = 4 \text{ kips} \quad \text{[to the right]}$$

Determine the footing size. (Do not design the reinforcement.)

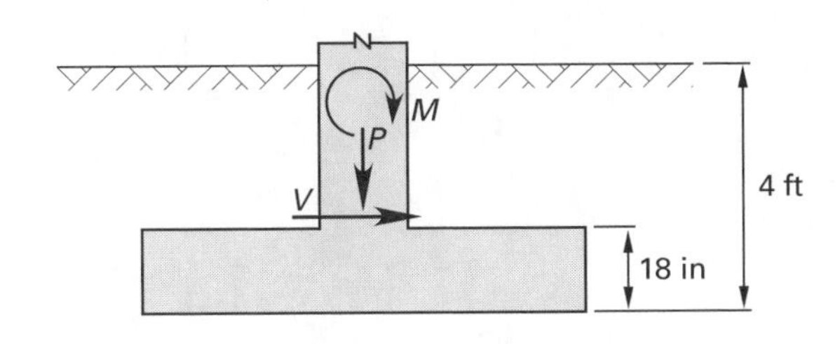

2. An 8 in thick concrete wall carries a service load of 22 kips/ft. As shown, the footing for the wall is located 5 ft below grade. 3000 psi concrete is used in the footing. At this depth, the allowable gross soil pressure is 5000 lbf/ft^2. The soil has a unit weight of 120 lbf/ft^3. No. 4 bars are to be used for reinforcement. Take the ultimate load as 1.5 times the service load. Size the footing and determine the steel requirements. Do not check development length.

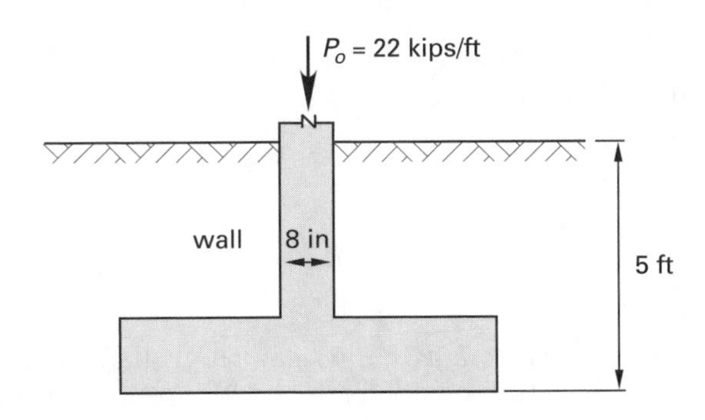

3. (*Time limit: one hour*) A 16 in thick, 10 ft square footing on grade supports a 12 in square column as shown. The net allowable soil pressure is 3000 lbf/ft^2.

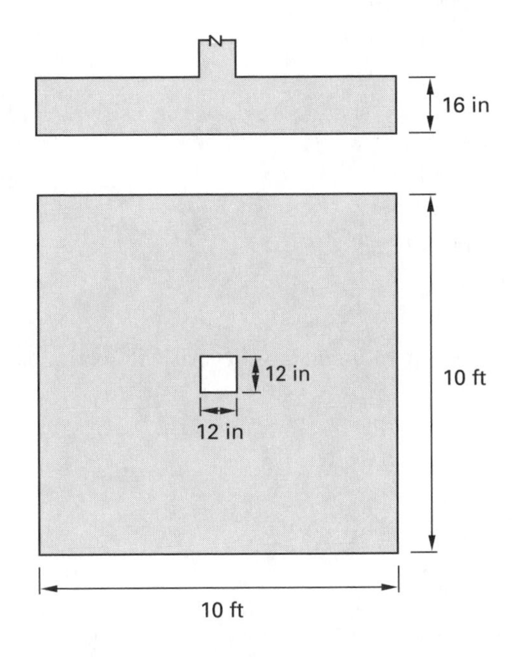

(a) If the column is not subjected to a moment, the maximum service load that the footing can transmit safely to the soil is most nearly

(A) 200 kips
(B) 250 kips
(C) 300 kips
(D) 350 kips

Structural

(b) If the column has a total axial service load (dead plus live) of 200 kips, the maximum moment about the centroid of the footing print that the footing can support is most nearly

(A) 140 ft-kips
(B) 170 ft-kips
(C) 190 ft-kips
(D) 220 ft-kips

(c) The area of the critical section resisting punching shear is most nearly

(A) 680 in^2
(B) 960 in^2
(C) 1200 in^2
(D) 1400 in^2

For parts (d) through (f), assume the column is subjected to a factored axial force of 340 kips and no moment.

(d) The punching shear stress at the critical section is most nearly

(A) 200 lbf/in^2
(B) 250 lbf/in^2
(C) 280 lbf/in^2
(D) 320 lbf/in^2

(e) The one-way shear stress at the critical section is most nearly

(A) 60 lbf/in^2
(B) 80 lbf/in^2
(C) 100 lbf/in^2
(D) 120 lbf/in^2

(f) The design moment at the critical section is most nearly

(A) 280 ft-kips
(B) 300 ft-kips
(C) 340 ft-kips
(D) 440 ft-kips

In parts (g) through (j), assume the loading on the footing consists of a factored load of 340 kips plus a moment of 150 ft-kips.

(g) The maximum punching shear stress at the critical section is most nearly

(A) 220 lbf/in^2
(B) 250 lbf/in^2
(C) 360 lbf/in^2
(D) 410 lbf/in^2

(h) The maximum one-way shear stress at the critical section is most nearly

(A) 86 lbf/in^2
(B) 97 lbf/in^2
(C) 100 lbf/in^2
(D) 120 lbf/in^2

(i) The design moment at the critical section is most nearly

(A) 350 ft-kips
(B) 380 ft-kips
(C) 410 ft-kips
(D) 510 ft-kips

(j) The number of no. 6 bars needed as reinforcement for the moment of part (i) is most nearly

(A) 12
(B) 14
(C) 19
(D) 24

4. (*Time limit: one hour*) A 10 ft by 14 ft rectangular footing supports a 12 in by 16 in reinforced concrete column placed at its center. The strength of the concrete for the column and the footing are 5000 lbf/in^2 and 3000 lbf/in^2, respectively. All steel is f_y = 60,000 lbf/in^2. The relative orientation of the column with respect to the footing is shown.

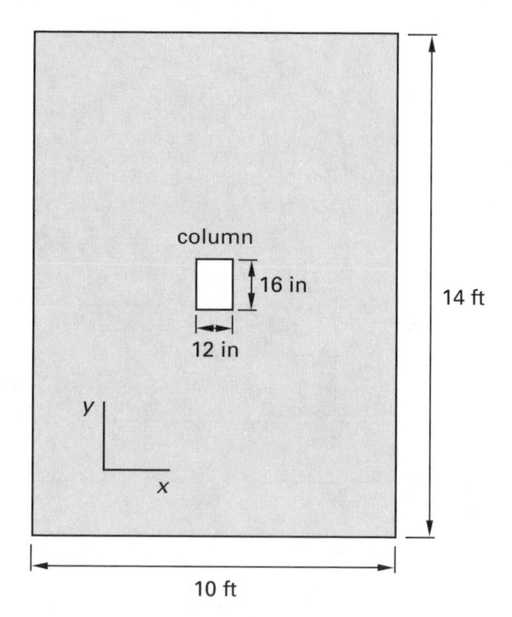

(a) Assume the column is subjected to a factored axial load of 650 kips. The minimum area of dowels needed to transfer the axial force from the column to the footing is most nearly

(A) 0 in^2
(B) 1.0 in^2
(C) 1.5 in^2
(D) 3.8 in^2

(b) Assume the column is subjected to axial compression only. If the dowels are no. 6 bars with 90° standard hooks, the minimum depth of the footing for which they will be fully anchored is most nearly

(A) 16 in
(B) 18 in
(C) 25 in
(D) 35 in

In parts (c) through (e), assume the footing is 24 in thick with an effective depth of 20 in.

(c) The minimum permitted amount of steel oriented parallel to the 10 ft side is most nearly

(A) 6.0 in^2
(B) 7.3 in^2
(C) 11 in^2
(D) 12 in^2

(d) As controlled by punching shear, the maximum factored axial load that can be resisted is most nearly

(A) 270 kips
(B) 480 kips
(C) 540 kips
(D) 610 kips

(e) As controlled by one-way shear, the maximum axial load that can be transferred by the footing is most nearly

(A) 300 kips
(B) 590 kips
(C) 670 kips
(D) 1200 kips

In parts (f) through (h), assume the column is subjected to a factored axial force of 400 kips and a factored moment of 600 ft-kips acting about the y axis.

(f) The punching shear stress at the critical section is most nearly

(A) 140 lbf/in^2
(B) 220 lbf/in^2
(C) 300 lbf/in^2
(D) 350 lbf/in^2

(g) The one-way shear stress at the critical section is most nearly

(A) 21 lbf/in^2
(B) 32 lbf/in^2
(C) 49 lbf/in^2
(D) 55 lbf/in^2

(h) The design moment for bars parallel to the x direction is most nearly

(A) 470 ft-kips
(B) 530 ft-kips
(C) 660 ft-kips
(D) 1100 ft-kips

(i) If the required steel area for bars parallel to the x direction (see illustration) is 12 in^2 for some applied load, the spacing of no. 6 bars parallel to the x direction in the region near the column is most nearly

(A) 4.0 in
(B) 4.5 in
(C) 5.2 in
(D) 6.0 in

(j) The available anchorage length for bars in the y direction is most nearly

(A) 69 in
(B) 74 in
(C) 78 in
(D) 84 in

5. A square footing is to carry a live load of 240,000 lbf that is transmitted through a 16 in square column. (The column dead load can be disregarded, although the footing dead load cannot.) The allowable soil pressure is 4000 lbf/ft^2, and the top of the footing is level with the surrounding soil surface. Material properties are f'_c = 3000 lbf/in^2 and f_y = 40,000 lbf/in^2. The footing thickness is 20 in. Design the footing.

6. An 18 in square column supports 200,000 lbf dead load and 145,000 lbf live load. The allowable soil pressure is 4000 lbf/ft^2. Material properties are f'_c = 3000 lbf/in^2 and f_y = 40,000 lbf/in^2. The column steel consists of ten no. 9 bars. Design the footing.

7. A 10 ft wide combined footing carries the service loads and moments shown. Loads and moments are transmitted through 18 in square columns. The footing weight can be disregarded. (a) What dimension L will result in a uniform soil pressure? (b) Draw the shear and moment diagrams. (c) Does the footing need a layer of top steel?

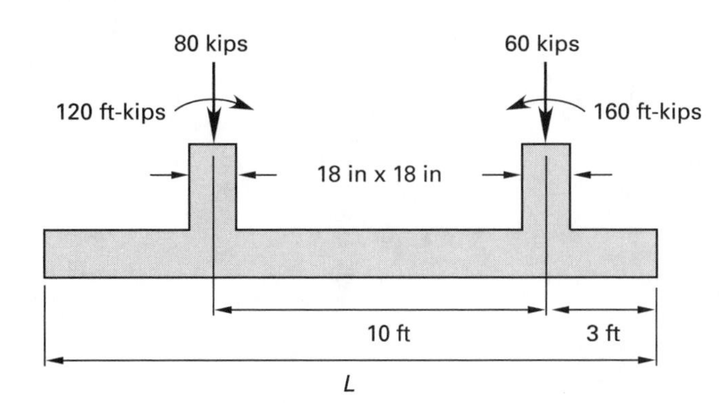

8. A 14 in square column carries a live load of 200 kips and a live moment of 100 ft-kips, as shown. The column self-weight can be disregarded. The allowable soil pressure is 6000 lbf/ft^2. The concrete's compressive strength is 3000 lbf/in^2. The soil specific weight is 110 lbf/ft^3. The top of the footing is even with the soil surface. (a) Size the footing. (Do not design the steel reinforcement.) (b) What are the maximum and minimum pressures along the footing base? (c) What thickness of footing would be required if shear reinforcement was to be completely eliminated? (d) What is the maximum moment the footing must resist?

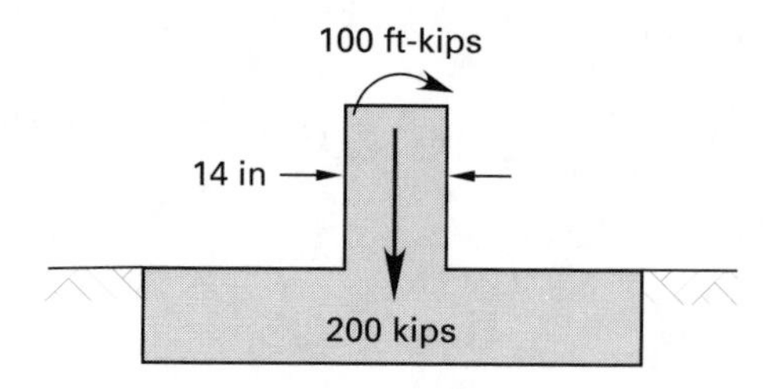

SOLUTIONS

1. The solution and answer remain the same whether ACI 318 Apps. B and C are used or ACI 318 Unified Design Method and Ch. 9 are used.

The service loads are not factored when sizing the footing.

$$P_s = 100 \text{ kips} + 58 \text{ kips} = 158 \text{ kips}$$
$$V_s = 2 \text{ kips} + 4 \text{ kips} = 6 \text{ kips}$$
$$H = 4 \text{ ft}$$
$$h = 1.5 \text{ ft}$$

The service moment is

$$M_s = 20 \text{ ft-kips} + 40 \text{ ft-kips} + (2 \text{ kips} + 4 \text{ kips})\left(\frac{18 \text{ in}}{12 \frac{\text{in}}{\text{ft}}}\right) = 69 \text{ ft-kips}$$

From Eq. 55.1,

$$2\,\frac{\text{kips}}{\text{ft}^2} = \frac{158 \text{ kips}}{A_f} + \left(0.15\,\frac{\text{kip}}{\text{ft}^3}\right)(1.5 \text{ ft}) + \left(0.11\,\frac{\text{kip}}{\text{ft}^3}\right)(4 \text{ ft} - 1.5 \text{ ft}) + \frac{(69 \text{ ft-kips})\left(\frac{B}{2}\right)}{I_f}$$

$$1.5\,\frac{\text{kips}}{\text{ft}^2} = \frac{158 \text{ kips}}{A_f} + \frac{(34.5 \text{ ft-kips})B}{I_f}$$

Since the footing is square,

$$A_f = B^2$$
$$I_f = \tfrac{1}{12}B^4$$

Therefore,

$$1.5\,\frac{\text{kips}}{\text{ft}^2} = \frac{158 \text{ kips}}{B^2} + \frac{414 \text{ ft-kips}}{B^3}$$

The above is most easily solved by trial substitution.

$$B = 11.4 \text{ ft} \quad [\text{use 11.5 ft by 11.5 ft}]$$

2. Use either **ACI 318 Apps. B and C** or **ACI 318 Ch. 9** (see p. 55-6) to solve this problem.

Solution Using ACI 318 Apps. B and C

Consider a unit length of footing. Determine the footing width. The service loads are not factored when sizing the footing.

Assume $h = 12$ in. Use Eq. 55.1.

$$q_a = \frac{P_s}{A} + \gamma_c h + \gamma_s (H - h)$$

$$5\ \frac{\text{kips}}{\text{ft}^2} = \frac{22\ \text{kips}}{B(1\ \text{ft})} + \left(0.15\ \frac{\text{kip}}{\text{ft}^3}\right)(1\ \text{ft}) + \left(0.12\ \frac{\text{kip}}{\text{ft}^3}\right)(5\ \text{ft} - 1\ \text{ft})$$

$$B = 5.03\ \text{ft}\quad [\text{use 5 ft}]$$

Use Eq. 55.3.

$$q_u = \frac{(1.5)\left(22\ \frac{\text{kips}}{\text{ft}}\right)}{5\ \text{ft}} = 6.6\ \text{kips/ft}^2$$

Use Eq. 55.6.

$$v_c = 2\sqrt{f'_c} = \frac{2\sqrt{3000\ \frac{\text{lbf}}{\text{in}^2}}\left(144\ \frac{\text{in}^2}{\text{ft}^2}\right)}{1000\ \frac{\text{lbf}}{\text{kip}}} = 15.77\ \text{kips/ft}^2$$

Obtain the required depth for shear.

$$d = \frac{q_u(B-t)}{2(q_u + \phi v_c)} = \frac{\left(6.6\ \frac{\text{kips}}{\text{ft}^2}\right)(5\ \text{ft} - 0.67\ \text{ft})\left(12\ \frac{\text{in}}{\text{ft}}\right)}{(2)\left(6.6\ \frac{\text{kips}}{\text{ft}^2} + (0.85)\left(15.77\ \frac{\text{kips}}{\text{ft}^2}\right)\right)} = \boxed{8.57\ \text{in}}$$

Using no. 4 bars in each direction,

$$h \geq 8.57\ \text{in} + \left(\frac{1}{2}\right)(0.5\ \text{in}) + 0.5\ \text{in} + 3\ \text{in} = \boxed{12.32\ \text{in}\quad [\text{use 13 in}]}$$

(The assumed value of h of 13 in is close to the h selected, so there is no need to revise the calculations for the width B.)

Determine the flexural steel.

$$d = 13\ \text{in} - 3\ \text{in} - 0.5\ \text{in} - 0.25\ \text{in} = 9.25\ \text{in}$$

$$t = \frac{8\ \text{in}}{12\ \frac{\text{in}}{\text{ft}}} = 0.67\ \text{ft}$$

The moment at the critical section is

$$M_u = \frac{q_u\left(\frac{B}{2} - \frac{t}{2}\right)^2}{2} = \frac{q_u(B-t)^2}{8} = \frac{\left(6.6\ \frac{\text{kips}}{\text{ft}^2}\right)(5\ \text{ft} - 0.67\ \text{ft})^2}{8} = 15.47\ \text{ft-kips/ft}$$

Assume $\lambda = 0.1d = (0.1)(9.25\ \text{in}) = 0.925$ in.

Assume the section is tension controlled. The area of steel is

$$A_s = \frac{M_u}{\phi f_y (d - \lambda)} = \frac{\left(15.47\ \frac{\text{ft-kips}}{\text{ft}}\right)\left(12\ \frac{\text{in}}{\text{ft}}\right)}{(0.9)\left(60\ \frac{\text{kips}}{\text{in}^2}\right)(9.25\ \text{in} - 0.925\ \text{in})} = 0.413\ \text{in}^2/\text{ft}$$

Check λ.

$$A_c = \frac{A_s f_y}{0.85 f'_c} = \frac{(0.413\ \text{in}^2)\left(60\ \frac{\text{kips}}{\text{in}^2}\right)}{(0.85)\left(3\ \frac{\text{kips}}{\text{in}^2}\right)} = 9.72\ \text{in}^2$$

$$\lambda = \frac{9.72\ \text{in}^2}{(12\ \text{in})(2)} = 0.405\ \text{in}$$

Check whether the section is tension controlled.

$$c = \frac{(0.404\ \text{in})(2)}{0.85} = 0.95\ \text{in}$$

$$\epsilon_t = \left(\frac{d-c}{c}\right)0.003 = \left(\frac{9.25\ \text{in} - 0.95\ \text{in}}{0.95\ \text{in}}\right)(0.003) = 0.026 > 0.005$$

Therefore, the section is tension controlled.

The revised A_s is

$$A_s = \frac{\left(15.47\ \frac{\text{ft-kips}}{\text{ft}}\right)\left(12\ \frac{\text{in}}{\text{ft}}\right)}{(0.9)\left(60\ \frac{\text{kips}}{\text{in}^2}\right)(9.25\ \text{in} - 0.405\ \text{in})} = 0.39\ \text{in}^2/\text{ft}$$

The minimum steel is

$$A_{s,\text{min}} = (0.0018)(12\ \text{in})(12\ \text{in}) = 0.259\ \text{in}^2/\text{ft}\quad [\text{does not control}]$$

The number of no. 4 bars per foot is

$$\frac{A_s}{A_b} = \frac{0.39 \ \frac{\text{in}^2}{\text{ft}}}{0.2 \ \text{in}^2} = 2.0$$

The spacing of no. 4 bars is

$$\frac{12 \ \text{in}}{2.0} = 6 \ \text{in}$$

The longitudinal (temperature) steel is

$$(0.0018)(5 \ \text{ft})\left(12 \ \frac{\text{in}}{\text{ft}}\right)(12 \ \text{in}) = 1.30 \ \text{in}^2$$

Use eight no. 4 bars.

Space the outside bars to provide 3 in of cover.

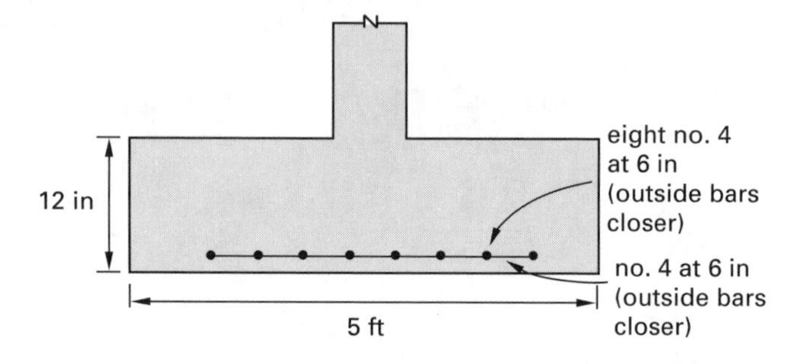

(The next step would be to check development length and the need for hooked bars.)

Solution Using ACI 318 Ch. 9

The ϕ factor for shear is 0.75. Therefore, the required depth of the footing for shear resistance is

$$\begin{aligned} d &= \frac{q_u(B-t)}{2(q_u + \phi v_c)} \\ &= \frac{\left(6.6 \ \frac{\text{kips}}{\text{ft}^2}\right)(5 \ \text{ft} - 0.67 \ \text{ft})\left(12 \ \frac{\text{in}}{\text{ft}}\right)}{(2)\left(6.6 \ \frac{\text{kips}}{\text{ft}^2} + (0.75)\left(15.77 \ \frac{\text{kips}}{\text{ft}^2}\right)\right)} \\ &= \boxed{9.305 \ \text{in}} \end{aligned}$$

Using no. 4 bars in each direction,

$$\begin{aligned} h &\geq 9.305 \ \text{in} + \left(\frac{1}{2}\right)(0.5 \ \text{in}) + 0.5 \ \text{in} + 3 \ \text{in} \\ &= \boxed{13.06 \ \text{in} \quad [\text{use 13 in}]} \end{aligned}$$

(The assumed value of h of 13 in is close to the h selected, so there is no need to revise the calculations for the width B.)

Determine the flexural steel.

$$d = 13 \ \text{in} - 3 \ \text{in} - 0.5 \ \text{in} - 0.25 \ \text{in} = 9.25 \ \text{in}$$

$$t = \frac{8 \ \text{in}}{12 \ \frac{\text{in}}{\text{ft}}} = 0.67 \ \text{ft}$$

The moment at the critical section is

$$\begin{aligned} M_u &= \frac{q_u\left(\frac{B}{2} - \frac{t}{2}\right)^2}{2} = \frac{q_u(B-t)^2}{8} \\ &= \frac{\left(6.6 \ \frac{\text{kips}}{\text{ft}^2}\right)(5 \ \text{ft} - 0.67 \ \text{ft})^2}{8} \\ &= 15.47 \ \text{ft-kips/ft} \end{aligned}$$

Assume $\lambda = 0.1d = (0.1)(9.25 \ \text{in}) = 0.925 \ \text{in}$.

Assume the section is tension controlled. The area of steel is

$$\begin{aligned} A_s &= \frac{M_u}{\phi f_y(d-\lambda)} \\ &= \frac{\left(15.47 \ \frac{\text{ft-kips}}{\text{ft}}\right)\left(12 \ \frac{\text{in}}{\text{ft}}\right)}{(0.9)\left(60 \ \frac{\text{kips}}{\text{in}^2}\right)(9.25 \ \text{in} - 0.925 \ \text{in})} \\ &= 0.413 \ \text{in}^2/\text{ft} \end{aligned}$$

Check λ.

$$A_c = \frac{A_s f_y}{0.85 f'_c} = \frac{(0.413 \ \text{in}^2)\left(60 \ \frac{\text{kips}}{\text{in}^2}\right)}{(0.85)\left(3 \ \frac{\text{kips}}{\text{in}^2}\right)} = 9.72 \ \text{in}^2$$

$$\lambda = \frac{9.72 \ \text{in}^2}{(12 \ \text{in})(2)} = 0.405 \ \text{in}$$

Check whether the section is tension controlled.

$$c = \frac{(0.404 \ \text{in})(2)}{0.85} = 0.95 \ \text{in}$$

$$\begin{aligned} \epsilon_t &= \left(\frac{d-c}{c}\right)0.003 = \left(\frac{9.25 \ \text{in} - 0.95 \ \text{in}}{0.95 \ \text{in}}\right)(0.003) \\ &= 0.026 > 0.005 \end{aligned}$$

Therefore, the section is tension controlled.

The revised A_s is

$$\begin{aligned} A_s &= \frac{\left(15.47 \ \frac{\text{ft-kips}}{\text{ft}}\right)\left(12 \ \frac{\text{in}}{\text{ft}}\right)}{(0.9)\left(60 \ \frac{\text{kips}}{\text{in}^2}\right)(9.25 \ \text{in} - 0.405 \ \text{in})} \\ &= 0.39 \ \text{in}^2/\text{ft} \end{aligned}$$

The minimum steel is

$$A_{s,\min} = (0.0018)(12\ \text{in})(12\ \text{in}) = 0.259\ \text{in}^2/\text{ft}$$

[does not control]

The number of no. 4 bars per foot is

$$\frac{A_s}{A_b} = \frac{0.39\ \frac{\text{in}^2}{\text{ft}}}{0.2\ \text{in}^2} = 2.0$$

The spacing of no. 4 bars is

$$\frac{12\ \text{in}}{2.0} = 6\ \text{in}$$

The longitudinal (temperature) steel is

$$(0.0018)(5\ \text{ft})\left(12\ \frac{\text{in}}{\text{ft}}\right)(12\ \text{in}) = 1.30\ \text{in}^2$$

Use eight no. 4 bars.

Space the outside bars to provide 3 in of cover.

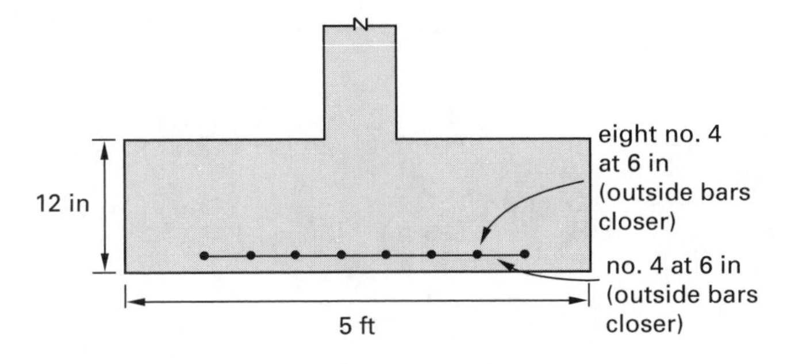

(The next step would be to check development length and the need for hooked bars.)

3. The solution and answers are the same whether ACI 318 App. C or ACI 318 Ch. 9 is used.

(a)

$$P = \left(3\ \frac{\text{kips}}{\text{ft}^2}\right)(10\ \text{ft})(10\ \text{ft}) = \boxed{300\ \text{kips}}$$

The answer is (C).

(b) $I_f = \left(\frac{1}{12}\right)(LB^3) = \left(\frac{1}{12}\right)(10\ \text{ft})^4 = 833.3\ \text{ft}^4$

Use Eq. 55.1. Disregard the correction for burial.

$$q_a = \frac{P_s}{A_f} + \gamma_c h + \gamma_s(H-h) + \frac{M_s\left(\frac{B}{2}\right)}{I_f}$$

$$3\ \frac{\text{kips}}{\text{ft}^2} = \frac{200\ \text{kips}}{100\ \text{ft}^2} + \frac{M_s\left(\frac{10\ \text{ft}}{2}\right)}{833.3\ \text{ft}^4}$$

$$M_s = \boxed{166.7\ \text{ft-kips}}$$

The answer is (B).

(c) Leaving room for bars and cover, assume

$$d \approx h - 4\ \text{in} = 16\ \text{in} - 4\ \text{in} = 12\ \text{in}$$

$$b_1 = b_2 = \text{column size} + 2\left(\frac{d}{2}\right) = 12\ \text{in} + (2)\left(\frac{12\ \text{in}}{2}\right) = 24\ \text{in}$$

$$A_p = (4)(24\ \text{in})(12\ \text{in}) = \boxed{1152\ \text{in}^2}$$

The answer is (C).

(d) Use Eq. 55.14.

$$R = \frac{P_u b_1 b_2}{A_f} = \frac{(340\ \text{kips})(24\ \text{in})^2}{(120\ \text{in})^2} = 13.6\ \text{kips}$$

$$v_u = \frac{P_u - R}{A_p} = \frac{340\ \text{kips} - 13.6\ \text{kips}}{1152\ \text{in}^2} = 0.283\ \text{kip/in}^2 \quad \boxed{(283\ \text{lbf/in}^2)}$$

The answer is (C).

(e) The distance from the critical sections to the edge is

$$e = \frac{10\ \text{ft} - 1\ \text{ft}}{2} - 1\ \text{ft} = 3.5\ \text{ft}$$

$$q_u = \frac{340\ \text{kips}}{100\ \text{ft}^2} = 3.4\ \text{kips/ft}^2$$

$$v_u = \frac{\left(3.4\ \frac{\text{kips}}{\text{ft}^2}\right)(3.5\ \text{ft})\left(1000\ \frac{\text{lbf}}{\text{kip}}\right)}{(1\ \text{ft})\left(144\ \frac{\text{in}^2}{\text{ft}^2}\right)} = \boxed{82.64\ \text{lbf/in}^2}$$

The answer is (B).

(f) The critical section for flexure is at the face of the column, at 4.5 ft from the edge.

$$M_u = \left(\frac{340\ \text{kips}}{(10\ \text{ft})(10\ \text{ft})}\right)(10\ \text{ft})(4.5\ \text{ft})\left(\frac{4.5}{2}\ \text{ft}\right) = \boxed{344.3\ \text{ft-kips}}$$

The answer is (C).

(g) From Eq. 55.16,

$$J = \left(\frac{(12\ \text{in})(24\ \text{in})^3}{6}\right) \times \left(1 + \left(\frac{12\ \text{in}}{24\ \text{in}}\right)^2 + (3)\left(\frac{24\ \text{in}}{24\ \text{in}}\right)\right) = 117{,}504\ \text{in}^4$$

$$\gamma_v = 1 - \frac{1}{1 + \frac{2}{3}\sqrt{\frac{24\ \text{in}}{24\ \text{in}}}} = 0.4$$

Structural

Use Eq. 55.13.

$$v_u = \frac{340 \text{ kips} - 13.6 \text{ kips}}{1152 \text{ in}^2} + \frac{(0.4)(150 \text{ ft-kips})\left(12 \ \frac{\text{in}}{\text{ft}}\right)(0.5)(24 \text{ in})}{117{,}504 \text{ in}^4}$$

$$= 0.283 \ \frac{\text{kip}}{\text{in}^2} + 0.074 \ \frac{\text{kip}}{\text{in}^2}$$

$$= 0.356 \text{ kip/in}^2 \quad \boxed{(356 \text{ lbf/in}^2)}$$

The answer is (C).

(h) Determine the pressure distribution. Use Eq. 55.1.

$$q = \frac{340 \text{ kips}}{100 \text{ ft}^2} \pm \frac{(150 \text{ ft-lbf})\left(\frac{10 \text{ ft}}{2}\right)}{833 \text{ ft}^4}$$

$$= 3.4 \ \frac{\text{kips}}{\text{ft}^2} \pm 0.9 \ \frac{\text{kip}}{\text{ft}^2} = (4.3 \text{ kips/ft}^2, \ 2.5 \text{ kips/ft}^2)$$

At the critical section,

$$q = 4.3 \ \frac{\text{kips}}{\text{ft}^2} - \left(\frac{3.5 \text{ ft}}{10 \text{ ft}}\right)\left(4.3 \ \frac{\text{kips}}{\text{ft}^2} - 2.5 \ \frac{\text{kips}}{\text{ft}^2}\right)$$

$$= 3.67 \text{ kips/ft}^2$$

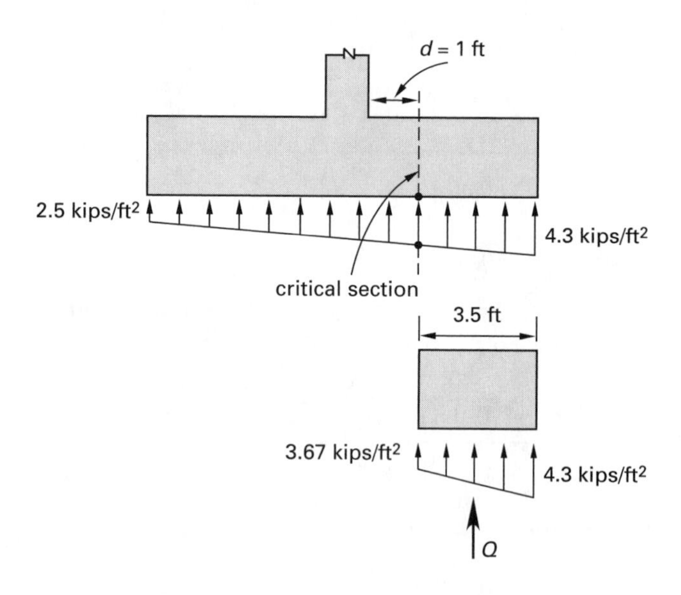

$$Q = \left(4.3 \ \frac{\text{kips}}{\text{ft}^2} + 3.67 \ \frac{\text{kips}}{\text{ft}^2}\right)\left(\frac{3.5 \text{ ft}}{2}\right)(10 \text{ ft})$$

$$= 139.5 \text{ kips}$$

$$v_u = \frac{(139.5 \text{ kips})\left(1000 \ \frac{\text{lbf}}{\text{kip}}\right)}{(10 \text{ ft})(1 \text{ ft})\left(144 \ \frac{\text{in}^2}{\text{ft}^2}\right)} = \boxed{96.86 \text{ lbf/in}^2}$$

The answer is (B).

(i) From the free-body diagram,

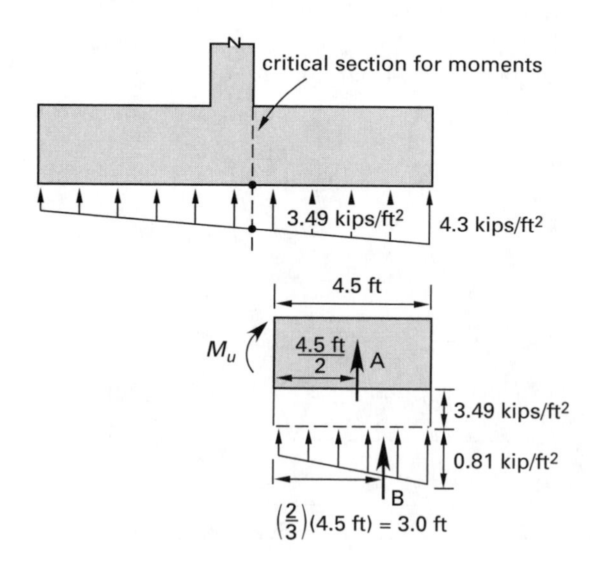

$$A = \left(3.49 \ \frac{\text{kips}}{\text{ft}^2}\right)(10 \text{ ft})(4.5 \text{ ft}) = 157.05 \text{ kips}$$

$$B = \left(0.81 \ \frac{\text{kips}}{\text{ft}^2}\right)(10 \text{ ft})(4.5 \text{ ft})\left(\tfrac{1}{2}\right) = 18.23 \text{ kips}$$

$$M_u = (157.05 \text{ kips})\left(\frac{4.5 \text{ ft}}{2}\right) + (18.23 \text{ kips})\left(\tfrac{2}{3}\right)(4.5 \text{ ft})$$

$$= \boxed{408.05 \text{ ft-kips}}$$

The answer is (C).

(j) Assume $\lambda = 0.1d$.

$$A_s = \frac{M_u}{\phi f_y (d - \lambda)}$$

$$= \frac{(408.05 \text{ ft-kips})\left(12 \ \frac{\text{in}}{\text{ft}}\right)}{(0.9)\left(60 \ \frac{\text{kips}}{\text{in}^2}\right)(12 \text{ in} - 1.2 \text{ in})}$$

$$= 8.40 \text{ in}^2$$

The number of no. 6 bars is

$$\frac{A_s}{A_b} = \frac{8.4 \text{ in}^2}{0.44 \text{ in}^2} = \boxed{19.09}$$

A second iteration from the beginning would refine this value.

The answer is (C).

4. Use either **ACI 318 App. C** or **ACI 318 Ch. 9** (see p. 55-12) to solve this problem.

Solution Using ACI 318 App. C

(a) $A_c = (16 \text{ in})(12 \text{ in}) = 192 \text{ in}^2$

The factored load is given as $P_u = 650$ kips.

The bearing strength of the column is

$$\begin{aligned}\phi(0.85f'_c)A_c &= (0.70)(0.85)\left(5\ \frac{\text{kips}}{\text{in}^2}\right)(192 \text{ in}^2)\\ &= 571.2 \text{ kips}\end{aligned}$$

For the bearing strength of the footing, which uses a different concrete, the α factor may be used.

The aspect ratio of the column is 16 in/12 in = 1.333. The longest footing side length that can be used in the calculation of A_{ff} is

$$(1.333)(10 \text{ ft})\left(12\ \frac{\text{in}}{\text{ft}}\right) = 160 \text{ in}$$

$$\alpha = \sqrt{\frac{A_{ff}}{A_c}} = \sqrt{\frac{(160 \text{ in})(120 \text{ in})}{192 \text{ in}^2}} = 10 > 2.0$$

Use $\alpha = 2.0$. The bearing strength of the footing is

$$\begin{aligned}\phi\alpha(0.85f'_c)A_c &= (0.70)(2.0)(0.85)\left(3\ \frac{\text{kips}}{\text{in}^2}\right)(192 \text{ in}^2)\\ &= 685.4 \text{ kips}\end{aligned}$$

The smaller value between the bearing strength of the column and the bearing strength of the footing is the controlling strength. Therefore, $\phi P_{n\text{-}b} = 571.2$ kips. Since the $P_u > \phi P_{n,b}$, the excess that must be carried by the dowels is

$$650 \text{ kips} - 571.2 \text{ kips} = 78.8 \text{ kips}$$

Design the dowel steel. The dowel steel area reduces the concrete bearing surface area by the same amount. Subtract the concrete bearing strength from the steel yield strength.

$$78.8 \text{ kips} = \phi P_{n,\text{steel}} = \phi A_s(f_y - 0.85f'_{c,\text{column}})$$

$$\begin{aligned}A_s &= \frac{\phi P_{n,\text{steel}}}{\phi(f_y - 0.85f'_{c,\text{column}})}\\ &= \frac{78.8 \text{ kips}}{(0.70)\left(60\ \frac{\text{kips}}{\text{in}^2} - (0.85)\left(5\ \frac{\text{kips}}{\text{in}^2}\right)\right)}\\ &= 2.019 \text{ in}^2\end{aligned}$$

Check the minimum requirement.

$$A_{db,\min} = 0.005A_c = (0.005)(192 \text{ in}^2) = 0.96 \text{ in}^2$$
$$2.019 \text{ in}^2 > 0.96 \text{ in}^2 \quad \text{(ok)}$$

Use $A_s = 2 \text{ in}^2$.

The answer is (C).

(b) Hooks are ineffective for compression. The development length of a bar in compression does not include the bend length. Use Eq. 55.31.

$$\begin{aligned}l_d &= 0.02d_b\left(\frac{f_y}{\sqrt{f'_c}}\right) = \frac{(0.02)(0.75 \text{ in})\left(60{,}000\ \frac{\text{lbf}}{\text{in}^2}\right)}{\sqrt{3000\ \frac{\text{lbf}}{\text{in}^2}}}\\ &= 16.43 \text{ in} \quad \text{[controls]}\\ &= 0.0003d_bf_y = (0.0003)(0.75 \text{ in})\left(60{,}000\ \frac{\text{kips}}{\text{in}^2}\right)\\ &= 13.5 \text{ in}\end{aligned}$$

Since the bar is hooked, it is necessary to add the radius of the bend to the cover and the steel layer depth.

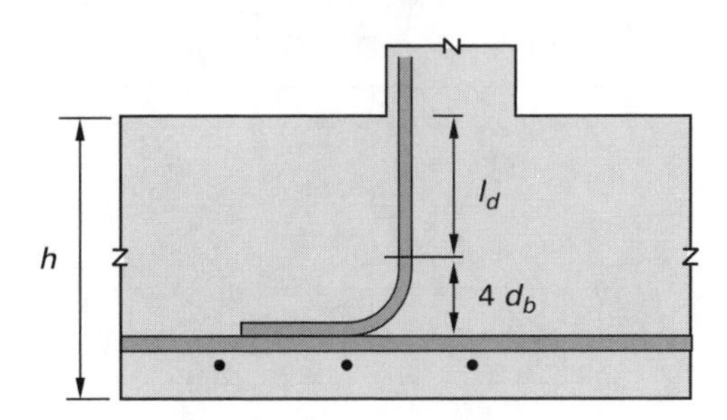

For no. 8 reinforcement,

$$h = (16.43 \text{ in}) + (4)(0.75 \text{ in}) + 2 \text{ in} + 3 \text{ in} = 24.43 \text{ in}$$

The answer is (C).

(c) The minimum steel permitted is

$$(0.0018)(14 \text{ ft})\left(12\ \frac{\text{in}}{\text{ft}}\right)(20 \text{ in}) = 6.05 \text{ in}^2$$

The answer is (A).

(d)
$$b_1 = 12 \text{ in} + 20 \text{ in} = 32 \text{ in}$$
$$b_2 = 16 \text{ in} + 20 \text{ in} = 36 \text{ in}$$

Use Eq. 55.12.

$$\begin{aligned}A_p &= 2(b_1 + b_2)d = (2)(32 \text{ in} + 36 \text{ in})(20 \text{ in})\\ &= 2720 \text{ in}^2\end{aligned}$$

Structural

Use Eq. 55.14.

$$R = \frac{P_u(32\text{ in})(36\text{ in})}{(120\text{ in})(168\text{ in})} = 0.0571P_u$$

Use Eq. 55.13.

$$v_u = \frac{P_u - 0.0571P_u}{2720\text{ in}^2} = \left(3.466 \times 10^{-4}\right) P_u$$

The allowable punching shear stress is given by Eq. 55.17.

$$\phi v_c = (0.85)(4)\sqrt{3000\ \frac{\text{lbf}}{\text{in}^2}} = 186\text{ lbf/in}^2$$

Equating v_u to ϕv_c,

$$P_u = \frac{186\ \frac{\text{lbf}}{\text{in}^2}}{3.466 \times 10^{-4}} = 5.37 \times 10^5\text{ lbf} \quad \boxed{(540\text{ kips})}$$

The answer is (C).

(e) The pressure q_u is given by

$$q_u = \frac{P_u}{A_f} = \frac{P_u}{140\text{ ft}^2} = 0.0071P_u$$

$$e = \frac{\dfrac{(14\text{ ft})\left(12\ \frac{\text{in}}{\text{ft}}\right) - 16\text{ in}}{2} - 20\text{ in}}{12\ \frac{\text{in}}{\text{ft}}} = 4.67\text{ ft}$$

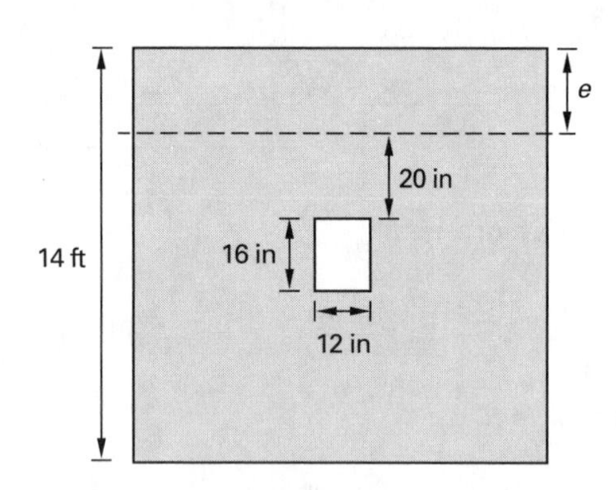

Use Eq. 55.11.

$$v_u = \frac{q_u e}{d} = \frac{(0.0071P_u)(4.67\text{ ft})}{1.67\text{ ft}} = 0.02P_u$$

The allowable concrete stress, using Eqs. 55.5 and 55.6, is

$$\phi v_c = (0.85)(2)\sqrt{f'_c} = \frac{(0.85)(2)\sqrt{3000\ \frac{\text{lbf}}{\text{in}^2}}\left(144\ \frac{\text{in}^2}{\text{ft}^2}\right)}{1000\ \frac{\text{lbf}}{\text{kip}}}$$

$$= 13.41\text{ kips/ft}^2$$

Equating v_u to ϕv_c,

$$P_u = \frac{13.41\ \frac{\text{kips}}{\text{ft}^2}}{0.02\text{ ft}^2} = \boxed{670.4\text{ kips}}$$

The answer is (C).

(f) From Eq. 55.16,

$$J = \left(\frac{db_1^3}{6}\right)\left(1 + \left(\frac{d}{b_1}\right)^2 + \frac{3b_2}{b_1}\right)$$

$$= \left(\frac{(20\text{ in})(32\text{ in})^3}{6}\right)\left(1 + \left(\frac{20\text{ in}}{32\text{ in}}\right)^2 + (3)\left(\frac{36\text{ in}}{32\text{ in}}\right)\right)$$

$$= 520{,}530\text{ in}^4$$

Use Eq. 55.14.

$$R = \frac{P_u b_1 b_2}{A_f} = \frac{(400\text{ kips})(32\text{ in})(36\text{ in})}{(140\text{ ft}^2)\left(144\ \frac{\text{in}^2}{\text{ft}^2}\right)} = 22.86\text{ kips}$$

$$\gamma_v = 1 - \frac{1}{1 + \frac{2}{3}\sqrt{\frac{b_1}{b_2}}} = 1 - \frac{1}{1 + \frac{2}{3}\sqrt{\frac{32\text{ in}}{36\text{ in}}}} = 0.386$$

Use Eq. 55.13.

$$v_u = \frac{P_u - R}{A_p} + \frac{\gamma_u M_u(0.5b_1)}{J}$$

$$= \frac{400\text{ kips} - 22.86\text{ kips}}{2720\text{ in}^2} + \frac{(0.386)(600\text{ ft-kips})\left(12\ \frac{\text{in}}{\text{ft}}\right)(0.5)(32\text{ in})}{520{,}530\text{ in}^4}$$

$$= 0.224\text{ kip/in}^2 \quad \boxed{(224\text{ lbf/in}^2)}$$

The answer is (B).

(g) Determine the pressure distribution.

$$I_f = \left(\tfrac{1}{12}\right)(14\text{ ft})(10\text{ ft})^3 = 1166.7\text{ ft}^4$$

From Eq. 55.1,

$$q_u = \frac{400\text{ kips}}{140\text{ ft}^2} \pm \frac{(600\text{ ft-kips})(5\text{ ft})}{1166.7\text{ ft}^4}$$

$$= 2.86\ \frac{\text{kips}}{\text{ft}^2} \pm 2.57\ \frac{\text{kips}}{\text{ft}^2}$$

$$= \left(5.43\text{ kips/ft}^2,\ 0.29\text{ kip/ft}^2\right)$$

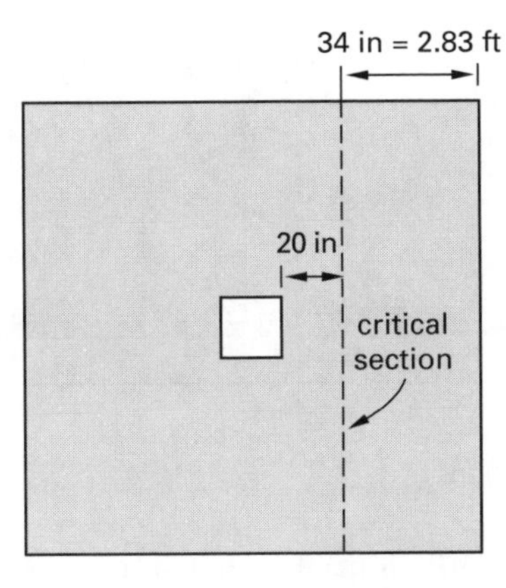

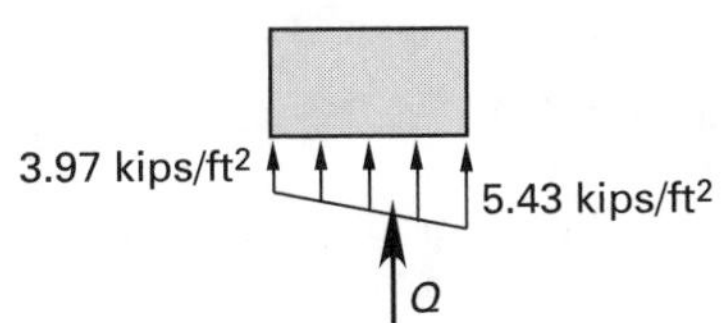

$$Q = \frac{\left(5.43\ \frac{\text{kips}}{\text{ft}^2} + 3.97\ \frac{\text{kips}}{\text{ft}^2}\right)(14\ \text{ft})(2.83\ \text{ft})}{2}$$

$$= 186.21\ \text{kips}$$

$$v_u = \frac{186.21\ \text{kips}}{(14\ \text{ft})\left(12\ \frac{\text{in}}{\text{ft}}\right)(20\ \text{in})}$$

$$= 0.0554\ \text{kip/in}^2 \quad \boxed{(55.4\ \text{lbf/in}^2)}$$

The answer is (D).

(h)

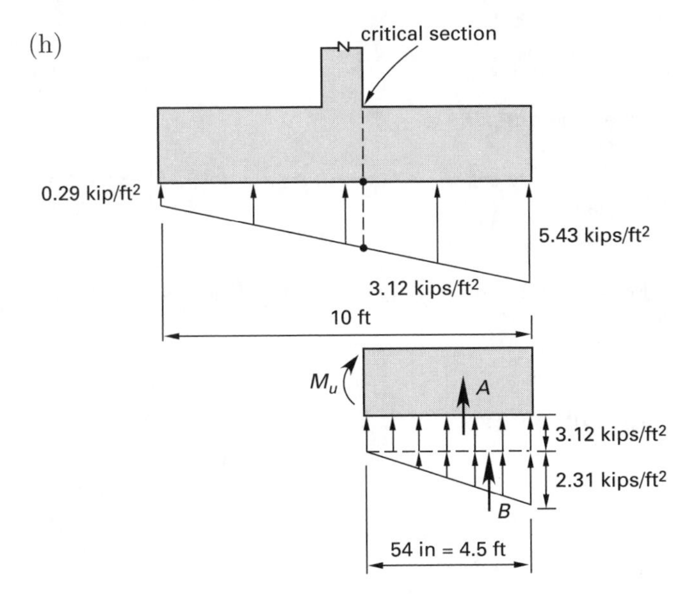

$$A = \left(3.12\ \frac{\text{kips}}{\text{ft}^2}\right)(14\ \text{ft})(4.5\ \text{ft})$$

$$= 196.6\ \text{kips}$$

$$B = \left(\tfrac{1}{2}\right)\left(2.31\ \frac{\text{kips}}{\text{ft}^2}\right)(14\ \text{ft})(4.5\ \text{ft})$$

$$= 72.77\ \text{kips}$$

$$M_u = (196.6\ \text{kips})\left(\frac{4.5\ \text{ft}}{2}\right) + \left(\tfrac{2}{3}\right)(72.77\ \text{kips})(4.5\ \text{ft})$$

$$= \boxed{660.7\ \text{ft-kips}}$$

The answer is (C).

(i)

$$\beta = \frac{14\ \text{ft}}{10\ \text{ft}} = 1.4$$

Use Eq. 55.25.

$$A_1 = A_{sd}\left(\frac{2}{\beta+1}\right) = (12\ \text{in}^2)\left(\frac{2}{1.4+1}\right)$$

$$= 10\ \text{in}^2$$

The number of no. 6 bars is

$$\frac{A_s}{A_b} = \frac{10\ \text{in}^2}{0.44} = 22.73 \quad (23\ \text{bars})$$

The width of the band is 10 ft.

The spacing is

$$s = \frac{(10\ \text{ft})\left(12\ \frac{\text{in}}{\text{ft}}\right)}{23} = \boxed{5.2\ \text{in}}$$

The answer is (C).

(j)

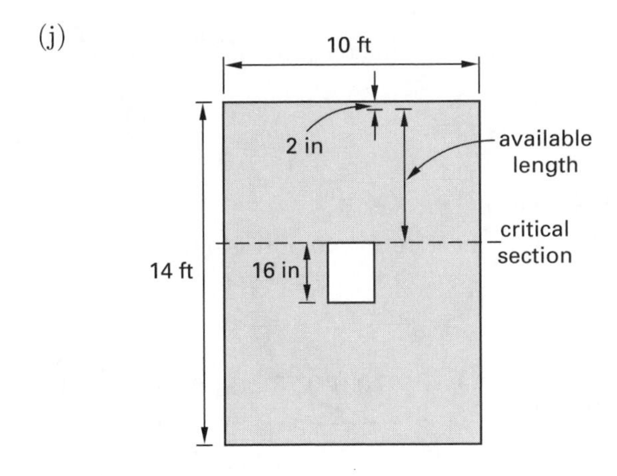

Use a 2 in tip cover.

$$\text{length} = \frac{(14\ \text{ft})\left(12\ \frac{\text{in}}{\text{ft}}\right) - 16\ \text{in}}{2} - 2\ \text{in} = \boxed{74\ \text{in}}$$

The answer is (B).

Solution Using ACI 318 Ch. 9

(a) The bearing strength of the column is

$$\phi(0.85f'_c)A_c = (0.65)(0.85)\left(5\ \frac{\text{kips}}{\text{in}^2}\right)(192\ \text{in}^2)$$
$$= 530.4\ \text{kips}$$

Use $\alpha = 2.0$. The bearing strength of the footing is

$$\phi\alpha(0.85f'_c)A_c = (0.65)(2.0)(0.85)\left(3\ \frac{\text{kips}}{\text{in}^2}\right)(192\ \text{in}^2)$$
$$= 636.5\ \text{kips}$$

Therefore, $\phi P_{n,b} = 530.4$ kips.

Since the $P_u > \phi P_{n,b}$, the excess that must be carried by the dowels is

$$650\ \text{kips} - 530.4\ \text{kips} = 119.6\ \text{kips}$$

Design the dowel steel.

$$119.6\ \text{kips} = \phi P_{n,\text{steel}} = \phi A_s(f_y - 0.85f'_{c,\text{column}})$$

$$A_s = \frac{\phi P_{n,\text{steel}}}{\phi(f_y - 0.85f'_{c,\text{column}})}$$
$$= \frac{119.6\ \text{kips}}{(0.65)\left(60\ \frac{\text{kips}}{\text{in}^2} - (0.85)\left(5\ \frac{\text{kips}}{\text{in}^2}\right)\right)}$$
$$= 3.30\ \text{in}^2$$

Check the minimum requirement.

$$A_{db,\text{min}} = 0.005A_c = (0.005)(192\ \text{in}^2) = 0.96\ \text{in}^2$$

Use $A_s = 3.3\ \text{in}^2$.

The answer is (D).

(b) Hooks are ineffective for compression. The development length of a bar in compression does not include the bend length. Use Eq. 55.31.

$$l_d = 0.02d_b\left(\frac{f_y}{\sqrt{f'_c}}\right) = \frac{(0.02)(0.75\ \text{in})\left(60{,}000\ \frac{\text{lbf}}{\text{in}^2}\right)}{\sqrt{3000\ \frac{\text{lbf}}{\text{in}^2}}}$$
$$= 16.43\ \text{in}\quad [\text{controls}]$$
$$= 0.0003d_bf_y = (0.0003)(0.75\ \text{in})\left(60{,}000\ \frac{\text{kips}}{\text{in}^2}\right)$$
$$= 13.5\ \text{in}$$

Since the bar is hooked, it is necessary to add the radius of the bend to the cover and the steel layer depth.

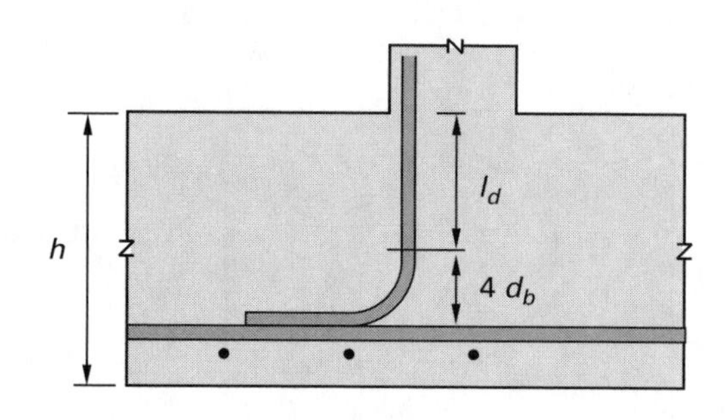

For no. 8 reinforcement,

$$h = (16.43\ \text{in}) + (4)(0.75\ \text{in}) + 2\ \text{in} + 3\ \text{in} = \boxed{24.43\ \text{in}}$$

The answer is (C).

(c) The minimum steel permitted is

$$(0.0018)(14\ \text{ft})\left(12\ \frac{\text{in}}{\text{ft}}\right)(20\ \text{in}) = \boxed{6.05\ \text{in}^2}$$

The answer is (A).

(d) The value for v_u is the same as when ACI 318 App. C is used.

$$v_u = (3.466\times10^{-4})P_u$$

The allowable punching shear stress is given by Eq. 55.17.

$$\phi v_c = (0.75)(4)\sqrt{3000\ \frac{\text{lbf}}{\text{in}^2}} = 164.3\ \text{lbf/in}^2$$

Equating v_u to ϕv_c,

$$P_u = \frac{164.3\ \frac{\text{lbf}}{\text{in}^2}}{3.466\times10^{-4}} = 4.74\times10^5\ \text{lbf}\quad \boxed{(474\ \text{kips})}$$

The answer is (B).

(e) The value for v_u is the same as when ACI 318 App. C is used.

$$v_u = 0.02P_u$$

The allowable concrete stress, using Eqs. 55.5 and 55.6, is

$$\phi v_c = (0.75)(2)\sqrt{f'_c} = \frac{(0.75)(2)\sqrt{3000\ \frac{\text{lbf}}{\text{in}^2}}\left(144\ \frac{\text{in}^2}{\text{ft}^2}\right)}{1000\ \frac{\text{lbf}}{\text{kip}}}$$
$$= 11.83\ \text{kips/ft}^2$$

Equating v_u to ϕv_c,

$$P_u = \frac{11.83\ \frac{\text{kips}}{\text{ft}^2}}{0.02\ \text{ft}^2} = \boxed{591.5\ \text{kips}}$$

The answer is (B).

(f) From Eq. 55.16,

$$J = \left(\frac{db_1^3}{6}\right)\left(1 + \left(\frac{d}{b_1}\right)^2 + \frac{3b_2}{b_1}\right)$$
$$= \left(\frac{(20 \text{ in})(32 \text{ in})^3}{6}\right)\left(1 + \left(\frac{20 \text{ in}}{32 \text{ in}}\right)^2 + (3)\left(\frac{36 \text{ in}}{32 \text{ in}}\right)\right)$$
$$= 520{,}530 \text{ in}^4$$

Use Eq. 55.14.

$$R = \frac{P_u b_1 b_2}{A_f} = \frac{(400 \text{ kips})(32 \text{ in})(36 \text{ in})}{(140 \text{ ft}^2)\left(144 \frac{\text{in}^2}{\text{ft}^2}\right)} = 22.86 \text{ kips}$$

$$\gamma_v = 1 - \frac{1}{1 + \frac{2}{3}\sqrt{\frac{b_1}{b_2}}} = 1 - \frac{1}{1 + \frac{2}{3}\sqrt{\frac{32 \text{ in}}{36 \text{ in}}}} = 0.386$$

Use Eq. 55.13.

$$v_u = \frac{P_u - R}{A_p} + \frac{\gamma_u M_u (0.5 b_1)}{J}$$
$$= \frac{400 \text{ kips} - 22.86 \text{ kips}}{2720 \text{ in}^2}$$
$$+ \frac{(0.386)(600 \text{ ft-kips})\left(12 \frac{\text{in}}{\text{ft}}\right)(0.5)(32 \text{ in})}{520{,}530 \text{ in}^4}$$
$$= 0.224 \text{ kip/in}^2 \quad \boxed{(224 \text{ lbf/in}^2)}$$

The answer is (B).

(g) Determine the pressure distribution.

$$I_f = \left(\tfrac{1}{12}\right)(14 \text{ ft})(10 \text{ ft})^3 = 1166.7 \text{ ft}^4$$

From Eq. 55.1,

$$q_u = \frac{400 \text{ kips}}{140 \text{ ft}^2} \pm \frac{(600 \text{ ft-kips})(5 \text{ ft})}{1166.7 \text{ ft}^4}$$
$$= 2.86 \frac{\text{kips}}{\text{ft}^2} \pm 2.57 \frac{\text{kips}}{\text{ft}^2}$$
$$= \left(5.43 \text{ kips/ft}^2,\ 0.29 \text{ kip/ft}^2\right)$$

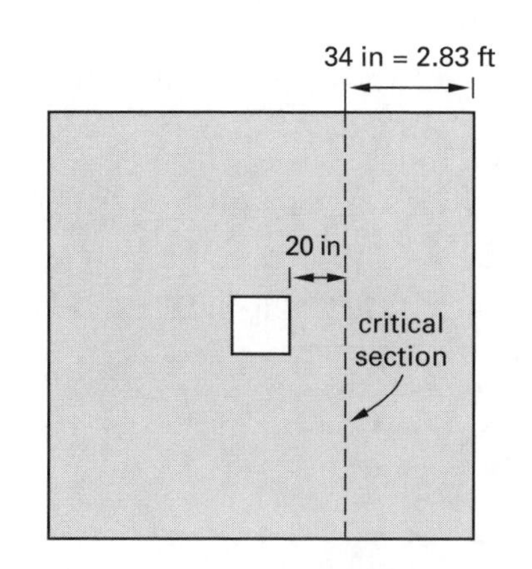

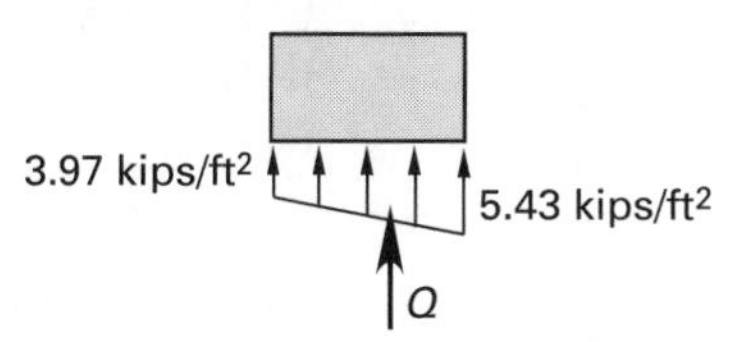

$$Q = \frac{\left(5.43 \frac{\text{kips}}{\text{ft}^2} + 3.97 \frac{\text{kips}}{\text{ft}^2}\right)(14 \text{ ft})(2.83 \text{ ft})}{2}$$
$$= 186.21 \text{ kips}$$

$$v_u = \frac{186.21 \text{ kips}}{(14 \text{ ft})\left(12 \frac{\text{in}}{\text{ft}}\right)(20 \text{ in})}$$
$$= 0.0554 \text{ kip/in}^2 \quad \boxed{(55.4 \text{ lbf/in}^2)}$$

The answer is (D).

(h)

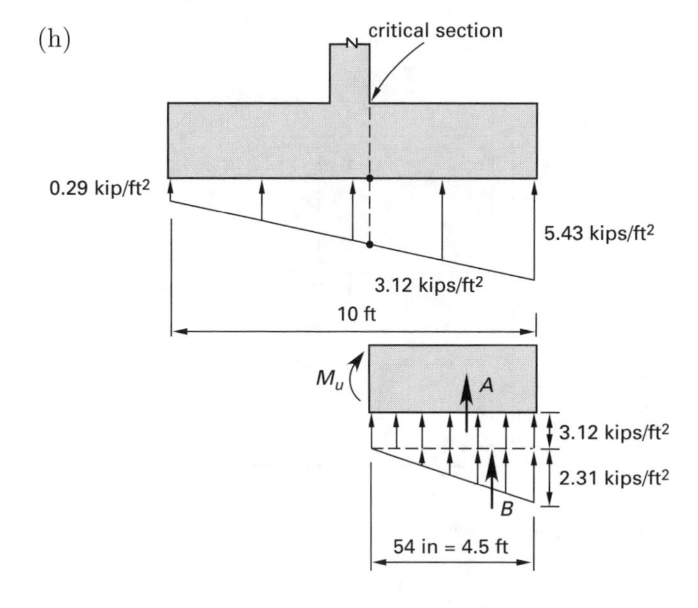

$$A = \left(3.12 \frac{\text{kips}}{\text{ft}^2}\right)(14 \text{ ft})(4.5 \text{ ft})$$
$$= 196.6 \text{ kips}$$

$$B = \left(\tfrac{1}{2}\right)\left(2.31\ \frac{\text{kips}}{\text{ft}^2}\right)(14\ \text{ft})(4.5\ \text{ft})$$
$$= 72.77\ \text{kips}$$

$$M_u = (196.6\ \text{kips})\left(\frac{4.5\ \text{ft}}{2}\right) + \left(\tfrac{2}{3}\right)(72.77\ \text{kips})(4.5\ \text{ft})$$
$$= \boxed{660.7\ \text{ft-kips}}$$

The answer is (C).

(i)
$$\beta = \frac{14\ \text{ft}}{10\ \text{ft}} = 1.4$$

Use Eq. 55.25.

$$A_1 = A_{sd}\left(\frac{2}{\beta+1}\right) = (12\ \text{in}^2)\left(\frac{2}{1.4+1}\right)$$
$$= 10\ \text{in}^2$$

The number of no. 6 bars is

$$\frac{A_s}{A_b} = \frac{10\ \text{in}^2}{0.44} = 22.73 \quad (23\ \text{bars})$$

The width of the band is 10 ft.

The spacing is

$$s = \frac{(10\ \text{ft})\left(12\ \frac{\text{in}}{\text{ft}}\right)}{23} = \boxed{5.2\ \text{in}}$$

The answer is (C).

(j)

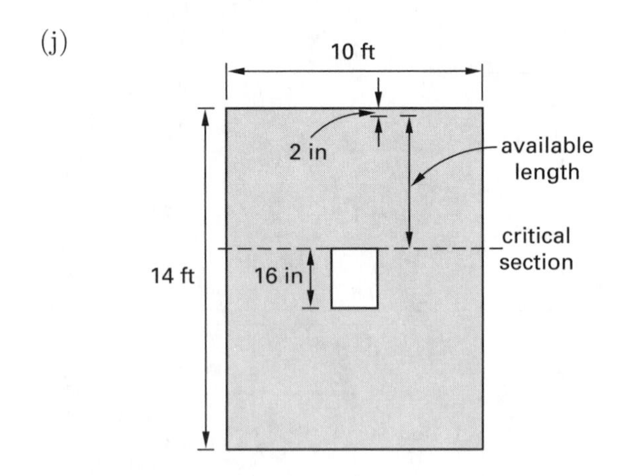

Use a 2 in tip cover.

$$\text{length} = \frac{(14\ \text{ft})\left(12\ \frac{\text{in}}{\text{ft}}\right) - 16\ \text{in}}{2} - 2\ \text{in} = \boxed{74\ \text{in}}$$

The answer is (B).

5. Use either **ACI 318 Apps. B and C** or **ACI 318 Ch. 9 and Unified Design Method** (see p. 55-16) to solve this problem.

Solution Using ACI 318 Apps. B and C

steps 1 and 2: The unfactored (net) allowable soil pressure is

$$q_{a,\text{net}} = 4000\ \frac{\text{lbf}}{\text{ft}^2} - \left(\frac{20\ \text{in}}{12\ \frac{\text{in}}{\text{ft}}}\right)\left(150\ \frac{\text{lbf}}{\text{ft}^3}\right)$$
$$= 3750\ \text{lbf/ft}^2$$

The required footing area is

$$A_f = \frac{240{,}000\ \text{lbf}}{3750\ \frac{\text{lbf}}{\text{ft}^2}} = 64.0\ \text{ft}^2$$

$$\boxed{\text{Use } (8.0\ \text{ft})(8.0\ \text{ft})} = 64.0\ \text{ft}^2.$$

$$q'_{\text{actual}} = \frac{240{,}000\ \text{lbf}}{64.0\ \text{ft}^2} = 3750\ \text{lbf/ft}^2$$

Assume a depth to reinforcement of

$$\boxed{d = 16\ \text{in.}}$$

The critical area is

$$(8.0\ \text{ft})^2 - \left(\frac{16\ \text{in} + 16\ \text{in}}{12\ \frac{\text{in}}{\text{ft}}}\right)^2 = 56.89\ \text{ft}^2$$

The critical perimeter is

$$\frac{(4)(16\ \text{in} + 16\ \text{in})}{12\ \frac{\text{in}}{\text{ft}}} = 10.67\ \text{ft}$$

step 3: Assuming all of the load is live, the required ultimate shear is

$$V_u = (1.7)\left(3750\ \frac{\text{lbf}}{\text{ft}^2}\right)(56.89\ \text{ft}^2)$$
$$= 362{,}674\ \text{lbf}$$

step 4: The allowable shear load is

$$\beta_c = \frac{16\ \text{in}}{16\ \text{in}} = 1 \quad [\text{must be a minimum of 2}]$$

Use Eq. 55.17.

$$V_c = \left(\left(2 + \tfrac{4}{2}\right)\sqrt{3000\ \frac{\text{lbf}}{\text{in}^2}}\right)\left(144\ \frac{\text{in}^2}{\text{ft}^2}\right) \times \left((10.67\ \text{ft})\left(\frac{16\ \text{in}}{12\ \frac{\text{in}}{\text{ft}}}\right)\right)$$
$$= 448{,}834\ \text{lbf}$$

step 5: Use Eq. 55.5.

$$\phi V_c = (0.85)(448{,}834 \text{ lbf}) = 381{,}509 \text{ lbf}$$
$$V_u < \phi V_c \quad \text{[acceptable]}$$

Square footings should also be checked for one-way shear.

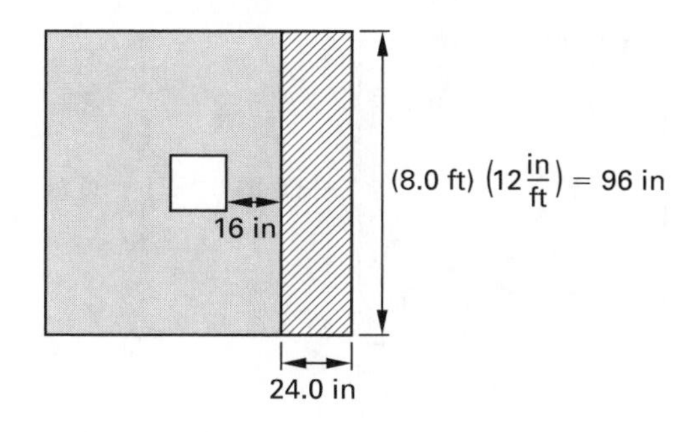

$$\text{critical area} = \frac{(24 \text{ in})(96 \text{ in})}{144 \ \frac{\text{in}^2}{\text{ft}^2}} = 16.0 \text{ ft}^2$$

$$V_u = (1.7)\left(3750 \ \frac{\text{lbf}}{\text{ft}^2}\right)(16.0 \text{ ft}^2) = 102{,}000 \text{ lbf}$$

The allowable shear, using Eq. 55.6, is

$$V_c = \left(2\sqrt{3000 \ \frac{\text{lbf}}{\text{in}^2}}\right)\left(144 \ \frac{\text{in}^2}{\text{ft}^2}\right) \times \left((8.0 \text{ ft})\left(\frac{16 \text{ in}}{12 \ \frac{\text{in}}{\text{ft}}}\right)\right) = 168{,}260 \text{ lbf}$$

$$\phi V_c = (0.85)(168{,}260 \text{ lbf}) = 143{,}021 \text{ lbf}$$
$$V_u < \phi V_c \quad \text{[acceptable]}$$

step 6:

$$b_m = \frac{8.0 \text{ ft} - \frac{16 \text{ in}}{12 \ \frac{\text{in}}{\text{ft}}}}{2} = 3.33 \text{ ft} \quad [40.0 \text{ in}]$$

From Eq. 55.23,

$$M_u = \frac{\left((1.7)\left(3750 \ \frac{\text{lbf}}{\text{ft}^2}\right)\right)(8.0 \text{ ft})(3.33 \text{ ft})^2}{2} = 282{,}767 \text{ ft-lbf}$$

step 7: Solve for the required flexural steel from Eq. 50.23.

$$M_n = \frac{M_u}{\phi} = \frac{282{,}767 \text{ ft-lbf}}{0.9} = 314{,}186 \text{ ft-lbf} = \rho b d^2 f_y \left(1 - \frac{\rho f_y}{1.7 f'_c}\right)$$

$$\rho(8.0 \text{ ft})\left(12 \ \frac{\text{in}}{\text{ft}}\right)(16 \text{ in})^2\left(40{,}000 \ \frac{\text{lbf}}{\text{in}^2}\right) \times \left(1 - \frac{\rho\left(40{,}000 \ \frac{\text{lbf}}{\text{in}^2}\right)}{(1.7)\left(3000 \ \frac{\text{lbf}}{\text{in}^2}\right)}\right) = (314{,}186 \text{ ft-lbf}) \times \left(12 \ \frac{\text{in}}{\text{ft}}\right)$$

$$\rho = 0.00396$$

step 8:

$$A_s = (0.00396)(16 \text{ in})(8.0 \text{ ft})\left(12 \ \frac{\text{in}}{\text{ft}}\right) = 6.08 \text{ in}^2$$

If it is not necessary to know the reinforcement ratio, it is easier to use Eq. 55.24.

$$\lambda = 0.1d = (0.1)(16 \text{ in}) = 1.6 \text{ in}$$

$$A_s = \frac{M_u}{\phi f_y (d - \lambda)} = \frac{(282{,}767 \text{ ft-lbf})\left(12 \ \frac{\text{in}}{\text{ft}}\right)}{(0.9)\left(40{,}000 \ \frac{\text{lbf}}{\text{in}^2}\right)(16 \text{ in} - 1.6 \text{ in})} = 6.55 \text{ in}^2$$

Use eleven no. 7 bars (6.6 in^2).

step 9: Minimum spacing is not a problem.

step 10: Assuming straight bars with no hooks, from Eq. 55.27,

$$l_d = \text{maximum}\begin{cases} \dfrac{(3)(0.875 \text{ in})\left(40{,}000 \ \frac{\text{lbf}}{\text{in}^2}\right)(1)}{50\sqrt{3000 \ \frac{\text{lbf}}{\text{in}^2}}} = 38.3 \text{ in} \\ 12 \text{ in} \end{cases}$$

$$38.3 \text{ in} < b_m = 40.0 \text{ in} \quad \text{[acceptable]}$$

Structural

Solution Using ACI 318 Ch. 9 and Unified Design Method

steps 1 and 2: The unfactored (net) allowable soil pressure is

$$q_{a,\text{net}} = 4000\ \frac{\text{lbf}}{\text{ft}^2} - \left(\frac{20\ \text{in}}{12\ \frac{\text{in}}{\text{ft}}}\right)\left(150\ \frac{\text{lbf}}{\text{ft}^3}\right)$$
$$= 3750\ \text{lbf/ft}^2$$

The required footing area is

$$A_f = \frac{240{,}000\ \text{lbf}}{3750\ \frac{\text{lbf}}{\text{ft}^2}} = 64.0\ \text{ft}^2$$

Use (8.0 ft)(8.0 ft) = 64.0 ft².

$$q'_{\text{actual}} = \frac{240{,}000\ \text{lbf}}{64.0\ \text{ft}^2} = 3750\ \text{lbf/ft}^2$$

Assume a depth to reinforcement of

$d = 16$ in.

The critical area is

$$(8.0\ \text{ft})^2 - \left(\frac{16\ \text{in} + 16\ \text{in}}{12\ \frac{\text{in}}{\text{ft}}}\right)^2 = 56.89\ \text{ft}^2$$

The critical perimeter is

$$\frac{(4)(16\ \text{in} + 16\ \text{in})}{12\ \frac{\text{in}}{\text{ft}}} = 10.67\ \text{ft}$$

step 3: Assuming all of the load is live, the required ultimate shear is

$$V_u = (1.6)\left(3750\ \frac{\text{lbf}}{\text{ft}^2}\right)(56.89\ \text{ft}^2)$$
$$= 341{,}340\ \text{lbf}$$

step 4: The allowable shear load is

$$\beta_c = \frac{16\ \text{in}}{16\ \text{in}} = 1 \quad \text{[must be a minimum of 2]}$$

Use Eq. 55.17.

$$V_c = \left(\left(2 + \tfrac{4}{2}\right)\sqrt{3000\ \frac{\text{lbf}}{\text{in}^2}}\right)\left(144\ \frac{\text{in}^2}{\text{ft}^2}\right) \times \left((10.67\ \text{ft})\left(\frac{16\ \text{in}}{12\ \frac{\text{in}}{\text{ft}}}\right)\right)$$
$$= 448{,}834\ \text{lbf}$$

step 5: Use Eq. 55.5.

$$\phi V_c = (0.75)(448{,}834\ \text{lbf}) = 381{,}509\ \text{lbf}$$
$$V_u < \phi V_c \quad \text{[acceptable]}$$

Square footings should also be checked for one-way shear.

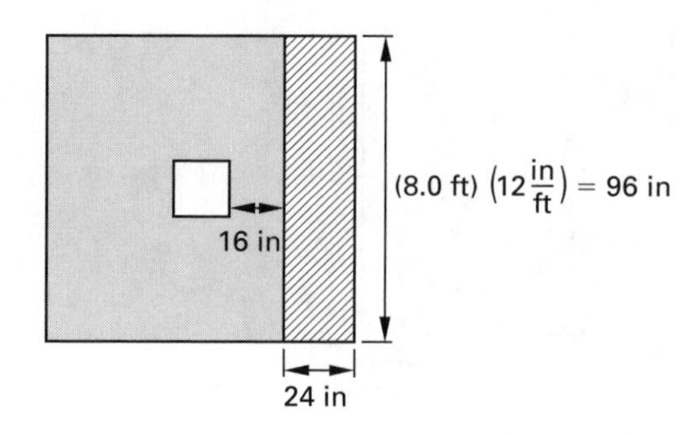

$$\text{critical area} = \frac{(24\ \text{in})(96\ \text{in})}{144\ \frac{\text{in}^2}{\text{ft}^2}} = 16.0\ \text{ft}^2$$

$$V_u = (1.6)\left(3750\ \frac{\text{lbf}}{\text{ft}^2}\right)(16.0\ \text{ft}^2)$$
$$= 102{,}000\ \text{lbf}$$

The allowable shear, using Eq. 55.6, is

$$V_c = \left(2\sqrt{3000\ \frac{\text{lbf}}{\text{in}^2}}\right)\left(144\ \frac{\text{in}^2}{\text{ft}^2}\right) \times \left((8.0\ \text{ft})\left(\frac{16\ \text{in}}{12\ \frac{\text{in}}{\text{ft}}}\right)\right)$$
$$= 168{,}260\ \text{lbf}$$

$$\phi V_c = (0.75)(168{,}260\ \text{lbf}) = 126{,}195\ \text{lbf}$$
$$V_u < \phi V_c \quad \text{[acceptable]}$$

step 6:
$$b_m = \frac{8.0\ \text{ft} - \frac{16\ \text{in}}{12\ \frac{\text{in}}{\text{ft}}}}{2} = 3.33\ \text{ft} \quad [40.0\ \text{in}]$$

From Eq. 55.23,

$$M_u = \frac{\left((1.6)\left(3750\ \frac{\text{lbf}}{\text{ft}^2}\right)\right)(8.0\ \text{ft})(3.33\ \text{ft})^2}{2}$$
$$= 266{,}134\ \text{ft-lbf}$$

step 7: Solve for the required flexural steel from Eq. 50.32, assuming the section is tension controlled.

$$M_n = \frac{M_u}{\phi} = \frac{266{,}134 \text{ ft-lbf}}{0.9} = 295{,}704 \text{ ft-lbf} = \rho b d^2 f_y \left(1 - \frac{\rho f_y}{1.7 f'_c}\right)$$

$$\rho(8.0 \text{ ft})\left(12\ \frac{\text{in}}{\text{ft}}\right)(16 \text{ in})^2\left(40{,}000\ \frac{\text{lbf}}{\text{in}^2}\right) \times \left(1 - \frac{\rho\left(40{,}000\ \frac{\text{lbf}}{\text{in}^2}\right)}{(1.7)\left(3000\ \frac{\text{lbf}}{\text{in}^2}\right)}\right) = (295{,}704 \text{ ft-lbf}) \times \left(12\ \frac{\text{in}}{\text{ft}}\right)$$

$$\rho = 0.00351$$

The reinforcement ratio corresponding to $\epsilon_t = 0.005$ is

$$\rho = \frac{0.85\beta_1 t'_c}{f_y}\left(\frac{0.003}{0.003 + 0.005}\right) = 0.02$$

Since $\rho = 0.00351 < 0.02$, the section is tension controlled.

step 8: $A_s = (0.00351)(16 \text{ in})(8.0 \text{ ft})\left(12\ \frac{\text{in}}{\text{ft}}\right) = 5.39 \text{ in}^2$

(If it is not necessary to know the reinforcement ratio, it is easier to use Eq. 55.24.)

$$\lambda = 0.1d = (0.1)(16 \text{ in}) = 1.6 \text{ in}$$

In this case, the section is tension controlled.

$$A_s = \frac{M_u}{\phi f_y (d - \lambda)} = \frac{(266{,}134 \text{ ft-lbf})\left(12\ \frac{\text{in}}{\text{ft}}\right)}{(0.9)\left(40{,}000\ \frac{\text{lbf}}{\text{in}^2}\right)(16 \text{ in} - 1.6 \text{ in})} = 6.16 \text{ in}^2$$

Use 14 no. 6 bars (6.16 in^2).

step 9: Minimum spacing is not a problem.

step 10: Assuming straight bars with no hooks, from Eq. 55.27,

$$l_d = \text{maximum}\left\{\begin{array}{l} \dfrac{(3)(0.75 \text{ in})\left(40{,}000\ \frac{\text{lbf}}{\text{in}^2}\right)(1)}{50\sqrt{3000\ \frac{\text{lbf}}{\text{in}^2}}} = 32.8 \text{ in} \\ 12 \text{ in} \end{array}\right\}$$

$$32.8 \text{ in} < b_m = 40.0 \text{ in} \quad \text{[acceptable]}$$

6. Use either **ACI 318 Apps. B and C** or **ACI 318 Ch. 9 and Unified Design Method** (see p. 55-19) to solve this problem.

Solution Using ACI 318 Apps. B and C

This is similar to Prob. 5 except for the information about the column steel.

steps 1 and 2: Assume a 25 in thick footing. The approximate footing size is

$$A = \frac{200{,}000 \text{ lbf} + 145{,}000 \text{ lbf}}{4000\ \frac{\text{lbf}}{\text{ft}^2} - \left(\frac{25 \text{ in}}{12\ \frac{\text{in}}{\text{ft}}}\right)\left(150\ \frac{\text{lbf}}{\text{ft}^3}\right)} = 93.6 \text{ ft}^2$$

Try a 10 ft by 10 ft footing (100 ft^2).

(9.75 ft by 9.75 ft could also be used.)

$$q_{\text{actual}} = \frac{(1.4)(200{,}000 \text{ lbf}) + (1.7)(145{,}000 \text{ lbf})}{100 \text{ ft}^2} = 5265 \text{ lbf/ft}^2$$

Assume a depth to reinforcement of $d = 21$ in.

The critical area is

$$(10 \text{ ft})^2 - \left(\frac{18 \text{ in} + 21 \text{ in}}{12\ \frac{\text{in}}{\text{ft}}}\right)^2 = 89.44 \text{ ft}^2$$

The critical perimeter is

$$(4)\left(\frac{18 \text{ in} + 21 \text{ in}}{12\ \frac{\text{in}}{\text{ft}}}\right) = 13 \text{ ft}$$

Structural

step 3: The ultimate shear is

$$V_u = (1.4)(200{,}000 \text{ lbf}) + (1.7)(145{,}000 \text{ lbf}) = 526{,}500 \text{ lbf}$$

step 4: The allowable two-way shear is

$$\beta_c = 2 \quad \text{[minimum value]}$$

$$\phi V_c = \left(\left(2 + \tfrac{4}{2}\right)(0.85)\sqrt{3000 \ \frac{\text{lbf}}{\text{ft}^2}} \right)\left(144 \ \frac{\text{in}^2}{\text{ft}^2}\right) \times \left((13 \text{ ft}) \left(\frac{21 \text{ in}}{12 \ \frac{\text{in}}{\text{ft}}} \right) \right) = 610{,}075 \text{ lbf}$$

step 5: Since $V_u < \phi V_c$, depth and size are acceptable. (The one-way shear should also be checked.)

step 6:
$$b_m = \frac{10 \text{ ft} - \dfrac{18 \text{ in}}{12 \ \frac{\text{in}}{\text{ft}}}}{2} = 4.25 \text{ ft} \quad [51.0 \text{ in}]$$

Use Eq. 55.23.

$$M_u = \frac{\left(5265 \ \frac{\text{lbf}}{\text{ft}^2}\right)(10 \text{ ft})(4.25 \text{ ft})^2}{2} = 475{,}495 \text{ ft-lbf}$$

step 7:
$$M_n = \frac{M_u}{\phi} = \frac{(475{,}495 \text{ ft-lbf})\left(12 \ \frac{\text{in}}{\text{ft}}\right)}{0.90} = 6.34 \times 10^6 \text{ in-lbf}$$

From Eq. 50.32,

$$M_n = \rho b d^2 f_y \left(1 - \frac{\rho f_y}{1.7 f'_c}\right)$$

$$6.34 \times 10^6 \text{ in-lbf} = \rho \left((10 \text{ ft}) \left(12 \ \frac{\text{in}}{\text{ft}}\right) \right)(21 \text{ in})^2 \times \left(40{,}000 \ \frac{\text{lbf}}{\text{in}^2}\right) \times \left(1 - \frac{\rho \left(40{,}000 \ \frac{\text{lbf}}{\text{in}^2}\right)}{(1.7)\left(3000 \ \frac{\text{lbf}}{\text{in}^2}\right)} \right)$$
$$= (2.117 \times 10^9 \text{ lbf})\rho(1 - 7.843\rho)$$

$$\rho = 0.00307$$

step 8:
$$A_{st} = (0.00307)(21 \text{ in})(10 \text{ ft})\left(12 \ \frac{\text{in}}{\text{ft}}\right) = 7.74 \text{ in}^2$$

Use ten no. 8 bars (7.9 in²).

step 9: Minimum spacing is not a problem.

step 10: Assuming straight bars with no hooks, and using Eq. 55.31,

$$l_d = \text{maximum} \begin{cases} \dfrac{(3)(1 \text{ in})\left(40{,}000 \ \frac{\text{lbf}}{\text{in}^2}\right)(1)}{50\sqrt{3000 \ \frac{\text{lbf}}{\text{in}^2}}} = 43.8 \text{ in} \\ 12 \text{ in} \end{cases}$$

Since 43.8 in $< b_m = 51.0$ in, it is acceptable.

Finally, since column steel was specified, check the need for dowel bars.

The gross column area is

$$A_g = (18 \text{ in})^2 = 324 \text{ in}^2$$

The bearing strength is

$$f_{\text{bearing}} = (0.85)\left(3000 \ \frac{\text{lbf}}{\text{in}^2}\right) = 2550 \text{ lbf/in}^2$$

The bearing capacity of concrete is

$$\phi f_{\text{bearing}} A_g = (0.70)\left(2550 \ \frac{\text{lbf}}{\text{in}^2}\right)(324 \text{ in}^2) = 578{,}340 \text{ lbf}$$

Since 578,340 lbf > 526,500 lbf ($\phi P_n > P_u$), only the minimum dowel steel is required.

$$A_{st,\text{min}} = (0.005)(18 \text{ in})(18 \text{ in}) = 1.62 \text{ in}$$

Use four no. 6 bars (1.76 in²).

Check the development length of the dowels in compression. From Eq. 55.31,

$$l_d = \text{maximum} \begin{cases} \dfrac{(0.02)(0.75 \text{ in})\left(40{,}000 \ \frac{\text{lbf}}{\text{in}^2}\right)}{\sqrt{3000 \ \frac{\text{lbf}}{\text{in}^2}}} = 10.95 \text{ in} \\ (0.0003)(0.75 \text{ in})\left(40{,}000 \ \frac{\text{lbf}}{\text{in}^2}\right) = 9.0 \text{ in} \\ 8 \text{ in} \end{cases}$$

$$l_d = 10.95 \text{ in} \quad \text{[controls]}$$

Since 10.95 in $< d = 21$ in, it is acceptable.

Solution Using ACI 318 Ch. 9 and Unified Design Method

steps 1 and 2: Assume a 25 in thick, 10 ft by 10 ft footing.

$$q_{\text{actual}} = \frac{(1.2)(200{,}000\ \text{lbf}) + (1.6)(145{,}000\ \text{lbf})}{100\ \text{ft}^2} = 4720\ \text{lbf/ft}^2$$

step 3: The ultimate shear is

$$V_u = (1.2)(200{,}000\ \text{lbf}) + (1.6)(145{,}000\ \text{lbf}) = 472{,}000\ \text{lbf}$$

step 4: The allowable two-way shear is

$$\beta_c = 2 \quad \text{[minimum value]}$$

$$\phi V_c = \left(\left(2+\tfrac{4}{2}\right)(0.75)\sqrt{3000\ \frac{\text{lbf}}{\text{ft}^2}}\right)\left(144\ \frac{\text{in}^2}{\text{ft}^2}\right) \times \left((13\ \text{ft})\left(\frac{21\ \text{in}}{12\ \frac{\text{in}}{\text{ft}}}\right)\right) = 538{,}301\ \text{lbf}$$

step 5: Since $V_u < \phi V_c$, depth and size are acceptable. (The one-way shear should also be checked.)

step 6:

$$b_m = \frac{10\ \text{ft} - \frac{18\ \text{in}}{12\ \frac{\text{in}}{\text{ft}}}}{2} = 4.25\ \text{ft} \quad [51.0\ \text{in}]$$

Use Eq. 55.23.

$$M_u = \frac{\left(4720\ \frac{\text{lbf}}{\text{ft}^2}\right)(10\ \text{ft})(4.25\ \text{ft})^2}{2} = 426{,}275\ \text{ft-lbf}$$

step 7: Assume the section is tension controlled.

$$M_n = \frac{M_u}{\phi} = \frac{(426{,}275\ \text{ft-lbf})\left(12\ \frac{\text{in}}{\text{ft}}\right)}{0.90} = 5.68 \times 10^6\ \text{in-lbf}$$

From Eq. 50.32,

$$M_n = \rho b d^2 f_y \left(1 - \frac{\rho f_y}{1.7 f'_c}\right)$$

Solve for ρ.

$$\rho = 0.00274$$

With this low reinforcement ratio, the section is going to be tension controlled.

step 8:

$$A_{\text{st}} = (0.00274)(21\ \text{in})(10\ \text{ft})\left(12\ \frac{\text{in}}{\text{ft}}\right) = 6.90\ \text{in}^2$$

Use 12 no. 7 bars (7.2 in^2).

step 9: Minimum spacing is not a problem.

step 10: Assuming straight bars with no hooks, and using Eq. 55.31,

$$l_d = \text{maximum}\begin{cases} \dfrac{(3)(0.875\ \text{in})\left(40{,}000\ \frac{\text{lbf}}{\text{in}^2}\right)(1)}{40\sqrt{3000\ \frac{\text{lbf}}{\text{in}^2}}} = 47.9\ \text{in} \\ 12\ \text{in} \end{cases}$$

Since 47.9 in < b_m = 51.0 in, it is acceptable.

Finally, since column steel was specified, check the need for dowel bars.

The bearing capacity of concrete is

$$\phi f_{\text{bearing}} A_g = (0.65)\left(2550\ \frac{\text{lbf}}{\text{in}^2}\right)(324\ \text{in}^2) = 537{,}030\ \text{lbf}$$

Since 537,030 lbf > 472,000 lbf ($\phi P_n > P_u$), only the minimum dowel steel is required.

$$A_{\text{st,min}} = (0.005)(18\ \text{in})(18\ \text{in}) = 1.62\ \text{in}$$

Use four no. 6 bars (1.76 in^2).

Check the development length of the dowels in compression. From Eq. 55.31,

$$l_d = \text{maximum}\begin{cases} \dfrac{(0.02)(0.75\ \text{in})\left(40{,}000\ \frac{\text{lbf}}{\text{in}^2}\right)}{\sqrt{3000\ \frac{\text{lbf}}{\text{in}^2}}} = 10.95\ \text{in} \\ (0.0003)(0.75\ \text{in})\left(40{,}000\ \frac{\text{lbf}}{\text{in}^2}\right) = 9.0\ \text{in} \\ 8\ \text{in} \end{cases}$$

$$l_d = 10.95\ \text{in} \quad \text{[controls]}$$

Since 10.95 in < d = 21 in, it is acceptable.

Structural

7. The solution and answers remain the same whether ACI 318 Apps. B and C are used or ACI 318 Unified Design Method and Ch. 9 are used.

(a) Find the centroid of the load group.

$$\overline{x} = \frac{(3 \text{ ft})(60 \text{ kips}) + (13 \text{ ft})(80 \text{ kips})}{60 \text{ kips} + 80 \text{ kips}}$$
$$= 8.71 \text{ ft} \quad \text{[from right end]}$$

Calculate the eccentricities that would produce the same moments.

$$\epsilon_1 = \frac{M}{P} = \frac{120 \text{ ft-kips}}{80 \text{ kips}} = 1.5 \text{ ft}$$
$$\epsilon_2 = \frac{160 \text{ ft-kips}}{60 \text{ kips}} = 2.67 \text{ ft}$$

Move the loads.

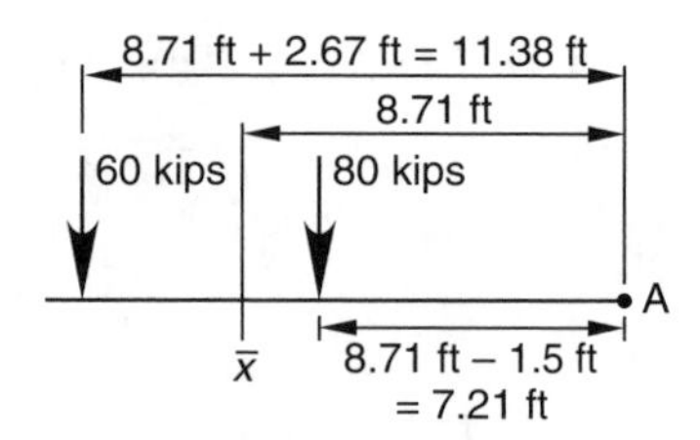

The total net vertical reaction must be 60 kips + 80 kips = 140 kips. This can be assumed to be at 1/2L if it is uniform.

$$\sum M_A = (60 \text{ kips})(11.38 \text{ ft}) + (80 \text{ kips})(7.21 \text{ ft}) - (140 \text{ kips})\left(\frac{L}{2}\right) = 0$$

$$L = \boxed{17.99 \text{ ft} \quad \text{[round to 18 ft]}}$$

Alternative Solution

Sum moments about the left free end.

$$(80 \text{ kips})(L - 13 \text{ ft}) + 120 \text{ ft-kips} + (60 \text{ kips})(L - 3 \text{ ft}) - 160 \text{ ft-kips} - (140 \text{ kips})\left(\frac{L}{2}\right) = 0$$

$$L = \boxed{18 \text{ ft}}$$

(b)
$$w = \text{soil pressure} = \frac{140 \text{ kips}}{18.0 \text{ ft}}$$
$$= 7.78 \text{ kips/ft} \quad (8 \text{ kips/ft})$$

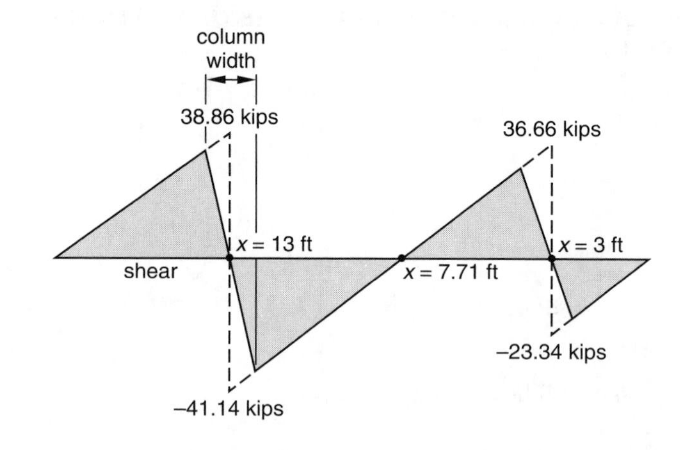

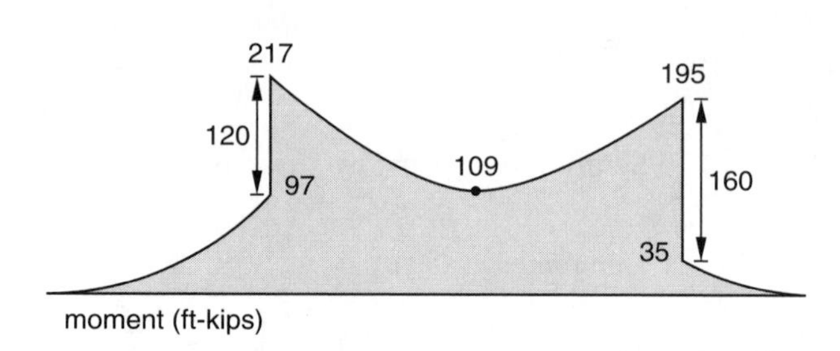

(c) No top steel is needed because the entire footing is in single-curvature bending.

8. Use either **ACI 318 App. C** or **ACI 318 Ch. 9** (see p. 55-22) to solve this problem.

Solution Using ACI 318 App. C

(a)
$$\epsilon = \frac{100 \text{ ft-kips}}{200 \text{ kips}} = 0.5 \text{ ft}$$

Assume a footing thickness of 24 in.

The net allowable soil pressure is

$$p_{a,\text{net}} = 6000 \ \frac{\text{lbf}}{\text{ft}^2} - \left(\frac{24 \text{ in}}{12 \ \frac{\text{in}}{\text{ft}}}\right)\left(150 \ \frac{\text{lbf}}{\text{ft}^3}\right) = 5700 \text{ lbf/ft}^2$$

For a footing with no moment,

$$A = \frac{200{,}000 \text{ lbf}}{5700 \ \frac{\text{lbf}}{\text{ft}^2}} = 35.09 \text{ ft}^2$$

$$B = \sqrt{35.09 \text{ ft}^2} = 5.92 \text{ ft} \quad \text{[round to 6 ft]}$$

$$q_{\max} = 5700 \ \frac{\text{lbf}}{\text{ft}^2} = \left(\frac{200{,}000}{6L}\right)\left(1 + \frac{(6)(0.5)}{L}\right)$$

$$L = 8.03 \text{ ft}$$

$$\boxed{\text{Use 6 ft by 8.25 ft footing.}}$$

(b)

$$q_{max} = \left(\frac{200{,}000\ \text{lbf}}{(6\ \text{ft})(8.25\ \text{ft})}\right)\left(1+\frac{(6\ \text{ft})(0.5)}{8.25\ \text{ft}}\right) + \left(\frac{24\ \text{in}}{12\ \frac{\text{in}}{\text{ft}}}\right)\left(150\ \frac{\text{lbf}}{\text{ft}^3}\right) = \boxed{5810\ \text{lbf/ft}^2}$$

$$q_{min} = \left(\frac{200{,}000\ \text{lbf}}{(6\ \text{ft})(8.25\ \text{ft})}\right)\left(1-\frac{(6\ \text{ft})(0.5)}{8.25\ \text{ft}}\right) + \left(\frac{24\ \text{in}}{12\ \frac{\text{in}}{\text{ft}}}\right)\left(150\ \frac{\text{lbf}}{\text{ft}^3}\right) = \boxed{2871\ \text{lbf/ft}^2}$$

(c) Disregarding dead load and factoring the live load, the maximum and minimum ultimate pressures are

$$q_{max} = (1.7)\left(5810\ \frac{\text{lbf}}{\text{ft}^2}\right) = 9877\ \text{lbf/ft}^2$$

$$q_{min} = (1.7)\left(2871\ \frac{\text{lbf}}{\text{ft}^2}\right) = 4881\ \text{lbf/ft}^2$$

Assume $d = 20$ in. The critical ultimate soil pressure is 20 in from the face of the column. The distance from the far (left) edge of the footing is

$$\frac{(8.25\ \text{ft})\left(12\ \frac{\text{in}}{\text{ft}}\right)}{2} + \frac{14\ \text{in}}{2\ \text{in}} + 20\ \text{in} = 76.5\ \text{in}$$

At the critical line, the soil pressure is

$$4881\ \frac{\text{lbf}}{\text{ft}^2} + \left(9877\ \frac{\text{lbf}}{\text{ft}^2} - 4881\ \frac{\text{lbf}}{\text{ft}^2}\right) \times \left(\frac{76.5\ \text{in}}{(8.25\ \text{ft})\left(12\ \frac{\text{in}}{\text{ft}}\right)}\right) = 8742\ \text{lbf/ft}^2$$

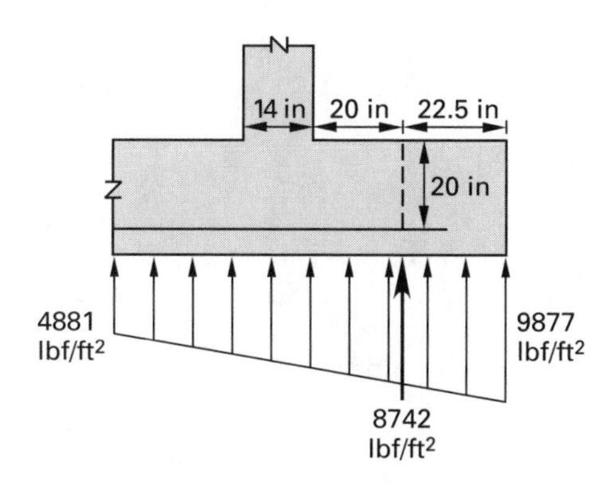

The required ultimate shear at the critical line is given by Eq. 55.13.

$$V_u = \left(\frac{22.5\ \text{in}}{12\ \frac{\text{in}}{\text{ft}}}\right)\left(8742\ \frac{\text{lbf}}{\text{ft}^2}\right) + \left(\tfrac{1}{2}\right)\left(\frac{22.5\ \text{in}}{12\ \frac{\text{in}}{\text{ft}}}\right)\left(9877\ \frac{\text{lbf}}{\text{ft}^2} - 8742\ \frac{\text{lbf}}{\text{ft}^2}\right) = 17{,}455\ \text{lbf}\quad \text{[per foot of width]}$$

The ultimate shear strength is calculated from Eq. 55.6.

$$\phi v_c = (2)(0.85)\sqrt{3000\ \frac{\text{lbf}}{\text{in}^2}} = 93.1\ \text{lbf/in}^2$$

The depth required is

$$d = \frac{(17{,}455\ \text{lbf})\left(12\ \frac{\text{in}}{\text{ft}}\right)}{\left(93.1\ \frac{\text{lbf}}{\text{in}^2}\right)\left(144\ \frac{\text{in}^2}{\text{ft}^2}\right)(1\ \text{ft})} = \boxed{15.6\ \text{in}\quad (16\ \text{in})}$$

(d) The pressure at the critical flexure section is

$$4881\ \frac{\text{lbf}}{\text{ft}^2} + \left(9877\ \frac{\text{lbf}}{\text{ft}^2} - 4881\ \frac{\text{lbf}}{\text{ft}^2}\right) \times \left(\frac{(8.25\ \text{ft})\left(12\ \frac{\text{in}}{\text{ft}}\right) - 20\ \text{in} - 22.5\ \text{in}}{(8.25\ \text{ft})\left(12\ \frac{\text{in}}{\text{ft}}\right)}\right) = 7732\ \text{lbf/ft}^2$$

The moment per unit length at the critical line is given by Eq. 55.23.

$$\frac{M_u}{L} = \frac{q_u l^2}{2} = \left(\tfrac{1}{2}\right)\left(7732\ \frac{\text{lbf}}{\text{ft}^2}\right)\left(\frac{42.5\ \text{in}}{12\ \frac{\text{in}}{\text{ft}}}\right)^2 + \left(\tfrac{1}{2}\right)\left(\frac{42.5\ \text{in}}{12\ \frac{\text{in}}{\text{ft}}}\right)\left(9877\ \frac{\text{lbf}}{\text{ft}^2} - 7732\ \frac{\text{lbf}}{\text{ft}^2}\right) \times \left(\tfrac{2}{3}\right)\left(\frac{42.5\ \text{in}}{12\ \frac{\text{in}}{\text{ft}}}\right) = \boxed{57{,}461\ \text{ft-lbf per foot of footing}}$$

Structural

Solution Using ACI 318 Ch. 9

(a)

$$\epsilon = \frac{100 \text{ ft-kips}}{200 \text{ kips}} = 0.5 \text{ ft}$$

Assume a footing thickness of 24 in.

The net allowable soil pressure is

$$p_{a,\text{net}} = 6000 \ \frac{\text{lbf}}{\text{ft}^2} - \left(\frac{24 \text{ in}}{12 \ \frac{\text{in}}{\text{ft}}}\right)\left(150 \ \frac{\text{lbf}}{\text{ft}^3}\right) = 5700 \text{ lbf/ft}^2$$

For a footing with no moment,

$$A = \frac{200{,}000 \text{ lbf}}{5700 \ \frac{\text{lbf}}{\text{ft}^2}} = 35.09 \text{ ft}^2$$

$$B = \sqrt{35.09 \text{ ft}^2} = 5.92 \text{ ft} \quad [\text{round to 6 ft}]$$

$$q_{\text{max}} = 5700 \ \frac{\text{lbf}}{\text{ft}^2} = \left(\frac{200{,}000}{6L}\right)\left(1 + \frac{(6)(0.5)}{L}\right)$$

$$L = 8.03 \text{ ft}$$

Use 6 ft by 8.25 ft footing.

(b)

$$q_{\text{max}} = \left(\frac{200{,}000 \text{ lbf}}{(6 \text{ ft})(8.25 \text{ ft})}\right)\left(1 + \frac{(6 \text{ ft})(0.5)}{8.25 \text{ ft}}\right) + \left(\frac{24 \text{ in}}{12 \ \frac{\text{in}}{\text{ft}}}\right)\left(150 \ \frac{\text{lbf}}{\text{ft}^3}\right) = 5810 \text{ lbf/ft}^2$$

$$q_{\text{min}} = \left(\frac{200{,}000 \text{ lbf}}{(6 \text{ ft})(8.25 \text{ ft})}\right)\left(1 - \frac{(6 \text{ ft})(0.5)}{8.25 \text{ ft}}\right) + \left(\frac{24 \text{ in}}{12 \ \frac{\text{in}}{\text{ft}}}\right)\left(150 \ \frac{\text{lbf}}{\text{ft}^3}\right) = 2871 \text{ lbf/ft}^2$$

(c) Disregarding dead load and factoring the live load, the maximum and minimum ultimate pressures are

$$q_{\text{max}} = (1.6)\left(5810 \ \frac{\text{lbf}}{\text{ft}^2}\right) = 9296 \text{ lbf/ft}^2$$

$$q_{\text{min}} = (1.6)\left(2871 \ \frac{\text{lbf}}{\text{ft}^2}\right) = 4594 \text{ lbf/ft}^2$$

Assume $d = 20$ in. The critical ultimate soil pressure is 20 in from the face of the column. The distance from the far (left) edge of the footing is

$$\frac{(8.25 \text{ ft})\left(12 \ \frac{\text{in}}{\text{ft}}\right)}{2} + \frac{14 \text{ in}}{2 \text{ in}} + 20 \text{ in} = 76.5 \text{ in}$$

At the critical line, the soil pressure is

$$4594 \ \frac{\text{lbf}}{\text{ft}^2} + \left(9296 \ \frac{\text{lbf}}{\text{ft}^2} - 4594 \ \frac{\text{lbf}}{\text{ft}^2}\right) \times \left(\frac{76.5 \text{ in}}{(8.25 \text{ ft})\left(12 \ \frac{\text{in}}{\text{ft}}\right)}\right) = 8227 \text{ lbf/ft}^2$$

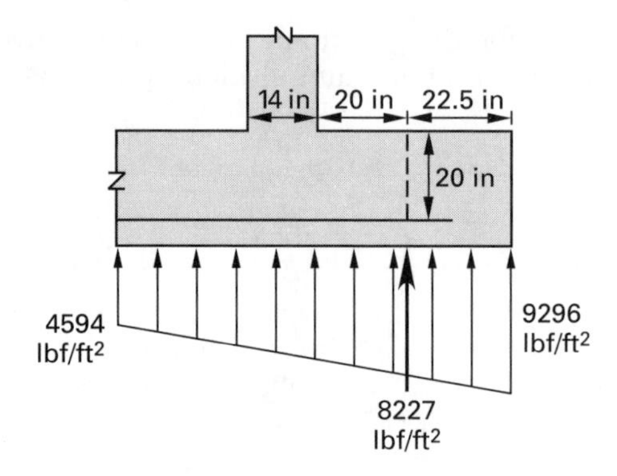

The required ultimate shear at the critical line is given by Eq. 55.13.

$$V_u = \left(\frac{22.5 \text{ in}}{12 \ \frac{\text{in}}{\text{ft}}}\right)\left(8227 \ \frac{\text{lbf}}{\text{ft}^2}\right) + \left(\tfrac{1}{2}\right)\left(\frac{22.5 \text{ in}}{12 \ \frac{\text{in}}{\text{ft}}}\right)\left(9296 \ \frac{\text{lbf}}{\text{ft}^2} - 8227 \ \frac{\text{lbf}}{\text{ft}^2}\right) = 16{,}428 \text{ lbf} \quad [\text{per foot of width}]$$

The ultimate shear strength is calculated from Eq. 55.6.

$$\phi v_c = (2)(0.75)\sqrt{3000 \ \frac{\text{lbf}}{\text{in}^2}} = 82.2 \text{ lbf/in}^2$$

The depth required is

$$d = \frac{(16{,}428 \text{ lbf})\left(12 \ \frac{\text{in}}{\text{ft}}\right)}{\left(82.2 \ \frac{\text{lbf}}{\text{in}^2}\right)\left(144 \ \frac{\text{in}^2}{\text{ft}^2}\right)(1 \text{ ft})} = 16.7 \text{ in} \quad (18 \text{ in})$$

(d) The pressure at the critical flexure section is

$$4594\ \frac{\text{lbf}}{\text{ft}^2} + \left(9296\ \frac{\text{lbf}}{\text{ft}^2} - 4594\ \frac{\text{lbf}}{\text{ft}^2}\right) \times \left(\frac{(8.25\ \text{ft})\left(12\ \frac{\text{in}}{\text{ft}}\right) - 20\ \text{in} - 22.5\ \text{in}}{(8.25\ \text{ft})\left(12\ \frac{\text{in}}{\text{ft}}\right)}\right) = 7277\ \text{lbf/ft}^2$$

The moment per unit length at the critical line is given by Eq. 55.23.

$$\begin{aligned}\frac{M_u}{L} &= \frac{q_u l^2}{2} \\ &= \left(\tfrac{1}{2}\right)\left(7277\ \frac{\text{lbf}}{\text{ft}^2}\right)\left(\frac{42.5\ \text{in}}{12\ \frac{\text{in}}{\text{ft}}}\right)^2 \\ &\quad + \left(\tfrac{1}{2}\right)\left(\frac{42.5\ \text{in}}{12\ \frac{\text{in}}{\text{ft}}}\right)\left(9296\ \frac{\text{lbf}}{\text{ft}^2} - 7277\ \frac{\text{lbf}}{\text{ft}^2}\right) \\ &\quad \times \left(\tfrac{2}{3}\right)\left(\frac{42.5\ \text{in}}{12\ \frac{\text{in}}{\text{ft}}}\right) \\ &= \boxed{54{,}081\ \text{ft-lbf per foot of footing}}\end{aligned}$$

Structural

56 Pretensioned Concrete

PRACTICE PROBLEMS

1. Calculate the nominal moment strength of the cross section of the prestressed beam shown. $f'_c = 5000$ psi, $f_{pu} = 270{,}000$ psi (stress-relieved strands), and $f_{py} = 0.85 f_{pu}$.

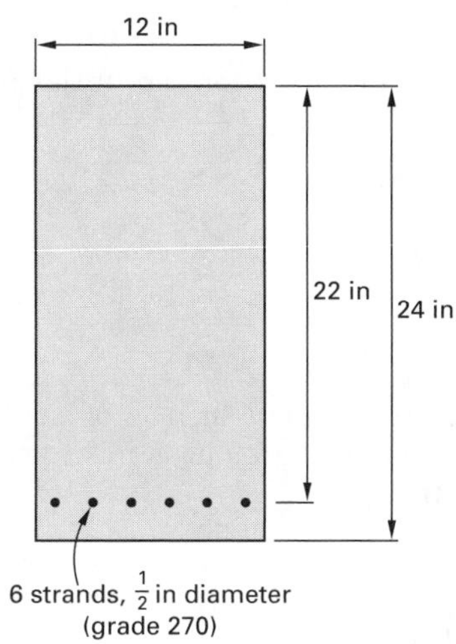

2. The cross section of the simply supported pretensioned concrete beam is shown. The superimposed live and dead loads (in excess of beam weight) are as indicated. The beam has straight cables with initial prestress of 216 kips and final prestress after losses of 140 kips. The concrete is uncracked. The creep coefficient is 2.0, and the modulus of elasticity is 4×10^6 psi. (a) Calculate the centerline deflection of the beam immediately after the cables are cut. (b) Calculate the centerline deflection of the beam five years after the cables are cut.

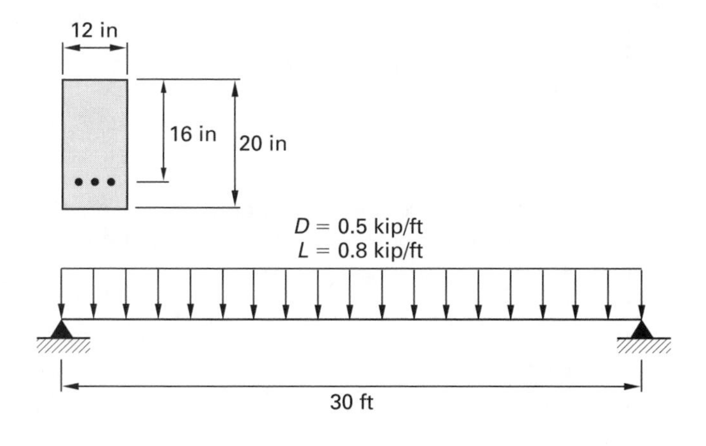

3. *(Time limit: one hour)* The 12 in by 24 in prestressed beam shown carries a dead load of 3 kips/ft (which includes the beam's self-weight). The beam length is 20 ft. The tensile force in the prestressing steel is 250 kips.

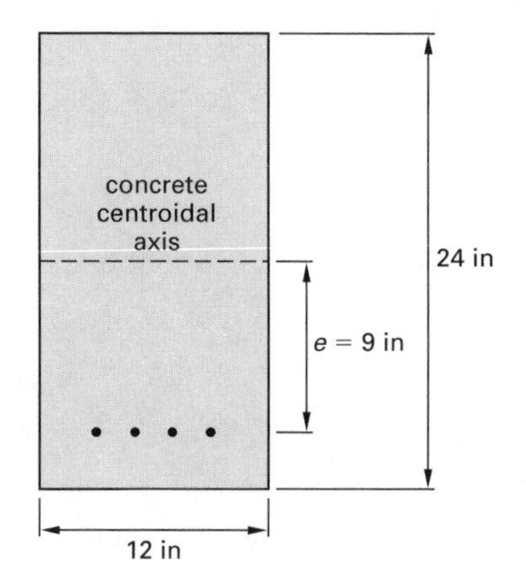

(a) The moment of inertia of the section of the beam shown is most nearly

(A) 13,000 in^4
(B) 14,000 in^4
(C) 15,000 in^4
(D) 16,000 in^4

(b) The area of the cross section of the beam is most nearly

(A) 250 in^2
(B) 290 in^2
(C) 300 in^2
(D) 320 in^2

(c) The maximum bending moment due to dead load is most nearly

(A) 100 ft-kips
(B) 130 ft-kips
(C) 150 ft-kips
(D) 180 ft-kips

(d) The axial compression stress at the top fibers at the midspan of the beam is most nearly

(A) 0.470 ksi
(B) 0.570 ksi
(C) 0.770 ksi
(D) 0.870 ksi

(e) The axial compression stress at the bottom fibers at the midspan of the beam is most nearly

(A) 0.470 ksi
(B) 0.570 ksi
(C) 0.770 ksi
(D) 0.870 ksi

(f) The axial compression stress at the top fibers at the ends of the beam is most nearly

(A) 0.470 ksi
(B) 0.570 ksi
(C) 0.770 ksi
(D) 0.870 ksi

(g) The total stress at the top fibers at the midspan of the beam is most nearly

(A) 0.48 ksi
(B) 0.87 ksi
(C) 1.6 ksi
(D) 2.0 ksi

(h) The total stress at the bottom fibers at the midspan of the beam is most nearly

(A) 0.87 ksi
(B) 1.3 ksi
(C) 2.0 ksi
(D) 2.3 ksi

(i) The total stress at the top fibers at the ends of the beam is most nearly

(A) 0.48 ksi
(B) 0.87 ksi
(C) 1.1 ksi
(D) 1.6 ksi

(j) The total stress at the bottom fibers at the ends of the beam is most nearly

(A) 0.87 ksi
(B) 2.0 ksi
(C) 2.8 ksi
(D) 3.8 ksi

4. *(Time limit: one hour)* The cross section for a 12 m simply supported beam is shown. The initial tendon prestress is 1.10 GPa. $f'_c = 34.5$ MPa, and $f_{pu} = 1.725$ GPa. The total area of prestressing steel is 774 mm^2.

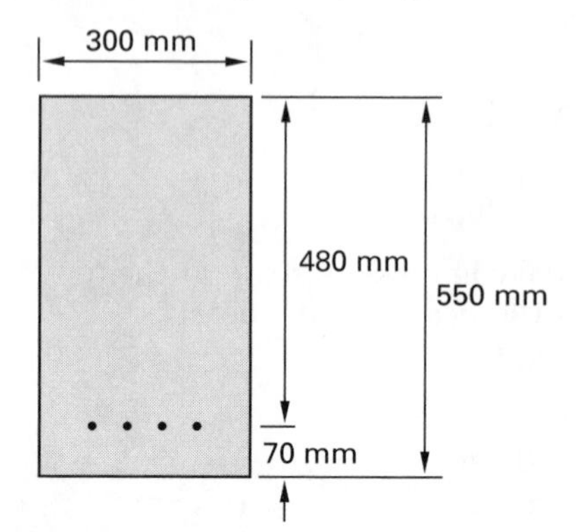

(a) The eccentricity for this section of the beam is most nearly

(A) 70 mm
(B) 210 mm
(C) 480 mm
(D) 550 mm

(b) The maximum bending moment due to the beam dead load is most nearly

(A) 60 kN·m
(B) 63 kN·m
(C) 70 kN·m
(D) 73 kN·m

(c) The initial force used for the initial prestressing is most nearly

(A) 800 kN
(B) 810 kN
(C) 840 kN
(D) 850 kN

(d) The concrete stress in the top of the beam at midspan after the tendons are cut is most nearly

(A) 1.8 MPa
(B) 4.6 MPa
(C) 5.2 MPa
(D) 12 MPa

(e) The concrete stress in the bottom of the beam at midspan after the tendons are cut is most nearly

(A) 4.6 MPa
(B) 5.2 MPa
(C) 10 MPa
(D) 12 MPa

(f) Assuming the losses in the tendons are 18%, the concrete stress at the top of the beam at midspan is most nearly

(A) 0.62 MPa
(B) 1.8 MPa
(C) 4.6 MPa
(D) 5.2 MPa

(g) Assuming the losses in the tendons are 18%, the concrete stress in the bottom of the beam at midspan is most nearly

(A) 5.2 MPa
(B) 4.6 MPa
(C) 9.1 MPa
(D) 9.8 MPa

(h) Assuming the allowable stress is $0.45f'_c$ in compression, the maximum allowable bending moment at midspan is most nearly

(A) 250 kN·m
(B) 290 kN·m
(C) 310 kN·m
(D) 340 kN·m

(i) Assuming the allowable stress is $0.5\sqrt{f'_c}$ in tension, the maximum allowable bending moment at midspan is most nearly

(A) 250 kN·m
(B) 270 kN·m
(C) 300 kN·m
(D) 330 kN·m

(j) Using the answers from parts (h) and (i), the maximum allowable uniform live load that the beam can support in addition to its own weight is most nearly

(A) 8.1 kN/m
(B) 8.6 kN/m
(C) 9.1 kN/m
(D) 10 kN/m

5. A prestressed hollow box cross section is shown. The prestressing is applied through twelve 1/2 in diameter grade 270 seven-wire strand tendons. For these strands, $f_{py} = 0.95f_{pu}$, and the effective prestress after losses is f_{se} = 148 ksi. The specified strength of the concrete is 5000 psi. (a) Determine the flexural strength of the section. (b) Determine the cracking moment. (c) Does the relationship between the cracking moment and the flexural capacity satisfy the requirements of the ACI Code?

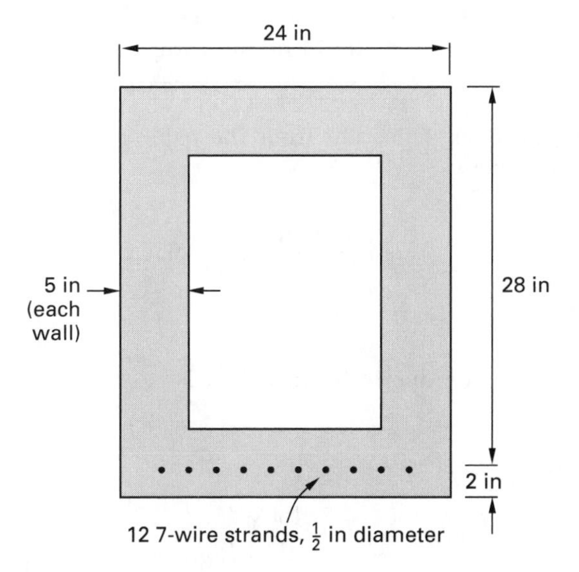

6. *(Time limit: one hour)* A pre-tensioned girder is constructed with 30 bonded strands. The girder spans 100 ft and is supported on simple supports. It carries a uniform live load of 540 lbf/ft as well as its own weight of 150 lbf/ft. The modular ratio is 7, f'_c = 5000 psi, f'_{ci} = 3800 psi, f_{pu} = 250 ksi, f_y/f_{pu} = 0.85, and A_p = 0.144 in^2/strand. It is not necessary to check the tendon stress. (a) If the effective prestress before losses is 80% of f_{pu}, is the stress in the concrete acceptable immediately after transfer, before any live load is added? (b) If the effective prestress after losses is 60% of f_{pu}, is the concrete stress acceptable under service loading?

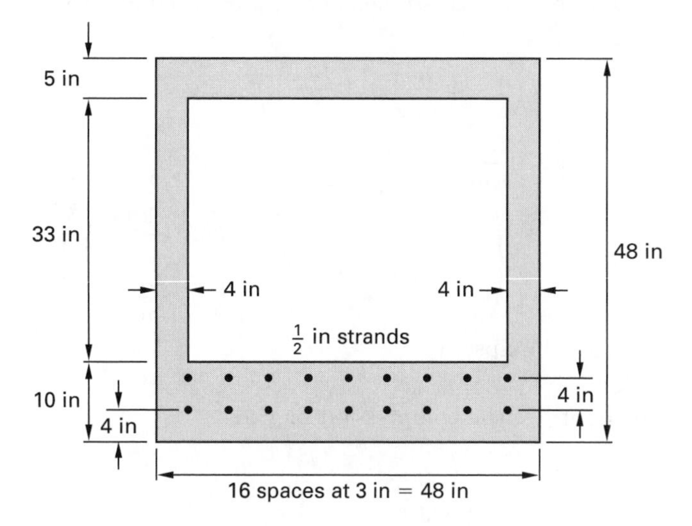

SOLUTIONS

1. $\gamma_p = 0.40$ for $\frac{f_{py}}{f_{pu}} = 0.85$ for stress-relieved strand

$$\beta_1 = 0.80 \text{ for } f'_c = 5000 \text{ lbf/in}^2$$

$$\rho_p = \frac{A_{ps}}{bd_p} = \frac{(6)(0.153 \text{ in}^2)}{(12 \text{ in})(22 \text{ in})} = 0.00348$$

For a fully prestressed member, Eq. 56.3 reduces to

$$\begin{aligned} f_{ps} &= f_{pu}\left(1 - \left(\frac{\gamma_p}{\beta_1}\right)\rho_p\left(\frac{f_{pu}}{f'_c}\right)\right) \\ &= \left(270 \ \frac{\text{kips}}{\text{in}^2}\right) \\ &\quad \times \left(1 - \left(\frac{0.40}{0.80}\right)(0.00348)\left(\frac{270 \ \frac{\text{kips}}{\text{in}^2}}{5 \ \frac{\text{kips}}{\text{in}^2}}\right)\right) \\ &= 245 \text{ kips/in}^2 \end{aligned}$$

The depth of the compression block is

$$\begin{aligned} a &= \frac{A_{ps}f_{ps}}{0.85bf'_c} \\ &= \frac{(6)(0.153 \text{ in}^2)\left(245 \ \frac{\text{kips}}{\text{in}^2}\right)}{(0.85)(12 \text{ in})\left(5 \ \frac{\text{kips}}{\text{in}^2}\right)} \\ &= 4.41 \text{ in} \end{aligned}$$

The nominal moment strength is

$$\begin{aligned} M_n &= A_{ps}f_{ps}\left(d_p - \frac{a}{2}\right) \\ &= (0.918 \text{ in}^2)\left(245 \ \frac{\text{kips}}{\text{in}^2}\right)\left(22 \text{ in} - \frac{4.41 \text{ in}}{2}\right) \\ &= 4452 \text{ in-kip} \end{aligned}$$

$$M_n = \frac{4452 \text{ in-kip}}{12 \ \frac{\text{in}}{\text{ft}}} = \boxed{371 \text{ ft-kips}}$$

2. (a) The gross moment of inertia is

$$I_g = \frac{bh^3}{12} = \frac{(12 \text{ in})(20 \text{ in})^3}{12} = 8000 \text{ in}^4$$

The eccentricity is

$$e = 16 \text{ in} - 10 \text{ in} = 6 \text{ in}$$

The loads are not factored when calculating deflections. The beam's dead weight is

$$w = \frac{(12 \text{ in})(20 \text{ in})\left(150 \ \frac{\text{lbf}}{\text{ft}^3}\right)}{144 \ \frac{\text{in}^2}{\text{ft}^2}} = 250 \text{ lbf/ft}$$

The deflection due to the pretensioning cables is

$$\begin{aligned} \delta_1 &= \frac{-PeL^2}{8E_cI} \\ &= \frac{-(216{,}000 \text{ lbf})(6 \text{ in})\left((30 \text{ ft})\left(12 \ \frac{\text{in}}{\text{ft}}\right)\right)^2}{(8)\left(4 \times 10^6 \ \frac{\text{lbf}}{\text{in}^2}\right)(8000 \text{ in}^4)} \\ &= -0.656 \text{ in} \quad \text{[upward]} \end{aligned}$$

The deflection due to the beam's own weight is

$$\begin{aligned} \delta_2 = \frac{5wL^4}{384E_cI} &= \frac{(5)\left(\frac{250 \ \frac{\text{lbf}}{\text{ft}}}{12 \ \frac{\text{in}}{\text{ft}}}\right)\left((30 \text{ ft})\left(12 \ \frac{\text{in}}{\text{ft}}\right)\right)^4}{(384)\left(4 \times 10^6 \ \frac{\text{lbf}}{\text{in}^2}\right)(8000 \text{ in}^4)} \\ &= 0.142 \text{ in} \quad \text{[downward]} \end{aligned}$$

The total deflection is

$$\begin{aligned} \delta &= -0.656 \text{ in} + 0.142 \text{ in} \\ &= \boxed{-0.514 \text{ in} \quad \text{[upward]}} \end{aligned}$$

(b) The 5 yr deflection from the cable can be found proportionally.

$$\begin{aligned} \delta_1 &= (2)\left(\frac{140 \text{ kips}}{216 \text{ kips}}\right)(-0.656 \text{ in}) \\ &= -0.85 \text{ in} \end{aligned}$$

The deflection due to beam weight is

$$\begin{aligned} \delta_2 &= \frac{5wL^4C_c}{384E_cI} \\ &= \frac{(5)\left(\frac{500 \ \frac{\text{lbf}}{\text{ft}} + 250 \ \frac{\text{lbf}}{\text{ft}}}{12 \ \frac{\text{in}}{\text{ft}}}\right)\left((30 \text{ ft})\left(12 \ \frac{\text{in}}{\text{ft}}\right)\right)^4(2)}{(384)\left(4 \times 10^6 \ \frac{\text{lbf}}{\text{in}^2}\right)(8000 \text{ in}^4)} \\ &= 0.854 \text{ in} \quad \text{[downward]} \end{aligned}$$

The deflection due to the live load is found proportionally.

$$\left(\frac{800\ \frac{\text{lbf}}{\text{ft}}}{250\ \frac{\text{lbf}}{\text{ft}}}\right)(0.142\text{ in}) = 0.454\text{ in} \quad [\text{downward}]$$

The total deflection after 5 yr is

$$\begin{aligned}\delta &= -0.85\text{ in} + 0.854\text{ in} + 0.454\text{ in}\\ &= \boxed{0.458\text{ in} \qquad [\text{downward}]}\end{aligned}$$

Note that this assumes a Class U section. This should be verified by computing extreme fiber stress, f_t, at service loads.

3. (a) $I = \frac{bh^3}{12} = \frac{(12\text{ in})(24\text{ in})^3}{12} = \boxed{13{,}824\text{ in}^4}$

The answer is (B).

(b) $A = (12\text{ in})(24\text{ in}) = \boxed{288\text{ in}^2}$

The answer is (B).

(c) $M_{\text{max}} = \frac{wL^2}{8} = \frac{\left(3\ \frac{\text{kips}}{\text{ft}}\right)(20\text{ ft})^2}{8} = 150\text{ ft-kips}$

The answer is (C).

(d) $f = \frac{-P}{A} = \frac{-250\text{ kips}}{288\text{ in}^2} = \boxed{-0.868\text{ kip/in}^2}$

The answer is (D).

(e) $f = \frac{-P}{A} = \frac{-250\text{ kips}}{288\text{ in}^2} = \boxed{-0.868\text{ kip/in}^2}$

The answer is (D).

(f) $f = \frac{-P}{A} = \frac{-250\text{ kips}}{288\text{ in}^2} = \boxed{-0.868\text{ kip/in}^2}$

The answer is (D).

(g) Use Eq. 56.6.

$$\begin{aligned}f_{\text{top}} &= -\frac{P}{A} + \frac{Pec}{I} - \frac{Mc}{I}\\ &= -\frac{250\text{ kips}}{288\text{ in}^2}\\ &\quad + \frac{(250\text{ kips})(9\text{ in})(12\text{ in})}{13{,}824\text{ in}^4}\\ &\quad - \frac{(12\text{ ft})(150\text{ ft-kips})\left(12\ \frac{\text{in}}{\text{ft}}\right)}{13{,}824\text{ in}^4}\\ &= -0.868\ \frac{\text{kip}}{\text{in}^2} + 1.953\ \frac{\text{kips}}{\text{in}^2} - 1.562\ \frac{\text{kips}}{\text{in}^2}\\ &= \boxed{-0.477\text{ kip/in}^2}\end{aligned}$$

The answer is (A).

(h)
$$\begin{aligned}f_{\text{bottom}} &= -\frac{P}{A} - \frac{Pec}{I} + \frac{Mc}{I}\\ &= -0.868\ \frac{\text{kip}}{\text{in}^2} - 1.953\ \frac{\text{kips}}{\text{in}^2} + 1.562\ \frac{\text{kips}}{\text{in}^2}\\ &= \boxed{-1.259\text{ kips/in}^2}\end{aligned}$$

The answer is (B).

(i)
$$\begin{aligned}f_{\text{top}} &= -\frac{P}{A} + \frac{Pec}{I}\\ &= -0.868\ \frac{\text{kip}}{\text{in}^2} + 1.953\ \frac{\text{kips}}{\text{in}^2}\\ &= \boxed{1.085\text{ kips/in}^2}\end{aligned}$$

The answer is (C).

(j)
$$\begin{aligned}f_{\text{bottom}} &= -\frac{P}{A} - \frac{Pec}{I}\\ &= -0.868\ \frac{\text{kip}}{\text{in}^2} - 1.953\ \frac{\text{kips}}{\text{in}^2}\\ &= \boxed{-2.821\text{ kips/in}^2}\end{aligned}$$

The answer is (C).

Note that (h) shows the critical midspan section to be Class U, which justifies the use of the gross moment of inertia in stress calculations.

4. (a) The eccentricity is

$$e = \frac{550\text{ mm}}{2} - 70\text{ mm} = \boxed{205\text{ mm}}$$

The answer is (B).

(b) The beam's own loading is

$$\begin{aligned}w &= \frac{(300\text{ mm})(550\text{ mm})\left(23.5\ \frac{\text{kN}}{\text{m}^3}\right)}{\left(1000\ \frac{\text{mm}}{\text{m}}\right)^2}\\ &= 3.877\text{ kN/m}\end{aligned}$$

$$\begin{aligned}M_{\text{max}} &= \frac{wL^2}{8} = \frac{\left(3.877\ \frac{\text{kN}}{\text{m}}\right)(12\text{ m})^2}{8}\\ &= \boxed{69.78\text{ kN·m}}\end{aligned}$$

The answer is (C).

Structural

(c) $$P_{\text{initial}} = \frac{(1.1\text{ GPa})\left(10^9\ \frac{\text{Pa}}{\text{GPa}}\right)(774\text{ mm}^2)}{\left(10^6\ \frac{\text{mm}^2}{\text{m}^2}\right)\left(1000\ \frac{\text{N}}{\text{kN}}\right)} = \boxed{851.4\text{ kN}}$$

The answer is (D).

(d) $$c = \frac{500\text{ mm}}{2} = 275\text{ mm}$$

$$A = (300\text{ mm})(550\text{ mm}) = 165\,000\text{ mm}^2$$

$$I = \frac{bh^3}{12} = \frac{(300\text{ mm})(550\text{ mm})^3}{12} = 4.159 \times 10^9\text{ mm}^4$$

$$\begin{aligned} f_{\text{top}} &= -\frac{P}{A} + \frac{Pec}{I} - \frac{Mc}{I} \\ &= -\frac{851\,400\text{ N}}{165\,000\text{ mm}^2} + \frac{(851\,400\text{ N})(205\text{ mm})(275\text{ mm})}{4.159 \times 10^9\text{ mm}^4} \\ &\quad - \frac{(69.78\text{ kN·m})(275\text{ mm}) \times \left(1000\ \frac{\text{mm}}{\text{m}}\right)\left(1000\ \frac{\text{N}}{\text{kN}}\right)}{4.159 \times 10^9\text{ mm}^4} \\ &= -5.16\text{ MPa} + 11.54\text{ MPa} - 4.61\text{ MPa} \\ &= \boxed{1.77\text{ MPa}} \end{aligned}$$

The answer is (A).

(e) $$\begin{aligned} f_{\text{bottom}} &= -\frac{P}{A} - \frac{Pec}{I} + \frac{Mc}{I} \\ &= -5.16\text{ MPa} - 11.54\text{ MPa} + 4.61\text{ MPa} \\ &= \boxed{-12.09\text{ MPa}} \end{aligned}$$

The answer is (D).

(f) After an 18% loss, there is 82% remaining.

$$\begin{aligned} f_{\text{top}} &= (-5.16\text{ MPa})(0.82) + (11.54\text{ MPa})(0.82) \\ &\quad - 4.61\text{ MPa} \\ &= \boxed{0.618\text{ MPa}} \end{aligned}$$

The answer is (A).

(g) $$\begin{aligned} f_{\text{bottom}} &= (-5.16\text{ MPa})(0.82) \\ &\quad - (11.54\text{ MPa})(0.82) + 4.61\text{ MPa} \\ &= \boxed{-9.08\text{ MPa}} \end{aligned}$$

The answer is (C).

(h) $$f_{\text{top}} = -\frac{P}{A} + \frac{Pec}{I} - \frac{Mc}{I}$$

$$\begin{aligned} (-0.45)(34.5\text{ MPa}) &= -(0.82)\left(\frac{851\,400\text{ N}}{165\,000\text{ mm}^2}\right) \\ &\quad + \frac{(0.82)(851\,400\text{ N}) \times (205\text{ mm})(275\text{ mm})}{4.159 \times 10^9\text{ mm}^4} \\ &\quad - \frac{M(275\text{ mm}) \times \left(1000\ \frac{\text{mm}}{\text{m}}\right)\left(1000\ \frac{\text{N}}{\text{kN}}\right)}{4.159 \times 10^9\text{ mm}^4} \\ M &= \boxed{313.9\text{ kN·m}} \end{aligned}$$

The answer is (C).

(i) $$f_{\text{bottom}} = -\frac{P}{A} - \frac{Pec}{I} + \frac{Mc}{I}$$

$$\begin{aligned} 0.5\sqrt{34.5\text{ MPa}} &= -(0.82)\left(\frac{851\,400\text{ N}}{165\,000\text{ mm}^2}\right) \\ &\quad - \frac{(0.82)(851\,400\text{ N}) \times (205\text{ mm})(275\text{ mm})}{4.159 \times 10^9\text{ mm}^4} \\ &\quad + \frac{M(275)\left(1000\ \frac{\text{mm}}{\text{m}}\right)\left(1000\ \frac{\text{N}}{\text{kN}}\right)}{4.159 \times 10^9\text{ mm}^4} \\ M &= \boxed{251.5\text{ kN·m}} \end{aligned}$$

The answer is (A).

(j) From parts (h) and (i), the maximum bending moment is 251.5 kN·m.

$$M_{\text{max}} = \frac{wL^2}{8}$$

$$251.5\text{ kN·m} = \frac{w(12\text{ m})^2}{8}$$

$$w = 13.972\text{ kN/m}$$

The beam weight is 3.833 kN/m.

The allowable uniform load is

$$13.972\ \frac{\text{kN}}{\text{m}} - 3.877\ \frac{\text{kN}}{\text{m}} = \boxed{10.095\ \text{kN/m}}$$

The answer is (D).

Note the tensile stress limitation in (i) makes this a Class U section.

5. (a) The total area of the prestressing steel is

$$A_{\text{ps}} = (12)(0.153\ \text{in}^2) = 1.836\ \text{in}^2$$

The prestressing steel ratio is

$$\rho_p = \frac{A_{\text{ps}}}{bd_p} = \frac{1.836\ \text{in}^2}{(24\ \text{in})(28\ \text{in})} = 0.0028 \quad \text{[round up]}$$

Since $f_{\text{py}}/f_{\text{pu}} > 0.9$, $\gamma_p = 0.28$ and $\beta_1 = 0.8$. The stress in the tendons at the attainment of the nominal moment strength is

$$\begin{aligned} f_{\text{ps}} &= f_{\text{pu}}\left(1 - \left(\frac{\gamma_p}{\beta_1}\right)\left(\rho_p\left(\frac{f_{\text{pu}}}{f_c'}\right) + \left(\frac{d}{d_p}\right)(\omega - \omega')\right)\right) \\ &= \left(270\ \frac{\text{kips}}{\text{in}^2}\right) \\ &\quad \times \left(1 - \left(\frac{0.28}{0.80}\right)\left((0.0028)\left(\frac{270\ \frac{\text{kips}}{\text{in}^2}}{5\ \frac{\text{kips}}{\text{in}^2}}\right)\right)\right) \\ &= 255.7\ \text{kips/in}^2 \end{aligned}$$

The nominal moment strength is computed in the same manner as for a standard reinforced section. The total tension at maximum stress is

$$\begin{aligned} T = A_{\text{ps}}f_{\text{ps}} &= (1.836\ \text{in}^2)\left(255.71\ \frac{\text{kips}}{\text{in}^2}\right) \\ &= 469.5\ \text{kips} \end{aligned}$$

The area of concrete required to balance this force is

$$\begin{aligned} A_c = \frac{T}{0.85f_c'} &= \frac{469.5\ \text{kips}}{(0.85)\left(5\ \frac{\text{kips}}{\text{in}^2}\right)} \\ &= 110.5\ \text{in}^2 \end{aligned}$$

Assume the depth of the equivalent rectangular stress block is within the 5 in thickness.

$$\begin{aligned} a = \frac{A_c}{b} &= \frac{110.5\ \text{in}^2}{24\ \text{in}} \\ &= 4.60\ \text{in} \end{aligned}$$

The assumption is valid. The centroid of the compressed area is at $\lambda = a/2$.

$$\begin{aligned} M_n = T(d - \lambda) &= \frac{(469.5\ \text{kips})\left(28\ \text{in} - \frac{4.60\ \text{in}}{2}\right)}{12\ \frac{\text{in}}{\text{ft}}} \\ &= 1006\ \text{ft-kips} \end{aligned}$$

Check to see if the section is tension controlled.

$$\begin{aligned} \epsilon_t &= \left(\frac{d - c}{c}\right)(0.003) = \left(\frac{28\ \text{in} - 5.75\ \text{in}}{5.75\ \text{in}}\right)(0.003) \\ &= 0.0116 > 0.005 \end{aligned}$$

Therefore, the section is tension controlled.

The design moment capacity is

$$\phi M_n = (0.9)(1006\ \text{ft-kips}) = \boxed{905\ \text{ft-kips}}$$

(b) The cracking moment, M_{cr}, is computed as the value that leads to a maximum tensile stress equal to the modulus of rupture. The gross moment of inertia may be used in the calculation.

$$A = (30\ \text{in})(24\ \text{in}) - (20\ \text{in})(14\ \text{in}) = 440\ \text{in}^2$$

$$\begin{aligned} I &= \frac{b_1h_1^3}{12} - \frac{b_2h_2^3}{12} \\ &= \left(\tfrac{1}{12}\right)\left((24\ \text{in})(30\ \text{in})^3 - (14\ \text{in})(20\ \text{in})^3\right) \\ &= 44{,}667\ \text{in}^4 \end{aligned}$$

$$\begin{aligned} y &= \text{distance from neutral axis to tension fiber} \\ &= 15\ \text{in} \end{aligned}$$

$$\begin{aligned} P = f_{\text{se}}A_{\text{ps}} &= \left(148\ \frac{\text{kips}}{\text{in}^2}\right)(1.836\ \text{in}^2) \\ &= 271.7\ \text{kips} \end{aligned}$$

$$e = \text{eccentricity of tendons} = 13\ \text{in}$$

$$\begin{aligned} f_r = 7.5\sqrt{f_c'} &= 7.5\sqrt{5000\ \frac{\text{lbf}}{\text{in}^2}} \\ &= 530\ \text{lbf/in}^2 \end{aligned}$$

$$f_c = \frac{-P}{A} + \frac{Pey}{I} - \frac{M_sy}{I}$$

$$\begin{aligned} 0.53\ \frac{\text{kips}}{\text{in}^2} &= \frac{-271.7\ \text{kips}}{440\ \text{in}^2} \\ &\quad + \frac{(271.7\ \text{kips})(13\ \text{in})(-15\ \text{in})}{44{,}667\ \text{in}^4} \\ &\quad - \frac{M_{\text{cr}}(-15\ \text{in})}{44{,}667\ \text{in}} \end{aligned}$$

$$M_{\text{cr}} = 6949.1\ \text{in-kips}$$

$$\begin{aligned} M_{\text{cr}} &= \frac{6949.1\ \text{in-kips}}{12\ \frac{\text{in}}{\text{ft}}} \\ &= \boxed{579.1\ \text{ft-kips}} \end{aligned}$$

(c) ACI 318 requires that the design moment, ϕM_n, exceed the cracking moment by at least 20%.

$$\frac{\phi M_n}{M_{cr}} = \frac{905 \text{ ft-kips}}{579 \text{ ft-kips}}$$

$$= 1.56 \quad \boxed{[56\% \text{ increase, so OK}]}$$

6. (a) *step 1:* $w_D = 150$ lbf/ft [given]

$$\textit{step 2:}\ M_D = \frac{wL^2}{8} = \frac{\left(150 \frac{\text{lbf}}{\text{ft}}\right)(100 \text{ ft})^2}{8}$$

$$= 187{,}500 \text{ ft-lbf}$$

step 3: Use the top as the reference. Neglect the tension holes.

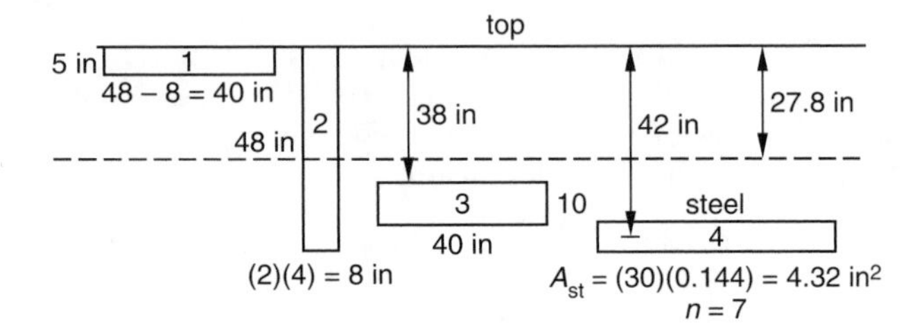

$$\overline{x} = \frac{\sum A_i \overline{x}_i}{\sum A_i}$$

$$= \frac{\begin{array}{c}(5 \text{ in})(40 \text{ in})(2.5 \text{ in}) + (48 \text{ in})(8 \text{ in})(24 \text{ in}) \\ + (10 \text{ in})(40 \text{ in})(38 \text{ in} + 5 \text{ in}) \\ + (4.32 \text{ in}^2)(7)(42 \text{ in})\end{array}}{\begin{array}{c}(5 \text{ in})(40 \text{ in}) + (48 \text{ in})(8 \text{ in}) \\ + (10 \text{ in})(40 \text{ in}) + (44.32 \text{ in}^2)(7)\end{array}}$$

$$= \frac{28{,}186 \text{ in}^3}{1014.2 \text{ in}^2} = 27.8 \text{ in} \quad [\text{from top}]$$

step 4: Assume the entire beam is in compression. Set up a table to calculate the moment of inertia.

section	$I_c = \frac{bh^3}{12}$ (in⁴)	A (in²)	d (in)	Ad^2 (in⁴)
1	416.7	200	25.3	128,018
2	73,728	384	3.8	5545
3	3333.3	400	15.2	92,416
4	0	30.24	14.2	6098
totals	77,478			232,077

$$I = 77{,}478 \text{ in}^4 + 232{,}077 \text{ in}^4 = 309{,}555 \text{ in}^4$$

step 5: $e = 42 \text{ in} - 27.8 \text{ in} = 14.2 \text{ in}$

step 6: Disregarding elastic shortening losses immediately after transfer (conservative assumption), and disregarding the moment caused by the beam weight at the end of the transfer zone, the initial prestress is

$$f_s = \text{initial prestress} - \text{losses}$$

$$= (0.8)\left(250{,}000 \frac{\text{lbf}}{\text{in}^2}\right) - 0$$

$$= 200{,}000 \text{ lbf/in}^2$$

$$P = \left(200{,}000 \frac{\text{lbf}}{\text{in}^2}\right)\left(0.144 \frac{\text{in}^2}{\text{strand}}\right) \times (30 \text{ strands})$$

$$= 864{,}000 \text{ lbf}$$

$$A_c = 200 \text{ in}^2 + 384 \text{ in}^2 + 400 \text{ in}^2 + (7)(4.32 \text{ in}^2)$$

$$= 1014 \text{ in}^2$$

$$c_{top} = 27.8 \text{ in} \quad [\text{for top fibers}]$$

$$c_{bottom} = 48 \text{ in} - 27.8 \text{ in} = 20.2 \text{ in} \quad [\text{for bottom fibers}]$$

$$f_{top} = -\frac{864{,}000 \text{ lbf}}{1014 \text{ in}} + \frac{(864{,}000 \text{ lbf})(14.2 \text{ in})(27.8 \text{ in})}{309{,}555 \text{ in}^4}$$

$$= 250 \text{ lbf/in}^2 \quad [\text{tension in concrete}]$$

$$f_{bottom} = -\frac{864{,}000 \text{ lbf}}{1014 \text{ in}^2} - \frac{(864{,}000 \text{ lbf})(14.2 \text{ in})(20.2 \text{ in})}{309{,}555 \text{ in}^4}$$

$$= -1653.6 \text{ lbf/in}^2 \quad \left[\begin{array}{c}\text{compression in} \\ \text{concrete}\end{array}\right]$$

step 7: The stress due to dead load is

$$f_{b,top} = \frac{Mc}{I}$$

$$= \frac{(187{,}500 \text{ ft-lbf})\left(12 \frac{\text{in}}{\text{ft}}\right)(27.8 \text{ in})}{309{,}555 \text{ in}^4}$$

$$= 202.1 \text{ lbf/in}^2 \quad \left[\begin{array}{c}\text{compression in} \\ \text{concrete}\end{array}\right]$$

$$f_{b,bottom} = \frac{(187{,}500 \text{ ft-lbf})\left(12 \frac{\text{in}}{\text{ft}}\right)(20.2 \text{ in})}{309{,}555 \text{ in}^4}$$

$$= 146.8 \text{ lbf/in}^2 \quad [\text{tension in concrete}]$$

step 8:

$$f_{top} = 250 \frac{\text{lbf}}{\text{in}^2} - 202.1 \frac{\text{lbf}}{\text{in}^2}$$

$$= 48 \text{ lbf/in}^2 \quad [\text{tension}]$$

$$f_{bottom} = -1653 \frac{\text{lbf}}{\text{in}^2} + 146.8 \frac{\text{lbf}}{\text{in}^2}$$

$$= -1506 \text{ lbf/in}^2 \quad [\text{compression}]$$

step 9: $(0.60)\left(3800\ \frac{\text{lbf}}{\text{in}^2}\right) = 2280\ \text{lbf/in}^2$

$$> 1506\ \text{lbf/in}^2$$

$$3\sqrt{3800\ \frac{\text{lbf}}{\text{in}^2}} = 185\ \text{lbf/in}^2 > 48\ \text{lbf/in}^2$$

The stresses are acceptable.

(b) The prestress from step 6 is reduced.

$$\frac{(0.6)\left(250{,}000\ \frac{\text{lbf}}{\text{in}^2}\right)}{(0.8)\left(250{,}000\ \frac{\text{lbf}}{\text{in}^2}\right)} = 0.75 \text{ of step 6 prestress}$$

Recalculate the stresses with the lower prestress.

$$f_{\text{top}} = (0.75)\left(250\ \frac{\text{lbf}}{\text{in}^2}\right) = 188\ \text{lbf/in}^2 \quad [\text{tension}]$$

$$f_{\text{bottom}} = (0.75)\left(-1653\ \frac{\text{lbf}}{\text{in}^2}\right) = -1240\ \text{lbf/in}^2 \quad [\text{compression}]$$

$$M_{D+L} = \frac{wL^2}{8} = \frac{\left(150\ \frac{\text{lbf}}{\text{ft}} + 540\ \frac{\text{lbf}}{\text{ft}}\right)(100\ \text{ft})^2}{8} = 862{,}500\ \text{ft-lbf}$$

$$f_{\text{top}} = \frac{(862{,}500\ \text{ft-lbf})\left(12\ \frac{\text{in}}{\text{ft}}\right)(27.8\ \text{in})}{309{,}555\ \text{in}^4} = -929.5\ \text{lbf/in}^2 \quad [\text{compression}]$$

$$f_{\text{bottom}} = \frac{(862{,}500\ \text{ft-lbf})\left(12\ \frac{\text{in}}{\text{ft}}\right)(20.2\ \text{in})}{309{,}555\ \text{in}^4} = 675.4\ \text{lbf/in}^2 \quad [\text{tension}]$$

$$f_{\text{top,total}} = 188\ \frac{\text{lbf}}{\text{in}^2} - 202.1\ \frac{\text{lbf}}{\text{in}^2} - 929.5\ \frac{\text{lbf}}{\text{in}^2} = -944\ \text{lbf/in}^2 \quad [\text{compression}]$$

$$f_{\text{bottom,total}} = -1240\ \frac{\text{lbf}}{\text{in}^2} + 146.8\ \frac{\text{lbf}}{\text{in}^2} + 675.4\ \frac{\text{lbf}}{\text{in}^2} = -418\ \text{lbf/in}^2 \quad [\text{compression}]$$

$$(0.45)\left(5000\ \frac{\text{lbf}}{\text{in}^2}\right) = 2250\ \frac{\text{lbf}}{\text{in}^2} > 1240\ \frac{\text{lbf}}{\text{in}^2} - 146.8\ \frac{\text{lbf}}{\text{in}^2} = 1093\ \text{lbf/in}^2$$

$$(0.6)\left(5000\ \frac{\text{lbf}}{\text{in}^2}\right) = 3000\ \text{lbf/in}^2 > 741\ \text{lbf/in}^2$$

The compressive stresses are acceptable. (There is no tensile stress.)

57 Composite Concrete and Steel Bridge Girders

PRACTICE PROBLEMS

1. *(Time limit: one hour)* A simple bridge span is shown. The deck of the bridge consists of a 7 in thick concrete slab supported by 88 ft W 24×94 steel beams. Normal-weight concrete is used. The concrete strength is 3000 psi. Use AASHTO specifications.

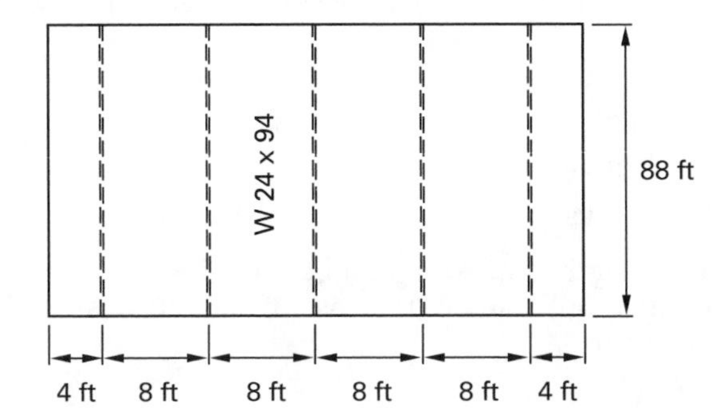

(a) The effective width for an interior girder is most nearly

(A) 42 in
(B) 84 in
(C) 96 in
(D) 260 in

(b) The effective width for an exterior girder is most nearly

(A) 42 in
(B) 48 in
(C) 88 in
(D) 98 in

(c) The modular ratio is most nearly

(A) 6
(B) 7
(C) 8
(D) 9

(d) The equivalent width (steel) of the effective width for an interior girder is most nearly

(A) 8.3 in
(B) 9.3 in
(C) 10 in
(D) 11 in

(e) The centroidal distance of the transformed section measured from the bottom of the steel beam is most nearly

(A) 21 in
(B) 22 in
(C) 23 in
(D) 24 in

(f) The moment of inertia of the transformed section about the centroidal axis is most nearly

(A) 7700 in^4
(B) 8100 in^4
(C) 8300 in^4
(D) 8500 in^4

(g) The section modulus with respect to the top fiber is most nearly

(A) 910 in^3
(B) 930 in^3
(C) 950 in^3
(D) 990 in^3

(h) The section modulus with respect to the bottom fiber is most nearly

(A) 290 in^3
(B) 330 in^3
(C) 380 in^3
(D) 420 in^3

(i) If a 1 in by 8 in plate is welded to the bottom flange of the steel beam, the centroid of the transformed section will

(A) move up.
(B) move down.
(C) not move.
(D) move only after the concrete has failed (i.e., is 100% cracked).

(j) The addition of the 1 in by 8 in steel plate to the bottom flange will

(A) increase the moment of inertia about the x-axis.
(B) decrease the moment of inertia about the x-axis.
(C) have no effect on the moment of inertia about the y-axis.
(D) decrease the moment of inertia about the y-axis.

2. *(Time limit: one hour)* The steel support beams described in Prob. 1 are reinforced with a 1 in by 8 in steel plate welded to the bottom flange. After curing, the live load moment is 620 ft-kip. Use AASHTO specifications to evaluate an interior girder with a span length of 39.3 ft.

(a) The moment of inertia of the transformed section of the composite section about the centroidal axis is most nearly

(A) 11,000 in^4
(B) 12,000 in^4
(C) 13,000 in^4
(D) 14,000 in^4

(b) The centroidal distance of the noncomposite section measured from the bottom of the steel plate is most nearly

(A) 8.3 in
(B) 10 in
(C) 11 in
(D) 12 in

(c) The moment of inertia of the noncomposite section about the centroidal axis is most nearly

(A) 3700 in^4
(B) 3800 in^4
(C) 3900 in^4
(D) 4000 in^4

(d) The bending stress at the top steel fibers due to noncomposite dead load in the case of unshored construction is most nearly

(A) 4.3 ksi
(B) 6.9 ksi
(C) 7.7 ksi
(D) 8.5 ksi

(e) The bending stress at the bottom fibers due to noncomposite dead load in the case of unshored construction is most nearly

(A) 5.3 ksi
(B) 6.2 ksi
(C) 7.4 ksi
(D) 8.9 ksi

(f) The additional bending stress at the top fibers of the concrete slab due to live moment in the case of unshored construction is most nearly

(A) 0.5 ksi
(B) 0.6 ksi
(C) 0.7 ksi
(D) 0.8 ksi

(g) The additional bending stress at the top fibers of the steel beam due to live moment in the case of unshored construction is most nearly

(A) 1.1 ksi
(B) 1.9 ksi
(C) 2.4 ksi
(D) 3.8 ksi

(h) The additional bending stress at the bottom fibers of the steel plate due to live moment in the case of unshored construction is most nearly

(A) 14 ksi
(B) 15 ksi
(C) 20 ksi
(D) 22 ksi

(i) The bending stress at the top fibers of the concrete slab due to all loads in the case of shored construction is most nearly

(A) 0.2 ksi
(B) 0.9 ksi
(C) 1.3 ksi
(D) 2.3 ksi

(j) The bending stress at the bottom fibers of the steel plate due to all loads in the case of shored construction is most nearly

(A) 11 ksi
(B) 14 ksi
(C) 16 ksi
(D) 18 ksi

3. Calculate the ultimate moment capacity of the composite section shown. All pieces are A36 steel. The beam is W 460×113 (metric). The concrete compressive strength is 20.7 MPa.

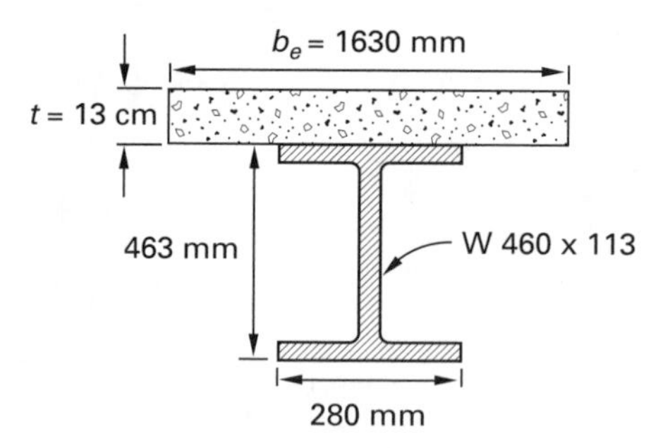

4. The beam described in Prob. 3 is changed such that the equivalent width of the concrete slab is 130 cm. The material properties are unchanged. Determine the ultimate moment capacity.

SOLUTIONS

1. (a) $$\frac{L}{4} = \frac{(88\text{ ft})\left(12\ \frac{\text{in}}{\text{ft}}\right)}{4} = 264\text{ in}$$

$$b_o = (8\text{ ft})\left(12\ \frac{\text{in}}{\text{ft}}\right) = 96\text{ in}$$

$$12t = (12)(7\text{ in}) = \boxed{84\text{ in}\quad[\text{controls}]}$$

The answer is (B).

(b) $$\frac{L}{12} = \frac{(88\text{ ft})\left(12\ \frac{\text{in}}{\text{ft}}\right)}{12} = 88\text{ in}$$

$$\tfrac{1}{2}b_o = \frac{(8\text{ ft})\left(12\ \frac{\text{in}}{\text{ft}}\right)}{2} = 48\text{ in}$$

$$6t = (6)(7\text{ in}) = \boxed{42\text{ in}\qquad[\text{controls}]}$$

The answer is (A).

(c) Assume normal-weight concrete has a specific weight of 145 lbf/ft^3.

$$\begin{aligned} n &= \frac{E_s}{E_c} \\ &= \frac{29{,}000{,}000\ \frac{\text{lbf}}{\text{in}^2}}{\left(145\ \frac{\text{lbf}}{\text{ft}^3}\right)^{1.5} 33\sqrt{3000\ \frac{\text{lbf}}{\text{in}^2}}} \\ &= \boxed{9.19} \end{aligned}$$

(Same as Table 57.1.)

The answer is (D).

(d) The equivalent width is

$$\frac{b_e}{n} = \frac{84\text{ in}}{9} = \boxed{9.33\text{ in}}$$

The answer is (B).

(e) $$\begin{aligned} y_b &= \frac{\sum A_i y_i}{\sum A_i} \\ &= \frac{(27.7\text{ in}^2)\left(\frac{24.31\text{ in}}{2}\right) + (7)(9.33\text{ in})\left(24.31\text{ in} + \frac{7\text{ in}}{2}\right)}{27.7\text{ in}^2 + (9.33\text{ in})(7\text{ in})} \\ &= \boxed{23.15\text{ in}} \end{aligned}$$

The answer is (C).

(f)

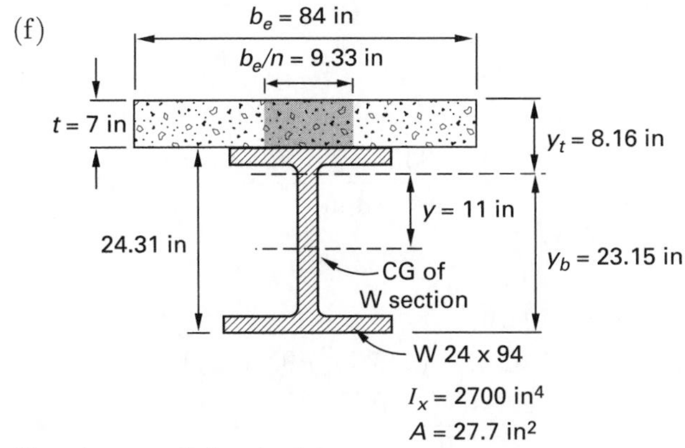

Use the parallel axis theorem.

$$\begin{aligned} I &= I_x + Ad^2 \\ &= 2700\text{ in}^4 + (27.7\text{ in}^2)(11\text{ in})^2 + \frac{(9.33\text{ in})(7\text{ in})^3}{12} \\ &\quad + (7\text{ in})(9.33\text{ in})(4.66\text{ in})^2 \\ &= \boxed{7733.8\text{ in}^4} \end{aligned}$$

The answer is (A).

(g) $$S_t = \frac{7733.6\text{in}^4}{8.16\text{ in}} = \boxed{947.74\text{ in}^3}$$

The answer is (C).

(h) $$S_b = \frac{7733.6\text{ in}^4}{23.15\text{ in}} = \boxed{334.1\text{ in}^3}$$

The answer is (B).

(i) The centroid of the transformed section will move down when a steel plate is added to the bottom flange.

The answer is (B).

(j) The moment of inertia about the x- and y-axes both will increase.

The answer is (A).

2. (a)

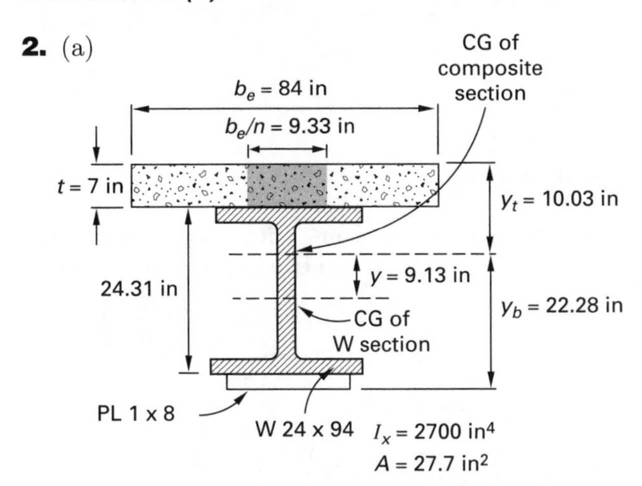

Structural

$$y_b = \frac{\sum A_i y_i}{\sum A_i}$$

$$= \frac{(27.7\ \text{in}^2)(13.16\ \text{in}) + (8\ \text{in}^2)(0.5\ \text{in}) + (9.33\ \text{in})(7\ \text{in})(28.81\ \text{in})}{27.7\ \text{in}^2 + 8\ \text{in}^2 + 65.31\ \text{in}^2}$$

$$= \frac{2250.11\ \text{in}^3}{101.01\ \text{in}^2} = 22.28\ \text{in}$$

$$I = I_x + Ad^2$$

$$= 2700\ \text{in}^4 + (27.7\ \text{in}^2)(9.13\ \text{in})^2 + \frac{(8\ \text{in})(1\ \text{in})^3}{12} + (8\ \text{in}^2)(21.78\ \text{in})^2 + \frac{(9.33\ \text{in})(7\ \text{in})^3}{12} + (9.33\ \text{in})(7\ \text{in})(6.53\ \text{in})^2$$

$$= \boxed{11{,}856\ \text{in}^4}$$

The answer is (B).

(b)
$$y_b = \frac{\sum A_i y_i}{\sum A_i} = \frac{(8\ \text{in}^2)(0.5\ \text{in}) + (27.7\ \text{in}^2)(13.16\ \text{in})}{8\ \text{in}^2 + 27.7\ \text{in}^2} = \boxed{10.32\ \text{in}}$$

The answer is (B).

(c)

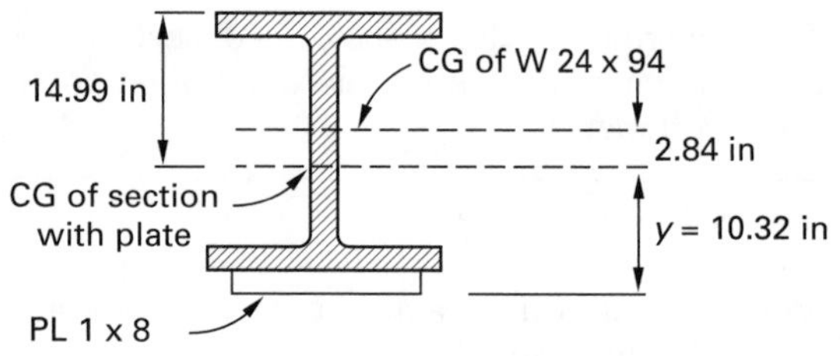

$$I = I_x + Ad^2$$

$$= 2700\ \text{in}^4 + (27.7\ \text{in}^2)(2.84\ \text{in})^2 + \frac{(8\ \text{in})(1\ \text{in})^3}{12} + (8\ \text{in}^2)(9.82\ \text{in})^2$$

$$= \boxed{3695.5\ \text{in}^4}$$

The answer is (A).

(d) For construction without temporary shoring, the noncomposite dead load is carried by the steel section alone.

$$w_{\text{slab}} = \frac{(7\ \text{in})(8\ \text{ft})\left(0.150\ \frac{\text{kip}}{\text{ft}^3}\right)}{12\ \frac{\text{in}}{\text{ft}}} = 0.70\ \text{kip/ft}$$

$$w_{\text{beam and plate}} = 0.121\ \text{kip/ft}$$

$$w_D = 0.70\ \frac{\text{kip}}{\text{ft}} + 0.121\ \frac{\text{kip}}{\text{ft}} = 0.821\ \text{kip/ft}$$

$$M_D = \frac{w_D L^2}{8} = \frac{\left(0.821\ \frac{\text{kip}}{\text{ft}}\right)(39.3\ \text{ft})^2}{8} = 158.5\ \text{ft-kips}$$

$$c = \frac{24.31\ \text{in}}{2} + 2.84\ \text{in} = 14.99\ \text{in}$$

$$f_{\text{top}} = \frac{M_D}{S_{t,\text{steel}}} = \frac{(158.50\ \text{ft-kips})\left(12\ \frac{\text{in}}{\text{ft}}\right)}{\frac{3695.5\ \text{in}^4}{14.99\ \text{in}}} = \boxed{7.71\ \text{kips/in}^2}$$

The answer is (C).

(e)
$$c = 10.32\ \text{in} \quad [\text{part (b)}]$$

$$f_{\text{bottom}} = \frac{M_D}{S_{b,\text{steel}}} = \frac{(158.50\ \text{ft-kips})\left(12\ \frac{\text{in}}{\text{ft}}\right)}{\frac{3695.5\ \text{in}^4}{10.32\ \text{in}}} = \boxed{5.31\ \text{kips/in}^2}$$

The answer is (A).

(f)
$$c = 1\ \text{in} + 24.31\ \text{in} + 7\ \text{in} - 22.28\ \text{in} = 10.03\ \text{in}$$

$$f_{\text{top,concrete}} = \frac{M_L}{S_{t,\text{composite}}} = \frac{(620\ \text{ft-kips})\left(12\ \frac{\text{in}}{\text{ft}}\right)}{\frac{(9)(11{,}856\ \text{in}^4)}{10.03\ \text{in}}} = \boxed{0.70\ \text{kip/in}^2}$$

The answer is (C).

(g)

$$c = 1 \text{ in} + 24.31 \text{ in} - 22.28 \text{ in} = 3.03 \text{ in}$$

$$\begin{aligned} f_{\text{top,steel}} &= \frac{M_L}{S_{t,\text{composite}}} \\ &= \frac{(620 \text{ ft-kips})\left(12 \ \frac{\text{in}}{\text{ft}}\right)}{\frac{11{,}856 \text{ in}^4}{3.03 \text{ in}}} \\ &= \boxed{1.90 \text{ kips/in}^2} \end{aligned}$$

The answer is (B).

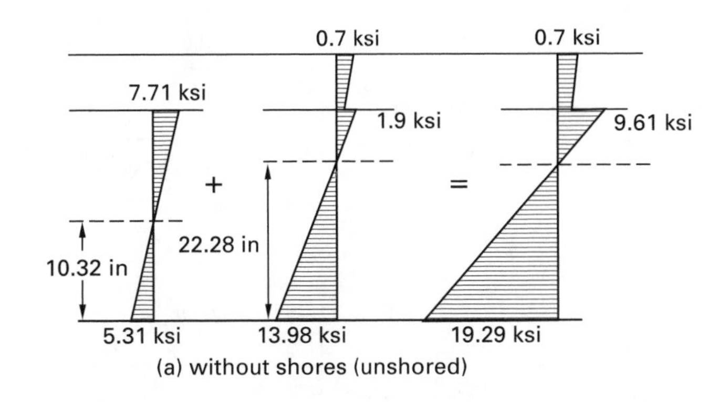

(a) without shores (unshored)

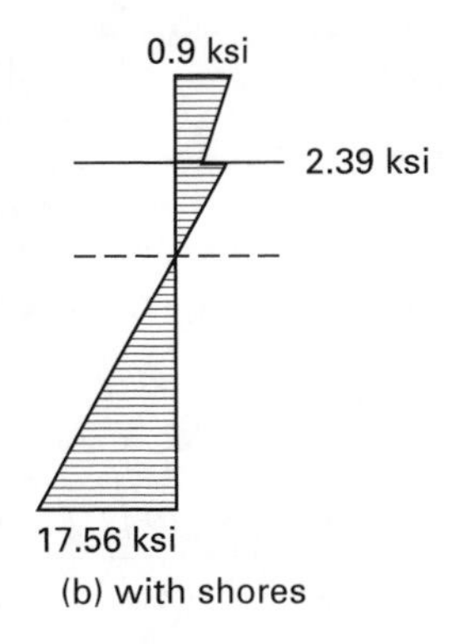

(b) with shores

(h)

$$c = 22.28 \text{ in} \quad \text{[part (a)]}$$

$$\begin{aligned} f_{\text{bottom,steel}} &= \frac{M_L}{S_{b,\text{composite}}} \\ &= \frac{(620 \text{ ft-kips})\left(12 \ \frac{\text{in}}{\text{ft}}\right)}{\frac{11{,}856 \text{ in}^4}{22.28 \text{ in}}} \\ &= \boxed{13.98 \text{ kips/in}^2} \end{aligned}$$

The answer is (A).

(i) With temporary shoring, the steel beam and the wet concrete will be supported by temporary falsework until the concrete has cured.

$$\begin{aligned} f_{\text{top}} &= \frac{M_L + M_D}{S_{t,\text{composite}}} \\ &= \frac{(620 \text{ ft-lbf} + 158.50 \text{ ft-lbf})\left(12 \ \frac{\text{in}}{\text{ft}}\right)}{\frac{(9)(11{,}856 \text{ in}^4)}{10.03}} \\ &= \boxed{0.9 \text{ kip/in}^2} \end{aligned}$$

The answer is (B).

(j)

$$\begin{aligned} f_{\text{bottom}} &= \frac{M_L + M_D}{S_{b,\text{composite}}} \\ &= \frac{(620 \text{ ft-lbf} + 158.50 \text{ ft-lbf})\left(12 \ \frac{\text{in}}{\text{ft}}\right)}{\frac{11{,}856 \text{ in}^4}{22.28 \text{ in}}} \\ &= \boxed{17.56 \text{ kips/in}^2} \end{aligned}$$

The answer is (D).

3.

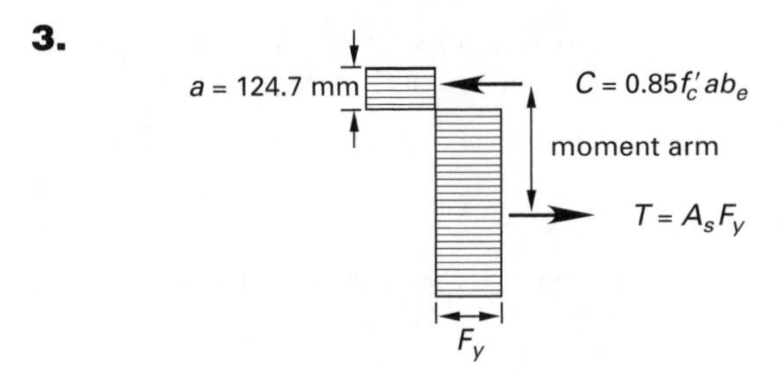

Calculate the depth, a, of the stress block.

$$\begin{aligned} a &= \frac{A_s F_y}{0.85 f'_c b_e} \\ &= \frac{(14\,400 \text{ mm}^2)(248.3 \text{ MPa})}{(0.85)(20.7 \text{ MPa})(1630 \text{ mm})} \\ &= 124.7 \text{ mm} \end{aligned}$$

Since $a < t$, the slab is adequate.

$$\begin{aligned} C &= 0.85 f'_c a b_e \\ &= \frac{(0.85)(20.7 \text{ MPa})(124.7 \text{ mm})(1630 \text{ mm})}{\left(1000 \ \frac{\text{mm}}{\text{m}}\right)^2} \\ &= 3.58 \text{ MN} \end{aligned}$$

The moment arm is

$$\begin{aligned} \frac{d}{2} + t - \frac{a}{2} &= \frac{463 \text{ mm}}{2} + 130 \text{ mm} - \frac{124.7 \text{ mm}}{2} \\ &= 299.2 \text{ mm} \end{aligned}$$

Structural

The ultimate composite moment capacity is

$$M_u = C(\text{moment arm}) = \frac{(3.58 \text{ MN})(299.3 \text{ mm})}{1000 \ \frac{\text{mm}}{\text{m}}} = \boxed{1.07 \text{ MN·m}}$$

4.

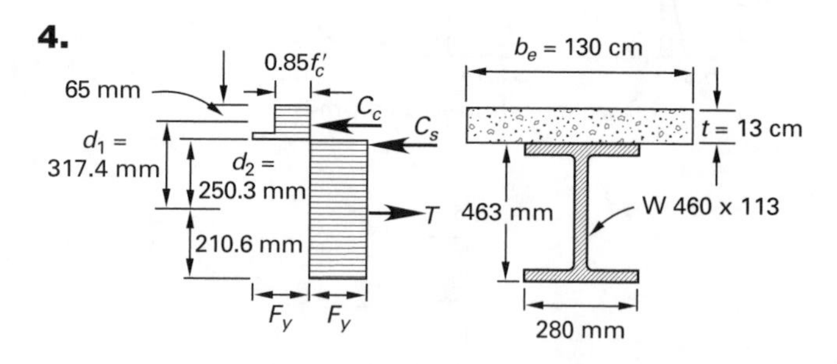

Calculate the depth, a, of the stress block.

$$a = \frac{A_sF_y}{0.85f'_cb_e} = \frac{(14\,400 \text{ mm}^2)(248.3 \text{ MPa})}{(0.85)(20.7 \text{ MPa})(1300 \text{ mm})} = 156.3 \text{ mm}$$

Since $a > t$, the slab is inadequate. A portion of the compressive stress will be carried by the steel beam.

The compressive stress in the concrete is

$$C_c = 0.85f'_cb_et = (0.85)(20.7 \text{ MPa})(1.3 \text{ m})(0.13 \text{ m}) = 2.97 \text{ MN}$$

However, the total tensile force is equal to the total compressive force.

$$T = C_c + C_s$$

The tensile force can also be calculated from the stress carried by the steel.

$$T = A_sF_y - C_s$$

Equate the two expressions for T and solve for C_s.

$$C_s = \frac{A_sF_y - C_c}{2} = \frac{\frac{(14\,400 \text{ mm}^2)(248.3 \text{ MPa})}{\left(1000 \ \frac{\text{mm}}{\text{m}}\right)^2} - 2.97 \text{ MN}}{2} = 0.30 \text{ MN}$$

Calculate the portion of the top flange needed to carry C_s.

$$d_f = \frac{(0.30 \text{ MN})\left(1000 \ \frac{\text{mm}}{\text{m}}\right)}{(248.3 \text{ MPa})(0.280 \text{ m})} = 4.3 \text{ mm}$$

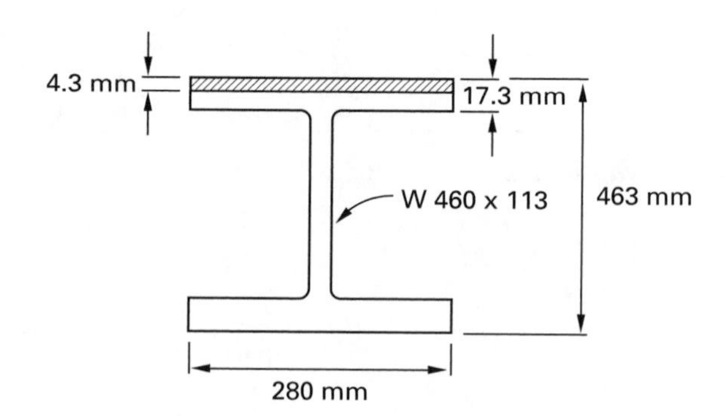

The height of the centroid of the tension portion of the steel beam, from the bottom, is

$$y = \frac{(14\,400 \text{ mm}^2)\left(\frac{463 \text{ mm}}{2}\right) - (4.3 \text{ mm})(280 \text{ mm})\left(463 \text{ mm} - \frac{4.3 \text{ mm}}{2}\right)}{14\,400 \text{ mm}^2 - (280 \text{ mm})(4.3 \text{ mm})} = 210.57 \text{ mm}$$

The ultimate moment capacity of the composite section is

$$M_u = C_cd_1 + C_sd_2 = \frac{(2.97 \text{ MN})(317.4 \text{ mm}) + (0.30 \text{ MN})(250.3 \text{ mm})}{1000 \ \frac{\text{mm}}{\text{m}}} = \boxed{1.02 \text{ MN·m}}$$

58 Structural Steel: Introduction

PRACTICE PROBLEMS

There are no problems in this book corresponding to Ch. 58 of the *Civil Engineering Reference Manual*.

59 Structural Steel: Beams

PRACTICE PROBLEMS

1. A 25 ft beam of A36 steel is simply supported at its left end and 7 ft from its right end. A load of 3000 lbf/ft is uniformly distributed over its entire length. Lateral support is provided only at the reactions. The beam's dead weight is initially estimated as 60 lbf/ft. Choose an economical W shape for this application. (Do not check shear stress.)

2. A beam spans 24 ft and has a deflection limited to $L/300$. The compression flange has complete lateral bracing. The beam carries a uniformly distributed load of 800 lbf/ft over its entire length, excluding the beam's self-weight. The beam's dead weight is initially estimated as 50 lbf/ft. Select an economical A36 W shape.

3. Select an A36 W shape with lateral support at $5^1/_2$ ft intervals to span 20 ft and carry a uniformly distributed load of 1 kip/ft. The load includes a uniform dead load allowance of 40 lbf/ft. The maximum deflection is limited to $L/240$.

4. (*Time limit: one hour*) The steel beam shown carries a load of 1.4 kips/ft, excluding self-weight, over its entire span. Assume lateral bracing only at the reaction points.

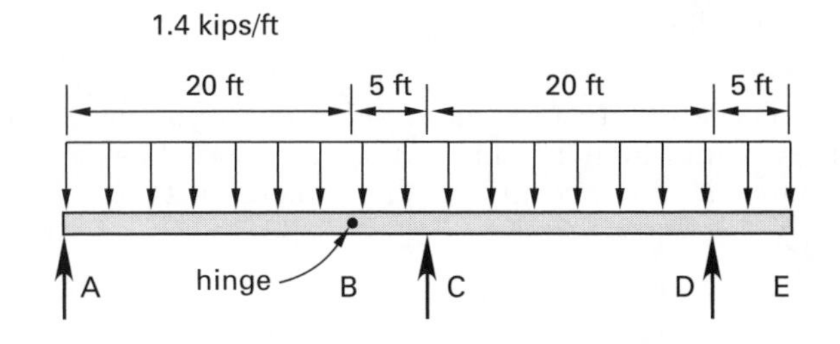

(Disregard the weight of the beam in the following questions.)

(a) The reaction at A is most nearly

(A) 12 kips
(B) 14 kips
(C) 16 kips
(D) 18 kips

(b) The reaction at C is most nearly

(A) 36 kips
(B) 37 kips
(C) 39 kips
(D) 42 kips

(c) The reaction at D is most nearly

(A) 13 kips
(B) 18 kips
(C) 19 kips
(D) 22 kips

(d) The absolute maximum value of shear is most nearly

(A) 12 kips
(B) 14 kips
(C) 18 kips
(D) 21 kips

(e) The absolute maximum shear occurs closest to

(A) support A
(B) hinge B
(C) support C
(D) support D

(f) The absolute maximum value of moment is most nearly

(A) 70 ft-kips
(B) 79 ft-kips
(C) 88 ft-kips
(D) 98 ft-kips

(g) The absolute maximum moment occurs

(A) halfway between A and B
(B) at support C
(C) halfway between C and D
(D) at support D

(h) Considering the unbraced length $L_b = 25$ ft, what is the lightest W 18 section of A36 steel that can be used from a consideration of the moments?

(A) W 18×76
(B) W 18×65
(C) W 18×60
(D) W 18×55

(i) For the lightest beam chosen in part (h), the actual maximum shear stress (ignoring the weight of the beam) is most nearly

(A) 2.0 ksi
(B) 2.4 ksi
(C) 2.8 ksi
(D) 3.4 ksi

(j) If the beam chosen in part (h) is constructed of 42 ksi steel, the maximum unbraced length that permits the development of $0.66F_y$ bending stress is most nearly

(A) 78 in
(B) 84 in
(C) 89 in
(D) 96 in

5. (*Time limit: one hour*) A beam is constructed from a W shape and a C shape, both of A36 steel, as shown. Two equal loads remain 3 ft apart as they move across a 40 ft span. Sufficient lateral support is provided so that the allowable bending stress is $0.6F_y$. The beam is not subjected to impact loading.

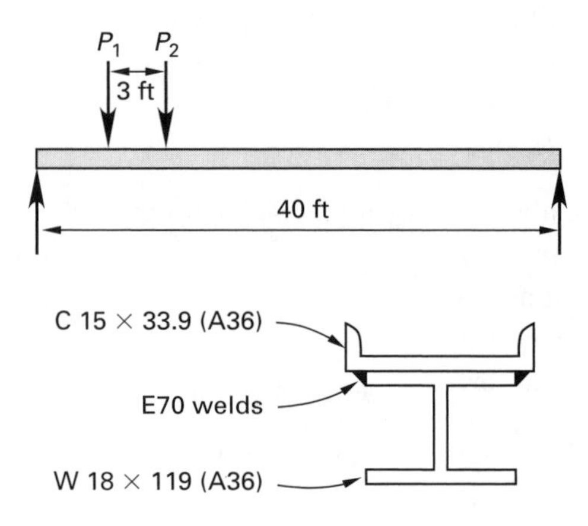

(a) The distance of the neutral axis of the beam cross section measured from the bottom of the section is most nearly

(A) 10.0 in
(B) 11.8 in
(C) 12.3 in
(D) 12.9 in

(b) The distance to the extreme compression fiber from the neutral axis is most nearly

(A) 10.6 in
(B) 11.3 in
(C) 12.0 in
(D) 12.4 in

(c) The distance to the extreme tension fiber from the neutral axis is most nearly

(A) 11.0 in
(B) 11.8 in
(C) 12.3 in
(D) 12.9 in

(d) The moment of inertia of the section (strong-axis bending) is most nearly

(A) 1950 in^4
(B) 2390 in^4
(C) 3020 in^4
(D) 3160 in^4

(e) For maximum moment to occur in the beam under load P_2, the position of the loads will be such that their resultant will be located at a distance (measured from the left support) equal to

(A) 18.8 ft
(B) 19.3 ft
(C) 19.8 ft
(D) 20.0 ft

(f) The maximum moment obtained as a function of each unknown moving load of P (in kips) is most nearly

(A) $16.0P$ ft-kips
(B) $17.1P$ ft-kips
(C) $18.5P$ ft-kips
(D) $19.1P$ ft-kips

(g) Considering moment only, the maximum allowable value of P is most nearly

(A) 24.9 kips
(B) 25.3 kips
(C) 25.8 kips
(D) 26.5 kips

(h) The absolute maximum value of the actual shear stress (anywhere in the beam) corresponding to the value of P obtained in part (g) is most nearly

(A) 3.3 ksi
(B) 3.9 ksi
(C) 4.5 ksi
(D) 5.1 ksi

(i) The minimum beam bearing length required at the reaction point from local web yielding criteria is most nearly

(A) less than zero
(B) 0.60 in
(C) 1.1 in
(D) 2.3 in

(j) The minimum beam bearing length required at the reaction point from web crippling criteria is most nearly

(A) less than zero
(B) 0.65 in
(C) 1.0 in
(D) 2.2 in

6. An A36 steel beam is loaded as shown. Lateral support is provided only at the three reaction points. The beam carries a uniform load of 1400 lbf/ft over its entire span. (a) Determine the reactions. (b) Draw the shear and moment diagrams. (c) Choose the lightest W 14 beam that is capable of developing a bending stress of $0.60F_y$. (d) Specify the maximum unbraced length for the beam chosen.

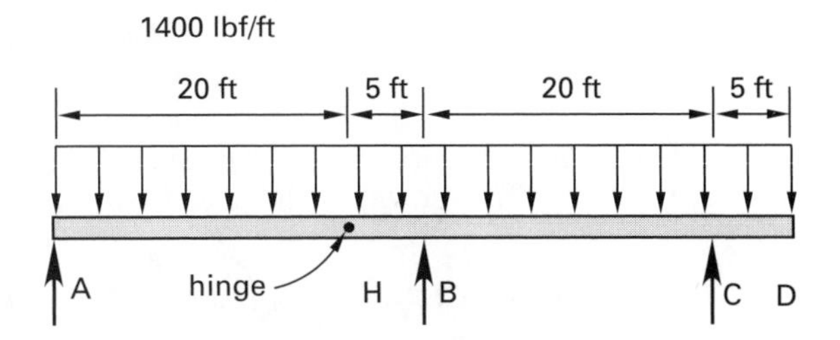

SOLUTIONS

1. Assume the weight of the beam is 60 lbf/ft.

The shear and moment diagrams are as shown.

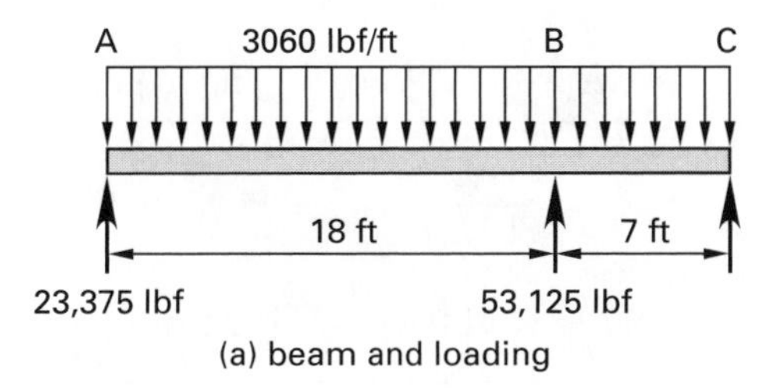

(a) beam and loading

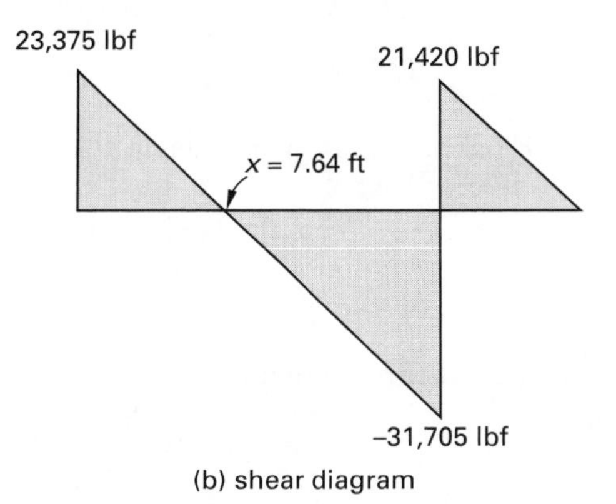

(b) shear diagram

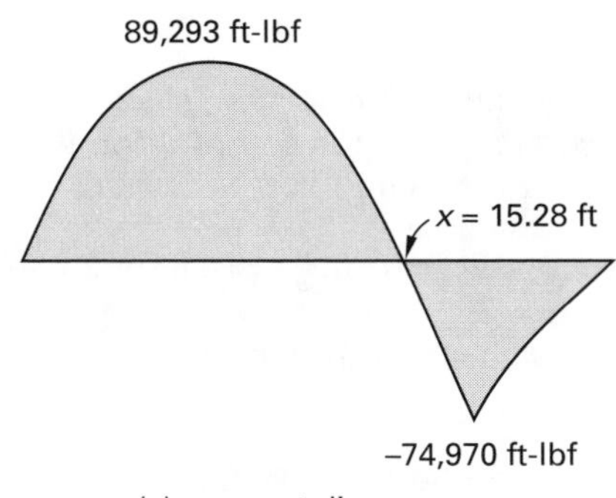

(c) moment diagram

Since M_{max} does not occur at the end of the cantilever span, C_b does not apply.

From the beam selection chart, choose W 14 × 43 to meet the requirements of 89.3 ft-kips with an unbraced length of 18 ft.

(Notice that $L_b > L_u$. This is acceptable at lower stress levels.)

2. *step 1:* Assume the beam weight is 50 lbf/ft.

$$w = \frac{800 \ \frac{\text{lbf}}{\text{ft}} + 50 \ \frac{\text{lbf}}{\text{ft}}}{1000 \ \frac{\text{lbf}}{\text{kip}}} = 0.85 \text{ kip/ft}$$

step 2: The maximum deflection is

$$y = \frac{L}{300} = \frac{(24\text{ ft})\left(12\ \frac{\text{in}}{\text{ft}}\right)}{300} = 0.96\text{ in}$$

step 3: Use App. 44.A, case 9. The required moment of inertia is

$$I_x = \frac{5wL^4}{384Ey} = \frac{(5)\left(0.85\ \frac{\text{kip}}{\text{ft}}\right)(24\text{ ft})^4\left(12\ \frac{\text{in}}{\text{ft}}\right)^3}{(384)\left(29{,}000\ \frac{\text{kips}}{\text{in}^2}\right)(0.96\text{ in})} = 227.9\text{ in}^4$$

step 4: From the moment of inertia selection table, choose W 14 × 26.

$$I_x = 245\text{ in}^4$$
$$S_x = 35.3\text{ in}^3$$

step 5: The maximum moment is

$$M_{\text{max}} = \frac{wL^2}{8} = \frac{\left(0.85\ \frac{\text{kip}}{\text{ft}}\right)(24\text{ ft})^2}{8} = 61.2\text{ ft-kips}$$

step 6: The allowable stress (since the section is compact and the compression flange has full lateral bracing) is

$$F_b = 0.66F_y = (0.66)\left(36\ \frac{\text{kips}}{\text{in}^2}\right) = 23.76\text{ kips/in}^2$$

step 7: The actual stress is

$$f_b = \frac{M_{\text{max}}}{S_x} = \frac{(61.2\text{ ft-kips})\left(12\ \frac{\text{in}}{\text{ft}}\right)}{35.3\text{ in}^3} = 20.80\text{ kips/in}^2$$

The actual stress is less than the allowable stress, so the section is acceptable.

Use W 14 × 26.

3. *step 1:* The reactions are each

$$R = \frac{wL}{2} = \frac{\left(1\ \frac{\text{kip}}{\text{ft}}\right)(20\text{ ft})}{2} = 10\text{ kips}$$

step 2: The maximum moment is

$$M_{\text{max}} = \frac{wL^2}{8} = \frac{\left(1\ \frac{\text{kip}}{\text{ft}}\right)(20\text{ ft})^2}{8} = 50\text{ ft-kips}$$

step 3: From the AISC beam selection chart, try W 14 × 22 to meet 50 ft-kips and $L_b = 5.5$ ft.

$$t_w = 0.23\text{ in}$$
$$d = 13.74\text{ in}$$
$$I = 199\text{ in}^4$$

step 4: Check for shear stress.

The maximum shear force is

$$V = R = 10\text{ kips}$$

See *Civil Engineering Reference Manual* p. 59-4. The allowable shear stress is

$$F_v = 0.4F_y = (0.4)\left(36\ \frac{\text{kips}}{\text{in}^2}\right) = 14.4\text{ kips/in}^2$$

The actual shear stress is

$$f_v = \frac{V}{dt_w} = \frac{10\text{ kips}}{(13.74\text{ in})(0.23\text{ in})} = 3.164\text{ kips/in}^2$$

$$f_v < F_v \quad \text{[OK]}$$

(Beam charts are based on maximum allowable bending stress, so it is not necessary to check bending stress.)

step 5: Check deflection. The allowable deflection is

$$y_{\text{allowable}} = \frac{L}{240} = \frac{(20\text{ ft})\left(12\ \frac{\text{in}}{\text{ft}}\right)}{240} = 1\text{ in}$$

The actual deflection is

$$y = \frac{5wL^4}{384EI} = \frac{(5)\left(1\ \frac{\text{kip}}{\text{ft}}\right)(20\text{ ft})^4\left(12\ \frac{\text{in}}{\text{ft}}\right)^3}{(384)\left(29{,}000\ \frac{\text{kips}}{\text{in}^2}\right)(199\text{ in}^4)} = 0.624\text{ in}$$

$$y < y_{\text{allowable}} \quad \text{[OK]}$$

Use W 14 × 22.

4. (a) Ignoring the weight of the beam, find the reactions.

Taking clockwise moments as positive,

$$\sum M_{\text{hinge to left}} = R_A(20 \text{ ft}) - \left(1.4 \ \frac{\text{kips}}{\text{ft}}\right)(20 \text{ ft})\left(\frac{20 \text{ ft}}{2}\right) = 0$$

$$R_A = \boxed{14 \text{ kips}}$$

The answer is (B).

(b)
$$\sum M_D = (14 \text{ kips})(45 \text{ ft}) - \left(1.4 \ \frac{\text{kips}}{\text{ft}}\right)(45 \text{ ft})\left(\frac{45 \text{ ft}}{2}\right) + R_C(20 \text{ ft}) + \left(1.4 \ \frac{\text{kips}}{\text{ft}}\right)(5 \text{ ft})\left(\frac{5 \text{ ft}}{2}\right) = 0$$

$$R_C = \boxed{38.5 \text{ kips}}$$

The answer is (C).

(c)
$$R_D = \left(1.4 \ \frac{\text{kips}}{\text{ft}}\right)(50 \text{ ft}) - 14 \text{ kips} - 38.5 \text{ kips} = \boxed{17.5 \text{ kips}}$$

The answer is (B).

(d)

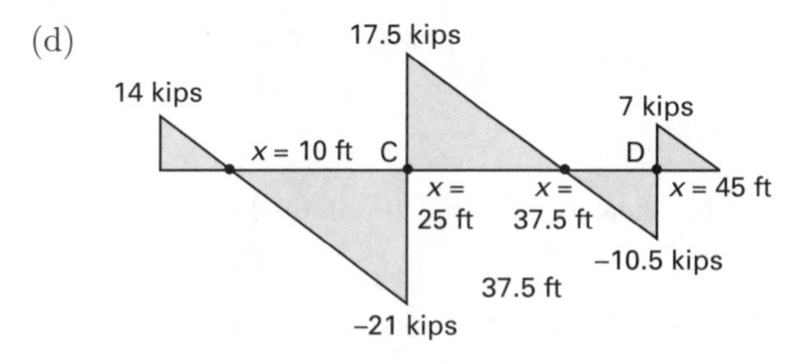

(a) shear diagram

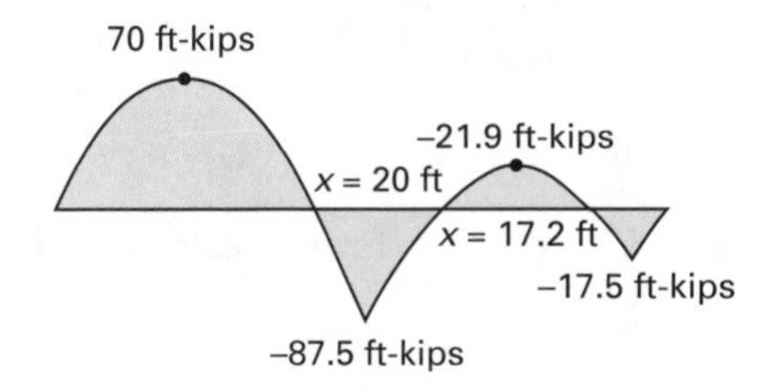

(b) moment diagram

From the shear diagram, the absolute maximum shear is $\boxed{21 \text{ kips.}}$

The answer is (D).

(e) From the shear diagram, the absolute maximum shear occurs $\boxed{\text{just to the left of support C.}}$

The answer is (C).

(f) From the moment diagram, the absolute maximum moment is $\boxed{87.5 \text{ ft-kips.}}$

The answer is (C).

(g) From the moment diagram, the absolute maximum moment occurs $\boxed{\text{at support C.}}$

The answer is (B).

(h) $L_u = 25$ ft; $M_{\text{max}} = 87.5$ ft-kips.

From the beam selection chart, $\boxed{\text{select W } 18{\times}60.}$

The allowable moment of 103.75 ft-kips far exceeds the actual maximum moment of 87.5 ft-kips (excluding beam weight). There is ample reserve capacity for the additional moment due to the weight of the beam.

The answer is (C).

(i) For W 18×60,

$$d = 18.24 \text{ in}$$
$$t_w = 0.415 \text{ in}$$

The actual shear stress is

$$f_v = \frac{V}{dt_w} = \frac{21 \text{ kips}}{(18.24 \text{ in})(0.415 \text{ in})} = \boxed{2.774 \text{ kips/in}^2}$$

The answer is (C).

(j) For W 18×60, $b_f = 7.555$ in and $d/A_f = 3.47$.

$$L_c = \frac{76b_f}{\sqrt{F_y}} = \frac{(76)(7.555 \text{ in})}{\sqrt{42 \ \frac{\text{kips}}{\text{in}^2}}} = 88.6 \text{ in}$$

$$L_c = \frac{20{,}000}{\left(\frac{d}{A_f}\right)F_y} = \frac{20{,}000}{(3.47)\left(42 \ \frac{\text{kips}}{\text{in}^2}\right)} = 137.2 \text{ in}$$

The smaller value controls. Therefore, $L_c = \boxed{88.6 \text{ in.}}$

The answer is (C).

5. (a) Locate the neutral axis (measured from the bottom of the beam). For W 18×119, $A = 35.1 \text{ in}^2$, $d = 18.97$ in, and $I_x = 2190 \text{ in}^4$. For C 15 × 33.9, $A = 9.96 \text{ in}^2$, $b_f = 3.4$ in, $\bar{x} = 0.787$ in, and $I_y = 8.13 \text{ in}^4$.

$$y = \frac{\sum A_i y_i}{\sum A_i}$$

$$= \frac{(35.1 \text{ in}^2)\left(\frac{18.97 \text{ in}}{2}\right) + (9.96 \text{ in}^2)(18.97 \text{ in} + 0.787 \text{ in})}{35.1 \text{ in}^2 + 9.96 \text{ in}^2}$$

$$= \boxed{11.76 \text{ in}}$$

The answer is (B).

(b) The distance to extreme compression fiber is

$$c = 18.97 \text{ in} + 3.4 \text{ in} - 11.76 \text{ in} = \boxed{10.61 \text{ in}}$$

The answer is (A).

(c) The distance to the extreme tension fiber is

$$c = y = \boxed{11.76 \text{ in}}$$

The answer is (B).

(d) The channel properties are found in the *AISC Manual*. The moment of inertia of the section is

$$I = I_{\text{W shape}} + I_{\text{channel}}$$

$$= 2190 \text{ in}^4 + (35.1 \text{ in}^2)\left(\frac{18.97 \text{ in}}{2} - 11.76 \text{ in}\right)^2 + 8.13 \text{ in}^4 + (9.96 \text{ in}^2)(18.97 \text{ in} + 0.787 \text{ in} - 11.76 \text{ in})^2$$

$$= \boxed{3016.8 \text{ in}^4}$$

The answer is (C).

(e) The moment is maximum when the loads are positioned as shown.

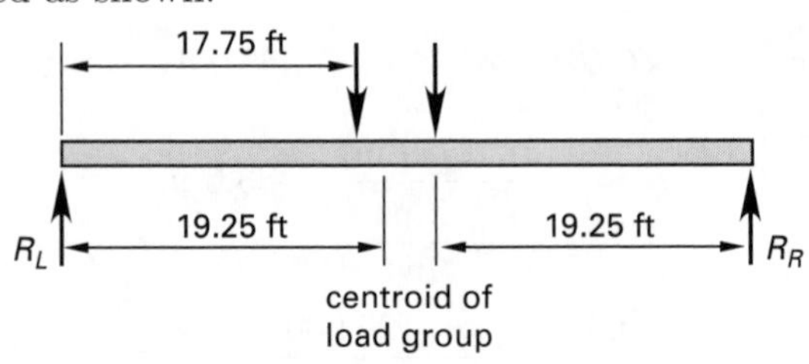

The maximum moment occurs when the resultant of the two loads is located 19.25 ft from the left end of the beam.

The answer is (B).

(f) (Note: This problem states that the beam is not subject to impact loading. But if it were, the load P would have been increased by 10 to 25%.)

Taking clockwise moments as positive,

$$\sum M_{\text{left support}} = P(17.75 \text{ ft}) + P(20.75 \text{ ft}) - R_R(40 \text{ ft}) = 0$$

$$R_R = 0.9625P$$

$$R_L = 2P - 0.9625P = 1.0375P$$

$$M_{\max} = (0.9625P)(19.25 \text{ ft})$$

$$= \boxed{18.53P \text{ ft-kips}}$$

The answer is (C).

(g) The allowable bending stress is

$$F_b = 0.6F_y = (0.6)\left(36 \frac{\text{kips}}{\text{in}^2}\right)$$

$$= 21.6 \text{ kips/in}^2$$

$$f = \frac{Mc}{I}$$

$$21.6 \frac{\text{kips}}{\text{in}^2} = \frac{(18.53P \text{ ft-kips})\left(12 \frac{\text{in}}{\text{ft}}\right)(11.76 \text{ in})}{3016.8 \text{ in}^4}$$

$$P = \boxed{24.92 \text{ kips}}$$

The answer is (A).

(h) Check the shear.

Placing the load group at one beam end,

$$V = R = 24.92 \text{ kips} + \frac{(24.92 \text{ kips})(40 \text{ ft} - 3 \text{ ft})}{40 \text{ ft}}$$

$$= 47.97 \text{ kips}$$

The area for shear (assuming the channel does not carry any shear) is

$$A_w = dt_w = (18.97 \text{ in})(0.655 \text{ in}) = 12.43 \text{ in}^2$$

The maximum shear stress is

$$f_v = \frac{V}{A_w} = \frac{47.97 \text{ kips}}{12.43 \text{ in}^2} = \boxed{3.86 \text{ kips/in}^2}$$

The answer is (B).

(i) From the *AISC Manual*, Part 2, Allowable Uniform Load Tables, for W 18 × 119,

$$R_1 = 68.1 \text{ kips}$$

$$R_2 = 15.6 \text{ kips/in}$$

From a consideration of web yielding,

$$N_{\text{min}} = \frac{R - R_1}{R_2} = \frac{47.97 \text{ kips} - 68.1 \text{ kips}}{15.6 \ \frac{\text{kips}}{\text{in}}}$$

$$= \boxed{-1.30 \text{ in} < 0}$$

The answer is (A).

(j) From the *AISC Manual*, Part 2, Allowable Uniform Load Tables, for W 18 × 119,

$$R_3 = 111 \text{ kips}$$
$$R_4 = 8.55 \text{ kips/in}$$

From a consideration of web crippling,

$$N_{\text{min}} = \frac{R - R_3}{R_4} = \frac{47.97 \text{ kips} - 111 \text{ kips}}{8.55 \ \frac{\text{kips}}{\text{in}}}$$

$$= \boxed{-7.38 \text{ in} < 0}$$

The answer is (A).

6. This is a determinate beam. Taking clockwise moments from the hinge to the left as positive,

(a) $$\sum M_H = 20R_{\text{A}} - \left(\tfrac{1}{2}\right)\left(1400 \ \frac{\text{lbf}}{\text{ft}}\right)(20 \text{ ft})^2 = 0$$

$$R_{\text{A}} = \boxed{14{,}000 \text{ lbf}}$$

$$V_H = 14{,}000 \text{ lbf} - (20 \text{ ft})\left(1400 \ \frac{\text{lbf}}{\text{ft}}\right) = -14{,}000 \text{ lbf}$$

$$R_{\text{B}} + R_{\text{C}} = (5 \text{ ft} + 20 \text{ ft} + 5 \text{ ft})\left(1400 \ \frac{\text{lbf}}{\text{ft}}\right) + 14{,}000 \text{ lbf} = 56{,}000 \text{ lbf}$$

$$\sum M_{\text{C}} = 20R_{\text{B}} - (25 \text{ ft})(14{,}000 \text{ lbf}) - \left(\tfrac{1}{2}\right)\left(1400 \ \frac{\text{lbf}}{\text{ft}}\right)(25 \text{ ft})^2 + \left(\tfrac{1}{2}\right)\left(1400 \ \frac{\text{lbf}}{\text{ft}}\right)(5 \text{ ft})^2 = 0$$

$$R_{\text{B}} = \boxed{38{,}500 \text{ lbf}}$$

$$R_{\text{C}} = 56{,}000 \text{ lbf} - 38{,}500 \text{ lbf} = \boxed{17{,}500 \text{ lbf}}$$

(b)

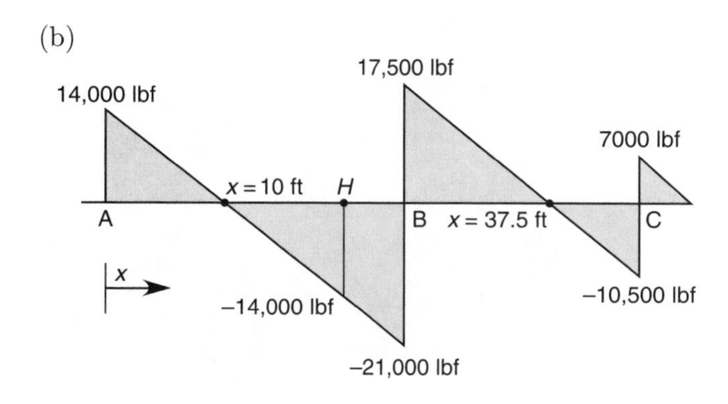

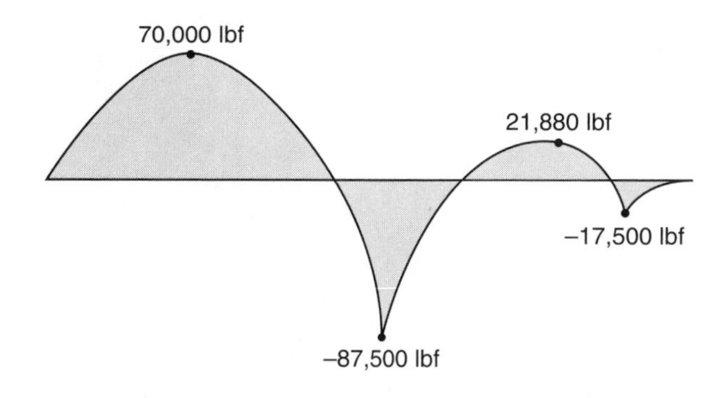

(c) $$M_{\text{max}} = \frac{87{,}500 \text{ ft-lbf}}{1000 \ \frac{\text{lbf}}{\text{kip}}} = 87.5 \text{ ft-kips}$$

Interpretation 1

Since there is lateral bracing only at points of support, $L_b = 25$ ft. From the allowable moment beam chart, $\boxed{\text{W 14} \times 74.}$

(W 14 × 53 and subsequent beams encountered in the chart show dashed sloping lines, indicating $F_b < 0.60F_y$. W 14 × 74 is the first beam encountered with a solid horizontal line at $L_b = 25$ ft.)

Interpretation 2

The flange is discontinuous at the hinge, so $L_b = 20$ ft.

From the allowable moment beam chart, $\boxed{\text{W 14} \times 61.}$

(d) $L_u = \boxed{\begin{array}{l}25.9 \text{ ft} \quad [\text{from table for W 14} \times 74] \\ (21.5 \text{ ft for W 14} \times 61)\end{array}}$

60 Structural Steel: Tension Members

PRACTICE PROBLEMS

1. A single-angle tension member (L$7 \times 4 \times {}^{5}/_{8}$) of A36 steel has two gage lines in its long leg and one in its short leg for ${}^{3}/_{4}$ in diameter bolts arranged as shown. Determine the tensile capacity of the member, neglecting block shear.

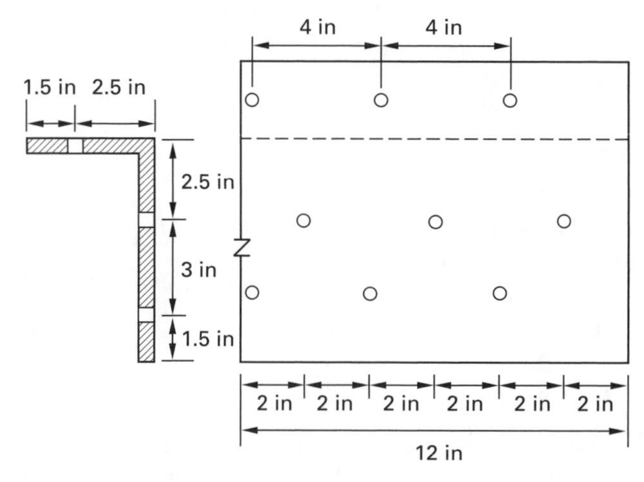

(A) 70 kips
(B) 110 kips
(C) 140 kips
(D) 180 kips

2. Design a plate eyebar of A36 steel to carry a tensile load of 300 kips.

3. (*Time limit: one hour*) For the truss shown, member DE is made of 2L4 × $3^{1}/_{2} \times {}^{1}/_{4}$, of A36 steel. All joints are assumed to be pin-connected with one gage line of ${}^{3}/_{4}$ in high-strength bolts through one leg of each angle. Each end of member DE is connected with four bolts on the gage line with an edge distance of $1^{1}/_{4}$ in and a $2^{1}/_{4}$ in pitch.

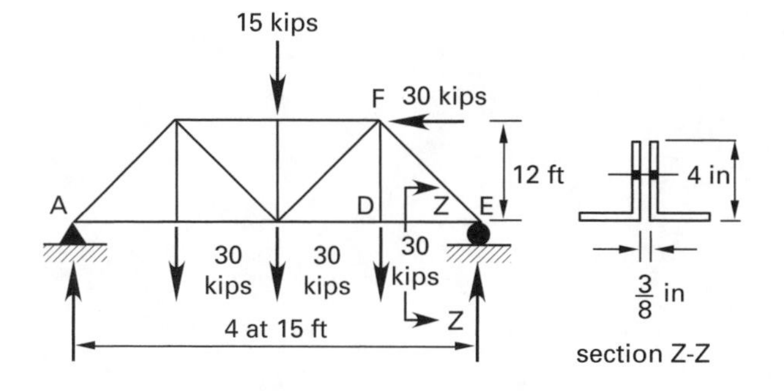

(a) The support reaction at E is most nearly

(A) 38 kips
(B) 47 kips
(C) 52 kips
(D) 58 kips

(b) The axial force in member DE is most nearly

(A) 38 kips
(B) 46 kips
(C) 52 kips
(D) 58 kips

(c) The effective net area of member DE is most nearly

(A) 1.7 in^2
(B) 2.4 in^2
(C) 2.7 in^2
(D) 2.9 in^2

(d) The tensile load capacity of member DE, based on the fracture criterion, is most nearly

(A) 58 kips
(B) 64 kips
(C) 69 kips
(D) 79 kips

(e) The tensile load capacity of member DE, based on the yielding criterion, is most nearly

(A) 58 kips
(B) 68 kips
(C) 75 kips
(D) 78 kips

(f) In the computation of the block shear strength of member DE, the area in shear is most nearly

(A) 0.8 in^2
(B) 1.4 in^2
(C) 2.5 in^2
(D) 3.0 in^2

(g) In the computation of the block shear strength of member DE, the area in tension is most nearly

(A) 0.5 in^2
(B) 0.9 in^2
(C) 1.3 in^2
(D) 2.8 in^2

(h) The tensile load capacity of member DE, based on the block shear criterion, is most nearly

(A) 55 kips
(B) 58 kips
(C) 68 kips
(D) 73 kips

(i) The tensile load capacity of member DE, based on yielding, fracture, and block shear criteria, is most nearly

(A) 55 kips
(B) 58 kips
(C) 68 kips
(D) 78 kips

(j) The maximum slenderness ratio of the member DE is most nearly

(A) 140
(B) 200
(C) 250
(D) 300

4. (*Time limit: one hour*) A 25 ft long W-shape member is to carry an axial tensile load of 420 kips. It is assumed that there will be two lines of holes for $^7/_8$ in diameter bolts in each flange. There will be at least three bolts in each line.

(For parts (a) through (e), assume A36 steel is used.)

(a) The required gross area based on the yielding criterion is most nearly

(A) 10 in^2
(B) 15 in^2
(C) 19 in^2
(D) 25 in^2

(b) The required effective net area based on the fracture criterion is most nearly

(A) 10 in^2
(B) 15 in^2
(C) 20 in^2
(D) 25 in^2

(c) The required net area based on the fracture criterion is most nearly

(A) 10 in^2
(B) 16 in^2
(C) 20 in^2
(D) 25 in^2

(d) The minimum radius of gyration required to satisfy the preferred slenderness limit for the member is most nearly

(A) 0.55 in
(B) 0.75 in
(C) 0.85 in
(D) 1.0 in

(e) Among the four choices listed, the lightest W 12 (A36) section suitable for this member is

(A) W 12 × 50
(B) W 12 × 65
(C) W 12 × 79
(D) W 12 × 106

(For parts (f) through (j), assume A572, grade-50 steel is used.)

(f) The required gross area based on the yielding criterion is most nearly

(A) 10 in^2
(B) 14 in^2
(C) 20 in^2
(D) 25 in^2

(g) The required net effective area based on the fracture criterion is most nearly

(A) 10 in^2
(B) 13 in^2
(C) 20 in^2
(D) 25 in^2

(h) The required net area based on the fracture criterion is most nearly

(A) 10 in^2
(B) 14 in^2
(C) 20 in^2
(D) 25 in^2

(i) The minimum radius of gyration required to satisfy the preferred slenderness limit for the member is most nearly

(A) 0.55 in
(B) 0.75 in
(C) 0.85 in
(D) 1.0 in

(j) The lightest 50 $kips/in^2$ W 12 section suitable for this member is

(A) W 12 × 50
(B) W 12 × 65
(C) W 12 × 79
(D) W 12 × 106

SOLUTIONS

1. The gross area of the member is $A_g = 6.48\ \text{in}^2$ (*AISC Manual*, Part 1).

The effective hole diameter includes a $^1/_8$ in allowance for clearance and manufacturing tolerances.

$$d_h = 0.75\ \text{in} + 0.125\ \text{in} = 0.875\ \text{in}$$

To calculate the net area, the net width is calculated first using Eq. 60.3.

$$b_n = b - \sum d_h + \sum \frac{s^2}{4g}$$

The net width of the member must be evaluated by paths ABCD and ABECD.

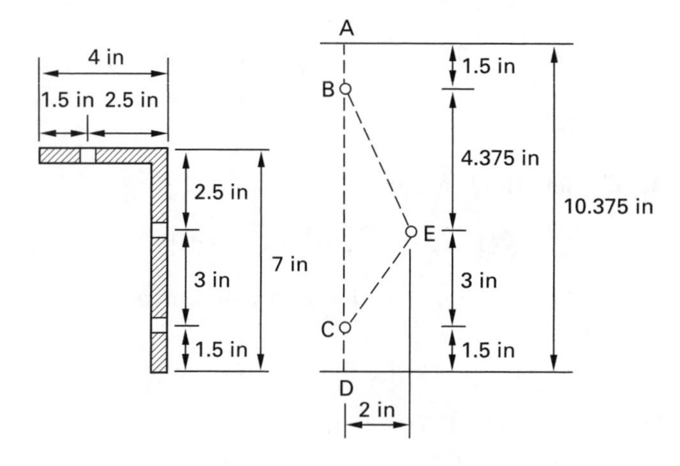

Path ABCD does not have any diagonal runs.

$$b_{n,\text{ABCD}} = 10.375\ \text{in} - (2)(0.875\ \text{in}) = 8.625\ \text{in}$$

Path ABECD has two diagonal runs, BE and EC, for which the quantity $s^2/4g$ must be calculated. The pitch, s, is shown as 2 in, and the gage values are 4.375 in and 3 in, respectively.

For diagonal BE,

$$\frac{s^2}{4g} = \frac{(2\ \text{in})^2}{(4)(4.375\ \text{in})} = 0.229\ \text{in}$$

For diagonal EC,

$$\frac{s^2}{4g} = \frac{(2\ \text{in})^2}{(4)(3\ \text{in})} = 0.333\ \text{in}$$

$$\begin{aligned} b_{n,\text{ABECD}} &= 10.375\ \text{in} - (3)(0.875\ \text{in}) + (1)(0.229\ \text{in}) \\ &\quad + (1)(0.333\ \text{in}) \\ &= 8.312\ \text{in} \quad \text{[controls]} \end{aligned}$$

Therefore, the critical net area is given by Eq. 60.4.

$$\begin{aligned} A_n &= b_{n,\text{ABECD}}t = (8.312\ \text{in})\left(\tfrac{5}{8}\ \text{in}\right) \\ &= 5.195\ \text{in}^2 < 0.85\%\, A_g \quad \text{[OK]} \end{aligned}$$

The effective net area is given by Eq. 60.6.

$$A_e = UA_n = (1)(5.195\ \text{in}^2) = 5.195\ \text{in}^2$$

The capacity of the member based on the gross section is given by Eq. 60.11.

$$\begin{aligned} P_t &= 0.60F_yA_g = (0.60)\left(36\ \frac{\text{kips}}{\text{in}^2}\right)(6.48\ \text{in}^2) \\ &= 140\ \text{kips} \end{aligned}$$

The capacity of the member based on the net section is given by Eq. 60.12.

$$\begin{aligned} P_t &= 0.50F_uA_e = (0.50)\left(58\ \frac{\text{kips}}{\text{in}^2}\right)(5.195\ \text{in}^2) \\ &= 150.7\ \text{kips} \end{aligned}$$

The tensile capacity is the smaller value, **140 kips.**

The answer is (C).

2. Eyebar design is covered in the AISC Specification D3.3.

step 1: The allowable stress on the net area of the eye is

$$\begin{aligned} F_{te} &= 0.45F_y = (0.45)\left(36\ \frac{\text{kips}}{\text{in}^2}\right) \\ &= 16.2\ \text{kips/in}^2 \end{aligned}$$

step 2: The allowable stress F_a on the bar is the minimum of the following two equations.

$$\begin{aligned} F_{tg} &= 0.60F_y = (0.60)\left(36\ \frac{\text{kips}}{\text{in}^2}\right) \\ &= 21.6\ \text{kips/in}^2 \quad \text{[controls]} \end{aligned}$$

$$\begin{aligned} F_t &= 0.50F_u = (0.50)\left(58\ \frac{\text{kips}}{\text{in}^2}\right) \\ &= 29.0\ \text{kips/in}^2 \end{aligned}$$

step 3: The bar cross-sectional area is

$$A = \frac{P}{F_a} = \frac{300\ \text{kips}}{21.6\ \frac{\text{kips}}{\text{in}^2}} = 13.89\ \text{in}^2$$

Try a $1^1/_2$ in thick plate. The width is

$$b = \frac{A}{t} = \frac{13.89\ \text{in}^2}{1.5\ \text{in}} = 9.26\ \text{in}$$

Use PL $1^1/_2 \times 9^1/_2$ for actual area of 14.25 in^2.

step 4: Check the b/t ratio.

$$b/t = \frac{9.5 \text{ in}}{1.5 \text{ in}} = 6.33 < 8 \quad \text{[acceptable]}$$

step 5: The net area at the eye is

$$A_n = \frac{P}{F_{te}} = \frac{300 \text{ kips}}{16.2 \ \frac{\text{kips}}{\text{in}^2}} = 18.52 \text{ in}^2$$

step 6: The eye width is

$$w = \frac{A_n}{2t} = \frac{18.52 \text{ in}^2}{(2)(1.5 \text{ in})} = 6.17 \text{ in}$$

The minimum value of w is

$$w_{\text{min}} = \tfrac{2}{3}b = \left(\tfrac{2}{3}\right)(9.5 \text{ in}) = 6.33 \text{ in}$$

The maximum value of w is

$$w_{\text{max}} = \tfrac{3}{4}b = \left(\tfrac{3}{4}\right)(9.5 \text{ in}) = 7.13 \text{ in}$$

Use $w = 6.5$ in.

step 7: Calculate the minimum pin diameter.

$$d_{\text{min}} = \tfrac{7}{8}b = \left(\tfrac{7}{8}\right)(9.5 \text{ in}) = 8.31 \text{ in}$$

step 8: Refer to ASD Ch. D, Sec. D3.3. The maximum hole diameter is

$$\begin{aligned} d_{h,\text{max}} &\approx d_{\text{min}} + \tfrac{1}{32} \text{ in} \\ &= 8.31 \text{ in} + 0.03125 \text{ in} \\ &= 8.34 \text{ in} \end{aligned}$$

Make the hole 8.5 in diameter.

The pin diameter is 8.5 in − 0.03 in = 8.47 in.

step 9: The minimum external diameter is

$$d_h + 2w = 8.5 \text{ in} + (2)(6.5 \text{ in}) = \boxed{21.5 \text{ in}}$$

step 10: The allowable bearing stress (from AISC Specification J8) is

$$\begin{aligned} F_p &= 0.90F_y = (0.90)\left(36 \ \frac{\text{kips}}{\text{in}^2}\right) \\ &= 32.4 \text{ kips/in}^2 \end{aligned}$$

The actual bearing stress is

$$\begin{aligned} f_p &= \frac{P}{dt} = \frac{300 \text{ kips}}{(8.47 \text{ in})(1.5 \text{ in})} \\ &= 23.613 \text{ kips/in}^2 < 32.4 \text{ kips/in}^2 \quad \text{[OK]} \end{aligned}$$

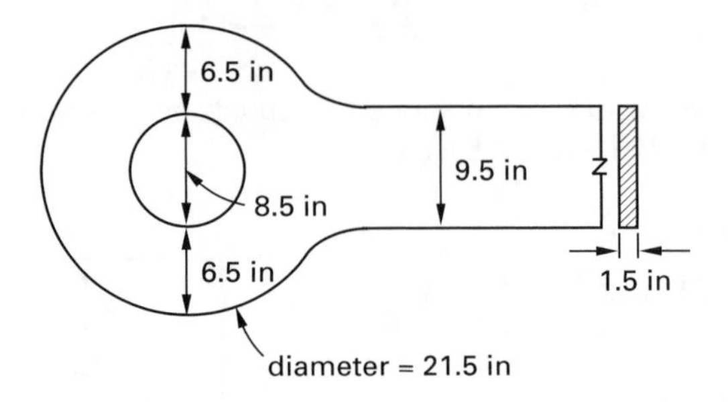

3. The length of member FE is

$$\sqrt{(15 \text{ ft})^2 + (12 \text{ ft})^2} = 19.21 \text{ ft}$$

(a) Find the support reaction E_y by summing moments about point A.

For $\sum M_A = 0$ (clockwise moment positive),

$$\begin{aligned} &(30 \text{ kips})(15 \text{ ft} + 30 \text{ ft} + 45 \text{ ft}) \\ &\quad + (15 \text{ kips})(30 \text{ ft}) - (30 \text{ kips})(12 \text{ ft}) \\ &\quad - E_y(60 \text{ ft}) = 0 \end{aligned}$$

$$E_y = \boxed{46.5 \text{ kips}}$$

The answer is (B).

(b) Find the axial force in member DE by applying the method of joints at E.

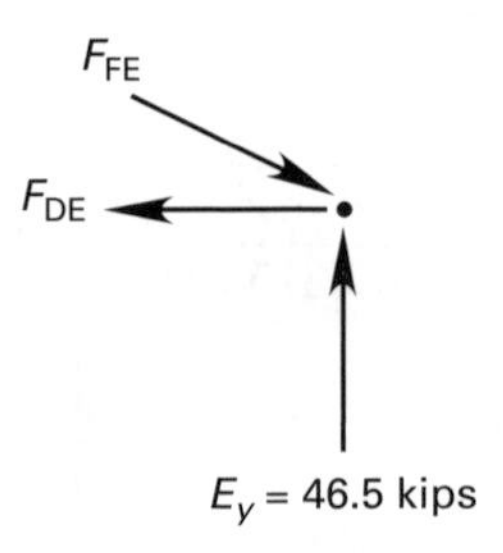

For $\sum F_y = 0$,

$$46.5 \text{ kips} - F_{\text{FE}}\left(\frac{12 \text{ ft}}{19.21 \text{ ft}}\right) = 0$$

$$F_{\text{FE}} = 74.44 \text{ kips}$$

For $\sum F_x = 0$,

$$F_{DE} - (74.44 \text{ kips})\left(\frac{15 \text{ ft}}{19.21 \text{ ft}}\right) = 0$$

$$F_{DE} = \boxed{58.13 \text{ kips}}$$

The answer is (D).

(c) The gross area of 2L4 × 3½ × ¼ is obtained from the *AISC Manual* Shape Table.

$$A_g = 3.63 \text{ in}^2$$

The angle thickness is $t = {}^1\!/_4$ in.

The diameter of the bolt hole is

$$d_h = \tfrac{3}{4} \text{ in} + \tfrac{1}{8} \text{ in} = {}^7\!/_8 \text{ in}$$

The net area is given by Eq. 60.5.

$$\begin{aligned} A_n &= A_g - d_h(2t) = 3.63 \text{ in}^2 - \left(\tfrac{7}{8} \text{ in}\right)(2)\left(\tfrac{1}{4} \text{ in}\right) \\ &= 3.19 \text{ in}^2 \end{aligned}$$

Since the connection is made to only one leg of the angles, from Table 60.1, the reduction coefficient is $U = 0.85$.

The effective net area is

$$A_e = UA_n = (0.85)(3.19 \ \in^2) = \boxed{2.71 \text{ in}^2}$$

The answer is (C).

(d) The tensile load capacity of member DE based on the fracture criterion (Eq. 60.12) is

$$\begin{aligned} P_t &= 0.5F_uA_e = (0.5)\left(58 \ \frac{\text{kips}}{\text{in}^2}\right)(2.71 \text{ in}^2) \\ &= \boxed{78.6 \text{ kips}} \end{aligned}$$

The answer is (D).

(e) The tensile load capacity of member DE based on the yielding criterion (Eq. 60.11) is

$$\begin{aligned} P_t &= 0.6F_yA_g = (0.6)\left(36 \ \frac{\text{kips}}{\text{in}^2}\right)(3.63 \text{ in}^2) \\ &= \boxed{78.4 \text{ kips}} \end{aligned}$$

The answer is (D).

(f) For block shear strength computation of member DE, the area in shear is

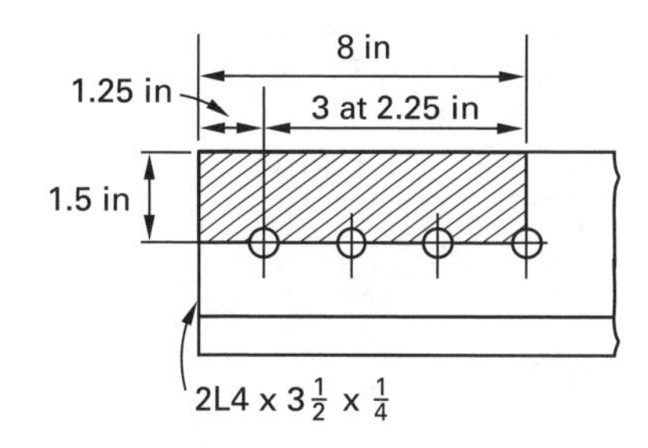

$$\begin{aligned} A_v &= 2(\text{length in shear} \times \text{thickness}) \\ &= (2)\left(8 \text{ in} - (3.5)\left(\tfrac{7}{8} \text{ in}\right)\right)\left(\tfrac{1}{4} \text{ in}\right) = \boxed{2.47 \text{ in}^2} \end{aligned}$$

The answer is (C).

(g) For block shear strength computation of member DE, the area in tension is

$$\begin{aligned} A_t &= 2(\text{length in tension} \times \text{thickness}) \\ &= (2)\left(1.5 \text{ in} - (0.5)\left(\tfrac{7}{8} \text{ in}\right)\right)\left(\tfrac{1}{4} \text{ in}\right) = \boxed{0.53 \text{ in}^2} \end{aligned}$$

The answer is (A).

(h) The tensile load capacity of member DE is based on the block shear criterion (Eq. 60.13).

$$\begin{aligned} P_t &= A_v(0.3F_u) + A_t(0.5F_u) \\ &= (2.47 \text{ in}^2)(0.3)\left(58 \ \frac{\text{kips}}{\text{in}^2}\right) \\ &\quad + (0.53 \text{ in}^2)\left((0.5)\left(58 \ \frac{\text{kips}}{\text{in}^2}\right)\right) \\ &= \boxed{58.35 \text{ kips}} \end{aligned}$$

The answer is (B).

(i) The tensile load capacity of member DE, based on yielding, fracture, and block shear criteria, is the smallest of the three values. The smallest P_t is $\boxed{58.35 \text{ kips.}}$

The answer is (B).

(j) The maximum slenderness ratio of member DE is

$$\begin{aligned} l &= (15 \text{ ft})\left(12 \ \frac{\text{in}}{\text{ft}}\right) = 180 \text{ in} \\ r_{\min} &= r_x = 1.27 \text{ in} \\ \frac{l}{r_{\min}} &= \frac{180 \text{ in}}{1.27 \text{ in}} = \boxed{141.7} \end{aligned}$$

The answer is (A).

Structural

4. The axial tensile load to be carried by the W-shape member is $P_t = 420$ kips.

(a) The required gross area based on the yielding criterion can be found from Eq. 60.11.

$$A_g = \frac{P_t}{0.6F_y} = \frac{420 \text{ kips}}{(0.6)\left(36 \ \frac{\text{kips}}{\text{in}^2}\right)} = \boxed{19.44 \text{ in}^2}$$

The answer is (C).

(b) The required effective net area based on the fracture criterion from Eq. 60.12 is

$$A_e = \frac{P_t}{0.5F_u} = \frac{420 \text{ kips}}{(0.5)\left(58 \ \frac{\text{kips}}{\text{in}^2}\right)} = \boxed{14.48 \text{ in}^2}$$

The answer is (B).

(c) From Table 60.1, the reduction coefficient is $U = 0.90$.

The required net area is based on the fracture criterion from Eq. 60.6.

$$A_n = \frac{A_e}{U} = \frac{14.48 \text{ in}^2}{0.90} = \boxed{16.1 \text{ in}^2}$$

The answer is (B).

(d) The minimum radius of gyration required to satisfy the preferred slenderness ratio of 300 is

$$r_{\min} = \frac{l}{300} = \frac{(25 \text{ ft})\left(12 \ \frac{\text{in}}{\text{ft}}\right)}{300} = \boxed{1 \text{ in}}$$

The answer is (D).

(e) From the Shape Tables of the *AISC Manual*,

W 12 × 79 is the lightest suitable section of the four choices listed.

$$\begin{aligned} t_f &= 0.735 \text{ in} \\ r_{\min} &= 3.05 \text{ in} > \text{required } 1 \text{ in} \\ A_g &= 23.2 \text{ in}^2 > \text{required } 19.44 \text{ in}^2 \quad \text{[yielding]} \\ &> \text{required } 16.1 \text{ in}^2 + (4)(0.735 \text{ in})(1 \text{ in}) \\ &= 19.04 \text{ in}^2 \quad \text{[fracture]} \end{aligned}$$

The answer is (C).

For parts (f) through (j), A572 grade-50 steel is used.

$$\begin{aligned} F_y &= 50 \text{ kips/in}^2 \\ F_u &= 65 \text{ kips/in}^2 \end{aligned}$$

(f) The required gross area is based on the yielding criterion from Eq. 60.11.

$$A_g = \frac{P_t}{0.6F_y} = \frac{420 \text{ kips}}{(0.6)\left(50 \ \frac{\text{kips}}{\text{in}^2}\right)} = 14.0 \text{ in}^2$$

The answer is (B).

(g) The required effective net area is based on the fracture criterion from Eq. 60.12.

$$A_e = \frac{P_t}{0.5F_u} = \frac{420 \text{ kips}}{(0.5)\left(65 \ \frac{\text{kips}}{\text{in}^2}\right)} = \boxed{12.92 \text{ in}^2}$$

The answer is (B).

(h) The reduction coefficient is $U = 0.90$.

The required net area based on the fracture criterion from Eq. 60.6 is

$$A_n = \frac{A_e}{U} = \frac{12.92 \text{ in}^2}{0.90} = \boxed{14.36 \text{ in}^2}$$

The answer is (B).

(i) The minimum radius of gyration required to satisfy the preferred slenderness ratio of 300 is

$$r_{\min} = \frac{l}{300} = \frac{(25 \text{ ft})\left(12 \ \frac{\text{in}}{\text{ft}}\right)}{300} = \boxed{1 \text{ in}}$$

The answer is (D).

(j) From the Shape Tables of the *AISC Manual*,

W 12 × 65 is the lightest suitable section of the four listed.

$$\begin{aligned} t_f &= 0.605 \text{ in} \\ r_{\min} &= 3.02 \text{ in} > \text{required } 1 \text{ in} \\ A_g &= 19.1 \text{ in}^2 > \text{required } 14.0 \text{ in}^2 \quad \text{[yielding]} \\ &> \text{required } 14.36 \text{ in}^2 + (4)(0.605 \text{ in})(1 \text{ in}) \\ &= 16.78 \text{ in}^2 \quad \text{[fracture]} \end{aligned}$$

The answer is (B).

61 Structural Steel: Compression Members

PRACTICE PROBLEMS

1. (*Time limit: one hour*) The two-story, unbraced frame of A36 steel shown is one of several placed on 30 ft centers. All columns are pinned (rotation free and translation fixed) in the weak axis direction (i.e., perpendicular to the plane of the frame). The following loadings are applied.

roof dead load: 4 in thick reinforced concrete slab (unit weight = 150 lbf/ft^3) and roofing material (weight = 6 lbf/ft^2)
roof live load: 30 lbf/ft^2
floor dead load: 6 in thick reinforced concrete slab
floor live load: 90 lbf/ft^2

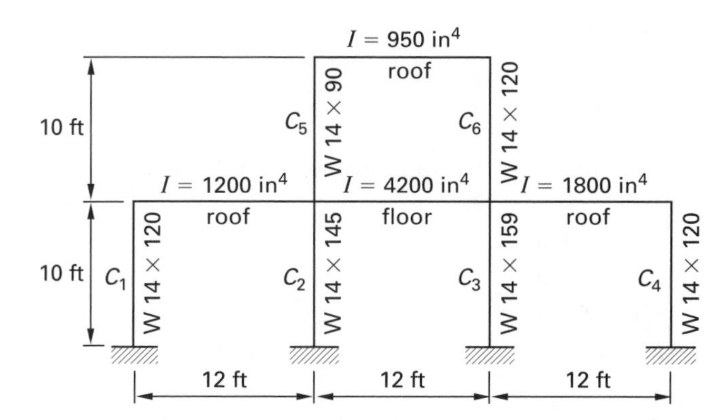

(a) Based on the framing member information provided, the effective length factor K_x of column C_2 for buckling about the strong axis is most nearly

(A) 0.70
(B) 0.85
(C) 1.0
(D) 1.2

(b) Based on the framing member information provided, the effective length factor K_x of column C_5 for buckling about the strong axis is most nearly

(A) 0.78
(B) 0.95
(C) 1.3
(D) 1.7

(c) The slenderness ratios of columns C_2 and C_5 with respect to the strong axis are most nearly

(A) 23 and 27, respectively
(B) 49 and 54, respectively
(C) 62 and 67, respectively
(C) 78 and 85, respectively

(d) The maximum slenderness ratio for column C_2 is most nearly

(A) 24
(B) 30
(C) 48
(D) 63

(e) The maximum slenderness ratio for column C_5 is most nearly

(A) 28
(B) 32
(C) 45
(D) 68

(f) The allowable axial compressive load on column C_2 is most nearly

(A) 580 kips
(B) 680 kips
(C) 850 kips
(D) 920 kips

(g) The allowable axial compressive load on column C_5 is most nearly

(A) 520 kips
(B) 650 kips
(C) 790 kips
(D) 940 kips

(h) The computed axial compressive load on column C_5 is most nearly

(A) 16 kips
(B) 78 kips
(C) 180 kips
(D) 220 kips

(i) The computed axial compressive load on column C_2 is most nearly

(A) 63 kips
(B) 78 kips
(C) 180 kips
(D) 220 kips

(j) Considering axial loads only,

(A) both columns C_2 and C_5 are adequate.
(B) both columns C_2 and C_5 are inadequate.
(C) column C_2 is adequate, but column C_5 is inadequate.
(D) column C_5 is adequate, but column C_2 is inadequate.

2. (*Time limit: one hour*) A 20 ft high column, pinned at both ends, is constructed from an A36 channel and an A36 wide flange shape as shown. The column is constrained in the x-plane by bracing at midheight.

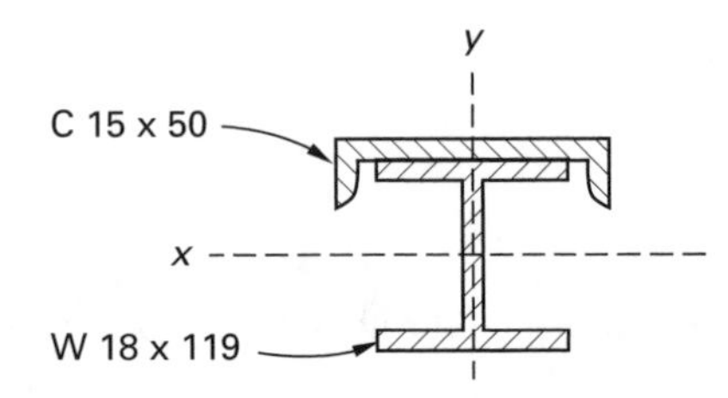

(a) The cross-sectional area of the column is most nearly

(A) 35 in^2
(B) 42 in^2
(C) 50 in^2
(D) 55 in^2

(b) The centroidal moment of inertia of the column for bending about the x-axis is most nearly

(A) 2700 in^4
(B) 2900 in^4
(C) 3100 in^4
(D) 3300 in^4

(c) The centroidal moment of inertia of the column for bending about the y-axis is most nearly

(A) 570 in^4
(B) 660 in^4
(C) 790 in^4
(D) 820 in^4

(d) The radius of gyration of the column with respect to the x-axis is most nearly

(A) 4.9 in
(B) 5.7 in
(C) 6.9 in
(D) 7.9 in

(e) The radius of gyration of the column with respect to the y-axis is most nearly

(A) 2.9 in
(B) 3.1 in
(C) 3.6 in
(D) 4.1 in

(f) The slenderness ratio of the column with respect to the x-axis is most nearly

(A) 28
(B) 30
(C) 33
(D) 41

(g) The slenderness ratio of the column with respect to the y-axis is most nearly

(A) 28
(B) 30
(C) 33
(D) 41

(h) The maximum slenderness ratio for the column is most nearly

(A) 28
(B) 30
(C) 33
(D) 41

(i) The allowable compressive stress for the column is most nearly

(A) 12 ksi
(B) 15 ksi
(C) 18 ksi
(D) 20 ksi

(j) The allowable axial compressive load on the column is most nearly

(A) 590 kips
(B) 680 kips
(C) 980 kips
(D) 1200 kips

3. Select the lightest W section of A36 steel to serve as a column 30 ft long and to carry an axial compressive load of 160 kips. The member is pinned at its top and bottom. The member's self-weight is estimated as 2000 lbf. It is supported at midheight in its weak direction.

(A) W 8 × 40
(B) W 10 × 39
(C) W 12 × 45
(D) W 14 × 48

4. Select A36 interior columns of the frame shown. Assume columns are braced in the plane perpendicular to the frame and built in at the bottom. Sidesway is uninhibited. Girders are W 12×96.

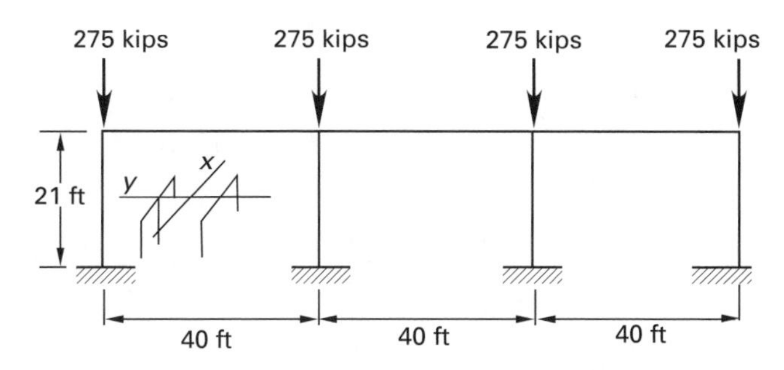

(A) W 12 × 40
(B) W 12 × 45
(C) W 12 × 53
(D) W 12 × 65

5. (*Time limit: one hour:*) The truss shown is constructed of a variety of members, all of which are assumed to be pin-ended. Determine whether all members are adequate. If members are not adequate, specify replacements with the same nominal depth. Assume that the top chord is braced laterally at every joint.

top and bottom chords:	W 8 × 28 (A36)
diagonal bracing:	L 3 1/2 × 3 1/2 × 5/16 (pair)
vertical members:	L 4 × 4 × 3/8 (pair)

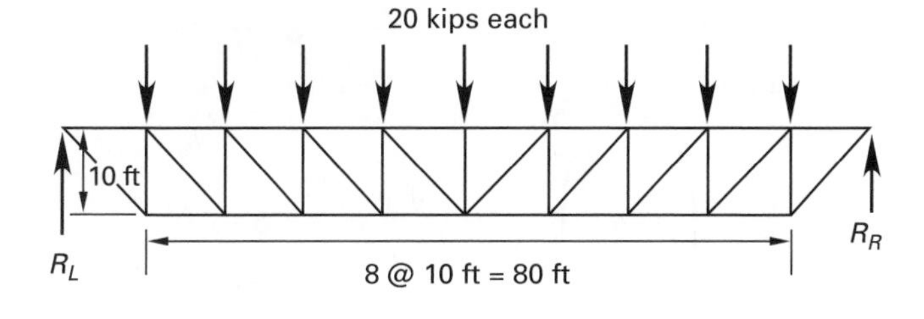

SOLUTIONS

1. From the shape tables of the *AISC Manual*, for column C_2 (W 14 × 145),

$$A = 42.7\ \text{in}^2$$
$$I_x = 1710\ \text{in}^4$$
$$r_x = 6.33\ \text{in}$$
$$r_y = 3.98\ \text{in}$$

For column C_5 (W 14 × 90),

$$A = 26.5\ \text{in}^2$$
$$I_x = 999\ \text{in}^4$$
$$r_x = 6.14\ \text{in}$$
$$r_y = 3.70\ \text{in}$$

For both columns C_2 and C_5,

$$L_x = L_y = 10\ \text{ft}$$
$$K_y = 1.0\quad [\text{pinned at both ends}]$$

(a) For column C_2, $G_{\text{bottom}} = 1.0$ (rigid connection at footing).

$$G_{\text{top}} = \frac{\sum\left(\frac{I}{L}\right)_{\text{columns}}}{\sum\left(\frac{I}{L}\right)_{\text{beams}}} = \frac{\frac{1710\ \text{in}^4}{10\ \text{ft}} + \frac{999\ \text{in}^4}{10\ \text{ft}}}{\frac{1200\ \text{in}^4}{12\ \text{ft}} + \frac{4200\ \text{in}^4}{12\ \text{ft}}}$$
$$= 0.6$$

From the alignment chart (Fig. 61.2(b)) for uninhibited sidesway, $K_x \approx \boxed{1.2.}$

The answer is (D).

(b) For column C_5, $G_{\text{bottom}} = 0.6$ (same as G_{top} for column C_2).

$$G_{\text{top}} = \frac{\sum\left(\frac{I}{L}\right)_{\text{columns}}}{\sum\left(\frac{I}{L}\right)_{\text{beams}}} = \frac{\frac{999\ \text{in}^4}{10\ \text{ft}}}{\frac{950\ \text{in}^4}{12\ \text{ft}}}$$
$$= 1.262\quad [\text{say } 1.3]$$

From the alignment chart for uninhibited sidesway, $K_x \approx \boxed{1.3.}$

The answer is (C).

(c) Calculate the slenderness ratio with respect to the strong axis. For column C_2,

$$\frac{K_x L_x}{r_x} = \frac{(1.2)(10\ \text{ft})\left(12\ \frac{\text{in}}{\text{ft}}\right)}{6.33\ \text{in}}$$
$$= \boxed{22.74}$$

Structural

For column C_5,

$$\frac{K_xL_x}{r_x} = \frac{(1.3)(10\text{ ft})\left(12\ \frac{\text{in}}{\text{ft}}\right)}{6.14\text{ in}} = \boxed{25.41}$$

The answer is (A).

(d) For column C_2,

$$\frac{K_yL_y}{r_y} = \frac{(1)(10\text{ ft})\left(12\ \frac{\text{in}}{\text{ft}}\right)}{3.98\text{ in}} = 30.15$$

From part (c),

$$\frac{K_xL_x}{r_x} = 22.74$$

Therefore, the maximum slenderness ratio is $\boxed{30.15.}$

The answer is (B).

(e) For column C_5,

$$\frac{K_yL_y}{r_y} = \frac{(1)(10\text{ ft})\left(12\ \frac{\text{in}}{\text{ft}}\right)}{3.70\text{ in}} = 32.43$$

From part (c),

$$\frac{K_xL_x}{r_x} = 25.41$$

Therefore, the maximum slenderness ratio is $\boxed{32.43.}$

The answer is (B).

(f) For column C_2, $\text{SR}_{\text{max}} = 30.15$. From Table C-36 of the *AISC Manual*, $F_a = 19.93$ ksi. The allowable axial compressive load is

$$P_{\text{allowable}} = F_aA = \left(19.93\ \frac{\text{kips}}{\text{in}^2}\right)(42.7\text{ in}^2) = \boxed{851\text{ kips}}$$

The answer is (C).

(g) For column C_5, $\text{SR}_{\text{max}} = 32.43$. From Table C-36 of the *AISC Manual*, $F_a = 19.77$ ksi. The allowable axial compressive load is

$$P_{\text{allowable}} = F_aA = \left(19.77\ \frac{\text{kips}}{\text{in}^2}\right)(26.5\text{ in}^2) = \boxed{524\text{ kips}}$$

The answer is (A).

(h) The tributary area of the roof to one side (only) of column C_5 is

$$(30\text{ ft})\left(\frac{12\text{ ft}}{2}\right) = 180\text{ ft}^2$$

The loadings are as follows.

roof DL:
- roofing material: 6 lbf/ft^2
- concrete slab: $\left(\frac{4\text{ in}}{12\ \frac{\text{in}}{\text{ft}}}\right)\left(150\ \frac{\text{lbf}}{\text{ft}^3}\right) = 50\text{ lbf/ft}^2$

roof LL: 30 lbf/ft^2

total: 86 lbf/ft^2

The total axial compressive load on column C_5 including the self-weight of 90 lbf/ft is

$$\begin{aligned}&(\text{tributary area})(\text{total loading}) + \text{column weight}\\ &= (180\text{ ft}^2)\left(86\ \frac{\text{lbf}}{\text{ft}^2}\right) + \left(90\ \frac{\text{lbf}}{\text{ft}}\right)(10\text{ ft})\\ &= \boxed{16{,}380\text{ lbf (16.38 kips)}}\end{aligned}$$

The answer is (A).

(i) The tributary area of the second story floor on column C_2 is

$$(30\text{ ft})\left(\frac{12\text{ ft}}{2}\right) = 180\text{ ft}^2$$

The loadings are as follows.

floor DL:
- concrete slab: $\left(\frac{6\text{ in}}{12\ \frac{\text{in}}{\text{ft}}}\right)\left(150\ \frac{\text{lbf}}{\text{ft}^3}\right) = 75\text{ lbf/ft}^2$

floor LL: 90 lbf/ft^2

total: 165 lbf/ft^2

The tributary area of the first story roof to the left of column C_2 is 180 ft^2. The total axial compressive load on column C_2 is

$$\begin{aligned}&(\text{tributary area})(\text{total load}) + \text{column weight}\\ &\quad + \text{total axial compressive load on column } C_5\\ &= (180\text{ ft}^2)\left(165\ \frac{\text{lbf}}{\text{ft}^2}\right) + (180\text{ ft}^2)\left(86\ \frac{\text{lbf}}{\text{ft}^2}\right)\\ &\quad + \left(145\ \frac{\text{lbf}}{\text{ft}}\right)(10\text{ ft}) + 16{,}380\text{ lbf}\\ &= \boxed{63{,}010\text{ lbf}\quad(63\text{ kips})}\end{aligned}$$

The answer is (A).

(j) Considering axial loads only,

$$\text{column } C_5: \ P = 16.38 \text{ kips} < P_{\text{allowable}} = 524 \text{ kips} \quad \text{[OK]}$$

$$\text{column } C_2: \ P = 77.23 \text{ kips} < P_{\text{allowable}} = 851 \text{ kips} \quad \text{[OK]}$$

The answer is (A).

2. From the *AISC Manual* shape tables for W 18×119,

$$A = 35.1 \text{ in}^2$$
$$d = 18.97 \text{ in}$$
$$I_x = 2190 \text{ in}^4$$
$$I_y = 253 \text{ in}^4$$

For C 15×50,

$$A = 14.7 \text{ in}^2$$
$$I_x = 404 \text{ in}^4$$
$$I_y = 11.0 \text{ in}^4$$
$$t_w = 0.716 \text{ in}$$
$$\overline{x} = 0.798 \text{ in}$$

(a) The cross-sectional area of the column is

$$\sum A_i = 35.1 \text{ in}^2 + 14.7 \text{ in}^2 = \boxed{49.8 \text{ in}^2}$$

The answer is (C).

(b) To calculate the moment of inertia, the location of centroid needs to be determined first.

$$\overline{y} = \frac{\sum A_i \overline{y}_i}{\sum A_i} = \frac{(35.1 \text{ in}^2)\left(\frac{18.97 \text{ in}}{2}\right) + (14.7 \text{ in}^2)\big(18.97 \text{ in} + (0.716 \text{ in} - 0.798 \text{ in})\big)}{35.1 \text{ in}^2 + 14.7 \text{ in}^2} = 12.26 \text{ in}$$

From the parallel axis theorem, the moment of inertia about the strong axis is

$$\begin{aligned} I_x &= 2190 \text{ in}^4 + (35.1 \text{ in}^2)\left(\frac{18.97 \text{ in}}{2} - 12.26 \text{ in}\right)^2 \\ &\quad + 11.0 \text{ in}^4 + (14.7 \text{ in}^2) \\ &\quad \times \big(18.97 \text{ in} + (0.716 \text{ in} - 0.798 \text{ in} - 12.26 \text{ in})\big)^2 \\ &= \boxed{3117 \text{ in}^4} \end{aligned}$$

The answer is (C).

(c) The moment of inertia about the weak axis is

$$I_y = 253 \text{ in}^4 + 404 \text{ in}^4 = \boxed{657 \text{ in}^4}$$

The answer is (B).

(d) The radius of gyration with respect to the x-axis is

$$r_x = \sqrt{\frac{I_x}{A}} = \sqrt{\frac{3117 \text{ in}^4}{49.8 \text{ in}^2}} = \boxed{7.91 \text{ in}}$$

The answer is (D).

(e) The radius of gyration with respect to the y-axis is

$$r_y = \sqrt{\frac{I_y}{A}} = \sqrt{\frac{657 \text{ in}^4}{49.8 \text{ in}^2}} = \boxed{3.63 \text{ in}}$$

The answer is (C).

(f) For buckling about the x-axis, from Table 62.1, $K_x = 1$ (both ends pinned). The unbraced length, L_x, is 20 ft. The slenderness ratio of the column with respect to the x-axis is

$$\frac{K_x L_x}{r_x} = \frac{(1)(20 \text{ ft})\left(12 \ \frac{\text{in}}{\text{ft}}\right)}{7.91 \text{ in}} = \boxed{30.34}$$

The answer is (B).

(g) For buckling about the y-axis, from Table 62.1, $K_y = 1$ (both ends pinned). The unbraced length, L_y, is 10 ft. The slenderness ratio of the column with respect to the y-axis is

$$\frac{K_y L_y}{r_y} = \frac{(1)(10 \text{ ft})\left(12 \ \frac{\text{in}}{\text{ft}}\right)}{3.63 \text{ in}} = \boxed{33.06}$$

The answer is (C).

(h) The maximum slenderness ratio is $\boxed{33.06.}$

The answer is (C).

Structural

(i) From the *AISC Manual*, Part 3, Table C-36, the allowable compressive stress for $KL/r = 33.06$ is $F_a =$

$$\boxed{19.73 \text{ ksi.}}$$

The answer is (D).

(j) The allowable axial compressive load is given by Eq. 61.11.

$$P_a = F_a A = \left(19.73 \ \frac{\text{kips}}{\text{in}^2}\right)(49.8 \text{ in}^2) = \boxed{982 \text{ kips}}$$

The answer is (C).

3. Assume the column weight will be about 2 kips. The column load, then, is 160 kips + 2 kips = 162 kips.

The effective length factors are $K_x = K_y = K = 1.0$.

The unbraced lengths are $L_x = 30$ ft and $L_y = 15$ ft.

$$L_{e,\text{min}} = KL_y = (1.0)(15 \text{ ft}) = 15 \text{ ft}$$

From the *AISC Manual*, Part 3, column tables, try W 8×40.

$$P = 169 \text{ kips} \quad [\text{for } KL_y = 15 \text{ ft}]$$
$$\frac{r_x}{r_y} = 1.73$$

Since $1.73 < 30/15 = 2.0$, the strong axis buckling controls. The equivalent unbraced length for the strong axis is

$$\frac{30 \text{ ft}}{1.73} = 17.3 \text{ ft}$$

For W 8×40, P = 153 kips (for L = 17.3 ft). This is not enough, so try W 10×39.

$$P = 162 \text{ kips} \quad [\text{for } KL_y = 15 \text{ ft}]$$
$$\frac{r_x}{r_y} = 2.16$$
$$r_y = 1.98 \text{ in}$$
$$A = 11.5 \text{ in}^2$$

Since $2.16 > 2.0$, the weak axis buckling controls. Therefore, the strong axis buckling load does not have to be checked.

The answer is (B).

4. This solution is found through trial and error. Other shapes may be more economical. Each column carries 275 kips. Neglect self-weight. From the *AISC Manual*, Part 3, column tables, try W 12×65, which has an allowable load for buckling about the minor axis (by interpolation) of

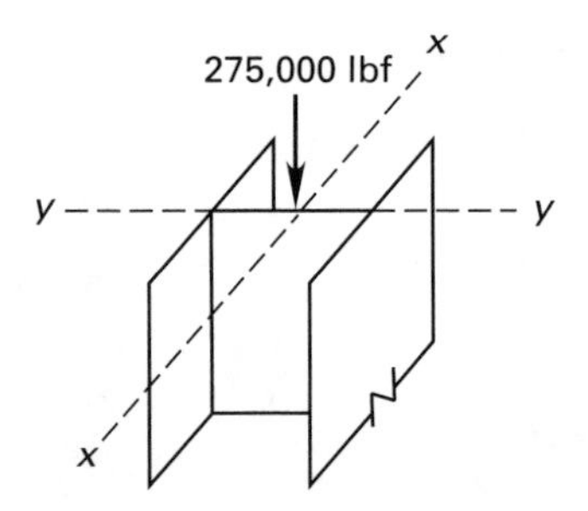

$$P_{21} = \tfrac{1}{2}(P_{20} + P_{22}) = \left(\tfrac{1}{2}\right)(294 \text{ kips} + 277 \text{ kips}) = 285.5 \text{ kips}$$

From the *AISC Manual*, Part 1, Shape Tables, for the W 12×65,

$$I_x = 533 \text{ in}^4$$
$$A = 19.1 \text{ in}^2$$
$$\frac{r_x}{r_y} = 1.75$$

The column is braced against bending about the major axis by the 40 ft girders. The effective length factor K in this problem is most easily evaluated by use of the alignment charts (Fig. 62.2). For the W 12×96 horizontal beams, $I = 833 \text{ in}^4$. For an interior column from Eq. 61.4,

$$G_{\text{top}} = \frac{\sum\left(\frac{I}{L}\right)_{\text{columns}}}{\sum\left(\frac{I}{L}\right)_{\text{beams}}} = \frac{\frac{533 \text{ in}^4}{21 \text{ ft}}}{\frac{833 \text{ in}^4}{40 \text{ ft}} + \frac{833 \text{ in}^4}{40 \text{ ft}}} = 0.609$$

Since the bottom end is built in, $G_{\text{bottom}} = 1.0$. From the alignment chart (sidesway permitted), $K_x = 1.25$. The effective length for major axis bending is

$$K_x L_x = (1.25)(21 \text{ ft}) = 26.25 \text{ ft}$$

The table is for minor axis bending only, however. So to use the column tables, the equivalent effective length for minor axis bending is calculated from Eq. 61.13.

$$K_x L_x' = \frac{K_x L_x}{\frac{r_x}{r_y}} = \frac{26.25 \text{ ft}}{1.75} = 15 \text{ ft}$$

Since 15 ft < 21 ft, the effective length for minor axis bending is critical. W 12×65 is acceptable.

Check the actual compressive stress.

$$f = \frac{P}{A} = \frac{275 \text{ kips}}{19.1 \text{ in}^2} = 14.4 \text{ kips/in}^2$$

The slenderness ratios with respect to x- and y-axes are

$$\frac{K_xL_x}{r_x} = \frac{(1.25)(21\text{ ft})\left(12\ \frac{\text{in}}{\text{ft}}\right)}{5.28\text{ in}} = 59.7$$

$$\frac{K_yL_y}{r_y} = \frac{(1)(21\text{ ft})\left(12\ \frac{\text{in}}{\text{ft}}\right)}{3.02\text{ in}} = 83.5$$

The controlling slenderness ratio is 83.5.

From Table C-36 of the *AISC Manual*, $F_a = 14.96$ ksi.

$$14.96\text{ ksi} > 14.4\text{ ksi} \quad [\text{OK}]$$

The answer is (D).

5. Neglect self-weight.

$$R_L = R_R = \frac{(9)(20\text{ kips})}{2} = 90\text{ kips}$$

- Top chords: W 8 × 28

The force in horizontal members is proportional to the moment across that panel. The moment on the truss is maximum at center.

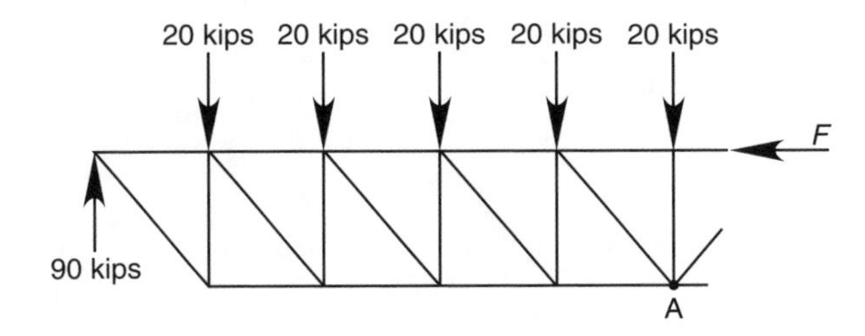

Taking clockwise moments as positive,

$$\sum M_A = (90\text{ kips})(50\text{ ft}) - (20\text{ kips})(10\text{ ft} + 20\text{ ft} + 30\text{ ft} + 40\text{ ft}) - 10F = 0$$

$$F = 250\text{ kips (compression)}$$

$$K_x = K_y = 1$$

From the column table, capacity = 132 kips at KL = 10 ft, which is inadequate.

Choose W 8 × 58 (capacity = 303 kips at KL = 10 ft). (From a practical standpoint, W 8 × 48 might also be chosen.)

- Bottom chords: W 8 × 28

Force is proportional to moment across the panel.

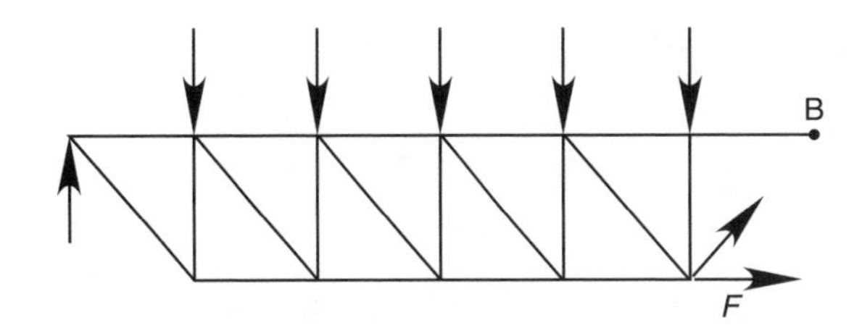

$$\sum M_B = (90\text{ kips})(60\text{ ft}) - (20\text{ ft})(10\text{ ft}+20\text{ ft}+30\text{ ft}+40\text{ ft}+50\text{ ft}) - 10F = 0$$

$$F = 240\text{ kips (tension)}$$

$$\text{SR} = 74.1 < 300 \quad [\text{acceptable}]$$

$$F_t = (0.6)(36\text{ ksi}) = 21.6\text{ ksi}$$

$$f_t = \frac{240\text{ kips}}{8.25\text{ in}^2} = 29.1\text{ ksi}$$

inadequate

The required area is

$$A = \frac{240\text{ kips}}{21.6\ \frac{\text{kips}}{\text{in}^2}} = 11.11\text{ in}^2$$

This disregards net section reductions. Try W 8 × 40 (11.7 in^2).

- Diagonal bracing: L 3 1/2 × 3 1/2 × 5/16

The force in the diagonals is proportional to shear across the panel. At the end diagonal, V = 90 kips.

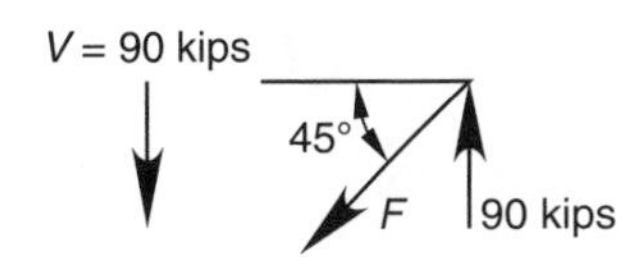

$$F = \sqrt{(90\text{ kips})^2 + (90\text{ kips})^2} = 127.3\text{ kips (tension)}$$

$$L = (1.41)(10\text{ ft}) = 14.1\text{ ft}$$

From the double-angle tables,

$$r_x = 1.08\text{ in}$$

$$A = 4.18\text{ in}^2$$

$$\text{SR} = \frac{(1)(14.1\text{ ft})\left(12\ \frac{\text{in}}{\text{ft}}\right)}{1.08} = 156.7 < 300 \quad [\text{acceptable}]$$

$$F_a = (0.6)(36\text{ ksi}) = 21.6\text{ ksi}$$

$$f_a = \frac{127.3\text{ kips}}{4.18\text{ in}^2} = 30.45\text{ ksi}$$

inadequate

Structural

The required area is

$$A = \frac{127.3 \text{ kips}}{21.6 \ \frac{\text{kips}}{\text{in}^2}} = 5.89 \text{ in}^2$$

The area of L $3^1/_2 \times 3^1/_2 \times {}^1/_2$ in pair is

$$(2)(3.25 \text{ in}^2) = 6.5 \text{ in}^2$$

This is sufficient if the angle is available.

- Vertical members: L $4 \times 4 \times {}^3/_8$

The force in the vertical member is proportional to vertical force in the adjacent diagonal.

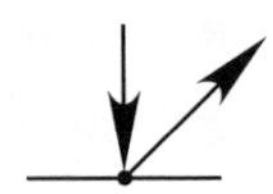

By inspection, at the first vertical member,

$$F = 90 \text{ kips (compression)}$$

From the Double-Angle Tables, the capacity = 76 kips.

$$\boxed{\text{inadequate}}$$

Try L $4 \times 4 \times {}^1/_2$ (capacity = 99 kips at $KL = 10$ ft).

62 Structural Steel: Beam-Columns

PRACTICE PROBLEMS

1. (*Time limit: one hour*) A column carries an axial load of 525 kips as shown. The beams framing into it apply a moment of 225 ft-kips. Bracing is provided to prevent bending about the weak axis, which can be disregarded. All members are grade-50 steel. (a) If the column is W 14×82 and the beams are W 14×53, determine if the column is adequate. (b) If the design is inadequate, specify an acceptable W 14 grade-50 column.

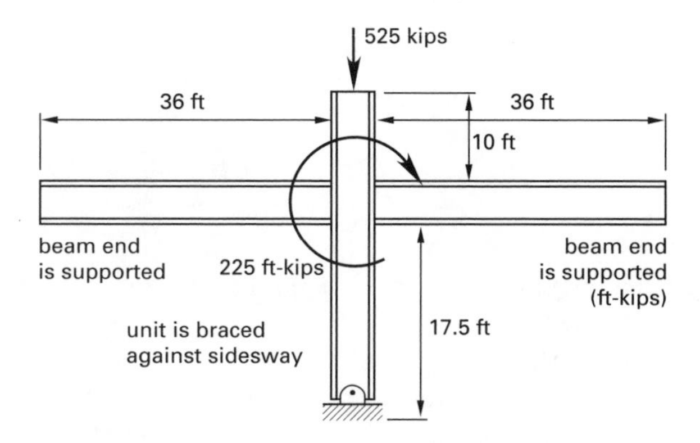

2. (*Time limit: one hour*) A tall column in an industrial storage building carries the load shown. The lateral wind loads are carried into a braced frame. The column is braced in its weak direction by a strut 24 ft from the bottom hinged end. There is no support of the column compression flange along its length. Select an economical A36 W 24 section based on stress and deflection criteria.

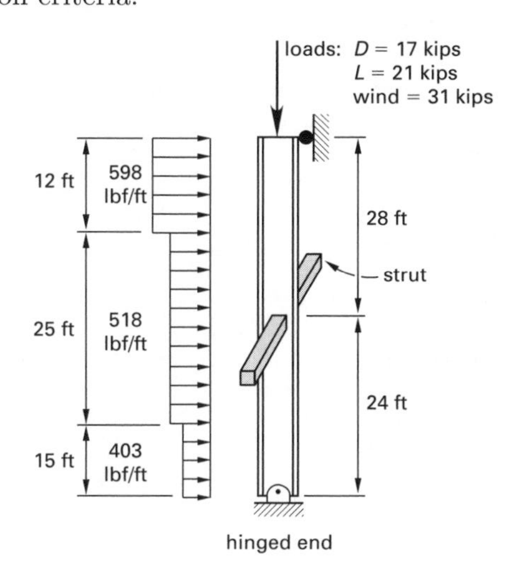

3. (*Time limit: one hour*) A 16 ft column is acted upon by an axial gravity load of 85 kips and a uniform wind load of w (in kips/ft) as shown. The W 12×58 column is made of A572 grade-50 steel. The lower end is built in. The upper end is not supported in the weak direction.

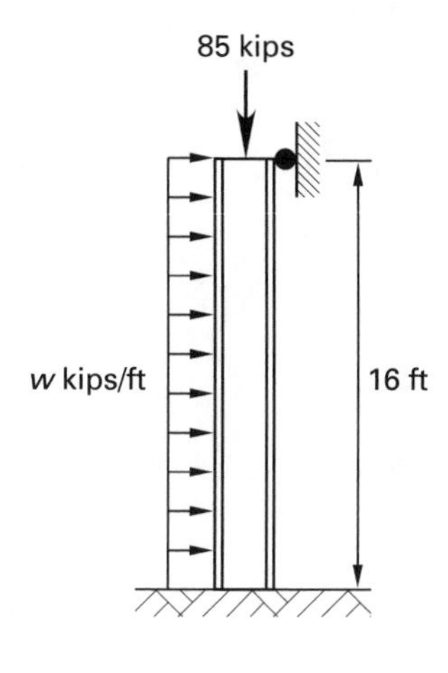

(a) The actual axial stress in the member is most nearly

(A) 2.5 ksi
(B) 4.0 ksi
(C) 5.0 ksi
(D) 6.5 ksi

(b) The critical slenderness ratio is most nearly

(A) 39
(B) 98
(C) 160
(D) 190

(c) The allowable axial stress is most nearly

(A) 3 ksi
(B) 6 ksi
(C) 9 ksi
(D) 15 ksi

(d) The allowable bending stress is most nearly

(A) 22 ksi
(B) 27 ksi
(C) 30 ksi
(D) 33 ksi

(e) The actual bending stress, assuming $w = 4$ kips/ft, is most nearly

(A) 2 ksi
(B) 8 ksi
(C) 12 ksi
(D) 20 ksi

(f) The beam-column is subjected to

(A) small compression and is adequate.
(B) large compression and is adequate.
(C) small compression and is inadequate.
(D) large compression and is inadequate.

For parts (g) through (j), assume lateral bracing (pinned support) is provided in the weak direction at mid-height as well as at the top of the member.

(g) The critical slenderness ratio is most nearly

(A) 39
(B) 98
(C) 160
(D) 190

(h) The allowable axial stress is most nearly

(A) 18 ksi
(B) 21 ksi
(C) 26 ksi
(D) 32 ksi

(i) The allowable bending stress with respect to the strong axis is most nearly

(A) 22 ksi
(B) 27 ksi
(C) 30 ksi
(D) 33 ksi

(j) The maximum lateral wind load w the member can safely carry is most nearly

(A) 5.5 kips/ft
(B) 6.8 kips/ft
(C) 7.5 kips/ft
(D) 9.5 kips/ft

4. (*Time limit: one hour*) A W 14 × 48 member of A36 steel carries the service loads shown. The member is part of a braced system, and has support in the weak direction at mid-height, but it has support only at top and bottom for the strong direction.

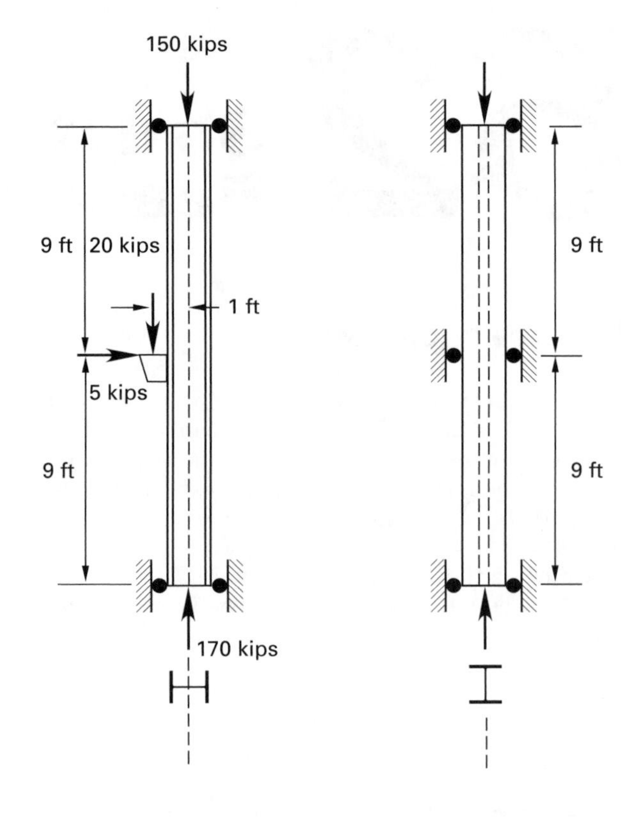

(a) The maximum primary bending moment about the strong axis occurs

(A) at the top.
(B) at one-quarter height from top.
(C) at mid-height.
(D) at three-quarter height from top.

(b) The maximum primary bending moment about the strong axis is most nearly

(A) 28 ft-kips
(B) 33 ft-kips
(C) 40 ft-kips
(D) 43 ft-kips

(c) The slenderness ratio with respect to buckling about the weak axis is most nearly

(A) 27
(B) 37
(C) 47
(D) 57

(d) The slenderness ratio with respect to buckling about the strong axis is most nearly

(A) 27
(B) 37
(C) 47
(D) 57

(e) The critical slenderness ratio is most nearly

(A) 27
(B) 37
(C) 47
(D) 57

(f) The allowable axial stress is most nearly

(A) 9 ksi
(B) 12 ksi
(C) 18 ksi
(D) 21 ksi

(g) The actual bending stress is most nearly

(A) 5.5 ksi
(B) 6.7 ksi
(C) 10 ksi
(D) 14 ksi

(h) The ratio of actual axial stress to allowable axial stress is most nearly

(A) 0.12
(B) 0.37
(C) 0.51
(D) 0.68

(i) The allowable bending stress is most nearly

(A) 16 ksi
(B) 18 ksi
(C) 22 ksi
(D) 24 ksi

(j) The beam-column is subjected to

(A) small compression and is adequate.
(B) large compression and is adequate.
(C) small compression and is inadequate.
(D) large compression and is inadequate.

SOLUTIONS

1. (a)

$$P = 525 \text{ kips}$$
$$M_x = 225 \text{ ft-kips}$$
$$M_y = 0$$

Evaluate the 17.5 ft lower column. For the grade-50, W 14 × 82 member,

$$\begin{aligned} F_y &= 50 \text{ ksi} \\ A &= 24.1 \text{ in}^2 \\ I_x &= 882 \text{ in}^4 \\ S_x &= 123 \text{ in}^3 \\ r_x &= 6.05 \text{ in} \\ L_c &= 9.1 \text{ ft} \\ L_u &= 20.2 \text{ ft} \end{aligned}$$

For W 14 × 53, $I_x = 541 \text{ in}^4$.

The axial stress is

$$f_a = \frac{P}{A} = \frac{525 \text{ kips}}{24.1 \text{ in}^2} = 21.8 \text{ kips/in}^2$$

Use the alignment chart to obtain K. With no sidesway, $G_{\text{A}} = 10$ (pinned), and

$$G_{\text{B}} = \frac{\sum\left(\frac{I}{L}\right)_{\text{columns}}}{\sum\left(\frac{I}{L}\right)_{\text{girders}}} = \frac{\frac{882 \text{ in}^4}{10 \text{ ft}} + \frac{882 \text{ in}^4}{17.5 \text{ ft}}}{\frac{541 \text{ in}^4}{36 \text{ ft}} + \frac{541 \text{ in}^4}{36 \text{ ft}}} = 4.6$$

$$K = 0.94$$

The problem stated that the weak axis bending can be disregarded. The maximum slenderness ratio is

$$\text{SR} = \frac{KL}{r_x} = \frac{(0.94)(17.5 \text{ ft})\left(12 \frac{\text{in}}{\text{ft}}\right)}{6.05 \text{ in}} = 32.6 \quad [\text{say } 33]$$

From the *AISC Manual*, Part 3, Table C-50, $F_a = 26.77$ ksi.

Since $F_a > f_a$, additional checking as a beam-column is required.

$$\frac{f_a}{F_a} = \frac{21.8 \text{ ksi}}{26.77 \text{ ksi}} = 0.814 > 0.15$$

So the axial compression is large.

$$\begin{aligned} f_{bx} &= \frac{M_x}{S_x} = \frac{(225 \text{ ft-kips})\left(12 \frac{\text{in}}{\text{ft}}\right)}{123 \text{ in}^3} \\ &= 22.0 \text{ ksi} \end{aligned}$$

Structural

Since $L_u > L_b > L_c$,

$$F_{bx} = 0.6F_y = (0.6)(50\text{ ksi}) = 30\text{ ksi}$$

$$\frac{f_a}{0.60F_y} + \frac{f_{bx}}{F_{bx}} + \frac{f_{by}}{F_{by}} = \frac{21.8\text{ ksi}}{(0.60)(50\text{ ksi})} + \frac{22.0\text{ ksi}}{30\text{ ksi}} + 0$$
$$= 1.46 > 1.0$$

The column is inadequate.

(b) *step 1:* Since the column is not known, conservatively estimate $K = 1$.

$$KL = (1)(17.5\text{ ft}) = 17.5\text{ ft} \quad [\text{say 18 ft}]$$

step 2: From the $KL = 18$ column, read $m = 1.7$. (Since $M_y = 0$ and a W14 shape is to be used, only one iteration will be required.)

step 3: (skip)

step 4:
$$\begin{aligned} P_{\text{eff}} &= P + M_x m + M_y m U \\ &= 525\text{ kips} + (225\text{ ft-kips})(1.7) + 0 \\ &= 907.5\text{ kips} \end{aligned}$$

step 5: Try W 14 × 132.

$$\text{column capacity} = 898\text{ kips at 18 ft}$$
[50 ksi steel]

Prove this is adequate.

For 50 ksi W 14 × 132,

$$\begin{aligned} A &= 38.8\text{ in}^2 \\ I_x &= 1530\text{ in}^4 \\ S_x &= 209\text{ in}^3 \\ r_x &= 6.28\text{ in} \\ L_c &= 13.2\text{ ft} \\ L_u &= 34.4\text{ ft} \\ B_x &= 0.186 \\ f_a &= \frac{P}{A} = \frac{525\text{ kips}}{38.8\text{ in}^2} \\ &= 13.5\text{ kips/in}^2 \end{aligned}$$

From the alignment chart, with $G_A = 10$,

$$G_B = \frac{\sum\left(\frac{I}{L}\right)_{\text{columns}}}{\sum\left(\frac{I}{L}\right)_{\text{girders}}} = \frac{\frac{1530\text{ in}^4}{10\text{ ft}} + \frac{1530\text{ in}^4}{17.5\text{ ft}}}{\frac{541\text{ in}^4}{36\text{ ft}} + \frac{541\text{ in}^4}{36\text{ ft}}} = 8.0$$

$K = 0.96$

$$\begin{aligned} \text{SR} &= \frac{KL}{r_x} = \frac{(0.96)(17.5\text{ ft})\left(12\ \frac{\text{in}}{\text{ft}}\right)}{6.28\text{ in}} \\ &= 32.1 \end{aligned}$$

$$F_a = 26.9\text{ ksi} \quad [\textit{AISC Manual}\text{, Table C-50}]$$

$$\frac{f_a}{F_a} = \frac{13.5\text{ ksi}}{26.9\text{ ksi}} = 0.502 > 0.15$$

$$\begin{aligned} f_{bx} &= \frac{M_x}{S_x} = \frac{(225\text{ ft-kips})\left(12\ \frac{\text{in}}{\text{ft}}\right)}{209\text{ in}^3} \\ &= 12.92\text{ kips/in}^2 \end{aligned}$$

Since $L_u > L_b > L_c$,

$$F_{bx} = 0.6F_y = (0.6)(50\text{ ksi}) = 30\text{ ksi}$$

$$\frac{f_a}{0.60F_y} + \frac{f_{bx}}{F_{bx}} + \frac{f_{by}}{F_{by}} = \frac{13.5\text{ ksi}}{(0.60)(50\text{ ksi})} + \frac{12.92\text{ ksi}}{30\text{ ksi}} + 0$$
$$= 0.88 < 1.0 \quad [\text{OK}]$$

One more check needs to be made. Although C_m could be calculated, choose $C_m = 1$ as a conservative shortcut. From the *AISC ASD Manual*, Table 8 (p. 5-122), $F'_e = 145.83$ ksi. Use Eq. 62.4.

$$\frac{f_a}{F_a} + \frac{C_{mx}f_{bx}}{\left(1 - \frac{f_a}{F'_{ex}}\right)F_{bx}} + \frac{C_{my}f_{by}}{\left(1 - \frac{f_a}{F'_{ey}}\right)F_{by}}$$
$$= \frac{13.5\text{ ksi}}{26.9\text{ ksi}} + \frac{(1)(12.92\text{ ksi})}{\left(1 - \frac{13.5\text{ ksi}}{145.83\text{ ksi}}\right)(30ksi)} + 0$$
$$= 0.976 < 1.0 \quad [\text{OK}]$$

The W 14 × 132 column is adequate.

2. All vertical loads act simultaneously with the wind load. The total axial load is

$$17\text{ kips} + 21\text{ kips} + 31\text{ kips} = 69\text{ kips}$$

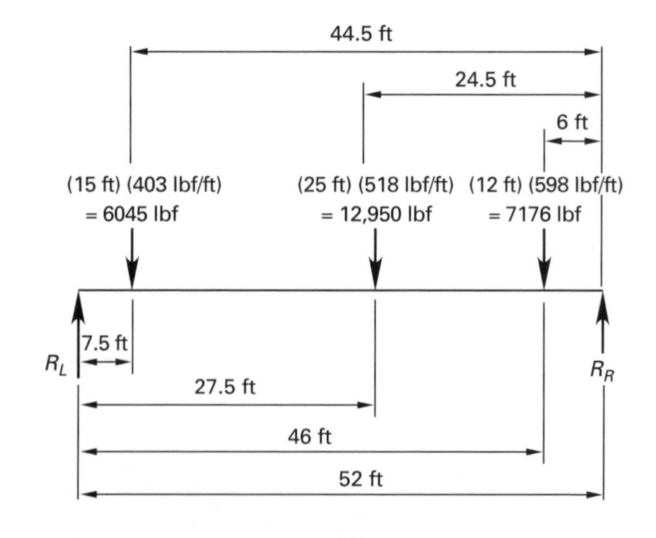

Find the reactions. Taking clockwise moments as positive,

$\sum M$ about R_L:

$$(6.045\,\text{kips})(7.5\,\text{ft}) + (12.95\,\text{kips})(27.5\,\text{ft}) + (7.176\,\text{kips})(46\,\text{ft}) - R_R(52\,\text{ft}) = 0$$

$$R_R = 14.07\,\text{kips}$$

$\sum M$ about R_R:

$$-(6.045\,\text{kips})(44.5\,\text{ft}) - (12.95\,\text{kips})(24.5\,\text{ft}) - (7.176\,\text{kips})(6\,\text{ft}) + R_L(52\,\text{ft}) = 0$$

$$R_L = 12.10\,\text{kips}$$

The shear diagram is

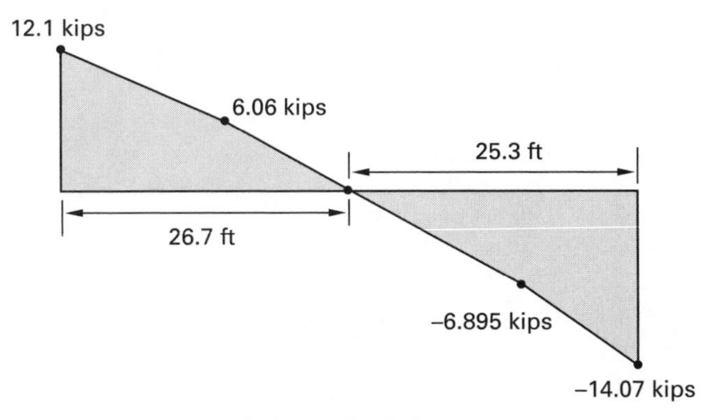

$M_{\max}$ at $x = 26.7$ ft from the left:

$$\left(\tfrac{1}{2}\right)(12.1\,\text{kips} - 6.06\,\text{kips})(15\,\text{ft}) + (6.06\,\text{kips})(15\,\text{ft}) + \left(\tfrac{1}{2}\right)(6.06\,\text{kips})(11.7\,\text{ft}) = 171.65\,\text{ft-kips}$$

Since a W 24 section is specified, the column tables in the *AISC Manual* cannot be used.

Try W 24 × 117.

$$A = 34.4\ \text{in}^2$$
$$d = 24.26\ \text{in}$$
$$b_f = 12.80\ \text{in}$$
$$\frac{d}{A_f} = 2.23\ \text{in}^{-1}$$
$$I_x = 3540\ \text{in}^4$$
$$S_x = 291\ \text{in}^3$$
$$r_x = 10.1\ \text{in}$$
$$r_y = 2.94\ \text{in}$$
$$L_c = 13.5\ \text{ft}$$
$$L_u = 20.8\ \text{ft}$$

Since both ends are free to rotate, $K = 1$. The slenderness ratio is

$$\text{SR} = \text{larger of}\left\{\begin{array}{l} \dfrac{K_xL_x}{r_x} = \dfrac{(1)(52\ \text{ft})\left(12\ \dfrac{\text{in}}{\text{ft}}\right)}{10.1\ \text{in}} = 61.8 \\[2ex] \dfrac{K_yL_y}{r_y} = \dfrac{(1)(28\ \text{ft})\left(12\ \dfrac{\text{in}}{\text{ft}}\right)}{2.94\ \text{in}} = 114.3 \end{array}\right\}$$

SR is 114.3.

$$F_a = 11.09\ \text{ksi} \quad \left[\begin{array}{c}\text{interpolated from the } \textit{AISC} \\ \textit{Manual}, \text{ Table C-36}\end{array}\right]$$

$$f_a = \frac{P}{A} = \frac{69\ \text{kips}}{34.4\ \text{in}^2} = 2.0\ \text{kips/in}^2$$

$$\frac{f_a}{F_a} = \frac{2.0\ \text{ksi}}{11.09\ \text{ksi}} = 0.18 > 0.15$$

The compression is large.

$$f_{bx} = \frac{M_x}{S_x} = \frac{(171.6\ \text{ft-kips})\left(12\ \dfrac{\text{in}}{\text{ft}}\right)}{291\ \text{in}^3} = 7.08\ \text{kips/in}^2$$

Use AISC Specification Eq. F1-8 as a shortcut to finding F_b. As the compression flange is not supported, $l_b = 52$ ft. Therefore, $l_b > L_u$. Use $C_b = 1$ since there is no lateral support.

$$F_{\text{bx}} = \frac{(12\times10^3)C_b}{l\left(\dfrac{d}{A_f}\right)} = \frac{(12\times10^3)(1)}{(52\ \text{ft})\left(12\ \dfrac{\text{in}}{\text{ft}}\right)\left(\dfrac{2.23}{\text{in}}\right)} = 8.62\ \text{ksi}$$

Using the one-third increase in allowable stress due to wind load in Eq. 62.7,

$$\frac{f_a}{0.60F_y} + \frac{f_{bx}}{F_{bx}} + \frac{f_{by}}{F_{by}} = \frac{2.0\ \text{ksi}}{(0.60)(36\ \text{ksi})} + \frac{7.08\ \text{ksi}}{8.62\ \text{ksi}} + 0 = 0.91 < \tfrac{4}{3} \quad \text{[OK]}$$

Check Eq. 62.4 for beam-column interaction. $K_xL_x/r_x = 61.8$ is used because x is the beam-action direction. (The member is not a beam in the y-direction.) From *AISC Manual* Table 8, $F'_e \approx 39$ ksi.

$$(39\ \text{ksi})\left(\tfrac{4}{3}\right) = 52\ \text{ksi}$$

From AISC Specification, Sec. H1, case c(ii), $C_m = 1.0$.

From Eq. 62.4,

$$\frac{f_a}{F_a} + \frac{C_{mx}f_{bx}}{\left(1 - \dfrac{f_a}{F'_{ex}}\right)F_{bx}} + \frac{C_{my}f_{by}}{\left(1 - \dfrac{f_a}{F'_{ey}}\right)F_{by}} = \frac{2\ \text{ksi}}{11.13\ \text{ksi}} + \frac{(1)(7.08\ \text{ksi})}{\left(1 - \dfrac{2\ \text{ksi}}{52\ \text{ksi}}\right)(8.62\ \text{ksi})} + 0 = 1.03 < \tfrac{4}{3} \quad \text{[OK]}$$

An exact deflection calculation is difficult because the transverse load is nonuniform. Assume 598 lbf/ft over the entire column (limiting case). For a uniformly distributed load,

$$\Delta_{max} = \frac{5wL^4}{384EI} = \frac{(5)\left(598\ \frac{\text{lbf}}{\text{ft}}\right)(52\ \text{ft})^4\left(12\ \frac{\text{in}}{\text{ft}}\right)^3}{(384)\left(29\times10^6\ \frac{\text{lbf}}{\text{in}^2}\right)(3540\ \text{in}^4)}$$
$$= 0.96\ \text{in} = L/650$$

$L/650$ is less than the traditional guidelines for deflection. So, $\boxed{\text{use W 24} \times 117.}$

3. (a) For W 12 × 58, $A = 17.0\ \text{in}^2$. The actual axial stress is

$$f_a = \frac{P}{A} = \frac{85\ \text{kips}}{17\ \text{in}^2} = \boxed{5.0\ \text{ksi}}$$

The answer is (C).

(b) For W 12 × 58, $r_x = 5.28$ in and $r_y = 2.51$ in. For buckling about the strong (x-x) axis, $K_x = 0.8$ (one end fixed, one end pinned). $L_x = 16$ ft. For buckling about the weak (y-y) axis, $K_y = 2.1$ (one end fixed and one end free). $L_y = 16$ ft.

$$\text{SR} = \text{larger of}\begin{Bmatrix} \frac{K_xL_x}{r_x} = \frac{(0.8)(16\ \text{ft})\left(12\ \frac{\text{in}}{\text{ft}}\right)}{5.28\ \text{in}} = 29.1 \\ \frac{K_yL_y}{r_y} = \frac{(2.1)(16\ \text{ft})\left(12\ \frac{\text{in}}{\text{ft}}\right)}{2.51\ \text{in}} = 160.64 \end{Bmatrix}$$

SR is $\boxed{160.64.}$

The answer is (C).

(c) From the *AISC Manual*, Part 3, Table C-50, for $KL/r = 160.64$, $F_a = \boxed{5.76\ \text{ksi.}}$

The answer is (B).

(d) The unbraced length $l_b = 16$ ft. For 50 ksi W 12 × 58, $L_c = 9.0$ ft and $L_u = 17.5$ ft. Therefore, $L_c < l_b < L_u$.

$$F_{bx} = 0.6F_y = (0.6)(50\ \text{ksi}) = \boxed{30\ \text{ksi}}$$

The answer is (C).

(e) The lateral load is $w = 4.0$ kips/ft. For a beam fixed at one end and supported at the other and carrying a uniformly distributed load (*AISC* beam diagram, case 12),

$$M_x = \frac{wL^2}{8} = \frac{\left(4\ \frac{\text{kips}}{\text{ft}}\right)(16\ \text{ft})^2}{8} = 128\ \text{ft-kips}$$

For a W 12 × 58, $S_x = 78.0\ \text{in}^3$. The bending stress is

$$f_{bx} = \frac{M_x}{S_x} = \frac{(128\ \text{ft-kips})\left(12\ \frac{\text{in}}{\text{ft}}\right)}{78\ \text{in}^3}$$
$$= \boxed{19.7\ \text{kips/in}^2\ \text{(ksi)}}$$

The answer is (D).

(f) The axial stress ratio is

$$\frac{f_a}{F_a} = \frac{5.0\ \text{ksi}}{5.76\ \text{ksi}} = 0.868 > 0.15$$

Therefore, the beam-column is subjected to large compression.

Check Eqs. 62.4 and 62.7 for beam-column interaction, using one-third increase in allowable stresses due to wind load. (See *AISC Manual* Sec. A5.2.) $K_xL_x/r_x = 29.1$ is used because x is the beam-action direction. (The member is not a beam in the y-direction.) From *AISC Manual* Table 8, $F'_e \approx 176$ ksi.

$$(176\ \text{ksi})\left(\tfrac{4}{3}\right) = 235\ \text{ksi}$$

From *AISC Specifications*, Sec. H1, case c(ii), $C_m = 1.0$. From Eq. 62.4,

$$\frac{f_a}{F_a} + \frac{C_{mx}f_{bx}}{\left(1 - \frac{f_a}{F'_{ex}}\right)F_{bx}} + \frac{C_{my}f_{by}}{\left(1 - \frac{f_a}{F'_{ey}}\right)F_{by}}$$
$$= \frac{5\ \text{ksi}}{5.76\ \text{ksi}} + \frac{(1)(19.7\ \text{ksi})}{\left(1 - \frac{5\ \text{ksi}}{235\ \text{ksi}}\right)(30\ \text{ksi})} + 0$$
$$= 1.53 > \tfrac{4}{3} \quad \text{[inadequate]}$$

The member is inadequate.

There is no need to check by using Eq. 62.7.

The answer is (D).

(g) For buckling about the strong (x-x) axis, $K_x = 0.8$ (one end fixed and the other end pinned). $L_x = 16$ ft. For buckling about the weak (y-y) axis, $K_y = 1.0$ (column with both ends pinned) and $L_y = 8$ ft.

From Eq. 62.9,

$$\text{SR} = \text{larger of} \left\{ \begin{array}{l} \dfrac{K_x L_x}{r_x} = \dfrac{(0.8)(16 \text{ ft})\left(12 \dfrac{\text{in}}{\text{ft}}\right)}{5.28 \text{ in}} = 29.1 \\ \dfrac{K_y L_y}{r_y} = \dfrac{(1)(8 \text{ ft})\left(12 \dfrac{\text{in}}{\text{ft}}\right)}{2.51 \text{ in}} = 38.25 \end{array} \right\}$$

SR is $\boxed{38.25.}$

The answer is (A).

(h) From the *AISC Manual*, Part 3, Table C-50 for $KL/r = 38.25$, $F_a = \boxed{26.07 \text{ ksi.}}$

The answer is (C).

(i) The unbraced length is $l_b = 16$ ft. For 50 ksi W 12 × 58, $L_c = 9.0$ ft and $L_u = 17.5$ ft. Therefore, $l_b < L_\mu$. From Chap. 59,

$$F_{bx} = 0.60F_y = (0.60)(50 \text{ ksi}) = \boxed{30 \text{ ksi}}$$

The answer is (C).

(j) The axial stress ratio is

$$\frac{f_a}{F_a} = \frac{5.0 \text{ ksi}}{26.07 \text{ ksi}} = 0.192 > 0.15$$

Therefore, the beam-column is subjected to large compression. Check Eqs. 62.5 and 62.7 for beam-column interaction. $K_x L_x / r_x = 29.1$ is used because x is the beam-action direction. (The member is not a beam in the y-direction.) From *AISC Manual* Table 8, $F'_e \approx$ 176 ksi.

$$(176 \text{ ksi})\left(\tfrac{4}{3}\right) = 235 \text{ ksi}$$

From *AISC Specifications*, Sec. H1, case c(ii), $C_m =$ 1.0.

From Eq. 62.4, using one-third increase in allowable stresses due to wind load,

$$\frac{f_a}{F_a} + \frac{C_{mx} f_{bx}}{\left(1 - \dfrac{f_a}{F'_{ex}}\right) F_{bx}} + \frac{C_{my} f_{by}}{\left(1 - \dfrac{f_a}{F'_{ey}}\right) F_{by}}$$
$$= \frac{5 \text{ ksi}}{26.07 \text{ ksi}} + \frac{(1) f_{bx}}{\left(1 - \dfrac{5 \text{ ksi}}{235 \text{ ksi}}\right)(30 \text{ ksi})} + 0$$
$$= \frac{4}{3}$$
$$f_{bx} = 33.52 \text{ ksi}$$

$$M_x = f_{bx} S_x = \frac{\left(33.52 \dfrac{\text{kips}}{\text{in}^2}\right)(78 \text{ in}^3)}{12 \dfrac{\text{in}}{\text{ft}}}$$
$$= 217.9 \text{ ft-kips}$$

From $M_x = wL^2/8$,

$$w = \frac{8M_x}{L^2} = \frac{(8)(217.9 \text{ ft-kips})}{(16 \text{ ft})^2}$$
$$= \boxed{6.81 \text{ kips/ft}}$$

From Eq. 62.7, using the one-third increase in allowable stress due to wind load,

$$\frac{f_a}{0.60F_y} + \frac{f_a}{F_{bx}} + \frac{f_{by}}{F_{by}} = \frac{5.0 \text{ ksi}}{(0.60)(50 \text{ ksi})} + \frac{f_{bx}}{30 \text{ ksi}} + 0$$
$$= \frac{4}{3}$$
$$f_{bx} = 35.0 \text{ ksi}$$

This value is larger than 33.52 ksi, so the values obtained from Eq. 62.4 do not control.

The answer is (B).

4. The notable features of this problem are that (a) the bracing is not at the same points for both directions and (b) the lateral transverse loading causes the primary bending moment.

For W 14 × 48, $A = 14.1 \text{ in}^2$, $r_x = 5.85$ in, $r_y =$ 1.91 in, $S_x = 70.3 \text{ in}^3$, $F'_y > 36$ ksi, $L_c = 8.5$ ft, and $L_u = 16.0$ ft.

(a) The maximum axial load, $P = 170$ kips, occurs in the lower half of the member. Determine the maximum primary bending moment. From the moment diagram,

$\boxed{\text{the maximum primary moment occurs at the mid-height of the member.}}$

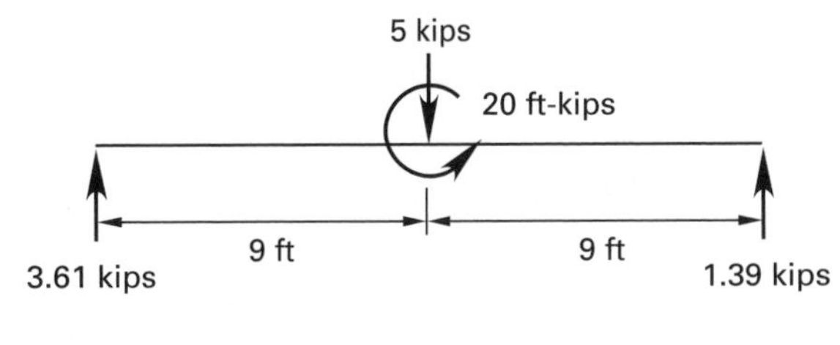

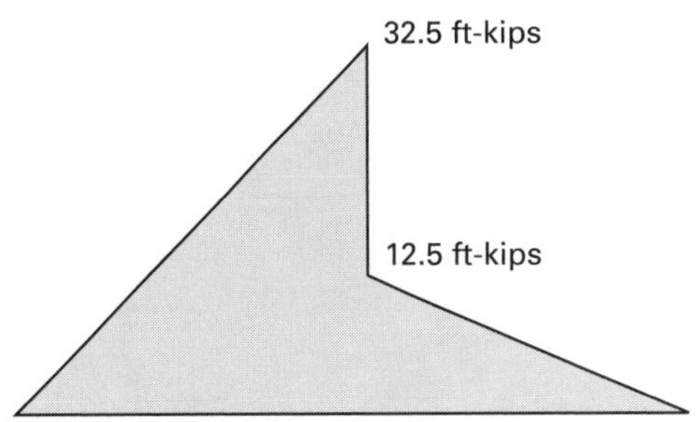

The answer is (C).

Structural

(b) From the moment diagram, the maximum primary bending moment is

$$M_x = \boxed{32.5 \text{ ft-kips}}$$

The moment about the weak axis is zero.

The answer is (B).

(c) For buckling about the weak (y-y) axis, $K_y = 1.0$ for both ends pinned. $L_y = 9$ ft.

$$\frac{K_y L_y}{r_y} = \frac{(1)(9 \text{ ft})\left(12 \ \frac{\text{in}}{\text{ft}}\right)}{1.91 \text{ in}} = \boxed{56.54 \quad [\text{use } 57]}$$

The answer is (D).

(d) For buckling about the strong (x-x) axis, from Table 62.1, $K_x = 1.0$ for both ends pinned. $L_x = 18$ ft.

$$\frac{K_x L_x}{r_x} = \frac{(1)(18 \text{ ft})\left(12 \ \frac{\text{in}}{\text{ft}}\right)}{5.85 \text{ in}} = \boxed{36.92 \quad [\text{use } 37]}$$

The answer is (B).

(e) The critical slenderness ratio is 57.

The answer is (D).

(f) From the *AISC Manual*, Table C-36, for $KL/r = 57$,

$$F_a = \boxed{17.71 \text{ ksi}}$$

The answer is (C).

(g) Calculate the bending stress.

$$f_{bx} = \frac{M_x}{S_x} = \frac{(32.5 \text{ ft-kips})\left(12 \ \frac{\text{in}}{\text{ft}}\right)}{70.3 \text{ in}^3} = \boxed{5.55 \text{ ksi}}$$

The answer is (A).

(h) Calculate the axial stress.

$$f_a = \frac{P}{A} = \frac{170 \text{ kips}}{14.1 \text{ in}^2} = 12.06 \text{ kips/in}^2$$

Calculate the ratio of actual axial stress to allowable axial stress.

$$\frac{f_a}{F_a} = \frac{12.06 \text{ ksi}}{17.71 \text{ ksi}} = \boxed{0.68}$$

The answer is (D).

(i) Calculate the allowable bending stress. The section is compact since $F_y' > F_y = 36$ ksi. The unsupported length of the compression flange is $l_b = 9$ ft. Since $L_c < l_b < L_u$,

$$F_{bx} = 0.60F_y = (0.60)(36 \text{ ksi}) = \boxed{21.6 \text{ ksi}}$$

The answer is (C).

(j) Since $f_a/F_a = 0.68 > 0.15$,

the member is subjected to large compression.

Both Eqs. 62.4 and 62.7 need to be checked.

Bending is about the strong axis only. From the *AISC Manual*, Table 8, for $Kl_b/r_b = 37$, $F_{ex}' = 109.08$ ksi. For columns in braced frames with ends unrestrained against rotation, $C_m = 1.0$ [*AISC Specifications* Sec. H1, case (ii)].

From Eq. 62.4,

$$\begin{aligned}
&\frac{f_a}{F_a} + \frac{C_{mx} f_{bx}}{\left(1 - \frac{f_a}{F_{ex}'}\right) F_{bx}} + \frac{C_{my} f_{by}}{\left(1 - \frac{f_a}{F_{ey}'}\right) F_{by}} \\
&= \frac{12.06 \text{ ksi}}{17.71 \text{ ksi}} + \frac{(1)(5.55 \text{ ksi})}{\left(1 - \frac{12.06 \text{ ksi}}{109.08 \text{ ksi}}\right)(21.6 \text{ ksi})} + 0 \\
&= 0.97 < 1.0 \quad [\text{OK}]
\end{aligned}$$

From Eq. 62.7,

$$\begin{aligned}
\frac{f_a}{0.60F_y} + \frac{f_{bx}}{F_{bx}} + \frac{f_{by}}{F_{by}} &= \frac{12.06 \text{ ksi}}{(0.60)(36 \text{ ksi})} + \frac{5.55 \text{ ksi}}{21.6 \text{ ksi}} + 0 \\
&= 0.82 < 1.0 \quad [\text{OK}]
\end{aligned}$$

The beam-column is adequate.

The answer is (B).

63 Structural Steel: Plate Girders

PRACTICE PROBLEMS

1. A simply supported plate girder spans 40 ft and is constructed with $^3/_8$ in A36 steel plates. The flanges are 16 in wide and the web is 48 in deep. A uniform but unknown load is carried across the girder's entire length. (a) Based on shear consideration only, determine the maximum uniform load (in kips/ft) that the beam can carry. (b) A36 web stiffeners are used in pairs and are placed every 4 ft. Design the web stiffeners located 4 ft from the supports.

2. A welded A36 plate girder with a 60 ft span has to carry a uniformly distributed load of 2 kips/ft, its own weight, and a concentrated load of 135 kips at midspan. Lateral support is provided for the compression flange at beam ends and at midspan. No intermediate stiffeners are to be used. The web-to-flange weld size is $w = {}^5/_{16}$ in. The length of bearing under the concentrated load is $N = 10$ in. (a) Design the girder cross section such that the overall girder depth does not exceed 60 in. (b) Design the bearing stiffeners.

3. (*Time limit: one hour*) A simply supported plate girder with a 60 ft span is composed of $1^1/_2$ in × 16 in flanges and a $^3/_8$ in × 64 in web. A36 steel is used. The girder supports a uniformly distributed service load of 2 kips/ft (including the girder weight) over the entire span and two concentrated service loads of 50 kips magnitude each, placed symmetrically one third of the span from each end. Lateral supports for the compression flange are provided at the beam ends and the concentrated load points. Except for bearing stiffeners at the supports and concentrated loads, no transverse stiffeners are provided.

(a) The actual value and the allowable limit of web depth-thickness ratio to prevent vertical buckling of the compression flange into the web are, respectively, most nearly

(A) 171 and 322
(B) 171 and 333
(C) 179 and 322
(D) 179 and 333

(b) The moment of inertia about the strong axis is most nearly

(A) 50,000 in^4
(B) 54,000 in^4
(C) 60,000 in^4
(D) 61,000 in^4

(c) The maximum service load shear is most nearly

(A) 85 kips
(B) 110 kips
(C) 130 kips
(D) 160 kips

(d) The maximum service load moment is most nearly

(A) 1600 ft-kips
(B) 1800 ft-kips
(C) 1900 ft-kips
(D) 2000 ft-kips

(e) The maximum computed bending stress due to service loads is most nearly

(A) 12 ksi
(B) 13 ksi
(C) 19 ksi
(D) 23 ksi

(f) The allowable bending stress in the compression flange (at the location of maximum computed bending stress) is most nearly

(A) 18.6 ksi
(B) 19.4 ksi
(C) 21.3 ksi
(D) 23.8 ksi

(g) The maximum computed shear stress due to service loads is most nearly

(A) 4.6 ksi
(B) 5.8 ksi
(C) 6.8 ksi
(D) 7.1 ksi

(h) The allowable shear stress is most nearly

(A) 1.6 ksi
(B) 2.9 ksi
(C) 4.5 ksi
(D) 8.2 ksi

(i) The plate girder is adequate for

(A) moment only
(B) shear only
(C) both moment and shear
(D) neither moment nor shear

(j) Assuming the web plate thickness is changed to $^5/_8$ in, the plate girder is adequate for

(A) moment only
(B) shear only
(C) both moment and shear
(D) neither moment nor shear

4. (*Time limit: one hour*) A simply supported welded plate girder having a 72 ft span carries a uniformly distributed load of 2 kips/ft and a concentrated load of 120 kips 24 ft from the left end. A572 grade-50 steel is used. The web is $^3/_8$ in thick and 48 in deep. Each flange is $1^1/_2$ in thick and 16 in wide. Bearing stiffeners are provided at the reaction points and at the concentrated load.

(a) The maximum shear force the plate girder can carry without any intermediate stiffener is most nearly

(A) 83 kips
(B) 91 kips
(C) 100 kips
(D) 120 kips

(b) The maximum allowable spacing for the first intermediate stiffener from the left end is most nearly

(A) 51 in
(B) 55 in
(C) 60 in
(D) 70 in

(c) If the first intermediate stiffener is placed 48 in from the left end, the maximum allowable spacing for the next stiffener is most nearly

(A) 50 in
(B) 55 in
(C) 60 in
(D) 70 in

(d) Assuming a single plate is placed 48 in from the left end, the minimum size required for the first intermediate stiffener from the left end is most nearly

(A) 1.3 in^2
(B) 1.8 in^2
(C) 2.3 in^2
(D) 2.8 in^2

(e) The maximum bending stress at the concentrated load is most nearly

(A) 18 ksi
(B) 24 ksi
(C) 29 ksi
(D) 32 ksi

(f) Assuming that intermediate stiffeners are provided every 4 ft throughout the girder span, the actual shear stress and the allowable tensile stress, respectively, at the concentrated load, are most nearly

(A) 4.6 ksi and 32 ksi
(B) 5.8 ksi and 29 ksi
(C) 6.5 ksi and 30 ksi
(D) 7.3 ksi and 26 ksi

(g) The bearing stiffener size that will satisfy the width-thickness ratio requirement is most nearly

(A) $^1/_4$ in × 6 in
(B) $^3/_8$ in × 7 in
(C) $^3/_8$ in × $7^1/_2$ in
(D) $^1/_2$ in × $6^1/_2$ in

(h) Assuming two matched $^1/_2$ in × 7 in bars are used for end bearing stiffening, the total radius of gyration of the bearing stiffener is most nearly

(A) 2.0 in
(B) 2.6 in
(C) 3.8 in
(D) 4.1 in

(i) The computed stress and the allowable stress, respectively, in the end bearing stiffener are, most nearly

(A) 7.6 ksi and 17.8 ksi
(B) 8.4 ksi and 21.3 ksi
(C) 10.1 ksi and 24.2 ksi
(D) 17.5 ksi and 29.3 ksi

(j) Assume that no intermediate stiffeners are to be used and only bearing stiffeners are to be provided at the beam ends and the concentrated load. The required minimum thickness of the web is most nearly

(A) $^7/_{16}$ in
(B) $^1/_2$ in
(C) $^9/_{16}$ in
(D) $^5/_8$ in

SOLUTIONS

1. (a) Assume adequate stiffeners as required to achieve $F_v = 0.40F_y$.

$$F_v = 0.4F_y = (0.4)(36 \text{ ksi}) = 14.4 \text{ ksi}$$

The shear is

$$V = \left(\tfrac{1}{2}\right)(40w) = 20w$$

The shear stress is given by Eq. 63.3.

$$f_v = \frac{V_{\text{max}}}{ht_w} = F_v$$

$$\frac{20w}{(48 \text{ in})(0.375 \text{ in})} = 14.4 \text{ kips/in}^2$$

$$w = 12.96 \text{ kips/ft}$$

The girder weight per foot is

$$(1 \text{ ft})\left(\left(\frac{\frac{3}{8} \text{ in}}{12 \frac{\text{in}}{\text{ft}}}\right)\left(\frac{16 \text{ in} + 16 \text{ in} + 48 \text{ in}}{12 \frac{\text{in}}{\text{ft}}}\right)\right) \times \left(\frac{490 \frac{\text{lbf}}{\text{ft}^3}}{1000 \frac{\text{lbf}}{\text{kip}}}\right) = 0.1 \text{ kip/ft}$$

$$w_{\text{load}} = 12.96 \frac{\text{kips}}{\text{ft}} - 0.1 \frac{\text{kip}}{\text{ft}} = \boxed{12.86 \text{ kips/ft}}$$

(b)

$$a = (4 \text{ ft})\left(12 \frac{\text{in}}{\text{ft}}\right) = 48 \text{ in}$$

$$\frac{a}{h} = \frac{48 \text{ in}}{48 \text{ in}} = 1$$

From Eq. 63.7 or 63.8,

$$k_v = 4.00 + \frac{5.34}{\left(\frac{a}{h}\right)^2} = 4.00 + \frac{5.34}{(1)^2} = 9.34$$

Assume $C_v < 0.8$. From Eq. 63.5,

$$C_v = \frac{45{,}000k_v}{F_y\left(\frac{h}{t_w}\right)^2} = \frac{\left(45{,}000 \frac{\text{kips}}{\text{in}^2}\right)(9.34)}{\left(36 \frac{\text{kips}}{\text{in}^2}\right)\left(\frac{48 \text{ in}}{0.375 \text{ in}}\right)^2} = 0.71$$

From Eq. 63.22, the steel area for stiffener pairs is

$$A_{\text{st}} = \left(\frac{1-C_v}{2}\right)\left(\frac{a}{h} - \frac{\left(\frac{a}{h}\right)^2}{\sqrt{1+\left(\frac{a}{h}\right)^2}}\right) \times \left(\frac{F_{y,\text{web}}}{F_{y,\text{stiffener}}}\right)Dht_w$$

$$= \left(\frac{1-0.71}{2}\right)\left(1 - \frac{(1)^2}{\sqrt{1+(1)^2}}\right)\left(\frac{36 \text{ ksi}}{36 \text{ ksi}}\right) \times (1)(48 \text{ in})\left(\tfrac{3}{8} \text{ in}\right)$$

$$= 0.76 \text{ in}^2$$

The total width of $^3/_8$ in thick plates is

$$\frac{0.76 \text{ in}^2}{\frac{3}{8} \text{ in}} = 2.0 \text{ in}$$

Check the minimum moment of inertia. From Eq. 63.25,

$$I_{\text{st}} \geq \left(\frac{h}{50}\right)^4 = \left(\frac{48 \text{ in}}{50}\right)^4 = 0.85 \text{ in}^4$$

From Eq. 63.24,

$$b_{\text{st}} = \left(\frac{12I_{\text{st}}}{t_{\text{st}}}\right)^{1/3} = \left(\frac{(12)(0.85 \text{ in}^4)}{\frac{3}{8} \text{ in}}\right)^{1/3} = 3.0 \text{ in}$$

Referring to *AISC Specification* Sec. G4, the length required is $h - w - 6t_w = 48 \text{ in} - \frac{5}{16} \text{ in} - (6)(\frac{3}{8} \text{ in}) = 45.4$ [use 46 in].

($^5/_{16}$ in is the minimum weld size required for the connection between the web and the flange.)

Use $^3/_8$ in × $1^1/_2$ in × 3 ft 10 in stiffeners on each side.

2. (a) *step 1:* Assume the weight of the plate girder will be approximately 350 lbf/ft (0.35 kip/ft). The distributed load, including the weight of the girder, is

$$w = 2 \frac{\text{kips}}{\text{ft}} + 0.35 \frac{\text{kip}}{\text{ft}} = 2.35 \text{ kips/ft}$$

The maximum shear is

$$V_{\text{max}} = \tfrac{1}{2}(wL + P) = \left(\tfrac{1}{2}\right)\left(\left(2.35 \frac{\text{kips}}{\text{ft}}\right)(60 \text{ ft}) + 135 \text{ kips}\right) = 138 \text{ kips}$$

The maximum moment is

$$M_{max} = \left(\frac{L}{4}\right)\left(\frac{wL}{2}+P\right)$$
$$= \left(\frac{60\text{ ft}}{4}\right) \times \left(\frac{\left(2.35\ \frac{\text{kips}}{\text{ft}}\right)(60\text{ ft})}{2} + 135\text{ kips}\right)$$
$$= 3083\text{ ft-kips}$$

Assume an allowable bending stress of

$$F_b = 0.6F_y = (0.6)(36\text{ ksi}) = 21.6\text{ ksi}$$

The required section modulus is

$$S_x = \frac{M_{max}}{F_b} = \frac{(3083\text{ ft-kips})\left(12\ \frac{\text{in}}{\text{ft}}\right)}{21.6\ \frac{\text{kips}}{\text{in}^2}}$$
$$= 1713\text{ in}^3$$

From the *AISC Manual*, Part 2, Table of Welded Plate Girders: Dimensions and Properties, try a girder having a $3/8$ in × 52 in web with $1^3/4$ in × 18 in flange plates. The overall depth is

$$52\text{ in} + (2)(1.75\text{ in}) = 55.5\text{ in} < 60\text{ in} \quad [\text{OK}]$$
$$S_x = 1800\text{ in}^3 > 1713\text{ in}^3$$

step 2: Determine the web required. Assumed web depth, h, is 52 in and web thickness, t_w, is 0.375 in.

$$A_w = ht_w = (52\text{ in})(0.375\text{ in})$$
$$= 19.5\text{ in}^2$$
$$\frac{h}{t_w} = \frac{52\text{ in}}{0.375\text{ in}} = 139$$

Find the minimum thickness of the web. From Eq. 63.2 or *AISC Manual*, Part 5, Table 5 of Numerical Values, $h/t_w = 322$. The minimum t_w is

$$\frac{52\text{ in}}{322} = 0.16\text{ in} < 0.375\text{ in} \quad [\text{OK}]$$

Since no intermediate stiffeners are to be used, from *AISC Specifications* Sec. F5, $h/t_w = 260$. The minimum t_w is

$$\frac{52\text{ in}}{260} = 0.2\text{ in} < 0.375\text{ in} \quad [\text{OK}]$$

step 3: For no reduction in flange stress, from *AISC Specifications* Sec. G2,

$$\frac{h}{t_w} \le \frac{970}{\sqrt{F_y}}$$
$$= \frac{970}{\sqrt{36\ \frac{\text{kips}}{\text{in}^2}}} = 162$$

The corresponding thickness of the web is

$$t_w = \frac{52\text{ in}}{162} = 0.32\text{ in} < 0.375\text{ in} \quad [\text{OK}]$$

Check the web shear stress. From the *AISC Manual*, Part 2, Table 2-36, interpolating for $h/t_w = 139$ and $a/h > 3$ (since no intermediate stiffeners are to be provided), $F_v = 4.27$ ksi. The allowable vertical shear is

$$V = F_vA_w = \left(4.27\ \frac{\text{kips}}{\text{in}^2}\right)(19.5\text{ in}^2)$$
$$= 83.3\text{ kips} < 138\text{ kips} \quad [\text{no good}]$$

Try a thicker web plate: $7/16$ in × 52 in.

$$A_w = \left(\tfrac{7}{16}\text{ in}\right)(52\text{ in}) = 22.75\text{ in}^2$$
$$\frac{h}{t_w} = \frac{52\text{ in}}{0.4375\text{ in}} = 119$$

Using *AISC Table* 1-36 with $h/t_w = 119$ and $a/h > 3$, $F_v = 5.9$ ksi. The allowable vertical shear is

$$V = F_vA_w = \left(5.9\ \frac{\text{kips}}{\text{in}^2}\right)(22.75\text{ in}^2)$$
$$= 134\text{ kips} < 138\text{ kips} \quad [\text{no good}]$$

Try a thicker web plate: $1/2$ in × 52 in.

$$A_w = \left(\tfrac{1}{2}\text{ in}\right)(52\text{ in}) = 26\text{ in}^2$$
$$\frac{h}{t_w} = \frac{52\text{ in}}{0.5\text{ in}} = 104$$

Using *AISC Table* 1-36 with $h/t_w = 104$ and $a/h > 3$, $F_v = 7.74$ ksi. The allowable vertical shear is

$$V = F_vA_w = \left(7.74\ \frac{\text{kips}}{\text{in}^2}\right)(26\text{ in}^2)$$
$$= 201\text{ kips} > 138\text{ kips} \quad [\text{OK}]$$

Determine the flange required. Using Eq. 63.12, the required flange area is

$$A_f = \frac{M_x}{F_b h} - \frac{t_w h}{6}$$
$$= \frac{(3083 \text{ ft-kips})\left(12 \frac{\text{in}}{\text{ft}}\right)}{\left(21.6 \frac{\text{kips}}{\text{in}^2}\right)(52 \text{ in})} - \frac{(0.5 \text{ in})(52 \text{ in})}{6}$$
$$= 28.6 \text{ in}^2$$

For a $1^3/_4$ in × 18 in flange plate,

$$A_f = 1^3/_4 \text{ in} \times 18 \text{ in}$$
$$= 31.5 \text{ in}^2 > 28.6 \text{ in}^2 \quad \text{[OK]}$$

step 4: Check the bending stress.

step 4.1: Calculate the gross moment of inertia.

section	dimensions (in)	A (in^2)	y (in)	Ay^2 (in^4)	I_o (in^4)	I_{gr} (in^4)
web	$\frac{1}{2} \times 52$	26	0	0	5859	5859
flange	$1\frac{3}{4} \times 18$	31.5	26.875	22,751	8	22,759
flange	$1\frac{3}{4} \times 18$	31.5	26.875	22,751	8	22,759
				total		51,377

The distance from the neutral axis to the extreme fiber is

$$c = \frac{52 \text{ in}}{2} + 1\tfrac{3}{4} \text{ in} = 27.75 \text{ in}$$

The section modulus of the plate girder cross section is

$$S_x = \frac{I}{c} = \frac{51{,}377 \text{ in}^4}{27.75 \text{ in}}$$
$$= 1851 \text{ in}^3$$

The maximum bending stress is

$$f_b = \frac{(3083 \text{ ft-kips})\left(12 \frac{\text{in}}{\text{ft}}\right)}{1851 \text{ in}^3}$$
$$= 20.0 \text{ kips/in}^2$$

step 4.2: Check for adequacy against local buckling using equations 63.14 and 63.15, and *AISC Specifications* Table B5-1. (Use Ftn. e.)

$$\frac{h}{t_w} = \frac{52 \text{ in}}{0.5 \text{ in}} = 104 > 70$$

$$k_c = \frac{4.05}{\left(\frac{h}{t_w}\right)^{0.46}} = \frac{4.05}{(104)^{0.46}}$$
$$= 0.48$$

$$\frac{95}{\sqrt{\frac{F_y}{k_c}}} = \frac{95}{\sqrt{\frac{36 \frac{\text{kips}}{\text{in}^2}}{0.48}}}$$
$$= 10.97$$

$$\frac{b_f}{2t_f} = \frac{18 \text{ in}}{(2)(1.75 \text{ in})} = 5.14$$

$$5.14 < 10.97 \quad \text{[OK]}$$

step 4.3: Calculate allowable bending stress based on lateral buckling criterion (*AISC Specification* Sec. F1). The unbraced length of the compression flange is

$$l_b = \left(\frac{60 \text{ ft}}{2}\right)\left(12 \frac{\text{in}}{\text{ft}}\right) = 360 \text{ in}$$

$$L_c = \frac{76 b_f}{\sqrt{F_y}}$$
$$= \frac{(76)(18 \text{ in})}{\sqrt{36 \frac{\text{kips}}{\text{in}^2}}}$$
$$= 228 \text{ in}$$

Since $l_b > L_c$, *AISC Specifications* Sec. F1.3 applies.

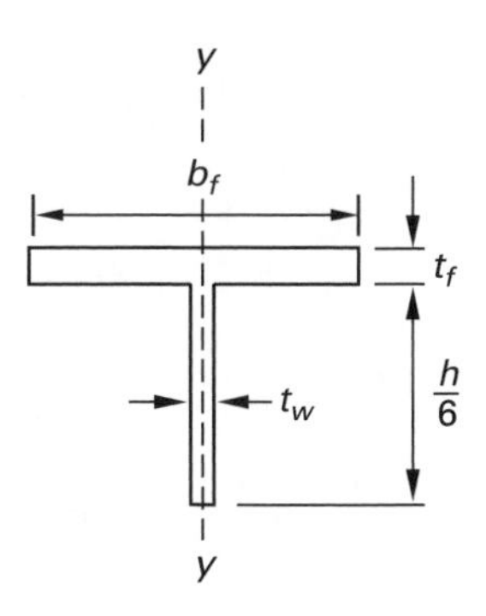

Determine the moment of inertia about the y-y axis of the flange plus $^1/_6$ web. Neglecting the web portion,

$$I_{oy} = \frac{t_f {b_f}^3}{12}$$
$$= \frac{(1.75 \text{ in})(18 \text{ in})^3}{12} = 850.5 \text{ in}^4$$

$$A_f + \tfrac{1}{6}A_w = 31.5\ \text{in}^2 + \left(\tfrac{1}{6}\right)(26\ \text{in}^2) = 35.8\ \text{in}^2$$

$$r_T = \sqrt{\frac{I_{oy}}{A_f + \frac{1}{6}A_w}} = \sqrt{\frac{850.5\ \text{in}^4}{31.5\ \text{in}^2 + \left(\frac{1}{6}\right)(26\ \text{in}^2)}} = 4.87\ \text{in}$$

$$\frac{l_b}{r_T} = \frac{360\ \text{in}}{4.87\ \text{in}} = 73.92$$

For the unbraced length of the compression flange, $M_1 = 0$ and $M_2 = M_{max} = 3083$ ft-kips, so $M_1/M_2 = 0$. From *AISC Manual*, Part 5, Table 6, $C_b = 1.75$.

Referring to the *AISC Manual*, Part 5, Table 5, for $F_y = 36$ ksi steel,

$$53\sqrt{C_b} = 53\sqrt{1.75} = 70.1$$
$$119\sqrt{C_b} = 119\sqrt{1.75} = 157.4$$
$$70.1 < 73.92 < 157.4$$

Using *AISC Specifications* Eq. F1-6,

$$F_b = \left(\frac{2}{3} - \frac{F_y\left(\frac{l}{r_T}\right)^2}{1530\times 10^3\,C_b}\right)F_y = \left(\frac{2}{3} - \frac{\left(36\ \frac{\text{kips}}{\text{in}^2}\right)(73.92)^2}{(1530\times 10^3)(1.75)}\right)\left(36\ \frac{\text{kips}}{\text{in}^2}\right) = 21.4\ \text{ksi}$$

Using *AISC Specifications* Eq. F1-8,

$$F_b = \frac{12\times 10^3 C_b}{l\left(\frac{d}{A_f}\right)} = \frac{(12\times 10^3)(1.75)}{(360\ \text{in})\left(\frac{55.5\ \text{in}}{31.5\ \text{in}^2}\right)} = 33.1\ \text{ksi}$$

However, F_b must not exceed $0.6F_y = (0.6) \times 36\ \text{ksi}) = 21.6$ ksi, so $F_b = 21.6$ ksi. Pick the larger of the two values from AISC Eqs. F1-6 and F1-8.

$$F_b = 21.6\ \text{ksi}$$
$$f_b = 20.0\ \text{ksi} < 21.6\ \text{ksi} \quad \text{[OK]}$$

For the web, one plate = **1/2 in × 52 in.**

For the flanges, two plates = **1 3/4 in × 18 in.**

(b) *step 5:* Design the bearing stiffeners.

step 5.1: Bearing stiffeners are required at unframed girder ends.

step 5.2: Check the bearing under concentrated load. Assume a web-to-flange weld size of $w = 5/16$ in and length of bearing of $N = 10$ in.

$$k = t_f + w = 1.75\ \text{in} + 0.3125\ \text{in} = 2.063\ \text{in}$$

Check local web yielding (*AISC Specifications* Eq. K1-2).

$$\frac{R}{t_w(N+5k)} = \frac{135\ \text{kips}}{(0.5\ \text{in})\big(10\ \text{in} + (5)(2.063\ \text{in})\big)} = 13.3\ \text{ksi}$$

$$13.3\ \text{ksi} < 0.66F_y = (0.66)(36\ \text{ksi}) = 23.76\ \text{ksi} \quad \text{[OK]}$$

So bearing stiffeners are not required.

Check web crippling (*AISC Specifications* Eq. K1-4). Bearing stiffeners are required if the concentrated load exceeds the value of R computed as

$$R = 67.5t_w^2\left(1 + 3\left(\frac{N}{d}\right)\left(\frac{t_w}{t_f}\right)^{1.5}\right)\sqrt{\frac{F_{yw}t_f}{t_w}}$$
$$= (67.5)(0.5\ \text{in})^2 \times\left(1 + (3)\left(\frac{10\ \text{in}}{55.5\ \text{in}}\right)\left(\frac{0.5\ \text{in}}{1.75\ \text{in}}\right)^{1.5}\right) \times\sqrt{\frac{\left(36\ \frac{\text{kips}}{\text{in}^2}\right)(1.75\ \text{in})}{0.5\ \text{in}}}$$
$$= 205\ \text{kips}$$

135 kips < 205 kips, so bearing stiffeners are not required.

Check sidesway web buckling (*AISC Specifications* Eq. K1.5).

$$d_c = d - 2k = 55.5\ \text{in} - (2)(2.063\ \text{in}) = 51.37\ \text{in}$$

Assume the loaded flange is restrained against rotation.

The longest unbraced length is

$$l = \left(\frac{60\ \text{ft}}{2}\right)\left(12\ \frac{\text{in}}{\text{ft}}\right) = 360\ \text{in}$$

$$\frac{\frac{d_c}{t_w}}{\frac{l}{b_f}} = \frac{\frac{51.37\ \text{in}}{0.5\ \text{in}}}{\frac{360\ \text{in}}{18\ \text{in}}} = 5.14 > 2.3$$

Bearing stiffeners are not required.

step 5.3: Determine the bearing stiffener size. At the end of the girder, try two PL $7^1/_2 \times {}^1/_2$ stiffeners.

Check the width-thickness ratio using Eq. 63.26.

$$\frac{b}{t} = \frac{7.5 \text{ in}}{0.5 \text{ in}} = 15 < \frac{95}{\sqrt{F_y}}$$
$$= \frac{95}{\sqrt{36 \; \frac{\text{kips}}{\text{in}^2}}} = 15.8 \quad [\text{OK}]$$

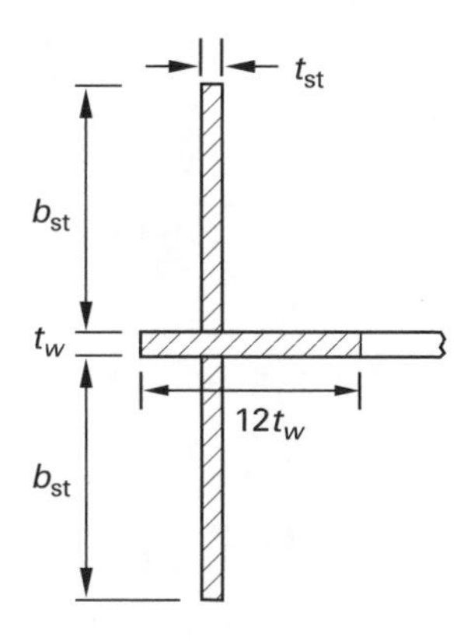

Check the compressive stress. For two stiffeners, one on either side of the web, the total stiffener length is

$$h = (2)\left(7\tfrac{1}{2} \text{ in}\right) + 0.5 \text{ in} = 15.5 \text{ in}$$
$$I = \frac{bh^3}{12} = \frac{(0.5 \text{ in})(15.5 \text{ in})^3}{12}$$
$$= 155.2 \text{ in}^4$$

$$A_{\text{eff}} = 2b_{\text{st}}t_{\text{st}} + 12t_w^2$$
$$= (2)(7.5 \text{ in})(0.5 \text{ in}) + (12)(0.5 \text{ in})^2$$
$$= 10.5 \text{ in}^2$$

$$r = \sqrt{\frac{I}{A_{\text{eff}}}} = \sqrt{\frac{155.2 \text{ in}^4}{10.5 \text{ in}^2}}$$
$$= 3.8 \text{ in}$$

$$kl = 0.75h = (0.75)(52 \text{ in}) = 39 \text{ in}$$
$$\frac{kl}{r} = \frac{39 \text{ in}}{3.8 \text{ in}} = 10.3$$

From the *AISC Manual*, Part 3, Table C-36, by interpolation, $F_a = 21.12$ ksi.

The reaction at the beam end is 138 kips.

$$f_a = \frac{138 \text{ kips}}{10.5 \text{ in}^2} = 13.14 \text{ ksi} < 21.12 \text{ ksi} \quad [\text{OK}]$$

Use (for bearing stiffeners) two $^1/_2$ in × $7^1/_2$ in × 4 ft $7^1/_4$ in bars, one on each side of the web, with close bearing on flanges receiving reactions.

3. (a) $$\frac{h}{t_w} = \frac{64 \text{ in}}{0.375 \text{ in}} = \boxed{170.76} \quad [\text{say } 171]$$

The allowable limit on a web depth-thickness ratio from the *AISC Manual* Part 5, Table 5 (or Eq. 63.2) is 322.

The answer is (A).

(b) The moment of inertia of the plate girder is

$$I = \sum \left(\frac{bh^3}{12} + Ad^2\right)$$
$$= (2)\left(\left(\tfrac{1}{12}\right)(16 \text{ in})(1.5 \text{ in})^3 + (16 \text{ in})(1.5 \text{ in})(32.75 \text{ in})^2\right)$$
$$+ \left(\tfrac{1}{12}\right)(0.375 \text{ in})(64 \text{ in})^3$$
$$= \boxed{59{,}684 \text{ in}^4}$$

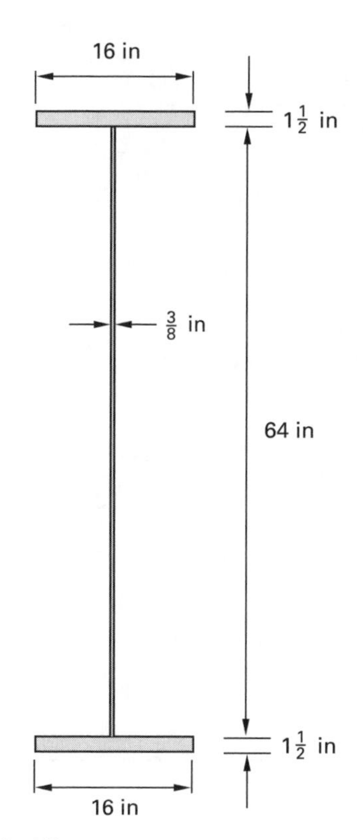

The answer is (C).

(c) The maximum service load shear is 110 kips.

The answer is (B).

(See the following illustration.)

(d) The maximum service load moment is

1900 ft-kips.

The answer is (C).

(e) $c = \frac{1}{2}(\text{total depth of plate girder}) = \left(\frac{1}{2}\right)(67 \text{ in})$
$= 33.5 \text{ in}$

The maximum bending stress is

$$f_b = \frac{Mc}{I} = \frac{(1900 \text{ ft-kips})\left(12 \frac{\text{in}}{\text{ft}}\right)(33.5 \text{ in})}{59{,}684 \text{ in}^4} = \boxed{12.8 \text{ kips/in}^2}$$

The answer is (B).

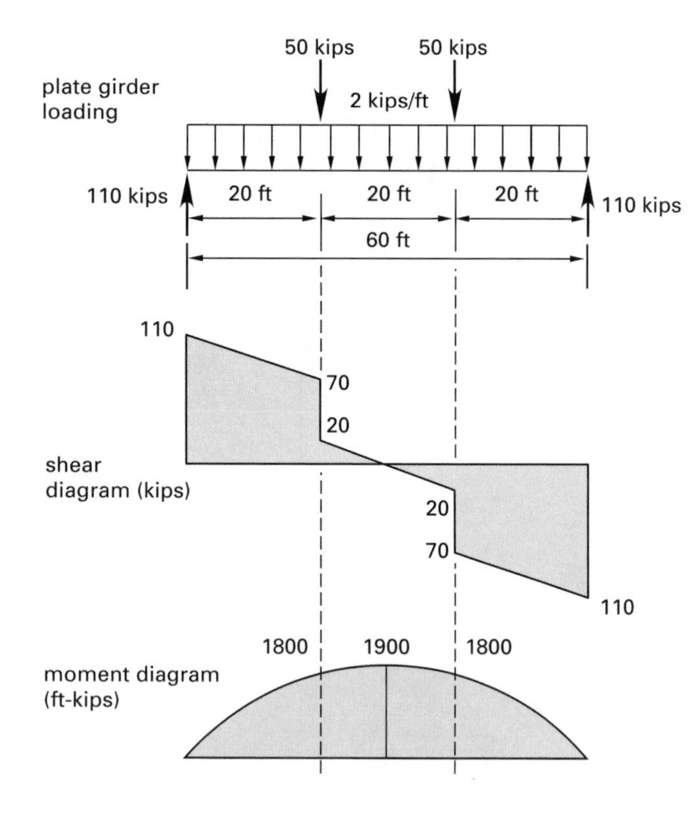

(f)
$$A_f = (16 \text{ in})(1.5 \text{ in}) = 24 \text{ in}^2$$
$$A_w = (64 \text{ in})(0.375 \text{ in}) = 24 \text{ in}^2$$

From the *AISC Specifications* F1.3, the moment of inertia of compression flange plus $^1/_6$ web about the Y-Y axis (neglecting the moment of inertia of the $^1/_6$ web part) is

$$I_{oy} = \frac{t_f b_f^3}{12} = \frac{(1.5 \text{ in})(16 \text{ in})^3}{12} = 512 \text{ in}^4$$

$$A_f + \tfrac{1}{6}A_w = 24 \text{ in}^2 + \left(\tfrac{1}{6}\right)(24 \text{ in}^2) = 28 \text{ in}^2$$

$$r_T = \sqrt{\frac{I_{oy}}{A_f + \frac{1}{6}A_w}} = \sqrt{\frac{512 \text{ in}^4}{28 \text{ in}^2}} = 4.276 \text{ in}$$

In the middle 20 ft panel of the plate girder, $M_{\max} =$ 1900 ft-kips and $M_1 = M_2 = 1800$ ft-kips, so $C_b = 1$.

$$\frac{l}{r_T} = \frac{(20 \text{ ft})\left(12 \frac{\text{in}}{\text{ft}}\right)}{4.276 \text{ in}} = 56.1$$

From the *AISC Manual*, Part 5, Table 5 of Numerical Values,

$$53\sqrt{C_b} = 53\sqrt{1} = 53$$
$$119\sqrt{C_b} = 119\sqrt{1} = 119$$
$$53 < 56.1 < 119$$

The allowable stress based upon the lateral buckling criterion, using AISC Eq. F1-6, is

$$F_b = \left(\frac{2}{3} - \frac{F_y\left(\frac{l}{r_T}\right)^2}{(1530 \times 10^3)C_b}\right)F_y = \left(\frac{2}{3} - \frac{(36 \text{ ksi})(56.1)^2}{(1530 \times 10^3)(1.0)}\right)(36 \text{ ksi}) = 21.33 \text{ ksi} < 0.6F_y = 21.6 \text{ ksi}$$

The allowable bending stress in the compression flange must be reduced if h/t_w exceeds $970/\sqrt{F_b}$.

$$\frac{h}{t_w} = 171 \quad \text{[from part (a)]}$$

$$\frac{970}{\sqrt{F_b}} = \frac{970}{\sqrt{21.3 \frac{\text{kips}}{\text{in}^2}}} = 210$$

Therefore, no reduction in flange stress is required.

$$F_b' = \boxed{21.33 \text{ ksi}}$$

The answer is (C).

(g) From Eq. 63.3, the maximum computed shear stress is

$$f_v = \frac{V_{\text{max}}}{ht_w} = \frac{110 \text{ kips}}{(64 \text{ in})(0.375 \text{ in})} = \boxed{4.58 \text{ kips/in}^2}$$

The answer is (A).

(h) For $h/t_w = 171$ and $a/h > 3$ (since no intermediate stiffeners are provided), interpolating from the *AISC Manual*, Part 2, Table 1-36, the allowable web shear stress is $F_v = \boxed{2.87 \text{ ksi.}}$

The answer is (B).

(i) From parts (e) and (f), $f_b < F_b$ (12.8 ksi < 19.4 ksi).

Therefore, the plate girder is adequate for moments.

From parts (g) and (h), $f_v > F_v$ (4.58 ksi > 2.87 ksi).

Therefore, the plate girder is inadequate for shear.

The answer is (A).

(j) If $t_w = 5/8$ in,

$$\frac{h}{t_w} = \frac{64 \text{ in}}{\frac{5}{8} \text{ in}} = 102.4$$

From the *AISC Manual*, Part 2, Table 1-36, the allowable web shear stress is $F_v = 7.96$ ksi (by interpolation). From Eq. 63.3, the maximum computed shear stress is

$$f_v = \frac{V_{\text{max}}}{ht_w} = \frac{110 \text{ kips}}{(64 \text{ in})(0.625 \text{ in})} = 2.75 \ \frac{\text{kips}}{\text{in}^2} < F_v \quad [\text{OK}]$$

Since the thinner web was adequate for moment, the thicker web will also be adequate.

The answer is (C).

4. (a)

$$A_w = ht_w = (48 \text{ in})\left(\tfrac{3}{8} \text{ in}\right) = 18 \text{ in}^2$$

$$\frac{h}{t_w} = \frac{48 \text{ in}}{0.375 \text{ in}} = 128$$

If no intermediate stiffeners are used, $a/h > 3$. From the *AISC Manual*, Part 2, Table 1-50 (by interpolation), $F_v = 5.08 \text{ ksi} = f_v$. The maximum allowable shear without intermediate stiffeners is obtained using Eq. 63.3.

$$V_{\text{allowable}} = F_v h t_w = \left(5.08 \ \frac{\text{kips}}{\text{in}^2}\right)(48 \text{ in})\left(\tfrac{3}{8} \text{ in}\right) = \boxed{91.4 \text{ kips}}$$

The answer is (B).

$$V_{\text{max}} = 152 \text{ kips} \quad [\text{at left support}]$$

(b)

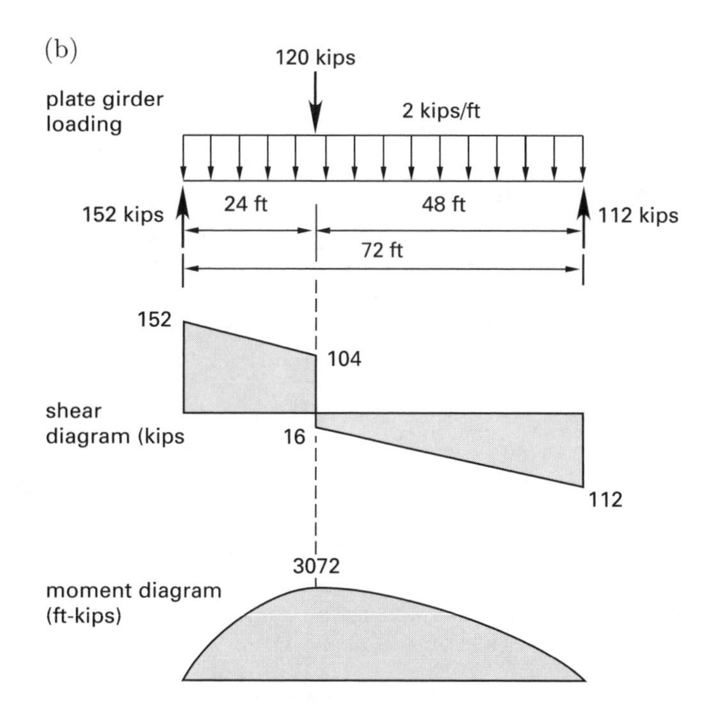

From Eq. 63.3,

$$f_v = \frac{V_{\text{max}}}{ht_w} = \frac{152 \text{ kips}}{(48 \text{ in})(0.375 \text{ in})} = 8.44 \text{ kips/in}^2$$

Set $F_v = f_v = 8.44$ ksi.

Tension field action is not permitted in the end panel (AISC Specification Section G4). From the *AISC Manual*, Part 2, Table 1-50 for $h/t_w = 128$ and $F_v = 8.44$ ksi (by interpolation), $a/h = 1.06$.

$$a = (1.06)(48 \text{ in}) = \boxed{50.9 \text{ in}}$$

The answer is (A).

(c) If the first intermediate stiffener is placed 48 in from the left end of the girder, the shear at that location is

$$V = 152 \text{ kips} - \left(2 \ \frac{\text{kips}}{\text{ft}}\right)\left(\frac{48 \text{ in}}{12 \ \frac{\text{in}}{\text{ft}}}\right) = 144 \text{ kips}$$

From Eq. 63.3,

$$f_v = \frac{V_{\text{max}}}{ht_w} = \frac{144 \text{ kips}}{(48 \text{ in})(0.375 \text{ in})} = 8.0 \text{ kips/in}^2$$

Set $F_v = f_v = 8.0$ ksi.

From the *AISC Manual*, Part 2, Table 1-50 for $h/t_w = 128$ and $F_v = 8.0$ ksi (by interpolation), $a/h = 1.14$.

$$a = (1.14)(48 \text{ in}) = \boxed{54.7 \text{ in}}$$

The answer is (B).

(d) The first intermediate stiffener is placed 48 in from left end. Therefore, $a = 48$ in.

$$\frac{a}{h} = \frac{48 \text{ in}}{48 \text{ in}} = 1$$

As in part (c), use $a/h = 1$ and $h/t_w = 128$ to interpolate $F_v = 8.9$ ksi from *AISC Manual*, Part 2, Table 1-50.

Since a part of Eq. 63.23 represents the fraction of the web area (tabulated in the *AISC Manual*) for the gross area of stiffeners and since the same grade of steel is used for all elements (flange, web, and stiffeners) in this problem, the equation can be rewritten as

$$A_{\text{st}} = (\text{fraction web area})(Dht_w)$$

For $h/t_w = 128$ and $a/h = 1$, from the *AISC Manual*, Part 2, Table 2-50 (by interpolation), the percentage web area is 7.14%. $D = 2.4$ for single plate stiffeners.

$$\begin{aligned} A_{\text{st}} &= (0.0714)(2.4)(48 \text{ in})(0.375 \text{ in}) \\ &= 3.084 \text{ in}^2 \quad [\text{without reduction}] \end{aligned}$$

The reduced area A_{st} is calculated with $f_v = 8.0$ ksi (calculated in Part (c)) and $F_v = 8.9$ ksi. From the *AISC Manual*, Part 2, Table 2-50, by interpolation, using Eq. 63.23,

$$\begin{aligned} A'_{\text{st}} &= \left(\frac{f_v}{F_v}\right) A_{\text{st}} = \left(\frac{8.0 \text{ ksi}}{8.9 \text{ ksi}}\right)(3.084 \text{ in}^2) \\ &= \boxed{2.77 \text{ in}^2} \end{aligned}$$

The answer is (D).

(e) From the moment diagram, the moment (maximum) at the concentrated load is $M_{\max} = 3072$ ft-kips.

From the *AISC Manual*, Part 2, Welded Plate Girder Dimensions and Properties, for 48 in × 3/8 in web and 16 in × 1 1/2 in flanges, $S_x = 1290 \text{ in}^3$. The maximum bending stress at the concentrated load is

$$\begin{aligned} f_b &= \frac{M_{\max}}{S_x} = \frac{(3072 \text{ ft-kips})\left(12 \frac{\text{in}}{\text{ft}}\right)}{1290 \text{ in}^3} \\ &= \boxed{28.6 \text{ ksi}} \end{aligned}$$

The answer is (C).

(f) From the shear diagram, the shear just to the left of the concentrated load is $V = 104$ kips. Using Eq. 63.3, the computed shear stress is

$$\begin{aligned} f_v &= \frac{V_{\max}}{ht_w} = \frac{104 \text{ kips}}{(48 \text{ in})(0.375 \text{ in})} \\ &= \boxed{5.8 \text{ ksi}} \end{aligned}$$

From part (d), $F_v = 8.9$ ksi and from part (e), $f_b = 28.6$ ksi. Using Eq. 63.21, the allowable tensile bending stress is

$$\begin{aligned} F_b &= \left(0.825 - 0.375\left(\frac{f_v}{F_v}\right)\right) F_y \\ &= \left(0.825 - (0.375)\left(\frac{5.78 \text{ ksi}}{8.9 \text{ ksi}}\right)\right) F_y = 0.58F_y \end{aligned}$$

(F_b cannot exceed $0.6F_y$.)

$$F_b = (0.58)(50 \text{ ksi}) = \boxed{29.07 \text{ ksi}}$$

The answer is (B).

(g) Using Eq. 63.26, the limiting width-thickness ratio for bearing stiffeners is

$$\frac{b_{\text{st}}}{t_{\text{st}}} = \frac{95}{\sqrt{F_y}} = \frac{95}{\sqrt{50 \frac{\text{kips}}{\text{in}^2}}} = 13.4$$

1/4 in × 6 in: $\frac{b}{t} = \frac{6 \text{ in}}{0.25 \text{ in}} = 24 > 13.4$ [no good]

3/8 in × 7 in: $\frac{b}{t} = \frac{7 \text{ in}}{0.375 \text{ in}} = 18.7 > 13.4$ [no good]

3/8 in × 7 1/2 in: $\frac{b}{t} = \frac{7.5 \text{ in}}{0.375 \text{ in}} = 20 > 13.4$ [no good]

$\boxed{\text{1/2 in} \times \text{6 1/2 in}}$: $\frac{b}{t} = \frac{6.5 \text{ in}}{0.5 \text{ in}} = 13 < 13.4$ [OK]

The answer is (D).

(h) $h = 7 \text{ in} + 7 \text{ in} + \frac{3}{8} \text{ in} = 14.375 \text{ in}$

$$\begin{aligned} I &= \frac{bh^3}{12} = \frac{(0.5 \text{ in})(14.375 \text{ in})^3}{12} = 123.77 \text{ in}^4 \\ A_{\text{eff}} &= 2A_{\text{st}} + 12t_w^2 \\ &= (2)((7 \text{ in})(0.5 \text{ in})) + (12)(0.375 \text{ in})^2 = 8.69 \text{ in}^2 \\ r &= \sqrt{\frac{I}{A_{\text{eff}}}} = \sqrt{\frac{123.77 \text{ in}^4}{8.69 \text{ in}^2}} \\ &= \boxed{3.77 \text{ in}} \end{aligned}$$

The answer is (C).

(i) The effective length of the bearing stiffener is

$$Kl = 0.75h = (0.75)(48 \text{ in}) = 36 \text{ in}$$

From part (h), the radius of gyration is $r = 3.77$ in.

$$\frac{Kl}{r} = \frac{36 \text{ in}}{3.77 \text{ in}} = 9.55$$

By interpolation, the allowable stress from the *AISC Manual*, Part 3, Table C-50, is $F_a = \boxed{29.3 \text{ ksi.}}$

The cross-sectional area of the "stiffener column" (see Fig. 63.3) for two end stiffeners is

$$\begin{aligned} A_c &= (2 \text{ stiffeners})(7 \text{ in})(0.5 \text{ in}) \\ &\quad + (12)(0.375 \text{ in})(0.375 \text{ in}) \\ &= 8.6875 \text{ in}^2 \end{aligned}$$

The actual stress is

$$f_a = \frac{V}{A_c} = \frac{152 \text{ kips}}{8.6875 \text{ in}^2} = \boxed{17.5 \text{ ksi}}$$

The answer is (D).

(j) For the left end panel,

$$a = (24 \text{ ft})\left(12 \frac{\text{in}}{\text{ft}}\right) = 288 \text{ in}$$

$$\frac{a}{h} = \frac{288 \text{ in}}{48 \text{ in}} = 6 > 3$$

For the right end panel,

$$a = (48 \text{ ft})\left(12 \frac{\text{in}}{\text{ft}}\right) = 576 \text{ in}$$

$$\frac{a}{h} = \frac{576 \text{ in}}{48 \text{ in}} = 12 > 3$$

If $t_w = 7/16$ in, then

$$\begin{aligned} \frac{h}{t_w} &= \frac{48 \text{ in}}{0.4375 \text{ in}} = 110 \\ A_w &= (48 \text{ in})(0.4375 \text{ in}) = 21 \text{ in}^2 \\ f_v &= \frac{V}{A_w} = \frac{152 \text{ kips}}{21 \text{ in}^2} = 7.24 \text{ kips/in}^2 \end{aligned}$$

From the *AISC Manual*, Part 2, Table 1-50, for $h/t_w = 110$ and $a/h > 3$, $F_v = 6.9$ ksi. Since $f_v > F_v$, this is inadequate.

Try $t_w = 1/2$ in.

$$\begin{aligned} \frac{h}{t_w} &= \frac{48 \text{ in}}{0.5 \text{ in}} = 96 \\ A_w &= (48 \text{ in})(0.5 \text{ in}) = 24 \text{ in}^2 \\ f_v &= \frac{V}{A_w} = \frac{152 \text{ kips}}{24 \text{ in}^2} = 6.33 \text{ kips/in}^2 \end{aligned}$$

From the *AISC Manual*, Part 2, Table 1-50, for $h/t_w = 96$ and $a/h > 3$ (by interpolation), $F_v = 9.1$ ksi. Since $f_v < F_v$, $\boxed{t_w = 1/2 \text{ in}}$ is adequate.

The answer is (B).

64 Structural Steel: Composite Steel Members

PRACTICE PROBLEMS

1. A cross section of a composite beam is shown. The steel beam is an A36 W 21 × 50, and the concrete strength is $f'_c = 4000$ psi. The modular ratio is 9. Assume shored construction and full composite action. Calculate the resisting moment of the beam.

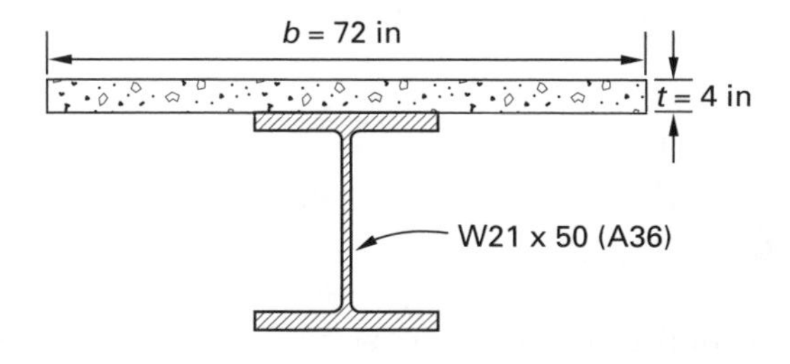

(A) 180 ft-kips
(B) 230 ft-kips
(C) 270 ft-kips
(D) 350 ft-kips

2. The floor framing system of an office building consists of A36 steel, simply supported beams of 40 ft span, spaced at 9 ft on center, that act compositely with a 4 in thick concrete slab ($f'_c = 3$ ksi). The unit weight of concrete is 150 lbf/ft^3, and the modular ratio is $n = 9$. No temporary shoring is to be used during construction. The loads to be applied after the concrete hardens are: floor live (service) load = 100 psf, and ceiling load = 10 psf. (a) Design an A36 interior beam such that the dead-load deflection is limited to 1 in and the live load deflection to $L/360$. (b) Design the shear connectors.

3. *(Time limit: one hour)* A building floor framing consists of A36 W 21 × 68 beams of 30 ft span placed at 12 ft center-to-center spacing. The beams are attached to a 4 in thick concrete slab ($f'_c = 3$ ksi, unit weight = 150 lbf/ft^3) through stud connectors. Assume shored construction, full composite action, and a modular ratio of 9. All loads applied after the concrete has hardened are uniformly distributed live loads. The following questions relate to an interior beam.

(a) The effective concrete flange width is most nearly

(A) 64 in
(B) 74 in
(C) 84 in
(D) 90 in

(b) The neutral axis of the transformed composite section is located at a distance (measured from the bottom of the steel beam) most nearly

(A) 18.9 in
(B) 20.5 in
(C) 21.7 in
(D) 22.4 in

(c) The moment of inertia of the transformed composite section is most nearly

(A) 3100 in^4
(B) 3600 in^4
(C) 4400 in^4
(D) 5100 in^4

(d) The allowable concrete stress is most nearly

(A) 0.85 ksi
(B) 0.97 ksi
(C) 1.1 ksi
(D) 1.3 ksi

(e) The allowable steel stress is most nearly

(A) 14 ksi
(B) 22 ksi
(C) 24 ksi
(D) 30 ksi

(f) The resisting moment of the composite beam is most nearly

(A) 240 ft-kips
(B) 380 ft-kips
(C) 410 ft-kips
(D) 520 ft-kips

(g) The maximum service dead load moment is most nearly

(A) 68 ft-kips
(B) 75 ft-kips
(C) 88 ft-kips
(D) 110 ft-kips

(h) The maximum service uniformly distributed live load capacity of the composite beam is most nearly

(A) 120 lbf/ft^2
(B) 180 lbf/ft^2
(C) 230 lbf/ft^2
(D) 270 lbf/ft^2

(i) For full composite action, the total horizontal shear V_h to be resisted at service loads between the point of maximum moment and the end of the beam is most nearly

(A) 260 kips
(B) 360 kips
(C) 440 kips
(D) 490 kips

(j) Assuming $^3/_4$ in dia. × 3 in stud connectors are used, the total number of studs required for the entire beam is most nearly

(A) 52
(B) 64
(C) 72
(D) 88

4. *(Time limit: one hour)* A floor framing system of composite construction (unshored) consists of simply supported, W 21 × 50 A36 steel beams of 40 ft span, equally spaced at 8 ft between beam centerlines, and a 3 in thick lightweight concrete slab (f'_c = 3000 psi, unit weight = 115 lbf/ft^3) on a 3 in deep formed steel deck with ribs running perpendicular to the beams. The average width of concrete rib (w_r) is 3 in. The steel beams are connected to the concrete slab by headed studs that have $^3/_4$ in diameters and 5 in lengths. The beams support a service dead load of 65 lbf/ft^2 due to slab and deck, and a live load of 100 lbf/ft^2 (applied after the concrete has hardened). Use a modular ratio of $n = 9$. The following questions relate to an interior beam.

(a) The moment of inertia of the composite section is most nearly

(A) 2600 in^4
(B) 3200 in^4
(C) 4100 in^4
(D) 5100 in^4

(b) Under combined service dead and live loads, the maximum computed concrete stress is most nearly

(A) 0.4 ksi
(B) 0.8 ksi
(C) 1.0 ksi
(D) 1.4 ksi

(c) Under combined service dead and live loads, the maximum beam stress at the bottom of the steel is most nearly

(A) 16 ksi
(B) 19 ksi
(C) 27 ksi
(D) 32 ksi

(d) Assuming full composite action, the total horizontal shear, V_h, to be resisted at service loads between the point of maximum moment and the point of zero moment is most nearly

(A) 270 kips
(B) 370 kips
(C) 410 kips
(D) 440 kips

(e) Assuming 1 stud per rib, the stud reduction factor is most nearly

(A) 0.41
(B) 0.46
(C) 0.51
(D) 0.57

(f) Assuming full composite action under service loads, the number of studs required between the beam ends is most nearly

(A) 70
(B) 82
(C) 94
(D) 98

(g) Assuming partial composite action (75%) under service loads, the total number of studs required between the beam ends is most nearly

(A) 52
(B) 62
(C) 72
(D) 82

(h) Assuming the ratio $V'_h/V_h = 0.75$, the effective section modulus, S_{eff}, is most nearly

(A) 110 in^3
(B) 150 in^3
(C) 180 in^3
(D) 220 in^3

(i) Considering full composite action, the service live load deflection is most nearly

(A) 0.34 in
(B) 0.49 in
(C) 0.58 in
(D) 0.68 in

(j) Assuming partial composite action (75%) and studs provided accordingly, the service live load deflection is most nearly

(A) 0.39 in
(B) 0.54 in
(C) 0.62 in
(D) 0.78 in

5. *(Time limit: one hour)* A portion of a bridge is supported by a simply supported plate girder. Bridge loads are applied after construction and are resisted by composite action. The girder is 80 ft long. It will not be shored during construction. Flange support is provided at 20 ft intervals. The girder is loaded by a moving 40 kip load and a 1 kip/ft dead load (which includes an allowable for the concrete). The concrete strength is 3000 psi. The modular ratio is 10. The yield stress for all steel plates is 36 ksi. Impact loading is negligible. Assume a worst-case stiffener placement. Determine if the plate girder is adequate.

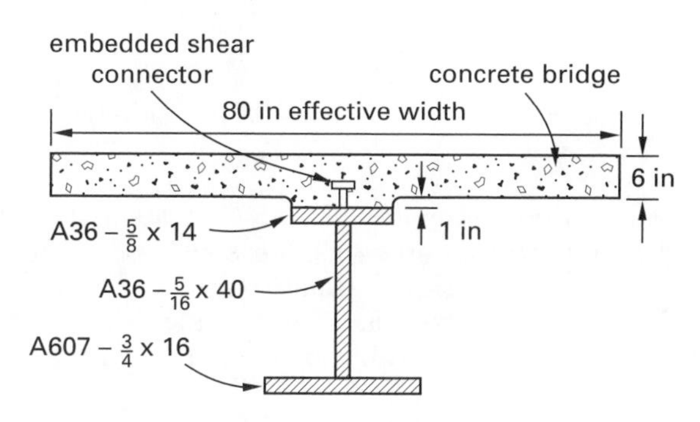

SOLUTIONS

(Refer to the *AISC Manual*, Part 2, Composite Design for Building Construction, General Notes, and Figs. 1 and 2, for notations used.)

1. The effective width of concrete slab is $b = 72$ in. The width of the equivalent steel area of the concrete slab is

$$\frac{b}{n} = \frac{72 \text{ in}}{9} = 8 \text{ in}$$

The concrete transformed area in compression is given by Eq. 64.4.

$$A_{\text{ctr}} = \left(\frac{b}{n}\right) t = (8 \text{ in})(4 \text{ in}) = 32 \text{ in}^2$$

The distance from the top of the steel beam to the centroid of the concrete area in compression is given by Eq. 64.7.

$$Y2 = \frac{t}{2} = \frac{4 \text{ in}}{2} = 2 \text{ in}$$

From the *AISC Manual*, for W 21 × 50, $d = 20.83$ in, $A_s = 14.7 \text{ in}^2$, and $I_s = 984 \text{ in}^4$.

The distance from the neutral axis of the transformed composite section (all steel) to the bottom of the steel beam is given by Eq. 64.3.

$$\begin{aligned}\overline{y}_b &= \frac{A_s\left(\frac{d}{2}\right) + A_{\text{ctr}}(d + Y2)}{A_s + A_{\text{ctr}}} \\ &= \frac{(14.7 \text{ in}^2)\left(\frac{20.83 \text{ in}}{2}\right) + (32 \text{ in}^2)(20.83 \text{ in} + 2 \text{ in})}{14.7 \text{ in}^2 + 32 \text{ in}^2} \\ &= 18.92 \text{ in}\end{aligned}$$

The moment of inertia of the transformed section is given by Eq. 64.5.

$$\begin{aligned}I_{\text{tr}} &= I_s + A_s\left(\overline{y}_b - \frac{d}{2}\right)^2 + \frac{\left(\frac{b}{n}\right)t^3}{12} \\ &\quad + A_{\text{ctr}}(d + Y2 - \overline{y}_b)^2 \\ &= 984 \text{ in}^4 + (14.7 \text{ in}^2)\left(18.92 \text{ in} - \left(\tfrac{1}{2}\right)(20.83 \text{ in})\right)^2 \\ &\quad + \frac{(8 \text{ in})(4 \text{ in})^3}{12} \\ &\quad + (32 \text{ in}^2)(20.83 \text{ in} + 2 \text{ in} - 18.92 \text{ in})^2 \\ &= 2579 \text{ in}^4\end{aligned}$$

(Alternatively, I_{tr} can be obtained by interpolation from the *AISC Manual*, Part 2, Composite Beam Selection Table, Transformed Section Properties.)

The allowable stresses are as follows.

For concrete, $0.45f'_c = (0.45)(4 \text{ ksi}) = 1.8$ ksi.

Structural

For steel, $0.66F_y = (0.66)(36 \text{ ksi}) = (0.66)(36 \text{ ksi}) = 23.8 \text{ ksi}$

Calculate the resisting moment based on the assumption that concrete attains the allowable stress.

$$\begin{aligned} y_t &= d + t - \bar{y}_b \\ &= 20.83 \text{ in} + 4 \text{ in} - 18.92 \text{ in} = 5.91 \text{ in} \end{aligned}$$

$$\begin{aligned} M_R &= \frac{n(0.45f'_c)I_{tr}}{y_t} \\ &= \frac{(9)\left(1.8 \frac{\text{kips}}{\text{in}^2}\right)(2579 \text{ in}^4)}{(5.91 \text{ in})\left(12 \frac{\text{in}}{\text{ft}}\right)} \\ &= 589 \text{ ft-kips} \end{aligned}$$

The resisting moment based on the assumption that steel attains the allowable stress is

$$\begin{aligned} M_R &= \frac{(0.66F_y)I_{tr}}{\bar{y}_b} \\ &= \frac{\left(23.8 \frac{\text{kips}}{\text{in}^2}\right)(2579 \text{ in}^4)}{(18.92 \text{ in})\left(12 \frac{\text{in}}{\text{ft}}\right)} \\ &= \boxed{270 \text{ ft-kips} \quad [\text{governs}]} \end{aligned}$$

The answer is (C).

2. (a) The weight of the 4 in slab is

$$\left(\frac{4 \text{ in}}{12 \frac{\text{in}}{\text{ft}}}\right)(9 \text{ ft})\left(150 \frac{\text{lbf}}{\text{ft}^3}\right) = 450 \text{ lbf/ft}$$

Assume a steel beam weight of 60 lbf/ft. The total dead load is

$$w_D = \frac{450 \frac{\text{lbf}}{\text{ft}} + 60 \frac{\text{lbf}}{\text{ft}}}{1000 \text{ lbf/kip}} = 0.51 \text{ kip/ft}$$

$$M_D = \frac{w_D L^2}{8} = \frac{\left(0.51 \frac{\text{kip}}{\text{ft}}\right)(40 \text{ ft})^2}{8} = 102 \text{ ft-kips}$$

After the concrete has hardened,

$$\begin{array}{c}\text{live load} \\ +\text{ceiling load}\end{array} = 100 \text{ psf} + 10 \text{ psf} = 110 \text{ psf}$$

$$w_L = \frac{\left(110 \frac{\text{lbf}}{\text{ft}^2}\right)(9 \text{ ft})}{1000 \frac{\text{lbf}}{\text{kip}}} = 0.99 \text{ kip/ft}$$

$$\begin{aligned} M_L &= \frac{w_L L^2}{8} = \frac{\left(0.99 \frac{\text{kip}}{\text{ft}}\right)(40 \text{ ft})^2}{8} \\ &= 198 \text{ ft-kips} \end{aligned}$$

The maximum moment is

$$\begin{aligned} M_{max} &= M_D + M_L = 102 \text{ ft-kips} + 198 \text{ ft-kips} \\ &= 300 \text{ ft-kips} \end{aligned}$$

The maximum shear is

$$\begin{aligned} V_{max} &= \frac{(w_D + w_L)L}{2} \\ &= \frac{\left(0.51 \frac{\text{kip}}{\text{ft}} + 0.99 \frac{\text{kip}}{\text{ft}}\right)(40 \text{ ft})}{2} \\ &= 30 \text{ kips} \end{aligned}$$

The effective width of the concrete slab is the smaller of

$$b = \frac{L}{4} = \frac{(40 \text{ ft})\left(12 \frac{\text{in}}{\text{ft}}\right)}{4} = 120 \text{ in}$$

$$b = s = (9 \text{ ft})\left(12 \frac{\text{in}}{\text{ft}}\right) = 108 \text{ in} \quad [\text{governs}]$$

To get an initial estimate of beam size, assume shored construction. (Equation 64.19 is easier to solve for S_{tr} than Eq. 64.24.) The approximate required section modulus for $F_y = 36$ ksi steel is determined as follows (from *AISC Manual*, Part 5, Table 1, Numerical Values).

$$F_b = 0.66F_y = 23.8 \text{ ksi}$$

$$\begin{aligned} S_{tr} &= \frac{M_{max}}{F_b} = \frac{(300 \text{ ft-kips})\left(12 \frac{\text{in}}{\text{ft}}\right)}{23.8 \frac{\text{kips}}{\text{in}^2}} \\ &= 151.3 \text{ in}^3 \end{aligned}$$

For M_D, assuming that the compression flange is adequately braced,

$$\begin{aligned} S_s &= \frac{M_D}{F_b} = \frac{(102 \text{ ft-kips})\left(12 \frac{\text{in}}{\text{ft}}\right)}{23.8 \frac{\text{kips}}{\text{in}^2}} \\ &= 51.5 \text{ in}^3 \end{aligned}$$

Select the section. Use Eq. 64.4.

$$Y2 = 2 \text{ in}$$

$$A_{ctr} = \left(\frac{b}{n}\right)t = \left(\frac{108 \text{ in}}{9}\right)(4 \text{ in}) = 48 \text{ in}^2$$

Enter the *AISC Manual*, Part 2, Composite Beam Selection Table, with the approximate required values of

$S_{\text{tr}} = 151.3 \text{ in}^4$. W 21 × 57 works, but select a larger beam to allow for unshored construction.

Select a W 21 × 62 beam.

$$S_{\text{tr}} = 173 \text{ in}^3$$
$$Y2 = 2 \text{ in}$$
$$A_{\text{ctr}} = 40 \text{ in}^2 < \text{maximum } A_{\text{ctr}} = 48 \text{ in}^2$$

From *AISC Manual*, Part 1, Dimension and Properties of W Shapes, for W 21×62,

$$d = 21.0 \text{ in}$$
$$A_s = 18.3 \text{ in}^2$$
$$I_s = 1330 \text{ in}^4$$
$$t_f = 0.615 \text{ in}$$
$$t_w = 0.400 \text{ in}$$
$$S_s = 127 \text{ in}^3$$

Calculate section properties. From the Composite Beam Selection Table, for $Y2 = 2$ in and $A_{\text{ctr}} = 48 \text{ in}^2$, by interpolation,

$$\overline{S}_{\text{tr}} = 174.2 \text{ in}^3$$
$$\overline{I}_{\text{tr}} = 3386 \text{ in}^4$$

Add the optional corrections from *AISC Manual*, Part 2, Eqs. 9 and 10.

$$t = 4 \text{ in}$$

$$\begin{aligned} I_{\text{tr}} &= \overline{I}_{\text{tr}} + \frac{A_{\text{ctr}}(t^2 - 1)}{12} \\ &= 3386 \text{ in}^4 + \frac{(48 \text{ in}^2)((4 \text{ in})^2 - 1)}{12} \\ &= 3446 \text{ in}^4 \end{aligned}$$

$$\begin{aligned} S_{\text{tr}} &= \left(\frac{I_{\text{tr}}}{\overline{I}_{\text{tr}}}\right)\overline{S}_{\text{tr}} \\ &= \left(\frac{3446 \text{ in}^4}{3386 \text{ in}^4}\right)(174.2 \text{ in}^3) = 177.3 \text{ in}^3 \end{aligned}$$

$$\begin{aligned} \overline{y}_b &= \frac{I_{\text{tr}}}{S_{\text{tr}}} = \frac{3446 \text{ in}^4}{177.3 \text{ in}^3} \\ &= 19.43 \text{ in} \end{aligned}$$

Check the concrete stress (unshored). Use Eq. 64.9.

$$\begin{aligned} S_t &= \frac{I_{\text{tr}}}{d + t - \overline{y}_b} \\ &= \frac{3446 \text{ in}^4}{21.0 \text{ in} + 4 \text{ in} - 19.43 \text{ in}} = 618.7 \text{ in}^3 \end{aligned}$$

$$\begin{aligned} f_c &= \frac{M_L}{nS_t} = \frac{(198 \text{ ft-kips})\left(12 \ \frac{\text{in}}{\text{ft}}\right)}{(9)(618.7 \text{ in}^3)} \\ &= 0.43 \ \frac{\text{kip}}{\text{in}^2} < 0.45f'_c = (0.45)(3 \text{ ksi}) \\ &= 1.35 \text{ ksi} \quad [\text{OK}] \end{aligned}$$

Check the steel stresses.

During construction (unshored), the steel beam carries all dead load. The beam weight per foot (62 lbf/ft) is close to the initial assumption (60 lbf/ft).

$$\begin{aligned} f_s &= \frac{M_D}{S_S} = \left(\frac{102 \text{ ft-kips}}{127 \text{ in}^3}\right)\left(12 \ \frac{\text{in}}{\text{ft}}\right) \\ &= 9.64 \text{ kips/in}^2 \quad [< F_b = 23.8 \text{ ksi, so OK}] \end{aligned}$$

Use Eq. 64.24 to check the loaded condition in full composite action.

$$\begin{aligned} f_s &= \frac{M_D}{S_S} + \frac{M_L}{S_{\text{tr}}} = \left(\frac{102 \text{ ft-kips}}{127 \text{ in}^3} + \frac{198 \text{ ft-kips}}{177.3 \text{ in}^3}\right) \\ &\quad \times \left(12 \ \frac{\text{in}}{\text{ft}}\right) \\ &= 23.04 \text{ kips/in}^2 \quad [< F_b = 23.8 \text{ ksi, so OK}] \end{aligned}$$

$$\begin{aligned} f_v &= \frac{V_{\text{max}}}{dt_w} = \frac{30 \text{ kips}}{(21.0 \text{ in})(0.400 \text{ in})} \\ &= 3.57 \text{ ksi} < 0.4F_y = 14.4 \text{ ksi} \quad [\text{OK}] \end{aligned}$$

Block shear does not control because the beam is not coped.

Check deflection.

$$\begin{aligned} \Delta_D &= \frac{5w_D L^4}{384EI_s} = \frac{(5)\left(0.51 \ \frac{\text{kip}}{\text{ft}}\right)(40 \text{ ft})^4\left(12 \ \frac{\text{in}}{\text{ft}}\right)^3}{(384)\left(29{,}000 \ \frac{\text{kips}}{\text{in}^2}\right)(1330 \text{ in}^4)} \\ &= 0.76 \text{ in} < 1 \text{ in} \quad [\text{OK}] \end{aligned}$$

$$\begin{aligned} \Delta_L &= \frac{5w_L L^4}{384EI_{\text{tr}}} = \frac{(5)\left(0.99 \ \frac{\text{kip}}{\text{ft}}\right)(40 \text{ ft})^4\left(12 \ \frac{\text{in}}{\text{ft}}\right)^3}{(384)\left(29{,}000 \ \frac{\text{kips}}{\text{in}^2}\right)(3446 \text{ in}^4)} \\ &= 0.57 \text{ in} < \frac{L}{360} = \frac{(40 \text{ ft})\left(12 \ \frac{\text{in}}{\text{ft}}\right)}{360} = 1.33 \text{ in} \ [\text{OK}] \end{aligned}$$

(b) Design the shear connectors. Try $^3/_4$ in diameter × 3 in studs.

The maximum stud diameter (unless located directly over the web) is

$$2.5t_f = (2.5)(0.615 \text{ in}) = 1.54 \text{ in} > \tfrac{3}{4} \text{ in} \quad [\text{OK}]$$

Evaluate for full composite action. Evaluate total horizontal shear from AISC Eq. I4-1.

$$\begin{aligned} V_h &= \frac{0.85f'_c A_c}{2} = \frac{(0.85)\left(3 \ \frac{\text{kips}}{\text{in}^2}\right)(108 \text{ in})(4 \text{ in})}{2} \\ &= 551 \text{ kips} \end{aligned}$$

From AISC Eq. I4-2 (Eq. 64.15),

$$V_h = \frac{A_sF_y}{2} = \frac{(18.3\ \text{in}^2)\left(36\ \frac{\text{kips}}{\text{in}^2}\right)}{2}$$
$$= 329\ \text{kips}\quad \text{[governs]}$$

(Alternatively, read the maximum value from the tabulated values as 329 kips.)

From *AISC Specifications* Table I4.1, the allowable horizontal shear for one $^3/_4$ in × 3 in stud is $q = 11.5$ kips. The number of shear connectors is given by Eq. 64.28.

$$N = \frac{V_h}{q} = \frac{329\ \text{kips}}{11.5\ \text{kips}}$$
$$= 28.6\quad \begin{bmatrix}\text{(say 30) on each side of the}\\ \text{point of maximum moment}\end{bmatrix}$$

The total number of shear connectors is

$$2N = (2)(30) = 60$$

Since the actual $S_{tr} = 177.3\ \text{in}^3$ exceeds the required $S_{tr} = 151.3\ \text{in}^3$, the number of shear connectors can be reduced considering partial composite action instead of full composite action.

For partial composite action, solve AISC Eq. I2-1 for V'_h using $S_{\text{eff}} = S_{\text{tr,required}} = 151.3\ \text{in}^3$. From Eq. 64.13,

$$V'_h = V_h\left(\frac{S_{\text{eff}} - S_s}{S_{\text{tr}} - S_s}\right)^2$$
$$= (329\ \text{kips})\left(\frac{151.3\ \text{in}^3 - 127\ \text{in}^3}{177.3\ \text{in}^3 - 127\ \text{in}^3}\right)^2 = 76.8\ \text{kips}$$
$$= 76.8\ \text{kips} > \left(\tfrac{1}{4}\right)(301\ \text{kips}) = 75.3\ \text{kips}\quad \text{[OK]}$$
$$N = \frac{V'_h}{q} = \frac{76.8\ \text{kips}}{11.5\ \text{kips}} = 6.7\quad \text{[say 8]}$$

The total number of shear connectors is

$$2N = (2)(8) = 16$$

Check the deflection with partial composite action.

$$V'_h = Nq = (8)(11.5\ \text{kips}) = 92\ \text{kips}$$

Using *AISC Specifications* Eq. I4-4, the effective moment of inertia is given by Eq. 64.17.

$$I_{\text{eff}} = I_s + \sqrt{\frac{V'_h}{V_h}}(I_{\text{tr}} - I_s)$$

$$I_{\text{eff}} = 1330\ \text{in}^4 + \sqrt{\frac{92\ \text{kips}}{329\ \text{kips}}}$$
$$\times (3446\ \text{in}^4 - 1330\ \text{in}^4)$$
$$= 2449\ \text{in}^4$$

Check the deflection.

$$\Delta_L\,(\text{partial}) = \frac{5w_LL^4}{384EI_{\text{eff}}}$$
$$= \frac{(5)\left(0.99\ \frac{\text{kip}}{\text{ft}}\right)(40\ \text{ft})^4\left(12\ \frac{\text{in}}{\text{ft}}\right)^3}{(384)\left(29{,}000\ \frac{\text{kips}}{\text{in}^2}\right)(2449\ \text{in}^4)}$$
$$= 0.80\ \text{in}$$

$$\frac{L}{360} = \frac{(40\ \text{ft})\left(12\ \frac{\text{in}}{\text{ft}}\right)}{360}$$
$$= 1.33\ \text{in} > 0.80\ \text{in}\quad \text{[OK]}$$

Use an *A36* W 21 × 62 beam with

16 $^3/_4$ in diameter × 3 in studs, 8 on each side of the midspan.

3. From the *AISC Manual*, for W 21 × 68,

$$d = 21.13\ \text{in}$$
$$A_s = 20\ \text{in}^2$$
$$I_s = 1480\ \text{in}^4$$

(a) The effective concrete flange width is the smaller of

$$b = \frac{L}{4} = \frac{(30\ \text{ft})\left(12\ \frac{\text{in}}{\text{ft}}\right)}{4} = \boxed{90\ \text{in}\quad \text{[governs]}}$$

$$b = s = (12\ \text{ft})\left(12\ \frac{\text{in}}{\text{ft}}\right) = 144\ \text{in}$$

The answer is (D).

(b) The width of the equivalent steel area of the concrete slab is

$$\frac{b}{n} = \frac{90\ \text{in}}{9} = 10\ \text{in}$$

The concrete transformed area in compression is given by Eq. 64.4.

$$A_{\text{ctr}} = \left(\frac{b}{n}\right)t = (10\ \text{in})(4\ \text{in}) = 40\ \text{in}^2$$

The distance from the top of the steel beam to the centroid of the concrete area in compression is given by Eq. 64.7.

$$Y2 = \frac{t}{2} = \frac{4\ \text{in}}{2} = 2\ \text{in}$$

The distance of the neutral axis of the transformed composite (all steel) section from the bottom of the steel beam is given by Eq. 64.3.

$$\overline{y}_b = \frac{A_s\left(\frac{d}{2}\right) + A_{\text{ctr}}(d + Y2)}{A_s + A_{\text{ctr}}}$$

$$= \frac{(20 \text{ in}^2)\left(\frac{21.13 \text{ in}}{2}\right) + (40 \text{ in}^2)(21.13 \text{ in} + 2 \text{ in})}{20 \text{ in}^2 + 40 \text{ in}^2}$$

$$= \boxed{18.94 \text{ in}}$$

The answer is (A).

(c) The moment of inertia of the transformed composite (all steel) section is given by Eq. 64.5.

$$I_{\text{tr}} = I_s + A_s\left(\overline{y}_b - \frac{d}{2}\right)^2 + \frac{\left(\frac{b}{n}\right)t^3}{12} + A_{\text{ctr}}(d + Y2 - \overline{y}_b)^2$$

$$= 1480 \text{ in}^4 + (20 \text{ in}^2)\left(18.94 \text{ in} - \left(\tfrac{1}{2}\right)(21.13 \text{ in})\right)^2 + \frac{(10 \text{ in})(4 \text{ in})^3}{12} + (40 \text{ in}^2)(21.13 \text{ in} + 2 \text{ in} - 18.94 \text{ in})^2$$

$$= \boxed{3638 \text{ in}^4}$$

The answer is (B).

(d) The allowable concrete stress is

$$0.45f'_c = (0.45)(3 \text{ ksi}) = \boxed{1.35 \text{ ksi}}$$

The answer is (D).

(e) The allowable steel stress is

$$0.66F_y = (0.66)(36 \text{ ksi}) = \boxed{23.8 \text{ ksi}}$$

The answer is (C).

(f) The resisting moment based on the assumption that concrete attains the allowable stress is found as follows.

$$y_t = d + t - \overline{y}_b = 21.13 \text{ in} + 4 \text{ in} - 18.94 \text{ in} = 6.19 \text{ in}$$

From $f = Mc/I$,

$$M_R = n(0.45f'_c)\left(\frac{I_{\text{tr}}}{y_t}\right) = \frac{(9)\left(1.35 \frac{\text{kips}}{\text{in}^2}\right)(3638 \text{ in}^4)}{\left(12 \frac{\text{in}}{\text{ft}}\right)(6.19 \text{ in})} = 595 \text{ ft-kips}$$

The resisting moment based on the assumption that steel attains the allowable stress is

$$M_R = \frac{(0.66F_y)I_{\text{tr}}}{\overline{y}_b} = \frac{\left(23.8 \frac{\text{kips}}{\text{in}^2}\right)(3638 \text{ in}^4)}{\left(12 \frac{\text{in}}{\text{ft}}\right)(18.94 \text{ in})} = \boxed{381 \text{ ft-kips} \quad \text{[governs]}}$$

The answer is (B).

(g) The dead load from the 4 in slab is

$$\left(\frac{4 \text{ in}}{12 \frac{\text{in}}{\text{ft}}}\right)\left(150 \frac{\text{lbf}}{\text{ft}^3}\right)(12 \text{ ft}) = 600 \text{ lbf/ft}$$

The W 21 × 68 steel beam contributes 68 lbf/ft.

$$w_D = \frac{600 \frac{\text{lbf}}{\text{ft}} + 68 \frac{\text{lbf}}{\text{ft}}}{1000 \frac{\text{lbf}}{\text{kip}}} = 0.668 \text{ kip/ft}$$

The maximum service dead load moment is

$$M_D = \frac{w_D L^2}{8} = \frac{\left(0.668 \frac{\text{kip}}{\text{ft}}\right)(30 \text{ ft})^2}{8} = \boxed{75.2 \text{ ft-kips}}$$

The answer is (B).

(h) The maximum allowable moment due to the service live load is

$$M_L = M_R - M_D = 381 \text{ ft-kips} - 75.2 \text{ ft-kips} = 305.8 \text{ ft-kips}$$

$$w_L = \frac{8M_L}{L^2} = \frac{(8)(305.8 \text{ ft-kips})}{(30 \text{ ft})^2} = 2.718 \text{ kips/ft}$$

The service live load capacity is

$$\frac{w_L}{s} = \frac{\left(2.718 \frac{\text{kips}}{\text{ft}}\right)\left(1000 \frac{\text{lbf}}{\text{kip}}\right)}{12 \text{ ft}} = \boxed{227 \text{ lbf/ft}^2}$$

The answer is (C).

Structural

(i) Find the total horizontal shear. Use AISC Eq. I4-1 (Eq. 64.14).

$$V_h = \frac{0.85 f'_c A_c}{2} = \frac{(0.85)\left(3\ \frac{\text{kips}}{\text{in}^2}\right)(90\ \text{in})(4\ \text{in})}{2} = 459\ \text{kips}$$

Use AISC Eq. I4-2 (Eq. 64.15).

$$V_h = \frac{A_s F_y}{2} = \frac{(20\ \text{in}^2)\left(36\ \frac{\text{kips}}{\text{in}^2}\right)}{2} = \boxed{360\ \text{kips} \quad [\text{governs}]}$$

The answer is (B).

(j) From *AISC Specifications* Table I4.1, the allowable horizontal shear for one 3/4 in dia. × 3 in stud is $q = 11.5$ kips.

The number of shear connectors required on each side of the point of maximum moment is given by Eq. 64.28.

$$N = \frac{V_h}{q} = \frac{360\ \text{kips}}{11.5\ \text{kips}} = 31.3 \quad [\text{say } 32]$$

The total number of shear connectors required for the entire beam is

$$2N = (2)(32) = \boxed{64}$$

The answer is (B).

4. The nominal rib height of the steel deck is $h_r = 3$ in.

The thickness of the concrete slab above the steel deck is $t_o = 3$ in.

A conservative estimate of the thickness of the concrete in compression is $t = t_o = 3$ in.

From the *AISC Manual*, for W 21 × 50,

$$d = 20.83\ \text{in}$$
$$A_s = 14.7\ \text{in}^2$$
$$I_s = 984\ \text{in}^4$$
$$S_s = 94.5\ \text{in}^3$$

(a) The effective concrete flange width is the smaller of

$$b = \frac{L}{4} = \frac{(40\ \text{ft})\left(12\ \frac{\text{in}}{\text{ft}}\right)}{4} = 120\ \text{in}$$

$$b = s = (8\ \text{ft})\left(12\ \frac{\text{in}}{\text{ft}}\right) = 96\ \text{in} \quad [\text{governs}]$$

From Eq. 64.8,

$$Y2 = h_r + \frac{t}{2} = 3\ \text{in} + \frac{3\ \text{in}}{2} = 4.5\ \text{in}$$

From Eq. 64.4,

$$A_{\text{ctr}} = \left(\frac{b}{n}\right)t = \left(\frac{96\ \text{in}}{9}\right)(3\ \text{in}) = 32\ \text{in}^2$$

The distance from the neutral axis of the transformed composite section (all steel) to the bottom of the steel beam is given by Eq. 64.3.

$$\bar{y}_b = \frac{A_s\left(\frac{d}{2}\right) + A_{\text{ctr}}(d + Y2)}{A_s + A_{\text{ctr}}} = \frac{(14.7\ \text{in}^2)\left(\frac{20.83\ \text{in}}{2}\right) + (32\ \text{in}^2)(20.83\ \text{in} + 4.5\ \text{in})}{14.7\ \text{in}^2 + 32\ \text{in}^2} = 20.63\ \text{in}$$

$$20.63\ \text{in} < d = 20.83\ \text{in}$$

Therefore, the neutral axis is below the steel deck.

The moment of inertia of the transformed composite section (all steel) is given by Eq. 64.5.

$$\begin{aligned} I_{\text{tr}} &= I_s + A_s\left(\bar{y}_b - \frac{d}{2}\right)^2 + \frac{\left(\frac{b}{n}\right)t^3}{12} + A_{\text{ctr}}(d + Y2 - \bar{y}_b)^2 \\ &= 984\ \text{in}^4 + (14.7\ \text{in}^2)\left(20.63\ \text{in} - \left(\tfrac{1}{2}\right)(20.83\ \text{in})\right)^2 + \frac{\left(\frac{96\ \text{in}}{9}\right)(3\ \text{in})^3}{12} + (32\ \text{in}^2)(20.83\ \text{in} + 4.5\ \text{in} - 20.63\ \text{in})^2 \\ &= \boxed{3249\ \text{in}^4} \end{aligned}$$

The answer is (B).

(b) Evaluate the construction loads. The slab and deck contribute

$$\left(65\ \frac{\text{lbf}}{\text{ft}^2}\right)(8\ \text{ft}) = 520\ \text{lbf/ft}$$

The W 21 × 50 steel beam contributes 50 lbf/ft.

The total dead load is

$$w_D = \frac{520\ \frac{\text{lbf}}{\text{ft}} + 50\ \frac{\text{lbf}}{\text{ft}}}{1000\ \frac{\text{lbf}}{\text{kip}}} = 0.57\ \text{kip/ft}$$

$$M_D = \frac{w_D L^2}{8} = \frac{\left(0.57\ \frac{\text{kip}}{\text{ft}}\right)(40\ \text{ft})^2}{8} = 114\ \text{ft-kips}$$

Evaluate the loads applied after the concrete has hardened. The live load is 100 psf.

$$w_L = \frac{\left(100\ \frac{\text{lbf}}{\text{ft}^2}\right)(8\ \text{ft})}{1000\ \frac{\text{lbf}}{\text{kip}}} = 0.8\ \text{kip/ft}$$

$$M_L = \frac{w_L L^2}{8} = \frac{\left(0.8\ \frac{\text{kip}}{\text{ft}}\right)(40\ \text{ft})^2}{8} = 160\ \text{ft-kips}$$

The distance from the neutral axis of the transformed composite section to the top of slab is

$$y_t = d + h_r + t_o - \bar{y}_b = 20.83\ \text{in} + 3\ \text{in} + 3\ \text{in} - 20.63\ \text{in} = 6.2\ \text{in}$$

The section modulus of the transformed composite cross section, referred to the top of concrete, is

$$S_t = \frac{I_{\text{tr}}}{y_t} = \frac{3249\ \text{in}^4}{6.2\ \text{in}} = 524\ \text{in}^3$$

The maximum stress in the concrete (unshored) is

$$f_c = \frac{M_L}{nS_t} = \frac{(160\ \text{ft-kips})\left(12\ \frac{\text{in}}{\text{ft}}\right)}{(9)(524\ \text{in}^3)} = \boxed{0.41\ \text{kip/in}^2\ \text{(ksi)}}$$

The answer is (A).

(c) The section modulus of the transformed composite cross section, referred to the bottom flange of steel beam, is

$$S_{\text{tr}} = \frac{I_{\text{tr}}}{\bar{y}_b} = \frac{3249\ \text{in}^4}{20.63\ \text{in}} = 157.5\ \text{in}^3$$

The maximum stress at the bottom of the steel beam under combined service dead and live loads is given by Eq. 64.24.

$$f_s = \frac{M_D}{S_s} + \frac{M_L}{S_{\text{tr}}} = \frac{(114\ \text{ft-kips})\left(12\ \frac{\text{in}}{\text{ft}}\right)}{94.5\ \text{in}^3} + \frac{(160\ \text{ft-kips})\left(12\ \frac{\text{in}}{\text{ft}}\right)}{157.5\ \text{in}^3} = \boxed{26.7\ \text{kips/in}^2\ \text{(ksi)}}$$

The answer is (C).

(d) Evaluate the total horizontal shear for full composite action. Use AISC Eq. I4-1 (Eq. 64.14).

$$V_h = \frac{0.85 f_c' A_c}{2} = \frac{(0.85)\left(3\ \frac{\text{kips}}{\text{in}^2}\right)(96\ \text{in})(3\ \text{in})}{2} = 367\ \text{kips}$$

Use AISC Eq. I4-2 (Eq. 64.15).

$$V_h = \frac{A_s F_y}{2} = \frac{(14.7\ \text{in}^2)\left(36\ \frac{\text{kips}}{\text{in}^2}\right)}{2} = \boxed{265\ \text{kips}\quad \text{[governs]}}$$

The answer is (A).

(e) Determine the stud reduction factor, R. From the *AISC Specification* Eq. I5-1,

N_r = number of stud connectors in one rib = 1
w_r = average width of concrete rib = 3 in
h_r = nominal rib height = 3 in
H_s = length of stud connector = 5 in

$$R = \left(\frac{0.85}{\sqrt{N_r}}\right)\left(\frac{w_r}{h_r}\right)\left(\frac{H_s}{h_r} - 1\right) = \left(\frac{0.85}{\sqrt{1}}\right)\left(\frac{3\ \text{in}}{3\ \text{in}}\right)\left(\frac{5\ \text{in}}{3\ \text{in}} - 1\right) = \boxed{0.57}$$

The answer is (D).

(f) From the *AISC Specifications* Table I4.1, the allowable horizontal shear for each $^3/_4$ in diameter by 3 in stud is $q = 11.5$ kips. From Table I4.2, a reduction factor of 0.86 is required for lightweight concrete with a density of 115 lbm/ft^3. The number of studs required

Structural

(for full composite action) to resist the horizontal shear V_h on each side of the point of maximum moment is

$$N = \frac{V_h}{qR} = \frac{265 \text{ kips}}{(0.86)(11.5 \text{ kips})(0.57)} = 47.0$$

The total number of studs required for the entire beam is

$$2N = (2)(47) = \boxed{94}$$

The answer is (C).

(g) 75% partial composite action means $V_h'/V_h = 0.75$.

$$V_h' = 0.75V_h = (0.75)(265 \text{ kips}) = 199 \text{ kips}$$

The stud strength is $q = 11.5$ kips.

The number of studs required to resist the horizontal shear V_h' on each side of the point of maximum moment is

$$N = \frac{V_h'}{qR} = \frac{199 \text{ kips}}{(0.86)(11.5 \text{ kips})(0.57)} = 35.3 \quad (36)$$

The total number of studs required for the entire beam is

$$2N = (2)(36) = \boxed{72}$$

The answer is (C).

(h) The effective composite section modulus (from *AISC Specifications* Eq. I2-1) is given by Eq. 64.13.

$$S_{\text{eff}} = S_s + \sqrt{\frac{V_h'}{V_h}}(S_{\text{tr}} - S_s)$$
$$= 94.5 \text{ in}^3 + \sqrt{0.75}(157.5 \text{ in}^3 - 94.5 \text{ in}^3)$$
$$= \boxed{149 \text{ in}^3}$$

The answer is (B).

(i) The service live load deflection for full composite action is

$$\Delta_L = \frac{5w_L L^4}{384EI_{\text{tr}}} = \frac{(5)\left(0.8 \frac{\text{kip}}{\text{ft}}\right)(40 \text{ ft})^4\left(12 \frac{\text{in}}{\text{ft}}\right)^3}{(384)\left(29{,}000 \frac{\text{kips}}{\text{in}^2}\right)(3249 \text{ in}^4)}$$
$$= \boxed{0.49 \text{ in}}$$

The answer is (B).

(j) The service live load deflection for partial (75%) composite action (i.e., $V_h'/V_h = 0.75$) from *AISC Specifications* Eq. I4-4 is given by Eq. 64.17.

$$I_{\text{eff}} = I_s + \sqrt{\frac{V_h'}{V_h}}(I_{\text{tr}} - I_s)$$
$$= 984 \text{ in}^4 + \sqrt{0.75}(3249 \text{ in}^2 - 984 \text{ in}^2)$$
$$= 2946 \text{ in}^4$$

$$\Delta_L \text{ (partial)} = \frac{5w_L L^4}{384EI_{\text{eff}}}$$
$$= \frac{(5)\left(0.8 \frac{\text{kip}}{\text{ft}}\right)(40 \text{ ft}^4)\left(12 \frac{\text{in}}{\text{ft}}\right)^3}{(384)\left(29{,}000 \frac{\text{kips}}{\text{in}^2}\right)(2946 \text{ in}^4)}$$
$$= \boxed{0.54 \text{ in}}$$

The answer is (B).

5. Find the moment of inertia of the plate girder with and without the concrete flange.

shape	A (in^2)	$\overline{y}$ (in)	$\overline{y}A$ (in^3)	I_c (in^4)
$\frac{5}{8} \times 14$	8.75	41.0625	359.297	0.2848
$\frac{5}{16} \times 40$	12.5	20.75	259.375	1666.67
$\frac{3}{4} \times 16$	12	0.375	4.5	0.5625
subtotals	33.25		623.172	1667.52
transformed concrete	48.0	45.375	2178	144
totals	81.25 in^2		2801.2 in^3	1811.52 in^4

Work with the steel acting alone before the concrete hardens.

$$\overline{y} = \frac{623.172 \text{ in}^3}{33.25 \text{ in}^2} = 18.74 \text{ in}$$

$$I = 1667.52 \text{ in}^4 + (12 \text{ in}^2)(18.74 \text{ in} - 0.375 \text{ in})^2 + (12.5 \text{ in}^2)(20.75 \text{ in} - 18.74 \text{ in})^2 + (8.75 \text{ in}^2)(41.0625 \text{ in} - 18.74 \text{ in})^2$$
$$= 10{,}125.4 \text{ in}^4$$

The distances to the extreme fibers are

$$c = 18.74 \text{ in (T)}$$
$$c = 22.64 \text{ in (C)}$$

The section modulus referred to the bottom of steel is

$$S = \frac{I}{c} = \frac{10{,}125.4 \text{ in}^4}{18.74 \text{ in}} = 540 \text{ in}^3$$

The maximum moment due to 1 kip/ft^2 dead load is

$$M_D = \frac{wL^2}{8} = \frac{\left(1 \frac{\text{kip}}{\text{ft}}\right)(80 \text{ ft})^2}{8} = 800 \text{ ft-kips}$$

The shear due to dead load is

$$V_D = \frac{wL}{2} = \frac{\left(1\ \frac{\text{kip}}{\text{ft}}\right)(80\ \text{ft})}{2} = 40\ \text{kips}$$

Check the width-thickness ratios of the compression elements.

Check the width-thickness ratio of the web.

$$\frac{h}{t} \leq \frac{14{,}000}{\sqrt{F_y(F_y + 16.5)}}$$

$$\frac{14{,}000}{\sqrt{(36\ \text{ksi})(36\ \text{ksi} + 16.5\ \text{ksi})}} = 322.0$$

$$\frac{h}{t} = \frac{40\ \text{in}}{\frac{5}{16}\ \text{in}} = 128$$

$$128 < 322 \quad \text{[acceptable]}$$

For the compression flange, considered stiffened along one edge,

$$k_c = \frac{4.05}{\left(\frac{h}{t}\right)^{0.46}} = \frac{4.05}{(128)^{0.46}} = 0.435$$

Typical values of k_c are in the 2.0 to 4.0 range. The very thin web contributes to this small value.

$$b = \left(\tfrac{1}{2}\right)(14\ \text{in}) = 7\ \text{in}$$

$$\frac{b}{t} = \frac{7\ \text{in}}{\frac{5}{8}\ \text{in}} = 11.2$$

From Eq. 63.14,

$$\frac{95}{\sqrt{\frac{F_{y,f}}{k_c}}} = \frac{95}{\sqrt{\frac{36\ \text{ksi}}{0.435}}} = 10.4$$

$$11.2 > 10.4 \quad \text{[not acceptable]}$$

Check the bending stress.

$$\frac{760}{\sqrt{F_b}} = \frac{760}{\sqrt{(0.60)(36\ \text{ksi})}} = 163.5$$

Calculate h/t for the web.

$$\frac{h}{t} = \frac{40\ \text{in}}{\frac{5}{16}\ \text{in}} = 128$$

Since $128 < 163.5$, F_b does not have to be reduced, and higher values may be justified. From *AISC Manual of Steel Construction* F1.3,

$$r_T = \sqrt{\frac{I_{y,\text{flange}} + I_{1/3\ \text{of web in compression}}}{A_{\text{flange}} + A_{1/3\ \text{of web in compression}}}}$$

Disregard the contribution of one third of the web to the moment of inertia. Approximately half of the web is in compression.

$$r_T = \sqrt{\frac{\left(\frac{1}{12}\right)\left(\frac{5}{8}\ \text{in}\right)(14\ \text{in})^3 + 0}{8.75\ \text{in}^2 + \left(\frac{5}{16}\ \text{in}\right)\left(\frac{22.0}{3}\ \text{in}\right)}} = 3.59\ \text{in}$$

$$\frac{l}{r_T} = \frac{(20\ \text{ft})\left(12\ \frac{\text{in}}{\text{ft}}\right)}{3.59\ \text{in}} = 66.8$$

The moment diagram due to the dead load is

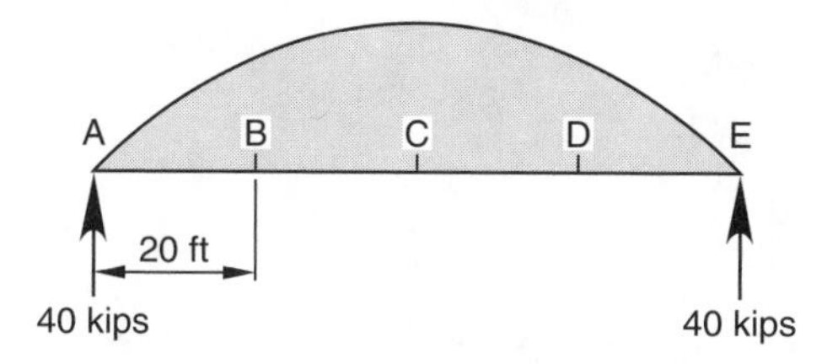

$C_b > 1$ in all 4 panels.

$$M_B = (40\ \text{kips})(20\ \text{ft}) - \left(\tfrac{1}{2}\right)\left(1\ \frac{\text{kip}}{\text{ft}}\right)(20\ \text{ft})^2 = 600\ \text{ft-kips}$$

$$M_C = 800\ \text{ft-kips}$$

$$C_{b,\text{span A-B}} = 1.75 + (1.05)\left(\frac{-0}{600\ \text{ft-kips}}\right) + (0.3)\left(\frac{-0}{600\ \text{ft-kips}}\right)^2 = 1.75$$

$$C_{b,\text{span B-C}} = 1.75 + (1.05)\left(\frac{-600}{800\ \text{ft-kips}}\right) + (0.3)\left(\frac{-600}{800\ \text{ft-kips}}\right)^2 = 1.13$$

Use $C_b = 1.13$ as the more conservative value.

From *AISC Manual of Steel Construction* F1-6,

$$\sqrt{\frac{102{,}000C_b}{F_y}} = \sqrt{\frac{(102{,}000)(1.13)}{36\ \text{ksi}}} = 56.6$$

$$\sqrt{\frac{510{,}000C_b}{F_y}} = \sqrt{\frac{(510{,}000)(1.13)}{36\ \text{ksi}}} = 126.5$$

Since $56.6 < 66.8 < 126.5$,

$$F_b = \left(\frac{2}{3} - \frac{F_y\left(\frac{l}{r_T}\right)^2}{(1530 \times 10^3)\,C_b}\right)F_y$$

$$= \left(\frac{2}{3} - \frac{(36)(66.8)^2}{(1530 \times 10^3)(1.13)}\right)(36\ \text{ksi}) = 20.7\ \text{ksi}$$

Structural

Check the bending stress.

$$f_b = \frac{Mc}{I} = \frac{(800 \text{ ft-kips})(22.64 \text{ in})\left(12 \frac{\text{in}}{\text{ft}}\right)}{10{,}125.4 \text{ in}^4} = 21.47 \text{ ksi}$$

$$21.47 > 20.7 \text{ ksi} \quad [\text{not acceptable}]$$

Check the shear stress.

For a worst-case stiffener placement,

$$k = 5.34$$

$$\frac{h}{t} = 128 > \frac{548}{\sqrt{36 \text{ ksi}}} = 91.3$$

$$F_v = \frac{83{,}150 \text{ ksi}}{(128)^2} = 5.08 \text{ ksi}$$

$$f_v = \frac{V}{A_{\text{web}}} = \frac{40 \text{ kips}}{(40 \text{ in})\left(\frac{5}{16} \text{ in}\right)} = \frac{40 \text{ kips}}{12.5 \text{ in}^2} = 3.2 \text{ ksi} < 5.08 \text{ ksi} \quad [\text{acceptable}]$$

After the concrete hardens,

$$\bar{y} = \frac{2801.2 \text{ in}^3}{81.25 \text{ in}^2} = 34.48 \text{ in}$$

All of the concrete is in compression. The distances to the extreme fibers are

$$c = \begin{Bmatrix} 6.895 \text{ in} & [\text{steel in compression}] \\ 13.895 \text{ in} & [\text{concrete in compression}] \end{Bmatrix} = 34.48 \text{ in} \quad [\text{steel in tension}]$$

$$\begin{aligned} I &= 1811.52 \text{ in}^4 + (8.75 \text{ in}^2)(41.0625 \text{ in} - 34.48 \text{ in})^2 \\ &\quad + (12.5 \text{ in}^2)(34.48 \text{ in} - 20.75 \text{ in})^2 \\ &\quad + (12 \text{ in}^2)(34.48 \text{ in} - 0.375 \text{ in})^2 \\ &\quad + (48 \text{ in}^2)(45.375 \text{ in} - 34.48 \text{ in})^2 \\ &= 24{,}202.5 \text{ in}^4 \end{aligned}$$

The transformed section modulus referred to the tension flange is

$$S = \frac{I}{c} = \frac{24{,}202.5 \text{ in}^4}{34.48 \text{ in}} = 701.9 \text{ in}^3$$

From the live load, the maximum shears and moments are

$$V_t = 40 \text{ kips} + \frac{\left(1 \frac{\text{kip}}{\text{ft}}\right)(80 \text{ ft})}{2} = 80 \text{ kips}$$

$$M_L = (20 \text{ ft})(40 \text{ kips}) = 800 \text{ ft-kips}$$

$$M_t = 800 \text{ ft-kips} + 800 \text{ ft-kips} = 1600 \text{ ft-kips}$$

The maximum S_{tr} allowed without shoring, per *AISC Manual of Steel Construction*, is

$$\begin{aligned} S_{\text{tr,max}} &= \left(1.35 + 0.35\left(\frac{M_L}{M_D}\right)\right) S \\ &= \left(1.35 + (0.35)\left(\frac{800 \text{ ft-kips}}{800 \text{ ft-kips}}\right)\right)(540 \text{ in}^3) \\ &= 918 \text{ in}^3 \end{aligned}$$

$$701.9 \text{ in}^3 < 918 \text{ in}^3 \quad [\text{acceptable}]$$

Check the shear stress.

$$f_v = \frac{V}{A_{\text{web}}} = \frac{80 \text{ kips}}{12.5 \text{ in}^2} = 6.4 \text{ ksi}$$

$$F_v = 5.08 \text{ ksi}$$

$$6.4 \text{ ksi} > 5.08 \text{ ksi} \quad [\text{not acceptable}]$$

Check the compressive stress in steel.

Since there is no shoring, the dead and live loads are superimposed.

$$\begin{aligned} f_b &= \frac{M_D c_{\text{steel}}}{I_{\text{steel}}} + \frac{M_L c_{\text{composite}}}{I_{\text{composite}}} \\ &= \frac{(800 \text{ ft-kips})\left(12 \frac{\text{in}}{\text{ft}}\right)(22.64 \text{ in})}{10{,}125.4 \text{ in}^4} \\ &\quad + \frac{(800 \text{ ft-kips})\left(12 \frac{\text{in}}{\text{ft}}\right)(6.895 \text{ in})}{24{,}202.5 \text{ in}^4} \\ &= 24.2 \text{ ksi} > 20.7 \text{ ksi} \quad [\text{not acceptable}] \end{aligned}$$

Check the tensile stress in steel.

$$F_{b,\text{tension}} = (0.6)(36 \text{ ksi}) = 21.6 \text{ ksi}$$

$$\begin{aligned} f_t &= \frac{(800 \text{ ft-kips})\left(12 \frac{\text{in}}{\text{ft}}\right)(18.74 \text{ in})}{10{,}125.4 \text{ in}^4} \\ &\quad + \frac{(800 \text{ ft-kips})\left(12 \frac{\text{in}}{\text{ft}}\right)(34.48 \text{ in})}{24{,}202.5 \text{ in}^4} \\ &= 31.4 \text{ ksi} > 21.6 \text{ ksi} \quad [\text{not acceptable}] \end{aligned}$$

Check the concrete compressive stress.

From *AISC Manual of Steel Construction* I2.2, $f_c < 0.45f'_c$. With 3000 psi concrete,

$$F_c = (0.45)(3000 \text{ ksi}) = 1.35 \text{ ksi}$$

Only the live load is applied after curing.

$$f_c = \frac{\frac{(800 \text{ ft-kips})\left(12 \frac{\text{in}}{\text{ft}}\right)(13.895 \text{ in})}{24{,}202.5 \text{ in}^4}}{10} = 0.55 \text{ ksi}$$

$$0.55 \text{ ksi} < 1.35 \text{ ksi} \quad [\text{acceptable}]$$

The plate girder is inadequate due to the excessive steel bending and shear stresses.

65 Structural Steel: Connectors

PRACTICE PROBLEMS

1. A bracing member made of a pair of $5 \times 3 \times 1/2$ angles to carry a service tensile load of 140 kips is attached to a W 12 column via a structural tee as shown. All connections are to be bearing-type and made with $7/8$ in diameter A325 bolts with threads in the plane of shear. A36 steel is used for all structural members. Neglecting block shear, determine the required number of bolts for (a) the connection between the bracing member and the structural tee and (b) the connection between the structural tee and the column.

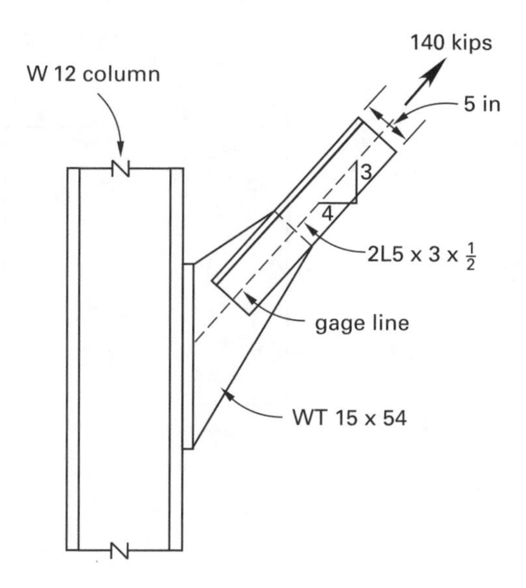

2. In the bracket-to-column connection shown, four $3/4$ in diameter bolts are used in each of the two vertical rows at 3 in pitch (for a total of eight bolts), such that the load P passes through the center of gravity of the bolt group. Determine the maximum value of the load P (assuming the connected parts do not govern) for (a) a slip-critical connection and (b) a bearing-type connection with threads excluded from the shear plane. Assume standard holes.

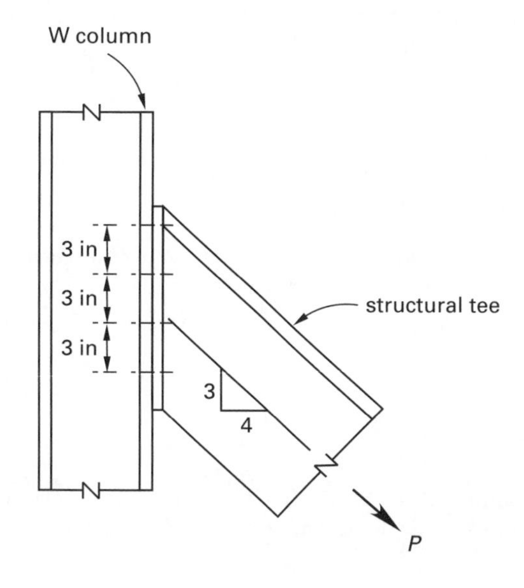

3. *(Time limit: one hour)* Two $5/8$ in plates are spliced as shown to form a member carrying an axial tensile force of unknown magnitude. Four $7/8$ in diameter bolts are used at center-to-center spacing of 3 in in both longitudinal and transverse directions. All holes are punched. The plate steel is A588, grade 50. Neglect block shear.

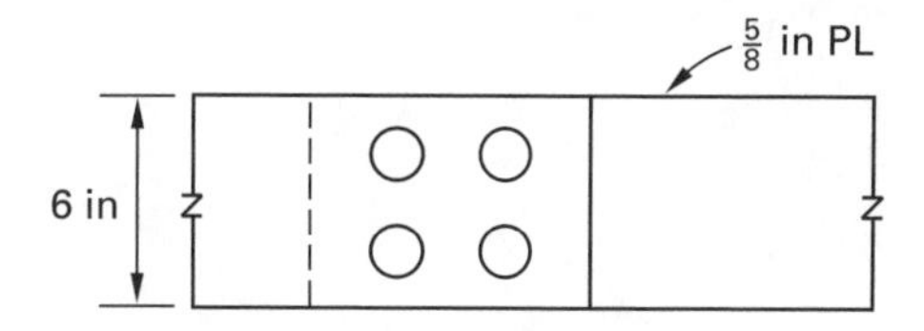

(a) The allowable tensile load for the member, based on the yielding criterion alone, is most nearly

(A) 68 kips
(B) 80 kips
(C) 88 kips
(D) 110 kips

Structural

(b) The allowable tensile load for the member, based on the fracture criterion alone, is most nearly

(A) 68 kips
(B) 80 kips
(C) 88 kips
(D) 110 kips

(c) Assuming that deformation around bolt holes is not a design consideration, the allowable bearing stress in the member is most nearly

(A) 76 ksi
(B) 84 ksi
(C) 95 ksi
(D) 105 ksi

(d) Assuming an A325SC slip-critical connection, the shear capacity of the connection is most nearly

(A) 35 kips
(B) 41 kips
(C) 56 kips
(D) 68 kips

(e) Assuming an A325SC slip-critical connection, the allowable tensile load capacity of the member is most nearly

(A) 41 kips
(B) 68 kips
(C) 80 kips
(D) 87 kips

(f) Assuming an A325N bearing-type connection with threads in the shear plane, the shear capacity of the four bolts is most nearly

(A) 35 kips
(B) 50 kips
(C) 72 kips
(D) 90 kips

(g) Assuming an A325N bearing-type connection with threads in the shear plane, the allowable tensile load capacity of the member is most nearly

(A) 41 kips
(B) 50 kips
(C) 80 kips
(D) 87 kips

(h) Assuming an A325X bearing-type connection with threads excluded from the shear plane, the shear capacity of the four bolts is most nearly

(A) 35 kips
(B) 41 kips
(C) 72 kips
(D) 90 kips

(i) Assuming an A3255X bearing-type connection with threads excluded from the shear plane, the maximum service tensile load the member can carry is most nearly

(A) 41 kips
(B) 72 kips
(C) 87 kips
(D) 90 kips

(j) The actual bearing stress in the member corresponding to its tensile capacity with a slip-critical connection is most nearly

(A) 17.3 kips/in^2
(B) 18.7 kips/in^2
(C) 19.9 kips/in^2
(D) 21.3 kips/in^2

4. *(Time limit: one hour)* In the bracket connection shown, all fasteners are $^3/_4$ in diameter bolts.

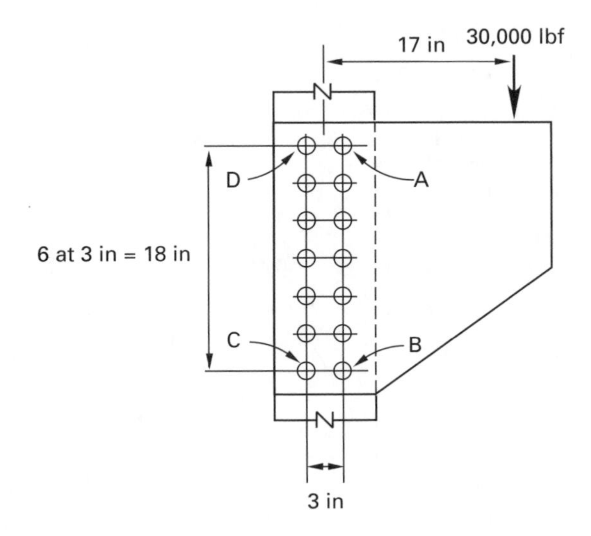

For parts (a) through (g), use a traditional elastic approach.

(a) The polar area moment of inertia of the bolt group is most nearly

(A) 190 in^4
(B) 240 in^4
(C) 290 in^4
(D) 300 in^4

(b) The torsional shear stress in bolts A, B, C, and D is most nearly

(A) 15,900 psi
(B) 18,700 psi
(C) 19,600 psi
(D) 20,300 psi

(c) The vertical component of the torsional shear stress in bolts A, B, C, and D is most nearly

(A) 2200 psi
(B) 2800 psi
(C) 3200 psi
(D) 3600 psi

(d) The horizontal component of the torsional shear stress in bolts A, B, C, and D is most nearly

(A) 12,500 psi
(B) 18,200 psi
(C) 18,900 psi
(D) 19,400 psi

(e) The direct vertical shear stress in the bolts is most nearly

(A) 4200 psi
(B) 4400 psi
(C) 4900 psi
(D) 5000 psi

(f) The total shear stress in bolts A and B is most nearly

(A) 16,200 psi
(B) 18,500 psi
(C) 21,000 psi
(D) 22,600 psi

(g) The total shear stress in bolts C and D is most nearly

(A) 15,200 psi
(B) 17,700 psi
(C) 18,900 psi
(D) 19,400 psi

For parts (h) through (j), use the *AISC* tables.

(h) Assuming A325SC bolts in standard holes, the allowable load for the bolt group is most nearly

(A) 20,800 lbf
(B) 25,600 lbf
(C) 26,900 lbf
(D) 30,400 lbf

(i) Assuming A325N bolts in standard holes, the allowable load for the bolt group is most nearly

(A) 22,900 lbf
(B) 27,300 lbf
(C) 37,700 lbf
(D) 38,400 lbf

(j) The connection is adequate for

(A) A325SC bolts only
(B) A325N bolts only
(C) both A325SC and A325N bolts
(D) neither A325SC nor A325N bolts

SOLUTIONS

1. (a) For the connection between the bracing member (double-angle) and the structural tee web, the bolts are in double shear. From the *AISC Manual*, Part 4, Table I-D, the capacity of one $^7/_8$ in diameter A325N bolt in double shear is 25.3 kips.

Bearing is on the web of the structural tee. For WT15 × 54, $t_w = 0.545$ in.

From the *AISC Manual*, Part 4, Table I-E, the capacity of one $^7/_8$ in diameter A325N bolt in bearing is 60.9 kips/in.

$$\left(60.9\ \frac{\text{kips}}{\text{in}}\right)(0.545\ \text{in}) = 33.2\ \text{kips}$$

Bolt shear controls.

The required number of bolts is

$$n = \frac{140\ \text{kips}}{25.3\ \text{kips}} = 5.53$$

Use six bolts with a 3 in pitch.

(b) For the connection between the structural tee flange and the column flange, bolts are in combined shear and tension. The design of this connection is a trial-and-error procedure.

The total shear to be resisted by the bolt group is

$$V = \left(\tfrac{3}{5}\right)(140\ \text{kips}) = 84\ \text{kips}$$

The total tension to be resisted by the bolt group is

$$T = \left(\tfrac{4}{5}\right)(140\ \text{kips}) = 112\ \text{kips}$$

The allowable shear in each bolt is 12.6 kips (*AISC Manual*, Part 4, Table I-D). Assuming the bolts are in shear alone, the required number of bolts is

$$n_1 = \frac{84\ \text{kips}}{12.6\ \text{kips}} = 6.6$$

The allowable tension in each bolt is 26.5 kips (*AISC Manual*, Part 4, Table I-A). Assuming bolts are in tension alone, the required number of bolts is

$$n_2 = \frac{112\ \text{kips}}{26.5\ \text{kips}} = 4.23$$

Try eight bolts, with four in each vertical row. Assume that the tensile load passes through the center of gravity of the bolt group. The allowable bolt shear stress is $F_v = 21$ ksi (*AISC Manual*, Part 4, Table I-D).

The average bolt shear stress is

$$f_v = \frac{V}{nA_b} = \frac{84\ \text{kips}}{(8)(0.6013\ \text{in}^2)} = 17.5\ \text{ksi} < 21\ \text{ksi} \quad [\text{OK}]$$

Structural

The allowable tensile stress in bolts is (*AISC Specifications* Table J3.3)

$$\begin{aligned}F_t &= \sqrt{(44)^2 - 4.39f_v^2}\\ &= \sqrt{(44)^2 - (4.39)(17.5\text{ ksi})^2}\\ &= 24.3\text{ ksi}\end{aligned}$$

Neglecting prying action, the actual tensile stress in bolts is

$$\begin{aligned}f_t &= \frac{T}{nA_b} = \frac{112\text{ kips}}{(8)(0.6013\text{ in}^2)}\\ &= 23.3\text{ ksi} < 24.3\text{ ksi}\quad[\text{OK}]\end{aligned}$$

> Use eight bolts, with four in each vertical row, at 3 in pitch.

(The prying force on these bolts should also be considered. See the *AISC Manual*, Part 4 example problem for design calculations.)

2. (a) Consider a slip-critical connection with eight $^3/_4$ in diameter A325 bolts. The tensile load on the bolt group is

$$P_t = \tfrac{4}{5}P = 0.8P$$

The shear load on the bolt group is

$$P_v = \tfrac{3}{5}P = 0.6P$$

The bolt area is $A_b = 0.4418\text{ in}^2$.

The actual shear stress per bolt is

$$f_v = \frac{0.6\text{P}}{(8)(0.4418\text{ in}^2)} = 0.17P$$

Neglecting prying action, the actual tensile stress per bolt is

$$f_t = \frac{0.8\text{P}}{(8)(0.4418\text{ in}^2)} = 0.226P$$

The allowable shear stress, F_v', is calculated from the tabulated allowable value of $F_v = 17$ ksi (*AISC Specifications* Table J3.2) and the reduction factor from Eq. 65.3 (with T_b = minimum pretension load = 28 kips from the *AISC Specifications* Table J3.7).

$$\begin{aligned}F_v' &= F_v\left(1 - \frac{f_tA_b}{T_b}\right)\\ &= (17\text{ ksi})\left(1 - \frac{(0.226P)(0.4418\text{ in}^2)}{28\text{ kips}}\right)\\ &= 17 - 0.061P\end{aligned}$$

Equating f_v to F_v',

$$\begin{aligned}0.17P &= 17 - 0.061P\\ P &= 73.6\text{ kips}\\ f_t &= \left(0.226\ \frac{1}{\text{in}^2}\right)(73.6\text{ kips}) = 16.6\text{ ksi}\end{aligned}$$

The allowable tensile stress is $F_t = 44$ ksi (*AISC Manual* Table J3.2).

$$\begin{aligned}f_t &< F_t\quad[\text{OK}]\\ P &= \boxed{73.6\text{ kips}}\end{aligned}$$

(b) Consider a bearing-type connection with eight $^3/_4$ in diameter A325X bolts. From part (a), $f_t = 0.226P$ and $f_v = 0.17P$. From the *AISC Specifications* Table J3.3, the allowable tensile stress is

$$\begin{aligned}F_t &= \sqrt{(44)^2 - 2.15f_v^2}\\ &= \sqrt{(44)^2 - (2.15)(0.17P)^2}\end{aligned}$$

Equating f_t to F_t,

$$0.226P = \sqrt{(44)^2 - (2.15)(0.17P)^2}$$

Squaring both sides,

$$\begin{aligned}0.1132P^2 &= 1936\\ P &= 130.8\text{ kips}\\ f_v &= \left(0.17\ \frac{1}{\text{in}^2}\right)(130.8 kips) = 22.2\text{ ksi}\end{aligned}$$

The allowable shear stress is $F_v = 30$ ksi (*AISC Manual* Table J3.2).

$$\begin{aligned}f_v &< F_v\quad[\text{OK}]\\ P &= \boxed{130.8\text{ kips}}\end{aligned}$$

3. For A588 grade-50 steel, $F_y = 50$ ksi and $F_u = 70$ ksi.

(a) The gross area of each plate is

$$A_g = bt = (6\text{ in})\left(\tfrac{5}{8}\text{ in}\right) = 3.75\text{ in}^2$$

The allowable tensile load on the member based on the yielding criterion is

$$\begin{aligned}P_t &= 0.6F_yA_g = (0.6)(50\text{ ksi})(3.75\text{ in}^2)\\ &= \boxed{112.5\text{ kips}}\end{aligned}$$

The answer is (D).

(b) For $^7/_8$ in diameter bolts, the standard hole diameter for the purpose of analysis or design is

$$\begin{aligned} d_h &= d_b + \tfrac{1}{8}\ \text{in} = \tfrac{7}{8}\ \text{in} + \tfrac{1}{8}\ \text{in} \\ &= 1\ \text{in} \end{aligned}$$

The net area is

$$\begin{aligned} A_n &= b_n t = \left(6\ \text{in} - (2)(1\ \text{in})\right)\left(\tfrac{5}{8}\ \text{in}\right) \\ &= 2.5\ \text{in}^2 \end{aligned}$$

Since $U = 1.0$, the effective net area is

$$\begin{aligned} A_e &= UA_n = (1.0)(2.5\ \text{in}^2) \\ &= 2.5\ \text{in}^2 \end{aligned}$$

The allowable tensile load on the member based on the fracture criterion is

$$\begin{aligned} P_t &= 0.5F_uA_e = (0.5)(70\ \text{ksi})(2.5\ \text{in}^2) \\ &= \boxed{87.5\ \text{kips}} \end{aligned}$$

The answer is (C).

(c) The allowable bearing stress in the member is [AISC Eq. J3-4]

$$\begin{aligned} F_p &= 1.5F_u = (1.5)(70\ \text{ksi}) \\ &= \boxed{105\ \text{ksi}} \end{aligned}$$

The answer is (D).

(d) For each A325SC $^7/_8$ in diameter bolt in single shear, the allowable shear is 10.2 kips (*AISC Manual*, Part 4, Table I-D). The shear capacity of the slip-critical connection is

$$\begin{aligned} \text{shear capacity} &= (4)(10.2\ \text{kips}) \\ &= \boxed{40.8\ \text{kips}} \end{aligned}$$

The answer is (B).

(e) The tensile load capacity is the smallest of the results of parts (a), (b), and (d) (i.e., smallest of 112.5 kips, 87.5 kips, and 40.8 kips), which is $\boxed{40.8\ \text{kips.}}$

The answer is (A).

(f) For each A325N $^7/_8$ in diameter bolt in single shear, the allowable shear is 12.6 kips (*AISC Manual*, Part 4, Table I-D). The shear capacity of the slip-critical connection is

$$(4)(12.6\ \text{kips}) = \boxed{50.4\ \text{kips}}$$

The answer is (B).

(g) The tensile load capacity is the smallest of the results of parts (a), (b), and (f) (i.e., the smallest of 112.5 kips, 87.5 kips, and 50.4 kips), which is $\boxed{50.4\ \text{kips.}}$

The answer is (B).

(h) For each A325X $^7/_8$ in diameter bolt in single shear, the allowable shear is 18.0 kips (*AISC Manual*, Part 4, Table I-D). The shear capacity of the slip-critical connection is

$$(4)(18.0\ \text{kips}) = \boxed{72\ \text{kips}}$$

The answer is (C).

(i) The tensile load capacity is the smallest of the results of parts (a), (b), and (h) (i.e., the smallest of 112.5 kips, 87.5 kips, and 72 kips), which is $\boxed{72\ \text{kips.}}$

The answer is (B).

(j) From part (d) with a slip-critical connection, the tensile capacity of the member is 40.8 kips. The load per bolt is

$$P = \frac{40.8\ \text{kips}}{4} = 10.2\ \text{kips}$$

The actual bearing stress is

$$\begin{aligned} f_p &= \frac{P}{dt} = \frac{10.2\ \text{kips}}{\left(\frac{7}{8}\ \text{in}\right)\left(\frac{5}{8}\ \text{in}\right)} \\ &= \boxed{18.65\ \text{kips/in}^2} \end{aligned}$$

The answer is (B).

4. Since the bolt group is symmetrical, the centroid is located between bolts 4 and 11.

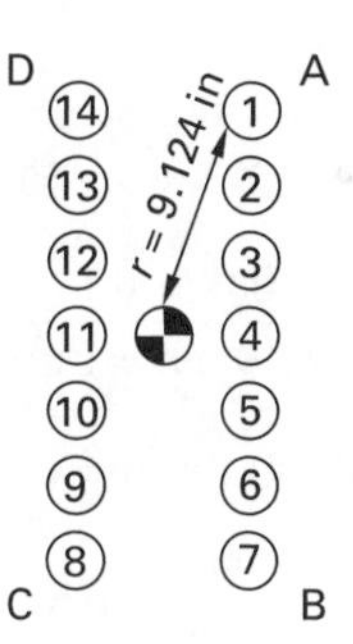

Structural

(a) For each bolt,

$$A_i = \left(\frac{\pi}{4}\right)(0.75 \text{ in})^2 = 0.4418 \text{ in}^2$$

$$J_i = \left(\frac{\pi}{32}\right)(0.75 \text{ in})^4 = 0.03106 \text{ in}^4$$

The solution is tabulated.

bolt	r (in)	r^2A (in^4)
3	3.354	4.97
2	6.185	16.90
1	9.124	36.78
	total	58.65
4	1.5	0.99

The total polar moment of inertia is

$$J = (14)(0.03106 \text{ in}^4) + (4)(58.65 \text{ in}^4) + (2)(0.99 \text{ in}^4)$$

$$= \boxed{237.0 \text{ in}^4}$$

The answer is (B).

(b) The torsional shear stress in bolts A, B, C, and D is

$$f_{v,t} = \frac{Tr}{J} = \frac{(30{,}000 \text{ lbf})(17 \text{ in})(9.124 \text{ in})}{237 \text{ in}^4}$$

$$= \boxed{19{,}634 \text{ lbf/in}^2 \quad (19{,}634 \text{ psi})}$$

The answer is (C).

(c) The vertical component of the torsional shear stress in bolts A, B, C, and D is

$$f_{v,t,y} = \left(\frac{1.5 \text{ in}}{9.124 \text{ in}}\right)(19{,}634 \text{ psi}) = \boxed{3228 \text{ psi}}$$

The answer is (C).

(d) The horizontal component of the torsional shear stress in bolts A, B, C, and D is

$$f_{v,t,x} = \left(\frac{9 \text{ in}}{9.124 \text{ in}}\right)(19{,}634 \text{ psi}) = \boxed{19{,}367 \text{ psi}}$$

The answer is (D).

(e) The direct vertical shear stress in the bolts is

$$f_{v,d} = \frac{30{,}000 \text{ lbf}}{(14)(0.4418 \text{ in}^2)}$$

$$= \boxed{4850 \text{ lbf/in}^2 \text{ (4850 psi)}}$$

The answer is (C).

(f) The total shear stress in bolts A and B is

$$\sqrt{(19{,}367 \text{ psi})^2 + (3228 \text{ psi} + 4850 \text{ psi})^2}$$

$$= \boxed{20{,}984 \text{ psi}}$$

The answer is (C).

(g) The total shear stress in bolts C and D is

$$\sqrt{(19{,}367 \text{ psi})^2 + (3228 \text{ psi} - 4850 \text{ psi})^2}$$

$$= \boxed{19{,}435 \text{ psi}}$$

The answer is (D).

(h) Use the *AISC Manual*, Part 4, Table XII with a total number of fasteners in one vertical row, $n = 7$, and bolt pitch, $b = 3$ in. For a lever arm (eccentricity of load), $l = 16$ in, and the coefficient C is 4.27. For $l = 18$ in, $C = 3.83$. By interpolation, for $l = 17$ in,

$$C = \frac{4.27 + 3.83}{2} = 4.05$$

From the *AISC Manual*, Part 4, Table I-D, for an A325SC $^3/_4$ in bolt in single shear (standard hole), the allowable shear load, r_v, is 7.51 kips. The allowable eccentric load on the fastener group is

$$P = Cr_v = (4.05)(7.51 \text{ kips})\left(1000 \frac{\text{lbf}}{\text{kip}}\right)$$

$$= \boxed{30{,}415 \text{ lbf}}$$

The answer is (D).

(i) From part (h), $C = 4.05$. From the *AISC Manual*, Part 4, Table I-D, for A325N $^3/_4$ in bolt in single shear (standard hole), the allowable shear load, r_v, is 9.3 kips. The allowable eccentric load on the fastener group is

$$P = Cr_v = (4.05)(9.3 \text{ kips})\left(1000 \frac{\text{lbf}}{\text{kip}}\right)$$

$$= \boxed{37{,}665 \text{ lbf}}$$

The answer is (C).

(j) The actual eccentric load is 30,000 lbf. The allowable load for the A325SC bolt group is 30,415 lbf, which is greater than 30,000 lbf. The allowable load for the A325N bolt group is 37,665 lbf, which is also greater than 30,000 lbf. The connection is adequate for both A325SC and A325N bolts.

The answer is (C).

66 Structural Steel: Welding

PRACTICE PROBLEMS

1. Determine the size of the fillet weld required to connect the 1 in thick plate bracket to the column flange as shown. A36 steel is used for both the bracket and the column. Assume SMAW process with E70XX electrodes.

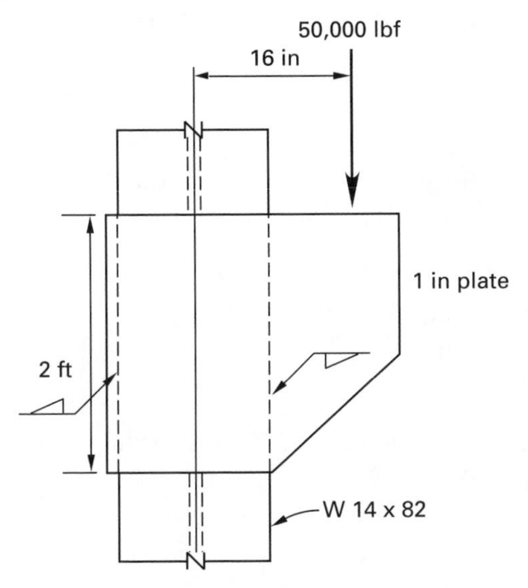

2. The connection between the column flange and plate is made with a $^3/_8$ in size fillet weld using E70 electrodes. Determine the capacity, P, of the connection if the process is (a) SMAW and (b) SAW.

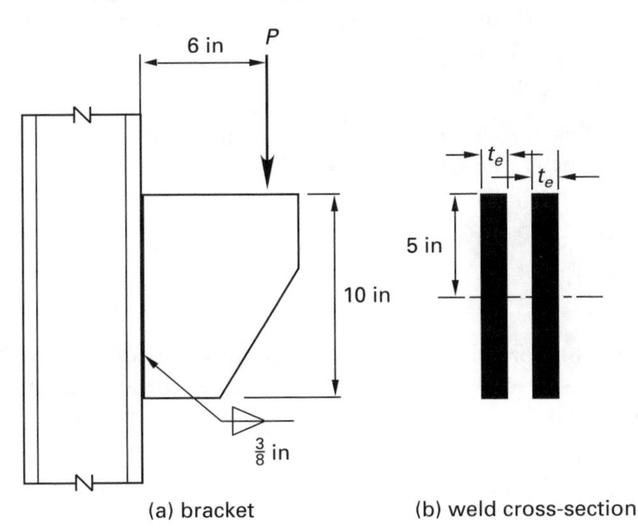

(a) bracket (b) weld cross-section

3. *(Time limit: one hour)* The fillet-welded lap joint shown is to be designed as a connection between the long leg of an L $5 \times 3 \times {}^1/_2$ tension member of A36 steel and a $^1/_2$ in thick gusset plate. The full strength of the angle is to be developed, minimizing the effect of eccentricity. E70 electrodes will be used. The gusset plate does not govern the design.

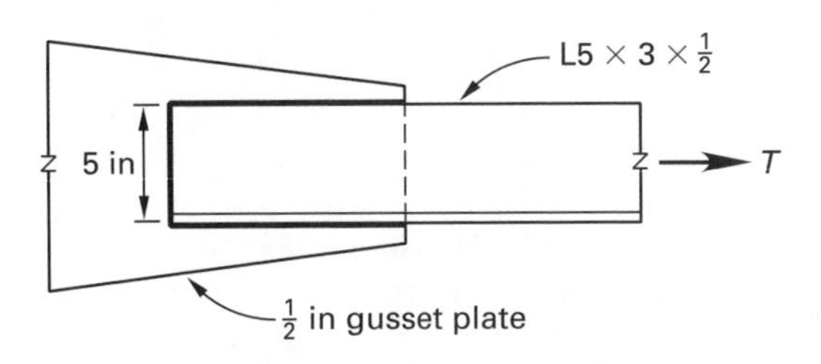

(a) The axial tensile capacity of the angle member is most nearly

(A) 72 kips
(B) 81 kips
(C) 89 kips
(D) 110 kips

(b) The minimum and maximum sizes of fillet weld that can be used are most nearly

(A) $^1/_8$ in and $^3/_8$ in
(B) $^3/_{16}$ in and $^7/_{16}$ in
(C) $^1/_4$ in and $^1/_2$ in
(D) $^5/_{16}$ in and $^5/_8$ in

For parts (c) through (g), assume an SMAW process with the minimum permissible weld size is used.

(c) The strength of the fillet weld per unit length is most nearly

(A) 2.0 kips/in
(B) 2.3 kips/in
(C) 2.8 kips/in
(D) 3.0 kips/in

(d) The load carried by the end (transverse) weld, P_2, is most nearly

(A) 11.4 kips
(B) 12.7 kips
(C) 13.9 kips
(D) 14.3 kips

(e) The load P_3 carried by the bottom longitudinal weld is most nearly

(A) 33 kips
(B) 40 kips
(C) 46 kips
(D) 51 kips

(f) The load P_1 carried by the top longitudinal weld is most nearly

(A) 17 kips
(B) 21 kips
(C) 26 kips
(D) 31 kips

(g) The required lengths of the top and bottom longitudinal welds are, respectively, most nearly

(A) 7 in and 17 in
(B) 8 in and 17 in
(C) 7 in and 18 in
(D) 8 in and 18 in

For parts (h) through (j), assume an SAW process with the minimum permissible weld size is used.

(h) The load, P_3, carried by the bottom longitudinal weld is most nearly

(A) 33 kips
(B) 38 kips
(C) 40 kips
(D) 43 kips

(i) The load, P_1, carried by the top longitudinal weld is most nearly

(A) 19 kips
(B) 21 kips
(C) 26 kips
(D) 31 kips

(j) The required lengths of the top and bottom longitudinal welds are, respectively, most nearly

(A) 5 in and 11 in
(B) 6 in and 11 in
(C) 5 in and 12 in
(D) 6 in and 12 in

4. *(Time limit: one hour)* The adequacy of a $^3/_8$ in fillet welded connection between a bracket and a column is to be checked. E70 electrodes are used. The column and the bracket do not control the design.

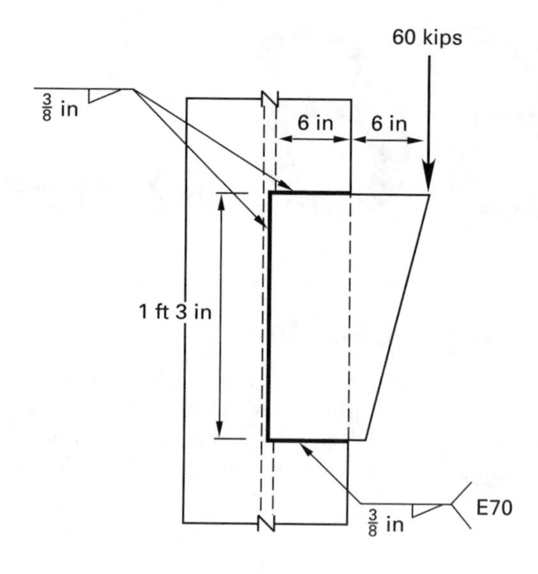

For parts (a) through (g), assume an effective throat thickness, t_e, of 1 in.

(a) The distance measured from the left end of the bracket plate to the centroid of the weld group is most nearly

(A) 1.1 in
(B) 1.3 in
(C) 1.9 in
(D) 2.1 in

(b) The polar moment of inertia of the weld group is most nearly

(A) 830 in^4
(B) 970 in^4
(C) 1100 in^4
(D) 1300 in^4

(c) The torsional moment acting on the weld group is most nearly

(A) 410 in-kip
(B) 470 in-kip
(C) 520 in-kip
(D) 640 in-kip

(d) The maximum stress in weld due to the torsional moment is most nearly

(A) 3.5 ksi
(B) 4.2 ksi
(C) 5.4 ksi
(D) 6.8 ksi

(e) The horizontal and vertical components of the maximum stress in weld due to the torsional moment are, respectively, most nearly

(A) 3.9 ksi and 3.4 ksi
(B) 4.6 ksi and 2.8 ksi
(C) 5.0 ksi and 2.1 ksi
(D) 5.7 ksi and 1.9 ksi

(f) The stress in weld due to direct shear is most nearly

(A) 1.7 ksi
(B) 2.2 ksi
(C) 2.9 ksi
(D) 3.2 ksi

(g) The resultant maximum stress (combination of torsional and direct shear stresses) in the weld group is most nearly

(A) 3.5 ksi
(B) 4.8 ksi
(C) 5.8 ksi
(D) 6.8 ksi

(h) Assuming that the SMAW process is used, the actual resistance per unit length of weld is most nearly

(A) 3.8 kips/in
(B) 4.7 kips/in
(C) 5.6 kips/in
(D) 6.8 kips/in

(i) Assuming that the SAW process is used, the shear resistance per unit length of weld is most nearly

(A) 3.8 kips/in
(B) 4.9 kips/in
(C) 6.3 kips/in
(D) 7.9 kips/in

(j) The connection is adequate for

(A) both SMAW and SAW processes
(B) neither SMAW nor SAW processes
(C) an SMAW process only
(D) an SAW process only

5. *(Time limit: one hour)* Two 4 × 3 × 1/2 angles are welded back to back to an intermediate wide-flange section that is bolted to a column as shown. All pieces are A36 steel. E70 electrodes are used to form the maximum weld size permitted. No threads are in the shear plane. (a) Specify the lengths and locations of the welds used to connect the angles to the intermediate section using a balanced configuration. (b) Determine if the bolts are adequate for a slip-critical-type connection. (c) Determine if the bolts are adequate for a bearing-type connection.

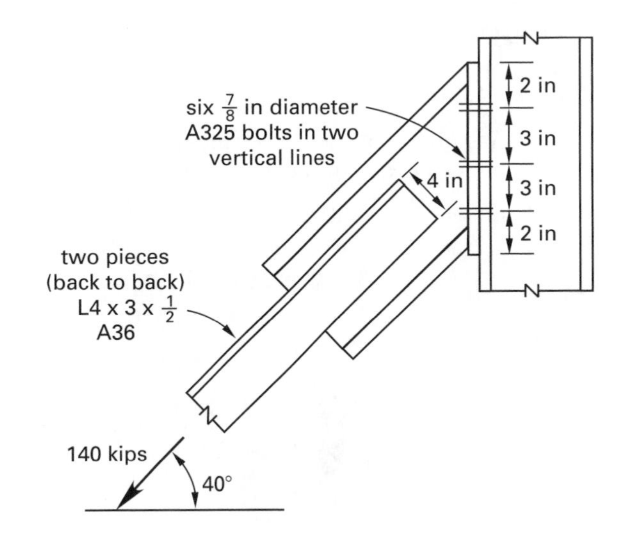

Structural

SOLUTIONS

1. *Elastic Analysis:*

step 1: Assume the weld has thickness t. (All dimensions are in inches.)

step 2: Locate the centroid by inspection.

$$x_c = y_c = 0$$

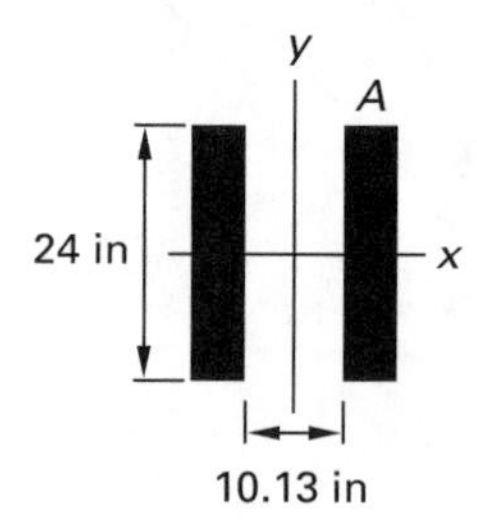

step 3: $I_x = (2)\left(\frac{t(24\text{ in})^3}{12}\right) = 2304t$

step 4: $b_f = 10.13\text{ in}$ [for $W\,14 \times 82$]

$$\begin{aligned} I_y &= (2)\left(\frac{24t^3}{12} + (24t)\left(\frac{10.13+t}{2}\right)^2\right) \\ &\approx 1232t \quad [\text{assuming } t^2 = t^3 = 0] \end{aligned}$$

step 5: $J = 2304t + 1232t = 3536t$

step 6: The maximum shear stress will occur at point A. The radial distance to point A is

$$r = \sqrt{\left(\frac{10.13\text{ in}}{2}\right)^2 + (12\text{ in})^2} = 13.03\text{ in}$$

step 7: The applied moment is

$$\begin{aligned} M &= Pe = (50{,}000\text{ lbf})(16\text{ in}) \\ &= 800{,}000\text{ in-lbf} \end{aligned}$$

step 8: Use Eq. 66.13. The torsional shear stress is

$$\begin{aligned} f_{v,t} &= \frac{Mc}{J} = \frac{(800{,}000\text{ in-lbf})(13.03\text{ in})}{3536t} \\ &= 2948/t \end{aligned}$$

$$f_{v,t,x} = \left(\frac{12\text{ in}}{13.03\text{ in}}\right)\left(\frac{2948}{t}\right) = 2715/t \quad [\text{to the right}]$$

$$\begin{aligned} f_{v,t,y} &= \left(\frac{10.13\text{ in}}{(2)(13.03\text{ in})}\right)\left(\frac{2948}{t}\right) \\ &= 1146/t \quad [\text{down}] \end{aligned}$$

step 9: Use Eq. 66.14. The direct shear stress is

$$\begin{aligned} f_{v,d} &= \frac{P}{A} = \frac{50{,}000\text{ lbf}}{(2)(24t)} \\ &= 1042/t \quad [\text{down}] \end{aligned}$$

step 10: Use Eq. 66.15. The resultant shear stress at point A is

$$\begin{aligned} f_v &= \left(\frac{1}{t}\right)\sqrt{(2715)^2 + (1146+1042)^2} \\ &= 3487/t \end{aligned}$$

step 11: Use Eq. 66.4. The allowable shear stress in the weld is

$$\begin{aligned} F_v &= 0.30F_{u,\text{rod}} = (0.30)\left(70{,}000\ \frac{\text{lbf}}{\text{in}^2}\right) \\ &= 21{,}000\text{ lbf/in}^2 \end{aligned}$$

Equating the actual and allowable stresses,

$$\begin{aligned} \frac{3487}{t} &= 21{,}000\text{ lbf/in}^2 \\ t &= 0.166\text{ in} \end{aligned}$$

step 12: Use Eq. 66.1. The weld size is

$$w = \frac{0.166\text{ in}}{0.707} = 0.24\text{ in}$$

The flange thickness of $W\,14 \times 82$ is 0.855 in, and the thickness of the bracket plate is 1 in. Refer to Table 66.1.

Since the thickness of the thicker part is over $^3/_4$ in, use a minimum weld size of $\boxed{^5/_{16}\text{ in.}}$

AISC Method:

Refer to the *AISC Manual*, Part 4, Table XIX (Eccentric Loads on Weld Groups).

$$\begin{aligned} l &= (2\text{ ft})\left(12\ \frac{\text{in}}{\text{ft}}\right) = 24\text{ in} \\ al &= 16\text{ in} \\ a &= \frac{al}{l} = \frac{16\text{ in}}{24\text{ in}} = 0.667 \\ kl &= b_f = 10.13\text{ in} \\ k &= \frac{kl}{l} = \frac{10.13\text{ in}}{24\text{ in}} = 0.422 \end{aligned}$$

Use double interpolation with $a = 0.666$ and $k = 0.422$ to get $C = 0.746$. $C_1 = 1.0$ for E70 electrodes. Use Eq. 66.16.

$$D = \frac{P}{CC_1 l} = \frac{50 \text{ kips}}{(0.746)(1)(24 \text{ in})}$$
$$= 2.79 \quad [\text{say } 3]$$
$$w = D\left(\tfrac{1}{16} \text{ in}\right) = 3/16 \text{ in}$$

From Table 66.1, since the thickness of the thicker part is over 3/4 in, the minimum weld size is $\boxed{5/16 \text{ in.}}$

2. *Elastic Method:*

This is an eccentrically loaded welded connection where the weld group is subjected to combined shear and bending stresses.

$$w = \tfrac{3}{8} \text{ in}$$
$$L = 10 \text{ in}$$

For E70 electrodes, $F_u = 70 \text{ kips/in}^2$.

(a) *SMAW Process:* (All forces are in kips. All dimensions are in inches.) Use Eq. 66.1.

$$t_e = 0.707w = (0.707)\left(\tfrac{3}{8} \text{ in}\right) = 0.265 \text{ in}$$

From Eq. 66.17, the shear stress is

$$f_v = \frac{P}{2Lt_e} = \frac{\text{P}}{(2)(10 \text{ in})(0.265 \text{ in})}$$
$$= 0.189P$$

The moment of inertia of the weld group is

$$I = \frac{2t_e L^3}{12} = \frac{(2)(0.265 \text{ in})(10 \text{ in})^3}{12}$$
$$= 44.167 \text{ in}^4$$

From Eq. 66.18, the maximum bending stress is

$$f_b = \frac{Mc}{I} = \frac{Pec}{I} = \frac{P(6 \text{ in})(5 \text{ in})}{44.167 \text{ in}^4}$$
$$= 0.679P$$

From Eq. 66.19, the maximum resultant stress is

$$f = \sqrt{f_v^2 + f_b^2}$$
$$= \sqrt{(0.189P)^2 + (0.679P)^2}$$
$$= 0.705P$$

The allowable shear stress for the weld is

$$F_v = 0.3F_u = (0.3)\left(70 \frac{\text{kips}}{\text{in}^2}\right) = 21 \text{ kips/in}^2$$

Equating the resultant stress to the allowable stress,

$$0.705P = 21 \text{ kips/in}^2$$
$$P = \boxed{29.8 \text{ kips}}$$

(b) *SAW Process:* (All forces are in kips. All dimensions are in inches.) Use Eq. 66.2.

$$t_e = w = \tfrac{3}{8} \text{ in} = 0.375 \text{ in}$$

From Eq. 66.17, the shear stress is

$$f_v = \frac{P}{2Lt_e} = \frac{P}{(2)(10 \text{ in})(0.375 \text{ in})}$$
$$= 0.133P$$

The moment of inertia of the weld group is

$$I = \frac{2t_e L^3}{12} = \frac{(2)(0.375 \text{ in})(10 \text{ in})^3}{12}$$
$$= 62.5 \text{ in}^4$$

From Eq. 66.18, the maximum bending stress is

$$f_b = \frac{Mc}{I} = \frac{Pec}{I} = \frac{P(6 \text{ in})(5 \text{ in})}{62.5 \text{ in}^4}$$
$$= 0.48P$$

From Eq. 66.19, the maximum resultant stress is

$$f = \sqrt{f_v^2 + f_b^2}$$
$$= \sqrt{(0.133P)^2 + (0.48P)^2}$$
$$= 0.498P$$

Use Eq. 66.4. The allowable shear stress for the weld is

$$F_v = 0.3F_u = (0.3)\left(70 \frac{\text{kips}}{\text{in}^2}\right) = 21 \text{ kips/in}^2$$

Equating the resultant stress to the allowable stress,

$$0.498P = 21 \text{ kips/in}^2$$
$$P = \boxed{42.2 \text{ kips}}$$

AISC Method:

Refer to *AISC Manual*, Part 4, Table XIX (Special Case: Load Not in Plane of Weld Group).

$$l = (2)(5 \text{ in}) = 10 \text{ in}$$
$$al = 6 \text{ in}$$
$$a = \frac{al}{l} = \frac{6 \text{ in}}{10 \text{ in}} = 0.6$$
$$k = 0$$
$$C = 0.673 \quad [\text{table value}]$$

$C_1 = 1.0$ for E70 electrodes.

Structural

(a) *SMAW Process:*

$$t_e = 0.707w = (0.707)\left(\frac{3}{8}\text{ in}\right) = 0.265\text{ in}$$

$$D = \frac{t_e}{\frac{1}{16}\text{ in}} = \frac{0.265\text{ in}}{\frac{1}{16}\text{ in}} = 4.24$$

$$P = CC_1Dl = (0.673)(1)(4.24)(10) = \boxed{28.5\text{ kips}}$$

(b) *SAW Process:*

$$t_e = w = \frac{3}{8}\text{ in}$$

$$D = \frac{t_e}{\frac{1}{16}\text{ in}} = \frac{\frac{3}{8}\text{ in}}{\frac{1}{16}\text{ in}} = 6$$

$$P = CC_1Dl = (0.673)(1)(6)(10) = \boxed{40.4\text{ kips}}$$

3. (a) For an A36 steel angle member, $F_y = 36\text{ kips/in}^2$ and $F_u = 58\text{ kips/in}^2$.

From the *AISC Manual*, Part 1, Shape Tables, for L5 × 3 × 1/2, $A_g = 3.75\text{ in}^2$.

From Table 60.2, $U = 1.00$.

By Eq. 60.7,

$$\begin{aligned} A_e &= UA_g = (1.00)(3.75\text{ in}^2) \\ &= 3.75\text{ in}^2 \end{aligned}$$

The tensile capacity is calculated using Eqs. 60.11 and 60.12.

The force to cause yielding is

$$\begin{aligned} P_t &= 0.6F_yA_g = (0.6)\left(36\ \frac{\text{kips}}{\text{in}^2}\right)(3.75\text{ in}^2) \\ &= \boxed{81\text{ kips}\quad[\text{controls}]} \end{aligned}$$

The force to cause fracture is

$$\begin{aligned} P_t &= 0.5F_uA_e = (0.5)\left(58\ \frac{\text{kips}}{\text{in}^2}\right)(3.75\text{ in}^2) \\ &= 108.75\text{ kips} \end{aligned}$$

The answer is (B).

(b) From Table 66.1, the minimum size fillet weld is $\boxed{3/16\text{ in}}$ for the thicker plate size of 1/2 in. The maximum size of the fillet weld is

$$\tfrac{1}{2}\text{ in} - \tfrac{1}{16}\text{ in} = \boxed{\tfrac{7}{16}\text{ in}}$$

The answer is (B).

(c) For an SMAW process with $F_{u,\text{rod}} = 70\text{ kips/in}^2$,

$$w = \tfrac{3}{16}\text{ in}\quad[\text{minimum size of weld}]$$

From Eqs. 66.1 and 66.5,

$$\begin{aligned} R_w &= (0.30)(0.707w)F_{u,\text{rod}} \\ &= (0.30)(0.707)\left(\tfrac{3}{16}\text{ in}\right)\left(70\ \frac{\text{kips}}{\text{in}^2}\right) \\ &= \boxed{2.78\text{ kips/in}} \end{aligned}$$

The answer is (C).

Refer to Fig. 66.4 for (d) through (j).

(d) $L_2 = d = 5\text{ in}$

From Eq. 66.8, the load carried by the end (transverse) weld is

$$P_2 = R_wL_2 = \left(2.78\ \frac{\text{kips}}{\text{in}}\right)(5\text{ in}) = \boxed{13.9\text{ kips}}$$

The answer is (C).

(e) From the *AISC Manual* Part 1, Shape Tables, for L5 × 3 × 1/2, $y = 1.75$ in.

From part (a), $T = 81$ kips. From Eq. 66.7, the load carried by the bottom longitudinal weld is

$$\begin{aligned} P_3 &= T\left(1 - \frac{y}{d}\right) - \frac{P_2}{2} \\ &= (81\text{ kips})\left(1 - \frac{1.75\text{ in}}{5\text{ in}}\right) - \frac{13.9\text{ kips}}{2} \\ &= \boxed{45.7\text{ kips}} \end{aligned}$$

The answer is (C).

(f) From Eq. 66.9, the load carried by the top longitudinal weld is

$$\begin{aligned} P_1 &= T - P_2 - P_3 \\ &= 81\text{ kips} - 13.9\text{ kips} - 45.7\text{ kips} = \boxed{21.4\text{ kips}} \end{aligned}$$

The answer is (B).

(g) From Eq. 66.10, the required length of the top longitudinal weld is

$$\begin{aligned} L_1 &= \frac{P_1}{R_w} = \frac{21.4\text{ kips}}{2.78\ \frac{\text{kips}}{\text{in}}} \\ &= \boxed{7.7\text{ in}} \end{aligned}$$

From Eq. 66.11, the required length of the bottom longitudinal weld is

$$L_3 = \frac{P_3}{R_w} = \frac{45.7 \text{ kips}}{2.78 \ \frac{\text{kips}}{\text{in}}} = \boxed{16.44 \text{ in}}$$

The answer is (B).

(h) For an SAW process with $F_{u,\text{rod}} = 70 \text{ kips/in}^2$,

$$w = \tfrac{3}{16} \text{ in} \quad [\text{minimum size of weld}] \le \tfrac{3}{8} \text{ in}$$

From Eq. 66.2,

$$t_e = w = \tfrac{3}{16} \text{ in}$$

From Eq. 66.5,

$$R_w = 0.30 t_e F_{u,\text{rod}} = (0.30)\left(\tfrac{3}{16} \text{ in}\right)\left(70 \ \frac{\text{kips}}{\text{in}^2}\right) = 3.94 \text{ kips/in}$$

By Eq. 66.8, the load carried by the end (transverse) weld is

$$P_2 = R_w L_2 = \left(3.94 \ \frac{\text{kips}}{\text{in}}\right)(5 \text{ in}) = 19.7 \text{ kips}$$

From Eq. 66.7 (from part (a), $T = 81$ kips), the load carried by the bottom longitudinal weld is

$$P_3 = T\left(1 - \frac{y}{d}\right) - \frac{P_2}{2} = (81 \text{ kips})\left(1 - \frac{1.75 \text{ in}}{5 \text{ in}}\right) - \frac{19.7 \text{ kips}}{2} = \boxed{42.8 \text{ kips}}$$

The answer is (D).

(i) From Eq. 66.9, the load carried by the top longitudinal weld is

$$P_1 = T - P_2 - P_3 = 81 \text{ kips} - 19.7 \text{ kips} - 42.8 \text{ kips} = \boxed{18.5 \text{ kips}}$$

The answer is (A).

(j) By Eq. 66.10, the required length of the top longitudinal weld is

$$L_1 = \frac{P_1}{R_w} = \frac{18.5 \text{ kips}}{3.94 \ \frac{\text{kips}}{\text{in}}} = \boxed{4.7 \text{ in}}$$

From Eq. 66.11, the required length of the bottom longitudinal weld is

$$L_3 = \frac{P_3}{R_w} = \frac{42.8 \text{ kips}}{3.94 \ \frac{\text{kips}}{\text{in}}} = \boxed{10.86 \text{ in}}$$

The answer is (A).

4. The weld group is subjected to combined shear and torsion.

(a) The location of the centroid of the weld group is obtained by summing moments about the 15 in side.

$$\overline{x} = \frac{t_e\big((2)(6 \text{ in})(3 \text{ in}) + (15 \text{ in})(0)\big)}{t_e\big((2)(6 \text{ in}) + 15 \text{ in}\big)} = \boxed{1.33 \text{ in}}$$

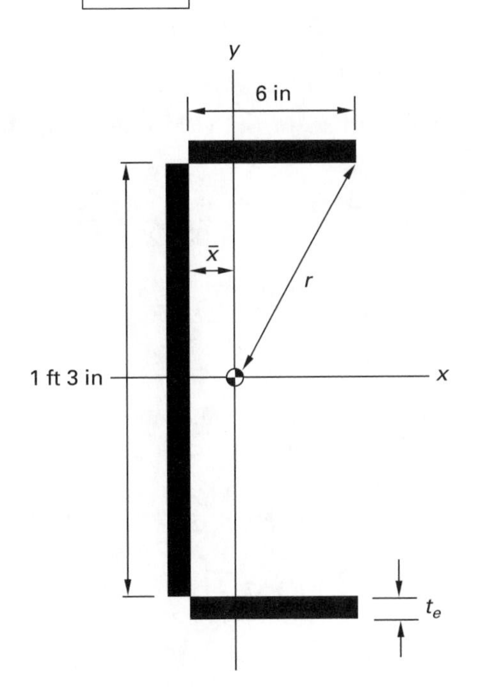

The answer is (B).

(b) The moment of inertia of the weld group about the x-axis (neglecting the moment of inertia of the top and bottom welds about their own centroidal axes) is

$$I_x = \left(\tfrac{1}{12}\right)(1 \text{ in})(15 \text{ in})^3 + (2)(1 \text{ in})(6 \text{ in})(7.5 \text{ in})^2 = 956.25 \text{ in}^4$$

The moment of inertia of the weld group about the y-axis is

$$I_y = (2)\left(\tfrac{1}{12}\right)(1 \text{ in})(6 \text{ in})^3 + (2)(1 \text{ in})(6 \text{ in})(1.67 \text{ in})^2 + (1 \text{ in})(15 \text{ in})(1.33 \text{ in})^2 = 96 \text{ in}^4$$

The polar moment of inertia of the weld group about its centroid is

$$J = I_x + I_y = 956.25 \text{ in}^4 + 96 \text{ in}^4 = \boxed{1052.25 \text{ in}^4}$$

The answer is (C).

(c) The torsional moment on the weld group is

$$\begin{aligned} M &= Pe = (60 \text{ kips})(6 \text{ in} + 6 \text{ in} - 1.33 \text{ in}) \\ &= \boxed{640 \text{ in-kips}} \end{aligned}$$

The answer is (D).

(d) The maximum radial distance is

$$\begin{aligned} x &= 6 \text{ in} - 1.33 \text{ in} = 4.67 \text{ in} \\ y &= \frac{15 \text{ in}}{2} = 7.5 \text{ in} \\ r &= \sqrt{x^2 + y^2} = \sqrt{(4.67 \text{ in})^2 + (7.5 \text{ in})^2} \\ &= 8.83 \text{ in} \end{aligned}$$

The maximum stress in the weld due to the torsional moment is

$$\begin{aligned} f_{v,t} &= \frac{Mr}{J} = \frac{(640 \text{ in-kips})(8.83 \text{ in})}{1052.25 \text{ in}^4} \\ &= \boxed{5.37 \text{ kips/in}^2} \end{aligned}$$

The answer is (C).

(e) The horizontal and vertical components of the maximum stress in the weld are

$$f_{v,t,x} = \left(\frac{7.5 \text{ in}}{8.83 \text{ in}}\right)\left(5.37 \, \frac{\text{kips}}{\text{in}^2}\right) = \boxed{4.56 \text{ kips/in}^2}$$

$$f_{v,t,y} = \left(\frac{4.67 \text{ in}}{8.83 \text{ in}}\right)\left(5.37 \, \frac{\text{kips}}{\text{in}^2}\right) = \boxed{2.84 \text{ kips/in}^2}$$

The answer is (B).

(f) The direct shear stress in the weld is

$$\begin{aligned} f_{v,d} &= \frac{P}{A} = \frac{60 \text{ kips}}{(27 \text{ in})(1 \text{ in})} \\ &= \boxed{2.22 \text{ kips/in}^2} \end{aligned}$$

The answer is (B).

(g) Use Eq. 66.15. The resultant maximum stress is

$$\begin{aligned} f_v &= \sqrt{\left(4.56 \, \frac{\text{kips}}{\text{in}^2}\right)^2 + \left(2.22 \, \frac{\text{kips}}{\text{in}^2} + 2.84 \, \frac{\text{kips}}{\text{in}^2}\right)^2} \\ &= \boxed{6.81 \text{ kips/in}^2} \end{aligned}$$

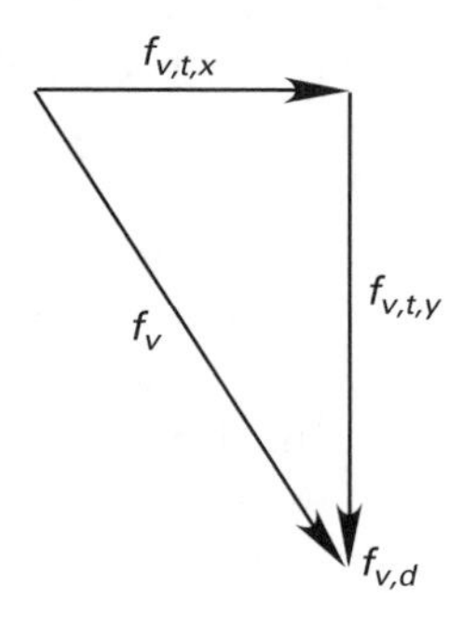

The answer is (D).

(h) For an SMAW process, with $w = 3/8$ in, from Eqs. 66.1 and 66.5, the shear strength of the weld per unit length is

$$\begin{aligned} R_w &= (0.30)(0.707w)F_{u,\text{rod}} \\ &= (0.30)(0.707)\left(\tfrac{3}{8} \text{ in}\right)\left(70 \, \frac{\text{kips}}{\text{in}^2}\right) = \boxed{5.57 \text{ kips/in}} \end{aligned}$$

The answer is (C).

(i) For an SAW process, with $w = 3/8$ in, from Eqs. 66.2 and 66.5, the shear strength of the weld per unit length is

$$\begin{aligned} R_w &= 0.30wF_{u,\text{rod}} = (0.30)\left(\tfrac{3}{8} \text{ in}\right)\left(70 \, \frac{\text{kips}}{\text{in}^2}\right) \\ &= \boxed{7.87 \text{ kips/in}} \end{aligned}$$

The answer is (D).

(j) Use Eq. 66.4. The actual weld shear strength is

$$F_v = 0.3F_{u,\text{rod}} = (0.3)\left(70 \, \frac{\text{kips}}{\text{in}^2}\right) = 21 \text{ kips/in}^2$$

Assuming $t_e = 1$ in, from part (g), the computed shear stress (i.e., the required shear resistance) is 6.81 kips/in^2.

Use Eq. 66.1. For a SMAW process with $w = 3/8$ in,

$$t_e = (0.707)\left(\tfrac{3}{8} \text{ in}\right) = 0.265 \text{ in}$$

For this actual weld size, the required shear resistance is

$$f_v = \left(6.81 \, \frac{\text{kips}}{\text{in}^2}\right)\left(\frac{1 \text{ in}}{0.265 \text{ in}}\right) = 25.7 \text{ kips/in}^2$$

Since $f_v > F_v$, SMAW is not adequate.

For a SAW process with $w = 3/8$ in,

$$t_e = \tfrac{3}{8} \text{ in} = 0.375 \text{ in}$$

For this actual weld size, the required shear resistance is

$$f_v = \left(6.81\ \frac{\text{kips}}{\text{in}^2}\right)\left(\frac{1\ \text{in}}{0.375\ \text{in}}\right) = 18.2\ \text{kips/in}^2$$

Since $f_v < F_v$, SAW is adequate.

The connection is **adequate only for an SAW process.**

The answer is (D).

5. Each angle carries (1/2)(140 kips) = 70 kips. Check the tensile stress.

$$A = 3.25\ \text{in}^2$$

$$F_a = (0.6)\left(36\ \frac{\text{kips}}{\text{in}^2}\right) = 21.6\ \text{kips/in}^2$$

$$f_a = \frac{70\ \text{kips}}{3.25\ \text{in}^2} = 21.5\ \frac{\text{kips}}{\text{in}^2} < 21.6\ \text{kips/in}^2 \quad \text{[acceptable]}$$

(a) Since thickness = 1/2 in, the maximum weld thickness is

$$w = \tfrac{1}{2}\ \text{in} - \tfrac{1}{16}\ \text{in} = \tfrac{7}{16}\ \text{in}$$

$$t_e = (0.707)\left(\tfrac{7}{16}\ \text{in}\right) = 0.309\ \text{in}$$

For the angles, the thickness is 0.50 in. For E70 electrodes,

$$R_w = (0.309\ \text{in})(0.3)\left(70\ \frac{\text{kips}}{\text{in}^2}\right) = 6.50\ \text{kips/in}$$

$$d = 4\ \text{in}$$

$$y = 1.33\ \text{in}$$

From Eq. 66.8,

$$P_2 = \left(6.5\ \frac{\text{kips}}{\text{in}}\right)(4\ \text{in}) = 26.0\ \text{kips}$$

From Eq. 66.7,

$$P_3 = (70\ \text{kips})\left(1 - \frac{1.33\ \text{in}}{4\ \text{in}}\right) - \frac{26.0\ \text{kips}}{2} = 33.7\ \text{kips}$$

$$L_3 = \frac{33.7\ \text{kips}}{6.5\ \frac{\text{kips}}{\text{in}}} = \boxed{5.18\ \text{in}} \quad \text{[round to } 5\tfrac{1}{4}\ \text{in]}$$

$$P_1 = 70\ \text{kips} - 26.0\ \text{kips} - 33.7\ \text{kips} = 10.3\ \text{kips}$$

$$L_1 = \frac{10.3\ \text{kips}}{6.5\ \frac{\text{kips}}{\text{in}}} = \boxed{1.58\ \text{in}} \quad \text{[round to } 1\tfrac{3}{4}\ \text{in]}$$

Check the minimum weld length.

$$L_1 \geq \left\{ \begin{array}{l} 4 \times \tfrac{7}{16}\ \text{in} = 1.75\ \text{in} \\ 1\tfrac{1}{2}\ \text{in} \end{array} \right\} = 1.75\ \text{in} \quad \text{[acceptable]}$$

(b)

$$P_x = (140\ \text{kips})(\cos 40^\circ) = 107.2\ \text{kips}$$

$$P_y = (140\ \text{kips})(\sin 40^\circ) = 90.0\ \text{kips}$$

For 7/8 in bolts,

$$A = \left(\frac{\pi}{4}\right)\left(\tfrac{7}{8}\ \text{in}\right)^2 = 0.6013\ \text{in}^2$$

The shear stress in each of the six bolts is

$$f_v = \frac{90\ \text{kips}}{(6)(0.6013\ \text{in}^2)} = 24.9\ \text{kips/in}^2$$

The tensile stress in each of the six bolts is

$$f_t = \frac{107.2\ \text{kips}}{(6)(0.6013\ \text{in}^2)} = 29.7\ \text{kips/in}^2$$

Assume a slip-critical type connection.

For A325 bolts in shear-only configurations, F_v = 17.0 kips/in^2.

From the *AISC Manual of Steel Construction*, Sec. J3.6 (for combined shear and tension),

$$F_v' = F_v\left(1 - \frac{f_t A_b}{T_b}\right)$$

Assume bolt tension = 39 kips (minimum preload for 7/8 in bolts from *AISC Manual of Steel Construction*, Table J3.7).

$$F_v' = \left(17.0\ \frac{\text{kips}}{\text{in}^2}\right)\left(1 - \frac{\left(29.7\ \frac{\text{kips}}{\text{in}^2}\right)(0.6013\ \text{in}^2)}{39\ \text{kips}}\right) = 9.22\ \text{kips/in}^2$$

Since 24.9 kips/in^2 > 9.22 kips/in^2,

it is not adequate.

(c) Assume a bearing-type connection.

From the *AISC Manual of Steel Construction*, Table J3.3,

$$F_t = \sqrt{(44)^2 - 2.15 f_v^2} = \sqrt{(44)^2 - (2.15)\left(24.9\ \frac{\text{kips}}{\text{in}^2}\right)^2} = 24.6\ \text{kips/in}^2$$

Since 29.7 kips/in^2 > 24.6 kips/in^2,

it is not adequate.

Check the bolt locations.

When prying action is absent, all bolts will be stressed equally. To avoid prying action, the line of action of the force (y = 1.33 in from the short side of the angle) should coincide with the bolt group centroid. This does not appear to be the case.

Structural

67 Structural Steel: Introduction to LRFD

PRACTICE PROBLEMS

There are no problems in this book corresponding to Ch. 67 of the *Civil Engineering Reference Manual*.

68 Properties of Masonry

PRACTICE PROBLEMS

There are no problems in this book corresponding to Ch. 68 of the *Civil Engineering Reference Manual*.

69 Masonry Walls

PRACTICE PROBLEMS

1. Determine the allowable vertical load of the masonry wall shown. Face shell mortar bedding is used.

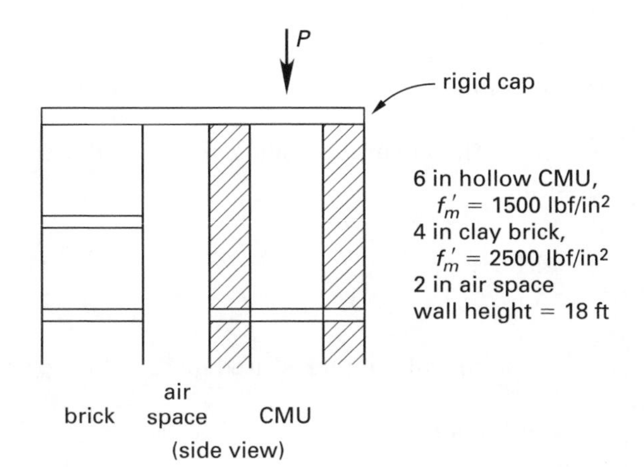

2. A 25 ft high, 10 in CMU wall is partially grouted and reinforced. The wall is nonloadbearing and is subjected to wind only. The wall is simply supported at top and bottom, and reinforcement is placed in the center of the wall.

$$f'_m = 2000 \text{ lbf/in}^2$$
$$n = 16.1$$

(a) With reinforcement spaced at 24 in, determine the largest reinforcing bars that can be used such that the neutral axis falls within the face shell of the unit.

(A) no. 4 bars
(B) no. 5 bars
(C) no. 6 bars
(D) no. 7 bars

For parts (b) through (f), assume the wall has no. 6 vertical bars spaced 32 in apart.

(b) Determine the location of the neutral axis.

(A) 1.0 in
(B) 1.2 in
(C) 1.7 in
(D) 2.0 in

(c) Determine the resisting moment assuming steel governs.

(A) 17,500 in-lbf/ft
(B) 20,000 in-lbf/ft
(C) 23,000 in-lbf/ft
(D) 24,500 in-lbf/ft

(d) Determine the allowable compressive stress.

(A) 670 lbf/in^2
(B) 890 lbf/in^2
(C) 1000 lbf/in^2
(D) 1140 lbf/in^2

(e) Determine the resisting moment of the wall.

(A) 17,000 in-lbf/ft
(B) 22,000 in-lbf/ft
(C) 23,000 in-lbf/ft
(D) 29,000 in-lbf/ft

(f) Determine the shear capacity.

(A) 2600 lbf/ft
(B) 3400 lbf/ft
(C) 3900 lbf/ft
(D) 4300 lbf/ft

(g) Determine the area of steel required to produce a balanced design condition.

(A) 0.15 in^2/ft
(B) 0.18 in^2/ft
(C) 0.21 in^2/ft
(D) 0.25 in^2/ft

(h) Determine the maximum moment on the wall if it is subjected to a 25 lbf/ft^2 wind load.

(A) 23,400 in-lbf/ft
(B) 24,200 in-lbf/ft
(C) 24,800 in-lbf/ft
(D) 25,100 in-lbf/ft

(i) Using $\rho = 0.004$ and $d = 4.81$ in, determine the wind-induced service moment the wall can withstand.

(A) 24,000 in-lbf/ft
(B) 32,000 in-lbf/ft
(C) 38,000 in-lbf/ft
(D) 47,000 in-lbf/ft

For part (j), the wall must resist a moment of 20,000 in-lbf/ft. Assume fully grouted masonry.

(j) With no. 5 bars spaced at 16 in within a 10 in thick wall, determine the steel stress.

(A) 14,000 lbf/in^2
(B) 16,000 lbf/in^2
(C) 18,000 lbf/in^2
(D) 20,000 lbf/in^2

3. An 8 in hollow unreinforced, ungrouted concrete masonry wall, 12 ft high, is simply supported at top and bottom. M-type mortar cement is used.

For parts (a) through (e), $f'_m = 1800$ lbf/in^2.

(a) Determine the allowable stresses.

(A) $F_a = 390$ lbf/in^2, $F_b = 600$ lbf/in^2
(B) $F_a = 320$ lbf/in^2, $F_b = 800$ lbf/in^2
(C) $F_a = 520$ lbf/in^2, $F_b = 600$ lbf/in^2
(D) $F_a = 390$ lbf/in^2, $F_b = 800$ lbf/in^2

(b) Determine the maximum vertical load the wall can be designed for, with a 2 in eccentricity.

(A) 6700 lbf/ft
(B) 7900 lbf/ft
(C) 10,000 lbf/ft
(D) 14,000 lbf/ft

(c) Determine the maximum wind load if the wall carries no vertical load. Neglect the self-weight of the wall.

(A) 9.4 lbf/ft^2
(B) 12 lbf/ft^2
(C) 16 lbf/ft^2
(D) 18 lbf/ft^2

For parts (d) and (e), the wall is subjected to an axial load of 5000 lbf/ft. There is no lateral load.

(d) Determine the allowable shear stress.

(A) 64 lbf/in^2
(B) 110 lbf/in^2
(C) 120 lbf/in^2
(D) 130 lbf/in^2

(e) Assuming the wall has no net tension, determine the maximum shear force the wall can be designed for.

(A) 900 lbf/ft
(B) 1000 lbf/ft
(C) 1100 lbf/ft
(D) 1300 lbf/ft

(f) Determine the minimum net area required to sustain an 18,000 lbf/ft axial load given that $f'_m = 1800$ lbf/in^2.

(A) 30 in^2/ft
(B) 39 in^2/ft
(C) 46 in^2/ft
(D) 52 in^2/ft

For parts (g) through (j), assume the wall is subjected to a vertical dead load, D, of 4500 lbf/ft ($e = 2.5$ in) and a lateral wind load, w, of 35 lbf/ft^2.

(g) Considering both wind and dead loads, determine the flexural compressive stress in the masonry.

(A) 120 lbf/in^2
(B) 140 lbf/in^2
(C) 160 lbf/in^2
(D) 180 lbf/in^2

(h) Considering both wind and dead loads, determine the masonry strength, f'_m, required.

(A) 890 lbf/in^2
(B) 950 lbf/in^2
(C) 1100 lbf/in^2
(D) 1500 lbf/in^2

(i) Determine the flexural tension capacity required.

(A) 6 lbf/in^2
(B) 10 lbf/in^2
(C) 14 lbf/in^2
(D) 15 lbf/in^2

(j) Determine the minimum required masonry strength, f'_m.

(A) 890 lbf/in^2
(B) 950 lbf/in^2
(C) 1100 lbf/in^2
(D) 1500 lbf/in^2

4. Design the joint reinforcement for a concrete masonry wall to carry a 20 lbf/ft^2 wind load. The wall spans 10 ft horizontally and is simply supported at both ends. Use allowable stress design. Assume the following.

- $f'_m = 1500$ lbf/in^2
- 8 in hollow units, laid with face shell mortar bedding, type-S mortar
- The wall will be built using horizontal joint reinforcement every other course (spaced at 16 in).
- The wall carries no axial loads.
- The masonry modulus of elasticity is $E_m = 2.08 \times 10^6$ lbf/in^2.
- The modulus of elasticity of steel is $E_s =$ 29,000,000 lbf/in^2.
- Steel stress governs the design.

SOLUTIONS

1. Determine section properties (assume face shell mortar bedding) of CMU.

$$A_n = 24 \text{ in}^2/\text{ft}$$
$$I = 139.3 \text{ in}^4/\text{ft}$$
$$r = 2.08 \text{ in}$$
$$\frac{h}{r} = \frac{(18 \text{ ft})\left(12 \frac{\text{in}}{\text{ft}}\right)}{2.08 \text{ in}} = 103.8$$

Since 103.8 is greater than 99, use Eq. 69.21.

$$F_a = \tfrac{1}{4}f'_m\left(\frac{70r}{h}\right)^2 = \left(\tfrac{1}{4}\right)\left(1500 \frac{\text{lbf}}{\text{in}^2}\right)\left(\frac{70}{103.8}\right)^2$$
$$= 170.5 \text{ lbf/in}^2$$
$$P_a = F_aA_n = \left(170.5 \frac{\text{lbf}}{\text{in}^2}\right)\left(24 \frac{\text{in}^2}{\text{ft}}\right) = 4092 \text{ lbf/ft}$$

Check buckling (Eqs. 69.18 and 69.19).

$$E_m = 900f'_m = (900)\left(1500 \frac{\text{lbf}}{\text{in}^2}\right)$$
$$= 1{,}350{,}000 \text{ lbf/in}^2$$
$$P_e = \left(\frac{\pi^2 EI}{h^2}\right)\left(1 - 0.577\frac{e}{r}\right)^3$$
$$= \frac{\pi^2\left(1{,}350{,}000 \frac{\text{lbf}}{\text{in}^2}\right)\left(139.3 \frac{\text{in}^4}{\text{ft}}\right)}{\left((18 \text{ ft})\left(12 \frac{\text{in}}{\text{ft}}\right)\right)^2}$$
$$= 39{,}781 \text{ lbf/ft}$$
$$P \le \tfrac{1}{4}P_e$$
$$4092 \text{ lbf/ft} \le \left(\tfrac{1}{4}\right)\left(39{,}781 \frac{\text{lbf}}{\text{ft}}\right) = 9945 \text{ lbf/ft} \quad \text{[OK]}$$
$$P = \boxed{4092 \text{ lbf/ft}}$$

2. (a) 10 in is the nominal thickness. The actual thickness is 10 in − 0.375 in = 9.625 in.

$$kd = t_f = 1.375 \text{ in}$$
$$d = \frac{9.625 \text{ in}}{2} = 4.81 \text{ in} \quad \text{[since fully grouted]}$$
$$k = \frac{t_f}{d} = \frac{1.375 \text{ in}}{4.81 \text{ in}} = 0.286$$

From Eq. 69.4,

$$k^2 + 2\rho nk - 2\rho n = 0$$
$$(0.286)^2 + 2\rho(16.1)(0.286) - 2\rho(16.1) = 0$$
$$\rho = 0.00356$$

$$A_s = \rho bd$$
$$= (0.00356)(24 \text{ in})(4.81 \text{ in})$$
$$= 0.411 \text{ in}^2 \quad \text{[for every 24 in]}$$

The area of a no. 6 bar is 0.44 in^2, which is too large.

The area of a no. 5 bar is 0.31 in^2, which is acceptable.

The answer is (B).

(b) From Table 69.2,

$$A_s = 0.166 \text{ in}^2/\text{ft}$$
$$\rho = \frac{A_s}{bd} = \frac{0.166 \text{ in}^2}{(12 \text{ in})(4.81 \text{ in})} = 0.00288$$
$$\rho n = (0.00288)(16.1) = 0.0464$$

From Eq. 69.4,

$$k = \sqrt{2\rho n + (\rho n)^2} - \rho n$$
$$= \sqrt{(2)(0.0464) + (0.0464)^2} - 0.0464 = 0.255$$
$$kd = (0.255)(4.81 \text{ in}) = \boxed{1.23 \text{ in}}$$

The answer is (B).

(c) Use Eq. 69.5,

$$j = 1 - \frac{k}{3} = 1 - \frac{0.255}{3} = 0.915$$
$$M_s = A_sF_sjd$$
$$= \left(0.166 \frac{\text{in}^2}{\text{ft}}\right)\left(\left(\tfrac{4}{3}\right)\left(24{,}000 \frac{\text{lbf}}{\text{in}^2}\right)\right)(0.913)(4.81 \text{ in})$$
$$= \boxed{23{,}378 \text{ in-lbf/ft}}$$

The answer is (C).

(Note the one third increase in allowable stress for wind load.)

(d) $$F_b = \tfrac{1}{3}f'_m\left(\tfrac{4}{3}\right) = \left(\tfrac{1}{3}\right)\left(2000 \frac{\text{lbf}}{\text{in}^2}\right)\left(\tfrac{4}{3}\right)$$
$$= \boxed{889 \text{ lbf/in}^2}$$

The answer is (B).

(e) $$M_m = \frac{F_bbd^2jk}{2}$$
$$= \frac{\left(889 \frac{\text{lbf}}{\text{in}^2}\right)\left(12 \frac{\text{in}}{\text{ft}}\right)(4.81 \text{ in})^2(0.915)(0.255)}{2}$$
$$= 28{,}790 \text{ in-lbf/ft}$$
$$M_R = \text{lesser of } M_s \text{ and } M_m = \boxed{23{,}378 \text{ in-lbf/ft}}$$

Therefore, steel governs.

The answer is (C).

Structural

(f)

$$V_R = F_v bd$$

$$F_v = \sqrt{f'_m} = \sqrt{2000\ \frac{\text{lbf}}{\text{in}^2}} = 44.7\ \text{lbf/in}^2$$

$$V_R = \left(\tfrac{4}{3}\right)\left(44.7\ \frac{\text{lbf}}{\text{in}^2}\right)\left(12\ \frac{\text{in}}{\text{ft}}\right)(4.81\ \text{in})$$

$$= \boxed{3440\ \text{lbf/ft}}$$

The answer is (B).

(g) From Eq. 69.10,

$$\rho_{\text{bal}} = \frac{nF_b}{2F_s\left(n + \frac{F_s}{F_b}\right)}$$

$$= \frac{(16.1)\left(889\ \frac{\text{lbf}}{\text{in}^2}\right)}{(2)\left(31{,}920\ \frac{\text{lbf}}{\text{in}^2}\right)\left(16.1 + \frac{31{,}920\ \frac{\text{lbf}}{\text{in}^2}}{889\ \frac{\text{lbf}}{\text{in}^2}}\right)}$$

$$= 0.0043$$

$$A_{s,\text{bal}} = \rho_{\text{bal}} bd = (0.0043)\left(12\ \frac{\text{in}}{\text{ft}}\right)(4.81\ \text{in})$$

$$= \boxed{0.248\ \text{in}^2/\text{ft}}$$

The answer is (D).

(h)

$$M = \frac{wl^2}{8} = \frac{\left(25\ \frac{\text{lbf}}{\text{ft}^2}\right)(25\ \text{ft})^2\left(12\ \frac{\text{in}}{\text{ft}}\right)}{8}$$

$$= \boxed{23{,}438\ \text{in-lbf/ft}}$$

The answer is (A).

(i)

$$\rho n = (0.004)(16.1) = 0.0644$$

$$k = \sqrt{2\rho n + (\rho n)^2} - \rho n$$

$$= \sqrt{(2)(0.0644) + (0.0644)^2} - 0.0644 = 0.300$$

$$j = 1 - \frac{k}{3} = 1 - \frac{0.300}{3} = 0.900$$

$$A_s = \rho bd = (0.004)\left(12\ \frac{\text{in}}{\text{ft}}\right)(4.81\ \text{in})$$

$$= 0.231\ \text{in}^2/\text{ft}$$

From Eq. 69.7,

$$M_s = A_s F_s jd$$

$$= \left(0.231\ \frac{\text{in}^2}{\text{ft}}\right)\left(31{,}920\ \frac{\text{lbf}}{\text{in}^2}\right)(0.900)(4.81\ \text{in})$$

$$= 31{,}920\ \text{in-lbf/ft}$$

The answer is (B).

(j) From Table 69.2,

$$A_s = 0.230\ \text{in}^2/\text{ft}$$

$$\rho = \frac{A_s}{bd} = \frac{0.230\ \frac{\text{in}^2}{\text{ft}}}{\left(12\ \frac{\text{in}}{\text{ft}}\right)(4.81\ \text{in})}$$

$$= 0.00398$$

$$\rho n = (0.00398)(16.1) = 0.0641$$

$$k = \sqrt{2\rho n + (\rho n)^2} - \rho n$$

$$= \sqrt{(2)(0.0641) + (0.0641)^2} - 0.0641 = 0.299$$

$$j = 1 - \frac{k}{3} = 1 - \frac{0.299}{3} = 0.900$$

$$f_s = \frac{M}{A_s jd} = \frac{20{,}000\ \frac{\text{in-lbf}}{\text{ft}}}{\left(0.230\ \frac{\text{in}^2}{\text{ft}}\right)(0.900)(4.81\ \text{in})}$$

$$= \boxed{20{,}090\ \text{lbf/in}^2}$$

The answer is (D).

3. (a) From App. 69.A,

$$r = 2.84\ \text{in}$$

$$S = 81\ \text{in}^3/\text{ft}$$

$$A_n = 30\ \text{in}^2/\text{ft}$$

$$I = 334\ \text{in}^4/\text{ft}$$

$$\frac{h}{r} = \frac{(12\ \text{ft})\left(12\ \frac{\text{in}}{\text{ft}}\right)}{2.84\ \text{in}} = 50.7$$

Since 50.7 is less than 99, use Eq. 69.20.

$$F_a = \tfrac{1}{4} f'_m \left(1 - \left(\frac{h}{140r}\right)^2\right)$$

$$= \left(\tfrac{1}{4}\right)\left(1800\ \frac{\text{lbf}}{\text{in}^2}\right)\left(1 - \left(\frac{50.7}{140}\right)^2\right)$$

$$= \boxed{391\ \text{lbf/in}^2}$$

$$F_b = \tfrac{1}{3} f'_m = \left(\tfrac{1}{3}\right)\left(1800\ \frac{\text{lbf}}{\text{in}^2}\right)$$

$$= \boxed{600\ \text{lbf/in}^2}$$

The answer is (A).

(b)

$$f_a = \frac{P}{A_n} = \frac{P}{30\ \frac{\text{in}^2}{\text{ft}}} = 0.033P$$

$$f_b = \frac{M}{S} = \frac{Pe}{S} = \frac{P(2\ \text{in})}{81\ \frac{\text{in}^3}{\text{ft}}} = 0.025P$$

From Eq. 69.17,

$$\frac{f_a}{F_a} + \frac{f_b}{F_b} \le 1$$

$$\frac{0.033P}{391\ \frac{\text{in}^2}{\text{ft}}} + \frac{0.025P}{600\ \frac{\text{in}^2}{\text{ft}}} = 1$$

$$P = 7932 \text{ lbf/ft}$$

Check buckling criteria.

$$E_m = 900f'_m = (900)\left(1800\ \frac{\text{lbf}}{\text{in}^2}\right) = 1{,}620{,}000 \text{ lbf/in}^2$$

$$P \le \tfrac{1}{4}P_e = \left(\tfrac{1}{4}\right)\left(\frac{\pi^2 EI}{h^2}\right)\left(1 - 0.577\left(\frac{e}{r}\right)\right)^3$$

$$= \left(\tfrac{1}{4}\right)\left(\frac{\pi^2\left(1{,}620{,}000\ \frac{\text{lbf}}{\text{in}^2}\right)\left(334\ \frac{\text{in}^4}{\text{ft}}\right)}{(12\text{ ft})^2\left(12\ \frac{\text{in}}{\text{ft}}\right)^2}\right) \times \left(1 - (0.577)\left(\frac{2\text{ in}}{2.84\text{ in}}\right)\right)^3$$

$$= 13{,}471 \text{ lbf/ft}$$

Buckling does not control. Therefore,

$$\boxed{P_{\text{max}} = 7932 \text{ lbf/ft.}}$$

The answer is (B).

(c)

$$M = \frac{wh^2}{8} = \frac{w\left((12\text{ ft})\left(12\ \frac{\text{in}}{\text{ft}}\right)\right)^2}{(8)\left(12\ \frac{\text{in}}{\text{ft}}\right)} = 216w \quad \text{[in in-lbf/ft]}$$

$$f_b = \frac{M}{S} = \frac{216w}{81\ \frac{\text{in}^3}{\text{ft}}} = 2.67w \quad \text{[in lbf/in}^2\text{]}$$

From Table 69.1,

$$F_b = \left(\tfrac{4}{3}\right)\left(25\ \frac{\text{lbf}}{\text{in}^2}\right) = 33.33 \text{ lbf/in}^2$$

Set f_b equal to F_b.

$$2.67w = 33.33\ \frac{\text{lbf}}{\text{in}^2}$$

$$w = \boxed{12.5 \text{ lbf/ft}^2}$$

The answer is (B).

(d) F_v is the least of the following.

- $1.5\sqrt{f'_m} = 1.5\sqrt{1800\ \frac{\text{lbf}}{\text{in}^2}} = 63.6 \text{ lbf/in}^2$
- $v + 0.45\left(\frac{N_v}{A_n}\right) = 37\ \frac{\text{lbf}}{\text{in}^2} + \frac{(0.45)\left(5000\ \frac{\text{lbf}}{\text{ft}}\right)}{30\ \frac{\text{in}^2}{\text{ft}}} = 112 \text{ lbf/in}^2$
- 120 lbf/in^2

The first criterion controls.

$$F_v = \boxed{63.6 \text{ lbf/in}^2}$$

The answer is (A).

(e) From Eq. 69.29,

$$f_v = \frac{3V}{2A_n}$$

Set f_v equal to F_v.

$$V = \frac{2A_nF_v}{3} = \frac{(2)\left(30\ \frac{\text{in}^2}{\text{ft}}\right)\left(63.6\ \frac{\text{lbf}}{\text{in}^2}\right)}{3} = \boxed{1272 \text{ lbf/ft}}$$

The answer is (D).

(f) From part (a), $F_a = 391 \text{ lbf/in}^2$.

$$F_a = \frac{P}{A_n}$$

$$391\ \frac{\text{lbf}}{\text{in}^2} = \frac{18{,}000\ \frac{\text{lbf}}{\text{ft}}}{A_n}$$

$$A_n = \boxed{46 \text{ in}^2\text{/ft}}$$

The answer is (C).

(g) At the mid-height of the wall,

$$M = \frac{Pe}{2} + \frac{wh^2}{8} = \frac{\left(4500\ \frac{\text{lbf}}{\text{ft}}\right)(2.5\text{ in})}{2} + \frac{\left(35\ \frac{\text{lbf}}{\text{ft}^2}\right)\left((12\text{ ft})\left(12\ \frac{\text{in}}{\text{ft}}\right)\right)^2}{(8)\left(12\ \frac{\text{in}}{\text{ft}}\right)} = 13{,}185 \text{ in-lbf/ft}$$

$$f_b = \frac{M}{S} = \frac{13{,}185\ \frac{\text{in-lbf}}{\text{ft}}}{81\ \frac{\text{in}^3}{\text{ft}}} = \boxed{163 \text{ lbf/in}^2}$$

The answer is (C).

(h)
$$F_a = \left(\tfrac{4}{3}\right)\tfrac{1}{4}f'_m\left(1-\left(\frac{h}{140r}\right)^2\right) = \left(\tfrac{4}{3}\right)\tfrac{1}{4}f'_m\left(1-\left(\frac{50.7}{140}\right)^2\right) = 0.290f'_m$$

$$f_a = \frac{P}{A} = \frac{4500\ \frac{\text{lbf}}{\text{ft}}}{30\ \frac{\text{in}^2}{\text{ft}}} = 150\ \text{lbf/in}^2$$

$$F_b = \left(\tfrac{4}{3}\right)\tfrac{1}{3}f'_m = 0.444f'_m$$

$$\frac{f_a}{F_a}+\frac{f_b}{F_b} = 1$$

$$\frac{150\ \frac{\text{lbf}}{\text{in}^2}}{0.290f'_m}+\frac{163\ \frac{\text{lbf}}{\text{in}^2}}{0.444f'_m} = 1$$

$$f'_m = \boxed{884\ \text{lbf/in}^2}$$

The answer is (A).

(i) For dead and wind loads,

$$-f_a + f_b = \tfrac{4}{3}F_t$$

$$-150\ \frac{\text{lbf}}{\text{in}^2} + 163\ \frac{\text{lbf}}{\text{in}^2} = \tfrac{4}{3}F_t$$

$$F_t = 9.75\ \text{lbf/in}^2$$

Considering dead load only, the moment at the top of the wall is

$$Pe = \left(4500\ \frac{\text{lbf}}{\text{ft}}\right)(2.5\ \text{in}) = 11{,}250\ \text{in-lbf/ft}$$

$$f_b = \frac{M}{S} = \frac{11{,}250\ \frac{\text{in-lbf}}{\text{ft}}}{81\ \frac{\text{in}^3}{\text{ft}}} = 139\ \text{lbf/in}^2$$

$$-f_a + f_b = F_t$$

$$-150\ \frac{\text{lbf}}{\text{in}^2} + 139\ \frac{\text{lbf}}{\text{in}^2} = -11\ \frac{\text{lbf}}{\text{in}^2} = F_t$$

There is no tension in the wall. The dead and wind load case controls.

$$F_{t,\text{reg}} = \boxed{9.75\ \text{lbf/in}^2}$$

The answer is (B).

(j) For dead and wind loads, from part (g), $f'_m = 884\ \text{lbf/in}^2$.

Check dead load only. From part (i),

$$f_a = 150\ \text{lbf/in}^2$$

$$F_a = \frac{0.290\ \frac{\text{lbf}}{\text{in}^2}}{\frac{4}{3}} = 0.218f'_m$$

From part (i),

$$f_b = 139\ \text{lbf/in}^2$$

$$F_b = \tfrac{1}{3}f'_m$$

$$\frac{f_a}{F_a}+\frac{f_b}{F_b} = 1$$

$$\frac{150\ \frac{\text{lbf}}{\text{in}^2}}{0.218f'_m}+\frac{139\ \frac{\text{lbf}}{\text{in}^2}}{0.333f'_m} = 1$$

$$f'_m = \boxed{1105\ \text{lbf/in}^2\quad [\text{governs}]}$$

The answer is (C).

4. The allowable tension in joint reinforcement due to wind loading is

$$F_s = \left(\tfrac{4}{3}\right)\left(30{,}000\ \frac{\text{lbf}}{\text{in}^2}\right) = 40{,}000\ \text{lbf/in}^2$$

The allowable masonry compressive stress due to flexure from wind loading is

$$F_b = \left(\tfrac{4}{3}\right)\left(\tfrac{1}{3}f'_m\right) = \left(\tfrac{4}{9}\right)\left(1500\ \frac{\text{lbf}}{\text{in}^2}\right) = 667\ \text{lbf/in}^2$$

$$n = \frac{E_s}{E_m} = \frac{29\ \text{ksi}}{2.08\ \text{ksi}} = 13.9$$

Determine the moment due to wind load. For simply supported conditions,

$$M = \frac{wl^2}{8} = \left(\frac{\left(20\ \frac{\text{lbf}}{\text{ft}^2}\right)(10\ \text{ft})^2}{8}\right)\left(12\ \frac{\text{in}}{\text{ft}}\right) = 3000\ \text{in-lbf/ft}$$

Determine the balanced reinforcement ratio, ρ_b.

$$\rho_b = \frac{nF_b}{2F_s\left(n+\frac{F_s}{F_b}\right)} = \frac{(13.9)\left(667\ \frac{\text{lbf}}{\text{in}^2}\right)}{(2)\left(40{,}000\ \frac{\text{lbf}}{\text{in}^2}\right)\left(13.9+\frac{40{,}000\ \frac{\text{lbf}}{\text{in}^2}}{667\ \frac{\text{lbf}}{\text{in}^2}}\right)} = 0.0016$$

If the actual steel ratio, ρ, is less than ρ_b, steel stress governs the design.

Try W2.1 (8 gage) horizontal joint reinforcement every other course.

The effective depth of reinforcement is

$$\begin{aligned} d &= \text{actual width of wall} - \tfrac{5}{8} \text{ in cover} \\ &= 7\tfrac{5}{8} \text{ in} - \tfrac{5}{8} \text{ in} = 7 \text{ in} \\ A_s &= 0.020 \text{ in}^2 \text{ per wire} \end{aligned}$$

(Note that since only one wire is in tension, this is the effective steel area.) With reinforcement every other course, the spacing is (2)(8 in) = 16 in.

$$\frac{0.020 \text{ in}^2}{\dfrac{16 \text{ in}}{12 \dfrac{\text{in}}{\text{ft}}}} = 0.015 \text{ in}^2/\text{ft}$$

$$\rho = \frac{A_s}{bd} = \frac{0.020 \text{ in}^2}{(16 \text{ in})(7 \text{ in})} = 0.00018$$

Since $\rho < \rho_b$, steel governs and the resisting moment of the wall is

$$M_r = A_s F_s j d$$

Determine j.

$$\begin{aligned} k &= \sqrt{2n\rho + (n\rho)^2} - n\rho \\ &= \sqrt{(2)(13.9)(0.00018) + \left((13.9)(0.00018)\right)^2} \\ &\quad - (13.9)(0.00018) \\ &= 0.068 \\ j &= 1 - \frac{k}{3} \\ &= 1 - \frac{0.068}{3} \\ &= 0.977 \\ M_r &= A_s F_s j d \\ &= \left(0.015 \, \frac{\text{in}^2}{\text{ft}}\right) \left(\tfrac{4}{3}\right) \left(40{,}000 \, \frac{\text{lbf}}{\text{in}^2}\right) (0.977)(7 \text{ in}) \\ &= 5471 \text{ in-lbf/ft} \end{aligned}$$

Since $M_r > M$, the design is adequate with W2.1 joint reinforcement every other course.

Structural

70 Masonry Columns

PRACTICE PROBLEMS

1. *(Time limit: one hour)* A 36 ft tall vertical concrete masonry column with the cross section shown is firmly connected to its foundation at the base but is unsupported at the top. A 30 lbf/ft^2 wind acts perpendicular to the column face uniformly over the entire length. $f'_m = 1500\ \text{lbf/in}^2$, $f_y = 60{,}000\ \text{lbf/in}^2$, and $n = 20$. Neglecting the axial stresses due to the dead load, determine if the stresses in the steel and masonry are acceptable.

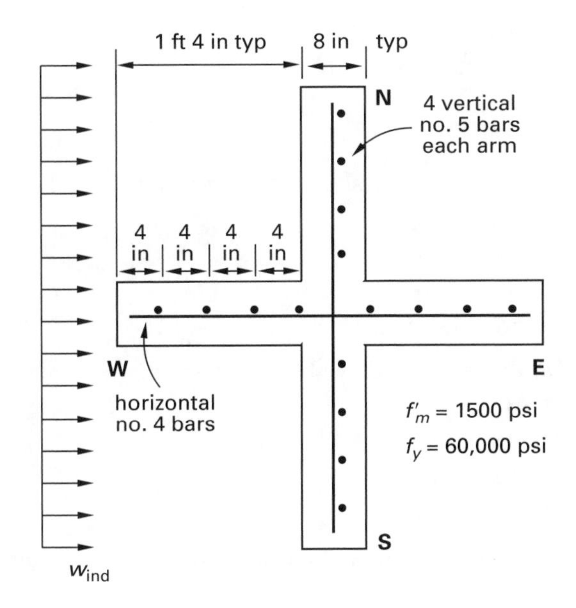

2. A 16 in square concrete masonry column with four vertical no. 6 bars and an effective height of 18 ft supports an axial load of 75,000 lbf. Determine the required masonry strength.

$$r = 4.53\ \text{in}$$

$$S = 635.8\ \text{in}^3$$

$$A_n = 244.1\ \text{in}^2$$

3. A 32 in by 16 in clay brick column supports a compressive load of 350,000 lbf.

$$e = 4\ \text{in}\quad \text{[along the major axis]}$$

$$f'_m = 3000\ \text{lbf/in}^2$$

$$\text{effective height} = 10\ \text{ft}$$

$$r = 9.1\ \text{in}$$

$$g = 0.8$$

Determine the required steel area.

4. A brick column, 24 in by 24 in, is reinforced with four no. 5 bars. Its effective height is 16 ft, and $f'_m = 3000\ \text{lbf/in}^2$. It has a compressive load with 6 in eccentricity in the x-direction. Assume $g = 0.6$.

(a) Determine the radius of gyration, r.

(A) 5.9 in
(B) 6.2 in
(C) 6.8 in
(D) 7.0 in

(b) Determine the reinforcement ratio, ρ_t.

(A) 0.002
(B) 0.003
(C) 0.004
(D) 0.005

(c) Determine the axial capacity.

(A) 120,000 lbf
(B) 150,000 lbf
(C) 180,000 lbf
(D) 212,000 lbf

(d) Determine the moment capacity in the x-direction.

(A) 870,000 in-lbf
(B) 930,000 in-lbf
(C) 1,270,000 in-lbf
(D) 1,520,000 in-lbf

(e) Determine the axial stress.

(A) 320 lbf/in^2
(B) 380 lbf/in^2
(C) 420 lbf/in^2
(D) 500 lbf/in^2

(f) Determine the allowable compressive force.

(A) 300,000 lbf
(B) 400,000 lbf
(C) 500,000 lbf
(D) 600,000 lbf

For parts (g) through (j), assume an additional eccentricity of 4 in in the y-direction.

(g) Determine P_y.

(A) 160,000 lbf
(B) 190,000 lbf
(C) 230,000 lbf
(D) 280,000 lbf

(h) Determine P_a.

(A) 300,000 lbf
(B) 330,000 lbf
(C) 390,000 lbf
(D) 420,000 lbf

(i) Determine the maximum biaxial load.

(A) 140,000 lbf
(B) 170,000 lbf
(C) 200,000 lbf
(D) 230,000 lbf

(j) Determine the moment capacity in the y-direction.

(A) 980,000 in-lbf
(B) 1,030,000 in-lbf
(C) 1,120,000 in-lbf
(D) 1,200,000 in-lbf

5. A 32 in square concrete masonry column is subjected to the following: a dead load, D, of 180,000 lbf; a live load, L, of 140,000 lbf; and design shear force of 30,000 lbf. The forces due to earthquake, E, are as follows.

$$\text{bending moment} = 140{,}000 \text{ ft-lbf}$$

$$\text{axial compressive force} = 45{,}000 \text{ lbf}$$

Use $f'_m = 1500$ lbf/in^2, $h = 10$ ft, $r = 9.1$ in, $g = 0.4$, $n = 14$, and $d = 23$ in.

(a) Determine the design eccentricity for dead and live loads.

(A) 2.7 in
(B) 3.2 in
(C) 3.8 in
(D) 4.6 in

(b) Determine the design eccentricity for dead and live loads and earthquake forces.

(A) 2.5 in
(B) 3.7 in
(C) 4.6 in
(D) 6.9 in

(c) Determine the design eccentricity for $0.9D + E$.

(A) 4.3 in
(B) 5.2 in
(C) 6.9 in
(D) 8.1 in

(d) Determine the maximum axial stress on the column.

(A) 180 lbf/in^2
(B) 250 lbf/in^2
(C) 320 lbf/in^2
(D) 365 lbf/in^2

(e) Determine the maximum allowable axial load on the column. Neglect the steel's contribution.

(A) 370,000 lbf
(B) 420,000 lbf
(C) 460,000 lbf
(D) 510,000 lbf

(f) Considering the three load combinations ($D + L$, $D + L + E$, and $0.9D + E$), determine the minimum steel area required.

(A) none
(B) 2 in^2
(C) 3 in^2
(D) 4 in^2

(g) Determine the design shear stress on the column.

(A) 35 lbf/in^2
(B) 41 lbf/in^2
(C) 50 lbf/in^2
(D) 57 lbf/in^2

(h) Determine the allowable shear stress.

(A) 39 lbf/in^2
(B) 48 lbf/in^2
(C) 52 lbf/in^2
(D) 57 lbf/in^2

(i) Determine the area of shear reinforcement required.

(A) none
(B) 0.11 in^2/ft
(C) 0.20 in^2/ft
(D) 0.27 in^2/ft

(j) If shear reinforcement is provided to resist the total design shear, determine the minimum spacing of no. 5 ties.

(A) 4 in o.c.
(B) 8 in o.c.
(C) 16 in o.c.
(D) 24 in o.c.

SOLUTIONS

1. The area projected to the wind per foot of column is

$$\left(1\ \frac{\text{ft}}{\text{ft}}\right)\left(\frac{16\ \text{in} + 8\ \text{in} + 16\ \text{in}}{12\ \frac{\text{in}}{\text{ft}}}\right) = 3.333\ \text{ft}^2/\text{ft}$$

The distributed wind load per foot is

$$w = \left(30\ \frac{\text{lbf}}{\text{ft}^2}\right)\left(3.333\ \frac{\text{ft}^2}{\text{ft}}\right) = 100\ \text{lbf/ft}$$

The pier acts like a uniformly loaded horizontal cantilever beam. The moment at the base is

$$\begin{aligned} M &= \tfrac{1}{2}wL^2 = \left(\tfrac{1}{2}\right)\left(100\ \frac{\text{lbf}}{\text{ft}}\right)(36\ \text{ft})^2 \\ &= 64{,}800\ \text{ft-lbf} \end{aligned}$$

The west half of the column will be in tension, and masonry cannot support tension. Assume the neutral axis will be in the east arm, placing the north-south steel in tension.

Disregard the effect of the horizontal no. 4 bars, which contribute to crack control. Use the most westerly end as the reference point to find the neutral axis.

The masonry in compression is A_1. (All dimensions are in inches.)

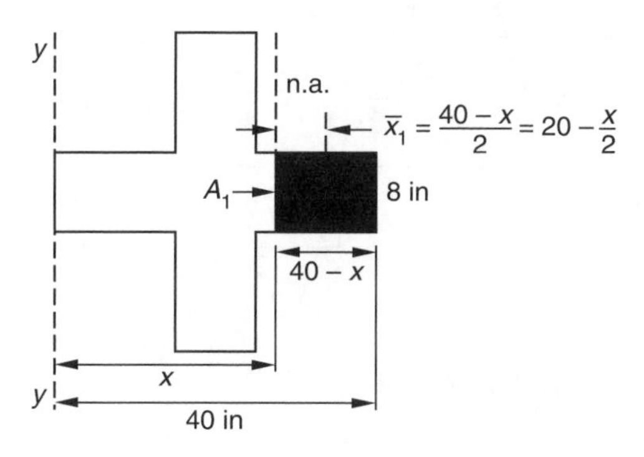

$$A_1 = (40 - x)(8) = 320 - 8x$$

$$\overline{x}_1 = \frac{40 - x}{2} = 20 - \frac{x}{2}$$

Note that x is measured from the y-y axis, but $\overline{x}$ is measured from the neutral axis.

Each steel bar contributes a transformed area of

$$nA_b = (20)(0.31\ \text{in}^2) = 6.2\ \text{in}^2$$

Assume three bars in the east arm are in compression.

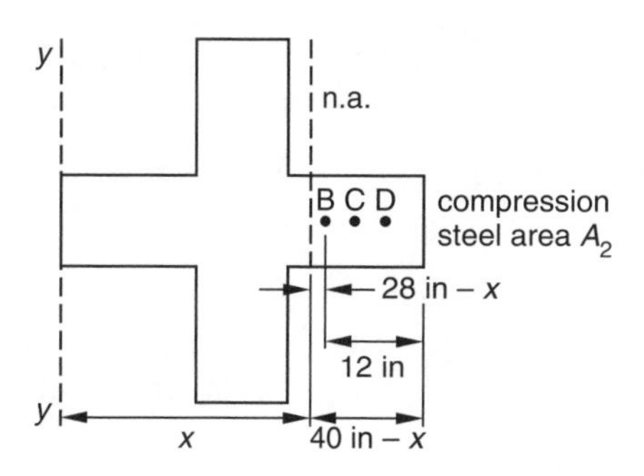

$$\begin{aligned} A_{2B} &= 6.2\ \text{in}^2 \\ \overline{x}_{2B} &= 40 - x - 12 \\ &= 28 - x \quad \text{[from neutral axis]} \\ A_{2C} &= 6.2\ \text{in}^2 \\ \overline{x}_{2C} &= 32 - x \quad \text{[from neutral axis]} \\ A_{2D} &= 6.2\ \text{in}^2 \\ \overline{x}_{2D} &= 36 - x \quad \text{[from neutral axis]} \end{aligned}$$

The steel in tension is

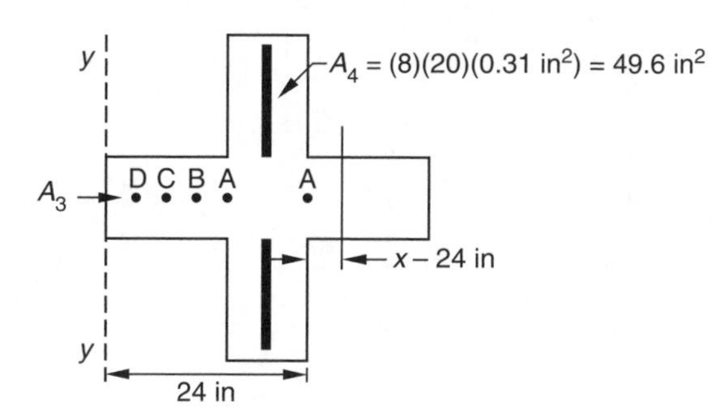

$$\begin{aligned} A_{2A} &= 6.2\ \text{in}^2 \\ \overline{x}_{2A} &= x - 24 \\ A_{3A} &= 6.2\ \text{in}^2 \\ \overline{x}_{3A} &= x - 16 \\ A_{3B} &= 6.2\ \text{in}^2 \\ \overline{x}_{3B} &= x - 12 \\ A_{3C} &= 6.2\ \text{in}^2 \\ \overline{x}_{3C} &= x - 8 \\ A_{3D} &= 6.2\ \text{in}^2 \\ \overline{x}_{3D} &= x - 4 \\ A_4 &= 49.6\ \text{in}^2 \\ \overline{x}_4 &= x - 20 \end{aligned}$$

The neutral axis is located where

$$\sum_{\text{compression}} A_i\overline{x}_i = \sum_{\text{tension}} A_i\overline{x}_i \quad (320 - 8x)\left(20 - \frac{x}{2}\right)$$
$$+ (6.2)\big((28 - x) + (32 - x) + (36 - x)\big)$$
$$= (6.2)\big((x - 24) + (x - 16) + (x - 12) + (x - 8) + (x - 4)\big) + (49.6)(x - 20)$$
$$= (6.2)(5x - 64) + (49.6)(x - 20)$$

$$4x^2 - 419.2x = -8384$$
$$x^2 - 104.8x = -2096$$
$$x = 26.91 \text{ in} \quad \left[\begin{array}{c}\text{round to 27.0;} \\ \text{location of neutral axis}\end{array}\right]$$

(The location of the neutral axis is not significantly affected by the initial assumption about the number of bars in compression.)

To find the centroidal moment of inertia, create a table.

element	A (in^2)	$\overline{x}$ (in)	$(\overline{x})^2$ (in^2)	I (in^4)
A_1				$\frac{bh^3}{3} = 5859$
A_{2A}	6.2	3	9	55.8
A_{2B}	6.2	1	1	6.2
A_{2C}	6.2	5	25	155.0
A_{2D}	6.2	9	81	502.2
A_{3A}	6.2	11	121	750.2
A_{3B}	6.2	15	225	1395.0
A_{3C}	6.2	19	361	2238.2
A_{3D}	6.2	23	529	3279.8
A_4	49.6	7	49	2430.2
			total	16,671 in^4

The extreme masonry stress is

$$f_m = \frac{Mc}{I} = \frac{(64{,}800 \text{ ft-lbf})\left(12 \frac{\text{in}}{\text{ft}}\right)(40 \text{ in} - 27 \text{ in})}{16{,}671 \text{ in}^4}$$
$$= 606.4 \text{ lbf/in}^2 \quad \text{(psi)}$$

The allowable bending stress due to wind loading is

$$\left(\tfrac{4}{3}\right)\left(\tfrac{1}{3}\right)\left(1500 \frac{\text{lbf}}{\text{in}^2}\right) = 667 \text{ lbf/in}^2 \quad \text{(psi)}$$

acceptable

Use bar A_{3D} to find the maximum steel stress.

$$f_{st} = \frac{(20)(64{,}800 \text{ ft-lbf})\left(12 \frac{\text{in}}{\text{ft}}\right)(23 \text{ in})}{16{,}671 \text{ in}^4}$$
$$= 21{,}456 \text{ lbf/in}^2 \quad \text{(psi)}$$

For grade-60 steel, f_{st} is limited to

$$\left(\tfrac{4}{3}\right)(24{,}000 \text{ psi}) = 32{,}000 \text{ psi}$$

acceptable

2.

$$\frac{h}{r} = \frac{(18 \text{ ft})\left(12 \frac{\text{in}}{\text{ft}}\right)}{4.53 \text{ in}} = 47.68$$

Since $h/r < 99$,

$$F_a = \tfrac{1}{4}f'_m\left(1 - \left(\frac{h}{140r}\right)^2\right)$$
$$= \frac{P_a}{A} \quad \text{[neglecting steel]}$$
$$= \tfrac{1}{4}f'_m\left(1 - \left(\frac{47.68}{140}\right)^2\right) = 0.221f'_m$$

$$f_a = \frac{P}{A} = \frac{75{,}000 \text{ lbf}}{244.1 \text{ in}^2} = 307.3 \text{ lbf/in}^2$$

$$\text{minimum } e = 0.1t = (0.1)(15.625 \text{ in}) = 1.56 \text{ in}$$

$$e_k = \frac{t}{6} = \frac{15.625 \text{ in}}{6} = 2.6 \text{ in}$$
$$e < e_k$$

The section is in compression.

$$f_b = \frac{M}{S} = \frac{Pe}{S}$$
$$= \frac{(75{,}000 \text{ lbf})(1.56 \text{ in})}{635.8 \text{ in}^3}$$
$$= 184.0 \text{ lbf/in}^2$$

$$F_b = \tfrac{1}{3}f'_m$$

$$\frac{f_a}{F_a} + \frac{f_b}{F_b} = 1$$

$$\frac{307.3 \frac{\text{lbf}}{\text{in}^2}}{0.221f'_m} + \frac{184.0 \frac{\text{lbf}}{\text{in}^2}}{0.333f'_m} = 1$$

$$\boxed{f'_m = 1943 \text{ lbf/in}^2}$$

3.

$$\frac{h}{r} = \frac{(10 \text{ ft})\left(12 \frac{\text{in}}{\text{ft}}\right)}{9.1 \text{ in}} = 13.19$$

Since $h/r < 99$,

$$F_a = \tfrac{1}{4}f'_m\left(1 - \left(\frac{h}{140r}\right)^2\right)$$
$$= \left(\tfrac{1}{4}\right)\left(3000 \frac{\text{lbf}}{\text{in}^2}\right)\left(1 - \left(\frac{13.19}{140}\right)^2\right)$$
$$= 743 \text{ lbf/in}^2$$

The maximum allowable load, P_a, is

$$P_a = F_a bt = \left(743\ \frac{\text{lbf}}{\text{in}^2}\right)(32\ \text{in})(16\ \text{in}) = 380{,}416\ \text{lbf}$$

Since 350,000 lbf < 380,416 lbf, the column is not overloaded.

Per *MSJC Code*, the minimum eccentricity is

$$\rho = 0.1t = (0.1)(32\ \text{in}) = 3.2\ \text{in} \quad [\text{use 4 in}]$$

$$F_b = \tfrac{1}{3}f'_m = \left(\tfrac{1}{3}\right)\left(3000\ \frac{\text{lbf}}{\text{in}^2}\right) = 1000\ \text{lbf/in}^2$$

$$\frac{P}{F_b bt} = \frac{350{,}000\ \text{lbf}}{\left(1000\ \frac{\text{lbf}}{\text{in}^2}\right)(16\ \text{in})(32\ \text{in})} = 0.684$$

$$\frac{Pe}{F_b bt^2} = \frac{(350{,}000\ \text{lbf})(4\ \text{in})}{\left(1000\ \frac{\text{lbf}}{\text{in}^2}\right)(16\ \text{in})(32\ \text{in})^2} = 0.0854$$

From App. 70.C, $n\rho_t = 0.15$.

$$E_m = 700f'_m = (700)\left(3000\ \frac{\text{lbf}}{\text{in}^2}\right) = 2{,}100{,}000\ \text{lbf/in}^2$$

$$n = \frac{E_s}{E_m} = \frac{29{,}000{,}000\ \frac{\text{lbf}}{\text{in}^2}}{2{,}100{,}000\ \frac{\text{lbf}}{\text{in}^2}} = 13.8$$

$$\rho_t = \frac{0.15}{n} = \frac{0.15}{13.8} = 0.0109$$

$$A_s = \rho_t bt = (0.0109)(16\ \text{in})(32\ \text{in}) = \boxed{5.58\ \text{in}^2}$$

(Note that 1/4 in diameter lateral ties are also required per *MSJC Code* 5.9.1.6.)

4. (a)
$$I = \frac{bt^3}{12} = \frac{(23.625\ \text{in})^4}{12} = 25{,}960\ \text{in}^4$$

$$A = bt = (23.625\ \text{in})^2 = 558\ \text{in}^2$$

$$r = \sqrt{\frac{I}{A}} = \sqrt{\frac{25{,}960\ \text{in}^4}{558\ \text{in}^2}} = \boxed{6.82\ \text{in}}$$

The answer is (C).

(b)
$$A_s = 4\pi r^2 = 4\pi\left(\tfrac{5}{16}\ \text{in}\right)^2 = 1.23\ \text{in}^2$$

$$\rho_t = \frac{A_s}{bt} = \frac{1.23\ \text{in}^2}{(23.625\ \text{in})(23.625\ \text{in})} = \boxed{0.0022}$$

The answer is (A).

(c)
$$\frac{e}{t} = \frac{6\ \text{in}}{23.625\ \text{in}} = 0.25$$

$$E_m = 700f'_m = (700)\left(3000\ \frac{\text{lbf}}{\text{in}^2}\right) = 2{,}100{,}000\ \text{lbf/in}^2$$

$$n = \frac{E_s}{E_m} = \frac{29{,}000{,}000\ \frac{\text{lbf}}{\text{in}^2}}{2{,}100{,}000\ \frac{\text{lbf}}{\text{in}^2}} = 13.8$$

$$n\rho_t = (13.8)(0.002) = 0.0276 \quad [\text{use 0.03}]$$

From App. 70.B,

$$\frac{P}{F_b bt} = 0.38$$

$$P = (0.38)\left(\tfrac{1}{3}\right)\left(3000\ \frac{\text{lbf}}{\text{in}^2}\right)(23.625\ \text{in})(23.625\ \text{in}) = \boxed{212{,}093\ \text{lbf}}$$

The answer is (D).

(d) $M = Pe = (212{,}093\ \text{lbf})(6\ \text{in}) = \boxed{1{,}272{,}558\ \text{in-lbf}}$

The answer is (C).

(e)
$$f_a = \frac{P}{A} = \frac{212{,}093\ \text{lbf}}{(23.625\ \text{in})^2} = \boxed{380\ \text{lbf/in}^2}$$

The answer is (B).

(f) Ignoring the steel,

$$P_a = 0.25f'_m A_n\left(1 - \left(\frac{h}{140r}\right)^2\right)$$
$$= (0.25)\left(3000\ \frac{\text{lbf}}{\text{in}^2}\right)(23.625\ \text{in})^2 \times\left(1 - \left(\frac{(16\ \text{ft})\left(12\ \frac{\text{in}}{\text{ft}}\right)}{(140)(6.82\ \text{in})}\right)^2\right)$$
$$= \boxed{401{,}678\ \text{lbf}}$$

The answer is (B).

(g)
$$\frac{e}{t} = \frac{4\ \text{in}}{23.625\ \text{in}} = 0.17$$
$$n\rho_t = 0.03$$

Structural

From App. 70.B,

$$\frac{P}{F_bbt} = 0.5$$

$$P = 0.5F_bbt = (0.5)\left(\tfrac{1}{3}\right)\left(3000\ \frac{\text{lbf}}{\text{in}^2}\right)(23.625\ \text{in})^2 = \boxed{279{,}070\ \text{lbf}}$$

The answer is (D).

(h) From Eq. 70.1,

$$\begin{aligned} P_a &= (0.25f'_mA_n + 0.65A_{\text{st}}F_s)\left(1-\left(\frac{h}{140r}\right)^2\right) \\ &= \left((0.25)\left(3000\ \frac{\text{lbf}}{\text{in}^2}\right)(558\ \text{in}^2) + (0.65)(1.23\ \text{in}^2)\left(24{,}000\ \frac{\text{lbf}}{\text{in}^2}\right)\right) \\ &\quad \times\left(1-\left(\frac{(16\ \text{ft})\left(12\ \frac{\text{in}}{\text{ft}}\right)}{(140)(6.82\ \text{in})}\right)^2\right) \\ &= \boxed{419{,}989\ \text{lbf}} \end{aligned}$$

The answer is (D).

(i) From Eq. 70.3,

$$\begin{aligned} \frac{1}{P_{\text{biaxial}}} &= \frac{1}{P_x} + \frac{1}{P_y} - \frac{1}{P_a} \\ &= \frac{1}{212{,}093\ \text{lbf}} + \frac{1}{279{,}070\ \text{lbf}} - \frac{1}{419{,}989\ \text{lbf}} \end{aligned}$$

$$P_{\text{biaxial}} = \boxed{168{,}998\ \text{lbf}}$$

The answer is (B).

(j)

$$M_y = P_ye_y = (279{,}070\ \text{lbf})(4\ \text{in}) = \boxed{1{,}116{,}280\ \text{in-lbf}}$$

The answer is (C).

5. (a)

$$e = \text{minimum} = 0.1t = (0.1)(32\ \text{in}) = \boxed{3.2\ \text{in}}$$

The answer is (B).

(b)

$$\begin{aligned} P &= 180{,}000\ \text{lbf} + 140{,}000\ \text{lbf} + 45{,}000\ \text{lbf} \\ &= 365{,}000\ \text{lbf} \end{aligned}$$

$$e = \frac{M}{P} = \left(\frac{140{,}000\ \text{ft-lbf}}{365{,}000\ \text{lbf}}\right)\left(12\ \frac{\text{in}}{\text{ft}}\right) = \boxed{4.6\ \text{in}}$$

The answer is (C).

(c)

$$\begin{aligned} P &= (0.9)(180{,}000\ \text{lbf}) + (45{,}000\ \text{lbf}) \\ &= 207{,}000\ \text{lbf} \end{aligned}$$

$$\rho = \frac{M}{P} = \left(\frac{140{,}000\ \text{ft-lbf}}{207{,}000\ \text{lbf}}\right)\left(12\ \frac{\text{in}}{\text{ft}}\right) = \boxed{8.1\ \text{in}}$$

The answer is (D).

(d)

$$\frac{P}{A} = \frac{180{,}000\ \text{lbf} + 140{,}000\ \text{lbf}}{(31.625\ \text{in})^2} = \boxed{320\ \text{lbf/in}^2}$$

The answer is (C).

(e) Use Eq. 70.1. Neglect steel.

$$\begin{aligned} P_a &= (0.25f'_mA_n + 0.65A_{\text{st}}F_s)\left(1-\left(\frac{h}{140r}\right)^2\right) \\ &= (0.25)\left(1500\ \frac{\text{lbf}}{\text{in}^2}\right)(31.625\ \text{in})^2 \\ &\quad \times\left(1-\left(\frac{(10\ \text{ft})\left(12\ \frac{\text{in}}{\text{ft}}\right)}{(140)(9.1\ \text{in})}\right)^2\right) \\ &= \boxed{371{,}725\ \text{lbf}} \end{aligned}$$

The answer is (A).

(f) For $D+L$,

$$\frac{P}{F_bbt} = \frac{320{,}000\ \text{lbf}}{\left(500\ \frac{\text{lbf}}{\text{in}^2}\right)(31.625\ \text{in})^2} = 0.64$$

$$\frac{Pe}{F_bbt^2} = \frac{(0.64)(3.2\ \text{in})}{31.625\ \text{in}} = 0.065$$

From App. 70.A, $n\rho_t = 0.06$.

For $D+L+E$, increase F_b by one third for load combinations including earthquake.

$$F_b = \left(\tfrac{4}{3}\right)\left(\tfrac{1}{3}\right)\left(1500\ \frac{\text{lbf}}{\text{in}^2}\right) = 667\ \text{lbf/in}^2$$

$$\frac{P}{F_bbt} = \frac{365{,}000\ \text{lbf}}{\left(667\ \frac{\text{lbf}}{\text{in}^2}\right)(31.625\ \text{in})^2} = 0.55$$

$$\frac{Pe}{F_bbt^2} = \frac{(0.55)(4.6\ \text{in})}{31.625\ \text{in}} = 0.080$$

From App. 70.A, $n\rho_t = 0.06$.

For $0.9D + E$,

$$\frac{P}{F_b bt} = \frac{207{,}000 \text{ lbf}}{\left(667 \ \frac{\text{lbf}}{\text{in}^2}\right)(31.625 \text{ in})^2} = 0.31$$

$$\frac{Pe}{F_b bt^2} = \frac{(0.31)(8.1 \text{ in})}{31.625 \text{ in}} = 0.079$$

From App. 70.A, $n\rho_t = 0$.

Use maximum steel of $n\rho_t = 0.06$.

$$\rho_t = \frac{n\rho_t}{n} = \frac{0.06}{14} = 0.0043$$

Since the code minimum of 0.0025 is less than 0.004, this is acceptable.

$$A_{\text{st}} = \rho_t bt = (0.0043)(31.625 \text{ in})^2 = \boxed{4.3 \text{ in}^2}$$

The answer is (D).

(g) $$f_v = \frac{V}{bd} = \frac{30{,}000 \text{ lbf}}{(31.625 \text{ in})(23 \text{ in})} = \boxed{41.2 \text{ lbf/in}^2}$$

The answer is (B).

(h) $$F_v = \left(\tfrac{4}{3}\right)\sqrt{f'_m} = \left(\tfrac{4}{3}\right)\sqrt{1500 \ \frac{\text{lbf}}{\text{in}^2}} = \boxed{51.6 \text{ lbf/in}^2}$$

$$51.6 \ \frac{\text{lbf}}{\text{in}^2} > 41.2 \ \frac{\text{lbf}}{\text{in}^2} \quad \text{[OK]}$$

The answer is (C).

(i) Since F_v is greater than f_v, $\boxed{\text{none is required.}}$

The answer is (A).

(j) Use Eq. 69.35.

$$s = \frac{A_v F_s d}{V} = \frac{2\pi\left(\tfrac{5}{16} \text{ in}\right)^2\left(24{,}000 \ \frac{\text{lbf}}{\text{in}^2}\right)(23 \text{ in})}{30{,}000 \text{ lbf}} = 11.3 \text{ in}$$

Use no. 5 ties spaced at 8 in.

The answer is (B).

Structural

71 Properties of Solid Bodies

PRACTICE PROBLEMS

1. A spoked flywheel has an outside diameter of 60 in (1500 mm). The rim thickness is 6 in (150 mm) and the width is 12 in (300 mm). The cylindrical hub has an outside diameter of 12 in (300 mm), a thickness of 3 in (75 mm), and a width of 12 in (300 mm). The rim and hub are connected by six equally spaced cylindrical radial spokes, each with a diameter of 4.25 in (110 mm). All parts of the flywheel are ductile cast iron with a density of 0.256 lbm/in^3 (7080 kg/m^3). What is the rotational mass moment of inertia of the entire flywheel?

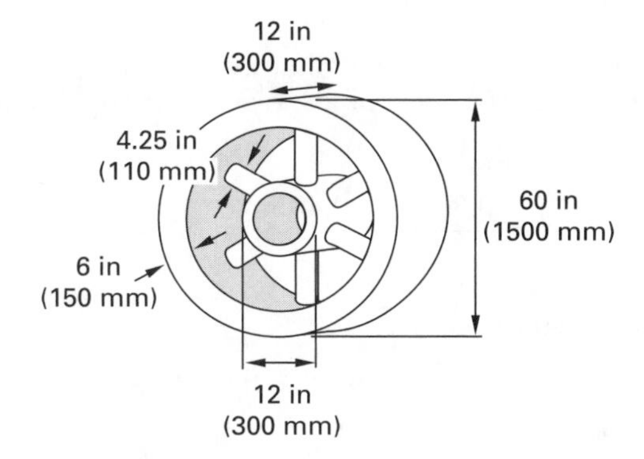

(A) 16,000 lbm-ft^2 (625 kg·m^2)
(B) 16,200 lbm-ft^2 (631 kg·m^2)
(C) 16,800 lbm-ft^2 (654 kg·m^2)
(D) 193,000 lbm-ft^2 (7500 kg·m^2)

SOLUTIONS

1. *Customary U.S. Solution*

From the hollow circular cylinder in App. 71.A, the mass moment of inertia of the rim is

$$\begin{aligned}I_{x,1} &= \left(\frac{\pi\rho L}{2}\right)(r_o^4 - r_i^4)\\ &= \left(\frac{\pi}{2}\right)\left(0.256\ \frac{\text{lbm}}{\text{in}^3}\right)(12\text{ in})\\ &\quad\times\left(\left(\frac{60\text{ in}}{2}\right)^4 - \left(\frac{60\text{ in} - 12\text{ in}}{2}\right)^4\right)\left(\frac{1\text{ ft}^2}{144\text{ in}^2}\right)\\ &= 16{,}025\text{ lbm-ft}^2\end{aligned}$$

From the hollow circular cylinder in App. 71.A, the mass moment of inertia of the hub is

$$\begin{aligned}I_{x,2} &= \left(\frac{\pi\rho L}{2}\right)(r_o^4 - r_i^4)\\ &= \left(\frac{\pi}{2}\right)\left(0.256\ \frac{\text{lbm}}{\text{in}^3}\right)(12\text{ in})\\ &\quad\times\left(\left(\frac{12\text{ in}}{2}\right)^4 - \left(\frac{12\text{ in} - 6\text{ in}}{2}\right)^4\right)\left(\frac{1\text{ ft}^2}{144\text{ in}^2}\right)\\ &= 42\text{ lbm-ft}^2\end{aligned}$$

The length of a cylindrical spoke is

$$L = 24\text{ in} - 6\text{ in} = 18\text{ in}$$

The mass of a spoke is

$$\begin{aligned}m &= \rho AL = \rho\pi r^2 L\\ &= \left(0.256\ \frac{\text{lbm}}{\text{in}^3}\right)\pi\left(\frac{4.25\text{ in}}{2}\right)^2(18\text{ in})\\ &= 65.37\text{ lbm}\end{aligned}$$

From the solid circular cylinder in App. 71.A, the mass moment of inertia of a spoke about its own centroidal axis is

$$\begin{aligned}I_z &= \frac{m(3r^2 + L^2)}{12}\\ &= \frac{(65.37\text{ lbm})\left((3)\left(\frac{4.25\text{ in}}{2}\right)^2 + (18\text{ in})^2\right)\left(\frac{1\text{ ft}^2}{144\text{ in}^2}\right)}{12}\\ &= 13\text{ lbm-ft}^2\end{aligned}$$

Use the parallel axis theorem, Eq. 71.12, to find the mass moment of inertia of a spoke about the axis of the flywheel.

$$d = \frac{12 \text{ in}}{2} + \frac{18 \text{ in}}{2} = 15 \text{ in}$$

$$\begin{aligned} I_{x,3} \text{ per spoke} &= I_z + md^2 \\ &= 13 \text{ lbm-ft}^2 + (65.37 \text{ lbm}) \left(\frac{15 \text{ in}}{12 \frac{\text{in}}{\text{ft}}} \right)^2 \\ &= 115 \text{ lbm-ft}^2 \end{aligned}$$

The total for six spokes is

$$\begin{aligned} I_{x,3} &= 6I_{x,3} \text{ per spoke} \\ &= (6)(115 \text{ lbm-ft}^2) = 690 \text{ lbm-ft}^2 \end{aligned}$$

Finally, the total rotational mass moment of inertia of the flywheel is

$$\begin{aligned} I &= I_{x,1} + I_{x,2} + I_{x,3} \\ &= 16{,}025 \text{ lbm-ft}^2 + 42 \text{ lbm-ft}^2 + 690 \text{ lbm-ft}^2 \\ &= \boxed{16{,}757 \text{ lbm-ft}^2} \end{aligned}$$

SI Solution

From the hollow circular cylinder in App. 71.A, the mass moment of inertia of the rim is

$$\begin{aligned} I_{x,1} &= \left(\frac{\pi \rho L}{2} \right) (r_o^4 - r_i^4) \\ &= \left(\frac{\pi \left(7080 \frac{\text{kg}}{\text{m}^3} \right) (0.3 \text{ m})}{2} \right) \\ &\quad \times \left(\left(\frac{1.5 \text{ m}}{2} \right)^4 - \left(\frac{1.5 \text{ m} - 0.30 \text{ m}}{2} \right)^4 \right) \\ &= 623.3 \text{ kg·m}^2 \end{aligned}$$

From the hollow circular cylinder in App. 71.A, the mass moment of inertia of the hub is

$$\begin{aligned} I_{x,2} &= \left(\frac{\pi \rho L}{2} \right) (r_o^4 - r_i^4) \\ &= \left(\frac{\pi \left(7080 \frac{\text{kg}}{\text{m}^3} \right) (0.3 \text{ m})}{2} \right) \\ &\quad \times \left(\left(\frac{0.3 \text{ m}}{2} \right)^4 - \left(\frac{0.3 \text{ m} - 0.15 \text{ m}}{2} \right)^4 \right) \\ &= 1.6 \text{ kg·m}^2 \end{aligned}$$

The length of a cylindrical spoke is

$$L = 0.6 \text{ m} - 0.15 \text{ m} = 0.45 \text{ m}$$

The mass of a spoke is

$$\begin{aligned} m &= \rho A L = \rho \pi r^2 L \\ &= \left(7080 \frac{\text{kg}}{\text{m}^3} \right) \pi \left(\frac{0.11 \text{ m}}{2} \right)^2 (0.45 \text{ m}) \\ &= 30.3 \text{ kg} \end{aligned}$$

From the solid circular cylinder in App. 71.A, the mass moment of inertia of a spoke about its own centroidal axis is

$$\begin{aligned} I_z &= \frac{m(3r^2 + L^2)}{12} \\ &= \frac{(30.3 \text{ kg}) \left((3) \left(\frac{0.11 \text{ m}}{2} \right)^2 + (0.45 \text{ m})^2 \right)}{12} \\ &= 0.53 \text{ kg·m}^2 \end{aligned}$$

Use the parallel axis theorem, Eq. 71.12, to find the mass moment of inertia of a spoke about the axis of the flywheel.

$$d = \frac{0.30 \text{ m}}{2} + \frac{0.45 \text{ m}}{2} = 0.375 \text{ m}$$

$$\begin{aligned} I_{x,3} \text{ per spoke} &= I_z + md^2 \\ &= 0.53 \text{ kg·m}^2 + (30.3 \text{ kg})(0.375 \text{ m})^2 \\ &= 4.8 \text{ kg·m}^2 \end{aligned}$$

The total for six spokes is

$$\begin{aligned} I_{x,3} &= 6I_{x3} \text{ per spoke} \\ &= (6)(4.8 \text{ kg·m}^2) = 28.8 \text{ kg·m}^2 \end{aligned}$$

Finally, the total rotational mass moment of inertia of the flywheel is

$$\begin{aligned} I &= I_{x,1} + I_{x,2} + I_{x,3} \\ &= 623.3 \text{ kg·m}^2 + 1.6 \text{ kg·m}^2 + 28.8 \text{ kg·m}^2 \\ &= \boxed{653.7 \text{ kg·m}^2} \end{aligned}$$

72 Kinematics

PRACTICE PROBLEMS

Linear Particle Motion

1. A particle moves horizontally according to $s = 2t^2 - 8t + 3$.

(a) When $t = 2$, what are the position, velocity, and acceleration?

(b) What are the linear displacement and total distance traveled between $t = 1$ and $t = 3$?

Uniform Acceleration

2. A jet plane acquires a speed of 180 mph (290 km/h) in 60 sec. What is its acceleration?

(A) 4.4 ft/sec^2 (1.3 m/s^2)
(B) 6.3 ft/sec^2 (1.9 m/s^2)
(C) 8.9 ft/sec^2 (2.7 m/s^2)
(D) 12 ft/sec^2 (3.6 m/s^2)

3. What is the acceleration of a train that increases its speed from 5 ft/sec to 20 ft/sec (1.5 m/s to 6 m/s) in 2 min?

(A) 0.13 ft/sec^2 (0.038 m/s^2)
(B) 0.39 ft/sec^2 (0.12 m/s^2)
(C) 0.82 ft/sec^2 (0.25 m/s^2)
(D) 1.3 ft/sec^2 (0.39 m/s^2)

4. A car traveling at 60 mph (100 km/h) applies its brakes and stops in 5 sec. What is its acceleration and distance traveled before stopping?

Projectile Motion

5. A projectile is fired at 45° from the horizontal with an initial velocity of 2700 ft/sec (820 m/s). Find the maximum altitude and range neglecting air friction.

6. A projectile is launched with an initial velocity of 900 ft/sec (270 m/s). The target is 12,000 ft (3600 m) away and 2000 ft (600 m) higher than the launch point. Air friction is to be neglected. At what angle should the projectile be launched?

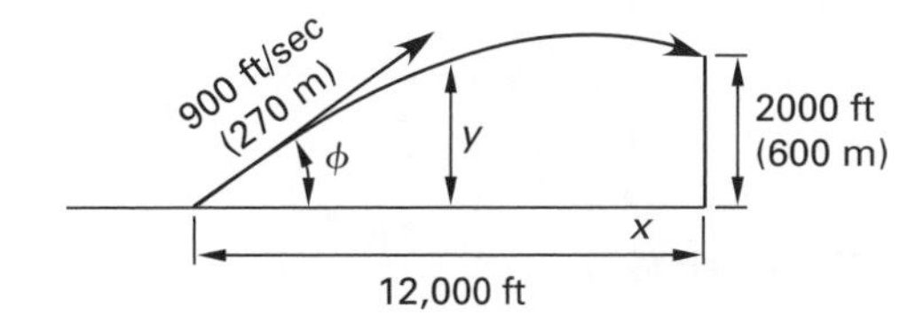

7. A baseball is hit at 60 ft/sec (20 m/s) and 36.87° from the horizontal. It strikes a fence 72 ft (22 m) away. Find the velocity components and the elevation above the origin at impact.

8. A bomb is dropped from a plane that is climbing at 30° and 600 ft/sec (180 m/s) while traveling at 12,000 ft (3600 m) altitude.

(a) What is the bomb's maximum altitude?

(A) 12,500 ft (3700 m)
(B) 13,000 ft (3900 m)
(C) 13,400 ft (4000 m)
(D) 14,200 ft (4300 m)

(b) How long will it take for the bomb to reach the ground from the release point?

(A) 24 sec
(B) 38 sec
(C) 43 sec
(D) 55 sec

Rotational Particle Motion

9. A point travels in a circle according to $\omega = 6t^2 - 10t$. At $t = 2$, the direction of motion is clockwise.

(a) What is the angular velocity at $t = 2$?

(A) 4
(B) 6
(C) 8
(D) 12

(b) What is the displacement between $t = 1$ and $t = 3$?

(A) 6 rad
(B) 10 rad
(C) 12 rad
(D) 15 rad

(c) What is the total angle turned through between $t = 1$ and $t = 3$? Assume $\theta(0) = 0$.

(A) 6 rad
(B) 8 rad
(C) 12 rad
(D) 15 rad

10. What is the linear speed of a point on the edge of a 14 in diameter (36 cm diameter) disk turning at 40 rpm?

(A) 2.4 ft/sec (0.75 m/s)
(B) 4.5 ft/sec (1.4 m/s)
(C) 7.8 ft/sec (2.3 m/s)
(D) 9.2 ft/sec (2.8 m/s)

11. What angular acceleration is required to increase an electric motor's speed from 1200 rpm to 3000 rpm in 10 sec?

(A) 10 rad/sec^2
(B) 14 rad/sec^2
(C) 18 rad/sec^2
(D) 25 rad/sec^2

12. An apparatus for determining the speed of a bullet consists of 2 paper disks mounted 5 ft (1.5 m) apart on a single horizontal shaft which is turning at 1750 rpm. A bullet pierces both disks at radius 6 in (15 cm) and an angle of 18° exists between each hole. What is the bullet velocity?

(A) 1700 ft/sec (510 m/s)
(B) 2100 ft/sec (630 m/s)
(C) 2400 ft/sec (720 m/s)
(D) 2900 ft/sec (880 m/s)

13. Disks B and C are in contact and rotate without slipping. A and B are splined together and rotate counterclockwise. Angular velocity and acceleration of disk C are 2 rad/sec (2 rad/s) and 6 rad/sec^2 (6 rad/s^2), respectively. What is the velocity and acceleration of point D?

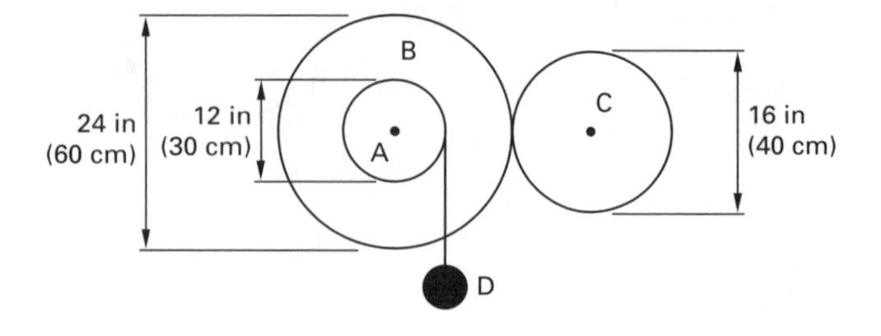

Relative Motion

14. The center of a wheel with an outer diameter of 24 in (610 mm) is moving at 28 mph (12.5 m/s). There is no slippage between the wheel and surface. A valve stem is mounted 6 in (150 mm) from the center. What are the velocity and direction of the valve stem when it is 45° from the horizontal?

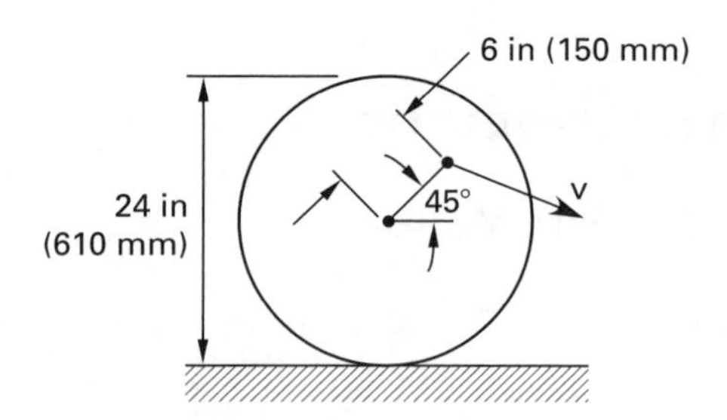

15. A balloon is 200 ft (60 m) above the ground and is rising at a constant 15 ft/sec (4.5 m/s). An automobile passes under it traveling along a straight and level road at 45 mph (72 km/h). How fast is the distance between them changing 1 sec later?

(A) 13 ft/sec (4.0 m/s)
(B) 34 ft/sec (10 m/s)
(C) 37 ft/sec (11 m/s)
(D) 51 ft/sec (15 m/s)

Rotation About a Fixed Axis

16. Find the velocities of points A and B with respect to point O if the wheel rolls without slipping. The axle to which the wheel is attached moves at 10 ft/sec (3 m/s) to the right.

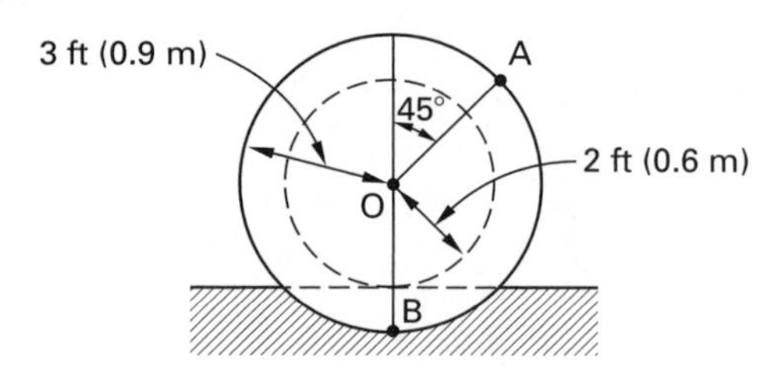

17. What is the velocity of point B with respect to point A in Prob. 16?

SOLUTIONS

1. *Customary U.S. Solution*

(a)

$$x(t) = 2t^2 - 8t + 3$$

$$\mathrm{v}(t) = \frac{dx}{dt} = 4t - 8$$

$$a(t) = \frac{d\mathrm{v}}{dt} = 4$$

At $t = 2$:

$$x = (2)(2)^2 - (8)(2) + 3 = \boxed{-5}$$

$$\mathrm{v} = (4)(2) - 8 = \boxed{0}$$

$$a = \boxed{4}$$

(b) At $t = 1$:

$$x = (2)(1)^2 - (8)(1) + 3 = \boxed{-3}$$

At $t = 3$:

$$x = (2)(3)^2 - (8)(3) + 3 = \boxed{-3}$$

displacement $= x_3 - x_1 = -3 - (-3) = 0$

total distance traveled from $t = 1$ to $t = 3$

$$= \int_1^3 |\mathrm{v}(t)|dt = \int_1^3 |4t - 8|dt$$

$$= \int_1^2 (8 - 4t)dt + \int_2^3 (4t - 8)dt$$

$$= 2 + 2 = \boxed{4}$$

SI Solution

(a)

$$x(t) = 2t^2 - 8t + 3$$

$$\mathrm{v}(t) = \frac{dx}{dt} = 4t - 8$$

$$a(t) = \frac{d\mathrm{v}}{dt} = 4$$

At $t = 2$:

$$x = (2)(2)^2 - (8)(2) + 3 = \boxed{-5}$$

$$\mathrm{v} = (4)(2) - 8 = \boxed{0}$$

$$a = \boxed{4}$$

(b) At $t = 1$:

$$x = (2)(1)^2 - (8)(1) + 3 = \boxed{-3}$$

At $t = 3$:

$$x = (2)(3)^2 - (8)(3) + 3 = \boxed{-3}$$

displacement $= x_3 - x_1 = -3 - (-3) = 0$

total distance traveled from $t = 1$ to $t = 3$

$$= \int_1^3 |\mathrm{v}(t)|dt = \int_1^3 |4t - 8|dt$$

$$= \int_1^2 (8 - 4t)dt + \int_2^3 (4t - 8)dt$$

$$= 2 + 2 = \boxed{4}$$

2. *Customary U.S. Solution*

$$a = \frac{\Delta \mathrm{v}}{\Delta t} = \frac{\left(180\ \frac{\text{mi}}{\text{hr}}\right)\left(5280\ \frac{\text{ft}}{\text{mi}}\right)\left(\frac{1\ \text{hr}}{3600\ \text{sec}}\right)}{60\ \text{sec}}$$

$$= \boxed{4.4\ \text{ft/sec}^2}$$

The answer is (A).

SI Solution

$$a = \frac{\Delta \mathrm{v}}{\Delta t} = \frac{\left(290\ \frac{\text{km}}{\text{h}}\right)\left(1000\ \frac{\text{m}}{\text{km}}\right)\left(\frac{1\ \text{h}}{3600\ \text{s}}\right)}{60\ \text{s}}$$

$$= \boxed{1.34\ \text{m/s}^2}$$

The answer is (A).

3. *Customary U.S. Solution*

$$a = \frac{\mathrm{v} - \mathrm{v}_0}{t} = \frac{20\ \frac{\text{ft}}{\text{sec}} - 5\ \frac{\text{ft}}{\text{sec}}}{(2\ \text{min})\left(60\ \frac{\text{sec}}{\text{min}}\right)}$$

$$= \boxed{0.125\ \text{ft/sec}^2}$$

The answer is (A).

SI Solution

$$a = \frac{v - v_0}{t} = \frac{6\ \frac{m}{s} - 1.5\ \frac{m}{s}}{(2\ \text{min})\left(60\ \frac{s}{\text{min}}\right)}$$

$$= \boxed{0.0375\ \text{m/s}^2}$$

The answer is (A).

4. *Customary U.S. Solution*

Assume uniform deceleration.

$$v_0 = \left(60\ \frac{\text{mi}}{\text{hr}}\right)\left(5280\ \frac{\text{ft}}{\text{mi}}\right)\left(\frac{1\ \text{hr}}{3600\ \text{sec}}\right) = 88\ \text{ft/sec}$$

$$a = \frac{v - v_0}{t} = \frac{0 - 88\ \frac{\text{ft}}{\text{sec}}}{5\ \text{sec}} = \boxed{-17.6\ \text{ft/sec}^2}$$

$$s = \frac{1}{2}t(v + v_0) = \frac{1}{2}tv_0$$

$$= \left(\frac{1}{2}\right)(5\ \text{sec})\left(88\ \frac{\text{ft}}{\text{sec}}\right) = \boxed{220\ \text{ft}}$$

SI Solution

Assume uniform deceleration.

$$v_0 = \left(100\ \frac{\text{km}}{\text{h}}\right)\left(1000\ \frac{\text{m}}{\text{km}}\right)\left(\frac{1\ \text{h}}{3600\ \text{s}}\right) = 27.78\ \text{m/s}$$

$$a = \frac{v - v_0}{t} = \frac{0 - 27.78\ \frac{m}{s}}{5\ \text{s}} = \boxed{-5.56\ \text{m/s}^2}$$

$$s = \frac{1}{2}t(v + v_0) = \frac{1}{2}tv_0$$

$$= \left(\frac{1}{2}\right)(5\ \text{s})\left(27.78\ \frac{m}{s}\right) = \boxed{69.5\ \text{m}}$$

5. *Customary U.S. Solution*

$$H = \frac{v_0^2 \sin^2\phi}{2g}$$

$$= \frac{\left(2700\ \frac{\text{ft}}{\text{sec}}\right)^2(\sin^2 45°)}{(2)\left(32.2\ \frac{\text{ft}}{\text{sec}^2}\right)} = \boxed{56{,}599\ \text{ft}}$$

$$R = \frac{v_0^2 \sin 2\phi}{g}$$

$$= \frac{\left(2700\ \frac{\text{ft}}{\text{sec}}\right)^2(\sin 90°)}{32.2\ \frac{\text{ft}}{\text{sec}^2}} = \boxed{226{,}398\ \text{ft}}$$

SI Solution

$$H = \frac{v_0^2 \sin^2\phi}{2g}$$

$$= \frac{\left(820\ \frac{m}{s}\right)^2(\sin^2 45°)}{(2)\left(9.81\ \frac{m}{s^2}\right)} = \boxed{17\,136\ \text{m}}$$

$$R = \frac{v_0^2 \sin 2\phi}{g}$$

$$= \frac{\left(820\ \frac{m}{s}\right)^2(\sin 90°)}{9.81\ \frac{m}{s^2}} = \boxed{68\,542\ \text{m}}$$

6. *Customary U.S. Solution*

From Table 72.2, the x-distance is

$$x = (v_0 \cos\phi)t$$

Solve for t.

$$t = \frac{x}{v_0 \cos\phi} = \frac{12{,}000\ \text{ft}}{\left(900\ \frac{\text{ft}}{\text{sec}}\right)(\cos\phi)} = \frac{13.33}{\cos\phi}$$

From Table 72.2, the y-distance is

$$y = (v_0 \sin\phi)t - \tfrac{1}{2}gt^2$$

Substitute t and the value for y.

$$2000\ \text{ft} = \left(900\ \frac{\text{ft}}{\text{sec}}\right)(\sin\phi)\left(\frac{13.33}{\cos\phi}\right) - \left(\tfrac{1}{2}\right)\left(32.2\ \frac{\text{ft}}{\text{sec}^2}\right)\left(\frac{13.33}{\cos\phi}\right)^2$$

Simplify.

$$1 = 6.0\tan\phi - \frac{1.43}{\cos^2\phi}$$

Use the identity.

$$\frac{1}{\cos^2\phi} = 1 + \tan^2\phi$$

$$1 = 6.0\tan\phi - 1.43 - 1.43\tan^2\phi$$

Simplify.

$$\tan^2\phi - 4.20\tan\phi + 1.70 = 0$$

Use the quadratic formula.

$$\tan\phi = \frac{4.20 \pm \sqrt{(4.20)^2 - (4)(1)(1.70)}}{(2)(1)}$$

$$= 3.75,\ 0.454 \quad \text{[radians]}$$

$$\phi = \tan^{-1}(3.75),\ \tan^{-1}(0.454)$$

$$= \boxed{75.1°,\ 24.4°}$$

SI Solution

From Table 72.2, the x-distance is

$$x = (\mathrm{v}_0 \cos\phi)t$$

Solve for t.

$$t = \frac{x}{\mathrm{v}_0 \cos\phi} = \frac{3600 \text{ m}}{\left(270 \ \frac{\text{m}}{\text{s}}\right)(\cos\phi)} = \frac{13.33}{\cos\phi}$$

From Table 72.2, the y-distance is

$$y = (\mathrm{v}_0 \sin\phi)t - \tfrac{1}{2}gt^2$$

Substitute t and the value of y.

$$600 \text{ m} = \left(270 \ \frac{\text{m}}{\text{s}}\right)(\sin\phi)\left(\frac{13.33}{\cos\phi}\right) - \left(\tfrac{1}{2}\right)\left(9.81 \ \frac{\text{m}}{\text{s}^2}\right)\left(\frac{13.33}{\cos\phi}\right)^2$$

Simplify.

$$1 = 6.0\tan\phi - \frac{1.45}{\cos^2\phi}$$

Use the identity.

$$\frac{1}{\cos^2\phi} = 1 + \tan^2\phi$$

$$1 = 6.0\tan\phi - 1.45 - 1.45\tan^2\phi$$

Simplify.

$$\tan^2\phi - 4.14\tan\phi + 1.69 = 0$$

Use the quadratic formula.

$$\begin{aligned}\tan\phi &= \frac{4.14 \pm \sqrt{(4.14)^2 - (4)(1)(1.69)}}{(2)(1)} \\ &= 3.68,\ 0.459 \quad \text{[radians]} \\ \phi &= \tan^{-1}(3.68),\ \tan^{-1}(0.459) \\ &= \boxed{74.8^\circ,\ 24.7^\circ}\end{aligned}$$

7. *Customary U.S. Solution*

Neglect air friction.

$$\begin{aligned}\mathrm{v}_x = \mathrm{v}_0\cos\phi &= \left(60 \ \frac{\text{ft}}{\text{sec}}\right)(\cos 36.87^\circ) \\ &= \boxed{48 \text{ ft/sec}}\end{aligned}$$

$$t = \frac{s}{\mathrm{v}_x} = \frac{72 \text{ ft}}{48 \ \frac{\text{ft}}{\text{sec}}} = 1.5 \text{ sec}$$

$$\begin{aligned}\mathrm{v}_y &= \mathrm{v}_0\sin\phi - gt \\ &= \left(60 \ \frac{\text{ft}}{\text{sec}}\right)(\sin 36.87^\circ) - \left(32.2 \ \frac{\text{ft}}{\text{sec}^2}\right)(1.5 \text{ sec}) \\ &= \boxed{-12.3 \text{ ft/sec}}\end{aligned}$$

$$\begin{aligned}y &= \mathrm{v}_0\sin\phi\, t - \frac{1}{2}gt^2 \\ &= \left(60 \ \frac{\text{ft}}{\text{sec}}\right)(\sin 36.87^\circ)(1.5 \text{ sec}) - \left(\frac{1}{2}\right)\left(32.2 \ \frac{\text{ft}}{\text{sec}^2}\right)(1.5 \text{ sec})^2 \\ &= \boxed{17.78 \text{ ft}}\end{aligned}$$

SI Solution

Neglect air friction.

$$\begin{aligned}\mathrm{v}_x = \mathrm{v}_0\cos\phi &= \left(20 \ \frac{\text{m}}{\text{s}}\right)(\cos 36.87^\circ) \\ &= \boxed{16 \text{ m/s}}\end{aligned}$$

$$t = \frac{s}{\mathrm{v}_x} = \frac{22 \text{ m}}{16 \ \frac{\text{m}}{\text{s}}} = 1.375 \text{ s}$$

$$\begin{aligned}\mathrm{v}_y &= \mathrm{v}_0\sin\phi - gt \\ &= \left(20 \ \frac{\text{m}}{\text{s}}\right)(\sin 36.87^\circ) - \left(9.81 \ \frac{\text{m}}{\text{s}^2}\right)(1.375 \text{ s}) \\ &= \boxed{-1.49 \text{ m/s}}\end{aligned}$$

$$\begin{aligned}y &= \mathrm{v}_0\sin\phi t - \frac{1}{2}gt^2 \\ &= \left(20 \ \frac{\text{m}}{\text{s}}\right)(\sin 36.87^\circ)(1.375 \text{ s}) - \left(\frac{1}{2}\right)\left(9.81 \ \frac{\text{m}}{\text{s}^2}\right)(1.375 \text{ s})^2 \\ &= \boxed{7.227 \text{ m}}\end{aligned}$$

8. *Customary U.S. Solution*

(a)

$$\begin{aligned}H &= z + \frac{\mathrm{v}_0^2\sin^2\phi}{2g} \\ &= 12{,}000 \text{ ft} + \frac{\left(600 \frac{\text{ft}}{\text{sec}}\right)^2(\sin^2 30^\circ)}{(2)\left(32.2 \ \frac{\text{ft}}{\text{sec}^2}\right)} \\ &= \boxed{13{,}398 \text{ ft}}\end{aligned}$$

The answer is (C).

Transportation

(b) Let t_1 be the time the bomb takes to reach the maximum altitude.

$$t_1 = \frac{1}{2}\left(\frac{2\text{v}_0 \sin\phi}{g}\right) = \frac{\left(600 \ \frac{\text{ft}}{\text{sec}}\right)(\sin 30°)}{32.2 \ \frac{\text{ft}}{\text{sec}^2}} = 9.32 \text{ sec}$$

Let t_2 be the time the bomb takes to fall from H.

$$t_2 = \sqrt{\frac{2H}{g}} = \sqrt{\frac{(2)(13{,}398 \text{ ft})}{32.2 \ \frac{\text{ft}}{\text{sec}^2}}} = 28.85 \text{ sec}$$

$$t = t_1 + t_2 = 9.32 \text{ sec} + 28.85 \text{ sec}$$
$$= \boxed{38.17 \text{ sec}}$$

The answer is (B).

SI Solution

(a)
$$H = z + \frac{\text{v}_0^2 \sin^2\phi}{2g}$$
$$= 3600 \text{ m} + \frac{\left(180 \frac{\text{m}}{\text{s}}\right)^2 (\sin^2 30°)}{(2)\left(9.81 \frac{\text{m}}{\text{s}^2}\right)}$$
$$= \boxed{4012.8 \text{ m}}$$

The answer is (C).

(b) Let t_1 be the time the bomb takes to reach the maximum altitude.

$$t_1 = \frac{1}{2}\left(\frac{2\text{ft}_0 \sin\phi}{g}\right) = \frac{\left(180 \ \frac{\text{m}}{\text{s}}\right)(\sin 30°)}{9.81 \ \frac{\text{m}}{\text{s}^2}} = 9.17 \text{ s}$$

Let t_2 be the time the bomb takes to fall from H.

$$t_2 = \sqrt{\frac{2H}{g}} = \sqrt{\frac{(2)(4012.8 \text{ m})}{9.81 \ \frac{\text{m}}{\text{s}^2}}} = 28.60 \text{ s}$$

$$t = t_1 + t_2 = 9.17 \text{ s} + 28.60 \text{ s}$$
$$= \boxed{37.77 \text{ s}}$$

The answer is (B).

9. *Customary U.S. Solution*

(a)
$$\omega(2) = (6)(2)^2 - (10)(2) = \boxed{4 \text{ clockwise}}$$

(b)
$$\theta = \theta(3) - \theta(1) = \int_1^3 \omega(t)dt$$
$$= \int_1^3 (6t^2 - 10t)dt = 2t^3 - 5t^2 + C\Big]_1^3$$
$$[C = 0 \text{ since } \theta(0) = 0]$$
$$= 2(3)^3 - 5(3)^2 - 2(1)^3 + 5(1)^2$$
$$= \boxed{12}$$

To find the total distance traveled, check for sign reversals in $\omega(t)$ over the interval $t = 1$ to $t = 3$.

$$6t^2 - 10t = 0$$
$$\text{sign reversal at } t = \frac{5}{3}$$

(c) The total angle turned is

$$\int_1^3 |\omega(t)|dt = \int_1^{\frac{5}{3}} (10t - 6t^2)dt + \int_{\frac{5}{3}}^3 (6t^2 - 10t)dt$$
$$= \boxed{15.26 \text{ rad}}$$

The answer is (D).

SI Solution

(a)
$$\omega(2) = (6)(2)^2 - (10)(2) = \boxed{4 \text{ clockwise}}$$

(b)
$$\theta = \theta(3) - \theta(1) = \int_1^3 \omega(t)dt$$
$$= \int_1^3 (6t^2 - 10t)dt = \boxed{12}$$

To find the total distance traveled, check for sign reversals in $\omega(t)$ over the interval $t = 1$ to $t = 3$.

$$6t^2 - 10t = 0$$
$$\text{sign reversal at } t = \frac{5}{3}$$

(c) The total angle turned is

$$\int_1^3 |\omega(t)|dt = \int_1^{\frac{5}{3}} (10t - 6t^2)dt + \int_{\frac{5}{3}}^3 (6t^2 - 10t)dt$$
$$= \boxed{15.26 \text{ rad}}$$

The answer is (D).

10. *Customary U.S. Solution*

$$\text{v} = r\omega = r2\pi f = d\pi f$$
$$= (14 \text{ in})\left(\frac{\text{ft}}{12 \text{ in}}\right)(\pi)\left(40 \ \frac{\text{rev}}{\text{min}}\right)\left(\frac{1 \text{ min}}{60 \text{ sec}}\right)$$
$$= \boxed{2.44 \text{ ft/sec}}$$

The answer is (A).

SI Solution

$$\text{v} = r\omega = r2\pi f = d\pi f$$
$$= (0.36 \text{ m})(\pi)\left(40 \ \frac{\text{rev}}{\text{min}}\right)\left(\frac{1 \text{ min}}{60 \text{ s}}\right)$$
$$= \boxed{0.754 \text{ m/s}}$$

The answer is (A).

11. *Customary U.S. Solution*

$$\omega_1 = 2\pi f_1 = \left(2\pi \frac{\text{rad}}{\text{rev}}\right)\left(1200 \frac{\text{rev}}{\text{min}}\right)\left(\frac{1 \text{ min}}{60 \text{ sec}}\right) = 125.66 \text{ rad/sec}$$

$$\omega_2 = \left(2\pi \frac{\text{rad}}{\text{rev}}\right)\left(3000 \frac{\text{rev}}{\text{min}}\right)\left(\frac{1 \text{ min}}{60 \text{ sec}}\right) = 314.16 \text{ rad/sec}$$

$$\alpha = \frac{\omega_2 - \omega_1}{\Delta t} = \frac{314.16 \frac{\text{rad}}{\text{sec}} - 125.66 \frac{\text{rad}}{\text{sec}}}{10 \text{ sec}} = \boxed{18.85 \text{ rad/sec}^2}$$

The answer is (C).

SI Solution

$$\omega_1 = 2\pi f_1 = \left(2\pi \frac{\text{rad}}{\text{rev}}\right)\left(1200 \frac{\text{rev}}{\text{min}}\right)\left(\frac{1 \text{ min}}{60 \text{ s}}\right) = 125.66 \text{ rad/s}$$

$$\omega_2 = \left(2\pi \frac{\text{rad}}{\text{rev}}\right)\left(3000 \frac{\text{rev}}{\text{min}}\right)\left(\frac{1 \text{ min}}{60 \text{ s}}\right) = 314.16 \text{ rad/s}$$

$$\alpha = \frac{\omega_2 - \omega_1}{\Delta t} = \frac{314.16 \frac{\text{rad}}{\text{s}} - 125.66 \frac{\text{rad}}{\text{s}}}{10 \text{ s}} = \boxed{18.85 \text{ rad/s}^2}$$

The answer is (C).

12. *Customary U.S. Solution*

$$f = \left(1750 \frac{\text{rev}}{\text{min}}\right)\left(\frac{1 \text{ min}}{60 \text{ sec}}\right) = 29.167 \text{ rev/sec}$$

$$\theta = (18^\circ)\left(\frac{1 \text{ rev}}{360^\circ}\right) = 0.05 \text{ rev}$$

$$t = \frac{\theta}{f} = \frac{0.05 \text{ rev}}{29.167 \frac{\text{rev}}{\text{sec}}} = 0.001714 \text{ sec}$$

$$\text{v} = \frac{s}{t} = \frac{5 \text{ ft}}{0.001714 \text{ sec}} = \boxed{2916.7 \text{ ft/sec (fps)}}$$

The answer is (D).

SI Solution

$$\omega = \left(1750 \frac{\text{rev}}{\text{min}}\right)\left(\frac{1 \text{ min}}{60 \text{ s}}\right)\left(\frac{2\pi \text{ rad}}{\text{rev}}\right) = 183.26 \text{ rad/s}$$

$$\theta = (18^\circ)\left(\frac{2\pi \text{ rad}}{360^\circ}\right) = 0.314 \text{ rad}$$

$$t = \frac{\theta}{\omega} = \frac{0.314 \text{ rad}}{183.26 \frac{\text{rad}}{\text{s}}} = 0.001\,713 \text{ s}$$

$$\text{v} = \frac{s}{t} = \frac{1.5 \text{ m}}{0.001\,713 \text{ s}} = \boxed{875.66 \text{ m/s}}$$

The answer is (D).

13. *Customary U.S. Solution*

$$\omega_B = \left(\frac{16}{24}\right)\omega_C = \left(\frac{16}{24}\right)\left(2 \frac{\text{rad}}{\text{sec}}\right) = 1.333 \text{ rad/sec}$$

$$\alpha_B = \left(\frac{16}{24}\right)\alpha_C = \left(\frac{16}{24}\right)\left(6 \frac{\text{rad}}{\text{sec}^2}\right) = 4 \text{ rad/sec}^2$$

Since A and B are splined together,

$$\omega_A = \omega_B$$

$$\alpha_A = \alpha_B$$

$$\text{v}_D = r_A\omega_A = (0.5 \text{ ft})\left(1.333 \frac{\text{rad}}{\text{sec}}\right) = \boxed{0.6665 \text{ ft/sec (fps)}}$$

$$a_D = r_A\alpha_A = (0.5 \text{ ft})\left(4 \frac{\text{rad}}{\text{sec}^2}\right) = \boxed{2 \text{ ft/sec}^2}$$

SI Solution

$$\omega_B = \left(\frac{40}{60}\right)\omega_C = \left(\frac{40}{60}\right)\left(2 \frac{\text{rad}}{\text{s}}\right) = 1.333 \text{ rad/s}$$

$$\alpha_B = \left(\frac{40}{60}\right)\alpha_C = \left(\frac{40}{60}\right)\left(6 \frac{\text{rad}}{\text{s}^2}\right) = 4 \text{ rad/s}^2$$

Since A and B are splined together,

$$\omega_A = \omega_B$$

$$\alpha_A = \alpha_B$$

$$\text{v}_D = r_A\omega_A = (0.15 \text{ m})\left(1.333 \frac{\text{rad}}{\text{s}}\right) = \boxed{0.2 \text{ m/s}}$$

$$a_D = r_A\alpha_A = (0.15 \text{ m})\left(4 \frac{\text{rad}}{\text{s}^2}\right) = \boxed{0.6 \text{ m/s}^2}$$

14. *Customary U.S. Solution*

The angular velocity of the wheel is

$$\omega = \frac{\mathrm{v_c}}{r} = \frac{\left(28\ \frac{\text{mi}}{\text{hr}}\right)\left(5280\ \frac{\text{ft}}{\text{mi}}\right)\left(12\ \frac{\text{in}}{\text{ft}}\right)\left(\frac{1\ \text{hr}}{3600\ \text{sec}}\right)}{12\ \text{in}} = 41.07\ \text{rad/sec}$$

The distance from the valve stem to the instant center of the point of contact of the wheel and the surface is determined from the law of cosines for the triangle defined by the valve stem, instant center, and center of wheel.

$$l^2 = (12\ \text{in})^2 + (6\ \text{in})^2 - (2)(12\ \text{in})(6\ \text{in})(\cos 135°)$$
$$l = 16.79\ \text{in}$$

Use Eq. 72.52.

$$\mathrm{v} = l\omega = (16.79\ \text{in})\left(\frac{1\ \text{ft}}{12\ \text{in}}\right)\left(41.07\ \frac{\text{rad}}{\text{sec}}\right) = \boxed{57.46\ \text{ft/sec}}$$

The direction is perpendicular to the direction of l.

SI Solution

The angular velocity of the wheel is

$$\omega = \frac{\mathrm{v_c}}{r} = \frac{12.5\ \frac{\text{m}}{\text{s}}}{0.305\ \text{m}} = 40.98\ \text{rad/s}$$

The distance from the valve stem to the instant center at the point of contact of the wheel and the surface is determined from the law of cosines for the triangle defined by the valve stem, instant center, and center of wheel.

$$l^2 = (0.305\ \text{m})^2 + (0.150\ \text{m})^2 - (2)(0.305\ \text{m})(0.150\ \text{m})(\cos 135°)$$
$$l = 0.425\ \text{m}$$

Use Eq. 72.52.

$$\mathrm{v} = l\omega = (0.425\ \text{m})\left(40.98\ \frac{\text{rad}}{\text{s}}\right) = \boxed{17.4\ \text{m/s}}$$

The direction is perpendicular to the direction of l.

15. *Customary U.S. Solution*

$$\mathrm{v_{car}} = \left(45\ \frac{\text{mi}}{\text{hr}}\right)\left(5280\ \frac{\text{ft}}{\text{mi}}\right)\left(\frac{1\ \text{hr}}{3600\ \text{sec}}\right) = 66\ \text{ft/sec}$$

The separation distance after 1 sec is

$$s(1) = \sqrt{\left(200\ \text{ft} + \left(15\ \frac{\text{ft}}{\text{sec}}\right)(1\ \text{sec})\right)^2 + \left(\left(66\ \frac{\text{ft}}{\text{sec}}\right)(1\ \text{sec})\right)^2} = 224.9\ \text{ft}$$

The separation velocity is the difference in components of the car's and balloon's velocities along a mutually parallel line. Use the separation vector as this line.

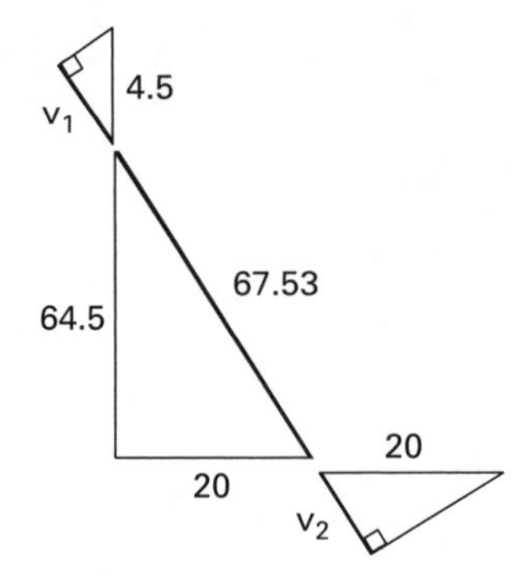

$$\mathrm{v_1} = \left(15\ \frac{\text{ft}}{\text{sec}}\right)\left(\frac{215}{224.9}\right) = 14.34\ \text{ft/sec}$$
$$\mathrm{v_2} = \left(66\ \frac{\text{ft}}{\text{sec}}\right)\left(\frac{66}{224.9}\right) = 19.37\ \text{ft/sec}$$
$$\Delta\mathrm{v} = \mathrm{v_1} + \mathrm{v_2} = 14.34\ \frac{\text{ft}}{\text{sec}} + 19.37\ \text{ft/sec} = \boxed{33.71\ \text{ft/sec}}$$

The answer is (B).

SI Solution

$$\mathrm{v_{car}} = \left(72\ \frac{\text{km}}{\text{h}}\right)\left(1000\ \frac{\text{m}}{\text{km}}\right)\left(\frac{1\ \text{h}}{3600\ \text{s}}\right) = 20\ \text{m/s}$$

The separation distance after 1 second is

$$s(1) = \sqrt{\left(60\ \text{m} + \left(4.5\ \frac{\text{m}}{\text{s}}\right)(1\ \text{s})\right)^2 + \left(\left(20\ \frac{\text{m}}{\text{s}}\right)(1\ \text{s})\right)^2} = 67.53\ \text{m}$$

The separation velocity is the difference in components of the car's and balloon's velocities along a mutually parallel line. Use the separation vector as this line.

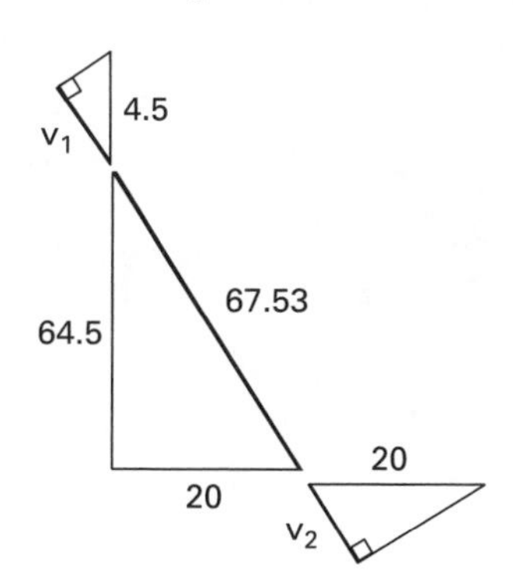

$$v_1 = \left(4.5\ \frac{\text{m}}{\text{s}}\right)\left(\frac{64.5}{67.53}\right) = 4.298\ \frac{\text{m}}{\text{s}}$$

$$v_2 = \left(20\ \frac{\text{m}}{\text{s}}\right)\left(\frac{20}{67.53}\right) = 5.923\ \frac{\text{m}}{\text{s}}$$

$$\Delta v = v_1 + v_2 = 4.298\ \frac{\text{m}}{\text{s}} + 5.923\ \frac{\text{m}}{\text{s}}$$

$$= \boxed{10.22\ \text{m/s}}$$

The answer is (B).

16. *Customary U.S. Solution*

$$\omega = \frac{v_0}{r_{\text{inner}}} = \frac{10\ \frac{\text{ft}}{\text{sec}}}{2\ \frac{\text{ft}}{\text{rad}}} = 5\ \text{rad/sec}$$

$$v_{A/O} = r_{\text{outer}}\,\omega = \left(3\ \frac{\text{ft}}{\text{rad}}\right)\left(5\ \frac{\text{rad}}{\text{sec}}\right)$$

$$= \boxed{\text{15 ft/sec, 45° below the horizontal, to the right}}$$

$$v_{B/O} = r_{\text{outer}}\,\omega = \boxed{\text{15 ft/sec, horizontal, to the left}}$$

SI Solution

$$\omega = \frac{v_0}{r_{\text{inner}}} = \frac{3\ \frac{\text{m}}{\text{s}}}{0.6\ \frac{\text{m}}{\text{rad}}} = 5\ \text{rad/s}$$

$$v_{A/O} = r_{\text{outer}}\,\omega = \left(0.9\ \frac{\text{m}}{\text{rad}}\right)\left(5\ \frac{\text{rad}}{\text{s}}\right)$$

$$= \boxed{\text{4.5 m/s, 45° below the horizontal, to the right}}$$

$$v_{B/O} = r_{\text{outer}}\,\omega = \boxed{\text{4.5 m/s, horizontal, to the left}}$$

17. *Customary U.S. Solution*

$$|AB|^2 = (3\ \text{ft})^2 + (3\ \text{ft})^2 - (2)(3\ \text{ft})^2(\cos 135°)$$
$$= 30.728\ \text{ft}^2$$
$$|AB| = 5.543\ \text{ft}$$

$$v_{B/A} = |AB|\omega = (5.543\ \text{ft})\left(5\ \frac{\text{rad}}{\text{sec}}\right)$$

$$= \boxed{\text{27.72 ft/sec, 22.5° above the horizontal, to the left}}$$

SI Solution

$$|AB|^2 = (0.9\ \text{m})^2 + (0.9\ \text{m})^2 - (2)(0.9\ \text{m})^2(\cos 135°)$$
$$= 2.7655\ \text{m}^2$$
$$|AB| = 1.663\ \text{m}$$

$$v_{B/A} = |AB|\omega = (1.663\ \text{m})\left(5\ \frac{\text{rad}}{\text{s}}\right)$$

$$= \boxed{\text{8.315 m/s, 22.5° above the horizontal, to the left}}$$

73 Kinetics

PRACTICE PROBLEMS

Centripetal and Centrifugal Forces

1. Calculate the superelevation in percent necessary on a curve with a 6000 ft (1800 m) radius so that at 60 mph (100 km/h) cars will not have to rely on friction to stay on the roadway.

(A) 4%
(B) 7%
(C) 11%
(D) 14%

2. What is the angle between the pole and the wire if the radius of the path is 4 ft (1.2 m)? The 8.05 lbm (3.65 kg) object is rotating at 20 ft/sec (6 m/s).

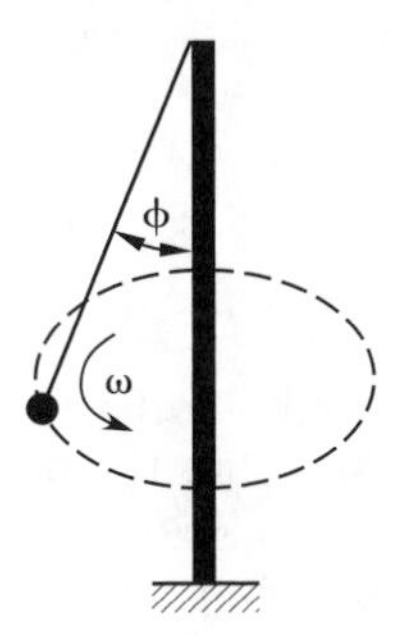

(A) 46°
(B) 58°
(C) 72°
(D) 79°

3. A 10 lbm (5 kg) mass is tied to a 2 ft (50 cm) string and whirled at 5 rev/sec horizontally to the ground.

(a) Find the centripetal acceleration.

(A) 1400 ft/sec^2 (350 m/s^2)
(B) 1600 ft/sec^2 (400 m/s^2)
(C) 1800 ft/sec^2 (450 m/s^2)
(D) 2000 ft/sec^2 (490 m/s^2)

(b) Find the centrifugal force.

(A) 490 lbf (1900 N)
(B) 610 lbf (2500 N)
(C) 750 lbf (2900 N)
(D) 820 lbf (3100 N)

(c) Find the centripetal force.

(A) 490 lbf (1900 N)
(B) 610 lbf (2500 N)
(C) 750 lbf (2900 N)
(D) 820 lbf (3100 N)

(d) Find the angular momentum.

(A) 39 ft-lbf-sec (39 J·s)
(B) 44 ft-lbf-sec (44 J·s)
(C) 53 ft-lbf-sec (53 J·s)
(D) 71 ft-lbf-sec (71 J·s)

Friction

4. A 100 lbm (50 kg) body has an initial velocity of 12.88 ft/sec (3.93 m/s) while moving on a plane with coefficient of friction of 0.2. What distance will it travel before coming to rest?

(A) 13 ft (3.9 m)
(B) 15 ft (4.5 m)
(C) 18 ft (5.4 m)
(D) 24 ft (7.2 m)

5. A box is dropped onto a conveyor belt moving at 10 ft/sec (3 m/s). The coefficient of friction between the box and the belt is 0.333. How long will it take before the box stops slipping on the belt?

(A) 0.48 sec
(B) 0.67 sec
(C) 0.93 sec
(D) 1.2 sec

6. A motorcycle and rider weigh 400 lbm (200 kg). They travel horizontally around the inside of a hollow, right-angle cylinder of 100 ft (30 m) inside diameter. What is the coefficient of friction that will allow a speed of 40 mph (60 km/h)?

(A) 0.3
(B) 0.4
(C) 0.5
(D) 0.6

7. When a force acts on a 10 lbm (5 kg) body initially at rest, a speed of 12 ft/sec (3.6 m/s) is attained in 36 ft (11 m). What is the force if the coefficient of friction is 0.25 and the acceleration is constant?

(A) 1.8 lbf (8.6 N)
(B) 2.4 lbf (13 N)
(C) 3.1 lbf (15 N)
(D) 4.6 lbf (22 N)

8. A 130 lbm (60 kg) block slides up a 22.62° incline with coefficient of friction of 0.1 and initial velocity of 30 ft/sec (9 m/s). How far will the block slide up the incline before coming to rest?

(A) 18 ft (5.4 m)
(B) 29 ft (8.7 m)
(C) 42 ft (13 m)
(D) 57 ft (17 m)

9. A 100 lbm (50 kg) block is acted upon by a 100 lbf (400 N) force while resting on a horizontal surface with coefficient of friction of 0.2. A 50 lbm (25 kg) block sits on top of the 100 lbm (50 kg) block. What is the minimum coefficient of friction between the 50 lbm and 100 lbm (25 kg and 50 kg) blocks for there to be no slipping?

Rigid Body Motion

10. A constant force of 20 lbf (90 N) is applied to a 100 lbm (50 kg) door supported on rollers at A and B.

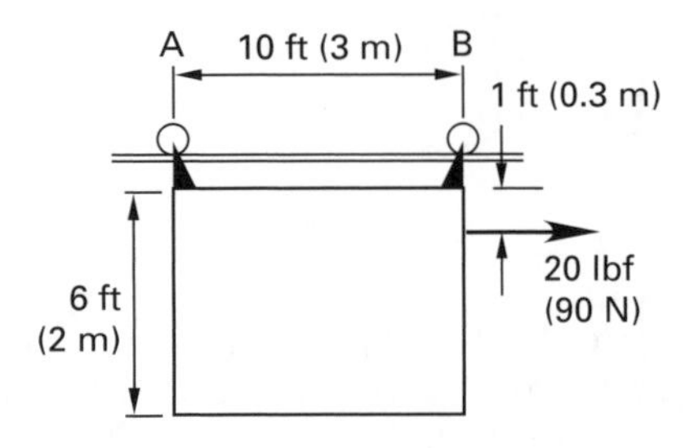

(a) What is the acceleration of the door?

(b) What are the reactions at A and B?

(c) Where would the 20 lbf (90 N) force have to be applied if the reactions at A and B were to be equal?

Constrained Motion

11. A solid sphere rolls without slipping down a 30° incline, starting from rest. What is its speed after 2 sec?

(A) 17 ft/sec (5.1 m/s)
(B) 23 ft/sec (7.0 m/s)
(C) 34 ft/sec (10 m/s)
(D) 86 ft/sec (26 m/s)

12. Object A weighs 10 lbm (5 kg) and rests on a frictionless plane with a 36.87° slope. Object B weighs 20 lbm (10 kg). What is the velocity of B 3 sec after release?

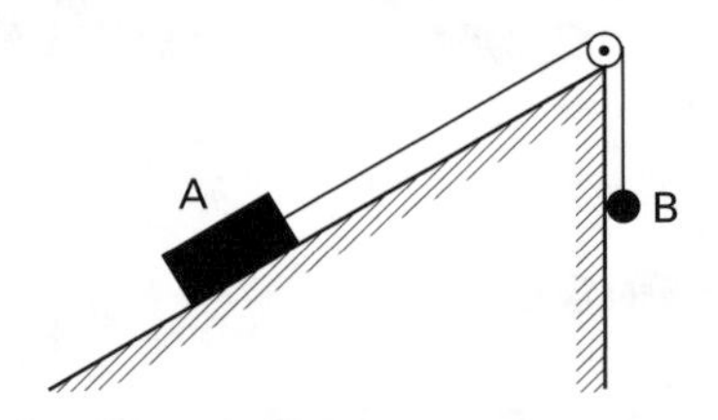

(A) 14 ft/sec (4.2 m/s)
(B) 22 ft/sec (6.6 m/s)
(C) 38 ft/sec (11 m/s)
(D) 45 ft/sec (14 m/s)

Impulse and Momentum

13. Sand drops at the rate of 560 lbm/min (250 kg/min) onto a conveyor belt moving with a velocity of 3.2 ft/sec (0.98 m/s). What force is required to keep the belt moving?

(A) 0.93 lbf (4.1 N)
(B) 1.2 lbf (5.3 N)
(C) 2.3 lbf (10 N)
(D) 4.8 lbf (14 N)

14. What is the impulse imparted to a 0.4 lbm (0.2 kg) baseball that approaches the batter at 90 ft/sec (30 m/s) and leaves at 130 ft/sec (40 m/s)?

(A) 1.8 lbf-sec (9.4 N·s)
(B) 2.7 lbf-sec (14 N·s)
(C) 4.1 lbf-sec (21 N·s)
(D) 8.9 lbf-sec (46 N·s)

15. At what velocity will a 1000 lbm (500 kg) gun mounted on wheels recoil if a 2.6 lbm (1.2 kg) projectile is propelled to 2100 ft/sec (650 m/s)?

(A) 3.7 ft/sec (1.1 m/s)
(B) 4.8 ft/sec (1.4 m/s)
(C) 5.5 ft/sec (1.6 m/s)
(D) 15 ft/sec (4.5 m/s)

16. A 0.15 lbm (60 g) bullet traveling 2300 ft/sec (700 m/s) embeds itself in a 9 lbm (4.5 kg) wooden block. What will be the block's initial velocity?

(A) 24 ft/sec (7.2 m/s)
(B) 38 ft/sec (9.2 m/s)
(C) 66 ft/sec (200 m/s)
(D) 110 ft/sec (330 m/s)

17. A nozzle discharges 40 gal/min (0.15 m^3/min) of water at 60 ft/sec (20 m/s). Find the total force required to hold a flat plate in front of the nozzle. Assume that there is no splashback.

(A) 10 lbf (50 N)
(B) 20 lbf (100 N)
(C) 40 lbf (200 N)
(D) 80 lbf (400 N)

18. In a water turbine, 100 gal/sec (0.4 m^3) impinge on a stationary blade. The water is turned through an angle of 160° and exits at 57 ft/sec (17 m/s). The water impinges at 60 ft/sec (20 m/s).

(a) Find the force exerted by the blade on the stream.

(b) What is the angle from the horizontal of the force?

Impacts

19. An electron collides with a hydrogen atom initially at rest. The electron's initial velocity is 1500 ft/sec (500 m/s). Its final velocity is 65 ft/sec (20.2 m/s) with a path 30° from its original path. Find the velocity of the hydrogen atom in the x-direction if it recoils 1.2° from the original path of the electron.

(A) 0.043 ft/sec (0.013 m/s)
(B) 0.79 ft/sec (0.26 m/s)
(C) 4.6 ft/sec (1.4 m/s)
(D) 53 ft/sec (16 m/s)

20. Two freight cars weighing 5 tons (5000 kg) each roll toward each other and couple. The left car has a velocity of 5 ft/sec (1.5 m/s) and the right car has velocity of 4 ft/sec (1.2 m/s) prior to the impact. What is the velocity of the two cars coupled together after the impact?

(A) 0 ft/sec (0 m/s)
(B) 0.5 ft/sec (0.15 m/s)
(C) 1.0 ft/sec (0.30 m/s)
(D) 2.0 ft/sec (0.60 m/s)

21. Two 2 lbm billiard balls collide as shown. What are the velocities after impact? Use $e = 0.8$.

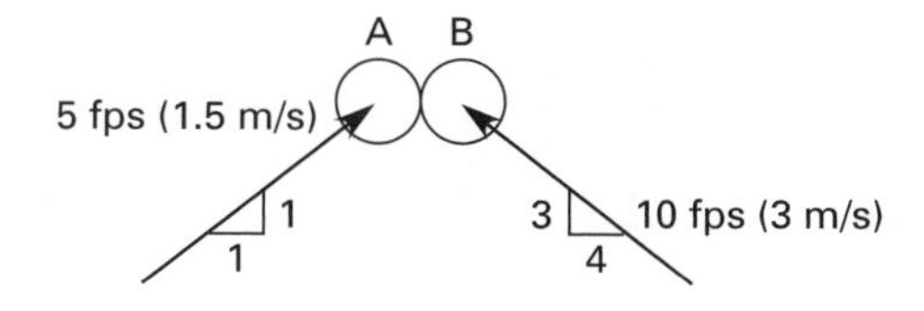

22. A 10 lbm (5 kg) pendulum is released from rest and strikes a 50 lbm (25 kg) block with $e = 0.7$. The block slides on a frictionless surface.

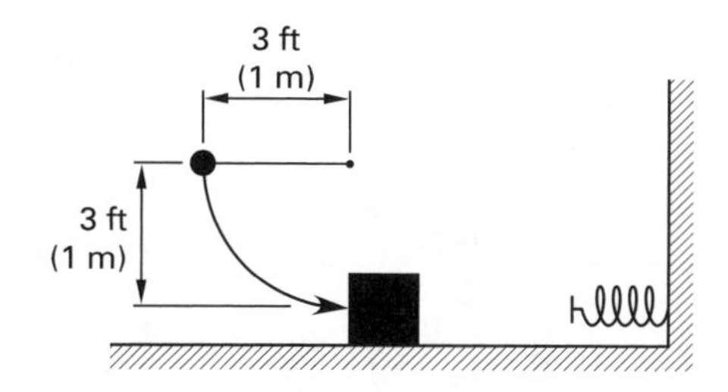

(a) Find the velocity of the pendulum at impact.

(A) 14 ft/sec (4.4 m/s)
(B) 26 ft/sec (7.8 m/s)
(C) 38 ft/sec (11 m/s)
(D) 51 ft/sec (15 m/s)

(b) Find the tension in the cord at impact.

(A) 10 lbf (50 N)
(B) 20 lbf (100 N)
(C) 30 lbf (150 N)
(D) 40 lbf (200 N)

(c) Find the block's velocity immediately after impact.

(A) 2.2 ft/sec (0.66 m/s)
(B) 3.9 ft/sec (1.3 m/s)
(C) 4.6 ft/sec (1.4 m/s)
(D) 5.8 ft/sec (1.7 m/s)

(d) Find the required spring constant to stop the block with less than 6 in (15 cm) deflection.

(A) 54 lbf/ft (1200 N/m)
(B) 62 lbf/ft (1400 N/m)
(C) 75 lbf/ft (1600 N/m)
(D) 96 lbf/ft (1800 N/m)

SOLUTIONS

1. *Customary U.S. Solution*

$$\mathrm{v}_t = \left(60\ \frac{\text{mi}}{\text{hr}}\right)\left(5280\ \frac{\text{ft}}{\text{mi}}\right)\left(\frac{1\ \text{hr}}{3600\ \text{sec}}\right) = 88\ \text{ft/sec}$$

$$\tan\phi = \frac{\mathrm{v}_t^2}{gr} = \frac{\left(88\ \frac{\text{ft}}{\text{sec}}\right)^2}{\left(32.2\ \frac{\text{ft}}{\text{sec}^2}\right)(6000\ \text{ft})} = \boxed{0.04\quad(4\%)}$$

The answer is (A).

SI Solution

$$\mathrm{v}_t = \left(100\ \frac{\text{km}}{\text{h}}\right)\left(1000\ \frac{\text{m}}{\text{km}}\right)\left(\frac{1\ \text{h}}{3600\ \text{s}}\right) = 27.78\ \text{m/s}$$

$$\tan\phi = \frac{\mathrm{v}_t^2}{gr} = \frac{\left(27.78\ \frac{\text{m}}{\text{s}}\right)^2}{\left(9.81\ \frac{\text{m}}{\text{s}^2}\right)(1800\ \text{m})} = \boxed{0.0437\ (4.37\%)}$$

The answer is (A).

2. *Customary U.S. Solution*

The net acceleration vector is directed along the string.

$$\begin{aligned}\phi &= \arctan\left(\frac{a_c}{a_n}\right) = \arctan\left(\frac{\mathrm{v}_t^2}{gr}\right)\\ &= \arctan\left(\frac{\left(20\ \frac{\text{ft}}{\text{sec}}\right)^2}{\left(32.2\ \frac{\text{ft}}{\text{sec}^2}\right)(4\ \text{ft})}\right)\\ &= \boxed{72.15^\circ\ \text{(independent of object's mass)}}\end{aligned}$$

The answer is (C).

SI Solution

The net acceleration vector is directed radially along the string.

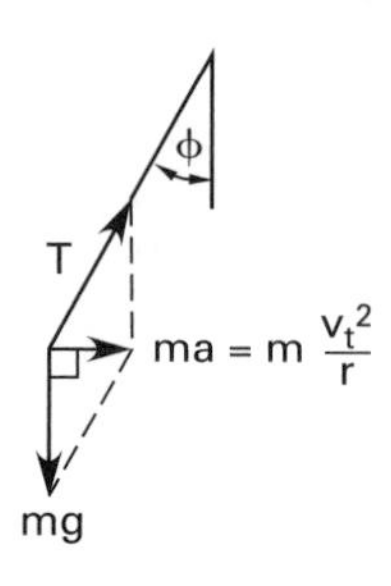

$$\frac{m\mathrm{v}_t^2}{r}\cos\phi = mg\sin\phi$$

$$\begin{aligned}\tan\phi &= \frac{\mathrm{v}_t^2}{rg}\\ \phi &= \arctan\left(\frac{\mathrm{v}_t^2}{rg}\right)\\ &= \arctan\left(\frac{\left(6\ \frac{\text{m}}{\text{s}}\right)^2}{\left(9.81\ \frac{\text{m}}{\text{s}^2}\right)(1.2\ \text{m})}\right)\\ &= \boxed{71.89^\circ\ \text{(independent of object's mass)}}\end{aligned}$$

The answer is (C).

3. *Customary U.S. Solution*

(a)
$$\begin{aligned}\mathrm{v}_t &= \omega r = 2\pi f r\\ &= \left(2\pi\ \frac{\text{rad}}{\text{rev}}\right)\left(5\ \frac{\text{rev}}{\text{sec}}\right)(2\ \text{ft}) = 62.83\ \text{ft/sec}\end{aligned}$$

$$a_n = \frac{\mathrm{v}_t^2}{r} = \frac{\left(62.83\ \frac{\text{ft}}{\text{sec}}\right)^2}{2\ \text{ft}} = \boxed{1974\ \text{ft/sec}^2}$$

The answer is (D).

(b)
$$F_{\text{centrifugal}} = \frac{ma_n}{g_c} = \frac{(10\ \text{lbm})\left(1974\ \frac{\text{ft}}{\text{sec}^2}\right)}{32.2\ \frac{\text{lbm-ft}}{\text{sec}^2\text{-lbf}}} = \boxed{613.04\ \text{lbf (directed outward)}}$$

The answer is (B).

(c)
$$|\mathbf{F}_{\text{centripetal}}| = |\mathbf{F}_{\text{centrifugal}}|$$

$$F_{\text{centripetal}} = \boxed{613.04\ \text{lbf (directed inward)}}$$

The answer is (B).

(d)
$$\begin{aligned}\mathbf{H} &= \frac{\mathbf{r}\times m\mathbf{v}_t}{g_c}\\ H &= \frac{rm\mathrm{v}_t}{g_c}\quad [\text{since } \mathbf{r}\perp\mathbf{v}_t]\\ &= \frac{(2\ \text{ft})(10\ \text{lbm})\left(62.83\ \frac{\text{ft}}{\text{sec}}\right)}{32.2\ \frac{\text{lbm-ft}}{\text{sec}^2\text{-lbf}}}\\ &= \boxed{39.02\ \text{ft-lbf-sec}}\end{aligned}$$

The answer is (A).

SI Solution

(a)
$$\mathrm{v}_t = \omega r = 2\pi f r = \left(2\pi\,\frac{\text{rad}}{\text{rev}}\right)\left(5\,\frac{\text{rev}}{\text{s}}\right)(0.5\text{ m}) = 15.708\text{ m/s}$$

$$a_n = \frac{\mathrm{v}_t^2}{r} = \frac{\left(15.708\,\frac{\text{m}}{\text{s}}\right)^2}{0.5\text{ m}} = \boxed{493.5\text{ m/s}^2}$$

The answer is (D).

(b)
$$F_{\text{centrifugal}} = m a_n = (5\text{ kg})\left(493.5\,\frac{\text{m}}{\text{s}^2}\right) = \boxed{2467.5\text{ N (directed outward)}}$$

The answer is (B).

(c)
$$|\mathbf{F}_{\text{centripetal}}| = |\mathbf{F}_{\text{centrifugal}}|$$

$$F_{\text{centripetal}} = \boxed{2467.5\text{ N (directed inward)}}$$

The answer is (B).

(d)
$$\mathrm{H} = \mathbf{r} \times m\mathbf{v}_t$$

$$H = rm\mathrm{v}_t \quad [\text{since } \mathbf{r} \perp \mathbf{v}_t]$$

$$= (0.5\text{ m})(5\text{ kg})\left(15.708\,\frac{\text{m}}{\text{s}}\right) = \boxed{39.27\text{ J·s}}$$

The answer is (A).

4. *Customary U.S. Solution*

$$a = fg = (0.2)\left(32.2\,\frac{\text{ft}}{\text{sec}^2}\right) = 6.44\text{ ft/sec}^2$$

$$s_{\text{skidding}} = \frac{\mathrm{v}^2}{2a} = \frac{\left(12.88\,\frac{\text{ft}}{\text{sec}}\right)^2}{(2)\left(6.44\,\frac{\text{ft}}{\text{sec}^2}\right)} = \boxed{12.88\text{ ft}}$$

The answer is (A).

SI Solution

$$a = fg = (0.2)\left(9.81\,\frac{\text{m}}{\text{s}^2}\right) = 1.962\text{ m/s}^2$$

$$s_{\text{skidding}} = \frac{\mathrm{v}^2}{2a} = \frac{\left(3.93\,\frac{\text{m}}{\text{s}}\right)^2}{(2)\left(1.962\,\frac{\text{m}}{\text{s}^2}\right)} = \boxed{3.936\text{ m}}$$

The answer is (A).

5. *Customary U.S. Solution*

$$a = fg = (0.333)\left(32.2\,\frac{\text{ft}}{\text{sec}^2}\right) = 10.72\text{ ft/sec}^2$$

$$\Delta t = \frac{\mathrm{v}}{a} = \frac{10\,\frac{\text{ft}}{\text{sec}}}{10.72\,\frac{\text{ft}}{\text{sec}^2}} = \boxed{0.933\text{ sec}}$$

The answer is (C).

SI Solution

$$a = fg = (0.333)\left(9.81\,\frac{\text{m}}{\text{s}^2}\right) = 3.267\text{ m/s}^2$$

$$\Delta t = \frac{\mathrm{v}}{a} = \frac{3\,\frac{\text{m}}{\text{s}}}{3.267\,\frac{\text{m}}{\text{s}^2}} = \boxed{0.918\text{ s}}$$

The answer is (C).

6. *Customary U.S. Solution*

$$\mathrm{v}_t = \left(40\,\frac{\text{mi}}{\text{hr}}\right)\left(5280\,\frac{\text{ft}}{\text{mi}}\right)\left(\frac{1\text{ hr}}{3600\text{ sec}}\right) = 58.67\text{ ft/sec}$$

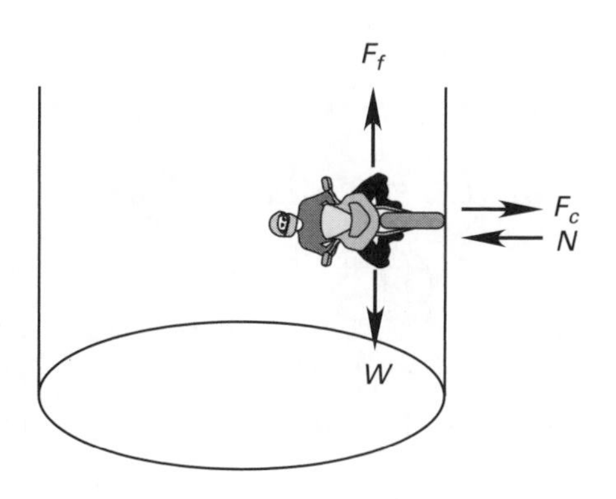

$$N = F_c = \frac{m a_n}{g_c} = \left(\frac{m}{g_c}\right)\left(\frac{\mathrm{v}_t^2}{r}\right)$$

$$W = F_f = fN$$

$$f = \frac{W}{N} = \frac{\frac{mg}{g_c}}{\frac{m\mathrm{v}_t^2}{g_c r}} = \frac{gr}{\mathrm{v}_t^2} = \frac{\left(32.2\,\frac{\text{ft}}{\text{sec}^2}\right)(50\text{ ft})}{\left(58.67\,\frac{\text{ft}}{\text{sec}}\right)^2} = \boxed{0.468}$$

The answer is (C).

Transportation

SI Solution

$$\mathrm{v}_t = \left(60\ \frac{\text{km}}{\text{h}}\right)\left(1000\ \frac{\text{m}}{\text{km}}\right)\left(\frac{1\ \text{h}}{3600\ \text{s}}\right) = 16.67\ \text{m/s}$$

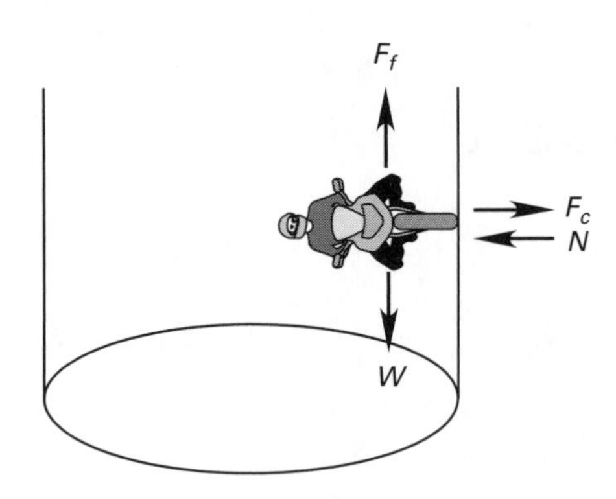

$$N = F_c = ma_n = m\left(\frac{\mathrm{v}_t^2}{r}\right)$$

$$W = F_f = fN$$

$$f = \frac{W}{N} = \frac{mg}{\frac{m\mathrm{v}_t^2}{r}} = \frac{gr}{\mathrm{v}_t^2} = \frac{\left(9.81\ \frac{\text{m}}{\text{s}^2}\right)(15\ \text{m})}{\left(16.67\ \frac{\text{m}}{\text{s}}\right)^2} = \boxed{0.530}$$

The answer is (C).

7. *Customary U.S. Solution*

$$a = \frac{\mathrm{v}^2}{2s} = \frac{\left(12\ \frac{\text{ft}}{\text{sec}}\right)^2}{(2)(36\ \text{ft})} = 2\ \text{ft/sec}^2$$

$$F_f = fN = fW = (0.25)(10\ \text{lbf}) = 2.5\ \text{lbf}$$

$$P = F_f + \frac{ma}{g_c} = 2.5\ \text{lbf} + \frac{(10\ \text{lbm})\left(2\ \frac{\text{ft}}{\text{sec}^2}\right)}{32.2\ \frac{\text{lbm-ft}}{\text{lbf-sec}^2}} = \boxed{3.12\ \text{lbf}}$$

The answer is (C).

SI Solution

$$a = \frac{\mathrm{v}^2}{2s} = \frac{\left(3.6\ \frac{\text{m}}{\text{s}}\right)^2}{(2)(11\ \text{m})} = 0.589\ \text{m/s}^2$$

$$F_f = fN = fW = (0.25)(5\ \text{kg})\left(9.81\ \frac{\text{m}}{\text{s}^2}\right) = 12.26\ \text{N}$$

$$P = F_f + ma = 12.26\ \text{N} + (5\ \text{kg})\left(0.589\ \frac{\text{m}}{\text{s}^2}\right) = \boxed{15.21\ \text{N}}$$

The answer is (C).

8. *Customary U.S. Solution*

$$a = g\sin\theta + fg\cos\theta = \left(32.2\ \frac{\text{ft}}{\text{sec}^2}\right)\left(\sin(22.62^\circ) + (0.1)(\cos 22.62^\circ)\right) = 15.36\ \text{ft/sec}^2$$

$$s = \frac{\mathrm{v}_0^2}{2a} = \frac{\left(30\ \frac{\text{ft}}{\text{sec}}\right)^2}{(2)\left(15.36\ \frac{\text{ft}}{\text{sec}^2}\right)} = \boxed{29.30\ \text{ft}}$$

The answer is (B).

SI Solution

$$a = g\sin\theta + fg\cos\theta = \left(9.81\ \frac{\text{m}}{\text{s}^2}\right)\left(\sin 22.62^\circ + (0.1)(\cos 22.62^\circ)\right) = 4.679\ \text{m/s}^2$$

$$s = \frac{\mathrm{v}_0^2}{2a} = \frac{\left(9\ \frac{\text{m}}{\text{s}}\right)^2}{(2)\left(4.679\ \frac{\text{m}}{\text{s}^2}\right)} = \boxed{8.656\ \text{m}}$$

The answer is (B).

9. *Customary U.S. Solution*

	block 2 = 50 lbm	
F = 100 lbf →	block 1 = 100 lbm	$f_{1,G}$ = 0.2
	ground, G	

$$F_{f(1,G)} = (W_1 + W_2)f_{1,G} = (100\ \text{lbf} + 50\ \text{lbf})(0.2) = 30\ \text{lbf}$$

[< F; blocks move together at acceleration = a, assuming no slipping]

$$a = \frac{(F - F_{f(1,G)})g_c}{m_1 + m_2} = \frac{(100\ \text{lbf} - 30\ \text{lbf})\left(32.2\ \frac{\text{lbm-ft}}{\text{lbf-sec}^2}\right)}{100\ \text{lbm} + 50\ \text{lbm}} = 15.03\ \text{ft/sec}^2$$

If block 2 is not slipping,

$$F_{f(1,2)} = f_{1,2}W_2 = \frac{m_2 a}{g_c}$$

$$f_{1,2} = \frac{a}{g} = \frac{15.03\ \frac{\text{ft}}{\text{sec}^2}}{32.2\ \frac{\text{ft}}{\text{sec}^2}} = \boxed{0.467}$$

SI Solution

block 2 = 25 kg
block 1 = 50 kg
F = 400 N
$f_{1,G} = 0.2$
ground, G

$$F_{f(1,G)} = (W_1 + W_2) f_{1,G}$$
$$= (50\text{ kg} + 25\text{ kg})\left(9.81\ \frac{\text{m}}{\text{s}^2}\right)(0.2)$$
$$= 147.2\text{ N}\ \left[\begin{array}{l} < F;\ \text{blocks move together at accel-} \\ \text{eration } a,\ \text{assuming no slipping} \end{array}\right]$$

$$a = \frac{F - F_{f(1,G)}}{m_1 + m_2}$$
$$= \frac{400\text{ N} - 147.2\text{ N}}{50\text{ kg} + 25\text{ kg}} = 3.371\text{ m/s}^2$$

If block 2 is not slipping,

$$F_{f(1,2)} = f_{1,2} W_2 = m_2 a$$
$$f_{1,2} = \frac{a}{g} = \frac{3.371\ \frac{\text{m}}{\text{s}^2}}{9.81\ \frac{\text{m}}{\text{s}^2}} = \boxed{0.344}$$

10. *Customary U.S. Solution*

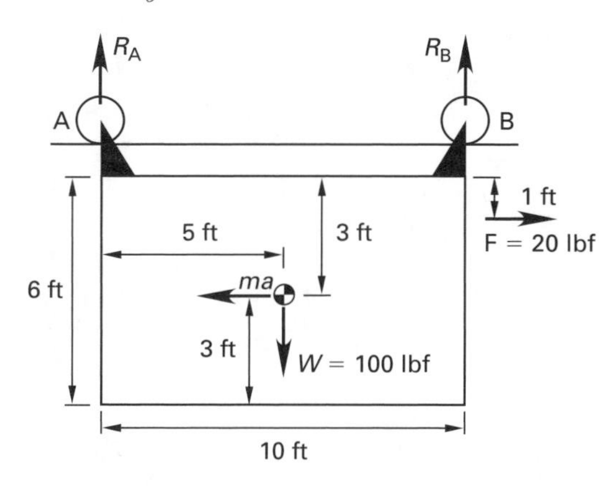

(a)

$$a_x = \frac{F_x g_c}{m} = \frac{(20\text{ lbf})\left(32.2\ \frac{\text{lbm-ft}}{\text{lbf-sec}^2}\right)}{100\text{ lbm}}$$
$$= \boxed{6.44\text{ ft/sec}^2}$$

(b) The inertial force, ma, is also 20 lbf.

$$\sum M_A = -(5\text{ ft})(100\text{ lbf}) + (10\text{ ft})R_\text{B} + (1\text{ ft})(20\text{ lbf}) - (3\text{ ft})(20\text{ lbf}) = 0$$
$$R_\text{B} = \boxed{54\text{ lbf (upward)}}$$
$$\sum F_y = R_\text{A} + 54\text{ lbf} - 100\text{ lbf} = 0$$
$$R_\text{A} = \boxed{46\text{ lbf (upward)}}$$

(c) Let y be the distance (positive upwards) from the center of gravity to F's line of action for $R_\text{A} = R_\text{B} = R$.

$$\sum M_\text{CG} = W(0) - (5\text{ ft})R + (5\text{ ft})R + y(20\text{ lbf}) = 0$$
$$y = 0$$

$$\boxed{\text{Apply the force in line with the center of gravity.}}$$

SI Solution

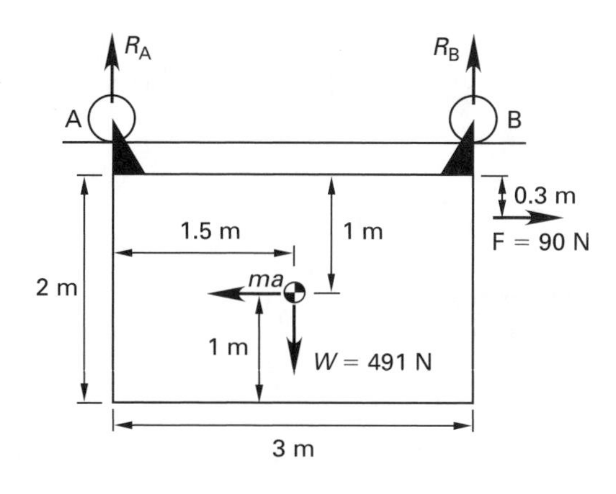

(a)

$$a_x = \frac{F_x}{m} = \frac{90\text{ N}}{50\text{ kg}} = \boxed{1.8\text{ m/s}^2}$$

(b) The inertial force is also 90 N.

$$\sum M_A = -(1.5\text{ m})(491\text{ N}) + (3\text{ m})R_\text{B} + (0.3\text{ m})(90\text{ N}) - (1\text{ m})(90\text{ N}) = 0$$
$$R_\text{B} = \boxed{266.5\text{ N (upward)}}$$
$$\sum F_y = R_\text{A} + 266.5\text{ N} - 491\text{ N} = 0$$
$$R_\text{A} = \boxed{224.5\text{ N (upward)}}$$

(c) Let y be the distance (positive upward) from the center of gravity to F's line of action for $R_A = R_\text{B} = R$.

$$\sum M_\text{CG} = W(0) - (1.5\text{ m})R + (1.5\text{ m})R + y(90\text{ N}) = 0$$
$$y = 0$$

$$\boxed{\text{Apply the force in line with the center of gravity.}}$$

11. *Customary U.S. Solution*

$$F_f = m\frac{g}{g_c}\sin\phi - \frac{ma}{g_c} \qquad \text{[Eq. 1]}$$

Transportation

For constrained motion,

$$F_f r = I_0 \frac{\alpha}{g_c} = \left(\frac{2}{5}mr^2\right)\left(\frac{a}{r}\right)\left(\frac{1}{g_c}\right) = \left(\frac{2}{5}\right)\left(\frac{mar}{g_c}\right)$$

$$F_f = \left(\frac{2}{5}\right)\left(\frac{ma}{g_c}\right) \quad \text{[Eq. 2]}$$

From Eq. 1 and Eq. 2, $mg\sin\phi - ma = (2/5)ma$.

$$a = \left(\frac{5}{7}\right) g\sin\phi = \left(\frac{5}{7}\right)\left(32.2\ \frac{\text{ft}}{\text{sec}^2}\right)(\sin 30°) = 11.5\ \text{ft/sec}^2$$

$$\text{v} = at = \left(11.5\ \frac{\text{ft}}{\text{sec}^2}\right)(2\ \text{sec}) = \boxed{23\ \text{ft/sec (fps)}}$$

The answer is (B).

SI Solution

$$F_f = mg\sin\phi - ma \quad \text{[Eq. 1]}$$

For constrained motion,

$$F_f r = I_0\alpha = \left(\frac{2}{5}mr^2\right)\left(\frac{a}{r}\right) = \frac{2}{5}mar$$

$$F_f = \frac{2}{5}ma \quad \text{[Eq. 2]}$$

From Eq. 1 and Eq. 2, $mg\sin\phi - ma = \frac{2}{5}ma.$

$$a = \left(\frac{5}{7}\right) g\sin\phi = \left(\frac{5}{7}\right)\left(9.81\ \frac{\text{m}}{\text{s}^2}\right)(\sin 30°) = 3.504\ \text{m/s}^2$$

$$\text{v} = at = \left(3.504\ \frac{\text{m}}{\text{s}^2}\right)(2\ \text{s}) = \boxed{7.007\ \text{m/s}}$$

The answer is (B).

12. *Customary U.S. Solution*

Let B's acceleration and velocity be a and v, positive upwards. Then A's acceleration and velocity are $-a$ and $-$v, respectively.

For B: $$T - \frac{m_B g}{g_c} = \frac{m_B a}{g_c}$$
$$T = \frac{m_B a + m_B g}{g_c}$$

For A: $$T - \frac{m_A g\sin\phi}{g_c} = \frac{m_A(-a)}{g_c}$$
$$T = \frac{m_A g\sin\phi - m_A a}{g_c}$$

$$a = \frac{m_A g\sin\phi - m_B g}{m_B + m_A}$$

$$= \frac{(10\ \text{lbm})\left(32.2\ \frac{\text{ft}}{\text{sec}^2}\right)(\sin 36.87°) - (20\ \text{lbm})\left(32.2\ \frac{\text{ft}}{\text{sec}^2}\right)}{20\ \text{lbm} + 10\ \text{lbm}}$$

$$= -15.03\ \text{ft/sec}^2$$

$$\text{v} = at = \left(-15.03\ \frac{\text{ft}}{\text{sec}^2}\right)(3\ \text{sec}) = -45.1\ \text{ft/sec} = \boxed{45.1\ \text{ft/sec (downward)}}$$

The answer is (D).

SI Solution

Let B's acceleration and velocity be a and v, positive upwards. Then A's acceleration and velocity are $-a$ and $-$v, respectively.

For B: $$T - m_B g = m_B a$$
$$T = m_B a + m_B g$$

For A: $$T - m_A g\sin\phi = m_A(-a)$$
$$T = m_A g\sin\phi - m_A a$$

$$a = \frac{m_A g\sin\phi - m_B g}{m_A + m_B}$$

$$= \frac{(5\ \text{kg})\left(9.81\ \frac{\text{m}}{\text{s}^2}\right)(\sin 36.87°) - (10\ \text{kg})\left(9.81\ \frac{\text{m}}{\text{s}^2}\right)}{5\ \text{kg} + 10\ \text{kg}}$$

$$= -4.578\ \text{m/s}^2$$

$$\text{v} = at = \left(-4.578\ \frac{\text{m}}{\text{s}^2}\right)(3\ \text{s}) = -13.7\ \text{m/s} = \boxed{13.7\ \text{m/s (downward)}}$$

The answer is (D).

13. *Customary U.S. Solution*

$$F = \frac{\dot{m}\Delta\text{v}}{g_c} = \frac{\left(560\ \frac{\text{lbm}}{\text{min}}\right)\left(\frac{1\ \text{min}}{60\ \text{sec}}\right)\left(3.2\ \frac{\text{ft}}{\text{sec}}\right)}{32.2\ \frac{\text{lbm-ft}}{\text{lbf-sec}^2}} = \boxed{0.9275\ \text{lbf}}$$

The answer is (A).

SI Solution

$$F = \dot{m}\Delta \mathrm{v} = \left(250\ \frac{\mathrm{kg}}{\mathrm{min}}\right)\left(\frac{1\ \mathrm{min}}{60\ \mathrm{s}}\right)\left(0.98\ \frac{\mathrm{m}}{\mathrm{s}}\right) = \boxed{4.083\ \mathrm{N}}$$

The answer is (A).

14. *Customary U.S. Solution*

$$|\mathbf{Imp}| = \frac{|\Delta\mathbf{p}|}{g_c} = \frac{|\mathbf{p}_2 - \mathbf{p}_1|}{g_c} = \frac{m\mathrm{v}_2 - (-m\mathrm{v}_1)}{g_c}$$

$$= \frac{(0.4\ \mathrm{lbm})\left(90\ \frac{\mathrm{ft}}{\mathrm{sec}} + 130\ \frac{\mathrm{ft}}{\mathrm{sec}}\right)}{32.2\ \frac{\mathrm{lbm\text{-}ft}}{\mathrm{lbf\text{-}sec}^2}} = \boxed{2.73\ \mathrm{lbf\text{-}sec}}$$

The answer is (B).

SI Solution

$$|\mathbf{Imp}| = |\Delta\mathbf{p}| = |\mathbf{p}_2 - \mathbf{p}_1| = m\mathrm{v}_2 - (-m\mathrm{v}_1)$$

$$= (0.2\ \mathrm{kg})\left(40\ \frac{\mathrm{m}}{\mathrm{s}} + 30\ \frac{\mathrm{m}}{\mathrm{s}}\right) = \boxed{14\ \mathrm{N{\cdot}s}}$$

The answer is (B).

15. *Customary U.S. Solution*

$$\Delta p_{\mathrm{gun}} = -\Delta p_{\mathrm{projectile}}$$

$$\frac{m_{\mathrm{gun}}\mathrm{v}_{\mathrm{gun}}}{g_c} = \frac{-m_{\mathrm{projectile}}\mathrm{v}_{\mathrm{projectile}}}{g_c}$$

$$\mathrm{v}_{\mathrm{gun}} = -\frac{m_{\mathrm{projectile}}\mathrm{v}_{\mathrm{projectile}}}{m_{\mathrm{gun}}} = \frac{-(2.6\ \mathrm{lbm})\left(2100\ \frac{\mathrm{ft}}{\mathrm{sec}}\right)}{1000\ \mathrm{lbm}}$$

$$= \boxed{-5.46\ \mathrm{ft/sec\ (opposite\ projectile)}}$$

The answer is (C).

SI Solution

$$\Delta p_{\mathrm{gun}} = -\Delta p_{\mathrm{projectile}}$$

$$m_{\mathrm{gun}}\mathrm{v}_{\mathrm{gun}} = -m_{\mathrm{projectile}}\mathrm{v}_{\mathrm{projectile}}$$

$$\mathrm{v}_{\mathrm{gun}} = -\frac{m_{\mathrm{projectile}}\mathrm{v}_{\mathrm{projectile}}}{m_{\mathrm{gun}}} = \frac{-(1.2\ \mathrm{kg})\left(650\ \frac{\mathrm{m}}{\mathrm{s}}\right)}{500\ \mathrm{kg}}$$

$$= \boxed{-1.56\ \mathrm{m/s\ (opposite\ projectile)}}$$

The answer is (C).

16. *Customary U.S. Solution*

In absence of external forces, $\Delta\mathbf{p} = 0$.

$$\frac{m_{\mathrm{bullet}}\mathrm{v}_{\mathrm{bullet}}}{g_c} = \frac{m_{(\mathrm{bullet+block})}\mathrm{v}_{(\mathrm{bullet+block})}}{g_c}$$

$$\mathrm{v}_{(\mathrm{bullet+block})} = \frac{m_{\mathrm{bullet}}\mathrm{v}_{\mathrm{bullet}}}{m_{(\mathrm{bullet+block})}} = \frac{(0.15\ \mathrm{lbm})\left(2300\ \frac{\mathrm{ft}}{\mathrm{sec}}\right)}{0.15\ \mathrm{lbm} + 9\ \mathrm{lbm}} = \boxed{37.7\ \mathrm{ft/sec}}$$

The answer is (B).

SI Solution

In absence of external forces, $\Delta\mathbf{p} = 0$.

$$m_{\mathrm{bullet}}\mathrm{v}_{\mathrm{bullet}} = m_{(\mathrm{bullet+block})}\mathrm{v}_{(\mathrm{bullet+block})}$$

$$\mathrm{v}_{(\mathrm{bullet+block})} = \frac{m_{\mathrm{bullet}}\mathrm{v}_{\mathrm{bullet}}}{m_{(\mathrm{bullet+block})}} = \frac{(0.06\ \mathrm{kg})\left(700\ \frac{\mathrm{m}}{\mathrm{s}}\right)}{4.5\ \mathrm{kg} + 0.06\ \mathrm{kg}} = \boxed{9.21\ \mathrm{m/s}}$$

The answer is (B).

17. *Customary U.S. Solution*

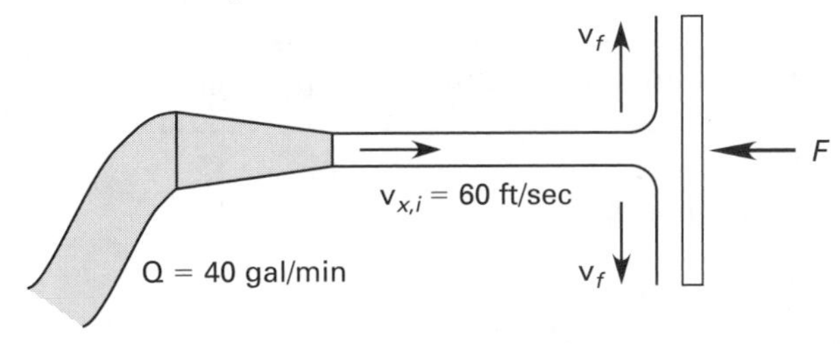

Neglecting gravity, water is turned through an angle of 90° in equal portions in all directions.

$$\dot{m} = \rho Q = \left(62.4\ \frac{\mathrm{lbm}}{\mathrm{ft}^3}\right)\left(40\ \frac{\mathrm{gal}}{\mathrm{min}}\right)\left(\frac{1\ \mathrm{min}}{60\ \mathrm{sec}}\right)\left(0.13368\ \frac{\mathrm{ft}^3}{\mathrm{gal}}\right) = 5.56\ \mathrm{lbm/sec}$$

For every direction other than along the x-axis, equal amounts of water are directed in opposite senses; no net force is applied in these directions.

$$F_x = \frac{\dot{m}\Delta\mathrm{v}_x}{g_c} = \frac{-\dot{m}\mathrm{v}_{x,i}}{g_c} = \frac{-\left(5.56\ \frac{\mathrm{lbm}}{\mathrm{sec}}\right)\left(60\ \frac{\mathrm{ft}}{\mathrm{sec}}\right)}{32.2\ \frac{\mathrm{lbm\text{-}ft}}{\mathrm{lbf\text{-}sec}^2}} = \boxed{10.36\ \mathrm{lbf\ (opposite\ water\ direction)}}$$

The answer is (A).

SI Solution

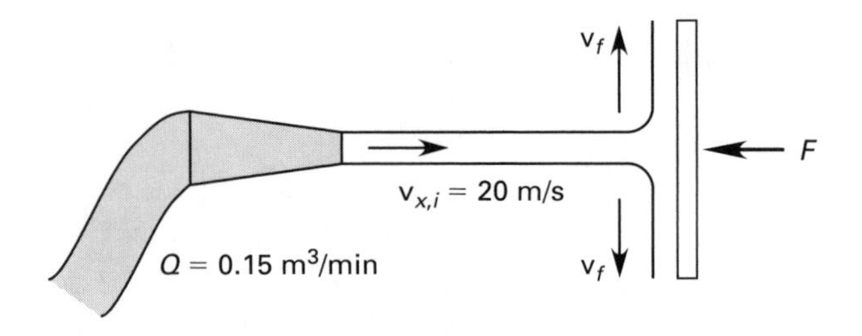

The mass flow rate is

$$\dot{m} = \rho Q = \left(1000\ \frac{\text{kg}}{\text{m}^3}\right)\left(0.15\ \frac{\text{m}^3}{\text{min}}\right)\left(\frac{1\ \text{min}}{60\ \text{s}}\right) = 2.5\ \text{kg/s}$$

Neglecting gravity, water is turned through an angle of 90° in equal portions in all directions. For every direction other than along the x-axis, equal amounts of water are directed in opposite senses; no net force is applied in these directions.

$$F_x = \dot{m}\Delta v_x = -\dot{m}v_{x,i} = -\left(2.5\ \frac{\text{kg}}{\text{s}}\right)\left(20\ \frac{\text{m}}{\text{s}}\right) = \boxed{50\ \text{N (opposite water direction)}}$$

The answer is (A).

18. *Customary U.S. Solution*

(a)
$$\dot{m} = \rho Q = \left(62.4\ \frac{\text{lbm}}{\text{ft}^3}\right)\left(100\ \frac{\text{gal}}{\text{sec}}\right)\left(0.13368\ \frac{\text{ft}^3}{\text{gal}}\right) = 834.16\ \text{lbm/sec}$$

Let inward directions be positive.

$$\dot{p}_{\text{out},x} = \dot{m}v_{\text{out}}\cos 160° = \left(834.16\ \frac{\text{lbm}}{\text{sec}}\right)\left(57\ \frac{\text{ft}}{\text{sec}}\right)(\cos 160°) = -44{,}680\ \frac{\text{lbm-ft}}{\text{sec}^2}$$

$$\dot{p}_{\text{out},y} = \dot{m}v_{\text{out}}\sin 160° = \left(834.16\ \frac{\text{lbm}}{\text{sec}}\right)\left(57\ \frac{\text{ft}}{\text{sec}}\right)(\sin 160°) = 16{,}262\ \text{lbm-ft/sec}^2$$

$$\dot{p}_{\text{in},x} = \dot{m}v_{\text{in}} = \left(834.16\ \frac{\text{lbm}}{\text{sec}}\right)\left(60\ \frac{\text{ft}}{\text{sec}}\right) = 50{,}050\ \text{lbm-ft/sec}^2$$

$$F_x = \frac{\Delta \dot{p}_x}{g_c} = \frac{\dot{p}_{\text{out},x} - \dot{p}_{\text{in},x}}{g_c} = \frac{-44{,}680\ \frac{\text{lbm-ft}}{\text{sec}^2} - 50{,}050\ \frac{\text{lbm-ft}}{\text{sec}^2}}{32.2\ \frac{\text{lbm-ft}}{\text{lbf-sec}^2}} = -2942\ \text{lbf}$$

$$F_y = \frac{\Delta \dot{p}_y}{g_c} = \frac{16{,}262\ \frac{\text{lbm-ft}}{\text{sec}^2} - 0}{32.2\ \frac{\text{lbm-ft}}{\text{lbf-sec}^2}} = 505.0\ \text{lbf}$$

$$F = \sqrt{{F_x}^2 + {F_y}^2} = \sqrt{(-2942\ \text{lbf})^2 + (505\ \text{lbf})^2} = \boxed{2985\ \text{lbf}}$$

(b)
$$\phi = \arctan\left(\frac{F_y}{F_x}\right) = \arctan\left(\frac{505\ \text{lbf}}{-2942\ \text{lbf}}\right) = \boxed{170.26°\ \text{from the horizontal, counterclockwise}}$$

SI Solution

(a)
$$\dot{m} = \rho Q = \left(1000\ \frac{\text{kg}}{\text{m}^3}\right)\left(0.4\ \frac{\text{m}^3}{\text{s}}\right) = 400\ \text{kg/s}$$

Let inward directions be positive.

$$\dot{p}_{\text{out},x} = \dot{m}v_{\text{out}}\cos 160° = \left(400\ \frac{\text{kg}}{\text{s}}\right)\left(17\ \frac{\text{m}}{\text{s}}\right)(\cos 160°) = -6390\ \text{N}$$

$$\dot{p}_{\text{out},y} = \dot{m}v_{\text{out}}\sin 160° = \left(400\ \frac{\text{kg}}{\text{s}}\right)\left(17\ \frac{\text{m}}{\text{s}}\right)(\sin 160°) = 2326\ \text{N}$$

$$\dot{p}_{\text{in},x} = \dot{m}v_{\text{in}} = \left(400\ \frac{\text{kg}}{\text{s}}\right)\left(20\ \frac{\text{m}}{\text{s}}\right) = 8000\ \text{N}$$

$$F_x = \Delta\dot{p}_x = \dot{p}_{\text{out},x} - \dot{p}_{\text{in},x} = -6390\ \text{N} - 8000\ \text{N} = -14\,390\ \text{N}$$

$$F_y = \Delta\dot{p}_y = 2326\ \text{N} - 0 = 2326\ \text{N}$$

$$F = \sqrt{F_x^2 + F_y^2} = \sqrt{(-14\,390\ \text{N})^2 + (2326\ \text{N})^2} = \boxed{14\,576\ \text{N}}$$

$$(b)\ \phi = \arctan\left(\frac{F_y}{F_x}\right) = \arctan\left(\frac{2326\text{ N}}{-14\,390\text{ N}}\right) = \boxed{170.82^\circ \text{ from the horizontal, counterclockwise}}$$

19. *Customary U.S. Solution*

Since the electron is deflected from its original path, the collision is oblique. The ratio of the masses is

$$\frac{m_H}{m_e} = \frac{1.007277u}{0.0005486u} = 1836$$

Kinetic energy may or may not be conserved in this collision; there is insufficient information to make the determination. Momentum is always conserved, regardless of the axis along which it is evaluated. Consider the original path of the electron to be parallel to the x-axis. Then, by conservation of momentum in the x-direction,

$$m_e v_{e,x} + m_H v_{H,x} = m_e v'_{e,x} + m_H v'_{H,x}$$

Recognizing that $v_{H,x} = 0$ and substituting the ratio of masses,

$$v_{e,x} = v'_e \cos\theta_e + 1836\, v'_{H,x}$$

$$1500\ \frac{\text{ft}}{\text{sec}} = \left(65\ \frac{\text{ft}}{\text{sec}}\right)\cos 30^\circ + (1836)(v'_{H,x})$$

$$v'_{H,x} = \boxed{0.786\text{ ft/sec}}$$

Notice that the x-component of v'_H was requested.

The answer is (B).

SI Solution

Since the electron is deflected from its original path, the collision is oblique. The ratio of the masses is

$$\frac{m_H}{m_e} = \frac{1.673\times10^{-27}\text{ kg}}{9.11\times10^{-31}\text{ kg}} = 1836$$

Kinetic energy may or may not be conserved in this collision; there is insufficient information to make the determination. Momentum is always conserved, regardless of the axis along which it is evaluated. Consider the original path of the electron to be parallel to the x-axis. Then, by conservation of momentum in the x-direction,

$$m_e v_{e,x} + m_H v_{H,x} = m_e v'_{e,x} + m_H v'_{H,x}$$

Recognizing that $v_{H,x} = 0$ and substituting the ratio of masses,

$$v_{e,x} = v'_e \cos\theta_e + 1836\, v'_{H,x}$$

$$500\ \frac{\text{m}}{\text{s}} = \left(20.2\ \frac{\text{m}}{\text{s}}\right)(\cos 30^\circ) + (1836)(v'_{H,x})$$

$$v'_{H,x} = \boxed{0.263\text{ m/s}}$$

Notice that the x-component of v'_H was requested.

The answer is (B).

20. *Customary U.S. Solution*

$$m_{\text{left}}v_{\text{left}} + m_{\text{right}}v_{\text{right}} = m_{\text{couple}}v_{\text{couple}}$$

$$v_{\text{couple}} = \frac{(5\text{ tons})\left(5\ \frac{\text{ft}}{\text{sec}}\right) + (5\text{ tons})\left(-4\ \frac{\text{ft}}{\text{sec}}\right)}{10\text{ tons}} = \boxed{\frac{1}{2}\text{ ft/sec (to the right)}}$$

The answer is (B).

SI Solution

$$m_{\text{left}}v_{\text{left}} + m_{\text{right}}v_{\text{right}} = m_{\text{couple}}v_{\text{couple}}$$

$$v_{\text{couple}} = \frac{(5000\text{ kg})\left(1.5\ \frac{\text{m}}{\text{s}}\right) + (5000\text{ kg})\left(-1.2\ \frac{\text{m}}{\text{s}}\right)}{10\,000\text{ kg}} = \boxed{0.15\text{ m/s (to the right)}}$$

The answer is (B).

21. *Customary U.S. Solution*

$$v_{Ay} = 5\sin 45^\circ = 3.536\text{ ft/sec}$$

$$v_{Ax} = 5\cos 45^\circ = 3.536\text{ ft/sec}$$

$$v_{By} = \left(10\ \frac{\text{ft}}{\text{sec}}\right)\left(\frac{3}{5}\right) = 6\text{ ft/sec}$$

$$v_{Bx} = \left(10\ \frac{\text{ft}}{\text{sec}}\right)\left(-\frac{4}{5}\right) = -8\text{ ft/sec}$$

The force of impact is in the x-direction only.

$$v'_{Ay} = v_{Ay}$$

$$v'_{By} = v_{By}$$

In the x-direction,

$$e = \frac{v'_{Ax} - v'_{Bx}}{v_{Bx} - v_{Ax}} = \frac{v'_{Ax} - v'_{Bx}}{-8\ \frac{\text{ft}}{\text{sec}} - 3.536\ \frac{\text{ft}}{\text{sec}}}$$

$$v'_{Ax} - v'_{Bx} = (0.8)\left(-8\ \frac{\text{ft}}{\text{sec}} - 3.536\ \frac{\text{ft}}{\text{sec}}\right) = -9.229\text{ ft/sec} \quad \text{[Eq. 1]}$$

$$m_A v_{Ax} + m_B v_{Bx} = m_A v'_{Ax} + m_B v'_{Bx}$$

Since $m_A = m_B$,

$$v'_{Ax} + v'_{Bx} = 3.536\ \frac{\text{ft}}{\text{sec}} - 8\ \frac{\text{ft}}{\text{sec}} = -4.464\text{ ft/sec} \quad \text{[Eq. 2]}$$

Transportation

Solving Eq. 1 and Eq. 2 simultaneously,

$$v'_{Ax} = -6.846 \text{ ft/sec}$$

$$v'_{Bx} = 2.382 \text{ ft/sec}$$

$$v'_A = \sqrt{(v'_{Ax})^2 + (v'_{Ay})^2}$$

$$= \sqrt{\left(3.536 \frac{\text{ft}}{\text{sec}}\right)^2 + \left(-6.846 \frac{\text{ft}}{\text{sec}}\right)^2} = \boxed{7.705 \text{ ft/sec}}$$

$$v'_B = \sqrt{\left(6 \frac{\text{ft}}{\text{sec}}\right)^2 + \left(2.382 \frac{\text{ft}}{\text{sec}}\right)^2} = \boxed{6.456 \text{ ft/sec}}$$

$$\phi_A = \arctan\left(\frac{3.536}{-6.846}\right) = \boxed{152.7°}$$

$$\phi_B = \arctan\left(\frac{6}{2.382}\right) = \boxed{68.3°}$$

SI Solution

$$v_{Ay} = 1.5 \sin 45° = 1.061 \text{ m/s}$$

$$v_{Ax} = 1.5 \cos 45° = 1.061 \text{ m/s}$$

$$v_{By} = \left(3 \frac{\text{m}}{\text{s}}\right)\left(\frac{3}{5}\right) = 1.8 \text{ m/s}$$

$$v_{Bx} = \left(3 \frac{\text{m}}{\text{s}}\right)\left(-\frac{4}{5}\right) = -2.4 \text{ m/s}$$

The force of impact is in the x-direction only.

$$v'_{Ay} = v_{Ay}$$

$$v'_{By} = v_{By}$$

In the x-direction,

$$e = \frac{v'_{Ax} - v'_{Bx}}{v_{Bx} - v_{Ax}} = \frac{v'_{Ax} - v'_{Bx}}{-2.4 \frac{\text{m}}{\text{s}} - 1.061 \frac{\text{m}}{\text{s}}}$$

$$v'_{Ax} - v'_{Bx} = (0.8)\left(-2.4 \frac{\text{m}}{\text{s}} - 1.061 \frac{\text{m}}{\text{s}}\right)$$

$$= -2.769 \text{ m/s} \quad \text{[Eq. 1]}$$

$$m_A v_{Ax} + m_B v_{Bx} = m_A v'_{Ax} + m_B v'_{Bx}$$

Since $m_A = m_B$,

$$v'_{Ax} + v'_{Bx} = 1.061 \frac{\text{m}}{\text{s}} - 2.4 \frac{\text{m}}{\text{s}} = -1.339 \text{ m/s} \quad \text{[Eq. 2]}$$

Solving Eq. 1 and Eq. 2 simultaneously,

$$v'_{Ax} = -2.054 \text{ m/s}$$

$$v'_{Bx} = 0.715 \text{ m/s}$$

$$v'_A = \sqrt{(v'_{Ax})^2 + (v'_{Ay})^2}$$

$$= \sqrt{\left(-2.054 \frac{\text{m}}{\text{s}}\right)^2 + \left(1.061 \frac{\text{m}}{\text{s}}\right)^2} = \boxed{2.312 \text{ m/s}}$$

$$v'_B = \sqrt{\left(0.715 \frac{\text{m}}{\text{s}}\right)^2 + \left(1.8 \frac{\text{m}}{\text{s}}\right)^2} = \boxed{1.937 \text{ m/s}}$$

$$\phi_A = \arctan\left(\frac{1.061}{-2.054}\right) = \boxed{152.7°}$$

$$\phi_B = \arctan\left(\frac{1.8}{0.715}\right) = \boxed{68.34°}$$

22. *Customary U.S. Solution*

(a) $mgh = \frac{1}{2}mv^2$

The answer is (A).

$$v = \sqrt{2gh} = \sqrt{(2)\left(32.2 \frac{\text{ft}}{\text{sec}^2}\right)(3 \text{ ft})} = \boxed{13.9 \text{ ft/sec}}$$

(b)

$$F_c = \frac{mv^2}{g_c r} = \frac{(10 \text{ lbm})\left(13.9 \frac{\text{ft}}{\text{sec}}\right)^2}{\left(32.2 \frac{\text{lbm-ft}}{\text{lbf} - \text{sec}^2}\right)(3 \text{ ft})} = 20 \text{ lbf}$$

$$T = W + F_c = 10 \text{ lbf} + 20 \text{ lbf} = \boxed{30 \text{ lbf}}$$

The answer is (C).

(c)

$$e = \frac{v_1' - v_2'}{v_2 - v_1} = \frac{v_1' - v_2'}{0 - 13.9 \frac{\text{ft}}{\text{sec}}}$$

$$v_1' - v_2' = (0.7)\left(-13.9 \frac{\text{ft}}{\text{sec}}\right) = -9.73 \text{ ft/sec} \quad \text{[Eq. 1]}$$

$$m_1 v_1 + m_2 v_2 = m_1 v_1' + m_2 v_2'$$

$$(10 \text{ lbm})v_1' + (50 \text{ lbm})v_2' = 139 \text{ ft/sec} \quad \text{[Eq. 2]}$$

Solving Eq. 1 and Eq. 2 simultaneously,

$$v_1' = -5.79 \text{ ft/sec}$$

$$v_2' = \boxed{3.94 \text{ ft/sec}}$$

The answer is (B).

(d) $\left(\frac{1}{2}\right)\left(\frac{m}{g_c}\right)\mathrm{v}^2 = \frac{1}{2}kx^2$

$$k = \frac{m\mathrm{v}^2}{g_c x^2} = \frac{(50\ \text{lbm})\left(3.94\ \frac{\text{ft}}{\text{sec}}\right)^2}{\left(32.2\ \frac{\text{lbm-ft}}{\text{lbf-sec}^2}\right)(0.5\ \text{ft})^2}$$

$$= \boxed{96.4\ \text{lbf/ft}}$$

The answer is (D).

SI Solution

(a) $mgh = \frac{1}{2}m\mathrm{v}^2$

$$\mathrm{v} = \sqrt{2gh} = \sqrt{(2)\left(9.81\ \frac{\text{m}}{\text{s}^2}\right)(1\ \text{m})} = \boxed{4.43\ \text{m/s}}$$

The answer is (A).

(b) $F_c = \frac{m\mathrm{v}^2}{r} = \frac{(5\ \text{kg})\left(4.43\ \frac{\text{m}}{\text{s}}\right)^2}{1\ \text{m}} = 98.1\ \text{N}$

$$T = W + F_c = (5\ \text{kg})\left(9.81\ \frac{\text{m}}{\text{s}^2}\right) + 98.1\ \text{N}$$

$$= \boxed{147\ \text{N}}$$

The answer is (C).

(c) $e = \frac{\mathrm{v}_1{}' - \mathrm{v}_2{}'}{\mathrm{v}_2 - \mathrm{v}_1} = \frac{\mathrm{v}_1{}' - \mathrm{v}_2{}'}{0 - 4.43\ \frac{\text{m}}{\text{s}}}$

$$\mathrm{v}_1{}' - \mathrm{v}_2{}' = (0.7)\left(-4.43\ \frac{\text{m}}{\text{s}}\right) = -3.1\ \text{m/s} \quad \text{[Eq. 1]}$$

$$m_1\mathrm{v}_1 + m_2\mathrm{v}_2 = m_1\mathrm{v}_1{}' + m_2\mathrm{v}_2{}'$$

$$(5\ \text{kg})\mathrm{v}_1{}' + (25\ \text{kg})\mathrm{v}_2{}' = 22.15\ \text{m/s} \quad \text{[Eq. 2]}$$

The answer is (B).

Solving Eq. 1 and Eq. 2 simultaneously,

$$\mathrm{v}_1{}' = -1.845\ \text{m/s}$$

$$\mathrm{v}_2{}' = \boxed{1.255\ \text{m/s}}$$

(d) $\frac{1}{2}m\mathrm{v}^2 = \frac{1}{2}kx^2$

$$k = \frac{m\mathrm{v}^2}{x^2} = \frac{(25\ \text{kg})\left(1.255\ \frac{\text{m}}{\text{s}}\right)^2}{(0.15\ \text{m})^2}$$

$$= \boxed{1750\ \text{N/m}}$$

The answer is (D).

Transportation

74 Roads and Highways: Capacity Analysis

PRACTICE PROBLEMS

Design Speeds/Volume Parameters

1. *(Time limit: one hour)* Commuter trains travel between stations spaced 1.0 mi apart. A train arrives every 5 min, and the trains can attain a maximum speed of 80 mph. Only one train is permitted between stations at a time. Each train has 5 cars, with a maximum capacity of 220 people per car. When the train stops, it must wait 1.0 min to allow for passenger movement. The uniform acceleration of the train is 5.5 ft/sec^2. Deceleration is 4.4 ft/sec^2. (a) What is the top speed of the train? (b) What is the maximum capacity of the line in people per hour? (c) What is the average train speed? (d) What is the average running speed? (e) In regard to the answer in part (d), explain whether the basis is time or spacing. (f) How much time would be saved between stations if the maximum speed was increased to 100 mph?

2. A rural two-lane highway passing through rolling terrain has a design speed of 60 mph in both directions. One particular segment of the highway is a 3% uphill grade lasting for 2 mi with 40% of that length unavailable for passing. There are no access points along the segment. The lane width is 11 ft. The shoulders are 2 ft wide on the uphill side and 6 ft wide on the downhill side. In the peak 60 min of travel, vehicular volume is near the maximum for level of service B. The traffic distribution is 5% trucks, 15% buses, and 80% cars. 70% of the traffic goes uphill during the peak hour, while the remaining 30% travels in the opposite direction.

(a) What is the highest possible directional flow rate for extended lengths of a level, two-way, two-lane highway?

(A) 1600 pcph each lane
(B) 1700 pcph each lane
(C) 3200 pcph both lanes combined
(D) 6400 pcph both lanes combined

(b) What is the uphill grade adjustment factor?

(A) 0.91
(B) 0.93
(C) 0.97
(D) 0.99

(c) What is the heavy vehicle adjustment factor going uphill?

(A) 0.50
(B) 0.62
(C) 0.71
(D) 0.84

(d) What is the passenger car equivalent volume for the uphill lane during the peak 15-min period?

(A) 1000 pcph
(B) 1150 pcph
(C) 1300 pcph
(D) 1450 pcph

(e) Based on the design speed, what is the uphill free flow speed?

(A) 51 mph
(B) 53 mph
(C) 55 mph
(D) 57 mph

(f) Assuming a free flow speed of 57 mph, what is the average uphill travel speed?

(A) 42 mph
(B) 50 mph
(C) 54 mph
(D) 56 mph

(g) What is the uphill percent time spent following?

(A) 45
(B) 60
(C) 75
(D) 90

(h) What treatments might be made to improve the level of service?

(A) adding a median and providing curbs along the lane edge
(B) adding a sidewalk, adding a traffic signal, and improving the pavement markings
(C) adding roadway lighting and reducing the speed limit
(D) realigning the roadway to improve sight distance and adding a passing lane

3. One lane of a two-lane highway was observed for an hour during the day. The following data were gathered.

average distance between front bumpers of successive cars	80 ft
spot mean speed	30 mph
space mean speed	29 mph

(a) What is the average headway?

(A) 1.9 sec/veh
(B) 2.7 sec/veh
(C) 2.8 sec/veh
(D) 42 sec/veh

(b) What is the density in vehicles per mile?

(A) 30 vpm
(B) 66 vpm
(C) 180 vpm
(D) 2000 vpm

(c) What is the traffic volume in vehicles per hour?

(A) 1300 vph
(B) 1350 vph
(C) 1900 vph
(D) 2900 vph

(d) What is the maximum (ideal) capacity in both directions sustainable for short distances?

(A) 2800 pcph
(B) 3200 pcph
(C) 4400 pcph
(D) 4800 pcph

(e) Which is a more accurate parameter of traffic capacity: volume or density? Why?

(A) Density is more accurate: Only density is a function of cars in a given length of roadway.
(B) Density is more accurate: Only density has units of time.
(C) Volume is more accurate: Only volume is a function of cars in a given length of roadway.
(D) Volume is more accurate: Only volume has units of time.

(f) What is the maximum (ideal) capacity of one lane of a four-lane freeway?

(A) 8800 pcph
(B) 4400 pcph
(C) 2400 pcph
(D) 1900 pcph

Intersection Design

4. The level intersection of First Street and Main Street, located in a suburb with negligible pedestrian and bicycle activity, is being investigated. The peak-hour factor on all approaches is 0.85. The signal cycle is 60 sec, two-phase. On Main Street, there is one 20 ft lane in each direction, and parking is prohibited. First Street is one-way westbound with two 11 ft lanes and parking lanes on both sides. There are 40 parking maneuvers per hour. Traffic travels equally in the two lanes of First Street. Turning on red is prohibited. All of the amber time is considered lost. There is no residual delay.

parameter	First (westbound)	Main (northbound)	Main (southbound)
green time	27 sec	27 sec	27 sec
amber time	3 sec	3 sec	3 sec
trucks	4%	5%	5%
buses	0%	0%	0%
left turns	10%	10%	0%
right turns	10%	0%	10%
volume	1240 vph	500 vph	350 vph

(a) What is the saturation flow rate for level of service E on First Street?

(A) 1900 vph
(B) 2700 vph
(C) 2900 vph
(D) 3800 vph

(b) What is the lane group capacity on First Street?

(A) 800 vph
(B) 1200 vph
(C) 1300 vph
(D) 2900 vph

(c) What is the maximum delay for level of service E?

(A) 35 sec
(B) 55 sec
(C) 60 sec
(D) 80 sec

(d) Determine the First Street lane group volume/capacity ratio. Will the intersection operation be satisfactory?

(A) 1.0; satisfactory
(B) 1.0; unsatisfactory
(C) 1.1; satisfactory
(D) 1.1; unsatisfactory

(e) Determine the maximum approach volume for First Street for satisfactory operation.

(A) 1130 vph
(B) 1260 vph
(C) 1330 vph
(D) 1480 vph

(f) What is the level of service for the given volume on First Street?

(A) C
(B) D
(C) E
(D) F

5. Two streets in a town of 250,000 people intersect at a stop sign: a 36 ft side street and a 44 ft arterial. There have been many complaints from the side street users that traffic signals are needed at the intersection. Use information from the *Manual on Uniform Traffic Control Devices.*

(a) What factors should be taken into consideration when deciding whether to signalize the intersection?

(A) warrants 1, 2, and 6
(B) warrants 1, 2, and 8
(C) warrants 1 through 7
(D) all warrants as applicable

(b) Give a logical step-by-step description of how the question should be investigated.

(A) Start with warrant 6, then work only the warrants needed.
(B) Work through the warrants for which adverse data is known.
(C) Work through the warrants in forward order.
(D) Follow warrants 10 and 11 for peak hour analysis, then review warrants for a stop sign.

(c) Explain the criteria for arriving at a recommendation.

(A) Compare actual conditions to the warrant criteria.
(B) Use the *Highway Capacity Manual* to design the signal.
(C) Install a stop beacon and see if the situation improves.
(D) Recommend further study by a traffic engineer.

(d) If a signal is required, what conditions would justify an actuated signal?

(A) Only pedestrian criteria are satisfied, and there are no turning movements.
(B) There are low, fluctuating, or unbalanced traffic volumes.
(C) The intersection has poor visibility.
(D) The traffic volumes are too low for a pretimed signal.

Economic Justification of Highway Safety Features

6. A benefit/cost analysis is to be performed to justify the installation of safety-related road improvements such as flexible barriers and breakaway poles. What records could be used to obtain an economic value for property damage prevented and human lives saved by the safety improvements?

(A) police records, telephone directories, and voting records
(B) school bus incident reports, municipal building permits, and zoning records
(C) hospital emergency room reports and paramedic files
(D) insurance records, state disability fund records, accepted federal standards (e.g., OSHA), and property damage reports on file with police

SOLUTIONS

1. (a) For the standard case in which the station spacing is great enough to allow the train to accelerate to maximum speed, the distance and time diagram is

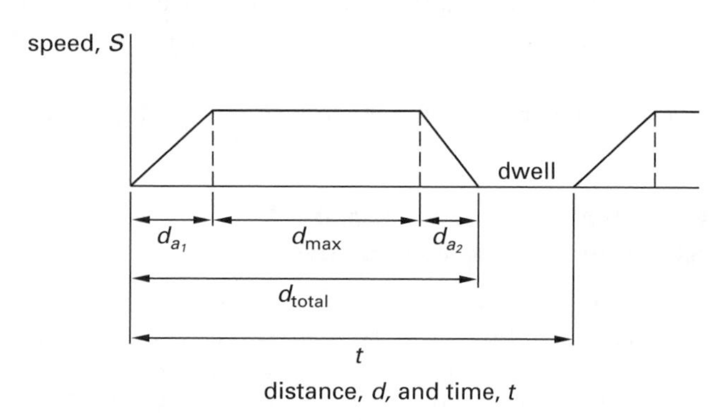

The first step is to check the acceleration and deceleration distance against the distance between stations to determine if the train can reach maximum speed. The acceleration distance is

$$d_{a_1} = \frac{S_{max}^2 - S_0^2}{2a_1} = \frac{\left(80\ \frac{\text{mi}}{\text{hr}}\right)^2 - \left(0\ \frac{\text{mi}}{\text{hr}}\right)^2}{(2)\left(5.5\ \frac{\text{ft}}{\text{sec}^2}\right)\left(0.68\ \frac{\text{sec}^2\text{-mi}^2}{\text{hr}^2\text{-ft}^2}\right)}$$

$$= 856 \text{ ft}$$

The deceleration distance is

$$d_{a_2} = \frac{S_0^2 - S_{max}^2}{2_{a_2}} = \frac{\left(0\ \frac{\text{mi}}{\text{hr}}\right)^2 - \left(80\ \frac{\text{mi}}{\text{hr}}\right)^2}{(2)\left(-4.4\ \frac{\text{ft}}{\text{sec}^2}\right)\left(0.68\ \frac{\text{sec}^2\text{-mi}^2}{\text{hr}^2\text{-ft}^2}\right)}$$

$$= 1070 \text{ ft}$$

The sum of the acceleration and deceleration distances is

$$d_{a_1} + d_{a_2} = 856 \text{ ft} + 1070 \text{ ft} = 1926 \text{ ft}$$

This distance is less than the distance between stations (5280 ft). Therefore, the train can reach maximum speed between stations.

$$\boxed{80 \text{ mph}}$$

(b)
$$\left(\frac{60\ \frac{\text{min}}{\text{hr}}}{5\ \frac{\text{min}}{\text{train}}}\right)\left(\frac{5 \text{ cars}}{1 \text{ train}}\right)\left(\frac{220 \text{ passengers}}{1 \text{ car}}\right)$$

$$= \boxed{13{,}200 \text{ passengers/hr}}$$

(c) The average travel speed between stations does not consider station dwell.

Subtracting the acceleration and deceleration distances from the total distance between two stations yields the distance that is traveled at constant maximum speed.

$$\begin{aligned} d_{max} &= d_{total} - d_{a_1} - d_{a_2} \\ &= 5280 \text{ ft} - 856 \text{ ft} - 1070 \text{ ft} \\ &= 3354 \text{ ft} \end{aligned}$$

The travel time between stations is the sum of times needed for acceleration, travel at constant maximum speed, and deceleration.

$$t_{a_1} = \frac{S_{max} - S_0}{a_1} = \frac{\left(80\ \frac{\text{mi}}{\text{hr}} - 0\ \frac{\text{mi}}{\text{hr}}\right)\left(1.467\ \frac{\text{hr-ft}}{\text{sec-mi}}\right)}{5.5\ \frac{\text{ft}}{\text{sec}^2}}$$

$$= 21.3 \text{ sec}$$

$$t_{max} = \frac{d_{max}}{S} = \frac{3354 \text{ ft}}{\left(80\ \frac{\text{mi}}{\text{hr}}\right)\left(1.467\ \frac{\text{hr-ft}}{\text{sec-mi}}\right)}$$

$$= 28.6 \text{ sec}$$

$$t_{a_2} = \frac{S_0 - S_{max}}{a_2} = \frac{\left(0\ \frac{\text{mi}}{\text{hr}} - 80\ \frac{\text{mi}}{\text{hr}}\right)\left(1.467\ \frac{\text{hr-ft}}{\text{sec-mi}}\right)}{-4.4\ \frac{\text{ft}}{\text{sec}^2}}$$

$$= 26.7 \text{ sec}$$

The average travel speed is

$$\begin{aligned} S_{travel,avg} &= \frac{d_{total}}{t_{a_1} + t_{max} + t_{a_2}} \\ &= \frac{1 \text{ mi}}{(21.3 \text{ sec} + 28.6 \text{ sec} + 26.7 \text{ sec})\left(\frac{1 \text{ hr}}{3600 \text{ sec}}\right)} \\ &= \boxed{47 \text{ mph}} \end{aligned}$$

(d) The average running speed includes the station dwell time. It is a parameter used to set schedules. The average running speed is

$$\begin{aligned} S_{running,avg} &= \frac{d_{total}}{t_{a_1} + t_{max} + t_{a_2} + t_{dwell}} \\ &= \frac{1 \text{ mi}}{(21.3 \text{ sec} + 28.6 \text{ sec} + 26.7 \text{ sec} + 60 \text{ sec})} \\ &\quad \times \left(\frac{1 \text{ hr}}{3600 \text{ sec}}\right) \\ &= \boxed{26.4 \text{ mph}} \end{aligned}$$

(e) The average travel speed over a specified distance during a time period is termed *space mean speed.*

(f) The acceleration distance for the increased maximum speed is

$$d_{a_1} = \frac{S_{\max}^2 - S_0^2}{2a_1} = \frac{\left(100\ \frac{\text{mi}}{\text{hr}}\right)^2 - \left(0\ \frac{\text{mi}}{\text{hr}}\right)^2}{(2)\left(5.5\ \frac{\text{ft}}{\text{sec}^2}\right)\left(0.68\ \frac{\text{sec}^2\text{-mi}^2}{\text{hr}^2\text{-ft}^2}\right)} = 1337\ \text{ft}$$

The deceleration distance is

$$d_{a_2} = \frac{S_0^2 - S_{\max}^2}{2a_2} = \frac{\left(0\ \frac{\text{mi}}{\text{hr}}\right)^2 - \left(100\ \frac{\text{mi}}{\text{hr}}\right)^2}{(2)\left(-4.4\ \frac{\text{ft}}{\text{sec}^2}\right)\left(0.68\ \frac{\text{sec}^2\text{-mi}^2}{\text{hr}^2\text{-ft}^2}\right)} = 1671\ \text{ft}$$

The distance traveled at constant maximum speed is the distance between stations minus the acceleration and deceleration distances.

$$\begin{aligned} d_{\max} &= d_{\text{total}} - d_{a_1} - d_{a_2} \\ &= 5280\ \text{ft} - 1337\ \text{ft} - 1671\ \text{ft} \\ &= 2272\ \text{ft} \end{aligned}$$

The travel time between stations is the sum of times for acceleration, travel at maximum speed, and deceleration.

$$t_{a_1} = \frac{S_{\max} - S_0}{a_1} = \frac{\left(100\ \frac{\text{mi}}{\text{hr}} - 0\ \frac{\text{mi}}{\text{hr}}\right)\left(1.467\ \frac{\text{hr-ft}}{\text{sec-mi}}\right)}{5.5\ \frac{\text{ft}}{\text{sec}^2}} = 26.7\ \text{sec}$$

$$t_{\max} = \frac{d_{\max}}{S_{\max}} = \frac{2272\ \text{ft}}{\left(100\ \frac{\text{mi}}{\text{hr}}\right)\left(1.467\ \frac{\text{hr-ft}}{\text{sec-mi}}\right)} = 15.5\ \text{sec}$$

$$t_{a_2} = \frac{S_0 - S_{\max}}{a_2} = \frac{\left(0\ \frac{\text{mi}}{\text{hr}} - 100\ \frac{\text{mi}}{\text{hr}}\right)\left(1.467\ \frac{\text{hr-ft}}{\text{sec-mi}}\right)}{-4.4\ \frac{\text{ft}}{\text{sec}^2}} = 33.3\ \text{sec}$$

$$t_{a_1} + t_{\max} + t_{a_2} = 26.7\ \text{sec} + 15.5\ \text{sec} + 33.3\ \text{sec} = 75.5\ \text{sec}$$

The time saved at 100 mph maximum speed versus 80 mph maximum speed is 76.6 sec − 75.5 sec = 1.1 sec.

2. (a) Refer to *HCM* p. 20-10. The combined (i.e., both ways) capacity is 3200 pcph. This would occur at LOS E.

The answer is (C).

(b) From *HCM* p. 20-14, "Any upgrade of three percent or more and a length of 0.6 mi or more must be analyzed as a specific upgrade." Relevant parameters for specific upgrades are different from those for general rolling terrain. From *HCM* p. 12-16, the total maximum volume at LOS B is 780 pcph in both directions. From *HCM* Exh. 20-13 with a two-way flow rate of 780 pcph in rolling terrain, the grade adjustment factor for a 3% grade and directional flow of (0.70)(780 pcph) = 546 pcph is 0.97.

The answer is (C).

(c) From *HCM* Exh. 20-15, the passenger car equivalent of trucks moving at 300–600 pcph is 5.9. Trucks and buses are grouped together. There are no RVs. From *HCM* Eq. 20-4, the heavy vehicle adjustment factor is

$$f_{\text{HV}} = \frac{1}{1 + P_T(E_T - 1) + P_R(E_R - 1)} = \frac{1}{1 + (0.15 + 0.05)(5.9 - 1)} = 0.505$$

The answer is (A).

(d) From *HCM* p. 12-16, the total maximum volume at LOS B is 780 pcph in both directions. The uphill volume is $V = (0.7)(780\ \text{pcph}) = 546\ \text{pcph}$. The default value of the peak hour factor (PHF) in rural areas is 0.88 (*HCM* p. 12-10).

From *HCM* Eq. 20-12,

$$\text{v}_d = \frac{V}{(\text{PHF})f_G f_{\text{HV}}} = \frac{546\ \frac{\text{pc}}{\text{hr}}}{(0.88)(0.97)(0.505)} = 1267\ \text{pc/hr}\quad (\text{pcph})$$

The answer is (C).

(e) The base free flow speed (BFFS) is taken as the design speed of 60 mph. From *HCM* Exh. 20-5, for a lane width of 11 ft and a shoulder of 2 ft, the lane and shoulder width adjustment factor is 3.0 mph. The

Transportation

adjustment for access point density is 0.0. From *HCM* Eq. 20-2,

$$\begin{aligned}\text{FFS} &= \text{BFFS} - f_{\text{LS}} - f_A \\ &= 60\ \text{mph} - 3.0\text{mph} - 0.0\text{mph} \\ &= \boxed{57\ \text{mph}}\end{aligned}$$

The answer is (D).

(f) The passenger car equivalent flow rate downhill is given by *HCM* Eq. 20-13.

$$\begin{aligned}v_o &= \frac{V}{(\text{PHF})f_G f_{\text{HV}}} = \frac{(0.30)(780\ \text{pcph})}{(0.88)(0.97)(0.505)} \\ &= 543\ \text{pcph}\end{aligned}$$

From *HCM* Exh. 20-19, the adjustment to travel speed for percentage of no-passing zones ($v_o = 600$ pcph, 40% no-passing zones, FFS = 55 mph) is $f_{np} = 1.1$ mph.

From *HCM* Eq. 20-15,

$$\begin{aligned}\text{ATS}_d &= \text{FFS}_d - 0.00776(v_d + v_o) - f_{np} \\ &= 57\ \text{mph} - (0.00776)(1266\ \text{pcph} + 543\ \text{pcph}) \\ &\quad - 1.1\ \text{mph} \\ &= \boxed{41.9\ \text{mph}}\end{aligned}$$

The answer is (A).

(g) Use *HCM* Exh. 20-21 to determine the percent time spent following coefficients.

For $v_o = 543$ pcph (600 pcph)

$$\begin{aligned}a &= -0.100 \\ b &= 0.413\end{aligned}$$

Use *HCM* Eq. 20-17 to find the base percent time spent following.

$$\begin{aligned}\text{BPTSF}_d &= 100\left(1 - e^{av_d^b}\right) \\ &= (100\%)\left(1 - e^{(-0.100)(1266)^{0.413}}\right) \\ &= 85.2\%\end{aligned}$$

Use *HCM* Exh. 20-20 with FFS = 55 mph, 40% no-passing zones, and $v_o = 543$ pcph (600 pcph) to find the no-passing zone adjustment factor for the directional segment.

$$f_{np} = 7.3\%$$

Use *HCM* Eq. 20-16.

$$\begin{aligned}\text{PTSF}_d &= \text{BPTSF}_d + f_{np} \\ &= 85.2\% + 7.3\% = \boxed{92.5\%}\end{aligned}$$

The answer is (D).

(h) Curbs, traffic signals, and reductions in speed will reduce the capacity. Improving the sight distance and reducing the percent of no-passing distance will allow the trucks to pull over and will improve the level of service.

The answer is (D).

3. Spot speed is an instantaneous value at a particular point and is not a function of the number of vehicles on the highway. Use space mean speed.

$$29\ \text{mph} = \left(29\ \frac{\text{mi}}{\text{hr}}\right)\left(\frac{5280\ \frac{\text{ft}}{\text{mi}}}{3600\ \frac{\text{sec}}{\text{hr}}}\right) = 42.53\ \text{ft/sec}$$

(a) From *HCM* Eq. 74.12,

$$\text{headway} = \frac{80\ \frac{\text{ft}}{\text{veh}}}{42.53\ \frac{\text{ft}}{\text{sec}}} = \boxed{1.88\ \text{sec/veh}}$$

The answer is (A).

$$\text{(b) density} = \frac{5280\ \frac{\text{ft}}{\text{mi}}}{80\ \frac{\text{ft}}{\text{veh}}} = \boxed{66\ \text{veh/mi (vpm)}}$$

The answer is (B).

(c) From *HCM* Eq. 74.13,

$$\text{volume} = \frac{3600\ \frac{\text{sec}}{\text{hr}}}{1.88\ \frac{\text{sec}}{\text{veh}}} = \boxed{1914\ \text{veh/hr (vph)}}$$

The answer is (C).

(d) From the *HCM* p. 20-3, the maximum stable capacity is $\boxed{\text{3200–3400 pcph (two directions).}}$

The answer is (B).

(e) Only volume has units of time. Density is a function of the number of cars in a given distance, but has nothing to do with speed. The cars could be stopped and still have a high density.

$$\boxed{\text{Capacity based on volume is more accurate.}}$$

The answer is (D).

(f) From *HCM* Exh. 13-3 or p. 13-4,

$$\boxed{2400\ \text{passenger cars/hr (pcph)}}$$

The free-flow speed has little effect on the maximum capacity.

The answer is (C).

4.

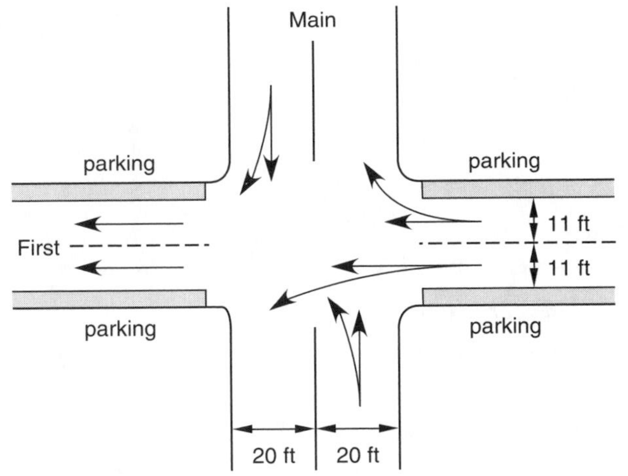

Since the lost time includes the amber time, the effective green time is the actual green time.

Treat First Street as one lane group.

lane width (HCM Exh. 16-7):

$$f_W = 1 + \frac{W - 12}{30} = 1 + \frac{11 - 12}{30} = 0.967$$

heavy vehicles (HCM Exh. 16-7):

$$f_{\text{HV}} = \frac{100}{100 + \%\text{HV}(E_T - 1)} = \frac{100}{100 + (4)(2 - 1)} = 0.962$$

grade (HCM Exh. 16-7):

$$\%G = 0$$
$$f_g = 1.0$$

parking (HCM Exh. 16-7):

$$f_P = \frac{N - 0.1 - \frac{18N_m}{3600}}{N} = \frac{2 - 0.1 - \frac{(18)(40)}{3600}}{2} = 0.85$$

bus blockage (HCM Exh. 16-7):

$$N_b = 0$$
$$f_{\text{bb}} = 1.0$$

type of area (HCM Exh. 16-7):
suburb

$$f_a = 1.0$$

lane utilization (HCM Exh. 16-7):
No information given, so uniform flow assumed.

$$f_{\text{LU}} = 1.0$$

left turns (HCM Exh. 16-7)
shared lanes

$$f_{\text{LT}} = \frac{1}{1.0 + 0.05P_{\text{LT}}} = \frac{1}{1.0 + (0.05)(0.10)} = 0.995$$

right turns (HCM Exh. 16-7)
shared lanes

$$f_{\text{RT}} = 1.0 - (0.15)P_{\text{RT}} = 1.0 - (0.15)(0.10) = 0.985$$

pedestrian bicycle blockage:
(negligible bicycle and pedestrian traffic)

$$f_{\text{Lpb}} = f_{\text{Rpb}} = 1.0$$

(a) Therefore, the saturation flow rate is given by *HCM* Eq. 74.26.

$$\begin{aligned} s &= s_o N f_w f_{\text{HV}} f_g f_p f_{\text{bb}} f_a f_{\text{LU}} f_{\text{RT}} f_{\text{LT}} f_{\text{Lpb}} f_{\text{Rpb}} \\ &= \left(1900\ \frac{\text{vph}}{\text{lane}}\right)(2\ \text{lanes})(0.967)(0.962)(1.000)(0.85) \\ &\quad \times (1.000)(1.00)(0.985)(0.995)(1.00)(1.00) \\ &= \boxed{2944\ \text{vph}} \end{aligned}$$

The answer is (C).

(b) The lane group capacity [*HCM* Eq. 16-6] is

$$\begin{aligned} c = s(\text{effective green ratio}) &= \left(2944\ \frac{\text{veh}}{\text{hr}}\right)\left(\frac{27\ \text{sec}}{60\ \text{sec}}\right) \\ &= \boxed{1325\ \text{vph}} \end{aligned}$$

The answer is (C).

(c) Maximum delay for level of service E is 80 sec [*HCM* Exh. 16-2].

The answer is (D).

(d)

$$v_p = \frac{v}{\text{PHF}} = \frac{1240\ \frac{\text{veh}}{\text{hr}}}{0.85} = 1459\ \text{vph}$$

$$\frac{v}{c} = \frac{1459\ \frac{\text{veh}}{\text{hr}}}{1325\ \frac{\text{veh}}{\text{hr}}} = \boxed{1.1}$$

The First Street approach is oversaturated and the operation will be unsatisfactory.

The answer is (D).

(e) With v_p at capacity, $v_p = c$.

$$v = c(\text{PHF}) = \left(1325 \ \frac{\text{veh}}{\text{hr}}\right)(0.85)$$

$$= \boxed{1126 \text{ vph} \quad (1130 \text{ vph})}$$

The answer is (A).

(f) The level of service is found by the intersection approach delay, found by *HCM* Eq. 16-9.

$$\begin{aligned} d &= d_1(\text{PF}) + d_2 + d_3 \\ &= 0.50C \left(\frac{\left(1-\frac{g}{C}\right)^2}{1-X\left(\frac{g}{C}\right)}\right)\left(\frac{(1-P)f_{\text{PA}}}{1-\frac{g}{C}}\right) \\ &\quad + 900T\left((X-1)+\sqrt{(X-1)^2+\frac{8kIX}{cT}}\right)+0 \\ &= (0.50)(60 \text{ sec})\left(\frac{\left(1-\frac{27 \text{ sec}}{60 \text{ sec}}\right)^2}{(1-60 \text{ sec})\left(\frac{27 \text{ sec}}{60 \text{ sec}}\right)}\right) \\ &\quad \times \left(\frac{\left(1-\frac{27 \text{ sec}}{60 \text{ sec}}\right)(1.0)}{1-\frac{27 \text{ sec}}{60 \text{ sec}}}\right) \\ &\quad + (900)(0.25)\left(\begin{array}{l}(1.1-1) \\ + \sqrt{\begin{array}{l}(1.1-1)^2 \\ + \frac{(8)(0.50)(1.0)(1.1)}{(1325)(0.25)}\end{array}}\end{array}\right) \\ &\quad + 0 \\ &= \boxed{56 \text{ sec}} \end{aligned}$$

From *HCM* Exh. 16-2 the level of service is E.

The answer is (C).

5. (a) Although the geometric design of the intersection is specified, no information about any aspect of intersection performance is given. All warrants are theoretically applicable in this instance.

The answer is (D).

(b) All aspects of intersection performance should be evaluated.

The answer is (C).

(c) Signalization is an art, not a science. The warrants are necessary but not sufficient conditions for signalization. Therefore, a strict comparison of warrants to actual performance is only part of the procedure.

The answer is (D).

(d) Traffic-actuated controllers should be considered when

- there are low, fluctuating, or unbalanced traffic volumes
- there are high side street flows and delays during peak hours only
- only the accident or crosswalk warrant is satisfied
- only one direction of the two-way traffic is to be controlled

The answer is (B).

6. Accident data can be obtained from the following.

- police records (property damage only)
- municipal records
- commercial insurance records
- state disability funds
- expected remaining earnings estimates
- accepted federal standards (e.g., OSHA)

The answer is (D).

75 Vehicle Dynamics and Accident Analysis

PRACTICE PROBLEMS

Collisions

1. Two cars are moving at 60 mph in the same direction and in the same lane. The cars are separated by one car length (20 ft) for each 10 mph. The coefficient of friction (skidding) between the tires and the roadway is 0.6. The reaction time is assumed to be 0.5 sec.

(a) If the lead car hits a parked truck, what is the speed of the second car when it hits the first (stationary) car?

(A) 13 mph
(B) 24 mph
(C) 38 mph
(D) 47 mph

(b) If the lead car hits a parked truck and comes to an abrupt halt, at what speed does the rule of one car length of spacing for every 10 mph of speed become unsafe?

(A) 23 mph
(B) 33 mph
(C) 55 mph
(D) 85 mph

(c) At 60 mph, what should the rule actually be?

(A) 30 ft for every 10 mph
(B) 40 ft for every 10 mph
(C) 50 ft for every 10 mph
(D) 60 ft for every 10 mph

2. You have been hired by the owner of a damaged house to investigate a car crash that caused the damage. From the police report, you learn that the car was traveling down a 3% grade at an unknown speed. The skid marks are 185 ft (56 m) long, and the pavement was dry at the time of the accident. The police report estimates that the initial speed of the car was 25 mph (40 km/h). From an inspection, you know that the car's tires were new. The house owner does not believe the estimate of initial speed. You have the following data from tests performed on dry, level roadways.

initial speed in mph (km/h):	30 (50)	40 (60)	50 (800)	60 (100)
coefficient of friction:	0.59	0.51	0.45	0.35

(a) Find the minimum initial speed of the car.

(A) 49 mph (78 kph)
(B) 50 mph (81 kph)
(C) 56 mph (89 kph)
(D) 74 mph (120 kph)

(b) If the police report was mistaken in assuming the road surface was dry, what was the minimum initial speed of the car?

(A) 39 mph (63 kph)
(B) 41 mph (66 kph)
(C) 43 mph (69 kph)
(D) 45 mph (72 kph)

3. You are hired as a consultant to give expert evidence in a court action arising from a vehicle collision. The two vehicles collided head-on while traveling on a curve tangent with a 4% grade. Vehicle A skidded 195 ft (59.4 m) downhill before colliding with vehicle B. Vehicle B skidded 142 ft (43.3 m). The police report estimates that both vehicles were traveling at 25 mph (40 kph) prior to the collision, based on vehicle deformation. Assume a coefficient of friction of 0.48.

(a) What was the speed of vehicle A prior to the application of brakes?

(A) 55.7 mph (89.6 kph)
(B) 56.6 mph (90.8 kph)
(C) 60.5 mph (97.4 kph)
(D) 61.5 mph (98.9 kph)

(b) What was the speed of vehicle B prior to the application of the brakes?

(A) 53.3 mph (85.6 kph)
(B) 56.6 mph (90.8 kph)
(C) 60.5 mph (97.4 kph)
(D) 61.5 mph (98.9 kph)

(c) Which of your assumptions (impact speed or coefficient of friction) produces the greatest possible error in your initial speed calculations?

(A) The impact speed is more sensitive to error.
(B) The impact speed and coefficient of friction are equally sensitive to error.
(C) The impact speed is more sensitive to error over 50 mph (80 kph).
(D) The coefficient of friction is more sensitive to error.

Accident Data Analysis

4. *(Time limit: one hour)* Four intersections and five segments of highway have been evaluated using prior years' accident data. (a) Calculate the number of accidents per million vehicles for each intersection. Rank the intersections in order of highest to lowest need for improvement. (b) For the highway segments, calculate the number of accidents per year per million vehicle miles. Rank the segments in order of highest to lowest need for improvement. (c) Explain why calculations in (a) and (b) should be done. (d) Assume that the intersections varied in terms of the fractions of accidents that were fatal, injury, and property damage only. How would you compare the safety needs at those intersections?

intersection	ADT	accidents/year
A	820	4
B	1200	5
C	1070	7
D	1400	6

highway segment	ADT	accidents/year	length, mi (km)
1	1900	1	1.50 (2.41)
2	2000	14	1.35 (2.17)
3	5500	18	4.50 (7.24)
4	3000	11	0.53 (0.85)
5	4000	30	2.48 (4.00)

SOLUTIONS

1. (a) $$\left(60\ \frac{\text{mi}}{\text{hr}}\right)\left(\frac{5280\ \frac{\text{ft}}{\text{mi}}}{3600\ \frac{\text{sec}}{\text{hr}}}\right) = 88\ \text{ft/sec}$$

$$\text{initial separation} = (20\ \text{ft})(6) = 120\ \text{ft}$$

$$\text{separation after reaction time} = 120\ \text{ft} - (0.5\ \text{sec})\left(88\ \frac{\text{ft}}{\text{sec}}\right) = 76\ \text{ft}$$

From Newton's second law,

$$\begin{aligned} F_f &= \text{frictional retarding force} = ma \\ &= w\mu = mg\mu \\ ma &= mg\mu \\ a &= g\mu = \left(32.2\ \frac{\text{ft}}{\text{sec}^2}\right)(0.6) = 19.32\ \text{ft/sec}^2 \\ \text{v}_o &= 88\ \text{ft/sec} \\ s &= 76\ \text{ft} \\ a &= -19.32\ \text{ft/sec}^2 \\ \text{v} &= \sqrt{\text{v}_o^2 + 2as} \\ &= \sqrt{\left(88\ \frac{\text{ft}}{\text{sec}}\right)^2 + (2)\left(-19.32\ \frac{\text{ft}}{\text{sec}^2}\right)(76\ \text{ft})} \\ &= \boxed{69.3\ \text{ft/sec}\quad(47.3\ \text{mph})} \end{aligned}$$

The answer is (D).

(b) The rule is one car length (20 ft) per 10 mph. Convert 10 mph to ft/sec.

$$\left(10\ \frac{\text{mi}}{\text{hr}}\right)\left(\frac{5280\ \frac{\text{ft}}{\text{mi}}}{3600\ \frac{\text{sec}}{\text{hr}}}\right) = 14.67\ \text{ft/sec}$$

Working backward, knowing that terminal velocity must be zero,

$$0 = \sqrt{\begin{array}{l}\text{v}_o^2 + (2)\left(-19.32\ \dfrac{\text{ft}}{\text{sec}^2}\right) \\ \quad\times\left((20\ \text{ft})\left(\dfrac{\text{v}_o}{14.67\ \frac{\text{ft}}{\text{sec}}}\right) - (0.5\ \text{sec})\text{v}_o\right)\end{array}}$$

$$\text{v}_o^2 - 33.36\text{v}_o = 0$$

$$\text{v}_o = \boxed{33.36\ \text{ft/sec}\quad(22.7\ \text{mph})}$$

The answer is (A).

(c) If a similar simple car-length rule is to apply at 60 mph,

$$0 = \left(88\ \frac{\text{ft}}{\text{sec}}\right)^2 - (2)\left(19.32\ \frac{\text{ft}}{\text{sec}^2}\right) \times \left(d\left(\frac{88\ \frac{\text{ft}}{\text{sec}}}{14.67\ \frac{\text{ft}}{\text{sec}}}\right) - (0.5\text{ sec})\left(88\ \frac{\text{ft}}{\text{sec}}\right)\right)$$

$$d = \boxed{40.7\text{ ft} \quad \begin{bmatrix}\text{approximately two car lengths}\\ \text{for every 10 mph}\end{bmatrix}}$$

The answer is (B).

2. (a) Use Eq. 75.24.

$$185\text{ ft} = \frac{v_{\text{mph}}^2}{30(f - 0.03)}$$

By trial and error, using the test values of f,

$$v \geq \boxed{49\text{ mph}}$$

The answer is (A).

(b) Use Table 75.1 for a wet road surface. By trial and error,

$$v \geq \boxed{41\text{ mph}}$$

The answer is (B).

3. (a) 25 mph = terminal speed at time of accident (not original speed).

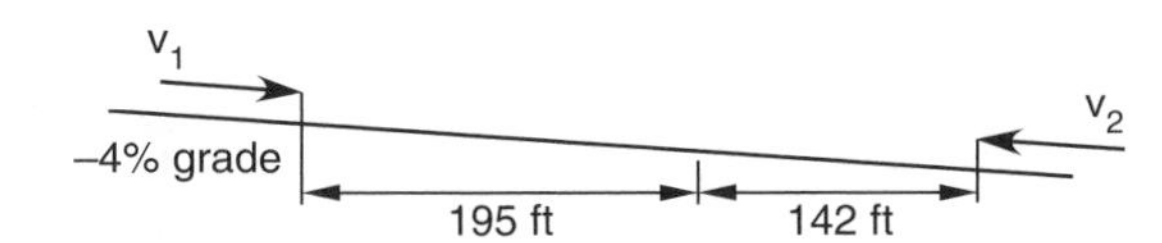

Vehicle A:

If there had been no collision, the skidding distance would have been

$$s_{v_1} = \frac{v_1^2}{(30)(0.48 - 0.04)} = \frac{v_1^2}{13.2} \quad \text{[Eq. 75.24]}$$

At 25 mph, the skidding distance would be

$$s_{25,1} = \frac{\left(25\ \frac{\text{mi}}{\text{hr}}\right)^2}{13.2} = 47.3\text{ ft}$$

The actual skid was 195 ft, and the car was still moving at 25 mph when the collision occurred. Determine the skid distance without a collision.

$$s_{v_1} = 195\text{ ft} + 47.35\text{ ft} = 242.35\text{ ft}$$

Then,

$$242.35\text{ ft} = \frac{v_1^2}{13.2}$$

$$v_1 = \boxed{56.6\text{ mph}}$$

The answer is (B).

Vehicle B:

(b)

$$s_{v_2} = \frac{v_2^2}{(30)(0.48 + 0.04)} = \frac{v_2^2}{15.6}$$

$$s_{25,2} = \frac{\left(25\ \frac{\text{mi}}{\text{hr}}\right)^2}{15.6} = 40.06\text{ ft}$$

$$s_{v_2} = 142\text{ ft} + 40.06\text{ ft} = 182.06\text{ ft}$$

$$182.06\text{ ft} = \frac{v_2^2}{15.6}$$

$$v_2 = \boxed{53.3\text{ mph}}$$

The answer is (A).

(c) The speed of impact and coefficient of friction are the only two variables. Vary each by 10% and see how velocities change.

If the speed at impact is (1+0.10)(25 mph) = 27.5 mph,

$$v_1 = 57.7\text{ mph} \quad [1.6\%\text{ change}]$$

If the coefficient of friction is (1 + 0.10)(0.48) = 0.528,

$$v_1 = 59.0\text{ mph} \quad [4.2\%\text{ change}]$$

$$\boxed{\text{The coefficient of friction is more sensitive to error.}}$$

The answer is (D).

4. (a) Use Eq. 75.25.

$$\text{A:} \quad \frac{\left(4\ \frac{1}{\text{yr}}\right)(1{,}000{,}000)}{\left(820\ \frac{1}{\text{day}}\right)\left(365\ \frac{\text{days}}{\text{yr}}\right)} = 13.4\text{ accidents/MEV}$$

$$\text{B:} \quad \frac{\left(5\ \frac{1}{\text{yr}}\right)(1{,}000{,}000)}{\left(1200\ \frac{1}{\text{day}}\right)\left(365\ \frac{\text{days}}{\text{yr}}\right)} = 11.4\text{ accidents/MEV}$$

Transportation

C: $$\frac{\left(7\ \frac{1}{\text{yr}}\right)(1{,}000{,}000)}{\left(1070\ \frac{1}{\text{day}}\right)\left(365\ \frac{\text{days}}{\text{yr}}\right)} = 17.9 \text{ accidents/MEV}$$

D: $$\frac{\left(6\ \frac{1}{\text{yr}}\right)(1{,}000{,}000)}{\left(1400\ \frac{1}{\text{day}}\right)\left(365\ \frac{\text{days}}{\text{yr}}\right)} = 11.7 \text{ accidents/MEV}$$

Ranking from worst to best,

C, A, D, B

(b) Use Eq. 75.26. (Use 10^6 instead of 10^8 because the ranking is per million vehicle miles, not 100 million.)

1: $$\frac{\left(1\ \frac{1}{\text{yr}}\right)(1{,}000{,}000)}{\left(1900\ \frac{1}{\text{day}}\right)\left(365\ \frac{\text{days}}{\text{yr}}\right)(1.5 \text{ mi})}$$
$$= 0.96 \text{ accidents/mi-yr}$$

2: $$\frac{\left(14\ \frac{1}{\text{yr}}\right)(1{,}000{,}000)}{\left(2000\ \frac{1}{\text{day}}\right)\left(365\ \frac{\text{days}}{\text{yr}}\right)(1.35 \text{ mi})}$$
$$= 14.2 \text{ accidents/mi-yr}$$

3: $$\frac{\left(18\ \frac{1}{\text{yr}}\right)(1{,}000{,}000)}{\left(5500\ \frac{1}{\text{day}}\right)\left(365\ \frac{\text{days}}{\text{yr}}\right)(4.5 \text{ mi})}$$
$$= 1.99 \text{ accidents/mi-yr}$$

4: $$\frac{\left(11\ \frac{1}{\text{yr}}\right)(1{,}000{,}000)}{\left(3000\ \frac{1}{\text{day}}\right)\left(365\ \frac{\text{days}}{\text{yr}}\right)(0.53 \text{ mi})}$$
$$= 18.95 \text{ accidents/mi-yr}$$

5: $$\frac{\left(30\ \frac{1}{\text{yr}}\right)(1{,}000{,}000)}{\left(4000\ \frac{1}{\text{day}}\right)\left(365\ \frac{\text{days}}{\text{yr}}\right)(2.48 \text{ mi})}$$
$$= 8.29 \text{ accidents/mi-yr}$$

Ranking from worst to best,

4, 2, 5, 3, 1

(c) Calculations are performed in order to rank highway improvement priorities. However, this assumes that all locations produce accidents of equal costs. Actually, some locations may produce a higher fraction of fatalities than others.

(d) The need for safety improvements at various intersections could be ranked by cumulative cost (i.e., value) of the accidents occurring over a fixed period of time.

76 Flexible Pavement Design

PRACTICE PROBLEMS

Marshall Mixture Testing

1. Results of Marshall mix testing are shown. An asphalt pavement is to be designed for medium traffic using aggregate with a $^{3}/_{4}$ in nominal size. What is the appropriate percentage of asphalt for the mix?

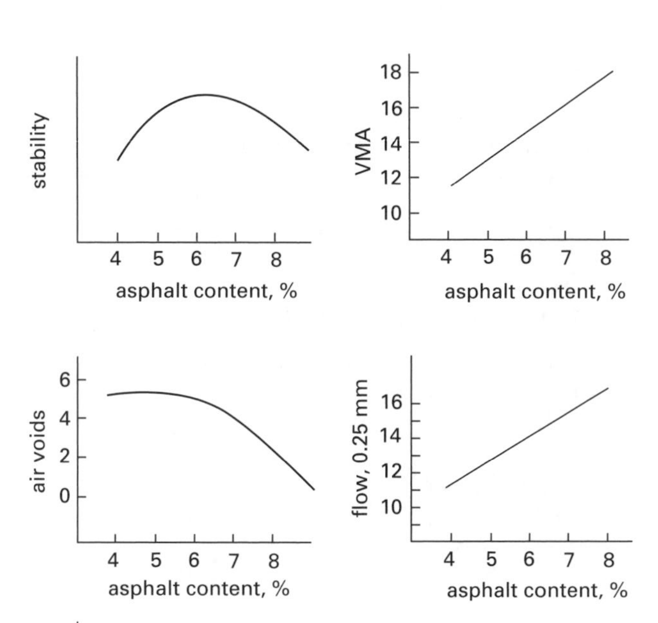

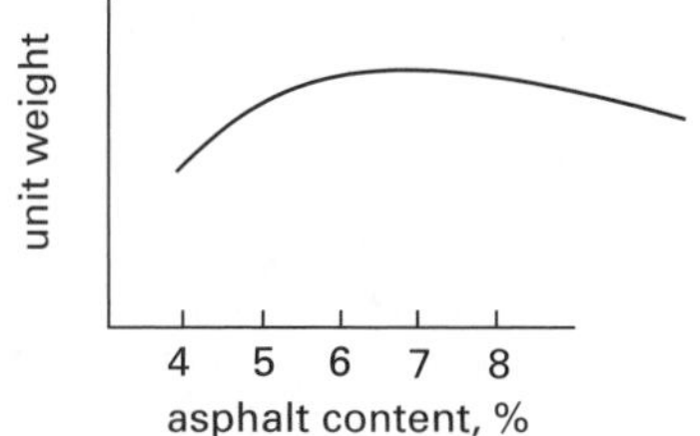

Answer: 6.8%

Mixture Properties

2. The mass of a core of asphalt concrete in air is 1150.0 g. The mass of the core in water is 498.3 g. What is the core's density in lbm/ft^3?

3. The specific gravity of a voidless mixture (i.e., the maximum theoretical specific gravity) of an asphalt concrete is 2.550. The components are specified as follows.

material	specific gravity	apparent specific gravity	by weight, %
asphalt cement	1.020	–	6.3
limestone dust	2.820	2.650	13.7
sand	2.650	2.905	30.4
gravel	2.650	2.873	49.6

(a) What is the air void content if the bulk specific gravity of the mixture is 2.340?

(A) 8.0%
(B) 8.2%
(C) 8.5%
(D) 8.6%

(b) What is the bulk specific gravity of the aggregate?

(A) 2.67
(B) 2.69
(C) 2.73
(D) 2.75

(c) What is the VMA of the mixture?

(A) 16%
(B) 17%
(C) 18%
(D) 19%

(d) What is the effective specific gravity of the aggregate?

(A) 2.80
(B) 2.82
(C) 2.84
(D) 2.86

(e) What is the asphalt absorption?

(A) 1.90%
(B) 2.05%
(C) 2.13%
(D) 2.18%

(f) What is the effective asphalt content of the mixture?

(A) 4.26%
(B) 4.50%
(C) 4.71%
(D) 5.03%

(g) What is the VFA of the mixture?

(A) 52.4%

(B) 53.8%

(C) 54.3%

(D) 54.6%

(h) What is the apparent specific gravity of the aggregate?

(A) 2.35

(B) 2.55

(C) 2.65

(D) 2.85

(i) How much would the percent air voids of the mixture change if the weight of the asphalt were increased by 2%?

(A) 0.05% decrease

(B) 0.14% decrease

(C) no change

(D) 0.14% increase

(j) How much would the VMA of the mixture change if the weight of the asphalt were increased by 2%?

(A) 0.1% decrease

(B) 0.2% decrease

(C) no change

(D) 0.1% increase

AASHTO Pavement Design

4. The following data apply to a particular road.

initial ADT	5000 vpd
growth rate	4%/yr
design period	20 yr
fraction of truck traffic	0.25

loadometer data

axle load (lbf)	number of axles out of 1250 vehicles
single-axle trucks	
2000	800
6000	922
14,000	1851
20,000	37
double-axle trucks	
16,000	417
20,000	608
34,000	84

(a) Assuming a structural number of 3, what is the 18 kip equivalent axle load (EAL)?

(A) 1000

(B) 1030

(C) 1070

(D) 1240

(b) Assuming a structural number of 3, calculate the 18 kip equivalent axle load per truck.

(A) 0.61

(B) 0.63

(C) 0.78

(D) 0.83

(c) Calculate the 20 yr AADT (i.e., the ADT in an average year) in one direction.

(A) 2700 vpd

(B) 3000 vpd

(C) 3700 vpd

(D) 4000 vpd

(d) Calculate the total number of trucks in one direction for the 20 yr design period.

(A) 5.2×10^6

(B) 6.0×10^6

(C) 6.8×10^6

(D) 7.3×10^6

(e) Calculate the total 18 kip single-axle load applications over the design period.

(A) 5.6×10^6

(B) 6.1×10^6

(C) 7.9×10^6

(D) 8.3×10^6

(f) Calculate the effective roadbed modulus for use in design, given the following data.

season	roadbed soil modulus (psi)
winter (Dec–Feb.)	16,000
spring (Mar–Apr.)	2000
summer (May–Sep.)	6000
fall (Oct–Nov.)	4000

(A) 2500 psi

(B) 2800 psi

(C) 3250 psi

(D) 3730 psi

(g) Using the AASHTO design charts, a standard deviation of 0.35, and a reliability of 0.9, calculate the required structural number for a rural divided highway.

(A) 4.7

(B) 5.2

(C) 5.6

(D) 6.7

(h) If the minimum plant-mix asphalt layer thickness is 4 in, calculate the appropriate crushed-stone base layer thickness if it is placed directly on the subgrade. Use $m = 1$ for all drainage coefficients.

(A) 24 in

(B) 26 in

(C) 28 in

(D) 30 in

(i) For a 6 in plant-mix asphalt surface placed over an 8 in crushed stone base, calculate the required subbase thickness for a subbase with a layer coefficient of 0.08. Use $m = 1$ for all drainage coefficients.

(A) 23 in
(B) 25 in
(C) 27 in
(D) 31 in

(j) Given the following minimum thicknesses and unit costs, design the most economical pavement section.

material	minimum thickness	unit cost
asphalt plant mix	4 in	\$0.17/in
crushed stone base	6 in	\$0.04/in
granular subbase	–	\$0.01/in

(A) AC, 4 in; base, 6 in; subbase, 28 in
(B) AC, 4 in; base, 6 in; subbase, 38 in
(C) AC, 6 in; base, 6 in; subbase, 28 in
(D) AC, 4 in; base, 8 in; subbase, 30 in

SOLUTIONS

1.

$$P_{\text{air voids}} = 7.0\%$$
$$P_{\text{stability}} = 6.5\%$$
$$P_{\text{unit weight}} = 7.0\%$$
$$P_{\text{ave}} = \frac{7.0\% + 6.5\% + 7.0\%}{3} = 6.83\%$$

Check that VMA $\geq$ 13.5% [OK]

Check that $8 < \text{flow} < 16$ [OK]

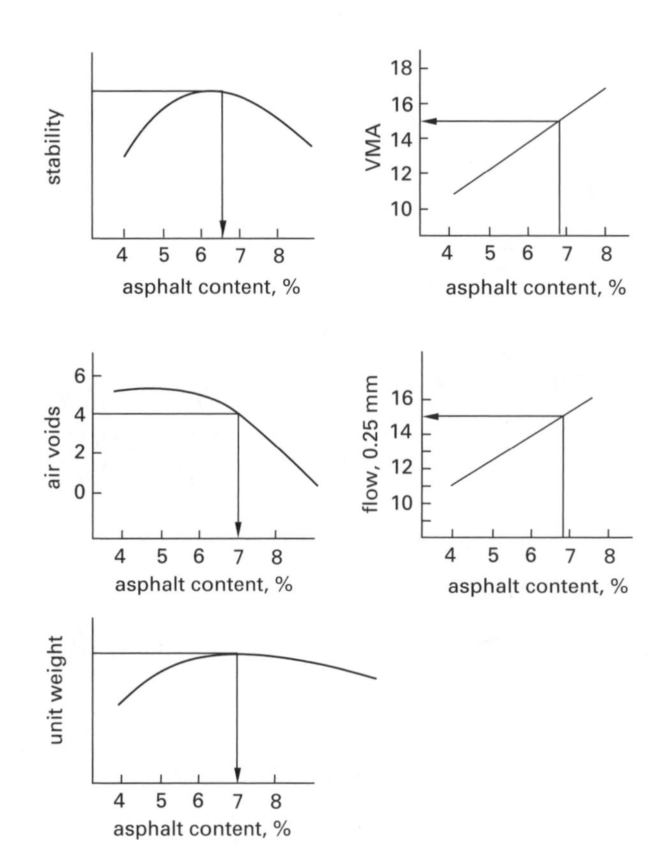

Select 6.8% AC.

2. Use Eq. 76.7.

$$\begin{aligned} \rho &= G\rho_{\text{H}_2\text{O}} \\ &= \left(\frac{m_{\text{air}}}{m_{\text{air}} - m_{\text{H}_2\text{O}}}\right)\rho_{\text{H}_2\text{O}} \\ &= \left(\frac{1150 \text{ g}}{1150 \text{ g} - 498.3 \text{ g}}\right)\left(62.4 \ \frac{\text{lbm}}{\text{ft}^3}\right) \\ &= \boxed{110.11 \text{ lbm/ft}^3} \end{aligned}$$

3. (a) Use Eq. 76.19. For the air voids,

$$\begin{aligned}\text{VTM} &= \frac{(100\%)(G_{\text{mm}} - G_{\text{mb}})}{G_{\text{mm}}} \\ &= \frac{(100\%)(2.550 - 2.340)}{2.550} \\ &= \boxed{8.23\%}\end{aligned}$$

The answer is (B).

(b) Use 76.6. For the aggregate bulk specific gravity,

$$\begin{aligned}G_{\text{sb}} &= \frac{P_{\text{LD}} + P_S + P_G}{\frac{P_{\text{LD}}}{G_{\text{LD}}} + \frac{P_S}{G_S} + \frac{P_G}{G_G}} \\ &= \frac{93.7\%}{\frac{13.7\%}{2.820} + \frac{30.4\%}{2.650} + \frac{49.6\%}{2.650}} \\ &= \boxed{2.674}\end{aligned}$$

The answer is (A).

(c) Use Eq. 76.18. For the VMA,

$$\begin{aligned}\text{VMA} &= 100\% - (100\%)\left(\frac{G_{\text{mb}}P_s}{G_{sb}}\right) \\ &= 100\% - (100\%)\left(\frac{(2.340)(1 - 0.063}{2.674}\right) \\ &= \boxed{18.000\%}\end{aligned}$$

The answer is (C).

(d) Use Eq. 76.13. For the effective specific gravity of aggregate,

$$\begin{aligned}G_{\text{se}} &= \frac{P_{\text{mm}} - P_b}{\frac{P_{\text{mm}}}{G_{\text{mm}}} - \frac{P_b}{G_b}} \\ &= \frac{100\% - 6.3\%}{\frac{100\%}{2.550} - \frac{6.3\%}{1.020}} \\ &= \boxed{2.836}\end{aligned}$$

The answer is (C).

(e) Use Eq. 76.16. For the asphalt absorption,

$$\begin{aligned}P_{\text{ba}} &= (100\%)\left(\frac{G_{\text{se}} - G_{\text{sb}}}{G_{\text{sb}}G_{\text{se}}}\right)G_b \\ &= (100\%)(1.020)\left(\frac{2.836 - 2.674}{(2.674)(2.836)}\right) \\ &= \boxed{2.179\%}\end{aligned}$$

The answer is (D).

(f) Use Eq. 76.17. For the effective asphalt content,

$$\begin{aligned}P_{\text{be}} &= P_b - \left(\frac{P_{\text{ba}}}{100\%}\right)P_s \\ &= 6.3\% - \left(\frac{2.179\%}{100\%}\right)(93.7\%) \\ &= \boxed{4.258\%}\end{aligned}$$

The answer is (A).

(g) Use Eq. 76.20. For the VFA,

$$\begin{aligned}\text{VFA} &= \frac{(100\%)(\text{VMA} - P_a)}{\text{VMA}} \\ &= \frac{(100\%)(18\% - 8.23\%)}{18\%} \\ &= \boxed{54.278\%}\end{aligned}$$

The answer is (C).

(h) Use Eq. 76.6. For the apparent specific gravity of aggregate,

$$\begin{aligned}G_{\text{sa}} &= \frac{P_{\text{LD}} + P_S + P_G}{\frac{P_{\text{LD}}}{G_{\text{LD}_\text{a}}} + \frac{P_S}{G_{S_\text{a}}} + \frac{P_G}{G_{G_\text{a}}}} \\ &= \frac{93.7\%}{\frac{13.7\%}{2.650} + \frac{30.4\%}{2.905} + \frac{49.6\%}{2.873}} \\ &= \boxed{2.848}\end{aligned}$$

The answer is (D).

(i) Calculate G_{mm}.

Assume a 100 lbm mix. The weight of AC is

$$(6.3 \text{ lbm})(1.02) = 6.426 \text{ lbm}$$

The new weight of the mix is 100.126 lbm.

$$\begin{aligned}P_b &= \frac{6.426 \text{ lbm}}{100.126 \text{ lbm}} \\ &= 6.4\%\end{aligned}$$

$$\begin{aligned}P_s &= \frac{93.7 \text{ lbm}}{100.126 \text{ lbm}} \\ &= 93.6\%\end{aligned}$$

$$G_{\text{mm}} = \frac{P_{\text{mm}}}{\frac{P_s}{G_{\text{se}}} + \frac{P_b}{G_b}} = \frac{100\%}{\frac{93.6\%}{2.836} + \frac{6.4\%}{1.02}} = 2.546$$

Use Eq. 76.19.

$$\begin{aligned} \text{VTM} &= \left(\frac{G_{\text{mm}} - G_{\text{mb}}}{G_{\text{mm}}}\right)(100\%) \\ &= \left(\frac{2.546 - 2.340}{2.546}\right)(100\%) \\ &= 8.087\% \end{aligned}$$

The change is

$$8.23\% - 8.09\% = \boxed{0.14\% \text{ decrease}}$$

The answer is (B).

(j) Use Eq. 76.18.

$$\begin{aligned} \text{VMA} &= 100\% - \frac{G_{\text{mb}} P_s}{G_{\text{sb}}} \\ &= 100\% - \frac{(2.340)(93.6\%)}{2.674} \\ &= 18.091\% \end{aligned}$$

The change is

$$18.091\% - 18.000\% = \boxed{0.091\% \text{ increase}}$$

The answer is (D).

4. (a) Use Apps. 76.A and 76.B. Use nonlinear interpolation for greatest accuracy.

axle load (kips)	equiv. factor	× no. axles	= ESALs
	single axle		
2000	0.0003	800	0.24
6000	0.017	922	15.67
14,000	0.399	1851	738.55
20,000	1.49	37	55.13
	double axle		
16,000	0.070	417	29.19
20,000	0.162	608	98.50
34,000	1.11	84	93.24
		total	1030.52

The answer is (B).

(b)

$$\frac{1030.52 \text{ ESALs}}{1250 \text{ trucks}} = \boxed{0.824 \text{ ESALs/truck}}$$

The answer is (D).

(c) The initial ADT in one direction is 5000 vpd/2 = 2500 vpd, which grows at 4% per year for 20 years.

From App. 85.B,

$$(F/A,\ 4\%,\ 20) = 29.7781$$

The AADT in one direction is

$$\frac{(29.7781)(2500 \text{ vpd})}{20 \text{ years}} = 3722 \text{ vpd}$$

The answer is (C).

(d)

$$(3722 \text{ vpd})(0.25) = 931 \text{ trucks/day}$$

The total number of trucks in one direction for a 20 yr period is

$$\begin{aligned} \text{AADT} &= \left(931\ \frac{\text{trucks}}{\text{day}}\right)(20 \text{ yr})\left(365\ \frac{\text{days}}{\text{yr}}\right) \\ &= \boxed{6{,}796{,}300 \text{ trucks}} \end{aligned}$$

The answer is (C).

(e)

$$\begin{aligned} &(6{,}796{,}300 \text{ trucks})\left(0.824\ \frac{\text{ESALs}}{\text{truck}}\right) \\ &= \boxed{5{,}600{,}151} \end{aligned}$$

The answer is (A).

(f) The effective M_R is calculated as shown.

Transportation

month	roadbed soil modulus, (psi) M_R	relative damage u_f
January	16000	0.021
February	16000	0.021
March	2000	2.6
April	2000	2.6
May	6000	0.2
June	6000	0.2
July	6000	0.2
August	6000	0.2
September	6000	0.2
October	4000	0.52
November	4000	0.52
December	16000	0.021
	summation: Σu_f =	7.30

average: $\bar{u}_f = \frac{\Sigma u_f}{n} = \frac{7.30}{12} = 0.61$

effective roadbed soil resilient modulus,
M_R (psi) = 3730 (corresponds to $\bar{u}_f$)

$$M_R = 3730 \text{ psi}$$

The answer is (D).

(g) Select $R = 0.9$, $S_o = 0.35$, and $\Delta\text{PSI} \approx 4 - 2.5 = 2$.

$$\text{SN} \approx 5.6$$

The answer is (C).

(h) Use Eq. 76.29. Use $a_1 = 0.44$ 1/in and $a_2 = 0.14$ 1/in.

$$5.6 = (4 \text{ in})\left(0.44 \, \frac{1}{\text{in}}\right) + D_2\left(0.14 \, \frac{1}{\text{in}}\right)$$

$$D_2 = 27.4 \text{ in} \approx \boxed{28 \text{ in}}$$

The answer is (C).

(i) Use Eq. 76.29.

$$5.6 = (6 \text{ in})\left(0.44 \, \frac{1}{\text{in}}\right) + (8 \text{ in})\left(0.14 \, \frac{1}{\text{in}}\right) + D_3\left(0.08 \, \frac{1}{\text{in}}\right)$$

$$D_3 = \boxed{23 \text{ in}}$$

The answer is (A).

(j) The total SN is 5.6.

For asphalt, select 4 in.

$$(4 \text{ in})\left(0.44 \, \frac{1}{\text{in}}\right) = 1.76 = \text{SN}_1^*$$

For aggregate, select 6 in.

$$(4 \text{ in})\left(0.44 \, \frac{1}{\text{in}}\right) + (6 \text{ in})\left(0.14 \, \frac{1}{\text{in}}\right) = 2.60 = \text{SN}_2^*$$

The required D_3 is

$$\frac{5.6 \text{ in} - 2.6 \text{ in}}{0.08} = 37.5 \text{ in} \quad (38 \text{ in subbase})$$

4 in AC
6 in base
38 in subbase

The answer is (B).

77 Rigid Pavement Design

PRACTICE PROBLEMS

Concrete Paving Mixtures

1. *(Time limit: one hour)* A concrete mixture is made up of materials with the following properties.

material	absolute volume fraction	bulk specific gravity (oven dry)	absorption (%)
sand	0.25	2.60	1.0
gravel	0.50	2.70	1.5
cement	0.10	3.15	–
water	0.15	–	–

(a) What is the volumetric water-cement ratio?

(A) 1.2:1
(B) 1.4:1
(C) 1.5:1
(D) 1.7:1

(b) What is the gravimetric water-cement ratio?

(A) 0.3:1
(B) 0.5:1
(C) 0.7:1
(D) 0.9:1

(c) What is the water-cement ratio?

(A) 4.8 gal/sack
(B) 5.4 gal/sack
(C) 5.7 gal/sack
(D) 6.0 gal/sack

(d) If the moisture content of the sand is 3.0%, what is the weight of free water in the sand used for a 1 ft^3 batch?

(A) 0.7 lbf excess
(B) 0.8 lbf excess
(C) 0.9 lbf excess
(D) 0.8 lbf deficit

(e) If the moisture content is 0.5%, what is the weight of free water in the gravel for a 1 ft^3 batch?

(A) 0.7 lbf excess
(B) 0.8 lbf excess
(C) 0.9 lbf excess
(D) 0.8 lbf deficit

(f) Using the results in parts (d) and (e), calculate the weight of the "added water."

(A) 8.7 lbf
(B) 8.9 lbf
(C) 9.1 lbf
(D) 9.4 lbf

(g) What is the weight of the materials in 1 ft^3 of the mix?

(A) 144 lbf
(B) 148 lbf
(C) 155 lbf
(D) 161 lbf

(h) If 0.25 ft^3 of the mixture weighs 37 lbf, what percent air is in the mixture?

(A) 3.2%
(B) 4.5%
(C) 4.9%
(D) 5.3%

(i) Another concrete mixture has the proportions of cement to sand and gravel of 1:2.5:3.5 by weight, and the water-cement ratio is 6 gal/sack. The specific gravities of the components are unchanged. What is the relative proportion by weight of water in the mixture?

(A) 0.5
(B) 0.7
(C) 0.9
(D) 1.1

(j) What is the volume of all ingredients in a one-sack batch of the mixture described in part (i)?

(A) 3.5 ft^3
(B) 3.8 ft^3
(C) 4.2 ft^3
(D) 4.7 ft^3

2. How much water must be added to the following materials to obtain 6 gal net of mixing water?

material	weight (lbf)	moisture content (%)	absorption capacity (%)
damp gravel	75.0	3.4	1.50
damp sand	46.0	4.5	0.90

Transportation

3. The batch weights for a concrete mixture include 1344 lbf damp sand with a moisture content of 4.24%, and 2260 lbf of damp gravel with a moisture content of 2.10%. The absorption capacity of the sand is 0.90%, and the absorption capacity of the gravel is 1.50%. 220 lbf of water are added to the mixture. What is the net mixing water?

Rigid Pavement Design

4. *(Time limit: one hour)* A jointed, reinforced concrete pavement using doweled joints and asphalt shoulders without dowels is being designed with the AASHTO rigid pavement procedure. A reliability of 90% is needed. The change in pavement serviceability index is specified as 2.0. The slab length is 50 ft. Two 12 ft lanes are tied together. The pavement is to be constructed in an environment with average drainage. The subgrade is a sandy clay with a CBR of 7. The friction factor, F, is 1.8. At 28 days, the concrete has a compressive strength of 4000 psi with a modulus of rupture of 700 psi. The traffic in the design lane is projected to be 30×10^6 ESALs. An 8 in cement-treated subbase will be used. Grade-40 steel and $^1/_2$ in tie bars will be used.

(a) What subgrade k-value would you use in the design?

(A) 50 lbf/in^3
(B) 100 lbf/in^3
(C) 150 lbf/in^3
(D) 200 lbf/in^3

(b) What is the effective k-value you would use in the design?

(A) 400 lbf/in^3
(B) 500 lbf/in^3
(C) 600 lbf/in^3
(D) 700 lbf/in^3

(c) Estimate the concrete modulus of elasticity, E_c, for use in the design.

(A) 3.0×10^6 lbf/in^2
(B) 3.2×10^6 lbf/in^2
(C) 3.6×10^6 lbf/in^2
(D) 3.8×10^6 lbf/in^2

(d) What is the required slab thickness?

(A) 10.0 in
(B) 11.0 in
(C) 12.0 in
(D) 13.0 in

(e) What size dowel bars are to be used?

(A) no. 8
(B) no. 10
(C) no. 12
(D) no. 14

(f) How much longitudinal steel is required, expressed in terms of percentage of area?

(A) 0.15%
(B) 0.18%
(C) 0.20%
(D) 0.24%

(g) How much transverse steel is required, expressed in terms of percentage of area?

(A) 0.05%
(B) 0.07%
(C) 0.10%
(D) 0.14%

(h) At what distance from the free longitudinal edge should the first tie bar be placed?

(A) 10 ft
(B) 12 ft
(C) 14 ft
(D) 18 ft

(i) What is the required increase in slab thickness if traffic is decreased by 40% and all other original factors remain unchanged?

(A) 0.5 in
(B) 1.0 in
(C) 1.5 in
(D) 2.0 in

(j) By how many inches must the pavement thickness be increased if the subbase is untreated and all other original factors remain unchanged?

(A) 0.5 in
(B) 1.0 in
(C) 1.5 in
(D) 2.0 in

SOLUTIONS

1. (a) The water-cement ratio by volume is

$$\frac{w}{c} = \frac{0.15}{0.10} = \boxed{1.5\ (1.5{:}1)}$$

The answer is (C).

(b) Work with 1 cm^3 of concrete. The water-cement ratio by weight is found as follows.

Convert to weight using specific gravity.

The weight of the cement is

$$(3.15)(0.10\ \text{cm}^3)\left(1\ \frac{\text{g}}{\text{cm}^3}\right) = 0.315\ \text{g}$$

The weight of the water is

$$(1.0)(0.15\ \text{cm}^3)\left(1\ \frac{\text{g}}{\text{cm}^3}\right) = 0.15\ \text{g}$$

Therefore,

$$\frac{w}{c} = \frac{0.15\ \text{g}}{0.315\ \text{g}} = \boxed{0.476\ (0.476{:}1)}$$

The answer is (B).

(c) The standard cement sack weighs 94 lbf. The water-cement ratio in gal/sack is

$$\left(\frac{94\ \frac{\text{lbf}}{\text{sack}}}{8.34\ \frac{\text{lbf}}{\text{gal}}}\right)(0.476) = \boxed{5.36\ \text{gal/sack}}$$

The answer is (B).

(d) With 3% moisture content in sand, the free water in 1 ft^3 of the concrete mixture is

$$\begin{aligned}\text{weight of sand} &= (0.25\ \text{ft}^3)(2.60)\left(62.4\ \frac{\text{lbf}}{\text{ft}^3}\right)\\ &= 40.56\ \text{lbf}\\ \text{free water} &= (40.56\ \text{lbf})(0.03 - 0.01)\\ &= \boxed{0.81\ \text{lbf excess}}\end{aligned}$$

The answer is (B).

(e) With a moisture content of 0.5%, the free water in gravel is

$$\begin{aligned}\text{weight gravel} &= (0.5\ \text{ft}^3)(2.70)\left(62.4\ \frac{\text{lbf}}{\text{ft}^3}\right) = 84.24\ \text{lbf}\\ \text{free water} &= (84.24\ \text{lbf})(0.005 - 0.015)\\ &= \boxed{-0.84\ \text{lbf (deficit)}}\end{aligned}$$

The answer is (D).

(f) Calculate the added water.

The required mixing water is

$$(0.15\ \text{ft}^3)\left(62.4\ \frac{\text{lbf}}{\text{ft}^3}\right) = 9.36\ \text{lbf}$$

The net water is

$$9.36\ \text{lbf} - 0.81\ \text{lbf} + 0.84\ \text{lbf} = \boxed{9.39\ \text{lbf}}$$

The answer is (D).

(g) The weight of 1 ft^3 is

ingredient	calculation	weight (lbf)
sand (dry)		40.56
gravel (dry)		84.24
water (net)		9.39
cement	$(0.1\ \text{ft}^3)(3.15)\left(62.4\ \frac{\text{lbf}}{\text{ft}^3}\right) =$	19.66
absorbed water	$(40.56\ \text{lbf})(0.01) + (84.24\ \text{lbf})(0.015) =$	1.67
	total	$\boxed{155.52\ \text{lbf}}$

The answer is (C).

(h) The air content is

$$\begin{aligned}\frac{0.25\ \text{ft}^3}{37\ \text{lbf}} &= \frac{x}{155.52\ \text{lbf}}\\ x &= 1.051\ \text{ft}^3\\ \%\text{air} &= \frac{1.051\ \text{ft}^3 - 1.000\ \text{ft}^3}{1.051\ \text{ft}^3}\\ &= \boxed{4.85\%}\end{aligned}$$

The answer is (C).

(i) The proportion of water is

$$\frac{w}{c} = \left(6\ \frac{\text{gal}}{\text{sack}}\right)\left(8.34\ \frac{\text{lbf}}{\text{gal}}\right)\left(\frac{1\ \text{sack}}{94\ \text{lbf}}\right) = \boxed{0.532}$$

The answer is (A).

(j) For a one-sack batch,

ingredient	weight (lbf)	volume $\text{ft}^3) = \frac{\text{weight}}{(\text{SG})\gamma_w}$
cement	94 lbf	$\frac{94}{(3.15)(62.4)} = 0.478$
water	(94 lbf)(0.532) = 50 lbf	$\frac{50}{(1.00)(62.4)} = 0.801$
sand	(94 lbf)(2.5) = 235 lbf	$\frac{235}{(2.60)(62.4)} = 1.448$
gravel	(94 lbf)(3.5) = 329 lbf	$\frac{329}{(2.70)(62.4)} = 1.953$
	total	$\boxed{4.680\ \text{ft}^3}$

The answer is (D).

2. The net mixing water required is

$$(6 \text{ gal})\left(8.34 \ \frac{\text{lbf}}{\text{gal}}\right) = 50.04 \text{ lbf}$$

The free water in gravel is

$$\left(\frac{75 \text{ lbf}}{1.034}\right)(0.034 - 0.015) = 1.38 \text{ lbf}$$

The free water in sand is

$$\left(\frac{46 \text{ lbf}}{1.045}\right)(0.045 - 0.009) = 1.58 \text{ lbf}$$

The total free water is 1.38 lbf + 1.58 lbf = 2.96 lbf.

The added water is

$$50.04 \text{ lbf} - 2.96 \text{ lbf} = \boxed{47.08 \text{ lbf}}$$

3. The free water is

$$\left(\frac{1344 \text{ lbf}}{1.0424}\right)(0.0424 - 0.009) + \left(\frac{2260 \text{ lbf}}{1.0210}\right)(0.021 - 0.015) = 56.34 \text{ lbf}$$

The net mixing water is

$$220 \text{ lbf} + 56.34 \text{ lbf} = \boxed{276.34 \text{ lbf}}$$

4. (a) From Fig. 77.1 with CBR = 7, subgrade $k = \boxed{150 \text{ lbf/in}^3.}$

The answer is (C).

(b) From Table 77.7, interpolating for $k_{\text{subgrade}} = 150$, the effective k is

$$\left(\tfrac{1}{2}\right)\left(520 \ \frac{\text{lbf}}{\text{in}^3} + 830 \ \frac{\text{lbf}}{\text{in}^3}\right) = \boxed{675 \text{ lbf/in}^3}$$

The answer is (D).

(c) Use Eq. 77.2.

$$\begin{aligned} E_c &= 57{,}000\sqrt{f'_c} \\ &= 57{,}000\sqrt{4000} \\ &= \boxed{3.6 \times 10^6 \text{ lbf/in}^2} \end{aligned}$$

The answer is (C).

(d) Find the slab thickness.

$$\begin{aligned} \text{CBR} &= 7 \\ k_{\text{subgrade}} &= 150 \text{ lbf/in}^3 \\ k_{\text{effective}} &= 675 \text{ lbf/in}^3 \\ E_c &= 3.6 \times 10^6 \text{ lbf/in}^2 \\ J &= 4 \quad \text{[doweled shoulders]} \\ C_d &= 1.0 \quad \text{[average drainage]} \\ R &= 90\% \quad \text{[reliability]} \\ S_o &= 0.35 \quad \text{[typical rigid pavement]} \\ \Delta\text{PSI} &= 2.0 \end{aligned}$$

The slab thickness, D, is $\boxed{12.0 \text{ in.}}$

The answer is (C).

(e) The dowel bar diameter is

$$\frac{D}{8} \approx \frac{12.0 \text{ in}}{8} = 1.50 \text{ in} = \boxed{1^5/_8 \text{ in} \quad \text{[or no. 14]}}$$

(There is no no. 12 bar.)

The answer is (D).

(f) With grade-40 steel, the working stress is

$$f_s = 0.75 f_y = (0.75)\left(40 \ \frac{\text{kips}}{\text{in}^2}\right) = 30 \text{ kips/in}^2$$

The friction factor is given as $F = 1.8$.

The percentage of longitudinal steel is given by Eq. 77.3.

$$P_s = \frac{LF(100\%)}{2f_s} = \frac{(50 \text{ ft})(1.8)(100\%)}{(2)\left(30{,}000 \ \frac{\text{lbf}}{\text{in}^2}\right)} = \boxed{0.15\%}$$

The answer is (A).

(g) Since the two lanes are tied, both must be considered for horizontal steel. The width is 24 ft.

From Eq. 77.3, the area of transverse steel is

$$P_t = \frac{LF(100\%)}{2f_s} = \frac{(24 \text{ ft})(1.8)(100\%)}{(2)\left(30{,}000 \ \frac{\text{lbf}}{\text{in}^2}\right)} = \boxed{0.072\%}$$

The answer is (B).

(h) From Fig. 77.3, the distance is $\boxed{14 \text{ in.}}$

The answer is (C).

(i) The new traffic volume is

$$(30 \times 10^6)(1 - 0.40) = 18 \times 10^6$$

The new thickness is

$$D = 11 \text{ in}$$

The increase is

$$12 \text{ in} - 11 \text{ in} = \boxed{1.0 \text{ in}}$$

The answer is (B).

(j)

$$\begin{aligned}
\text{CBR} &= 7 \\
k_{\text{subgrade}} &= 150 \text{ lbf/in}^3 \\
k_{\text{effective}} &= 205 \text{ lbf/in}^3 \\
&\quad \left[\begin{array}{c}\text{Table 77.6, double interpolation} \\ \text{for } k = 150 \text{ and } t = 8 \text{ in}\end{array}\right] \\
E_c &= 3.6 \times 10^6 \text{ lbf/in}^2 \\
S_c' &= 700 \text{ lbf/in}^2 \\
J &= 4 \\
C_d &= 1.0 \\
(\text{match line} &\approx 75) \\
\Delta\text{PSI} &= 2 \\
R &= 90\% \\
S_o &= 0.35 \\
\text{ESAL} &= 30 \times 10^6 \\
D &\approx 12.5 \text{ in}
\end{aligned}$$

The thickness increase is

$$12.5 \text{ in} - 12.0 \text{ in} = \boxed{0.5 \text{ in}}$$

The answer is (A).

Transportation

78 Plane Surveying

PRACTICE PROBLEMS

Errors

1. Four level circuits were run over four different routes to determine the elevation of a benchmark. The observed elevations and probable errors for each circuit are as given. What is the most probable value of the benchmark's elevation?

route 1: 745.08 ± 0.03

route 2: 745.22 ± 0.01

route 3: 745.45 ± 0.09

route 4: 745.17 ± 0.05

Leveling

2. A level is placed where the height of the instrument above the datum is 294.43 ft. The rod is held on a point 5000 ft away, and a rod reading of 4.63 ft is recorded. (a) What is the effect of the earth's curvature on the rod reading? (b) What is the effect of atmospheric refraction on the rod reading?

3. Complete the survey notes from a leveling session.

station	+ BS	HI	−FS	elevation
BMA	5.54			100.60
1+00			2.88	
2+00			5.56	
3+00			4.37	
4+00			2.22	
TP1			7.89	
	6.34			
5+00			10.89	
6+00			11.12	
7+00			10.45	
8+00			12.61	
TP2			9.32	
	1.45			
9+00			5.82	
10+00			6.99	
11+00			5.34	
12+00			4.98	
TP3			5.67	
	2.38			
13+00			6.24	
14+00			6.76	
BMB			4.90	

4. A transit with an interval factor of 100 and an instrument factor of 1.0 is used to take the stadia sights recorded. The height of the instrument above the ground was 4.8 ft. The ground elevation where the instrument was set up was 297.8 ft. (a) What is the horizontal distance between points A and B? (b) What is the elevation of point B?

point	azimuth	rod interval ($R_2 - R_1$)	middle hair reading	vertical angle
A	42.17°	3.22	5.7	−6.3°
B	222.17°	2.60	10.9	+4.17°

5. Recorded data from a leveling circuit is shown. (a) What is the elevation of BM11? (b) What is the elevation of BM12?

station	backsight	foresight	elevation
BM10	4.64		179.65
TP1	5.80	5.06	
TP2	2.25	5.02	
BM11	6.02	5.85	
TP3	8.96	4.34	
TP4	8.06	3.22	
TP5	9.45	3.71	
TP6	12.32	2.02	
BM12		1.98	

Traverses

6. The balanced latitudes and departure of the legs of a closed traverse were determined as shown. What is the traverse area?

leg	latitude	departure
AB	N 350	E 0
BC	N 550	E 600
CD	S 250	E 1200
DE	S 750	E 200
EF	S 550	W 1100
FA	N 650	W 900

7. Use the compass rule to balance and adjust (to the nearest 0.1) the traverse.

leg	bearing	length
AB	N	500
BC	N 45.00° E	848.6
CD	S 69.45° E	854.4
DE	S 11.32° E	1019.8
EF	S 79.70° W	1118.0
FA	N 54.10° W	656.8

8. A five-leg closed traverse is taped and scoped in the field, but obstructions make it impossible to collect all readings. It is known that the general direction of leg EA is east-west. Complete the field notes.

leg	north azimuth	distance
AB	106.22°	1081.3
BC	195.23°	1589.5
CD	247.12°	1293.7
DE	332.37°	______
EA	______	1737.9

Photogrammetry

9. *(Time limit: one hour)* A camera with a focal length of 6.0 in and a picture size of 12 in by 12 in is used to photograph an area from an altitude of 4000 ft above sea level.

(a) If the average ground elevation is 650 ft, what is the average scale of the photo?

(A) 1 in = 558 ft
(B) 1 in = 600 ft
(C) 1:6000
(D) 1:6500

(b) What is the ground area covered by a single photograph?

(A) 1030 ac
(B) 1500 ac
(C) 2000 ac
(D) 2070 ac

(c) If overlap and sidelap are 60% and 20% respectively, what is the ground spacing between exposures?

(A) 2670 ft
(B) 2672 ft
(C) 2678 ft
(D) 2778 ft

(d) With the overlap as given in part (c), what is the ground spacing between flight lines?

(A) 5236 ft
(B) 5336 ft
(C) 5346 ft
(D) 5356 ft

(e) If the aircraft flies at a ground speed of 300 mph along the flight line outlined, what is the time interval, to the nearest 0.1 sec, between exposures along the flight line?

(A) 6.0 sec
(B) 6.1 sec
(C) 6.2 sec
(D) 8.1 sec

(f) Two points at an elevation of 665 ft appear on a photograph. The distance between these points is 89 mm on the photograph. What is the ground distance, in ft, between the two points?

(A) 1557 ft
(B) 1647 ft
(C) 1737 ft
(D) 1947 ft

(g) At what elevation should the plane be flown to obtain photography at an average scale of 1:10,000 if the terrain to be photographed lies in the range of 1000 to 1200 ft above sea level?

(A) 6000 ft
(B) 6050 ft
(C) 6100 ft
(D) 6150 ft

(h) A pair of images lie 5.68 in apart on a photo taken with a lens whose focal length is 12 in. A map with a scale of 1 in = 1680 ft shows the same points 0.54 in apart. What is the scale of the photo?

(A) 1 in = 160 ft
(B) 1 in = 200 ft
(C) 1:1500
(D) 1:2000

(i) For part (h), at what elevation was the plane flying if the elevation of the points is 889 ft?

(A) 2000 ft
(B) 2500 ft
(C) 2800 ft
(D) 3000 ft

(j) For part (h), at what elevation was the plane flying if the same two points are 0.98 in apart on the map and the ground elevation was 889 ft?

(A) 4400 ft
(B) 4500 ft
(C) 4600 ft
(D) 4700 ft

SOLUTIONS

1. $(0.03)^2 + (0.01)^2 + (0.09)^2 + (0.05)^2 = 0.0116$

The weights to be applied are

$$\frac{0.0116}{(0.03)^2} = 12.89$$
$$\frac{0.0116}{(0.01)^2} = 116.0$$
$$\frac{0.0116}{(0.09)^2} = 1.43$$
$$\frac{0.0116}{(0.05)^2} = 4.64$$

The most probable elevation is

$$745 + \frac{(12.89)(0.08) + (116)(0.22) + (1.43)(0.45) + (4.64)(0.17)}{12.89 + 116 + 1.43 + 4.64} = \boxed{745.21}$$

2. (a) Use Eq. 78.12.

$$\begin{aligned} h_c &= (2.4 \times 10^{-8})x^2 \\ &= \left(2.4 \times 10^{-8}\ \frac{1}{\text{ft}}\right)(5000\ \text{ft})^2 \\ &= \boxed{0.6\ \text{ft}} \end{aligned}$$

(b) Use Eq. 78.13.

$$\begin{aligned} h_r &= (3.0 \times 10^{-9})x^2 \\ &= \left(3.0 \times 10^{-9}\ \frac{1}{\text{ft}}\right)(5000\ \text{ft})^2 \\ &= \boxed{0.075\ \text{ft}} \end{aligned}$$

3.

station	+ BS	HI	− FS	elevation
BMA	5.54	106.14		100.60
1+00			2.88	103.26
2+00			5.56	100.58
3+00			4.37	101.77
4+00			2.22	103.92
TP1	6.34		7.89	98.25
5+00		104.59	10.89	93.70
6+00			11.12	93.47
7+00			10.45	94.14
8+00			12.61	91.98
TP2	1.45		9.32	95.27
9+00		96.72	5.82	90.90
10+00			6.99	89.73
11+00			5.34	91.38
12+00			4.98	91.74
TP3	2.38		5.67	91.05
13+00		93.43	6.24	87.19
14+00			6.76	86.67
BMB			4.90	88.53
total BS	15.71	total TP	27.78	

Check.

$$100.6 + 15.71 - 27.78 = 88.53 \quad \text{[OK]}$$

4. (a) Use Eq. 78.10.

$$\begin{aligned} H_{\text{scope-A}} &= (100)(3.22\ \text{ft})\left(\cos^2(-6.3°)\right) \\ &\quad + (1)\left(\cos(-6.3°)\right) \\ &= 319.1\ \text{ft} \\ H_{\text{scope-B}} &= (100)(2.6\ \text{ft})(\cos^2 4.17°) \\ &\quad + (1\ \text{ft})(\cos 4.17°) \\ &= 259.6\ \text{ft} \end{aligned}$$

Since the azimuths differ by 180°, the telescope is in line with points A and B.

$$\text{AB} = 319.1\ \text{ft} + 259.6\ \text{ft} = \boxed{578.7\ \text{ft}}$$

(b) Use Eq. 78.11.

$$\begin{aligned} V_{scope-B} &= \left(\tfrac{1}{2}\right)(100)(2.60\ \text{ft})\left(\sin(2)(4.17°)\right) \\ &\quad + (1\ \text{ft})(\sin 4.17°) \\ &= 18.9\ \text{ft} \\ \text{elevation B} &= 297.8\ \text{ft} + 18.9\ \text{ft} + 4.8\ \text{ft} - 10.9\ \text{ft} \\ &= \boxed{310.6\ \text{ft}} \end{aligned}$$

5. (a)

$$\begin{aligned} \text{elevation BM11} &= 179.65 + \sum(\text{BS}) - \sum(\text{FS}) \\ &= 179.65 + 4.64 + 5.80 + 2.25 \\ &\quad - (5.06 + 5.02 + 5.85) \\ &= \boxed{176.41} \end{aligned}$$

(b)

$$\begin{aligned} \text{elevation BM12} &= 179.65 + 57.5 - 31.2 \\ &= \boxed{205.95} \end{aligned}$$

6. While it is not necessary, convert the latitudes and departures into (x, y) coordinates. Assume point A is at (0,0). Then,

$$\text{B}_x = 0 + 0 = 0$$
$$\text{B}_y = 0 + 350 = 350$$

point	(x, y)
A	(0,0)
B	(0,350)
C	(600,900)
D	(1800,650)
E	(2000,− 100)
F	(900,− 650)

Transportation

Use Eq. 78.33.

$$\frac{0}{0} \times \frac{0}{350} \times \frac{600}{900} \times \frac{1800}{650} \times \frac{2000}{-100} \times \frac{900}{-650} \times \frac{0}{0}$$

$$\sum \text{full line products} = -1{,}090{,}000$$

$$\sum \text{dotted line products} = 3{,}040{,}000$$

$$\text{area} = \tfrac{1}{2}\left| -1{,}090{,}000 - 3{,}040{,}000 \right|$$

$$= \boxed{2{,}065{,}000}$$

7.

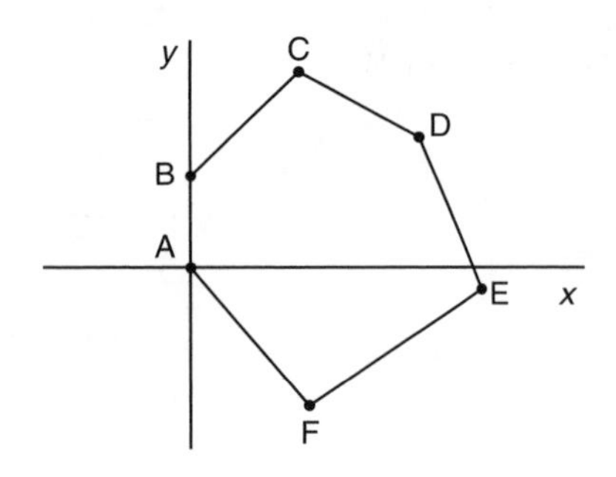

step 1: The interior angles are

$$\begin{aligned}
&\text{A: } 90° - 54.10° + 90° = 125.90° \\
&\text{B: } 180° - 45° = 135° \\
&\text{C: } 45° + 69.45° = 114.45° \\
&\text{D: } 90° - 69.45° + 90° + 11.32° = 121.87° \\
&\text{E: } 90° - 11.32° + 90° - 79.70° = 88.98° \\
&\text{F: } 79.70° + 54.10° = 133.8°
\end{aligned}$$

step 2: The sum of the interior angles is 720°. For a six-sided polygon, the sum of the interior angles should be

$$(6 - 2)(180°) = 720°$$

Since the sum of the angles is exactly 720°, no angle balancing is necessary. However, there may be a closure error.

For leg AB,

$$\text{latitude} = 500$$

$$\text{departure} = 0$$

For leg BC,

$$\text{latitude} = (848.6)(\cos 45°) = 600$$

$$\text{departure} = (848.6)(\sin 45°) = 600$$

Similarly, the following table is prepared (rounding to the nearest 0.1).

leg	latitude	departure
AB	+500.0	0
BC	+600.0	+600.0
CD	−299.9	+800.0
DE	−1000.0	+200.2
EF	−199.9	−1100.0
FA	+385.1	−532.0
totals	−14.7	−31.8

The total perimeter length is 4997.6 (round to 5000).

The latitude correction to line AB is

$$\left(\frac{500}{5000}\right)(14.7) = 1.47 \quad \text{[round to nearest 0.1 to get 1.5]}$$

The departure correction to line AB is

$$\left(\frac{500}{5000}\right)(31.8) = 3.18 \quad \text{[round to 3.2]}$$

The corrected latitude and departure of leg AB is

$$\text{latitude: } 500 + 1.5 = 501.5$$

$$\text{departure: } 0 + 3.2 = 3.2$$

Similarly, the following table is prepared.

leg	latitude	departure
AB	501.5	3.2
BC	602.5	605.4
CD	−297.4	805.4
DE	−997.0	206.7
EF	−196.6	−1092.9
FA	387.0	−527.8
totals	0.0	0.0

The angles and lengths calculated from these latitudes and departures will not be the same as the given angles and lengths. Therefore, corrected angles and lengths must be calculated.

For leg AB, the corrected angle is

$$\arctan\left(\frac{3.2}{501.5}\right) = 0.37°$$

The corrected bearing is N0.37°E.

The corrected length is

$$\sqrt{(3.2)^2 + (501.5)^2} = 501.5$$

The following table is similarly prepared.

leg	corrected bearing	corrected length
AB	N0.37°E	501.5
BC	N45.14°E	854.1
CD	S69.73°E	858.6
DE	S11.71°E	1018.2
EF	S79.80°W	1110.4
FA	N53.75°W	654.5

8. The plot of the available data shows two possible locations for point E. However, only one is essentially west of point A. Point E′ is ruled out since E′-A is not easterly.

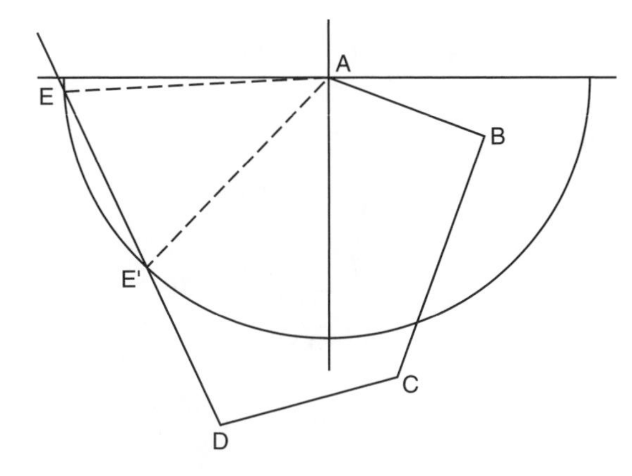

The latitudes and departures of the first three legs are

leg	latitude	departure
AB	−302.0	1038.3
BC	−1533.7	−417.6
CD	−503.0	−1191.9
totals	−2338.7	−571.2

The slope of line DE is

$$-\tan(332.37° - 270°) = -1.91$$

The coordinates of point D are (−571.2, −2338.7). The equation of line DE is

$$\begin{aligned} y &= mx + b \\ -2338.7 &= (-1.91)(-571.2) + b \\ b &= -3429.7 \\ y &= -1.91x - 3429.7 \end{aligned}$$

The equation of a circle centered at point A with a radius of 1737.9 is

$$x^2 + y^2 = (1737.9)^2$$

Since point E is on both the line and the circle,

$$\begin{aligned} x^2 + (-1.91x - 3429.7)^2 &= (1737.9)^2 \\ x^2 + 3.65x^2 + 13{,}101.5x + (3429.7)^2 &= (1737.9)^2 \\ x^2 + 2817.5x &= -1.88 \times 10^6 \end{aligned}$$

This has solutions $x = -1085.2$ and -1732.4. The larger (negative) x is -1732.4.

From the line equation,

$$y = -(1.91)(-1732.4) - 3429.7 = -120.8$$

The angle of line EA is

$$\arctan\left(\frac{120.8}{1732.4}\right) \approx 4.0°$$

The north azimuth of line EA is

$$90° - 4.0° = \boxed{86°}$$

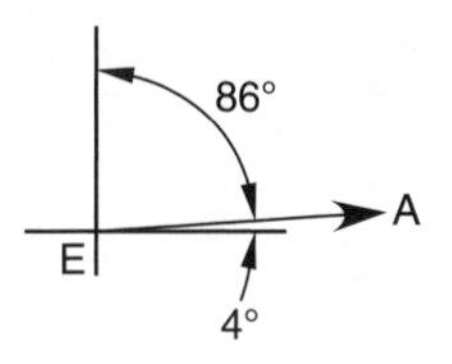

The length of line DE is

$$\sqrt{(-1732.4 + 571.2)^2 + (-120.8 + 2338.7)^2} = \boxed{2503.5}$$

9. (a) Use Eq. 78.39.

$$\begin{aligned} S_E &= \frac{f}{H - h} = \frac{6 \text{ in}}{4000 \text{ ft} - 650 \text{ ft}} \\ &= \boxed{1 \text{ in} = 558 \text{ ft}} \end{aligned}$$

The answer is (A).

(b) The ground area is

$$\frac{\left((12 \text{ in})\left(558 \ \frac{\text{ft}}{\text{in}}\right)\right)^2}{43{,}560 \ \frac{\text{ft}^2}{\text{ac}}} = \boxed{1030 \text{ ac}}$$

The answer is (A).

(c)

$$\text{overlap} = (0.40)(12\ \text{in})\left(558\ \frac{\text{ft}}{\text{in}}\right) = \boxed{2678\ \text{ft}}$$

The answer is (C).

(d)

$$\text{sidelap} = (0.80)(12\ \text{in})\left(558\ \frac{\text{ft}}{\text{in}}\right) = \boxed{5356\ \text{ft}}$$

The answer is (D).

(e)

$$t = \frac{(2678\ \text{ft})\left(3600\ \frac{\text{sec}}{\text{hr}}\right)}{\left(300\ \frac{\text{mi}}{\text{hr}}\right)\left(5280\ \frac{\text{ft}}{\text{mi}}\right)} = \boxed{6.09\ \text{sec}}$$

The answer is (B).

(f) Use Eq. 78.39.

$$S = \frac{f}{H - 665} = \frac{6\ \text{in}}{4000\ \text{ft} - 665\ \text{ft}} = \frac{1\ \text{in}}{555.8\ \text{ft}}$$

$$S = \frac{\text{photo}}{\text{true}}$$

$$\text{true} = \frac{\text{photo}}{S} = \frac{89\ \text{mm}}{\left(25.4\ \frac{\text{mm}}{\text{in}}\right)\left(\frac{1\ \text{in}}{555.8\ \text{ft}}\right)} = \boxed{1947\ \text{ft}}$$

The answer is (D).

(g)

$$f = \frac{6\ \text{in}}{12\ \frac{\text{in}}{\text{ft}}} = 0.5\ \text{ft}$$

Use Eq. 78.39.

$$S = \frac{f}{H - h}$$

$$\frac{1}{10{,}000} = \frac{0.5\ \text{ft}}{H - 1100\ \text{ft}}$$

$$H = \boxed{6100\ \text{ft}}$$

The answer is (C).

(h)

$$\text{real distance} = \frac{\text{map distance}}{\text{scale}} = \frac{0.54\ \text{in}}{\frac{1\ \text{in}}{1680\ \text{ft}}} = 907.2\ \text{ft}$$

$$\text{scale} = \frac{\text{map distance}}{\text{real distance}} = \frac{5.68\ \text{in}}{907.2\ \text{ft}} = \boxed{1\ \text{in} = 159.7\ \text{ft}}$$

The answer is (A).

(i)

$$\frac{1\ \text{in}}{159.7\ \text{ft}} = \frac{12\ \text{in}}{H - 889\ \text{ft}}$$

$$H = \boxed{2805.7\ \text{ft}}$$

The answer is (C).

(j)

$$\text{real distance} = \frac{\text{map distance}}{\text{scale}} = \frac{0.98\ \text{in}}{\frac{1\ \text{in}}{1680\ \text{ft}}} = 1646.4\ \text{ft}$$

$$\text{scale} = \frac{5.68\ \text{in}}{1646.4\ \text{ft}} = \frac{1\ \text{in}}{289.9\ \text{ft}}$$

$$\frac{1\ \text{in}}{289.9\ \text{ft}} = \frac{12\ \text{in}}{H - 889\ \text{ft}}$$

$$H = \boxed{4367.3\ \text{ft}}$$

The answer is (A).

79 Horizontal, Compound, Vertical, and Spiral Curves

PRACTICE PROBLEMS

Vertical Sag Curves

1. *(Time limit: one hour)* A sag vertical curve is shown.

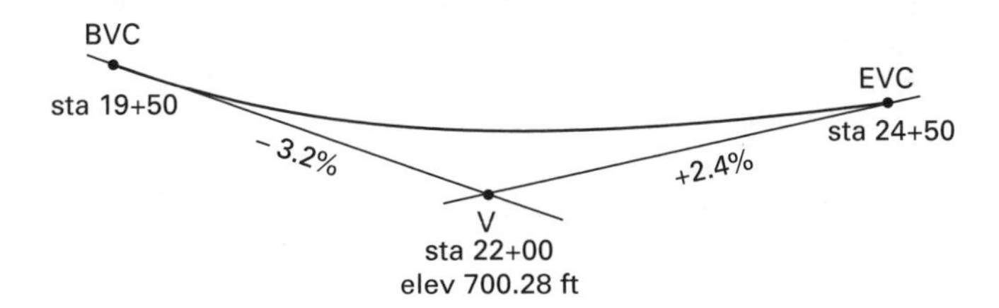

(a) At what station is the low point located?

(A) sta 22+15
(B) sta 22+36
(C) sta 22+68
(D) sta 22+79

(b) What is the elevation of the low point?

(A) 701 ft
(B) 702 ft
(C) 703 ft
(D) 704 ft

(c) What is the elevation of sta 21+25.04?

(A) 701 ft
(B) 704 ft
(C) 706 ft
(D) 708 ft

(d) Based on comfort, what is the highest design speed for this curve?

(A) 60 mph
(B) 64 mph
(C) 68 mph
(D) 72 mph

(e) What is the rate of grade change per station?

(A) 1.0%/sta
(B) 1.1%/sta
(C) 1.4%/sta
(D) 1.6%/sta

(f) If 20 ft of clearance are required, the lowest elevation that the bottom of an overhead bridge deck may have at sta 23+16 is most nearly

(A) 724 ft
(B) 726 ft
(C) 728 ft
(D) 730 ft

(g) What would the curve length have to be if the bridge deck in part (f) were kept at the same elevation, but 22 ft (6.7 m) of clearance were required? Assume that the stationing and elevation of the BVC are kept the same.

(A) 660 ft
(B) 680 ft
(C) 700 ft
(D) 710 ft

(h) For the curve solved in part (g), how much additional excavation is required at sta 22+00?

(A) 1.0 ft
(B) 1.9 ft
(C) 2.5 ft
(D) 2.9 ft

(i) What is the AASHTO K-value for the longer curve?

(A) 120.8 ft%
(B) 121.8 ft%
(C) 122.8 ft%
(D) 123.8 ft%

(j) At 65 mph, stopping sight distance for the longer curve is approximately equal to

(A) 270 ft
(B) 520 ft
(C) 650 ft
(D) 880 ft

2. A falling grade of 4% meets a rising grade of 5% in a sag curve. At the start of the curve, the elevation is 123.06 ft at sta 4034+20. At sta 4040+20, there is an overpass with an underside elevation of 134.06 ft. If the curve is designed to afford a clearance of 15 ft under the overpass, calculate the required length.

3. *(Time limit: one hour)* Two highways are planned, each running perpendicular to the other. The geometric details of the underpass have been determined. The design process is continuing for the perpendicular overpass, which must maintain a 25 ft clearance (measured between highway surface centerlines). (a) What is the minimum station where the new highway can be constructed? (b) Locate the maximum station where the new highway can be constructed. (c) What is the station of the lowest point on the existing curve? (d) What is the elevation of the lowest point on the existing curve?

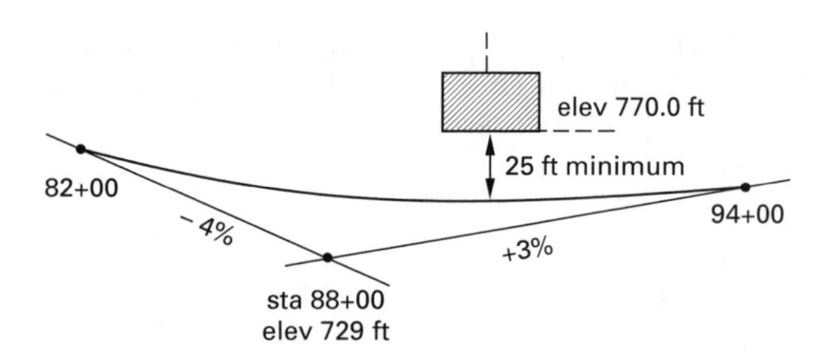

4. *(Time limit: one hour)* A new highway can go either over or under (perpendicular to) an existing highway at sta 75+00. The elevation of the existing highway at sta 75+00 is 510.00 ft. 22 ft of clearance (between the centerlines of the two road surfaces) are required if the new highway goes over the existing highway. 20 ft of clearance are required if the new highway goes under the old highway. (a) What is the longest vertical curve that will pass under the existing highway? (b) What is the shortest vertical curve that will pass over the existing highway?

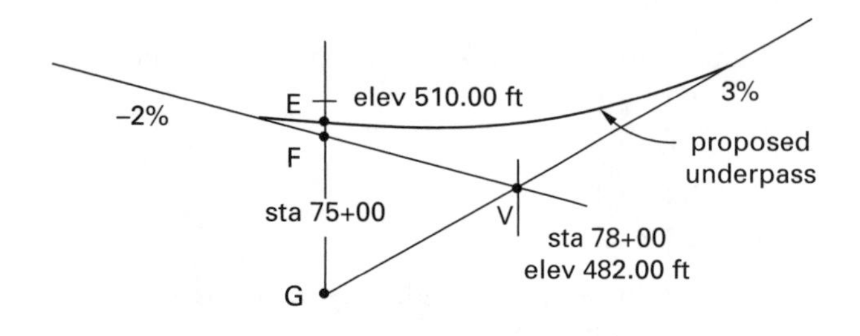

5. *(Time limit: one hour)* The grades on a vertical curve intersect at sta 105+00 and elevation 350 ft. A drain grating exists at sta 105+00 and elevation 358.30 ft. (a) If a curve length of 1900 ft is used, what is the elevation of the BVC? (b) If a length of 1900 ft is used, what is the elevation at sta 106+40? (c) If a curve length of 1900 ft is used, what is the elevation of the lowest point on the curve? (d) If a curve length of 1900 ft is used, what is the sta of the BVC? (e) If a curve length of 1900 ft is used, what is the station of the EVC? (f) What is the minimum continuous curve length such that no point on the curve will be lower than the drain grating?

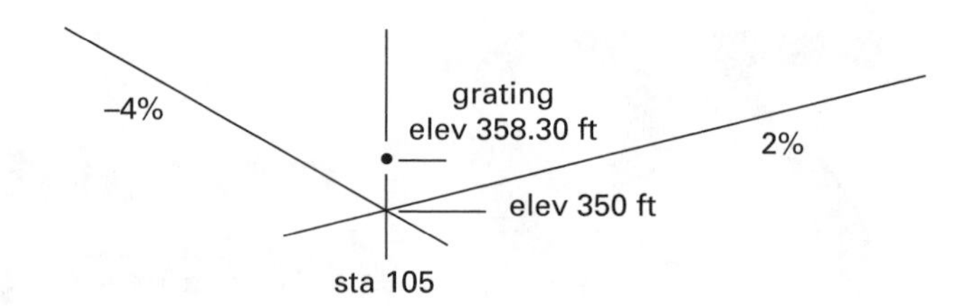

Vertical Crest Curves

6. An existing length of road consists of a rising gradient of 1 in 20 followed by a vertical parabolic crest curve 300 ft long, and then a falling gradient of 1 in 40. The curve joins both gradients tangentially, and the elevation of the highest point on the curve is 173.07 ft. Visibility is to be improved over this stretch of road by replacing the curve with another parabolic curve 600 ft long. (a) Find the depth of excavation required at the midpoint of the curve. (b) Tabulate the elevations of points at 100 ft intervals along the new curve.

7. *(Time limit: one hour)* An equal-tangent vertical crest curve has the following geometric properties.

BVC sta	110+00
PVI sta	116+00
PVI elev	1262
EVC	122+00
ascending grade	+3%
descending grade	−2%

(a) Find the centerline roadway elevation of sta 110+50. (b) Find the centerline roadway elevation of sta 115+50. (c) The road continues on to a horizontal curve to the right. Superelevation is 8%. Assume the curve is in full superelevation throughout. The road is 60 ft wide at all points. The horizontal curve is supported by three pile bents symmetrical about the curve's PVI. The pile bents are located at stations 115+50, 116+00, and 116+50. Each pile bent consists of seven piles symmetrical about the road centerline, placed 10 ft apart. The tops of the piles are 3.5 ft below the roadway. What are the elevations of the pile heads at stations 115+50, 116+00, and 116+50?

Horizontal Curves

8. Three points (A, B, and C) were selected on the centerline of an existing horizontal circular curve. The instrument was set horizontally at point B. Readings were taken on a vertical staff at points A and C. The instrument had constants of 100 and 0. The instrument axis was 4.7 ft above the road at point B. (a) What is the radius of the curve? (b) What is the line-of-sight gradient AB? (c) What is the line-of-sight gradient BC?

staff at	horizontal bearing	stadia readings (ft)		
A	0.00°	10.038	6.221	2.403
C	211.14°	7.236	5.778	4.320

9. *(Time limit: one hour)* Two straight tangents of a two-lane highway intersect at point B. The tangents have azimuths of 270° and 110°, respectively. They are to be joined by a circular curve that must pass through a point D, 350 ft from point B. The azimuth of line BD is 260°.

(a) The required radius is most nearly

(A) 7280 ft
(B) 7330 ft
(C) 7650 ft
(D) 7900 ft

(b) The tangent lengths are most nearly

(A) 1280 ft
(B) 1300 ft
(C) 1340 ft
(D) 1380 ft

(c) The length of curve is most nearly

(A) 2000 ft
(B) 2500 ft
(C) 3000 ft
(D) 3500 ft

(d) The central angle for a 500 ft arc is most nearly

(A) 1°
(B) 2°
(C) 3°
(D) 4°

(e) If the curve is laid out by the tangent-offset method, the tangent distance required to place the stake located an arc distance of 100 ft from the PC is most nearly

(A) 80 ft
(B) 90 ft
(C) 100 ft
(D) 200 ft

(f) Assuming that snow and ice are not normally encountered on the highway, the maximum design speed for this curve without relying on friction is most nearly

(A) 55 mph
(B) 65 mph
(C) 75 mph
(D) 100 mph

(g) Assuming a design speed of 70 mph, the passing sight distance for this curve is most nearly

(A) 2000 ft
(B) 2500 ft
(C) 3000 ft
(D) 3500 ft

(h) What would the maximum superelevation rate be if snow and ice were commonly encountered on the freeway?

(A) 0.02
(B) 0.04
(C) 0.06
(D) 0.10

(i) Assume lane widths are 12 ft and the crown cross-slope is most nearly 0.02 ft/ft. The transition rate is 1:200. A vehicle is traveling out of the curve. Over what distance (from the PC or PT) should all superelevation be removed from the alignment and the crown be fully developed?

(A) 130 ft
(B) 140 ft
(C) 170 ft
(D) 190 ft

(j) What is the approximate minimum stopping sight distance for this curve for a vehicle traveling at 70 mph?

(A) 730 ft
(B) 800 ft
(C) 850 ft
(D) 900 ft

10. *(Time limit: one hour)* An existing 6° horizontal circular curve connects a PC and PT_1 as shown. It is desired to avoid having vehicles pass too close to a historical monument, so a proposal has been made to relocate the curve 120 ft back. The PC will remain the same. (a) What is the PC station? (b) What is the radius of curvature of the new curve? (c) What is the stationing of the new curve's PT?

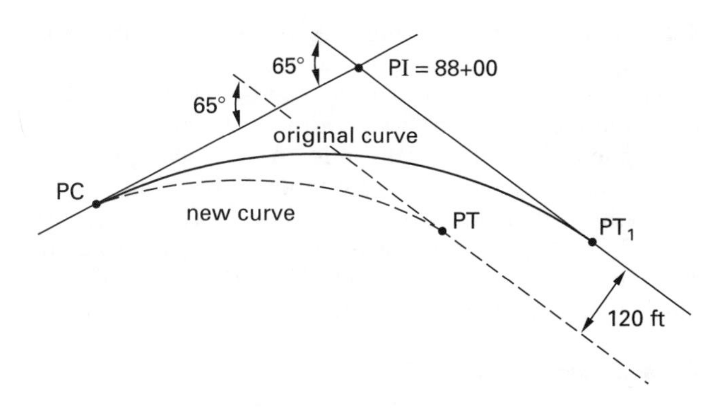

11. The tangent of the horizontal curve shown is relocated 10 ft west of its original position. The PC remains the same. (a) What is the original curve radius? (b) What is the new curve radius?

Transportation

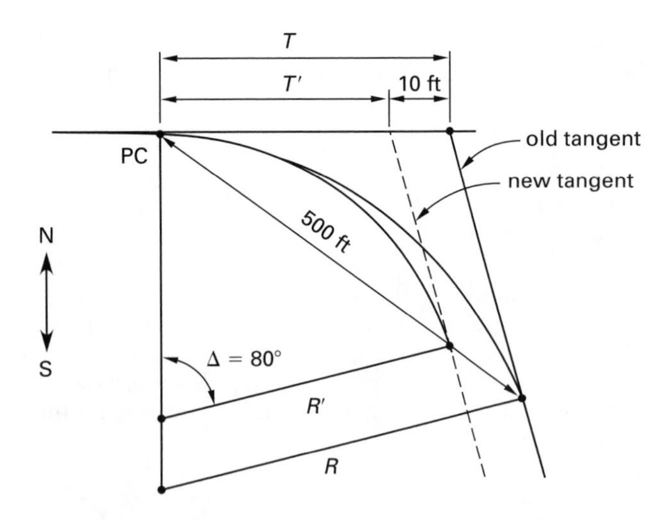

12. *(Time limit: one hour)* The tangent of a horizontal curve is parallel to a railroad line 150 ft away. The two curve tangents intersect at sta 182+27.52. Other information is provided in the illustration. Determine the (a) station of the PC, (b) the station of the PT, (c) the middle ordinate, (d) the external ordinate, and (e) the station of the railroad line's intersection with the highway.

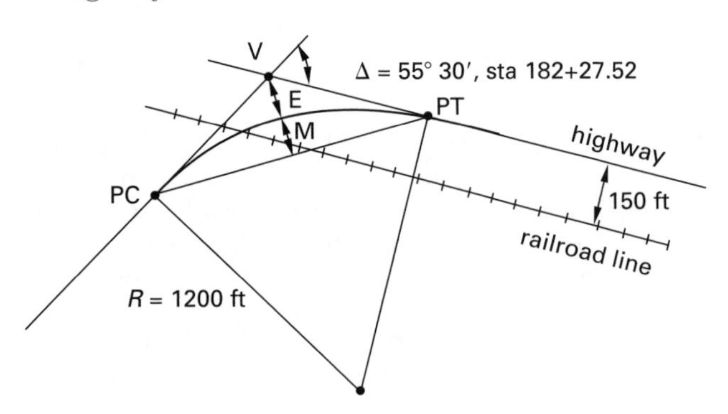

13. *(Time limit: one hour)* A freeway is being constructed to pass close to an existing first-order USGS survey monument. The freeway can be on either side of the monument. The freeway has a 75 ft right-of-way on each side of the curve centerline. A 1 ft clearance from the monument (which is considered to be a dimensionless point) to the edge of the right-of-way is required. The clearance is radial, measured along a line from the curve center to the monument. Find the range of radii of circular arcs that will provide the required clearance.

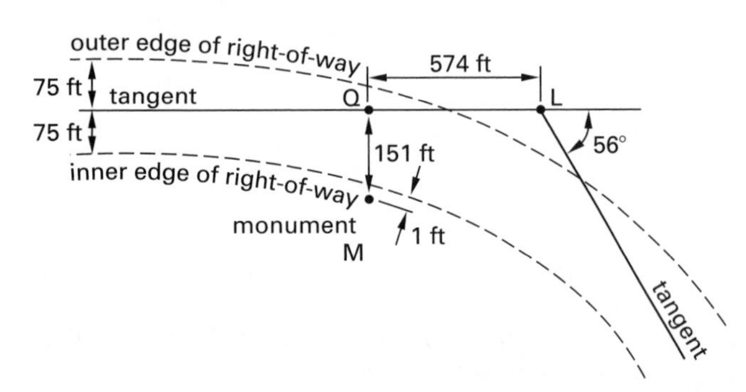

Reverse Curves

14. *(Time limit: one hour)* A historical monument must be avoided near a proposed highway. A reverse curve is being considered. Both curves will have the same curvature. What is the station of the (a) PC, (b) PRC, and (c) PT? (d) What is the deflection angle from the PC to the midpoint (MPC) of the first curve? (e) What are the disadvantages associated with reverse curves?

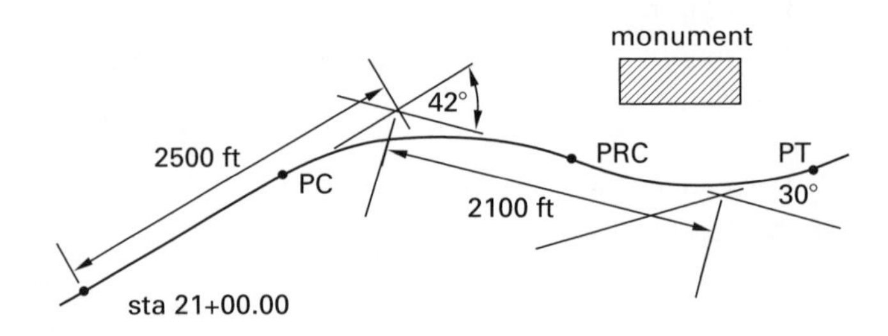

Sight Distance

15. *(Time limit: one hour)* An unsignalized blind intersection is shown. It is desired to prohibit parking near the intersection in order to provide adequate sight distance into the intersection. Vehicles travel on the major street at 35 mph or slower 85% of the time. PIEV time is 1.0 sec for the average driver. Vehicles are 20 ft long. Drivers' eyes are 5 ft behind the front bumpers and from the far side of the vehicle. Assume wet pavement and worn tires. Vehicles accelerate according to the data given. (a) Determine the length, L, needed to allow adequate sight distance. (b) How many 24 ft parallel parking places will be lost?

distance traveled, ft	time, sec
10	2.3
30	3.7
50	6.2
80	8.5
100	11.8

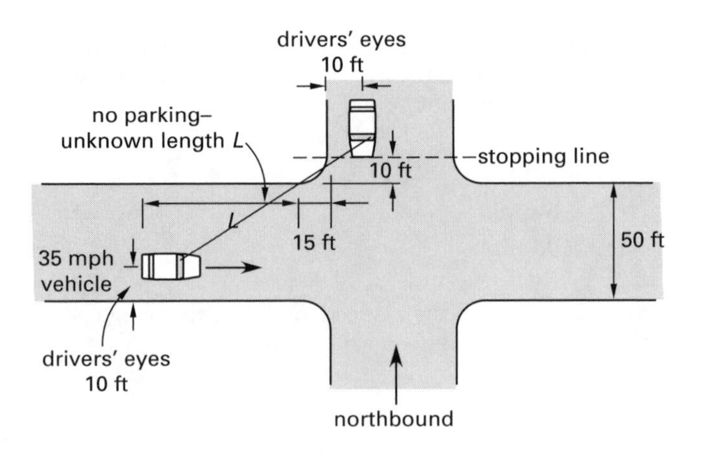

SOLUTIONS

1. (a) Use Eq. 79.46.

$$R = \frac{G_2 - G_1}{L} = \frac{2.4\% - (-3.2\%)}{5 \text{ sta}} = 1.12\%/\text{sta}$$

The low point is given by Eq. 79.49.

$$x = \frac{-G_1}{R} = \frac{3.2\%}{1.12 \frac{\%}{\text{sta}}} = 2.857 \text{ sta}$$

The low point's station is

$$(19{+}50) + 2.857 \text{ sta} = \boxed{\text{sta } 22{+}35.71}$$

The answer is (B).

(b) The elevation of the BVC is

$$700.28 \text{ ft} + (3.2\%)(2.5 \text{ sta}) = 708.28 \text{ ft}$$

The elevation of the low point is

$$\begin{aligned} y &= \frac{Rx^2}{2} + G_1 d + \text{elev}_{\text{BVC}} \\ &= \frac{\left(1.12 \frac{\%}{\text{sta}}\right)(2.857 \text{ sta})^2}{2} \\ &\quad + (-3.2\%)(2.857 \text{ sta}) + 708.28 \text{ ft} \\ &= \boxed{703.71 \text{ ft}} \end{aligned}$$

The answer is (D).

(c)

$$\begin{aligned} x &= (21{+}25.04) - (19{+}50) = 1.7504 \text{ sta} \\ y &= \frac{Rx^2}{2} + G_1 x + \text{elev}_{\text{BVC}} \\ &= \frac{\left(1.12 \frac{\%}{\text{sta}}\right)(1.75 \text{ sta})^2}{2} \\ &\quad + (-3.2\%)(1.75 \text{ sta}) + 708.28 \text{ ft} \\ &= \boxed{704.4 \text{ ft}} \end{aligned}$$

The answer is (B).

(d) The highest comfort speed is found from Eq. 79.58.

$$\begin{aligned} L &= \frac{|(G_2 - G_1)|\text{v}^2}{46.5} \\ \text{v} &= \sqrt{\frac{(500 \text{ ft})(46.5)}{|(2.4\% - (-3.2\%)|}} \\ &= \boxed{64.43 \text{ mph}} \end{aligned}$$

The answer is (B).

(e) $R = \boxed{1.12\%/\text{sta}}$

The answer is (B).

(f) At sta 23+16,

$$\begin{aligned} x &= (23{+}16) - (19{+}50) = 3.66 \text{ sta} \\ y &= \frac{Rx^2}{2} + G_1 x + \text{elev}_{\text{BVC}} \\ &= \frac{\left(1.12 \frac{\%}{\text{sta}}\right)(3.66 \text{ sta})^2}{2} \\ &\quad + (-3.2\%)(3.66 \text{ sta}) + 708.28 \text{ ft} \\ &= 704.07 \text{ ft} \end{aligned}$$

20 ft of clearance are needed.

$$\begin{aligned} \text{elevation} &= y + 20 \text{ ft} = 704.07 \text{ ft} + 20 \text{ ft} \\ &= \boxed{724.07 \text{ ft}} \end{aligned}$$

The answer is (A).

(g) At sta 23+16,

$$\begin{aligned} x &= (23{+}16) - (19{+}50) = 3.66 \text{ sta} \\ R &= \frac{2.4\% - (-3.2\%)}{L} = \frac{5.6\%}{L} \\ y &= \frac{Rx^2}{2} + G_1 x + \text{elev}_{\text{BVC}} \\ 724.07 \text{ ft} - 22 \text{ ft} &= \frac{\left(\frac{5.6}{L} \frac{\%}{\text{sta}}\right)(3.66 \text{ sta})^2}{2} \\ &\quad + (-3.2\%)(3.66 \text{ sta}) + 708.28 \text{ ft} \\ L &= \boxed{6.82 \text{ sta} \quad (680 \text{ ft})} \end{aligned}$$

The answer is (B).

(h) For curve 1,

$$\begin{aligned} y &= \frac{Rx^2}{2} + G_1 x + \text{elev}_{\text{BVC}} \\ &= \frac{\left(1.12 \frac{\%}{\text{sta}}\right)(2.5 \text{ sta})^2}{2} \\ &\quad + (-3.2\%)(2.5 \text{ sta}) + 708.28 \text{ ft} \\ &= 703.78 \text{ ft} \end{aligned}$$

For curve 2,

$$\begin{aligned} y &= \frac{Rx^2}{2} + G_1 x + \text{elev}_{\text{BVC}} \\ &= \frac{\left(\frac{5.6}{6.82} \frac{\%}{\text{sta}}\right)(2.5 \text{ sta})^2}{2} \\ &\quad + (-3.2\%)(2.5 \text{ sta}) + 708.28 \text{ ft} \\ &= 702.84 \text{ ft} \end{aligned}$$

The additional excavation needed is

$$703.78\text{ ft} - 702.84\text{ ft} = \boxed{0.94\text{ ft}}$$

The answer is (A).

(i) From Eq. 79.57,

$$K = \frac{L}{A} = \frac{682\text{ ft}}{|-3.2\% - 2.4\%|} = \frac{682\text{ ft}}{5.6\%} = \boxed{121.8\text{ ft}/\%}$$

The answer is (B).

(j) Use Table 79.2 (AASHTO Green Book Exh. 3-1). For a design speed of 65 mph, the design stopping distance is $\boxed{645\text{ ft.}}$

The answer is (C).

2.

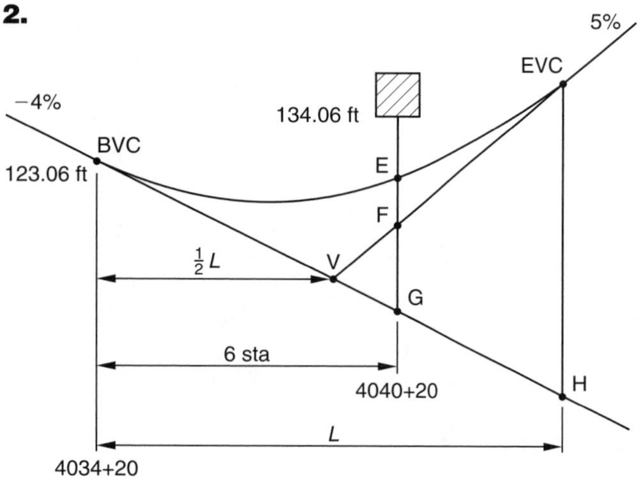

−4%
5%
EVC
134.06 ft
BVC
123.06 ft
E
F
$\frac{1}{2}L$
V
G
6 sta
4040+20
H
L
4034+20

step 1: The elevation at point E is

$$134.06\text{ ft} - 15\text{ ft} = 119.06\text{ ft}$$

step 2: The elevation at point G is

$$123.06\text{ ft} - \left(4\ \frac{\text{ft}}{\text{sta}}\right)(4040.2\text{ sta} - 4034.2\text{ sta}) = 99.06\text{ ft}$$

$$\text{distance EG} = 119.06\text{ ft} - 99.06\text{ ft} = 20\text{ ft}$$

step 3: ∠FVG is a 5% − (−4%) = 9% divergence.

$$\text{distance EVC-H} = (9)\tfrac{1}{2}L = 4.5L$$

step 4: Measuring all x distances from BVC, and since parabolic offsets from a line (BVC-H in this case) are proportional to x^2,

$$\frac{\text{EG}}{(6\text{ sta})^2} = \frac{\text{EVC-H}}{L^2}$$

$$\frac{20\text{ ft}}{(6\text{ sta})^2} = \frac{4.5L}{L^2}$$

$$L = \boxed{8.1\text{ sta}}$$

3. Define the curve mathematically.

$$L = 94\text{ sta} - 82\text{ sta} = 12\text{sta}$$

$$\text{elevation}_{\text{BVC}} = 729\text{ ft} + (88\text{ sta} - 82\text{ sta})\left(4\ \frac{\text{ft}}{\text{sta}}\right) = 729\text{ ft} + 24\text{ ft} = 753\text{ ft}$$

The rate of grade per station is

$$R = \frac{G_2 - G_1}{L} = \frac{3\% - (-4\%)}{12\text{ sta}} = \frac{7}{12}\%/\text{sta}$$

$$y = \left(\frac{7}{24}\right)x^2 - 4x + 753\text{ ft}$$

The points in question have an elevation of 770 ft − 25 ft = 745 ft. Solve for x.

$$745\text{ ft} = \left(\frac{7}{24}\right)x^2 - 4x + 753\text{ ft}$$

$$x = 11.29\text{ sta},\ 2.43\text{sta}$$

(a) $82{+}2.43 = \boxed{\text{sta } 84{+}43}$

(b) $82{+}11.29 = \boxed{\text{sta } 93{+}29}$

(c) The turning point is at

$$x = \frac{-(-4\%)}{\frac{7}{12}\frac{\%}{\text{sta}}} = 6.86\text{ ft}$$

$$\text{location} = (82{+}00) + (6{+}86) = \boxed{\text{sta } 88{+}86}$$

(d) Putting x back into the equation for y,

$$y = \left(\frac{7}{24}\frac{\%}{\text{sta}}\right)(6.86\text{ sta})^2 - (4\%)(6.86\text{ sta}) + 753\text{ ft} = \boxed{739.29\text{ ft}}$$

4. (a) Underpass—20 ft clearance:

step 1: $\text{elev}_E = 510 \text{ ft} - 20 \text{ ft} = 490 \text{ ft}$

step 2:

$$\begin{aligned}\text{elev}_G &= \text{elev}_V - (\text{slope})(x)\\ &= 482 \text{ ft} - (3\%)(3 \text{ sta})\\ &= 473 \text{ ft}\\ \text{EG} &= \text{elev}_E - \text{elev}_G\\ &= 490 \text{ ft} - 473 \text{ ft} = 17 \text{ ft}\end{aligned}$$

step 3:

$$\begin{aligned}\text{EF} &= \text{elev}_E - \text{elev}_V + \Delta\text{elev}_{\text{V-F}}\\ &= 490 \text{ ft} - 482 \text{ ft} - (2\%)(3 \text{ sta})\\ &= 2 \text{ ft}\end{aligned}$$

step 4: The offset from the grade line is proportional to the square of the distance from the BVC (or EVC) for a parabolic curve, since for a parabola,

$$y = ax^2$$

$$\frac{y_1}{x_1^2} = \frac{y_2}{x_2^2}$$

$$\frac{17 \text{ ft}}{\left(\frac{L}{2} + 3 \text{ sta}\right)^2} = \frac{2 \text{ ft}}{\left(\frac{L}{2} - 3 \text{ sta}\right)^2}$$

$$L = \boxed{12.2 \text{ sta}}$$

(b) Overpass—22 ft clearance:

step 1: $\text{elev}_E = 510 \text{ ft} + 22 \text{ ft} = 532 \text{ ft}$

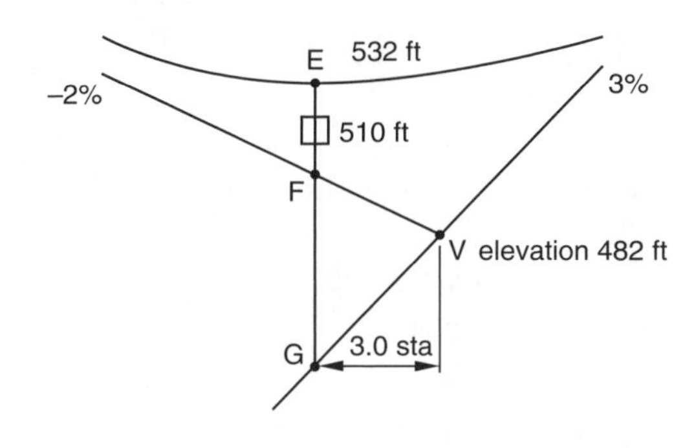

step 2:

$$\begin{aligned}\text{EG} &= 532 \text{ ft} - 482 \text{ ft} + (3 \text{ sta})(3\%)\\ &= 59 \text{ ft}\end{aligned}$$

step 3:

$$\begin{aligned}\text{EF} &= 532 \text{ ft} - 482 \text{ ft} - (3 \text{ sta})(2\%)\\ &= 44 \text{ ft}\end{aligned}$$

step 4:

$$\frac{59 \text{ ft}}{\left(\frac{L}{2} + 3 \text{ sta}\right)^2} = \frac{44 \text{ ft}}{\left(\frac{L}{2} - 3 \text{ sta}\right)^2}$$

$$L = \boxed{82.0 \text{ sta}}$$

5. (a)

$$\begin{aligned}R &= \frac{2\% - (-4\%)}{19 \text{ sta}}\\ &= 0.316\%/\text{sta}\\ \text{distance from BVC to V} &= \frac{19 \text{ sta}}{2}\\ &= 9.5 \text{ sta}\\ \text{elev}_{\text{BVC}} &= 350 \text{ ft} + (4\%)(9.5 \text{ sta})\\ &= \boxed{388 \text{ ft}}\end{aligned}$$

(b)

$$y = \left(\frac{0.316}{2}\right)x^2 - 4x + 388 \text{ ft}$$

The number of stations from BVC to sta 106.40 is 106.40 – (105 – 9.5) = 10.9 sta.

$$y = \left(\frac{0.316}{2} \frac{\%}{\text{sta}}\right)(10.9 \text{ sta})^2 - (4\%)(10.9 \text{ sta}) + 388 \text{ ft}$$

$$= \boxed{363.17 \text{ ft}}$$

(c) Use Eq. 79.48. The minimum point occurs at

$$x = \frac{-(-4\%)}{0.316 \frac{\%}{\text{sta}}} = 12.66 \text{ sta}$$

$$y = \left(\frac{0.316}{2} \frac{\%}{\text{sta}}\right)(12.66 \text{ sta})^2 - (4\%)(12.66 \text{ sta}) + 388 \text{ ft}$$

$$= \boxed{362.68 \text{ ft}}$$

(d) and (e) The stations of BVC and EVC are

$$105 \text{ sta} \pm \frac{19 \text{ sta}}{2} = \boxed{\text{sta } 95+50, \text{ sta } 114+50}$$

The road must go through the grating, so the low point is at the vertex.

(f) With a symmetrical (equal tangent) parabolic curve, the lowest point will not be at the vertex unless $G_1 = G_2$. Therefore, the curve cannot be equal tangent.

Each tangent of an unequal curve can be thought of as an equal-tangent curve with one of the grades equal to zero (i.e., $g = 0\%$).

BVC to V:

$$\begin{aligned}G_1 &= -4\%\\ G_2 &= 0\\ L_{\text{BVC-V}} &= \text{sta}_V - \text{sta}_{\text{BVC}} = 105 - \text{sta}_{\text{BVC}}\\ R &= \frac{0 - (-4\%)}{L_{\text{BVC-V}}} = \frac{4\%}{L_{\text{BVC-V}}}\end{aligned}$$

The minimum elevation is 358.30 ft at sta 105. Use Eq. 79.48.

$$x = \frac{-(-4\%)}{\frac{4\%}{L_{\text{BVC-V}}}} = L_{\text{BVC-V}}$$

The elevation of BVC is

$$\begin{aligned}
\text{elev}_\text{V} + |G_1|x &= 350 \text{ ft} + 4L_{\text{BVC-V}} \\
358.30 \text{ ft} &= \frac{4L^2}{2L} - 4L + 350 + 4L \\
8.30 \text{ ft} &= 2L \\
L_{\text{BVC-V}} &= 4.15 \text{ sta} \\
\text{location}_\text{BVC} &= 105 \text{ sta} - 4.15 \text{ sta} \\
&= 100.85 \text{ sta} \quad (\text{sta } 100+85) \\
\text{elevation}_\text{BVC} &= 350 \text{ ft} + (4\%)(4.15 \text{ sta}) \\
&= 366.60 \text{ ft}
\end{aligned}$$

V to EVC:

(Work with a mirror image to avoid $x = 0$ problems.)

$$\begin{aligned}
G_1 &= -2\% \\
G_2 &= 0 \\
L_{\text{V-EVC}} &= \text{sta}_\text{V} - \text{sta}_\text{EVC} \\
&= 105 - \text{sta}_\text{EVC} \\
R &= \frac{0 - (-2\%)}{L_{\text{V-EVC}}} = \frac{2\%}{L_{\text{V-EVC}}} \\
x &= \frac{-(-2\%)}{\frac{2\%}{L_{\text{V-EVC}}}} = L_{\text{V-EVC}} \\
\text{elev}_\text{EVC} &= 350 \text{ ft} + 2L_{\text{V-EVC}} \\
358.30 \text{ ft} &= \left(\frac{2}{2L}\right)L^2 - 2L + 350 \text{ ft} + 2L \\
L &= 8.30 \text{ sta}
\end{aligned}$$

Working with the actual curve,

$$\begin{aligned}
\text{location}_\text{EVC} &= 105 \text{ sta} + 8.30 \text{ sta} \\
&= 113.30 \text{ sta} \quad (\text{sta } 113+30) \\
\text{elev}_\text{EVC} &= 350 \text{ ft} + \left(2 \ \frac{\text{ft}}{\text{sta}}\right)(8.30 \text{ sta}) = 366.6 \text{ ft} \\
L_{\text{BVC-EVC}} &= 4.15 \text{ sta} + 8.30 \text{ sta} = \boxed{12.45 \text{ sta}}
\end{aligned}$$

This is the same answer that you would get without recognizing that the curve was not symmetrical. However, the premise would be wrong, and the curve equation $y(x)$ would be incorrect.

6. (a) Use Eq. 79.46.

$$R_1 = \frac{-2.5\% - 5\%}{3 \text{ sta}} = -2.5\%/\text{sta}$$

Use Eq. 79.48.

$$x = \frac{-5\%}{-2.5 \ \frac{\%}{\text{sta}}} = 2 \text{ sta} \quad [\text{distance to turning point}]$$

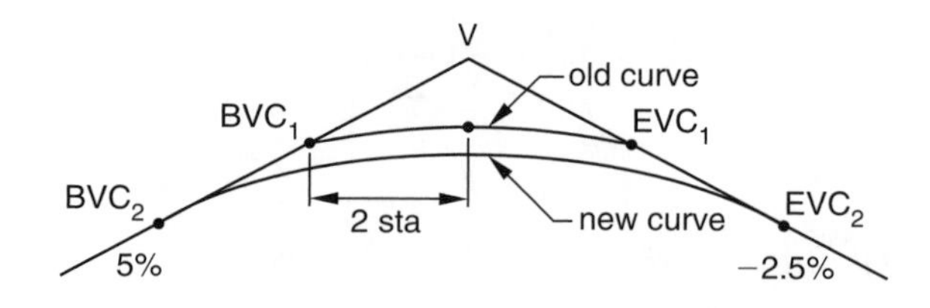

$$\begin{aligned}
173.07 \text{ ft} &= \left(\frac{-2.5 \ \frac{\%}{\text{sta}}}{2}\right)(2 \text{ sta})^2 \\
&\quad + \left(5 \ \frac{\text{ft}}{\text{sta}}\right)(2 \text{ sta}) + \text{elev BVC}_1 \\
\text{elev BVC}_1 &= 168.07 \text{ ft}
\end{aligned}$$

The equation of curve 1 is

$$y_1 = -1.25x^2 + 5x + 168.07 \text{ ft}$$

The elevation at the midpoint of curve 1 is

$$\begin{aligned}
y &= \left(-1.25 \ \frac{\%}{\text{sta}}\right)(1.5 \text{ sta})^2 + (5\%)(1.5 \text{ sta}) \\
&\quad + 168.07 \text{ ft} \\
&= 172.76 \text{ ft}
\end{aligned}$$

BVC_1 and BVC_2 are separated by 150 ft (1.5 sta). The elevation at BVC_2 is

$$168.07 \text{ ft} - \left(5 \ \frac{\text{ft}}{\text{sta}}\right)(1.5 \text{ sta}) = 160.57 \text{ ft}$$

$$\begin{aligned}
R_2 &= \frac{-2.5\% - 5\%}{6 \text{ sta}} = -1.25\%/\text{sta} \\
y_2 &= -0.625x^2 + 5x + 160.57 \text{ ft}
\end{aligned}$$

The elevation at the midpoint of curve 2 is

$$\begin{aligned}
y_2 &= \left(-0.625 \ \frac{\%}{\text{sta}}\right)(3 \text{ sta})^2 + (5\%)(3 \text{ sta}) + 160.57 \text{ ft} \\
&= 169.95 \text{ ft}
\end{aligned}$$

The midpoint cut is

$$172.76 \text{ ft} - 169.95 \text{ ft} = \boxed{2.81 \text{ ft}}$$

(Equation 79.49 can be used to solve for M directly for each curve. The difference in M values is 2.8 ft.)

(b) Measuring x from BVC_2,

$$\boxed{\begin{aligned} y_0 &= 160.57 \text{ ft} \\ y_1 &= 164.95 \text{ ft} \\ y_2 &= 168.07 \text{ ft} \\ y_3 &= 169.95 \text{ ft} \\ y_4 &= 170.57 \text{ ft} \\ y_5 &= 169.95 \text{ ft} \\ y_6 &= 168.07 \text{ ft} \end{aligned}}$$

7.

elevation 1262 ft
PVI
BVC
EVC
+3%
−2%
sta 110
sta 116
sta 122

(a)

$$\begin{aligned} \text{elev}_{\text{BVC}} &= 1262 \text{ ft} - (3\%)(116 \text{ sta} - 110 \text{ sta}) \\ &= 1244 \text{ ft} \\ L &= 122 \text{ sta} - 110 \text{ sta} = 12 \text{ sta} \\ R &= \frac{-2\% - 3\%}{12 \text{ sta}} = -0.417\%/\text{sta} \end{aligned}$$

For the first half-station $x = 0.5$ sta,

$$\begin{aligned} y &= \left(\frac{-0.417\%}{2 \text{ sta}}\right)(0.5 \text{ sta})^2 + (3\%)(0.5 \text{ sta}) + 1244 \text{ ft} \\ &= \boxed{1245.6 \text{ ft}} \end{aligned}$$

(b)

$$\begin{aligned} x &= 5.5 \text{ sta} \\ y_{115.5} &= \left(\frac{-0.417 \frac{\%}{\text{sta}}}{2}\right)(5.5 \text{ sta})^2 + (3\%)(5.5 \text{ sta}) \\ &\quad + 1244 \text{ ft} \\ &= \boxed{1254.2 \text{ ft}} \end{aligned}$$

(c) Label the piles A through G. D is the centerline pile.

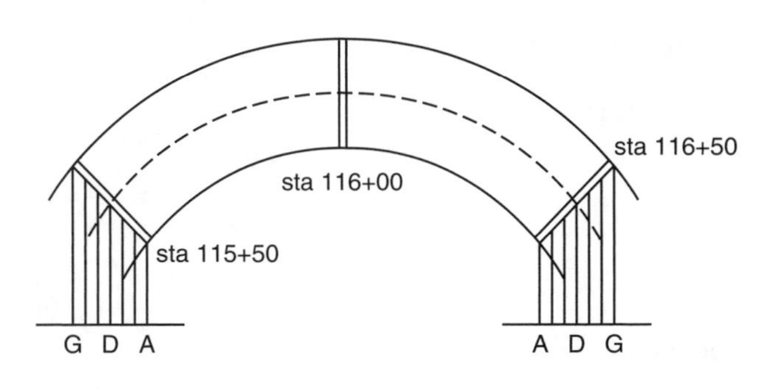

The elevation of the centerline pile (pile D) is found from part (b), less 3.5 ft.

This is an 8% superelevation, so for each 10 ft run, the rise is 0.8 ft.

For the first bent at sta 115+50, the top of pile D is at

$$\begin{aligned} &\text{D:} \quad 1254.2 \text{ ft} - 3.5 \text{ ft} = 1250.7 \text{ ft} \\ &\text{E:} \quad 1250.7 \text{ ft} + 0.8 \text{ ft} = 1251.5 \text{ ft} \\ &\text{F:} \quad 1250.7 \text{ ft} + (2)(0.8 \text{ ft}) = 1252.3 \text{ ft} \\ &\text{G:} \quad 1250.7 \text{ ft} + (3)(0.8 \text{ ft}) = 1253.1 \text{ ft} \\ &\text{C:} \quad 1250.7 \text{ ft} - 0.8 \text{ ft} = 1249.9 \text{ ft} \\ &\text{B:} \quad 1250.7 \text{ ft} - (2)(0.8 \text{ ft}) = 1249.1 \text{ ft} \\ &\text{A:} \quad 1250.7 \text{ ft} - (3)(0.8 \text{ ft}) = 1248.3 \text{ ft} \end{aligned}$$

bent at station	A	B	C	D	E	F	G
115 + 50	1248.3	1249.1	1249.9	1250.7	1251.5	1252.3	1253.1
116	1248.6	1249.4	1250.2	1251.0	1251.8	1252.6	1253.4
116 + 50	1248.8	1249.6	1250.4	1251.2	1252.0	1252.8	1253.6

8. This is essentially a horizontal curve.

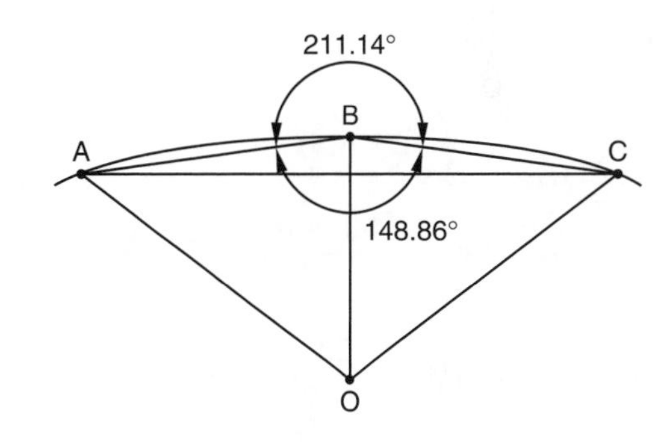

(a) The 211.14° angle is shown. The interior angle is 360° − 211.14° = 148.86°.

The distances AB and BC can be found from the stadia readings.

$$\begin{aligned} \text{AB} &= (100)(10.038 \text{ ft} - 2.403 \text{ ft}) + 0 = 763.5 \text{ ft} \\ \text{BC} &= (100)(7.236 \text{ ft} - 4.320 \text{ ft}) = 291.6 \text{ ft} \end{aligned}$$

Use the law of cosines to calculate distance AC.

$$\begin{aligned} (\text{AC})^2 &= (763.5 \text{ ft})^2 + (291.6 \text{ ft})^2 \\ &\quad - (2)(763.5 \text{ ft})(291.6 \text{ ft})(\cos 148.86°) \\ \text{AC} &= 1024.2 \text{ ft} \end{aligned}$$

Angles BAC and BCA can be found from the law of sines.

$$\frac{\sin \angle \text{BAC}}{291.6 \text{ ft}} = \frac{\sin \angle \text{BCA}}{763.5 \text{ ft}} = \frac{\sin 148.86°}{1024.2 \text{ ft}}$$

$$\angle \text{BAC} = 8.47°$$

$$\angle \text{BCA} = 22.67°$$

($\angle$BCA is not needed for the solution but can be used for checking.)

Check.

$$\angle \text{BAC} + \angle \text{BCA} + \angle \text{ABC} = 8.47° + 22.67° + 148.86° = 180° \quad \text{[check]}$$

$$\text{DC} = \left(\tfrac{1}{2}\right)(291.6 \text{ ft}) = 145.8 \text{ ft}$$

$$\angle \text{BOC} = 2\angle \text{BAC}$$

From the deflection angle principles,

$$\angle \text{DOC} = \tfrac{1}{2}\angle \text{BOC} = \angle \text{BAC}$$

$$\angle \text{DOC} = \angle \text{BAC} = 8.47°$$

$$R = \text{OC} = \frac{145.8 \text{ ft}}{\sin 8.47°} = \boxed{989.9 \text{ ft}}$$

(b)

$$\text{grade AB} = \frac{6.221 \text{ ft} - 4.7 \text{ ft}}{763.5 \text{ ft}} = \boxed{+0.00199 \text{ ft/ft}}$$

(c)

$$\text{grade BC} = \frac{4.7 \text{ ft} - 5.778 \text{ ft}}{291.6 \text{ ft}} = \boxed{-0.00370 \text{ ft/ft}}$$

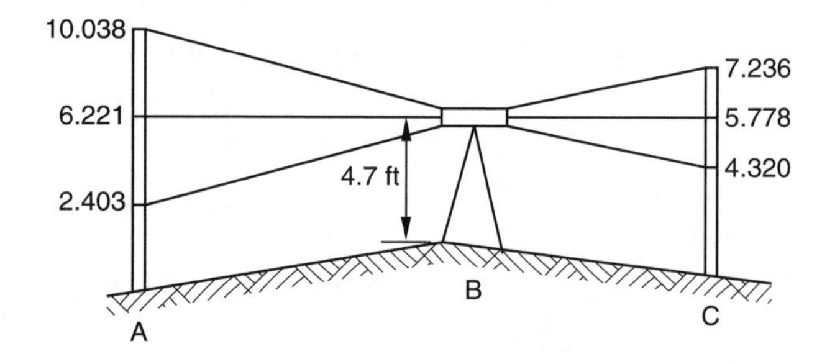

9.

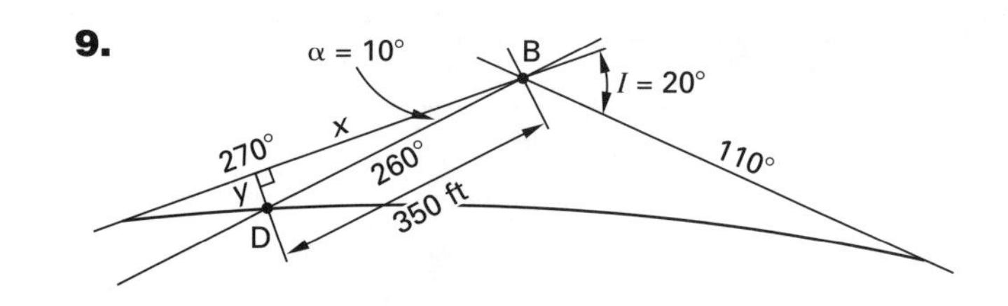

(a) Refer to Fig. 79.6.

$$m = 350 \text{ ft}$$

$$\alpha = 10°$$

$$\cos \alpha = \frac{x}{350 \text{ ft}}$$

$$x = 344.68 \text{ ft}$$

$$\sin \alpha = \frac{y}{350 \text{ ft}}$$

$$y = 60.78 \text{ ft}$$

$$\gamma = 90° - \frac{I}{2} - \alpha = 90° - \frac{20°}{2} - 10° = 70°$$

$$\phi = 180° - \arcsin\left(\frac{\sin 70°}{\cos 10°}\right) = 107.41°$$

$$\theta = 180° - \gamma - \phi = 180° - 70° - 107.41° = 2.59°$$

$$\frac{\sin \theta}{m} = \frac{\sin \phi \cos\left(\frac{I}{2}\right)}{R}$$

$$\frac{\sin 2.59°}{350 \text{ ft}} = \frac{(\sin 107.41°)\left(\cos\left(\frac{20°}{2}\right)\right)}{R}$$

$$R = \boxed{7278.21 \text{ ft}}$$

The answer is (A).

(b) Use Eq. 79.4.

$$T = R \tan\left(\frac{I}{2}\right) = (7278.21 \text{ ft}) \tan\left(\frac{20°}{2}\right) = \boxed{1283.34 \text{ ft}}$$

The answer is (A).

(c) Use Eqs. 79.8 and 79.10.

$$L = \frac{100IR}{5729.6} = \frac{(100)(20°)(7278.21 \text{ ft})}{5729.6} = \boxed{2540.56 \text{ ft}}$$

The answer is (B).

(d) Use Eq. 79.14. To set out a 500 ft arc, the deflection angle would be

$$\beta = \frac{(360°)(\text{arc length})}{2\pi R} = \frac{(360°)(500 \text{ ft})}{2\pi(7278.21 \text{ ft})} = \boxed{3.936°}$$

The answer is (D).

(e) Use Eq. 79.8. To lay out the first stake at a 100 ft arc distance,

$$\beta = \frac{5729.6}{R} = \frac{5729.6}{7278.21 \text{ ft}} = 0.7872°$$

$$\alpha = \frac{\beta}{2} = \frac{0.7872°}{2} = 0.3936°$$

$$\begin{aligned}\text{tangent distance} &= \text{NQ}\cos\alpha = 2Rg\sin\alpha\cos\alpha \\ &= (2)(7278.21 \text{ ft})(\sin 0.3936°) \\ &\quad \times (\cos 0.3936°) \\ &= \boxed{99.99 \text{ ft}}\end{aligned}$$

The answer is (C).

(f) When snow and ice are not present, the superelevation is generally limited to 0.08–0.12. Use Eq. 79.35 with 0.10.

$$\tan\phi_{\text{max}} = \frac{\text{v}^2}{gR}$$

$$\begin{aligned}\text{v} &= \sqrt{\tan\phi g R} \\ &= \sqrt{(0.1)\left(32.2 \frac{\text{ft}}{\text{sec}^2}\right)(7278.21 \text{ ft})} \\ &= 153.09 \text{ ft/sec}\end{aligned}$$

$$\text{v} = \frac{\left(153.09 \frac{\text{ft}}{\text{sec}}\right)\left(3600 \frac{\text{sec}}{\text{hr}}\right)}{5280 \frac{\text{ft}}{\text{mi}}} = \boxed{104 \text{ mph}}$$

The answer is (D).

(g) Use AASHTO Green Book Exh. 3-7. Given a design speed of 70 mph, the passing sight distance is $\boxed{2480 \text{ ft.}}$

The answer is (B).

(h) The maximum superelevation rate in areas encountering snow and ice is typically $\boxed{0.06\text{–}0.08.}$

The answer is (C).

(i) Remove all of the superelevation by $^2/_3L$ and develop the crown on the tangent runout. Use Eq. 79.40.

$$\text{SRR} = \frac{1}{200}$$

$$T_R = \frac{wp}{\text{SRR}} = \frac{(12 \text{ ft})\left(0.02 \frac{\text{ft}}{\text{ft}}\right)}{\frac{1}{200}} = 48 \text{ ft}$$

Use Eq. 79.41.

$$L = \frac{we}{\text{SRR}}$$

$$\tfrac{2}{3}L = \tfrac{2}{3}\left(\frac{we}{\text{SRR}}\right) = \left(\tfrac{2}{3}\right)(12 \text{ ft})\frac{\left(0.06 \frac{\text{ft}}{\text{ft}}\right)}{\frac{1}{200}} = 96 \text{ ft}$$

$$48 \text{ ft} + 96 \text{ ft} = \boxed{144 \text{ ft}}$$

The answer is (B).

(j) From Table 79.2, the minimum stopping sight distance is $\boxed{727.6 \text{ ft.}}$

The answer is (A).

10. (a) Calculate the radius.

$$R = \left(\frac{360°}{6°}\right)\left(\frac{100 \text{ ft}}{2\pi}\right) = 954.93 \text{ ft}$$

Calculate the original back tangent length.

$$T = (954.93 \text{ ft})\tan\left(\frac{65°}{2}\right) = 608.36 \text{ ft}$$

$$\begin{aligned}\text{location of PC} &= \text{sta PVI} - T \\ &= (88{+}00) - (6{+}8.36) \\ &= \boxed{\text{sta } 81{+}91.64}\end{aligned}$$

(b) The back tangent length decreases.

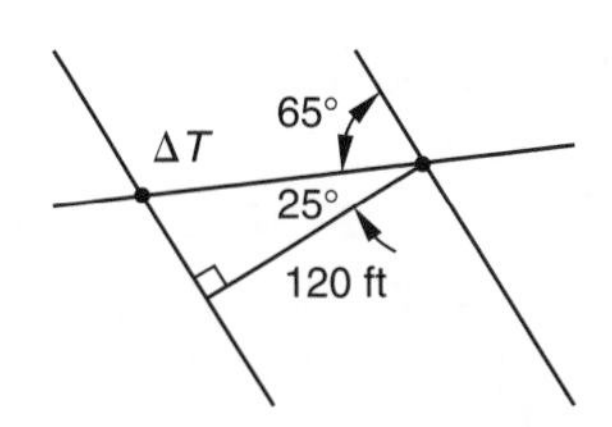

$$\Delta T = \frac{120 \text{ ft}}{\cos 25°} = 132.41 \text{ ft}$$

Transportation

The new back tangent is 608.36 ft − 132.41 ft = 475.95 ft.

$$R_{\text{new}} = \frac{475.95\text{ ft}}{\tan\left(\frac{65^\circ}{2}\right)} = \boxed{747.09\text{ ft}}$$

(c) The length of the curve is

$$L = (747.09\text{ ft})(65^\circ)\left(\frac{2\pi}{360^\circ}\right) = 847.55\text{ ft}$$

PT is located at

$$\begin{aligned}\text{sta PC} + L &= (81{+}91.64) + (8{+}47.55)\\ &= \boxed{\text{sta } 90{+}39.19}\end{aligned}$$

11.

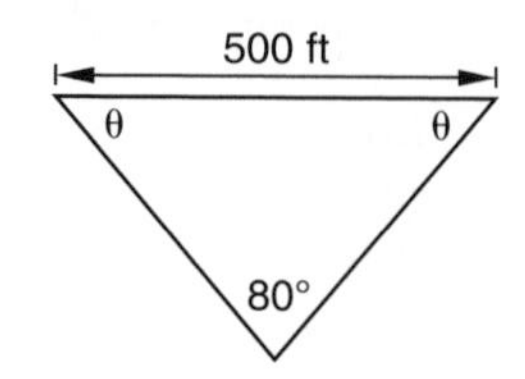

$$\theta = \frac{180^\circ - 80^\circ}{2} = 50^\circ$$

(a) From the sine rule,

$$\frac{500\text{ ft}}{\sin 80^\circ} = \frac{R}{\sin 50^\circ}$$

$$R = \boxed{388.9\text{ ft}}$$

(b) $T = (388.9\text{ ft})\tan\left(\frac{80^\circ}{2}\right) = 326.3\text{ ft}$

$$T' = T - 10\text{ ft} = 326.3\text{ ft} - 10\text{ ft} = 316.3\text{ ft}$$

$$R' = \frac{316.3\text{ ft}}{\tan 40^\circ} = \boxed{377.0\text{ ft}}$$

12. (a) $T = (1200\text{ ft})\tan\left(\frac{55.5^\circ}{2}\right) = 631.35\text{ ft}$

$$\begin{aligned}\text{location of PC} &= \text{sta } 182{+}27.52 - \text{sta } 6{+}31.35\\ &= \boxed{\text{sta } 175{+}96.17}\end{aligned}$$

(b) $L = (1200\text{ ft})(55.5^\circ)\left(\frac{2\pi}{360^\circ}\right) = 1162.39\text{ ft}$

$$\begin{aligned}\text{location of PT} &= \text{sta PC} + L\\ &= \text{sta } 175{+}96.17 + \text{sta } 11{+}62.39\\ &= \boxed{\text{sta } 187{+}58.56}\end{aligned}$$

(c)

$$\begin{aligned}M &= (1200\text{ ft})\left(1 - \cos\frac{55.5^\circ}{2}\right)\\ &= \boxed{138.0\text{ ft}}\end{aligned}$$

(d) From Eq. 79.5,

$$\begin{aligned}E &= R\tan\left(\frac{I}{2}\right)\tan\left(\frac{I}{4}\right)\\ &= (1200\text{ ft})\tan\left(\frac{55.5^\circ}{2}\right)\tan\left(\frac{55.5^\circ}{4}\right)\\ &= \boxed{155.95\text{ ft}}\end{aligned}$$

(e)

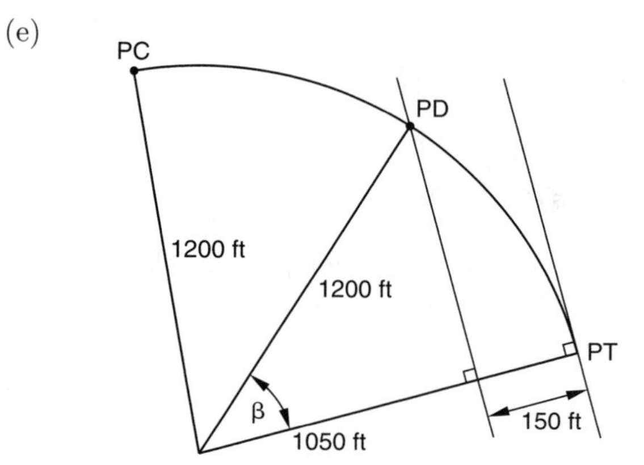

$$\beta = \arccos\left(\frac{1050\text{ ft}}{1200\text{ ft}}\right) = 28.96^\circ$$

The curve length from PD to PT is

$$\begin{aligned}(1200\beta)\left(\frac{2\pi}{360^\circ}\right) &= (1200\text{ ft})(28.96^\circ)\left(\frac{2\pi}{360^\circ}\right)\\ &= 606.54\text{ ft}\\ \text{sta}_{\text{PD}} &= \text{sta}_{\text{PT}} - \text{arc length}\\ &= 18{,}758.56\text{ ft} - 606.54\text{ ft}\\ &= 18{,}152.02\text{ ft}\\ &= \boxed{\text{sta } 181{+}52.02}\end{aligned}$$

13. Refer to the following diagram. Notice that the 1 ft clearance is not in line with segment QM.

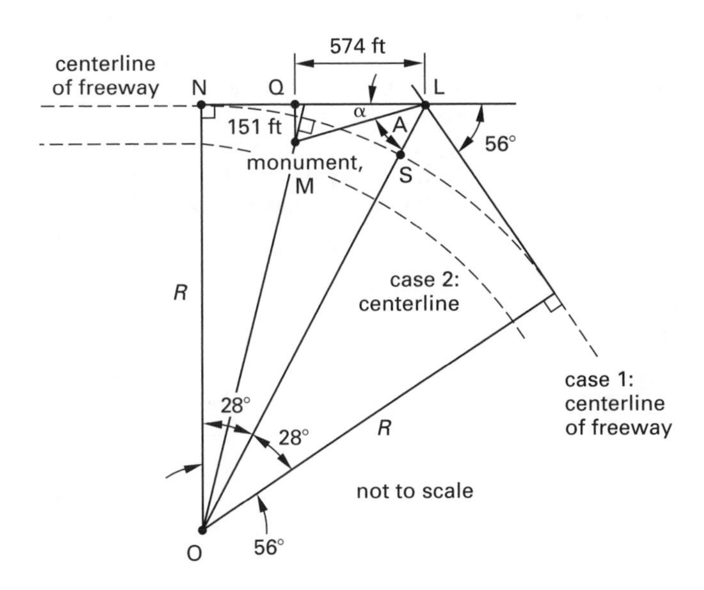

First, determine the unknown angles and distances.

$$\alpha = \arctan\left(\frac{\text{QM}}{\text{QL}}\right) = \arctan\left(\frac{151 \text{ ft}}{574 \text{ ft}}\right) = 14.739°$$

For triangle LNO,

$$\angle \text{L} = 180° - 90° - 28° = 62°$$

$$\angle \text{A} = 62° - \alpha = 62° - 14.739° = 47.261°$$

$$\text{LM} = \sqrt{(\text{QL})^2 + (\text{QM})^2} = \sqrt{(574 \text{ ft})^2 + (151 \text{ ft})^2} = 593.53 \text{ ft}$$

From Eq. 79.5,

$$\text{LS} = R\tan\left(\frac{I}{2}\right)\tan\left(\frac{I}{4}\right) = R\tan\left(\frac{56°}{2}\right)\tan\left(\frac{56°}{4}\right) = 0.13257R$$

$$c = \text{distance OL} = R + \text{LS} = 1.13257R$$

Case 1: The monument is between the freeway centerline and point O. Work with triangle OML.

$$(\text{OM})^2 = (\text{LM})^2 + (\text{OL})^2 - 2(\text{LM})(\text{OL})\cos A$$

$$\text{LM} = 593.53 \text{ ft}$$

$$\text{OL} = 1.13257R$$

$$\text{A} = 47.261°$$

$$\text{OM} = R - \tfrac{1}{2} \text{ road width} - \text{clearance} = R - \frac{150 \text{ ft}}{2} - 1 \text{ ft} = R - 76 \text{ ft}$$

$$(R - 76 \text{ ft})^2 = (593.53 \text{ ft})^2 + (1.13257R)^2 - (2)(593.53 \text{ ft})(1.13257R)\cos 47.261°$$

By trial and error, equation solver, or other appropriate means,

$$\boxed{R_1 = 2108.35 \text{ ft}}$$

Case 2: The freeway is between the monument and point O.

$$\text{OM} = R_2 + 76 \text{ ft}$$

$$(R + 76 \text{ ft})^2 = (593.53 \text{ ft})^2 + (1.13257R)^2 - (2)(593.53 \text{ ft})(1.13257R)\cos 47.261°$$

$$\boxed{R_2 = 3405.01 \text{ ft}}$$

14.

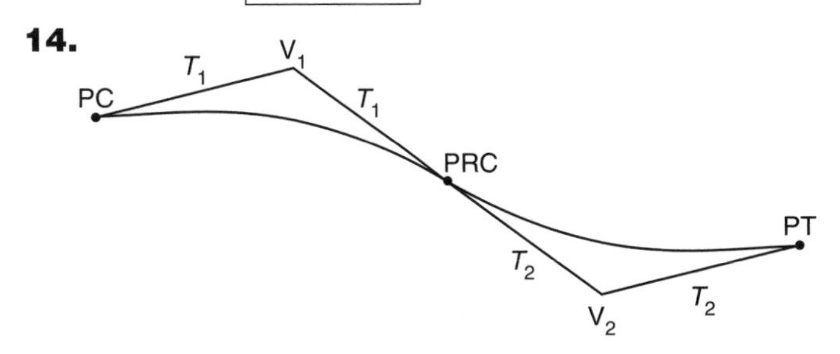

(a) Since the two intersection angles are not the same, the tangents are not the same.

$$T_1 + T_2 = 2100 \text{ ft}$$

$$T_1 = R\tan\left(\frac{42°}{2}\right) = 0.384R$$

$$T_2 = R\tan\left(\frac{30°}{2}\right) = 0.268R$$

$$(0.384 + 0.268)R = 2100 \text{ ft}$$

$$R = 3220.9 \text{ ft}$$

$$T_1 = (0.384)(3220.9 \text{ ft}) = 1236.8 \text{ ft}$$

$$T_2 = (0.268)(3220.9 \text{ ft}) = 863.2 \text{ ft}$$

$$L_1 = (3220.9 \text{ ft})(42°)\left(\frac{2\pi}{360°}\right) = 2361.04 \text{ ft}$$

$$L_2 = (3220.9 \text{ ft})(30°)\left(\frac{2\pi}{360°}\right) = 1686.46 \text{ ft}$$

Station of V_1 is 21 + 25 = 46+00.

(a)

$$\text{sta PC} = (46{+}00) - (12{+}36.8) = \boxed{\text{sta } 33{+}63.20}$$

(b)

$$\text{sta PRC} = \text{sta PC} + L_1 = (33{+}63.20) + (23{+}61.04) = \boxed{\text{sta } 57{+}24.24}$$

Transportation

(c)
$$\begin{aligned} \text{sta PT} &= \text{sta PRC} + L_2 \\ &= (57+24.24) + (16+86.46) \\ &= \boxed{\text{sta } 74+10.70} \end{aligned}$$

(d) $\text{deflection angle} = \left(\tfrac{1}{2}\right)\left(\left(\tfrac{1}{2}\right)(42°)\right) = \boxed{10.5°}$

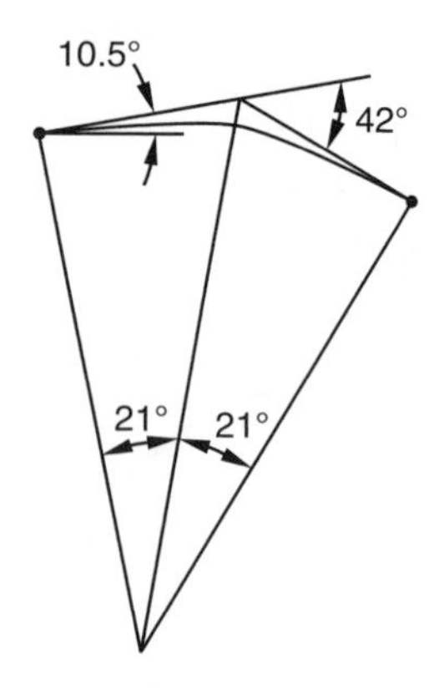

(e) The hazards are

- driving off the road due to distraction or inattention
- tipping over (whipping) at high speeds
- difficulty in getting proper superelevation

15. Since the angles (with respect to an E-W line) made by the diagonal between drivers' eyes are relatively acute (i.e., less than 15°), the diagonal and E-W distances are essentially the same. The offset and other distances are not used in this first analysis. At 35 mph, the stopping sight distance is given by Eq. 79.43. Assume the worst case (wet pavement, worn tires). From Table 75.1,

$$f = 0.34$$

(a) *Case 1:* An eastbound car sees a southbound car appear suddenly in front and skids to a stop.

If sight is into the intersection only (that is, drivers cannot see each other until the other is into the intersection), add 2.3 sec (the time to accelerate 10 ft from a stop into the intersection).

$$\begin{aligned} S &= (1.47)(1 \text{ sec} + 2.3 \text{ sec})(35 \text{ mph}) + \frac{(35 \text{ mph})^2}{(30)(0.34)} \\ &= 289.9 \text{ ft} \end{aligned}$$

Case 2: After entering the intersection, the southbound car sees the eastbound car (whose driver is not watching) and accelerates across the intersection.

Starting from a stop, to clear the intersection completely, the southbound car will have to travel 10 ft + 50 ft + 20 ft = 80 ft, where 20 ft is the car length. From the acceleration data, this will take 8.5 sec. By the time the vehicle has traveled the first 10 ft and has become vulnerable, 2.3 sec will have elapsed. The time that the southbound vehicle is vulnerable to being hit is

$$t = 8.5 \text{ sec} - 2.3 \text{ sec} = 6.2 \text{ sec}$$

The distance traveled by the eastbound car at 35 mph is

$$(1.47)(6.2 \text{ sec})(35 \text{ mph}) \approx 319 \text{ ft}$$

Since 319 ft > 289.9 ft,

$$L \approx \boxed{319 \text{ ft}}$$

(b) The calculation of the number of parking spaces lost must take into consideration the length of interior parking places (say 24 ft), the length of the end parking place (say 22 ft), a no-parking clear zone at the corner (say 15 ft), the southbound lane width (say 12 ft) and position of the car, and perhaps the distance between the drivers' eyes and the front/side of the vehicle (say 5 ft). Too many assumptions are required to integrate all of the factors accurately.

Simplistically, the number of 24 ft spaces lost is approximately

$$\begin{aligned} &1 + \frac{319 \text{ ft} - 22 \text{ ft} - 15 \text{ ft} - 12 \text{ ft} + 5 \text{ ft} + 5 \text{ ft}}{24 \text{ ft}} \\ &= \boxed{12.7 \text{ spaces} \quad (13 \text{ spaces})} \end{aligned}$$

80 Construction Earthwork, Staking, and Surveying

PRACTICE PROBLEMS

There are no problems in this book corresponding to Ch. 80 of the *Civil Engineering Reference Manual*.

81 Project Management

PRACTICE PROBLEMS

1. The activities that constitute a project are listed. The project starts at $t = 0$. (a) Draw the critical path network. (b) Indicate the critical path. (c) What is the earliest finish? (d) What is the latest finish? (e) What is the slack along the critical path? (f) What is the float along the critical path?

activity	predecessors	successors	duration
start	–	A	0
A	start	B,C,D	7
B	A	G	6
C	A	E,F	5
D	A	G	2
E	C	H	13
F	C	H,I	4
G	D,B	I	18
H	E,F	finish	7
I	F,G	finish	5
finish	H,I	–	0

2. PERT activities constituting a short project are listed with their characteristic completion times. If the project starts at $t = 15$, what is the probability that the project will be completed on or before $t = 42$?

activity	predecessors	successors	t_{min}	t_{likely}	t_{max}
start	–	A	0	0	0
A	start	B,D	1	2	5
B	A	C	7	9	20
C	B	D	5	12	18
D	A,C	finish	2	4	7
finish	D	–	0	0	0

3. Listed is a set of activities and sequence requirements to start a warehouse construction project. Prepare a CPM project diagram.

activity	letter code	code of immediate predecessor
move-in	A	
job layout	B	A
excavations	C	B
make-up forms	D	A
shop drawing, order rebar	E	A
erect forms	F	C,D
rough in plumbing	G	F
install rebar	H	E,F
pour, finish concrete	I	G,H

4. Activities constituting a bridge construction project are listed. (a) Draw the CPM network showing the critical path. (b) Compute ES, EF, LS, and LF for each of the activities. Assume a target time for completing the project that, for the bridge, is 3 days after the EF time.

job number		immediate predecessors	time (days)
a	start		0
b		a	4
c		b	2
d		c	4
e		d	6
f		c	1
g		f	2
h		f	3
i		d	2
j		d,g	4
k		i,j,h	10
l		k	3
m		l	1
n		l	2
o		l	3
p		e	2
q		p	1
r		c	1
s		o,t	2
t		m,n	3
u		t	1
v		q,r	2
w		v	5
x	finish	s,u,w	0

5. Prepare a PERT diagram for the activities and sequence requirements given in Prob. 4.

6. Listed is a set of activities, sequence requirements, and estimated activity times required for the renewal of a pipeline.

(a) Prepare a PERT project diagram.

activity	letter code	code of immediate predecessor	activity time requirement (days)
assemble crew for job	A		10
use old line to build inventory	B	D	28
measure and sketch old line	C	A	2
develop materials list	D	C	1
erect scaffold	E	D	2
procure pipe	F	D	30
procure valves	G	D	45
deactivate old line	H	B	1
remove old line	I	E,H	6
prefabricate new pipe	J	F	5
place valves	K	I,G	1
place new pipe	L	I,J	6
weld pipe	M	L	2
connect valves	N	K,M	1
insulate	O	N	4
pressure test	P	K	1
remove scaffold	Q	O,P	1
clean up and turn over to operating crew	R	O,Q	1

(b) There is additional information in the form of optimistic, most likely, and pessimistic time estimates for the project. Compute the expected mean time, t_m, and the variance, σ^2, for the activities. Which activities have the greatest uncertainty in their completion schedules?

activity code	optimistic time ($t_{\min}$)	most likely time (t_{ml})	pessimistic time ($t_{\max}$)
A	8	10	12
B	26	26.5	36
C	1	2	3
D	0.5	1	1.5
E	1.5	1.63	4
F	28	28	40
G	40	42.5	60
H	1	1	1
I	4	6	8
J	4	4.5	8
K	0.5	0.9	2
L	5	5.25	10
M	1	2	3
N	0.5	1	1.5
O	3	3.75	6
P	1	1	1
Q	1	1	1
R	1	1	1

(c) Suppose that, due to penalties in the contract, each day the pipeline renewal project can be shortened is worth \$100. Which of the following possibilities would you follow and why?

- Shorten t_{m} of activity B by 4 days at a cost of \$100.
- Shorten t_{ml} of activity G by 5 days at a cost of \$50.
- Shorten t_{m} of activity O by 2 days at a cost of \$150.
- Shorten t_{m} of activity O by 2 days by drawing resources from activity N, thereby lengthening its t_e by 2 days.

SOLUTIONS

1. (a) The critical path network diagram is as follows.

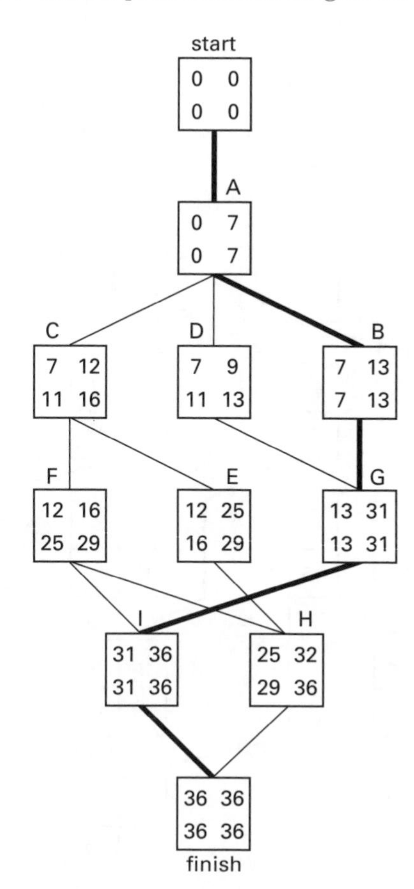

ES (earliest start) Rule: The earliest start time for an activity leaving a particular node is equal to the largest of the earliest finish times for all activities entering the node.

LF (latest finish) Rule: The latest finish time for an activity entering a particular node is equal to the smallest of the latest start times for all activities leaving the node.

The activity is critical if the earliest start equals the latest start.

(b) The critical path is A-B-G-I.

(c) The earliest finish is 36.

(d) The latest finish is 36.

(e) The slack along the critical path is 0.

(f) The float along the critical path is 0.

2. From Eq. 81.1,

$$\mu = \frac{t_{\text{minimum}} + 4t_{\text{likely}} + t_{\text{maximum}}}{6}$$

From Eq. 81.2,

$$\sigma^2 = \left(\frac{t_{\text{maximum}} - t_{\text{minimum}}}{6}\right)^2$$

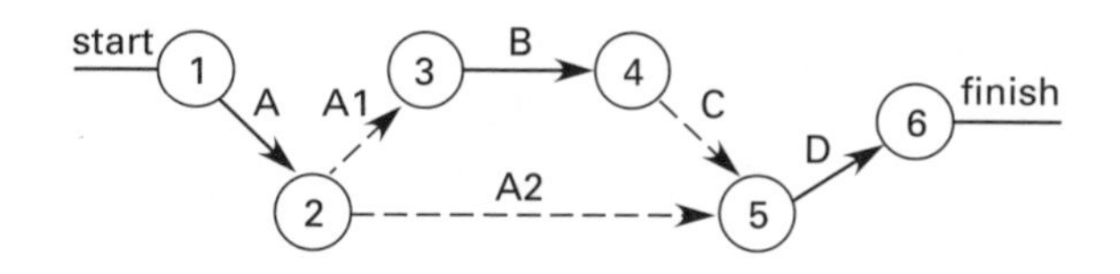

The critical path is A-A1-B-C-D.

The following probability calculations assume that all activities are independent. Use the following theorems for the sum of independent random variables and use the normal distribution for T (project time).

$$\mu_{\text{total}} = t_{\text{A}} + t_{\text{B}} + t_{\text{C}} + t_{\text{D}}$$
$$\sigma^2_{\text{total}} = \sigma^2_{\text{A}} + \sigma^2_{\text{B}} + \sigma^2_{\text{C}} + \sigma^2_{\text{D}}$$

The variance is 10.52778, and the standard deviation is 3.244654. (See table below.)

PERT Analysis for Prob. 2

activity				earliest	latest	earliest	latest	slack
no.	name	exp. time	variance	start	start	finish	finish	LS − ES
1	A	+2.33333	+0.44444	15	15	+17.3333	+17.3333	0
2	A1	0	0	+17.3333	+17.3333	+17.3333	+17.3333	0
3	A2	0	0	+17.3333	+39.6667	+17.3333	+39.6667	+22.3333
4	B	+10.5000	+4.69444	+17.3333	+17.3333	+27.8333	+27.8333	0
5	C	+11.8333	+4.69444	+27.8333	+27.8333	+39.6667	+39.6667	0
6	D	+4.16667	+0.69444	+39.6667	+39.6667	+43.8333	+43.8333	0
expected completion time = 43.83333								

Systems, Mgmt., and Professional

$$\mu_{\text{total}} = 43.83333$$
$$\sigma^2_{\text{total}} = 10.52778$$
$$\sigma_{\text{total}} = 3.244654$$
$$z = \frac{t - \mu_{\text{total}}}{\sigma} = \left| \frac{42 - 43.83333}{3.244654} \right| = 0.565$$

From the normal table, the probability of finishing for $T \leq 42$ is 0.286037 (28.6%).

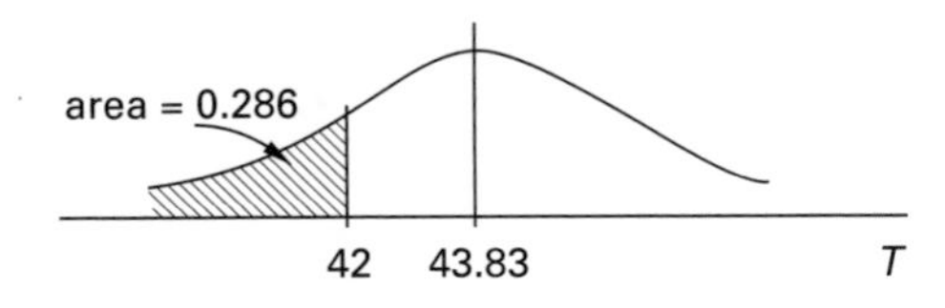

3.

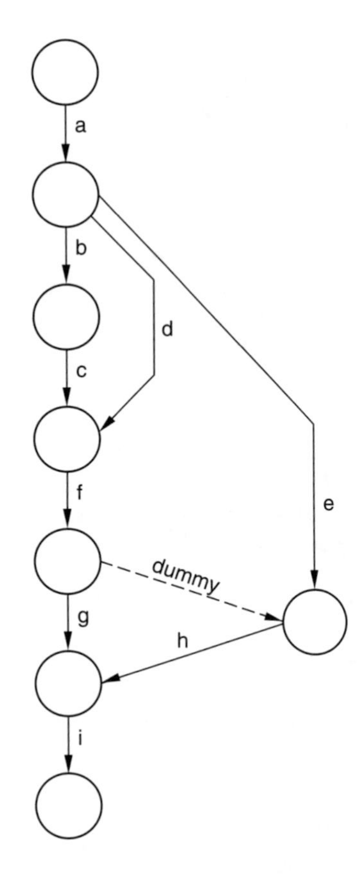

(Notice that a dummy node cannot be replaced with any other activity letter without duplicating that activity.)

4.

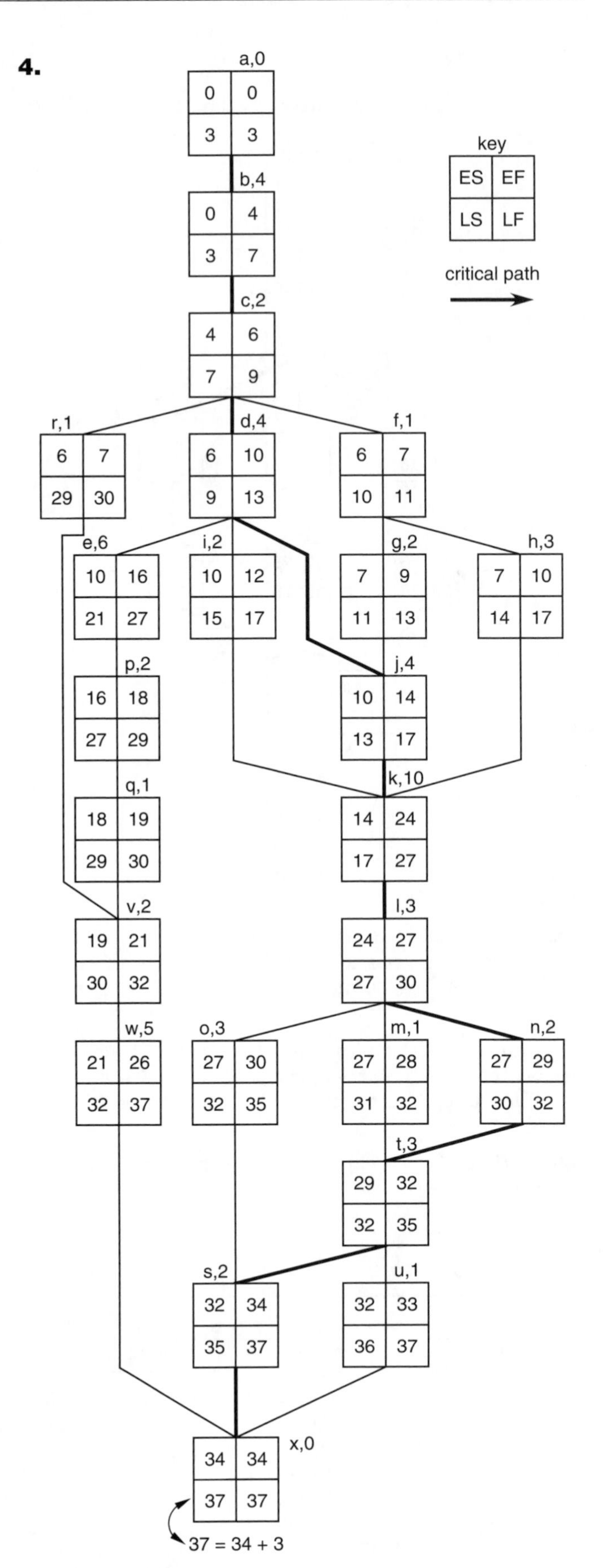

5.

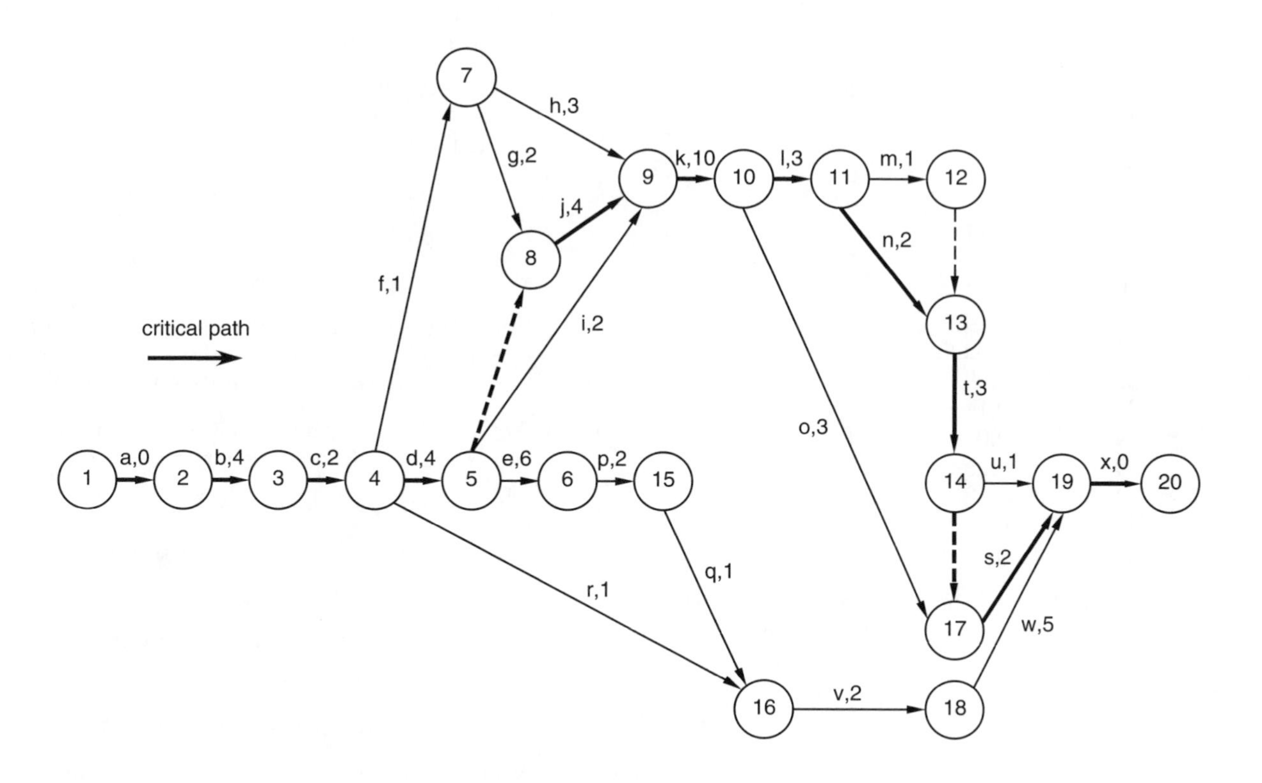

6. (a)

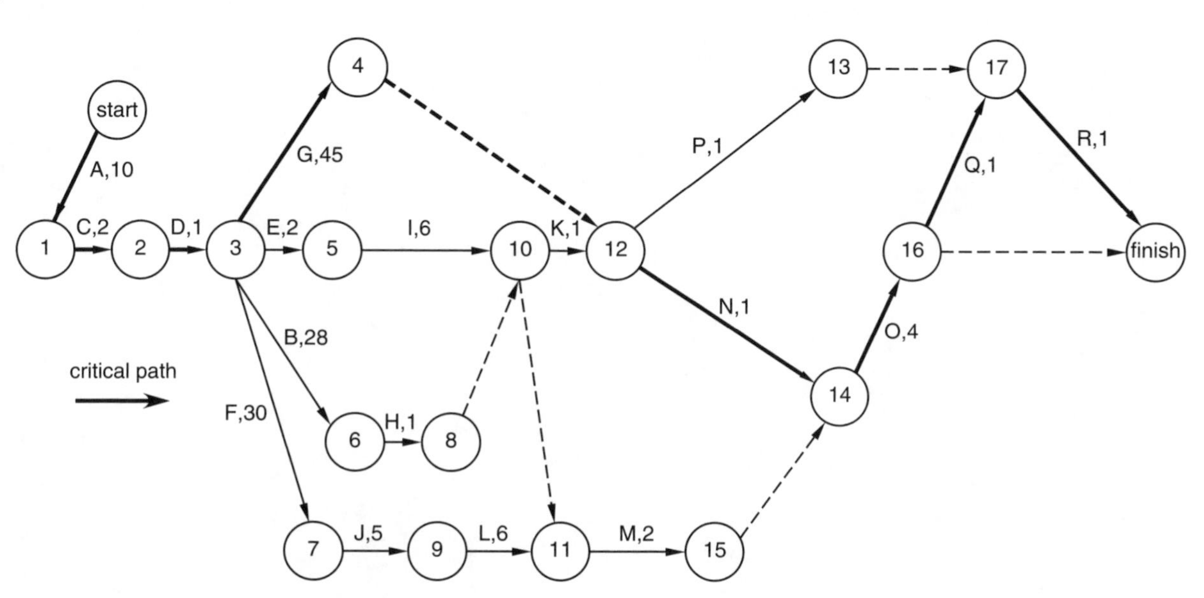

Systems, Mgmt., and Professional

(b)

activity	variance	meantime
A	0.44	10
B	2.77	28
C	0.11	2
D	0.027	1
E	0.17	2
F	4.0	30
G	11.11	45
H	0	1
I	0.44	6
J	0.44	5
K	0.6225	1
L	0.69	6
M	0.11	2
N	0.027	1
O	0.25	4
P	0	1
Q	0	1
R	0	1

Activities B, F, and G have the three largest variances, so these activities have the greatest uncertainties.

(c)

- Activity B is not on the critical path, so shortening it will not shorten the project.
- Activity G is on the critical path, so it should be shortened. Shortening it 5 days may change the critical path, however.

  $$\text{path D-G-K: } 1 + 45 = 46 \text{ days}$$

  $$\text{path D-E-I-K: } 1 + 2 + 6 = 9 \text{ days}$$

  $$\text{path D-B-H-I-K: } 1 + 28 + 1 + 6 = 36 \text{ days}$$

  Since the shortened path D-G-K has a length of 46 days − 5 days = 41 days and is still the longest path from D to K, it should be shortened.
- Activity O is on the critical path, but the cost of the crash schedule exceeds the benefit.
- The current path length is

  $$\text{N-O-Q-R: } 1 + 4 + 1 = 6 \text{ days}$$

  The new paths would be

  $$\text{N-P-Q-R: } 3 + 1 + 1 = 5 \text{ days}$$

  $$\text{N-O-Q-R: } 3 + 2 + 1 = 6 \text{ days}$$

  $$\text{N-O-R: } (3 + 2) + (4 - 2) = 7 \text{ days}$$

  N-O-R would become part of the critical path and would be longer than the original critical path. Therefore, it is not acceptable.

82 Construction and Jobsite Safety

PRACTICE PROBLEMS

There are no problems in this book corresponding to Ch. 82 of the *Civil Engineering Reference Manual*.

83 Electrical Systems and Equipment

PRACTICE PROBLEMS

1. Calculate the full-load phase current drawn by a 440 V (rms) 20 hp (per phase) induction motor having a full-load efficiency of 86% and a full-load power factor of 76%.

2. A 200 hp, three-phase, four-pole, 60 Hz, 440 V (rms) squirrel-cage induction motor operates at full load with an efficiency of 85%, power factor of 91%, and 3% slip.

(a) Find the speed in rpm.

(b) Find the torque developed.

(c) Find the line current.

3. A factory's induction motor load draws 550 kW at 82% power factor. What size synchronous motor is required to carry 250 hp and raise the power factor to 95%? The line voltage is 220 V (rms).

4. The nameplate of an induction motor lists 960 rpm as the full-load speed. For what frequency was the motor designed?

SOLUTIONS

1.
$$I_{\text{phase}} = \frac{P_{\text{phase}}}{\eta V \cos\phi} = \frac{(20\ \text{hp})\left(0.7457\times10^3\ \frac{\text{W}}{\text{hp}}\right)}{(0.86)(440\ \text{V})(0.76)} = \boxed{51.86\ \text{A}}$$

2. (a)
$$n_r = \left(\frac{(2)\left(60\ \frac{\text{sec}}{\text{min}}\right) f}{p}\right)(1-s)$$
$$= \left(\frac{(2)\left(60\ \frac{\text{sec}}{\text{min}}\right)(60\ \text{Hz})}{4}\right)(1-0.03)$$
$$= \boxed{1746\ \text{rpm}}$$

(b)
$$T = \frac{P}{\omega} = \frac{(200\ \text{hp})\left(550\ \frac{\text{ft-lbf}}{\text{hp-sec}}\right)}{2\pi\left(1746\ \frac{\text{rev}}{\text{min}}\right)\left(\frac{1\ \text{min}}{60\ \text{sec}}\right)}$$
$$= \boxed{602\ \text{ft-lbf}}$$

(c)
$$I_l = \left(\frac{1}{\sqrt{3}}\right)\left(\frac{P}{\eta V_l \cos\phi}\right)$$
$$= \left(\frac{1}{\sqrt{3}}\right)\left(\frac{(200\ \text{hp})\left(0.7457\times10^3\ \frac{\text{W}}{\text{hp}}\right)}{(0.85)(440\ \text{V})(0.91)}\right)$$
$$= \boxed{253\ \text{A}}$$

3.

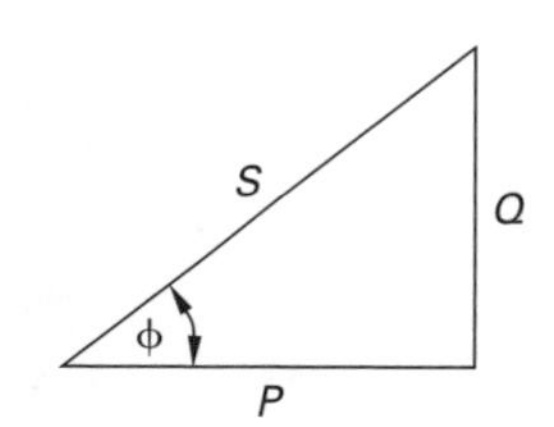

The original power angle is

$$\phi_i = \arccos 0.82 = 34.92°$$

$$P_1 = 550 \text{ kW}$$

$$Q_i = P_1 \tan\phi_i = (550 \text{ kW})(\tan 34.92°) = 384.0 \text{ kVAR}$$

The new conditions are

$$P_2 = (250 \text{ hp})\left(0.7457 \,\frac{\text{kW}}{\text{hp}}\right) = 186.4 \text{ kW}$$

$$\phi_f = \arccos 0.95 = 18.19°$$

Since both motors perform real work,

$$P_f = P_1 + P_2 = 550 \text{ kW} + 186.4 \text{ kW} = 736.4 \text{ kW}$$

The new reactive power is

$$\begin{aligned} Q_f &= P_f \tan\phi_f = (736.4 \text{ kW})(\tan 18.19°) \\ &= 242.0 \text{ kVAR} \end{aligned}$$

The change in reactive power is

$$\Delta Q = 384.0 \text{ kVAR} - 242.0 \text{ kVAR} = 142 \text{ kVAR}$$

Synchronous motors used for power factor correction are rated by apparent power.

$$\begin{aligned} S &= \sqrt{(\Delta P)^2 + (\Delta Q)^2} \\ &= \sqrt{(186.4 \text{ kW})^2 + (142 \text{ kVAR})^2} = \boxed{234.3 \text{ kVA}} \end{aligned}$$

4. $f = \dfrac{pn_s}{(20)\left(60\,\frac{\text{sec}}{\text{min}}\right)} = \dfrac{pn}{(2)\left(60\,\frac{\text{sec}}{\text{min}}\right)(1-s)}$

The slip and number of poles are unknown. Assume $s = 0$ and $p = 4$.

$$f = \frac{(4)\left(960\,\frac{\text{rev}}{\text{min}}\right)}{(2)\left(60\,\frac{\text{sec}}{\text{min}}\right)} = 32 \text{ Hz}$$

This is not close to anything in commercial use. Try $p = 6$.

$$f = \frac{(6)\left(960\,\frac{\text{rev}}{\text{min}}\right)}{(2)\left(60\,\frac{\text{sec}}{\text{min}}\right)} = 48 \text{ Hz}$$

With a 4% slip, $f = 50$ Hz.

$$\boxed{\text{50 Hz (European)}}$$

84 Instrumentation and Measurements

PRACTICE PROBLEMS

There are no problems in this book corresponding to Ch. 84 of the *Civil Engineering Reference Manual*.

85 Engineering Economic Analysis

PRACTICE PROBLEMS

1. At 6% effective annual interest, how much will be accumulated if $1000 is invested for ten years?

2. At 6% effective annual interest, what is the present worth of $2000 that becomes available in four years?

3. At 6% effective annual interest, how much should be invested to accumulate $2000 in 20 years?

4. At 6% effective annual interest, what year-end annual amount deposited over seven years is equivalent to $500 invested now?

5. At 6% effective annual interest, what will be the accumulated amount at the end of ten years if $50 is invested at the end of each year for ten years?

6. At 6% effective annual interest, how much should be deposited at the start of each year for ten years (a total of 10 deposits) in order to empty the fund by drawing out $200 at the end of each year for ten years (a total of 10 withdrawals)?

7. At 6% effective annual interest, how much should be deposited at the start of each year for five years to accumulate $2000 on the date of the last deposit?

8. At 6% effective annual interest, how much will be accumulated in ten years if three payments of $100 are deposited every other year for four years, with the first payment occurring at $t = 0$?

9. $500 is compounded monthly at a 6% nominal annual interest rate. How much will have accumulated in five years?

10. What is the effective annual rate of return on an $80 investment that pays back $120 in seven years?

11. A new machine will cost $17,000 and will have a resale value of $14,000 after five years. Special tooling will cost $5000. The tooling will have a resale value of $2500 after five years. Maintenance will be $2000 per year. The effective annual interest rate is 6%. What will be the average annual cost of ownership during the next five years?

12. An old covered wooden bridge can be strengthened at a cost of $9000, or it can be replaced for $40,000. The present salvage value of the old bridge is $13,000. It is estimated that the reinforced bridge will last for 20 years, will have an annual cost of $500, and will have a salvage value of $10,000 at the end of 20 years. The estimated salvage value of the new bridge after 25 years is $15,000. Maintenance for the new bridge would cost $100 annually. The effective annual interest rate is 8%. Which is the best alternative?

13. A firm expects to receive $32,000 each year for 15 years from sales of a product. An initial investment of $150,000 will be required to manufacture the product. Expenses will run $7530 per year. Salvage value is zero, and straight-line depreciation is used. The income tax rate is 48%. What is the after-tax rate of return?

14. A public works project has initial costs of $1,000,000, benefits of $1,500,000, and disbenefits of $300,000. (a) What is the benefit/cost ratio? (b) What is the excess of benefits over costs?

15. A speculator in land pays $14,000 for property that he expects to hold for ten years. $1000 is spent in renovation, and a monthly rent of $75 is collected from the tenants. (Use the year-end convention.) Taxes are $150 per year, and maintenance costs are $250 per year. What must be the sale price in ten years to realize a 10% rate of return?

16. What is the effective annual interest rate for a payment plan of 30 equal payments of $89.30 per month when a lump sum payment of $2000 would have been an outright purchase?

17. A depreciable item is purchased for $500,000. The salvage value at the end of 25 years is estimated at $100,000. What is the depreciation in each of the first three years using the (a) straight line, (b) sum-of-the-years' digits, and (c) double-declining balance methods?

18. Equipment that is purchased for $12,000 now is expected to be sold after ten years for $2000. The estimated maintenance is $1000 for the first year, but it is expected to increase $200 each year thereafter. The effective annual interest rate is 10%. What are the (a) present worth and (b) annual cost?

19. A new grain combine with a 20-year life can remove seven pounds of rocks from its harvest per hour. Any rocks left in its output hopper will cause $25,000 damage in subsequent processes. Several investments are available to increase the rock-removal capacity, as listed in the table. The effective annual interest rate is 10%. What should be done?

rock removal rate	probability of exceeding rock removal rate	required investment to achieve removal rate
7	0.15	0
8	0.10	$15,000
9	0.07	$20,000
10	0.03	$30,000

20. (*Time limit: one hour*) A mechanism that costs $10,000 has operating costs and salvage values as given. An effective annual interest rate of 20% is to be used.

year	operating cost	salvage value
1	$2000	$8000
2	$3000	$7000
3	$4000	$6000
4	$5000	$5000
5	$6000	$4000

(a) What is the economic life of the mechanism? (b) Assuming that the mechanism has been owned and operated for four years already, what is the cost of owning and operating the mechanism for one more year?

21. (*Time limit: one hour*) A salesperson intends to purchase a car for $50,000 for personal use, driving 15,000 miles per year. Insurance for personal use costs $2000 per year, and maintenance costs $1500 per year. The car gets 15 miles per gallon, and gasoline costs $1.50 per gallon. The resale value after five years will be $10,000. The salesperson's employer has asked that the car be used for business driving of 50,000 miles per year and has offered a reimbursement of $0.30 per mile. Using the car for business would increase the insurance cost to $3000 per year and maintenance to $2000 per year. The salvage value after five years would be reduced to $5000. If the employer purchased a car for the salesperson to use, the initial cost would be the same, but insurance, maintenance, and salvage would be $2500, $2000, and $8000, respectively. The salesperson's effective annual interest rate is 10%. (a) Is the reimbursement offer adequate? (b) With a reimbursement of $0.30 per mile, how many miles must the car be driven per year to justify the employer buying the car for the salesperson to use?

22. (*Time limit: one hour*) Alternatives A and B are being evaluated. The effective annual interest rate is 10%. What alternative is economically superior?

	alternative A	alternative B
first cost	$80,000	$35,000
life	20 years	10 years
salvage value	$7000	0
annual costs		
years 1–5	$1000	$3000
years 6–10	$1500	$4000
years 11–20	$2000	0
additional cost		
year 10	$5000	0

23. (*Time limit: one hour*) A car is needed for three years. Plans A and B for acquiring the car are being evaluated. An effective annual interest rate of 10% is to be used. Which plan is economically superior?

plan A: lease the car for $0.25/mile (all inclusive)

plan B: purchase the car for $30,000
keep the car for three years
sell the car after three years for $7200
pay $0.14 per mile for oil and gas
pay other costs of $500 per year

24. (*Time limit: one hour*) Two methods are being considered to meet strict air pollution control requirements over the next ten years. Method A uses equipment with a life of ten years. Method B uses equipment with a life of five years that will be replaced with new equipment with an additional life of five years. Capacities of the two methods are different, but operating costs do not depend on the throughput. Operation is 24 hours per day, 365 days per year. The effective annual interest rate for this evaluation is 7%.

	method A	method B	
	years 1–10	years 1–5	years 6–10
installation cost	$13,000	$6000	$7000
equipment cost	$10,000	$2000	$2200
operating cost per hour	$10.50	$8.00	$8.00
salvage value	$5000	$2000	$2000
capacity (tons/yr)	50	20	20
life	10 years	5 years	5 years

(a) What is the uniform annual cost per ton for each method? (b) Over what range of throughput (in tons/yr) does each method have the minimum cost?

25. (*Time limit: one hour*) A transit district has asked for your assistance in determining the proper fare for its bus system. An effective annual interest rate of 7% is to be used. The following additional information was compiled for your study.

cost per bus	$60,000
bus life	20 years
salvage value	$10,000
miles driven per year	37,440
number of passengers per year	80,000
operating cost	$1.00 per mile in the first year, increasing $0.10 per mile each year thereafter

(a) If the fare is to remain constant for the next 20 years, what is the break-even fare per passenger? (b) If the transit district decides to set the per-passenger fare at $0.35 for the first year, by what amount should the per-passenger fare go up each year thereafter such that the district can break even in 20 years? (c) If the transit district decides to set the per-passenger fare at $0.35 for the first year and the per-passenger fare goes up $0.05 each year thereafter, what additional governmental subsidy (per passenger) is needed for the district to break even in 20 years?

26. Make a recommendation to your client to accept one of the following alternatives. Use the present worth comparison method. (Initial costs are the same.)

(A) a 25 year annuity paying $4800 at the end of each year, where the interest rate is a nominal 12% per annum

(B) a 25 year annuity paying $1200 every quarter at 12% nominal annual interest

27. A firm has two alternatives for improvement of its existing production line. The data are as follows.

	alternative A	alternative B
initial installment cost	$1500	$2500
annual operating cost	$800	$650
service life	5 years	8 years
salvage value	0	0

Determine the best alternative using an interest rate of 15%.

28. Two mutually exclusive alternatives requiring different investments are being considered. The life of both alternatives is estimated at 20 years with no salvage values. The minimum rate of return that is considered acceptable is 4%. Which alternative is best?

	alternative A	alternative B
investment required	$70,000	$40,000
net income per year	$5620	$4075
rate of return on total investment	5%	8%

29. Compare the costs of two plant renovation schemes, A and B. Assume equal lives of 25 years, no salvage values, and interest at 25%. Make the comparison on the basis of (a) present worth, (b) capitalized cost, and (c) annual cost.

	alternative A	alternative B
first cost	$20,000	$25,000
annual expenditure	$3000	$2500

30. With interest at 8%, obtain the solutions to the following to the nearest dollar. (a) A machine costs $18,000 and has a salvage value of $2000. It has a useful life of 8 years. What is its book value at the end of 5 years using straight line depreciation? (b) Using data from part (a), find the depreciation in the first three years using the sinking fund method. (c) Repeat part (a) using double declining balance depreciation to find the first five years' depreciation.

31. A chemical pump motor unit is purchased for $14,000. The estimated life is 8 years, after which it will be sold for $1800. Find the depreciation in the first two years by the sum-of-the-years' digits method. Calculate the after-tax depreciation recovery using 15% interest with 52% income tax.

32. A soda ash plant has the water effluent from processing equipment treated in a large settling basin. The settling basin eventually discharges into a river that runs alongside the basin. Recently enacted environmental regulations require all rainfall on the plant to be diverted and treated in the settling basin. A heavy rainfall will cause the entire basin to overflow. An uncontrolled overflow will cause environmental damage and heavy fines. The construction of additional height on the existing basic walls is under consideration.

Data on the costs of construction and expected costs for environmental cleanup and fines are shown. Data on 50 typical winters have been collected. The soda ash plant management considers 12% to be their minimum rate of return, and it is felt that after 15 years the plant will be closed. The company wants to select the alternative that minimizes its total expected costs.

additional basin height (ft)	number of winters with basin overflow	expense for environmental clean up per year	construction cost
0	24	$550,000	0
5	14	$600,000	$600,000
10	8	$650,000	$710,000
15	3	$700,000	$900,000
20	1	$800,000	$1,000,000
	50		

Systems, Mgmt., and Professional

33. A wood processing plant installed a waste gas scrubber at a cost of $30,000 to remove pollutants from the exhaust discharged into the atmosphere. The scrubber has no salvage value and will cost $18,700 to operate next year, with operating costs expected to increase at the rate of $1200 per year thereafter. When should the company consider replacing the scrubber? Money can be borrowed at 12%.

34. Two alternative piping schemes are being considered by a water treatment facility. On the basis of a 10-year life and an interest rate of 12%, determine the number of hours of operation for which the two installations will be equivalent.

	alternative A	alternative B
pipe diameter	4 in	6 in
head loss for required flow	48 ft	26 ft
size motor required	20 hp	7 hp
energy cost per hour of operation	$ 0.30	$ 0.10
cost of motor installed	$3600	$2800
cost of pipes and fittings	$3050	$5010
salvage value at end of 10 years	$200	$280

35. An 88% learning curve is used with an item whose first production time was 6 weeks. How long will it take to produce the fourth item? How long will it take to produce the sixth through fourteenth items?

36. *(Time limit: one hour)* A company is considering two alternatives, only one of which can be selected.

alternative	initial investment	salvage value	annual net profit	life
A	$120,000	$15,000	$57,000	5 yr
B	$170,000	$20,000	$67,000	5 yr

The net profit is after operating and maintenance costs, but before taxes. The company pays 45% of its year-end profit as income taxes. Use straight line depreciation. Do not use investment tax credit. Find the best alternative if the company's minimum attractive rate of return is 15%.

37. *(Time limit: one hour)* A company is considering the purchase of equipment to expand its capacity. The equipment cost is $300,000. The equipment is needed for 5 years, after which it will be sold for $50,000. The company's before-tax cash flow will be improved $90,000 annually by the purchase of the asset.

The corporate tax rate if 48%, and straight line depreciation will be used. The company will take an investment tax credit of 6.67%. What is the after-tax rate of return associated with this equipment purchase?

38. *(Time limit: one hour)* A 120-room hotel is purchased for $2,500,000. A 25-year loan is available for 12%. A study was conducted to determine the various occupancy rates.

occupancy	probability
65% full	0.40
70%	0.30
75%	0.20
80%	0.10

The operating costs of the hotel are as follows.

taxes and insurance	$20,000 annually
maintenance	$50,000 annually
operating	$200,000 annually

The life of the hotel is figured to be 25 years when operating 365 days per year. The salvage value after 25 years is $500,000.

Neglect tax credit and income taxes. Determine the average rate that should be charged per room per night to return 15% of the initial cost each year.

39. *(Time limit: one hour)* A company is insured for $3,500,000 against fire. The insurance rate is $0.69/1000. The insurance company will decrease the rate to $0.47/1000 if fire sprinklers are installed. The initial cost of the sprinklers is $7500. Annual costs are $200; additional taxes are $100 annually. The system life is 25 years. What is the rate of return on this investment?

40. *(Time limit: one hour)* Heat losses through the walls in an existing building cost a company $1,300,000 per year. This amount is considered excessive, and two alternatives are being evaluated. Neither of the alternatives will increase the life of the existing building beyond the current expected life of 6 years, and neither of the alternatives will produce a salvage value.

Alternative A: Do nothing. Continue with current losses.

Alternative B: Spend $2,000,000 immediately to upgrade the building and reduce the loss by 80%. This alternative will require annual maintenance of $150,000.

Alternative C: Spend $1,200,000 immediately. Repeat the $1,200,000 expenditure 3 years from now. Heat loss the first year will be reduced 80%. Due to deterioration, the reduction will be 55% and 20% in the second and third years. (The pattern is repeated starting after the second expenditure.) There are no maintenance costs.

All energy and maintenance costs are considered expenses for tax purposes. The company's tax rate is 48%, and straight line depreciation is used. 15% is regarded as the effective annual interest rate. Evaluate each alternative on an after-tax basis, and recommend the best alternative.

41. *(Time limit: one hour)* You have been asked to determine if a 7-year-old machine should be replaced. Give a full explanation for your recommendation. Base your decision on a before-tax interest rate of 15%.

The existing machine is presumed to have a 10-year life. It has been depreciated on a straight line basis from its original value of \$1,250,000 to a current book value of \$620,000. Its ultimate salvage value was assumed to be \$350,000 for purposes of depreciation. Its present salvage value is estimated at \$400,000, and this is not expected to change over the next 3 years. The current operating costs are not expected to change from \$200,000 per year.

A new machine costs \$800,000, with operating costs of \$40,000 the first year, and increasing by \$30,000 each year thereafter. The new machine has an expected life of 10 years. The salvage value depends on the year the new machine is retired.

year retired	salvage
1	\$600,000
2	\$500,000
3	\$450,000
4	\$400,000
5	\$350,000
6	\$300,000
7	\$250,000
8	\$200,000
9	\$150,000
10	\$100,000

SOLUTIONS

1.

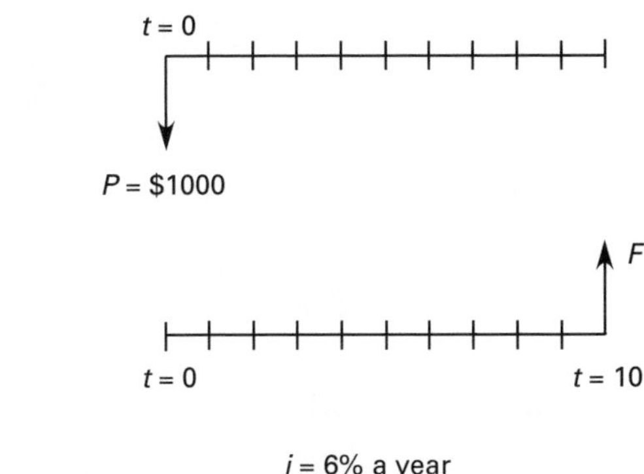

By the formula from Table 85.1,

$$F = P(1+i)^n = (\$1000)(1+0.06)^{10} = \boxed{\$1790.85}$$

By the factor converting P to F, $(F/P, i, n) = 1.7908$ for $i = 6\%$ a year and $n = 10$ years.

$$F = P(F/P, 6\%, 10) = (\$1000)(1.7908) = \boxed{\$1790.80}$$

2.

F = \$2000

t = 0 t = 10

t = 0 t = 4

P

i = 6% a year

By the formula from Table 85.1,

$$P = \frac{F}{(1+i)^n} = \frac{\$2000}{(1+0.06)^4} = \boxed{\$1584.19}$$

From the factor converting F to P, $(P/F, i, n) = 0.7921$ for $i = 6\%$ a year and $n = 4$ years.

$$\text{P} = F(P/F, 6\%, 4) = (\$2000)(0.7921) = \boxed{\$1584.20}$$

3.

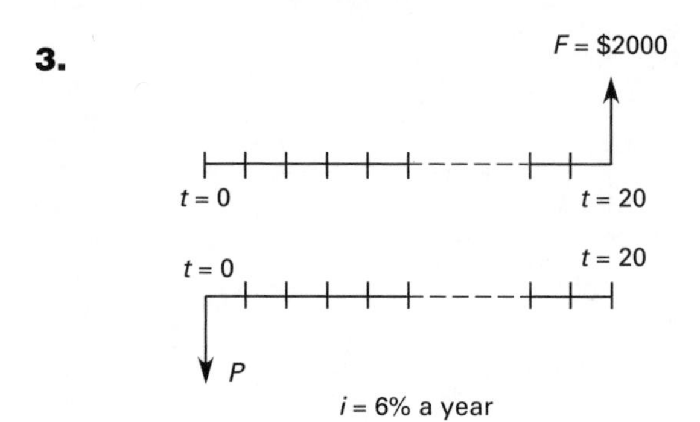

By the formula from Table 85.1,

$$P = \frac{F}{(1+i)^n} = \frac{\$2000}{(1+0.06)^{20}} = \boxed{\$623.61}$$

From the factor converting F to P, $(P/F, i, n) = 0.3118$ for $i = 6\%$ a year and $n = 20$ years.

$$P = F(P/F, 6\%, 20) = (\$2000)(0.3118) = \boxed{\$623.60}$$

4.

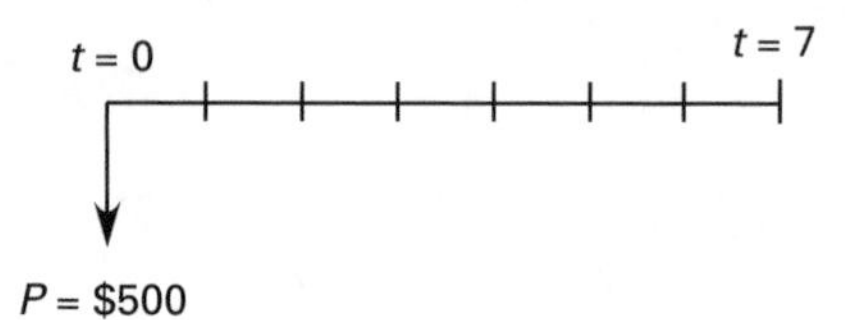

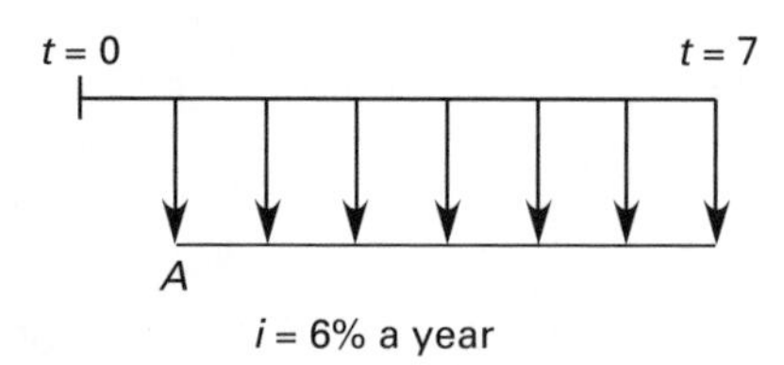

By the formula from Table 85.1,

$$A = P\left(\frac{i(1+i)^n}{(1+i)^n - 1}\right) = (\$500)\left(\frac{(0.06)(1+0.06)^7}{(1+0.06)^7 - 1}\right) = \boxed{\$89.57}$$

By the factor converting P to A, $(A/P, i, n) = 0.17914$ for $i = 6\%$ a year and $n = 7$ years.

$$A = P(A/P, 6\%, 7) = (\$500)(0.17914) = \boxed{\$89.57}$$

5.

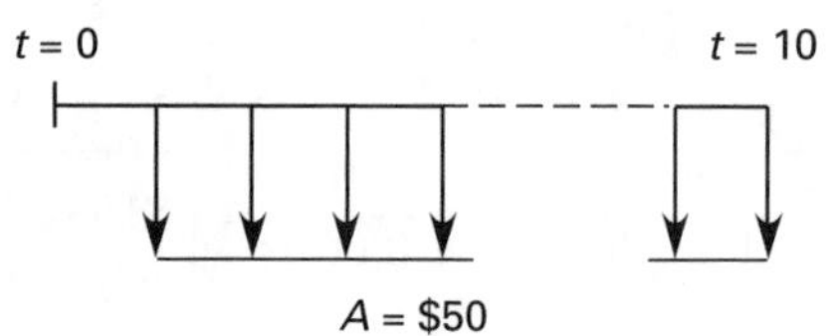

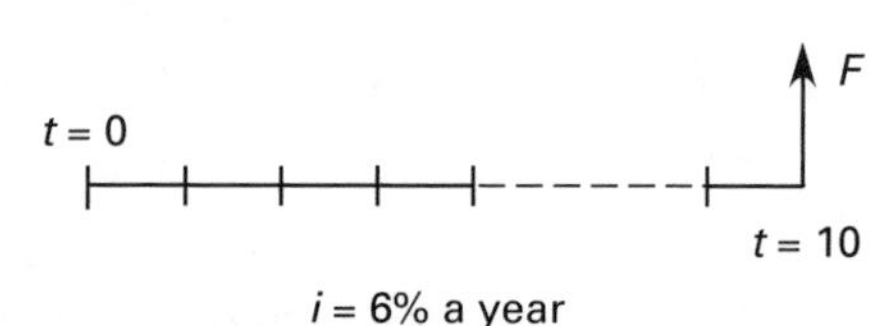

By the formula from Table 85.1,

$$F = A\left(\frac{(1+i)^n - 1}{i}\right) = (\$50)\left(\frac{(1+0.06)^{10} - 1}{0.06}\right) = \boxed{\$659.04}$$

By the factor converting A to F, $(F/A, i, n) = 13.181$ for $i = 6\%$ a year and $n = 10$ years.

$$F = A(F/A, 6\%, 10) = (\$50)(13.181) = \boxed{\$659.05}$$

6.

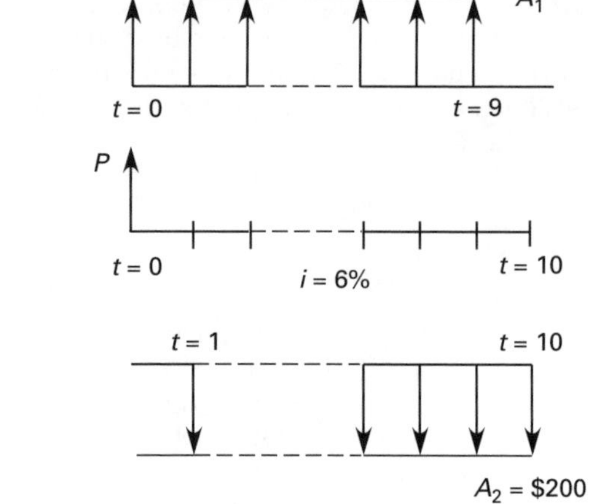

By the formula from Table 85.1, for each cash flow diagram,

$$\begin{aligned} P &= A_1 + (A_1)\left(\frac{(1+0.06)^9 - 1}{(0.06)(1+0.06)^9}\right) \\ &= A_2\left(\frac{(1+0.06)^{10} - 1}{(0.06)(1+0.06)^{10}}\right) \end{aligned}$$

Therefore, for $A_2 = \$200$,

$$\begin{aligned} A_1 + (A_1)\left(\frac{(1+0.06)^9 - 1}{(0.06)(1+0.06)^9}\right) &= (\$200)\left(\frac{(1+0.06)^{10} - 1}{(0.06)(1+0.06)^{10}}\right) \\ 7.80A_1 &= \$1472.02 \\ A_1 &= \boxed{\$188.72} \end{aligned}$$

By the factor converting A to P,

$$\begin{aligned} (P/A, 6\%, 9) &= 6.8017 \\ (P/A, 6\%, 10) &= 7.3601 \\ A_1 + A_1(6.8017) &= (\$200)(7.3601) \\ 7.8017A_1 &= \$1472.02 \\ A_1 &= \frac{\$1472.02}{7.8017} = \boxed{\$188.68} \end{aligned}$$

7.

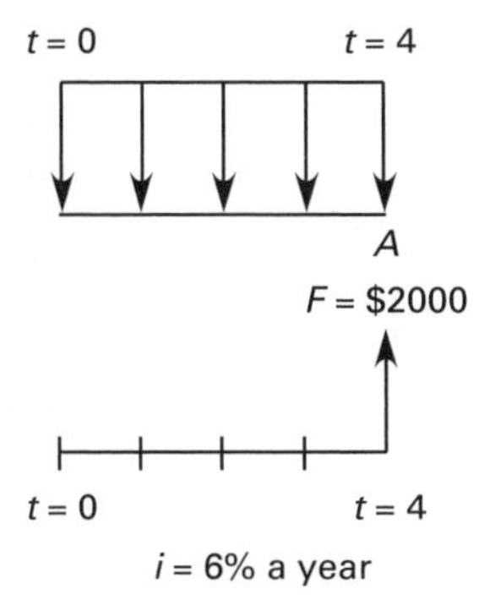

By the formula from Table 85.1,

$$F = A\left(\frac{(1+i)^n - 1}{i}\right)$$

Since the deposits start at the start of each year, $n = 4 + 1$, for a total of five deposits.

$$\begin{aligned} F &= A\left(\frac{(1+i)^{n+1} - 1}{i}\right) = \$2000 \\ &= A\left(\frac{(1+0.06)^5 - 1}{0.06}\right) \\ \$2000 &= 5.6371A \\ A &= \frac{\$2000}{5.6371} = \boxed{\$354.79} \end{aligned}$$

By the factor converting P and A to F,

$$\begin{aligned} F &= A\big((F/P, 6\%, 4) + (F/A, 6\%, 4)\big) \\ \$2000 &= A(1.2625 + 4.3746) \\ A &= \boxed{\$354.79} \end{aligned}$$

8.

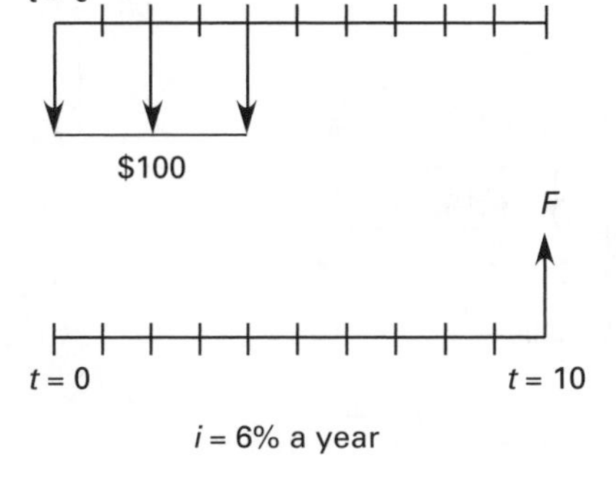

By the formula from Table 85.1, $F = P(1+i)^n$. If each deposit is considered as P, each will accumulate interest for periods of 10, 8, and 6 years.

Therefore,

$$\begin{aligned} F &= (\$100)(1+0.06)^{10} + (\$100)(1+0.06)^8 \\ &\quad + (\$100)(1+0.06)^6 \\ &= (\$100)(1.7908 + 1.5938 + 1.4185) \\ &= (\$100)(4.8031) \\ &= \boxed{\$480.31} \end{aligned}$$

By the factor converting P to F,

$$\begin{aligned} (F/P, i, n) &= 1.7908 \text{ for } i = 6\% \;\; and\ n = 10 \\ &= 1.5938 \text{ for } i = 6\% \text{ and } n = 8 \\ &= 1.4185 \text{ for } i = 6\% \text{ and } n = 6 \end{aligned}$$

By summation,

$$\begin{aligned} F &= (\$100)(1.7908 + 1.5938 + 1.4185) \\ &= (\$100)(4.8031) \\ &= \boxed{\$480.31} \end{aligned}$$

9.

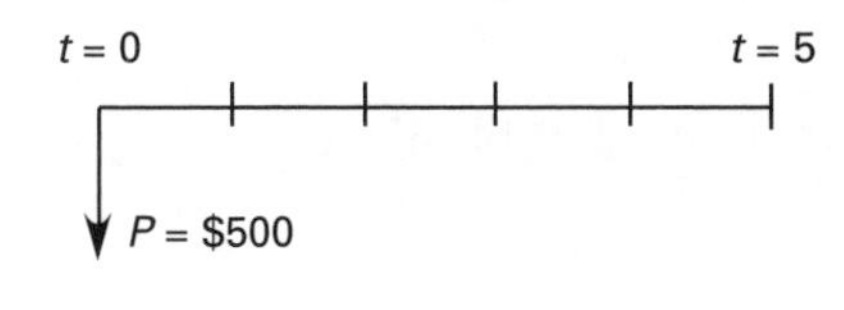

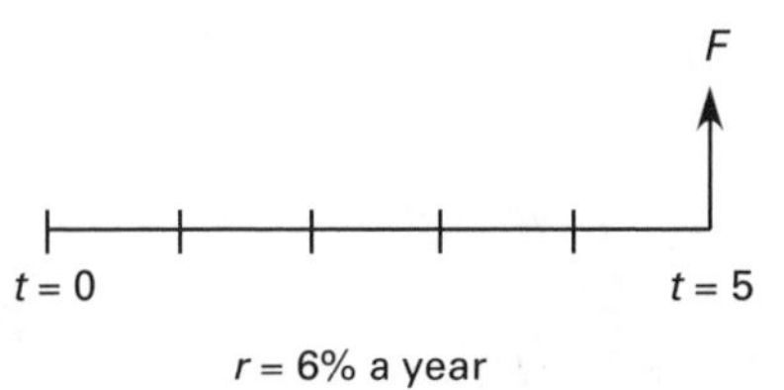

Since the deposit is compounded monthly, the effective interest rate should be calculated as shown by Eq. 85.54.

$$\begin{aligned} i &= \left(1 + \frac{r}{k}\right)^k - 1 = \left(1 + \frac{0.06}{12}\right)^{12} - 1 \\ &= 0.061677 \quad (6.1677\%) \end{aligned}$$

By the formula from Table 85.1,

$$F = P(1+i)^n = (\$500)(1 + 0.061677)^5 = \boxed{\$674.42}$$

To use a table of factors, interpolation is required.

$i\%$	factor F/P
6	1.3382
6.1677	desired
7	1.4026

Systems, Mgmt., and Professional

$$i = \left(\frac{6.1677 - 6}{7 - 6}\right)(1.4026 - 1.3382)$$
$$= 0.0108$$

Therefore,

$$F/P = 1.3382 + 0.0108 = 1.3490$$
$$F = P(F/P, 6.1677\%, 5) = (\$500)(1.3490)$$
$$= \boxed{\$674.50}$$

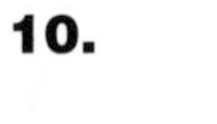

10.

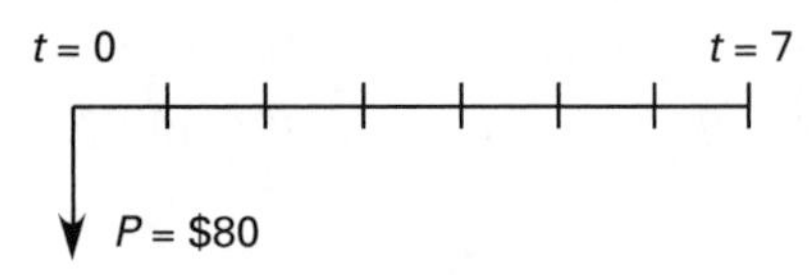

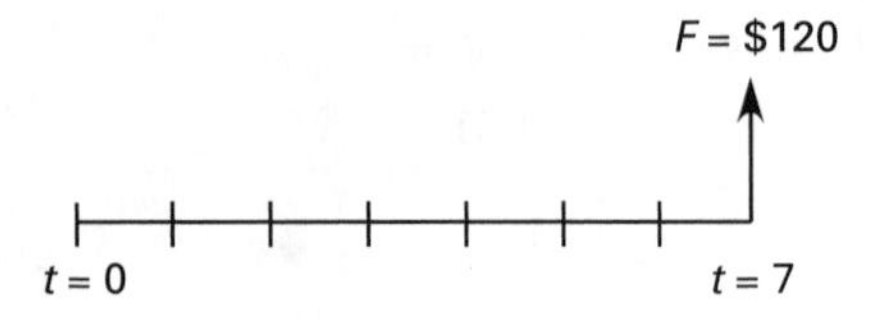

By the formula from Table 85.1,

$$F = P(1+i)^n$$

Therefore,

$$(1+i)^n = F/P$$
$$i = (F/P)^{1/n} - 1 = \left(\frac{\$120}{\$80}\right)^{1/7} - 1$$
$$= 0.059 \approx \boxed{6\%}$$

By the factor coverting P to F,

$$F = P(F/P, i\%, 7)$$
$$(F/P, i\%, 7) = F/P = \frac{\$120}{\$80} = 1.5$$

By checking the interest tables,

$$(F/P, i\%, 7) = 1.4071 \text{ for } i = 5\%$$
$$= 1.5036 \text{ for } i = 6\%$$
$$= 1.6058 \text{ for } i = 7\%$$

Therefore, $i \approx \boxed{6\%.}$

11.

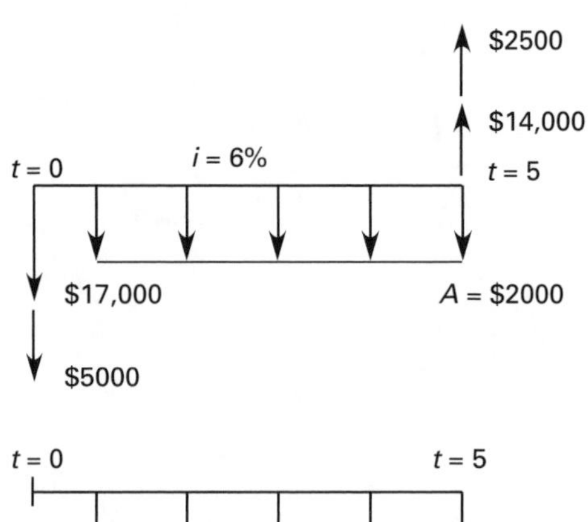

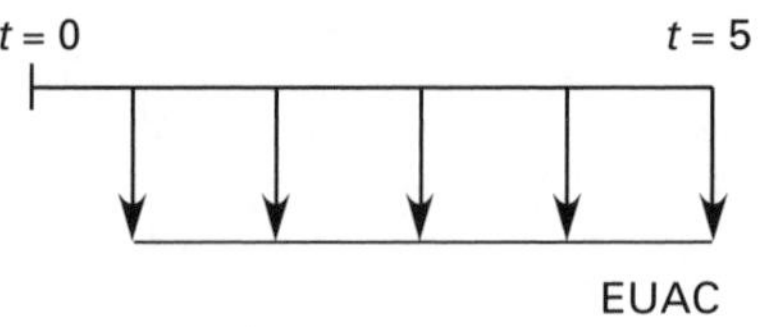

Annual cost of ownership, EUAC, can be obtained by the factors converting P to A and F to A.

$$P = \$17{,}000 + \$5000$$
$$= \$22{,}000$$
$$F = \$14{,}000 + \$2500$$
$$= \$16{,}500$$
$$\text{EUAC} = A + P(A/P, 6\%, 5) - F(A/F, 6\%, 5)$$
$$(A/P, 6\%, 5) = 0.23740$$
$$(A/F, 6\%, 5) = 0.17740$$
$$\text{EUAC} = \$2000 + (\$22{,}000)(0.23740) - (\$16{,}500)(0.17740)$$
$$= \boxed{\$4295.70}$$

12.

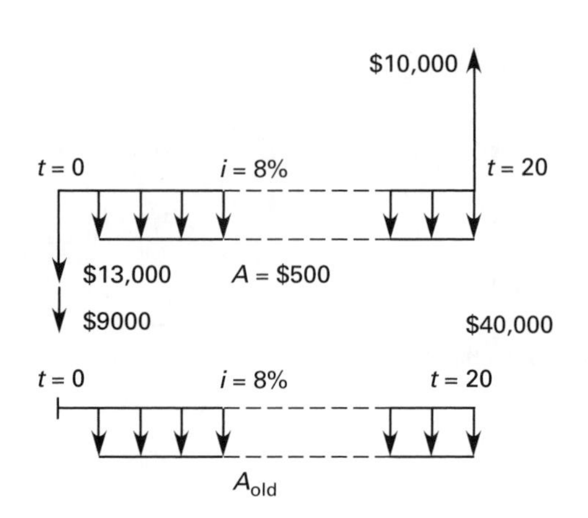

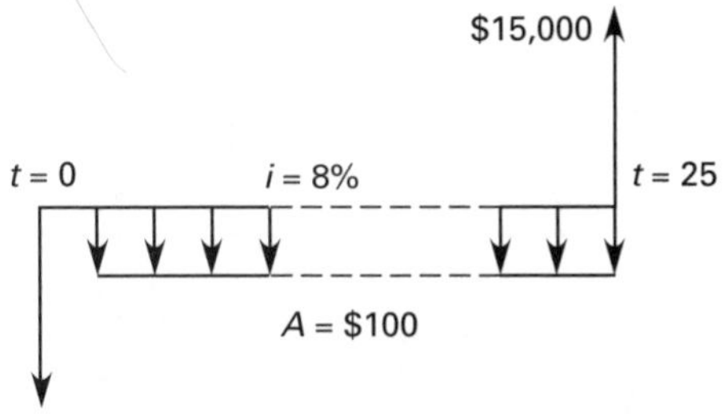

$15,000
t = 0   i = 8%   t = 25
A = $100

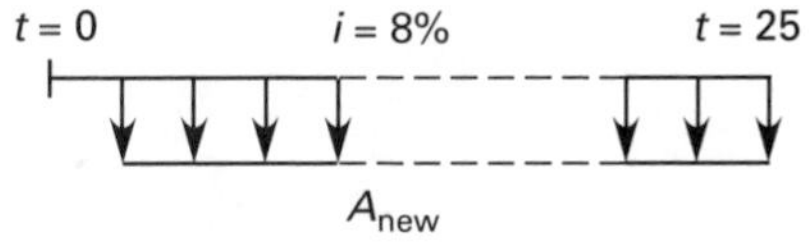

t = 0   i = 8%   t = 25
A_{new}

Consider the salvage value as a benefit lost (cost).

$$\begin{aligned} \text{EUAC}_{\text{old}} &= \$500 + (\$22{,}000)(A/P, 8\%, 20) \\ &\quad - (\$10{,}000)(A/F, 8\%, 20) \\ (A/P, 8\%, 20) &= 0.1019 \\ (A/F, 8\%, 20) &= 0.0219 \\ \text{EUAC}_{\text{old}} &= \$500 + (\$22{,}000)(0.1019) \\ &\quad - (\$10{,}000)(0.0219) \\ &= \$2522.80 \end{aligned}$$

Similarly,

$$\begin{aligned} \text{EUAC}_{\text{new}} &= \$100 + (\$40{,}000)(A/P, 8\%, 25) \\ &\quad - (\$15{,}000)(A/F, 8\%, 25) \\ (A/P, 8\%, 25) &= 0.0937 \\ (A/F, 8\%, 25) &= 0.0137 \\ \text{EUAC}_{\text{new}} &= \$100 + (\$40{,}000)(0.0937) \\ &\quad - (\$15{,}000)(0.0137) \\ &= \$3642.50 \end{aligned}$$

Therefore, the new bridge is going to be more costly.

The best alternative is to strengthen the old bridge.

13.

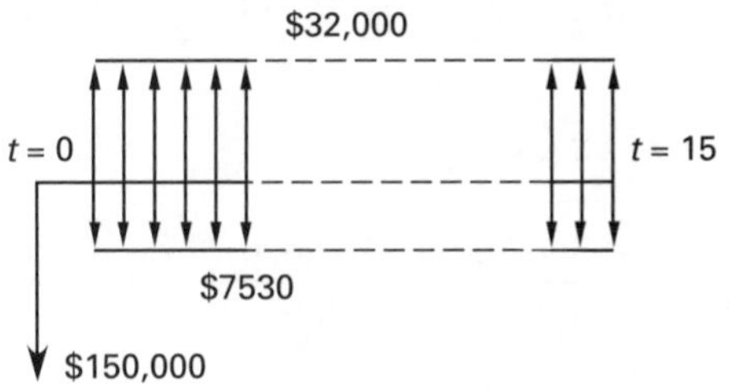

$32,000
t = 0   t = 15
$7530
$150,000

The annual depreciation is

$$\begin{aligned} D &= \frac{C - S_n}{n} = \frac{\$150{,}000}{15} \\ &= \$10{,}000/\text{year} \end{aligned}$$

The taxable income is

$$\$32{,}000 - \$7530 - \$10{,}000 = \$14{,}470/\text{year}$$

Taxes paid are

$$(\$14{,}470)(0.48) = \$6945.60/\text{year}$$

The after-tax cash flow is

$$\$24{,}470 - \$6945.60 = \$17{,}524.40$$

The present worth of the alternate is zero when evaluated at its ROR.

$$0 = -\$150{,}000 + (\$17{,}524.40)(P/A, i\%, 15)$$

Therefore,

$$(P/A, i\%, 15) = \frac{\$150{,}000}{\$17{,}524.40} = 8.55949$$

By checking the tables, this factor matches $i = 8\%$.

$$\boxed{\text{ROR} = 8\%}$$

14. The conventional benefit/cost ratio is

$$B/C = \frac{B - D}{D}$$

(a) The benefit/cost ratio will be

$$B/C = \frac{\$1{,}500{,}000 - \$300{,}000}{\$1{,}000{,}000} = \boxed{1.2}$$

(b) The excess of benefits over cost is $\boxed{\$200{,}000.}$

Systems, Mgmt., and Professional

15.

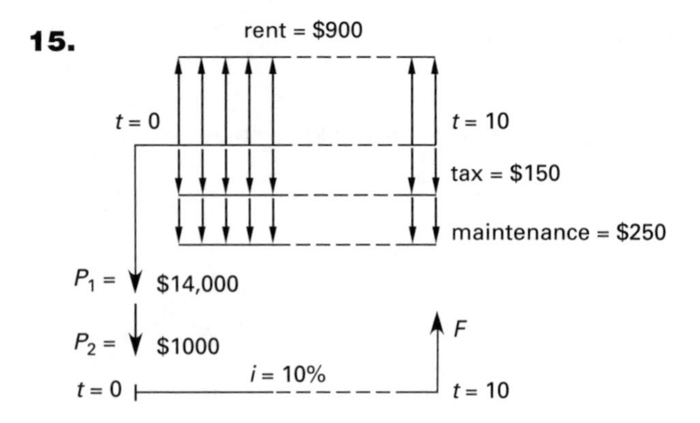

The annual rent is

$$(\$75)\left(12\ \frac{\text{months}}{\text{year}}\right) = \$900$$

$$P = P_1 + P_2 = \$15{,}000$$
$$A_1 = -\$900$$
$$A_2 = \$250 + \$150 = \$400$$

By the factors converting P to F and A to F,

$$F = (\$15{,}000)(F/P, 10\%, 10) + (\$400)(F/A, 10\%, 10) - (\$900)(F/A, 10\%, 10)$$

$$(F/P, 10\%, 10) = 2.5937$$
$$(F/A, 10\%, 10) = 15.937$$
$$F = (\$15{,}000)(2.5937) + (\$400)(15.937) - (\$900)(15.937) = \boxed{\$30{,}937}$$

16.

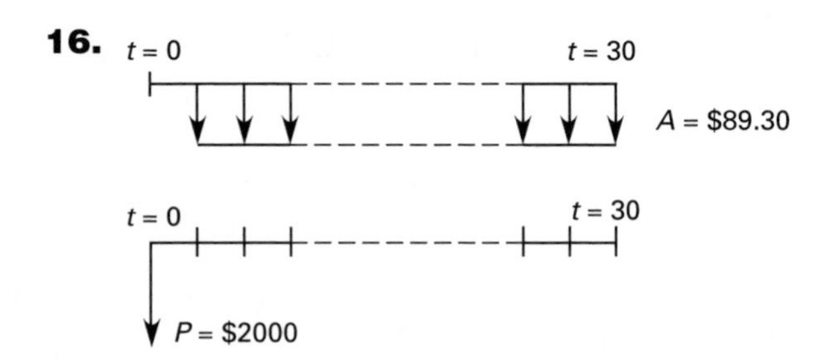

By the formula relating P to A,

$$P = A\left(\frac{(1+i)^n - 1}{i(1+i)^n}\right)$$

$$\frac{(1+i)^{30} - 1}{i(1+i)^{30}} = \frac{\$2000}{\$89.30} = 22.30$$

By trial and error,

$i(\%)$	$(1+i)^{30}$	$\frac{(1+i)^{30}-1}{i(1+i)^{30}}$
10	17.45	9.42
6	5.74	13.76
4	3.24	17.28
2	1.81	22.37

2% per month is close.

$$i = (1 + 0.02)^{12} - 1 = \boxed{0.2682\ \ (26.82\%)}$$

17. (a) Use the straight line method, Eq. 85.25.

$$D = \frac{C - S_n}{n}$$

Each year depreciation will remain the same.

$$D = \frac{\$500{,}000 - \$100{,}000}{25} = \boxed{\$16{,}000}$$

(b) Sum-of-the-years' digits (SOYD) can be calculated as shown by Eq. 85.28.

$$D_j = \frac{(C - S_n)(n - j + 1)}{T}$$

Use Eq. 85.27.

$$T = \tfrac{1}{2}n(n+1) = \left(\tfrac{1}{2}\right)(25)(25+1) = 325$$

$$D_1 = \frac{(\$500{,}000 - \$100{,}000)(25 - 1 + 1)}{325} = \boxed{\$30{,}769}$$

$$D_2 = \frac{(\$500{,}000 - \$100{,}000)(25 - 2 + 1)}{325} = \boxed{\$29{,}538}$$

$$D_3 = \frac{(\$500{,}000 - \$100{,}000)(25 - 3 + 1)}{325} = \boxed{\$28{,}308}$$

(c) The double declining balance (DDB) method can be used. By Eq. 85.32,

$$D_j = dC(1 - d)^{j-1}$$

Use Eq. 85.31.

$$d = \frac{2}{n} = \tfrac{2}{25}$$

$$D_1 = \left(\tfrac{2}{25}\right)(\$500{,}000)\left(1 - \tfrac{2}{25}\right)^0 = \boxed{\$40{,}000}$$

$$D_2 = \left(\tfrac{2}{25}\right)(\$500{,}000)\left(1 - \tfrac{2}{25}\right)^1 = \boxed{\$36{,}800}$$

$$D_3 = \left(\tfrac{2}{25}\right)(\$500{,}000)\left(1 - \tfrac{2}{25}\right)^2 = \boxed{\$33{,}856}$$

18.

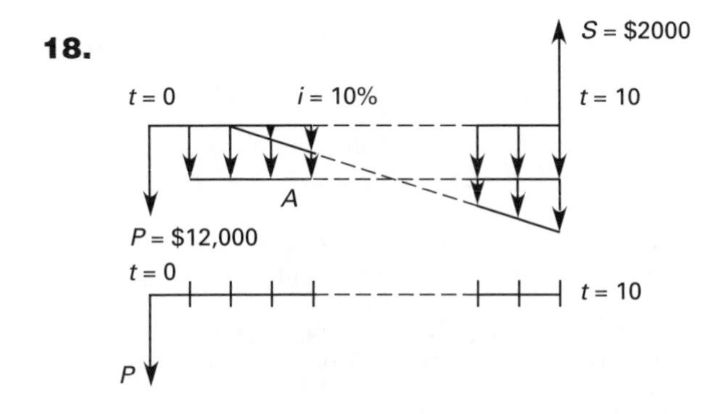

(a) $A = \$1000$ and $G = \$200$ for $t = n - 1 = 9$ years.

$$\begin{aligned} F &= S = \$2000 \\ P &= \$12{,}000 + A(P/A, 10\%, 10) + G(P/G, 10\%, 10) \\ &\quad - F(P/F, 10\%, 10) \\ &= \$12{,}000 + (\$1000)(6.1446) + (\$200)(22.8913) \\ &\quad - (\$2000)(0.3855) \\ &= \boxed{\$21{,}952} \end{aligned}$$

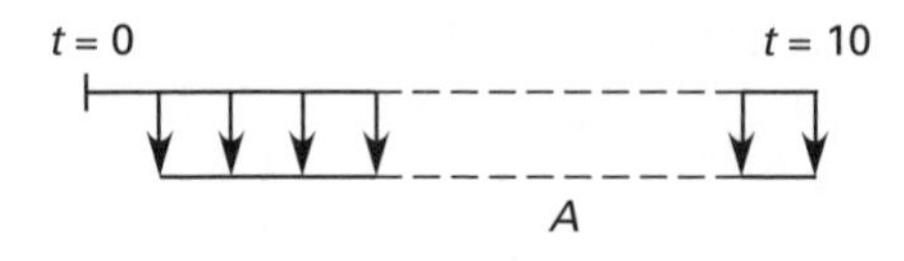

(b)
$$\begin{aligned} A &= (\$12{,}000)(A/P, 10\%, 10) + \$1000 \\ &\quad + (\$200)(A/G, 10\%, 10) \\ &\quad - (\$2000)(A/F, 10\%, 10) \\ &= (\$12{,}000)(0.1627) + \$1000 + (\$200)(3.7255) \\ &\quad - (\$2000)(0.0627) \\ &= \boxed{\$3572.70} \end{aligned}$$

19. An increase in rock removal capacity can be achieved by a 20-year loan (investment). Different cases available can be compared by equivalent uniform annual cost (EUAC).

$$\begin{aligned} \text{EUAC} &= \text{annual loan cost} \\ &\quad + \text{expected annual damage} \\ &= \text{cost } (A/P, 10\%, 20) \\ &\quad + (\$25{,}000)(\text{probability}) \\ (A/P, 10\%, 20) &= 0.1175 \end{aligned}$$

A table can be prepared for different cases.

rock removal rate	cost ($)	annual loan cost ($)	expected annual damage ($)	EUAC ($)
7	0	0	3750	3750.00
8	15,000	1761.90	2500	4261.90
9	20,000	2349.20	1750	4099.20
10	30,000	3523.80	750	4273.80

It is cheapest to do nothing.

20. Calculate the cost of owning and operating for each year.

$$\begin{aligned} A_1 &= (\$10{,}000)(A/P, 20\%, 1) + \$2000 \\ &\quad - (\$8000)(A/F, 20\%, 1) \\ (A/P, 20\%, 1) &= 1.2 \\ (A/F, 20\%, 1) &= 1.0 \\ A_1 &= (\$10{,}000)(1.2) + \$2000 - (\$8000)(1.0) \\ &= \$6000 \\ A_2 &= (\$10{,}000)(A/P, 20\%, 2) + \$2000 \\ &\quad + (\$1000)(A/G, 20\%, 2) \\ &\quad - (\$7000)(A/F, 20\%, 2) \\ (A/P, 20\%, 2) &= 0.6545 \\ (A/G, 20\%, 2) &= 0.4545 \\ (A/F, 20\%, 2) &= 0.4545 \\ A_2 &= (\$10{,}000)(0.6545) + \$2000 \\ &\quad + (\$1000)(0.4545) - (\$7000)(0.4545) \\ &= \$5818 \\ A_3 &= (\$10{,}000)(A/P, 20\%, 3) + \$2000 \\ &\quad + (\$1000)(A/G, 20\%, 3) \\ &\quad - (\$6000)(A/F, 20\%, 3) \\ (A/P, 20\%, 3) &= 0.4747 \\ (A/G, 20\%, 3) &= 0.8791 \\ (A/F, 20\%, 3) &= 0.2747 \end{aligned}$$

$$\begin{aligned} A_3 &= (\$10{,}000)(0.4747) + \$2000 \\ &\quad + (\$1000)(0.8791) \\ &\quad - (\$6000)(0.2747) \\ &= \$5977.90 \end{aligned}$$

$$\begin{aligned} A_4 &= (\$10{,}000)(A/P, 20\%, 4) \\ &\quad + \$2000 + (\$1000)(A/G, 20\%, 4) \\ &\quad - (\$5000)(A/F, 20\%, 4) \end{aligned}$$

$$(A/P, 20\%, 4) = 0.3863$$
$$(A/G, 20\%, 4) = 1.2762$$
$$(A/F, 20\%, 4) = 0.1863$$

$$\begin{aligned} A_4 &= (\$10{,}000)(0.3863) + \$2000 \\ &\quad + (\$1000)(1.2762) - (\$5000)(0.1863) \\ &= \$6207.70 \end{aligned}$$

$$\begin{aligned} A_5 &= (\$10{,}000)(A/P, 20\%, 5) + \$2000 \\ &\quad + (\$1000)(A/G, 20\%, 5) \\ &\quad - (\$4000)(A/F, 20\%, 5) \end{aligned}$$

$$(A/P, 20\%, 5) = 0.3344$$
$$(A/G, 20\%, 5) = 1.6405$$
$$(A/F, 20\%, 5) = 0.1344$$

$$\begin{aligned} A_5 &= (\$10{,}000)(0.3344) + \$2000 \\ &\quad + (\$1000)(1.6405) - (\$4000)(0.1344) \\ &= \$6446.90 \end{aligned}$$

(a) Since the annual owning and operating cost is smallest after two years of operation, it is advantageous to sell the mechanism after the second year.

The economic life is two years.

(b) After four years of operation, the owning and operating cost of the mechanism for one more year will be

$$A = \$6000 + (\$5000)(1 + i) - \$4000$$
$$i = 0.2 \quad (20\%)$$
$$A = \$6000 + (\$5000)(1.2) - \$4000$$
$$= \boxed{\$8000}$$

21. To find out if the reimbursement is adequate, calculate the business-related expense.

Charge the company for business travel.

$$\text{insurance: } \$3000 - \$2000 = \$1000$$
$$\text{maintenance: } \$2000 - \$1500 = \$500$$
$$\text{drop in salvage value: } \$10{,}000 - \$5000 = \$5000$$

The annual portion of the drop in salvage value is

$$A = (\$5000)(A/F, 10\%, 5)$$
$$(A/F, 10\%, 5) = 0.1638$$
$$A = (\$5000)(0.1638) = \$819/\text{year}$$

(a) The annual cost of gas is

$$\left(\frac{50{,}000 \text{ mi}}{15 \ \frac{\text{mi}}{\text{gal}}}\right)\left(\frac{\$1.50}{\text{gal}}\right) = \$5000$$

$$\begin{aligned} \text{EUAC per mile} &= \frac{\$1000 + \$500 + \$819 + \$5000}{50{,}000 \text{ mi}} \\ &= \boxed{\$0.14638/\text{mi}} \end{aligned}$$

Since the reimbursement per mile was \$0.30 and since $\$0.30 > \$0.14638$, the reimbursement is adequate.

(b) Next, determine (with reimbursement) how many miles the car must be driven to break even.

If the car is driven M miles per year,

$$\begin{aligned} \left(\frac{\$0.30}{1 \text{ mi}}\right) M &= (\$50{,}000)(A/P, 10\%, 5) + \$2500 \\ &\quad + \$2000 - (\$8000)(A/F, 10\%, 5) \\ &\quad + \left(\frac{M}{15 \ \frac{\text{mi}}{\text{gal}}}\right)(\$1.50) \end{aligned}$$

$$(A/P, 10\%, 5) = 0.2638$$
$$(A/F, 10\%, 5) = 0.1638$$

$$\begin{aligned} 0.3M &= (\$50{,}000)(0.2638) + \$2500 + \$2000 \\ &\quad - (\$8000)(0.1638) + 0.1M \end{aligned}$$
$$0.2M = \$16{,}379.60$$
$$M = \frac{\$16{,}379.60}{0.2 \ \frac{\$}{\text{mi}}} = \boxed{81{,}898 \text{ mi}}$$

22.

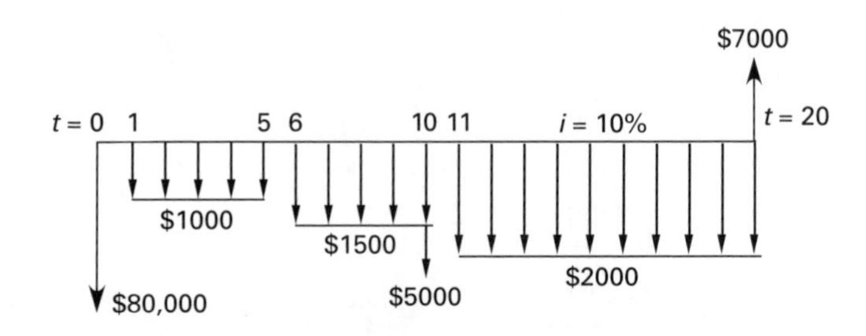

$$\begin{aligned} P_A = \$80{,}000 &+ (\$1000)(P/A, 10\%, 5) \\ &+ (\$1500)(P/A, 10\%, 5)(P/F, 10\%, 5) \\ &+ (\$2000)(P/A, 10\%, 10)(P/F, 10\%, 10) \\ &+ (\$5000)(P/F, 10\%, 10) \\ &- (\$7000)(P/F, 10\%, 20) \end{aligned}$$

$$\begin{aligned} (P/A, 10\%, 5) &= 3.7908 \\ (P/F, 10\%, 5) &= 0.6209 \\ (P/A, 10\%, 10) &= 6.1446 \\ (P/F, 10\%, 10) &= 0.3855 \\ (P/F, 10\%, 20) &= 0.1486 \\ P_A &= \$80{,}000 + (\$1000)(3.7908) \\ &\quad + (\$1500)(3.7908)(0.6209) \\ &\quad + (\$2000)(6.1446)(0.3855) \\ &\quad + (\$5000)(0.3855) - (\$7000)(0.1486) \\ &= \$92{,}946.15 \end{aligned}$$

Since the lives are different, compare by EUAC.

$$\begin{aligned} \text{EUAC}(A) &= (\$92{,}946.14)(A/P, 10\%, 20) \\ &= (\$92{,}946.14)(0.1175) = \$10{,}921 \end{aligned}$$

Similarly, evaluate alternative B.

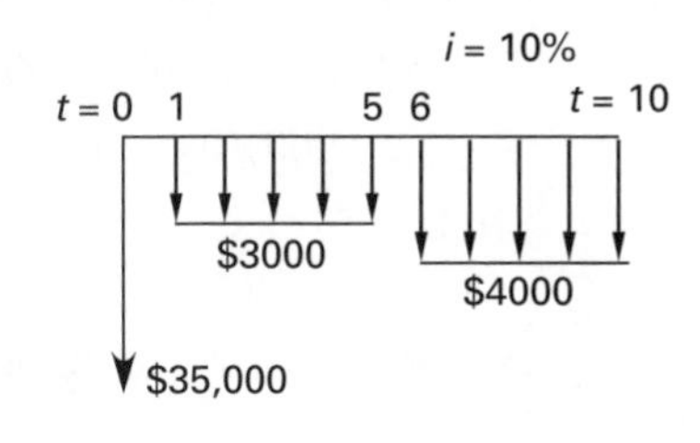

$$\begin{aligned} P_B &= \$35{,}000 + (\$3000)(P/A, 10\%, 5) \\ &\quad + (\$4000)(P/A, 10\%, 5)(P/F, 10\%, 5) \\ (P/A, 10\%, 5) &= 3.7908 \\ (P/F, 10\%, 5) &= 0.6209 \\ P_B &= \$35{,}000 + (\$3000)(3.7908) \\ &\quad + (\$4000)(3.7908)(0.6209) \\ &= \$55{,}787.23 \\ \text{EUAC}(B) &= (\$55{,}787.23)(A/P, 10\%, 10) \\ &= (\$55{,}787.23)(0.1627) = \$9077 \end{aligned}$$

Since $\text{EUAC}(B) < \text{EUAC}(A)$,

alternative B is economically superior.

23. For both cases, if the annual cost is compared with a total annual mileage of M,

$$\begin{aligned} A_A &= \$0.25M \\ A_B &= (\$30{,}000)(A/P, 10\%, 3) + \$0.14M \\ &\quad + \$500 - (\$7200)(A/F, 10\%, 3) \\ (A/P, 10\%, 3) &= 0.4021 \\ (A/F, 10\%, 3) &= 0.3021 \\ A_B &= (\$30{,}000)(0.4021) + \$0.14M + \$500 \\ &\quad - (\$7200)(0.3021) \\ &= \$12{,}063 + \$0.14M \\ &\quad + \$500 - \$2175.12 \\ &= \$10{,}387.88 + \$0.14M \end{aligned}$$

For an equal annual cost $A_A = A_B$,

$$\$0.25M = \$10{,}387.88 + \$0.14M$$

An annual mileage would be $M = 94{,}435$ mi.

For an annual mileage less than that, $A_A < A_B$.

Plan A is economically superior until that mileage is exceeded.

24. (a) Method A:

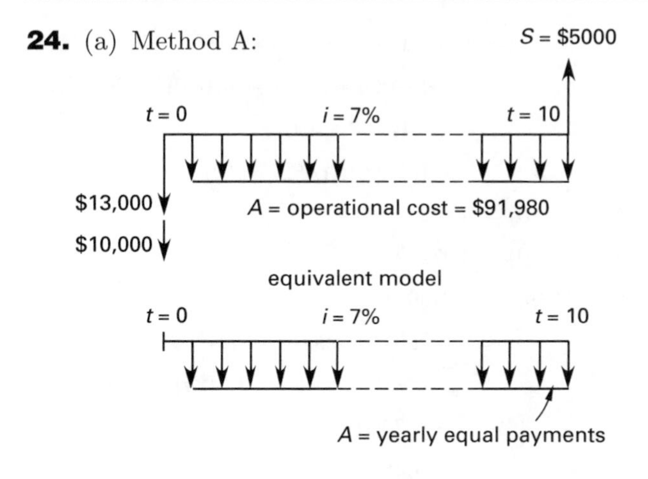

24 hr/day
365 days/yr
total of (24)(365) = 8760 hr/yr
$10.50 operational cost/hr
total of (8760)($10.50) = $91,980 operational cost/yr

$$\begin{aligned} A &= \$91{,}980 + (\$23{,}000)(A/P, 7\%, 10) \\ &\quad - (\$5000)(A/F, 7\%, 10) \\ (A/P, 7\%, 10) &= 0.1424 \\ (A/F, 7\%, 10) &= 0.0724 \\ A &= \$91{,}980 + (\$23{,}000)(0.1424) \\ &\quad - (\$5000)(0.0724) \\ &= \$94{,}893.20/\text{yr} \end{aligned}$$

Therefore, the uniform annual cost per ton each year will be

$$\frac{\$94{,}893.30}{50 \text{ ton}} = \boxed{\$1897.87}$$

Method B:

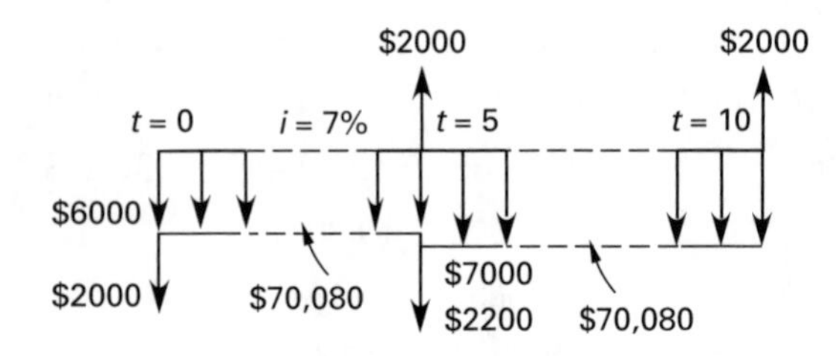

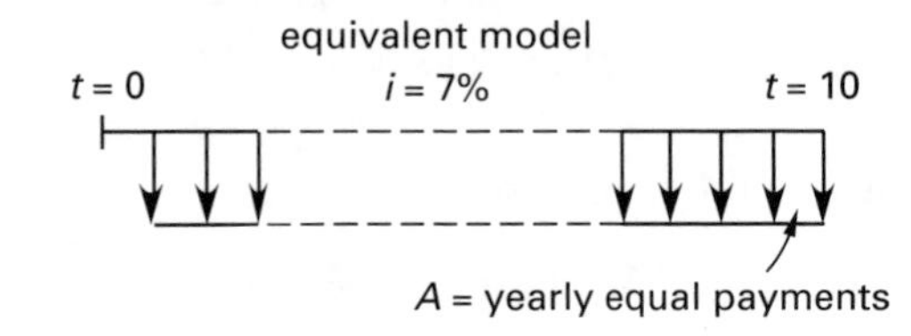

8760 hr/yr
\$8 operational cost/hr
total of \$70,080 operational cost/yr

$$\begin{aligned} A &= \$70{,}080 + (\$6000 + \$2000)(A/P, 7\%, 10) \\ &\quad + (\$7000 + \$2200 - \$2000)(P/F, 7\%, 5) \\ &\quad \times (A/P, 7\%, 10) \\ &\quad - (\$2000)(A/F, 7\%, 10) \end{aligned}$$

$$(A/P, 7\%, 10) = 0.1424$$

$$(A/F, 7\%, 10) = 0.0724$$

$$(P/F, 7\%, 5) = 0.7130$$

$$\begin{aligned} A &= \$70{,}080 + (\$8000)(0.1424) \\ &\quad + (\$7200)(0.7130)(0.1424) \\ &\quad - (\$2000)(0.0724) \\ &= \$71{,}805.42/\text{yr} \end{aligned}$$

Therefore, the uniform annual cost per ton each year will be

$$\frac{\$71{,}805.42}{20 \text{ ton}} = \boxed{\$3590.27}$$

(b)

tons/yr	cost of using A	cost of using B	cheapest
0–20	\$94,893 (1x)	\$71,805 (1x)	B
20–40	\$94,893 (1x)	\$143,610 (2x)	A
40–50	\$94,893 (1x)	\$215,415 (3x)	A
50–60	\$189,786 (2x)	\$215,415 (3x)	A
60–80	\$189,786 (2x)	\$287,220 (4x)	A

25.

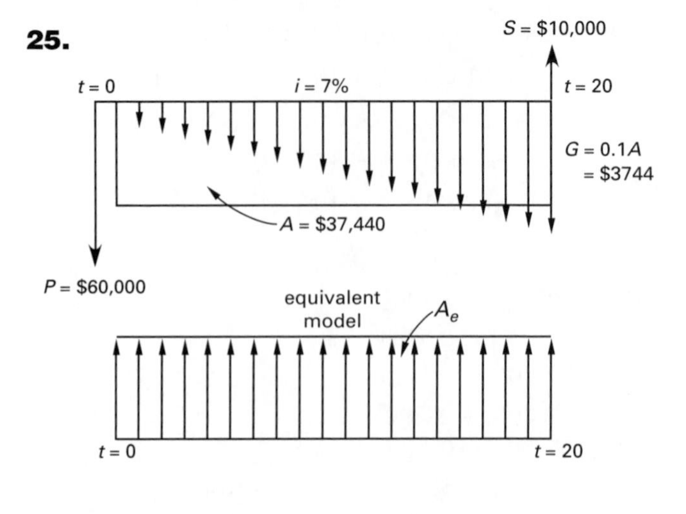

$$\begin{aligned} A_e &= (\$60{,}000)(A/P, 7\%, 20) + A \\ &\quad + G(P/G, 7\%, 20)(A/P, 7\%, 20) \\ &\quad - (\$10{,}000)(A/F, 7\%, 20) \end{aligned}$$

$$(A/P, 7\%, 20) = 0.0944$$

$$A = (37{,}440 \text{ mi})\left(\frac{\$1.0}{\text{mi}}\right) = \$37{,}440$$

$$G = 0.1A = (0.1)(\$37{,}440) = \$3744$$

$$(P/G, 7\%, 20) = 77.5091$$

$$(A/F, 7\%, 20) = 0.0244$$

$$\begin{aligned} A_e &= (\$60{,}000)(0.0944) + \$37{,}440 \\ &\quad + (\$3744)(77.509)(0.0944) \\ &\quad - (\$10{,}000)(0.0244) \\ &= \$70{,}253.78 \end{aligned}$$

(a) With 80,000 passengers a year, the break-even fare per passenger would be

$$\text{fare} = \frac{A_e}{80{,}000} = \frac{\$70{,}253.78}{80{,}000} = \boxed{\$0.878/\text{passenger}}$$

(b)

$$\begin{aligned} \$0.878 &= \$0.35 + G(A/G, 7\%, 20) \\ G &= \frac{\$0.878 - \$0.35}{7.3163} \\ &= \boxed{\$0.072 \text{ increase per year}} \end{aligned}$$

(c) As in part (b), the subsidy should be

$$\text{subsidy} = \text{cost} - \text{revenue}$$

$$\begin{aligned} P &= \$0.878 - \big(\$0.35 + (\$0.05)(A/G, 7\%, 20)\big) \\ &= \$0.878 - \big(\$0.35 + (\$0.05)(7.3163)\big) \\ &= \boxed{\$0.162} \end{aligned}$$

26.

$$\begin{aligned} P(\text{A}) &= (\$4800)(P/A, 12\%, 25) \\ &= (\$4800)(7.8431) \\ &= \$37{,}646.88 \end{aligned}$$

(4 quarters)(25 years) = 100 compounding periods

$$\begin{aligned} P(\text{B}) &= (\$1200)(P/A, 3\%, 100) \\ &= (\$1200)(31.5989) \\ &= \$37{,}918.68 \end{aligned}$$

Alternative B is economically superior.

27.

$$\begin{aligned} \text{EUAC(A)} &= (\$1500)(A/P, 15\%, 5) + \$800 \\ &= (\$1500)(0.2983) + \$800 \\ &= \$1247.45 \\ \text{EUAC(B)} &= (\$2500)(A/P, 15\%, 8) + \$650 \\ &= (\$2500)(0.2229) + \$650 \\ &= \$1207.25 \end{aligned}$$

Alternative B is economically superior.

28. The data given imply that both investments return 4% or more. However, the increased investment of \$30,000 may not be cost effective. Do an incremental analysis.

incremental cost = \$70,000 – \$40,000 = \$30,000

incremental income = \$5620 – \$4075 = \$1545

$$0 = -\$30{,}000 + (\$1545)(P/A, i\%, 20)$$

$$(P/A, i\%, 20) = 19.417$$

$$i \approx 0.25\% < 4\%$$

Alternative B is economically superior.

(The same conclusion could be reached by taking the present worths of both alternatives at 4%.)

29. (a)

$$\begin{aligned} P(\text{A}) &= (-\$3000)(P/A, 25\%, 25) - \$20{,}000 \\ &= (-\$3000)(3.9849)\$20{,}000 \\ &= -\$31{,}954.70 \\ P(\text{B}) &= (-\$2500)(3.9849)\$25{,}000 \\ &= -\$34{,}962.25 \end{aligned}$$

A is better.

(b)

$$\text{CC(A)} = \$20{,}000 + \frac{\$3000}{0.25} = \$32{,}000$$

$$\text{CC(B)} = \$25{,}000 + \frac{\$2500}{0.25} = \$35{,}000$$

A is better.

(c)

$$\begin{aligned} \text{EUAC(A)} &= (\$20{,}000)(A/P, 25\%, 25) + \$3000 \\ &= (\$20{,}000)(0.2509) + \$3000 \\ &= \$8018.00 \\ \text{EUAC(B)} &= (\$25{,}000)(0.2509) + \$2500 \\ &= \$8772.50 \end{aligned}$$

A is better.

30. (a)

$$\text{BV} = \$18{,}000 - (5)\left(\frac{\$18{,}000 - \$2000}{8}\right) = \boxed{\$8000}$$

(b) With the sinking fund method, the basis is

$$\begin{aligned} &(\$18{,}000 - \$2000)(A/F, 8\%, 8) \\ &= (\$18{,}000 - \$2000)(0.0940) \\ &= \$1504 \end{aligned}$$

$$D_1 = (\$1504)(1.000) = \boxed{\$1504}$$

$$D_2 = (\$1504)(1.0800) = \boxed{\$1624}$$

$$D_3 = (\$1504)(1.0800)^2 = (\$1504)(1.1664) = \boxed{\$1754}$$

(c)

$$D_1 = \left(\tfrac{2}{8}\right)(\$18{,}000) = \boxed{\$4500}$$

$$D_2 = \left(\tfrac{2}{8}\right)(\$18{,}000 - \$4500) = \boxed{\$3375}$$

$$D_3 = \left(\tfrac{2}{8}\right)(\$18{,}000 - \$4500 - \$3375) = \boxed{\$2531}$$

$$D_4 = \left(\tfrac{2}{8}\right)(\$18{,}000 - \$4500 - \$3375 - \$2531) = \boxed{\$1898}$$

$$D_5 = \left(\tfrac{2}{8}\right)(\$18{,}000 - \$4500 - \$3375 - \$2531 - \$1898) = \boxed{\$1424}$$

$$\text{BV} = \$18{,}000 - \$4500 - \$3375 - \$2531 - \$1898 - \$1424 = \boxed{\$4272}$$

31.

$$T = \left(\tfrac{1}{2}\right)(8)(9) = 36$$
$$D_1 = \left(\tfrac{8}{36}\right)(\$14{,}000 - \$1800) = \$2711$$
$$\Delta D = \left(\tfrac{1}{36}\right)(\$14{,}000 - \$1800) = \$339$$
$$D_2 = \$2711 - \$339 = \$2372$$
$$\begin{aligned}\text{DR} &= (0.52)(\$2711)(P/A, 15\%, 8) \\ &\quad - (0.52)(\$339)(P/G, 15\%, 8) \\ &= (0.52)(\$2711)(4.4873) \\ &\quad - (0.52)(\$339)(12.4807) \\ &= \boxed{\$4125.74}\end{aligned}$$

32.

$$\begin{aligned}\text{EUAC}_{5\text{ ft}} &= (\$600{,}000)(A/P, 12\%, 15) \\ &\quad + \left(\tfrac{14}{50}\right)(\$600{,}000) + \left(\tfrac{8}{50}\right)(\$650{,}000) \\ &\quad + \left(\tfrac{3}{50}\right)(\$700{,}000) + \left(\tfrac{1}{50}\right)(\$800{,}000) \\ &= \$418{,}080\end{aligned}$$

$$\begin{aligned}\text{EUAC}_{10\text{ ft}} &= (\$710{,}000)(A/P, 12\%, 15) \\ &\quad + \left(\tfrac{8}{50}\right)(\$650{,}000) + \left(\tfrac{3}{50}\right)(\$700{,}000) \\ &\quad + \left(\tfrac{1}{50}\right)(\$800{,}000) \\ &= \$266{,}228\end{aligned}$$

$$\begin{aligned}\text{EUAC}_{15\text{ ft}} &= (\$900{,}000)(A/P, 12\%, 15) \\ &\quad + \left(\tfrac{3}{50}\right)(\$700{,}000) + \left(\tfrac{1}{50}\right)(\$800{,}000) \\ &= \$190{,}120\end{aligned}$$

$$\begin{aligned}\text{EUAC}_{20\text{ ft}} &= (\$1{,}000{,}000)(A/P, 12\%, 15) \\ &\quad + \left(\tfrac{1}{50}\right)(\$800{,}000) \\ &= \$162{,}800\end{aligned}$$

$$\boxed{\text{Build to 20 ft.}}$$

33. Assume replacement after 1 year.

$$\begin{aligned}\text{EUAC}(1) &= (\$30{,}000)(A/P, 12\%, 1) + \$18{,}700 \\ &= (\$30{,}000)(1.12) + \$18{,}700 = \$52{,}300\end{aligned}$$

Assume replacement after 2 years.

$$\begin{aligned}\text{EUAC}(2) &= (\$30{,}000)(A/P, 12\%, 2) \\ &\quad + \$18{,}700 + (\$1200)(A/G, 12\%, 2) \\ &= (\$30{,}000)(0.5917) + \$18{,}700 \\ &\quad + (\$1200)(0.4717) = \$37{,}017\end{aligned}$$

Assume replacement after 3 years.

$$\begin{aligned}\text{EUAC}(3) &= (\$30{,}000)(A/P, 12\%, 3) \\ &\quad + \$18{,}700 + (\$1200)(A/G, 12\%, 3) \\ &= (\$30{,}000)(0.4163) + \$18{,}700 \\ &\quad + (\$1200)(0.9246) = \$32{,}299\end{aligned}$$

Similarly, calculate to obtain the numbers in the following table.

years in service	EUAC
1	\$52,300
2	\$37,017
3	\$32,299
4	\$30,207
5	\$29,152
6	\$28,602
7	\$28,335
8	\$28,234
9	\$28,240
10	\$28,312

$$\boxed{\text{Replace after 8 yr.}}$$

34. Assume the head and horsepower data are already reflected in the hourly operating costs.

Let N = no. of hours operated each year.

$$\begin{aligned}\text{EUAC(A)} &= (\$3600 + \$3050)(A/P, 12\%, 10) \\ &\quad - (\$200)(A/F, 12\%, 10) + 0.30N \\ &= (\$6650)(0.1770) - (\$200)(0.0570) + 0.30N \\ &= 1165.65 + 0.30N\end{aligned}$$

$$\begin{aligned}\text{EUAC(B)} &= (\$2800 + \$5010)(A/P, 12\%, 10) \\ &\quad - (\$280)(A/F, 12\%, 10) + 0.10N \\ &= (\$7810)(0.1770) - (\$280)(0.0570) + 0.10N \\ &= 1366.41 + 0.10N\end{aligned}$$

$$\text{EUAC(A)} = \text{EUAC(B)}$$
$$1165.65 + 0.30N = 1366.41 + 0.10N$$
$$N = \boxed{1003.8 \text{ hr}}$$

35. (a) From Eq. 85.78,

$$\begin{aligned}\frac{T_2}{T_1} &= 0.88 = 2^{-b} \\ \log 0.88 &= -b \log 2 \\ -0.0555 &= -(0.3010)b \\ b &= 0.1843\end{aligned}$$

$$T_4 = (6)(4)^{-0.1843} = \boxed{4.65 \text{ wk}}$$

(b) From Eq. 85.79,

$$\begin{aligned} T_{6\text{-}14} &= \left(\frac{6}{1-0.1843}\right) \\ &\quad\times \left(\left(14+\tfrac{1}{2}\right)^{1-0.1843} - \left(6-\tfrac{1}{2}\right)^{1-0.1843}\right) \\ &= \left(\frac{6}{0.8157}\right)(8.857-4.017) \\ &= \boxed{35.6 \text{ wk}} \end{aligned}$$

36. First check that both alternatives have an ROR greater than the MARR. Work in thousands of dollars.

Evaluate alternative A.

$$\begin{aligned} P(\text{A}) &= -\$120 + (\$15)(P/F, i\%, 5) \\ &\quad + (\$57)(P/A, i\%, 5)(1-0.45) \\ &\quad + \left(\frac{\$120-\$15}{5}\right)(P/A, i\%, 5)(0.45) \\ &= -\$120 + (\$15)(P/F, i\%, 5) \\ &\quad + (\$40.8)(P/A, i\%, 5) \end{aligned}$$

Try 15%.

$$\begin{aligned} P(\text{A}) &= -\$120 + (\$15)(0.4972) + (\$40.8)(3.3522) \\ &= \$24.23 \end{aligned}$$

Try 25%.

$$\begin{aligned} P(\text{A}) &= -\$120 + (\$15)(0.3277) + (\$40.8)(2.6893) \\ &= -\$5.36 \end{aligned}$$

Since $P(\text{A})$ goes through 0,

$$(\text{ROR})_\text{A} > \text{MARR} = 15\%$$

Next, evaluate alternative B.

$$\begin{aligned} P(\text{B}) &= -\$170 + (\$20)(P/F, i\%, 5) \\ &\quad + (\$67)(P/A, i\%, 5)(1-0.45) \\ &\quad + \left(\frac{\$170-\$20}{5}\right)(P/A, i\%, 5)(0.45) \\ &= -\$170 + (\$20)(P/F, i\%, 5) \\ &\quad + (\$50.35)(P/A, i\%, 5) \end{aligned}$$

Try 15%.

$$\begin{aligned} P(\text{B}) &= -\$170 + (\$20)(0.4972) + (\$50.35)(3.352) \\ &= \$8.72 \end{aligned}$$

Since $P(\text{B}) > 0$ and will decrease as i increases,

$$(\text{ROR})_\text{B} > 15\%$$

ROR > MARR for both alternatives.

Do an incremental analysis to see if it is worthwhile to invest the extra $170 − $120 = $50.

$$\begin{aligned} P(\text{B}-\text{A}) &= -\$50 + (\$20-\$15)(P/F, i\%, 5) \\ &\quad + (\$50.35-\$40.8)(P/A, i\%, 5) \end{aligned}$$

Try 15%.

$$\begin{aligned} P(\text{B}-\text{A}) &= -\$50 + (\$5)(0.4972) \\ &\quad + (\$9.55)(3.3522) \\ &= -\$15.50 \end{aligned}$$

Since $P(\text{B}-\text{A}) < 0$ and would become more negative as i increases, the ROR of the added investment is < 15%.

$$\boxed{\text{Alternative A is superior.}}$$

37. Use the year-end convention with the tax credit. The purchase is made at $t = 0$. However, the tax credit is received at $t = 1$ and must be multiplied by $(P/F, i\%, 1)$.

(Note that 0.0667 is actually 2/3 of 10%.)

$$\begin{aligned} P &= -\$300{,}000 + (0.0667)(\$300{,}000)(P/F, i\%, 1) \\ &\quad + (\$90{,}000)(P/A, i\%, 5)(1-0.48) \\ &\quad + \left(\frac{\$300{,}000-\$50{,}000}{5}\right)(P/A, i\%, 5)(0.48) \\ &\quad + (\$50{,}000)(P/F, i\%, 5) \\ &= -\$300{,}000 + (\$20{,}000)(P/F, i\%, 1) \\ &\quad + (\$46{,}800)(P/A, i\%, 5) \\ &\quad + (\$24{,}000)(P/A, i\%, 5) \\ &\quad + (\$50{,}000)(P/F, i\%, 5) \end{aligned}$$

By trial and error,

i	P
10%	$17,616
15%	−$20,412
12%	$1448
13%	−$6142
$12\frac{1}{4}$%	−$479

$$\boxed{i \text{ is between } 12\% \text{ and } 12^1/_4\%.}$$

38. Assume loan payments are made at the end of each year. Find the annual payment.

$$\begin{aligned} \text{payment} &= (\$2{,}500{,}000)(A/P, 12\%, 25) \\ &= (\$2{,}500{,}000)(0.1275) \\ &= \$318{,}750 \\ \text{distributed profit} &= (0.15)(\$2{,}500{,}000) \\ &= \$375{,}000 \end{aligned}$$

After paying all expenses and distributing the 15% profit, the remainder should be 0.

$$\begin{aligned}0 = \text{EUAC} &= \$20{,}000 + \$50{,}000 + \$200{,}000 \\ &\quad + \$375{,}000 + \$318{,}750 - \text{annual receipts} \\ &\quad - (\$500{,}000)(A/F, 15\%, 25) \\ &= \$963{,}750 - \text{annual receipts} \\ &\quad - (\$500{,}000)(0.0047)\end{aligned}$$

This calculation assumes $i = 15\%$, which equals the desired return. However, this assumption only affects the salvage calculation, and since the number is so small, the analysis is not sensitive to the assumption.

$$\text{annual receipts} = \$961{,}400$$

The average daily receipts are

$$\frac{\$961{,}400}{365} = \$2634$$

Use the expected value approach. The average occupancy is

$$\begin{aligned}&(0.40)(0.65) + (0.30)(0.70) + (0.20)(0.75) \\ &\quad + (0.10)(0.80) = 0.70\end{aligned}$$

The average number of rooms occupied each night is

$$(0.70)(120 \text{ rooms}) = 84 \text{ rooms}$$

The minimum required average daily rate per room is

$$\frac{\$2634}{84} = \boxed{\$31.36}$$

39.

$$\begin{aligned}\text{annual savings} &= \left(\frac{0.69 - 0.47}{1000}\right)(\$3{,}500{,}000) = \$770 \\ P &= -\$7500 + (\$770 - \$200 - \$100) \\ &\quad \times (P/A, i\%, 25) = 0 \\ (P/A, i\%, 25) &= 15.957\end{aligned}$$

Searching the tables and interpolating,

$$i \approx \boxed{3.75\%}$$

40. Work in millions of dollars.

$$\begin{aligned}P(\text{A}) &= -(\$1.3)(1 - 0.48)(P/A, 15\%, 6) \\ &= (\$1.3)(0.52)(3.7845) \\ &= -\$2.56 \quad [\text{millions}]\end{aligned}$$

Since this is an after-tax analysis and since the salvage value was mentioned, assume that the improvements can be depreciated.

Use straight line depreciation.

Evaluate alternative B.

$$\begin{aligned}D_j &= \tfrac{2}{6} = 0.333 \\ P(\text{B}) &= -\$2 - (\$0.20)(\$1.3)(1 - 0.48)(P/A, 15\%, 6) \\ &\quad - (\$0.15)(1 - 0.48)(P/A, 15\%, 6) \\ &\quad + (\$0.333)(0.48)(P/A, 15\%, 6) \\ &= -\$2 - (\$0.20)(\$1.3)(0.52)(3.7845) \\ &\quad - (\$0.15)(0.52)(3.7845) \\ &\quad + (\$0.333)(0.48)(3.7845) \\ &= -\$2.206 \quad [\text{millions}]\end{aligned}$$

Next, evaluate alternative C.

$$\begin{aligned}D_j &= \frac{1.2}{3} = 0.4 \\ P(\text{C}) &= -(\$1.2)\big(1 + (P/F, 15\%, 3)\big) \\ &\quad - (\$0.20)(\$1.3)(1 - 0.48) \\ &\quad \times \big((P/F, 15\%, 1) + (P/F, 15\%, 4)\big) \\ &\quad - (\$0.45)(\$1.3)(1 - 0.48) \\ &\quad \times \big((P/F, 15\%, 2) + (P/F, 15\%, 5)\big) \\ &\quad - (\$0.80)(\$1.3)(1 - 0.48) \\ &\quad \times \big((P/F, 15\%, 3) + (P/F, 15\%, 6)\big) \\ &\quad + (\$0.4)(\$0.48)(P/A, 15\%, 6) \\ &= -(\$1.2)(1.6575) \\ &\quad - (\$0.20)(\$1.3)(0.52)(0.8696 + 0.5718) \\ &\quad - (\$0.45)(\$1.3)(0.52)(0.7561 + 0.4972) \\ &\quad - (\$0.80)(\$1.3)(0.52)(0.6575 + 0.4323) \\ &\quad + (\$0.4)(0.48)(3.7845) \\ &= -\$2.436 \quad [\text{millions}]\end{aligned}$$

$$\boxed{\text{Alternative B is superior.}}$$

41. This is a replacement study. Since production capacity and efficiency are not a problem with the defender, the only question is when to bring in the challenger.

Since this is a before-tax problem, depreciation is not a factor, nor is book value.

The cost of keeping the defender one more year is

$$\begin{aligned}\text{EUAC(defender)} &= \$200{,}000 + (0.15)(\$400{,}000) \\ &= \$260{,}000\end{aligned}$$

For the challenger,

$$\begin{aligned}\text{EUAC(challenger)}&\\ &= (\$800{,}000)(A/P, 15\%, 10) + \$40{,}000\\ &\quad + (\$30{,}000)(A/G, 15\%, 10)\\ &\quad - (\$100{,}000)(A/F, 15\%, 10)\\ &= (\$800{,}000)(0.1993) + \$40{,}000\\ &\quad + (\$30{,}000)(3.3832)\\ &\quad - (\$100{,}000)(0.0493)\\ &= \$296{,}006\end{aligned}$$

Since the defender is cheaper, keep it. The same analysis next year will give identical answers. Therefore, keep the defender for the next 3 years, at which time the decision to buy the challenger will be automatic.

Having determined that it is less expensive to keep the defender than to maintain the challenger for 10 years, determine whether the challenger is less expensive if retired before 10 years.

If retired in 9 years,

$$\begin{aligned}\text{EUAC(challenger)} &= (\$800{,}000)(A/P, 15\%, 9) + \$40{,}000\\ &\quad + (\$30{,}000)(A/G, 15\%, 9)\\ &\quad - (\$150{,}000)(A/F, 15\%, 9)\\ &= (\$800{,}000)(0.2096)\\ &\quad + \$40{,}000 + (\$30{,}000)(3.0922)\\ &\quad - (\$150{,}000)(0.0596)\\ &= \$291{,}506\end{aligned}$$

Similar calculations yield the following results for all the retirement dates.

n	EUAC
10	$296,000
9	$291,506
8	$287,179
7	$283,214
6	$280,016
5	$278,419
4	$279,909
3	$288,013
2	$313,483
1	$360,000

Since none of these equivalent uniform annual costs is less than that of the defender, it is not economical to buy and keep the challenger for any length of time.

Keep the defender.

Systems, Mgmt., and Professional

86 Engineering Law

PRACTICE PROBLEMS

1. List the different forms of company ownership. What are the advantages and disadvantages of each?

2. Define the requirements for a contract to be enforceable.

3. What standard features should a written contract include?

4. Describe the ways a consulting fee can be structured.

5. What is a retainer fee?

SOLUTIONS

1. The three different forms of company ownership are the (1) sole proprietorship, (2) partnership, and (3) corporation.

A *sole proprietor* is his or her own boss. This satisfies the proprietor's ego and facilitates quick decisions, but unless the proprietor is trained in business, the company will usually operate without the benefit of expert or mitigating advice. The sole proprietor also personally assumes all the debts and liabilities of the company. A sole proprietorship is terminated upon the death of the proprietor.

A *partnership* increases the capitalization and the knowledge base beyond that of a proprietorship, but offers little else in the way of improvement. In fact, the partnership creates an additional disadvantage of one partner's possible irresponsible actions creating debts and liabilities for the remaining partners.

A *corporation* has sizable capitalization (provided by the stockholders) and a vast knowledge base (provided by the board of directors). It keeps the company and owner liability separate. It also survives the death of any employee, officer, or director. Its major disadvantage is the administrative work required to establish and maintain the corporate structure.

2. To be legal, a contract must contain an *offer*, some form of *consideration* (which does not have to be equitable), and an *acceptance* by both parties. To be enforceable, the contract must be voluntarily entered into, both parties must be competent and of legal age, and the contract cannot be for illegal activities.

3. A written contract will identify both parties, state the purpose of the contract and the obligations of the parties, give specific details of the obligations (including relevant dates and deadlines), specify the consideration, state the boilerplate clauses to clarify the contract terms, and leave places for signatures.

4. A consultant will either charge a fixed fee, a variable fee, or some combination of the two. A one-time fixed fee is known as a *lump-sum fee*. In a *cost plus fixed fee* contract, the consultant will also pass on certain costs to the client. Some charges to the client may depend on other factors, such as the salary of the consultant's staff, the number of days the consultant works, or the eventual cost or value of an item being designed by the consultant.

5. A *retainer* is a (usually) nonreturnable advance paid by the client to the consultant. While the retainer may be intended to cover the consultant's initial expenses until the first big billing is sent out, there does not need to be any rational basis for the retainer. Often, a small retainer is used by the consultant to qualify the client (i.e., to make sure the client is not just shopping around and getting free initial consultations) and as a security deposit (to make sure the client does not change consultants after work begins).

87 Engineering Ethics

PRACTICE PROBLEMS

(Note: Each problem has two parts. Determine whether the situation is (or can be) permitted legally. Then, determine whether the situation is permitted ethically.)

1. (a) Was it legal and/or ethical for an engineer to sign and seal plans that were not prepared by him or prepared under his responsible direction, supervision, or control?

(b) Was it legal and/or ethical for an engineer to sign and seal plans that were not prepared by him but were prepared under his responsible direction, supervision, and control?

2. Under what conditions would it be legal and/or ethical for an engineer to rely on the information (e.g., elevations and amounts of cuts and fills) furnished by a grading contractor?

3. Was it legal and/or ethical for an engineer to alter the soils report prepared by another engineer for his client?

4. Under what conditions would it be legal and/or ethical for an engineer to assign work called for in his contract to another engineer?

5. A licensed professional engineer was convicted of a felony totally unrelated to his consulting engineering practice.

(a) What actions would you recommend be taken by the state registration board?

(b) What actions would you recommend be taken by the professional or technical society (e.g., ASCE, ASME, IEEE, NSPE, etc.)?

6. An engineer came across some work of a predecessor. After verifying the validity and correctness of all assumptions and calculations, the engineer used the work. Under what conditions would such use be legal and/or ethical?

7. A building contractor made it a policy to provide cellular car telephones to the engineers of projects he was working on. Under what conditions could the engineers accept the telephones?

8. An engineer designed a tilt-up slab building for a client. The design engineer sent the design out to another engineer for checking. The checking engineer himself sent the plans to a concrete contractor for review. The concrete contractor made suggestions that were incorporated into the recommendations of the checking contractor. These recommendations were subsequently incorporated into the plans by the original design engineer. What steps must be taken to keep the design process legal and/or ethical?

9. A consulting engineer registered his corporation as "John Williams, P.E. and Associates, Inc." even though he had no associates. Under what conditions would this name be legal and/or ethical?

10. When it became known that a chemical plant was planning on producing a toxic product, an engineer wrote to the local newspaper condemning the action. Under what conditions would this action be legal and/or ethical?

11. An engineer signed a contract with a client. The fee the client agreed to pay was based on the engineer's estimate of time required. The engineer was able to complete the contract satisfactorily in half the time he expected. Under what conditions would it be legal and/or ethical for the engineer to keep the full fee?

12. After working on a project for a client, the engineer was asked by a competitor of the client to perform design services. Under what conditions would it be legal and/or ethical for the engineer to work for the competitor?

13. Two engineers submitted bids to a prospective client for a design project. The client told engineer A how much engineer B had bid and invited engineer A to beat the amount. Under what conditions could engineer A legally/ethically submit a lower bid?

14. A registered civil engineer specializing in well-drilling, irrigation pipelines, and farmhouse sanitary systems took a booth at a county fair located in a farming town. By a random drawing, the engineer's booth was located next to a hog-breeder's booth, complete with live (prize) hogs. The engineer gave away helium balloons with his name and phone number to all visitors to the booth. Did the engineer violate any laws/ethical guidelines?

15. While in a developing country supervising construction of a project he designed, an engineer discovered his client's project manager was treating local workers in an unsafe and inhuman (but for that country, legal) manner. When he objected, the client told the engineer to mind his own business. Later, the local workers asked the engineer to participate in a walkout and strike with them.

(a) What legal/ethical positions should the engineer take?

(b) Should it have made any difference if the engineer had or had not yet accepted any money from the client?

16. While working for a client, an engineer learns confidential knowledge of a proprietary production process being used by the client's chemical plant. The process is clearly destructive to the environment, but the client will not listen to the objections of the engineer. To inform the proper authorities will require the engineer to release information that he gained in confidence. Is it legal and/or ethical for the engineer to expose the client?

17. While working for an engineering design firm, an engineer was moonlighting as a soils engineer. At night, the engineer used the facilities of his employer to analyze and plot the results of soils tests. He then used his employer's computers and word processors to write his reports. The equipment, computers, and word processors would otherwise be unused. Under what conditions could the engineer's actions be considered legal and/or ethical?

SOLUTIONS

Introduction to the Answers

Case studies in law and ethics can be interpreted in many ways. The problems presented are simple thumbnail outlines. In most real cases, there will be more facts to influence a determination than are presented in the case scenarios. In some cases, a state may have specific laws affecting the determination; in other cases, prior case law will have been established.

The determination of whether an action is legal can be made in two ways. The obvious interpretation of an illegal action is one that violates a specific law or statute. An action can also be *found to be illegal* if it is judged in court to be a breach of a written, verbal, or implied contract. Both of these approaches are used in the following solutions.

These answers have been developed to teach legal and ethical principals. While being realistic, they are not necessarily based on actual incidents or prior case law.

1. (a) Stamping plans for someone else is illegal. The registration laws of all states permit a registered engineer to stamp/sign/seal only plans that were prepared by him personally or were prepared under his direction, supervision, or control. This is sometimes called being in *responsible charge*. The stamping/signing/sealing, for a fee or gratis, of plans produced by another person, whether that person is registered or not and whether that person is an engineer or not, is illegal.

(b) The act is unethical. An illegal act, being a concealed act, is intrinsically unethical. In addition, stamping/signing/sealing plans that have not been checked violates the rule contained in all ethical codes that requires an engineer to protect the public.

2. Unless the engineer and contractor worked together such that the engineer had personal knowledge that the information was correct, accepting the contractor's information is illegal. Not only would using unverified data violate the state's registration law (for the same reason that stamping/signing/sealing unverified plans in problem 1 was illegal), but the engineer's contract clause dealing with assignment of work to others would probably be violated.

The act is unethical. An illegal act, being a concealed act, is intrinsically unethical. In addition, using unverified data violates the rule contained in all ethical codes that requires an engineer to protect the client.

3. It is illegal to alter a report to bring it "more into line" with what the client wants unless the alterations represent actual, verified changed conditions. Even when the alterations are warranted, however, use of the unverified remainder of the report is a violation of the state registration law requiring an engineer only to stamp/sign/seal plans developed by or under him. Furthermore, this would be a case of fraudulent misrepresentation unless the originating engineer's name was removed from the report.

Unless the engineer who wrote the original report has given permission for the modification, altering the report would be unethical.

4. Assignment of engineering work is legal (1) if the engineer's contract permitted assignment, (2) all prerequisites (i.e., notifying the client) were met, and (3) the work was performed under the direction of another licensed engineer.

Assignment of work is ethical (1) if it is not illegal, (2) if it is done with the awareness of the client, and (3) if the assignor has determined that the assignee is competent in the area of the assignment.

5. (a) The registration laws of many states require a hearing to be held when a licensee is found guilty of unrelated, but nevertheless unforgivable, felonies (e.g., moral turpitude). The specific action (e.g., suspension, revocation of license, public censure, etc.) taken depends on the customs of the state's registration board.

(b) By convention, it is not the responsibility of technical and professional organizations to monitor or judge the personal actions of their members. Such organizations do not have the authority to discipline members (other than to revoke membership), nor are they immune from possible retaliatory libel/slander lawsuits if they publicly censure a member.

6. The action is legal because, by verifying all the assumptions and checking all the calculations, the engineer effectively does the work himself. Very few engineering procedures are truly original; the fact that someone else's effort guided the analysis does not make the action illegal.

The action is probably ethical, particularly if the client and the predecessor are aware of what has happened (although it is not necessary for the predecessor to be told). It is unclear to what extent (if at all) the predecessor should be credited. There could be other extenuating circumstances that would make referring to the original work unethical.

7. Gifts, per se, are not illegal. Unless accepting the phones violates some public policy or other law, or is in some way an illegal bribe to induce the engineer to favor the contractor, it is probably legal to accept the phones.

Ethical acceptance of the phones requires (among other considerations) that (1) the phones be required for the job, (2) the phones be used for business only, (3) the phones are returned to the contractor at the end of the job, and (4) the contractor's and engineer's clients know and approve of the transaction.

8. There are two issues here: (1) the assignment and (2) the incorporation of work done by another. To avoid a breach, the contracts of both the design and checking engineers must permit the assignments. To avoid a violation of the state registration law requiring engineers to be in responsible charge of the work they stamp/sign/seal, both the design and checking engineers must verify the validity of the changes.

To be ethical, the actions must be legal and all parties (including the design engineer's client) must be aware that the assignments have occurred and that the changes have been made.

9. The name is probably legal. If the name was accepted by the state's corporation registrar, it is a legally formatted name. However, some states have engineering registration laws that restrict what an engineering corporation may be named. For example, all individuals listed in the name (e.g., "Cooper, Williams, and Somerset—Consulting Engineers") may need to be registered. Whether having "Associates" in the name is legal depends on the state.

Using the name is unethical. It misleads the public and represents unfair competition with other engineers running one-person offices.

10. Unless the engineeer's accusation is known to be false or exaggerated, or the engineer has signed an agreement (e.g., confidentiality, non-disclosure, etc.) with his employer forbidding the disclosure, the letter to the newspaper is probably not illegal.

The action is probably unethical. (If the letter to the newspaper is unsigned it is a concealed action and is definitely unethical.) While whistle-blowing to protect the public is implicitly an ethical procedure, unless the engineer is reasonably certain that manufacture of the toxic product represents a hazard to the public, he has a responsibility to the employer. Even then, the engineer should exhaust all possible remedies to render the manufacture nonhazardous before blowing the whistle. Of course, the engineer may quit working for the chemical plant and be as critical as the law allows without violating engineer-employer ethical considerations.

11. Unless the engineer's payment was explicitly linked in the contract to the amount of time spent on the job, taking the full fee would not be illegal or a breach of the contract.

An engineer has an obligation to be fair in estimates of cost, particularly when the engineer knows no one else is providing a competitive bid. Taking the full fee would be ethical if the original estimate was arrived at logically and was not meant to deceive or take advantage of the client. An engineer is permitted to take advantage of economies of scale, state-of-the-art techniques, and break-through methods. (Similarly, when a job costs more than the estimate, the engineer may be ethically bound to stick with the original estimate.)

12. In the absence of a nondisclosure or noncompetition agreement or similar contract clause, working for the competitor is probably legal.

Working for both clients is unethical. Even if both clients know and approve, it is difficult for the engineer not to "cross-pollinate" his work and improve one client's position with knowledge and insights gained at the expense of the other client. Furthermore, the mere appearance of a conflict of interest of this type is a violation of most ethical codes.

13. In the absence of a sealed-bid provision mandated by a public agency and requiring all bids to be opened at once (and the award going to the lowest bidder), the action is probably legal.

It is unethical for an engineer to undercut the price of another engineer. Not only does this violate a standard of behavior expected of professionals, it unfairly benefits one engineer because a similar chance is not given to the other engineer. Even if both engineers are bidding openly against each other (in an auction format), the client must understand that a lower price means reduced service. Each reduction in price is an incentive to the engineer to reduce the quality or quantity of service.

14. It is generally legal for an engineer to advertise his services. Unless the state has relevant laws, the engineer probably did not engage in illegal actions.

Most ethical codes prohibit unprofessional advertising. The unfortunate location due to a random drawing might be excusable, but the engineer should probably refuse to participate. In any case, the balloons are a form of unprofessional advertising, and as such, are unethical.

15. (a) As stated in the scenario statement, the client's actions are legal for that country. The fact that the actions might be illegal in another country is irrelevant. Whether or not the strike is legal depends on the industry and the laws of the land. Some or all occupations (e.g., police and medical personnel) may be forbidden to strike. Assuming the engineer's contract does not prohibit his own participation, the engineer should determine the legality of the strike before making a decision to participate.

If the client's actions are inhuman, the engineer has an ethical obligation to withdraw from the project. Not doing so associates the profession of engineering with human misery.

(b) The engineer has a contract to complete the project for the client. (It is assumed that the contract between the engineer and client was negotiated in good faith, that the engineer had no knowledge of the work conditions prior to signing, and that the client did not falsely induce the engineer to sign.) Regardless of the reason for withdrawing, the engineer is breaching his contract. In the absence of proof of illegal actions by the client, withdrawal by the engineer requires a return of all fees received. Even if no fees have been received, withdrawal exposes the engineer to other delay-related claims by the client.

16. A contract for an illegal action cannot be enforced. Therefore, any confidentiality or nondisclosure agreement that the engineer has signed is unenforceable if the production process is illegal, uses illegal chemicals, or violates laws protecting the environment. If the production process is not illegal, it is not legal for the engineer to expose the client.

Society and the public are at the top of the hierarchy of an engineer's responsibilities. Obligations to the public take precedence over the client. If the production process is illegal, it would be ethical to expose the client.

17. It is probably legal for the engineer to use the facilities, particularly if the employer is aware of the use. (The question of whether the engineer is trespassing or violating a company policy cannot be answered without additional information.)

Moonlighting, in general, is not ethical. Most ethical codes prohibit running an engineering consulting business while receiving a salary from another employer. The rationale is that the moonlighting engineer is able to offer services at a much lower price, placing other consulting engineers at a competitive disadvantage. The use of someone else's equipment only compounds the problem. Since the engineer does not have to pay for using the equipment, he does not have to charge his clients for it. This places him at an unfair competitive advantage compared to other consultants who have invested heavily in equipment.

88 Engineering Licensing in the United States

PRACTICE PROBLEMS

There are no problems in this book corresponding to Ch. 88 of the *Civil Engineering Reference Manual*.

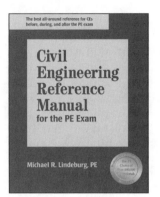

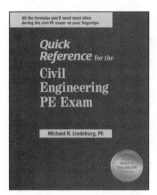

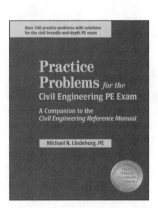

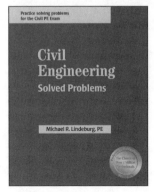